Geometry of Isotropic Convex Bodies

Mathematical
Surveys
and
Monographs

Volume 196

Geometry of Isotropic Convex Bodies

Silouanos Brazitikos
Apostolos Giannopoulos
Petros Valettas
Beatrice-Helen Vritsiou

American Mathematical Society
Providence, Rhode Island

EDITORIAL COMMITTEE

Ralph L. Cohen, Chair
Robert Guralnick
Michael A. Singer
Benjamin Sudakov
Michael I. Weinstein

2010 *Mathematics Subject Classification.* Primary 52Axx, 46Bxx, 60Dxx, 28Axx.

For additional information and updates on this book, visit
www.ams.org/bookpages/surv-196

Library of Congress Cataloging-in-Publication Data

Brazitikos, Silouanos, 1990– author.
  Geometry of isotropic convex bodies / Silouanos Brazitikos, Apostolos Giannopoulos, Petros Valettas, Beatrice-Helen Vritsiou.
    pages cm. – (Mathematical surveys and monographs ; volume 196)
  Includes bibliographical references and indexes.
  ISBN 978-1-4704-1456-6 (alk. paper)
  1. Convex geometry. 2. Banach lattices. I. Giannopoulos, Apostolos, 1963– author. II. Valettas, Petros, 1982– author. III. Vritsiou, Beatrice-Helen, author. IV. Title.

QA331.5.B73 2014
516.3′62–dc23
                                                                              2013041914

Copying and reprinting. Individual readers of this publication, and nonprofit libraries acting for them, are permitted to make fair use of the material, such as to copy a chapter for use in teaching or research. Permission is granted to quote brief passages from this publication in reviews, provided the customary acknowledgment of the source is given.

Republication, systematic copying, or multiple reproduction of any material in this publication is permitted only under license from the American Mathematical Society. Requests for such permission should be addressed to the Acquisitions Department, American Mathematical Society, 201 Charles Street, Providence, Rhode Island 02904-2294 USA. Requests can also be made by e-mail to reprint-permission@ams.org.

© 2014 by the American Mathematical Society. All rights reserved.
The American Mathematical Society retains all rights
except those granted to the United States Government.
Printed in the United States of America.

∞ The paper used in this book is acid-free and falls within the guidelines
established to ensure permanence and durability.
Visit the AMS home page at http://www.ams.org/

10 9 8 7 6 5 4 3 2 1   19 18 17 16 15 14

Contents

Preface	ix
Chapter 1. Background from asymptotic convex geometry	1
1.1. Convex bodies	1
1.2. Brunn–Minkowski inequality	4
1.3. Applications of the Brunn-Minkowski inequality	8
1.4. Mixed volumes	12
1.5. Classical positions of convex bodies	16
1.6. Brascamp-Lieb inequality and its reverse form	22
1.7. Concentration of measure	25
1.8. Entropy estimates	34
1.9. Gaussian and sub-Gaussian processes	38
1.10. Dvoretzky type theorems	43
1.11. The ℓ-position and Pisier's inequality	50
1.12. Milman's low M^*-estimate and the quotient of subspace theorem	52
1.13. Bourgain-Milman inequality and the M-position	55
1.14. Notes and references	58
Chapter 2. Isotropic log-concave measures	63
2.1. Log-concave probability measures	63
2.2. Inequalities for log-concave functions	66
2.3. Isotropic log-concave measures	72
2.4. ψ_α-estimates	78
2.5. Convex bodies associated with log-concave functions	84
2.6. Further reading	94
2.7. Notes and references	100
Chapter 3. Hyperplane conjecture and Bourgain's upper bound	103
3.1. Hyperplane conjecture	104
3.2. Geometry of isotropic convex bodies	108
3.3. Bourgain's upper bound for the isotropic constant	116
3.4. The ψ_2-case	123
3.5. Further reading	128
3.6. Notes and references	134
Chapter 4. Partial answers	139
4.1. Unconditional convex bodies	139
4.2. Classes with uniformly bounded isotropic constant	144
4.3. The isotropic constant of Schatten classes	150
4.4. Bodies with few vertices or few facets	155

4.5.	Further reading	161
4.6.	Notes and references	170

Chapter 5. L_q-centroid bodies and concentration of mass 173
5.1.	L_q-centroid bodies	174
5.2.	Paouris' inequality	182
5.3.	Small ball probability estimates	190
5.4.	A short proof of Paouris' deviation inequality	197
5.5.	Further reading	202
5.6.	Notes and references	209

Chapter 6. Bodies with maximal isotropic constant 213
6.1.	Symmetrization of isotropic convex bodies	214
6.2.	Reduction to bounded volume ratio	223
6.3.	Regular isotropic convex bodies	227
6.4.	Reduction to negative moments	231
6.5.	Reduction to $I_1(K, Z_q^\circ(K))$	234
6.6.	Further reading	239
6.7.	Notes and references	242

Chapter 7. Logarithmic Laplace transform and the isomorphic slicing problem 243
7.1.	Klartag's first approach to the isomorphic slicing problem	244
7.2.	Logarithmic Laplace transform and convex perturbations	249
7.3.	Klartag's solution to the isomorphic slicing problem	251
7.4.	Isotropic position and the reverse Santaló inequality	254
7.5.	Volume radius of the centroid bodies	256
7.6.	Notes and references	270

Chapter 8. Tail estimates for linear functionals 271
8.1.	Covering numbers of the centroid bodies	273
8.2.	Volume radius of the ψ_2-body	284
8.3.	Distribution of the ψ_2-norm	292
8.4.	Super-Gaussian directions	298
8.5.	ψ_α-estimates for marginals of isotropic log-concave measures	301
8.6.	Further reading	304
8.7.	Notes and references	310

Chapter 9. M and M^*-estimates 313
9.1.	Mean width in the isotropic case	313
9.2.	Estimates for $M(K)$ in the isotropic case	322
9.3.	Further reading	330
9.4.	Notes and references	332

Chapter 10. Approximating the covariance matrix 333
10.1.	Optimal estimate	334
10.2.	Further reading	349
10.3.	Notes and references	354

Chapter 11. Random polytopes in isotropic convex bodies 357
11.1.	Lower bound for the expected volume radius	358

11.2.	Linear number of points	363
11.3.	Asymptotic shape	367
11.4.	Isotropic constant	377
11.5.	Further reading	381
11.6.	Notes and references	387

Chapter 12. Central limit problem and the thin shell conjecture — 389
- 12.1. From the thin shell estimate to Gaussian marginals — 391
- 12.2. The log-concave case — 397
- 12.3. The thin shell conjecture — 402
- 12.4. The thin shell conjecture in the unconditional case — 407
- 12.5. Thin shell conjecture and the hyperplane conjecture — 415
- 12.6. Notes and references — 422

Chapter 13. The thin shell estimate — 425
- 13.1. The method of proof and Fleury's estimate — 427
- 13.2. The thin shell estimate of Guédon and E. Milman — 436
- 13.3. Notes and references — 458

Chapter 14. Kannan-Lovász-Simonovits conjecture — 461
- 14.1. Isoperimetric constants for log-concave probability measures — 462
- 14.2. Equivalence of the isoperimetric constants — 473
- 14.3. Stability of the Cheeger constant — 477
- 14.4. The conjecture and the first lower bounds — 480
- 14.5. Poincaré constant in the unconditional case — 485
- 14.6. KLS-conjecture and the thin shell conjecture — 486
- 14.7. Further reading — 505
- 14.8. Notes and references — 509

Chapter 15. Infimum convolution inequalities and concentration — 511
- 15.1. Property (τ) — 512
- 15.2. Infimum convolution conjecture — 523
- 15.3. Concentration inequalities — 527
- 15.4. Comparison of weak and strong moments — 533
- 15.5. Further reading — 535
- 15.6. Notes and references — 546

Chapter 16. Information theory and the hyperplane conjecture — 549
- 16.1. Entropy gap and the isotropic constant — 550
- 16.2. Entropy jumps for log-concave random vectors with spectral gap — 552
- 16.3. Further reading — 559
- 16.4. Notes and references — 562

Bibliography — 565

Subject Index — 585

Author Index — 591

Preface

Asymptotic convex geometry may be described as the study of convex bodies from a geometric and analytic point of view, with an emphasis on the dependence of various parameters on the dimension. This theory stands at the intersection of classical convex geometry and the local theory of Banach spaces, but it is also closely linked to many other fields, such as probability theory, partial differential equations, Riemannian geometry, harmonic analysis and combinatorics. The aim of this book is to introduce a number of basic questions regarding the distribution of volume in high-dimensional convex bodies and to provide an up to date account of the progress that has been made in the last fifteen years. It is now understood that the convexity assumption forces most of the volume of a body to be concentrated in some canonical way and the main question is whether, under some natural normalization, the answer to many fundamental questions should be independent of the dimension.

One such normalization, that in many cases facilitates the study of volume distribution, is the *isotropic position*. A convex body K in $\mathbb{R}^n$ is called isotropic if it has volume 1, barycenter at the origin, and its inertia matrix is a multiple of the identity: there exists a constant $L_K > 0$ such that

$$\int_K \langle x, \theta \rangle^2 dx = L_K^2$$

for every θ in the Euclidean unit sphere S^{n-1}. It is easily verified that the affine class of any convex body K contains a unique, up to orthogonal transformations, isotropic convex body; this is the isotropic position of K. A first example of the role and significance of the isotropic position may be given through the *hyperplane conjecture* (or *slicing problem*), which is one of the main problems in the asymptotic theory of convex bodies, and asks if there exists an absolute constant $c > 0$ such that $\max_{\theta \in S^{n-1}} |K \cap \theta^\perp| \geqslant c$ for every convex body K of volume 1 in $\mathbb{R}^n$ that has barycenter at the origin. This question was posed by Bourgain [99], who was interested in finding L_p-bounds for maximal operators defined in terms of arbitrary convex bodies. It is not so hard to check that answering his question affirmatively is equivalent to the following statement:

Isotropic constant conjecture. *There exists an absolute constant $C > 0$ such that*

$$L_n := \max\{L_K : K \text{ is isotropic in } \mathbb{R}^n\} \leqslant C.$$

This problem became well-known due to an article of V. Milman and Pajor which remains a classical reference on the subject. Around the same time, K. Ball showed in his PhD Thesis that the notion of the isotropic constant and the conjecture can be reformulated in the language of logarithmically-concave (or log-concave for short) measures; however, without the problem becoming essentially more general. Let us note here that a finite Borel measure μ on $\mathbb{R}^n$ is called log-concave if, for any $\lambda \in (0,1)$ and any compact subsets A, B of $\mathbb{R}^n$, we have

$$\mu(\lambda A + (1-\lambda)B) \geqslant \mu(A)^\lambda \mu(B)^{1-\lambda};$$

note also that the indicator function of a convex body is the density (with respect to the Lebesgue measure) of a compactly supported log-concave measure, but that not all log-concave measures are compactly supported. Isotropic convex bodies now form a genuine subclass of isotropic log-concave measures, but several properties and results that (may) hold for this subclass, including the boundedness or not of the isotropic constants, immediately translate in the setting of log-concave measures. Around 1990, Bourgain obtained the upper bound $L_n \leqslant c\sqrt[4]{n}\log n$ and, in 2006, this estimate was improved by Klartag to $L_n \leqslant c\sqrt[4]{n}$.

The problem remains open and has become the starting point for many other questions and challenging conjectures in high-dimensional geometry, one of those being the *central limit problem*. The latter in the asymptotic theory of convex bodies means the task to identify those high-dimensional distributions which have approximately Gaussian marginals. It is a question inspired by a general fact that has appeared more than once in the literature and states that, if μ is an isotropic probability measure on $\mathbb{R}^n$ which satisfies the thin shell condition

$$\mu\left(\big|\,\|x\|_2 - \sqrt{n}\,\big| \geqslant \varepsilon\right) \leqslant \varepsilon$$

for some $\varepsilon \in (0,1)$, then, for all directions θ in a subset A of S^{n-1} with $\sigma(A) \geqslant 1 - \exp(-c_1\sqrt{n})$, one has

$$|\mu\left(\{x : \langle x, \theta\rangle \leqslant t\}\right) - \Phi(t)| \leqslant c_2(\varepsilon + n^{-\alpha}) \qquad \text{for all } t \in \mathbb{R},$$

where $\Phi(t)$ is the standard Gaussian distribution function and $c_1, c_2, \alpha > 0$ are absolute constants. Thus, the central limit problem is reduced to the question of identifying those high-dimensional distributions that satisfy a thin shell condition. It was the work of Anttila, Ball and Perissinaki that made this type of statement widely known in the context of isotropic convex bodies or, more generally, log-concave distributions. One of the main results in this area, first proved by Klartag in a breakthrough work, states that the assumption of log-concavity guarantees a thin shell bound, and hence an affirmative answer to the central limit problem. In fact, the following quantitative conjecture has been proposed.

Thin shell conjecture. *There exists an absolute constant $C > 0$ such that, for any $n \geqslant 1$ and any isotropic log-concave measure μ on $\mathbb{R}^n$, one has*

$$\sigma_\mu^2 := \int_{\mathbb{R}^n} \left(\|x\|_2 - \sqrt{n}\right)^2 d\mu(x) \leqslant C^2.$$

A third conjecture concerns the *Cheeger constant* Is_μ of an isotropic log-concave measure μ which is defined as the best constant $\kappa \geqslant 0$ such that

$$\mu^+(A) \geqslant \kappa \min\{\mu(A), 1 - \mu(A)\}$$

for every Borel subset A of $\mathbb{R}^n$, and where

$$\mu^+(A) := \liminf_{\varepsilon \to 0^+} \frac{\mu(A_\varepsilon) - \mu(A)}{\varepsilon}$$

is the Minkowski content of A (also, $A_\varepsilon := \{x : \mathrm{dist}(x, A) < \varepsilon\}$ is the ε-extension of A).

Kannan-Lovász-Simonovits conjecture. *There exists an absolute constant $c > 0$ such that*

$$\mathrm{Is}_n := \min\{\mathrm{Is}_\mu : \mu \text{ is isotropic log-concave measure on } \mathbb{R}^n\} \geqslant c.$$

Another way to formulate this conjecture is to ask if there exists an absolute constant $c > 0$ such that, for every isotropic log-concave measure μ on $\mathbb{R}^n$ and for every smooth function φ with $\int_{\mathbb{R}^n} \varphi \, d\mu = 0$, one has

$$c \int_{\mathbb{R}^n} \varphi^2 d\mu \leqslant \int_{\mathbb{R}^n} \|\nabla\varphi\|_2^2 d\mu.$$

We then say that μ satisfies the *Poincaré inequality* with constant $c > 0$. The equivalence of the two formulations can be seen by checking that

$$\mathrm{Is}_\mu^2 \simeq \inf_\mu \inf_\varphi \frac{\int \|\nabla\varphi\|_2^2 d\mu}{\int \varphi^2 d\mu}.$$

In this book we discuss these three conjectures and what is currently known about them, as well as other problems that are related to and arise from them. We now give a brief account of the contents of every chapter; more details can be found in the introduction of each individual chapter. In Chapters 2–4, we present the hyperplane conjecture and the first attempts to an answer. This presentation is given in the more general setting of logarithmically concave probability measures, which are introduced in Chapter 2 along with their main concentration properties. Some of these properties follow immediately from the Brunn-Minkowski inequality (more precisely, from Borell's lemma) and can be expressed in the form of reverse Hölder inequalities for seminorms: if $f : \mathbb{R}^n \to \mathbb{R}$ is a seminorm, then, for every log-concave probability measure μ on $\mathbb{R}^n$, one has $\|f\|_{\psi_1(\mu)} \leqslant c\|f\|_{L_1(\mu)}$, where

$$\|f\|_{\psi_\alpha(\mu)} = \inf\left\{t > 0 : \int \exp((|f|/t)^\alpha) \, d\mu \leqslant 2\right\}$$

is the Orlicz ψ_α-norm of f with respect to μ, $\alpha \in [1,2]$. Isotropic log-concave measures are the log-concave probability measures μ that have barycenter at the origin and satisfy the isotropic condition

$$\int_{\mathbb{R}^n} \langle x, \theta \rangle^2 d\mu(x) = 1$$

for every $\theta \in S^{n-1}$. The isotropic constant of a measure μ in this class is defined as

$$L_\mu := \left(\sup_{x \in \mathbb{R}^n} f(x)\right)^{1/n} \simeq (f(0))^{1/n},$$

where f is the log-concave density of μ. K. Ball introduced a family of convex bodies $K_p(\mu)$, $p \geqslant 1$, that can be associated with a given log-concave measure μ and showed that these bodies allow us to reduce the study of log-concave measures to that of convex bodies, but also enable us to use tools from the broader class of measures to tackle problems that have naturally, or merely initially, been formulated for bodies. A first example of their use, as mentioned above, is the fact that studying the magnitude of the isotropic constant of log-concave measures is completely equivalent to the respective task inside the more restricted class of convex bodies.

The isotropic constant conjecture is discussed in detail in Chapter 3; it reads that there exists an absolute constant $C > 0$ such that
$$L_\mu \leqslant C$$
for every $n \geqslant 1$ and every log-concave measure μ on $\mathbb{R}^n$. In order to understand its equivalence to the hyperplane conjecture we formulated above, we recall that

$$\max\{L_K : K \text{ is an isotropic convex body in } \mathbb{R}^n\}$$
$$\simeq \sup\{L_\mu : \mu \text{ is an isotropic log-concave measure on } \mathbb{R}^n\},$$

and then we have to explain the relation of the moments of inertia of a centered convex body to the volume of its hyperplane sections passing through the origin. In particular, in Section 3.1.2 we show that, if K is an isotropic convex body in $\mathbb{R}^n$, then for every $\theta \in S^{n-1}$ we have
$$\frac{c_1}{L_K} \leqslant |K \cap \theta^\perp| \leqslant \frac{c_2}{L_K},$$
where $c_1, c_2 > 0$ are absolute constants, and, thus, all hyperplane sections through the barycenter of K have approximately the same volume, this volume being large enough if and only if L_K is small enough. The hyperplane conjecture is also equivalent to the asymptotic versions of several classical problems in convex geometry. We discuss two of them: Sylvester's problem on the expected volume of a random simplex contained in a convex body and the Busemann-Petty problem. In Sections 3.3 and 3.4 we discuss Bourgain's upper bound $L_K \leqslant C\sqrt[4]{n}\log n$ for the isotropic constant of convex bodies K in $\mathbb{R}^n$. We describe two proofs of Bourgain's result. A key observation is that, if K is an isotropic convex body in $\mathbb{R}^n$, then, as we saw above for every log-concave probability measure μ, one has $\|\langle \cdot, \theta\rangle\|_{\psi_1(K)} \leqslant C\|\langle \cdot, \theta\rangle\|_{L_1(K)} \leqslant CL_K$ for all $\theta \in S^{n-1}$, where $C > 0$ is an absolute constant. In fact, Alesker's theorem shows that one has a stronger ψ_2-estimate for the function $f(x) = \|x\|_2$: one has $\|f\|_{\psi_2(K)} \leqslant C\|f\|_{L_2(K)} \leqslant C\sqrt{n}L_K$. Markov's inequality then implies exponential concentration of the mass of K in a strip of width CL_K and normal concentration in a ball of radius $C\sqrt{n}L_K$.

Chapter 4 is devoted to some partial affirmative answers to the hyperplane conjecture that were obtained soon after the problem became known. In order to make this statement more precise, we say that a class $\mathcal{C}$ of centered convex bodies satisfies the hyperplane conjecture uniformly if there exists a positive constant C such that $L_K \leqslant C$ for all $K \in \mathcal{C}$. The hyperplane conjecture has been verified for several important classes of convex bodies. A first example is the class of unconditional convex bodies; these are the centrally symmetric convex bodies K in $\mathbb{R}^n$ that have a position that is symmetric with respect to the standard coordinate subspaces, namely they have a position $\tilde{K}$ such that, if $(x_1, \ldots, x_n)$ belongs to $\tilde{K}$, then $(\epsilon_1 x_1, \ldots, \epsilon_n x_n)$ also belongs to $\tilde{K}$ for every $(\epsilon_1, \ldots, \epsilon_n) \in \{-1, 1\}^n$. The class of unconditional convex bodies will appear often in this book, mainly as a model for results or conjectures regarding the general cases. In this chapter, we also describe uniform bounds for the isotropic constants of some other classes of convex bodies and we give simple geometric proofs of the best known estimates for the isotropic constants of polytopes with N vertices or polyhedra with N facets, estimates that are logarithmic in N.

In Chapters 5–7, we discuss more recent approaches to the slicing problem and some very useful tools that have been developed for these approaches as well as for related problems in the theory. Bourgain's approach exploited the ψ_1-information we have for the behavior of the linear functionals $x \mapsto \langle x, \theta \rangle$ on an isotropic convex body. The aim to understand the distribution of linear functionals in an isotropic convex body or, more precisely, the behavior of their L_q-norms with respect to the uniform measure on the body, has been furthered by the introduction of the family of L_q-centroid bodies of a convex body K of volume 1 or, more generally, of a log-concave probability measure μ. For every $q \geqslant 1$, the L_q-centroid body $Z_q(K)$ of K or, respectively, the L_q-centroid body $Z_q(\mu)$ of μ is defined through its support function, which is given by

$$h_{Z_q(K)}(y) := \|\langle \cdot, y \rangle\|_{L_q(K)} = \left(\int_K |\langle x, y \rangle|^q dx \right)^{1/q},$$

$$\text{or by} \quad h_{Z_q(\mu)}(y) := \|\langle \cdot, y \rangle\|_{L_q(\mu)} = \left(\int |\langle x, y \rangle|^q d\mu(x) \right)^{1/q},$$

respectively, for every vector y. Note that, according to our normalization, a convex body K of volume 1 in $\mathbb{R}^n$ is isotropic if and only if it is centered and $Z_2(K) = L_K B_2^n$ and, respectively, a log-concave probability measure μ on $\mathbb{R}^n$ is isotropic if and only if it is centered and $Z_2(\mu) = B_2^n$. The development of an asymptotic theory for this family of bodies, and for their behavior as q increases from 2 up to the dimension n, was initiated by Paouris and has proved to be a very fruitful idea.

In Chapter 5 we present the basic properties of the family $\{L_q(\mu) : q \geqslant 2\}$ of the centroid bodies of a centered log-concave probability measure μ on $\mathbb{R}^n$ and prove some fundamental formulas. The first main application of this theory is a striking and very useful deviation inequality of Paouris: for every isotropic log-concave probability measure μ on $\mathbb{R}^n$ one has

$$\mu(\{x \in \mathbb{R}^n : \|x\|_2 \geqslant ct\sqrt{n}\}) \leqslant \exp\left(-t\sqrt{n}\right)$$

for every $t \geqslant 1$, where $c > 0$ is an absolute constant. This is a consequence of the following statement: there exists an absolute constant $C_1 > 0$ such that, if μ is an isotropic log-concave measure on $\mathbb{R}^n$, then

(1) $$I_q(\mu) \leqslant C_1 I_2(\mu)$$

for every $q \leqslant \sqrt{n}$, where $I_q(\mu)$ is defined by

$$I_q(\mu) = \left(\int_{\mathbb{R}^n} \|x\|_2^q d\mu(x) \right)^{1/q}$$

for all $0 \neq q > -n$. Paouris has, moreover, proved an extension to this theorem which we also present: there exists an absolute constant c_2 such that, if μ is an isotropic log-concave measure on $\mathbb{R}^n$, then for any $1 \leqslant q \leqslant c_2\sqrt{n}$ one has

$$I_{-q}(\mu) \simeq I_q(\mu).$$

In particular, this shows that, for all $1 \leqslant q \leqslant c_2\sqrt{n}$, one has $I_q(\mu) \leqslant C I_2(\mu)$, where $C > 0$ is an absolute constant. Using the extended result one can derive a small ball probability estimate: for every isotropic log-concave measure μ on $\mathbb{R}^n$ and for any $0 < \varepsilon < \varepsilon_0$, one has

$$\mu(\{x \in \mathbb{R}^n : \|x\|_2 < \varepsilon\sqrt{n}\}) \leqslant \varepsilon^{c_3\sqrt{n}},$$

where $\varepsilon_0, c_3 > 0$ are absolute constants. In a few words, the main results of Paouris imply that for any isotropic log-concave measure one has

$$\mu(\{x : c\sqrt{n} \leqslant \|x\|_2 \leqslant C\sqrt{n}\}) \geqslant 1 - \exp(-\sqrt{n}).$$

This is a rough version of the thin shell estimate, that is often enough for the applications. In fact, as we will explain in Chapter 13, a way to obtain a thin shell estimate is to prove a more precise version of (1), with the constant C_1 being, for example, of the form $1 + cq/\sqrt{n}$ for some absolute constant $c > 0$ and for as large $q \in [1, \sqrt{n}]$ as possible.

In Chapters 6 and 7 we discuss some recent approaches to and reductions of the hyperplane conjecture. Chapter 6 deals with properties that bodies with maximal isotropic constant have, namely bodies whose isotropic constant is equal to or very close to L_n. It turns out that the isotropic position of such bodies is closely related to their M-position and this enables one to establish several interesting facts: for example, a reduction of the hyperplane conjecture, due to Bourgain, Klartag and V. Milman, to the question of boundedness of the isotropic constant of a restricted class of convex bodies, those that have volume ratio bounded by an absolute constant. Next, we give two more reductions of the conjecture to the study of parameters that can be associated with any isotropic convex body. The proofs of these reductions rely heavily on the existence of convex bodies with maximal isotropic constant whose isotropic position is not only closely related to their M-position, but is also compatible with regular covering estimates. The first of these reductions is a continuation of the work of Paouris on the behavior of the negative moments of the Euclidean norm with respect to an isotropic measure μ on $\mathbb{R}^n$. As we mentioned above, we already know that $I_{-q}(\mu) \simeq I_2(\mu) = \sqrt{n}$ for $0 < q \leqslant \sqrt{n}$, however, the behavior of the negative moments $I_{-q}(\mu)$ for $q > \sqrt{n}$ is not known at all and, in fact, our current knowledge does not exclude the possibility that the moments stay constant for all positive q up to $n - 1$. Dafnis and Paouris have actually proved that this question is equivalent to the hyperplane conjecture: they introduce a parameter that, for each $\delta \geqslant 1$, is given by

$$q_{-c}(\mu, \delta) := \max\{1 \leqslant q \leqslant n - 1 : I_{-q}(\mu) \geqslant \delta^{-1} I_2(\mu) = \delta^{-1}\sqrt{n}\},$$

and they establish that

$$L_n \leqslant C\delta \sup_\mu \sqrt{\frac{n}{q_{-c}(\mu, \delta)}} \log^2\left(\frac{en}{q_{-c}(\mu, \delta)}\right)$$

for every $\delta \geqslant 1$; additionally, they show that, if the hyperplane conjecture is correct, then we must have $q_{-c}(\mu, \delta_0) = n - 1$ for some $\delta_0 \simeq 1$, for every isotropic log-concave measure μ on $\mathbb{R}^n$. The other reduction is a work of Giannopoulos, Paouris and Vritsiou, based on the study of the parameter

$$I_1(K, Z_q^\circ(K)) = \int_K h_{Z_q(K)}(x)dx = \int_K \|\langle \cdot, x \rangle\|_{L_q(K)}dx,$$

and can be viewed as a continuation of Bourgain's initial approach that led to the upper bound $L_K \leqslant c\sqrt[4]{n}\log n$. Roughly speaking, this last reduction can be formulated as follows: given $q \geqslant 2$ and $\frac{1}{2} \leqslant s \leqslant 1$, an upper bound of the form

$$I_1(K, Z_q^\circ(K)) \leqslant C_1 q^s \sqrt{n} L_K^2 \quad \text{for all bodies } K \text{ in isotropic position}$$

leads to the estimate
$$L_n \leqslant \frac{C_2 \sqrt[4]{n} \log^2 n}{q^{\frac{1-s}{2}}}.$$

Bourgain's estimate is (almost) recovered by choosing $q = 2$, however, the behavior of $I_1(K, Z_q^\circ(K))$ may allow one to use $s < 1$ along with large values of q to obtain improved bounds if possible.

In Chapter 7 we first discuss Klartag's solution to the isomorphic slicing problem, an isomorphic variation of the hyperplane conjecture that asks whether, given any convex body, we can find another convex body, with absolutely bounded isotropic constant, that is geometrically close to the first body. Klartag's method relies on properties of the logarithmic Laplace transform of the uniform measure on a convex body. In general, given a finite Borel measure μ on $\mathbb{R}^n$, the logarithmic Laplace transform of μ is given by

$$\Lambda_\mu(\xi) := \log\left(\frac{1}{\mu(\mathbb{R}^n)} \int_{\mathbb{R}^n} e^{\langle \xi, x \rangle} d\mu(x)\right).$$

Klartag proved that, if K is a convex body in $\mathbb{R}^n$, then, for every $\varepsilon \in (0,1)$, we can find a centered convex body $T \subset \mathbb{R}^n$ and a point $x \in \mathbb{R}^n$ such that $\frac{1}{1+\varepsilon}T \subseteq K + x \subseteq (1+\varepsilon)T$ and
$$L_T \leqslant C/\sqrt{\varepsilon},$$

where $C > 0$ is an absolute constant. Most remarkably, by combining this fact with the deviation inequality of Paouris, one may also deduce the currently best known upper bound for the isotropic constant, which is that

$$L_\mu \leqslant C' n^{1/4}$$

for every isotropic log-concave measure μ on $\mathbb{R}^n$. The logarithmic Laplace transform is another important tool of the theory that, since it was first employed in the setting of isotropic convex bodies and log-concave measures, has proved to be extremely useful given its various and interesting applications; these include Klartag's solution to the isomorphic slicing problem, that we already mentioned, as well as an alternative approach of Klartag and E. Milman that combines the advantages of both the logarithmic Laplace transform and the extensive theory of the L_q-centroid bodies, and occupies the second part of Chapter 7. Klartag and E. Milman looked for lower bounds for the volume radius of the L_q-centroid bodies of an isotropic log-concave measure μ. Through a delicate analysis of the logarithmic Laplace transform of μ, they showed that

$$(2) \qquad |Z_q(\mu)|^{1/n} \geqslant c_1 \sqrt{q/n}$$

for all $q \leqslant \sqrt{n}$, where $c_1 > 0$ is an absolute constant. Apart from being interesting on its own, this result leads again to the estimate $L_\mu \leqslant c_2 \sqrt[4]{n}$. It is also plausible that (2) can hold for larger values of $q \in [1, n]$ as well; this is the content of a recent work of Vritsiou that is also discussed in the chapter. She showed that (2) holds for every q up to a variant of the parameter $q_{-c}(\mu, \delta)$ of Dafnis and Paouris, which, as we previously mentioned, could be of the order of n (in fact, recall that the hyperplane conjecture is correct if and only if $q_{-c}(\mu, \delta_0)$ is of the order of n for some $\delta_0 \simeq 1$ and every isotropic log-concave measure μ on $\mathbb{R}^n$). However, even a small improvement to the estimates we currently have for $q_{-c}(\mu, \delta)$ and its variant could permit one to extend the range of q with which the method of Klartag and Milman can be applied, and also improve on the currently known bounds for the isotropic constant problem. Other applications of the logarithmic Laplace transform are discussed in some of the following chapters, the most important of these appearing in Chapters 12 and 15.

In Chapters 8–11, we deviate a little from those lines of results that are directly related to the hyperplane conjecture and the other two main conjectures of the theory so as to look at different applications of the tools that were developed in the previous part. Chapters 8 and 9 are devoted to some open questions, whose study so far has already shed more light on various geometric properties of convex bodies and log-concave measures. The first question was originally posed by V. Milman in the framework of convex bodies: it asks if there exists an absolute constant $C > 0$ such that every centered convex body K of volume 1 has at least one sub-Gaussian direction with constant C. Following some positive results for special classes of convex bodies, Klartag was the first to prove the existence of "almost sub-Gaussian" directions for any isotropic convex body. More precisely, using again properties of the logarithmic Laplace transform, he proved that for every log-concave probability measure μ on $\mathbb{R}^n$ there exists $\theta \in S^{n-1}$ such that

$$\mu(\{x : |\langle x, \theta \rangle| \geqslant ct\|\langle \cdot, \theta \rangle\|_2\}) \leqslant e^{-\frac{t^2}{(\log(t+1))^{2\alpha}}},$$

for all $1 \leqslant t \leqslant \sqrt{n}\log^\alpha n$, where $\alpha = 3$. We describe the best known estimate, due to Giannopoulos, Paouris and Valettas, according to which one can always have $\alpha = 1/2$. The main idea is to define the symmetric convex set $\Psi_2(\mu)$ whose support function is $h_{\Psi_2(\mu)}(\theta) = \|\langle \cdot, \theta \rangle\|_{\psi_2}$ and to estimate its volume. One can show that for every centered log-concave probability measure μ in $\mathbb{R}^n$ one has

$$c_1 \leqslant \left(\frac{|\Psi_2(\mu)|}{|Z_2(\mu)|} \right)^{1/n} \leqslant c_2 \sqrt{\log n},$$

where $c_1, c_2 > 0$ are absolute constants. An immediate consequence is the existence of at least one sub-Gaussian direction for μ with constant $b = O(\sqrt{\log n})$. The main tool in the proof of this result is estimates for the covering numbers $N(Z_q(K), sB_2^n)$. An even more interesting question is to determine the distribution of the function $\theta \mapsto \|\langle \cdot, \theta \rangle\|_{\psi_2}$ on the unit sphere; that is, to understand whether most of the directions have ψ_2-norm that is, say, logarithmic in the dimension.

In Chapter 9 we discuss the questions of obtaining an upper bound for the mean width

$$w(K) := \int_{S^{n-1}} h_K(x)\, d\sigma(x),$$

that is, the L_1-norm of the support function of K with respect to the Haar measure on the sphere, as well as the respective L_1-norm of the Minkowski functional of K,

$$M(K) := \int_{S^{n-1}} \|x\|_K\, d\sigma(x),$$

when K is an isotropic convex body. We present some non-trivial but non-optimal estimates. We also discuss the same questions for the L_q-centroid bodies of an isotropic log-concave measure. Answering these questions requires a deeper understanding of the behavior of linear functionals and of the local structure of the centroid bodies; this would bring new insights to the reductions of the hyperplane conjecture that were discussed in the previous chapters.

Chapters 10 and 11 contain applications of the theory of L_q-centroid bodies and of the main inequalities of Paouris to random matrices and random polytopes.

In Chapter 10 we discuss a question of Kannan, Lovász and Simonovits on the approximation of the covariance matrix of a log-concave measure. If K is an isotropic convex body in $\mathbb{R}^n$, then one has

$$I = \frac{1}{L_K^2} \int_K x \otimes x \, dx,$$

where I is the identity operator. Given $\varepsilon \in (0,1)$, the question is to find N_0, as small as possible, for which the following holds true: if $N \geqslant N_0$, then N independent random points $x_1, \ldots, x_N$ that are uniformly distributed in K must have, with probability greater than $1 - \varepsilon$, the property that

$$(1-\varepsilon)L_K^2 \leqslant \frac{1}{N} \sum_{i=1}^N \langle x_i, \theta \rangle^2 \leqslant (1+\varepsilon)L_K^2$$

for every $\theta \in S^{n-1}$. The question had its origin in the problem of finding a fast algorithm for the computation of the volume of a given convex body, and Kannan, Lovász and Simonovits proved that one can take $N_0 = C(\varepsilon)n^2$ for some constant $C(\varepsilon) > 0$ depending only on ε. This was improved to $N_0 = C(\varepsilon)n(\log n)^3$ by Bourgain and to $N_0 = C(\varepsilon)n(\log n)^2$ by Rudelson. It was finally proved by Adamczak, Litvak, Pajor and Tomczak-Jaegermann that the best estimate for N_0 is $C(\varepsilon)n$. We describe the history and the solution of the problem.

In Chapter 11 we discuss the asymptotic shape of the random polytope $K_N := \operatorname{conv}\{\pm x_1, \ldots, \pm x_N\}$ that is spanned by N independent random points $x_1, \ldots, x_N$ uniformly distributed in an isotropic convex body K in $\mathbb{R}^n$. The literature on the approximation of convex bodies by random polytopes is very rich, but the main point here is that N is fixed in the range $[n, e^n]$ and we are interested in estimates which do not depend on the affine class of a convex body K. Some basic tasks in this spirit are: to determine the asymptotic behavior of the volume radius $|K|^{1/n}$, to understand the typical "asymptotic shape" of K_N and to estimate the isotropic constant of K_N. The same questions can be formulated and studied more generally if we assume that we have N independent copies $X_1, \ldots, X_N$ of an isotropic log-concave random vector X. A general, and rather precise, description was obtained by Dafnis, Giannopoulos and Tsolomitis: given any isotropic log-concave measure μ on $\mathbb{R}^n$ and any $n \leqslant N \leqslant \exp(n)$, the random polytope K_N defined by N i.i.d. random points $X_1, \ldots, X_N$ which are distributed according to μ satisfies, with high probability, the next two conditions: (i) $K_N \supseteq c\, Z_{\log(N/n)}(\mu)$ and (ii) for every $\alpha > 1$ and $q \geqslant 1$,

$$\mathbb{E}\left[\sigma(\{\theta : h_{K_N}(\theta) \geqslant \alpha h_{Z_q(\mu)}(\theta)\})\right] \leqslant N\alpha^{-q}.$$

Using this description of the shape of K_N and the theory of centroid bodies which was developed in the previous chapters, one can determine the volume radius and the quermassintegrals of a random K_N, at least in the range $n \leqslant N \leqslant \exp(\sqrt{n})$. A question concerning the isotropic constant of K_N can be made precise in the following way: one would like to show that, with probability tending to 1 as $n \to \infty$, the isotropic constant of the random polytope $K_N := \operatorname{conv}\{\pm x_1, \ldots, \pm x_N\}$ is bounded by CL_K where $C > 0$ is a constant independent of K, n and N. We describe a method that was initiated by Klartag and Kozma when dealing with the class of Gaussian random polytopes. Variants of the method also work in the cases that the vertices x_j of K_N are distributed according to the uniform measure on an isotropic convex body which is either ψ_2 (with constant b) or unconditional. The general case remains open.

Chapters 12–14 provide an exposition of our state of knowledge on the thin shell and Kannan-Lovász-Simonovits (or KLS for short) conjectures. Historical and other information about the thin shell conjecture and its connections with the central limit problem is given in Chapter 12. We present the work of Anttila, Ball and Perissinaki and various central limit theorems for isotropic convex bodies which would follow from thin shell estimates. This question has been studied by many authors and has been verified in some special cases. Klartag was the first to give a positive answer in full generality. In fact, aside from the immediate consequence of a general thin shell estimate that, as we mentioned again earlier in the Introduction, is that most one-dimensional marginals are close to Gaussian distributions, Klartag also established normal approximation for multidimensional marginal distributions. In Section 12.4 we give an account of Klartag's positive answer to the thin shell conjecture for the class of unconditional isotropic log-concave random vectors, which is one of the special cases for which this question was fully verified. Klartag proved that if K is an unconditional isotropic convex body in $\mathbb{R}^n$, then

$$\sigma_K^2 := \mathbb{E}_{\mu_K}\left(\|x\|_2 - \sqrt{n}\right)^2 \leqslant C^2,$$

where $C \leqslant 4$ is an absolute positive constant. We also describe a result of Eldan and Klartag which shows that the thin shell conjecture is stronger than the hyperplane conjecture and implies it; more precisely, they prove that $L_n \leqslant C\sigma_n$ where

$$\sigma_n := \max\{\sigma_\mu : \mu \text{ is isotropic log-concave measure on } \mathbb{R}^n\},$$

and, hence, any estimate one establishes for the former conjecture immediately holds for the latter too. Chapter 13 is then devoted to a complete proof of the currently best known estimate for the thin shell conjecture, $\sigma_n \leqslant Cn^{1/3}$, which is due to Guédon and E. Milman.

Chapter 14 is devoted to the Kannan-Lovász-Simonovits conjecture. We first introduce various isoperimetric constants which provide information on the interplay between a log-concave probability measure μ and the underlying Euclidean metric (the Cheeger constant Is_μ, the Poincaré constant Poin_μ, the exponential concentration constant Exp_μ and the first moment concentration constant FM_μ) and we discuss their relation. Complementing classical results of Maz'ya, Cheeger, Gromov, V. Milman, Buser, Ledoux and others, E. Milman established the equivalence of all four constants in the log-concave setting: one has

$$\mathrm{Is}_\mu \simeq \sqrt{\mathrm{Poin}_\mu} \simeq \mathrm{Exp}_\mu \simeq \mathrm{FM}_\mu$$

for every log-concave probability measure, where $a \simeq b$ means that $c_1 a \leqslant b \leqslant c_2 b$ for some absolute constants $c_1, c_2 > 0$. As an application, E. Milman obtained stability results for the Cheeger constant of convex bodies. Loosely speaking, if K and T are two convex bodies in $\mathbb{R}^n$ and if $|K| \simeq |T| \simeq |K \cap T|$, then $\mathrm{Is}_K \simeq \mathrm{Is}_T$. We introduce the KLS-conjecture in Section 14.4 and we present the first general lower bounds for Is_μ in the isotropic log-concave case. From the work of Kannan, Lovász and Simonovits and Bobkov one has that $\sqrt{n}\mathrm{Is}_\mu \geqslant c$, where $c > 0$ is an absolute constant. Actually, Bobkov proved that

$$\sqrt[4]{n}\sqrt{\sigma_\mu}\mathrm{Is}_\mu \geqslant c;$$

this provides a direct link between the KLS-conjecture and the thin shell conjecture: combined with the thin shell estimate of Guédon and E. Milman his result leads to the bound $n^{5/12}\mathrm{Is}_\mu \geqslant c$. In Section 14.5 we describe Klartag's logarithmic in the dimension lower bound for the Poincaré constant Poin_K of an unconditional isotropic convex body K in $\mathbb{R}^n$; one has $\mathrm{Is}_K \simeq \sqrt{\mathrm{Poin}_K} \geqslant \frac{c}{\log n}$, where $c > 0$ is an absolute positive constant. We close this discussion with a result of Eldan which, again, connects the thin shell conjecture with the KLS-conjecture: there exists an absolute constant $C > 0$ such that

$$\frac{1}{\mathrm{Is}_n^2} \leqslant C \log n \sum_{k=1}^n \frac{\sigma_k^2}{k}.$$

Taking into account the result of Guédon and E. Milman, one gets the currently best known bound for Is_n: $\mathrm{Is}_n^{-1} \leqslant C n^{1/3} \log n$.

In the last two chapters of the book we are concerned with two more approaches to the main questions in this theory. Chapter 15 is devoted to a probabilistic approach and related conjectures of Latała and Wojtaszczyk on the geometry of log-concave measures. The starting point is an *infimum convolution inequality* which was first introduced by Maurey when he gave a simple proof of Talagrand's two level concentration inequality for the product exponential measure. In general, if μ is a probability measure and φ is a non-negative measurable function on $\mathbb{R}^n$, one says that the pair (μ, φ) has *property* (τ) if, for every bounded measurable function f on $\mathbb{R}^n$,

$$\left(\int_{\mathbb{R}^n} e^{f\Box\varphi} d\mu\right) \left(\int_{\mathbb{R}^n} e^{-f} d\mu\right) \leqslant 1,$$

where $f\Box\varphi$ is the infimum convolution of f and φ, defined by

$$(f\Box\varphi)(x) = \inf\{f(x-y) + \varphi(y) : y \in \mathbb{R}^n\}.$$

That the property (τ) is satisfied by a pair (μ, φ) is directly related to concentration properties of the measure μ since the former property implies that, for every measurable $A \subseteq \mathbb{R}^n$ and every $t > 0$, we have

$$\mu(x \notin A + B_\varphi(t)) \leqslant (\mu(A))^{-1} e^{-t},$$

where $B_\varphi(t) = \{\varphi \leqslant t\}$. Therefore, given a measure μ it makes sense to ask for the optimal cost function φ for which we have that (μ, φ) has property (τ). The first main observation is that, if we restrict ourselves to even probability measures μ and convex cost functions φ, then the (pointwise) largest candidate for a cost function is the Cramer transform Λ_μ^* of μ; this is the Legendre transform of the logarithmic Laplace transform of μ. In the setting of log-concave probability measures, the conjecture Latała and Wojtaszczyk formulate is that the pair (μ, Λ_μ^*) always has property (τ). A detailed analysis shows that this conjecture would imply an affirmative answer to most of the conjectures addressed in this book: among them, the thin shell conjecture as well as the hyperplane conjecture. The problems that are raised through this approach are very interesting and challenging. An affirmative answer has been given for some rather restricted classes of measures: even log-concave product measures, uniform distributions on ℓ_p^n-balls and rotationally invariant log-concave measures.

In the last chapter we give an account of K. Ball's information theoretic approach, which is based on the study of the Shannon entropy $\mathrm{Ent}(X) = -\int_{\mathbb{R}^n} f \log f$ of an isotropic random vector X with density f. It is known that, among all isotropic random vectors, the standard Gaussian random vector G has the largest entropy, and the main observation is that comparing the entropy gap $\mathrm{Ent}\left(\frac{X+Y}{\sqrt{2}}\right) - \mathrm{Ent}(X)$ (with Y being an independent copy of X) to $\mathrm{Ent}(G) - \mathrm{Ent}(X)$ provides a link between the KLS-conjecture and the hyperplane conjecture. A first result of this type was obtained by Ball, Barthe and Naor for one-dimensional distributions. The main result of this chapter is a recent high-dimensional analogue for isotropic log-concave random vectors, which is due to Ball and Nguyen: if X is an isotropic log-concave random vector in $\mathbb{R}^n$ and its density f satisfies the Poincaré inequality with constant $\kappa > 0$, then

$$\mathrm{Ent}\left(\frac{X+Y}{\sqrt{2}}\right) - \mathrm{Ent}(X) \geqslant \frac{\kappa}{4(1+\kappa)}\left(\mathrm{Ent}(G) - \mathrm{Ent}(X)\right),$$

where G is a standard Gaussian random vector in $\mathbb{R}^n$. In addition, Ball and Nguyen show that this implies $L_X \leqslant e^{17/\kappa}$. Thus, for each individual isotropic log-concave distribution X, a lower bound for the Poincaré constant implies a bound for the isotropic constant.

The book is primarily addressed to readers who are familiar with the basic theory of convex bodies and the asymptotic theory of finite dimensional normed spaces as these are developed in the books of Milman and Schechtman and of Pisier. Nevertheless, we have included an introductory chapter where all the prerequisites are described; short proofs are also provided for the most important results that are used in the sequel.

This book grew out of our working seminar in the last fifteen years. Among the main topics that were discussed in our meetings were the developments on the basic questions addressed in the text. A large part of the material forms the basis of PhD and MSc theses that were written at the University of Athens and the University of Crete. We are grateful to Nikos Dafnis, Dimitris Gatzouras, Marianna Hartzoulaki, Labrini Hioni, Lefteris Markessinis, Nikos Markoulakis, Grigoris Paouris, Eirini Perissinaki, Pantelis Stavrakakis and Antonis Tsolomitis for their active participation in our seminar, for collaborating with us at various stages, for numerous discussions around the subject of this book and for their friendship over the years.

We are very grateful to Sergei Gelfand for many kind reminders regarding this project and for believing that we would be able to complete it. We are also grateful to Christine Thivierge and Luann Cole for their precious help in the preparation of this book. Finally, we would like to acknowledge partial support from the ARISTEIA II programme of the General Secretariat of Research and Technology of Greece during the final stage of this project.

Athens, February 2014

CHAPTER 1

Background from asymptotic convex geometry

In this introductory chapter we survey the prerequisites from the theory of convex bodies and the asymptotic theory of finite dimensional normed spaces. Short proofs are provided for the most important results that are used in the sequel. Basic references on convex geometry are the monographs by Schneider [**463**] and Gruber [**237**]. The asymptotic theory of finite dimensional normed spaces is presented in the books by V. Milman and Schechtman [**387**], Pisier [**430**] and Tomczak-Jaegermann [**493**].

The books of Rockafellar [**442**], Bogachev [**92**], Brezis [**121**] and Feller [**169**], [**170**] are very useful sources of information on facts from convex analysis, functional analysis and probability theory that are being used throughout this book.

The first four sections of this chapter contain background material from classical convexity: the Brunn-Minkowski inequality and its functional forms, mixed volumes and classical geometric inequalities.

Section 1.5 introduces three classical positions of convex bodies: John's position, the minimal mean width position and the minimal surface area position. All of them arise as solutions of extremal problems and can be characterized as satisfying an isotropic condition with respect to an appropriate measure. This relates them to the Brascamp-Lieb inequality and its reverse. In Section 1.6 we discuss Barthe's proof of these inequalities and their applications to geometric problems; an example is K. Ball's sharp reverse isoperimetric inequality.

Section 1.7 introduces the concept of measure concentration and the main examples of metric probability spaces that will be used in this book: the sphere, the Gauss space and the discrete cube. The next two sections survey basic probabilistic tools that we will use: covering numbers and basic inequalities for them, Gaussian and sub-Gaussian processes and bounds for the expectation of their supremum.

The last sections of the chapter give a brief synopsis of the major results of asymptotic convex geometry: Dvoretzky type theorems, the notion of volume ratio and Kashin's theorem, the ℓ-position and Pisier's inequality on the Rademacher projection, the MM^*-estimate, Milman's low M^*-estimate and the quotient of subspace theorem. Finally, we present the reverse Santaló inequality and the reverse Brunn-Minkowski inequality; during this discussion M-ellipsoids and their basic properties are also introduced.

1.1. Convex bodies

We work in $\mathbb{R}^n$, which is equipped with a Euclidean structure $\langle \cdot, \cdot \rangle$. We denote by $\|\cdot\|_2$ the corresponding Euclidean norm, and write B_2^n for the Euclidean unit ball and S^{n-1} for the unit sphere. Volume is denoted by $|\cdot|$. We write ω_n for the volume of B_2^n and σ for the rotationally invariant probability measure on S^{n-1}.

We also denote the Haar measure on $O(n)$ by ν. The Grassmann manifold $G_{n,i}$ of i-dimensional subspaces of $\mathbb{R}^n$ is equipped with the Haar probability measure $\nu_{n,i}$. Let $i \leqslant n$ and $F \in G_{n,i}$. We will denote the orthogonal projection from $\mathbb{R}^n$ onto F by P_F. We also define $B_F = B_2^n \cap F$ and $S_F = S^{n-1} \cap F$.

We say that a subset A of $\mathbb{R}^n$ is convex if $(1-\lambda)x + \lambda y \in A$ for any $x, y \in A$ and any $\lambda \in [0, 1]$. The Minkowski sum of two sets $A, B \subset \mathbb{R}^n$ is defined by

$$A + B = \{a + b : a \in A, b \in B\},$$

and for every $\lambda \in \mathbb{R}$ we set

$$\lambda A = \{\lambda a : a \in A\}.$$

Note that both operations preserve convexity; also, A is convex if and only if $\lambda A + (1-\lambda)A = A$ for every $\lambda \in (0,1)$. We denote by $\tilde{\mathcal{K}}_n$ the convex cone (under Minkowski addition and multiplication by nonnegative real numbers) of all non-empty, compact convex subsets of $\mathbb{R}^n$. In this book we are mainly interested in convex bodies.

1.1.1. Convex bodies

DEFINITION 1.1.1. A *convex body* is a convex subset K of $\mathbb{R}^n$ which is compact and has non-empty interior. The class of convex bodies in $\mathbb{R}^n$ is denoted by $\mathcal{K}_n$.

We say that $K \in \mathcal{K}_n$ is a *symmetric convex body* if $x \in K$ implies that $-x \in K$. We also say that K is *centered* if the *barycenter*

$$\mathrm{bar}(K) = \frac{1}{|K|} \int_K x \, dx$$

of K is at the origin.

DEFINITION 1.1.2. The *support function* of a convex body K (and more generally of a compact convex set) in $\mathbb{R}^n$ is defined by

$$h_K(x) = \sup\{\langle x, y \rangle : y \in K\}$$

for all $x \in \mathbb{R}^n$.

One may check that h_K is positively homogeneous and convex. Note that if K and T are two convex bodies in $\mathbb{R}^n$ then $K \subseteq T$ if and only if $h_K \leqslant h_T$. Given $u \in S^{n-1}$, the quantity $h_K(u) + h_K(-u)$ is the *width* of K in the direction of u.

A compact set K in $\mathbb{R}^n$ will be called *star-shaped* at 0 if it contains the origin in its interior and every line through 0 meets K in a line segment. For such a set, the *radial function* ρ_K is defined for all $x \neq 0$ by

$$\rho_K(x) = \max\{\lambda > 0 : \lambda x \in K\}.$$

If ρ_K is continuous then we say that K is a *star body*. The volume of a star body K can be expressed in polar coordinates as

$$|K| = \omega_n \int_{S^{n-1}} \rho_K^n(\theta) \, d\sigma(\theta).$$

DEFINITION 1.1.3. Let K be a convex body in $\mathbb{R}^n$ with 0 in the interior of K. The *polar body* of K is the set

$$K^\circ = \left\{ y \in \mathbb{R}^n : \sup_{x \in K} \langle x, y \rangle \leqslant 1 \right\}.$$

It is easily checked that K° is a convex body, the mapping $K \mapsto K^\circ$ reverses order, and $K^{\circ\circ} = K$ for every K in the class $\mathcal{K}_n^{(0)}$ of convex bodies which contain 0 in their interior. Note that $(K \cap T)^\circ = \operatorname{conv}(K^\circ \cup T^\circ)$ for all $K, T \in \mathcal{K}_n^{(0)}$.

The natural topology on the space of convex bodies in $\mathbb{R}^n$ is induced by the Hausdorff metric δ^H: More generally, if $K, T \in \tilde{\mathcal{K}}_n$ then we define

$$\delta^H(K,T) = \max\left\{\max_{x \in K} \min_{y \in T} \|x-y\|_2, \max_{x \in T} \min_{y \in K} \|x-y\|_2\right\}.$$

Equivalently,

$$\delta^H(K,T) = \inf\{\delta \geqslant 0 : K \subseteq T + \delta B_2^n \text{ and } T \subseteq K + \delta B_2^n\}$$
$$= \max\{|h_K(u) - h_T(u)| : u \in S^{n-1}\}.$$

This shows that the embedding $K \mapsto h_K$ from $\mathcal{K}_n$ to the space $C(S^{n-1})$ of continuous functions on the sphere is an isometry between $(\mathcal{K}_n, \delta^H)$ and a subset of $C(S^{n-1})$ endowed with the supremum norm. Note that this mapping is positively linear (mapping Minkowski addition to sum of functions) and order-preserving (between inclusion and point-wise inequality). The *Blaschke selection theorem* provides a very useful compactness principle.

THEOREM 1.1.4 (Blaschke). *Let $\{K_j\}$ be a sequence of compact convex sets in $\mathbb{R}^n$. Assume that there exists $R > 0$ such that $K_j \subseteq RB_2^n$ for all j. Then, $\{K_j\}$ has a subsequence which converges to some $K \in \tilde{\mathcal{K}}_n$ with respect to δ^H.*

1.1.2. Symmetric convex bodies

Let K be a symmetric convex body in $\mathbb{R}^n$. The function

$$\|x\|_K = \min\{\lambda \geqslant 0 : x \in \lambda K\}$$

is a norm on $\mathbb{R}^n$. We denote the normed space $(\mathbb{R}^n, \|\cdot\|_K)$ by X_K. Conversely, if $X = (\mathbb{R}^n, \|\cdot\|)$ is a normed space, then its unit ball $K_X = \{x \in \mathbb{R}^n : \|x\| \leqslant 1\}$ is a symmetric convex body in $\mathbb{R}^n$.

The dual norm $\|\cdot\|_*$ of $\|\cdot\|$ is defined by

$$\|y\|_* = \max\{|\langle x,y\rangle| : \|x\| \leqslant 1\}.$$

From the definition it is clear that $|\langle x,y\rangle| \leqslant \|y\|_*\|x\|$ for all $x, y \in \mathbb{R}^n$. If $X^* = (\mathbb{R}^n, \|\cdot\|_*)$ is the dual space of X, then $K_{X^*} = K_X^\circ$. We will use the notation $\|\cdot\|_{K^\circ}$ or $\|\cdot\|_*$, and $\|\cdot\|_K$ or $\|\cdot\|$.

If K and T are two convex bodies in $\mathbb{R}^n$ that contain the origin in their interior, their *geometric distance* $d_G(K,T)$ is defined by

$$d_G(K,T) = \inf\{ab : a, b > 0, K \subseteq bT \text{ and } T \subseteq aK\}.$$

The natural distance between two n-dimensional normed spaces X_K and X_T is the Banach-Mazur distance

$$d_{\mathrm{BM}}(X_K, X_T) = \inf_{A \in GL(n)} \|A : X_K \to X_T\| \|A^{-1} : X_T \to X_K\|.$$

From the definition of the geometric distance we see that

$$d_{\mathrm{BM}}(X_K, X_T) = \inf\{d_G(K, A(T)) : A \in GL(n)\}.$$

In other words, the Banach-Mazur distance $d_{\mathrm{BM}}(X_K, X_T)$ is the smallest positive real λ for which we may find $A \in GL(n)$ such that $K \subseteq A(T) \subseteq \lambda K$. It is clear that

$d_{\mathrm{BM}}(X_K, X_T) \geqslant 1$ with equality if and only if X_K are X_T isometrically isomorphic. Note that $d_{\mathrm{BM}}(X, Z) \leqslant d_{\mathrm{BM}}(X, Y) d_{\mathrm{BM}}(Y, Z)$ for any triple of n-dimensional normed spaces.

If K and T are symmetric convex bodies in $\mathbb{R}^n$ we set $d_{\mathrm{BM}}(K,T) = d_{\mathrm{BM}}(X_K.X_T)$. The definition of the Banach-Mazur distance can be extended to the class of not necessarily symmetric convex bodies as follows: if $K, T \in \mathcal{K}_n$ then we set

$$d_{\mathrm{BM}}(K, T) = \inf\{\lambda > 0 : K - z \subseteq A(T - w) \subseteq \lambda(K - z)\},$$

where the infimum is over all $z, w \in \mathbb{R}^n$ and all $A \in GL(n)$. In the sequel, we usually denote d_{BM} simply by d. Also, the distance from an n-dimensional normed space X to ℓ_2^n will be denoted by $d_X \,(= d(X, \ell_2^n))$, and similarly we set $d_K = d(K, B_2^n)\, (= d_{\mathrm{BM}}(K, B_2^n))$.

1.2. Brunn–Minkowski inequality

The Brunn-Minkowski inequality relates Minkowski addition and volume in $\mathbb{R}^n$.

THEOREM 1.2.1 (Brunn-Minkowski). *Let K and T be two non-empty compact subsets of $\mathbb{R}^n$. Then,*

(1.2.1) $$|K + T|^{1/n} \geqslant |K|^{1/n} + |T|^{1/n}.$$

Theorem 1.2.1 expresses the fact that volume is a concave function with respect to Minkowski addition. For this reason we also write it in the following form: If K and T are non-empty compact subsets of $\mathbb{R}^n$ then for every $\lambda \in (0, 1)$ we have

(1.2.2) $$|\lambda K + (1-\lambda)T|^{1/n} \geqslant \lambda |K|^{1/n} + (1-\lambda)|T|^{1/n}.$$

From (1.2.2) and the arithmetic-geometric means inequality we get

(1.2.3) $$|\lambda K + (1-\lambda)T| \geqslant |K|^\lambda |T|^{1-\lambda}.$$

This form of the Brunn-Minkowski inequality has the advantage of being dimension free. In fact, one can show that it is equivalent to (1.2.1) in the sense that knowing (1.2.3) for all K, T and λ we can then show that the stronger inequality (1.2.1) holds true.

1.2.1. Brunn's principle

There are many interesting proofs of the Brunn-Minkowski inequality. The first one, in chronological order, was restricted to the class of convex bodies and it was based on Brunn's concavity principle.

THEOREM 1.2.2 (Brunn). *Let K be a convex body in $\mathbb{R}^n$ and let F be a k-dimensional subspace. Then, the function $f : F^\perp \to \mathbb{R}$ defined by $f(x) = |K \cap (F + x)|^{1/k}$ is concave on its support.*

For the proof of Brunn's principle we introduce *Steiner symmetrization*. For any convex body K in $\mathbb{R}^n$ and any $\theta \in S^{n-1}$ we consider the set $S_\theta(K)$ consisting of all points of the form $x + \lambda \theta$, where x is in the projection $P_{\theta^\perp}(K)$ of K onto $\theta^\perp$ and $|\lambda| \leqslant \frac{1}{2} \times \mathrm{length}[(x + \mathbb{R}\theta) \cap K]$. In other words, we obtain $S_\theta(K)$ by sliding its chords so that their midpoint will be on $\theta^\perp$ and take the union of all resulting chords. The set $S_\theta(K)$ is the Steiner symmetrization of K in the direction of θ.

From the definition one can check a number of basic properties of Steiner symmetrization which are summarized below:

(i) Steiner symmetrization preserves convexity: if K is a convex body then $S_\theta(K)$ is also a convex body.

(ii) $S_\theta(K)$ can be described as follows:
$$S_\theta(K) = \left\{ x + \frac{t_1 - t_2}{2}\theta : x \in P_{\theta^\perp}K, x + t_1\theta \in K, x + t_2\theta \in K \right\}.$$

(iii) Steiner symmetrization preserves volume: $|S_\theta(K)| = |K|$.

(iv) If K_1 and K_2 are two convex bodies then, for every $\lambda \in (0,1)$,
$$S_\theta(\lambda K_1 + (1-\lambda)K_2) \supseteq \lambda S_\theta(K_1) + (1-\lambda)S_\theta(K_2).$$

A very useful fact is that, given a convex body K in $\mathbb{R}^n$, we can find a sequence $\{\theta_j\}$ of directions so that applying successive Steiner symmetrizations with respect to θ_j we obtain a sequence of convex bodies which converges to a Euclidean ball in the Hausdorff metric. More generally, if F is a k-dimensional subspace of $\mathbb{R}^n$, $1 \leqslant k \leqslant n$, it is a well known fact, which goes back to Steiner and Schwarz, that for every convex body K one can find a sequence of successive Steiner symmetrizations in directions $\theta \in F$ so that the limiting convex body $\tilde{K}$ has the following property: For every $x \in F^\perp$, $\tilde{K} \cap (F + x)$ is a ball with center at x and radius $r(x)$ such that $|\tilde{K} \cap (F+x)| = |K \cap (F+x)|$.

Then, the proof of Theorem 1.2.2 is easily completed: using the convexity of $\tilde{K}$ we see that the function r is concave on its support, and hence f is also concave. $\square$

Proof of Theorem 1.2.1. Brunn's concavity principle implies the Brunn-Minkowski inequality for convex bodies as follows. If K and T are convex bodies in $\mathbb{R}^n$, we define
$$K_1 = K \times \{0\} \quad \text{and} \quad T_1 = T \times \{1\}$$
in $\mathbb{R}^{n+1}$ and consider their convex hull L. If
$$L(t) = \{x \in \mathbb{R}^n : (x,t) \in L\} \quad (t \in [0,1])$$
we easily check that $L(0) = K$, $L(1) = T$ and
$$L(1/2) = \frac{K+T}{2}.$$
Then, Brunn's concavity principle for $F = \mathbb{R}^n$ shows that
$$\left|\frac{K+T}{2}\right|^{1/n} \geqslant \frac{1}{2}|K|^{1/n} + \frac{1}{2}|T|^{1/n},$$
and (1.2.1) is proved. $\square$

1.2.2. Prékopa-Leindler inequality

We describe one more proof of the Brunn-Minkowski inequality, using an inequality of Prékopa and Leindler.

THEOREM 1.2.3 (Prékopa-Leindler). *Let $f, g, h : \mathbb{R}^n \to \mathbb{R}^+$ be measurable functions, and let $\lambda \in (0,1)$. We assume that f and g are integrable, and for every $x, y \in \mathbb{R}^n$*

(1.2.4) $$h(\lambda x + (1-\lambda)y) \geqslant f(x)^\lambda g(y)^{1-\lambda}.$$

Then,
$$\int_{\mathbb{R}^n} h \geqslant \left(\int_{\mathbb{R}^n} f\right)^\lambda \left(\int_{\mathbb{R}^n} g\right)^{1-\lambda}.$$

Proof. We present a proof which uses induction on the dimension n. We first work in dimension one: we may assume that f and g are continuous and strictly positive, and we define $x, y : (0, 1) \to \mathbb{R}$ by the equations
$$\int_{-\infty}^{x(t)} f = t \int_{\mathbb{R}} f \quad \text{and} \quad \int_{-\infty}^{y(t)} g = t \int_{\mathbb{R}} g.$$

In view of our assumptions, x and y are differentiable, and for every $t \in (0, 1)$ we have
$$x'(t) f(x(t)) = \int_{\mathbb{R}} f \quad \text{and} \quad y'(t) g(y(t)) = \int_{\mathbb{R}} g.$$

We now define $z : (0, 1) \to \mathbb{R}$ by
$$z(t) = \lambda x(t) + (1 - \lambda) y(t).$$

Since x and y are strictly increasing, z is also strictly increasing, and the arithmetic-geometric means inequality shows that
$$z'(t) = \lambda x'(t) + (1 - \lambda) y'(t) \geqslant (x'(t))^\lambda (y'(t))^{1-\lambda}.$$

Hence, we can estimate the integral of h making the change of variables $s = z(t)$, as follows:
$$\int_{\mathbb{R}} h = \int_0^1 h(z(t)) z'(t) dt$$
$$\geqslant \int_0^1 h(\lambda x(t) + (1-\lambda) y(t)) (x'(t))^\lambda (y'(t))^{1-\lambda} dt$$
$$\geqslant \int_0^1 f^\lambda(x(t)) g^{1-\lambda}(y(t)) \left(\frac{\int f}{f(x(t))}\right)^\lambda \left(\frac{\int g}{g(y(t))}\right)^{1-\lambda} dt$$
$$= \left(\int_{\mathbb{R}} f\right)^\lambda \left(\int_{\mathbb{R}} g\right)^{1-\lambda}.$$

Next, we assume that $n \geqslant 2$ and the theorem has been proved in all dimensions $k \in \{1, \ldots, n-1\}$. Let f, g and h be as in the theorem. For every $s \in \mathbb{R}$ we define $h_s : \mathbb{R}^{n-1} \to \mathbb{R}^+$ setting $h_s(w) = h(w, s)$, and $f_s, g_s : \mathbb{R}^{n-1} \to \mathbb{R}^+$ in an analogous way. From (1.2.4) it follows that if $x, y \in \mathbb{R}^{n-1}$ and $s_0, s_1 \in \mathbb{R}$ then
$$h_{\lambda s_1 + (1-\lambda) s_0}(\lambda x + (1-\lambda) y) \geqslant f_{s_1}(x)^\lambda g_{s_0}(y)^{1-\lambda},$$
and our inductive hypothesis gives
$$H(\lambda s_1 + (1-\lambda) s_0) := \int_{\mathbb{R}^{n-1}} h_{\lambda s_1 + (1-\lambda) s_0} \geqslant \left(\int_{\mathbb{R}^{n-1}} f_{s_1}\right)^\lambda \left(\int_{\mathbb{R}^{n-1}} g_{s_0}\right)^{1-\lambda}$$
$$=: F^\lambda(s_1) G^{1-\lambda}(s_0).$$

Applying the inductive hypothesis once again, this time with $n = 1$, to the functions F, G and H, we get
$$\int_{\mathbb{R}^n} h = \int_{\mathbb{R}} H \geqslant \left(\int_{\mathbb{R}} F\right)^\lambda \left(\int_{\mathbb{R}} G\right)^{1-\lambda} = \left(\int_{\mathbb{R}^n} f\right)^\lambda \left(\int_{\mathbb{R}^n} g\right)^{1-\lambda}.$$

This completes the proof. $\square$

The dimension free version (1.2.3) of the Brunn-Minkowski inequality is a simple consequence of the Prékopa-Leindler inequality. We consider two non-empty compact subsets K and T of $\mathbb{R}^n$ and, given $\lambda \in (0,1)$, we define $f = \mathbf{1}_K$, $g = \mathbf{1}_T$ and $h = \mathbf{1}_{\lambda K + (1-\lambda)T}$, where $\mathbf{1}_A$ denotes the indicator function of a set A. We check that the assumptions of Theorem 1.2.3 are satisfied, therefore

$$|\lambda K + (1-\lambda)T| = \int_{\mathbb{R}^n} h \geqslant \left(\int_{\mathbb{R}^n} f\right)^\lambda \left(\int_{\mathbb{R}^n} g\right)^{1-\lambda} = |K|^\lambda |T|^{1-\lambda}.$$

1.2.3. Knothe map

We fix an orthonormal basis $\{e_1, \ldots, e_n\}$ in $\mathbb{R}^n$, and consider two open convex bodies K and T. The properties of the *Knothe map* from K to T with respect to the given coordinate system are described in the following theorem.

THEOREM 1.2.4 (Knothe). *There exists a map $\phi : K \to T$ with the following properties:*

(i) *ϕ is triangular: the i-th coordinate function of ϕ depends only on $x_1, \ldots, x_i$. That is,*
$$\phi(x_1, \ldots, x_n) = (\phi_1(x_1), \phi_2(x_1, x_2), \ldots, \phi_n(x_1, \ldots, x_n)).$$

(ii) *The partial derivatives $\frac{\partial \phi_i}{\partial x_i}$ are positive on K, and the Jacobian determinant J_ϕ of ϕ is constant. More precisely, for every $x \in K$,*
$$J_\phi(x) = \prod_{i=1}^n \frac{\partial \phi_i}{\partial x_i}(x) = \frac{|T|}{|K|}.$$

Proof. For each $i = 1, \ldots, n$ and $s = (s_1, \ldots, s_i) \in \mathbb{R}^i$ we consider the section
$$K_s = \{y \in \mathbb{R}^{n-i} : (s,y) \in K\}$$
of K (similarly for T). We shall define a one to one and onto map $\phi : K \to T$ as follows.

Let $x = (x_1, \ldots, x_n) \in K$. Then, $K_{x_1} \neq \emptyset$ and we can define $\phi_1(x) = \phi_1(x_1)$ by
$$\frac{1}{|K|} \int_{-\infty}^{x_1} |K_{s_1}|_{n-1} ds_1 = \frac{1}{|T|} \int_{-\infty}^{\phi_1(x_1)} |T_{t_1}|_{n-1} dt_1.$$
In other words, we move in the direction of e_1 until we "catch" a percentage of T which is equal to the percentage of K occupied by $K \cap \{s = (s_1, \ldots, s_n) : s_1 \leqslant x_1\}$. Note that ϕ_1 is defined on K but $\phi_1(x)$ depends only on the first coordinate of $x \in K$. Also,
$$\frac{\partial \phi_1}{\partial x_1}(x) = \frac{|T|}{|K|} \frac{|K_{x_1}|_{n-1}}{|T_{\phi_1(x_1)}|_{n-1}}.$$
We continue by induction. Assume that we have defined $\phi_1(x) = \phi_1(x_1)$, $\phi_2(x) = \phi_2(x_1, x_2)$ and $\phi_{j-1}(x) = \phi_{j-1}(x_1, \ldots, x_{j-1})$ for some $j \geqslant 2$. If $x = (x_1, \ldots, x_n) \in K$ then $K_{(x_1, \ldots, x_{j-1})} \neq \emptyset$, and we define $\phi_j(x) = \phi_j(x_1, \ldots, x_j)$ by

$$\frac{|T_{(\phi_1(x_1), \ldots, \phi_{j-1}(x_1, \ldots, x_{j-1}))}|_{n-j+1}}{|K_{(x_1, \ldots, x_{j-1})}|_{n-j+1}} \int_{-\infty}^{x_j} |K_{(x_1, \ldots, x_{j-1}, s_j)}|_{n-j} ds_j$$
$$= \int_{-\infty}^{\phi_j(x_1, \ldots, x_j)} |T_{(\phi_1(x_1), \ldots, \phi_{j-1}(x_1, \ldots, x_{j-1}), t_j)}|_{n-j} dt_j.$$

It is clear that
$$\frac{\partial \phi_j}{\partial x_j}(x) = \frac{|T_{(\phi_1(x),\ldots,\phi_{j-1}(x))}|_{n-j+1}}{|K_{(x_1,\ldots,x_{j-1})}|_{n-j+1}} \frac{|K_{(x_1,\ldots,x_j)}|_{n-j}}{|T_{(\phi_1(x),\ldots,\phi_j(x))}|_{n-j}}.$$

Continuing in this way, we obtain a map $\phi = (\phi_1, \ldots, \phi_n) : K \to T$. It is easy to check that ϕ is one to one and onto. Note that
$$\frac{\partial \phi_n}{\partial x_n}(x) = \frac{|T_{(\phi_1(x),\ldots,\phi_{n-1}(x))}|_1}{|K_{(x_1,\ldots,x_{n-1})}|_1}.$$

By construction, ϕ has properties (i) and (ii). $\square$

REMARK 1.2.5. Observe that each choice of coordinate system in $\mathbb{R}^n$ produces a different Knothe map from K onto T. Using the Knothe map one can give one more proof of the Brunn-Minkowski inequality for convex bodies. We may clearly assume that K and T are open. Consider the Knothe map $\phi : K \to T$. It is clear that
$$(I + \phi)(K) \subseteq K + \phi(K) = K + T,$$
and hence, employing property (ii) of ϕ and the arithmetic-geometric means inequality, we write
$$|K + T| \geqslant \int_{(I+\phi)(K)} dx = \int_K |J_{I+\phi}(x)| \, dx$$
$$= \int_K \prod_{j=1}^n \left(1 + \frac{\partial \phi_j}{\partial x_j}(x)\right) dx \geqslant \int_K \left(1 + \left(\prod_{j=1}^n \frac{\partial \phi_j}{\partial x_j}(x)\right)^{1/n}\right)^n dx$$
$$= |K|\left(1 + \left(\frac{|T|}{|K|}\right)^{1/n}\right)^n$$
$$= \left(|K|^{1/n} + |T|^{1/n}\right)^n.$$

1.3. Applications of the Brunn-Minkowski inequality

In this section we collect a few very important geometric inequalities that will be frequently used in this book. Most of them are consequences of the Brunn-Minkowski inequality (and more will appear in subsequent chapters).

1.3.1. An inequality of Rogers and Shephard

The *difference body* of a convex body K is the symmetric convex body
$$K - K = \{x - y \mid x, y \in K\}.$$

From the Brunn-Minkowski inequality it is clear that $|K-K| \geqslant 2^n|K|$ with equality if and only if K has a center of symmetry. Rogers and Shephard gave a sharp upper bound for the volume of the difference body.

THEOREM 1.3.1 (Rogers-Shephard). *Let K be a convex body in $\mathbb{R}^n$. Then,*
$$|K - K| \leqslant \binom{2n}{n}|K|.$$

Proof. We consider the function $f(x) = |K \cap (x+K)|^{1/n}$; by the Brunn-Minkowski inequality this is a concave function supported on $K-K$. Note that every $x \in K-K$ can be written in the form $x = r\theta$, where $\theta \in S^{n-1}$ and $0 \leqslant r \leqslant \rho_{K-K}(\theta)$. We define a second function $g : K-K \to [0,\infty)$ by $g(r\theta) = f(0)(1 - r/\rho_{K-K}(\theta))$. Then, g is linear on the interval $[0, \rho_{K-K}(\theta)\theta]$, it vanishes on the boundary of $K-K$, and $g(0) = f(0)$. Since f is concave, we see that $f \geqslant g$ on $K-K$. Therefore,

$$\int_{K-K} |K \cap (x+K)| dx = \int_{K-K} f^n(x) dx \geqslant \int_{K-K} g^n(x) dx$$

$$= [f(0)]^n n\omega_n \int_{S^{n-1}} \int_0^{\rho_{K-K}(\theta)} r^{n-1}(1 - r/\rho_{K-K}(\theta))^n dr d\sigma(\theta)$$

$$= n\omega_n |K| \int_{S^{n-1}} \rho_{K-K}^n(\theta) \, d\sigma(\theta) \int_0^1 t^{n-1}(1-t)^n dt$$

$$= |K||K-K| \frac{n\Gamma(n)\Gamma(n+1)}{\Gamma(2n+1)}$$

$$= \binom{2n}{n}^{-1} |K||K-K|.$$

On the other hand, Fubini's theorem gives

$$\int_{K-K} |K \cap (x+K)| dx = \int_{\mathbb{R}^n} |K \cap (x+K)| dx$$

$$= \int_{\mathbb{R}^n} \int_{\mathbb{R}^n} \mathbf{1}_K(y) \mathbf{1}_{x+K}(y) dy dx$$

$$= \int_{\mathbb{R}^n} \mathbf{1}_K(y) \left(\int_{\mathbb{R}^n} \mathbf{1}_{y-K}(x) dx \right) dy$$

$$= \int_K |y - K| dy = |K|^2.$$

Combining the above, we conclude the proof. □

Theorem 1.3.1 will be used in the following way: we always have

$$|K - K|^{1/n} \leqslant 4|K|^{1/n},$$

and hence every convex body which contains the origin is contained in a symmetric convex body with the same more or less *volume radius* (for any convex body A in $\mathbb{R}^n$, its volume radius is defined as the radius $\mathrm{v.rad}(A) := (|A|/|B_2^n|)^{1/n}$ of a Euclidean ball that has the same volume as A).

Rogers and Shephard also proved that, when the barycenter of K is at 0, then

$$|K \cap (-K)| \geqslant 2^{-n}|K|.$$

This result implies that every convex body contains a convex body which has a center of symmetry and the same more or less volume radius. Let us also note that Milman and Pajor obtained the following generalization:

THEOREM 1.3.2 (Milman-Pajor). *Let K and L be two convex bodies in $\mathbb{R}^n$ with barycenter at the origin. Then,*

$$|K||L| \leqslant |K+L||K \cap (-L)|.$$

1.3.2. Borell's lemma

Borell's lemma states that if the intersection of $A \cap K$ of a convex body K with a symmetric convex set A captures more than half of the volume of K, then the percentage of K which stays outside tA, $t > 1$ decreases exponentially with respect to t as $t \to \infty$, at a rate which is independent from the body K and the dimension n.

THEOREM 1.3.3 (Borell). *Let K be a convex body of volume 1 in $\mathbb{R}^n$, and let A be a closed, convex and symmetric set such that $|K \cap A| = \delta > \frac{1}{2}$. Then, for every $t > 1$ we have*

$$|K \cap (tA)^c| \leqslant \delta \left(\frac{1-\delta}{\delta}\right)^{\frac{t+1}{2}}.$$

Proof. Observe that

$$A^c \supseteq \frac{2}{t+1}(tA)^c + \frac{t-1}{t+1}A.$$

Otherwise, there exists $a \in A$ which can be written in the form $a = \frac{2}{t+1}y + \frac{t-1}{t+1}a_1$ for some $a_1 \in A$ and $y \notin tA$, and then, using the convexity and symmetry of A, we can write

$$\frac{1}{t}y = \frac{t+1}{2t}a + \frac{t-1}{2t}(-a_1) \in A,$$

which implies that $y \in tA$, a contradiction. Since K is convex, we then have

$$A^c \cap K \supseteq \frac{2}{t+1}((tA)^c \cap K) + \frac{t-1}{t+1}(A \cap K).$$

From the Brunn-Minkowski inequality we get

$$1 - \delta = |A^c \cap K| \geqslant |(tA)^c \cap K|^{\frac{2}{t+1}}|A \cap K|^{\frac{t-1}{t+1}} = |(tA)^c \cap K|^{\frac{2}{t+1}}\delta^{\frac{t-1}{t+1}},$$

and the result follows. $\square$

1.3.3. The isoperimetric inequality for convex bodies

The *surface area* (or Minkowski content) $\partial(K)$ of a convex body K is defined by

$$\partial(K) = \lim_{t \to 0^+} \frac{|K + tB_2^n| - |K|}{t}.$$

It is a well-known fact that among all convex bodies of a given volume the ball has minimal surface area. This is an immediate consequence of the Brunn-Minkowski inequality: If K is a convex body in $\mathbb{R}^n$ and if we write $|K| = |rB_2^n|$ for some $r > 0$, then for every $t > 0$

$$|K + tB_2^n|^{1/n} \geqslant |K|^{1/n} + t|B_2^n|^{1/n} = (r+t)|B_2^n|^{1/n}.$$

It follows that the surface area $\partial(K)$ of K satisfies

$$\partial(K) = \lim_{t \to 0^+} \frac{|K + tB_2^n| - |K|}{t} \geqslant \lim_{t \to 0^+} \frac{(r+t)^n - r^n}{t}|B_2^n|$$
$$= nr^{n-1}|B_2^n|,$$

which shows that

(1.3.1) $$\partial(K) \geqslant n|B_2^n|^{\frac{1}{n}}|K|^{\frac{n-1}{n}}$$

with equality if $K = rB_2^n$. The question of uniqueness in the equality case is more delicate.

Actually, the argument above gives a stronger statement: if $|K| = |rB_2^n|$ then
$$|K + tB_2^n| \geqslant |rB_2^n + tB_2^n|$$
for every $t > 0$. If we fix the volume of K then, *for every $t > 0$*, the t-extension
$$K_t = \{y \mid d(y, K) \leqslant t\}$$
of K has minimal volume if K is a ball.

1.3.4. Blaschke-Santaló inequality

Let K be a symmetric convex body in $\mathbb{R}^n$. Recall that the polar body of K is the symmetric convex body
$$K^\circ = \{y \in \mathbb{R}^n \mid \forall x \in K \ |\langle x, y \rangle| \leqslant 1\}.$$
The *volume product* $s(K)$ of K is defined by
$$s(K) := |K||K^\circ|.$$
Since $(TK)^\circ = (T^{-1})^*(K^\circ)$ for every $T \in GL(n)$, we readily see that $s(K) = s(TK)$. So, the volume product is an invariant of the linear class of K. The Blaschke-Santaló inequality states that $s(K)$ is maximized when K is an ellipsoid.

THEOREM 1.3.4 (Blaschke-Santaló). *Let K be a symmetric convex body in $\mathbb{R}^n$. Then, $|K||K^\circ| \leqslant \omega_n^2$.*

Meyer and Pajor gave a very simple proof of this fact which is based on Steiner symmetrization. The main step is to show that if K is a symmetric convex body in $\mathbb{R}^n$ and if $K_1 = S_\theta(K)$ is the Steiner symmetrization of K in the direction of any $\theta \in S^{n-1}$, then

(1.3.2) $$|K^\circ| \leqslant |(K_1)^\circ|.$$

Since Steiner symmetrization preserves volume, this implies that
$$s(K) \leqslant s(K_1),$$
and then the theorem follows by applying a suitable sequence of Steiner symmetrizations to K.

For the proof of (1.3.2) we may clearly assume that $\theta^\perp = \mathbb{R}^{n-1}$ and then we write
$$S_\theta(K) = K_1 = \left\{ \left(x, \frac{t_1 - t_2}{2} \right) : x \in P_{\theta^\perp} K, (x, t_1) \in K, (x, t_2) \in K \right\}.$$
For every $A \subseteq \mathbb{R}^n$ we write
$$A(t) = \{x \in \mathbb{R}^{n-1} : (x, t) \in A\}.$$
With this notation we show that, for every $s \in \mathbb{R}$,
$$\frac{K^\circ(s) + K^\circ(-s)}{2} \subseteq (K_1)^\circ(s).$$
Then, we apply the Brunn-Minkowski inequality to get
$$|(K_1)^\circ(s)| \geqslant |K^\circ(s)|^{\frac{1}{2}} |K^\circ(-s)|^{\frac{1}{2}},$$
and since $|K^\circ(s)| = |K^\circ(-s)|$ by the symmetry of K°, we see that
$$|(K_1)^\circ(s)| \geqslant |K^\circ(s)|$$

for every $s \in \mathbb{R}$. Integrating with respect to s we have

$$|(K_1)^\circ| = \int_{-\infty}^{+\infty} |(K_1)^\circ(s)| ds \geqslant \int_{-\infty}^{+\infty} |K^\circ(s)| ds = |K^\circ|$$

and (1.3.2) is proved. □

1.3.5. Urysohn's inequality

The *mean width* $w(K)$ of a convex body K in $\mathbb{R}^n$ is defined by

$$w(K) = \int_{S^{n-1}} h_K(u) \, d\sigma(u),$$

where h_K is the support function of K. A classical inequality of Urysohn states that for fixed volume, Euclidean ball has minimal mean width.

THEOREM 1.3.5 (Urysohn). *Let K be a convex body in $\mathbb{R}^n$. Then,*

$$w(K) \geqslant \left(\frac{|K|}{|B_2^n|}\right)^{1/n}.$$

A simple proof of this fact may be given with the method of Steiner symmetrization. The main step is to show that

$$w(S_\theta(K)) \leqslant w(K)$$

for every $\theta \in S^{n-1}$. Urysohn's inequality then follows by applying a suitable sequence of Steiner symmetrizations to K.

We can give a second proof by averaging K using orthogonal transformations. One easily checks that $K_N = \frac{1}{N}\sum_{i=1}^N U_i(K)$ has the same mean width as K. Clearly, by taking U_i to be a net in $O(n)$ one may make sure that K_N converges to a multiple of the Euclidean ball, and thus it must converge to $w(K)B_2^n$. But, by the Brunn-Minkowski inequality, the volume of K_N is greater than the volume of K, and we get the inequality in the limit.

1.4. Mixed volumes

In this section we introduce mixed volumes and survey some of the fundamental results, formulas and inequalities, that will be used in this book.

1.4.1. Minkowski's theorem

Recall that $\tilde{\mathcal{K}}_n$ denotes the convex cone of all non-empty, compact convex subsets of $\mathbb{R}^n$. Minkowski's fundamental theorem on mixed volumes states that there exists a function $V : (\tilde{\mathcal{K}}_n)^n \to \mathbb{R}^+$ satisfying the following conditions:

(i) V has volume as its diagonal: if $K \in \tilde{\mathcal{K}}_n$ then $V(K, \ldots, K) = |K|$.
(ii) V is positive linear in each of its arguments: if $K_1, \ldots, K_i^{(1)}, K_i^{(2)}, \ldots, K_n \in \tilde{\mathcal{K}}_n$ and $t_1, t_2 \geqslant 0$, then

$$V(K_1, \ldots, t_1 K_i^{(1)} + t_2 K_i^{(2)}, \ldots, K_n) = \sum_{j=1}^{2} t_j V(K_1, \ldots, K_i^{(j)}, \ldots, K_n).$$

(iii) V is symmetric: if $K_1, \ldots, K_n \in \tilde{\mathcal{K}}_n$ and σ is any permutation of the indices, then
$$V(K_{\sigma(1)}, \ldots, K_{\sigma(n)}) = V(K_1, \ldots, K_n).$$

The value $V(K_1, \ldots, K_n)$ is called the *mixed volume* of $K_1, \ldots, K_n$. It follows that if $K_1, \ldots, K_m \in \tilde{\mathcal{K}}_n, m \in \mathbb{N}$, then the volume of $t_1 K_1 + \cdots + t_m K_m$ is a homogeneous polynomial of degree n in $t_i > 0$. That is,
$$|t_1 K_1 + \cdots + t_m K_m| = \sum_{1 \leqslant i_1, \ldots, i_n \leqslant m} V(K_{i_1}, \ldots, K_{i_n}) t_{i_1} \cdots t_{i_n}.$$

In particular, if $K, L \in \tilde{\mathcal{K}}_n$ then the function $|K + tL|$ is a polynomial in $t \in [0, \infty)$:

(1.4.1) $$|K + tL| = \sum_{j=0}^{n} \binom{n}{j} V_j(K, L) \, t^j,$$

where $V_j(K, L) = V(K; n-j, L; j)$ is the j-th mixed volume of K and L. Here and elsewhere we use the notation $L; j$ for $L, \ldots, L$ j-times.

From the last formula we obtain
$$V_1(K, L) = \frac{1}{n} \lim_{t \to 0^+} \frac{|K + tL| - |K|}{t},$$

which together with the classical Brunn-Minkowski inequality $|K+tL|^{1/n} \geqslant |K|^{1/n} + t|L|^{1/n}$ implies that
$$V_1(K, L) \geqslant |K|^{\frac{n-1}{n}} |L|^{1/n}$$

for all $K, L \in \tilde{\mathcal{K}}_n$. This is *Minkowski's first inequality*.

1.4.2. Steiner's formula and quermassintegrals

Steiner's formula may be viewed as a special case of (1.4.1). The volume of $K + tB_2^n$, $t > 0$, can be expanded as a polynomial in t:
$$|K + tB_2^n| = \sum_{j=0}^{n} \binom{n}{j} W_j(K) t^j,$$

where
$$W_j(K) = V_j(K, B_2^n) = V(K; n-j, B_2^n; j)$$

is the *j-th quermassintegral* of K. The quermassintegrals W_j inherit properties of mixed volumes: they are monotone, continuous with respect to the Hausdorff metric, and homogeneous of degree $n - j$.

It is easy to see that the surface area of K is given by
$$\partial(K) = n W_1(K).$$

Kubota's integral formula expresses the quermassintegral $W_j(K)$ as an average of the volumes of $(n-j)$-dimensional projections of K:
$$W_j(K) = \frac{\omega_n}{\omega_{n-j}} \int_{G_{n,n-j}} |P_F(K)| d\nu_{n,n-j}(F).$$

Applying this formula for $j = n - 1$ we see that
$$W_{n-1}(K) = \omega_n w(K).$$

It will be convenient for us to work with a normalized variant of $W_{n-j}(K)$: for every $1 \leqslant j \leqslant n$ we set

$$Q_j(K) = \left(\frac{1}{\omega_j} \int_{G_{n,j}} |P_F(K)| \, d\nu_{n,j}(F) \right)^{1/j}.$$

Note that $Q_1(K) = w(K)$. Kubota's formula shows that

$$Q_j(K) = \left(\frac{W_{n-j}(K)}{\omega_n} \right)^{1/j}.$$

1.4.3. Mixed area measures

We fix $K \in \tilde{\mathcal{K}}_n$, and for every $L \in \tilde{\mathcal{K}}_n$ we define $f(h_L) = V_1(K, L)$. We extend f linearly on the subspace $D(S^{n-1}) = \mathrm{span}\{h_L|_{S^{n-1}}, L \in \tilde{\mathcal{K}}_n\}$ of $C(S^{n-1})$. From the additivity of V_1 with respect to L and the fact that $h_{L_1+L_2} = h_{L_1} + h_{L_2}$ whenever $L_1, L_2 \in \tilde{\mathcal{K}}_n$, f is a well-defined positive functional on $D(S^{n-1})$, and hence it extends to a positive functional on $C(S^{n-1})$. By the Riesz representation theorem, we can find a Borel measure σ_K on S^{n-1} for which

$$V_1(K, L) = \frac{1}{n} \int_{S^{n-1}} h_L(u) \, d\sigma_K(u), \qquad L \in \mathcal{K}_n.$$

σ_K is called the *surface area measure* of K. Equivalently, if B is a Borel subset of S^{n-1} then $\sigma_K(B)$ is the $(n-1)$-dimensional surface measure of the set of all boundary points of K at which there exists an exterior normal in B (for a polytope K with facets $F_1, \ldots, F_m$ having exterior normals $u_1, \ldots, u_m$ respectively, σ_K is the measure supported by $\{u_1, \ldots, u_m\}$ with $\sigma_K(\{u_j\}) = |F_j|, j = 1, \ldots, m$). As a consequence of the integral representation for $V_1(K, L)$ we also see that $L_1 \subseteq L_2$ implies $V_1(K, L_1) \leqslant V_1(K, L_2)$.

More generally, the *mixed area measures* were introduced by Alexandrov and may be viewed as a local generalization of the mixed volumes. For any $(n-1)$-tuple $\mathcal{C} = (K_1, \ldots, K_{n-1})$ of elements of $\tilde{\mathcal{K}}_n$, the Riesz representation theorem guarantees the existence of a Borel measure $S(\mathcal{C}, \cdot)$ on the unit sphere S^{n-1} such that

$$V(L, K_1, \ldots, K_{n-1}) = \frac{1}{n} \int_{S^{n-1}} h_L(u) \, dS(\mathcal{C}, u)$$

for every $L \in \tilde{\mathcal{K}}_n$. The local analogue of Minkowski's theorem is

$$\sigma_{\sum_{i=1}^m t_i K_i}(B) = \sum_{1 \leqslant i_1, \ldots, i_n \leqslant m} S(K_{i_1}, \ldots, K_{i_{n-1}}, \omega) t_{i_1} \ldots t_{i_n}$$

for all Borel sets $B \subseteq S^{n-1}$ and all $t_i > 0$, $K_i \in \tilde{\mathcal{K}}_n$, $m \in \mathbb{N}$.

The *j-th area measure* of K is defined by $S_j(K, \cdot) = S(K; j, B_2^n; n-j-1, \cdot)$, $j = 0, 1, \ldots, n-1$. It follows that the quermassintegrals of K can be represented by

$$W_j(K) = \frac{1}{n} \int_{S^{n-1}} h_K(u) \, dS_{n-j-1}(K, u), \qquad j = 0, 1, \ldots, n-1$$

or alternatively,

$$W_j(K) = \frac{1}{n} \int_{S^{n-1}} dS_{n-j}(K, u), \qquad i = 1, \ldots, n.$$

1.4.4. The Alexandrov-Fenchel inequalities

The *Alexandrov-Fenchel inequality* generalizes the Brunn-Minkowski inequality and its consequences. It states that if $K, L, K_3, \ldots, K_n \in \hat{\mathcal{K}}_n$, then

$$V(K, L, K_3, \ldots, K_n)^2 \geqslant V(K, K, K_3, \ldots, K_n) V(L, L, K_3, \ldots, K_n).$$

From this inequality one can recover the Brunn-Minkowski inequality as well as the following generalization for the quermassintegrals:

$$W_j(K+L)^{\frac{1}{n-j}} \geqslant W_j(K)^{\frac{1}{n-j}} + W_j(L)^{\frac{1}{n-j}}, \qquad j = 0, \ldots, n-1$$

for any pair of convex bodies in $\mathbb{R}^n$.

Steiner's formula and the Brunn-Minkowski inequality show that

$$\sum_{j=0}^{n} \binom{n}{j} \frac{W_j(K)}{|B_2^n|} t^j = \frac{|K + tB_2^n|}{|B_2^n|} \geqslant \left(\left(\frac{|K|}{|B_2^n|} \right)^{1/n} + t \right)^n$$

$$= \sum_{j=0}^{n} \binom{n}{j} \left(\frac{|K|}{|B_2^n|} \right)^{\frac{n-j}{n}} t^j$$

for every $t > 0$. Since the first and the last term are equal on both sides of this inequality, we must have

$$\frac{W_1(K)}{|B_2^n|} \geqslant \left(\frac{|K|}{|B_2^n|} \right)^{\frac{n-1}{n}}$$

which is the isoperimetric inequality for convex bodies, and

$$w(K) = \frac{W_{n-1}(K)}{|B_2^n|} \geqslant \left(\frac{|K|}{|B_2^n|} \right)^{\frac{1}{n}},$$

which is Urysohn's inequality. Both inequalities are special cases of the set of *Alexandrov inequalities*

$$\left(\frac{W_i(K)}{|B_2^n|} \right)^{\frac{1}{n-i}} \geqslant \left(\frac{W_j(K)}{|B_2^n|} \right)^{\frac{1}{n-j}} \qquad n > i > j \geqslant 0.$$

This implies that the sequence $(W_0(K), \ldots, W_n(K))$ is log-concave: we have

$$W_j^{k-i} \geqslant W_i^{k-j} W_k^{j-i}$$

if $0 \leqslant i < j < k \leqslant n$. From these inequalities one can check that $Q_j(K)$ is a decreasing function of j.

1.4.5. Projection bodies

Minkowski's existence theorem states that if $u_1, \ldots, u_m$ are distinct unit vectors with the origin in the interior of their convex hull, and if $\gamma_1, \ldots, \gamma_m$ are positive real numbers with $\sum_{j=1}^{m} \gamma_j u_j = 0$, then there exists a polytope K having $u_1, \ldots, u_m$ as its (only) normal vectors and satisfying $|F_j| = \gamma_j, j = 1, \ldots, m$, where F_j is the facet of K corresponding to u_j. By approximation one obtains that a finite Borel measure μ on S^{n-1} is the surface area measure of some $K \in \mathcal{K}_n$ if and only if μ is not concentrated on any great subsphere of S^{n-1}, and

$$\int_{S^{n-1}} u \, d\mu(u) = 0.$$

Note that the second condition is always satisfied if μ is an even measure.

If $K \in \mathcal{K}_n$, the *projection body* ΠK of K is the symmetric convex body whose support function is defined by
$$h_{\Pi K}(\theta) = |P_\theta(K)|, \qquad \theta \in S^{n-1}.$$
One has the integral representation
$$|P_\theta(K)| = \frac{1}{2} \int_{S^{n-1}} |\langle u, \theta \rangle| \, d\sigma_K(u),$$
which is easily verified in the case of a polytope, and extends to any $K \in \mathcal{K}_n$ by approximation (*Cauchy's formula*). It follows that the support function of a projection body is given by
$$h_{\Pi K}(\theta) = \int_{S^{n-1}} |\langle u, \theta \rangle| \, d\mu(u)$$
for some finite even Borel measure on S^{n-1}. This also shows that all projection bodies are indeed convex. The integral representation of $h_{\Pi K}$ also shows that ΠK is a zonoid: a Hausdorff limit of a sequence of finite Minkowski sums of line segments. Minkowski's existence theorem implies that, conversely, every zonoid is the projection body of some symmetric convex body in $\mathbb{R}^n$. Moreover, if we denote by $\mathcal{Z}$ the class of zonoids, *Alexandrov's uniqueness theorem* shows that the Minkowski map $\Pi : \mathcal{C}_n \to \mathcal{Z}$ with $K \to \Pi K$, is injective. Note also that $\mathcal{Z}$ is invariant under invertible linear transformations and closed in the Hausdorff metric.

1.5. Classical positions of convex bodies

The family of positions of a convex body K in $\mathbb{R}^n$ is the class $\{z + T(K) : z \in \mathbb{R}^n, T \in GL(n)\}$. The right choice of a position is often quite important for the study of affinely invariant quantities. For example, let K be a symmetric convex body in $\mathbb{R}^n$ and consider the volume product $s(K) = |K||K^\circ|$. The Blaschke-Santaló inequality (Theorem 1.3.4) asserts that $s(K)$ is maximized if and only if K is an ellipsoid (note that $s(K)$ is invariant under $GL(n)$). On the other hand, a simple application of Hölder's inequality shows that, for every symmetric convex body A in $\mathbb{R}^n$,
$$\frac{|A|}{|B_2^n|} = \int_{S^{n-1}} \|\theta\|_A^{-n} d\sigma(\theta) \geqslant w(A^\circ)^{-n}.$$
This implies that
$$\left(\frac{s(B_2^n)}{s(K)} \right)^{1/n} \leqslant \min_{T \in GL(n)} w(TK)w((TK)^\circ).$$
Therefore, in order to obtain a "reverse Blaschke-Santaló inequality" it is useful to study the quantity
$$\max_K \min_{T \in GL(n)} w(TK)w((TK)^\circ),$$
or equivalently, to study the position $\tilde{K}$ of K which minimizes $w(TK)w((TK)^\circ)$ over all $T \in GL(n)$.

In this section we introduce three classical positions of convex bodies. All of them arise as solutions of extremal problems of the following type: we normalize the volume of K to be 1 and ask for the maximum or minimum of $f(TK)$ over all $T \in SL(n)$, where f is some functional on convex bodies (in the example above, f is the product of the mean widths of a body and its polar). An interesting feature of this procedure, which was put forward in [**204**], is that a simple variational method leads to a geometric description of the extremal position, and that in many cases this position satisfies an isotropic condition for an appropriate measure on S^{n-1}.

DEFINITION 1.5.1. A Borel measure μ on S^{n-1} is called *isotropic* if

$$\int_{S^{n-1}} \langle x, \theta \rangle^2 d\mu(x) = \frac{\mu(S^{n-1})}{n} \tag{1.5.1}$$

for every $\theta \in S^{n-1}$. We will make frequent use of the next standard lemma.

LEMMA 1.5.2. *Let μ be a Borel measure on S^{n-1}. The following are equivalent:*
  (i) *μ is isotropic.*
  (ii) *For every $i, j = 1, \ldots, n$,*

$$\int_{S^{n-1}} \phi_i \phi_j d\mu(\phi) = \frac{\mu(S^{n-1})}{n} \delta_{i,j}. \tag{1.5.2}$$

  (iii) *For every linear transformation $T : \mathbb{R}^n \to \mathbb{R}^n$ (we will write $T \in L(\mathbb{R}^n)$),*

$$\int_{S^{n-1}} \langle \phi, T\phi \rangle d\mu(\phi) = \frac{\operatorname{tr}(T)}{n} \mu(S^{n-1}). \tag{1.5.3}$$

Proof. Setting $\theta = e_i$ and $\theta = \frac{e_i + e_j}{\sqrt{2}}$ in (1.5.1) we get (1.5.2). On observing that if $T = (t_{ij})_{i,j=1}^n$ then $\langle \phi, T\phi \rangle = \sum_{i,j=1}^n t_{ij} \phi_i \phi_j$, we readily see that (1.5.2) implies (1.5.3). Finally, note that applying (1.5.3) with $T(\phi) = \langle \phi, \theta \rangle \theta$ we get (1.5.1). □

1.5.1. John's position

Given a convex body K in $\mathbb{R}^n$, we consider the family $\mathcal{E}(K)$ of the *ellipsoids* which are contained in K. An ellipsoid in $\mathbb{R}^n$ is a convex body of the form

$$\mathcal{E} = \left\{ x \in \mathbb{R}^n : \sum_{i=1}^n \frac{\langle x, v_i \rangle^2}{\alpha_i^2} \leqslant 1 \right\},$$

where $\{v_i\}_{i \leqslant n}$ is an orthonormal basis of $\mathbb{R}^n$, and $\alpha_1, \ldots, \alpha_n$ are positive reals (the directions and lengths of the semiaxes of $\mathcal{E}$, respectively). It is easy to check that $\mathcal{E} = T(B_2^n)$, where T is the linear transformation of $\mathbb{R}^n$ defined by $T(v_i) = \alpha_i v_i$, $i = 1, \ldots, n$. Therefore, the volume of $\mathcal{E}$ is equal to

$$|\mathcal{E}| = |B_2^n| \prod_{i=1}^n \alpha_i.$$

The *volume ratio* of K is the quantity

$$vr(K) = \inf \left\{ \left(\frac{|K|}{|\mathcal{E}|} \right)^{1/n} : \mathcal{E} \subseteq K \right\},$$

where the infimum is taken over all the ellipsoids which are contained in K. One can show that there is a unique ellipsoid $\mathcal{E}$ of maximal volume which is contained in K. We will say that $\mathcal{E}$ is the *maximal volume ellipsoid* of K. Moreover, there is a unique ellipsoid $\mathcal{E}$ which contains K and has minimal volume (the *minimal volume ellipsoid* of K).

Assume that B_2^n is the maximal volume ellipsoid of K; then we say that K is in John's position. We will say that $x \in \mathbb{R}^n$ is a *contact point* of K and B_2^n if

$$\|x\|_2 = \|x\|_K = \|x\|_{K^\circ} = 1.$$

John's theorem describes the distribution of contact points on the unit sphere S^{n-1}.

THEOREM 1.5.3 (John). *If B_2^n is the maximal volume ellipsoid of the symmetric convex body K in $\mathbb{R}^n$, then there exist contact points $x_1, \ldots, x_m$ of K and B_2^n, and positive real numbers $c_1, \ldots, c_m$ such that*
$$x = \sum_{j=1}^{m} c_j \langle x, x_j \rangle x_j$$
for every $x \in \mathbb{R}^n$.

REMARK 1.5.4. Theorem 1.5.3 says that the identity operator I of $\mathbb{R}^n$ can be represented in the form

(1.5.4) $$I = \sum_{j=1}^{m} c_j x_j \otimes x_j,$$

where $x_j \otimes x_j$ is the projection in the direction of x_j: $(x_j \otimes x_j)(x) = \langle x, x_j \rangle x_j$. Note that for every $x \in \mathbb{R}^n$
$$\|x\|_2^2 = \langle x, x \rangle = \sum_{j=1}^{m} c_j \langle x, x_j \rangle^2.$$

Also, if we choose $x = e_i$, $i = 1, \ldots, n$, where $\{e_i\}$ is the standard orthonormal basis of $\mathbb{R}^n$, we have
$$n = \sum_{i=1}^{n} \|e_i\|_2^2 = \sum_{i=1}^{n} \sum_{j=1}^{m} c_j \langle e_i, x_j \rangle^2 = \sum_{j=1}^{m} c_j \sum_{i=1}^{n} \langle e_i, x_j \rangle^2$$
$$= \sum_{j=1}^{m} c_j \|x_j\|_2^2 = \sum_{j=1}^{m} c_j.$$

Sketch of the proof of Theorem 1.5.3. We follow the presentation of K. Ball from [**39**]; Remark 1.5.4 shows that a necessary condition for a representation of the form (1.5.4) is that $\sum_{j=1}^{m} \frac{c_j}{n} = 1$. Our purpose is then to show that I/n is in the convex hull of the set of all matrices that have the form $x \otimes x$ for some contact point x of K and B_2^n. To this end, we define
$$\mathcal{C} = \{x \otimes x : \|x\|_2 = \|x\|_K = 1\},$$
and show that $I/n \in \mathrm{conv}(\mathcal{C})$. Note that $\mathrm{conv}(\mathcal{C})$ is a non-empty compact convex subset of $\mathbb{R}^{n^2}$.

Assume that $I/n \notin \mathrm{conv}(\mathcal{C})$. Using the Hahn-Banach theorem, one can prove the next lemma.

LEMMA 1.5.5. *If $I/n \notin \mathrm{conv}(\mathcal{C})$, there exist $s > 0$ and B symmetric with $\mathrm{tr}(B) = 0$, such that*
$$\langle B, x \otimes x \rangle \geqslant s$$
for every $x \otimes x \in \mathcal{C}$.

Then, we consider $\delta > 0$ small enough, and define the ellipsoid
$$\mathcal{E}_\delta = \{x \in \mathbb{R}^n : \langle (I + \delta B)x, x \rangle \leqslant 1\}.$$
The next step is to show that $\mathcal{E}_\delta \subseteq K$ if δ is small enough.

LEMMA 1.5.6. *There exists $\delta_0 > 0$ such that $\mathcal{E}_\delta \subseteq K$ for every $0 < \delta < \delta_0$.*

This leads to a contradiction. Choose $\delta > 0$ so small that $I + \delta B$ is positive definite and the ellipsoid $\mathcal{E}_\delta$ is contained in K. Since B_2^n is the maximal volume ellipsoid of K, we have $|\mathcal{E}_\delta| \leqslant |B_2^n|$. On the other hand, if $I + \delta B = S_\delta^2$, then

$$|\mathcal{E}_\delta| = |S_\delta^{-1}(B_2^n)| = |B_2^n|/\sqrt{\det(I + \delta B)}.$$

It follows that $\det(I + \delta B) \geqslant 1$. By the arithmetic-geometric means inequality

$$[\det(I + \delta B)]^{\frac{1}{n}} \leqslant \frac{\operatorname{tr}(I + \delta B)}{n} = 1 + \delta \frac{\operatorname{tr}(B)}{n} = 1,$$

because $\operatorname{tr}(B) = 0$. This means that we have equality in the arithmetic-geometric means inequality, and this implies that $I + \delta B$ is a multiple of the identity: $I + \delta B = \mu I$. But then, B is a multiple of the identity, and since $\operatorname{tr}(B) = 0$ we get $B = 0$.

This contradicts Lemma 1.5.5, because $\langle Bx, x \rangle \geqslant s > 0$ for all $x \otimes x \in \mathcal{C}$. Therefore, $I/n \in \operatorname{conv}(\mathcal{C})$. To finish the proof, we have to show that $\|x_j\|_2 = \|x_j\|_K = \|x_j\|_{K^\circ} = 1$ for all j. Since $x_j \in S^{n-1}$, we have

$$1 = \langle x_j, x_j \rangle \leqslant \|x_j\|_K \|x_j\|_{K^\circ} = \|x_j\|_{K^\circ}, \qquad j = 1, \ldots, m.$$

On the other hand, at each x_j, K and B_2^n have the same supporting hyperplane with normal vector x_j. Therefore, for every $x \in K$ we have $\langle x, x_j \rangle \leqslant 1$, and by the symmetry of K, $|\langle x, x_j \rangle| \leqslant 1$. It follows that $\|x_j\|_K = \|x_j\|_{K^\circ} = \|x_j\|_2 = 1$ for all $j = 1, \ldots, m$. □

Theorem 1.5.3 implies

$$\sum_{i=1}^m c_i \langle x_i, \theta \rangle^2 = 1$$

for every $\theta \in S^{n-1}$. In the terminology of Definition 1.5.1 the measure μ on S^{n-1} that gives mass c_i to the point x_i, $i = 1, \ldots, m$, is isotropic. In this sense, John's position is an *isotropic position*. Conversely, one can show that if K is a symmetric convex body in $\mathbb{R}^n$ which contains the Euclidean unit ball B_2^n and if there exists an isotropic Borel measure μ on S^{n-1} which is supported by the contact points of K and B_2^n, then B_2^n is the maximal volume ellipsoid of K.

A well-known consequence of Theorem 1.5.3 (which is usually called *John's theorem*) states that if K is a symmetric convex body in $\mathbb{R}^n$ and $\mathcal{E}$ is the maximal volume ellipsoid of K, then $K \subseteq \sqrt{n}\mathcal{E}$. This is equivalent to the next proposition.

THEOREM 1.5.7 (John). *If B_2^n is the maximal volume ellipsoid of K, then $K \subset \sqrt{n}B_2^n$.*

Proof. Consider the representation of the identity

$$x = \sum_{j=1}^m c_j \langle x, x_j \rangle x_j$$

of Theorem 1.5.3. We will use the fact that $\|x_j\|_K = \|x_j\|_{K^\circ} = \|x_j\|_2 = 1$, $j = 1, \ldots, m$. For every $x \in K$ we have

$$\|x\|_2^2 = \sum_{j=1}^m c_j \langle x, x_j \rangle^2 \leqslant \sum_{j=1}^m c_j = n.$$

This shows that $\|x\|_2 \leqslant \sqrt{n}$. Therefore, $B_2^n \subseteq K \subseteq \sqrt{n}B_2^n$. □

John's theorem can be extended to the case of not necessarily symmetric convex bodies.

THEOREM 1.5.8. *Let K be a convex body in $\mathbb{R}^n$ such that B_2^n is the ellipsoid of maximal volume inscribed in K. We can find contact points $x_1, \ldots, x_m$ of K and B_2^n, and positive reals $c_1, \ldots, c_m$, such that: $\sum_{j=1}^m c_j x_j = 0$ and*

$$I = \sum_{j=1}^m c_j x_j \otimes x_j.$$

Löwner's position is dual to John's position. We say that a convex body K is in Löwner's position if the ellipsoid of minimal volume containing K is the Euclidean unit ball B_2^n. By John's theorem, K is in Löwner's position if and only if $K \subseteq B_2^n$ and there exist $x_1, \ldots, x_m \in \mathrm{bd}(K) \cap S^{n-1}$ and positive real numbers $c_1, \ldots, c_m$ such that the measure μ on S^{n-1} which is supported by $\{x_1, \ldots, x_m\}$ and gives mass c_j to $\{x_j\}$, $j = 1, \ldots, m$, is isotropic.

1.5.2. Dvoretzky-Rogers lemmas

Assume that B_2^n is the ellipsoid of maximal volume contained in the symmetric convex body K. Starting from John's representation of the identity $I = \sum_{j=1}^m c_j x_j \otimes x_j$ of Theorem 1.5.3, Dvoretzky and Rogers obtained precise information on the distribution of the contact points x_j on the Euclidean unit sphere. There are several results of this type, known as "Dvoretzky-Rogers lemmas". An example is given by Theorem 1.5.10 for which we sketch a proof.

LEMMA 1.5.9. *If B_2^n is the maximal volume ellipsoid of the symmetric convex body K then for every $T \in L(\mathbb{R}^n)$ there exists a contact point y of K and B_2^n which satisfies*

$$\langle y, Ty \rangle \geqslant \frac{\mathrm{tr}(T)}{n}.$$

Proof. We have

$$\mathrm{tr}\, T = \langle T, I \rangle = \sum_{j=1}^m c_j \langle T, x_j \otimes x_j \rangle.$$

Since $\sum_{j=1}^m c_j = n$, we may clearly find y among the x_j's which satisfies

$$\langle y, Ty \rangle = \langle T, y \otimes y \rangle \geqslant \frac{\mathrm{tr}(T)}{n}.$$

THEOREM 1.5.10 (Dvoretzky-Rogers). *If B_2^n is the maximal volume ellipsoid of the symmetric convex body K then there exists an orthonormal sequence $z_1, \ldots, z_n$ in $\mathbb{R}^n$ such that*

$$\left(\frac{n-i+1}{n}\right)^{1/2} \leqslant \|z_i\| \leqslant \|z_i\|_2 = 1$$

for all $1 \leqslant i \leqslant n$.

Proof. We define the z_i's inductively; z_1 can be any contact point of K and B_2^n. Assume that $z_1, \ldots, z_k$ have been chosen for some $k < n$.

We set $F_k = \mathrm{span}\{z_1, \ldots, z_k\}$. Then, $\mathrm{tr}(P_{F_k^\perp}) = n - k$, and applying Lemma 1.5.9 we find a contact point y_{k+1} with

$$\|P_{F_k^\perp} y_{k+1}\|_2^2 = \langle y_{k+1}, P_{F_k^\perp} y_{k+1} \rangle \geqslant \frac{n-k}{n}.$$

It follows that $\|P_{F_k} y_{k+1}\| \leqslant \|P_{F_k} y_{k+1}\|_2 \leqslant \sqrt{k/n}$.

We define $z_{k+1} = P_{F_k^\perp} y_{k+1}/\|P_{F_k^\perp} y_{k+1}\|_2$. Then,

$$1 = \|z_{k+1}\|_2 \geqslant \|z_{k+1}\| \geqslant |\langle y_{k+1}, z_{k+1}\rangle| = \left\|P_{F_k^\perp} y_{k+1}\right\|_2 \geqslant \left(\frac{n-k}{n}\right)^{1/2}.$$

The next corollary of Theorem 1.5.10 will play an important role in the proof of Dvoretzky theorem (see Section 1.10.1).

COROLLARY 1.5.11. *Assume that B_2^n is the maximal volume ellipsoid of the symmetric convex body K. If $k = \lfloor n/2 \rfloor + 1$, then we can find orthonormal vectors $z_1, \ldots, z_k$ such that*

$$\frac{1}{\sqrt{2}} \leqslant \|z_j\| \leqslant 1$$

for all $j = 1, \ldots, k$.

1.5.3. Minimal mean width position

Let K be a convex body in $\mathbb{R}^n$ (without loss of generality we may assume that $0 \in \mathrm{int}\,(K)$). Recall that the mean width of K is the quantity

$$w(K) = \int_{S^{n-1}} h_K(x)\, d\sigma(x).$$

We say that K has *minimal mean width* if $w(K) \leqslant w(TK)$ for every $T \in SL(n)$.

We assume for simplicity that h_K is twice continuously differentiable (we then say that K is *smooth enough*) and we consider the measure ν_K on S^{n-1} with density h_K with respect to σ. The next theorem characterizes the minimal mean width position.

THEOREM 1.5.12 (Giannopoulos-Milman). *A smooth enough convex body K in $\mathbb{R}^n$ has minimal mean width if and only if*

$$\int_{S^{n-1}} h_K(x)\langle x, \theta\rangle^2 d\sigma(x) = \frac{w(K)}{n}$$

for every $\theta \in S^{n-1}$ (equivalently, if ν_K is isotropic). Moreover, this minimal mean width position is unique up to orthogonal transformations.

1.5.4. Minimal surface area position

Let K be a convex body of volume 1 in $\mathbb{R}^n$. As in the previous subsection, we consider the problem to find the minimum of the surface area $\partial(TK)$ over all $T \in SL(n)$. This minimum is attained for some T_0; we denote it by ∂_K and we call it the *minimal surface invariant* of K. We say that K has minimal surface area if $\partial(K) = \partial_K |K|^{\frac{n-1}{n}}$.

Recall the definition of the area measure σ_K of K from Section 1.4.3. It is defined on S^{n-1} and corresponds to the usual surface measure on K via the Gauss map: For every Borel $A \subseteq S^{n-1}$, we have

$$\sigma_K(A) = \nu\left(\{x \in \mathrm{bd}(K) : \text{the outer normal to } K \text{ at } x \text{ is in } A\}\right),$$

where ν is the $(n-1)$-dimensional surface measure on K. We obviously have $\partial(K) = \sigma_K(S^{n-1})$.

A characterization of the minimal surface position through the area measure was given by Petty.

THEOREM 1.5.13 (Petty). *Let K be a convex body of volume 1 in $\mathbb{R}^n$. Then, $\partial(K) = \partial_K$ if and only if σ_K is isotropic. Moreover, this minimal surface area position is unique up to orthogonal transformations.*

1.6. Brascamp-Lieb inequality and its reverse form

1.6.1. Brascamp-Lieb inequality

The Brascamp-Lieb inequality estimates the norm of the multilinear operator $I : L^{p_1}(\mathbb{R}) \times \cdots \times L^{p_m}(\mathbb{R}) \to \mathbb{R}$ defined by

$$I(f_1, \ldots, f_m) = \int_{\mathbb{R}^n} \prod_{j=1}^m f_j(\langle u_j, x\rangle)\, dx,$$

where $m \geqslant n$, $p_1, \ldots, p_m \geqslant 1$ with $\frac{1}{p_1} + \cdots + \frac{1}{p_m} = n$, and $u_1, \ldots, u_m \in \mathbb{R}^n$. Brascamp and Lieb proved that the norm of I is the supremum D of

$$\frac{I(g_1, \ldots, g_m)}{\prod_{j=1}^m \|g_j\|_{p_j}}$$

over all centered Gaussian functions $g_1, \ldots, g_m$, i.e. over all functions of the form $g_j(t) = e^{-\lambda_j t^2}$, $\lambda_j > 0$. This fact is a generalization of Young's convolution inequality $\|f * g\|_r \leqslant C_{p,q} \|f\|_p \|g\|_q$ for all $f \in L^p(\mathbb{R})$ and $g \in L^q(\mathbb{R})$, where $p, q, r \geqslant 1$ and $1/p + 1/q = 1 + 1/r$.

If we set $c_j = 1/p_j$ and replace f_j by $f_j^{c_j}$ then we can state the Brascamp-Lieb inequality in the following form.

THEOREM 1.6.1 (Brascamp-Lieb). *Let $m \geqslant n$, and let $u_1, \ldots, u_m \in \mathbb{R}^n$ and $c_1, \ldots, c_m > 0$ with $c_1 + \cdots + c_m = n$. Then,*

$$\int_{\mathbb{R}^n} \prod_{j=1}^m f_j^{c_j}(\langle x, u_j\rangle)\, dx \leqslant D \prod_{j=1}^m \left(\int_{\mathbb{R}} f_j\right)^{c_j}$$

for all integrable functions $f_j : \mathbb{R} \to [0, \infty)$.

Direct computation of the ratio $I(g_1, \ldots, g_m)/\prod_{j=1}^m \|g_j\|_{p_j}$ for Gaussian functions $g_j(t) = e^{-\lambda_j t^2}$ shows that $D = 1/\sqrt{F}$ where

$$F = \inf\left\{\frac{\det\left(\sum_{j=1}^m c_j \lambda_j u_j \otimes u_j\right)}{\prod_{j=1}^m \lambda_j^{c_j}} \;\Big|\; \lambda_j > 0\right\}.$$

1.6.2. Barthe's proof

A reverse form of Theorem 1.6.1 was proved by Barthe.

THEOREM 1.6.2 (Barthe). *Let $m \geqslant n$, $c_1, \ldots, c_m > 0$ with $c_1 + \cdots + c_m = n$, and $u_1, \ldots, u_m \in \mathbb{R}^n$. If $h_1, \ldots, h_m : \mathbb{R} \to [0, \infty)$ are measurable functions, we set*

$$K(h_1, \ldots, h_m) = \int_{\mathbb{R}^n}^* \sup\left\{\prod_{j=1}^m h_j^{c_j}(\theta_j) \;\Big|\; \theta_j \in \mathbb{R}, \; x = \sum_{j=1}^m \theta_j c_j u_j\right\} dx,$$

where $\int^*$ denotes the outer integral. Then,

$$\inf\left\{K(h_1, \ldots, h_m) \;\Big|\; \int_{\mathbb{R}} h_j = 1, \; j = 1, \ldots, m\right\} = \sqrt{F}.$$

1.6. BRASCAMP-LIEB INEQUALITY AND ITS REVERSE FORM

Although the Brascamp-Lieb inequality and its reverse form do not play a central role in this book, we sketch Barthe's argument which is very elegant, short and related in spirit with arguments that appear in subsequent chapters. A remarkable fact is that it gives a new direct proof of the Brascamp-Lieb inequality as well.

The first observation is that
$$\inf\left\{K(h_1,\ldots,h_m) \mid \int_\mathbb{R} h_j = 1,\ j=1,\ldots,m\right\} \leqslant \sqrt{F}.$$
This follows by direct computation with Gaussian functions. The main step in Barthe's argument is the following proposition.

PROPOSITION 1.6.3 (Barthe). *Let $f_1,\ldots,f_m : \mathbb{R} \to [0,\infty)$ and $h_1,\ldots,h_m : \mathbb{R} \to [0,\infty)$ be integrable functions with*
$$\int_\mathbb{R} f_j(t)\,dt = \int_\mathbb{R} h_j(t)\,dt = 1, \qquad j=1,\ldots,m.$$
Then,
$$F \cdot I(f_1,\ldots,f_m) \leqslant K(h_1,\ldots,h_m).$$

Proof. We may assume that f_j, h_j are continuous and strictly positive, and also that F is finite and positive. The key idea is to use a transportation of measure argument, which resembles the proof of the Prékopa-Leindler inequality in Section 1.2b. For every $j=1,\ldots,m$ we define $T_j : \mathbb{R} \to \mathbb{R}$ by the equation
$$\int_{-\infty}^{T_j(t)} h_j(s)\,ds = \int_{-\infty}^{t} f_j(s)\,ds.$$
Observe that each T_j is strictly increasing, 1-1 and onto, and
$$T_j'(t) h_j(T_j(t)) = f_j(t), \qquad t \in \mathbb{R}.$$
Then we define $W : \mathbb{R}^n \to \mathbb{R}^n$ by
$$W(y) = \sum_{j=1}^m c_j T_j(\langle y, u_j \rangle) u_j.$$
We check that $J_W(y) = \sum_{j=1}^m c_j T_j'(\langle y, u_j \rangle) u_j \otimes u_j$ and this implies that $\langle [J_W(y)](v), v \rangle > 0$ if $v \neq 0$, which shows that W is injective. We define
$$m(x) = \sup\left\{\prod_{j=1}^m h_j^{c_j}(\theta_j) \mid x = \sum_{j=1}^m \theta_j c_j u_j\right\}.$$
It is clear that
$$m(W(y)) \geqslant \prod_{j=1}^m h_j^{c_j}(T_j(\langle y, u_j \rangle))$$
for every $y \in \mathbb{R}^n$, and hence
$$\int_{\mathbb{R}^n} m(x)\,dx \geqslant \int_{W(\mathbb{R}^n)} m(x)\,dx$$
$$= \int_{\mathbb{R}^n} m(W(y)) |\det J_W(y)|\,dy$$
$$\geqslant \int_{\mathbb{R}^n} \prod_{j=1}^m h_j^{c_j}(T_j(\langle y, u_j \rangle)) \det\left(\sum_{j=1}^m c_j T_j'(\langle y, u_j \rangle) u_j \otimes u_j\right) dy.$$

By the definition of the constant F we have

$$\det\left(\sum_{j=1}^{m} c_j T'_j(\langle y, u_j\rangle) u_j \otimes u_j\right) \geqslant F \cdot \prod_{j=1}^{m} \left(T'_j(\langle y, u_j\rangle)\right)^{c_j}.$$

So, we can write

$$\int_{\mathbb{R}^n} m(x)dx \geqslant F \cdot \int_{\mathbb{R}^n} \prod_{j=1}^{m} h_j^{c_j}(T_j(\langle y, u_j\rangle)) \cdot \prod_{j=1}^{m} \left(T'_j(\langle y, u_j\rangle)\right)^{c_j} dy$$

$$= F \cdot \int_{\mathbb{R}^n} \prod_{j=1}^{m} f_j^{c_j}(\langle y, u_j\rangle) dy$$

$$= F \cdot I(f_1, \ldots, f_m).$$

This proves that $F \cdot I(f_1, \ldots, f_m) \leqslant K(h_1, \ldots, h_m)$. □

Proof of Theorems 1.6.1 and 1.6.2. We have

$$\sup\left\{I(f_1, \ldots, f_m) \mid \int_{\mathbb{R}} f_j = 1, \ j = 1, \ldots, m\right\} \geqslant D = \frac{1}{\sqrt{F}}.$$

On the other hand, Proposition 1.6.3 shows that

$$\frac{1}{\sqrt{F}} \leqslant \sup\left\{I(f_1, \ldots, f_m) \mid \int_{\mathbb{R}} f_j = 1\right\}$$

$$\leqslant \frac{1}{F} \cdot \inf\left\{K(h_1, \ldots, h_m) \mid \int_{\mathbb{R}} h_j = 1\right\} \leqslant \frac{1}{\sqrt{F}}.$$

Then, we must have equality everywhere, and this ends the proof. □

1.6.3. Reverse isoperimetric inequality

The calculation of the constant $F = F(\{u_j\}, \{c_j\})$ in Theorems 1.6.1 and 1.6.2 is not an easy task. An important observation of Ball is that its value is equal to 1 if the vectors u_j satisfy John's representation of the identity, i.e. if they behave like an orthogonal basis with weights equal to c_j.

THEOREM 1.6.4 (Ball). *Let $u_1, \ldots, u_m \in S^{n-1}$ and $c_1, \ldots, c_m > 0$ such that*

$$I = \sum_{j=1}^{m} c_j u_j \otimes u_j.$$

Then, the constant $F = F(\{u_j\}, \{c_j\})$ in Theorems 1.6.1 and 1.6.2 is equal to 1.

A well-known application of the Brascamp-Lieb inequality in this context is Ball's reverse isoperimetric inequality. We ask for the best constant $\partial(n)$ for which every symmetric convex body K in $\mathbb{R}^n$ has a position $\tilde{K}$ satisfying

$$\partial(\tilde{K}) \leqslant \partial(n)|\tilde{K}|^{(n-1)/n}.$$

The natural position of K is the minimal surface area position which was discussed in Section 1.5d. However, Ball's solution of the problem employs John's position. Assume that B_2^n is the maximal volume ellipsoid of K. Then,

$$\partial(K) = \lim_{t \to 0^+} \frac{|K + tB_2^n| - |K|}{t} \leqslant \lim_{t \to 0^+} \frac{|K + tK| - |K|}{t} = n|K|.$$

Then, Ball proves that among all bodies in John's position the cube has maximal volume.

THEOREM 1.6.5. *Let $Q_n = [-1,1]^n$ be the unit cube in $\mathbb{R}^n$. If K is a symmetric convex body in $\mathbb{R}^n$, and if K is in John's position, then $|K| \leqslant 2^n = |Q_n|$.*

For the proof we use John's representation of the identity, where the u_j's are contact points of K and B_2^n. Observe that
$$K \subseteq M := \{x : |\langle x, u_j \rangle| \leqslant 1, \ j = 1, \ldots, m\}.$$
Therefore,
$$|K| \leqslant |M| = \int_{\mathbb{R}^n} \prod_{j=1}^m \mathbf{1}_{[-1,1]}^{c_j}(\langle x, u_j\rangle) dx$$
$$\leqslant \prod_{j=1}^m \left(\int_{\mathbb{R}} \mathbf{1}_{[-1,1]}(t) dt \right)^{c_j} = 2^{\sum_{j=1}^m c_j} = 2^n,$$
where we used the Brascamp-Lieb inequality and applied Theorem 1.6.4 and the fact that $\sum_{j=1}^m c_j = n$. □

Theorem 1.6.5 shows that $\partial(K) \leqslant n|K| \leqslant 2n|K|^{(n-1)/n}$, and since K was arbitrary, $\partial(n) \leqslant 2n$. The example of the cube shows that, actually, $\partial(n) = 2n$.

Theorem 1.6.5 shows that the cube has maximal volume ratio among all symmetric convex bodies. In the general case, one can show that the simplex Δ_n is the extremal convex body (this was also proved by Ball).

1.7. Concentration of measure

It was for the purposes of geometric functional analysis that concentration of measure was initially understood to be very useful and was developed as a method; the starting point was V. Milman's proof of Dvoretzky theorem. As it was soon realized, it could be and it has been very well adapted to the needs of probability theory, asymptotic combinatorics and complexity as well. In this section we introduce the main examples of metric probability spaces that will be used in this book.

1.7.1. Metric probability spaces

Let (X, d) be a metric space. If μ is a probability measure on the Borel σ-algebra $\mathcal{B}(X)$ of (X, d), then the triple (X, d, μ) is called a *metric probability space*.

Typical examples of metric probability spaces include:
1. *The Euclidean sphere S^{n-1}*. We consider the unit sphere $S^{n-1} = \{x \in \mathbb{R}^n : \|x\|_2 = 1\}$ equipped with the geodesic metric ρ defined as follows: if $x, y \in S^{n-1}$ then $\rho(x, y)$ is the convex angle $\widehat{xoy}$ in the plane determined by the origin o and x, y. The sphere S^{n-1} becomes a probability space with the unique rotationally invariant measure σ: for any Borel set $A \subseteq S^{n-1}$ we set
$$\sigma(A) := \frac{|C(A)|}{|B_2^n|},$$

where
$$C(A) := \{sx : x \in A \text{ and } 0 \leqslant s \leqslant 1\}.$$
One can check that ρ is indeed a metric, and $\|x - y\|_2 = 2\sin\frac{\rho(x,y)}{2}$, therefore
$$\frac{2}{\pi}\rho(x,y) \leqslant \|x-y\|_2 \leqslant \rho(x,y).$$

2. *Gauss space.* We consider the measure γ_n on $\mathbb{R}^n$ with density
$$g_n(x) = (2\pi)^{-n/2} e^{-\|x\|_2^2/2}.$$
In other words, if A is a Borel subset of $\mathbb{R}^n$ then
$$\gamma_n(A) = \frac{1}{(2\pi)^{n/2}} \int_A e^{-\|x\|_2^2/2} dx.$$
The metric probability space $(\mathbb{R}^n, \|\cdot\|_2, \gamma_n)$ is called the n-dimensional Gauss space.

The standard Gaussian measure γ_n has two important properties: it is a product measure, more precisely $\gamma_n = \gamma_1 \otimes \cdots \otimes \gamma_1$, and it is invariant under orthogonal transformations, that is, if $U \in O(n)$ and A is a Borel subset of $\mathbb{R}^n$ then $\gamma_n(U(A)) = \gamma_n(A)$.

3. *The discrete cube.* We consider the set $E_2^n = \{-1,1\}^n$, which we identify with the set of vertices of the unit cube $Q_n = [-1,1]^n$ in $\mathbb{R}^n$. We equip E_2^n with the uniform probability measure μ_n which gives mass 2^{-n} to each point, and with the Hamming metric
$$d_n(x,y) = \frac{1}{n}\text{card}\{i \leqslant n : x_i \neq y_i\} = \frac{1}{2n}\sum_{i=1}^n |x_i - y_i|.$$

DEFINITION 1.7.1. Let (X, d, μ) be a metric probability space. For each non-empty $A \in \mathcal{B}(X)$ and any $t > 0$, the *t-extension* of A is the set
$$A_t = \{x \in X : d(x, A) < t\}.$$
The *concentration function* of (X, d, μ) is defined on $(0, \infty)$ by
$$\alpha_\mu(t) := \sup\{1 - \mu(A_t) : \mu(A) \geqslant 1/2\}.$$
The function α_μ is obviously decreasing and one can check that for every metric probability space (X, d, μ) one has
$$\lim_{t\to\infty} \alpha_\mu(t) = 0.$$
Roughly speaking, we say that we have measure concentration on the metric probability space (X, d, μ) if $\alpha_\mu(t)$ decreases fast to zero as $t \to \infty$. More precisely:

1. We say that μ has *normal concentration* on (X, d) if there exist constants $C, c > 0$ such that, for every $t > 0$,
$$\alpha_\mu(t) \leqslant Ce^{-ct^2}.$$
We shall say that a family (X_n, d_n, μ_n) is a *Lévy family* if for any $t > 0$
$$\alpha_{\mu_n}(t \cdot \text{diam}(X_n)) \to_{n\to\infty} 0$$
and we shall say that it is a *normal Lévy family* with constants c, C if for any $t > 0$
$$\alpha_{\mu_n}(t) \leqslant Ce^{-ct^2 n}.$$
We can check that this is the case for the examples we mentioned above: the sphere, the discrete cube and the Gauss space. Indeed, in the terminology of Definition 1.7.1, the following estimates hold:

(i) For the sphere (S^{n-1}, ρ, σ) one has
$$\alpha_\sigma(t) \leqslant \sqrt{\pi/8}\exp(-t^2 n/2).$$
(ii) For the Gauss space $(\mathbb{R}^n, \|\cdot\|_2, \gamma_n)$ one has
$$\alpha_{\gamma_n}(t) \leqslant \frac{1}{2}\exp(-t^2/2).$$
(iii) For the discrete cube (E_2^n, d_n, μ_n) one has
$$\alpha_{\mu_n}(t) \leqslant \frac{1}{2}\exp(-2t^2 n).$$

In addition,

(iv) For the family of the orthogonal groups $(SO(n), \rho_n, \mu_n)$ equipped with the Hilbert-Schmidt metric and the Haar probability measure one has
$$\alpha_{\rho_n}(t) \leqslant \sqrt{\pi/8}\exp(-t^2 n/8).$$
(v) All homogeneous spaces of $SO(n)$ inherit the property of forming Lévy families. In particular, any family of Stiefel manifolds W_{n,k_n} or any family of Grassman manifolds G_{n,k_n}, $n = 1, 2, \ldots$, $1 \leqslant k_n \leqslant n$, is a Lévy family with the same constants as $SO(n)$.

2. We say that μ has *exponential concentration* on (X, d) if there exist constants $C, c > 0$ such that, for every $t > 0$,
$$\alpha_\mu(t) \leqslant Ce^{-ct}.$$

1.7.2. Concentration of measure and Lipschitz functions

Many of the applications of measure concentration follow directly from the next theorem.

THEOREM 1.7.2. *Let (X, d, μ) be a metric probability space. If $f : X \to \mathbb{R}$ is a Lipschitz function with constant 1, i.e. $|f(x) - f(y)| \leqslant d(x, y)$ for all $x, y \in X$, then*
$$\mu(\{x \in X : |f(x) - \mathrm{med}(f)| > t\}) \leqslant 2\alpha_\mu(t),$$
where $\mathrm{med}(f)$ is a Lévy mean of f.

Note. Given an arbitrary function $g : X \to \mathbb{R}$, a real number $\mathrm{med}(g)$ is called a Lévy mean of g if we have
$$\mu(\{g \geqslant \mathrm{med}(g)\}) \geqslant 1/2 \text{ and } \mu(\{g \leqslant \mathrm{med}(g)\}) \geqslant 1/2.$$
Observe that there are cases in which there are more than one numbers having this property.

Proof of Theorem 1.7.2. We set $A = \{x : f(x) \geqslant \mathrm{med}(f)\}$ and $B = \{x : f(x) \leqslant \mathrm{med}(f)\}$. For every $y \in A_t$ there exists $x \in A$ with $d(x, y) \leqslant t$, and thus
$$f(y) = f(y) - f(x) + f(x) \geqslant -d(y, x) + \mathrm{med}(f) \geqslant \mathrm{med}(f) - t$$
because f is 1-Lipschitz. Similarly, if $y \in B_t$ then there exists $x \in B$ with $d(x, y) \leqslant t$, and thus
$$f(y) = f(y) - f(x) + f(x) \leqslant d(y, x) + \mathrm{med}(f) \leqslant \mathrm{med}(f) + t.$$
It follows that if $y \in A_t \cap B_t$ then $|f(x) - \mathrm{med}(f)| \leqslant t$. In other words,
$$\{x \in X : |f(x) - \mathrm{med}(f)| > t\} \subseteq (A_t \cap B_t)^c = A_t^c \cup B_t^c.$$

From the definition of the concentration function we have $\mu(A_t) \geq 1 - \alpha_\mu(t)$ and $\mu(B_t) \geq 1 - \alpha_\mu(t)$. It follows that
$$\mu(\{|f - \operatorname{med}(f)| > t\}) \leq (1 - \mu(A_t)) + (1 - \mu(B_t)) \leq 2\alpha_\mu(t),$$
as claimed. □

When the concentration function of the space decreases fast, Theorem 1.7.2 shows that 1-Lipschitz functions are "almost constant" on "almost all of the space". We will use this fact very often, in the following form: if $f : X \to \mathbb{R}$ is a Lipschitz function with constant $\|f\|_{\operatorname{Lip}}$, i.e. if $|f(x) - f(y)| \leq \|f\|_{\operatorname{Lip}} d(x, y)$ for all $x, y \in X$, then
$$\mu(\{x \in X : |f(x) - \operatorname{med}(f)| > t\}) \leq 2\alpha_\mu(t/\|f\|_{\operatorname{Lip}}),$$
where $\operatorname{med}(f)$ is a Lévy mean of f.

Converse statements are also valid.

THEOREM 1.7.3. *Let (X, d, μ) be a metric probability space. Assume that for some $\eta > 0$ and some $t > 0$ one has*
$$\mu(\{x \in X : |f(x) - \operatorname{med}(f)| > t\}) \leq \eta$$
for every 1-Lipschitz function $f : X \to \mathbb{R}$. Then, $\alpha_\mu(t) \leq \eta$.

Proof. Let A be a Borel subset of X with $\mu(A) \geq 1/2$. We consider the function $f(x) = d(x, A)$. Then, f is 1-Lipschitz and $\operatorname{med}(f) = 0$ because f is non-negative and $\mu(\{x : f(x) = 0\}) \geq 1/2$. From the assumption we get
$$\mu(\{x \in X : d(x, A) > t\}) \leq \eta,$$
that is $1 - \mu(A_t) \leq \eta$. It follows that $\alpha_\mu(t) \leq \eta$. □

The next theorem shows that we can draw the same conclusion if we replace the Lévy mean by the expectation; this is often easier to compute.

THEOREM 1.7.4. *Let (X, d, μ) be a metric probability space. Assume that there is some function $\alpha : [0, \infty) \to [0, \infty)$ such that, for every bounded 1-Lipschitz function $f : (X, d) \to \mathbb{R}$ and for every $t > 0$, one has*
$$\mu\left(\left\{x : f(x) \geq \int f \, d\mu + t\right\}\right) \leq \alpha(t).$$
Then, for every Borel set $A \subseteq X$ with $\mu(A) > 0$ and for every $t > 0$,
$$1 - \mu(A_t) \leq \alpha(\mu(A)t).$$
In particular,
$$\alpha_\mu(t) \leq \alpha(t/2), \qquad t > 0.$$

Proof. We fix $A \in \mathcal{B}(X)$ with $\mu(A) > 0$ and $t > 0$. We consider the function $f(x) = \min\{d(x, A), t\}$. Note that $\|f\|_{\operatorname{Lip}} \leq 1$ and
$$\int f \, d\mu \leq (1 - \mu(A))t.$$
From the assumption we have
$$1 - \mu(A_t) = \mu(\{f \geq t\}) \leq \mu\left(\left\{x : f(x) \geq \int f \, d\mu + \mu(A)t\right\}\right)$$
$$\leq \alpha(\mu(A)t).$$

In particular, if $\mu(A) \geqslant 1/2$ then we have $\int f\, d\mu \leqslant t/2$ and this gives

$$1 - \mu(A_t) = \mu(\{f \geqslant t\}) \leqslant \mu\left(\left\{x : f(x) \geqslant \int f\, d\mu + t/2\right\}\right)$$
$$\leqslant \alpha(t/2).$$

We thus conclude that $\alpha_\mu(t) \leqslant \alpha(t/2)$. $\square$

1.7.3. Isoperimetric problems and concentration of measure

In this subsection we discuss the isoperimetric problem, which can be formulated for an arbitrary metric probability space. In the next subsections we will see that for the typical examples of metric probability spaces that we saw in Subsection 1.7.1 the solution to the isoperimetric problem is known.

DEFINITION 1.7.5. Let (X, d) be a metric space and let μ be a (not necessarily finite) measure on the Borel σ-algebra $\mathcal{B}(X)$. The surface area (or *Minkowski content*) of a non-empty $A \in \mathcal{B}(X)$ is defined by

$$\mu^+(A) = \liminf_{t \to 0^+} \frac{\mu(A_t \setminus A)}{t},$$

where A_t is the t-extension of A. If $\mu(A) < \infty$ (which is certainly true if (X, d, μ) is a metric probability space) then

$$\mu^+(A) = \liminf_{t \to 0^+} \frac{\mu(A_t) - \mu(A)}{t}.$$

Given a metric probability space, one can now state the isoperimetric problem in one of the following ways:

(i) Given $0 < \alpha < 1$ and $t > 0$, find

$$\inf\{\mu(A_t) : A \in \mathcal{B}(X), \mu(A) \geqslant \alpha\}$$

and identify (if they exist) those sets A at which this infimum is attained.

(ii) Given $0 < \alpha < 1$ and $t > 0$, find

$$\inf\{\mu^+(A) : A \in \mathcal{B}(X), \mu(A) \geqslant \alpha\}$$

and identify (if they exist) those sets A at which this infimum is attained.

Obviously, there is no reason why the answer to the first question shouldn't vary with t. Nevertheless, in all classical examples, the minimizers turn out to be independent of t and quite symmetric subsets of X, therefore we can easily compute the measure of their t-extension as well as their surface area. Note that the examples of spaces that we discuss next, for which the solution to the isoperimetric problem is known, are also the cases which will be important for this book.

1.7.4. The spherical isoperimetric inequality

The isoperimetric problem for S^{n-1} is formulated as follows. Let $\alpha \in (0,1)$ and $t > 0$. Among all Borel subsets A of the sphere which satisfy $\sigma(A) = \alpha$, determine the ones for which the measure $\sigma(A_t)$ of their t-extension is minimal.

The answer to this question is given by the next theorem.

THEOREM 1.7.6 (Lévy). *Let $\alpha \in (0,1)$ and let $B(x,r)$ be a ball of radius $r > 0$ in S^{n-1} such that $\sigma(B(x,r)) = \alpha$. Then, for every $A \subseteq S^{n-1}$ with $\sigma(A) = \alpha$ and every $t > 0$, we have*
$$\sigma(A_t) \geqslant \sigma(B(x,r)_t) = \sigma(B(x,r+t)).$$

In other words, for any given value of α and any $t > 0$, the spherical caps of measure α provide the solution to the isoperimetric problem. A proof of the spherical isoperimetric inequality can be given with spherical symmetrization and induction on the dimension. Let us consider the special case $\alpha = 1/2$. If $\sigma(A) = 1/2$ and $t > 0$, then we can estimate the size of A_t using the isoperimetric inequality: from Theorem 1.7.6 we have

$$(1.7.1) \qquad \sigma(A_t) \geqslant \sigma\left(B\left(x, \frac{\pi}{2} + t\right)\right)$$

for every $t > 0$ and $x \in S^{n-1}$. Then, starting from (1.7.1) and computing the measure of a cap $B(x, \pi/2 + t)$ we obtain the following inequality.

THEOREM 1.7.7. *Let $A \subseteq S^{n+1}$ with $\sigma(A) = 1/2$ and let $t > 0$. Then,*
$$(1.7.2) \qquad \sigma(A_t) \geqslant 1 - \sqrt{\pi/8}\exp(-t^2 n/2).$$

The proof of Theorem 1.7.7 is heavily based on the spherical isoperimetric inequality. However, in the applications, we do not really need the precise solution to the isoperimetric problem; we only need an inequality providing a similar estimate to that in (1.7.2) (even with constants that are "much" larger than $\sqrt{\pi/8}$ and 2, as long as they are independent of the dimension). It turns out that one can give a very simple proof of an analogous exponential estimate using the Brunn-Minkowski inequality. The key point is the following result of Arias-de-Reyna, Ball and Villa.

LEMMA 1.7.8 (Arias de Reyna-Ball-Villa). *Consider the probability measure $\mu(A) = |A|/|B_2^n|$ on the Euclidean unit ball B_2^n. If A, B are subsets of B_2^n with $\mu(A) \geqslant \alpha$, $\mu(B) \geqslant \alpha$, and if $\rho(A,B) = \inf\{\|a-b\|_2 : a \in A, b \in B\} = \rho > 0$, then*
$$\alpha \leqslant \exp(-\rho^2 n/8).$$

Proof. We may assume that A and B are closed. By the Brunn-Minkowski inequality, $\mu(\frac{A+B}{2}) \geqslant \alpha$. On the other hand, the parallelogram law shows that if $a \in A, b \in B$ then
$$\|a+b\|_2^2 = 2\|a\|_2^2 + 2\|b\|_2^2 - \|a-b\|_2^2 \leqslant 4 - \rho^2.$$
It follows that $\frac{A+B}{2} \subseteq \sqrt{1 - \frac{\rho^2}{4}} B_2^n$, hence
$$\mu\left(\frac{A+B}{2}\right) \leqslant \left(1 - \frac{\rho^2}{4}\right)^{n/2} \leqslant \exp(-\rho^2 n/8).$$

Proof of Theorem 1.7.7. Assume that $A \subseteq S^{n-1}$ with $\sigma(A) = 1/2$. Let $t > 0$ and define $B = (A_t)^c \subseteq S^{n-1}$. We fix $\lambda \in (0,1)$ and consider the subsets
$$\tilde{A} = \bigcup\{tA : \lambda \leqslant t \leqslant 1\} \quad \text{and} \quad \tilde{B} = \bigcup\{tB : \lambda \leqslant t \leqslant 1\}$$
of B_2^n. These are disjoint with distance $\simeq \lambda t$. Lemma 1.7.8 shows that $\mu(\tilde{B}) \leqslant \exp(-c\lambda^2 t^2 n/8)$, and since $\mu(\tilde{B}) = (1 - \lambda^n)\sigma(B)$ we obtain
$$\sigma(A_t) \geqslant 1 - \frac{1}{1-\lambda^n}\exp(-c\lambda^2 t^2 n/8).$$
We conclude the proof by choosing $\lambda = 1/2$. $\square$

From Theorem 1.7.2 it follows that if $g : S^{n-1} \to \mathbb{R}$ is Lipschitz continuous, then
$$\sigma(\{\theta : |g(\theta) - \operatorname{med}(g)| \geq t\}) \leq 2\exp(-(n-1)t^2/2\|g\|_{\operatorname{Lip}}^2)$$
for every $t > 0$. We will often use the fact that, in this deviation inequality, one can replace the Lévy mean of g by the expectation of g (for a proof of this assertion in the setting of the Gaussian measure, see Theorem 15.1.11 and Corollary 15.1.12 in Chapter 15).

THEOREM 1.7.9. *If $g : S^{n-1} \to \mathbb{R}$ is a Lipschitz continuous function, then*
$$\text{(1.7.3)} \quad \sigma(\{\theta : |g(\theta) - \mathbb{E}_\sigma(g)| \geq t\}) \leq 2\exp(-(n-1)t^2/2\|g\|_{\operatorname{Lip}}^2)$$
for all $t > 0$.

Another useful fact is that if $g : \mathbb{R}^n \to \mathbb{R}$ is a norm then the Lévy mean $\operatorname{med}(g)$ and the expectation $\mathbb{E}_\sigma(g)$ of g on the sphere S^{n-1} are comparable: one has
$$\frac{1}{2}\operatorname{med}(g) \leq \mathbb{E}_\sigma(g) \leq c\operatorname{med}(g)$$
where $c > 0$ is an absolute constant.

1.7.5. Isoperimetric inequality in the Gauss space

The isoperimetric inequality in the "Gauss space" is the following statement.

THEOREM 1.7.10 (Borell, Sudakov-Tsirelson). *Let $\alpha \in (0,1)$ and $\theta \in S^{n-1}$ and let $H = \{x \in \mathbb{R}^n : \langle x, \theta \rangle \leq \lambda\}$ be a half-space in $\mathbb{R}^n$ with $\gamma_n(H) = \alpha$. Then, for every $t > 0$ and every Borel $A \subseteq \mathbb{R}^n$ with $\gamma_n(A) = \alpha$, we have*
$$\gamma_n(A_t) \geq \gamma_n(H_t).$$

COROLLARY 1.7.11. *If $\gamma_n(A) \geq 1/2$ then, for every $t > 0$,*
$$\text{(1.7.4)} \quad 1 - \gamma_n(A_t) \leq \frac{1}{2}\exp(-t^2/2).$$

Proof. From Theorem 1.7.10 we know that
$$1 - \gamma_n(A_t) \leq 1 - \gamma_n(H_t)$$
where H is a half-space of measure $1/2$. Since γ_n is invariant under orthogonal transformations, we may assume that $H = \{x \in \mathbb{R}^n : x_1 \leq 0\}$, and then it follows that
$$1 - \gamma_n(H_t) = \frac{1}{\sqrt{2\pi}} \int_t^\infty e^{-s^2/2} ds.$$
Differentiation shows that the function
$$F(x) = e^{x^2/2} \int_x^\infty e^{-s^2/2} ds$$
is decreasing on $[0, +\infty)$. The fact that $F(t) \leq F(0)$ completes the proof. $\square$

As in the case of the sphere, the proof of the approximate isoperimetric inequality (1.7.4) requires knowing the exact solution of the Gaussian isoperimetric problem. However, there are also simple, direct proofs which do not assume the isoperimetric inequality in the Gauss space; an example is the following proof by Maurey that makes use of the Prékopa-Leindler inequality.

THEOREM 1.7.12 (Maurey). *Let A be a non-empty Borel subset of $\mathbb{R}^n$. Then,*

$$\int_{\mathbb{R}^n} e^{d(x,A)^2/4} d\gamma_n(x) \leqslant \frac{1}{\gamma_n(A)},$$

where $d(x, A) = \inf\{\|x - y\|_2 : y \in A\}$. Therefore, if $\gamma_n(A) \geqslant \frac{1}{2}$ we have

$$1 - \gamma_n(A_t) \leqslant 2\exp(-t^2/4)$$

for every $t > 0$.

Proof. Consider the functions

$$f(x) = e^{d(x,A)^2/4} g_n(x), \quad h(x) = \mathbf{1}_A(x) g_n(x), \quad m(x) = g_n(x),$$

where g_n is the density of the gaussian measure γ_n. For every $x \in \mathbb{R}^n$ and $y \in A$ we see that

$$f(x)h(y) \leqslant \left(m\left(\frac{x+y}{2}\right)\right)^2,$$

using the parallelogram law and the fact that $d(x, A) \leqslant \|x - y\|_2$. Since $h(y) = 0$ whenever $y \notin A$, this implies that f, h, m satisfy the assumptions of the Prékopa-Leindler inequality with $\lambda = 1/2$. Therefore,

$$\left(\int_{\mathbb{R}^n} e^{d(x,A)^2/4} d\gamma_n(x)\right) \gamma_n(A) = \left(\int_{\mathbb{R}^n} f\right)\left(\int_{\mathbb{R}^n} h\right) \leqslant \left(\int_{\mathbb{R}^n} m\right)^2 = 1.$$

This proves the first assertion of the theorem. For the second one, observe that if $\gamma_n(A) \geqslant \frac{1}{2}$ then

$$e^{t^2/4} \gamma_n(\{x : d(x, A) \geqslant t\}) \leqslant \int_{\mathbb{R}^n} e^{d(x,A)^2/4} d\gamma_n(x) \leqslant \frac{1}{\gamma_n(A)} \leqslant 2.$$

This shows that $\gamma_n(A_t^c) \leqslant 2\exp(-t^2/4)$. □

1.7.6. Isoperimetric inequality in the discrete cube

The solution to the isoperimetric problem for E_2^n is given by by the d_n-balls (the so-called Hamming balls of E_2^n) in the case where N is the cardinality of some d_n ball. A combinatorial proof of this fact was given by Harper (the general minimizers for the isoperimetric problem are also known). Based on this information one can give an estimate for the concentration function of E_2^n.

THEOREM 1.7.13 (Harper). *If $\mu_n(A) \geqslant 1/2$ and $t > 0$, then*

$$\mu_n(A_t^c) \leqslant \frac{1}{2} \exp(-2t^2 n).$$

We will present a direct proof of an only slightly worse exponential estimate in Theorem 1.7.16. The proof is based on the following theorem of Talagrand.

THEOREM 1.7.14 (Talagrand). *Let A be a non-empty subset of E_2^n. We consider its convex hull $\mathrm{conv}(A)$ and for every $x \in E_2^n$ we define*

$$\phi_A(x) = \min\{\|x - y\|_2 : y \in \mathrm{conv}(A)\}.$$

Then,

$$\int_{E_2^n} \exp(\phi_A^2(x)/8) d\mu_n(x) \leqslant \frac{1}{\mu_n(A)}.$$

For every non-empty subset A of E_2^n, the function ϕ_A of Theorem 1.7.14 and the function
$$d_n(x, A) = \min\left\{\frac{1}{2n}\sum_{i=1}^n |x_i - y_i| : y \in A\right\}$$
which measures the distance from x to A are related as follows.

LEMMA 1.7.15. *For every non-empty $A \subseteq E_2^n$ and every $x \in E_2^n$,*
$$2\sqrt{n}d_n(x, A) \leqslant \phi_A(x).$$

Proof. Let $x \in E_2^n$. For every $y \in A$ we have
$$(1.7.5) \qquad \langle x - y, x \rangle = \sum_{i=1}^n x_i(x_i - y_i) = 2nd_n(x,y) \geqslant 2nd_n(x, A).$$

From (1.7.5) we see that, for every $y \in \mathrm{conv}(A)$,
$$\sqrt{n}\|x - y\|_2 \geqslant \langle x - y, x \rangle \geqslant 2nd_n(x, A).$$

This proves the lemma. $\square$

Combining the above we get the approximate isoperimetric inequality for E_2^n:

THEOREM 1.7.16. *Let $A \subseteq E_2^n$ with $\mu_n(A) \geqslant 1/2$. Then, for every $t > 0$, we have*
$$\mu_n(A_t) \geqslant 1 - 2\exp(-t^2 n/2).$$

Proof. If $x \notin A_t$, then $d_n(x, A) \geqslant t$ and Lemma 1.7.15 shows that $\phi_A(x) \geqslant 2t\sqrt{n}$. But, from Theorem 1.7.14 we have
$$e^{t^2 n/2}\mu_n(\{x : \phi_A(x) \geqslant 2t\sqrt{n}\}) \leqslant \int_{E_2^n} \exp(\phi_A^2(x)/8)d\mu_n(x) \leqslant \frac{1}{\mu_n(A)} \leqslant 2,$$
and this gives
$$\mu_n(A_t^c) \leqslant \mu_n(\{x : \phi_A(x) \geqslant 2t\sqrt{n}\}) \leqslant 2\exp(-t^2 n/2),$$
whence we are done. $\square$

1.7.7. Kahane-Khintchine inequality

The Rademacher functions $r_i : [0, 1] \to \mathbb{R}$, $i \geqslant 1$, are defined by
$$r_i(t) = \mathrm{sign}\,\sin(\pi 2^i t/2).$$
They are ± 1-valued (if we ignore a set of measure zero) independent random variables on $[0, 1]$ and they form an orthonormal sequence in $L^2[0,1]$. An equivalent way to define them is to consider $E_2 = \prod_{i=1}^\infty \{-1, 1\}$ endowed with the standard product measure and to define, for every $\epsilon = (\epsilon_i)_{i=1}^\infty$,
$$r_i(\epsilon) = \epsilon_i.$$

The classical Khintchine inequality states that for every $p > 0$ there exist constants $A_p, B_p > 0$ such that, for every $n \geqslant 1$ and any n-tuple of real numbers $a_1, \ldots, a_n$,
$$(1.7.6) \qquad A_p\left(\sum_{i=1}^n a_i^2\right)^{1/2} \leqslant \left(\int_{E_2^n} \left|\sum_{i=1}^n a_i \epsilon_i\right|^p d\mu_n(\epsilon)\right)^{1/p} \leqslant B_p\left(\sum_{i=1}^n a_i^2\right)^{1/2}.$$

Since
$$\Big(\sum_{i=1}^n a_i^2\Big)^{1/2} = \Big(\int_{E_2^n} \Big|\sum_{i=1}^n a_i \epsilon_i\Big|^2 d\mu_n(\epsilon)\Big)^{1/2},$$
an equivalent way to state Khintchine's inequality is the following.

THEOREM 1.7.17 (Khintchine). *For every $p > 0$ there exist $A_p, B_p > 0$ such that for every $n \geqslant 1$ and any $a = (a_1, \ldots, a_n) \in \ell_2^n$,*

$$(1.7.7) \quad A_p \Big\| \sum_{i=1}^n a_i \epsilon_i \Big\|_{L_2(E_2^n)} \leqslant \Big\| \sum_{i=1}^n a_i \epsilon_i \Big\|_{L_p(E_2^n)} \leqslant B_p \Big\| \sum_{i=1}^n a_i \epsilon_i \Big\|_{L_2(E_2^n)}.$$

Let A_p^*, B_p^* denote the best constants for which the statement of Theorem 1.7.17 is valid. From Hölder's inequality it is clear that $A_p^* = 1$ if $p \geqslant 2$ and $B_p^* = 1$ if $0 < p \leqslant 2$. The exact values of A_p^* and B_p^* have been determined by Szarek ($A_1^* = 1/\sqrt{2}$) and Haagerup (for all p). What is particularly important is to know the behavior of B_p^* as $p \to \infty$; the order of growth of B_p^* is $O(\sqrt{p})$ for large p.

Kahane's inequality generalizes Khintchine's inequality.

THEOREM 1.7.18 (Kahane). *There exists $C > 0$ such that for every normed space X, for any $n \geqslant 1$, for any $x_1, \ldots, x_n \in X$ and any $p \geqslant 1$,*

$$(1.7.8) \quad \Big(\mathbb{E}\Big\|\sum_{i=1}^n \epsilon_i x_i\Big\|^p\Big)^{1/p} \leqslant C\sqrt{p}\, \mathbb{E}\Big\|\sum_{i=1}^n \epsilon_i x_i\Big\|.$$

1.8. Entropy estimates

1.8.1. Covering numbers

Let A and B be two convex bodies in $\mathbb{R}^n$. The *covering number* $N(A, B)$ of A by B is the least number of translates of B that are needed in order to cover A:

$$N(A, B) = \min\{N \in \mathbb{N} \mid \exists\, x_1, \ldots, x_N \in \mathbb{R}^n : A \subseteq \cup(x_i + B)\}.$$

Sometimes, we require that the centers x_i belong to A; then we set

$$\overline{N}(A, B) = \min\{N \in \mathbb{N} \mid \exists\, x_1, \ldots, x_N \in A : A \subseteq \cup(x_i + B)\}.$$

Some of the basic properties of covering numbers are listed below; we will be using them very often in this book:

(i) For all convex A, B and every invertible linear operator $T : \mathbb{R}^n \to \mathbb{R}^n$ we have $N(A, B) = N(T(A), T(B))$.
(ii) For all convex A, B and C we have $N(A, B) \leqslant N(A, C)N(C, B)$.
(iii) For all convex A, B and C we have $N(A + C, B + C) \leqslant N(A, B)$.
(iv) For all convex A and B we have $\overline{N}(A, (A - A) \cap B) = \overline{N}(A, B)$. In particular we have that: for centrally symmetric A, $\overline{N}(A, 2A \cap B) = \overline{N}(A, B)$ and $N(A, 2(B \cap A)) \leqslant N(A, B)$.
(v) For all convex A we have $N(A, RB_2^n \cap (A - A)) = N(A, RB_2^n)$.

The *packing number* $M(A, B)$ of two centrally symmetric convex bodies A and B is the maximal cardinality of a B-separated set in A:

$$M(A, B) = \max\{N : \exists\, x_1, \ldots, x_N \in A \text{ s.t. } \forall j \neq i, x_j \notin (x_i + B)\}.$$

Equivalently, we ask that if $i \neq j$ then $\|x_i - x_j\|_B > 1$. The packing number is closely related to the covering number of A by B: one can check that
$$M(A, 2B) \leqslant N(A, B) \leqslant \overline{N}(A, B) \leqslant M(A, B).$$

1.8.2. Sudakov inequality and its dual

Let K be a symmetric convex body in $\mathbb{R}^n$. It will be very useful to have sharp estimates for the covering numbers $N(K, tB_2^n)$ and $N(B_2^n, tK)$. A well-known bound for $N(K, tB_2^n)$ is given by *Sudakov inequality* (see [**481**]).

THEOREM 1.8.1 (Sudakov). *Let K be a symmetric convex body in $\mathbb{R}^n$. For every $t > 0$,*
$$\log N(K, tB_2^n) \leqslant cn\left(\frac{w(K)}{t}\right)^2,$$
where $c > 0$ is an absolute constant.

The simplest way to prove Sudakov inequality is from the so-called *dual Sudakov inequality* which was proved by Pajor and Tomczak-Jaegermann.

THEOREM 1.8.2 (Pajor-Tomczak). *Let K be a symmetric convex body in $\mathbb{R}^n$. For every $t > 0$,*
$$\log N(B_2^n, tK) \leqslant cn\left(\frac{w(K^\circ)}{t}\right)^2,$$
where $c > 0$ is an absolute constant.

We present a simple proof of Theorem 1.8.2 which is due to Talagrand; his argument makes use of the Gaussian measure γ_n. We need a simple lemma.

LEMMA 1.8.3. *Let K be a symmetric convex body in $\mathbb{R}^n$. For every $z \in \mathbb{R}^n$ we have $\gamma_n(K + z) \geqslant \exp(-\|z\|_2^2/2)\gamma_n(K)$.* □

Proof of Theorem 1.8.2. Let $x_1, \ldots, x_N$ be a maximal (tK)-separated set of points in B_2^n. Then, the sets $x_i + \frac{t}{2}K$ have disjoint interiors, and hence, for every $\lambda > 0$ the sets $\lambda x_i + \frac{\lambda t}{2}K$ have disjoint interiors. Since γ_n is a probability measure, we have
$$\sum_{i=1}^N \gamma_n\left(\lambda x_i + \frac{\lambda t}{2}K\right) = \gamma_n\left(\bigcup_{i=1}^N \left(\lambda x_i + \frac{\lambda t}{2}K\right)\right) \leqslant 1.$$
Note that $\|\lambda x_i\|_2 \leqslant \lambda$; then, Lemma 1.8.3 shows that
$$\gamma_n\left(\lambda x_i + \frac{\lambda t}{2}K\right) \geqslant \exp(-\lambda^2/2)\gamma_n\left(\frac{\lambda t}{2}K\right) \qquad i = 1, \ldots, N.$$
Consequently, for every $\lambda > 0$ we have that
$$N(B_2^n, tK) \leqslant N \leqslant \frac{\exp(\lambda^2/2)}{\gamma_n\left(\frac{\lambda t}{2}K\right)}.$$
It remains to choose $\lambda > 0$ in an optimal way: integration in polar coordinates shows that
$$\int_{\mathbb{R}^n} \|x\|_K \gamma_n(dx) \leqslant c\sqrt{n} \int_{S^{n-1}} \|\theta\|_K d\sigma(\theta) = c\sqrt{n}\, w(K^\circ),$$
and applying Markov's inequality we get
$$\gamma_n(\|x\|_K \geqslant \lambda t/2) \leqslant \frac{2}{\lambda t}\int_{\mathbb{R}^n} \|x\|_K \gamma_n(dx) \leqslant \frac{2c\sqrt{n}}{\lambda t}w(K^\circ),$$

or equivalently,
$$1 - \gamma_n\left(\frac{\lambda t}{2}K\right) \leqslant \frac{2c\sqrt{n}}{\lambda t}w(K^\circ).$$
For $\lambda = 4c\sqrt{n}w(K^\circ)/t$ we get
$$\gamma_n\left(2c\sqrt{n}w(K^\circ)K\right) \geqslant \frac{1}{2},$$
and hence
$$N(B_2^n, tK) \leqslant 2\exp\left(8c^2nw^2(K^\circ)/t^2\right),$$
as claimed. □

Tomczak-Jaegermann observed that one can deduce Theorem 1.8.1 from Theorem 1.8.2 and vice versa.

THEOREM 1.8.4. *Let K be a symmetric convex body in $\mathbb{R}^n$. If*
$$B = \sup_{t>0} t\left(\log N(K, tB_2^n)\right)^{1/2},$$
and
$$A := \sup_{t>0} t\left(\log \overline{N}(B_2^n, tK^\circ)\right)^{1/2} \leqslant c\sqrt{n}w(K),$$
then $B \leqslant 10A$. In particular, we have Sudakov inequality: for every $t > 0$,
$$\log N(K, tB_2^n) \leqslant cn\left(\frac{w(K)}{t}\right)^2,$$
where $c > 0$ is an absolute constant.

Proof. We first observe that $2K \cap \left(\frac{t^2}{2}K^\circ\right) \subseteq tB_2^n$. It follows that
$$\overline{N}(K, tB_2^n) \leqslant \overline{N}\left(K, (2K) \cap \left(\frac{t^2}{2}K^\circ\right)\right).$$
Next, we observe that
$$\overline{N}\left(K, (2K) \cap \left(\frac{t^2}{2}K^\circ\right)\right) = \overline{N}\left(K, \frac{t^2}{2}K^\circ\right).$$
Using the above we see that
$$N(K, tB_2^n) \leqslant \overline{N}(K, tB_2^n) \leqslant \overline{N}\left(K, \frac{t^2}{2}K^\circ\right)$$
$$\leqslant N(K, \frac{t^2}{4}K^\circ) \leqslant N(K, 2tB_2^n)N\left(B_2^n, \frac{t}{8}K^\circ\right).$$
We write
$$t^2 \log N(K, tB_2^n) \leqslant \frac{1}{4}(2t)^2 \log N(K, 2tB_2^n) + 64(t/8)^2 \log N\left(B_2^n, \frac{t}{8}K^\circ\right)$$
$$\leqslant \frac{1}{4}(2t)^2 \log N(K, 2tB_2^n) + 64A^2,$$
and taking sup over all $t > 0$ we arrive at $3B^2 \leqslant 256A^2$.

We can now prove Sudakov inequality: for every $t > 0$ we have
$$t^2 \log N(K, tB_2^n) \leqslant 100A^2 \leqslant cnw^2(K),$$
where $c > 0$ is an absolute constant. □

Theorem 1.8.1 holds true for not necessarily symmetric convex bodies as well.

PROPOSITION 1.8.5. *Let K be a convex body in $\mathbb{R}^n$. For every $t > 0$,*
$$N(K, tB_2^n) \leqslant \exp\left(cn(w(K)/t)^2\right),$$
where $c > 0$ is an absolute constant.

Proof. Consider the difference body $K - K$ of K. Then,
$$w(K - K) = \int_{S^{n-1}} h_{K-K}(u) d\sigma(u) = \int_{S^{n-1}} [h_K(u) + h_{-K}(u)] d\sigma(u)$$
$$= \int_{S^{n-1}} [h_K(u) + h_K(-u)] d\sigma(u) = 2w(K).$$

Since there is a translate of K which is contained in $K - K$, Sudakov's inequality gives
$$t^2 \log N(K, tB_2^n) \leqslant t^2 \log N(K - K, tB_2^n) \leqslant cnw^2(K - K) = 4cnw^2(K),$$
which proves the theorem. □

1.8.3. Duality of entropy

The duality of entropy numbers conjecture asserts that if X, Y are Banach spaces, if $T : X \to Y$ is a compact operator and if $N(T, t)$ denotes the covering number $N(T(B_X), tB_Y)$, then
$$a^{-1} \log N(T, bt) \leqslant \log N(T^*, t) \leqslant a \log N(T, b^{-1}t)$$
for every $t > 0$, where $a, b > 0$ are absolute constants, and T^* is the adjoint operator of T. This conjecture has been verified only under strong assumptions for both spaces X and Y. In the case where one of the two spaces is a Hilbert space, the conjecture is equivalent to the following statement about covering numbers of convex bodies: There exist two constants $\alpha, \beta > 0$ such that
$$\log N(B_2^n, \beta K^\circ) \leqslant \alpha \log N(K, B_2^n)$$
for every symmetric convex body K in $\mathbb{R}^n$.

This case was settled by Artstein-Avidan, Milman and Szarek.

THEOREM 1.8.6 (Artstein-Milman-Szarek). *There exist two absolute constants α and $\beta > 0$ such that for any dimension n and any symmetric convex body K in $\mathbb{R}^n$, one has*
$$N(B_2^n, \alpha^{-1} K^\circ)^{1/\beta} \leqslant N(K, B_2^n) \leqslant N(B_2^n, \alpha K^\circ)^\beta$$

Theorem 1.8.6 establishes a strong connection between the geometry of a set and its polar. Observe that since the theorem is true for any K, we actually have that, for any $t > 0$,
$$\beta^{-1} \log N(B_2^n, \alpha^{-1} t K^\circ) \leqslant \log N(K, t B_2^n) \leqslant \beta \log N(B_2^n, \alpha t K^\circ).$$

A weaker but general duality inequality has been proved by König and Milman. Using the reverse Santaló and Brunn-Minkowski inequalities they showed that

(1.8.1) $$c^{-1} N(K_2^\circ, K_1^\circ)^{1/n} \leqslant N(K_1, K_2)^{1/n} \leqslant c N(K_2^\circ, K_1^\circ)^{1/n}$$

for every pair of symmetric convex bodies K_1 and K_2 in $\mathbb{R}^n$.

1.9. Gaussian and sub-Gaussian processes

1.9.1. Sub-Gaussian processes

Let (T,d) be a metric space and let $\mathcal{Y} = (Y_t)_{t \in T}$ be a family of real valued random variables, with indices from T, on a probability space $(\Omega, \mathcal{A}, P)$. We say that the process $\mathcal{Y} = (Y_t)_{t \in T}$ is *sub-Gaussian* with respect to d if $\mathbb{E}(Y_t) = 0$ for all $t \in T$ and, for all $t, s \in T$ and every $u > 0$,

$$\mathbb{P}(|Y_t - Y_s| \geqslant u) \leqslant 2 \exp\left(-\frac{u^2}{d^2(t,s)}\right).$$

A typical example is given by the discrete cube $E_2^n = \{-1, 1\}^n$, equipped with the uniform probability measure. Write $\epsilon = (\epsilon_1, \ldots, \epsilon_n)$ for the points of E_2^n and consider the *Rademacher functions* $r_i : E_2^n \to \{-1, 1\}$, $1 \leqslant i \leqslant n$, defined by $r_i(\epsilon) = \epsilon_i$.

For every $t = (t_1, \ldots, t_n) \in T \subseteq \mathbb{R}^n$ we define

$$Y_t(\epsilon) = \langle t, \epsilon \rangle = t_1 r_1(\epsilon) + \cdots + t_n r_n(\epsilon).$$

From Khintchine's inequality it follows that

$$\mathbb{P}\left(\epsilon \in E_2^n : |t_1 r_1(\epsilon) + \cdots + t_n r_n(\epsilon)| \geqslant u\right) \leqslant 2 \exp\left(-u^2/2(t_1^2 + \cdots + t_n^2)\right)$$

for every $u > 0$. This shows that $\mathcal{Y} = (Y_t)_{t \in \mathbb{R}^n}$ is sub-Gaussian with respect to the Euclidean metric.

A second example is given by Gaussian processes. We write g for a standard Gaussian random variable and $G = (g_1, \ldots, g_n)$ for the standard Gaussian random vector in $\mathbb{R}^n$. The distribution of G is the Gaussian measure γ_n, with density $(2\pi)^{-n/2} \exp(-\|x\|_2^2/2)$.

Let T be a non-empty set. A family $\mathcal{Z} = (Z_t)_{t \in T}$ of real valued random variables on $(\Omega, \mathcal{A}, P)$ is called a *Gaussian process* if, for any $a_1, \ldots, a_m \in \mathbb{R}$ and any $Z_{t_1}, \ldots, Z_{t_m} \in \mathcal{Z}$, the linear combination $a_1 Z_{t_1} + \cdots + a_m Z_{t_m}$ is a Gaussian random variable with mean 0. We may view $\mathcal{Z}$ as a subset of $L^2(\Omega)$, and then it induces on T the metric

$$d(t,s) = \|Z_t - Z_s\|_{L^2(\Omega)}.$$

By the definition of a Gaussian process, for every $t, s \in T$, $Z_t - Z_s$ is a Gaussian random variable with mean 0 and variance $\mathbb{E}(Z_t - Z_s)^2 = d^2(t,s)$. Consequently, for every $u > 0$ we have

$$\mathbb{P}(|Z_t - Z_s| \geqslant u) = \frac{2}{d(t,s)\sqrt{2\pi}} \int_u^\infty \exp\left(-\frac{r^2}{2d^2(t,s)}\right) dr \leqslant 2 \exp\left(-\frac{u^2}{d^2(t,s)}\right),$$

which implies that $\mathcal{Z}$ is sub-Gaussian with respect to the metric d it induces to T.

EXAMPLES 1.9.1. (i) If $g_1, \ldots, g_N$ are independent standard Gaussian random variables on $(\Omega, \mathcal{A}, P)$, then $\mathcal{Z} = \{g_1, \ldots, g_N\}$ is a Gaussian process.

(ii) Consider n independent standard Gaussian random variables $g_1, \ldots, g_n$. For every non-empty $T \subseteq \mathbb{R}^n$ we define a process $\mathcal{Z} = (Z_t)_{t \in T}$, by

$$Z_t(\omega) = \langle t, G(\omega) \rangle = \left\langle t, \sum_{i=1}^n g_i(\omega) e_i \right\rangle = \sum_{i=1}^n \langle t, e_i \rangle g_i(\omega),$$

where $\{e_1, \ldots, e_n\}$ is an orthonormal basis of $\mathbb{R}^n$ and $G = (g_1, \ldots, g_n)$. Then, $\mathcal{Z}$ is a Gaussian process and the induced metric is the Euclidean metric on $\mathbb{R}^n$: for all $t, s \in T$,
$$d(t,s) = \|Z_t - Z_s\|_{L^2(\Omega)} = \|t - s\|_2.$$

DEFINITION 1.9.2. Let (T, d) be a metric space and let $\mathcal{Y} = (Y_t)_{t \in T}$ be a sub-Gaussian process with respect to d. We define
$$\mathbb{E} \sup_{t \in T} Y_t = \sup \left\{ \mathbb{E} \max_{t \in F} Y_t : F \subseteq T, 0 < |F| < \infty \right\}.$$

An important question is to obtain sharp upper bounds for the expectation $\mathbb{E} \sup Y_t$ in terms of the geometry of (T, d); in the next two subsections we discuss a number of important related results, which will be used several times in this book.

REMARK 1.9.3. A basic observation, which connects this question with convex geometric analysis, is that if we consider a convex body K in $\mathbb{R}^n$ and the Gaussian process $\mathcal{Z} = (Z_t)_{t \in K}$, where $Z_t(\omega) = \langle t, G(\omega) \rangle$, that was defined in Example 1.9.1 (ii), then

(1.9.1)
$$\mathbb{E} \sup_{t \in K} Z_t = \mathbb{E} \sup_{t \in K} \langle t, G \rangle = \mathbb{E}\, h_K(G)$$
$$= \frac{1}{(2\pi)^{n/2}} \int_{\mathbb{R}^n} h_K(x) e^{-|x|^2/2} dx \simeq \sqrt{n} \int_{S^{n-1}} h_K(\theta) \sigma(d\theta) \simeq \sqrt{n}\, w(K).$$

We just used the fact that the distribution of G is the standard Gaussian measure on $\mathbb{R}^n$ and integration in polar coordinates.

1.9.2. Metric entropy — the case of Gaussian processes

Let (T, d) be a metric space. For every $\varepsilon > 0$ we define
$$N(T, d, \varepsilon) = \min \left\{ N : \text{there exist } t_1, \ldots, t_N \in T : T \subseteq \bigcup_{i=1}^{N} B(t_i, \varepsilon) \right\},$$
where $B(t, \varepsilon) = \{s \in T : d(t,s) < \varepsilon\}$. The function $\varepsilon \mapsto \log N(T, d, \varepsilon)$ is the *metric entropy function* of T.

Consider as an example the Gaussian process $\mathcal{Z} = \{g_1, \ldots, g_N\}$, $N \geq 2$. We easily check that $\|g_i - g_j\|_2 = \sqrt{2}$ if $i \neq j$, and hence, $N(\varepsilon) = N$ if $0 < \varepsilon \leq \sqrt{2}$ and $N(\varepsilon) = 1$ if $\varepsilon > \sqrt{2}$. Also, using the fact that g_i are independent we may check that
$$\mathbb{E} \max_{1 \leq i \leq N} g_i \simeq \sqrt{\log N}.$$

Let $\mathcal{Z} = (Z_t)_{t \in T}$ be a Gaussian process. We view T as a metric space with the induced metric d. The next theorem gives upper and lower bounds for $\mathbb{E} \sup Z_t$ in terms of the metric entropy function of (T, d).

THEOREM 1.9.4 (Sudakov-Dudley). *There exist constants $c_1, c_2 > 0$ with the following property: if $\mathcal{Z} = (Z_t)_{t \in T}$ is a Gaussian process and d is the induced metric, then*
$$c_1 \sup_{\varepsilon > 0} \varepsilon \sqrt{\log N(T, d, \varepsilon)} \leq \mathbb{E} \sup_{t \in T} Z_t \leq c_2 \int_0^\infty \sqrt{\log N(T, d, \varepsilon)}\, d\varepsilon.$$

The left hand side inequality is Sudakov's inequality, while the right hand side inequality is Dudley's inequality. In the example of $\mathcal{Z} = \{g_1, \ldots, g_N\}$, both bounds give the right order of $\mathbb{E} \sup g_i$. The proof of Sudakov's inequality is based on a classical comparison lemma of Slepian. We will discuss Dudley's inequality in the more general context of sub-Gaussian processes.

THEOREM 1.9.5 (Slepian). *If $(X_1, \ldots, X_N)$ and $(Y_1, \ldots, Y_N)$ are two N-tuples of Gaussian random variables with mean 0 which satisfy the condition*
$$\|X_i - X_j\|_2 \leqslant \|Y_i - Y_j\|_2$$
for all $i \neq j$, then
$$\mathbb{E} \max_{i \leqslant N} X_i \leqslant \mathbb{E} \max_{i \leqslant N} Y_i.$$

Proof of Sudakov's inequality. We use Slepian's lemma as follows: Let $\mathcal{Z} = (Z_t)_{t \in T}$ be a Gaussian process and let d be the induced metric. Given $\varepsilon > 0$ we consider a subset $\{t_1, \ldots, t_N\}$ of T which is maximal with respect to the condition "$d(t,s) \geqslant \varepsilon$ if $t \neq s$". Then $T \subseteq \bigcup_{i=1}^N B(t_i, \varepsilon)$, which implies $N(T, d, \varepsilon) \leqslant N$.

If $\delta = \min \|Z_{t_i} - Z_{t_j}\|_2$, we consider the N-tuple $\left(\frac{\delta g_1}{\sqrt{2}}, \ldots, \frac{\delta g_N}{\sqrt{2}}\right)$, where g_i are independent standard Gaussian random variables. If $i \neq j$ then
$$\left\|\frac{\delta g_i}{\sqrt{2}} - \frac{\delta g_j}{\sqrt{2}}\right\|_2 = \delta \leqslant \|Z_{t_i} - Z_{t_j}\|_2,$$
so we can apply Slepian's lemma. It follows that
$$\mathbb{E} \sup_{t \in T} Z_t \geqslant \mathbb{E} \max_{i \leqslant N} Z_{t_i} \geqslant \frac{\delta}{\sqrt{2}} \mathbb{E} \max_{i \leqslant N} g_i \geqslant c_1 \varepsilon \sqrt{\log N}.$$
Thus, $\mathbb{E} \sup_{t \in T} Z_t \geqslant c_1 \sup_{\varepsilon > 0} \varepsilon \sqrt{\log N(T, d, \varepsilon)}$. $\square$

1.9.3. Dudley's bound for sub-Gaussian processes

Dudley's inequality is more generally valid for sub-Gaussian processes. The proof uses a successive approximation argument which we briefly describe:

Proof of Dudley's inequality. We assume that (T, d) is a metric space and $\mathcal{Y} = (Y_t)_{t \in T}$ is a process such that $\mathbb{E}(Y_t) = 0$ for all $t \in T$ and, for all $t, s \in T$ and every $u > 0$,
$$\mathbb{P}(|Y_t - Y_s| \geqslant u) \leqslant 2 \exp\left(-\frac{u^2}{d^2(t,s)}\right).$$
We consider a non-empty finite subset F of T and fix $t_0 \in F$. We set $R = \max\{d(t, t_0) : t \in F\}$ and $r_k = R/2^k$ for all $k \geqslant 0$.

We define $A_0 = \{t_0\}$ and for every $k \geqslant 1$ we find $A_k \subseteq F$ with cardinality $|A_k| = N(F, d, r_k)$ such that $F \subseteq \bigcup_{t \in A_k} B(t, r_k)$. Finally, for every $t \in F$ and $k \geqslant 0$ we choose $\pi_k(t) \in A_k$ with the property $d(t, \pi_k(t)) \leqslant r_k$. Since F is finite, for every $t \in F$ we eventually have $\pi_k(t) = t$. Note also that
$$d(\pi_k(t), \pi_{k-1}(t)) \leqslant r_k + r_{k-1} = 3r_k.$$
For every $t \in F$ we write
$$Y_t - Y_{t_0} = \sum_{k=1}^{\infty} \left(Y_{\pi_k(t)} - Y_{\pi_{k-1}(t)}\right)$$

and, using the fact that $\mathbb{E}(Y_{t_0}) = 0$,
$$\mathbb{E}\max_{t\in F} Y_t = \mathbb{E}\max_{t\in F}(Y_t - Y_{t_0}) = \int_0^\infty P\left(\max_{t\in F}(Y_t - Y_{t_0}) \geq u\right)du.$$

We fix $\alpha_k > 0$ (which will be suitably chosen) with $S := \sum \alpha_k < \infty$ and set $B_k = \{(w,z) \in A_k \times A_{k-1} : d(w,z) \leq 3r_k\}$. Using the sub-Gaussian assumption we write

$$P\left(\max_{t\in F}(Y_t - Y_{t_0}) \geq uS\right) \leq P\left(\sum_{k=1}^\infty \max_{t\in F}(Y_{\pi_k(t)} - Y_{\pi_{k-1}(t)}) \geq \sum_{k=1}^\infty u\alpha_k\right)$$
$$\leq \sum_{k=1}^\infty P\left(\max_{(w,z)\in B_k}(Y_w - Y_z) \geq u\alpha_k\right)$$
$$\leq \sum_{k=1}^\infty \exp(-u^2\alpha_k^2/9r_k^2)|A_k|\cdot|A_{k-1}|.$$

We now choose $\alpha_k = 3r_k\sqrt{\log(2^k|A_k|^2)}$. For every $u \geq 1$ and every k we have
$$\exp(-u^2\alpha_k^2/9r_k^2)|A_k|\cdot|A_{k-1}| \leq |A_k|^2\left(2^k|A_k|^2\right)^{-u^2} \leq 2^{-u^2 k},$$
and hence
$$P\left(\max_{t\in F}(Y_t - Y_{t_0}) \geq uS\right) \leq \sum_{k=1}^\infty 2^{-u^2 k} \leq c2^{-u^2}.$$

This shows that
$$\mathbb{E}\max_{t\in F} Y_t = S\int_0^\infty P\left(\max_{t\in F}(Y_t - Y_{t_0}) \geq uS\right)du$$
$$\leq S + S\int_1^\infty c2^{-u^2}du \leq c'S.$$

Going back to the definition of S we see that
$$S = \sum_{k=1}^\infty \frac{3R}{2^k}\left(k\sqrt{\log 2} + \sqrt{\log N(F,d,R/2^k)}\right)$$
$$\simeq \sum_{k=1}^\infty \frac{R}{2^k}\sqrt{\log N(F,d,R/2^k)}$$
$$\simeq \int_0^\infty \sqrt{\log N(F,d,\varepsilon)}d\varepsilon.$$

Since $N(F,d,\varepsilon) \leq N(T,d,\varepsilon)$, the proof is complete. $\square$

1.9.4. Majorizing measures

Dudley's bound is not always sharp, as one can see by the following example: Consider an infinite sequence $\{g_n\}$ of independent standard Gaussian random variables, fix $a = (a_n) \in \ell_2$ and define the ellipsoid
$$\mathcal{E} = \left\{t = (t_n) \in \ell_2 : \sum_{n=1}^\infty t_n^2/a_n^2 \leq 1\right\}.$$

If we set $Z_t = \sum_n t_n g_n$, then $\mathcal{Z} = (Z_t)_{t\in\mathcal{E}}$ is a Gaussian process and

$$\mathbb{E}\sup_{t\in\mathcal{E}} Z_t \simeq \left(\sum_{n=1}^\infty a_n^2\right)^{1/2} < \infty.$$

On the other hand, one can choose $a \in \ell_2$ so that "Dudley's integral" will diverge.

Starting with the argument that we presented in the previous subsection one may check that the following (genuinely better) version of Dudley's inequality can be obtained along the same lines (see [**491**, Section 1.2]).

THEOREM 1.9.6. *Let (T,d) be a metric space and let $\{Y_t\}_{t\in T}$ be a sub-Gaussian process with respect to d. Assume that $\{T_n\}_{n\geqslant 0}$ is a sequence of subsets of T such that $|T_0| = 1$ and $|T_n| \leqslant 2^{2^n}$ for all $n \geqslant 1$. Then,*

$$\mathbb{E}\sup_{t\in T} Y_t \leqslant C \sup_{t\in T} \sum_{n=0}^\infty 2^{n/2} d(t, T_n),$$

where $C > 0$ is an absolute constant.

Talagrand's majorizing measure theorem, which we describe below, shows that the upper bound of Theorem 1.9.6 provides the correct estimate for $\mathbb{E}\sup_{t\in T} Z_t$ for every Gaussian process $\mathcal{Z} = (Z_t)_{t\in T}$.

Let (T,d) be a metric space and let $\{\mathcal{A}_n\}_{n=0}^\infty$ be an increasing sequence of partitions of T; the term *increasing* means that $\mathcal{A}_{n+1}$ is a refinement of $\mathcal{A}_n$ for every $n \geqslant 0$. We say that $\{\mathcal{A}_n\}$ is *admissible* if $\mathcal{A}_0 = \{T\}$ and $|\mathcal{A}_n| \leqslant 2^{2^n}$ for every $n \geqslant 1$. Given a partition P of T, for any $t \in T$ we denote by $P(t)$ the set from P which contains t.

Now, let $\{Y_t\}_{t\in T}$ be a sub-Gaussian process with respect to d. Let $\{\mathcal{A}_n\}$ be an admissible sequence of partitions of T. Given n, consider a subset T_n of T which contains exactly one point from each set in the partition $\mathcal{A}_n$. Then, Theorem 1.9.6 shows that

$$\mathbb{E}\sup_{t\in T} Y_t \leqslant C \sup_{t\in T} \sum_{n=0}^\infty 2^{n/2} d(t, T_n) \leqslant C \sup_{t\in T} \sum_{n=0}^\infty 2^{n/2} \mathrm{diam}(\mathcal{A}_n(t)).$$

The next theorem shows that in the case of Gaussian processes this bound is optimal.

THEOREM 1.9.7 (Talagrand). *There exists an absolute constant $c_0 > 0$ with the following property: if $\{Z_t\}_{t\in T}$ is a Gaussian process, then there exists an admissible sequence $\{\mathcal{A}_n\}$ of partitions of T such that*

$$\mathbb{E}\sup_{t\in T} Z_t \geqslant c_0 \sup_{t\in T} \sum_{n=0}^\infty 2^{n/2} \mathrm{diam}(\mathcal{A}_n(t)).$$

In other words,

$$\mathbb{E}\sup_{t\in T} Z_t \simeq \inf_{\{\mathcal{A}_n\}} \sup_{t\in T} \sum_{n=0}^\infty 2^{n/2} \mathrm{diam}(\mathcal{A}_n(t)).$$

A direct consequence of the above is the following comparison theorem.

THEOREM 1.9.8 (Talagrand). *Let $\mathcal{Z} = (Z_t)_{t \in T}$ be a Gaussian process and let d be the induced metric. If the process $\mathcal{Y} = (Y_t)_{t \in T}$ is sub-Gaussian with respect to d, then*
$$\mathbb{E} \sup_{t \in T} Y_t \leqslant C \cdot \mathbb{E} \sup_{t \in T} Z_t,$$
where $C > 0$ is an absolute constant.

1.10. Dvoretzky type theorems

Dvoretzky theorem states that every high-dimensional normed space has a subspace of "large dimension" which is C-isomorphic to a Euclidean space, where C is an absolute constant. We use the terminology "Dvoretzky-type theorems" to refer to a wide family of results concerning the existence of large nice substructures inside normed spaces of sufficiently high dimension. One of the most crucial and important aspects of the theory is to find concrete estimates for the different parameters we are interested in, such as the dimension of the substructures in relation to the dimension of the whole space; there are many theorems which provide such estimates (or in some cases even asymptotic formulas) depending on the various parameters which appear in this type of results. Another important topic is to find the optimal dependences.

The starting point for the original Dvoretzky theorem has been a lemma of Dvoretzky and Rogers which shows that, given a symmetric convex body K whose maximal volume ellipsoid is B_2^n, we can find a k-dimensional subspace E of $\mathbb{R}^n$, with $k \simeq \sqrt{n}$, such that $B_2^n \cap E \subseteq K \cap E \subseteq 2Q_k$, where we write Q_k for the unit cube in E with respect to a suitable coordinate system. Therefore, we have

THEOREM 1.10.1 (Dvoretzky-Rogers). *Assume that B_2^n is the maximal volume ellipsoid of the symmetric convex body K. There exist $k \simeq \sqrt{n}$ and orthonormal vectors $z_1, \ldots, z_k$ in $\mathbb{R}^n$ such that for all $a_1, \ldots, a_k \in \mathbb{R}$,*
$$\frac{1}{2} \max_{i \leqslant k} |a_i| \leqslant \left\| \sum_{i=1}^k a_i z_i \right\| \leqslant \left(\sum_{i=1}^k a_i^2 \right)^{1/2}.$$

Grothendieck asked whether it is possible to replace Q_k by $B_2^n \cap E$ in the above statement, and still have k increase to infinity with n. Dvoretzky theorem provides an affirmative answer to this question. The best known version is the following

THEOREM 1.10.2 (Dvoretzky-Milman). *Let X be an n-dimensional normed space and $\varepsilon \in (0, 1)$. There exist an integer $k \geqslant c\varepsilon^2 \log n$ and a k-dimensional subspace F of X which satisfies $d(F, \ell_2^k) \leqslant 1 + \varepsilon$.*

In geometric terms this can be stated as follows: if K is a symmetric convex body in $\mathbb{R}^n$, then for every $\varepsilon \in (0, 1)$ we can find $k \geqslant c\varepsilon^2 \log n$, a subspace $F \in G_{n,k}$ and an ellipsoid $\mathcal{E}$ in F so that
$$\mathcal{E} \subseteq K \cap F \subseteq (1 + \varepsilon)\mathcal{E}.$$

The example of ℓ_∞^n shows that the logarithmic dependence of k on n is the best possible if we fix small values of ε. The exact way that all three parameters, n, ε and k, are related to each other has not been settled yet. It seems reasonable that ℓ_∞^n represents the worst case, and if this proved to be true it would imply that, for fixed k and ε, every n-dimensional normed space has a k-dimensional subspace which is $(1+\varepsilon)$-isomorphic to ℓ_2^k, provided that $n \geqslant c(k)\varepsilon^{-\frac{k-1}{2}}$. The problem is of great interest even for small values of k.

1.10.1. Proof of Dvoretzky theorem

Vitali Milman's proof of Theorem 1.10.2 utilizes the concentration of measure on S^{n-1}. We start with an n-dimensional normed space X, and we assume without loss of generality that B_2^n is the ellipsoid of maximal volume inscribed in the unit ball K of X. Observe that the function $r : S^{n-1} \to \mathbb{R}$ defined by $r(x) = \|x\|$, where $\|\cdot\|$ is the norm of X, is Lipschitz continuous with constant 1. If L_r is the Lévy median of r, Theorems 1.7.7 and 1.7.2 show that for every $t \in (0, 1)$

$$\sigma\left(x \in S^{n-1} : |r(x) - L_r| \geqslant tL_r\right) \leqslant \exp(-ct^2 L_r^2 n),$$

where $c > 0$ is an absolute constant. The idea is that, since the function $r(x) = \|x\|$ is almost constant and equal to L_r on a subset of the sphere whose measure is practically equal to 1, one can extract a subsphere on the whole of which r will be almost equal to L_r; this is done by a discretization argument via nets of spheres.

THEOREM 1.10.3 (Milman). *Let $X = (\mathbb{R}^n, \|\cdot\|)$ and assume that $\|x\| \leqslant \|x\|_2$ for all $x \in \mathbb{R}^n$. For every $\varepsilon \in (0,1)$ we can find $k \geqslant c\varepsilon^2 L_r^2 n$ and a k-dimensional subspace F of $\mathbb{R}^n$ such that*

$$(1+\varepsilon)^{-1/2} L_r \|x\|_2 \leqslant \|x\| \leqslant L_r (1+\varepsilon)^{1/2} \|x\|_2$$

for every $x \in F$. □

If $Y = (F, \|\cdot\|)$, it is clear that $d(Y, \ell_2^k) \leqslant 1 + \varepsilon$, and what remains to do is to give a lower bound for L_r. To this end, it is easier to work with the expectation

$$M = M(X) = \int_{S^{n-1}} \|x\| \, d\sigma(x),$$

of the norm on the sphere, and then a rather simple computation, based on measure concentration, shows that $L_r \simeq M$.

Finally, we make full use of the fact that B_2^n is the ellipsoid of maximal volume inscribed in K. By the Dvoretzky-Rogers lemma (Corollary 1.5.11), we can find an orthonormal basis $\{v_1, \ldots, v_n\}$ with $\|v_i\| \geqslant 1/2$ for all $i \leqslant n/2$. One may check that

$$M = \int_{S^{n-1}} \|\sum_{i=1}^n a_i v_i\| \, d\sigma(a) = \int_{S^{n-1}} \int_{E_2^n} \|\sum_{i=1}^n \epsilon_i a_i v_i\| \, d\epsilon d\sigma(a)$$

$$\geqslant \int_{S^{n-1}} \max_{1 \leqslant i \leqslant n} \|a_i v_i\| \, d\sigma(a) \geqslant \frac{1}{2} \int_{S^{n-1}} \max_{1 \leqslant i \leqslant n/2} |a_i| \, d\sigma(a) \geqslant c\sqrt{\log n/n},$$

where $c > 0$ is an absolute constant. Going back to Theorem 1.10.3 we conclude the proof of Theorem 1.10.2. □

1.10.2. The critical dimension

Let $X = (\mathbb{R}^n, \|\cdot\|)$ be a normed space and let a, b be the smallest positive constants so that $a^{-1}\|x\|_2 \leqslant \|x\| \leqslant b\|x\|_2$ for all $x \in \mathbb{R}^n$. The proof of Dvoretzky theorem in the previous section shows that a subspace E of X with $\dim E = \lfloor c\varepsilon^2 n(M/b)^2 \rfloor$ is $(1+\varepsilon)$-Euclidean with high probability. This inspires the following definition:

1.10. DVORETZKY TYPE THEOREMS

DEFINITION 1.10.4. *Let X be an n-dimensional normed space. We set $k(X)$ to be the largest positive integer $k \leqslant n$ for which*

$$\nu_{n,k}\left(\left\{E_k \in G_{n,k} : \tfrac{1}{2}M\|x\|_2 \leqslant \|x\| \leqslant 2M\|x\|_2, x \in E_k\right\}\right) \geqslant 1 - \frac{k}{n+k}.$$

In other words, $k(X)$ is the largest possible dimension $k \leqslant n$ such that the distance of most of the k-dimensional subspaces of X from the Euclidean space is at most 4. Note that the presence of M in the definition corresponds to the right normalization, since the average of $M(E)$ over $G_{n,k}$ is equal to M for all $1 \leqslant k \leqslant n$.

Starting from the proof of Dvoretzky theorem and using the equivalence $L_r \simeq M$ one can check that $k(X) \geqslant cn(M/b)^2$. Milman and Schechtman observed that an inverse inequality also holds true.

THEOREM 1.10.5 (Milman-Schechtman). *For every n-dimensional normed space X one has*

$$k(X) \leqslant 8n(M/b)^2.$$

Proof. We fix orthogonal subspaces $E_1, \ldots, E_t$ of dimension $k(X)$ such that $E = \sum_{i=1}^{t} E_i$ has dimension $n \geqslant m > n - k(X)$, and we write $\mathbb{R}^n = \sum_{i=1}^{t} E_i + E^\perp$. We may also expand $E^\perp$ to a $k(X)$-dimensional subspace E_{t+1} of $\mathbb{R}^n$ in such a way that $\dim(E_t \cap E_{t+1}) = k(X) + m - n$. By the definition of $k(X)$, most orthogonal images of each E_i are 4-Euclidean, and we can find $U \in O(n)$ such that

$$\tfrac{1}{2}M\|x\|_2 \leqslant \|x\| \leqslant 2M\|x\|_2, \qquad x \in U(E_i)$$

for all $i = 1, \ldots, t, t+1$. Every $x \in \mathbb{R}^n$ can be written in the form $x = \sum_{i=1}^{t} x_i + y$, where $x_i \in U(E_i)$ and $y \in U(E^\perp) \subset U(E_{t+1})$. Since the x_i are orthogonal, we get

$$\|x\| \leqslant 2M \sum_{i=1}^{t} \|x_i\|_2 + 2M\|y\|_2 \leqslant 2M\sqrt{t+1}\|x\|_2.$$

This shows that $b \leqslant 2M\sqrt{t+1}$, and since $t = \lfloor n/k(X) \rfloor$ we conclude that $k(X) \leqslant 8n(M/b)^2$. □

Combining all the above, we arrive at an *asymptotic formula* for $k(X)$:

(1.10.1) $$k(X) \simeq n(M/b)^2$$

for every n-dimensional normed space X. In the case of the classical spaces ℓ_p^n, $1 \leqslant p \leqslant \infty$, we can use this formula to compute the order of $k_p := k(\ell_p^n)$ as a function of p and n.

THEOREM 1.10.6. *If $1 \leqslant p \leqslant 2$ then $k_p \simeq n$, whereas if $2 < q < \infty$ then $c_1 n^{2/q} \leqslant k_q \leqslant c_2(q) n^{2/q}$, with $c_2(q) \simeq q$.*

When $p = \infty$, we have $k_p \geqslant c \log n$ from Dvoretzky theorem. This estimate turns out to be sharp.

THEOREM 1.10.7. $k_\infty \simeq \log n$.

The idea of the proof of Theorem 1.10.7 gives a more general result.

PROPOSITION 1.10.8. *If P is a polytope with m facets in $\mathbb{R}^k$, and if $B_2^k \subseteq P \subseteq aB_2^k$, then*

$$m \geqslant \exp\left(\frac{k}{2a^2}\right).$$

Proof. We can write
$$P = \{x \in \mathbb{R}^k : \langle x, v_j \rangle \leqslant 1, \, j \leqslant m\}$$
for some vectors $v_j \in \mathbb{R}^k$. Since $B_2^k \subseteq P$, we must have $\|v_j\|_2 \leqslant 1$ for every $j = 1, \ldots, m$. From our other assumption, that $P \subseteq aB_2^k$, it follows that for every $\theta \in S^{n-1}$ there exists $j \leqslant m$ for which $\langle \theta, v_j \rangle \geqslant 1/a$.

We set $u_j = v_j/\|v_j\|_2$, $j = 1, \ldots, m$. Since $\|v_j\|_2 \leqslant 1$,
$$\{\theta \in S^{k-1} : \langle \theta, v_j \rangle \geqslant 1/a\} \subseteq \{\theta \in S^{k-1} : \langle \theta, u_j \rangle \geqslant 1/a\},$$
and hence
$$(1.10.2) \qquad S^{k-1} \subset \bigcup_{j=1}^m \{\theta \in S^{k-1} : \langle \theta, u_j \rangle \geqslant 1/a\}.$$

For each j, $\{\theta \in S^{k-1} : \langle \theta, u_j \rangle \geqslant 1/a\}$ is a cap in S^{k-1}, centered at u_j and with angular radius $2\arcsin\frac{1}{2a}$. Using the next lemma we can estimate its measure.

LEMMA 1.10.9. *For every $u \in S^{k-1}$ and for every $\varepsilon \in (0,1)$ we set $C(u,\varepsilon) := \{\theta \in S^{k-1} : \langle u, \theta \rangle \geqslant \varepsilon\}$. Then,*
$$\sigma(C(u,\varepsilon)) \leqslant \exp(-\varepsilon^2 k/2).$$

Proof. The measure σ of $C(u,\varepsilon)$ is equal to the percentage of B_2^k which is occupied by the *spherical cone* which corresponds to $C(u,\varepsilon)$. Observe that this cone is contained in a Euclidean ball of radius $(1-\varepsilon^2)^{1/2}$, and hence
$$\sigma(C(u,\varepsilon)) \leqslant (1-\varepsilon^2)^{k/2} \leqslant \exp(-\varepsilon^2 k/2),$$
as claimed. $\square$

Now, applying (1.10.2) and Lemma 1.10.9 with $\varepsilon = 1/a$, we get
$$1 = \sigma(S^{k-1}) \leqslant m\sigma(C(u,\varepsilon)) \leqslant m\exp(-k/2a^2),$$
and we complete the proof of Proposition 1.10.8. $\square$

Proof of Theorem 1.10.7. We assume that for some $k \in \mathbb{N}$ there exists a k-dimensional subspace of ℓ_∞^n such that $d(F, \ell_2^k) \leqslant 4$. Then, there exists an ellipsoid $\mathcal{E}$ in F such that $\mathcal{E} \subset Q_n \cap F \subset 4\mathcal{E}$. The cube Q_n has $2n$ facets, which implies that $P := Q_n \cap F$ has $m \leqslant 2n$ facets. Thus, applying a linear transformation, we find a polytope $P_1 = T(P) \subset \mathbb{R}^k$ with m facets, which satisfies
$$B_2^k \subset P_1 \subset 4B_2^k.$$
Proposition 1.10.8 shows that $2n \geqslant m \geqslant \exp(k/32)$, which gives
$$k \leqslant 32\log(2n).$$
It follows that $k_\infty \leqslant 32\log(2n)$, and hence $k_\infty \simeq \log n$. $\square$

Recall that, if $X = (\mathbb{R}^n, \|\cdot\|)$ is an n-dimensional normed space, then the dual norm is defined by
$$\|x\|_* = \sup\{|\langle x,y \rangle| : \|y\| \leqslant 1\}.$$
If $a^{-1}\|x\|_2 \leqslant \|x\| \leqslant b\|x\|_2$ for all x, it is clear that $b^{-1}\|x\|_2 \leqslant \|x\|_* \leqslant a^{-1}\|x\|_2$ for all x. Thus, if we define
$$k_*(X) = k(X^*) \quad \text{and} \quad M^*(X) = M(X^*),$$
we have
$$k_* \simeq n(M^*/a)^2.$$

On the other hand, a trivial application of the Cauchy-Schwarz inequality shows that
$$MM^* \geqslant \left(\int_{S^{n-1}} \|x\|_*^{1/2}\|x\|^{1/2} d\sigma(x)\right)^2 \geqslant \left(\int_{S^{n-1}} |\langle x,x\rangle|^{1/2} d\sigma(x)\right)^2 = 1.$$

This gives
$$kk_* \geqslant cn^2 \frac{(MM^*)^2}{(ab)^2} \geqslant \frac{cn^2}{(ab)^2}.$$

Using John's theorem, we can always bring the unit ball of X to a position such that $ab \leqslant \sqrt{n}$. This immediately proves the next fact.

THEOREM 1.10.10. *For every n-dimensional normed space X there exists a Euclidean structure for which one has*
$$k(X)k_*(X) \geqslant cn,$$
where $c > 0$ is an absolute constant.

1.10.3. Volume ratio and Kashin's theorem

Let K be a symmetric convex body in $\mathbb{R}^n$. Recall that the *volume ratio* of K is the quantity
$$vr(K) = \inf\left\{\left(\frac{|K|}{|\mathcal{E}|}\right)^{1/n} : \mathcal{E} \subseteq K\right\},$$
where the infimum is taken over all the ellipsoids which are contained in K. It is easy to check that the volume ratio is invariant under invertible linear transformations of $\mathbb{R}^n$.

EXAMPLE 1.10.11. Let K be a symmetric convex body in $\mathbb{R}^n$. If $\|\cdot\|$ is the norm induced by K, then
$$\int_{\mathbb{R}^n} e^{-\|x\|^p} dx = \int_{\mathbb{R}^n} \int_{\|x\|}^{\infty} pt^{p-1} e^{-t^p} dt dx = \int_0^{\infty} pt^{p-1} e^{-t^p} |\{x : \|x\| \leqslant t\}| dt$$
$$= |K| \int_0^{\infty} pt^{n+p-1} e^{-t^p} dt = |K|\Gamma\left(\frac{n}{p} + 1\right).$$

If we choose $K = B_p^n$, $1 \leqslant p < \infty$, we see that
$$\int_{\mathbb{R}^n} e^{-\|x\|^p} dx = \left(2\int_0^{\infty} e^{-t^p} dt\right)^n = [2\Gamma(1/p+1)]^n.$$

Therefore,
$$|B_p^n| = \frac{[2\Gamma(\frac{1}{p}+1)]^n}{\Gamma(\frac{n}{p}+1)}.$$

Observe that if $1 \leqslant p \leqslant 2$ then the maximal volume ellipsoid of B_p^n is $n^{\frac{1}{2}-\frac{1}{p}} B_2^n$. It follows that
$$vr(B_p^n) = \frac{2\Gamma(\frac{1}{p}+1)[\Gamma(\frac{n}{2}+1)]^{\frac{1}{n}}}{n^{\frac{1}{2}-\frac{1}{p}}[\Gamma(\frac{n}{p}+1)]^{\frac{1}{n}}\sqrt{\pi}} \leqslant C,$$
where $C > 0$ is an absolute constant. We say that the unit balls of ℓ_p^n, $1 \leqslant p \leqslant 2$ have *uniformly bounded volume ratio*.

The next theorem asserts that, if a body K has bounded volume ratio, then the space X_K contains subspaces F of dimension proportional to n which have bounded Banach-Mazur distance from $\ell_2^{\dim F}$. This fact was first proved by Kashin in the case of ℓ_1^n, and later Szarek and Tomczak-Jaegermann introduced the notion of volume ratio and proved the following

THEOREM 1.10.12 (Kashin, Szarek-Tomczak). *Let K be a symmetric convex body in $\mathbb{R}^n$ such that $B_2^n \subseteq K$ and $|K| = \alpha^n |B_2^n|$ for some $\alpha > 1$. Given $1 \leqslant k \leqslant n$, a random subspace $E \in G_{n,k}$ satisfies the following:*

$$B_E \subseteq K \cap E \subseteq (c\alpha)^{\frac{n}{n-k}} B_E,$$

with probability greater than $1 - e^{-n}$, where $c > 0$ is an absolute constant.

Proof. Since $B_2^n \subseteq K$, we have $\|x\| \leqslant \|x\|_2$ for every $x \in \mathbb{R}^n$. Let $k \leqslant n$. We may write

$$\int_{G_{n,k}} \int_{S_E} \|x\|^{-n} d\sigma_E(x) d\nu_{n,k}(E) = \int_{S^{n-1}} \|x\|^{-n} d\sigma(x) = \frac{|K|}{|B_2^n|} = \alpha^n.$$

From Markov's inequality, the measure of the set of $E \in G_{n,k}$ which satisfy

$$\int_{S_E} \|x\|^{-n} d\sigma_E(x) \leqslant (e\alpha)^n$$

is greater than $1 - e^{-n}$. Let E be such a subspace. Then, applying Markov's inequality again, we see that for every $r \in (0, 1)$

(1.10.3) $$\sigma_E\{x \in S_E : \|x\| < r\} \leqslant (er\alpha)^n.$$

We will use the following simple lemma.

LEMMA 1.10.13. *If $x \in S^{k-1}$ and $0 < t < 1$, then $\sigma_k(B(x,t)) \geqslant (t/3)^k$.*

Proof. There exists a t-net $\mathcal{N}$ of S^{k-1} of cardinality $|\mathcal{N}| \leqslant \left(1 + \frac{2}{t}\right)^k$. Since $S^{k-1} \subseteq \bigcup_{x \in \mathcal{N}} B(x,t)$ we must have $\sigma_k(B(x,t))|\mathcal{N}| \geqslant 1$. This implies that

$$\sigma_k(B(x,t)) \geqslant \left(\frac{t}{t+2}\right)^k \geqslant \left(\frac{t}{3}\right)^k,$$

and the lemma is proved. $\square$

We now return to the proof of Theorem 1.10.12. Let $E \in G_{n,k}$ satisfy (1.10.3). Fix $x \in S_E$. From the lemma we see that if $(er\alpha)^n < (r/6)^k$ then

$$C(x, r/2) \cap \{y \in S_E : \|y\| \geqslant r\} \neq \emptyset.$$

Then, we may find $y \in S_E$ with $\|x - y\|_2 \leqslant r/2$ and $\|y\| \geqslant r$. By the triangle inequality we get

$$\|x\| \geqslant \|y\| - \|x - y\| \geqslant r - \|x - y\|_2 \geqslant r/2.$$

Since $x \in S_E$ was arbitrary, this shows that

$$B_E \subseteq K \cap E \subseteq \frac{2}{r} B_E.$$

It remains to choose an optimal r: we want

$$e^n 6^k \alpha^n r^{n-k} < 1,$$

which gives $r_{\max} = (6e\alpha)^{-\frac{n}{n-k}}$. $\square$

1.10. DVORETZKY TYPE THEOREMS

REMARK 1.10.14. Theorem 1.10.12 says, for example, that if $1 \leqslant p \leqslant 2$ and $\lambda \in (0,1)$, then ℓ_p^n has subspaces F of dimension $k = \lfloor \lambda n \rfloor + 1$ with $d(F, \ell_2^k) \leqslant C_1^{\frac{1}{1-\lambda}}$, where $C_1 > 0$ is an absolute constant. Of course the estimate is bad when $\lambda \to 1$, but the distance remains uniformly bounded as long as, say, $\lambda \leqslant 1/2$.

The next result is a "global formulation" of the volume ratio theorem.

THEOREM 1.10.15. *Let K be a convex body in $\mathbb{R}^n$ such that $B_2^n \subseteq K$ and $|K| = \alpha^n |B_2^n|$ for some $\alpha > 1$. There exists $U \in O(n)$ with the property*
$$B_2^n \subset K \cap U(K) \subset c\alpha^2 B_2^n,$$
where $c > 0$ is an absolute constant.

Proof. Note that
$$\|x\|_{K \cap U(K)} = \max\{\|Ux\|, \|x\|\} \geqslant \frac{\|Ux\| + \|x\|}{2}$$
for all $U \in O(n)$ and $x \in \mathbb{R}^n$. Since $B_2^n \subset K \cap U(K)$ for every $U \in O(n)$, the theorem will follow if we find $U \in O(n)$ such that
$$N_U(\theta) := \frac{\|U\theta\| + \|\theta\|}{2} \geqslant \frac{1}{c\alpha^2}$$
for all $\theta \in S^{n-1}$. We have
$$\int_{O(n)} \int_{S^{n-1}} \frac{1}{\|U\theta\|^n \|\theta\|^n} d\sigma(\theta) d\nu(U) = \int_{S^{n-1}} \left(\int_{O(n)} \frac{1}{\|U\theta\|^n} d\nu(U) \right) \frac{1}{\|\theta\|^n} d\sigma(\theta)$$
$$= \int_{S^{n-1}} \left(\int_{S^{n-1}} \frac{1}{\|\phi\|^n} d\sigma(\phi) \right) \frac{1}{\|\theta\|^n} d\sigma(\theta)$$
$$= \left(\int_{S^{n-1}} \frac{1}{\|\theta\|^n} d\sigma(\theta) \right)^2$$
$$= \alpha^{2n}.$$

Therefore, we can find $U \in O(n)$ which satisfies
$$\int_{S^{n-1}} \left(\frac{2}{\|U\theta\| + \|\theta\|} \right)^{2n} d\sigma(\theta) \leqslant \int_{S^{n-1}} \frac{1}{\|U\theta\|^n \|\theta\|^n} d\sigma(\theta) \leqslant \alpha^{2n}.$$

Let $\theta \in S^{n-1}$ and set $N_U(\theta) = t$. If $\phi \in S^{n-1}$ and $\|\theta - \phi\|_2 \leqslant t$, then the fact that N_U is a norm with Lipschitz constant 1 gives
$$N_U(\phi) \leqslant N_U(\theta) + N_U(\phi - \theta) \leqslant t + \|\phi - \theta\|_2 \leqslant 2t.$$
On the other hand, $\sigma(B(\theta, t)) \geqslant (t/3)^n$, and hence
$$\left(\frac{t}{3} \right)^n \frac{1}{(2t)^{2n}} \leqslant \sigma(B(\theta, t)) \frac{1}{(2t)^{2n}} \leqslant \int_{S^{n-1}} \left(\frac{1}{N_U(\phi)} \right)^{2n} d\sigma(\phi) \leqslant \alpha^{2n}.$$

It is now clear that $t \geqslant 1/(c\alpha^2)$ for some absolute constant $c > 0$. This completes the proof. $\square$

REMARK 1.10.16. It is worth observing that the proofs of the two theorems proceed along the same lines. This is an instance of a much more general principle: local statements (like Theorem 1.10.12), which describe the geometry of proportional sections of a convex body K, frequently have analogous global statements (like Theorem 1.10.15), which relate K to its orthogonal images.

In the case of ℓ_1^n we get a very interesting application of Theorem 1.10.15.

THEOREM 1.10.17. *There exist vectors $y_1, \ldots, y_{2n} \in S^{n-1}$ such that*
$$c\sqrt{n}\|x\|_2 \leqslant \sum_{j=1}^{2n} |\langle x, y_j\rangle| \leqslant 2\sqrt{n}\|x\|_2$$
for every $x \in \mathbb{R}^n$, where $c > 0$ is an absolute constant.

Proof. The maximal volume ellipsoid of B_1^n is $n^{-1/2}B_2^n$, and its volume ratio is bounded by an absolute constant $C > 0$. From the proof of Theorem 1.10.15, we can find $U \in O(n)$ with the property
$$2\sqrt{n} \geqslant \|\theta\|_1 + \|U\theta\|_1 \geqslant \frac{\sqrt{n}}{C_1^2}$$
for every $\theta \in S^{n-1}$, where $C_1 > 0$ is an absolute constant. We set $y_i = e_i$ and $y_{n+i} = U^*(e_i)$, $i = 1, \ldots, n$. Then,
$$2\sqrt{n} \geqslant \sum_{j=1}^{2n} |\langle \theta, y_j\rangle| \geqslant \frac{\sqrt{n}}{C_1^2}$$
for every $\theta \in S^{n-1}$, and this completes the proof. $\square$

1.11. The ℓ-position and Pisier's inequality

1.11.1. ℓ-position

Let X be an n-dimensional normed space, and let α be a norm on the space $L(\ell_2^n, X)$ of linear operators $u : \ell_2^n \to X$. The *trace dual norm* is defined on $L(X^*, \ell_2^n)$ by
$$\alpha^*(v) = \sup\{\operatorname{tr}(vu) : \alpha(u) \leqslant 1\}.$$
The next lemma of Lewis applies to any pair of trace dual norms.

THEOREM 1.11.1. *For any norm α on $L(\ell_2^n, X)$, there exists $u : \ell_2^n \to X$ such that $\alpha(u) = 1$ and $\alpha^*(u^{-1}) = n$.*

The ℓ-*norm* on $L(\ell_2^n, X)$ was defined by Figiel and Tomczak-Jaegermann: Let $\{g_1, \ldots, g_n\}$ be a sequence of independent standard Gaussian random variables on some probability space, and let $\{e_1, \ldots, e_n\}$ be the standard orthonormal basis of $\mathbb{R}^n$. If $u : \ell_2^n \to X$, the ℓ-norm of u is defined by
$$\ell(u) = \left(\mathbb{E}\,\Big\|\sum_{i=1}^n g_i u(e_i)\Big\|^2\right)^{1/2}.$$
A standard computation gives
$$(1.11.1) \qquad \ell(u) \simeq \sqrt{n}\,w(u^*(K^\circ)),$$
where K is the unit ball of X. This formula connects the ℓ-norm to the mean width. It is more instructive to replace the Gaussians by the Rademacher functions $r_i : E_2^n \to \{-1, 1\}$ defined by $r_i(\epsilon) = \epsilon_i$, where $E_2^n = \{-1, 1\}^n$ is viewed as a

probability space with the uniform measure. An inequality of Maurey and Pisier shows that

$$\ell(u) \simeq \left(\int_{E_2^n} \Big\| \sum_{i=1}^n r_i(\epsilon) u(e_i) \Big\|^2 d\epsilon \right)^{1/2}$$

up to a $\sqrt{\log n}$-term.

Consider the Walsh functions $w_A(\varepsilon) = \prod_{i \in A} r_i(\varepsilon)$, where $A \subseteq \{1, \ldots, n\}$. It is not hard to see that every function $f : E_2^n \to X$ is uniquely represented in the form

$$f(\epsilon) = \sum_A w_A(\epsilon) x_A,$$

for some $x_A \in X$. The space of all functions $f : E_2^n \to X$ becomes a Banach space with the norm

$$\|f\|_{L_2(X)} = \left(\int_{E_2^n} \|f(\epsilon)\|^2 d\epsilon \right)^{1/2}$$

The *Rademacher projection* $R_n : L_2(X) \to L_2(X)$ is the operator sending $f = \sum w_A x_A$ to the function $R_n f := \sum_{i=1}^n r_i x_{\{i\}}$. Write $\mathrm{Rad}(X)$ for the operator norm of R_n. Figiel and Tomczak-Jaegermann proved the following.

THEOREM 1.11.2 (Figiel-Tomczak). *Let X be an n-dimensional normed space. There exists $u : \ell_2^n \to X$ such that*

$$\ell(u)\ell((u^{-1})^*) \leqslant n \mathrm{Rad}(X).$$

Let us briefly sketch the proof. From Theorem 1.11.1 we can find an isomorphism $u : \ell_2^n \to X$ such that $\ell(u)\ell^*(u^{-1}) = n$. On the other hand,

$$\ell\left((u^{-1})^*\right) = \left(\int_{E_2^n} \Big\| \sum_{i=1}^n r_i(\epsilon)(u^{-1})^*(e_i) \Big\|_*^2 d\epsilon \right)^{1/2}.$$

There exists a function $\phi : E_2^n \to X$, which can be represented in the form $\phi = \sum_A w_A x_A$ and has norm $\|\phi\|_{L_2(X)} = 1$, such that

$$\ell((u^{-1})^*) = \Big\langle \sum_{i=1}^n r_i(u^{-1})^*(e_i), \phi \Big\rangle = \sum_{i=1}^n \langle (u^{-1})^*(e_i), x_{\{i\}} \rangle.$$

If we define $v : \ell_2^n \to X$ by $v(e_i) = x_{\{i\}}$, we easily check that

$$\ell((u^{-1})^*) = \mathrm{tr}(u^{-1}v) \leqslant \ell^*(u^{-1})\ell(v).$$

Finally, observing that

$$\ell(v) = \|R_n(\phi)\|_{L_2(X)} \leqslant \mathrm{Rad}(X)\|\phi\|_{L_2(X)} = \mathrm{Rad}(X),$$

we get

$$\ell(u)\ell((u^{-1})^*) \leqslant \ell(u)\ell^*(u^{-1})\mathrm{Rad}(X) = n\mathrm{Rad}(X).$$

1.11.2. Pisier's inequality and the MM^*-estimate

Pisier gave a sharp estimate for $\mathrm{Rad}(X)$ in terms of the Banach-Mazur distance $d(X, \ell_2^n)$.

THEOREM 1.11.3 (Pisier). *Let X be an n-dimensional normed space. Then,*
$$\mathrm{Rad}(X) \leqslant c \log[d(X, \ell_2^n) + 1],$$
where $c > 0$ is an absolute constant.

Combined with the results of Lewis, Figiel and Tomczak-Jaegermann, Theorem 1.11.3 leads to the following statement (where we also use relation (1.11.1)).

THEOREM 1.11.4 (MM^*-estimate). *Let K be a symmetric convex body in $\mathbb{R}^n$. There exists a position $\tilde{K}$ of K for which*
$$w(\tilde{K}) w(\tilde{K}^\circ) \leqslant c \log[d(X_K, \ell_2^n) + 1],$$
where $c > 0$ is an absolute constant.

Computing the volume of $\tilde{K}$ in polar coordinates and using Hölder's inequality, we check that $w(\tilde{K}^\circ)^{-1} \leqslant c_2 \sqrt{n} |\tilde{K}|^{1/n}$. It follows that
$$w(\tilde{K}) \leqslant c\sqrt{n} \log n |\tilde{K}|^{1/n}.$$

Normalizing the volume we obtain the following *reverse Urysohn inequality*.

THEOREM 1.11.5. *If K is a symmetric convex body in $\mathbb{R}^n$, there exists a linear image $\tilde{K}$ of K with volume $|\tilde{K}| = 1$ and mean width*
$$w(\tilde{K}) \leqslant c\sqrt{n} \log n,$$
where $c > 0$ is an absolute constant.

Moreover, a simple argument based on the Rogers-Shephard inequality shows that we can remove the assumption of symmetry.

1.12. Milman's low M^*-estimate and the quotient of subspace theorem

1.12.1. Low M^*-estimate

Milman's low M^*-estimate is the first step towards a general theory of sections and projections of symmetric convex bodies in $\mathbb{R}^n$ onto subspaces with dimension proportional to n. In geometric terms, it says that for fixed $\lambda \in (0,1)$, the diameter of a random $\lfloor \lambda n \rfloor$-dimensional section of a body K is controlled by its mean width
$$w(K) = \int_{S^{n-1}} h_K(x) \, d\sigma(x)$$
up to a function depending only on λ.

THEOREM 1.12.1 (Milman). *There exists a function $f : (0,1) \to [0, \infty)$ with the following property: for every $\lambda \in (0,1)$ and every n-dimensional normed space X, a random subspace $H \in G_{n, \lfloor \lambda n \rfloor}$ satisfies*
$$\|x\| \geqslant \frac{f(\lambda)}{M^*} \|x\|_2 \quad \text{for every } x \in H,$$
where $M^* := w(K_X)$ is the mean width of the unit ball K_X of X.

1.12. MILMAN'S LOW M^*-ESTIMATE AND THE QUOTIENT OF SUBSPACE THEOREM

The precise dependence on λ has been established in a series of papers. Theorem 1.12.1 was originally proved by Milman, who also gave a second proof using the isoperimetric inequality on S^{n-1}, with a bound of the form $f(\lambda) \geqslant c(1-\lambda)$. Pajor and Tomczak-Jaegermann later showed that one can take $f(\lambda) \geqslant c\sqrt{1-\lambda}$. Finally, Gordon proved that the theorem holds true with

$$f(\lambda) \geqslant \sqrt{1-\lambda}\left(1 + O\left(\frac{1}{(1-\lambda)n}\right)\right).$$

Geometrically speaking, Theorem 1.12.1 says that for a random $\lfloor \lambda n \rfloor$-dimensional section of K_X we have

$$K_X \cap E \subseteq \frac{w(K_X)}{f(\lambda)} B_2^n \cap E,$$

that is, a random section does not have the same diameter as K_X but rather has radius $w(K_X)$, which is roughly the level r at which half of the supporting hyperplanes of rB_2^n cut the body K_X.

The dual formulation of the theorem has an interesting geometric interpretation too. A random λn-dimensional projection of K_X contains a ball of radius of the order of $1/M$. More precisely, for a random $E \in G_{n,\lfloor \lambda n \rfloor}$ we have

$$P_E(K_X) \supseteq \frac{f(\lambda)}{M} B_2^n \cap E.$$

Sketch of proof. We sketch Milman's proof of the inequality which gives linear dependence on λ. Consider the set $A = \{y \in S^{n-1} : \|y\|_* \leqslant 2M^*\}$. We obviously have $\sigma(A) \geqslant \frac{1}{2}$.

CLAIM 1.12.2. *For every $\lambda \in (0,1)$ there exists a subspace E of dimension $k = \lfloor \lambda n \rfloor$ such that*
$$E \cap S^{n-1} \subseteq A_{(\frac{\pi}{2} - \delta)},$$
where A_ε is the ε-extension of A on the sphere and $\delta \geqslant c(1-\lambda)$.

Proof. We have $\sigma(A_{\pi/4}) \geqslant 1 - c\sqrt{n} \int_0^{\pi/4} \sin^{n-2} t\, dt$, and integration over $G_{n,k}$ shows that a random $E \in G_{n,k}$ satisfies

$$\sigma_k(A_{\pi/4} \cap E) \geqslant 1 - c\sqrt{n} \int_0^{\pi/4} \sin^{n-2} t\, dt.$$

On the other hand, for every $x \in S^{n-1} \cap E$ we have

$$\sigma_k\left(B\left(x, \frac{\pi}{4} - \delta\right)\right) \simeq \sqrt{k} \int_0^{\frac{\pi}{4} - \delta} \sin^{k-2} t\, dt.$$

This implies that, if

(1.12.1) $$\sqrt{\lambda} \int_0^{\frac{\pi}{4} - \delta} \sin^{k-2} t\, dt \simeq \int_0^{\frac{\pi}{4}} \sin^{n-2} t\, dt,$$

then $A_{\pi/4} \cap B(x, \frac{\pi}{4} - \delta) \neq \emptyset$, and hence $x \in A_{\frac{\pi}{2} - \delta}$. Analyzing condition (1.12.1), we see that we can choose $\delta \geqslant c(1-\lambda)$. $\square$

We complete the proof of Theorem 1.12.1 as follows. Let $x \in S^{n-1} \cap E$. There exists $y \in A$ such that

$$\sin \delta \leqslant |\langle x, y \rangle| \leqslant \|y\|_* \|x\| \leqslant 2M^* \|x\|,$$

and since $\sin \delta \geqslant \frac{2}{\pi} \delta \geqslant c'(1-\lambda)$, the theorem follows. $\square$

1.12.2. Quotient of subspace theorem

Milman's *quotient of subspace theorem* states the following: by performing two operations in an n-dimensional space, first selecting a subspace and then a quotient of it, we can always arrive at a new space of dimension proportional to n which is close to being Euclidean (independently of n).

In order to interpret this in the language of convex bodies, observe that if K is the unit ball of $X = (\mathbb{R}^n, \|\cdot\|)$ and if $G \subseteq E \subseteq X$ then E/G is isometrically isomorphic to the subspace $F := E \cap G^\perp$ equipped with the norm induced by $P_F(K \cap E)$. We write $QS(X)$ for the class of all quotient spaces of subspaces of X; a space $Y \in QS(X)$ is of the form E/G where $G \subset E \subset X$. It is useful to note that $QS(X)$ is the same as the class $SQ(X)$ of all subspaces of quotient spaces of X. Indeed, if $F \subset E \subset \mathbb{R}^n$ one sees that $P_F(K \cap E) = (P_{F+E^\perp}(K)) \cap F$. This implies the following very useful property: if $Y \in QS(X)$ then every subspace or quotient space of Y also belongs to $QS(X)$ and $QS(Y) \subseteq QS(X)$. Thus, every iteration of the operation of choosing a quotient of a subspace leads to an element of $QS(X)$.

THEOREM 1.12.3 (Milman). *Let X be an n-dimensional normed space. For every $1 \leqslant k \leqslant n$ there exists $Y \in QS(X)$ with $\dim(Y) = n - k$ and*

$$(1.12.2) \qquad d(Y, \ell_2^{n-k}) \leqslant C \frac{n}{k} \log\left(\frac{Cn}{k}\right),$$

where $C > 0$ is an absolute constant.

Geometrically, the quotient of subspace theorem asserts that for every centrally symmetric convex body K in $\mathbb{R}^n$ and any $\alpha \in (0,1)$ we can find subspaces $F \subseteq E$ with $\dim(F) \geqslant \alpha n$ and an ellipsoid $\mathcal{E}$ in F such that

$$\mathcal{E} \subset P_F(K \cap E) \subset c(1-\alpha)^{-1}|\log(c(1-\alpha))|\mathcal{E}.$$

The proof of the theorem is based on the low M^*-estimate and an iteration procedure which makes essential use of the ℓ-position.

Proof. We may assume that K_X is in ℓ-position: then, by Theorem 1.11.4 we have $M(X)M^*(X) \leqslant c\log[d(X, \ell_2^n) + 1]$.

Step 1. Let $\lambda \in (0,1)$. We show that there exist a subspace E of X with $\dim(E) \geqslant \lambda n$ and a subspace F of E^* with $\dim(F) = k \geqslant \lambda^2 n$, such that $d(F, \ell_2^k) \leqslant c(1-\lambda)^{-1} \log[d(X, \ell_2^n) + 1]$.

The proof of this fact follows from a double application of the low M^*-estimate. By Theorem 1.12.1 a random λn-dimensional subspace E of X satisfies

$$\frac{c_1\sqrt{1-\lambda}}{M^*(X)}\|x\|_2 \leqslant \|x\| \leqslant b\|x\|_2, \qquad x \in E.$$

Moreover, since this is true for a random $E \in G_{n,\lambda n}$, we may also assume that $M(E) \leqslant c_2 M(X)$. Repeating the same argument for E^*, we may find a subspace F of E^* with $\dim(F) = k \geqslant \lambda^2 n$ and

$$\frac{c_3\sqrt{1-\lambda}}{M(X)}\|x\|_2 \leqslant \frac{c_1\sqrt{1-\lambda}}{M^*(E^*)}\|x\|_2 \leqslant \|x\|_{E^*} \leqslant \frac{M^*(X)}{c_1\sqrt{1-\lambda}}\|x\|_2$$

for every $x \in F$. Since K_X is in ℓ-position, we obtain

$$d(F, \ell_2^k) \leqslant c_4(1-\lambda)^{-1}M(X)M^*(X) \leqslant c(1-\lambda)^{-1}\log[d(X, \ell_2^n) + 1].$$

Step 2. Denote by $QS(X)$ the class of all quotient spaces of a subspace of X and define a function $f : (0,1) \to [0,\infty)$ by
$$f(\alpha) = \inf\{d(F, \ell_2^k) : F \in QS(X), \dim F \geqslant \alpha n\}.$$
Then, what we have really proved in Step 1 is the estimate
$$f(\lambda^2 \alpha) \leqslant c(1-\lambda)^{-1} \log f(\alpha).$$
An iteration lemma allows us to deduce that
$$f(\alpha) \leqslant c(1-\alpha)^{-1}|\log(1-\alpha)|,$$
and thus conclude the proof. $\square$

1.13. Bourgain-Milman inequality and the M-position

1.13.1. The Bourgain-Milman inequality

In Subsection 1.3.4 we saw that, for every symmetric convex body K in $\mathbb{R}^n$, the volume product $s(K) = |K||K^\circ|$ is less than or equal to the volume product $s(B_2^n)$; this is the Blaschke-Santaló inequality. In the opposite direction, a well-known conjecture of Mahler states that $s(K) \geqslant s(Q_n) = 4^n/n!$ for every symmetric convex body K, where $Q_n = [-1,1]^n$ is the n-dimensional cube. This has been verified for some classes of bodies, e.g. zonoids and 1-unconditional bodies but in general Mahler's question remains open. However, the Bourgain–Milman inequality does provide an "affirmative" answer to it in the asymptotic sense: for every symmetric convex body K in $\mathbb{R}^n$, the n-th root $s(K)^{1/n}$ of the volume product is of the order of $1/n$. Note that for many applications, some of which we will see in the rest of this book, knowing the order of the n-th root of the volume product suffices.

THEOREM 1.13.1 (Bourgain-Milman). *There exists an absolute constant $c > 0$ such that*
$$\left(\frac{s(K)}{s(B_2^n)}\right)^{1/n} \geqslant c$$
for every symmetric convex body K in $\mathbb{R}^n$.

The original proof of the Bourgain-Milman inequality (also called "reverse Santaló inequality") employed a dimension descending procedure which was based on Milman's quotient of subspace theorem. Later, Milman offered a second approach, introducing an "isomorphic symmetrization" technique. This symmetrization scheme, which we describe below, is closer to classical convexity, much more geometric in nature, and does preserve dimension unlike the procedure in the original proof of the reverse Santaló inequality; however, it is a symmetrization scheme which is in many ways different from the classical symmetrizations. At each step of the inductive procedure Milman's proof employs, none of the natural parameters of the body is being preserved, but the ones which are of interest remain under control. The MM^*-estimate is again crucial for the proof.

Since $s(K)$ is an affine invariant, we may start from a position of K which satisfies the inequality $M(K)M^*(K) \leqslant c\log[d(X_K, \ell_2^n)+1]$. We may also normalize so that $M(K) = 1$. We define
$$\lambda_1 = M^*(K)a_1 \quad \text{and} \quad \lambda_1' = M(K)a_1$$

for some $a_1 > 1$ which will be suitably chosen, and consider a new body
$$K_1 := \mathrm{conv}\left((K \cap \lambda_1 B_2^n) \cup \frac{1}{\lambda_1'} B_2^n\right).$$
Sudakov's inequality and elementary properties of covering numbers show that
$$|K_1| \geqslant |K \cap \lambda_1 B_2^n| \geqslant |K|/N(K, \lambda_1 B_2^n) \geqslant |K| \exp(-cn/a_1^2).$$
In an analogous way, using the dual Sudakov inequality one can show that
$$|K_1| \leqslant |\mathrm{conv}(K \cup (1/\lambda_1') B_2^n)| \leqslant |K| \exp(cn/a_1^2).$$
By the way K_1 is defined, one can apply the same reasoning with K_1°, and finally obtain that
$$\exp(-c/a_1^2) \leqslant \left(\frac{s(K_1)}{s(K)}\right)^{1/n} \leqslant \exp(c/a_1^2).$$
By construction, for the new body K_1 we have $d(X_{K_1}, \ell_2^n) \leqslant M(K)M^*(K)a_1^2$ and, since $s(K_1)$ is a linear invariant, we may also assume that K_1 is in (a suitably normalized) ℓ-position, so that $M(K_1)M^*(K_1) \leqslant c \log[d(X_{K_1}, \ell_2^n)+1]$ and $M(K_1) = 1$. If we set $\lambda_2 = M^*(K_1)a_2$, $\lambda_2' = M(K_1)a_2$ and define
$$K_2 = \mathrm{conv}\left((K_1 \cap \lambda_2 B_2^n) \cup \frac{1}{\lambda_2'} B_2^n\right)$$
for some $a_2 > 1$, we obtain
$$\exp(-c/a_2^2) \leqslant \left(\frac{s(K_2)}{s(K_1)}\right)^{1/n} \leqslant \exp(c/a_2^2).$$
We now iterate this procedure, choosing $a_1 = \log n$, $a_2 = \log \log n, \ldots, a_t = \log^{(t)} n$ – the t-iterated logarithm of n, and stop the procedure at the first t for which $a_t < 2$. It is easy to check that $d(X_{K_t}, \ell_2^n) \leqslant C$, therefore
$$\frac{1}{C} \leqslant \left(\frac{s(K_t)}{s(B_2^n)}\right)^{1/n} \leqslant C.$$
On the other hand,
$$c_1 \leqslant \exp\left(-c\left(\frac{1}{a_1^2} + \cdots + \frac{1}{a_t^2}\right)\right) \leqslant \left(\frac{s(K)}{s(K_t)}\right)^{1/n} \leqslant \exp\left(c\left(\frac{1}{a_1^2} + \cdots + \frac{1}{a_t^2}\right)\right),$$
which proves the theorem (observe that the series $\frac{1}{a_1^2} + \cdots + \frac{1}{a_t^2} + \cdots$ remains bounded by an absolute constant, irrespective of the final number of steps). □

1.13.2. M-position

The existence of an "M-ellipsoid" associated with any centered convex body K in $\mathbb{R}^n$ was first proved by Milman.

THEOREM 1.13.2 (Milman). *There exists an absolute constant $c > 0$ such that the following holds: given a symmetric convex body K in $\mathbb{R}^n$ we can find an ellipsoid $\mathcal{E}_K$ satisfying $|K| = |\mathcal{E}_K|$ and*

(1.13.1) $$\frac{1}{c}|\mathcal{E}_K + T|^{1/n} \leqslant |K + T|^{1/n} \leqslant c|\mathcal{E}_K + T|^{1/n},$$
$$\frac{1}{c}|\mathcal{E}_K^\circ + T|^{1/n} \leqslant |K^\circ + T|^{1/n} \leqslant c|\mathcal{E}_K^\circ + T|^{1/n}$$

for every convex body T in $\mathbb{R}^n$.

Sketch of proof. We use the same sequence of bodies as in the proof of Theorem 1.13.1 that we described above. For every s, we check that

$$\exp(-cn/a_s^2) \leqslant \frac{|K_s + T|}{|K_{s-1} + T|} \leqslant \exp(cn/a_s^2),$$

for every convex body T, and the same holds true with K_s°. After t steps, we arrive at a body K_t whose geometric distance from an ellipsoid $\mathcal{E}$ is bounded by an absolute constant c; if we normalize so that $|K_t| = |\mathcal{E}|$, then K_t and $\mathcal{E}$ satisfy estimates like the ones in (1.13.1). Our volume estimates show that $|K_t|^{1/n} \simeq |K|^{1/n}$ up to an absolute constant. If we define $\mathcal{E}_K = \rho\mathcal{E}$ where $\rho > 0$ is such that $|\mathcal{E}_K| = |K|$, then $\rho \simeq 1$ and the result follows. □

The existence of M-ellipsoids can be equivalently established by introducing the M-*position* of a convex body. To any given symmetric convex body K in $\mathbb{R}^n$ we can apply a linear transformation and find a position $\tilde{K} = u_K(K)$ of volume $|\tilde{K}| = |K|$ such that (1.13.1) is satisfied with $\mathcal{E}_K$ a multiple of B_2^n. This is the so-called M-position of K. It follows then that for every pair of convex bodies K_1 and K_2 in $\mathbb{R}^n$ and for all $t_1, t_2 > 0$,

$$(1.13.2) \qquad |t_1\tilde{K}_1 + t_2\tilde{K}_2|^{1/n} \leqslant c' \left(t_1|\tilde{K}_1|^{1/n} + t_2|\tilde{K}_2|^{1/n} \right),$$

where $c' > 0$ is an absolute constant, and that (1.13.2) remains true if we replace $\tilde{K}_1$ or $\tilde{K}_2$ (or both) by their polars. This statement is Milman's *reverse Brunn-Minkowski inequality*.

To define the M-position of a convex body, we can alternatively use covering numbers. Recall that the covering number $N(A, B)$ of a body A by a second body B is the least integer N for which there exist N translates of B whose union covers A. In a similar way as above, we can show that there exists an absolute constant $\beta > 0$ such that every centered convex body K in $\mathbb{R}^n$ has a linear image $\tilde{K}$ which satisfies $|\tilde{K}| = |B_2^n|$ and

$$(1.13.3) \qquad \max\{N(\tilde{K}, B_2^n), N(B_2^n, \tilde{K}), N(\tilde{K}^\circ, B_2^n), N(B_2^n, \tilde{K}^\circ)\} \leqslant \exp(\beta n).$$

We say that a convex body K which satisfies (1.13.3) is in M-position with constant β. If K_1 and K_2 are two such convex bodies, there is a standard way to show that they and their polar bodies satisfy the reverse Brunn–Minkowski inequality.

Pisier has proposed a different approach to these results, which allows one to acquire a whole family of M-ellipsoids along with more detailed information on how the corresponding covering numbers behave. The precise statement is as follows.

THEOREM 1.13.3 (Pisier). *For every $0 < \alpha < 2$ and every symmetric convex body K in $\mathbb{R}^n$, there exists a linear image $\tilde{K}$ of K such that*

$$\max\{N(\tilde{K}, tB_2^n), N(B_2^n, t\tilde{K}), N(\tilde{K}^\circ, tB_2^n), N(B_2^n, t\tilde{K}^\circ)\} \leqslant \exp\left(\frac{c(\alpha)n}{t^\alpha}\right)$$

for every $t \geqslant 1$, where $c(\alpha)$ is a constant depending only on α, with $c(\alpha) = O\big((2-\alpha)^{-\alpha/2}\big)$ as $\alpha \to 2$.

We say that a body $\tilde{K}$ which satisfies the conclusion of Theorem 1.13.3 is in M-*position of order α* (or α-*regular M-position*).

We close this section with a useful observation about the M-position.

PROPOSITION 1.13.4. *If K is in M-position with constant β then, given any $\lambda \in (0,1)$, a random orthogonal projection $P_E(K)$ onto a $\lfloor \lambda n \rfloor$-dimensional subspace E has volume ratio bounded by a constant $C(\beta, \lambda)$.*

Proof. Note that (1.13.3) implies

$$|\operatorname{conv}(K^\circ \cup B_2^n)|^{1/n} \leqslant C|B_2^n|^{1/n},$$

where C depends on β. In other words, $W = \operatorname{conv}(K^\circ \cup B_2^n)$ has bounded volume ratio, and thus Theorem 1.10.12 shows that, for a random $E \in G_{n,\lfloor \lambda n \rfloor}$,

$$K^\circ \cap E \subseteq W \cap E \subseteq C(\beta, \lambda) B_E.$$

By duality, this means that $P_E(K)$ contains a ball rB_E of radius $r \geqslant 1/C(\beta, \lambda)$. Since

$$|P_E(K)| \leqslant N(P_E(K), B_E)|B_E| \leqslant N(K, B_2^n)|B_E| \leqslant \exp(\beta n)|B_E|,$$

we arrive at a bound of the desired order for $\bigl(|P_E(K)|/|rB_E|\bigr)^{1/\lfloor \lambda n \rfloor}$. $\square$

1.14. Notes and references

Basic references

A thorough exposition of the theory of convex bodies can be found in the classical monographs by Schneider [**463**] and Gruber [**237**]. The books of Bonnesen and Fenchel [**94**], Burago and Zalgaller [**125**], Gardner [**191**] and Groemer [**228**] are very useful sources of additional information from different perspectives.

The asymptotic theory of finite dimensional normed spaces is presented in the books by V. Milman and Schechtman [**387**], Pisier [**430**] and Tomczak-Jaegermann [**493**]. Additional information can be found in the survey articles of Ball [**39**], Giannopoulos and V. Milman [**206**] and [**207**], Lindenstrauss [**329**], Lindenstrauss and V. Milman [**330**], V. Milman [**383**] and Pisier [**428**].

Convex bodies and mixed volumes

The tools that we will use from the theory of mixed volumes are described in Section 1.4. We refer to the books of Schneider [**463**] and Gruber [**237**] for proofs, historical information and detailed references. Minkowski's theory of mixed volumes appeared mainly in [**390**] and [**392**]; one can trace its roots in the works of Steiner [**478**] and Brunn [**122**], [**123**]. The theory of area measures was developed by Alexandrov in a series of works (see also Fenchel and Jessen [**172**]). In particular, the Alexandrov-Fenchel inequality appears in [5] and [6] (Fenchel sketched an alternative proof in [**171**]). Minkowski's existence theorem appears in [**390**] (see also [7]).

Brunn-Minkowski inequality

The Brunn-Minkowski inequality has its origin in the work of Brunn [**122**], [**123**] who discovered it in dimensions $n = 2, 3$. Minkowski established the n-dimensional case and characterized the case of equality in [**391**]. Blaschke gave a proof using Schwarz symmetrization in [**68**]. Lusternik [**342**] first extended the inequality to the class of compact sets. An alternative proof of Lusternik's result was obtained by Henstock and Macbeath in [**254**], and by Hadwiger and Ohmann [**244**], who also clarified Lusternik's conditions for equality. Several applications, analogues and variants of the Brunn-Minkowski inequality are discussed at length in the very informative survey paper of Gardner [**190**].

Theorem 1.2.3 is usually attributed to Prékopa and Leindler; it was proved in [**320**] and [**435**] (see also [**434**]). The survey article of Das Gupta [**155**] provides detailed information on the historical background and on related results by other groups of authors. The proof that we present in the text is more or less similar to the one given in [**430**].

The proof of the existence of Knothe's map is from [**289**].

Applications of the Brunn-Minkowski inequality

The inequality of Rogers and Shephard on the volume of the difference body was proved in [**443**]. The slightly different proof of Theorem 1.3.1 that we describe is due to Chakerian [**136**]. Other variants of the proof as well as extensions of this result can be found in [**444**], [**445**]. Theorem 1.3.2 is from [**386**].

Borell's Lema 1.3.3 appears in [**96**] and holds true in the more general setting of log-concave probability measures. It will play a very important role in this book, and we will discuss several of its applications.

The Blaschke-Santaló inequality was proved by Blaschke [**69**] in dimension $n = 3$ and by Santaló [**451**] in all dimensions. The simple proof that we describe appears in the article [**364**] of Meyer and Pajor, and in the PhD Thesis of Ball [**30**] (see also [**365**] for the not necessarily symmetric case). A well-known conjecture of Mahler states that, conversely, $s(K) \geqslant 4^n/n!$ for every centrally symmetric convex body K (with one of the minimizers being the n-dimensional cube), while in the not necessarily symmetric case one would expect that $s(K) \geqslant (n+1)^{n+1}/(n!)^2$ (with the minimum being attained at an n-dimensional regular simplex). The Bourgain-Milman inequality (which is presented in Section 1.13.1) verifies this conjecture in the asymptotic sense: for every centrally symmetric convex body K in $\mathbb{R}^n$ one has $s(K)^{1/n}$ is of the order of $1/n$.

Urysohn's inequality appears in [**494**].

Classical positions of convex bodies

John's theorem appears in [**260**]; the representation of the identity as a sum of rank one projections defined by contact points is from the same paper. Our sketch of the proof follows Ball's presentation in [**39**]. The isotropic characterization of John's position is due to Ball, see [**38**].

The Dvoretzky-Rogers lemma was proved in [**162**] and was used in the proof of the fact that every infinite dimensional Banach space X contains an unconditionally convergent series that is not absolutely convergent. It was the starting point for a question of Grothendieck that led to Dvoretzky theorem and it is used in the proof of Dvoretzky theorem.

Theorem 1.5.12 is due to Giannopoulos and Milman [**204**]. Theorem 1.5.13 is due to Petty [**422**] (see also [**215**] for a second proof and some applications to sharp inequalities for the volume of projection bodies and their polars). For a comparison of various classical positions of convex bodies see [**453**], [**351**] and [**352**].

Brascamp-Lieb inequality and its reverse form

The original proof of the Brascamp-Lieb inequality [**114**] was based on a general rearrangement inequality of Brascamp, Lieb and Luttinger[**115**] which states that if f^* is the symmetric decreasing rearrangement of a Borel measurable function f with level sets of finite measure, then
$$I(f_1, \ldots, f_m) \leqslant I(f_1^*, \ldots, f_m^*).$$
Then, Brascamp and Lieb used a generalized version of this inequality for functions of several variables and the fact that radial functions in high dimensions behave like Gaussian functions. The proof of the inequality and of its reverse form that we present in the text is due to Barthe [**52**] (see also [**53**] and [**51**]).

Theorem 1.6.4, the normalized form of the Brascamp-Lieb inequality, is due to Ball (see e.g. [**34**]) who applied it to obtain sharp volume estimates. Theorem 1.6.5 and the reverse isoperimetric inequality are also due to Ball [**37**]. Note that the reverse Brascamp-Lieb inequality plays a similar role if one is interested in dual statements: for example consider the outer volume ratio $\mathrm{ovr}(K) = \inf\left(|\mathcal{E}|/|K|\right)^{1/n}$, where the infimum is taken over all ellipsoids containing K. Then, $\mathrm{ovr}(K) \leqslant \mathrm{ovr}(\Delta_n)$ for every convex body K in $\mathbb{R}^n$. In the symmetric case the extremal body is the cross-polytope (the unit ball of ℓ_1^n). For a proof of these results see [**51**]).

Concentration of measure

General references on concentration, from various viewpoints, are the books of Ledoux [**316**], Ledoux and Talagrand [**318**], Gromov [**231**], and the articles of Ball [**39**], Gromov [**230**], Milman [**379**], [**381**], [**383**], Schechtman [**454**].

The solution of the isoperimetric problem on the sphere is given by Paul Lévy in [**321**] and by Schmidt [**459**]. For a proof using spherical symmetrization see [**174**]. Lemma 1.7.8, which leads to a very simple proof of the approximate isoperimetric inequality for the sphere, is due to Arias de Reyna, Ball and Villa [**20**].

The Gaussian isoperimetric inequality was discovered by Sudakov-Tsirelson [**483**] who used the isoperimetric theorem on the sphere and the observation that projections of uniform measures on N-dimensional spheres of radius $\sqrt{N}$ when projected to $\mathbb{R}^n$ approximate Gaussian measure as $N \to \infty$ (this is known as "Poincaré's lemma"; see [**480**] and [**313**]). The same result was also proved by Borell [**98**] who also obtained a Brunn-Minkowski inequality in Gauss space, and by Erhard [**168**] who developed a rearrangement of sets argument in Gauss space. Bobkov [**75**] proved an isoperimetric inequality on the discrete cube from which he also derived the Gaussian isoperimetric inequality (see also [**73**] and [**74**]). Theorem 1.7.12, which establishes the approximate isoperimetric inequality in Gauss space as a direct application of the Prékopa-Leindler inequality, is due to Maurey [**354**].

The isoperimetric problem for the discrete cube was solved by Harper in [**248**]. Theorem 1.7.14 is due to Talagrand [**488**].

Khintchine's appears in [**270**] and it was first stated in this form by Littlewood [**332**]. The best constants A_p^* and B_p^* were found by Szarek [**484**] who showed that $A_1^* = 1/\sqrt{2}$, and by Haagerup [**243**] who determined the best constants for all p. Kahane's inequality appears in [**265**] with constant proportional to p as $p \to \infty$. The optimal dependence $\sqrt{p}$ is due to Kwapien see [**304**].

Entropy estimates

Sudakov's inequality [**481**] in its original form gives a lower bound for the expectation of the supremum of a Gaussian process; this form is presented in Section 1.7. Theorem 1.8.1 is a direct application to the covering numbers of a convex body that follows from Sudakov's inequality once the geometric translation is done. The original proof of the dual statement, Theorem 1.8.2, is due to Pajor and Tomczak-Jaegermann [**404**]. The argument that allows one to pass from Sudakov's inequality to its dual and vice versa is due to Tomczak-Jaegermann [**492**] and is based on Theorem 1.8.4. We first present Talagrand's proof of the dual Sudakov inequality (see [**318**]) and following a similar route we deduce Sudakov's inequality using Tomczak's argument.

The duality of entropy theorem (Theorem 1.8.6) is due to Artstein-Avidan, Milman and Szarek [**24**]. The inequality of König and Milman was proved in [**299**].

Gaussian and sub-Gaussian processes

Theorem 1.9.4 combines the bounds of Sudakov and Dudley. Dudley proved the upper bound in [**158**] and conjectured the lower bound that was later proved by Sudakov in [**481**]. Theorem 1.9.5 was proved in [**468**] (see also [**481**], [**27**] and [**173**]).

Theorem 1.9.7 and Theorem 1.9.8 are due to Talagrand (see [**491**] and [**487**] for the original proof).

Dvoretzky theorem

Theorem 1.10.2 appears in [**160**] and [**161**]. Milman's proof (with the estimate $n(M/b)^2$) is from [**374**]. The definition of the critical dimension and Theorem 1.10.5 are from [**388**].

The "volume-ratio theorem" was first proved by Kashin [**269**] in the case of ℓ_1^n, and later Szarek and Tomczak-Jaegermann introduced the notion of volume ratio (in [**485**] and [**486**]) and proved Theorem 1.10.12.

The ℓ-position and Pisier's inequality

Theorem 1.11.1 appears in [**323**]. The ℓ-norm was introduced by Figiel and Tomczak-Jaegermann in [**175**] and Theorem 1.11.2 appears in the same paper. The first proof of Pisier's inequality (Theorem 1.11.3) appeared in [**426**]; see also [**427**], [**428**] and [**430**].

Low M^*-estimate and Milman's quotient of subspace theorem

Milman's first proof of the low M^*-estimate was using Urysohn's inequality and appears in [**375**]. Milman's second proof from [**376**], which makes use of the isoperimetric inequality on S^{n-1}, is the one that we sketch in the text. Afterwards, Pajor and Tomczak-Jaegermann obtained the asymptotically optimal version in [**405**]. Finally, Gordon [**222**] proved that the theorem holds true with

$$f(\lambda) \geqslant \sqrt{1-\lambda}\left(1 + O\left(\frac{1}{(1-\lambda)n}\right)\right).$$

The quotient of subspace theorem is due to V. Milman [**377**].

Bourgain-Milman inequality and M-position

Mahler's conjecture appears in [**349**] and [**350**] in connection with some questions from the geometry of numbers. The conjecture has been verified for some classes of bodies: for the class of 1-unconditional convex bodies by Saint-Raymond [**449**] (the equality cases were clarified by Meyer [**362**] and Reisner [**440**]) and for the class of zonoids by Reisner in [**438**] and [**439**]. A short proof of Mahler's conjecture for zonoids was also given in [**223**].

The reverse Brunn-Minkowski inequality was proved by Milman in [**378**]. The Bourgain-Milman inequality is from [**112**]. Milman introduced the method of isomorphic symmetrization in [**380**]. The proof of the reverse Santaló inequality that we present in this section comes from that paper.

Kuperberg's proof of the reverse Santaló inequality appears in [**302**]. The proof of Nazarov can be found in [**398**].

Theorem 1.13.3 is from [**429**]. See also Pisier's book [**430**, Chapter 7].

CHAPTER 2

Isotropic log-concave measures

The class of logarithmically concave probability measures is introduced in Section 2.1. These are the Borel probability measures on $\mathbb{R}^n$ which satisfy

$$\mu((1-\lambda)A + \lambda B) \geqslant \mu(A)^{1-\lambda}\mu(B)^\lambda$$

for any compact sets A, B in $\mathbb{R}^n$ and any $\lambda \in (0,1)$. In Section 2.2 we collect a number of useful inequalities for log-concave functions and log-concave probability measures; these are used frequently throughout this book.

Isotropic log-concave measures are defined in Section 2.3. These are the log-concave probability measures μ, with barycenter at the origin, which satisfy the isotropic condition

$$\int_{\mathbb{R}^n} \langle x, \theta \rangle^2 d\mu(x) = 1$$

for every $\theta \in S^{n-1}$. The isotropic constant of a measure μ in this class is defined as

$$L_\mu := \left(\sup_{x \in \mathbb{R}^n} f(x) \right)^{1/n} \simeq (f(0))^{1/n},$$

where f is the log-concave density of μ. We discuss in parallel the class of isotropic convex bodies and how the isotropic constant of a convex body or, in general, an arbitrary finite log-concave measure is defined.

The last two sections of this chapter introduce some fundamental tools. In Section 2.4 we study the main concentration properties of log-concave probability measures that follow immediately from the Brunn-Minkowski inequality (more precisely, from Borell's lemma) and express them in the form of reverse Hölder inequalities for seminorms. In Section 2.5 we introduce the family of convex bodies $K_p(\mu)$, $p > 0$, associated with a given log-concave probability measure μ. The bodies $K_p(\mu)$ were introduced by K. Ball and allow us to reduce the study of log-concave measures to that of convex bodies. We establish their convexity and their main properties. As a first example of their use, we show that studying the magnitude of the isotropic constant of log-concave probability measures can be reduced to the same task inside the more restricted class of convex bodies.

2.1. Log-concave probability measures

We denote by $\mathcal{P}_n$ the class of all Borel probability measures on $\mathbb{R}^n$ which are absolutely continuous with respect to the Lebesgue measure. The density of a measure $\mu \in \mathcal{P}_n$ is denoted by f_μ.

We say that a measure $\mu \in \mathcal{P}_n$ has center of mass (or *barycenter*) at $x_0 \in \mathbb{R}^n$, and we write $x_0 = \mathrm{bar}(\mu)$, if
$$\int_{\mathbb{R}^n} \langle x, \theta \rangle \, d\mu(x) = \langle x_0, \theta \rangle$$
for all $\theta \in S^{n-1}$. Equivalently, if
$$x_0 = \int_{\mathbb{R}^n} x \, d\mu(x).$$
The subclass $\mathcal{CP}_n$ of $\mathcal{P}_n$ consists of all *centered* $\mu \in \mathcal{P}_n$. These are the measures $\mu \in \mathcal{P}_n$ that have barycenter at the origin; so, $\mu \in \mathcal{CP}_n$ if
$$\int_{\mathbb{R}^n} \langle x, \theta \rangle d\mu(x) = 0$$
for all $\theta \in S^{n-1}$.

The subclass $\mathcal{SP}_n$ of $\mathcal{P}_n$ consists of all *even* (or *symmetric*) measures $\mu \in \mathcal{P}_n$; μ is called even (or symmetric) if $\mu(A) = \mu(-A)$ for every Borel subset A of $\mathbb{R}^n$.

Let $f : \mathbb{R}^n \to [0, \infty)$ be an integrable function with finite, positive integral. As in the case of measures, the barycenter of f is defined as
$$\mathrm{bar}(f) = \frac{\int_{\mathbb{R}^n} x f(x) \, dx}{\int_{\mathbb{R}^n} f(x) \, dx}.$$
In particular, f has barycenter (or center of mass) at the origin if
$$\int_{\mathbb{R}^n} \langle x, \theta \rangle f(x) \, dx = 0$$
for all $\theta \in S^{n-1}$. If so, we will say that f is *centered*.

DEFINITION 2.1.1. (i) A measure $\mu \in \mathcal{P}_n$ is called *log-concave* if for all compact subsets A, B of $\mathbb{R}^n$ and all $0 < \lambda < 1$ we have
$$\mu((1-\lambda)A + \lambda B) \geqslant \mu(A)^{1-\lambda} \mu(B)^{\lambda}.$$
(ii) A function $f : \mathbb{R}^n \to [0, \infty)$ is called *log-concave* if
$$f((1-\lambda)x + \lambda y) \geqslant f(x)^{1-\lambda} f(y)^{\lambda}$$
for all $x, y \in \mathbb{R}^n$ and any $0 < \lambda < 1$.

Let $f : \mathbb{R}^n \to [0, \infty)$ be a log-concave function with $\int_{\mathbb{R}^n} f(x)\,dx = 1$ (then we say that f is a *log-concave density*). From the Prékopa-Leindler inequality it follows that the measure μ with density f is log-concave: to see this, consider two compact sets A, B in $\mathbb{R}^n$ and any $\lambda \in (0,1)$. Then, the functions $w = \mathbf{1}_A f$, $g = \mathbf{1}_B f$ and $h = \mathbf{1}_{(1-\lambda)A + \lambda B} f$ satisfy
$$h((1-\lambda)x + \lambda y) \geqslant w(x)^{1-\lambda} g(y)^{\lambda}$$
for all $x, y \in \mathbb{R}^n$, and hence, Theorem 1.2.3 shows that
$$\mu((1-\lambda)A + \lambda B) = \int_{\mathbb{R}^n} h \geqslant \left(\int_{\mathbb{R}^n} w \right)^{1-\lambda} \left(\int_{\mathbb{R}^n} g \right)^{\lambda}$$
$$= \mu(A)^{1-\lambda} \mu(B)^{\lambda}.$$
The next theorem of Borell shows that, conversely, any non-degenerate log-concave probability measure on $\mathbb{R}^n$ belongs to the class $\mathcal{P}_n$ and has a log-concave density.

THEOREM 2.1.2 (Borell). *Let μ be a log-concave probability measure on $\mathbb{R}^n$ such that $\mu(H) < 1$ for any hyperplane H. Then, μ is absolutely continuous with respect to the Lebesgue measure and has a log-concave density f, that is $d\mu(x) = f(x)\,dx$.*

EXAMPLES 2.1.3. (i) Let K be a convex body of volume 1 in $\mathbb{R}^n$. We define a probability measure μ_K on $\mathbb{R}^n$, setting
$$\mu_K(A) = |K \cap A| = \int_A \mathbf{1}_K(x)dx$$
for every Borel $A \subseteq \mathbb{R}^n$. From the convexity of K we easily check that $\mathbf{1}_K$ is a log-concave function, and hence μ_K is a log-concave probability measure.

(ii) For every $c > 0$, the function $f_c(x) = \exp(-c\|x\|_2^2)$ is even and log-concave on $\mathbb{R}^n$: note that the Euclidean norm is a convex function, and the function $t \mapsto ct^2$ is also convex, thus their composition $c\|x\|_2^2 = -\log f(x)$ is an even convex function. It follows that, for every $c > 0$, the measure
$$\gamma_{n,c}(A) = \frac{1}{I(c)} \int_A \exp(-c\|x\|_2^2)dx,$$
where $I(c) = \int_{\mathbb{R}^n} \exp(-c\|x\|_2^2)dx$, is a log-concave probability measure. In particular, this holds true for the standard Gaussian measure γ_n.

(iii) Let K be a convex body in $\mathbb{R}^n$ with $0 \in \mathrm{int}(K)$ and let p_K denote the Minkowski functional of K. For any $s \geqslant 1$ we have
$$c(s, K) := \int_{\mathbb{R}^n} e^{-(p_K(x))^s} dx < \infty$$
and the measure $\mu_{s,K}$ with density $\frac{1}{c(s,K)} \exp(-(p_K(x))^s)$ is a log-concave probability measure on $\mathbb{R}^n$.

REMARK 2.1.4. Some basic properties of log-concave probability measures are the following:

(i) If μ is a log-concave probability measure on $\mathbb{R}^n$ and $T: \mathbb{R}^n \to \mathbb{R}^m$ is an affine transformation then $\mu \circ T^{-1}$ is a log-concave probability measure on $\mathbb{R}^m$.

(ii) If μ_i are log-concave probability measures on $\mathbb{R}^{n_i}$, $i = 1, \ldots, k$, then $\mu_1 \otimes \cdots \otimes \mu_k$ is a log-concave probability measure on $\mathbb{R}^{n_1} \times \cdots \times \mathbb{R}^{n_k}$.

(iii) If μ and ν are log-concave probability measures on $\mathbb{R}^n$ then their convolution $\mu * \nu$ (which is defined by
$$\int_{\mathbb{R}^n} h(x)\, d(\mu * \nu)(x) = \int_{\mathbb{R}^n} \int_{\mathbb{R}^n} h(x+y)\, d\mu(x)\, d\nu(y)$$
for any non-negative Borel measurable function h on $\mathbb{R}^n$) is a log-concave probability measure on $\mathbb{R}^n$. To see this, note that $\mu * \nu$ is the image of $\mu \times \nu$ under the affine transformation $T(x, y) = x + y$.

(iv) If $\{\mu_k\}_{k=1}^\infty$ is a sequence of log-concave probability measures on $\mathbb{R}^n$ that converges weakly to a measure μ, then μ is also a log-concave probability measure on $\mathbb{R}^n$. First note that by weak convergence μ is also a probability measure. Let A and B be non-empty compact subsets of $\mathbb{R}^n$ and, given $\lambda \in (0, 1)$, set $C = (1-\lambda)A + \lambda B$. Choosing $t_m = 1/m$, we have that the t_m extensions of A, B and C are compact sets with non-empty interior, which is exactly the "open t_m extension" of A, B and

C respectively, say $\tilde{A}_{t_m}$, $\tilde{B}_{t_m}$ and $\tilde{C}_{t_m}$. By weak convergence and the log-concavity of μ_k, we have that

$$\mu(C_{t_m}) \geqslant \limsup_{m \to \infty} \mu_k(C_{t_m}) \geqslant \limsup_{m \to \infty} \mu_k(A_{t_m})^{1-\lambda} \mu_k(B_{t_m})^\lambda$$
$$\geqslant \liminf_{m \to \infty} \mu_k(A_{t_m})^{1-\lambda} \mu_k(B_{t_m})^\lambda \geqslant \mu(\tilde{A}_{t_m})^{1-\lambda} \mu(\tilde{B}_{t_m})^\lambda$$
$$\geqslant \mu(A)^{1-\lambda} \mu(B)^\lambda.$$

Since $C = \bigcap_{m=1}^\infty C_{t_m}$ is compact, we have $\mu(C_{t_m}) \to \mu(C)$. Therefore,
$$\mu((1-\lambda)A + \lambda B) = \mu(C) \geqslant \mu(A)^{1-\lambda} \mu(B)^\lambda.$$

2.1.1. s-concave measures

In this short subsection we briefly introduce s-concave measures. Given $s \in [-\infty, \infty]$ we say that a measure μ on $\mathbb{R}^n$ is s-concave if

(2.1.1) $$\mu((1-\lambda)A + \lambda B) \geqslant ((1-\lambda)\mu^s(A) + \lambda \mu^s(B))^{1/s}$$

for all compact subsets A, B of $\mathbb{R}^n$ with $\mu(A)\mu(B) > 0$ and all $\lambda \in (0,1)$. The limiting cases are defined appropriately. For $s = 0$ the right hand side in (2.1.1) becomes $\mu(A)^{1-\lambda}\mu(B)^\lambda$ (therefore, 0-concave measures are the log-concave measures). In the case $s = -\infty$ the right hand side in (2.1.1) becomes $\min\{\mu(A), \mu(B)\}$ and in the case $s = \infty$ it is $\max\{\mu(A), \mu(B)\}$. Note that if μ is s-concave and $t \leqslant s$ then μ is t-concave.

Let $\gamma \in [-\infty, \infty]$. A function $f : \mathbb{R}^n \to [0, \infty)$ is called γ-concave if

(2.1.2) $$f((1-\lambda)x + \lambda y) \geqslant ((1-\lambda)f^\gamma(x) + \lambda f^\gamma(y))^{1/\gamma}$$

for all $x, y \in \mathbb{R}^n$ with $f(x)f(y) > 0$ and all $\lambda \in (0,1)$. Again, we define the cases $\gamma = 0, \pm\infty$ appropriately. The $-\infty$-concave functions are the quasi-concave functions: those functions f for which the set $\{f \geqslant t\}$ is convex for all $t \in \mathbb{R}$. Borell [**96**] studied the relation between s-concave probability measures and γ-concave functions (with a slightly different definition). The next theorem generalizes Theorem 2.1.2.

THEOREM 2.1.5 (Borell). *Let μ be a measure on $\mathbb{R}^n$ and let F be the affine subspace spanned by the support $\mathrm{supp}(\mu)$ of μ. If $\dim(F) = d$ then for every $-\infty \leqslant s \leqslant 1/d$ we have that μ is s-concave if and only if it has a non-negative density $\psi \in L^1_{\mathrm{loc}}(\mathbb{R}^n, dx)$ and ψ is γ-concave, where $\gamma = \frac{s}{1-sd} \in [-1/d, +\infty]$. If $s > 1/d$ then μ is s-concave if and only if μ is a Dirac measure.*

Finally, we say that a measure μ with density ψ is s-affine if ψ^γ (or $\log \psi$ in the case $s = \gamma = 0$) is affine on its convex support, where $\gamma = \frac{s}{1-sd}$.

2.2. Inequalities for log-concave functions

The first two results in this section establish two basic properties of log-concave functions. We first show that every integrable log-concave function $f : \mathbb{R}^n \to [0, \infty)$ has finite moments of all orders; this is a consequence of the fact that $f(x)$ decays exponentially as $\|x\|_2 \to \infty$.

LEMMA 2.2.1. *Let $f : \mathbb{R}^n \to [0, \infty)$ be a log-concave function with finite, positive integral. Then, there exist constants $A, B > 0$ such that $f(x) \leqslant Ae^{-B\|x\|_2}$ for all $x \in \mathbb{R}^n$. In particular, f has finite moments of all orders.*

Proof. Since $\int f > 0$, we can find $t \in (0,1)$ such that the set $C := \{x : f(x) > t\}$ has positive Lebesgue measure. Note that C is convex because f is log-concave. Since C has positive volume, its affine hull has dimension n, and hence C contains an affinely independent set $\{x_i\}_{i \leqslant n+1}$. By convexity, C contains the simplex $S = \mathrm{conv}\{x_i\}_{i \leqslant n+1}$; in particular, C has non-empty interior. Let $x_0 \in C$ and $r > 0$ such that $x_0 + rB_2^n \subseteq C$. Working with $f_1(\cdot) = f(\cdot + x_0)$ if needed, we may assume that $rB_2^n \subseteq C$.

We set $K = \{x : f(x) > t/e\}$. Then, Markov's inequality and the monotonicity of volume show that $0 < |K| < \infty$. Using the fact that K is convex, it has finite volume and contains rB_2^n, we see that it is bounded. So, we can find $R > 0$ such that $K \subset \frac{R}{2}B_2^n$. Then, for every x with $\|x\|_2 > R$ we have $R\frac{x}{\|x\|_2} \notin K$, and hence $f(Rx/\|x\|_2) \leqslant t/e$, while $r\frac{x}{\|x\|_2} \in C$, which shows that $f(rx/\|x\|_2) \geqslant t$. Moreover, we may write

$$\frac{Rx}{\|x\|_2} = \frac{\|x\|_2 - R}{\|x\|_2 - r} \frac{rx}{\|x\|_2} + \frac{R - r}{\|x\|_2 - r} x.$$

Since f is log-concave, we get

$$\frac{t}{e} \geqslant f\left(R\frac{x}{\|x\|_2}\right) \geqslant f\left(r\frac{x}{\|x\|_2}\right)^{\frac{\|x\|_2 - R}{\|x\|_2 - r}} f(x)^{\frac{R-r}{\|x\|_2 - r}} \geqslant t^{\frac{\|x\|_2 - R}{\|x\|_2 - r}} f(x)^{\frac{R-r}{\|x\|_2 - r}}.$$

It follows that

$$f(x) \leqslant t e^{-\frac{\|x\|_2 - r}{R - r}} < e^{-\|x\|_2/R}$$

for every $x \in \mathbb{R}^n$ with $\|x\|_2 > R$. On the other hand, for every $x \in RB_2^n$ and for every $y \in \frac{x}{2} + \frac{r}{2}B_2^n$ we have by the log-concavity of f that

$$f(y) \geqslant \sqrt{f(x)f(2y-x)} \geqslant \sqrt{t}\sqrt{f(x)};$$

this combined with the integrability of f shows that there exists $M > 0$ such that $f(x) \leqslant M$ for every $x \in RB_2^n$. So, we can clearly find two constants $A, B > 0$, which depend on f, so that $f(x) \leqslant Ae^{-B\|x\|_2}$ for every $x \in \mathbb{R}^n$. $\square$

The second result, which is due to Fradelizi, shows that the value of a centered log-concave function at the origin is comparable to its maximum (up to a constant depending on the dimension). Here, we present a different proof. Note that, if f is assumed even, then $f(0) = \|f\|_\infty$.

THEOREM 2.2.2 (Fradelizi). *Let $f : \mathbb{R}^n \to [0, \infty)$ be a centered log-concave function. Then,*

$$f(0) \leqslant \|f\|_\infty \leqslant e^n f(0).$$

Proof. We may assume that f is strictly positive, continuously differentiable and that $\int_{\mathbb{R}^n} f(y)dy = 1$. From Jensen's inequality and using the assumption that f is centered we have

$$(2.2.1) \qquad \log f(0) = \log f\left(\int_{\mathbb{R}^n} yf(y)dy\right) \geqslant \int_{\mathbb{R}^n} f(y) \log f(y)dy.$$

Let $x \in \mathbb{R}^n$. Using the fact that f is log-concave we have that

$$(2.2.2) \qquad -\log f(x) \geqslant -\log f(y) + \langle x - y, \nabla(-\log f)(y)\rangle.$$

Multiplying both terms of the last inequality by $f(y)$, and then integrating with respect to y, we get

$$(2.2.3) \quad -\log f(x) \geqslant -\int_{\mathbb{R}^n} f(y) \log f(y) dy + \int_{\mathbb{R}^n} \langle x - y, -\nabla f(y) \rangle \, dy$$

$$\geqslant -\int_{\mathbb{R}^n} f(y) \log f(y) dy - n,$$

where the last inequality follows if we integrate by parts (and since $f(y)$ decays exponentially as $\|y\|_2 \to \infty$). Combining (2.2.1) and (2.2.3) we get

$$\log f(0) \geqslant \int_{\mathbb{R}^n} f(y) \log f(y) dx \geqslant \log f(x) - n,$$

for every $x \in \mathbb{R}^n$. Taking the supremum over all x we get the result. $\square$

Next, we prove two technical inequalities for log-concave functions. The first one is a reverse Hölder inequality.

THEOREM 2.2.3. *Let $f : [0, \infty) \to [0, \infty)$ be a log-concave function with $f(0) > 0$. Then, the function*

$$G(p) := \left(\frac{1}{f(0)\Gamma(p)} \int_0^\infty f(x) x^{p-1} \, dx \right)^{1/p}$$

is a decreasing function of p on $(0, \infty)$.

Proof. Without loss of generality we may assume that $f(0) = 1$, otherwise we work with the log-concave function $f_1 = \frac{1}{f(0)} f$. Let $p > 0$. Applying the change of variables $y = cx$ we see that for every $c > 0$ one has

$$\int_0^\infty e^{-cx} x^{p-1} dx = \frac{1}{c^p} \int_0^\infty e^{-x} x^{p-1} dx = \frac{\Gamma(p)}{c^p}.$$

Thus, if we choose $c_p = \frac{1}{G(p)}$ we have

$$(2.2.4) \quad \int_0^\infty e^{-c_p x} x^{p-1} dx = \int_0^\infty f(x) x^{p-1} dx.$$

In particular, it cannot be that $e^{-c_p x} < f(x)$ for every $x \in (0, +\infty)$, hence the set $\{x > 0 : e^{-c_p x} \geqslant f(x)\}$ is non-empty and

$$x_0 := \inf\{x > 0 : e^{-c_p x} \geqslant f(x)\} \in [0, +\infty).$$

Obviously then

$$(2.2.5) \quad e^{-c_p x} < f(x) \text{ for every } 0 < x < x_0,$$

whereas for $x > x_0$ we can find $y \in [x_0, x) \cap \{y' > 0 : e^{-c_p y'} \geqslant f(y')\}$ and we can write

$$(2.2.6) \quad e^{-c_p y} \geqslant f(y) \geqslant f(x)^{\frac{y}{x}} f(0)^{1 - \frac{y}{x}} = f(x)^{\frac{y}{x}}$$

by the log-concavity of f, which means that $f(x) \leqslant (e^{-c_p y})^{\frac{x}{y}} = e^{-c_p x}$. It follows that

$$(2.2.7) \quad \int_x^\infty f(t) t^{p-1} dt \leqslant \int_x^\infty e^{-c_p t} t^{p-1} dt$$

for every $x > x_0$. On the other hand, by (2.2.5) we see that for $x \leqslant x_0$

$$\int_0^x f(t)t^{p-1}dt \geqslant \int_0^x e^{-c_p t}t^{p-1}dt,$$

and thus (2.2.4) implies (2.2.7) for every $x \leqslant x_0$ as well.

Let us consider $q > p$. Then, by Fubini's theorem and by (2.2.4)

$$\int_0^\infty f(x)x^{q-1}dx = \int_0^\infty f(x)x^{p-1}\int_0^x (q-p)t^{q-p-1}dt\,dx$$
$$= \int_0^\infty (q-p)t^{q-p-1}\int_t^\infty f(x)x^{p-1}dx\,dt$$
$$\leqslant \int_0^\infty (q-p)t^{q-p-1}\int_t^\infty e^{-c_p x}x^{p-1}dx\,dt$$
$$= \int_0^\infty e^{-c_p x}x^{q-1}dx = \frac{\Gamma(q)}{c_p^q}.$$

We conclude that

$$G(q) = \left(\frac{1}{\Gamma(q)}\int_0^\infty f(x)x^{q-1}\,dx\right)^{1/q} \leqslant \frac{1}{c_p} = G(p),$$

which was our claim. $\square$

The inequality that follows goes in the opposite direction.

LEMMA 2.2.4. *Let $f : [0, \infty) \to [0, \infty)$ be a log-concave function. Then,*

$$F(p) := \left(\frac{p}{\|f\|_\infty}\int_0^\infty x^{p-1}f(x)\,dx\right)^{1/p}$$

is an increasing function of p on $(0, \infty)$.

Proof. Without loss of generality we may assume that $\|f\|_\infty = 1$. Then, for any $0 < p < q$ and $\alpha > 0$, we may write

$$\frac{F(q)^q}{q} = \int_0^\infty x^{q-1}f(x)\,dx = \int_0^\alpha x^{q-1}f(x)\,dx + \int_\alpha^\infty x^{q-1}f(x)\,dx$$
$$\geqslant \int_0^\alpha x^{q-1}f(x)\,dx + \alpha^{q-p}\int_\alpha^\infty x^{p-1}f(x)\,dx$$
$$= \alpha^{q-p}\frac{F(p)^p}{p} - \alpha^q\int_0^1 (x^{p-1} - x^{q-1})f(\alpha x)\,dx$$
$$\geqslant \alpha^{q-p}\frac{F(p)^p}{p} - \alpha^q\left(\frac{1}{p} - \frac{1}{q}\right).$$

The choice $\alpha = F(p)$ minimizes the right hand side and gives the result. $\square$

A well-known inequality of Lyapunov (see, for example, [**246**]) asserts that, for every measurable function f, the function $p \mapsto \log\|f\|_p^p$, $-\infty < p < \infty$, is convex. The next result is an inverse Lyapunov inequality of Borell, which will be used in Chapter 13.

THEOREM 2.2.5 (Borell). *Let $f : [0, \infty) \to [0, \infty)$ be a log-concave function. Then, the function*

$$\Psi_f(p) = \frac{\int_0^\infty x^p f(x)\,dx}{\Gamma(p+1)}$$

is log-concave on $[0, \infty)$.

Proof. We may assume that $f > 0$ on $[0, \infty)$. It is enough to show that if $q > p \geqslant 0$ then $\Psi_f^2\left(\frac{p+q}{2}\right) \geqslant \Psi_f(p)\Psi_f(q)$. We consider functions of the form $g(x) = \alpha e^{-\beta x}$, where $\alpha, \beta > 0$. Note that

$$\Psi_g(r) := \frac{\int_0^\infty x^r \alpha e^{-\beta x}\, dx}{\Gamma(r+1)} = \frac{\alpha}{\beta^{r+1}} \tag{2.2.8}$$

for all $r \geqslant 0$. We choose $\alpha, \beta > 0$ so that $g(x) = \alpha e^{-\beta x}$ satisfies

$$\int_0^\infty x^r f(x)\,dx = \int_0^\infty x^r g(x)\,dx, \quad \text{for } r = p, q. \tag{2.2.9}$$

We first show that f and g intersect more than once. It is clear that one cannot have $f > g$ or $f < g$ on $[0, \infty)$, so there exists $x_1 > 0$ so that $f(x_1) = g(x_1)$. Assume that this the only point of intersection of f and g and, without loss of generality, assume that $g < f$ on $[0, x_1)$ and $g > f$ on (x_1, ∞). Then, using (2.2.9) as in the proof of Theorem 2.2.3 we easily check that

$$\int_y^\infty x^p g(x)\,dx > \int_y^\infty x^p f(x)\,dx$$

for all $y > 0$. We write

$$\int_0^\infty x^q f(x)\,dx = \int_0^\infty x^p f(x)\left(\int_0^x (q-p)s^{q-p-1}\,ds\right)dx$$
$$= \int_0^\infty (q-p)s^{q-p-1}\left(\int_s^\infty x^p f(x)\,dx\right)ds$$
$$< \int_0^\infty (q-p)s^{q-p-1}\left(\int_s^\infty x^p g(x)\,dx\right)ds$$
$$= \int_0^\infty x^q g(x)\,dx,$$

which is a contradiction. We have checked that there exist $b > a > 0$ so that $f(a) = g(a)$ and $f(b) = g(b)$, and by the log-concavity of $f - g$ (note that g is log-affine) we must have $f \geqslant g$ on $[a, b]$ and $f \leqslant g$ on $[0, a]$ and $[b, \infty)$. We set $r = (q-p)/2$; examining the sign of the integrand we see that

$$\int_0^\infty (x^r - a^r)(x^r - b^r)x^p(g(x) - f(x))\,dx \geqslant 0,$$

which implies

$$\int_0^\infty x^{2r+p}(g(x) - f(x))\,dx + (ab)^r \int_0^\infty x^p(g(x) - f(x))\,dx$$
$$- (a^r + b^r)\int_0^\infty x^{r+p}(g(x) - f(x))\,dx \geqslant 0.$$

Note that $2r + p = q$, and hence by the choice of g the first two integrals vanish. Therefore,

$$\int_0^\infty x^{\frac{p+q}{2}} f(x)\,dx = \int_0^\infty x^{r+p} f(x)\,dx \geqslant \int_0^\infty x^{r+p} g(x)\,dx = \int_0^\infty x^{\frac{p+q}{2}} g(x)\,dx,$$

which shows that $\Psi_f^2\left(\frac{p+q}{2}\right) \geqslant \Psi_g^2\left(\frac{p+q}{2}\right)$. From (2.2.8) we check that
$$\Psi_g^2\left(\frac{p+q}{2}\right) = \Psi_g(p)\Psi_g(q) = \Psi_f(p)\Psi_f(q),$$
which finally gives $\Psi_f^2\left(\frac{p+q}{2}\right) \geqslant \Psi_f(p)\Psi_f(q)$. □

The next result is known as Grünbaum's lemma. It states that if μ is a centered log-concave probability measure on $\mathbb{R}^n$ then every hyperplane through the origin defines two half-spaces of more or less the same measure.

LEMMA 2.2.6 (Grünbaum). *Let μ be a centered log-concave probability measure on $\mathbb{R}^n$. Then,*
$$\frac{1}{e} \leqslant \mu(\{x : \langle x, \theta \rangle \geqslant 0\}) \leqslant 1 - \frac{1}{e}$$
for every $\theta \in S^{n-1}$.

Proof. Without loss of generality we may assume that, for some $M > 0$,
$$\mu(\{x : |\langle x, \theta \rangle| > M\}) = 0.$$
The general case then follows by approximating a general log-concave measure by measures which have this property in the direction of θ.

Let $G(t) = \mu(\{x : \langle x, \theta \rangle \leqslant t\})$. Then, G is a log-concave increasing function and we have $G(t) = 0$ for $t \leqslant -M$ and $G(t) = 1$ for $t \geqslant M$. Since μ is centered, we have
$$\int_{-M}^{M} tG'(t)dt = 0,$$
and applying integration by parts we see that
$$\int_{-M}^{M} G(t)dt = M.$$
We want to prove that
$$G(0) \geqslant \frac{1}{e}$$
(that $G(0) \leqslant 1 - 1/e$ as well will then follow by replacing θ with $-\theta$ and repeating the argument). Observe that $\log G$ is a concave function, therefore
$$G(t) \leqslant G(0)e^{\alpha t}$$
with $\alpha = G'(0)/G(0)$. We may choose M large enough so that $1/\alpha < M$. Then, using that $G(t) \leqslant G(0)e^{\alpha t}$ if $t \leqslant 1/\alpha$ and that, trivially, $G(t) \leqslant 1$ if $t > 1/\alpha$, we can write
$$M = \int_{-M}^{M} G(t)dt \leqslant \int_{-\infty}^{1/\alpha} G(0)e^{\alpha t}dt + \int_{1/\alpha}^{M} 1\,dt = \frac{eG(0)}{\alpha} + M - \frac{1}{\alpha}.$$
We conclude that $G(0) \geqslant 1/e$ as claimed. □

REMARKS 2.2.7. We close this section with two remarks on s-concave and log-concave functions that will be useful and are of independent interest. The first one is related to Lemma 2.2.1.

(i) *If $s > 0$ and $f : \mathbb{R}^n \to [0, \infty)$ is an s-concave function with finite, positive integral then f is compactly supported.*

We briefly sketch an argument for this statement (for more details see Bobkov [81]). Let $K = \{x : f(x) > 0\}$. Using the integrability of f and arguing as in Lemma 2.2.1 one can show that f is bounded. Without loss of generality we assume that $\alpha := \sup(f)$ is attained at $x_0 = 0$. Using again the integrability and the s-concavity of f we may find $R > 0$ so that $f(x) \leqslant \alpha/2$ whenever $\|x\|_2 \geqslant R$. The function $g(x) = \left(\frac{f(x)}{\alpha}\right)^{1/s}$ is concave and satisfies: (i) $\lim_{x \to 0} g(x) = 1$ and (ii) $g(x) \leqslant 2^{-1/s}$ if $\|x\|_2 \geqslant R$. Let $y \in \mathbb{R}^n$ with $\|y\|_2 = R$ and assume that $\lambda y \in K$ for some $\lambda \geqslant 1$. We write $y = \left(1 - \frac{1}{\lambda}\right) 0 + \frac{1}{\lambda}(\lambda y)$ and using the concavity of g we get

$$2^{-1/s} \geqslant g(y) \geqslant 1 - \frac{1}{\lambda} + \frac{g(\lambda y)}{\lambda} > 1 - \frac{1}{\lambda}.$$

It follows that
$$\lambda \leqslant \lambda_0 := \frac{1}{1 - 2^{-1/s}}$$
and hence $K \subseteq (\lambda_0 R) B_2^n$.

(ii) *The class of log-concave probability measures is the smallest class of probability measures that contains uniform measures on convex bodies and is closed under affine images and weak limits.*

We will show that every log-concave probability measure μ on $\mathbb{R}^n$ is the weak limit of a sequence of projections of uniform measures of convex bodies. This can be seen as follows: starting with a log-concave density $f : \mathbb{R}^n \to \mathbb{R}^+$, for every $m \geqslant 1$ we define

$$f_m(x) = (1 + \log f(x)/m)_+^m,$$

where $a_+ = \max\{a, 0\}$. Note that $f_m^{1/m}$ is concave on its support A_m, $0 \leqslant f_m \leqslant f$ and this implies that

$$c_m := \int_{\mathbb{R}^n} f_m(x) dx \to \int_{\mathbb{R}^n} f(x) dx = 1.$$

The previous remark shows, for every m, that the closure of A_m is compact.

Given $m \geqslant 1$ consider the measure μ_m with density $c_m^{-1} f_m$. We set

$$K_m = \left\{(x, y) \in \mathbb{R}^n \times \mathbb{R}^m : x \in \overline{A_m}, \|y\|_2 \leqslant (f_m(x)/c_m \omega_m)^{1/m}\right\}.$$

One can check that K_m is a convex body of volume 1. If μ_{K_m} is the Lebesgue measure on K_m, one can also check that $\mu_m = \mu_{K_m} \circ P^{-1}$ where P is the orthogonal projection onto the first n coordinates. Finally, observe that $c_m^{-1} f_m \to f$ pointwise.

2.3. Isotropic log-concave measures

We first define the isotropic position of a centered convex body K and the isotropic constant L_K as an invariant of the linear class of K. In the next subsections we give a more general definition, in the setting of log-concave measures.

2.3.1. Isotropic position of a convex body

DEFINITION 2.3.1. A convex body K in $\mathbb{R}^n$ is called *isotropic* if it has volume $|K| = 1$, it is centered (i.e. its barycenter is at the origin), and there is a constant $\alpha > 0$ such that

(2.3.1) $$\int_K \langle x, y \rangle^2 dx = \alpha^2 \|y\|_2^2$$

for all $y \in \mathbb{R}^n$. Note that if K satisfies the *isotropic condition* (2.3.1) then

$$\int_K \|x\|_2^2 dx = \sum_{i=1}^n \int_K \langle x, e_i \rangle^2 dx = n\alpha^2,$$

where $x_j = \langle x, e_j \rangle$ are the coordinates of x with respect to some orthonormal basis $\{e_1, \ldots, e_n\}$ of $\mathbb{R}^n$. One can easily check that if K is an isotropic convex body in $\mathbb{R}^n$ then $U(K)$ is also isotropic for every $U \in O(n)$.

REMARK 2.3.2. It is not hard to check that the isotropic condition (2.3.1) is equivalent to each one of the following statements:

(i) For every $i, j = 1, \ldots, n$,

(2.3.2) $$\int_K x_i x_j dx = \alpha^2 \delta_{ij}.$$

(ii) For every $T \in L(\mathbb{R}^n)$,

(2.3.3) $$\int_K \langle x, Tx \rangle dx = \alpha^2 (\operatorname{tr} T).$$

To see this, we first assume that K is isotropic; setting $y = e_i$, $y = e_j$ and $y = e_i + e_j$ in (2.3.1) we get (2.3.2). Also, observing that if $T = (t_{ij})_{i,j=1}^n$ then $\langle x, T(x) \rangle = \sum_{i,j=1}^n t_{ij} x_i x_j$, we readily see that (2.3.2) implies (2.3.3). Finally, note that applying (2.3.3) with $T(x) = \langle x, y \rangle y$ we get the isotropic condition (2.3.1).

2.3.2. Existence

The next proposition shows that every centered convex body has a linear image which satisfies the isotropic condition.

PROPOSITION 2.3.3. *Let K be a centered convex body in $\mathbb{R}^n$. There exists $T \in GL(n)$ such that $T(K)$ is isotropic.*

Proof. The operator $M \in L(\mathbb{R}^n)$ defined by $M(y) = \int_K \langle x, y \rangle x \, dx$ is symmetric and positive definite; therefore, it has a symmetric and positive definite square root S. Consider the linear image $\tilde{K} = S^{-1}(K)$ of K. Then, for every $y \in \mathbb{R}^n$ we have

$$\int_{\tilde{K}} \langle x, y \rangle^2 dx = |\det S|^{-1} \int_K \langle S^{-1} x, y \rangle^2 dx$$
$$= |\det S|^{-1} \int_K \langle x, S^{-1} y \rangle^2 dx$$
$$= |\det S|^{-1} \left\langle \int_K \langle x, S^{-1} y \rangle x \, dx, S^{-1} y \right\rangle$$
$$= |\det S|^{-1} \langle MS^{-1}y, S^{-1}y \rangle = |\det S|^{-1} \|y\|_2^2.$$

Normalizing the volume of $\tilde{K}$ we get the result. $\square$

Proposition 2.3.3 shows that every centered convex body K in $\mathbb{R}^n$ has a *position* $\tilde{K}$ which is isotropic. We say that $\tilde{K}$ is an *isotropic position* of K. The next theorem shows that the isotropic position of a convex body is uniquely determined (if we ignore orthogonal transformations) and arises as a solution of a minimization problem.

THEOREM 2.3.4. *Let K be a centered convex body of volume 1 in $\mathbb{R}^n$. Define*

$$(2.3.4) \qquad B(K) = \inf\left\{\int_{TK} \|x\|_2^2 dx : T \in SL(n)\right\}.$$

Then, a position K_1 of K is isotropic if and only if

$$(2.3.5) \qquad \int_{K_1} \|x\|_2^2 dx = B(K).$$

If K_1 and K_2 are isotropic positions of K then there exists $U \in O(n)$ such that $K_2 = U(K_1)$.

Proof. Fix an isotropic position K_1 of K. Remark 2.3.2 shows that there exists $\alpha > 0$ such that

$$\int_{K_1} \langle x, Tx \rangle dx = \alpha^2 (\operatorname{tr} T)$$

for every $T \in L(\mathbb{R}^n)$. Then, for every $T \in SL(n)$ we have

$$(2.3.6) \qquad \int_{TK_1} \|x\|_2^2 dx = \int_{K_1} \|Tx\|_2^2 dx = \int_{K_1} \langle x, T^*Tx \rangle dx$$

$$= \alpha^2 \operatorname{tr}(T^*T) \geqslant n\alpha^2 = \int_{K_1} \|x\|_2^2 dx,$$

where we have used the arithmetic-geometric means inequality in the form

$$\operatorname{tr}(T^*T) \geqslant n[\det(T^*T)]^{1/n}.$$

This shows that K_1 satisfies (2.3.5). In particular, the infimum in (2.3.4) is a minimum.

Note also that if we have equality in (2.3.6) then $T^*T = I$, and hence $T \in O(n)$. This shows that any other position $\tilde{K}$ of K which satisfies (2.3.5) is an orthogonal image of K_1, therefore it is isotropic.

Finally, if K_2 is some other isotropic position of K then the first part of the proof shows that K_2 satisfies (2.3.5). By the previous step, we must have $K_2 = U(K_1)$ for some $U \in O(n)$. $\square$

REMARK 2.3.5. An alternative way to see that, if K is a solution of the minimization problem above, then K is isotropic is to use the following simple variational argument. Let $T \in L(\mathbb{R}^n)$. For small $\varepsilon > 0$, $I + \varepsilon T$ is invertible, and hence $(I + \varepsilon T)/[\det(I + \varepsilon T)]^{1/n}$ preserves volumes. Consequently,

$$\int_K \|x\|_2^2 dx \leqslant \int_K \frac{\|x + \varepsilon Tx\|_2^2}{[\det(I + \varepsilon T)]^{2/n}} dx.$$

Note that $\|x + \varepsilon Tx\|_2^2 = \|x\|_2^2 + 2\varepsilon \langle x, Tx \rangle + O_{T,K}(\varepsilon^2)$ and $[\det(I + \varepsilon T)]^{2/n} = 1 + 2\varepsilon \frac{\operatorname{tr} T}{n} + O_T(\varepsilon^2)$. Therefore, letting $\varepsilon \to 0^+$, we see that

$$\frac{\operatorname{tr} T}{n} \int_K \|x\|_2^2 dx \leqslant \int_K \langle x, Tx \rangle dx.$$

Since T was arbitrary, the same inequality holds with $-T$ instead of T, therefore

$$\frac{\operatorname{tr} T}{n} \int_K \|x\|_2^2 dx = \int_K \langle x, Tx \rangle dx$$

for all $T \in L(\mathbb{R}^n)$. This condition implies that K is isotropic.

DEFINITION 2.3.6. The preceding discussion shows that, for every centered convex body K in $\mathbb{R}^n$, the constant

$$L_K^2 = \frac{1}{n} \min\left\{ \frac{1}{|TK|^{1+\frac{2}{n}}} \int_{TK} \|x\|_2^2 dx \mid T \in GL(n) \right\}$$

is well-defined and depends only on the linear class of K. Also, if $\tilde{K}$ is an isotropic position of K then for all $\theta \in S^{n-1}$ we have

$$\int_{\tilde{K}} \langle x, \theta \rangle^2 dx = L_K^2.$$

The constant L_K is called the *isotropic constant* of K.

2.3.3. Isotropic log-concave measures

DEFINITION 2.3.7. Generalizing the definition of an isotropic convex body, we say that a probability measure $\mu \in \mathcal{P}_n$ is *isotropic* if it is centered and satisfies the isotropic condition

$$\int_{\mathbb{R}^n} \langle x, \theta \rangle^2 \, d\mu(x) = 1$$

for all $\theta \in S^{n-1}$. As in Remark 2.3.2, we easily check that for a centered measure $\mu \in \mathcal{P}_n$ the following are equivalent:

(a) μ is isotropic.

(b) For any $T \in L(\mathbb{R}^n)$ one has

$$\int_{\mathbb{R}^n} \langle x, Tx \rangle \, d\mu(x) = \mathrm{tr}(T).$$

(c) We have $\int_{\mathbb{R}^n} x_i x_j \, d\mu(x) = \delta_{ij}$ for all $i, j = 1, \ldots, n$.

REMARK 2.3.8. If μ is isotropic, then

$$\int_{\mathbb{R}^n} \|x\|_2^2 \, d\mu(x) = n.$$

Also, for any $T \in L(\mathbb{R}^n)$ we have

$$\int_{\mathbb{R}^n} \|Tx\|_2^2 \, d\mu(x) = \|T\|_{\mathrm{HS}}^2,$$

where $\|T\|_{\mathrm{HS}} = \left(\sum_{j=1}^n \|T(e_j)\|_2^2 \right)^{1/2}$ is the Hilbert-Schmidt norm of T.

Following the proof of Proposition 2.3.3, we can check that every non-degenerate centered measure $\mu \in \mathcal{P}_n$ has an isotropic image $\nu = \mu \circ S$, where $S: \mathbb{R}^n \to \mathbb{R}^n$ is a linear map. We define an operator $T: \mathbb{R}^n \to \mathbb{R}^n$ by setting

$$Ty := \int \langle x, y \rangle x \, d\mu(x),$$

we observe that T is symmetric and positive definite, and we set $\nu = \mu \circ S$ where $S \in GL(n)$ is symmetric, positive definite and satisfies $T = S^2$. Then, we easily check that for any $y \in \mathbb{R}^n$

$$\int \langle x, y \rangle^2 \, d\nu(x) = \|y\|_2^2.$$

Moreover, if μ is centered, we see that ν has the same property.

DEFINITION 2.3.9. Let $f : \mathbb{R}^n \to [0, \infty)$ be a log-concave density; that is, f is log-concave and $\int_{\mathbb{R}^n} f = 1$. Then, f is called *isotropic* if its barycenter is at the origin and
$$\int_{\mathbb{R}^n} \langle x, \theta \rangle^2 f(x)\, dx = 1$$
for all $\theta \in S^{n-1}$. As before, we can check that a centered, log-concave density f is isotropic if and only if one of the following conditions holds:

(i) For any $T \in L(\mathbb{R}^n)$ we have
$$\int_{\mathbb{R}^n} \langle x, Tx \rangle f(x)\, dx = \operatorname{tr}(T).$$

(ii) We have
$$\int_{\mathbb{R}^n} x_i x_j f(x)\, dx = \delta_{ij}, \quad \text{for all } i, j = 1, \ldots, n.$$

Again, if f is isotropic then $\int_{\mathbb{R}^n} \|x\|_2^2 f(x)\, dx = n$, and more generally,
$$\int_{\mathbb{R}^n} \|Tx\|_2^2 f(x)\, dx = \|T\|_{\mathrm{HS}}^2$$
for any $T \in L(\mathbb{R}^n)$.

It is easily checked that every log-concave $f : \mathbb{R}^n \to [0, \infty)$ with finite, positive integral has an isotropic image: there exists an affine isomorphism $S : \mathbb{R}^n \to \mathbb{R}^n$ and a positive number a such that $af \circ S$ is isotropic.

Finally, observe that a log-concave probability measure μ on $\mathbb{R}^n$ which is not supported by a hyperplane is isotropic if and only if its density f_μ is an isotropic log-concave function.

REMARK 2.3.10. It is useful to compare the definition of an isotropic convex body (Definition 2.3.1) with the definition of an isotropic log-concave measure. Note that a convex body K of volume 1 in $\mathbb{R}^n$ is isotropic if and only if the function $f_K := L_K^n \mathbf{1}_{\frac{1}{L_K} K}$ is an isotropic log-concave density.

DEFINITION 2.3.11 (General definition of the isotropic constant). Let f be a log-concave function with finite, positive integral. Then, we can define its *inertia* – or *covariance* – matrix $\operatorname{Cov}(f)$ as the matrix with entries
$$[\operatorname{Cov}(f)]_{ij} := \frac{\int_{\mathbb{R}^n} x_i x_j f(x)\, dx}{\int_{\mathbb{R}^n} f(x)\, dx} - \frac{\int_{\mathbb{R}^n} x_i f(x)\, dx}{\int_{\mathbb{R}^n} f(x)\, dx} \frac{\int_{\mathbb{R}^n} x_j f(x)\, dx}{\int_{\mathbb{R}^n} f(x)\, dx}.$$

Note that if f is isotropic then $\operatorname{Cov}(f)$ is the identity matrix.

Given a log-concave function f with finite, positive integral, we define its *isotropic constant* as

(2.3.7) $$L_f := \left(\frac{\sup_{x \in \mathbb{R}^n} f(x)}{\int_{\mathbb{R}^n} f(x)\, dx} \right)^{\frac{1}{n}} [\det \operatorname{Cov}(f)]^{\frac{1}{2n}}.$$

Also, given a finite, non-degenerate log-concave measure μ on $\mathbb{R}^n$ with density f_μ with respect to the Lebesgue measure, we define its isotropic constant by $L_\mu := L_{f_\mu}$, namely by

(2.3.8) $$L_\mu := \left(\frac{\|\mu\|_\infty}{\int_{\mathbb{R}^n} f_\mu(x)\, dx} \right)^{\frac{1}{n}} [\det \operatorname{Cov}(\mu)]^{\frac{1}{2n}},$$

where
$$\|\mu\|_\infty := \sup_{x \in \mathbb{R}^n} f_\mu(x)$$
and $\mathrm{Cov}(\mu) := \mathrm{Cov}(f_\mu)$.

With the above definition it is easy to check that the isotropic constant L_μ is an affine invariant; we have $L_\mu = L_{a\mu \circ A}$, $L_f = L_{af \circ A}$ for every invertible affine transformation A of $\mathbb{R}^n$ and every positive number a. Furthermore, we observe that

(i) Definition 2.3.11 is consistent with our previous definition (Definition 2.3.6) of the isotropic constant of a convex body, in the sense that $L_{\mathbf{1}_K} = L_K$; an easy way to see this is to assume that K is in isotropic position and then trivially observe that $\|\mathbf{1}_K\|_\infty = 1$, $\int \mathbf{1}_K(x)\,dx = 1$ and $\mathrm{Cov}(\mathbf{1}_K) = L_K^2 I$.

(ii) If μ is an isotropic log-concave measure on $\mathbb{R}^n$ then $\int f_\mu = 1$ and $\mathrm{Cov}(\mu) = I$, which shows that $L_\mu = \|\mu\|_\infty^{1/n}$. In addition, since μ is centered by definition, we have from Theorem 2.2.2 that $L_\mu \simeq (f_\mu(0))^{1/n}$; in the sequel we will make frequent use of this fact.

We could also prove a characterization of the isotropic constant which is completely analogous to the one in Theorem 2.3.4: if $f : \mathbb{R}^n \to [0, \infty)$ is a log-concave density, then

$$nL_f^2 = \inf_{\substack{T \in SL(n) \\ y \in \mathbb{R}^n}} \left(\sup_{x \in \mathbb{R}^n} f(x)\right)^{2/n} \int_{\mathbb{R}^n} \|Tx + y\|_2^2 f(x)\,dx.$$

The next proposition shows that the isotropic constants of all log-concave measures are uniformly bounded from below by a constant $c > 0$ which is independent of the dimension.

PROPOSITION 2.3.12. *Let* $f : \mathbb{R}^n \to [0, \infty)$ *be an isotropic log-concave density. Then*
$$L_f = \|f\|_\infty^{1/n} \geqslant c,$$
where $c > 0$ is an absolute constant.

Proof. Since f is isotropic, we may write

$$n = \int \|x\|_2^2 f(x)\,dx = \int_{\mathbb{R}^n} \left(\int_0^{\|x\|_2^2} \mathbf{1}\,dt\right) f(x)\,dx$$
$$= \int_0^\infty \int_{\mathbb{R}^n} \mathbf{1}_{\{x : \|x\|_2^2 \geqslant t\}}(x) f(x)\,dx\,dt$$
$$= \int_0^\infty \int_{\mathbb{R}^n \setminus \sqrt{t} B_2^n} f(x)\,dx\,dt$$
$$= \int_0^\infty \left(1 - \int_{\sqrt{t} B_2^n} f(x)dx\right) dt$$
$$\geqslant \int_0^{(\omega_n \|f\|_\infty)^{-2/n}} [1 - \omega_n \|f\|_\infty t^{n/2}]\,dt$$
$$= (\omega_n \|f\|_\infty)^{-2/n} \frac{n}{n+2}.$$

Since $\omega_n^{-1/n} \simeq \sqrt{n}$, we get $\|f\|_\infty^{1/n} \geqslant c$ for some absolute constant $c > 0$. $\square$

2.3.4. Isotropic random vectors

Let $(\Omega, \mathcal{A}, \mathbb{P})$ be a probability space. A *random vector* $X : \Omega \to \mathbb{R}^n$ will be called log-concave if its distribution

$$\mu(A) := \mathbb{P}(X \in A) = \mathbb{P}(\{\omega \in \Omega : X(\omega) \in A\})$$

is a log-concave probability measure on $\mathbb{R}^n$. We will say that X is an isotropic random vector if μ is isotropic and we will write the isotropic conditions in the form

$$\mathbb{E}(X) = 0 \quad \text{and} \quad \mathbb{E}(X \otimes X) = I.$$

The first equation is equivalent to the fact that μ is centered and the second one is equivalent to the fact that $\mathrm{Cov}(\mu) = I$.

We will adopt this terminology and notation mostly in the second part of this book.

2.4. ψ_α-estimates

Let $(\Omega, \mathcal{A}, \mu)$ be a probability space. Let $\Phi : \mathbb{R} \to [0, +\infty)$ be an even convex function satisfying $\Phi(0) = 0$ and $\lim_{x \to \infty} \Phi(x) = +\infty$ (we say that Φ is an Orlicz function). The Orlicz space $L_\Phi(\mu)$ that corresponds to Φ consists of all the $\mathcal{A}$-measurable functions f for which there is a constant $\kappa > 0$ such that $\int_\Omega \Phi(f/\kappa) d\mu < \infty$. The norm of any such function f is defined to be the infimum of all $\kappa > 0$ such that $\int_\Omega \Phi(f/\kappa) d\mu \leqslant 1$.

One can check that $L_\Phi(\mu) \subseteq L_1(\mu)$: if a measurable function f has finite $\Phi(\mu)$-norm then f is integrable with respect to μ. To see this, note first that, since Φ is convex and $\Phi(0) = (0)$, the function $t \mapsto \frac{\Phi(t)}{t}$ is increasing. This implies that $\Phi(t) \geqslant \frac{\Phi(t_0)}{t_0} \cdot t$ for all $t > t_0$ where t_0 is any positive number such that $\Phi(t_0) > 0$. Then, given any $\kappa > \|f\|_{\Phi(\mu)}$ we may write

$$\frac{1}{\kappa} \mathbb{E}_\mu(|f|) = \mathbb{E}_\mu\left(\frac{|f|}{\kappa} \cdot \mathbf{1}_{\{|f| \leqslant t_0 \kappa\}}\right) + \mathbb{E}_\mu\left(\frac{|f|}{\kappa} \cdot \mathbf{1}_{\{|f| > t_0 \kappa\}}\right)$$

$$\leqslant t_0 + \frac{t_0}{\Phi(t_0)} \mathbb{E}_\mu(\Phi(|f|/\kappa)) \leqslant t_0 \cdot [1 + (\Phi(t_0))^{-1}] < +\infty.$$

Next, we may also apply Jensen's inequality with the convex function Φ to get

$$\Phi(\mathbb{E}_\mu(|f|/\kappa)) \leqslant \mathbb{E}_\mu(\Phi(|f|/\kappa)) \leqslant 1$$

for every $\kappa > \|f\|_{\Phi(\mu)}$. Therefore,

$$\mathbb{E}_\mu(|f|) \leqslant \Phi_*^{-1}(1) \cdot \|f\|_{\Phi(\mu)}$$

where $\Phi_*^{-1}(1) = \inf\{s > 0 : \Phi(t) > 1 \text{ for all } t \geqslant s\}$.

The family of ψ_α-norms, which is a subclass of Orlicz norms, will play a central role in this book.

DEFINITION 2.4.1 (ψ_α-norm). Let $(\Omega, \mathcal{A}, \mu)$ be a probability space and let $f : \Omega \to \mathbb{R}$ be an $\mathcal{A}$-measurable function. For any $\alpha \geqslant 1$ we define the ψ_α-norm of f as follows:

$$\|f\|_{\psi_\alpha} := \inf\left\{t > 0 : \int_\Omega \exp\left(\frac{|f(\omega)|}{t}\right)^\alpha d\mu(\omega) \leqslant 2\right\},$$

provided that the set on the right hand side is non-empty. Note that the ψ_α-norm is exactly the Orlicz norm corresponding to the function $t \in \mathbb{R} \to e^{|t|^\alpha} - 1$.

The next lemma gives an equivalent expression for the ψ_α-norm in terms of the L_q-norms.

LEMMA 2.4.2. *Let $(\Omega, \mathcal{A}, \mu)$ be a probability space. Let $\alpha \geqslant 1$ and let $f : \Omega \to \mathbb{R}$ be an $\mathcal{A}$-measurable function. Then,*
$$\|f\|_{\psi_\alpha} \simeq \sup_{p \geqslant \alpha} \frac{\|f\|_{L_p(\mu)}}{p^{1/\alpha}},$$
up to some absolute constants.

Proof. First we show that there exists an absolute constant $C > 0$ such that for any $p \geqslant \alpha$ we have
$$\|f\|_p \leqslant C p^{1/\alpha} \|f\|_{\psi_\alpha}.$$
Indeed, we set $A = \|f\|_{\psi_\alpha}$ and using the elementary inequality $1 + \frac{t^k}{k!} \leqslant e^t$, valid for any $t > 0$, we obtain
$$1 + \int_\Omega \frac{|f(\omega)|^{k\alpha}}{k! A^{k\alpha}} \, d\mu \leqslant \int_\Omega \exp(|f|/A)^\alpha \, d\mu = 2,$$
which implies
$$\int_\Omega |f|^{k\alpha} \, d\mu \leqslant k! A^{k\alpha}.$$
Let $p \geqslant \alpha$. There exists a unique $k \in \mathbb{N}$ such that $k\alpha \leqslant p < (k+1)\alpha$. Then, using Hölder's inequality and Stirling's approximation we get
$$\|f\|_p \leqslant \|f\|_{(k+1)\alpha} \leqslant [(k+1)!]^{\frac{1}{(k+1)\alpha}} A \leqslant (2k)^{1/\alpha} A$$
$$\leqslant \left(\frac{2p}{\alpha}\right)^{1/\alpha} A \leqslant 2p^{1/\alpha} A.$$
Conversely, if $\gamma := \sup_{p \geqslant \alpha} \frac{\|f\|_p}{p^{1/\alpha}}$, then $\int_\Omega |f|^p \, d\mu \leqslant \gamma^p p^{p/\alpha}$ for all $p \geqslant \alpha$. Then, we fix $c > 0$ (which will be suitably defined) and write
$$\int_\Omega \exp(|f|/c\gamma)^\alpha = 1 + \sum_{k=1}^\infty \frac{1}{(c\gamma)^{k\alpha} k!} \int_\Omega |f|^{\alpha k} \, d\mu \leqslant 1 + \sum_{k=1}^\infty \frac{(k\alpha)^k}{k! c^{k\alpha}}$$
$$\leqslant 1 + \sum_{k=1}^\infty \left(\frac{e\alpha}{c^\alpha}\right)^k,$$
where we have used the elementary inequality $k! \geqslant (k/e)^k$. If we choose $c_\alpha := (2e\alpha)^{1/\alpha} \leqslant 2e \cdot e^{1/e} =: c$, then we have $\|f\|_{\psi_\alpha} \leqslant c_\alpha \gamma \leqslant c\gamma$. $\square$

DEFINITION 2.4.3. Let $\mu \in \mathcal{P}_n$, $\alpha \geqslant 1$ and $\theta \in S^{n-1}$. We say that μ satisfies a ψ_α-estimate with constant $b_\alpha = b_\alpha(\theta)$ in the direction of θ if we have
$$\|\langle \cdot, \theta \rangle\|_{\psi_\alpha} \leqslant b_\alpha \|\langle \cdot, \theta \rangle\|_2.$$
We say that μ is a ψ_α-measure with constant $B_\alpha > 0$ if
$$\sup_{\theta \in S^{n-1}} \frac{\|\langle \cdot, \theta \rangle\|_{\psi_\alpha}}{\|\langle \cdot, \theta \rangle\|_2} \leqslant B_\alpha.$$

Using Lemma 2.4.2 we see that μ satisfies a ψ_α-estimate with constant $b'_\alpha \simeq b_\alpha$ in the direction of $\theta \in S^{n-1}$ if

$$\|\langle \cdot, \theta \rangle\|_q \leqslant b_\alpha q^{1/\alpha} \|\langle \cdot, \theta \rangle\|_2$$

for all $q \geqslant \alpha$.

The next lemma gives one more useful description of the ψ_α-norm.

LEMMA 2.4.4. *Let $\mu \in \mathcal{P}_n$ and let $\alpha \geqslant 1$ and $\theta \in S^{n-1}$.*
  (i) *If μ satisfies a ψ_α-estimate with constant b in the direction of θ then for all $t > 0$ we have $\mu(\{x : |\langle x, \theta \rangle| \geqslant t \|\langle \cdot, \theta \rangle\|_2 \}) \leqslant 2e^{-t^\alpha/b^\alpha}$.*
  (ii) *If we have $\mu(\{x : |\langle x, \theta \rangle| \geqslant t \|\langle \cdot, \theta \rangle\|_2 \}) \leqslant 2e^{-t^\alpha/b^\alpha}$ for some $b > 0$ and for all $t > 0$ then μ satisfies a ψ_α-estimate with constant $\leqslant cb$ in the direction of θ, where $c > 0$ is an absolute constant.*

Proof. The first assertion is a direct application of Markov's inequality. For the second, it suffices to prove that

$$\left(\int_{\mathbb{R}^n} |\langle x, \theta \rangle|^p \, d\mu(x) \right)^{1/p} \leqslant c_1 b p^{1/\alpha} \|\langle \cdot, \theta \rangle\|_2,$$

for any $p \geqslant \alpha$, where $c_1 > 0$ is an absolute constant. We write

$$\int_{\mathbb{R}^n} |\langle x, \theta \rangle|^p \, d\mu(x) = \int_0^\infty p t^{p-1} \mu(\{x : |\langle x, \theta \rangle| \geqslant t\}) \, dt$$

$$= \|\langle \cdot, \theta \rangle\|_2^p \int_0^\infty p t^{p-1} \mu(\{x : |\langle x, \theta \rangle| \geqslant t \|\langle \cdot, \theta \rangle\|_2 \}) \, dt$$

$$\leqslant 2 \|\langle \cdot, \theta \rangle\|_2^p \int_0^\infty p t^{p-1} e^{-t^\alpha/b^\alpha} \, dt,$$

using the tail estimate from (ii). Making the change of variables $s = (t/b)^\alpha$, we arrive at

$$\int_{\mathbb{R}^n} |\langle x, \theta \rangle|^p \, d\mu(x) \leqslant 2(b \|\langle \cdot, \theta \rangle\|_2)^p \int_0^\infty \frac{p}{\alpha} s^{p/\alpha - 1} e^{-s} \, ds$$

$$= 2(b \|\langle \cdot, \theta \rangle\|_2)^p \Gamma\left(\frac{p}{\alpha} + 1 \right).$$

Using Stirling's formula, we get the result. □

Borell's lemma holds true in the more general context of log-concave probability measures.

LEMMA 2.4.5. *Let μ be a log-concave measure on $\mathcal{P}_n$. Then, for any symmetric convex set A in $\mathbb{R}^n$ with $\mu(A) = \alpha \in (0,1)$ and any $t > 1$ we have*

$$(2.4.1) \qquad 1 - \mu(tA) \leqslant \alpha \left(\frac{1-\alpha}{\alpha} \right)^{\frac{t+1}{2}}.$$

Proof. Using the symmetry and convexity of A we check that

$$\frac{2}{t+1}(\mathbb{R}^n \setminus (tA)) + \frac{t-1}{t+1} A \subseteq \mathbb{R}^n \setminus A.$$

for every $t > 1$. Then, we apply the log-concavity of μ to get the result. □

Note. The right hand side of (2.4.1) can be written in the form

$$(2.4.2) \qquad \frac{(1-\alpha)^{\frac{t+1}{2}}}{\alpha^{\frac{t-1}{2}}} < \frac{(1-\alpha)^{\frac{t-1}{2}}}{\alpha^{\frac{t-1}{2}}} = \left(\frac{1}{\alpha} - 1\right)^{\frac{t-1}{2}}.$$

Using Borell's lemma we see that there exists an absolute constant $C > 0$ such that every log-concave measure $\mu \in \mathcal{P}_n$ is a ψ_1-measure with constant C.

THEOREM 2.4.6. *Let $\mu \in \mathcal{P}_n$ be log-concave. If $f : \mathbb{R}^n \to \mathbb{R}$ is a seminorm then, for any $q > p \geqslant 1$, we have*

$$\left(\int_{\mathbb{R}^n} |f|^p \, d\mu\right)^{1/p} \leqslant \left(\int_{\mathbb{R}^n} |f|^q \, d\mu\right)^{1/q} \leqslant c\frac{q}{p} \left(\int_{\mathbb{R}^n} |f|^p \, d\mu\right)^{1/p},$$

where $c > 0$ is an absolute constant.

Proof. We write $\|f\|_p^p := \int |f|^p \, d\mu$. Then, the set

$$A = \{x \in \mathbb{R}^n : |f(x)| \leqslant 3\|f\|_p\}$$

is symmetric and convex. Also, for any $t > 0$ we get

$$tA = \{x \in \mathbb{R}^n : |f(x)| \leqslant 3t\|f\|_p\},$$

while $\mu(A) \geqslant 1 - 3^{-p}$. So, in our case $\frac{1}{\alpha} - 1 \leqslant \frac{3^{-p}}{1-3^{-p}} \leqslant e^{-p/2}$. Using (2.4.2) we see that

$$\mu(x : |f(x)| \geqslant 3t\|f\|_p) \leqslant e^{-c_1 p(t-1)}$$

for any $t > 1$, with $c_1 = \frac{1}{4}$. Now, we write

$$\int_{\mathbb{R}^n} |f|^q \, d\mu = \int_0^\infty q s^{q-1} \mu(\{x : |f(x)| \geqslant s\}) \, ds$$

$$\leqslant (3\|f\|_p)^q + (3\|f\|_p)^q \int_1^\infty q t^{q-1} e^{-c_1 p(t-1)} \, dt$$

$$\leqslant (3\|f\|_p)^q + e^{c_1 p}(3\|f\|_p)^q \int_0^\infty q t^{q-1} e^{-c_1 p t} \, dt$$

$$\leqslant (3\|f\|_p)^q + e^{c_1 p} \left(\frac{3\|f\|_p}{c_1 p}\right)^q \Gamma(q+1).$$

Stirling's formula and the fact that $(a+b)^{1/q} \leqslant a^{1/q} + b^{1/q}$ for all $a, b > 0$ and $q \geqslant 1$, imply that $\|f\|_{L^q(\mu)} \leqslant c\frac{q}{p} \|f\|_{L^p(\mu)}$. □

REMARKS 2.4.7. (a) For every $\theta \in S^{n-1}$ the function $x \mapsto |\langle x, \theta\rangle|$, $x \in \mathbb{R}^n$, satisfies the hypothesis of Theorem 2.4.6. Therefore,

$$\|\langle \cdot, \theta\rangle\|_q \leqslant cq \|\langle \cdot, \theta\rangle\|_1$$

for all $\theta \in S^{n-1}$ and $q \geqslant 1$, where $c > 0$ is an absolute constant. It follows that

$$\|\langle \cdot, \theta\rangle\|_{\psi_1} \leqslant c \|\langle \cdot, \theta\rangle\|_1$$

for all $\theta \in S^{n-1}$. This ψ_1-estimate will play an important role in our future work, starting from the next chapter.

(b) Using the fact that the n-dimensional Gaussian measure is a log-concave probability measure we get that any seminorm f satisfies the conclusion of Theorem 2.4.6. On the other hand, integrating in polar coordinates we get

$$\left(\int_{\mathbb{R}^n} |f(x)|^q \, d\gamma_n(x)\right)^{1/q} \simeq \sqrt{n+q} \left(\int_{S^{n-1}} |f(\theta)|^q \, d\sigma(\theta)\right)^{1/q},$$

for any $q \geqslant 1$. Combining these estimates we obtain:

$$\left(\int_{S^{n-1}} |f|^q \, d\sigma\right)^{1/q} \leqslant c\frac{q}{p}\sqrt{\frac{n+p}{n+q}} \left(\int_{S^{n-1}} |f|^p \, d\sigma\right)^{1/p}$$

for any $1 \leqslant p \leqslant q$, where $c > 0$ is an absolute constant.

The next result provides a small ball estimate for log-concave probability measures.

THEOREM 2.4.8 (Latała). *Let μ be a log-concave probability measure on $\mathbb{R}^n$. For any norm $\|\cdot\|$ on $\mathbb{R}^n$ and any $0 \leqslant t \leqslant 1$ one has*

(2.4.3) $$\mu(\{x : \|x\| \leqslant t\mathbb{E}_\mu(\|x\|)\}) \leqslant Ct,$$

where $C > 0$ is an absolute constant.

PROOF. Let K be the unit ball of $(\mathbb{R}^n, \|\cdot\|)$. We define

$$\alpha = \inf\left\{s > 0 : \mu(sK) \geqslant \frac{2}{3}\right\}.$$

Then $\mu(\alpha K) \leqslant 2/3$ and applying Borell's lemma we see that

$$\mathbb{E}_\mu(\|x\|) \leqslant \alpha + \alpha \int_1^\infty [1 - \mu(s\alpha K)] \, ds$$

$$\leqslant \alpha + \alpha \int_1^\infty \mu(\alpha K) \left(\frac{1 - \mu(\alpha K)}{\mu(\alpha K)}\right)^{\frac{s+1}{2}} ds$$

$$\leqslant \alpha + \frac{2\alpha}{3\sqrt{2}} \int_1^\infty 2^{-s/2} \, ds \leqslant c\alpha$$

where $c > 0$ is an absolute constant. We also have

$$1 - \mu(3\alpha K) \leqslant \mu(\alpha K) \left(\frac{1 - \mu(\alpha K)}{\mu(\alpha K)}\right)^2 \leqslant \frac{1}{6},$$

which implies

$$\mu(\{x : \alpha \leqslant \|x\| < 3\alpha\}) \geqslant \frac{1}{6}.$$

We fix $k \in \mathbb{N}$ and for any $u \geqslant \frac{\alpha}{2k}$ we define

$$T(u) = \left\{x : u - \frac{\alpha}{2k} \leqslant \|x\| < u + \frac{\alpha}{2k}\right\}.$$

Since

$$\{x : \alpha \leqslant \|x\| < 3\alpha\} = \bigcup_{s=1}^{2k} T\left(\alpha + \frac{(2s-1)\alpha}{2k}\right),$$

we may find $u_0 > \alpha$ so that $\mu(T(u_0)) \geqslant \frac{1}{12k}$. Note that

(2.4.4) $$\lambda T(u) + (1-\lambda)\frac{\alpha}{2k}K \subseteq T(\lambda u)$$

for every $0 < \lambda < 1$, by the triangle inequality.

Claim. For every $k \in \mathbb{N}$ we have
$$\mu\left(\frac{\alpha}{2k}K\right) \leqslant \frac{48}{k}.$$

To see this, assume the contrary and then observe that for any $0 < \lambda < \frac{1}{2}$ the log-concavity of μ and (2.4.4) imply
$$\mu(T(\lambda u_0)) \geqslant \mu(T(u_0))^\lambda \mu\left(\frac{\alpha}{2k}K\right)^{1-\lambda} \geqslant \left(\frac{1}{12k}\right)^\lambda \left(\frac{48}{k}\right)^{1-\lambda} > \frac{2}{k}.$$

This shows that $\mu(T(u)) > \frac{2}{k}$ for all $u < u_0/2$ (and hence, for all $u \leqslant \alpha/2$). We get a contradiction if we observe that the sets $T(s\alpha/k)$, $1 \leqslant s \leqslant k/2$ are disjoint, they have measure greater than $\frac{2}{k}$, and they are also disjoint from $\frac{\alpha}{2k}K$.

Now, let $0 < t \leqslant 1/2$. We may find $k \in \mathbb{N}$ so that $\frac{1}{4k} \leqslant t \leqslant \frac{1}{2k}$, and then
$$\mu(\{x : \|x\| \leqslant t\alpha\}) \leqslant \mu\left(\left\{x : \|x\| \leqslant \frac{\alpha}{2k}\right\}\right) = \mu\left(\frac{\alpha}{2k}K\right) \leqslant \frac{48}{k} \leqslant 192t.$$

Since $\mathbb{E}_\mu(\|x\|) \leqslant c\alpha$ the theorem follows. $\square$

A consequence of Theorem 2.4.8 is the next Kahane-Khintchine inequality for negative exponents.

THEOREM 2.4.9. *Let μ be a log-concave probability measure on $\mathbb{R}^n$. For any norm $\|\cdot\|$ on $\mathbb{R}^n$ and any $-1 < q < 0$ one has*

(2.4.5) $$\mathbb{E}_\mu(\|x\|) \leqslant \frac{C}{1+q}\left(\mathbb{E}_\mu(\|x\|^q)\right)^{1/q},$$

where $C > 0$ is an absolute constant.

Proof. We may assume that $\mathbb{E}_\mu(\|x\|) = 1$. We set $p = -q$ and using integration by parts and Theorem 2.4.8 we write
$$\mathbb{E}_\mu(\|x\|^q) = \mathbb{E}_\mu\left(\frac{1}{\|x\|^p}\right) = p\int_0^\infty t^{p-1}\mu\left(\left\{\frac{1}{\|x\|} \geqslant t\right\}\right)$$
$$\leqslant 1 + C_1 p \int_1^\infty t^{p-2}dt = 1 + \frac{C_1 p}{1-p} \leqslant \frac{1+C_1 p}{1-p} \leqslant \frac{e^{C_1 p}}{1-p}.$$

It follows that
$$\left(\mathbb{E}_\mu(\|x\|^q)\right)^{1/q} \geqslant \left(\frac{1-p}{e^{C_1 p}}\right)^{1/p} = e^{-C_1}(1-p)(1-p)^{1/p-1} \geqslant \frac{1-p}{C}$$

because $(1-p)^{1-1/p}$ is bounded on $(0,1)$. $\square$

We close this section with an observation that allows us to replace the expectation $\mathbb{E}_\mu$ by the median med_μ and vice versa in various functional inequalities involving Orlicz spaces $L_\Phi(\mu)$. We restrict our attention to functions Φ such that $\Phi(s) < \Phi(t)$ for all $0 \leqslant s < t$; all the main examples belong to this category.

LEMMA 2.4.10. *Let μ be a Borel probability measure on $\mathbb{R}^n$, let Φ be an Orlicz function that is strictly increasing on $\mathbb{R}^+$ and write $L_\Phi(\mu)$ for the corresponding Orlicz space. For every $f \in L_\Phi(\mu)$ we have*
$$\frac{1}{2}\|f - \mathbb{E}_\mu(f)\|_{L_\Phi(\mu)} \leqslant \|f - \text{med}_\mu(f)\|_{L_\Phi(\mu)} \leqslant 3\|f - \mathbb{E}_\mu(f)\|_{L_\Phi(\mu)}.$$

Proof. Observe that the constant function 1 has norm $\|1\|_{L_\Phi(\mu)} = 1/\Phi^{-1}(1)$, where Φ^{-1} is the inverse of the restriction of Φ to $\mathbb{R}^+$. Note that, since this restriction is convex, strictly increasing and $\Phi(0) = 0$, we have that $\Phi^{-1} : \mathbb{R}^+ \to \mathbb{R}^+$ is also strictly increasing, that it is concave and that $\Phi^{-1}(0) = 0$, hence the function $\frac{\Phi^{-1}(t)}{t} : (0, +\infty) \to (0, +\infty)$ is decreasing.

Applying Jensen's inequality twice, we see that

$$|\mathbb{E}_\mu(f) - \mathrm{med}_\mu(f)| \leqslant \mathbb{E}_\mu(|f - \mathrm{med}_\mu(f)|) \leqslant \Phi^{-1}(1) \cdot \|f - \mathrm{med}_\mu(f)\|_{L_\Phi(\mu)}.$$

Hence

$$\|f - \mathbb{E}_\mu(f)\|_{L_\Phi(\mu)} \leqslant \|f - \mathrm{med}_\mu(f)\|_{L_\Phi(\mu)} + \frac{1}{\Phi^{-1}(1)} \cdot |\mathbb{E}_\mu(f) - \mathrm{med}_\mu(f)|$$

$$\leqslant 2\|f - \mathrm{med}_\mu(f)\|_{L_\Phi(\mu)}.$$

Next, we assume that $\mathrm{med}_\mu(f) > \mathbb{E}_\mu(f)$ (because if the reverse inequality holds we can work with $-f$, and if $\mathrm{med}_\mu(f) = \mathbb{E}_\mu(f)$ we have nothing to prove). Markov's inequality implies

$$\frac{1}{2} \leqslant \mu(\{f \geqslant \mathrm{med}_\mu(f)\}) \leqslant \mu(\{|f - \mathbb{E}_\mu(f)| \geqslant \mathrm{med}_\mu(f) - \mathbb{E}_\mu(f)\})$$

$$\leqslant \left[\Phi\left(\frac{\mathrm{med}_\mu(f) - \mathbb{E}_\mu(f)}{\|f - \mathbb{E}_\mu f\|_{L_\Phi(\mu)}}\right)\right]^{-1}.$$

It follows that

$$|\mathrm{med}_\mu(f) - \mathbb{E}_\mu(f)| \leqslant \Phi^{-1}(2) \cdot \|f - \mathbb{E}_\mu(f)\|_{L_\Phi(\mu)},$$

and an application of the triangle inequality gives

$$\|f - \mathrm{med}_\mu(f)\|_{L_\Phi(\mu)} \leqslant \|f - \mathbb{E}_\mu(f)\|_{L_\Phi(\mu)} + \frac{1}{\Phi^{-1}(1)} \cdot |\mathbb{E}_\mu(f) - \mathrm{med}_\mu(f)|$$

$$\leqslant \left(1 + \frac{\Phi^{-1}(2)}{\Phi^{-1}(1)}\right) \|f - \mathbb{E}_\mu(f)\|_{L_\Phi(\mu)} \leqslant 3\|f - \mathbb{E}_\mu(f)\|_{L_\Phi(\mu)}$$

because $\frac{\Phi^{-1}(2)}{2} \leqslant \Phi^{-1}(1)$. This completes the proof. $\square$

2.5. Convex bodies associated with log-concave functions

In this section we associate a family of convex sets $K_p(f)$ with any given log-concave function f; we prove that these are convex bodies and we describe some of their basic properties. The bodies $K_p(f)$ were introduced by K. Ball (see [**32**]) who also established their convexity. They will play an important role in the sequel as they allow us to reduce the study of log-concave measures to that of convex bodies.

DEFINITION 2.5.1 (Ball). Let $f : \mathbb{R}^n \to [0, \infty)$ be a measurable function such that $f(0) > 0$. For any $p > 0$ we define the set $K_p(f)$ as follows:

$$K_p(f) = \left\{x \in \mathbb{R}^n : \int_0^\infty f(rx) r^{p-1} \, dr \geqslant \frac{f(0)}{p}\right\}.$$

From the definition it follows that the radial function of $K_p(f)$ is given by

(2.5.1) $$\rho_{K_p(f)}(x) = \left(\frac{1}{f(0)} \int_0^\infty p r^{p-1} f(rx) \, dr\right)^{1/p}$$

2.5. CONVEX BODIES ASSOCIATED WITH LOG-CONCAVE FUNCTIONS

for $x \neq 0$. If μ is a measure on $\mathbb{R}^n$ which is absolutely continuous with respect to the Lebesgue measure, and if f_μ is the density of μ and $f_\mu(0) > 0$, then we define

$$K_p(\mu) := K_p(f_\mu) = \left\{ x : \int_0^\infty r^{p-1} f_\mu(rx) \, dr \geqslant \frac{f_\mu(0)}{p} \right\},$$

LEMMA 2.5.2. *Let K be a convex body in $\mathbb{R}^n$ with $0 \in \mathrm{int}(K)$. Then, we have $K_p(\mathbf{1}_K) = K$ for all $p > 0$.*

Proof. For every $\theta \in S^{n-1}$ we have

$$\rho^p_{K_p(\mathbf{1}_K)}(\theta) = \frac{1}{\mathbf{1}_K(0)} \int_0^{+\infty} p r^{p-1} \mathbf{1}_K(r\theta) \, dr$$
$$= \int_0^{\rho_K(\theta)} p r^{p-1} \, dr = \rho_K^p(\theta).$$

It follows that $K_p(\mathbf{1}_K) = K$. □

The next proposition describes some basic properties of the sets $K_p(f)$.

PROPOSITION 2.5.3. *Let $f, g : \mathbb{R}^n \to [0, \infty)$ be two integrable functions with $f(0) = g(0) > 0$, and set*

$$m = \inf\left\{ \frac{f(x)}{g(x)} : g(x) > 0 \right\} \quad \text{and} \quad M^{-1} = \inf\left\{ \frac{g(x)}{f(x)} : f(x) > 0 \right\}.$$

Let V be a star body and write $\|\cdot\|_V$ for its gauge function. Then, for every $p > 0$ we have the following:

(i) $0 \in K_p(f)$.
(ii) $K_p(f)$ *is a star-shaped set.*
(iii) $K_p(f)$ *is symmetric if f is even.*
(iv) $m^{1/p} K_p(g) \subseteq K_p(f) \subseteq M^{1/p} K_p(g)$.
(v) *For any $\theta \in S^{n-1}$ we have*

$$\int_{K_{n+1}(f)} \langle x, \theta \rangle \, dx = \frac{1}{f(0)} \int_{\mathbb{R}^n} \langle x, \theta \rangle f(x) \, dx.$$

Thus, f is centered if and only if $K_{n+1}(f)$ is centered.
(vi) *For any $\theta \in S^{n-1}$ and $p > 0$ we have*

$$\int_{K_{n+p}(f)} |\langle x, \theta \rangle|^p \, dx = \frac{1}{f(0)} \int_{\mathbb{R}^n} |\langle x, \theta \rangle|^p f(x) \, dx.$$

(vii) *If $p > -n$ then*

(2.5.2) $$\int_{K_{n+p}(f)} \|x\|_V^p \, dx = \frac{1}{f(0)} \int_{\mathbb{R}^n} \|x\|_V^p f(x) \, dx.$$

Proof. The first three statements (i), (ii) and (iii) can be easily checked. To prove (iv) we compare the radial functions of $K_p(f)$ and $K_p(g)$. We have

$$\rho^p_{K_p(f)}(x) = \frac{1}{f(0)} \int_0^\infty p r^{p-1} f(rx) \, dr \leqslant \frac{M}{g(0)} \int_0^\infty p r^{p-1} g(rx) \, dr$$
$$= (M^{1/p} \rho_{K_p(g)}(x))^p,$$

and similarly $(m^{1/p} \rho_{K_p(g)}(x))^p \leqslant \rho^p_{K_p(f)}(x)$.

(v) Integrating in polar coordinates we see that, for any $\theta \in S^{n-1}$,

$$\int_{K_{n+1}(f)} \langle x, \theta \rangle dx = n\omega_n \int_{S^{n-1}} \langle \phi, \theta \rangle \int_0^{\rho_{K_{n+1}(f)}(\phi)} r^n dr d\sigma(\phi)$$

$$= \frac{n\omega_n}{f(0)} \int_{S^{n-1}} \langle \phi, \theta \rangle \int_0^\infty r^n f(r\phi) dr d\sigma(\phi)$$

$$= \frac{1}{f(0)} \int_{\mathbb{R}^n} \langle x, \theta \rangle f(x) dx.$$

So, if f is centered then $K_{n+1}(f)$ is also centered.

(vi) The same argument shows that, for every $p > -n$ and $\theta \in S^{n-1}$,

$$\int_{K_{n+p}(f)} |\langle x, \theta \rangle|^p dx = n\omega_n \int_{S^{n-1}} |\langle \phi, \theta \rangle|^p \int_0^{\rho_{K_{n+p}(f)}(\phi)} r^{n+p-1} dr d\sigma(\phi)$$

$$= \frac{n\omega_n}{f(0)} \int_{S^{n-1}} |\langle \phi, \theta \rangle|^p \int_0^\infty r^{n+p-1} f(r\phi) dr d\sigma(\phi)$$

$$= \frac{1}{f(0)} \int_{\mathbb{R}^n} |\langle x, \theta \rangle|^p f(x) dx.$$

(vii) Working in the same manner we see that for $p > -n$,

$$\int_{K_{n+p}(f)} \|x\|_V^p dx = n\omega_n \int_{S^{n-1}} \|\phi\|_V^p \int_0^{\rho_{K_{n+p}(f)}(\phi)} r^{n+p-1} dr d\sigma(\phi)$$

$$= \frac{n\omega_n}{f(0)} \int_{S^{n-1}} \|\phi\|_V^p \int_0^\infty r^{n+p-1} f(r\phi) dr d\sigma(\phi)$$

$$= \frac{1}{f(0)} \int_{\mathbb{R}^n} \|x\|_V^p f(x) dx.$$

We have thus checked (i)–(vii). $\square$

Assuming the log-concavity of f we can prove that the sets $K_p(f)$, $p > 0$, are convex. To this end, we use a three functions inequality in the spirit of the Prékopa-Leindler inequality. We first introduce some notation: Let $0 \leqslant \lambda \leqslant 1$ and $\gamma \in \mathbb{R}$. If (a, b) is a pair of positive real numbers we define their mean of order γ with coefficient λ by

$$M_\gamma^\lambda(a, b) = (\lambda a^\gamma + (1-\lambda) b^\gamma)^{1/\gamma}.$$

In the case $\gamma = 0$ we agree that $M_0^\lambda = a^\lambda b^{1-\lambda}$. We also extend the definition to pairs (a, b) with $ab = 0$ setting $M_\gamma^\lambda = 0$ in this case.

THEOREM 2.5.4. *Let $\gamma > 0$ and $\lambda, \mu > 0$ with $\lambda + \mu = 1$. Let $w, g, h : \mathbb{R}^+ \to \mathbb{R}^+$ be integrable functions such that*

$$h(M_{-\gamma}^\lambda(r, s)) \geqslant w(r)^{\frac{\lambda s^\gamma}{\lambda s^\gamma + \mu r^\gamma}} g(s)^{\frac{\mu r^\gamma}{\lambda s^\gamma + \mu r^\gamma}}$$

for every pair $(r, s) \in \mathbb{R}^+ \times \mathbb{R}^+$. Then,

$$\int_0^\infty h \geqslant M_{-\gamma}^\lambda \left(\int_0^\infty w, \int_0^\infty g \right).$$

Proof. We may assume that w and g are continuous and strictly positive. We define $r, s : [0, 1] \to \mathbb{R}^+$ by the equations
$$\int_0^{r(t)} w = t \int_0^\infty w \quad \text{and} \quad \int_0^{s(t)} g = t \int_0^\infty g.$$
Then, r and s are differentiable, and for every $t \in (0,1)$ we have
$$r'(t)w(r(t)) = \int_0^\infty w \quad \text{and} \quad s'(t)g(s(t)) = \int_0^\infty g.$$
Next, we define $z : [0,1] \to \mathbb{R}^+$ by
$$z(t) = M_{-\gamma}^\lambda(r(t), s(t)).$$
Note that
$$\begin{aligned}
z' &= z^{\gamma+1}\left(\frac{\lambda r'}{r^{\gamma+1}} + \frac{\mu s'}{s^{\gamma+1}}\right) \\
&= \lambda \frac{\int w}{w(r)} \left(\frac{s^\gamma}{\lambda s^\gamma + \mu r^\gamma}\right)^{\frac{\gamma+1}{\gamma}} + \mu \frac{\int g}{g(s)} \left(\frac{r^\gamma}{\lambda s^\gamma + \mu r^\gamma}\right)^{\frac{\gamma+1}{\gamma}} \\
&= \frac{\lambda s^\gamma}{\lambda s^\gamma + \mu r^\gamma}\left(\frac{\int w}{w(r)}\left(\frac{s^\gamma}{\lambda s^\gamma + \mu r^\gamma}\right)^{\frac{1}{\gamma}}\right) + \frac{\mu r^\gamma}{\lambda s^\gamma + \mu r^\gamma}\left(\frac{\int g}{g(s)}\left(\frac{r^\gamma}{\lambda s^\gamma + \mu r^\gamma}\right)^{\frac{1}{\gamma}}\right) \\
&\geqslant \left(\frac{\int w}{w(r)}\left(\frac{s^\gamma}{\lambda s^\gamma + \mu r^\gamma}\right)^{\frac{1}{\gamma}}\right)^{\frac{\lambda s^\gamma}{\lambda s^\gamma + \mu r^\gamma}} \left(\frac{\int g}{g(s)}\left(\frac{r^\gamma}{\lambda s^\gamma + \mu r^\gamma}\right)^{\frac{1}{\gamma}}\right)^{\frac{\mu r^\gamma}{\lambda s^\gamma + \mu r^\gamma}},
\end{aligned}$$
by the arithmetic-geometric means inequality. Now, making a change of variables we write
$$\begin{aligned}
\int_0^\infty h &= \int_0^1 h(z)z'\,dz \\
&\geqslant \int_0^1 M_0^{\frac{\lambda s^\gamma}{\lambda s^\gamma + \mu r^\gamma}}\left(\left(\frac{s^\gamma}{\lambda s^\gamma + \mu r^\gamma}\right)^{\frac{1}{\gamma}}\int_0^\infty w, \left(\frac{r^\gamma}{\lambda s^\gamma + \mu r^\gamma}\right)^{\frac{1}{\gamma}}\int_0^\infty g\right).
\end{aligned}$$
Since $M_0^\delta(a,b) \geqslant M_{-\gamma}^\delta(a,b)$ we finally get
$$\begin{aligned}
\int_0^\infty h &\geqslant \int_0^1 M_{-\gamma}^{\frac{\lambda s^\gamma}{\lambda s^\gamma + \mu r^\gamma}}\left(\left(\frac{s^\gamma}{\lambda s^\gamma + \mu r^\gamma}\right)^{\frac{1}{\gamma}}\int_0^\infty w, \left(\frac{r^\gamma}{\lambda s^\gamma + \mu r^\gamma}\right)^{\frac{1}{\gamma}}\int_0^\infty g\right) \\
&= \int_0^1 M_{-\gamma}^\lambda\left(\int_0^\infty w, \int_0^\infty g\right) = M_{-\gamma}^\lambda\left(\int_0^\infty w, \int_0^\infty g\right).
\end{aligned}$$
This completes the proof. $\square$

THEOREM 2.5.5 (Ball). *Let $f : \mathbb{R}^n \to [0, \infty)$ be a log-concave function such that $f(0) > 0$. For every $p > 0$, $K_p(f)$ is a convex set.*

Proof. Let $p > 0$. Let $x, y \in K_p(f)$; then,
$$p \int_0^\infty f(rx)r^{p-1}\,dr \geqslant f(0) \quad \text{and} \quad p \int_0^\infty f(ry)r^{p-1}\,dr \geqslant f(0).$$
Let $\lambda, \mu > 0$ with $\lambda + \mu = 1$. We set $\gamma = 1/p$ and define $w, g, h : \mathbb{R}^+ \to \mathbb{R}^+$ by
$$w(r) = f(r^\gamma x), \qquad g(s) = f(s^\gamma y), \qquad h(t) = f(t^\gamma(\lambda x + \mu y)).$$

Since f is log-concave, for every pair $(r,s) \in \mathbb{R}^+ \times \mathbb{R}^+$ we have

$$h(M^\lambda_{-\gamma}(r,s)) = f\Big(\frac{1}{\lambda r^{-\gamma} + \mu s^{-\gamma}}(\lambda x + \mu y)\Big)$$

$$= f\Big(\frac{\lambda s^\gamma}{\lambda s^\gamma + \mu r^\gamma} r^\gamma x + \frac{\mu r^\gamma}{\lambda s^\gamma + \mu r^\gamma} s^\gamma y\Big)$$

$$\geq w(r)^{\frac{\lambda s^\gamma}{\lambda s^\gamma + \mu r^\gamma}} g(s)^{\frac{\mu r^\gamma}{\lambda s^\gamma + \mu r^\gamma}}.$$

Using the previous theorem we get

$$\Big(\int_0^\infty f(r^\gamma(\lambda x + \mu y))dr\Big)^{-\gamma} \leq \lambda \Big(\int_0^\infty f(r^\gamma x)dr\Big)^{-\gamma} + \mu \Big(\int_0^\infty f(r^\gamma y)dr\Big)^{-\gamma}.$$

The change of variables $t = r^\gamma$ shows that

$$\Big(p\int_0^\infty r^{p-1} f(r(\lambda x + \mu y))dr\Big)^{-1/p}$$

$$\leq \lambda \Big(p\int_0^\infty r^{p-1} f(rx)dr\Big)^{-1/p} + \mu \Big(p\int_0^\infty r^{p-1} f(ry)dr\Big)^{-1/p}$$

$$\leq \lambda (f(0))^{-1/p} + \mu (f(0))^{-1/p} = (f(0))^{-1/p}.$$

This shows that

$$p\int_0^\infty r^{p-1} f(r(\lambda x + \mu y))dr \geq f(0),$$

and hence $\lambda x + \mu y \in K_p(f)$. $\square$

It remains to show that the convex sets $K_p(f)$, $p > 0$, are indeed convex bodies, namely they are compact and have non-empty interior, whenever the log-concave function f has finite, positive integral.

LEMMA 2.5.6. *For every measurable function $f : \mathbb{R}^n \to [0,\infty)$ such that $f(0) > 0$ we have*

$$|K_n(f)| = \frac{1}{f(0)} \int_{\mathbb{R}^n} f(x)dx.$$

In particular, if f is log-concave and has finite positive integral then, using Theorem 2.5.5 as well, we see that $K_n(f)$ is a convex body.

Proof. Working as in the proof of Proposition 2.5.3 (vi) we can write

$$|K_n(f)| = \int_{K_n(f)} \mathbf{1}\, dx$$

$$= n\omega_n \int_{S^{n-1}} \int_0^{\rho_{K_n(f)}(\phi)} r^{n-1} dr d\sigma(\phi)$$

$$= \frac{n\omega_n}{f(0)} \int_{S^{n-1}} \int_0^\infty r^{n-1} f(r\phi) dr d\sigma(\phi)$$

$$= \frac{1}{f(0)} \int_{\mathbb{R}^n} f(x)dx$$

using (2.5.1) and integration in polar coordinates. $\square$

Now we can show that, if f is log-concave with finite, positive integral, then all sets $K_p(f)$ have non-empty interior; inclusion relations between any two of them and estimates for their volumes are summarized in the next two propositions.

2.5. CONVEX BODIES ASSOCIATED WITH LOG-CONCAVE FUNCTIONS

PROPOSITION 2.5.7. *Let $f : \mathbb{R}^n \to [0, \infty)$ be a log-concave function such that $f(0) > 0$.*

(i) *If $0 < p \leqslant q$, then*

(2.5.3)
$$\frac{\Gamma(p+1)^{\frac{1}{p}}}{\Gamma(q+1)^{\frac{1}{q}}} K_q(f) \subseteq K_p(f) \subseteq \left(\frac{\|f\|_\infty}{f(0)}\right)^{\frac{1}{p}-\frac{1}{q}} K_q(f).$$

(ii) *If f has its barycenter at the origin then, for every $0 < p \leqslant q$,*

(2.5.4)
$$\frac{\Gamma(p+1)^{\frac{1}{p}}}{\Gamma(q+1)^{\frac{1}{q}}} K_q(f) \subseteq K_p(f) \subseteq e^{\frac{n}{p}-\frac{n}{q}} K_q(f).$$

Proof. Observe that (ii) is a direct consequence of (i) if we use Theorem 2.2.2: we know that if $\operatorname{bar}(f) = 0$ then $f(0) \leqslant \|f\|_\infty \leqslant e^n f(0)$. Therefore,

$$\left(\frac{\|f\|_\infty}{f(0)}\right)^{\frac{1}{p}-\frac{1}{q}} \leqslant e^{\frac{n}{p}-\frac{n}{q}}.$$

For the right inclusion in (2.5.3) we use Lemma 2.2.4: for every $x \neq 0$ the function

$$F(p) := \left(\frac{p}{\|f\|_\infty} \int_0^\infty r^{p-1} f(rx)\, dr\right)^{1/p}$$

is an increasing function of p on $(0, \infty)$. Then, for $q \geqslant p > 0$,

$$\begin{aligned}
\rho_{K_q(f)}(x) &= \left(\frac{q}{f(0)} \int_0^\infty r^{q-1} f(rx)\, dr\right)^{1/q} \\
&= \left(\frac{\|f\|_\infty}{f(0)}\right)^{1/q} \left(\frac{q}{\|f\|_\infty} \int_0^\infty r^{q-1} f(rx)\, dr\right)^{1/q} \\
&= \left(\frac{\|f\|_\infty}{f(0)}\right)^{1/q} F(q) \geqslant \left(\frac{\|f\|_\infty}{f(0)}\right)^{1/q} F(p) \\
&= \left(\frac{\|f\|_\infty}{f(0)}\right)^{1/q-1/p} \left(\frac{\|f\|_\infty}{f(0)}\right)^{1/p} F(p) \\
&= \left(\frac{\|f\|_\infty}{f(0)}\right)^{1/q-1/p} \rho_{K_p(f)}(x).
\end{aligned}$$

For the left inclusion in (2.5.3) we use Theorem 2.2.3: for every $x \neq 0$ the function

$$G(p) := \left(\frac{1}{f(0)\Gamma(p)} \int_0^\infty r^{p-1} f(rx)\, dr\right)^{1/p}$$

is a decreasing function of p on $(0, \infty)$. Then, for $q \geqslant p > 0$,

$$\begin{aligned}
\rho_{K_q(f)}(x) &= \left(\frac{q}{f(0)} \int_0^\infty r^{q-1} f(rx)\, dr\right)^{1/q} \\
&= \Gamma(q+1)^{1/q} \left(\frac{1}{f(0)\Gamma(q)} \int_0^\infty r^{q-1} f(rx)\, dr\right)^{1/q} \\
&= \Gamma(q+1)^{1/q} G(q) \\
&\leqslant \frac{\Gamma(q+1)^{1/q}}{\Gamma(p+1)^{1/p}} \Gamma(p+1)^{1/p} G(p) \\
&= \frac{\Gamma(q+1)^{1/q}}{\Gamma(p+1)^{1/p}} \rho_{K_p(f)}(x),
\end{aligned}$$

and the proof is complete. $\square$

Let f be a centered log-concave density. Knowing that

(2.5.5) $$|K_n(f)|\, f(0) = \int_{\mathbb{R}^n} f(x)\, dx = 1$$

by Lemma 2.5.6, we can use the inclusions of Proposition 2.5.7 in order to estimate the volume of $K_p(f)$ for every $p > 0$.

PROPOSITION 2.5.8. *Let $f: \mathbb{R}^n \to [0, \infty)$ be a log-concave density with barycenter at the origin. Then, for every $p > 0$ we have*

(2.5.6) $$e^{-1} \leqslant f(0)^{\frac{1}{n}+\frac{1}{p}} |K_{n+p}(f)|^{\frac{1}{n}+\frac{1}{p}} \leqslant e\frac{n+p}{n},$$

while for $-n < p < 0$ we have

(2.5.7) $$e^{-1} \leqslant f(0)^{\frac{1}{-p}-\frac{1}{n}} |K_{n+p}(f)|^{\frac{1}{-p}-\frac{1}{n}} \leqslant e.$$

Proof. To prove (2.5.6), we first use (2.5.4) to get

$$e^{\frac{n^2}{n+p}-n} |K_n(f)| \leqslant |K_{n+p}(f)| \leqslant \frac{\Gamma(n+p+1)^{\frac{n}{n+p}}}{\Gamma(n+1)} |K_n(f)|.$$

Thus, combining this inequality with (2.5.5) we see that, for every $p > 0$,

$$\frac{e^{-\frac{np}{n+p}}}{f(0)} \leqslant |K_{n+p}(f)| \leqslant (\Gamma(n+p+1))^{\frac{n}{n+p}} \frac{1}{n!\, f(0)},$$

and hence

$$\frac{1}{e} \leqslant f(0)^{\frac{1}{n}+\frac{1}{p}} |K_{n+p}(f)|^{\frac{1}{n}+\frac{1}{p}} \leqslant \frac{(\Gamma(n+p+1))^{\frac{1}{p}}}{(n!)^{\frac{n+p}{np}}}.$$

Using the bounds

$$\frac{(\Gamma(n+p+1))^{\frac{1}{p}}}{(n!)^{\frac{n+p}{np}}} \leqslant (n+p) \frac{(n!)^{\frac{1}{p}}}{(n!)^{\frac{n+p}{np}}} = \frac{n+p}{(n!)^{\frac{1}{n}}} \leqslant e \frac{n+p}{n},$$

we conclude the proof of (2.5.6). We verify (2.5.7) in a similar manner, using also the inequality

$$\frac{\Gamma(q+1)^{\frac{1}{q}}}{\Gamma(p+1)^{\frac{1}{p}}} \leqslant e^{\frac{q}{p}-1}$$

that holds true for all $0 < p \leqslant q$ (see [**499**]). $\square$

2.5. CONVEX BODIES ASSOCIATED WITH LOG-CONCAVE FUNCTIONS

A crucial observation of K. Ball who introduced the bodies $K_p(f)$ is that they allow us to reduce the study of log-concave measures to that of convex bodies. A first example of this use is given in the next two propositions which show that providing bounds for the isotropic constant of log-concave probability measures reduces to just finding bounds for the more restricted class of convex bodies. We start with the symmetric case.

PROPOSITION 2.5.9 (Ball). *Let $f : \mathbb{R}^n \to [0, \infty)$ be an even log-concave function with finite, positive integral. Then, the body $T = K_{n+2}(f)$ is a symmetric convex body with*

$$c_1 L_f \leqslant L_T \leqslant c_2 L_f,$$

where $c_1, c_2 > 0$ are absolute constants. Furthermore, if f is isotropic, then $\overline{T}$ is an isotropic convex body.

Proof. Since f is even and log-concave, T is a symmetric convex body; we also have $f(x) \leqslant f(0)$ for all $x \in \mathbb{R}^n$. Hence, $f(0) > 0$. From Proposition 2.5.3 (vi)

$$\int_T \langle x, \theta \rangle^2 \, dx = \frac{1}{f(0)} \int_{\mathbb{R}^n} \langle x, \theta \rangle^2 f(x) \, dx,$$

and more generally

$$\int_T \langle x, \theta \rangle \langle x, \phi \rangle \, dx = \frac{1}{f(0)} \int_{\mathbb{R}^n} \langle x, \theta \rangle \langle x, \phi \rangle f(x) \, dx$$

for all $\theta, \phi \in S^{n-1}$. From the discussion in Section 2.3.3 it follows that $L_{\mathbf{1}_T} = L_T$ and

$$|T| \operatorname{Cov}(\mathbf{1}_T) = \frac{\int f}{f(0)} \operatorname{Cov}(f).$$

By the definition of the isotropic constant (Definition 2.3.11) we obtain

$$L_T = \frac{1}{|T|^{\frac{1}{2} + \frac{1}{n}}} \left(\frac{1}{f(0)} \int_{\mathbb{R}^n} f(x) \, dx \right)^{\frac{1}{2} + \frac{1}{n}} L_f.$$

On the other hand, (2.5.5) and the proof of Proposition 2.5.8 with $p = 2$ show that

$$|T|^{\frac{1}{2} + \frac{1}{n}} = |K_{n+2}(f)|^{\frac{1}{2} + \frac{1}{n}} \simeq |K_n(f)|^{\frac{1}{2} + \frac{1}{n}} = \left(\frac{1}{f(0)} \int_{\mathbb{R}^n} f(x) \, dx \right)^{\frac{1}{2} + \frac{1}{n}}.$$

This shows that $L_T \simeq L_f$. Finally, note that if f is isotropic then

$$\int_{\overline{T}} \langle x, \theta \rangle^2 \, dx = \frac{1}{|T|^{1 + \frac{2}{n}}} \int_T \langle x, \theta \rangle^2 \, dx = \frac{1}{f(0)|T|^{1 + \frac{2}{n}}}$$

for every $\theta \in S^{n-1}$, which shows that $\overline{T}$ is in isotropic position. □

The next proposition, which is taken from [**272**], shows that we can further reduce our study of the behavior of the isotropic constant to the class of symmetric convex bodies.

PROPOSITION 2.5.10 (Klartag). *For every convex body K we can find an origin-symmetric convex body T with the property that*

$$L_T \simeq L_K.$$

Proof. Without loss of generality, we may assume that K has volume 1 and barycenter at the origin. We define a function f supported on $K - K$ as follows:

$$f(x) = (\mathbf{1}_K * \mathbf{1}_{-K})(x) = \int_{\mathbb{R}^n} \mathbf{1}_K(y)\mathbf{1}_{-K}(x-y)\,dy = |K \cap (x+K)|.$$

By the Brunn-Minkowski inequality f is log-concave. Given that for every x

$$|K \cap (x+K)| = |-x + (K \cap (x+K))| = |(-x+K) \cap K|,$$

we also have that f is even. In addition, it is easy to check that $\int_{\mathbb{R}^n} f = 1$ and that $f(x) \leqslant f(0) = |K| = 1$ for every x, therefore

$$L_f = [\det \operatorname{Cov}(f)]^{\frac{1}{2n}}.$$

Note now that for every $z \in \mathbb{R}^n$,

$$\int_{\mathbb{R}^n} \langle x, z\rangle^2 f(x)dx = \int_{\mathbb{R}^n}\int_{\mathbb{R}^n} \langle y + (x-y), z\rangle^2 \mathbf{1}_K(y)\mathbf{1}_{-K}(x-y)\,dydx$$

$$= \int_K \langle y, z\rangle^2 \left(\int_{-K+y} 1\,dx\right) dy + \int_{\mathbb{R}^n} \mathbf{1}_K(y)\left(\int_{-K} \langle w, z\rangle^2 dw\right) dy$$

$$+ 2\int_{\mathbb{R}^n} \langle y, z\rangle \mathbf{1}_K(y)\left(\int_{\mathbb{R}^n} \langle x-y, z\rangle \mathbf{1}_{-K}(x-y)\,dx\right) dy$$

$$= \int_K \langle x, z\rangle^2 dx + \int_{-K} \langle x, z\rangle^2 dx,$$

because K is centered (hence so is $-K$). It follows that

$$\operatorname{Cov}(f) = \operatorname{Cov}(K) + \operatorname{Cov}(-K) = 2\operatorname{Cov}(K),$$

and hence

$$L_f = [\det \operatorname{Cov}(f)]^{\frac{1}{2n}} = \sqrt{2}\,[\det \operatorname{Cov}(K)]^{\frac{1}{2n}} = \sqrt{2}\,L_K.$$

It is now easy to check that the convex body $T := K_{n+2}(f)$ has the desired properties: T is symmetric because f is even, and $L_T \simeq L_f = \sqrt{2}\,L_K$. $\square$

Assuming that the function $f : \mathbb{R}^n \to [0, \infty)$ is centered, but not necessarily even, we choose to work with the body $K_{n+1}(f)$ instead of $K_{n+2}(f)$, since $K_{n+1}(f)$ has the advantage of being centered and "almost isotropic". This last notion can be defined as follows.

DEFINITION 2.5.11. Let K be a convex body of volume 1 in $\mathbb{R}^n$. We say that K is *almost isotropic with constant* $C > 0$ if for any $T \in GL(n)$ such that $T(K)$ is isotropic we have $d_G(T(B_2^n), B_2^n) \leqslant C$.

PROPOSITION 2.5.12. *Let $f : \mathbb{R}^n \to [0, \infty)$ be a centered log-concave function with finite, positive integral. Then, $T = K_{n+1}(f)$ is a centered convex body in $\mathbb{R}^n$ with*

$$c_1 L_f \leqslant L_T \leqslant c_2 L_f,$$

where $c_1, c_2 > 0$ are absolute constants. Furthermore, if f is isotropic then $\overline{K_{n+1}}(f)$ is almost isotropic with some absolute constant $C > 0$.

Proof. Note that, by Theorem 2.2.2, f being centered implies $f(0) > 0$; thus, $K_{n+1}(f)$ is well-defined and by Proposition 2.5.3 we know that it is a centered convex body. Without loss of generality we may assume that f is a log-concave density, otherwise we work with $f_1 = \frac{f}{\int f}$ using the fact that $K_{n+1}(\lambda f) = K_{n+1}(f)$ and $L_{\lambda f} = L_f$ for any $\lambda > 0$. By Proposition 2.5.3 we have

$$\int_T |\langle x, \theta \rangle| \, dx = \frac{1}{f(0)} \int_{\mathbb{R}^n} |\langle x, \theta \rangle| f(x) \, dx.$$

From Borell's lemma we see that for every $y \in \mathbb{R}^n$

$$\left(\frac{1}{|T|} \int_T \langle x, y \rangle^2 \, dx \right)^{1/2} \simeq \frac{1}{|T|} \int_T |\langle x, y \rangle| \, dx$$

$$= \frac{1}{f(0)|T|} \int_{\mathbb{R}^n} |\langle x, y \rangle| f(x) \, dx$$

$$\simeq \frac{1}{f(0)|T|} \left(\int_{\mathbb{R}^n} \langle x, y \rangle^2 f(x) \, dx \right)^{1/2},$$

which, combined with the fact that T and f are both centered, implies that there exist absolute constants $c_1, c_2 > 0$ such that as positive definite matrices

$$c_2 \operatorname{Cov}(\mathbf{1}_T) \leqslant (|T| f(0))^{-2} \operatorname{Cov}(f) \leqslant c_1 \operatorname{Cov}(\mathbf{1}_T).$$

Therefore

(2.5.8) $$[\det \operatorname{Cov}(\mathbf{1}_T)]^{1/n} \simeq (|T| f(0))^{-2} [\det \operatorname{Cov}(f)]^{1/n}.$$

From the definition of the isotropic constant it follows that

$$L_T = \frac{1}{|T|^{1/n}} [\det \operatorname{Cov}(T)]^{\frac{1}{2n}}$$

$$\simeq |T|^{-1/n} (f(0)|T|)^{-1} [\det \operatorname{Cov}(f)]^{\frac{1}{2n}}$$

$$\simeq (f(0)|T|)^{-1-\frac{1}{n}} L_f,$$

where we have also used the fact that one has $\|f\|_\infty^{1/n} \simeq f(0)^{1/n}$ by Theorem 2.2.2. Finally, applying Proposition 2.5.8 with $p = 1$ we get that

(2.5.9) $$e^{-1} \leqslant (f(0)|T|)^{1+\frac{1}{n}} \leqslant e \frac{n+1}{n} \leqslant 2e.$$

This proves the first assertion.

Next, assume that f is isotropic. By (2.5.8) and (2.5.9), and given that $\operatorname{Cov}(\lambda K) = \lambda^2 \operatorname{Cov}(K)$ for any convex body K in $\mathbb{R}^n$ and any $\lambda > 0$, we conclude that

$$c_3 f(0)^{2/n} \|y\|_2^2 \leqslant \langle \operatorname{Cov}(\overline{T})(y), y \rangle \leqslant c_4 f(0)^{2/n} \|y\|_2^2$$

for all $y \in \mathbb{R}^n$ and some absolute constants $c_3, c_4 > 0$. Let $S \in SL(n)$ be a symmetric and positive definite matrix such that $T_1 = S(\overline{T})$ is isotropic. From the isotropic condition we have

$$\int_{T_1} \langle x, y \rangle^2 \, dx = L_T^2 \|y\|_2^2$$

for all $y \in \mathbb{R}^n$, thus we can write

$$L_T^2 \|y\|_2^2 = \int_{T_1} \langle x, y \rangle^2 \, dx = \int_{\overline{T}} \langle Sx, y \rangle^2 \, dx$$

$$= \int_{\overline{T}} \langle x, Sy \rangle^2 \, dx \geqslant c_3 f(0)^{2/n} \|Sy\|_2^2.$$

This gives that
$$\|Sy\|_2 \leqslant c_3^{-1/2} L_T f(0)^{-1/n} \|y\|_2$$
for all $y \in \mathbb{R}^n$. Similarly, we see that $\|Sy\|_2 \geqslant c_4^{-1/2} L_T f(0)^{-1/n} \|y\|_2$ for all $y \in \mathbb{R}^n$. These two estimates show that $d_G(S(B_2^n), B_2^n) \leqslant C$ with $C := \sqrt{c_4/c_3}$. Therefore, $\overline{T} = \overline{K_{n+1}(f)}$ is almost isotropic with constant C, where $C > 0$ is independent of f or the dimension n. □

2.6. Further reading

2.6.1. Localization lemma

The localization lemma of Lovász and Simonovits (see [**338**]) is a useful tool that often reduces the question to obtain an inequality for all $1/n$-concave measures on $\mathbb{R}^n$ to the problem to check this inequality on the $1/n$-affine measures that are supported by a segment.

THEOREM 2.6.1 (localization lemma). *Let f and g be lower semi-continuous integrable functions on $\mathbb{R}^n$ such that*
$$\int_{\mathbb{R}^n} f(x)dx > 0 \quad \text{and} \quad \int_{\mathbb{R}^n} g(x)dx > 0.$$
Then, there exist $a, b \in \mathbb{R}^n$ and an affine function $\ell : [0,1] \to \mathbb{R}$ such that we simultaneously have
$$\int_0^1 f((1-t)a + tb)\ell(t)^{n-1}dt > 0 \quad \text{and} \quad \int_0^1 g((1-t)a + tb)\ell(t)^{n-1}dt > 0.$$

Fradelizi and Guédon [**183**] offered a new approach to this result.

THEOREM 2.6.2 (Fradelizi-Guédon). *Let K be a compact convex set in $\mathbb{R}^n$, let $-\infty < s \leqslant 1/2$ and let $f : K \to \mathbb{R}$ be an upper-semicontinuous function. We denote by P_f the set of all s-concave probability measures μ that are supported on K and satisfy $\int f \, d\mu \geqslant 0$. Then the extreme points of $\operatorname{conv}(P_f)$ are precisely:*
  (i) *The Dirac measures at points $x \in K$ such that $f(x) \geqslant 0$, and*
  (ii) *The probability measures ν that are s-affine, supported by a segment $[a,b] \subset K$, and such that $\int f \, d\nu = 0$ and $\int_a^x f \, d\nu > 0$ for all $x \in (a,b)$ or $\int_x^b f \, d\nu > 0$ for all $x \in (a,b)$.*

We will need the next lemma.

LEMMA 2.6.3. *Let K be a convex set in $\mathbb{R}^n$ and let $-\infty \leqslant \gamma \leqslant 1$. If $F : K \to \mathbb{R}^+$ is γ-concave and $G : K \to \mathbb{R}^+$ is γ-affine, then $(F-G)_+ : K \to \mathbb{R}^+$ is γ-concave.*

Proof. Assume that $-\infty < \gamma \leqslant 1$ and $\gamma \neq 0$. Assume that $x, y \in K$ satisfy $F(x) \geqslant G(x)$ and $F(y) \geqslant G(y)$. Then, for any $\lambda \in (0,1)$ we have
$$((1-\lambda)F^\gamma(x) + \lambda F^\gamma(y))^{1/\gamma} - ((1-\lambda)G^\gamma(x) + \lambda G^\gamma(y))^{1/\gamma}$$
$$\geqslant ((1-\lambda)(F-G)^\gamma(x) + \lambda(F-G)^\gamma(y))^{1/\gamma}$$
by Minkowski's inequality, and since F is γ-concave and G is γ-affine, we see that if $x, y \in \operatorname{supp}((F-G)_+)$ then
$$(F-G)((1-\lambda)x + \lambda y) \geqslant ((1-\lambda)(F-G)^\gamma(x) + \lambda(F-G)^\gamma(y))^{1/\gamma}.$$
Using the fact that log-concave functions can be approximated by γ_n-concave functions with $\gamma_n \downarrow 0$ we see that the assertion of the lemma holds true for log-concave functions too. The case $\gamma = -\infty$ is simple. □

Proof of Theorem 2.6.2. It is clear that every Dirac measure at a point $x \in K$ for which $f(x) \geqslant 0$ is an extreme point of $\operatorname{conv}(P_f)$, so we consider an extreme point ν of $\operatorname{conv}(P_f)$ which is not a Dirac measure, and we show that ν has the properties described in (ii).

Let F be the affine subspace spanned by $\operatorname{supp}(\nu)$. We first show that $\dim(F) = 1$. Suppose that this is not true. Then, we may find x_0 in the relative interior of $\operatorname{supp}(\nu)$ and a two dimensional subspace E of $\mathbb{R}^n$ such that $x_0 + E \subseteq F$. Given $\theta \in S_E$ we set

$$H_\theta = \{x \in F : \langle x - x_0, \theta \rangle = 0\}$$
$$H_\theta^+ = \{x \in F : \langle x - x_0, \theta \rangle \geqslant 0\}$$
$$H_\theta^- = \{x \in F : \langle x - x_0, \theta \rangle \leqslant 0\}.$$

Then, we define $\phi : S_E \to \mathbb{R}$ setting

$$\phi(\theta) = \int_{H_\theta^+} f\, d\nu - \frac{1}{2} \int f\, d\nu.$$

Note that $\nu(H_\theta) = 0$ for all $\theta \in S_E$, because $\dim(H_\theta) = 1$. This implies that ϕ is odd and continuous. Thus, there exists $\theta_0 \in S_E$ such that $\phi(\theta_0) = 0$. Since x_0 is a relative interior point of $\operatorname{supp}(\nu)$ we have $\nu(H_{\theta_0}^+) > 0$ and $\nu(H_\theta^-) > 0$; so, we may define the measures

$$\nu_1 = \frac{\nu|_{H_{\theta_0}^+}}{\nu(H_{\theta_0}^+)} \quad \text{and} \quad \nu_2 = \frac{\nu|_{H_{\theta_0}^-}}{\nu(H_{\theta_0}^-)}.$$

Then, $\nu = (\nu(H_{\theta_0}^+))\nu_1 + (\nu(H_{\theta_0}^-))\nu_2$, and this is a contradiction because ν is an extreme point of $\operatorname{conv}(P_f)$.

Now, we apply Borell's characterization of s-concave measures to conclude that ν is supported by a segment $[a, b]$, with a non-negative density $\psi \in L_{\text{loc}}^1(\mathbb{R}^n, dx)$ which is γ-concave, where $\gamma = \frac{s}{1-s} \in [-1, 1]$.

Next, we show that

$$\int f\, d\nu = 0 \text{ and } \int_a^x f\, d\nu > 0 \text{ for all } x \in (a, b) \text{ or } \int_x^b f\, d\nu > 0 \text{ for all } x \in (a, b).$$

First, observe that the function $x \mapsto \int_a^x f\, d\nu$ is continuous on (a, b). If $\int_a^c f\, d\nu = 0$ for some $c \in (a, b)$ then we may define $\nu_1 = \frac{\nu|_{[a,c]}}{\nu([a,c])}$ and $\nu_2 = \frac{\nu|_{[c,b]}}{\nu([c,b])}$ and conclude that $\nu = (\nu([a, c]))\nu_1 + (\nu([c, b]))\nu_2$ is not an extreme point of $\operatorname{conv}(P_f)$. This shows that

$$\int_a^x f\, d\nu > 0 \text{ for all } x \in (a, b) \text{ or } \int_x^b f\, d\nu > 0 \text{ for all } x \in (a, b).$$

Similarly, if $\int f\, d\nu > 0$ then we can find $c \in (a, b)$ such that

$$\int_a^c f\, d\nu = \frac{1}{2} \int f\, d\nu.$$

Then, we define ν_1 and ν_2 as above and we obtain a contradiction.

Finally, we show that ν is s-affine. Without loss of generality, we assume that $\int_a^x f\, d\nu > 0$ for all $x \in (a, b)$. We set $u = (b - a)/\|b - a\|_2$, we consider $c \in (a, b)$, and for $t \in \mathbb{R}$ we define $g_t : [a, b] \to \mathbb{R}$ by $g_t(x) = \frac{\psi(c)}{2}(1 + \gamma t \langle x - c, u \rangle)_+^{1/\gamma}$ (for $\gamma = 0$ we extend the definition appropriately). Note that g_t is γ-affine on $[a, b]$. We define two measures μ_t and ν_t, supported in $[a, b]$, with densities $(\psi - g_t)_+$ and $\min\{\psi, g_t\}$ respectively. Since $\gamma \in [-1, 1]$, ψ is γ-concave and g_t is γ-affine on $[a, b]$, we see that $(\psi - g_t)_+$ is γ-concave, and hence μ_t is s-concave. Since $\min\{\psi, g_t\}$ is also γ-concave, we see that ν_t is also s-concave. The function $t \mapsto \int f\, d\nu_t$ is continuous on $\mathbb{R}$, therefore

$$\lim_{t \to -\infty} \int f\, d\nu_t = \int_a^c f\, d\nu > 0 \quad \text{and} \quad \lim_{t \to \infty} \int f\, d\nu_t = \int_c^b f\, d\nu < 0$$

because $\int f\, d\nu = 0$. It follows that
$$\int f\, d\nu_{t_0} = 0$$
for some $t_0 \in \mathbb{R}$. Note that $\nu = \mu_{t_0} + \nu_{t_0}$, which implies that $\int f\, d\mu_{t_0} = 0$. We set $\lambda = \nu_{t_0}([a,b])$. Then, $0 < \lambda < 1$ and $\nu = (1-\lambda)\nu_1 + \lambda\nu_2$, where $\nu_1 = \frac{\mu_{t_0}}{1-\lambda}$ and $\nu_2 = \frac{\nu_{t_0}}{\lambda}$. Since ν_1 and ν_2 are in P_f, from the fact that ν is an extreme point of $\mathrm{conv}(P_f)$ it follows that $\nu_1 = \nu_2 = \mu$. This shows that $\psi = g_{t_0}/\lambda$, and hence ν is s-affine.

In order to complete the proof we need to show that, conversely, every probability measure ν that satisfies (ii) is an extreme point of $\mathrm{conv}(P_f)$. To this end, we define $F : [a,b] \to \mathbb{R}$ by $F(x) = \int_a^x f\, d\nu$, and without loss of generality we also assume that $F(x) > 0$ on (a,b). Let ψ be the γ-affine density of ν on $[a,b]$. Suppose that there exist $\lambda_i > 0$ with $\sum_{i=1}^k \lambda_i = 1$ and $\mu_i \neq \nu$ in P_f such that
$$\nu = \sum_{i=1}^k \lambda_i \mu_i.$$
Since $\mu_i \in P_f$, for every $i = 1,\ldots,k$ we have $\int f\, d\mu_i \geqslant 0$. Since
$$0 = \int f\, d\nu = \sum_{i=1}^k \lambda_i \int f\, d\mu_i,$$
we must have $\int f\, d\mu_i = 0$ for all $1 \leqslant i \leqslant k$. Let ψ_i denote the γ-concave density of μ_i, and write $\rho_i = \psi_i/\psi$ for the density of μ_i with respect to ν. We observe that ρ_i is the quotient of a γ-concave function and a γ-affine function; therefore, ρ_i is a quasi-concave, non-negative continuous function on a segment contained in $[a,b]$. It follows that it is monotone or it is first increasing and then decreasing on its support. Since $\sum_{i=1}^k \lambda_i \rho_i \equiv 1$ on $[a,b]$, at least one of the functions ρ_i which are non-zero at a has to be decreasing in a neighborhood of a. Let ρ_j be one such function. Then, ρ_j is decreasing on its support $[a,c]$, where $c \in (a,b)$. We write
$$\int f\, d\mu_j = 0 = \int_a^c \rho_j \, dF = \rho_j(c) F(c) - \int_a^c F\, d\rho_j.$$
Since $\rho_j(c)F(c) \geqslant 0$, ρ_j is decreasing and $F > 0$ on (a,c), we must have that ρ_j is constant on $[a,c]$ and $F(c) = 0$. Then, $c = b$ and $\rho_j \equiv 1$ on $[a,b]$ (because ρ_j is the density of μ_j with respect to ν, and both μ_j and ν are probability measures). It follows that $\mu_j = \nu$, which is a contradiction. $\square$

As a consequence of Theorem 2.6.2 we get

THEOREM 2.6.4. *Let K be a compact convex set in $\mathbb{R}^n$, let $-\infty < s \leqslant 1/2$ and let $\mathcal{P}(K)$ denote the set of all probability measures supported on K. Let $f : K \to \mathbb{R}$ be an upper-semicontinuous function. We denote by P_f the set of all s-concave probability measures μ that are supported on K and satisfy $\int f\, d\mu \geqslant 0$. If $\Phi : \mathcal{P}(K) \to \mathbb{R}$ is a convex upper semi-continuous function, then $\sup\{\Phi(\mu) : \mu \in P_f\}$ is attained at some extreme point of $\mathrm{supp}(P_f)$, i.e. some Dirac measures at a point $x \in K$ such that $f(x) \geqslant 0$ or some probability measures ν which is s-affine, supported by a segment $[a,b] \subset K$, and such that $\int f\, d\nu = 0$ and $\int_a^x f\, d\nu > 0$ for all $x \in (a,b)$ or $\int_x^b f\, d\nu > 0$ for all $x \in (a,b)$.*

Proof. We use the fact (see [**96**]) that the set of s-concave probability measures that are supported in K is w^*-compact. Since f is upper semi-continuous, one can check that P_f is w^*-closed, and hence w^*-compact. By the Krein-Milman theorem, $\sup\{\Phi(\mu) : \mu \in P_f\}$ is attained at some $\nu \in \mathrm{Ext}(\overline{\mathrm{conv}}^{w^*}(P_f)) \subseteq \mathrm{Ext}(\mathrm{conv}(P_f))$. $\square$

A corollary of Theorem 2.6.4 is the next very useful geometric version of Theorem 2.6.1.

THEOREM 2.6.5. *Let $f_1, f_2, f_3, f_4 : \mathbb{R}^n \to \mathbb{R}^+$ such that $f_1^\alpha f_2^\beta \leqslant f_3^\alpha f_4^\beta$ for some $\alpha, \beta > 0$. We also assume that f_1, f_2 are upper semi-continuous and f_3, f_4 are lower semi-continuous, and that for some $-\infty \leqslant s \leqslant 1/2$ and every s-affine probability measure ν supported on a segment $[a, b]$ in $\mathbb{R}^n$,*

$$\left(\int f_1 \, d\nu\right)^\alpha \left(\int f_2 \, d\nu\right)^\beta \leqslant \left(\int f_3 \, d\nu\right)^\alpha \left(\int f_4 \, d\nu\right)^\beta.$$

Then, for every s-concave probability measure μ on $\mathbb{R}^n$,

$$\left(\int f_1 \, d\mu\right)^\alpha \left(\int f_2 \, d\mu\right)^\beta \leqslant \left(\int f_3 \, d\mu\right)^\alpha \left(\int f_4 \, d\mu\right)^\beta.$$

Proof. We may assume that $f_3 > 0$ on $\mathbb{R}^n$. Let μ be an s-concave compactly supported probability measure on $\mathbb{R}^n$ and write $K = \mathrm{supp}(\mu)$. We define

$$f = f_1 - \left(\frac{\int f_1 \, d\mu}{\int f_3 \, d\mu}\right) f_3$$

and

$$\Phi(\nu) = \left(\frac{\int f_1 \, d\mu}{\int f_3 \, d\mu}\right)^{\alpha/\beta} \left(\int f_2 \, d\nu\right) - \left(\int f_4 \, d\nu\right)$$

for all $\nu \in \mathcal{P}(K)$. Using the assumptions on f_i one can check that f and Φ are upper semi-continuous and Φ is affine. Since $\mu \in P_f$, Theorem 2.6.4 shows that there exists an extreme point ν of $\mathrm{conv}(P_f)$, which is either a Dirac measure or an s-affine measure supported by a segment $[a, b]$, such that $\Phi(\mu) \leqslant \Phi(\nu)$ and $\int f \, d\nu \geqslant 0$. Then,

$$\left(\frac{\int f_1 \, d\mu}{\int f_3 \, d\mu}\right)^{\alpha/\beta} \left(\int f_2 \, d\mu\right) - \left(\int f_4 \, d\mu\right) \leqslant \left(\frac{\int f_1 \, d\mu}{\int f_3 \, d\mu}\right)^{\alpha/\beta} \left(\int f_2 \, d\nu\right) - \left(\int f_4 \, d\nu\right)$$

$$\leqslant \left(\frac{\int f_1 \, d\nu}{\int f_3 \, d\nu}\right)^{\alpha/\beta} \left(\int f_2 \, d\nu\right) - \left(\int f_4 \, d\nu\right)$$

$$\leqslant 0.$$

An approximation argument shows that the same is true for every s-concave probability measure on $\mathbb{R}^n$. □

2.6.2. Distribution of polynomials on convex bodies

A special case of Theorem 2.4.6 (see Remark 2.4.7 (a)) asserts that if K is a convex body of volume 1 in $\mathbb{R}^n$ then

$$\|\langle \cdot, \theta \rangle\|_{L^q(K)} \leqslant c_1 q \|\langle \cdot, \theta \rangle\|_{L^1(K)}$$

for all $\theta \in S^{n-1}$ and $q \geqslant 1$. Equivalently,

$$\|\langle \cdot, \theta \rangle\|_{L^{\psi_1}(K)} \leqslant c_2 \|\langle \cdot, \theta \rangle\|_{L^1(K)}$$

for all $\theta \in S^{n-1}$. Note that a linear functional $x \mapsto \langle x, \theta \rangle = \sum_{i=1}^n \theta_i x_i$ is a polynomial of degree $d = 1$. V. Milman asked whether an analogous result holds true for polynomials of higher degree on high-dimensional convex bodies. Bourgain gave an affirmative answer.

THEOREM 2.6.6 (Bourgain). *Let d be a positive integer. For every $p > 1$ there exists a constant $C(d, p) > 0$ with the following property: For any $n \geqslant 1$, any convex body K of volume 1 in $\mathbb{R}^n$ and any polynomial $f : \mathbb{R}^n \to \mathbb{R}$ of degree d, we have:*

$$\left(\int_K |f(x)|^p \, dx\right)^{1/p} \leqslant C(d, p) \int_K |f(x)| \, dx.$$

For the proof of Theorem 2.6.6, Bourgain established a ψ-version of the above inequality for a suitable ψ-Orlicz function.

THEOREM 2.6.7. *Let d be a positive integer. For any $n \geqslant 1$, any convex body K of volume 1 in $\mathbb{R}^n$ and any polynomial $f : \mathbb{R}^n \to \mathbb{R}$ of degree d, we have:*

$$\|f\|_{L^\psi(K)} \leqslant C\|f\|_{L^1(K)},$$

where L^ψ is the Orlicz space that corresponds to the function $\psi(t) = e^{t^{c/d}} - 1$ and $c, C > 0$ are absolute constants.

Afterwards, Bobkov used a localization technique of Kannan, Lovász and Simonovits to prove that one can have $c = 1$ in Theorem 2.6.7. Finally, Carbery and Wright determined the sharp constants between the L_p and L_q norms of polynomials over convex bodies in $\mathbb{R}^n$. We present Bourgain's argument. The main tool is the Knothe map and a distributional inequality for polynomials of one variable. We begin with a simple lemma.

LEMMA 2.6.8. *There exists a constant $c > 0$ such that for any polynomial $p(t) = \sum_{j=0}^{d} a_j t^j$ of one variable we have*

$$\int_0^1 |p(t)|^{-\frac{1}{2d}} dt \leqslant c\|p\|_\infty^{-\frac{1}{2d}},$$

where $\|p\|_\infty = \sup_{t \in [0,1]} |p(t)|$.

Proof. Let $z_1, \ldots, z_d$ be the roots of p (in the complex plane) and write $p(t) = \alpha \prod_{j=1}^{d}(t - z_j)$. Note that

(2.6.1) $$\int_0^1 \frac{1}{\sqrt{|t-z|}} dt \leqslant \frac{c}{\sqrt{1+|z|}}$$

for any $z \in \mathbb{C}$, where $c = 4$. To see this we distinguish two cases:

(i) If $|z| \geqslant 1$ then we may write:

$$\int_0^1 |t-z|^{-1/2} dt \leqslant \int_0^1 (|z|-t)^{-1/2} dt \leqslant 2\sqrt{2}(1+|z|)^{-1/2}.$$

(ii) If $0 \leqslant |z| < 1$ then we write:

$$\int_0^1 |t-z|^{-1/2} dt \leqslant \int_0^{|z|} (|z|-t)^{-1/2} dt + \int_{|z|}^1 (t-|z|)^{-1/2} dt$$
$$= 2|z|^{1/2} + 2(1-|z|)^{1/2} \leqslant 2\sqrt{2} \leqslant 4(1+|z|)^{-1/2}.$$

Thus, from Hölder's inequality we obtain:

$$\int_0^1 |p(t)|^{-\frac{1}{2d}} dt = \int_0^1 |\alpha|^{-\frac{1}{2d}} \prod_{i=1}^{d} |t-z_i|^{-\frac{1}{2d}}$$
$$\leqslant |\alpha|^{-\frac{1}{2d}} \left(\prod_{i=1}^{d} \int_0^1 |t-z_i|^{-1/2} dt \right)^{1/d}$$
$$\leqslant c \left(|\alpha| \prod_{i=1}^{d} (1+|z_i|) \right)^{-\frac{1}{2d}} \leqslant c\|p\|_\infty^{-\frac{1}{2d}},$$

where we have used the fact that $\|p\|_\infty \leqslant |\alpha| \prod_{i=1}^{d}(1+|z_i|)$ and (2.6.1). □

A direct application of Markov's inequality shows that:

LEMMA 2.6.9. *There is a constant $c_1 > 1$ such that, for any polynomial $p(t) = \sum_{j=0}^{d} a_j t^j$ of one variable,*

$$\mathrm{Prob}(\{t \in [0,1] : |p(t)| \geqslant c_1^{-d}\|p\|_\infty\}) \geqslant \frac{1}{2}.$$

2.6. FURTHER READING

Proof of Theorem 2.6.7. By homogeneity we may assume that $\int_K |f(x)|\,dx = 1$. To bound $\|f\|_{L^\psi(K)}$ one needs to establish the corresponding distributional inequality:

$$|\{x \in K : |f(x)| \geq s\}| \leq e^{-(s/C)^{c/d}},$$

for any $0 < s < \|f\|_\infty := \sup_{x \in K} |f(x)|$. Define $K_s := \{x \in K : |f(x)| \geq s\}$ and consider the Knothe map $\phi : K_s \to K$ corresponding to the sets K_s and K. For fixed $x \in K_s$ we define a polynomial (in t) by

$$p_x(t) = f(tx + (1-t)\phi(x)).$$

Note that p_x is of degree at most d. By construction we have

$$\sup_{t \in [0,1]} |p_x(t)| \geq |p_x(1)| = |f(x)| \geq s.$$

Direct application of Lemma 2.6.9 yields

$$\operatorname{Prob}(\{t \in [0,1] : |p_x(t)| \geq c_1^{-d}s\}) \geq 1/2.$$

Since $x \in K_s$ was arbitrary, we conclude that

$$\int_{K_s} \operatorname{Prob}(\{t \in [0,1] : |p_x(t)| \geq c_1^{-d}s\})\,dx \geq \frac{1}{2}|K_s|.$$

Next, we apply Fubini's theorem to get

$$\int_0^1 |\{x \in K_s : |p_x(t)| \geq c_1^{-d}s\}|\,dt \geq \frac{1}{2}|K_s|,$$

and this implies that

$$\int_0^{2/3} |\{x \in K_s : |p_x(t)| \geq c_1^{-d}s\}|\,dt \geq \frac{1}{6}|K_s|.$$

Thus, we can find $0 \leq t_s \leq 2/3$ with

$$|\{x \in K_s : |p_x(t_s)| \geq c_1^{-d}s\}| \geq \frac{1}{4}|K_s|.$$

Because of the properties of the Knothe map, the map $x \mapsto t_sx + (1-t_s)\phi(x)$ is one-to-one on K_s, and maps the set $T := \{x \in K_s : |p_x(t_s)| \geq c_1^{-d}s\}$ on a set T' of measure

$$|T'| = \int_T |J_{t_sI+(1-t_s)\phi}(x)|\,dx$$

$$= \int_T \prod_{i=1}^n \left(t_s + (1-t_s)\frac{\partial \phi_i}{\partial x_i}(x)\right)\,dx$$

$$\geq \int_T \left(\prod_{i=1}^n \frac{\partial \phi_i}{\partial x_i}(x)\right)^{1-t_s}\,dx$$

$$= \int_T |J_\phi(x)|^{1-t_s}\,dx = (|K|/|K_s|)^{1-t_s}|T|,$$

where J_ϕ is the determinant of the Jacobian of ϕ. Since $|K| = 1$ and $|T| \geq \frac{1}{4}|K_s|$ we deduce that $|T'| > \frac{1}{4}|K_s|^{t_s}$. By construction we have

$$T' \subseteq \{t_sx + (1-t_s)\phi(x) : x \in K_s, |f(t_sx + (1-t_s)\phi(x))| \geq c_1^{-d}s\}$$
$$\subseteq \{y \in K : |f(y)| \geq c_1^{-d}s\} = K_{c_1^{-d}s},$$

so the previous estimate gives:

(2.6.2) $$|K_{c_1^{-d}s}| \geq \frac{1}{4}|K_s|^{t_s} \geq \frac{1}{4}|K_s|^{2/3},$$

where we have used the fact that $|K_s| \leq 1$ and $t_s \leq 2/3$. An iteration argument, which is based on (2.6.2), will give the result: Assuming that s is sufficiently large, we choose a positive integer m such that

(2.6.3) $$c_1^{md} \leq s/4^4 < c_1^{(m+1)d}.$$

Then, applying (2.6.2) m times we get

$$|K_{c_1^{-m}d_s}| \geq \left(\frac{1}{4}\right)^{\sum_{j=0}^{m-1}(2/3)^j} |K_s|^{(2/3)^m} \geq \frac{1}{4^3}|K_s|^{(2/3)^m}.$$

Since $K_{c_1^{-m}d} \subseteq K_{4^4}$, by Markov's inequality and the previous estimate we get that

$$\frac{1}{4^3}|K_s|^{(2/3)^m} \leq |K_{4^4}| \leq \frac{1}{4^4}\int_K |f| = \frac{1}{4^4}.$$

It follows that

$$|K_s| \leq 4^{-(3/2)^m} \leq e^{-(s/C)^{c/d}},$$

for some absolute constants $C, c > 0$, taking into account the right hand side inequality in (2.6.3). $\square$

2.7. Notes and references

Log-concave probability measures and log-concave functions

Theorems 2.1.2 and 2.1.5 are due to Borell (see [**96**] and [**97**]). The fact that every integrable log-concave function $f : \mathbb{R}^n \to [0, \infty)$ decays exponentially as $\|x\|_2 \to \infty$ appears in the article of Gromov and V. Milman [**233**] and in Bourgain's paper [**99**] in the case where $f = \mathbf{1}_K$ for a symmetric convex body K of volume 1 in $\mathbb{R}^n$. Our presentation of Lemma 2.2.1 follows [**276**, Lemma 2.1].

Fradelizi's inequality (Theorem 2.2.2) is from [**182**]. We present a different proof of the inequality $\|f\|_\infty \leq e^n f(0)$ that is simpler but does not clarify the cases of equality.

Theorem 2.2.3 is a generalization of [**273**, Lemma 2.6]; see [**353**] and [**384**] for a similar result in the case where f is, additionally, assumed decreasing. Related inequalities and more information can be found in [**67**] and [**95**]. The idea of the proof of Lemma 2.2.4 is from [**384**, Lemma 2.1]; a special case was proved by Hensley in [**253**].

Theorem 2.2.5 is due to Borell [**95**]; see also [**49**].

Grünbaum's lemma can be found in [**238**]. The argument that we present for Lemma 2.2.6 is from [**339**].

Isotropic log-concave measures

The isotropic position of a convex body appears in the work of Bourgain [**99**]; the extension of the definition to the setting of log-concave measures can be found in the PhD Thesis of Ball [**30**] (see also [**32**]). A deep study of isotropic convex bodies was done by V. Milman and Pajor in [**384**] which is a classical reference on the subject. Another source of information are the lecture notes [**200**]. The general definition of the isotropic constant that we use in this book (Definition 2.3.11) appears in Klartag's paper [**273**].

An isomorphic theory for high dimensional log-concave measures, analogous to the corresponding theory for convex bodies, has been initiated by Klartag and Milman in [**286**]; this direction is beyond the scope of this book (see e.g. [**446**]).

ψ_α-estimates

The general theory of Orlicz spaces is developed in the books of Krasnosel'skii and Rutickii [**300**], Lindenstrauss and Tzafriri [**331**], Rao and Ren [**436**], [**437**]. Khintchine type inequalities for linear functionals on convex bodies appear in the article of Gromov and V. Milman [**233**]; the role of ψ_1-estimates is emphasized in the book of V. Milman and Schechtman [**387**] and is discussed in depth in the article of V. Milman and Pajor [**384**] (see Chapter 3, Section 1). Theorem 2.4.8 is due to Latała [**306**]. For the Kahane-Khintchine inequality for negative exponents (Theorem 2.4.9) see [**239**].

Convex bodies associated with log-concave functions

The family of convex bodies $K_p(f)$, where f is a log-concave function f, were introduced by Ball (see [**32**]) who also established their convexity in the case $p \geqslant 1$. The argument that we present works for all $p > 0$ (it can be found e.g. in [**51**]). In a different language, Ball's bodies are also discussed in [**384**]. Proposition 2.5.7 and Proposition 2.5.8 are consequences of the basic inequalities of Section 2.2; in this form they can be found in [**414**]. Theorem 2.5.9 is one of the main observations of Ball in [**32**]. It reduces the study of the magnitude of the isotropic constants of log-concave probability measures to that of convex bodies. The fact that one can further reduce the study of the behavior of the isotropic constant to the class of symmetric convex bodies was observed by Klartag in [**272**].

An extension of this theory to the setting of s-concave measures was initiated by Bobkov in [**81**] and [**82**].

Localization lemma

Theorem 2.6.1, the localization lemma of Lovász and Simonovits, was proved in [**338**]. It is often applied in the form of Theorem 2.6.5. In Section 2.6.1 we present the approach of Fradelizi and Guédon who proved Theorem 2.6.2 and Theorem 2.6.4 in [**183**] (see also [**184**]).

Distribution of polynomials on convex bodies

Khintchine type inequalities for polynomials of higher degree on high-dimensional convex bodies were first established by Bourgain in [**101**]. We present his argument in Section 2.6.2. Later, Bobkov [**77**] using a localization technique of Kannan, Lovász and Simonovits [**266**] (based on the "localization lemma of Lovász and Simonovits [**338**]) proved that one can have $c = 1$ in Theorem 2.6.7. Nazarov, Sodin and Volberg proved a "geometric Kannan-Lovász-Simonovits lemma" stating that $\mu(F_t^c) \leqslant \mu(F^c)^{\frac{t+1}{2}}$, $t > 0$, for every log-concave probability measure μ and any Borel set F in $\mathbb{R}^n$, where F_t is the t-dilation of F defined by

$$F_t = \left\{ x \in \mathbb{R}^n : \text{ there exists an interval } I \text{ s. t. } x \in I \text{ and } |I| < \frac{t+1}{2}|F \cap I| \right\}.$$

In [**400**] they used this inequality to obtain sharp large and small deviations, as well as Khintchine-type inequalities for functions that satisfy a Remez type inequality. Using the original localization lemma of Lovász and Simonovits, Carbery and Wright (see [**133**]) determined the sharp constants between the L_p and L_q norms of polynomials over convex bodies in $\mathbb{R}^n$.

CHAPTER 3

Hyperplane conjecture and Bourgain's upper bound

In this chapter we introduce the main question of this book: the isotropic constant conjecture, which states that there exists an absolute constant $C > 0$ such that
$$L_\mu \leqslant C$$
for every $n \geqslant 1$ and every log-concave measure μ on $\mathbb{R}^n$. The question appears for the first time in the work of Bourgain [99] on high-dimensional maximal functions associated with arbitrary convex bodies. The conjecture was stated in this form in the well-known article of V. Milman and Pajor [384] and in the PhD Thesis of K. Ball.

In Section 3.1 we introduce the so-called hyperplane conjecture (or slicing problem) which asks if every centered convex body of volume 1 has a hyperplane section through the origin whose volume is greater than an absolute constant $c > 0$. We show that this is an equivalent formulation of the isotropic constant conjecture; actually, this was Bourgain's version of the problem. In order to show the equivalence, we explain the relation between the moments of inertia of a centered convex body and the volume of its hyperplane sections passing through the origin. In particular, we see that, if K is an isotropic convex body in $\mathbb{R}^n$, then for every $\theta \in S^{n-1}$ we have
$$\frac{c_1}{L_K} \leqslant |K \cap \theta^\perp| \leqslant \frac{c_2}{L_K},$$
where $c_1, c_2 > 0$ are absolute constants. In other words, all hyperplane sections through the barycenter of K have approximately the same volume, and boundedness of its isotropic constant shows that they are large enough.

Trying to understand the slicing problem leads to a number of deep questions about the distribution of volume on high dimensional isotropic convex bodies. Thus, the study of their geometric properties is naturally a central topic in this book. In Section 3.2 we obtain some first information on some of their basic geometric parameters; their inradius and circumradius, the behavior of their covering numbers by Euclidean balls, the ψ_α-behavior of linear functionals on them. A key observation is that, if K is an isotropic convex body in $\mathbb{R}^n$, then, as a consequence of the Brunn-Minkowski inequality, one has
$$\|\langle \cdot, \theta \rangle\|_{\psi_1} \leqslant C L_K$$
for all $\theta \in S^{n-1}$, where $C > 0$ is an absolute constant. In fact, Alesker's theorem shows that one has a stronger ψ_2-estimate for the function $f(x) = \|x\|_2$:
$$\|f\|_{\psi_2} \leqslant C\|f\|_2 \leqslant C\sqrt{n} L_K.$$

Markov's inequality then implies exponential concentration of the mass of K in a strip of width CL_K and normal concentration in a ball of radius $C\sqrt{n}L_K$.

The results of Section 3.2 play a key role in Bourgain's upper bound $L_K \leqslant C\sqrt[4]{n}\log n$ for the isotropic constant of convex bodies K in $\mathbb{R}^n$. In Section 3.3 we describe two proofs of Bourgain's result. The first one is closer to Bourgain's original argument which is based on a "reduction to small diameter" followed by a direct use of Talagrand's comparison theorem. The second one, which is due to Dar, is based on the ψ_1-behavior of linear functionals on convex bodies, which was the main ingredient in Bourgain's approach too, but it is more elementary, since it involves simpler entropy considerations. Section 3.4 is devoted to Bourgain's work on the behavior of the isotropic constant of ψ_2-bodies with constant b: in this case, his proof gives a bound $O(b\log(b+1))$ which does not depend on the dimension.

The hyperplane conjecture is equivalent to the "asymptotic versions" of several classical problems in convex geometry. In Section 3.5 we discuss two of them: Sylvester's problem on the expected volume of a random simplex in a convex body and the Busemann-Petty problem on volume comparison between two symmetric convex bodies K_1 and K_2 in $\mathbb{R}^n$ which satisfy $|K_1 \cap \theta^\perp| \leqslant |K_2 \cap \theta^\perp|$ for all $\theta \in S^{n-1}$.

3.1. Hyperplane conjecture

3.1.1. The isotropic constant conjecture

The main question that we discuss in this book asks if there exists a uniform upper bound, independent of the dimension, for the isotropic constants of all log-concave measures.

CONJECTURE 3.1.1 (isotropic constant). *There exists an absolute constant $C > 0$ such that*
$$L_K \leqslant C$$
for every $n \geqslant 1$ and every convex body K in $\mathbb{R}^n$. Equivalently, if K is an isotropic convex body in $\mathbb{R}^n$, then
$$\int_K \langle x, \theta \rangle^2 dx \leqslant C^2$$
for every $\theta \in S^{n-1}$.

More generally, there exists an absolute constant $C > 0$ such that
$$L_\mu \leqslant C$$
for every $n \geqslant 1$ and every non-degenerate log-concave measure μ on $\mathbb{R}^n$. Equivalently, if $f : \mathbb{R}^n \to [0, \infty)$ is an isotropic log-concave density, then
$$f(0)^{1/n} \leqslant C,$$
where $C > 0$ is an absolute constant. Recall that, from Proposition 2.5.12, this more general conjecture is in fact equivalent to the isotropic constant conjecture for convex bodies.

The origin of Conjecture 3.1.1 is the so-called *hyperplane conjecture* (or *slicing problem*) which asks if every centered convex body of volume 1 has a hyperplane section through the origin whose volume is greater than an absolute constant $c > 0$. The connection with the isotropic constant is presented in the next subsection.

3.1.2. Moments of inertia and maximal hyperplane sections

We first explain the relation of the moments of inertia of a centered convex body with the volume of its hyperplane sections passing through the origin. In the isotropic case, the main conclusion is the following:

THEOREM 3.1.2. *Let K be an isotropic convex body in $\mathbb{R}^n$. For every $\theta \in S^{n-1}$ we have*
$$\frac{c_1}{L_K} \leqslant |K \cap \theta^\perp| \leqslant \frac{c_2}{L_K},$$
where $c_1, c_2 > 0$ are absolute constants.

The theorem will follow from a series of observations. We first observe that hyperplane sections through the barycenter are, up to an absolute constant, maximal.

PROPOSITION 3.1.3. *Let K be a centered convex body of volume 1 in $\mathbb{R}^n$. Let $\theta \in S^{n-1}$ and consider the function*
$$f(t) = f_{K,\theta}(t) = |K \cap \{\langle x, \theta \rangle = t\}|, \qquad t \in \mathbb{R}.$$
Then,
$$\|f\|_\infty \leqslant ef(0) = e|K \cap \theta^\perp|.$$

Proof. The Brunn-Minkowski inequality implies that f is a log-concave density on $\mathbb{R}$ and the assumption that K is centered shows that $\mathrm{bar}(f) = 0$. Then, the result follows from Theorem 2.2.2 in dimension 1. $\square$

The basic observation in this subsection is that for every $\theta \in S^{n-1}$ the volume of the $(n-1)$-dimensional section $|K \cap \theta^\perp|$ of K is closely related to the L_q-norms
$$\|\langle \cdot, \theta \rangle\|_q = \|\langle \cdot, \theta \rangle\|_{L^q(K)} = \left(\int_K |\langle x, \theta \rangle|^q dx\right)^{1/q}$$
of the linear functional $x \mapsto \langle x, \theta \rangle$. This connection becomes clear once we write $\|\langle \cdot, \theta \rangle\|_q^q$ in the form
$$\int_K |\langle x, \theta \rangle|^q dx = \int_\mathbb{R} |t|^q f(t) dt.$$
We more or less follow the exposition of V. Milman and Pajor.

PROPOSITION 3.1.4. *Let K be a centered convex body of volume 1 in $\mathbb{R}^n$. For every $\theta \in S^{n-1}$ and every $q > 0$,*

(3.1.1) $$\left(\int_K |\langle x, \theta \rangle|^q dx\right)^{1/q} \geqslant \frac{1}{2e(q+1)^{1/q}} \frac{1}{|K \cap \theta^\perp|}.$$

Proof. We first apply Lemma 2.2.4 to the functions $f \cdot \mathbf{1}_{\{t \geqslant 0\}}$ and $f \cdot \mathbf{1}_{\{t \leqslant 0\}}$; we have
$$F(q+1) = \left(\frac{q+1}{\|f\|_\infty} \int_0^\infty t^q f(t)\, dt\right)^{1/(q+1)} \geqslant F(1) = \frac{1}{\|f\|_\infty} \int_0^\infty f(t) dt,$$
and similarly
$$\left(\frac{q+1}{\|f\|_\infty} \int_{-\infty}^0 |t|^q f(t)\, dt\right)^{1/(q+1)} \geqslant \frac{1}{\|f\|_\infty} \int_{-\infty}^0 f(t) dt.$$

Adding these two inequalities and using the fact that $a^s + b^s \leqslant 2^{1-s}(a+b)^s$ for all $a, b > 0$ and $0 < s < 1$, as well as the fact that $\int_{-\infty}^{\infty} f = |K| = 1$, we write

$$\frac{1}{\|f\|_{\infty}} \leqslant \left(\frac{q+1}{\|f\|_{\infty}} \int_0^{\infty} t^q f(t)\,dt\right)^{1/(q+1)} + \left(\frac{q+1}{\|f\|_{\infty}} \int_{-\infty}^0 |t|^q f(t)\,dt\right)^{1/(q+1)}$$

$$\leqslant 2^{q/(q+1)} \left(\frac{q+1}{\|f\|_{\infty}} \int_{-\infty}^{\infty} |t|^q f(t)\,dt\right)^{1/(q+1)}.$$

This shows that

$$\int_K |\langle x, \theta\rangle|^q dx = \int_{-\infty}^{\infty} |t|^q f(t)\,dt \geqslant \frac{1}{\|f\|_{\infty}^q 2^q (q+1)}$$

and the result follows from the fact that

$$\frac{1}{\|f\|_{\infty}} \geqslant \frac{1}{ef(0)} = \frac{1}{e|K \cap \theta^{\perp}|}$$

by Proposition 3.1.3. $\square$

Note. In the symmetric case we have $f(0) = \|f\|_{\infty}$ and the lower bound of Proposition 3.1.4 takes the form

$$\|\langle \cdot, \theta\rangle\|_q \geqslant \frac{1}{2(q+1)^{1/q}} \frac{1}{|K \cap \theta^{\perp}|}.$$

In the case $q = 2$ this was proved by Hensley in [**253**] (see also [**252**]). It is also useful to note that the constant $\frac{1}{2e(q+1)^{1/q}}$ in (3.1.1) is bounded from below by an absolute positive constant (independent of $q > 0$).

Next, we see that analogous upper bounds for $\|\langle \cdot, \theta\rangle\|_q$ hold true.

PROPOSITION 3.1.5. *Let K be a centered convex body of volume 1 in $\mathbb{R}^n$. For every $\theta \in S^{n-1}$ and every $q \geqslant 1$,*

(3.1.2) $$\left(\int_K |\langle x, \theta\rangle|^q dx\right)^{1/q} \leqslant \frac{cq}{|K \cap \theta^{\perp}|},$$

where $c > 0$ is an absolute constant.

Proof. We apply Theorem 2.2.3 as in the previous proof: we have

$$G(q+1) := \left(\frac{1}{f(0)\Gamma(q+1)} \int_0^{\infty} t^q f(t)\,dt\right)^{1/(q+1)} \leqslant G(1) = \frac{1}{f(0)} \int_0^{\infty} f(t)\,dt$$

and

$$\left(\frac{1}{f(0)\Gamma(q+1)} \int_{-\infty}^0 |t|^q f(t)\,dt\right)^{1/(q+1)} \leqslant \frac{1}{f(0)} \int_{-\infty}^0 f(t)\,dt.$$

Adding these two inequalities and using the fact that $a^s + b^s \geqslant (a+b)^s$ for all $a, b > 0$ and $0 < s < 1$, as well as the fact that $\int_{-\infty}^{\infty} f = |K| = 1$, we write

$$\frac{1}{f(0)} \geqslant \left(\frac{1}{f(0)\Gamma(q+1)} \int_0^{\infty} t^q f(t)\,dt\right)^{1/(q+1)}$$

$$+ \left(\frac{1}{f(0)\Gamma(q+1)} \int_{-\infty}^0 |t|^q f(t)\,dt\right)^{1/(q+1)}$$

$$\geqslant \left(\frac{1}{f(0)\Gamma(q+1)} \int_{-\infty}^{\infty} |t|^q f(t)\,dt\right)^{1/(q+1)}.$$

This shows that
$$\int_K |\langle x,\theta\rangle|^q dx = \int_{-\infty}^{\infty} |t|^q f(t)\,dt \leq \frac{\Gamma(q+1)}{f(0)^q}$$
and the result follows since $\Gamma(q+1)^{1/q} \leq cq$ by Stirling's formula. □

Assume that K is isotropic. Then, Propositions 3.1.4 and 3.1.5 imply Theorem 3.1.2. Choosing $q = 2$ we see that all $(n-1)$-dimensional sections $K \cap \theta^\perp$ of K have "the same volume":

Proof of Theorem 3.1.2. Let K be an isotropic convex body in $\mathbb{R}^n$. For every $\theta \in S^{n-1}$ we have $\|\langle \cdot, \theta \rangle\|_2 = L_K$. From (3.1.1) and (3.1.2) we see that
$$\frac{c_1}{L_K} \leq |K \cap \theta^\perp| \leq \frac{c_2}{L_K},$$
where $c_1, c_2 > 0$ are absolute constants. □

3.1.3. The hyperplane conjecture

Theorem 3.1.2 establishes the connection of the isotropic constant conjecture with the hyperplane conjecture (sometimes also called the slicing problem).

CONJECTURE 3.1.6 (hyperplane conjecture). There exists an absolute constant $c > 0$ with the following property: for every $n \geq 1$ and every centered convex body K of volume 1 in $\mathbb{R}^n$ there exists $\theta \in S^{n-1}$ such that
$$|K \cap \theta^\perp| \geq c. \tag{3.1.3}$$

It is now not hard to see that the two conjectures are equivalent. Assume that the slicing problem has an affirmative answer. If K is isotropic, Theorem 3.1.2 shows that *all* sections $K \cap \theta^\perp$ have volume bounded by c_2/L_K from above. Since (3.1.3) must be true for at least one $\theta \in S^{n-1}$, we get $L_K \leq c_2/c$.

Conversely, if there exists an absolute bound C for the isotropic constant, then the slicing conjecture follows. A simple way to see this is through the *Binet ellipsoid of inertia*. This is the ellipsoid $\mathcal{E}_B(K)$ defined by
$$\|y\|^2_{\mathcal{E}_B(K)} = \int_K \langle x,y\rangle^2 dx = \langle My,y\rangle, \tag{3.1.4}$$
where $M = M(K)$ is the matrix of inertia of K, with entries $m_{ij} = \int_K x_i x_j dx$.

The next proposition shows that the volume of $\mathcal{E}_B(K)$ is invariant under the action of $SL(n)$.

PROPOSITION 3.1.7. *Let K be a centered convex body of volume 1 in $\mathbb{R}^n$. Then,*
$$|\mathcal{E}_B(K)| = \omega_n L_K^{-n}.$$

Proof. Recall that if K is an isotropic convex body in $\mathbb{R}^n$ then $\mathcal{E}_B(K) = L_K^{-1} B_2^n$, and hence $|\mathcal{E}_B(K)| = \omega_n L_K^{-n}$. It is easily checked that if $T \in SL(n)$ then $M(T(K)) = TM(K)T^*$, and hence $|\det M(K)| = |\det M(T(K))|$. Since
$$\|y\|^2_{\mathcal{E}_B(T(K))} = \langle M(T(K))y,y\rangle,$$
we have $\mathcal{E}_B(T(K)) = S^{-1}(B_2^n)$ where $S^2 = M(T(K))$. It follows that
$$|\mathcal{E}_B(T(K))| = \omega_n |\det M(T(K))|^{-1/2} = \omega_n|\det M(K)|^{-1/2} = |\mathcal{E}_B(K)|$$
for every $T \in SL(n)$. □

COROLLARY 3.1.8. *Let K be a centered convex body of volume 1 in $\mathbb{R}^n$. There exists $\theta \in S^{n-1}$ such that*
$$\int_K \langle x, \theta \rangle^2 dx \leqslant L_K^2.$$

Proof. Note that
$$L_K^{-n} = |\mathcal{E}_B(K)|/\omega_n = \int_{S^{n-1}} \|\theta\|_{\mathcal{E}_B(K)}^{-n} \sigma(d\theta).$$
It follows that $\min_{\theta \in S^{n-1}} \|\theta\|_{\mathcal{E}_B(K)} \leqslant L_K$. □

We can now show that the isotropic constant conjecture implies a positive answer to the hyperplane conjecture. Assume that K is a centered convex body of volume 1 in $\mathbb{R}^n$. According to corollary 3.1.8, there exists $\theta \in S^{n-1}$ such that
$$\int_K \langle x, \theta \rangle^2 dx \leqslant L_K^2 \leqslant C^2.$$
Now, Proposition 3.1.4 shows that
$$|K \cap \theta^\perp| \geqslant c := \frac{1}{2\sqrt{3}eC}.$$

3.2. Geometry of isotropic convex bodies

Our main goal is to understand how volume is distributed in high dimensional isotropic convex bodies. Thus, the study of geometric properties of these bodies is naturally a central topic in this book. In this section we provide some first information in this direction; all the results will be used quite often in the sequel.

3.2.1. Radius and inradius

Recall that the inradius $r(K)$ of a convex body K in $\mathbb{R}^n$ with $0 \in \text{int}(K)$ is the largest $r > 0$ for which $rB_2^n \subseteq K$ while the circumradius $R(K) := \max\{\|x\|_2 : x \in K\}$ of K is the smallest $R > 0$ for which $K \subseteq RB_2^n$. It is not hard to see that the inradius and the circumradius of an isotropic convex body K in $\mathbb{R}^n$ satisfy the bounds

(3.2.1) $$c_1 L_K \leqslant r(K) \leqslant R(K) \leqslant c_2 n L_K,$$

where $c_1, c_2 > 0$ are absolute constants. The following simple argument proves the right hand side inequality: given $\theta \in S^{n-1}$, one knows that

(3.2.2) $$|K \cap \theta^\perp| \simeq \frac{1}{L_K}.$$

Consider $x_\theta \in K$ such that $\langle x_\theta, \theta \rangle = h_K(\theta)$ and let $C(\theta)$ denote the cone $\text{conv}(K \cap \theta^\perp, x_\theta)$. Then, $C(\theta) \subseteq K$, and hence
$$1 = |K| \geqslant |C(\theta)| = \frac{|K \cap \theta^\perp| h_K(\theta)}{n}.$$
It follows that $h_K(\theta) \leqslant c_2 n L_K$, and since $\theta \in S^{n-1}$ was arbitrary, this gives $R(K) \leqslant c_2 n L_K$. For the left hand side inequality we use Grünbaum's Lemma 2.2.6: since K is centered, for every $\theta \in S^{n-1}$ we have $|\{x \in K : \langle x, \theta \rangle \geqslant 0\}| \geqslant e^{-1}$. This implies that
$$e^{-1} \leqslant \|f_{K,\theta}\|_\infty h_K(\theta) \leqslant e|K \cap \theta^\perp| h_K(\theta),$$

where $f_{K,\theta}(t) = |K \cap \{\langle x,\theta\rangle = t\}|$. Now, using (3.2.2) we see that $h_K(\theta) \geqslant c_1 L_K$, and since θ was arbitrary, this gives $r(K) \geqslant c_1 L_K$. In the symmetric case one actually has the bound $r(K) \geqslant L_K$, because $|\langle x,\theta\rangle| \leqslant h_K(\theta)$ for all $x \in K$, and hence
$$h_K(\theta) \geqslant \left(\int_K \langle x,\theta\rangle^2 dx\right)^{1/2} = L_K$$
for every $\theta \in S^{n-1}$.

The argument of Kannan, Lovász and Simonovits which is presented below, results in a better constant for the upper bound in (3.2.1).

THEOREM 3.2.1. *Let K be an isotropic convex body in $\mathbb{R}^n$. Then, $R(K) \leqslant (n+1)L_K$.*

Proof. Let $x \in K$. We define $h : S^{n-1} \to \mathbb{R}$ by
$$h(u) = \max\{t \geqslant 0 : x + tu \in K\}.$$
We can express the volume of K as
$$1 = |K| = n\omega_n \int_{S^{n-1}} \int_0^{h(u)} t^{n-1} dt d\sigma(u) = \omega_n \int_{S^{n-1}} h^n(u) d\sigma(u).$$
Since K is isotropic, for every $\theta \in S^{n-1}$ we may write
$$L_K^2 = \int_K \langle y,\theta\rangle^2 dy$$
$$= n\omega_n \int_{S^{n-1}} \int_0^{h(u)} t^{n-1} \langle x+tu,\theta\rangle^2 dt d\sigma(u)$$
$$= n\omega_n \int_{S^{n-1}} \int_0^{h(u)} \left(t^{n-1}\langle x,\theta\rangle^2 + 2t^n \langle x,\theta\rangle\langle u,\theta\rangle + t^{n+1}\langle u,\theta\rangle^2\right) dt d\sigma(u)$$
$$= n\omega_n \int_{S^{n-1}} \left(\frac{h^n(u)}{n}\langle x,\theta\rangle^2 + \frac{2h^{n+1}(u)}{n+1}\langle x,\theta\rangle\langle u,\theta\rangle + \frac{h^{n+2}(u)}{n+2}\langle u,\theta\rangle^2\right) d\sigma(u)$$
$$= n\omega_n \int_{S^{n-1}} \left(\frac{h^n(u)}{n(n+1)^2}\langle x,\theta\rangle^2 + h^n(u)\left(\frac{h(u)\langle u,\theta\rangle}{\sqrt{n+2}} + \frac{\sqrt{n+2}\langle x,\theta\rangle}{n+1}\right)^2\right) d\sigma(u)$$
$$\geqslant \frac{\langle x,\theta\rangle^2}{(n+1)^2} \omega_n \int_{S^{n-1}} h^n(u) d\sigma(u) = \frac{\langle x,\theta\rangle^2}{(n+1)^2}.$$
This means that, for every $x \in K$ and $\theta \in S^{n-1}$,
$$|\langle x,\theta\rangle| \leqslant (n+1)L_K.$$
Therefore,
$$\|x\|_2 = \max_{\theta \in S^{n-1}} |\langle x,\theta\rangle| \leqslant (n+1)L_K.$$
Since $x \in K$ was arbitrary, the proof is complete. $\square$

3.2.2. Entropy estimates

The next theorem (proved by Milman-Pajor in the symmetric case, and by Hartzoulaki in the form below) gives an estimate for the covering numbers $N(K, tB_2^n)$ of an isotropic convex body K in $\mathbb{R}^n$ in terms of the quantity
$$I_1(K) = \int_K \|x\|_2 dx.$$

Note that
$$I_1(K) \leq \left(\int_K \|x\|_2^2 dx\right)^{1/2} = \sqrt{n} L_K.$$

THEOREM 3.2.2. *Let K be an isotropic convex body in $\mathbb{R}^n$. For every $t > 0$ we have*
$$N(K, tB_2^n) \leq 2\exp\left(\frac{4(n+1)I_1(K)}{t}\right) \leq 2\exp\left(\frac{6n^{3/2}L_K}{t}\right).$$

Proof. Consider the Minkowski functional $p_K(x) = \inf\{\lambda > 0 : x \in \lambda K\}$. It is clear that p_K is subadditive and positively homogeneous. We define a Borel probability measure on $\mathbb{R}^n$ by
$$\mu(A) = \frac{1}{c_K} \int_A e^{-p_K(x)} dx,$$
where $c_K = \int_{\mathbb{R}^n} \exp(-p_K(x)) dx$. Let $\{x_1, \ldots, x_N\}$ be a subset of K which is maximal with respect to the condition
$$i \neq j \implies \|x_i - x_j\|_2 \geq t.$$
Then, the balls $x_i + (t/2)B_2^n$ have disjoint interiors, and $K \subseteq \bigcup_{i \leq N}(x_i + tB_2^n)$. Consequently, $N(K, tB_2^n) \leq N$.

We choose $b > 0$ so that $\mu(bB_2^n) \geq 1/2$. If we set $y_i = (2b/t)x_i$, then
$$\mu(y_i + bB_2^n) = \frac{1}{c_K} \int_{bB_2^n} e^{-p_K(x+y_i)} dx \geq \frac{1}{c_K} \int_{bB_2^n} e^{-p_K(x)} e^{-p_K(y_i)} dx$$
$$= e^{-p_K(y_i)} \frac{1}{c_K} \int_{bB_2^n} e^{-p_K(x)} dx = e^{-\frac{2b}{t} p_K(x_i)} \mu(bB_2^n)$$
$$\geq e^{-2b/t} \mu(bB_2^n),$$
since $p_K(x_i) \leq 1$ for all $i = 1, \ldots, N$. The balls $y_i + bB_2^n$ have disjoint interiors, therefore
$$Ne^{-2b/t} \mu(bB_2^n) \leq \sum_{i=1}^N \mu(y_i + bB_2^n) = \mu\left(\bigcup_{i=1}^N (y_i + bB_2^n)\right) \leq 1.$$
It follows that
$$N(K, tB_2^n) \leq e^{2b/t}(\mu(bB_2^n))^{-1} \leq 2e^{2b/t}.$$
What remains is to estimate b. We first compute the constant
$$c_K = \int_{\mathbb{R}^n} e^{-p_K(x)} dx = \int_{\mathbb{R}^n} \int_{p_K(x)}^\infty e^{-s} ds dx = \int_0^\infty e^{-s} |\{x : p_K(x) \leq s\}| ds$$
$$= \int_0^\infty |sK| e^{-s} ds = \int_0^\infty s^n e^{-s} ds = n!.$$
It follows that
$$J := \int_{\mathbb{R}^n} \|x\|_2 d\mu(x) = \frac{1}{c_K} \int_{\mathbb{R}^n} \|x\|_2 \int_{p_K(x)}^\infty e^{-s} ds dx$$
$$= \frac{1}{n!} \int_0^\infty s^{n+1} e^{-s} ds \cdot \int_K \|x\|_2 dx = (n+1)I_1(K).$$
From Markov's inequality, $\mu(\{x \in \mathbb{R}^n : \|x\|_2 > 2J\}) \leq 1/2$, which shows that $\mu(2JB_2^n) \geq 1/2$. If we choose $b = 2J$, we get
$$N(K, tB_2^n) \leq 2\exp(4J/t) \leq 2\exp(4(n+1)I_1(K)/t) \leq 2\exp(6n^{3/2}L_K/t),$$
which is the assertion of the theorem. □

REMARK 3.2.3. Note that $|K| = 1$, and hence $|K| = |r_n B_2^n|$ for some $r_n > 0$ with $r_n \simeq \sqrt{n}$. Therefore,
$$N(K, t(r_n B_2^n)) \leqslant 2 \exp\left(\frac{cL_K n}{t}\right)$$
for all $t > 0$, where $c > 0$ is an absolute constant. This proves that if $L_K \leqslant C$ then K is in "M-position of order 1" (recall that the standard definition of the M-position requires bounds for the covering numbers $N(r_n B_2^n, tK)$ as well). In particular, if one had $L_K \leqslant C$ for all isotropic convex bodies then every pair of isotropic bodies K and T would satisfy the reverse Brunn-Minkowski inequality. We will show that the converse is also true; this is an observation of Bourgain, Kartag and Milman.

THEOREM 3.2.4. *Assume that there exists a constant $A > 0$ such that for every $n \geqslant 1$ the following holds true: if K and T are isotropic convex bodies in $\mathbb{R}^n$ then*

(3.2.3) $$|K + T|^{1/n} \leqslant 2A = A\left(|K|^{1/n} + |T|^{1/n}\right).$$

Then, for every convex body K in $\mathbb{R}^n$ we have
$$L_K \leqslant cA^4,$$
where $c > 0$ is an absolute constant.

We first prove a simple lemma.

LEMMA 3.2.5. *Let K and T be two isotropic convex bodies in $\mathbb{R}^n$ and $\mathbb{R}^m$ respectively. Then, $W := (L_T/L_K)^{\frac{m}{n+m}} K \times (L_K/L_T)^{\frac{n}{n+m}} T$ is an isotropic convex body in $\mathbb{R}^{n+m}$, and*
$$L_{K \times T} = L_K^{\frac{n}{n+m}} L_T^{\frac{m}{n+m}}.$$

Proof. Let E be the subspace spanned by the first n standard unit vectors in $\mathbb{R}^{n+m}$. We define $W = aK \times bT$ for some a, b to be specified in a while, and we ask that $a^n b^m = 1$ so that $|W| = 1$. If we write M for the operator $M_W \in L(\mathbb{R}^{n+m})$ defined by
$$M(z) = \int_W \langle w, z \rangle w \, dw$$
then it is clear that if $z \in E$ then $M(z) \in E$. Also,
$$\langle M(z), z \rangle = \int_W \langle w, z \rangle^2 dw = b^m \int_{aK} \langle x, z \rangle^2 dx = b^m a^{n+2} L_K^2 \|z\|_2^2 = a^2 L_K^2 \|z\|_2^2.$$
The same argument shows that if $z \in E^\perp$ then $M(z) \in E^\perp$ and
$$\langle M(z), z \rangle = b^2 L_T^2 \|z\|_2^2.$$
Since M acts as a multiple of the identity on both E and $E^\perp$, we see that W will be isotropic provided that
$$aL_K = bL_T.$$
Since $a^n b^m = 1$ this condition gives
$$a = \left(\frac{L_T}{L_K}\right)^{\frac{m}{n+m}} \quad \text{and} \quad b = \left(\frac{L_K}{L_T}\right)^{\frac{n}{n+m}}.$$
It remains to observe that $K \times T$ and W belong to the same linear class, and hence we have $L_{K \times T} = L_W = aL_K$ with a as above. $\square$

Proof of Theorem 3.2.4. Let $D_s = r_s B_2^s$ be the Euclidean ball of volume 1 in $\mathbb{R}^s$. Let K be an isotropic convex body in $\mathbb{R}^n$. According to Lemma 3.2.5, the body $W = (L_{B_2^n}/L_K)^{1/2} K \times (L_K/L_{B_2^n})^{1/2} D_n$ is an isotropic convex body in $\mathbb{R}^{2n}$ (and $L_W = (L_K L_{B_2^n})^{1/2}$). Applying (3.2.3) to W and D_{2n} we get

$$|W + D_{2n}|^{\frac{1}{2n}} \leqslant 2A.$$

On the other hand,

$$W + D_{2n} \supseteq \{\mathbf{0}_n\} \times (L_K/L_{B_2^n})^{1/2} D_n + D_{2n} \supseteq c_1 D_n \times (L_K/L_{B_2^n})^{1/2} D_n$$

(where $\mathbf{0}_n$ is the origin in $\mathbb{R}^n$ and $c_1 = \inf_n (r_{2n}/r_n) > 0$), which implies

$$|W + D_{2n}|^{\frac{1}{2n}} \geqslant c_1^{1/2} \left(\frac{L_K}{L_{B_2^n}}\right)^{1/4}.$$

It follows that

$$\left(\frac{L_K}{L_{B_2^n}}\right)^{1/4} c_1^{1/2} \leqslant 2A,$$

and hence $L_K \leqslant cA^4 L_{B_2^n}$ for some absolute constant $c > 0$. Finally, a simple computation shows that

$$(3.2.4) \qquad L_{B_2^n}^2 = \frac{1}{n} \int_{r_n B_2^n} \|x\|_2^2 dx = \frac{1}{n} \frac{n \omega_n}{n+2} r_n^{n+2} = \frac{\omega_n^{-2/n}}{n+2},$$

which implies that $L_{B_2^n} \simeq 1$ and completes the proof. $\square$

3.2.3. ψ_1-estimates for linear functionals

Propositions 3.1.4 and 3.1.5 state that if K is a centered convex body of volume 1 in $\mathbb{R}^n$ then, for every $\theta \in S^{n-1}$ and $q \geqslant 1$,

$$\frac{c_1}{|K \cap \theta^\perp|} \leqslant \left(\int_K |\langle x, \theta \rangle|^q dx\right)^{1/q} \leqslant \frac{c_2 q}{|K \cap \theta^\perp|},$$

where $c_1, c_2 > 0$ are absolute constants. In the isotropic case the result takes the following form.

THEOREM 3.2.6. *Let K be an isotropic convex body in $\mathbb{R}^n$. For every $q \geqslant 1$ and every $\theta \in S^{n-1}$ we have*

$$(3.2.5) \qquad c_1 L_K \leqslant \left(\int_K |\langle x, \theta \rangle|^q dx\right)^{1/q} \leqslant c_2 q L_K,$$

where $c_1, c_2 > 0$ are absolute constants. $\square$

One can say more, using the results of Section 2.4. By Theorem 2.4.6 we get the following

THEOREM 3.2.7. *Let K be a convex body of volume 1 in $\mathbb{R}^n$. For every $\theta \in S^{n-1}$ and any $q > p \geqslant 1$ we have*

$$\|\langle \cdot, \theta \rangle\|_{L^p(K)} \leqslant \|\langle \cdot, \theta \rangle\|_{L^q(K)} \leqslant c\frac{q}{p} \|\langle \cdot, \theta \rangle\|_{L^p(K)},$$

where $c > 0$ is an absolute constant. Note also that the two inequalities continue to hold if we replace $|\langle \cdot, \theta \rangle|$ by any seminorm on $\mathbb{R}^n$. $\square$

Given a convex body K of volume 1 in $\mathbb{R}^n$, the Orlicz norm $\|f\|_{L^{\psi_\alpha}(K)}$, $\alpha \geqslant 1$, of a bounded measurable function $f : K \to \mathbb{R}$ is defined by

$$\|f\|_{L^{\psi_\alpha}(K)} = \inf\left\{t > 0 \;\Big|\; \int_K \exp((|f(x)|/t)^\alpha) dx \leqslant 2\right\}.$$

Recall also that, by Lemma 2.4.2, for every $f : K \to \mathbb{R}$ we have

$$\|f\|_{\psi_\alpha} \simeq \sup\left\{\frac{\|f\|_q}{q^{1/\alpha}} : q \geqslant \alpha\right\}.$$

We will see that in the setting of centered convex bodies this supremum can be estimated over the interval $q \in [\alpha, n]$ when f is a linear functional. Our starting point is the next lemma (observed by Paouris).

LEMMA 3.2.8. *Let K be a centered convex body of volume 1 in $\mathbb{R}^n$. Then, for every $\theta \in S^{n-1}$ and every $q \geqslant 1$,*

$$\int_K |\langle x, \theta\rangle|^q dx \geqslant \frac{\Gamma(q+1)\Gamma(n)}{2e\Gamma(q+n+1)} \max\left\{h_K^q(\theta), h_K^q(-\theta)\right\}.$$

Proof. Consider the function $f_\theta(t) = |K \cap \{\langle x, \theta\rangle = t\}|$. By Brunn's principle, $f_\theta^{1/(n-1)}$ is concave. It follows that

$$f_\theta(t) \geqslant \left(1 - \frac{t}{h_K(\theta)}\right)^{n-1} f_\theta(0)$$

for all $t \in [0, h_K(\theta)]$. Therefore,

$$\int_K |\langle x, \theta\rangle|^q dx = \int_0^{h_K(\theta)} t^q f_\theta(t) dt + \int_0^{h_K(-\theta)} t^q f_{-\theta}(t) dt$$

$$\geqslant \int_0^{h_K(\theta)} t^q \left(1 - \frac{t}{h_K(\theta)}\right)^{n-1} f_\theta(0) dt$$

$$+ \int_0^{h_K(-\theta)} t^q \left(1 - \frac{t}{h_K(-\theta)}\right)^{n-1} f_\theta(0) dt$$

$$= f_\theta(0) \left(h_K^{q+1}(\theta) + h_K^{q+1}(-\theta)\right) \int_0^1 s^q (1-s)^{n-1} ds$$

$$= \frac{\Gamma(q+1)\Gamma(n)}{\Gamma(q+n+1)} f_\theta(0) \left(h_K^{q+1}(\theta) + h_K^{q+1}(-\theta)\right)$$

$$\geqslant \frac{\Gamma(q+1)\Gamma(n)}{\Gamma(q+n+1)} f_\theta(0) \frac{h_K(\theta) + h_K(-\theta)}{2} \max\left\{h_K^q(\theta), h_K^q(-\theta)\right\}.$$

Since K has its barycenter at the origin, we have $\|f_\theta\|_\infty \leqslant e f_\theta(0)$, and hence

$$1 = |K| = \int_{-h_K(-\theta)}^{h_K(\theta)} f_\theta(t) dt \leqslant e (h_K(\theta) + h_K(-\theta)) f_\theta(0).$$

This completes the proof. $\square$

COROLLARY 3.2.9. *Let K be a centered convex body of volume 1 in $\mathbb{R}^n$. For every $\theta \in S^{n-1}$ and every $q \geqslant n$,*

$$\|\langle \cdot, \theta\rangle\|_q \simeq \max\{h_K(\theta), h_K(-\theta)\}.$$

Proof. From Lemma 3.2.8 we easily see that $\|\langle\cdot,\theta\rangle\|_n \simeq \max\{h_K(\theta), h_K(-\theta)\}$, and the assertion follows. □

As an immediate consequence we get:

PROPOSITION 3.2.10. *Let K be a centered convex body of volume 1 in $\mathbb{R}^n$. Then, for every $\theta \in S^{n-1}$,*
$$\|\langle\cdot,\theta\rangle\|_{\psi_\alpha} \simeq \sup\left\{\frac{\|\langle\cdot,\theta\rangle\|_q}{q^{1/\alpha}} : \alpha \leqslant q \leqslant \max\{n,\alpha\}\right\}.$$

DEFINITION 3.2.11. Let K be a convex body of volume 1 in $\mathbb{R}^n$, and let $\alpha \geqslant 1$ and $\theta \in S^{n-1}$. As in Definition 2.4.3, we say that K satisfies a ψ_α-estimate with constant $b_\alpha = b_\alpha(\theta)$ in the direction of θ if we have
$$\|\langle\cdot,\theta\rangle\|_{\psi_\alpha} \leqslant b_\alpha \|\langle\cdot,\theta\rangle\|_2.$$
We say that K is a ψ_α-body with constant $B_\alpha > 0$ if
$$\sup_{\theta \in S^{n-1}} \frac{\|\langle\cdot,\theta\rangle\|_{\psi_\alpha}}{\|\langle\cdot,\theta\rangle\|_2} \leqslant B_\alpha.$$

Let us fix $\alpha = 1$. Theorem 3.2.6 shows that
$$\|\langle\cdot,\theta\rangle\|_q \leqslant CqL_K$$
for all $q \geqslant 1$. In other words, K is a ψ_1-body with constant C for some absolute constant $C > 0$.

PROPOSITION 3.2.12. *There exists an absolute constant $C > 0$ such that for every isotropic convex body K in $\mathbb{R}^n$ and every $\theta \in S^{n-1}$ one has*
$$\|\langle\cdot,\theta\rangle\|_{\psi_1} \leqslant CL_K,$$
and hence for every $t > 0$
$$|\{x \in K : |\langle x,\theta\rangle| \geqslant CL_K t\}| \leqslant 2e^{-t}.$$

Proof. We apply Markov's inequality to get
$$2 \geqslant \int_K \exp(|\langle x,\theta\rangle|/CL_K)dx \geqslant \int_{\{x:|\langle x,\theta\rangle|\geqslant CL_K t\}} \exp(|\langle x,\theta\rangle|/CL_K)dx$$
$$\geqslant e^t |\{x \in K : |\langle x,\theta\rangle| \geqslant CL_K t\}|$$
as claimed. □

REMARK 3.2.13. The linear dependence on q in Theorem 3.2.6 cannot be improved in general. If K is a cone in the direction of θ (in other words, if $|K \cap \{\langle x,\theta\rangle = t\}|^{\frac{1}{n-1}}$, $t > 0$, is linear on its support) then we easily check that
$$\|\langle\cdot,\theta\rangle\|_q \simeq q\|\langle\cdot,\theta\rangle\|_1.$$

REMARK 3.2.14. Note that
$$\|f\|_{\psi_2} \leqslant \sqrt{\|f\|_{\psi_1}\|f\|_\infty}$$
for every bounded measurable function on a probability space. If K is isotropic then, taking into account the fact that $\|\langle\cdot,\theta\rangle\|_\infty \leqslant R(K) = O(nL_K)$ for every $\theta \in S^{n-1}$, we get
$$\|\langle\cdot,\theta\rangle\|_{\psi_2} \leqslant \sqrt{(cL_K)R(K)} \leqslant c\sqrt{n}L_K.$$

3.2.4. Alesker's deviation estimate

Alesker's theorem asserts that the Euclidean norm on isotropic convex bodies satisfies a ψ_2-estimate.

THEOREM 3.2.15 (Alesker). *Let K be an isotropic convex body in $\mathbb{R}^n$. If $f(x) = \|x\|_2$, then*
$$\|f\|_{L^{\psi_2}(K)} \leqslant c\sqrt{n}L_K,$$
where $c > 0$ is an absolute constant.

We will use the following lemma.

LEMMA 3.2.16. *Let K be a centered convex body of volume 1 in $\mathbb{R}^n$. For every $q \geqslant 1$,*
$$\left(\int_{S^{n-1}} \int_K |\langle x,\theta\rangle|^q dx d\sigma(\theta)\right)^{1/q} \simeq \frac{\sqrt{q}}{\sqrt{n+q}} \left(\int_K \|x\|_2^q dx\right)^{1/q}.$$

Proof. For every $q \geqslant 1$ and $x \in \mathbb{R}^n$ we check that
$$(3.2.6) \qquad \left(\int_{S^{n-1}} |\langle x,\theta\rangle|^q d\sigma(\theta)\right)^{1/q} \simeq \frac{\sqrt{q}}{\sqrt{n+q}} \|x\|_2.$$

To see this, using polar coordinates we first see that
$$\int_{B_2^n} |\langle x,y\rangle|^q dy = n\omega_n \int_0^1 r^{n+q-1} dr \int_{S^{n-1}} |\langle x,\theta\rangle|^q d\sigma(\theta)$$
$$= \frac{n\omega_n}{n+q} \int_{S^{n-1}} |\langle x,\theta\rangle|^q d\sigma(\theta).$$

But we can also write the left hand side as
$$\int_{B_2^n} |\langle x,y\rangle|^q dy = \|x\|_2^q \int_{B_2^n} \left|\left\langle \frac{x}{\|x\|_2}, y\right\rangle\right|^q dy$$
$$= \|x\|_2^q \int_{B_2^n} |\langle e_1, y\rangle|^q dy$$
$$= 2\omega_{n-1} \|x\|_2^q \int_0^1 t^q (1-t^2)^{(n-1)/2} dt$$
$$= \omega_{n-1} \|x\|_2^q \frac{\Gamma\left(\frac{q+1}{2}\right) \Gamma\left(\frac{n+1}{2}\right)}{\Gamma\left(\frac{n+q+2}{2}\right)}.$$

Comparing the two expressions and using Stirling's formula we get (3.2.6). Then, a simple application of Fubini's Theorem gives the result. $\square$

Proof of Theorem 3.2.15. It suffices to prove that $\|f\|_q \leqslant c_1 \sqrt{q} \|f\|_2$ for every $q > 1$ or, equivalently,
$$(3.2.7) \qquad \left(\int_K \|x\|_2^q dx\right)^{1/q} \leqslant c_1 \sqrt{q} \sqrt{n} L_K$$

for some absolute constant $c_1 > 0$. We know that for every $\theta \in S^{n-1}$
$$\int_K |\langle x,\theta\rangle|^q dx \leqslant c_2^q q^q L_K^q.$$

Integrating on the sphere we get
$$\int_{S^{n-1}} \int_K |\langle x, \theta \rangle|^q dx d\sigma(\theta) \leqslant c_2^q q^q L_K^q.$$
Taking Lemma 3.2.16 into account we see that
$$\left(\int_K \|x\|_2^q dx \right)^{1/q} \leqslant c_3 q \sqrt{\frac{n+q}{q}} L_K \leqslant c_4 \sqrt{q} \sqrt{n} L_K,$$
provided that $q \leqslant n$. On the other hand, if $q > n$, using the fact that $R(K) \leqslant (n+1)L_K$, we get
$$\left(\int_K \|x\|_2^q dx \right)^{1/q} \leqslant c_5 n L_K \leqslant c_5 \sqrt{q} \sqrt{n} L_K.$$
Combining the above we see that there is an absolute constant $c_1 > 0$ such that (3.2.7) holds true for all $q > 1$. □

COROLLARY 3.2.17. *There exists an absolute constant $c > 0$ such that: if K is an isotropic convex body in $\mathbb{R}^n$ then*
$$|\{x \in K : \|x\|_2 \geqslant c\sqrt{n} L_K s\}| \leqslant 2\exp(-s^2)$$
for every $s > 0$.

Proof. From Theorem 3.2.15 and from the description of the ψ_2-norm given by Lemma 2.4.2, we have
$$\int_K \exp\left(\frac{\|x\|_2^2}{c^2 n L_K^2} \right) dx \leqslant 2$$
for some absolute constant c. The result follows from Markov's inequality. □

3.3. Bourgain's upper bound for the isotropic constant

Recall that the isotropic constant conjecture asks if $L_K \leqslant C$ for every $n \geqslant 1$ and every isotropic convex body K in $\mathbb{R}^n$. A lower bound is available and quite simple.

PROPOSITION 3.3.1. *For every isotropic convex body K in $\mathbb{R}^n$ we have*
$$L_K \geqslant L_{B_2^n} \geqslant c,$$
where $c > 0$ is an absolute constant.

Proof. If $r_n = \omega_n^{-1/n}$, then $|r_n B_2^n| = 1$ and $r_n B_2^n$ is isotropic. Let K be an isotropic convex body. Observe that $\|x\|_2 > r_n$ on $K \setminus r_n B_2^n$ and $\|x\|_2 \leqslant r_n$ on $r_n B_2^n \setminus K$. Since these two sets have the same Lebesgue measure, it follows that
$$nL_K^2 = \int_K \|x\|_2^2 dx = \int_{K \cap r_n B_2^n} \|x\|_2^2 dx + \int_{K \setminus r_n B_2^n} \|x\|_2^2 dx$$
$$\geqslant \int_{K \cap r_n B_2^n} \|x\|_2^2 dx + \int_{r_n B_2^n \setminus K} \|x\|_2^2 dx = \int_{r_n B_2^n} \|x\|_2^2 dx$$
$$= nL_{B_2^n}^2.$$
By (3.2.4) we know that $L_{B_2^n} \simeq 1$, therefore $L_K \geqslant L_{B_2^n} \geqslant c$ where $c > 0$ is an absolute constant. □

In view of Proposition 3.3.1, what the hyperplane conjecture states is that the isotropic constants of all convex bodies are (uniformly with respect to the dimension

3.3. BOURGAIN'S UPPER BOUND FOR THE ISOTROPIC CONSTANT

n) of the order of 1. A first upper bound for the isotropic constant follows rather easily from our bounds for the inradius of isotropic convex bodies.

PROPOSITION 3.3.2. *Let K be a convex body in $\mathbb{R}^n$. Then $L_K \leqslant c\sqrt{n}$.*

Proof. We may assume that K is in isotropic position. Then, $|K| = 1$ and by (3.2.1) we know that $K \supseteq c_1 L_K B_2^n$ for some absolute constant $c_1 > 0$. It follows that
$$\omega_n (c_1 L_K)^n \leqslant |K| = 1.$$
Since $\omega_n^{-1/n} \simeq \sqrt{n}$, the upper bound $L_K \leqslant c\sqrt{n}$ follows. $\square$

In this section we give two proofs of Bourgain's bound $L_K = O(\sqrt[4]{n} \log n)$ (see [**101**]). The first one is closer to Bourgain's original argument which is based on a "reduction to small diameter" followed by a direct use of Talagrand's comparison theorem. The second one, which is due to Dar [**152**], is based on the ψ_1-behavior of linear functionals on convex bodies, which was the main ingredient in Bourgain's approach too, but it is more elementary, since it involves simpler entropy considerations.

3.3.1. Reduction to small diameter

The next proposition reduces the question of whether the hyperplane conjecture is true for all bodies to whether it is true for isotropic convex bodies with "small diameter".

PROPOSITION 3.3.3 (Bourgain). *Let K be an isotropic convex body in $\mathbb{R}^n$. Then, there exists an isotropic convex body Q in $\mathbb{R}^n$ with $L_Q \simeq L_K$ and $R(Q) \leqslant c\sqrt{n} L_Q$, where $c > 0$ is an absolute constant.*

For the proof we need a simple observation about the stability of the isotropic constant.

LEMMA 3.3.4. *Let K be a convex body of volume 1 in $\mathbb{R}^n$. Assume that for some constant $L > 0$ and some $a, b > 0$ we have*
$$a^{-2} L^2 \leqslant \int_K \langle x, \theta \rangle^2 dx \leqslant b^2 L^2$$
for all $\theta \in S^{n-1}$. Then,
$$a^{-1} L \leqslant L_K \leqslant bL.$$

Proof. Consider the Binet ellipsoid $\mathcal{E}_B(K)$ which is defined by
$$\|\theta\|_{\mathcal{E}_B(K)}^2 = \int_K \langle x, \theta \rangle^2 dx.$$
Our assumption implies that
$$\frac{1}{bL} \omega_n^{1/n} \leqslant |\mathcal{E}_B(K)|^{1/n} \leqslant \frac{a}{L} \omega_n^{1/n}.$$
On the other hand,
$$|\mathcal{E}_B(K)|^{1/n} = L_K^{-1} \omega_n^{1/n}$$
from Proposition 3.1.7. Combining the above we conclude the proof. $\square$

Proof of Proposition 3.3.3. Since K is isotropic, we have
$$\int_K \|x\|_2^2 dx = nL_K^2.$$
For every $r > 0$ we define
$$K_r = \{x \in K : \|x\|_2 \leqslant r\sqrt{n}L_K\}.$$
Markov's inequality shows that $|K_r| \geqslant 1 - r^{-2}$. Then, using the Cauchy-Schwarz inequality and the fact that $\|\langle \cdot, \theta \rangle\|_4 \leqslant C \|\langle \cdot, \theta \rangle\|_2 = CL_K$ for every $\theta \in S^{n-1}$, we write
$$\int_{K_r} \langle x, \theta \rangle^2 dx = \int_K \langle x, \theta \rangle^2 dx - \int_{K \setminus K_r} \langle x, \theta \rangle^2 dx$$
$$\geqslant L_K^2 - |K \setminus K_r|^{1/2} \left(\int_K \langle x, \theta \rangle^4 dx \right)^{1/2}$$
$$\geqslant L_K^2 - r^{-1} C^2 L_K^2.$$
If we choose $r = 2C^2$, we have
$$\frac{L_K^2}{2} \leqslant \int_{K_r} \langle x, \theta \rangle^2 dx \leqslant L_K^2$$
for all $\theta \in S^{n-1}$. We find $a \geqslant 1$ with $a^n \leqslant c_1 := 1/(1 - r^{-2})$, so that $W = aK_r$ has volume $|W| = 1$. Then,

(3.3.1) $$\frac{L_K^2}{2} \|y\|_2^2 \leqslant \int_W \langle x, y \rangle^2 dx \leqslant c_1^2 L_K^2 \|y\|_2^2$$

for all $y \in \mathbb{R}^n$, and Lemma 3.3.4 shows that $L_W \simeq L_K$.

Choose $T \in SL(n)$ so that $Q = T(W)$ is isotropic. From (3.3.1), for every $\theta \in S^{n-1}$ we have
$$L_Q^2 = \int_Q \langle x, \theta \rangle^2 dx = \int_W \langle x, T^*\theta \rangle^2 dx \simeq \|T^*(\theta)\|_2^2 L_K^2.$$
Since $L_Q = L_W \simeq L_K$, we obtain
$$c_2 \leqslant \|T^*(\theta)\|_2 \leqslant c_3$$
for every $\theta \in S^{n-1}$. Therefore,
$$R(Q) \leqslant \|T : \ell_2^n \to \ell_2^n\| R(W) \leqslant c_3 ar\sqrt{n} L_K \leqslant c_4 \sqrt{n} L_K \leqslant c\sqrt{n} L_Q,$$
where $c > 0$ is an absolute constant. $\square$

3.3.2. Bourgain's upper bound via the majorizing measure theorem

For isotropic convex bodies of small diameter, Bourgain's upper bound can be obtained by an almost direct application of Talagrand's comparison theorem (Theorem 1.9.8). This leads to the next

THEOREM 3.3.5 (Bourgain). *Let K be an isotropic convex body in $\mathbb{R}^n$. Then*
$$L_K \leqslant c\sqrt[4]{n} \log n,$$
where $c > 0$ is an absolute constant.

Proof. From Proposition 3.3.3 we can find an isotropic convex body Q in $\mathbb{R}^n$ with $L_Q \simeq L_K$ and $R(Q) \leqslant c\sqrt{n}L_Q$. Recall that

$$\int_Q \langle x, Tx \rangle dx = (\operatorname{tr} T) \cdot L_Q^2$$

for every $T \in L(\mathbb{R}^n)$. Then, for every symmetric positive definite $T \in SL(n)$ we have

$$nL_Q^2 \leqslant (\operatorname{tr} T) \cdot L_Q^2 = \int_Q \langle x, Tx \rangle dx \leqslant \int_Q \max_{y \in TQ} \langle y, x \rangle dx.$$

Since $R(Q) \leqslant c\sqrt{n}L_Q$, Remark 3.2.14 shows that

$$\left\| \frac{\langle \cdot, y \rangle}{c_1 \sqrt[4]{n} L_Q \|y\|_2} \right\|_{\psi_2} \leqslant 1$$

for every $y \neq 0$, where $c_1 > 0$ is an absolute constant. It follows that

$$|\{x \in Q : |\langle x, y \rangle| \geqslant c_1 \sqrt[4]{n} L_Q t\}| \leqslant 2\exp(-t^2/\|y\|_2^2)$$

for every $y \neq 0$ and every $t > 0$.

We define a process $\mathcal{X} = (X_y)_{y \in TQ}$, where $X_y : Q \to \mathbb{R}$ is given by

$$X_y(x) = \frac{\langle x, y \rangle}{c_1 \sqrt[4]{n} L_Q}.$$

Then, for every $y \neq z \in TQ$ and every $t > 0$, we have

$$\operatorname{Prob}(|X_y - X_z| \geqslant t) = |\{x \in Q : |\langle y - z, x \rangle| \geqslant c_1 \sqrt[4]{n} L_Q t\}|$$
$$\leqslant 2\exp\left(-t^2/\|y - z\|_2^2\right),$$

that is, $\mathcal{X}$ is sub-Gaussian with respect to the Euclidean metric on TQ.

Let $g_1, \ldots, g_n$ be independent standard Gaussian random variables on some probability space Ω, and consider the Gaussian process $\mathcal{Z} = (Z_y)_{y \in TQ}$ with

$$Z_y(\omega) = \langle G(\omega), y \rangle,$$

where $G = (g_1, \ldots, g_n)$. Then, $\mathcal{X}$ and $\mathcal{Z}$ satisfy the assumptions of Talagrand's comparison theorem. Therefore,

$$\mathbb{E} \sup_{y \in TQ} X_y \leqslant C \cdot \mathbb{E} \sup_{y \in TQ} Z_y,$$

where $C > 0$ is an absolute constant. A simple computation (see Remark 1.9.3) shows that

$$\mathbb{E} \sup_{y \in TQ} Z_y \simeq \sqrt{n} w(TQ).$$

Therefore,

$$nL_Q^2 \leqslant c_1 \sqrt[4]{n} L_Q \cdot \mathbb{E} \sup_{y \in TQ} X_y \leqslant c_2 C \sqrt[4]{n} L_Q \cdot \sqrt{n} w(TQ).$$

In other words,

$$L_Q \leqslant c_3 w(TQ)/\sqrt[4]{n},$$

where $c_3 > 0$ is an absolute constant. The "reverse Urysohn inequality" (Theorem 1.11.5) shows that there exists a symmetric positive definite $T \in SL(n)$ for which $w(TQ) = O(\sqrt{n} \log n)$. This completes the proof. $\square$

3.3.3. Dar's argument

Next we present Dar's argument for Bourgain's bound, starting with the tools that we will use in the proof.

Expectation of the maximum of linear functionals on convex bodies

Let K be an isotropic convex body in $\mathbb{R}^n$, which satisfies the ψ_α-estimate

$$\|\langle\cdot,\theta\rangle\|_{\psi_\alpha} \leqslant b\|\langle\cdot,\theta\rangle\|_2 = bL_K$$

for all $\theta \in S^{n-1}$, where $\alpha \in [1,2]$ and $b > 0$.

PROPOSITION 3.3.6. *Let $N \geqslant 2$ and let $\theta_1,\ldots,\theta_N \in S^{n-1}$. Then,*

$$\int_K \max_{1\leqslant i\leqslant N} |\langle x,\theta_i\rangle|\,dx \leqslant CbL_K(\log N)^{1/\alpha},$$

where $C > 0$ is an absolute constant.

Proof. By the definition of the ψ_α-norm and Markov's inequality, for every $t > 0$ we have

$$\operatorname{Prob}\left(\left\{x \in K : \max_{1\leqslant i\leqslant N} |\langle x,\theta_i\rangle| \geqslant t\right\}\right) \leqslant \sum_{i=1}^{N} \operatorname{Prob}\left(\{x \in K : |\langle x,\theta_i\rangle| \geqslant t\}\right)$$
$$\leqslant 2N\exp\left(-(t/bL_K)^\alpha\right).$$

Then, given $A > 0$ we may write

$$\int_K \max_{1\leqslant i\leqslant N} |\langle x,\theta_i\rangle|\,dx = \int_0^\infty \operatorname{Prob}\left(\left\{x \in K : \max_{1\leqslant i\leqslant N} |\langle x,\theta_i\rangle| \geqslant t\right\}\right) dt$$
$$\leqslant A + \int_A^\infty \operatorname{Prob}\left(\left\{x \in K : \max_{1\leqslant i\leqslant N} |\langle x,\theta_i\rangle| \geqslant t\right\}\right) dt$$
$$\leqslant A + 2N\int_A^\infty \exp\left(-(t/bL_K)^\alpha\right) dt.$$

Choosing $A = 4bL_K(\log N)^{1/\alpha}$ we get

$$\int_A^\infty \exp\left(-(t/bL_K)^\alpha\right) dt = 4bL_K(\log N)^{1/\alpha}\int_1^\infty \exp(-4^\alpha s^\alpha \log N)ds$$
$$\leqslant 4bL_K(\log N)^{1/\alpha}\int_1^\infty \exp(-4s\log N)ds$$
$$\leqslant 4bL_K(\log N)^{1/\alpha}\exp(-2\log N)\int_1^\infty e^{-s}ds$$
$$\leqslant 4bL_K(\log N)^{1/\alpha}N^{-2},$$

where we have used the fact that

$$\exp(-4s\log N) \leqslant \exp(-2\log N)\cdot e^{-s}$$

is valid for all $s \geqslant 1$. It follows that

$$\int_K \max_{1\leqslant i\leqslant N} |\langle x,\theta_i\rangle|\,dx \leqslant CbL_K(\log N)^{1/\alpha}$$

with $C = 8$. □

Dudley-Fernique decomposition

Let K be a convex body in $\mathbb{R}^n$. We assume that $0 \in K$ and write R for the circumradius of K. For every $j \in \mathbb{N}$ we may find a subset N_j of K such that

$$|N_j| = \overline{N}(K, (R/2^j) B_2^n)$$

and

$$K \subseteq \bigcup_{y \in N_j} (y + (R/2^j) B_2^n).$$

Sudakov's inequality shows that

(3.3.2) $$\log |N_j| \leqslant cn \left(\frac{2^j w(K)}{R} \right)^2.$$

We define $N_0 = \{0\}$ and

$$W_j = N_j - N_{j-1} = \{y - y' \mid y \in N_j, y' \in N_{j-1}\}$$

for every $j \geqslant 1$.

LEMMA 3.3.7. *For every $x \in K$ and any $m \in \mathbb{N}$ we can find $z_j \in W_j \cap (3R/2^j) B_2^n$, $j = 1, \ldots, m$, and $w_m \in (R/2^m) B_2^n$ such that*

$$x = z_1 + \cdots + z_m + w_m.$$

Proof. Let $x \in K$. By the definition of N_j, we can find $y_j \in N_j$, $j = 1, \ldots, m$, such that

$$\|x - y_j\|_2 \leqslant \frac{R}{2^j}.$$

We write

$$x = (y_1 - 0) + (y_2 - y_1) + \cdots + (y_m - y_{m-1}) + (x - y_m).$$

We set $y_0 = 0$ and $w_m = x - y_m$, $z_j = y_j - y_{j-1}$ for $j = 1, \ldots, m$. Then, $\|w_m\|_2 = \|x - y_m\|_2 \leqslant R/2^m$, and $z_j \in N_j - N_{j-1} = W_j$. Also,

$$\|z_j\|_2 \leqslant \|x - y_j\|_2 + \|x - y_{j-1}\|_2 \leqslant \frac{R}{2^j} + \frac{R}{2^{j-1}} = \frac{3R}{2^j}.$$

Finally, $x = z_1 + \cdots + z_m + w_m$ as claimed. □

We set $Z_j = W_j \cap (3R/2^j) B_2^n$. Then, taking (3.3.2) into account, we may state the following.

THEOREM 3.3.8. *Let K be a convex body in $\mathbb{R}^n$, with $0 \in K$ and circumradius equal to R. There exist $Z_j \subseteq (3R/2^j) B_2^n$, $j \in \mathbb{N}$, with cardinality*

$$\log |Z_j| \leqslant cn \left(\frac{2^j w(K)}{R} \right)^2,$$

which satisfy the following: for every $x \in K$ and any $m \in \mathbb{N}$ we can find $z_j \in Z_j$, $j = 1, \ldots, m$, and $w_m \in (R/2^m) B_2^n$ such that $x = z_1 + \cdots + z_m + w_m$. □

We are now ready to give Dar's version of the proof of Bourgain's bound.

THEOREM 3.3.9 (Bourgain). *If K is an isotropic convex body in $\mathbb{R}^n$ then*

$$L_K \leqslant c \sqrt[4]{n} \log n,$$

where $c > 0$ is an absolute constant.

Proof. There exists a symmetric and positive definite $T \in SL(n)$ such that
$$w(TK) \leqslant c\sqrt{n}\log n.$$
We write
$$nL_K^2 = \int_K \|x\|_2^2 dx \leqslant \frac{\mathrm{tr}T}{n}\int_K \|x\|_2^2 = \int_K \langle x, Tx\rangle dx.$$
Therefore,
$$nL_K^2 \leqslant \int_K \max_{y \in TK}|\langle y, x\rangle| dx.$$
We now use Theorem 3.3.8 for TK. If R is the circumradius of TK, for every $m \in \mathbb{N}$ and every $1 \leqslant j \leqslant m$ we can find $Z_j \subset (3R/2^j)B_2^n$ such that
$$\log|Z_j| \leqslant cn\left(\frac{w(TK)2^j}{R}\right)^2,$$
and so that every $y \in TK$ can be written in the form $y = z_1 + \cdots + z_m + w_m$ with $z_j \in Z_j$ and $w_m \in (R/2^m)B_2^n$. This implies that
$$\max_{y \in TK}|\langle y, x\rangle| \leqslant \sum_{j=1}^m \max_{z \in Z_j}|\langle z, x\rangle| + \max_{w \in (R/2^m)B_2^n}|\langle w, x\rangle|$$
$$\leqslant \sum_{j=1}^m \frac{3R}{2^j}\max_{z \in Z_j}|\langle \overline{z}, x\rangle| + \frac{R}{2^m}\|x\|_2,$$
where $\overline{z}$ denotes the unit vector in the direction of z. Noting that $\int_K \|x\|_2 dx \leqslant \sqrt{n}L_K$ and using the above, we see that
$$nL_K^2 \leqslant \sum_{j=1}^m \frac{3R}{2^j}\int_K \max_{z \in Z_j}|\langle \overline{z}, x\rangle|dx + \frac{R}{2^m}\int_K \|x\|_2 dx$$
$$\leqslant \sum_{j=1}^m \frac{3R}{2^j}\int_K \max_{z \in Z_j}|\langle \overline{z}, x\rangle|dx + \frac{R}{2^m}\sqrt{n}L_K.$$
From Proposition 3.3.6 (with $\alpha = 1$) we get
$$(3.3.3) \qquad nL_K^2 \leqslant \sum_{j=1}^m \frac{3R}{2^j}c_1 nL_K\left(\frac{w(TK)2^j}{R}\right)^2 + \frac{R}{2^m}\sqrt{n}L_K.$$
The sum on the right is bounded by
$$c_2 L_K n w^2(TK)\frac{2^m}{R}.$$
Solving the equation
$$\frac{nw^2(TK)2^s}{R} = \frac{R\sqrt{n}}{2^s}$$
(where s here can be non-integer), we see that the optimal (integer) value of m satisfies the "equation"
$$\frac{R}{2^m} \simeq \sqrt[4]{n}w(TK).$$
Going back to (3.3.3), we obtain
$$nL_K^2 \leqslant c_3\sqrt{n}\sqrt[4]{n}w(TK)L_K.$$
Since $w(TK) \leqslant c_4\sqrt{n}\log n$, we get the result. $\square$

Assuming that K satisfies a ψ_α-estimate for some $\alpha \in [1, 2)$ and repeating the argument above, we get:

THEOREM 3.3.10. *Let K be an isotropic convex body in $\mathbb{R}^n$. Assume that there exist $1 \leqslant \alpha < 2$ and $B_\alpha > 0$ such that K satisfies the ψ_α-estimate*
$$\|\langle \cdot, \theta \rangle\|_{\psi_\alpha} \leqslant B_\alpha \|\langle \cdot, \theta \rangle\|_2 = B_\alpha L_K$$
for every $\theta \in S^{n-1}$. Then
$$L_K \leqslant C B_\alpha^{\frac{\alpha}{2}} (2-\alpha)^{-\frac{\alpha}{2}} n^{\frac{1}{2}-\frac{\alpha}{4}} \log n,$$
where $C > 0$ is an absolute constant. □

3.4. The ψ_2-case

Let us assume that K is a ψ_2-body with constant b. Adjusting the first argument of the previous section, we would just obtain the bound $L_K \leqslant cb \log n$. Bourgain showed in [**104**] that one can improve this estimate to $L_K \leqslant cb \log(1 + b)$. The exact statement is the following:

THEOREM 3.4.1 (Bourgain). *Let K be an isotropic symmetric convex body in $\mathbb{R}^n$ which is a ψ_2-body with constant $b \geqslant 1$. Then one has*
$$L_K \leqslant Cb \log(1 + b),$$
where $C > 0$ is an absolute constant.

We start with the following lemma.

LEMMA 3.4.2. *Let K be an isotropic symmetric convex body in $\mathbb{R}^n$ which is a ψ_2-body with constant $b \geqslant 1$. Let $\|\cdot\|$ be a norm on $\mathbb{R}^n$, let $\{v_1, \ldots, v_n\}$ be any orthonormal basis of $\mathbb{R}^n$ and let $\{u_1, \ldots, u_n\}$ be arbitrary vectors in $\mathbb{R}^n$. Then,*
$$\frac{1}{bL_K} \int_K \left\| \sum_{i=1}^n \langle x, v_i \rangle u_i \right\| dx \leqslant c_1 \int_\Omega \left\| \sum_{i=1}^n g_i(\omega) u_i \right\| d\omega,$$
where $(g_i)_{i=1}^n$ are independent standard Gaussian random variables defined on some probability space Ω and $c_1 > 0$ is an absolute constant.

Proof. Let $\|\cdot\|_*$ denote the dual norm of $\|\cdot\|$ and consider $T = \{t \in \mathbb{R}^n : \|t\|_* \leqslant 1\}$ as an index set for the process $\mathcal{X} = (X_t)_{t \in T}$, where $X_t : K \to \mathbb{R}$ is given by
$$X_t(x) = \frac{1}{bL_K} \sum_{i=1}^n \langle x, v_i \rangle \langle u_i, t \rangle.$$

We consider the pseudo-metric
$$d(t, t') := \left(\sum_{i=1}^n |\langle u_i, t - t' \rangle|^2 \right)^{1/2}$$
on T. Since K is a ψ_2-body, we have that $\|\langle \cdot, y \rangle\|_{\psi_2} \leqslant b \|\langle \cdot, y \rangle\|_2 = bL_K \|y\|_2$ for all $y \in \mathbb{R}^n$. Applying this to $y_t - y_{t'}$ where $y_t = \frac{1}{bL_K} \sum_{i=1}^n \langle u_i, t \rangle v_i$ we get
$$\|X_t - X_{t'}\|_{\psi_2} = \|\langle \cdot, y_t - y_{t'} \rangle\|_{\psi_2}$$
$$\leqslant bL_K \|y_t - y_{t'}\|_2 = \left(\sum_{i=1}^n |\langle u_i, t - t' \rangle|^2 \right)^{1/2} \equiv d(t, t').$$

In other words, $(X_t)_{t \in T}$ is sub-Gaussian with respect to d. Consider the Gaussian process $(Y_t)_{t \in T}$ with $Y_t = \sum_{i=1}^{n} g_i(\omega) \langle u_i, t \rangle$, $t \in T$. Since

$$\|Y_t - Y_{t'}\|_2 = d(t, t'),$$

we may apply Talagrand's comparison theorem (Theorem 1.9.8) to conclude that

$$\int_K \sup_{t \in T} |X_t(x)| \, dx \leqslant c_1 \int_\Omega \sup_{t \in T} |Y_t(\omega)| \, d\omega,$$

where $c_1 > 0$ is an absolute constant. This proves the lemma. $\square$

As in the proof of Proposition 3.3.3 we consider the convex body

$$K_1 = K \cap (rL_K \sqrt{n} B_2^n),$$

where $r > 0$ is an absolute constant which is chosen large enough so that $|K_1| \geqslant 1 - \frac{1}{r^2}$ and

(3.4.1) $$\int_{K_1} |\langle x, \theta \rangle|^2 \, dx > \frac{1}{2} L_K^2,$$

for all $\theta \in S^{n-1}$. We denote by $\|\cdot\| = \|\cdot\|_{K_1}$ the norm induced by K_1 and write $\|\cdot\|_* = \|\cdot\|_{K_1^\circ}$ for its dual norm. By the definition of K_1 we have

$$\|x\| \geqslant \frac{1}{rL_K \sqrt{n}} \|x\|_2 \quad \text{and} \quad \|x\|_* \leqslant rL_K \sqrt{n} \|x\|_2$$

for all $x \in \mathbb{R}^n$. With this notation in mind we prove the following lemma.

LEMMA 3.4.3. *Let E be a subspace of $\mathbb{R}^n$ with $\dim E > n/2$ such that*

$$\|x\|_* \geqslant \rho L_K \sqrt{n} \|x\|_2$$

for all $x \in E$ and some $\rho \in (0, 1/2)$. Then, for any $\delta \in (0, 1)$, there exists a subspace F of E satisfying

(3.4.2) $$\dim F > (1 - \delta) \dim E$$

and

(3.4.3) $$\|x\|_* \geqslant c\delta^{3/2} \left(\log \frac{1}{\rho}\right)^{-1} \frac{L_K}{b} \sqrt{n} \|x\|_2$$

for all $x \in F$.

Proof. Let $\dim E = m$. Note that the Banach-Mazur distance of $(E, \|\cdot\|_*)$ from the Euclidean space ℓ_2^m is less than or equal to r/ρ. We use the ℓ-position of K_1: there exists an operator $u : \ell_2^m \to (K_1 \cap E, \|\cdot\|)$ such that

(3.4.4) $$\int_{\mathbb{R}^m} \|u(x)\| \, d\gamma_m(x) \leqslant Cm \log(r/\rho) \quad \text{and} \quad \int_{\mathbb{R}^m} \|(u^{-1})^*(x)\| \, d\gamma_m(x) \leqslant c.$$

We may also assume that there exists an orthonormal basis $\{v_1, \ldots, v_m\}$ of E and positive numbers $0 < \lambda_1 \leqslant \cdots \leqslant \lambda_m$ such that $u(v_i) = \lambda_i v_i$, $1 \leqslant i \leqslant m$. Then, (3.4.4) can be written as follows:

(3.4.5) $$w(u(K_1 \cap E)) \leqslant C\sqrt{m} \log(r/\rho) \leqslant C'\sqrt{m} \log(1/\rho),$$

and

(3.4.6) $$M(u(K_1 \cap E)) \leqslant c/\sqrt{m}.$$

Now we use the low M^*-estimate in dual form: using (3.4.6) we see that for any fixed $\varepsilon \in (0, 1/2)$ there exists a subspace E_1 of E with $\dim E_1 > (1-\varepsilon)m$ such that

$$\|x\|_* = h_{K_1 \cap E}(x) = h_{u(K_1 \cap E)}((u^{-1})^*(x)) \geqslant \frac{c\sqrt{\varepsilon}}{M(u(K_1 \cap E))}\|(u^{-1})^*(x)\|_2$$

$$\geqslant c_1\sqrt{\varepsilon m}\|(u^{-1})^*(x)\|_2 = c_1\sqrt{\varepsilon m}\left(\sum_{i=1}^m \lambda_i^{-2}\langle x, v_i\rangle^2\right)^{1/2}$$

for all $x \in E_1$. Next, we use Lemma 3.4.2 as follows: we expand $\{v_1, \ldots, v_m\}$ to an orthonormal basis $\{v_1, \ldots, v_m, v_{m+1}, \ldots, v_n\}$ of $\mathbb{R}^n$ and we set $u_i = \lambda_i v_i$ for $i \leqslant m$ and $u_i = 0$ otherwise; since $K_1 \subseteq K$, we can then write

$$\frac{1}{bL_K}\int_{K_1}\left\|\sum_{i=1}^m \lambda_i \langle x, v_i\rangle v_i\right\|_* dx \leqslant \frac{1}{bL_K}\int_K \left\|\sum_{i=1}^m \lambda_i\langle x, v_i\rangle v_i\right\|_* dx$$

$$\leqslant c_2 \int_\Omega \left\|\sum_{i=1}^m \lambda_i g_i(\omega)v_i\right\|_* d\omega \leqslant C_1 m\log(1/\rho),$$

taking also (3.4.5) into account. On the other hand, using (3.4.1) we have

$$\int_{K_1}\left\|\sum_{i=1}^m \lambda_i\langle x, v_i\rangle v_i\right\|_* dx \geqslant \int_{K_1} \|x\| \cdot \left\|\sum_{i=1}^m \lambda_i\langle x, v_i\rangle v_i\right\|_* dx$$

$$\geqslant \int_{K_1} \left\langle x, \sum_{i=1}^m \lambda_i\langle x, v_i\rangle v_i\right\rangle dx \geqslant \sum_{i=1}^m \lambda_i \frac{L_K^2}{2}.$$

It follows that

(3.4.7) $$\lfloor \varepsilon m\rfloor \lambda_{\lfloor(1-\varepsilon)m\rfloor} \leqslant \sum_{i=1}^m \lambda_i \leqslant c_3 b \frac{m\log(1/\rho)}{L_K}.$$

Now, we further restrict x in $F = E_1 \cap \mathrm{span}\{v_i : i \leqslant (1-\varepsilon)m\}$ to obtain

$$\|x\|_* \geqslant c\sqrt{\varepsilon m}\left(\sum_{i\leqslant(1-\varepsilon)m} \lambda_i^{-2}\langle x, v_i\rangle^2\right)^{1/2}$$

$$\geqslant c'\frac{\varepsilon^{3/2}m^{3/2}}{b\,m\log(1/\rho)}L_K\left(\sum_{i\leqslant(1-\varepsilon)m}\langle x, v_i\rangle^2\right)^{1/2}$$

$$\geqslant c''\varepsilon^{3/2}\left(\log\frac{1}{\rho}\right)^{-1}\frac{L_K}{b}\sqrt{n}\|x\|_2,$$

where we have used the assumption that $m > n/2$ and (3.4.7). Finally, note that $\dim F > (1-2\varepsilon)m$; so, setting $\varepsilon = \delta/2$ we get the assertion. $\square$

Proof of Theorem 3.4.1. By (3.4.1) we have

$$\|\theta\|_* = h_{K_1}(\theta) \geqslant \left(\frac{1}{|K_1|}\int_{K_1}|\langle x, \theta\rangle|^2 dx\right)^{1/2} \geqslant \frac{L_K}{\sqrt{2}}$$

for all $\theta \in S^{n-1}$, and thus, when $E = \mathbb{R}^n$, the assumption of Lemma 3.4.3 is satisfied with $\rho = \rho_0 \simeq \frac{1}{\sqrt{n}}$. Using Lemma 3.4.3 repeatedly, we perform a flag construction

$$E_0 = \mathbb{R}^n \supseteq E_1 \supseteq \cdots \supseteq E_s \supseteq E_{s+1} \supseteq \cdots$$

of subspaces E_s with $\dim E_s = m_s > n/2$ as follows. Assume that $\|x\|_* \geq \rho_s L_K \sqrt{n} \|x\|_2$ for all $x \in E_s$ and some $\rho_s \in (0, e^{-2})$. We choose $\delta_{s+1} = (\log \frac{1}{\rho_s})^{-2}$ and define E_{s+1} so that the conclusion of Lemma 3.4.3 is satisfied with $F = E_{s+1}$. By (3.4.2) we have

$$\dim E_{s+1} = m_{s+1} > \left(1 - \frac{1}{(\log \frac{1}{\rho_s})^2}\right) m_s$$

and by (3.4.3)

$$\rho_{s+1} \simeq \left(\log \frac{1}{\rho_s}\right)^{-4} b^{-1}.$$

We end this procedure at the first s_0 for which $\rho_{s_0} \geq (c_1 b/c)^{-2}$, where $c \in (0,1)$ is the constant in (3.4.3) and c_1 is a large enough constant such that

$$4 \log \log t \leq \frac{1}{8} \log t \quad \text{for every } t \geq c_1^2.$$

Note that such an s_0 exists since for every integer s such that $\rho_s < (c_1 b/c)^{-2}$ we have that

$$\frac{1}{\rho_s} \geq (c_1 b/c)^2 \geq \max\{c_1^2, (b/c)^2\},$$

and hence

$$(3.4.8) \qquad \frac{\log \frac{1}{\rho_{s+1}}}{\log \frac{1}{\rho_s}} = \frac{\log\left(c^{-1} b \left(\log \frac{1}{\rho_s}\right)^4\right)}{\log \frac{1}{\rho_s}} = \frac{\log(b/c) + 4 \log \log \frac{1}{\rho_s}}{\log \frac{1}{\rho_s}} \leq \frac{5}{8};$$

thus, it cannot be that $\log(\rho_s^{-1}) > 2 \log(c_1 b/c)$ for every integer s. Because of (3.4.8), we also see that

$$\sum_{j < s_0} \left(\log \frac{1}{\rho_j}\right)^{-k} \leq \frac{1}{1 - \left(\frac{5}{8}\right)^k} \left(\log \frac{1}{\rho_{s_0 - 1}}\right)^{-k} < \frac{8}{3} \left(2 \log(c_1 b/c)\right)^{-k} < \frac{1}{2}$$

for every integer $k \geq 1$. A first consequence is that

$$\dim E_{s_0} \geq \left[\prod_{j < s_0} \left(1 - \frac{1}{(\log \frac{1}{\rho_j})^2}\right)\right] n \geq \left(1 - \sum_{j < s_0} \frac{1}{(\log \frac{1}{\rho_j})^2}\right) n > \frac{n}{2}.$$

We now use the classical Rogers-Shephard inequality and an iteration argument. For any $t < s_0$ we have

$$|K_1^\circ \cap E_t| \leq |K_1^\circ \cap E_{t+1}| \cdot |P_{E_{t+1}^\perp \cap E_t}(K_1^\circ \cap E_t)|.$$

By the assumption $\|x\|_* \geq \rho_t L_K \sqrt{n} \|x\|_2$ for all $x \in E_t$ we have

$$K_1^\circ \cap E_t \subseteq \frac{1}{\rho_t L_K \sqrt{n}} B_{E_t},$$

thus we get

$$|P_{E_t \cap E_{t+1}^\perp}(K_1^\circ \cap E_t)| \leq \left(\frac{C}{\rho_t L_K \sqrt{n} \sqrt{m_t - m_{t+1}}}\right)^{m_t - m_{t+1}}.$$

Iterating the previous estimates we obtain

$$
\begin{aligned}
|K_1^\circ| &\leqslant |K_1^\circ \cap E_{s_0}| \prod_{1\leqslant t\leqslant s_0} |P_{E_{t-1}\cap E_t^\perp}(K_1^\circ \cap E_{t-1})| \\
&\leqslant |K_1^\circ \cap E_{s_0}| \prod_{1\leqslant t\leqslant s_0} \left(\frac{C}{\rho_{t-1}\sqrt{n}L_K\sqrt{m_{t-1}-m_t}}\right)^{m_{t-1}-m_t} \\
&\leqslant |K_1^\circ \cap E_{s_0}| \left(\frac{c}{nL_K}\right)^{n-m_{s_0}} \prod_{t<s_0}\left(\frac{1}{\rho_t}\right)^{n(\log \rho_t^{-1})^{-2}} \prod_{t<s_0}\left(\frac{n}{m_t-m_{t+1}}\right)^{\frac{m_t-m_{t+1}}{2}} \\
&\leqslant |K_1^\circ \cap E_{s_0}| C^n \left(\frac{1}{nL_K}\right)^{n-m_{s_0}},
\end{aligned}
$$

where we use (3.4.8) and we also use the fact that, everytime we construct a new subspace E_{t+1}, $t<s_0$, with $\dim E_{t+1} = m_{t+1} \geqslant (1-(\log \rho_t^{-1})^{-2})m_t = \dim E_t$, we can choose it so that

$$m_t - m_{t+1} \simeq \left(\log \frac{1}{\rho_t}\right)^{-2} m_t \simeq \left(\log \frac{1}{\rho_t}\right)^{-2} n.$$

On the other hand, by the reverse Santaló inequality we have

$$|K_1^\circ| \geqslant \left(\frac{c_2}{n}\right)^n |K_1|^{-1} \geqslant \left(\frac{c_2}{n}\right)^n,$$

since $|K_1| \leqslant 1$. Combining the above we get

$$(3.4.9) \qquad |K_1^\circ \cap E_{s_0}| \geqslant c^n n^{-m_{s_0}} L_K^{n-m_{s_0}}.$$

Now, we recall the argument of Lemma 3.4.3 and observe that for the subspace E_{s_0} we have, by Sudakov's inequality and (3.4.6), that

$$
\begin{aligned}
|\sqrt{m}(u(K_1 \cap E_{s_0}))^\circ| &\leqslant |B_{E_{s_0}}| N(\sqrt{m}(u(K_1 \cap E_{s_0}))^\circ, B_{E_{s_0}}) \\
&\leqslant |B_{E_{s_0}}| \exp(cm_{s_0}^2 M^2(u(K_1 \cap E_{s_0}))) \leqslant |B_{E_{s_0}}| C^{m_{s_0}}.
\end{aligned}
$$

It follows that
(3.4.10)
$$
\begin{aligned}
|K_1^\circ \cap E_{s_0}| &\leqslant |P_{E_{s_0}}(K_1^\circ)| = |(K_1 \cap E_{s_0})^\circ| = |\det u| m^{m/2} |(u(K_1 \cap E_{s_0}))^\circ| \\
&\leqslant |B_{E_{s_0}}| C^{m_{s_0}} \prod_{1\leqslant j\leqslant m_{s_0}} \frac{\lambda_j}{\sqrt{m_{s_0}}}.
\end{aligned}
$$

Then, from (3.4.9) and (3.4.10) we obtain

$$L_K^{n-m_{s_0}} \leqslant C^{m_{s_0}} \prod_{j\leqslant m_{s_0}} \lambda_j,$$

thus

$$L_K^{\frac{n-m_{s_0}}{m_{s_0}}} \leqslant \frac{C}{m_{s_0}} \sum_{j\leqslant m_{s_0}} \lambda_j.$$

Going back to Lemma 3.4.3 and substituting this last estimate in (3.4.7) we obtain

$$L_K^{n/m_{s_0}} \leqslant c_3 b \log \frac{1}{\rho_{s_0}} \leqslant 2c_3 b \log(c_1' b) \leqslant c_3' b \log(1+b).$$

Taking into account the facts that $L_K \geqslant L_{B_2^n} \geqslant c$ for some absolute constant $c > 0$ and that $1 < n/m_{s_0} < 2$ we conclude the proof. $\square$

Note. In the proof above, if $s_0 = 0$, that is if $\rho_0 \simeq \frac{1}{\sqrt{n}} \geqslant (c_1 b/c)^{-2}$, then we combine (3.4.6) and (3.4.7) from the proof of Lemma 3.4.3 with Sudakov's inequality to get

$$\frac{c_2}{n} \leqslant |K_1^\circ|^{1/n} \leqslant C\, \omega_n^{1/n} \left(\prod_{1 \leqslant j \leqslant n} \frac{\lambda_j}{\sqrt{n}} \right)^{1/n}$$

$$\leqslant \frac{C'}{n} \frac{1}{n} \sum_{j=1}^n \lambda_j \leqslant \frac{c_3}{n} \frac{b \log(1/\rho_0)}{L_K} \leqslant \frac{c_3'}{n} \frac{b \log(1+b)}{L_K},$$

which gives the result again.

3.5. Further reading

3.5.1. Sylvester's problem

Let K be a convex body of volume 1 in $\mathbb{R}^n$ and let $x_1, \ldots, x_{n+1}$ be random points which are independently and uniformly distributed in K. Their convex hull $\mathrm{conv}\{x_1, \ldots, x_{n+1}\}$ is a random simplex contained in K. For every $p > 0$ we define

$$m_p(K) = \left(\int_K \cdots \int_K |\mathrm{conv}\{x_1, \ldots, x_{n+1}\}|^p dx_{n+1} \cdots dx_1 \right)^{1/p}.$$

If we drop the assumption that $|K| = 1$ then we normalize as follows:

$$m_p(K) = \left(\frac{1}{|K|^{n+p+1}} \int_K \cdots \int_K |\mathrm{conv}\{x_1, \ldots, x_{n+1}\}|^p dx_{n+1} \cdots dx_1 \right)^{1/p}.$$

Then, $m_p(K)$ is invariant under non-degenerate affine transformations: If $T \in GL(n)$ and $u \in \mathbb{R}^n$, then $m_p(K) = m_p(TK + u)$ for all $p > 0$. The quantity $m_1(K)$ is the expectation of the normalized volume of a random simplex inside K.

Sylvester's problem is the following question: *Describe the affine classes of convex bodies for which $m_p(K)$ is minimized or maximized.* It is known that, for every $p > 0$,

$$m_p(K) \geqslant m_p(B_2^n)$$

with equality if and only if K is an ellipsoid. In the opposite direction, the problem is open if $n \geqslant 3$ (see the "Notes and references" for a discussion).

CONJECTURE 3.5.1 (simplex conjecture). *For every convex body K in $\mathbb{R}^n$,*

$$m_1(K) \leqslant m_1(\Delta_n)$$

where Δ_n is a simplex in $\mathbb{R}^n$.

The simplex conjecture has been verified only in the case $n = 2$. As we will see, Sylvester's problem is related to the isotropic constant problem: we will prove that if the simplex conjecture is correct, then $L_K \leqslant C$ for every convex body K.

To see this, we define the variant of $m_p(K)$

$$S_p(K) = \left(\frac{1}{|K|^{n+p}} \int_K \cdots \int_K |\mathrm{conv}\{0, x_1, \ldots, x_n\}|^p dx_n \cdots dx_1 \right)^{1/p}.$$

PROPOSITION 3.5.2. *Let K be a centered convex body of volume 1 in $\mathbb{R}^n$. Then, for every $p \geqslant 1$ we have*

$$S_p(K) \leqslant m_p(K) \leqslant (n+1) S_p(K).$$

Proof. For every $x \in K$ we define
$$S_p(K;x) = \left(\int_K \cdots \int_K |\text{conv}\{x, x_1, \ldots, x_n\}|^p dx_n \cdots dx_1\right)^{1/p}.$$

We know that $|\text{conv}\{x, x_1, \ldots, x_n\}| = |\det(\tilde{x}, \tilde{x}_1, \ldots, \tilde{x}_n)|/n!$ where $\tilde{z} = (z, 1) \in \mathbb{R}^{n+1}$ for every $z \in \mathbb{R}^n$, and we know that this determinant is an affine function of x. It follows that $|\text{conv}\{x, x_1, \ldots, x_n\}|^p$ is a convex function on K. Integrating with respect to $x_1, \ldots, x_n$ we see that $S_p^p(K;x)$ is also convex. Since 0 is the barycenter of K and $|K| = 1$, we conclude that
$$S_p^p(K;0) \leqslant \int_K S_p^p(K;x) dx,$$
which gives
$$S_p^p(K) \leqslant m_p^p(K).$$
For the right hand side inequality we use again that if $x_1, \ldots, x_{n+1} \in K$, then
$$|\text{conv}\{x_1, \ldots, x_{n+1}\}| = \frac{1}{n!}|\det(\tilde{x}_1, \ldots, \tilde{x}_{n+1})|$$
where $\tilde{x}_j = (x_j, 1) \in \mathbb{R}^{n+1}$, and that if we expand the determinant with respect to the column $(1, \ldots, 1)$ and apply the triangle inequality we get
$$|\text{conv}\{x_1, \ldots, x_{n+1}\}| \leqslant \sum_{j=1}^{n+1} |\text{conv}\{0, x_i : i \neq j\}|.$$

It follows that
$$m_p(K) = \left(\int_K \cdots \int_K |\text{conv}\{x_1, \ldots, x_{n+1}\}|^p dx_{n+1} \cdots dx_1\right)^{1/p}$$
$$\leqslant \left(\int_K \cdots \int_K \left(\sum_{j=1}^{n+1} |\text{conv}\{0, x_i : i \neq j\}|\right)^p dx_{n+1} \cdots dx_1\right)^{1/p}$$
$$\leqslant \sum_{j=1}^{n+1} \left(\int_K \cdots \int_K |\text{conv}\{0, x_i : i \neq j\}|^p dx_{n+1} \cdots dx_1\right)^{1/p}$$
$$= (n+1)S_p(K),$$
as claimed. $\square$

REMARK 3.5.3. The function $f_i : K \to \mathbb{R}$ defined by $x_i \mapsto |\det(x_1, \ldots, x_n)|$ for fixed x_j in K, $j \neq i$, is a seminorm, as is the function $g_i : K \to \mathbb{R}$ defined by
$$x_i \mapsto \int_K \cdots \int_K |\det(x_1, \ldots, x_n)| dx_{i+1} \cdots dx_n$$
for fixed x_j in K when $j < i$. By consecutive applications of Fubini's theorem and of Theorem 3.2.7, we obtain the following proposition.

PROPOSITION 3.5.4. *Let K be a convex body of volume 1 in $\mathbb{R}^n$. Then,*
$$S_2(K) \leqslant c^n S_1(K),$$
where $c > 0$ is an absolute constant. $\square$

Fix an orthonormal basis in $\mathbb{R}^n$ and write $M(K)$ for the matrix of inertia
$$[M(K)]_{ij} = \int_K x_i x_j dx$$
of K. The connection of $m_p(K)$ and $S_p(K)$ with the isotropic constant of K becomes clear by the next identity (known as *Blaschke's formula*).

PROPOSITION 3.5.5. *Let K be a centered convex body of volume 1 in $\mathbb{R}^n$. Then,*
$$S_2^2(K) = \frac{\det(M(K))}{n!}.$$

Proof. By definition,
$$S_2^2(K) = \int_K \cdots \int_K |\mathrm{conv}\{0, x_1, \ldots, x_n\}|^2 dx_n \cdots dx_1.$$
We write $x_i = (x_{ij}), j = 1, \ldots, n$. Then,
$$(n!)^2 S_2^2(K) = \int_K \cdots \int_K |\det(x_1, \ldots, x_n)|^2 dx_n \cdots dx_1,$$
and expanding the determinant we get
$$(n!)^2 S_2^2(K) = \int_K \cdots \int_K \left(\sum_\sigma \epsilon_\sigma \prod_{i=1}^n x_{i,\sigma(i)} \right) \left(\sum_\tau \epsilon_\tau \prod_{i=1}^n x_{i,\tau(i)} \right) dx_n \cdots dx_1$$
$$= \int_K \cdots \int_K \left(\sum_{\sigma,\tau} \epsilon_\sigma \epsilon_\tau \prod_{i=1}^n x_{i,\sigma(i)} x_{i,\tau(i)} \right) dx_n \cdots dx_1$$
$$= \int_K \cdots \int_K \left(\sum_{\sigma,\varphi} \epsilon_\varphi \prod_{i=1}^n x_{i,\sigma(i)} x_{i,\varphi(\sigma(i))} \right) dx_n \cdots dx_1$$
$$= \sum_{\sigma,\varphi} \epsilon_\varphi \prod_{i=1}^n \left(\int_K x_i x_{\varphi(i)} dx \right)$$
$$= n! \det(M(K)),$$
which completes the proof. □

REMARK 3.5.6. Let K be a centered convex body of volume 1 in $\mathbb{R}^n$. In the proof of Proposition 3.1.7 we used the fact that if $T \in SL(n)$ then $M(T(K)) = TM(K)T^*$, and hence $|\det M(K)| = |\det M(T(K))|$. Since $S_2(K)$ is invariant under invertible linear transformations, Proposition 3.5.5 also shows that if $T \in SL(n)$ then
$$\det(M(TK)) = \det(M(K)).$$
If we choose T so that TK will be isotropic, then $M(TK) = L_K^2 I$, and hence $\det(M(K)) = L_K^{2n}$. Thus, we have proved the following

THEOREM 3.5.7. *Let K be a centered convex body in $\mathbb{R}^n$. Then,*
$$L_K^{2n} = n! \, S_2^2(K).$$

As an easy corollary, we get a second proof of the bound $L_K \leqslant c\sqrt{n}$ for every convex body K in $\mathbb{R}^n$. Indeed, we may assume that K is in isotropic position; noting that $S_2(K)$ is obviously bounded by 1, we then get
$$L_K \leqslant \sqrt[2n]{n!} \leqslant c\sqrt{n}$$
for some absolute constant $c > 0$.

COROLLARY 3.5.8. *If the simplex conjecture is correct, then $L_K \leqslant C$ for every convex body K in $\mathbb{R}^n$.*

Proof. Consider the simplex
$$\Delta_n = \left\{ x \in \mathbb{R}^n : -\frac{1}{n+1} \leqslant x_i \leqslant \frac{n}{n+1}, \ \sum_{i=1}^n x_i \leqslant \frac{1}{n+1} \right\}.$$

Then, $\Delta'_n = (n!)^{1/n}\Delta_n$ has volume 1 and barycenter at the origin. A simple computation shows that
$$\int_{\Delta'_n} x_i^2 dx < \frac{(n!)^{1+\frac{2}{n}}}{(n+2)!},$$
and since $M(K)$ is symmetric and positive definite, Hadamard's inequality gives
$$S_2^2(\Delta'_n) = \frac{\det(M(\Delta'_n))}{n!} \leqslant \frac{1}{n!}\left(\frac{(n!)^{1+\frac{2}{n}}}{(n+2)!}\right)^n \leqslant \frac{1}{n!}.$$

Now, if K is an isotropic convex body in $\mathbb{R}^n$ we have assumed that $m_1(K) \leqslant m_1(\Delta'_n)$, and combining Propositions 3.5.2, 3.5.5 and Theorem 3.5.7 we obtain
$$\begin{aligned} L_K^n &= \sqrt{n!}\, S_2(K) \leqslant \sqrt{n!}\, c^n S_1(K) \\ &\leqslant \sqrt{n!}\, c^n m_1(K) \leqslant \sqrt{n!}\, c^n m_1(\Delta'_n) \\ &\leqslant \sqrt{n!}\, c^n (n+1) S_1(\Delta'_n) \leqslant \sqrt{n!}\, c^n(n+1) S_2(\Delta'_n) \\ &\leqslant (n+1)c^n. \end{aligned}$$

It follows that $L_K \leqslant 2c$. $\square$

3.5.2. The Busemann-Petty problem

Let K be a convex body in $\mathbb{R}^n$. Assume that $0 \in \operatorname{int}(K)$. For every $\theta \in S^{n-1}$ consider the section $K \cap \theta^\perp$ of K, and the normalized version of $S_1(K \cap \theta^\perp)$ which is defined by
$$S_1(K \cap \theta^\perp) = \frac{1}{|K \cap \theta^\perp|^n} \int_{K \cap \theta^\perp} \cdots \int_{K \cap \theta^\perp} |\operatorname{conv}\{0, x_1, \ldots, x_{n-1}\}| dx_{n-1} \cdots dx_1.$$

Busemann proved a formula which connects the volume of K with the areas of the $(n-1)$-dimensional sections $K \cap \theta^\perp$, $\theta \in S^{n-1}$.

THEOREM 3.5.9. *If K is a convex body in $\mathbb{R}^n$ with $0 \in \operatorname{int}(K)$, then*
$$|K|^{n-1} = \frac{n!\,\omega_n}{2}\int_{S^{n-1}} |K \cap \theta^\perp|^n S_1(K \cap \theta^\perp) d\sigma(\theta).$$

Sketch of the proof. Consider $\mathbb{R}^{n(n-1)} = \mathbb{R}^n \times \cdots \times \mathbb{R}^n$, and set $\tilde{K} = K \times \cdots \times K \subset \mathbb{R}^{n(n-1)}$. We will write the coordinates of any $x_i \in \mathbb{R}^n$ in the form $x_i = (x_i^1, \ldots, x_i^n)$. Then,
$$|K|^{n-1} = \int_K \cdots \int_K dx_{n-1} \cdots dx_1 = \int_{\tilde{K}} dx_{n-1}^1 \cdots dx_{n-1}^n \cdots dx_1^1 \cdots dx_1^n.$$

Given $x_1, \ldots, x_{n-1}$, let $a_1, \ldots, a_{n-1}$ be the solution of the linear system
$$x_i^n = \sum_{j=1}^{n-1} a_j x_i^j, \qquad i = 1, \ldots, n-1,$$

and consider the new coordinates $x_1^1, \ldots, x_1^{n-1}, a_1, \ldots, x_{n-1}^1, \ldots, x_{n-1}^{n-1}, a_{n-1}$ on $\mathbb{R}^{n(n-1)}$. Note that the Jacobian $J = \det[(x_i^j)_{i,j=1}^{n-1}]$ of this transformation does not vanish, with an exception of a set of measure 0.

If $b = (1 + \sum_{j=1}^{n-1} a_j^2)^{-1/2}$ then the unit vector θ normal to the hyperplane $x^n = a_1 x^1 + \cdots + a_{n-1} x^{n-1}$ is described either by
$$\theta^j = ba_j \ (1 \leqslant j \leqslant n-1) \quad \text{and} \quad \theta^n = -b,$$
or by
$$\theta^j = -ba_j \ (1 \leqslant j \leqslant n-1) \quad \text{and} \quad \theta^n = b.$$

If ϕ is the angle formed by θ and the x^n-axis, then $b = |\cos\phi|$, and hence, $d\theta = \frac{1}{b}d\theta^1 \cdots d\theta^{n-1}$. Again, we may ignore those hyperplanes that are parallel to the x^n-axis. We can then check that
$$\det[(\partial\theta^j/\partial a_i)_{i,j=1}^{n-1}] = b^{n+1},$$
which gives
$$d\theta = \frac{1}{b}d\theta^1 \cdots d\theta^{n-1} = b^n da_1 \cdots da_{n-1} = |\cos^n\phi| da_1 \cdots da_{n-1}.$$
Note also that for the hyperplane $x_j^n = a_1 x_j^1 + \cdots + a_{n-1} x_j^{n-1}$ we have
$$dx_j = \frac{1}{|\cos\phi|} dx_j^1 \cdots dx_j^{n-1}.$$
Taking all the above into account, we write
$$|K|^{n-1} = \int_{J^{-1}\tilde{K}} |J| dx_1^1 \cdots dx_1^{n-1} \cdots dx_{n-1}^1 \cdots dx_{n-1}^{n-1} da_1 \cdots da_{n-1}$$
$$= \frac{n\omega_n}{2} \int_{S^{n-1}} \int_{K\cap\theta^\perp} \cdots \int_{K\cap\theta^\perp} |J/\cos\phi| dx_1 \cdots dx_{n-1} d\sigma(\theta),$$
where $x_1, \ldots, x_{n-1}$ are considered as points of $\theta^\perp$, and where the last integral is divided by 2 because each such hyperplane, $\theta^\perp$, is counted twice, as $\theta^\perp$ and as $(-\theta)^\perp$. Observe that the projection of the simplex $\mathrm{conv}\{0, x_1, \ldots, x_{n-1}\}$ onto $x^n = 0$ has volume $|J|/(n-1)!$. Since all these points belong to $\theta^\perp$, we get
$$|\mathrm{conv}\{0, x_1, \ldots, x_{n-1}\}||\cos\phi| = \frac{|J|}{(n-1)!}.$$
It follows that
$$|K|^{n-1} = \frac{n!\,\omega_n}{2} \int_{S^{n-1}} \int_{K\cap\theta^\perp} \cdots \int_{K\cap\theta^\perp} |\mathrm{conv}\{0, x_1, \ldots, x_{n-1}\}| dx_{n-1} \cdots dx_1 d\sigma(\theta)$$
$$= \frac{n!\,\omega_n}{2} \int_{S^{n-1}} |K\cap\theta^\perp|^n S_1(K\cap\theta^\perp) d\sigma(\theta),$$
which was the assertion. □

If we assume that K is symmetric, then the results of Section 3.5.1 show that
$$\bigl(S_1(K\cap\theta^\perp)\bigr)^{\frac{1}{n-1}} \simeq \bigl(S_2(K\cap\theta^\perp)\bigr)^{\frac{1}{n-1}} \simeq \frac{L_{K\cap\theta^\perp}}{\sqrt{n}}.$$
Therefore, Theorem 3.5.9 has the following immediate consequences.

COROLLARY 3.5.10. *Let K be a symmetric convex body in $\mathbb{R}^n$. Then,*
$$|K|^{\frac{n-1}{n}} \simeq \left(\int_{S^{n-1}} L_{K\cap\theta^\perp}^{n-1} |K\cap\theta^\perp|^n d\sigma(\theta)\right)^{1/n}.$$

Given that the isotropic constant of any convex body is bounded from below by an absolute positive constant, we get:

COROLLARY 3.5.11. *Let K be a symmetric convex body in $\mathbb{R}^n$. Then,*
$$|K|^{\frac{n-1}{n}} \geqslant c_1 \left(\int_{S^{n-1}} |K\cap\theta^\perp|^n d\sigma(\theta)\right)^{1/n},$$
where $c_1 > 0$ is an absolute constant.

These last results bring us to the Busemann-Petty problem which was originally formulated as follows: *Assume that K_1 and K_2 are symmetric convex bodies in $\mathbb{R}^n$ and satisfy*
$$|K_1 \cap \theta^\perp| \leqslant |K_2 \cap \theta^\perp|$$
for all $\theta \in S^{n-1}$. Does it follow that $|K_1| \leqslant |K_2|$?

The answer is positive if $n \leqslant 4$ and negative for all higher dimensions (see the "Notes and references" for a discussion). What remains open is the asymptotic version of the problem.

CONJECTURE 3.5.12 (Busemann-Petty problem). *There exists an absolute constant $c > 0$ such that if K_1 and K_2 are symmetric convex bodies in $\mathbb{R}^n$ and $|K_1 \cap \theta^\perp| \leqslant |K_2 \cap \theta^\perp|$ for all $\theta \in S^{n-1}$ then $|K_1| \leqslant c|K_2|$.*

The results of this paragraph show that Conjecture 3.5.12 is equivalent to the boundedness of the isotropic constant. Recall that we can restrict ourselves to the class of symmetric convex bodies. Let us first assume that there is a constant $C > 0$ such that $L_W \leqslant C$ for every symmetric convex body W. If K_1 and K_2 satisfy
$$|K_1 \cap \theta^\perp| \leqslant |K_2 \cap \theta^\perp|$$
for all $\theta \in S^{n-1}$, then corollaries 3.5.10 and 3.5.11 show that
$$|K_1|^{n-1} \leqslant c^n C^{n-1} \int_{S^{n-1}} |K_1 \cap \theta^\perp|^n d\sigma(\theta)$$
$$\leqslant c^n C^{n-1} \int_{S^{n-1}} |K_2 \cap \theta^\perp|^n \sigma(d\theta)$$
$$\leqslant (c/c_1)^n C^{n-1} |K_2|^{n-1},$$
which gives
$$|K_1| \leqslant c_3 |K_2|$$
for an absolute constant $c_3 > 0$. Conversely, let us assume that Conjecture 3.5.12 is correct and let K be an isotropic symmetric convex body in $\mathbb{R}^n$. Choose $\theta_0 \in S^{n-1}$ so that
$$|K \cap \theta_0^\perp| = \max_{\theta \in S^{n-1}} |K \cap \theta^\perp|$$
and $r > 0$ so that $\omega_{n-1} r^{n-1} = |K \cap \theta_0^\perp|$. Then,
$$|K \cap \theta^\perp| \leqslant \omega_{n-1} r^{n-1} = |(rB_2^n) \cap \theta^\perp|$$
for all $\theta \in S^{n-1}$, therefore
$$|K|^{n-1} \leqslant c^{n-1} |rB_2^n|^{n-1} = \frac{c^{n-1} \omega_n^{n-1}}{\omega_{n-1}^n} |K \cap \theta_0^\perp|^n \leqslant c_1^n |K \cap \theta_0^\perp|^n,$$
for some absolute constant $c_1 > 0$. Since $|K| = 1$, we see that
$$|K \cap \theta_0^\perp| \geqslant 1/c_1.$$
Since K is isotropic, we have $|K \cap \theta^\perp| \simeq 1/L_K$ for every $\theta \in S^{n-1}$. It follows that $L_K \leqslant C$ for some absolute constant $C > 0$.

3.5.3. Duality and the functional $\phi(K)$

Let K be a symmetric convex body in $\mathbb{R}^n$. We define
$$\phi(K) = \frac{1}{|K||K^\circ|} \int_K \int_{K^\circ} \langle x, y \rangle^2 dy dx.$$

Observe that $\phi(TK) = \phi(K)$ for every $T \in GL(n)$ and that $\phi(K)$ is trivially bounded by 1 (since $|\langle x, y \rangle| \leqslant 1$ for all $x \in K$, $y \in K^\circ$).

PROPOSITION 3.5.13. *Let K be a symmetric convex body in $\mathbb{R}^n$. Then,*
$$L_K^2 L_{K^\circ}^2 \leqslant cn\phi(K).$$
In particular, $L_K L_{K^\circ} \leqslant c_1 \sqrt{n}$.

Proof. We may assume that K° is isotropic. Then,
$$\frac{1}{|K|^{1+2/n}|K^\circ|^{1+2/n}}\int_K\int_{K^\circ}\langle x,y\rangle^2 dydx = L_{K^\circ}^2\frac{1}{|K|^{1+2/n}}\int_K\|x\|_2^2 dx \geqslant nL_K^2 L_{K^\circ}^2.$$
Therefore,
$$nL_K^2 L_{K^\circ}^2 \leqslant \bigl(|K|\,|K^\circ|\bigr)^{-2/n}\phi(K),$$
and the result follows from the reverse Santaló inequality:
$$\bigl(|K|\,|K^\circ|\bigr)^{1/n} \geqslant c_2\omega_n^{2/n} \geqslant c_3/n,$$
where $c_3 > 0$ is an absolute constant. $\square$

The problem of obtaining a non-trivial upper bound for $\phi(K)$ is open. Alonso-Gutiérrez proved that
$$\phi(B_p^n) \leqslant \phi(B_2^n)$$
for all $1 \leqslant p \leqslant \infty$. It is conceivable that $\phi(K) \leqslant \phi(B_2^n)$ for every symmetric convex body K in $\mathbb{R}^n$. Since $\phi(B_2^n) \simeq 1/n$, this would imply an affirmative answer to the hyperplane conjecture (see the "Notes and references" for a discussion).

3.6. Notes and references

Hyperplane conjecture

The hyperplane conjecture was stated explicitly as a problem in the article of V. Milman and Pajor [384] and in the PhD Thesis of Ball [30]. The question appears for the first time in the work of Bourgain [99] on high-dimensional maximal functions associated with arbitrary convex bodies. Bourgain was interested in bounds for the L_p-norm of the maximal function
$$M_K f(x) = \sup\left\{\frac{1}{|tK|}\int_{tK}|f(x+y)|\,dy \mid t > 0\right\}$$
of $f \in L_{\mathrm{loc}}^1(\mathbb{R}^n)$, where K is a symmetric convex body in $\mathbb{R}^n$. Let $C_p(K)$ denote the best constant such that
$$\|M_K f\|_p \leqslant C_p(K)\|f\|_p$$
and $C_{1,1}(K)$ the best constant so that the weak type inequality
$$\|M_K f\|_{1,\infty} \leqslant C_{1,1}(K)\|f\|_1$$
is satisfied. Stein proved in [475] that if $K = B_2^n$ is the Euclidean unit ball then $C_p(B_2^n)$ is bounded independently of the dimension for all $p > 1$. Bourgain showed that there exists an absolute constant $C > 0$ (independent of n and K) such that
$$\|M_K\|_{L_2(\mathbb{R}^n)\to L_2(\mathbb{R}^n)} \leqslant C.$$
By the definition of M_K it is clear that in order to obtain a uniform bound on $\|M_K\|_{2\to 2}$ one can start with a suitable position $T(K)$ (where $T \in GL(n)$) of K. Bourgain used the isotropic position; the property that played an important role in his argument was that when K is isotropic then $L_K|K\cap\theta^\perp| \simeq 1$ for all $\theta \in S^{n-1}$. Bourgain mentions the fact that $L_K \geqslant c$ and asks whether a reverse inequality holds true.

The result for $\|M_K\|_{2\to 2}$ was generalized to all $p > 3/2$ by Bourgain [100] and, independently, by Carbery [132]. Afterwards, Müller [395] obtained dimension free maximal bounds for all $p > 1$, which however depend on L_K and on the maximal volume of hyperplane projections of K. In the case of the cube, Bourgain [105] showed that for every $p > 1$ there exists a constant $C_p > 0$ such that $C_p(B_\infty^n) \leqslant C_p$ for all n. Regarding the weak inequality, Stein and Strömberg [477] showed that $C_{1,1}(B_2^n) \leqslant cn\log n$ (see also [397]). In the case of the cube, Aldaz [4] and Aubrun [26] showed that there is no dimension free weak $(1,1)$ maximal inequality; one has $C_{1,1}(B_\infty^n) \geqslant c(\epsilon)(\log n)^{1-\epsilon}$ for all $\epsilon > 0$.

Geometry of isotropic convex bodies

Our presentation in Section 3.1 essentially follows V. Milman and Pajor [**384**].

The argument of Kannan, Lovász and Simonovits for the circumradius of an isotropic convex body (Theorem 3.2.1) is from [**266**].

Theorem 3.2.2 was proved by Milman and Pajor [**385**] in the centrally symmetric case. The argument that we present in the text for the not necessarily symmetric case is due to Hartzoulaki [**249**].

Theorem 3.2.4 stating that the hyperplane conjecture is equivalent to the fact that every pair of isotropic convex bodies K and T satisfies the reverse Brunn-Minkowski inequality with a uniformly bounded constant was observed by Bourgain, Klartag and Milman in [**107**].

Theorem 3.2.15 was proved by Alesker in [**8**]; the fact that the Euclidean norm on isotropic convex bodies satisfies a ψ_2-estimate is the starting point for Paouris' deviation estimate that will be presented in Chapter 5.

Bourgain's upper bound for the isotropic constant

The first proof of Bourgain's bound $L_K = O(\sqrt[4]{n} \log n)$ that we present in Section 3.3 is closer to Bourgain's original argument in [**101**]. We use his idea of "reduction to small diameter" followed by a direct use of Talagrand's comparison theorem. The second one, which is due to Dar [**152**], is based on the ψ_1-behavior of linear functionals on convex bodies, which was the main ingredient in Bourgain's approach too, but it is more elementary, since it involves simpler entropy considerations. The original argument of Bourgain was about symmetric convex bodies, but one could obtain the same bound with a slight modification of Dar's argument; this was observed in [**407**].

Theorem 3.4.1 which provides the bound $L_K \leqslant cb \log(1+b)$ for the isotropic constant of a ψ_2-body with constant b is due to Bourgain [**104**]. In Chapter 7 we will see that a linear in b upper bound holds true (this is a result of Klartag and E. Milman).

Sylvester's problem

For any convex body K in $\mathbb{R}^n$ and any $N \geqslant n+1$ and $p > 0$ we define

$$m_p(K,N) = \left(\frac{1}{|K|^{N+p+1}} \int_K \cdots \int_K |\operatorname{conv}\{x_1,\ldots,x_N\}|^p dx_{n+1} \cdots dx_1 \right)^{1/p}.$$

Note that this is an affinely invariant quantity and that $m_p(K) = m_p(K, n+1)$; this was the quantity that we discussed in Section 3.5.1. Sylvester's problem is the question to describe the affine classes of convex bodies for which $m_p(K, N)$ is minimized or maximized. Blaschke settled the problem in the case $n = 2$ and $N = 3$: the minimum is attained when K is a disc and the maximum when K is a triangle (see [**70**] and [**71**]). It is known that, for every $p > 0$ and any $N \geqslant n+1$,

$$m_p(K) \geqslant m_p(B_2^n)$$

with equality if and only if K is an ellipsoid (see e.g. Groemer [**225**] and [**226**] for the case $p \geqslant 1$ and Giannopoulos-Tsolomitis [**217**] for an extension to all $p > 0$).

The problem of the maximum is completely open in dimensions $n \geqslant 2$. In the planar case it was proved by Dalla and Larman in [**151**] that if T is a triangle then $m_p(K,N) \leqslant m_p(T,N)$ for all K and for all $p \geqslant 1$ and $N \geqslant 3$. Giannopoulos showed in [**198**] that in the case $p = 1$ one can have equality only if K is also a triangle. Saroglou showed in [**452**] that the same is true for all $p \geqslant 1$.

The Busemann-Petty problem

Theorem 3.5.9 appears in [**127**] and is known as Busemann's intesection formula. The Busemann-Petty problem was posed in [**128**], first in a list of ten problems concerning central sections of symmetric convex bodies in $\mathbb{R}^n$ and coming from questions in Minkowski geometry. It was originally formulated as follows: *Assume that K_1 and K_2 are symmetric convex bodies in $\mathbb{R}^n$ and satisfy*

$$|K_1 \cap \theta^\perp| \leqslant |K_2 \cap \theta^\perp|$$

for all $\theta \in S^{n-1}$. Does it follow that $|K_1| \leqslant |K_2|$? Using Theorem 3.5.9, Busemann was able to give a positive answer in the case where K_1 is an ellipsoid. He returned to the problem in [**129**]; he concentrated on the three-dimensional case and he gave examples showing that the assumptions of convexity and symmetry are both essential. Later, Hadwiger [**245**] and Giertz [**218**] independently gave a positive answer to the problem in a special case: when K_1 and K_2 are coaxial convex symmetric bodies of revolution in $\mathbb{R}^3$. The first breakthrough on the Busemann-Petty problem came in 1975. Larman and Rogers [**305**] chose $K_2 = B_2^n$ and proved that if $n \geqslant 12$ then there exist symmetric convex bodies K_1 which are arbitrarily small perturbations of B_2^n such that the pair (K_1, B_2^n) provides a negative answer to the problem. The proof is probabilistic in nature and is described by Pichorides [**423**] as follows: "we put $2N$ points $u_1, \ldots, u_N, -u_1, \ldots, -u_N$ on the surface of B_2^n and consider the $2N$ spherical caps $C(u_j), C(-u_j), j = 1, \ldots, N$, of angular radius ε centered at these points. A suitable dilation of the set B_2^n minus the union of these caps gives a convex body K_1 symmetric with respect to the origin such that $|K_1| = |B_2^n|$. The idea is to show that, if $n > 11$, some choice of $u_1, \ldots, u_N$ guarantees that $|K_1 \cap \theta^\perp| < |B_2^n \cap \theta^\perp|$ for all $\theta \in S^{n-1}$. The necessary computations to make this idea work are greatly simplified (it is probably more accurate to say "they become doable") if these caps do not overlap".

Ball [**32**] proved that if $Q_n = [-1/2, 1/2]^n$ is the cube of volume 1 in $\mathbb{R}^n$ then $|Q_n \cap \theta^\perp| \leqslant \sqrt{2}$ for all $\theta \in S^{n-1}$. We will discuss this result in Chapter 4. This bound leads to a concrete counterexample to the Busemann-Petty problem: Ball observed in [**33**] that, if $n \geqslant 10, K_1 = Q_n$ and K_2 is the ball of volume 1, then

$$|Q_n \cap \theta^\perp| \leqslant \sqrt{2} < \omega_{n-1}\omega_n^{-\frac{n-1}{n}} = |K_2 \cap \theta^\perp|$$

for all $\theta \in S^{n-1}$. Ball's counterexample is essentially different from the one of Larman and Rogers: the cube is far from being a perturbation of a ball. After Ball's example, Giannopoulos [**197**] and Bourgain [**102**] constructed counterexamples for $n \geqslant 7$, and Papadimitrakis [**420**] obtained counterexamples in dimensions $n = 6$ and $n = 5$.

The key notion that led to the final solution to the Busemann-Petty problem is Lutwak's definition of an *intersection body*. The intersection body IK of a symmetric convex body K in $\mathbb{R}^n$ is the symmetric convex body whose radial function is defined by

$$\rho_{IK}(\theta) = |K \cap \theta^\perp|, \quad \theta \in S^{n-1}.$$

Lutwak observed that intersection bodies are closely connected with the Busemann-Petty problem. Using simple facts from the theory of *dual mixed volumes* one can see that if K_1 is an intersection body then the answer to the Busemann-Petty problem is affirmative for K_1 and any symmetric convex body K_2. However, if K_2 is a symmetric convex body that is not an intersection body, then K_1 and a perturbation K_2 of K_1 provide a counterexample to the question (see [**343**], [**344**] and [**345**]). Therefore, the Busemann-Petty problem has an affirmative answer in $\mathbb{R}^n$ if and only if every symmetric convex body in $\mathbb{R}^n$ is an intersection body. Using this reduction of the problem, Gardner [**187**], [**188**] and Zhang [**506**] gave a negative answer to the problem for $n \geqslant 5$ by providing examples of non-intersection bodies in $\mathbb{R}^5$. Around the same time, Gardner [**189**] proved that every symmetric convex body in $\mathbb{R}^3$ is an intersection body, and hence the Busemann-Petty problem has an affirmative answer in dimension $n = 3$. For a few years it was believed that the problem has a negative answer in the remaining case $n = 4$. Through the work of Koldobsky who established a Fourier analytic characterization of intersection bodies in [**291**] it was understood that the case $n = 4$ was still open. Zhang [**508**], [**507**] proved that the answer in $\mathbb{R}^4$ is affirmative; around the same time, using the Fourier analytic approach, Gardner, Koldobsky and Schlumprecht [**192**] gave a unified solution to the problem in all dimensions (afterwards, an alternative proof was also given in [**57**]).

Thus, the answer to the Busemann-Petty problem is positive if $n \leqslant 4$ and negative for all higher dimensions (see the books of Gardner [**191**] and Koldobsky [**292**] for a complete discussion of the problem and its history).

Duality and the functional $\phi(K)$

The observation of Proposition 3.5.13 appears in [**199**]. The problem of obtaining a non-trivial upper bound for $\phi(K)$ is open even in the unconditional case. Alonso-Gutiérrez proved in [**10**] that
$$\phi(B_p^n) \leqslant \phi(B_2^n)$$
for all $1 \leqslant p \leqslant \infty$. It is conceivable that $\phi(K) \leqslant \phi(B_2^n)$ for every symmetric convex body K in $\mathbb{R}^n$. Since $\phi(B_2^n) \simeq 1/n$, this would imply an affirmative answer to the hyperplane conjecture (see Kuperberg [**302**] for a discussion).

Ball studied in [**33**] the quantity
$$\psi(K) = \int_K \int_{K^\circ} \langle x, y \rangle^2 dy\, dx.$$
He proved that if K is an unconditional convex body in $\mathbb{R}^n$ then $\psi(K) \leqslant \psi(B_2^n) = n\omega_n^2/(n+2)^2$ and conjectured that the same is true for every symmetric convex body. If true, this inequality would imply the Blaschke-Santaló inequality: in [**30**] it is shown that
$$\frac{n|K|^{1+\frac{2}{n}}|K^\circ|^{1+\frac{2}{n}}}{(n+2)^2 \omega_n^{4/n}} \leqslant \int_K \int_{K^\circ} \langle x, y \rangle^2 dy\, dx.$$

CHAPTER 4

Partial answers

In this chapter we discuss some partial affirmative answers to the hyperplane conjecture. In order to make this statement more precise, we say that a class $\mathcal{C}$ of symmetric convex bodies satisfies the hyperplane conjecture uniformly if there exists a positive constant C such that $L_K \leqslant C$ for all $K \in \mathcal{C}$. The hyperplane conjecture has been verified for several important classes of convex bodies. Most of the results that we present in the next sections were obtained soon after the problem was posed.

In Section 4.1 we look at the class of unconditional convex bodies; these are the symmetric convex bodies in $\mathbb{R}^n$ which have a position which is symmetric with respect to the standard coordinate subspaces. We present two arguments which provide a uniform bound for their isotropic constants. The first one is based on the Loomis-Whitney inequality, while the second one is due to Bobkov and Nazarov. The class of unconditional convex bodies will appear often in this book, mainly as a model for results or conjectures regarding the general cases. All the major questions about the distribution of volume on isotropic convex bodies have been studied for this class, and the positive results will appear in the next chapters. In fact, in some cases, observations about unconditional convex bodies were the starting point for studying the more challenging general questions in the broader setting of convex bodies or log-concave probability measures.

In Sections 4.2 and 4.3 we describe uniform bounds for the isotropic constants of some other classes of convex bodies: convex bodies whose polar bodies contain large affine cubes, the unit balls of 2-convex spaces with a given constant α, bodies with small diameter (in particular, the class of zonoids) and the unit balls of the Schatten classes.

In Section 4.4 we discuss geometric proofs of the best known estimates for the isotropic constants of polytopes with N vertices or polyhedra with N facets. The bounds are logarithmic in N, and hence they are logarithmic in the dimension n as long as the number of vertices or facets remains "small", i.e. polynomial in n.

In the last section of this chapter we deviate from the asymptotic point of view of this book and present some sharp estimates, due to Vaaler, Meyer, Pajor, Ball and Koldobsky, on the volume of k-dimensional sections of the ℓ_p^n-balls

$$B_p^n = \{x \in \mathbb{R}^n : |x_1|^p + \cdots + |x_n|^p \leqslant 1\}.$$

4.1. Unconditional convex bodies

In this section we study the case of symmetric convex bodies which generate a norm with unconditional basis. After a linear transformation, we may assume that the standard orthonormal basis $\{e_1, \ldots, e_n\}$ of $\mathbb{R}^n$ is an unconditional basis for $\|\cdot\|_K$.

That is, for every choice of real numbers $t_1, \ldots, t_n$ and every choice of signs $\varepsilon_i = \pm 1$,

$$\|\varepsilon_1 t_1 e_1 + \cdots + \varepsilon_n t_n e_n\|_K = \|t_1 e_1 + \cdots + t_n e_n\|_K.$$

Geometrically, this means that if $x = (x_1, \ldots, x_n) \in K$ then the whole rectangle $\prod_{i=1}^n [-|x_i|, |x_i|]$ is contained in K.

Note that the matrix of inertia of such a body is diagonal, therefore one can bring it to the isotropic position by a diagonal operator. This explains that for every unconditional convex body K in $\mathbb{R}^n$ there exists a linear image $\tilde{K}$ of K which has the following properties:

1. The volume of $\tilde{K}$ is equal to 1.
2. If $x = (x_1, \ldots, x_n) \in \tilde{K}$ then $\prod_{i=1}^n [-|x_i|, |x_i|] \subseteq \tilde{K}$.
3. For every $j = 1, \ldots, n$,

$$\int_{\tilde{K}} x_j^2 dx = L_K^2.$$

This last condition implies that $\tilde{K}$ is in isotropic position, because

$$\int_{\tilde{K}} x_i x_j dx = 0 \text{ for all } i \neq j$$

by Property 2.

In the rest of this section we assume that K has these three properties. It is not hard to prove that $L_K \leqslant C$ for some absolute constant $C > 0$. In order to see this, one may use the Loomis-Whitney inequality [337] (which even holds without the convexity assumption).

THEOREM 4.1.1 (Loomis-Whitney). *For every convex body K in $\mathbb{R}^n$,*

$$|K|^{n-1} \leqslant \prod_{j=1}^n |P_j(K)|,$$

where $P_j(K)$ is the orthogonal projection of K onto $e_j^\perp$.

Proof. We state and prove the functional form of the Loomis-Whitney inequality which is the following: Let $\pi_i(x) = (x_1, \ldots, x_{i-1}, x_{i+1}, \ldots, x_n)$ denote the orthogonal projection of x onto the hyperplane which is orthogonal to e_i. Then, for any $f_1, \ldots, f_n : \mathbb{R}^{n-1} \to \mathbb{R}^+$ with $f_j \in L^{n-1}(\mathbb{R}^{n-1})$ we have

(4.1.1) $$\int_{\mathbb{R}^n} \prod_{j=1}^n f_j(\pi_j(x)) \, dx \leqslant \prod_{j=1}^n \|f_j\|_{L^{n-1}(\mathbb{R}^{n-1})}.$$

Note that the geometric form of the Loomis-Whitney inequality can be obtained directly from its functional form if we set $f_j = \mathbf{1}_{P_j(K)}$, since for any $x \in K$ we have

$$\mathbf{1}_K(x) \leqslant \prod_{j=1}^n \mathbf{1}_{P_j(K)}(\pi_j(x)).$$

The proof of (4.1.1) is based on successive applications of Hölder's inequality. We describe the first two steps of the scheme of the proof. First, using Hölder's inequality, we write

$$\int_{\mathbb{R}^n} \prod_{j=1}^n f_j(\pi_j(x))\, dx$$

$$= \int_{\mathbb{R}^{n-1}} \left(\int_{\mathbb{R}} \prod_{j=1}^{n-1} f_j(\pi_j(x))\, dx_n \right) f_n(\pi_n(x))\, dx_{n-1} \cdots dx_1$$

$$\leqslant \int_{\mathbb{R}^{n-1}} \prod_{j=1}^{n-1} \left(\int_{\mathbb{R}} f_j^{n-1}(\pi_j(x))\, dx_n \right)^{\frac{1}{n-1}} f_n(\pi_n(x))\, dx_{n-1} \cdots dx_1.$$

Now we continue integrating with respect to x_{n-1}:

$$\int_{\mathbb{R}^{n-1}} \prod_{j=1}^{n-1} \left(\int_{\mathbb{R}} f_j^{n-1}(\pi_j(x))\, dx_n \right)^{\frac{1}{n-1}} f_n(\pi_n(x))\, dx_{n-1} \cdots dx_1$$

$$= \int_{\mathbb{R}^{n-2}} \left[\int_{\mathbb{R}} \prod_{j=1}^{n-2} \left(\int_{\mathbb{R}} [f_j(\pi_j(x))]^{n-1}\, dx_n \right)^{\frac{1}{n-1}} f_n(\pi_n(x))\, dx_{n-1} \right]$$

$$\times \left(\int_{\mathbb{R}} f_{n-1}^{n-1}(\pi_{n-1}(x))\, dx_n \right)^{\frac{1}{n-1}} dx_{n-2} \cdots dx_1$$

$$\leqslant \int_{\mathbb{R}^{n-2}} \left(\prod_{j=1}^{n-2} \int_{\mathbb{R}^2} f_j^{n-1}(\pi_j(x))\, dx_n dx_{n-1} \right)^{\frac{1}{n-1}}$$

$$\times \left(\int_{\mathbb{R}} f_n^{n-1}(\pi_n(x))\, dx_{n-1} \right)^{\frac{1}{n-1}} \left(\int_{\mathbb{R}} f_{n-1}^{n-1}(\pi_{n-1}(x))\, dx_n \right)^{\frac{1}{n-1}} dx_{n-2} \cdots dx_1,$$

where we have used Hölder's inequality once again. We proceed in the same way, integrating with respect to x_{n-2} and so on, and after n steps we obtain (4.1.1). □

THEOREM 4.1.2. *There exists an absolute constant $C > 0$ such that $L_K \leqslant C$ for every $n \geqslant 1$ and every isotropic unconditional convex body K in $\mathbb{R}^n$.*

Proof. Since K has Property 2, it is clear that

$$P_j(K) = K \cap e_j^\perp$$

for all $j = 1, \ldots, n$. From the Loomis-Whitney inequality we see that

$$|K \cap e_j^\perp| \geqslant 1$$

for some $j \leqslant n$. From Proposition 3.1.5,

$$L_K^2 = \int_K x_j^2\, dx \leqslant \frac{c}{|K \cap e_j^\perp|^2}$$

where $c > 0$ is an absolute constant. This proves that $L_K \leqslant C$, with $C = \sqrt{c}$. □

We shall give a second proof of Theorem 4.1.2, due to Bobkov and Nazarov [90]. It will be convenient to consider the normalized part

$$K^+ = 2K \cap \mathbb{R}_+^n$$

of K in $\mathbb{R}_+^n = [0,+\infty)^n$. In other words, if $x = (x_1,\ldots,x_n)$ is uniformly distributed in K, then $(2|x_1|,\ldots,2|x_n|)$ is uniformly distributed in K^+. It is easy to check that K^+ has the following three properties:

4. The volume of K^+ is equal to 1.
5. If $x = (x_1,\ldots,x_n) \in K^+$ and $0 \leqslant y_j \leqslant x_j$ for all $1 \leqslant j \leqslant n$, then $y = (y_1,\ldots,y_n) \in K^+$.
6. For every $j = 1,\ldots,n$,
$$\int_{K^+} x_j^2 dx = 4L_K^2.$$

THEOREM 4.1.3 (Bobkov-Nazarov). *Let K be an isotropic unconditional convex body in $\mathbb{R}^n$. Then, $L_K^2 \leqslant 1/2$.*

Proof. Property 5 shows that if $x = (x_1,\ldots,x_n) \in K^+$, then $\prod_{j=1}^n [0,x_j] \subseteq K^+$. It follows that
$$\prod_{j=1}^n x_j \leqslant |K^+| = 1$$
for every $x \in K^+$. Define
$$V = \left\{ x \in \mathbb{R}_+^n : \prod_{j=1}^n x_j \geqslant 1 \right\}.$$

Then, V is convex and the sets K^+ and V have disjoint interiors. So, there exists a hyperplane H of the form
$$\lambda_1 x_1 + \cdots + \lambda_n x_n = \alpha$$
with $\lambda_j > 0$, which touches V at a point $y = (y_1,\ldots,y_n)$ and separates it from K^+. Since H touches V we can choose the λ_j's so that $\alpha = n$ and $\prod_{j=1}^n \lambda_j = 1$ (we have $\prod_{j=1}^n y_j = 1$ and we can choose $\lambda_j = \prod_{i \neq j} y_i$).

Since H separates V from K^+, we have
$$K^+ \subseteq \{ x \in \mathbb{R}_+^n : \lambda_1 x_1 + \cdots + \lambda_n x_n \leqslant n \}.$$
The arithmetic-geometric means inequality gives
$$\prod_{j=1}^n \left(\int_{K^+} x_j dx \right)^{\lambda_j/n} \leqslant \int_{K^+} \frac{\lambda_1 x_1 + \cdots + \lambda_n x_n}{n} dx \leqslant 1.$$

Now, applying the standard reverse Hölder inequality for $f(x) = x_j$, we get
$$4L_K^2 = \int_{K^+} x_j^2 dx \leqslant 2 \left(\int_{K^+} x_j dx \right)^2$$
for all $j = 1,\ldots,n$ (for the exact constant 2 see [**90**]). If we set $d = \frac{\lambda_1 + \cdots + \lambda_n}{n}$, then
$$(2L_K^2)^d = \prod_{j=1}^n (2L_K^2)^{\lambda_j/n} \leqslant \prod_{j=1}^n \left(\int_{K^+} x_j dx \right)^{2\lambda_j/n}$$
$$= \left[\prod_{j=1}^n \left(\int_{K^+} x_j dx \right)^{\lambda_j/n} \right]^2 \leqslant 1,$$
which shows that $L_K^2 \leqslant 1/2$. □

Combining Theorem 4.1.3 with Proposition 3.1.4 we have:

COROLLARY 4.1.4. *Let K be an isotropic unconditional convex body in $\mathbb{R}^n$. Then,*
$$|K \cap \theta^\perp| \geqslant \frac{1}{\sqrt{6}}$$
for every $\theta \in S^{n-1}$.

Proof. Let $\theta \in S^{n-1}$. Proposition 3.1.4 shows that
$$L_K = \left(\int_K \langle x, \theta \rangle^2 dx \right)^{1/2} \geqslant \frac{1}{2\sqrt{3}} \frac{1}{|K \cap \theta^\perp|}.$$
Note that if K is symmetric, then the factor e is not needed in the statement of Proposition 3.1.4. $\square$

We close with some additional results of Bobkov and Nazarov on the geometry of isotropic unconditional convex bodies. The first one is a distributional inequality which will be very useful in subsequent chapters.

THEOREM 4.1.5. *Let K be an isotropic unconditional convex body in $\mathbb{R}^n$. Then,*
$$|\{x \in K^+ : x_1 \geqslant \alpha_1, \ldots, x_n \geqslant \alpha_n\}| \leqslant \left(1 - \frac{\alpha_1 + \cdots + \alpha_n}{\sqrt{6}n}\right)^n,$$
for all $(\alpha_1, \ldots, \alpha_n) \in K^+$.

Proof. We define a function $u : K^+ \to [0, \infty)$ by
$$u(\alpha_1, \ldots, \alpha_n) = |\{x \in K^+ : x_1 \geqslant \alpha_1, \ldots, x_n \geqslant \alpha_n\}|.$$
The Brunn-Minkowski inequality shows that the function $h = u^{\frac{1}{n}}$ is concave on K^+. Observe that $u(0) = 1$ and
$$(4.1.2) \qquad \frac{\partial u}{\partial \alpha_j}(0) = -|K \cap e_j^\perp| \leqslant -\frac{1}{\sqrt{6}},$$
where the last inequality comes from Corollary 4.1.4. Let $\alpha \in K^+$ and consider the function $h_\alpha : [0, 1] \to \mathbb{R}$ defined by $h_\alpha(t) = h(\alpha t)$. Note that
$$h_\alpha'(0) = \sum_{j=1}^n \alpha_j \frac{\partial h}{\partial \alpha_j}(0) = \sum_{i=1}^n \alpha_i \cdot \frac{1}{n} \frac{\partial u}{\partial \alpha_j}(0) \leqslant -\frac{\alpha_1 + \cdots + \alpha_n}{\sqrt{6}n}$$
by (4.1.2). Since h is concave, h_α is concave on $[0, 1]$. This implies that h_α' is decreasing on $[0, 1]$, and hence,
$$h(\alpha) - 1 = h_\alpha(1) - h_\alpha(0) \leqslant h_\alpha'(0) \leqslant -\frac{\alpha_1 + \cdots + \alpha_n}{\sqrt{6}n}$$
for all $\alpha \in K^+$. This proves the theorem. $\square$

As a direct consequence we get the following statement, which is valid for all $\alpha_j \geqslant 0$.

COROLLARY 4.1.6. *Let K be an isotropic unconditional convex body in $\mathbb{R}^n$. Then,*
$$|\{x \in K^+ : x_1 \geqslant \alpha_1, \ldots, x_n \geqslant \alpha_n\}| \leqslant \exp(-c(\alpha_1 + \cdots + \alpha_n)),$$
for all $\alpha_1, \ldots, \alpha_n \geqslant 0$, where $c = 1/\sqrt{6}$.

Proof. If $(\alpha_1, \ldots, \alpha_n) \in K^+$ we apply Theorem 4.1.5 and then just use the fact that $1 - x \leqslant e^{-x}$ for all $x \geqslant 0$. If not, then the left hand side is equal to zero. □

Another consequence of Theorem 4.1.5 is that an interior point $(\alpha_1, \ldots, \alpha_n)$ of K^+ necessarily satisfies
$$\alpha_1 + \cdots + \alpha_n < \sqrt{6}n.$$
This observation can be equivalently stated as follows.

PROPOSITION 4.1.7. *Let K be an isotropic unconditional convex body in $\mathbb{R}^n$. Then $K \subseteq \sqrt{3/2}nB_1^n$, where $B_1^n = \{x \in \mathbb{R}^n : |x_1| + \cdots + |x_n| \leqslant 1\}$ is the unit ball of ℓ_1^n.* □

One last observation is that, on the other hand, K contains a large cube:

PROPOSITION 4.1.8. *Let K be an isotropic unconditional convex body in $\mathbb{R}^n$. Then,*
$$K \supseteq [-L_K/\sqrt{2}, L_K/\sqrt{2}]^n.$$

Proof. The center of mass $v = (v_1, \ldots, v_n)$ of K^+ is in K^+, therefore the rectangle $[0, v_1] \times \cdots \times [0, v_n]$ is contained in K^+, where
$$v_j = \int_{K^+} x_j \, dx \geqslant \sqrt{2} L_K.$$
Then,
$$[-v_1/2, v_1/2] \times \cdots \times [-v_n/2, v_n/2] \subseteq K,$$
and this proves the proposition. Note that $L_K/\sqrt{2} \geqslant L_{B_2^n}/\sqrt{2} \geqslant 1/(2\sqrt{\pi e})$. □

4.2. Classes with uniformly bounded isotropic constant

Recall that a class $\mathcal{C}$ of symmetric convex bodies is said to satisfy the hyperplane conjecture uniformly if there exists a positive constant C such that $L_K \leqslant C$ for all $K \in \mathcal{C}$. In Section 4.1 we saw that such uniform estimates are available for the class of unconditional convex bodies. In this section we collect several other examples of classes of bodies for which the hyperplane conjecture is known to have an affirmative answer.

4.2.1. Convex bodies whose polar bodies contain large affine cubes

Let K be a symmetric convex body in $\mathbb{R}^n$. The *outer volume ratio* of K is defined by
$$\mathrm{ovr}(K) = \min \left\{ \left(\frac{|\mathcal{E}|}{|K|} \right)^{1/n} : \mathcal{E} \text{ is an ellipsoid and } K \subseteq \mathcal{E} \right\}.$$
Our starting point is a simple observation. Assume that
$$\mathrm{ovr}(K) = \alpha.$$
This means that there exists a position $\tilde{K}$ of K such that $\tilde{K} \subseteq B_2^n$ (which means that $\|x\|_2 \leqslant \|x\|_K$ for all $x \in \mathbb{R}^n$) and
$$|\tilde{K}|^{1/n} = \frac{1}{\alpha} |B_2^n|^{1/n} \geqslant \frac{c}{\alpha \sqrt{n}}.$$

4.2. CLASSES WITH UNIFORMLY BOUNDED ISOTROPIC CONSTANT

We use the fact that

(4.2.1) $$nL_K^2 \leqslant \frac{1}{|\tilde{K}|^{1+\frac{2}{n}}} \int_{\tilde{K}} \|x\|_2^2 dx$$

to write

$$nL_K^2 \leqslant \frac{\alpha^2 n}{c^2} \frac{1}{|\tilde{K}|} \int_{\tilde{K}} \|x\|_K^2 dx \leqslant \frac{\alpha^2 n}{c^2}.$$

This proves the following.

PROPOSITION 4.2.1. *If K is a symmetric convex body in $\mathbb{R}^n$ then*

$$L_K \leqslant c_1 \mathrm{ovr}(K)$$

where $c_1 > 0$ is an absolute constant.

The next lemma provides a bound for L_K which is stronger than the one in (4.2.1).

LEMMA 4.2.2. *Let K be a symmetric convex body of volume 1 in $\mathbb{R}^n$. Then,*

$$L_K \leqslant c \cdot \exp\left(\int_K \log\left(\prod_{i=1}^n |x_i|\right)^{1/n} dx\right),$$

where $c > 0$ is an absolute constant.

Proof. We may find $T \in SL(n)$ such that $T(K)$ is isotropic. From Proposition 3.1.4 we know that the reverse Hölder inequality

$$\left(\int_{T(K)} |\langle x, \theta \rangle|^p dx\right)^{1/p} \geqslant \frac{1}{c} L_K$$

holds true for all $\theta \in S^{n-1}$ and all $p > 0$. Then, for any $0 < p < 2$, we have that

$$\int_K \sum_{i=1}^n |x_i|^p \, dx = \int_{T(K)} \|T^{-1}x\|_p^p \, dx$$

$$= \sum_{i=1}^n \int_{T(K)} |\langle x, (T^{-1})^*(e_i) \rangle|^p \, dx$$

$$\geqslant \frac{1}{c^p} \sum_{i=1}^n \left(\int_{T(K)} |\langle x, (T^{-1})^*(e_i) \rangle|^2 \, dx\right)^{p/2}$$

$$\geqslant \frac{1}{c^p} L_K^p \sum_{i=1}^n \|(T^{-1})^*(e_i)\|_2^p.$$

Applying the arithmetic-geometric means inequality and Hadamard's inequality we get

$$\sum_{i=1}^n \|(T^{-1})^*(e_i)\|_2^p \geqslant n \left(\prod_{i=1}^n \|(T^{-1})^*(e_i)\|_2\right)^{p/n} \geqslant n |\det(T^{-1})^*|^{p/n} = n.$$

Combining the above we see that

(4.2.2) $$\left(\frac{1}{n} \int_K \sum_{i=1}^n |x_i|^p \, dx\right)^{1/p} \geqslant \frac{L_K}{c}.$$

Letting $p \to 0$ we get the lemma. $\square$

Using Lemma 4.2.2 one can prove the following analogue of Proposition 4.2.1.

PROPOSITION 4.2.3. *Let K be a symmetric convex body in $\mathbb{R}^n$. Assume that $K \subseteq A$, where $A = \operatorname{conv}\{\pm x_1, \ldots, \pm x_n\}$ for some $x_1, \ldots, x_n$ in $\mathbb{R}^n$. Then,*
$$L_K \leqslant C \left(\frac{|A|}{|K|}\right)^{1/n},$$
where $C > 0$ is an absolute constant.

Proof. Since L_K is affinely invariant, we may assume that $|K| = 1$ and that $A = \alpha B_1^n$, where B_1^n is the unit ball of ℓ_1^n. Then (4.2.2) with $p = 1$ gives
$$L_K \leqslant c \int_K \frac{1}{n} \left(\sum_{i=1}^n |x_i|\right) dx \leqslant \frac{c\alpha}{n},$$
since $K \subseteq A = \alpha B_1^n$. On the other hand, $|A|^{1/n} = \alpha |B_1^n|^{1/n} = \frac{2\alpha}{(n!)^{1/n}} \geqslant \frac{c_1 \alpha}{n}$. Combining the above we get
$$L_K \leqslant C|A|^{1/n},$$
and the result follows. $\square$

Using the next lemma of Lozanovskii [**340**] we will prove an extension of Proposition 4.2.3.

LEMMA 4.2.4. *Let W be an unconditional convex body in $\mathbb{R}^n$. There exists $T \in GL(n)$ such that*
$$B_\infty^n \subseteq T(W) \subseteq nB_1^n.$$

Note. In fact, we may avoid using the exact statement of Lemma 4.2.4: we have already seen that if W is an unconditional convex body then its isotropic position $T(W)$ satisfies $c_1 B_\infty^n \subseteq T(W) \subseteq c_2 n B_1^n$, and we can prove the next proposition with the same more or less argument.

PROPOSITION 4.2.5. *Let K be a symmetric convex body in $\mathbb{R}^n$ and let W be an unconditional convex body in $\mathbb{R}^n$ with $K \subset W$. Then,*
$$L_K \leqslant C \left(\frac{|W|}{|K|}\right)^{1/n},$$
where $C > 0$ is an absolute constant.

Proof. From Lozanovskii's lemma we can find $T \in GL(n)$ such that $T(B_\infty^n) \subseteq W \subseteq nT(B_1^n)$. We apply Proposition 4.2.3 with $A = nT(B_1^n)$ to get
$$L_K \leqslant C_1 \frac{n|T(B_1^n)|^{1/n}}{|K|^{1/n}} \leqslant C_2 \frac{|\det T|^{1/n}}{|K|^{1/n}},$$
using the fact that $|B_1^n|^{1/n} = \frac{2}{(n!)^{1/n}} \leqslant \frac{c}{n}$. On the other hand,
$$|\det T|^{1/n} = \frac{1}{2}|T(B_\infty^n)|^{1/n} \leqslant \frac{1}{2}|W|^{1/n}.$$
Combining the above we conclude the proof. $\square$

As a corollary we see again that unconditional convex bodies have bounded isotropic constant; but we also see that convex bodies which are "close" to the class of unconditional convex bodies (in the sense of Banach-Mazur distance) have bounded isotropic constant too.

COROLLARY 4.2.6. *Let K be a symmetric convex body in $\mathbb{R}^n$.*
  (i) *If K is an unconditional convex body then $L_K \leqslant C$.*
  (ii) *If $d_{\mathrm{BM}}(K, L) \leqslant \alpha$ for some unconditional convex body L, then $L_K \leqslant C\alpha$.*

4.2.2. 2-convex bodies

A second class of bodies with uniformly bounded isotropic constant is the class of 2-convex bodies.

DEFINITION 4.2.7. *Let K be a symmetric convex body in $\mathbb{R}^n$. The modulus of convexity of K is the function*

$$\delta_K(\varepsilon) = \inf\left\{ 1 - \left\| \frac{x+y}{2} \right\|_K : \|x\|_K, \|y\|_K \leqslant 1, \|x-y\|_K \geqslant \varepsilon \right\},$$

defined for $0 < \varepsilon \leqslant 2$. We say that K is 2-convex with constant α if for all $0 < \varepsilon \leqslant 2$ we have that

$$\delta_K(\varepsilon) \geqslant \alpha \varepsilon^2.$$

A basic example of 2-convex bodies is given by the unit balls of subspaces of L_p-spaces, $1 < p \leqslant 2$. In this case, α is of the order of $p-1$.

The next result of Klartag and E. Milman [283] shows that the hyperplane conjecture holds true for the class of symmetric convex bodies which are 2-convex with a given constant α.

PROPOSITION 4.2.8. *Let K be a symmetric convex body of volume 1 in $\mathbb{R}^n$, which is also 2-convex with constant α. Then*

$$L_K \leqslant c/\sqrt{\alpha},$$

where $c > 0$ is an absolute constant.

We will also show that every (isotropic) 2-convex body with constant α has bounded volume ratio (which depends only on α).

PROPOSITION 4.2.9. *Let K be an isotropic 2-convex body with constant α. Then,*

$$c\sqrt{\alpha}\sqrt{n} B_2^n \subset K,$$

where $c > 0$ is an absolute constant.

The proof of the propositions is reduced to the next two lemmas.

LEMMA 4.2.10. *Let K be a symmetric convex body of volume 1 in $\mathbb{R}^n$. Assume that K is 2-convex with constant α. For any $\theta \in S^{n-1}$ and any $t > 0$, we have*

$$|\{x \in K : \langle x, \theta \rangle > t\}| \leqslant \exp\left(-\alpha n \left(t/\|\theta\|_*\right)^2\right),$$

where $\|x\|_* = h_K(x)$.

Proof. Let $\|\cdot\| := \|\cdot\|_K$ be the norm induced by K. Consider the half-spaces

$$A(t) = \{x \in K : \langle x, \theta \rangle > t\} \text{ and } B = \{x \in K : \langle x, \theta \rangle < 0\}.$$

Now, let $x \in A(t)$ and $y \in B$. Observe that

$$\|x-y\| \, \|\theta\|_* \geqslant |\langle x-y, \theta \rangle| \geqslant t.$$

Since K is 2-convex, setting $\varepsilon = \frac{t}{\|\theta\|_*}$ in Definition 4.2.7 we see that

$$\frac{B+A(t)}{2} \subseteq \left(1 - \alpha\, (t/\|\theta\|_*)^2\right) K.$$

Then, the Brunn-Minkowski inequality shows that

$$\min\{|B|, |A(t)|\} \leqslant \left|\frac{B+A(t)}{2}\right| \leqslant \left(1 - \alpha\, (t/\|\theta\|_*)^2\right)^n$$
$$\leqslant \exp\left(-\alpha n\, (t/\|\theta\|_*)^2\right).$$

By the symmetry of K we have that $|B| = \frac{1}{2}$; so, $\min\{|B|, |A(t)|\} = |A(t)|$ and the lemma follows. $\square$

Now, we are able to prove the main lemma which entails Propositions 4.2.8, 4.2.9.

LEMMA 4.2.11. *Let K be an isotropic convex body in $\mathbb{R}^n$, which is 2-convex with constant α. Then,*

(4.2.3) $$c\sqrt{\alpha}\sqrt{n}L_K B_2^n \subset K,$$

where $c > 0$ is an absolute constant.

Proof. We fix $\theta \in S^{n-1}$ and for every $t \in \mathbb{R}$ we define

$$B(t) = K \cap \{x \in K : \langle x, \theta \rangle < t\},$$

and $f(t) = |B(t)|$. From Lemma 4.2.10 we have that

(4.2.4) $$f(t) \geqslant 1 - \exp\left(-\alpha n\, (t/\|\theta\|_*)^2\right)$$

for all $t > 0$. On the other hand, the Brunn-Minkowski inequality implies that $f'(t) = |K \cap \{\langle x, \theta \rangle = t\}|$ is a log-concave function, which, in our case, is also even because of the symmetry of K. So, f' attains its maximum at 0. It follows that

$$f(t) \leqslant f(0) + tf'(0) \leqslant \frac{1}{2} + \frac{c}{L_K}t,$$

because, by Theorem 3.1.2, we know that $f'(0) = |K \cap \theta^\perp| \simeq \frac{1}{L_K}$. Choosing $t_0 = L_K/4c$, we get from (4.2.4) that

$$1 - \exp\left(-\alpha n\, (t_0/\|\theta\|_*)^2\right) \leqslant f(t_0) \leqslant \frac{3}{4},$$

which implies

$$c'\sqrt{\alpha}\sqrt{n}L_K \leqslant \|\theta\|_* = h_K(\theta).$$

Since the direction $\theta \in S^{n-1}$ was arbitrary, the lemma follows. $\square$

Now, Proposition 4.2.8 is a direct consequence of Lemma 4.2.11: taking volumes on both sides of (4.2.3) we get $L_K \leqslant c/\sqrt{\alpha}$. Finally, since $L_K \geqslant c$ for an absolute constant $c > 0$, Proposition 4.2.9 also follows.

4.2.3. Zonoids

Recall that a zonoid is a limit of Minkowski sums of line segments in the Hausdorff metric. Equivalently (see [**93**]), a symmetric convex body Z is a zonoid if and only if its polar body is the unit ball of an n-dimensional subspace of an L_1-space; i.e. if there exists a positive measure μ (the supporting measure of Z) on S^{n-1} such that

$$h_Z(x) = \|x\|_{Z^\circ} = \frac{1}{2}\int_{S^{n-1}} |\langle x, y\rangle| d\mu(y).$$

The class of zonoids coincides with the class of projection bodies. Recall that the projection body ΠK of a convex body K is the symmetric convex body whose support function is defined by

$$h_{\Pi K}(\theta) = |P_\theta(K)| \qquad (\theta \in S^{n-1})$$

where $P_\theta(K)$ is the orthogonal projection of K onto $\theta^\perp$. From Cauchy's formula (see Section 1.4.5)

$$|P_\theta(K)| = \frac{1}{2}\int_{S^{n-1}} |\langle u, \theta\rangle|\, d\sigma_K(u)$$

it follows that the projection body of K is a zonoid whose supporting measure is σ_K. Minkowski's existence theorem implies that, conversely, every zonoid is the projection body of some symmetric convex body in $\mathbb{R}^n$. We shall see that three natural positions of a zonoid have small diameter.

1. *Lewis position.* A result of Lewis [**322**] shows that every zonotope Z has a linear image Z_1 (the "Lewis position" of Z) with the following property: there exist unit vectors $u_1, \ldots, u_m$ and positive real numbers $c_1, \ldots, c_m$ such that

$$h_{Z_1}(x) = \sum_{j=1}^m c_j |\langle x, u_j\rangle|$$

and

$$I = \sum_{j=1}^m c_j u_j \otimes u_j.$$

Using the Brascamp-Lieb inequality, Ball proved in [**37**] that, under these conditions,

$$|Z_1^\circ| \leqslant \frac{2^n}{n!} \quad \text{and} \quad B_2^n \subseteq \sqrt{n} Z_1^\circ.$$

The exact reverse Santaló inequality for zonoids implies that

$$|Z_1| \geqslant 2^n \quad \text{and} \quad Z_1 \subseteq \sqrt{n} B_2^n.$$

This shows that

(4.2.5) $$R(Z_1) \leqslant \frac{\sqrt{n}}{2}|Z_1|^{1/n}.$$

2. *Löwner position.* Assume that B_2^n is the ellipsoid of minimal volume containing a zonoid Z_2. Let Z_1 be the Lewis position of Z_2. Then,

$$\frac{|B_2^n|}{|Z_2|} \leqslant \frac{|\sqrt{n}B_2^n|}{|Z_1|}.$$

It follows that

(4.2.6) $$2R(Z_2) \leqslant 2 \leqslant |Z_1|^{1/n} \leqslant \sqrt{n}|Z_2|^{1/n}.$$

3. *Minimal mean width position.* Assume that $Z_3 = \Pi K$ is a zonoid of volume 1 which has minimal mean width. Then, Theorem 1.5.12 shows that the area measure σ_K is isotropic, i.e.

$$\int_{S^{n-1}} \langle u, \theta \rangle^2 d\sigma_K(u) = \frac{\partial(K)}{n}$$

for every $\theta \in S^{n-1}$, where $\partial(K)$ is the surface area of K. Moreover, Theorem 1.5.13 shows that K has minimal surface area. Now, an application of the Cauchy-Schwarz inequality shows that

$$h_{Z_3}(\theta) = \frac{1}{2} \int_{S^{n-1}} |\langle \theta, u \rangle| d\sigma_K(u) \leq \frac{\partial(K)}{2\sqrt{n}}$$

for every $\theta \in S^{n-1}$. We will use the following consequence of Theorem 1.5.13 (see [215]):

LEMMA 4.2.12. *If K has minimal surface area, then*

$$\partial(K) \leq n|\Pi K|^{1/n}.$$

It follows that $h_{Z_3}(\theta) \leq \sqrt{n}/2$ for every $\theta \in S^{n-1}$. In other words,

$$R(Z_3) \leq \frac{\sqrt{n}}{2} |Z_3|^{1/n}.$$

The preceding discussion shows that zonoids have positions with "small diameter". This implies that their isotropic constants are uniformly bounded, in fact they are smaller than 1. More precisely, we have the following statement.

THEOREM 4.2.13. *Let Z be a zonoid in Lewis or Löwner or minimal mean width position. Then,*

$$R(Z) \leq \frac{\sqrt{n}}{2} |Z|^{1/n}.$$

In particular,

$$nL_Z^2 \leq \frac{1}{|Z|^{1+\frac{2}{n}}} \int_Z \|x\|_2^2 dx \leq \frac{n}{4},$$

which implies $L_Z \leq 1/2$.

Note. Apparently, it is not known if all *isotropic* zonoids Z in $\mathbb{R}^n$ satisfy $R(Z) \leq C\sqrt{n}$ (for an absolute constant $C > 0$).

4.3. The isotropic constant of Schatten classes

In this section we describe the work of König, Meyer and Pajor [298] on the uniform boundedness of the isotropic constants of the unit balls of the Schatten classes. For every $n \times n$ matrix $T \in \mathcal{M}_n(\mathbb{R})$ we write $s(T) = (s_1(T), \ldots, s_n(T))$ for the sequence of eigenvalues of $\sqrt{T^*T}$ in decreasing order, and for every $1 \leq p < +\infty$ we define

$$\sigma_p(T) = \|s(T)\|_p = \left(\sum_{i=1}^n |s_i(T)|^p \right)^{1/p}.$$

In the case $p = +\infty$ we define

$$\sigma_\infty(T) = \|s(T)\|_\infty = \max_{1 \leq i \leq n} s_i(T).$$

Note that $\sigma_2(T) = \|T\|_{\mathrm{HS}}$ (the Hilbert-Schmidt norm of T) and $\sigma_\infty(T) = \|T\|$ (the operator norm of $T: \ell_2^n \to \ell_2^n$). We denote by S_p^n the Schatten class of $n \times n$ matrices equipped with the σ_p-norm and consider its unit ball

$$B_{\mathbb{R}}(S_p^n) = \{T \in \mathcal{M}_n(\mathbb{R}): \sigma_p(T) \leqslant 1\}.$$

More generally, we may consider any *unitarily invariant* norm $\mathcal{N}$ on $\mathcal{M}_n(\mathbb{R})$ (a norm which satisfies $\mathcal{N}(USV) = \mathcal{N}(S)$ for all $S \in \mathcal{M}_n(\mathbb{R})$ and all $U, V \in O(n)$). It is known that to each unitarily invariant norm $\mathcal{N}$ on $\mathcal{M}_n(\mathbb{R})$ one can associate a 1-symmetric norm τ on $\mathbb{R}^n$ so that

(4.3.1) $\qquad \mathcal{N}(T) = \tau(s_1(T), \ldots, s_n(T)) \quad$ for all $T \in \mathcal{M}_n(\mathbb{R})$

and $\tau(x_1, \ldots, x_n) = \mathcal{N}(X)$, where X is a diagonal matrix with diagonal entries $(x_1, \ldots, x_n)$. In the example of $\mathcal{N}(T) = \sigma_p(T)$ one clearly has that the corresponding 1-symmetric norm is $\tau_p(x) = \|x\|_p$.

For the statement of the next lemma we need one more definition. We say that a function $F: \mathbb{R}^n \to \mathbb{R}$ is *symmetric* if for every permutation π of $\{1, \ldots, n\}$, we have

$$F(x_1, \ldots, x_n) = F(x_{\pi(1)}, \ldots, x_{\pi(n)}).$$

LEMMA 4.3.1. *For any symmetric function $F: \mathbb{R}^n \to \mathbb{R}$ and any unitarily invariant norm $\mathcal{N}$ on $\mathcal{M}_n(\mathbb{R})$ we have:*

(4.3.2) $\qquad \displaystyle\int_{B_{\mathbb{R}}(\mathcal{N})} F(s_1(T), \ldots, s_n(T)) \, dT = c_n \int_{B_\tau} F(x) f_n(x) dx,$

where $B_{\mathcal{N}}$ and B_τ are the unit balls of $\mathcal{N}$ and τ respectively, and

(4.3.3) $\quad c_n = \dfrac{n! \left(\prod_{k=1}^n \omega_k\right)^2}{4^n} \quad$ and $\quad f_n(x) = f_n(x_1, \ldots, x_n) = \displaystyle\prod_{1 \leqslant j < i \leqslant n} |x_i^2 - x_j^2|.$

Moreover, if F is also positively homogeneous of degree k, then for any $p > 0$ we have

(4.3.4) $\qquad \displaystyle\int_{B_{\mathbb{R}}(\mathcal{N})} F(s_1(T), \ldots, s_n(T)) \, dT = \dfrac{c_n}{\Gamma\left(1 + \frac{n^2+k}{p}\right)} \int_{\mathbb{R}^n} F(x) e^{-\tau(x)^p} f_n(x) dx.$

We will note provide the proof; since every $T \in \mathcal{M}_n(\mathbb{R})$ can be written in the form $T = UDV$, where D is a diagonal matrix with non-negative diagonal entries and $U, V \in O(n)$, (4.3.2) follows from a formula of Saint-Raymond in [**450**]. For (4.3.4) we write

$$\int_{\mathbb{R}^n} F(x) e^{-\tau(x)^p} f_n(x) dx = \int_{\mathbb{R}^n} F(x) f_n(x) \left(\int_{\tau(x)^p}^\infty e^{-t} dt\right) dx$$

$$= \int_0^\infty e^{-t} \left(\int_{\{x:\, \tau(x) \leqslant t^{1/p}\}} F(x) f_n(x) dx\right) dt$$

$$= \Gamma\left(1 + \frac{n^2 + k}{p}\right) \int_{B_\tau} F(x) f_n(x) dx.$$

and the result follows from (4.3.2).

We set $u = \displaystyle\sum_{i=1}^n e_i$, where $\{e_i\}_{i \leqslant n}$ is the standard orthonormal basis of $\mathbb{R}^n$. Without loss of generality we may assume that $\tau(e_i) = 1$, $1 \leqslant i \leqslant n$. From

Lozanovskii's theorem (see [**340**]) we have

$$\frac{1}{\tau(u)} B_\infty^n \subset B_\tau \subset \frac{n}{\tau(u)} B_1^n, \tag{4.3.5}$$

and hence

$$\frac{1}{\tau(u)} B_\mathbb{R}(S_\infty^n) \subset B_\mathcal{N} \subset \frac{n}{\tau(u)} B_\mathbb{R}(S_1^n). \tag{4.3.6}$$

From (4.3.5) we have

$$|B_\tau|^{\frac{1}{n}} \simeq \frac{1}{\tau(u)},$$

while from (4.3.6) we see that

$$\frac{1}{\tau(u)} |B_\mathbb{R}(S_\infty^n)|^{\frac{1}{n^2}} \leqslant |B_\mathbb{R}(\mathcal{N})|^{\frac{1}{n^2}} \leqslant \frac{n}{\tau(u)} |B_\mathbb{R}(S_1^n)|^{\frac{1}{n^2}}.$$

Saint-Raymond [**450**] has computed the volume radius of $B_\mathbb{R}(S_1^n)$ and $B_\mathbb{R}(S_\infty^n)$: one has

$$|B_\mathbb{R}(S_\infty^n)|^{1/n^2} \simeq n |B_\mathbb{R}(S_1^n)|^{1/n^2} \simeq \frac{1}{\sqrt{n}}.$$

Combining the above, we get:

LEMMA 4.3.2. *For any unitarily invariant norm $\mathcal{N}$ on $\mathcal{M}_n(\mathbb{R})$ with the property that the corresponding 1-symmetric norm τ satisfies $\tau(e_i) = 1$ for all $1 \leqslant i \leqslant n$, one has*

$$|B_\mathbb{R}(\mathcal{N})|^{1/n^2} \simeq \frac{1}{\tau(u)\sqrt{n}}. \tag{4.3.7}$$

In particular,

$$|B_\mathbb{R}(S_p^n)|^{1/n^2} \simeq n^{-\frac{1}{2} - \frac{1}{p}} \tag{4.3.8}$$

for every $1 \leqslant p \leqslant \infty$. □

We denote by $L_\mathbb{R}(\mathcal{N})$ the isotropic constant of $B_\mathbb{R}(\mathcal{N})$. Lemma 4.3.1 and (4.3.7) show that

$$L_\mathbb{R}(\mathcal{N})^2 \simeq \frac{n \tau^2(u)}{n^2 |B_\mathbb{R}(\mathcal{N})|} \int_{B_\mathbb{R}(\mathcal{N})} \sigma_2^2(T) dT = \frac{\tau^2(u)}{n} \frac{\int_{B_\tau} \left(\sum_{i=1}^n x_i^2 \right) f_n(x) dx}{\int_{B_\tau} f_n(x) dx}. \tag{4.3.9}$$

We will use this formula in order to estimate the isotropic constant of S_p^n.

LEMMA 4.3.3. *Let $1 \leqslant p \leqslant \infty$ and let $M_{n,p}$ denote the measure with density*

$$f(x) = f_{n,p}(x_1, \ldots, x_n) = \mathbf{1}_{\{x_1 \geqslant 0, \ldots, x_n \geqslant 0\}} f_n(x) e^{-\sum_{i=1}^n x_i^p}$$

with respect to the Lebesgue measure. We write $M_{n,p}(h)$ for the expectation of h with respect to $M_{n,p}$. If we denote by $L_\mathbb{R}(n,p)$ the square of the isotropic constant of $B_\mathbb{R}(S_p^n)$ then

$$L_\mathbb{R}(n,p) \sim n^{-\frac{2}{p}} \frac{M_{n,p}(x_1^2)}{M_{n,p}(1)}. \tag{4.3.10}$$

Proof. From (4.3.9), (4.3.8) and Lemma 4.3.1 we get

$$L_{\mathbb{R}}(n,p) \simeq n^{\frac{2}{p}-1} \frac{\int_{B_p^n \cap \mathbb{R}_+^n} \left(\sum_{i=1}^n x_i^2\right) f_n(x) dx}{\int_{B_p^n \cap \mathbb{R}_+^n} f_n(x) dx}$$

$$\simeq n^{\frac{2}{p}-1} \frac{\Gamma\left(1+\frac{n^2}{p}\right)}{\Gamma\left(1+\frac{n^2+2}{p}\right)} \frac{M_{n,p}\left(\sum_{i=1}^n x_i^2\right)}{M_{n,p}(1)}$$

$$= n^{\frac{2}{p}} \frac{\Gamma\left(1+\frac{n^2}{p}\right)}{\Gamma\left(1+\frac{n^2+2}{p}\right)} \frac{M_{n,p}(x_1^2)}{M_{n,p}(1)}.$$

A simple computation shows that

$$\frac{\Gamma\left(1+\frac{n^2}{p}\right)}{\Gamma\left(1+\frac{n^2+2}{p}\right)} \simeq n^{-\frac{4}{p}},$$

and (4.3.10) follows. $\square$

PROPOSITION 4.3.4. *For any $p \geqslant 1$ and $n \geqslant 1$,*

$$\frac{M_{n,p}(x_1^2)}{M_{n,p}(1)} \leqslant Cn^{2/p},$$

where $C > 0$ is an absolute constant.

Proof. We fix $x_2 \geqslant 0, \ldots, x_n \geqslant 0$ and set $x = (x_1, \ldots, x_n)$. Given $a > 0$, we define a function $\phi = \phi_{p,n,a} : [0, \infty) \to \mathbb{R}$ by

$$\phi(x_1) = x_1^a f_n(x).$$

Assume that $x_2 < x_3 < \cdots < x_n$, and also, $x_1 \in (x_m, x_{m+1})$ for some $2 \leqslant m \leqslant n$. Then, we write

$$\phi(x_1) = g(x_2, \ldots, x_n) x_1^a e^{-x_1^p} \prod_{i=1}^m (x_1^2 - x_i^2) \prod_{j=m+1}^n (x_j^2 - x_1^2),$$

where

$$g(x_2, \ldots, x_n) = e^{-\sum_{i=2}^n x_i^p} \prod_{2 \leqslant j < i \leqslant n} |x_i^2 - x_j^2|.$$

Then, ϕ is differentiable on the intervals $(0, x_2), (x_2, x_3), \ldots, (x_{n-1}, x_n), (x_n, \infty)$, with derivative equal to

$$\phi'(x_1) = f(x) \left(a x_1^{a-1} - p x_1^{p+a-1} + 2 x_1^{a+1} \left(\sum_{i=2}^m \frac{1}{x_1^2 - x_i^2} - \sum_{j=m+1}^n \frac{1}{x_j^2 - x_1^2} \right) \right)$$

$$= f(x) \left(a x_1^{a-1} - p x_1^{p+a-1} + 2 x_1^{a+1} \left(\sum_{i=2}^n \frac{1}{x_1^2 - x_i^2} \right) \right).$$

We also observe that for $a > 0$, ϕ' is continuous on all the open intervals (x_{i-1}, x_i) or $(0, x_2)$ and (x_n, ∞), and has integrable singularities at their endpoints; it follows that for almost all $(x_2, \ldots, x_n)$ (with respect to the Lebesgue measure on $\mathbb{R}^{n-1}$) we have

$$\int_0^\infty \phi'(x_1) \, dx_1 = \lim_{x_1 \to \infty} \phi(x_1) - \phi(0) = 0.$$

Integrating with respect to $(x_2, \ldots, x_n) \in [0, \infty)^{n-1}$ we get

$$aM_{n,p}(x_1^{a-1}) - pM_{n,p}(x_1^{a+p-1}) + M_{n,p}\left(2x_1^{a+1} \sum_{i=2}^{n} \frac{1}{x_1^2 - x_i^2}\right) = 0.$$

By symmetry and linearity, for all $2 \leqslant i \leqslant n$ we have

$$M_{n,p}\left(\frac{x_1^{a+1}}{x_1^2 - x_i^2}\right) = M_{n,p}\left(\frac{x_1^{a+1}}{x_1^2 - x_2^2}\right) = M_{n,p}\left(\frac{x_2^{a+1}}{x_2^2 - x_1^2}\right) = -M_{n,p}\left(\frac{x_2^{a+1}}{x_1^2 - x_2^2}\right),$$

therefore,

$$M_{n,p}\left(2x_1^{a+1} \sum_{i=2}^{n} \frac{1}{x_1^2 - x_i^2}\right) = (n-1)M_{n,p}\left(\frac{2x_1^{a+1}}{x_1^2 - x_2^2}\right)$$

$$= (n-1)M_{n,p}\left(\frac{x_1^{a+1} - x_2^{a+1}}{x_1^2 - x_2^2}\right).$$

It follows that

$$aM_{n,p}(x_1^{a-1}) - pM_{n,p}(x_1^{a+p-1}) + (n-1)M_{n,p}\left(\frac{x_1^{a+1} - x_2^{a+1}}{x_1^2 - x_2^2}\right) = 0.$$

Next, observe that if $1 \leqslant a \leqslant 3$ then for all $x_1, x_2 > 0$ we have

$$\frac{a+1}{4}(x_1^{a-1} + x_2^{a-1}) \leqslant \frac{x_1^{a+1} - x_2^{a+1}}{x_1^2 - x_2^2} \leqslant x_1^{a-1} + x_2^{a-1},$$

and hence

$$\frac{a+1}{2}M_{n,p}(x_1^{a-1}) \leqslant M_{n,p}\left(\frac{x_1^{a+1} - x_2^{a+1}}{x_1^2 - x_2^2}\right) \leqslant 2M_{n,p}(x_1^{a-1}).$$

Combining the above, we conclude that

$$\left(\frac{(n-1)(a+1) + 2a}{2p}\right)M_{n,p}(x_1^{a-1}) \leqslant M_{n,p}(x_1^{a+p-1}) \leqslant \frac{a + 2(n-1)}{p}M_{n,p}(x_1^{a-1}).$$

Setting $a = 1$ we obtain

(4.3.11) $$\frac{n}{p} \leqslant \frac{M_{n,p}(x_1^p)}{M_{n,p}(1)} \leqslant \frac{2n-1}{p},$$

and when $a = p + 1$ for some $0 < p \leqslant 2$, we have

(4.3.12) $$\frac{(n-1)(p+2) + 2(p+1)}{2p} \leqslant \frac{M_{n,p}(x_1^{2p})}{M_{n,p}(x_1^p)} \leqslant \frac{p + 1 + 2(n-1)}{p}.$$

If $1 \leqslant p \leqslant 2$, then Hölder's inequality gives

$$M_{n,p}(x_1^2)^{\frac{1}{2}} \leqslant M_{n,p}(x_1^p)^{\frac{p-1}{p}} M_{n,p}(x_1^{2p})^{\frac{2-p}{2p}},$$

thus,

$$\frac{M_{n,p}(x_1^2)}{M_{n,p}(1)} \leqslant \left(\frac{M_{n,p}(x_1^p)}{M_{n,p}(1)}\right)^{\frac{2p-2}{p}} \left(\frac{M_{n,p}(x_1^{2p})}{M_{n,p}(1)}\right)^{\frac{2-p}{p}}.$$

This last inequality, combined with (4.3.11) and (4.3.12) gives us

$$\frac{M_{n,p}(x_1^2)}{M_{n,p}(1)} = \frac{M_{n,p}(x_1^p)}{M_{n,p}(1)} \left(\frac{M_{n,p}(x_1^{2p})}{M_{n,p}(x_1^p)}\right)^{\frac{2-p}{p}}$$

$$\leqslant \frac{2n-1}{p} \left(\frac{p+1+2(n-1)}{p}\right)^{\frac{2-p}{p}}$$

$$\leqslant (2n+1)^{2/p} \leqslant 9n^{2/p}.$$

If $p \geqslant 2$, then Hölder's inequality and (4.3.11) imply

$$\frac{M_{n,p}(x_1^2)}{M_{n,p}(1)} \leqslant \left(\frac{M_{n,p}(x_1^p)}{M_{n,p}(1)}\right)^{\frac{2}{p}} \leqslant \left(\frac{2n-1}{p}\right)^{2/p} \leqslant n^{2/p}.$$

In both cases, $M_{n,p}(x_1^2)/M_{n,p}(1) \leqslant Cn^{2/p}$. □

Combining Proposition 4.3.4 with (4.3.10) we obtain a uniform upper bound for $L_{B(S_p^n)}$.

THEOREM 4.3.5 (König-Meyer-Pajor). *There exists an absolute constant $C > 0$ such that, for all $n \geqslant 1$ and $p \in [1, +\infty]$,*

$$L_{B(S_p^n)} \leqslant C.$$

Note. It should be mentioned that Dar had obtained (before the work of König, Meyer and Pajor) the almost optimal estimate $L_{B(S_p^n)} \leqslant C\sqrt{\log n}$ in [**154**] (see also [**152**] for the case $p = 1$).

4.4. Bodies with few vertices or few facets

In this section we discuss upper bounds for the isotropic constant of polytopes

$$K = \text{conv}\{y_1, \ldots, y_N\}$$

in $\mathbb{R}^n$ with a "small number of vertices" (assume, for example, that N is polynomial in the dimension n) or polyhedra

$$K = \{x \in \mathbb{R}^n : \langle x, w_i \rangle \leqslant 1, i = 1, \ldots, N\}$$

where $w_1, \ldots, w_N$ are vectors spanning $\mathbb{R}^n$, with a "small number of facets (assume, again, that N is polynomial in n). In the symmetric case, Junge obtained in [**262**] upper bounds which are logarithmic in N:

(i) If K is the unit ball of an n-dimensional subspace of an N-dimensional normed space with an unconditional basis then $L_K \leqslant c\sqrt{\log(1 + N/n)}$.

(ii) If K is the unit ball of an n-dimensional quotient of an N-dimensional normed space with an unconditional basis then $L_K \leqslant c \log N$.

Upper bounds for the isotropic constants of subspaces and quotients of L_p and related spaces were obtained by Ball in [**35**] and Junge in [**261**]. An alternative approach to these results was presented by E. Milman in [**366**]; he developed a technique, using dual mixed-volumes, to recover, strengthen and generalize these estimates.

The results of Junge and E. Milman imply that the isotropic constant of a symmetric polytope with $2N$ vertices is bounded by $c \log N$. Here, we present a more recent geometric approach of Alonso-Gutiérrez [**11**] which works for not necessarily symmetric polytopes as well. Let us also mention that a result from [**13**] states that if K is a polytope with N vertices then $L_K \leqslant c\sqrt{N/n}$; this is weaker than the previous results for most N, however it provides a sharp estimate as long as N is proportional to n.

4.4.1. Bodies with few vertices

THEOREM 4.4.1. *Let $K = \operatorname{conv}\{y_1, \ldots, y_N\}$ be a polytope with N vertices in $\mathbb{R}^n$. Then,*

$$L_K \leqslant c \log N,$$

where $c > 0$ is an absolute constant.

Proof. Assume that $K = \operatorname{conv}\{y_1, \ldots, y_N\}$ is in the isotropic position. For any symmetric and positive definite $T \in GL(n)$ we have

$$\operatorname{tr}(T) L_K^2 = \int_K \langle T(x), x \rangle dx \leqslant \int_K \max_{z \in K} |\langle T(z), x \rangle| dx = \int_K \max_{1 \leqslant i \leqslant N} |\langle T(y_i), x \rangle| dx$$
$$\leqslant \max_{1 \leqslant i \leqslant N} \|T(y_i)\|_2 \int_K \max_{1 \leqslant i \leqslant N} |\langle u_i, x \rangle| dx$$
$$\leqslant c_1 L_K (\log N) \max_{1 \leqslant i \leqslant N} \|T(y_i)\|_2,$$

where $u_i = T(y_i)/\|T(y_i)\|_2$ and we have also used Proposition 3.3.6.

It follows that

(4.4.1) $$n (\det T)^{1/n} L_K \leqslant c \log N \max_{i=1,\ldots,N} \|T(y_i)\|_2.$$

We now choose $w \in \mathbb{R}^n$ and $T \in GL(n)$ so that $w + T(K)$ is in Löwner's position, i.e. the Euclidean unit ball is the minimal volume ellipsoid of $w + T(K)$. From John's theorem, if we set $C = w + T(K)$ we can find $\lambda_1, \ldots, \lambda_m > 0$ and contact points $u_1, \ldots, u_m$ of C and B_2^n such that

(4.4.2) $$\sum_{j=1}^m \lambda_j u_j = 0 \quad \text{and} \quad I = \sum_{j=1}^m \lambda_j u_j \otimes u_j.$$

We use a result of Dar from [**153**].

LEMMA 4.4.2. *Let $t_1, \ldots, t_m > 0$ and $v_1, \ldots, v_m \in \mathbb{R}^n$ such that $\sum_{j=1}^m t_j = 1$ and*

(4.4.3) $$\sum_{j=1}^m t_j v_j = 0 \quad \text{and} \quad \sum_{j=1}^m t_j v_j \otimes v_j = A^2 |\operatorname{conv}\{v_1, \ldots, v_m\}|^{2/n} I.$$

Then,

$$A \leqslant \sqrt{n/e}.$$

Proof. If we consider the $n \times m$ matrix B with columns $\sqrt{t_j} v_j$ then (4.4.3) shows that

$$BB^* = A^2 |\operatorname{conv}\{v_1, \ldots, v_m\}|^{2/n} I.$$

Applying the Cauchy-Binet formula we get

$$A^{2n}|\operatorname{conv}\{v_1,\ldots,v_m\}|^2 = \det\left(A^2|\operatorname{conv}\{v_1,\ldots,v_m\}|^{2/n} I\right) = \det(BB^*)$$
$$= \sum_{1\leqslant i_1<\cdots<i_n\leqslant m} \det(B_{i_1\cdots i_n}B^*_{i_1\cdots i_n})$$
$$= \sum_{1\leqslant i_1<\cdots<i_n\leqslant m} \det(B_{i_1\cdots i_n})^2$$
$$= \sum_{1\leqslant i_1<\cdots<i_n\leqslant m} t_{i_1}\cdots t_{i_n} \det(v_{i_1},\ldots,v_{i_n})^2.$$

From (4.4.3) we see that $0 \in \operatorname{conv}\{v_1,\ldots,v_m\}$. Therefore, we have

$$|\det(v_{i_1},\ldots,v_{i_n})| = n!|\operatorname{conv}\{0,v_{i_1},\ldots,v_{i_n}\}| \leqslant n!|\operatorname{conv}\{v_1,\ldots,v_m\}|,$$

and hence, using Maclaurin's inequality we get

$$A^{2n} \leqslant (n!)^2 \left(\sum_{1\leqslant i_1<\cdots<i_n\leqslant m} t_{i_1}\cdots t_{i_n}\right)$$
$$\leqslant (n!)^2 \binom{m}{n}\left(\frac{1}{m}\sum_{j=1}^m t_j\right)^n = (n!)^2 \binom{m}{n}\frac{1}{m^n}$$
$$\leqslant (n!)^2 \frac{1}{n!} = n!.$$

It follows that $A \leqslant \sqrt{n/e}$. □

We apply Lemma 4.4.2 with $t_j = \lambda_j/n$ and $v_j = u_j$. From (4.4.2) it follows that $(t_j, v_j)_{j\leqslant m}$ satisfy (4.4.3) with

$$A = \frac{1}{\sqrt{n}|\operatorname{conv}\{v_1,\ldots,v_m\}|^{1/n}}.$$

Then, applying Lemma 4.4.2 we get

$$|\operatorname{conv}\{u_1,\ldots,u_m\}|^{1/n} = |\operatorname{conv}\{v_1,\ldots,v_m\}|^{1/n} \geqslant \frac{\sqrt{e}}{n}.$$

Since $w + T(K) = C \supseteq \operatorname{conv}\{u_1,\ldots,u_m\}$ we finally get:

LEMMA 4.4.3. *If $K = \operatorname{conv}\{y_1,\ldots,y_N\}$ is in isotropic position and $w + T(K)$ is in Löwner position, then*

$$|T(K)|^{1/n} = |w + T(K)|^{1/n} \geqslant \frac{\sqrt{e}}{n} \geqslant \frac{1}{n}.$$ □

Now, going back to (4.4.1) we get

$$L_K \leqslant n|T(K)|^{1/n}L_K = n\left(\det T\right)^{1/n} L_K \leqslant c\log N \max_{i=1,\ldots,N}\|T(y_i)\|_2$$

Observe that w is the centroid of $w + T(K)$ and, for each $1 \leqslant i \leqslant N$ we have $w + T(y_i) \in w + T(K)$ as well. Therefore, both w and $w + T(y_i)$ are in B_2^n, and this implies

$$\|T(y_i)\|_2 \leqslant \|w + T(y_i)\|_2 + \|w\|_2 \leqslant 2 \qquad (1 \leqslant i \leqslant N)$$

So,

$$L_K \leqslant c\log N \max_{i=1,\ldots,N}\|T(y_i)\|_2 \leqslant 2c\log N.$$ □

4.4.2. Bodies with few facets

Next, we discuss an upper bound (from [**11**]) for the isotropic constant L_K in the case where K is a polyhedron with N facets. So, we assume that
$$K = \{x \in \mathbb{R}^n : \langle x, w_i \rangle \leqslant 1, i = 1, \ldots, N\},$$
where $w_1, \ldots, w_N$ are vectors spanning $\mathbb{R}^n$.

For the proof of the result we need lower bounds for the volume of K. Carl-Pajor [**134**], and independently Gluskin [**220**], obtained a lower bound for the volume of a symmetric polyhedron $K = \{x \in \mathbb{R}^n : |\langle x, w_i \rangle| \leqslant 1, i = 1, \ldots, N\}$ in terms of $\max\{\|w_i\|_2 : 1 \leqslant i \leqslant N\}$:

THEOREM 4.4.4 (Carl-Pajor, Gluskin). *Let $w_1, \ldots, w_N$ be vectors spanning $\mathbb{R}^n$ with $\|w_i\|_2 \leqslant 1$ for all $1 \leqslant i \leqslant N$. Consider the symmetric convex body*
$$K = \{x \in \mathbb{R}^n : |\langle x, w_i \rangle| \leqslant 1, i = 1, \ldots, N\}.$$
Then,
$$|K|^{1/n} \geqslant \frac{c}{\sqrt{\log(1+N/n)}},$$
where $c > 0$ is an absolute constant.

We will give a proof of Theorem 4.4.4 using Sidák's lemma (see [**467**]). Observe that K is the intersection of the symmetric strips
$$P_i = \{x \in \mathbb{R}^n : |\langle x, w_i \rangle| \leqslant 1\} \qquad (i = 1, \ldots, N).$$
Sidák's lemma gives a lower bound for the n-dimensional Gaussian measure of such an intersection.

THEOREM 4.4.5 (Sidák). *If $P_1, \ldots, P_N$ are symmetric strips in $\mathbb{R}^n$ then*
$$\gamma_n(P_1 \cap \cdots \cap P_N) \geqslant \gamma_n(P_1) \cdots \gamma_n(P_N).$$

A simple proof of Theorem 4.4.5 can be based on the next observation.

PROPOSITION 4.4.6. *Let C be a symmetric closed convex set and let P be a symmetric strip in $\mathbb{R}^n$. Then,*
$$\gamma_n(C \cap P) \geqslant \gamma_n(C)\gamma_n(P).$$

Proof. We use induction on the dimension. If $n = 1$, then C and P are symmetric intervals on $\mathbb{R}$, and hence $C \cap P$ is either C or P. Since γ_1 is a probability measure, we obviously have
$$\gamma_1(C \cap P) = \min\{\gamma_1(C), \gamma_1(P)\} \geqslant \gamma_1(C)\gamma_1(P).$$
Assume that $n \geqslant 2$ and that the statement holds true for all dimensions less than n. Let C be a symmetric, closed and convex set and let P be a symmetric strip in $\mathbb{R}^n$. By the rotational invariance of γ_n we may assume that $P = \{x \in \mathbb{R}^n : |x_n| \leqslant t\}$ for some $t > 0$. Set $C_s = C \cap \{x_n = s\}$ and consider the even log-concave function $f(s) = \gamma_{n-1}(C_s)$. Then,
$$\gamma_n(C \cap P) = \int_{-t}^{t} \gamma_{n-1}(C_s) d\gamma_1(s) = \int_{-\infty}^{\infty} \mathbf{1}_{[-t,t]}(s) f(s) d\gamma_1(s)$$
$$= \int_{0}^{\infty} \gamma_1(\{s \in [-t,t] : f(s) \geqslant z\}) dz.$$

Since $\{f \geqslant z\}$ is symmetric and convex for any z, from the one-dimensional case of the theorem we have

$$\gamma_1(\{s \in [-t,t] : f(s) \geqslant z\}) = \gamma_1(\{f \geqslant z\} \cap [-t,t]) \geqslant \gamma_1(\{f \geqslant z\}) \cdot \gamma_1([-t,t]).$$

It follows that

$$\gamma_n(C \cap P) \geqslant \gamma_1([-t,t]) \int_0^\infty \gamma_1(\{f \geqslant z\})dz$$

$$= \gamma_1([-t,t]) \int_{-\infty}^\infty \gamma_{n-1}(C_s)ds = \gamma_1([-t,t])\gamma_n(C).$$

It is clear that $\gamma_n(P) = \gamma_1([-t,t])$, and this finally gives $\gamma_n(C \cap P) \geqslant \gamma_n(C)\gamma_n(P)$. □

Proof of Theorem 4.4.5. We assume that $N \geqslant 2$ and that the theorem holds true if the number of strips is less than N and consider N symmetric strips $P_1, \ldots, P_N$ in $\mathbb{R}^n$. Then, $C := P_1 \cap \cdots \cap P_{N-1}$ is symmetric, closed and convex, therefore

$$\gamma_n(P_1 \cap \cdots \cap P_{N-1} \cap P_N) = \gamma_n(C \cap P_N) \geqslant \gamma_n(C)\gamma_n(P_N)$$
$$= \gamma_n(P_1 \cap \cdots \cap P_{N-1})\gamma_n(P_N)$$

by Proposition 4.4.6. Applying the induction hypothesis for $P_1, \ldots, P_{N-1}$ we get the result. □

REMARK 4.4.7. A well-known open question (the *Gaussian correlation conjecture*) asks if for any pair of symmetric convex bodies K and C in $\mathbb{R}^n$ one has

$$\gamma_n(K \cap C) \geqslant \gamma_n(K)\gamma_n(C).$$

An affirmative answer has been given by Pitt [**431**] in dimension $n = 2$. Clearly, Sidák's lemma would be a direct consequence of this statement.

Besides Sidák's lemma, for the proof of Theorem 4.4.4 we will use the elementary inequality

$$(4.4.4) \qquad \frac{2}{\sqrt{2\pi}} \int_0^s e^{-t^2/2} dt \geqslant 1 - e^{-s^2/2} \qquad (s \geqslant 0).$$

Proof of Theorem 4.4.4. We fix $\alpha > 0$ and for every $1 \leqslant i \leqslant N$ we consider the symmetric strip

$$P_i = \{x \in \mathbb{R}^n : |\langle x, w_i \rangle| \leqslant \alpha\}.$$

Since $\|w_i\|_2 \leqslant 1$ for all i, the width of each P_i is at least 2α. Thus,

$$\gamma_n(P_i) \geqslant \frac{1}{\sqrt{2\pi}} \int_{-\alpha}^\alpha e^{-t^2/2} dt \geqslant 1 - e^{-\alpha^2/2}$$

by (4.4.4). Now, Theorem 4.4.5 shows that

$$\gamma_n(\alpha K) = \gamma_n\left(\bigcap_{i=1}^N P_i\right) \geqslant \left(1 - e^{-\alpha^2/2}\right)^N.$$

Note that $|\alpha K| \geqslant (2\pi)^{n/2}\gamma_n(\alpha K)$. So, if we choose $\alpha = 2\sqrt{\log(1 + N/n)}$ we have

$$2\sqrt{\log(1 + N/n)}|K|^{1/n} \geqslant \sqrt{2\pi}\left(1 - e^{-2\log(1+N/n)}\right)^{N/n}$$

$$= \sqrt{2\pi}\left(1 - \frac{1}{(1 + N/n)^2}\right)^{N/n} \geqslant c,$$

where $c > 0$ is an absolute constant. This shows that $|K|^{1/n} \geqslant c_1/\sqrt{\log(1 + N/n)}$. □

We will use Theorem 4.4.4 to give an upper bound for the isotropic constant of bodies with few facets.

THEOREM 4.4.8. *Let K be a polytope with N facets in $\mathbb{R}^n$. Then,*
$$L_K \leqslant c\sqrt{\log(1 + N/n)},$$
where $c > 0$ is an absolute constant.

Proof. We may assume that K is centered and that $K \cap (-K)$ is in John's position. Then, there exist $\{u_j\}_{j \leqslant m} \in \operatorname{bd}(K \cap (-K)) \cap S^{n-1}$ and positive real numbers $\{\lambda_j\}_{j \leqslant m}$ such that
$$I = \sum_{j=1}^{m} \lambda_j u_j \otimes u_j.$$

Since $K \cap (-K)$ is symmetric, we can change the sign of each u_j, and since $u_j \in \operatorname{bd}(K \cap (-K))$, we may then assume that $u_j \in \operatorname{bd}(K) \cap S^{n-1}$. In other words, we assume that $h_K(u_j) = h_{K \cap (-K)}(u_j) = 1$ for all $j = 1, \ldots, m$.

We will use the inequality
$$nL_K^2 \leqslant \frac{1}{|K|^{1+\frac{2}{n}}} \int_K \|x\|_2^2 \, dx$$
to obtain our bound. To this end, we first observe that

(4.4.5) $$\frac{1}{|K|} \int_K \langle x, u_i \rangle^2 \, dx \leqslant c_1 h_K(u_i)^2$$

where $c_1 > 0$ is an absolute constant (actually, it is proved in [**266**] that one can replace c_1 by $\frac{n}{n+2}$ in the right hand side). For the proof of (4.4.5) we recall that
$$\frac{1}{|K|} \int_K \langle x, u_i \rangle^2 \, dx \leqslant c_2 \frac{|K|^2}{|K \cap u_i^\perp|^2}$$
and then we use the facts that $|K| \leqslant e|K \cap \{\langle x, u_i \rangle \geqslant 0\}|$ by Grünbaum's lemma and $|K \cap \{\langle x, u_i \rangle \geqslant 0\}| \leqslant e h_K(u_i) |K \cap u^\perp|$ by Fradelizi's inequality (see Lemma 2.2.6 and Theorem 2.2.2 respectively). The assumption that K is centered is needed at this point. Using this observation and the fact that
$$\|x\|_2^2 = \sum_{j=1}^{m} \lambda_j \langle x, u_j \rangle^2$$
for all $x \in \mathbb{R}^n$ (which follows from John's representation of the identity), we get
$$\frac{1}{|K|} \int_K \|x\|_2^2 \, dx = \sum_{j=1}^{m} \lambda_j \left(\frac{1}{|K|} \int_K \langle x, u_j \rangle^2 \, dx \right) \leqslant c_1 \sum_{j=1}^{m} \lambda_j h_K^2(u_j)$$
$$= c_1 \sum_{j=1}^{m} \lambda_j = c_1 n.$$

In this way, we see that

(4.4.6) $$L_K^2 \leqslant \frac{c_1}{|K|^{2/n}} \leqslant \frac{c_1}{|K \cap (-K)|^{2/n}}.$$

Note that $K \cap (-K)$ is a polyhedron with $2N$-facets; it can be written in the form
$$K \cap (-K) = \{x \in \mathbb{R}^n : |\langle x, w_i \rangle| \leqslant 1,\, 1 \leqslant i \leqslant N\},$$
where $\|w_i\|_2 = 1$ because $K \cap (-K)$ is assumed to be in John's position. At this point, we use Theorem 4.4.4 to obtain the lower bound
$$|K \cap (-K)|^{\frac{1}{n}} \geqslant \frac{c}{\sqrt{\log(1 + N/n)}}.$$
and going back to (4.4.6) we get
$$L_K^2 \leqslant c_3 \log(1 + N/n)$$
as claimed. □

4.5. Further reading

In this section we describe some exact lower and upper bounds for the volume of k-dimensional sections of the ℓ_p^n-balls
$$B_p^n = \{x \in \mathbb{R}^n : |x_1|^p + \cdots + |x_n|^p \leqslant 1\}.$$

4.5.1. Vaaler's inequality and its extension by Meyer and Pajor

We start with a result of Vaaler [**495**]; he proved that every central section of the cube of volume 1 has volume greater than 1. This result was extended by Meyer and Pajor [**363**].

THEOREM 4.5.1 (Meyer-Pajor). *Let $B_p^n = \{x : \|x\|_p \leqslant 1\}$ denote the unit ball of ℓ_p^n, $1 \leqslant p \leqslant \infty$. For every $n \geqslant 1$ and every $F \in G_{n,k}$, $1 \leqslant k \leqslant n$, the function $h_F : [1, \infty] \to \mathbb{R}$ defined by*
$$h_F(p) := \frac{|B_p^n \cap F|}{|B_p^k|}$$
is increasing.

The monotonicity of $p \mapsto h_F(p)$ has the following consequence.

THEOREM 4.5.2 (Meyer-Pajor). *Let B_p^n be the unit ball of ℓ_p^n. Let $1 \leqslant k \leqslant n$ and $F \in G_{n,k}$.*
  (i) *If $1 \leqslant p \leqslant 2$ then $|B_p^n \cap F| \leqslant |B_p^k|$.*
  (ii) *If $2 \leqslant p \leqslant \infty$ then $|B_p^n \cap F| \geqslant |B_p^k|$.*

Proof. We observe that $h_F(2) = 1$ and use the fact that h_F is increasing. □

Note. Vaaler's result corresponds to the case $p = \infty$ of Theorem 4.5.2 (ii). If we consider the cube $C_n := \frac{1}{2} B_\infty^n$ of volume 1 in $\mathbb{R}^n$ then for every $1 \leqslant k \leqslant n$ we have
$$|C_n \cap F| \geqslant |C_k| = 1.$$

For the proof of Theorem 4.5.1 we introduce a partial order on the class of even log-concave probability measures. Given two such measures μ and ν we say that μ is *more peaked* than ν, and we write $\mu \succ \nu$, if
$$\mu(C) \geqslant \nu(C)$$
for every symmetric convex subset C of $\mathbb{R}^n$. A useful theorem of Kanter [**268**] asserts that $\succ$ is preserved by products of log-concave probability measures:

THEOREM 4.5.3 (Kanter). *Let μ_i, ν_i be even log-concave probability measures on $\mathbb{R}^{k_i}$, $i = 1, \ldots, m$, with $\mu_i \succ \nu_i$ for every $1 \leqslant i \leqslant m$. Then, $\mu_1 \otimes \cdots \otimes \mu_m \succ \nu_1 \otimes \cdots \otimes \nu_m$.*

For each $p \geq 1$ we consider the probability measure $\mu_{n,p}$ on $\mathbb{R}^n$ with density
$$f_{n,p}(x) = \exp(-\|\alpha_p x\|_p^p), \quad \text{where } \alpha_p = 2\Gamma(1+1/p).$$
Observe that
$$\mu_{n,p} = \mu_{1,p} \otimes \cdots \otimes \mu_{1,p}.$$
Meyer and Pajor proved and used the following fact.

PROPOSITION 4.5.4. *For all $1 \leq q \leq p \leq \infty$ we have $\mu_{n,p} \succ \mu_{n,q}$.*

Proof. Let $1 \leq q \leq p \leq \infty$. From Kanter's theorem it is enough to consider the case $n = 1$. According to the definition, we need to verify that
$$\mu_{1,p}([-x,x]) \geq \mu_{1,q}([-x,x])$$
for every $x > 0$. To see this, we observe that for the function
$$f(x) = \int_0^x (\exp(-(\alpha_p t)^p) - \exp(-(\alpha_q t)^q))dt$$
we have $f(0) = f(\infty) = 0$, f' has precisely one zero and is positive in a neighborhood of 0. Thus, $f(x) \geq 0$ for all x and the proposition follows. □

Let C be a symmetric convex body in $\mathbb{R}^n$ and let F be any k-dimensional subspace. The next lemma gives a convenient formula for $|C \cap F|$.

LEMMA 4.5.5. *Let $C = \{x : \|x\| \leq 1\}$ be a symmetric convex body in $\mathbb{R}^n$ and let $1 \leq k \leq n$. Consider any $F \in G_{n,k}$ and an orthonormal basis $\{u_1, \ldots, u_{n-k}\}$ of $F^\perp$. For any $0 < p \leq \infty$,*
$$\Gamma\left(1 + \frac{k}{p}\right)|C \cap F| = \lim_{\varepsilon \to 0^+} \frac{1}{(2\varepsilon)^{n-k}} \int_{F(\varepsilon)} \exp(-\|x\|^p)dx,$$
where
$$F(\varepsilon) = \{x \in \mathbb{R}^n : |\langle x, u_i \rangle| \leq \varepsilon \text{ for all } 1 \leq i \leq n-k\}.$$

Proof. For any $\varepsilon > 0$ we set
$$g(\varepsilon) = \frac{1}{(2\varepsilon)^{n-k}} \int_{F(\varepsilon)} \exp(-\|x\|^p)dx.$$
Using Fubini's theorem and a change of variables, we can rewrite this as follows:
$$(4.5.1) \quad g(\varepsilon) = \frac{1}{(2\varepsilon)^{n-k}} \int_{F(\varepsilon)} \int_{\|x\|^p}^\infty e^{-t} dt\, dx$$
$$= \int_0^\infty \frac{1}{(2\varepsilon t^{-1/p})^{n-k}} |C \cap F(\varepsilon t^{-1/p})| e^{-t} dt.$$
From the symmetry of C and the Brunn-Minkowski inequality we easily check that
$$(4.5.2) \quad \frac{1}{(2\delta)^{n-k}} |C \cap F(\delta)| \leq |C \cap F|$$
for every $\delta > 0$. Observe that
$$|C \cap F| = \lim_{\varepsilon \to 0^+} \frac{1}{(2\varepsilon)^{n-k}} |C \cap F(\varepsilon)|.$$
Therefore, using (4.5.1), (4.5.2) and the dominated convergence theorem, we may conclude the proof. □

Proof of Theorem 4.5.1. We apply Lemma 4.5.5 to the body $C = B_p^n$; we have
$$\Gamma\left(1 + \frac{k}{p}\right)|B_p^n \cap F| = \lim_{\varepsilon \to 0^+} \frac{1}{(2\varepsilon)^{n-k}} \int_{F(\varepsilon)} \exp(-\|x\|_p^p)dx,$$
and a change of variables shows that
$$|B_p^n \cap F| = \frac{\alpha_{k,p}}{\Gamma(1+k/p)} \lim_{\delta \to 0^+} \mu_{n,p}(F(\delta)).$$

Since $|B_p^k| = \alpha_{k,p}/\Gamma(1+k/p)$, this shows that
$$h_F(p) := \frac{|B_p^n \cap F|}{|B_p^k|} = \lim_{\delta \to 0^+} \mu_{n,p}(F(\delta)).$$

Now, the result follows from Proposition 4.5.4. □

4.5.2. Fourier transform and volume of hyperplane sections

In this section we describe Koldobsky's formula (see [**290**]) for the volume of central hyperplane sections of convex bodies in terms of the Fourier transform of their radial function. As an application we discuss the answer to the question to determine the central sections of the unit balls of the spaces ℓ_p^n, $1 \leqslant p \leqslant 2$, that have minimal volume.

We first recall the connection between the Fourier transform and the Radon transform. The Radon transform of an integrable function ϕ, which is also integrable on every hyperplane, is the function $\mathcal{R}\phi$ of $(\theta, t) \in S^{n-1} \times \mathbb{R}$ defined by
$$\mathcal{R}\phi(\theta, t) = \int_{\{x:\langle x,\theta\rangle = t\}} \phi(x)dx.$$

The connection between the two transforms is given by the next lemma.

LEMMA 4.5.6. *Fix $\theta \in S^{n-1}$. The Fourier transform of the function $t \mapsto g(t) = \mathcal{R}\phi(\theta, t)$ is the function $s \mapsto \widehat{\phi}(s\theta)$.*

Proof. Making the change of variables $t = \langle x, \theta \rangle$ we get

(4.5.3) $$\widehat{\phi}(s\theta) = \int_{\mathbb{R}^n} \phi(x) e^{-is\langle x,\theta\rangle} dx = \int_{-\infty}^{\infty} e^{-ist} \left(\int_{\{x:\langle x,\theta\rangle=t\}} \phi(x)dx \right) dt$$
$$= \int_{-\infty}^{\infty} g(t) e^{-ist} dt = \widehat{g}(s).$$

This proves the lemma. □

The spherical Radon transform $R: C(S^{n-1}) \to C(S^{n-1})$ is defined by
$$Rf(\theta) = \int_{S^{n-1} \cap \theta^\perp} f(x)dx$$
for any $f \in C(S^{n-1})$. It is directly related with the volume of central hyperplane sections of convex bodies: if K is a symmetric convex body in $\mathbb{R}^n$ then

(4.5.4) $$|K \cap \theta^\perp| = \frac{1}{n-1} R(\rho_K^{n-1})(\theta)$$

for all $\theta \in S^{n-1}$, where ρ_K is the radial function of K. A property of the spherical Radon transform that we will need is that it is self-adjoint (a simple proof can be found in [**228**, Lemma 1.3.3]). For any $f, g \in C(S^{n-1})$ one has
$$\int_{S^{n-1}} Rf(\theta) g(\theta) d\theta = \int_{S^{n-1}} f(\theta) Rg(\theta) d\theta.$$

The next lemma describes the Fourier transform of homogenous functions of degree $-n+1$ on $\mathbb{R}^n$.

LEMMA 4.5.7. *Let f be an even homogenous function of degree $-n+1$ on $\mathbb{R}^n$, which is continuous on S^{n-1}. Then, the Fourier transform of f is an even continuous function on $\mathbb{R}^n \setminus \{0\}$ which is homogeneous of degree -1 and satisfies*
$$\widehat{f}(\xi) = \pi \int_{S^{n-1} \cap \xi^\perp} f(\theta)\, d\theta = \pi Rf(\xi),$$
where the Radon transform is applied to $f|_{S^{n-1}}$.

Proof. Let ϕ be an even test function. Then,

$$(4.5.5) \qquad \langle \widehat{f}, \phi \rangle = \langle f, \widehat{\phi} \rangle = \int_{\mathbb{R}^n} f(x)\widehat{\phi}(x)dx = \int_{S^{n-1}} f(\theta)\left(\int_0^\infty \widehat{\phi}(t\theta)dt\right)d\theta.$$

From Lemma 4.5.6 we have that for every (fixed) $\theta \in S^{n-1}$ the Fourier transform of the function $t \mapsto \widehat{\phi}(t\theta)$ is the function

$$s \mapsto 2\pi \int_{\{x:\langle x,\theta\rangle=s\}} \phi(x)dx.$$

It follows that

$$2\int_0^\infty \widehat{\phi}(t\theta)\,dt = \int_{-\infty}^\infty \widehat{\phi}(t\theta)dt = (\widehat{\widehat{\phi}(t\theta)})(0) = 2\pi \int_{\{x:\langle x,\theta\rangle=0\}} \phi(x)dx.$$

Then,

$$\int_{S^{n-1}} f(\theta)\left(\int_0^\infty \widehat{\phi}(t\theta)dt\right)d\theta = \pi \int_{S^{n-1}} f(\theta)\left(\int_{S^{n-1}\cap\theta^\perp}\left(\int_0^\infty r^{n-2}\phi(r\xi)dr\right)d\xi\right)d\theta$$

$$= \pi \int_{S^{n-1}} f(\theta) R\left(\int_0^\infty r^{n-2}\phi(r\xi)dr\right)(\theta)\,d\theta.$$

We apply the self-adjointness of the spherical Radon transform for the functions f and $g(\xi) = \int_0^\infty r^{n-2}\phi(r\xi)dr$ to write

$$\pi \int_{S^{n-1}} f(\theta) R\left(\int_0^\infty r^{n-2}\phi(r\xi)dr\right)(\theta)\,d\theta = \pi \int_{S^{n-1}} \left(\int_0^\infty r^{n-2}\phi(r\xi)dr\right) Rf(\xi)d\xi$$

$$= \pi \int_{S^{n-1}} \left(\int_0^\infty r^{n-2}\phi(r\xi)dr\right)\left(\int_{S^{n-1}\cap\xi^\perp} f(\theta)d\theta\right)d\xi$$

$$= \pi \int_{\mathbb{R}^n} \frac{1}{\|x\|_2}\left(\int_{S^{n-1}\cap(x/\|x\|_2)^\perp} f(\theta)d\theta\right)\phi(x)dx$$

(for the last equality, use integration in polar coordinates with $x = r\xi$). This shows that

$$\widehat{f}(x) = \frac{\pi}{\|x\|_2}\int_{S^{n-1}\cap(x/\|x\|_2)^\perp} f(\theta)d\theta$$

for $x \neq 0$. In particular, $\widehat{f}(\xi) = \pi Rf(\xi)$ for all $\xi \in S^{n-1}$. $\square$

As a consequence we obtain the next theorem that relates the volume of central hyperplane sections to the Fourier transform.

THEOREM 4.5.8 (Koldobsky). *Let K be a symmetric convex body in $\mathbb{R}^n$. Then, for every $\xi \in S^{n-1}$,*

$$|K \cap \xi^\perp| = \frac{1}{\pi(n-1)}\left(\widehat{\|x\|_K^{-n+1}}\right)(\xi).$$

Proof. We apply Lemma 4.5.7 for the function $f = \|\cdot\|_K^{-n+1}$. From (4.5.4) we see that $\widehat{\|x\|_K^{-n+1}}$ is a continuous function on $\mathbb{R}^n \setminus \{0\}$ and

$$\left(\widehat{\|x\|_K^{-n+1}}\right)(\xi) = \pi R(\|\cdot\|_K^{-n+1})(\xi) = \pi(n-1)|K\cap\xi^\perp|$$

for all $\xi \in S^{n-1}$. $\square$

We are now going to use this formula in order to compute the volume of hyperplane sections of the unit ball of ℓ_p^n, $1 \leqslant p \leqslant 2$. We denote by γ_p the Fourier transform of the function $z \mapsto \exp(-|z|^p)$, $z \in \mathbb{R}$. For $0 < p \leqslant 2$, γ_p is (up to a constant) the density of the standard p-stable measure on $\mathbb{R}$, so γ_p is a non-negative function. Also, for every $p > 0$ we have that

$$\lim_{t\to\infty} t^{p+1}\gamma_p(t) = 2\Gamma(p+1)\sin(\pi p/2),$$

so γ_p decreases at infinity as $|t|^{-1-p}$. Simple calculations show that $\gamma_p(0) = 2\Gamma(1+1/p)$ and $\int_0^\infty \gamma_p(t)dt = \pi$.

LEMMA 4.5.9. *Fix $p > 0$, a non-negative integer n and β such that $-n < \beta < 0$. If $\xi = (\xi_1, \ldots, \xi_n) \in \mathbb{R}^n$ with $\xi_k \neq 0$ for all k, then*

$$\left(\widehat{\|x\|_p^\beta}\right)(\xi) = \frac{p}{\Gamma(-\beta/p)} \int_0^\infty t^{n+\beta-1} \prod_{k=1}^n \gamma_p(t\xi_k)\, dt.$$

Proof. By the definition of the Gamma function we get

$$(|x_1|^p + \cdots + |x_n|^p)^{\beta/p} = \frac{p}{\Gamma(-\beta/p)} \int_0^\infty y^{-1-\beta} \exp\left(-y^p\left(|x_1|^p + \cdots + |x_n|^p\right)\right) dy.$$

For every $y > 0$, the Fourier transform of the function

$$x \mapsto \exp\left(-y^p\left(|x_1|^p + \cdots + |x_n|^p\right)\right)$$

at any point $\xi \in \mathbb{R}^n$ is equal to $y^{-n} \prod_{k=1}^n \gamma_p(\xi_k/y)$. Making the change of variables $t = 1/y$ we get

$$\left(\widehat{\|x\|_p^\beta}\right)(\xi) = \frac{p}{\Gamma(-\beta/p)} \int_0^\infty y^{-n-\beta-1} \prod_{k=1}^n \gamma_p(\xi_k/y)\, dy$$

$$= \frac{p}{\Gamma(-\beta/p)} \int_0^\infty t^{n+\beta-1} \prod_{k=1}^n \gamma_p(t\xi_k)\, dt$$

as claimed. $\square$

Using Lemma 4.5.9 (with $\beta = -n+1$) and Theorem 4.5.8 we get the following.

COROLLARY 4.5.10. *For every $p \geq 1$ and $\xi \in S^{n-1}$,*

$$|B_p^n \cap \xi^\perp| = \frac{p}{\pi(n-1)\Gamma((n-1)/p)} \int_0^\infty \prod_{k=1}^n \gamma_p(t\xi_k)\, dt.$$

For the proof of the main theorem about minimal sections of the ℓ_p^n-balls for $1 \leq p \leq 2$ we need one more property of γ_p.

LEMMA 4.5.11. *For every $p \in (0,2)$, the function $\gamma_p(\sqrt{t})$ is log-convex on $(0,\infty)$.*

Proof. It is a well-known fact that there exists a measure μ on $[0,\infty)$ whose Laplace transform is equal to $\exp\left(-t^{p/2}\right)$. For this measure μ we have that

$$\exp(-|z|^p) = \int_0^\infty \exp(-uz^2)\, d\mu(u).$$

Calculating the Fourier transform of both sides of the latter equality (viewed as functions of the variable z) we see that, for every $t \in \mathbb{R}$,

$$\gamma_p(t) = \sqrt{2\pi} \int_0^\infty u^{-1/2} \exp\left(\frac{-t^2}{4u}\right) d\mu(u).$$

Now the fact that γ_p is log-convex follows from the Cauchy-Schwarz inequality applied to the functions $\exp(-t_1/8u)$ and $\exp(-t_2/(8u))$ and the measure $u^{-1/2}d\mu(u)$. $\square$

THEOREM 4.5.12 (Koldobsky). *For every $1 \leq p \leq 2$ and every $\xi \in S^{n-1}$ we have that*

$$\frac{p}{\pi(n-1)\Gamma((n-1)/p)} \int_0^\infty \gamma_p^n(t/\sqrt{n})\, dt \leq |B_p \cap \xi^\perp| \leq$$

$$\frac{p}{\pi(n-1)\Gamma((n-1)/p)} \gamma_p^{n-1}(0) \int_0^\infty \gamma_p(t)\, dt = \frac{2^{n-1}p\left(\Gamma(1+1/p)\right)^{n-1}}{(n-1)\Gamma((n-1)/p)},$$

with equality if and only if $|\xi_i| = 1/\sqrt{n}$ for every i for the lower bound and upper bound occurs if and only if one of the coordinates of ξ is equal to ± 1 and the others are equal to zero.

Proof. From Corollary 4.5.10 it suffices to prove that for every $\xi \in S^{n-1}$ and every $t > 0$

$$\gamma_p^n(t/\sqrt{n}) \leqslant \prod_{k=1}^n \gamma_p(t\xi_k) \leqslant \gamma_p^{n-1}(0)\gamma_p(t)$$

The left hand side inequality is an immediate consequence of Jensen's inequality applied to the convex function $t \mapsto \log \gamma_p(\sqrt{t})$ and the right hand side inequality is a consequence of Karamata's majorization inequality applied to the same function. □

4.5.3. Sections of the cube

Next we discuss upper bounds for the volume of the sections of the cube. The case of hyperplane sections was settled by Ball in [**31**].

THEOREM 4.5.13 (Ball). *Let $C_n := \frac{1}{2}B_\infty^n$ be the centered cube of volume 1 in $\mathbb{R}^n$. For any unit vector $\theta = (\theta_1, \ldots, \theta_n) \in S^{n-1}$ we have*

$$|C_n \cap \theta^\perp| \leqslant \sqrt{2}.$$

We have equality if and only if θ has exactly two non-zero coordinates of absolute value $\frac{1}{\sqrt{2}}$. In other words, the maximal hyperplane section of C_n is the product of C_{n-2} and the diagonal of C_2.

Proof. Let $\theta \in S^{n-1}$. We may assume that all $\theta_j \neq 0$ ($1 \leqslant j \leqslant n$), otherwise we are reduced to a similar lower-dimensional problem. Because of the symmetries of C_n we may also assume that $\theta_j > 0$ for all j. We distinguish two cases:

Case 1. Assume that $\theta_j \geqslant \frac{1}{\sqrt{2}}$ for some j. We can check that $P_{e_j^\perp}(C_n \cap \theta^\perp) = C_n \cap e_j^\perp$, which implies that

$$|C_n \cap \theta^\perp| = \frac{1}{\theta_j}|P_{e_j^\perp}(C_n \cap \theta^\perp)| = \frac{1}{\theta_j}|C_n \cap e_j^\perp| = \frac{1}{\theta_j} \leqslant \sqrt{2}.$$

Case 2. We assume that $0 < \theta_j \leqslant \frac{1}{\sqrt{2}}$ for all $j = 1, \ldots, n$. If we define

$$S(t) = |C_n \cap \{\langle x, \theta\rangle = t\}| \qquad (t \in \mathbb{R})$$

then, for every $u \in \mathbb{R}$,

$$\widehat{S}(u) = \int_\mathbb{R} S(t)e^{-2\pi i ut}dt = \int_{C_n} e^{-2\pi i u\langle x,\theta\rangle}dx = \prod_{j=1}^n \int_{-1/2}^{1/2} e^{-2\pi i u\theta_j x_j}dx_j$$

$$= \prod_{j=1}^n \frac{\sin(\pi\theta_j u)}{\pi\theta_j u}.$$

It follows that

$$S(t) = \int_{-\infty}^\infty \widehat{S}(u)e^{2\pi i ut}du = \int_{-\infty}^\infty e^{2\pi i ut}\prod_{j=1}^n \frac{\sin(\pi\theta_j u)}{\pi\theta_j u}\,du.$$

Since $\theta_1^2 + \cdots + \theta_n^2 = 1$, applying Hölder's inequality we may write

$$S(t) \leqslant \int_{-\infty}^\infty \prod_{j=1}^n \left|\frac{\sin(\pi\theta_j u)}{\pi\theta_j u}\right|du \leqslant \prod_{j=1}^n \left(\int_{-\infty}^\infty \left|\frac{\sin(\pi\theta_j u)}{\pi\theta_j u}\right|^{1/\theta_j^2}du\right)^{\theta_j^2}.$$

Note that $1/\theta_j^2 \geqslant 2$ for all j; then, making the change of variables $x = \theta_j u$, we see that it is enough to show the following: for every $s \geqslant 2$,

(4.5.6) $$\int_{-\infty}^{\infty} \left|\frac{\sin \pi x}{\pi x}\right|^s dx < \sqrt{\frac{2}{s}}$$

for all $s > 2$ (of course, this becomes an equality for $s = 2$). This will give

$$S(t) \leqslant \prod_{j=1}^{n} \left(\int_{-\infty}^{\infty} \left|\frac{\sin(\pi x)}{\pi x}\right|^{1/\theta_j^2} \frac{du}{\theta_j}\right)^{\theta_j^2} \leqslant \prod_{j=1}^{n} \left(\sqrt{2\theta_j^2} \frac{1}{\theta_j}\right)^{\theta_j^2}$$
$$= 2^{\frac{1}{2}\sum_{j=1}^n \theta_j^2} = \sqrt{2},$$

and, in particular,

$$|C_n \cap \theta^\perp| = S(0) \leqslant \sqrt{2}.$$

Proof of (4.5.6). The proof is based on the next lemma. Recall that the distribution function of a non-negative measurable function $f : (X, \mu) \to \mathbb{R}^+$ is defined by

$$F(y) := \mu(\{x \in X : f(x) \geqslant y\})$$

and is decreasing on $(0, \infty)$.

LEMMA 4.5.14 (Nazarov-Podkorytov). *Let $f, g : (X, \mu) \to \mathbb{R}^+$ be two measurable functions and let F, G be their distribution functions. Assume that $F(y)$ and $G(y)$ are finite for every $y > 0$ and that there exists $y_0 > 0$ such that $F(y) \leqslant G(y)$ on $(0, y_0)$ and $F(y) \geqslant G(y)$ on (y_0, ∞). Let $S = \{s > 0 : f^s - g^s \in L^1(X, \mu)\}$. Then, the function*

$$\varphi(s) = \frac{1}{s y_0^s} \int_X (f^s - g^s) \, d\mu$$

is increasing on S. In particular, if

$$\int_X (f^{s_0} - g^{s_0}) \, d\mu = 0$$

then

$$\int_X (f^s - g^s) \, d\mu \geqslant 0$$

for every $s > s_0$, with equality only if $F \equiv G$.

Proof. We first claim that if $f - g$ is integrable and $F(y), G(y)$ are finite for all $y > 0$, then

$$\int_X (f - g) \, d\mu = \int_0^\infty (F(y) - G(y)) \, dy.$$

To see this, define $h(x) = \min(f(x), g(x))$. Note that the distribution function H of h is bounded by F, and hence it is finite for all $y > 0$. Also $0 \leqslant f - h \leqslant |f - g|$, so $f - h$ is integrable. Applying Fubini's theorem for the characteristic function of $\{(x, y) \in X \times (0, \infty) : h(x) \leqslant y < f(x)\}$, we get

$$0 \leqslant \int_X (f - h) \, d\mu = \int_0^\infty (F(y) - H(y)) \, dy < \infty.$$

In the same way, we see that

$$0 \leqslant \int_X (g - h) \, d\mu = \int_0^\infty (G(y) - H(y)) \, dy < \infty,$$

and subtracting the two identities we get our claim.

Note that, for any $s > 0$, the distribution function of f^s is the function $y \mapsto F(y^{1/s})$. Thus,

$$\int_X (f^s - g^s) \, d\mu = \int_0^\infty (F(y^{1/s}) - G(y^{1/s})) \, dy = s \int_0^\infty y^{s-1} (F(y) - G(y)) \, dy.$$

It follows that if $s, s_0 \in S$ and $s > s_0$ then
$$\varphi(s) - \varphi(s_0) = \frac{1}{y_0} \int_0^\infty \left(\left(\frac{y}{y_0}\right)^{s-1} - \left(\frac{y}{y_0}\right)^{s_0-1} \right) (F(y) - G(y)) \, dy \geq 0,$$
where we use the fact that both factors of the integrand change their signs exactly at y_0. □

We continue with the **proof of (4.5.6)**. Using the identity $\int_{\mathbb{R}} e^{-\pi s x^2} dx = \frac{1}{2\sqrt{s}}$ we rewrite the inequality in the form
$$\int_0^\infty g^s < \int_0^\infty f^s,$$
where $f(x) = e^{-\pi x^2/2}$ and $g(x) = \left|\frac{\sin \pi x}{\pi x}\right|$. We have equality for $s = 2$ and we want to prove this inequality for $s > 2$. So, it is enough to check that the conditions of Lemma 4.5.14 are satisfied for
$$X = (0, \infty), \qquad d\mu = dx, \qquad s_0 = 2.$$
Observe that $\max(f, g) \leq 1$ on $(0, \infty)$, and hence $F(y) = G(y) = 0$ if $y \geq 1$. So, we may work with values $y \in (0, 1)$.

Since f is decreasing, we have
$$F(y) = f^{-1}(y) = \sqrt{\frac{2}{\pi} \log \frac{1}{y}}.$$
In order to estimate $G(y)$, we set $y_m = \max_{[m, m+1]}(g)$ for all $m \in \mathbb{N}$. Note that $\frac{1}{\pi(m+\frac{1}{2})} < y_m < \frac{1}{\pi m}$.

For $x \in (0, 1)$ we have
$$g(x) = \prod_{k=1}^\infty \left(1 - \frac{x^2}{k^2}\right) \leq \prod_{k=1}^\infty e^{-x^2/k^2} = e^{-(\pi x)^2/6} < e^{-\pi x^2/2} = f(x).$$
It follows that if $y \in (y_1, 1)$ then
$$G(y) = \mu(\{x \in (0,1) : g(x) > y\}) < \mu(\{x \in (0,1) : f(x) > y\}) \leq F(y).$$
This shows that $F - G > 0$ on $(y_1, 1)$.

Since
$$\int_0^\infty 2y(F(y) - G(y)) \, dy = \int_0^\infty (f(x)^2 - g(x)^2) \, dx = 0,$$
the difference $F - G$ must change sign at least once on $(0, \infty)$. We need to show that this happens only once, and it suffices to show that $F - G$ is increasing on $(0, y_1)$. Since F and G are continuous and decreasing, it is enough to check that $|G'(y)| > |F'(y)|$ on each interval (y_{m+1}, y_m). The derivative of G at $y \in (0, y_1)$, $y \neq y_m$, satisfies
$$|G'(y)| = \sum_{\{x > 0 : g(x) = y\}} \frac{1}{|g'(x)|}.$$
For $y \in (y_{m+1}, y_m)$ the equation $g(x) = y$ has one root in the interval $(0, 1)$ and two roots in each interval $(k, k+1)$, $1 \leq k \leq m$. In the first case we have
$$|g'(x)| = \frac{\sin \pi x - \pi x \cos \pi x}{\pi x^2} = \frac{1}{\pi x^2} \int_0^{\pi x} t \sin t \, dt \leq \frac{1}{\pi x^2} \int_0^{\pi x} t \, dt = \frac{\pi}{2}.$$
If the root $x \in (k, k+1)$, $k \geq 1$, then
$$|g'(x)| = \left|\frac{\cos \pi x}{x} - \frac{\sin \pi x}{\pi x^2}\right| \leq \frac{1}{x}\left(1 + \frac{|\sin \pi(x-k)|}{\pi x}\right) \leq \frac{1}{x}\left(1 + \frac{\pi(x-k)}{\pi k}\right) = \frac{1}{k}.$$

Therefore, if $y \in (y_{m+1}, y_m)$ we have

$$|G'(y)| \geqslant \frac{2}{\pi} + 2\sum_{k=1}^{m} k = \frac{2}{\pi} + m + m^2.$$

Recall that $F(y) = \sqrt{\frac{2}{\pi} \log \frac{1}{y}}$. Then,

$$\left|\frac{G'(y)}{F'(y)}\right| = |G'(y)| y\sqrt{2\pi \log \frac{1}{y}} \geqslant \left(\frac{2}{\pi} + m + m^2\right) y \sqrt{2\pi \log \frac{1}{y}}.$$

The function $y \mapsto y\sqrt{2\pi \log \frac{1}{y}}$ is increasing on $(0, y_1)$ because $y_1 < \frac{1}{\pi} < \frac{1}{\sqrt{e}}$. Since $y > y_{m+1} > \frac{1}{\pi(m+\frac{3}{2})}$, we see that

$$\left|\frac{G'(y)}{F'(y)}\right| = |G'(y)| > \frac{\frac{2}{\pi} + m + m^2}{\frac{3}{2} + m} \sqrt{\frac{2}{\pi} \log \pi(m+\tfrac{3}{2})} > \sqrt{\frac{2}{\pi} \log \frac{5\pi}{2}},$$

because $m \geqslant 1$ and $\frac{2}{\pi} + m^2 > \frac{1}{2} + 1 = \frac{3}{2}$.

It remains to check that this last quantity is greater than 1: this is true because $\log 5x > x$ on $[1, 2]$ (to check it, use the fact that $\log 5x$ is concave and compute at the endpoints 1 and 2); we apply this for $x = \pi/2$. □

Using the Brascamp-Lieb inequality, Ball [34] extended his result to the case of k-codimensional central sections of the cube.

THEOREM 4.5.15 (Ball). *Let $C_n := \frac{1}{2} B_\infty^n$ be the centered cube of volume 1 in $\mathbb{R}^n$. For any k-codimensional subspace F of $\mathbb{R}^n$ we have*

$$|C_n \cap F| \leqslant (\sqrt{2})^k.$$

Sketch of the proof. We assume that the result has been proved for the $(n-1)$-dimensional cube C_{n-1} and, as in the proof of Theorem 4.5.13, we distinguish two cases. If $F^\perp$ contains a unit vector θ with at least one coordinate $|\theta_j| > \frac{1}{\sqrt{2}}$ then a simple geometric argument, which is analogous to the one in the first part of the proof of Theorem 4.5.13, and the inductive hypothesis yield the bound $|C_n \cap F| \leqslant (\sqrt{2})^k$.

We assume that every unit vector θ in $F^\perp$ has all it coordinates $|\theta_j| \leqslant \frac{1}{\sqrt{2}}$. We set $P := P_{F^\perp}$, $a_i = \|P(e_i)\|_2$ and $u_i = \frac{P(e_i)}{a_i}$ for all $1 \leqslant i \leqslant n$. Observe that a_i is the i-th coordinate of the unit vector u_i in $F^\perp$, and hence $a_i \leqslant \frac{1}{\sqrt{2}}$ for all $1 \leqslant i \leqslant n$. We may also assume that $a_i > 0$ for all i, otherwise we are reduced to the case of the $(n-1)$-dimensional cube C_{n-1}. We define $f : F^\perp \to [0, \infty)$ by

$$f(z) = |C_n \cap (F + z)|.$$

Let $X_1, \ldots, X_n$ be independent random variables, uniformly distributed in $\left[-\frac{1}{2}, \frac{1}{2}\right]$ and consider the random vector $X = (X_1, \ldots, X_n)$. Then, f is the density of the random vector $P(X)$. The characteristic function $\phi : F^\perp \to \mathbb{R}$ of $P(X)$ is given by

$$\phi(\omega) = \int_{F^\perp} e^{i\langle z, \omega \rangle} f(z) dz = \mathbb{E}\left(\exp(i\langle P(X), \omega \rangle)\right)$$

$$= \mathbb{E}\left(\exp\left(i \sum_{j=1}^n X_j \langle P(e_j), \omega \rangle\right)\right) = \mathbb{E}\left(\exp\left(i \sum_{j=1}^n X_j a_j \langle u_j, \omega \rangle\right)\right)$$

$$= \prod_{j=1}^n \frac{2 \sin \frac{1}{2} a_j \langle u_j, \omega \rangle}{a_j \langle u_j, \omega \rangle}.$$

By the Fourier inversion formula we get

$$|C_n \cap F| = f(0) = \frac{1}{(2\pi)^k} \int_{F^\perp} \phi(\omega)\,d\omega = \frac{1}{(2\pi)^k} \int_{F^\perp} \prod_{j=1}^n \frac{2\sin\frac{1}{2}a_j\langle u_j, \omega\rangle}{a_j\langle u_j, \omega\rangle}\,d\omega$$

$$= \frac{1}{\pi^k} \int_{F^\perp} \prod_{j=1}^n \frac{\sin a_j\langle u_j, \omega\rangle}{a_j\langle u_j, \omega\rangle}\,d\omega \leqslant \frac{1}{\pi^k} \int_{F^\perp} \prod_{j=1}^n \left|\frac{\sin a_j\langle u_j, \omega\rangle}{a_j\langle u_j, \omega\rangle}\right|\,d\omega$$

$$= \frac{1}{\pi^k} \int_{F^\perp} \prod_{j=1}^n \phi_j(\langle u_j, \omega\rangle)\,d\omega,$$

where $\phi_j : \mathbb{R} \to [0, \infty)$ is defined by

$$\phi_j(t) = \left|\frac{\sin a_j t}{a_j t}\right|, \qquad 1 \leqslant j \leqslant n.$$

Now, observe that

$$I_{F^\perp} = \sum_{j=1}^n P(e_j) \otimes P(e_j) = \sum_{j=1}^n a_j^2 u_j \otimes u_j.$$

In particular,

$$\sum_{j=1}^n a_j^2 = k.$$

An application of the Brascamp-Lieb inequality gives

$$|C_n \cap F| \leqslant \frac{1}{\pi^k} \prod_{j=1}^n \left(\int_\mathbb{R} \phi_j(t)^{a_j^{-2}}\,dt\right)^{a_j^2}.$$

We set $p_j = a_j^{-2}$, $1 \leqslant j \leqslant n$, and apply (4.5.6) to get

$$|C_n \cap F| \leqslant \prod_{j=1}^n \left(\frac{1}{\pi}\int_{-\infty}^\infty \left|\frac{\sin a_j t}{a_j t}\right|^{p_j}\,dt\right)^{1/p_j} = \prod_{j=1}^n \left(\frac{1}{a_j\pi}\int_{-\infty}^\infty \left|\frac{\sin t}{t}\right|^{p_j}\,dt\right)^{1/p_j}$$

$$\leqslant \prod_{j=1}^n \left(\frac{1}{a_j}\sqrt{\frac{2}{p_j}}\right)^{1/p_j} = \prod_{j=1}^n (\sqrt{2})^{1/p_j} = (\sqrt{2})^{\sum_{j=1}^n a_j^2}$$

$$= (\sqrt{2})^k,$$

as claimed.

4.6. Notes and references

Unconditional convex bodies

The Loomis-Whitney inequality [**337**] was used by Schmuckenschläger in [**460**] for a proof of Theorem 4.1.2. The second proof that we present in this section, as well as the inclusions $K \subseteq \sqrt{3/2n}B_1^n$ and $K \supseteq [-L_K/\sqrt{2}, L_K/\sqrt{2}]^n$, come from the article of Bobkov and Nazarov [**90**].

Convex bodies whose polar bodies contain large affine cubes

The results of this section (Lemma 4.2.2, Proposition 4.2.3 and Proposition 4.2.5) are from the article of Milman and Pajor [**384**, Section 3].

2-convex bodies

Chapter 1.e of the second volume of the book of Lindenstrauss and Tzafriri [**331**] is a standard reference on uniform convexity and related notions. Sections, projections, and sections of projections of L_p-balls are basic examples of 2-convex bodies, with constants that depend only on p. Concentration properties of the uniform measure on 2-convex bodies were studied by Gromov and Milman [**234**], Bobkov and Ledoux [**86**], and by Schmuckenschläger who observed the boundedness of their isotropic constant in [**461**]. The proof of the main inequality that is used this section (Lemma 4.2.10) exploits the idea of the proof of Lemma 1.7.8 (which in turn has its roots in [**234**]) by Arias de Reyna, Ball and Villa and appears in the paper [**283**] of Klartag and E. Milman. Proposition 4.2.8 and Proposition 4.2.9 are from the same paper.

Zonoids

We refer to the articles of Bolker [**93**], Bourgain and Lindenstrauss [**108**], Goodey and Weil [**221**] for more information on zonoids. For the full proofs of the sharp bounds on the circumradius of a zonoid in Lewis position, Löwner position and minimal mean width position, the interested reader may also consult [**322**], [**37**] and [**215**]. It seems that it is not known whether a zonoid Z in $\mathbb{R}^n$ which is in the isotropic position satisfies $R(Z) \leqslant c\sqrt{n}$ for some absolute constant $c > 0$.

The isotropic constant of Schatten classes

Uniform boundedness of the isotropic constants of the unit balls of the Schatten classes was established by König, Meyer and Pajor in [**298**]. One of the main ingredients of the proof is a formula of Saint-Raymond from [**450**] (see Lemma 4.3.1). The proof of Proposition 4.3.4 is related to ideas of Aomoto [**19**] who studied Jacobi integrals associated to Selberg integrals (see also [**359**]). Before the work of König, Meyer and Pajor, Dar had obtained the estimate $L_{B(S_p^n)} \leqslant C\sqrt{\log n}$ in [**154**] (see also [**152**] for the case $p = 1$).

Bodies with few vertices or few facets

There are several articles that give bounds for the isotropic constant of polytopes; these bounds depend on the number of their vertices or facets. Some of them are formulated in an equivalent functional analytic language. Among them are the works of Ball [**35**], Junge [**261**] and [**262**] and E. Milman [**366**]. In this section we present a more geometric approach of Alonso-Gutiérrez [**11**]; besides the fact that it is closer to the setting of this book, its main advantage is that it addresses the technical questions that arise when one wants to work with not necessarily symmetric polytopes as well.

On the way we discuss well-known lower bounds for the volume of symmetric polytopes with few facets, obtained by Carl-Pajor [**134**], and independently by Gluskin [**220**] (also, in dual form, by Bárány and Füredi in [**46**]). The simple proof that we present here makes use of Sidák's lemma [**467**]; it appears in the survey article of Ball [**41**]. For related volume inequalities see also [**44**].

For the Gaussian correlation conjecture see Pitt [**431**], Schechtman, Schlumprecht and Zinn [**455**], Hargé [**247**] and Cordero-Erausquin [**141**].

Vaaler's inequality and its extension by Meyer and Pajor

Vaaler's inequality was proved in [**495**]. For the notion of peakedness (or dominance) see [**18**], [**466**], [**441**] and [**268**]. The main result of this section is due to Meyer and Pajor. It appeared in [**363**] and extends Vaaler's inequality.

Fourier transform and volume of hyperplane sections

Koldobsky's formula (Theorem 4.5.8) which expresses the volume of central hyperplane sections of convex bodies in terms of the Fourier transform of their radial function appears in [290]. Theorem 4.5.12 which identifies the central sections of the unit balls of the spaces ℓ_p^n, $1 \leqslant p \leqslant 2$, that have minimal volume is from the same article. The case $p = 1$ had been established by Meyer and Pajor in [363]. A complex version of the problem was studied by Koldobsky and Zymonopoulou [297]. The problem to find the maximal hyperplane section of B_q^n in the case $q > 2$ is open; Oleszkiewicz has shown [402] that the answer depends on both q and n.

The dual problem, concerning hyperplane projections of B_p^n, was studied by Barthe and Naor [60] and by Koldobsky, Ryabogin and Zvavitch [294] and [295]. We refer to the books of Koldobsky [292], Koldobsky and Yaskin [296] for a thorough presentation of the Fourier analytic approach and its applications.

Sections of the cube

Theorem 4.5.13 is Ball's well-known $\sqrt{2}$-bound for the volume of hyperplane sections of the unit cube. It was proved in [31]. The inequality

$$\int_{-\infty}^{\infty} \left|\frac{\sin \pi x}{\pi x}\right|^s dx < \sqrt{\frac{2}{s}}, \qquad s > 2$$

plays a key role in Ball's theorem. Here we present a proof of this inequality, which is due to Nazarov and Podkorytov [399]. A complex version of Ball's inequality was proved by Oleszkiewicz and Pelczynski in [403]. For a related interesting question regarding extremal slabs in the cube, see [59].

Using the Brascamp-Lieb inequality, Ball extended his result to the case of k-codimensional central sections of the cube. Theorem 4.5.15 is from [34].

CHAPTER 5

L_q-centroid bodies and concentration of mass

Bourgain's approach to the hyperplane conjecture, as described in Chapter 3, exploited the fact that the ψ_1-norm of any linear functional $x \mapsto \langle x, \theta \rangle$ on an isotropic convex body K in $\mathbb{R}^n$ is bounded by its isotropic constant L_K. In this chapter we introduce the family of L_q-centroid bodies of an isotropic convex body K (or, more generally, of an isotropic log-concave measure) in $\mathbb{R}^n$. For every $q \geqslant 1$, the L_q-centroid body $Z_q(K)$ of K is defined through its support function

$$h_{Z_q(K)}(y) := \|\langle \cdot, y \rangle\|_{L_q(K)} = \left(\int_K |\langle x, y \rangle|^q dx \right)^{1/q}.$$

Note that a convex body K of volume 1 in $\mathbb{R}^n$ is isotropic if and only if it is centered and $Z_2(K) = L_K B_2^n$. The development of an asymptotic theory for this family of bodies, and for their behavior as q increases from 2 to n, was initiated by Paouris and has proved to be a very fruitful idea.

In Section 5.1 we present the basic properties of the family $\{Z_q(\mu) : q \geqslant 2\}$ of the centroid bodies of a centered log-concave probability measure μ on $\mathbb{R}^n$ and prove some fundamental formulas. Two of them will play an important role in this and in subsequent chapters (recall that we write f_μ for the density of μ with respect to the Lebesgue measure):

(i) $f_\mu(0)^{1/n} |Z_n(\mu)|^{1/n} \simeq 1$.
(ii) For every $1 \leqslant k < n$ and for every $F \in G_{n,k}$ and $q \geqslant 1$, one has

$$P_F(Z_q(\mu)) = Z_q(\pi_F(\mu)),$$

where $\pi_F(\mu)$ is the marginal of μ with respect to F, defined by $\pi_F(\mu)(A) := \mu(P_F^{-1}(A))$ for all Borel subsets of F.

The first main application of the theory of L_q-centroid bodies is a striking, very useful, deviation inequality of Paouris: for every isotropic log-concave probability measure μ on $\mathbb{R}^n$ one has

$$\mu(\{x \in \mathbb{R}^n : \|x\|_2 \geqslant ct\sqrt{n}\}) \leqslant \exp\left(-t\sqrt{n}\right)$$

for every $t \geqslant 1$, where $c > 0$ is an absolute constant. The proof of this fact is given in Section 5.2. It is a consequence of the following statement: there exist absolute constants $c_1, c_2 > 0$ such that, if μ is an isotropic log-concave measure on $\mathbb{R}^n$, then

$$I_q(\mu) \leqslant c_2 I_2(\mu)$$

for every $q \leqslant c_1 \sqrt{n}$, where $I_q(\mu)$ is defined, for all $0 \neq q > -n$, by

$$I_q(\mu) = \left(\int_{\mathbb{R}^n} \|x\|_2^q d\mu(x) \right)^{1/q}.$$

In Section 5.3 we present a second theorem of Paouris which, in fact, extends the previous one. If μ is an isotropic log-concave measure on $\mathbb{R}^n$ then, for any $1 \leqslant q \leqslant c_3\sqrt{n}$, one has
$$I_{-q}(\mu) \simeq I_q(\mu).$$
In particular, this shows that for all $1 \leqslant q \leqslant c_3\sqrt{n}$ one has $I_q(\mu) \leqslant cI_2(\mu)$, where $c > 0$ is an absolute constant. Using this result one can derive a small ball probability estimate: for every isotropic log-concave measure μ on $\mathbb{R}^n$ and for any $0 < \varepsilon < \varepsilon_0$, one has
$$\mu(\{x \in \mathbb{R}^n : \|x\|_2 < \varepsilon\sqrt{n}\}) \leqslant \varepsilon^{c_4\sqrt{n}},$$
where $\varepsilon_0, c_4 > 0$ are absolute constants.

5.1. L_q-centroid bodies

5.1.1. Definition and first basic properties

DEFINITION 5.1.1. Let K be a convex body of volume 1 in $\mathbb{R}^n$. For every $q \geqslant 1$ we define the L_q-centroid body $Z_q(K)$ of K by its support function:
$$h_{Z_q(K)}(y) = \|\langle \cdot, y\rangle\|_{L^q(K)} = \left(\int_K |\langle x, y\rangle|^q dx\right)^{1/q}.$$

Observe that $Z_q(K)$ is always symmetric, and $Z_q(T(K)) = T(Z_q(K))$ for every $T \in GL(n)$ and $q \geqslant 1$. Isotropicity can be described through the L_2-centroid body: a centered convex body K of volume one is in the isotropic position if $Z_2(K)$ is a multiple of the Euclidean ball.

The definition is naturally extended to the setting of probability measures. Let $f : \mathbb{R}^n \to [0, \infty)$ be a log-concave function with $\int f = 1$. For any $q \geqslant 1$ we define the L_q-centroid body $Z_q(f)$ of f as the symmetric convex body whose support function is
$$h_{Z_q(f)}(y) := \left(\int_{\mathbb{R}^n} |\langle x, y\rangle|^q f(x)\, dx\right)^{1/q}.$$
More generally, if μ is a probability measure on $\mathbb{R}^n$ then we define
$$h_{Z_q(\mu)}(y) := \left(\int_{\mathbb{R}^n} |\langle x, y\rangle|^q\, d\mu(x)\right)^{1/q}.$$
Note that if μ has a density f_μ with respect to the Lebesgue measure, then $Z_q(\mu) = Z_q(f_\mu)$.

As in the case of convex bodies, $Z_q(\mu)$ is always symmetric, and $Z_q(T(\mu)) = T(Z_q(\mu))$ for every $T \in SL(n)$ and $q \geqslant 1$. A centered log-concave density f is isotropic if $Z_2(f) = B_2^n$.

Assume that K is a convex body of volume 1 in $\mathbb{R}^n$. From Hölder's inequality it is clear that
$$Z_1(K) \subseteq Z_p(K) \subseteq Z_q(K) \subseteq Z_\infty(K)$$
for every $1 \leqslant p \leqslant q \leqslant \infty$, where $Z_\infty(K) = \mathrm{conv}\{K, -K\}$.

Recall that, from Theorem 3.2.7, for all $y \in \mathbb{R}^n$ and all $q > p \geqslant 1$ we have
$$\|\langle \cdot, y\rangle\|_q \leqslant \frac{cq}{p}\|\langle \cdot, y\rangle\|_p.$$

Moreover, if we assume that K is centered, then Corollary 3.2.9 states that
$$\|\langle \cdot, y\rangle\|_{L^n(K)} \simeq \max\{h_K(y), h_K(-y)\}.$$
In the language of L_q-centroid bodies, these two results take the following form.

PROPOSITION 5.1.2. *Let K be a convex body of volume 1 in $\mathbb{R}^n$. Then, for all $1 \leqslant p < q$ we have*
$$Z_p(K) \subseteq Z_q(K) \subseteq \frac{c_1 q}{p} Z_p(K),$$
where $c_1 > 0$ is an absolute constant. If K is also centered, then
$$Z_q(K) \supseteq c_2 Z_\infty(K)$$
for all $q \geqslant n$, where $c_2 > 0$ is an absolute constant. □

A completely analogous fact is true for log-concave measures.

PROPOSITION 5.1.3. *Let μ be a log-concave probability measure, with density f, on $\mathbb{R}^n$. Then, for all $1 \leqslant p < q$ we have*
$$Z_p(f) \subseteq Z_q(f) \subseteq \frac{cq}{p} Z_p(f),$$
where $c > 0$ is an absolute constant. □

L_q-centroid bodies have appeared in the literature under a different normalization. If K is a convex body in $\mathbb{R}^n$ then, for every $1 \leqslant q < \infty$, the body $\Gamma_q(K)$ was defined by Lutwak and Zhang through its support function
$$h_{\Gamma_q(K)}(y) = \left(\frac{1}{c_{n,q}|K|} \int_K |\langle x, y \rangle|^q dx \right)^{1/q},$$
where
$$c_{n,q} = \frac{\omega_{n+q}}{\omega_2 \omega_n \omega_{q-1}}.$$
In other words, $Z_q(K) = c_{n,q}^{1/q} \Gamma_q(K)$ if $|K| = 1$. The normalization of $\Gamma_q(K)$ is chosen so that $\Gamma_q(B_2^n) = B_2^n$ for every q. Lutwak, Yang and Zhang have established the following L_q affine isoperimetric inequality.

THEOREM 5.1.4 (Lutwak-Yang-Zhang). *Let K be a convex body of volume 1 in $\mathbb{R}^n$. For every $q \geqslant 1$,*
$$|\Gamma_q(K)| \geqslant 1,$$
with equality if and only if K is a centered ellipsoid of volume 1. □

The systematic study of the L_q-centroid bodies from an asymptotic point of view started with the works of Paouris [410] and [411]. His main results were obtained a few years after; in this chapter we describe his results from [413] and [414] in the setting of log-concave probability measures.

5.1.2. L_q-centroid bodies of $K_p(\mu)$

Next, we discuss the close relation of the family of L_q-centroid bodies of a centered log-concave probability measure μ with the family of convex bodies $K_p(\mu)$. Our starting point is a consequence of Proposition 2.5.3. Recall that for every compact $A \subset \mathbb{R}^n$ with $|A| > 0$ we denote by $\overline{A}$ the set $A/|A|^{1/n}$.

PROPOSITION 5.1.5. *Let f be a centered log-concave density in $\mathbb{R}^n$. For every $p > 0$,*
$$Z_p(\overline{K_{n+p}}(f))|K_{n+p}(f)|^{\frac{1}{p}+\frac{1}{n}}f(0)^{1/p} = Z_p(f).$$

Proof. Let $p \geq 1$. From Proposition 2.5.3 (vi) we know that
$$\int_{K_{n+p}(f)} |\langle x, \theta \rangle|^p \, dx = \frac{1}{f(0)} \int_{\mathbb{R}^n} |\langle x, \theta \rangle|^p f(x) \, dx$$
for all $\theta \in S^{n-1}$. Since
$$\int_{K_{n+p}(f)} |\langle x, \theta \rangle|^p \, dx = |K_{n+p}|^{1+\frac{p}{n}} \int_{\overline{K_{n+p}}(f)} |\langle x, \theta \rangle|^p \, dx,$$
the result follows. $\square$

Now, we use Proposition 2.5.8. We know that for every $p > 0$ we have
$$e^{-1} \leq f(0)^{\frac{1}{p}+\frac{1}{n}}|K_{n+p}(f)|^{\frac{1}{p}+\frac{1}{n}} \leq e\frac{n+p}{n}.$$
Then, from Proposition 5.1.5 we get:

PROPOSITION 5.1.6. *Let f be a centered log-concave density in $\mathbb{R}^n$. For every $p > 0$,*
$$\frac{1}{e} Z_p(\overline{K_{n+p}}(f)) \subseteq f(0)^{1/n} Z_p(f) \subseteq e\frac{n+p}{n} Z_p(\overline{K_{n+p}}(f)).$$

In subsequent chapters we will also need to compare the L_q-centroid bodies of a centered log-concave density with those of the bodies $\overline{K_{n+1}}(f)$ and $\overline{K_{n+2}}(f)$.

THEOREM 5.1.7. *Let f be a centered log-concave density in $\mathbb{R}^n$. Then, for every $1 \leq q \leq n$, one has*

(5.1.1) $$c_1 f(0)^{1/n} Z_q(f) \subseteq Z_q(\overline{K_{n+1}}(f)) \subseteq c_2 f(0)^{1/n} Z_q(f)$$

and

(5.1.2) $$c_3 f(0)^{1/n} Z_q(f) \subseteq Z_q(\overline{K_{n+2}}(f)) \subseteq c_4 f(0)^{1/n} Z_q(f),$$

where $c_1, c_2, c_3, c_4 > 0$ are absolute constants.

Proof. We may assume that q is a positive integer. In view of Proposition 5.1.6 it suffices to compare $Z_q(\overline{K_{n+1}}(f))$ with $Z_q(\overline{K_{n+q}}(f))$. Using Proposition 2.5.7 (ii) we have that
$$e^{\frac{n}{n+q}-\frac{n}{n+1}}\left(\frac{|K_{n+1}(f)|}{|K_{n+q}(f)|}\right)^{1/n}\overline{K_{n+1}}(f) \subseteq \overline{K_{n+q}}(f)$$
$$\subseteq \frac{(\Gamma(n+q+1))^{\frac{1}{n+q}}}{(\Gamma(n+2))^{\frac{1}{n+1}}}\left(\frac{|K_{n+1}(f)|}{|K_{n+q}(f)|}\right)^{1/n}\overline{K_{n+1}}(f).$$

Taking volumes we obtain
$$\frac{(\Gamma(n+2))^{\frac{1}{n+1}}}{(\Gamma(n+q+1))^{\frac{1}{n+q}}} \leq \left(\frac{|K_{n+1}(f)|}{|K_{n+q}(f)|}\right)^{1/n} \leq e^{-\frac{n}{n+q}+\frac{n}{n+1}},$$
and hence
$$A^{-1}\overline{K_{n+1}}(f) \subseteq \overline{K_{n+q}}(f) \subseteq A\,\overline{K_{n+1}}(f),$$

where $A := e^{\frac{n}{n+1} - \frac{n}{n+q}} \frac{\Gamma(n+q+1)^{\frac{1}{n+q}}}{\Gamma(n+2)^{\frac{1}{n+1}}}$. It follows that

$$A^{-\frac{n+q}{q}} Z_q\left(\overline{K_{n+1}}(f)\right) \subseteq Z_q\left(\overline{K_{n+q}}(f)\right) \subseteq A^{\frac{n+q}{q}} Z_q\left(\overline{K_{n+1}}(f)\right),$$

with

$$A^{\frac{n+q}{q}} = \frac{(\Gamma(n+q+1))^{\frac{1}{n+q} \cdot \frac{n+q}{q}}}{(\Gamma(n+2))^{\frac{1}{n+1} \cdot \frac{n+q}{q}}} \cdot e^{(-\frac{n}{n+q} + \frac{n}{n+1}) \cdot \frac{n+q}{q}}$$

$$= \left(\frac{e^{\frac{n(q-1)}{n+1}}(n+2)\dots(n+q)}{(n+1)!^{\frac{q-1}{n+1}}}\right)^{\frac{1}{q}}$$

$$\leqslant \left(e^{\frac{(2n+1)(q-1)}{n+1}} \frac{(n+q)^{q-1}}{(n+1)^{q-1}}\right)^{\frac{1}{q}}$$

$$\leqslant e^2 \frac{n+q}{n+1}$$

(where we have used the fact that $(n+1)! \geqslant \left(\frac{n+1}{e}\right)^{n+1}$ and the simple bound $(n+2)\cdots(n+q) \leqslant (n+q)^{q-1}$). A similar argument proves the claim for $\overline{K_{n+2}}(f)$. □

5.1.3. Volume radius of $Z_n(f)$

In this short subsection we isolate a formula that follows immediately from the results of the previous subsection but will be very useful in the sequel.

PROPOSITION 5.1.8 (Paouris). *Let f be a centered log-concave density in $\mathbb{R}^n$. Then,*

(5.1.3) $$\frac{c_1}{f(0)^{1/n}} \leqslant |Z_n(f)|^{1/n} \leqslant \frac{c_2}{f(0)^{1/n}},$$

where $c_1, c_2 > 0$ are absolute constants.

Proof. Note that $K_{n+1}(f)$ has center of mass at the origin, hence from Proposition 5.1.2 we know that

$$Z_n(\overline{K_{n+1}}(f)) \simeq \operatorname{conv}\left\{\overline{K_{n+1}}(f), -\overline{K_{n+1}}(f)\right\}.$$

It follows that

$$|Z_n(\overline{K_{n+1}}(f))|^{1/n} \simeq 1.$$

Then, from Theorem 5.1.7 we get

$$f(0)^{1/n} |Z_n(f)|^{1/n} \simeq |Z_n(\overline{K_{n+1}}(f))|^{1/n} \simeq 1.$$

and the result follows. □

If μ is a centered log-concave probability measure on $\mathbb{R}^n$ with density f then Theorem 2.2.2 shows that $\|\mu\|_\infty^{1/n} \simeq f(0)^{1/n}$, therefore we can restate the result of Proposition 5.1.8 as follows.

COROLLARY 5.1.9. *Let μ be a log-concave probability measure on $\mathbb{R}^n$ with $\operatorname{bar}(\mu) = 0$. Then,*

$$|Z_n(\mu)|^{1/n} \simeq \frac{1}{\|\mu\|_\infty^{1/n}}.$$

5.1.4. Marginals and Projections

DEFINITION 5.1.10. Let $f : \mathbb{R}^n \to [0, \infty)$ be an integrable function. Let $1 \leqslant k < n$ be an integer and let $F \in G_{n,k}$. We define the *marginal* $\pi_F(f) : F \to [0, \infty)$ of f with respect to F by

$$\pi_F(f)(x) := \int_{x+F^\perp} f(y) dy. \tag{5.1.4}$$

More generally, for every $\mu \in \mathcal{P}_n$ we can define the marginal of μ with respect to a k-dimensional subspace F setting

$$\pi_F(\mu)(A) := \mu(P_F^{-1}(A))$$

for all Borel subsets of F. In the case where μ has a (log-concave) density f_μ then these two definitions coincide. We can see that

$$f_{\pi_F(\mu)} = \pi_F(f_\mu) \tag{5.1.5}$$

almost everywhere. Indeed; for every Borel subset A of F, we have that

$$\pi_F(\mu)(A) = \mu(P_F^{-1}(A)) = \int f_\mu(x) \mathbf{1}_A(P_F x) \, dx$$
$$= \int_F \int_{F^\perp} f_\mu(x+y) \mathbf{1}_A(x) \, dy \, dx,$$

by Fubini's theorem. Making a change of variables we get

$$\pi_F(\mu)(A) = \int_A \left(\int_{x+F^\perp} f_\mu(z) \, dz \right) dx = \int_A \pi_F(f_\mu)(x) \, dx.$$

In the next proposition we collect some properties of marginals.

PROPOSITION 5.1.11. *Let $f : \mathbb{R}^n \to [0, \infty)$ be an integrable function and let $1 \leqslant k < n$ and $F \in G_{n,k}$.*

(i) *If f is even then the same holds true for $\pi_F(f)$.*
(ii) *We have*

$$\int_F \pi_F(f)(x) \, dx = \int_{\mathbb{R}^n} f(x) \, dx.$$

(iii) *For any measurable function $g : F \to \mathbb{R}$ we have*

$$\int_{\mathbb{R}^n} g(P_F x) f(x) \, dx = \int_F g(x) \pi_F(f)(x) \, dx.$$

(iv) *For every $\theta \in S_F$,*

$$\int_F \langle x, \theta \rangle \pi_F(f)(x) dx = \int_{\mathbb{R}^n} \langle x, \theta \rangle f(x) dx. \tag{5.1.6}$$

In particular, if f is centered then, for every $F \in G_{n,k}$, $\pi_F(f)$ is centered.

(v) *For every $p > 0$ and $\theta \in S_F$,*

$$\int_{\mathbb{R}^n} |\langle x, \theta \rangle|^p f(x) dx = \int_F |\langle x, \theta \rangle|^p \pi_F(f)(x) dx.$$

In particular, if f is isotropic, the same holds true for $\pi_F(f)$.

(vi) *If f is log-concave, then $\pi_F(f)$ is log-concave.*

We note that the same formulas are valid for an arbitrary measure $\mu \in \mathcal{P}_n$.

Proof. The first property is clear. Facts (ii)-(v) are direct consequences of Fubini's Theorem. For the last assertion we use the log-concavity of μ to write

$$\begin{aligned}\pi_F(\mu)((1-\lambda)A+\lambda B) &= \mu(P_F^{-1}((1-\lambda)A+\lambda B)) \\ &= \mu((1-\lambda)P_F^{-1}(A)+\lambda P_F^{-1}(B)) \\ &\geqslant \mu(P_F^{-1}(A))^{1-\lambda}\mu(P_F^{-1}(B))^{\lambda} \\ &= (\pi_F(\mu)(A))^{1-\lambda}(\pi_F(\mu)(B))^{\lambda}\end{aligned}$$

for every pair of compact sets A and B in F and any $\lambda \in (0,1)$. $\square$

5.1.5. Projections of $Z_q(f)$

A simple but crucial observation of Paouris states that every projection of the L_q-centroid body of a density f coincides with the L_q-centroid body of the corresponding marginal of f. The proof is a direct consequence of Fubini's theorem, but this identity will play a very important role in the sequel.

THEOREM 5.1.12 (Paouris). *Let $f : \mathbb{R}^n \to [0,\infty)$ be a log-concave density on $\mathbb{R}^n$. For every $1 \leqslant k < n$ and for every $F \in G_{n,k}$ and $q \geqslant 1$, one has*

(5.1.7) $$P_F(Z_q(f)) = Z_q(\pi_F(f)).$$

Proof. This is clear from Proposition 5.1.11 (v): we saw that for every $q \geqslant 1$ and $\theta \in S_F$ we have

$$\int_{\mathbb{R}^n} |\langle x,\theta\rangle|^q f(x)dx = \int_F |\langle x,\theta\rangle|^q \pi_F(f)(x)dx.$$

This is equivalent to

$$h_{Z_q(f)}(\theta) = h_{Z_q(\pi_F f)}(\theta),$$

for all $\theta \in S_F$, and the result follows from the fact that $h_{P_F(Z_q(f))}(\theta) = h_{Z_q(f)}(\theta)$ for all $\theta \in S_F$ (recall that in general, if C is a convex body then $h_{P_F(C)}(\theta) = h_C(\theta)$ for all F and all $\theta \in S_F$). $\square$

Let f be a centered log-concave density on $\mathbb{R}^n$. Then, for every $F \in G_{n,k}$, the function $\pi_F(f)$ is a centered log-concave density on F. So, we may apply Proposition 5.1.8 for $\pi_F(f)$. It follows that

$$\frac{c_1}{[\pi_F(f)(0)]^{1/k}} \leqslant |Z_k(\pi_F(f))|^{1/k} \leqslant \frac{c_2}{[\pi_F(f)(0)]^{1/k}}.$$

This fact, combined with (5.1.7), proves the following.

THEOREM 5.1.13 (Paouris). *Let f be a centered log-concave density on $\mathbb{R}^n$. Then, for any $1 \leqslant k < n$ and $F \in G_{n,k}$, one has*

(5.1.8) $$c_1 \leqslant [\pi_F(f)(0)]^{\frac{1}{k}}|P_F(Z_k(f))|^{\frac{1}{k}} \leqslant c_2,$$

where $c_1, c_2 > 0$ are absolute constants. $\square$

Let K be a centered convex body of volume 1 in $\mathbb{R}^n$. If we choose $f = \mathbf{1}_K$, then observing that $\pi_F(f)(0) = |K \cap F^{\perp}|$ we obtain the following geometric inequality, which may be viewed as an "L_q-version of the Rogers-Shephard inequality".

THEOREM 5.1.14. *Let K be a centered convex body of volume 1 in $\mathbb{R}^n$. For every $1 \leqslant k < n$ and $F \in G_{n,k}$, one has*

$$(5.1.9) \qquad c_1 \leqslant |K \cap F^\perp|^{\frac{1}{k}} |P_F(Z_k(K))|^{\frac{1}{k}} \leqslant c_2,$$

where $c_1, c_2 > 0$ are absolute constants. □

The next proposition gives some very useful expressions for the volume of central sections of an isotropic convex body; these will be often used, for example, in Chapters 6 and 8.

PROPOSITION 5.1.15. *Let K be an isotropic convex body in $\mathbb{R}^n$. We denote by μ_K the isotropic log-concave measure with density $L_K^n \mathbf{1}_{\frac{K}{L_K}}$. Then, for every $1 \leqslant k < n$ and $F \in G_{n,k}$, the body $\overline{K_{k+1}}(\pi_F(\mu_K))$ is almost isotropic and*

$$(5.1.10) \qquad |K \cap F^\perp|^{1/k} \simeq \frac{L_{\overline{K_{k+1}}(\pi_F(\mu_K))}}{L_K},$$

Also, for all $1 \leqslant q \leqslant k$,

$$(5.1.11) \qquad Z_q\left(\overline{K_{k+1}}(\pi_F(\mu_K))\right) \simeq |K \cap F^\perp|^{1/k} P_F(Z_q(K)).$$

Proof. Fix $1 \leqslant k < n$ and $F \in G_{n,k}$. Let f_K be the density of μ_K. Since f_K is isotropic, Proposition 5.1.11 shows that $\pi_F(f_K)$ is isotropic. Hence, by Proposition 2.5.12 we get that $\overline{K_{k+1}}(\pi_F(f_K))$ is almost isotropic with some absolute constant $C > 0$. Using Theorem 5.1.7 (with $q = 2$) we get:

$$L_{\overline{K_{k+1}}(\pi_F(f_K))} = \left(\frac{|Z_2(\overline{K_{k+1}}(\pi_F(f_K)))|}{|B_F|}\right)^{1/k}$$

$$\simeq \pi_F(f_K)(0)^{1/k} \left(\frac{|Z_2(\pi_F(f_K))|}{|B_F|}\right)^{1/k}$$

$$= \pi_F(f_K)(0)^{1/k} \left(\frac{|P_F(Z_2(f_K))|}{|B_F|}\right)^{1/k},$$

where we have used the fact that $Z_2(\pi_F(f)) = P_F(Z_2(f))$ for any log-concave function f. Note that, since K is isotropic, we get

$$Z_2(f_K) = L_K^{-1} Z_2(K) = B_2^n \quad \text{and hence} \quad P_F(Z_2(f_K)) = B_F.$$

Moreover, we have

$$\pi_F(f_K)(0) = \int_{F^\perp} f_K(y)\, dy = L_K^n \left|\tfrac{1}{L_K} K \cap F^\perp\right| = L_K^k |K \cap F^\perp|.$$

Combining the above we conclude that

$$L_{\overline{K_{k+1}}(\pi_F(f_K))} \simeq L_K |K \cap F^\perp|^{1/k}.$$

The second assertion follows immediately from Theorem 5.1.7 and the equalities $\pi_F(\mu_K)(0)^{1/k} = L_K |K \cap F^\perp|^{1/k}$ and $Z_q(\pi_F(\mu_K)) = L_K^{-1} P_F(Z_q(K))$. □

5.1.6. Volume of L_q-centroid bodies

According to the normalization that we use in this book, the L_q-affine isoperimetric inequality of Lutwak, Yang and Zhang can be written in the following form.

PROPOSITION 5.1.16 (Lutwak-Yang-Zhang). *Let K be a convex body of volume 1 in $\mathbb{R}^n$. Then,*
$$|Z_q(K)|^{1/n} \geqslant |Z_q(\overline{B_2^n})|^{1/n} \geqslant c\sqrt{q/n}$$
for every $1 \leqslant q \leqslant n$, where $c > 0$ is an absolute constant.

Our goal in this section is to show that a reverse inequality holds true (up to the isotropic constant).

THEOREM 5.1.17 (Paouris). *If μ is an isotropic log-concave measure on $\mathbb{R}^n$, then for every $2 \leqslant q \leqslant n$ we have that*
$$(5.1.12) \qquad |Z_q(\mu)|^{1/n} \leqslant c\sqrt{q/n}.$$
Moreover, if K is a centered convex body of volume 1 in $\mathbb{R}^n$, then for every $2 \leqslant q \leqslant n$ we have that
$$(5.1.13) \qquad |Z_q(K)|^{1/n} \leqslant c\sqrt{q/n}\, L_K,$$
where $c > 0$ is an absolute constant.

For the proof we will use some facts from Section 1.4. Steiner's formula asserts that for every convex body C in $\mathbb{R}^n$ we have
$$|C + tB_2^n| = \sum_{k=0}^{n} \binom{n}{k} W_k(C) t^k$$
for all $t > 0$, where $W_k(C) = V_{n-k}(C) = V(C; n-k, B_2^n; k)$ is the k-th quermassintegral of C.

Also, the Alexandrov-Fenchel inequality implies the log-concavity of the sequence $(W_0(C), \ldots, W_n(C))$ and, in particular we have that
$$(5.1.14) \qquad \left(\frac{W_{n-i}(C)}{\omega_n}\right)^{1/i} \geqslant \left(\frac{W_{n-j}(C)}{\omega_n}\right)^{1/j},$$
for all $1 \leqslant i < j \leqslant n$.

We will also use Kubota's integral formula:
$$(5.1.15) \qquad W_{n-m}(C) = \frac{\omega_n}{\omega_m} \int_{G_{n,m}} |P_F(C)| \, d\nu_{n,m}(F), \quad (1 \leqslant m \leqslant n).$$

Proof of Theorem 5.1.17. It is enough to prove (5.1.12) for integer values of $1 \leqslant q \leqslant n-1$. Observe that for any $F \in G_{n,q}$ we have
$$|P_F(Z_q(\mu))|^{1/q} = |Z_q(\pi_F(\mu))|^{1/q} \leqslant \frac{c_1}{[f_{\pi_F(\mu)}(0)]^{1/q}} \leqslant c_2,$$
where we have used Theorem 5.1.12, Proposition 5.1.8 and Proposition 2.3.12, respectively, for the isotropic function $f_{\pi_F(\mu)} = \pi_F(f_\mu)$. Applying (5.1.15) we get
$$W_{n-q}(Z_q(\mu)) \leqslant \frac{\omega_n}{\omega_q} c_2^q.$$
Now, we apply (5.1.14) for $C = Z_q(\mu)$ with $j = n$ and $i = q$; this gives
$$W_{n-q}^{1/q}(Z_q(\mu)) \geqslant |Z_q(\mu)|^{1/n} \omega_n^{1/q - 1/n}.$$

Combining the above, we get
$$|Z_q(\mu)|^{1/n} \leqslant \frac{\omega_n^{1/n}}{\omega_q^{1/q}} c_2.$$

Since $\omega_k^{1/k} \simeq 1/\sqrt{k}$, we get (5.1.12). For the second assertion of the theorem we may assume that K is isotropic (because the volume of $Z_q(T(K))$ is the same for all $T \in SL(n)$). Consider the measure μ with density $f_\mu = L_K^n \mathbf{1}_{\frac{K}{L_K}}$. Then, μ is isotropic and $Z_q(\mu) = L_K^{-1} Z_q(K)$. Thus, the result follows immediately from (5.1.12). □

5.2. Paouris' inequality

In this section we present the proof of the following deviation inequality of Paouris.

THEOREM 5.2.1 (Paouris). *Let μ be an isotropic log-concave probability measure on $\mathbb{R}^n$. Then,*
$$(5.2.1) \qquad \mu(\{x \in \mathbb{R}^n : \|x\|_2 \geqslant ct\sqrt{n}\}) \leqslant \exp(-t\sqrt{n})$$
for every $t \geqslant 1$, where $c > 0$ is an absolute constant.

Let K be an isotropic convex body in $\mathbb{R}^n$. Then, (5.2.1) takes the form
$$(5.2.2) \qquad |\{x \in K : \|x\|_2 \geqslant ct\sqrt{n}L_K\}| \leqslant \exp(-t\sqrt{n})$$
for all $t \geqslant 1$. It is instructive to compare this dimension dependent estimate to the dimension free estimate
$$(5.2.3) \qquad |\{x \in K : \|x\|_2 \geqslant ct\sqrt{n}L_K\}| \leqslant \exp(-t^2)$$
obtained by Alesker. Since $R(K) \leqslant (n+1)L_K$, it is clear that both inequalities are meaningful in the range $1 \leqslant t \leqslant \sqrt{n}$. Since $t\sqrt{n} \geqslant t^2$ for these values of t, we see that (5.2.2) is stronger than (5.2.3); in fact, it is "much stronger" for "small" values of $t \geqslant 1$.

5.2.1. Reduction of the problem

We first show that the proof of the theorem can be reduced to the study of the moments of the function $x \mapsto \|x\|_2$. For every $q \geqslant 1$ we define
$$I_q(\mu) = \left(\int_{\mathbb{R}^n} \|x\|_2^q d\mu(x) \right)^{1/q}.$$

We have seen that, as a consequence of Borell's lemma, one has the following Khintchine type inequalities.

LEMMA 5.2.2. *For every $y \in \mathbb{R}^n$ and every $p, q \geqslant 1$ we have*
$$\|\langle \cdot, y \rangle\|_{L^{pq}(\mu)} \leqslant c_1 q \|\langle \cdot, y \rangle\|_{L^p(\mu)},$$
where $c_1 > 0$ is an absolute constant. Also, since $\|x\|_2$ is a norm, for every $p, q \geqslant 1$ we have
$$I_{pq}(\mu) \leqslant c_1 q I_p(\mu).$$
□

In particular, we obtain
$$I_q(\mu) \leqslant c_1 q I_2(\mu)$$
for all $q \geqslant 2$. Recall also that by Alesker's theorem (Theorem 3.2.15) we have
$$I_q(\mu) \leqslant c_2 \sqrt{q} I_2(\mu)$$
for all $2 \leqslant q \leqslant n$, where $c_2 > 0$ is an absolute constant. We will prove the following stronger estimate.

THEOREM 5.2.3 (Paouris). *There exist absolute constants $c_3, c_4 > 0$ with the following property: if μ is an isotropic log-concave measure on $\mathbb{R}^n$, then*

(5.2.4) $$I_q(\mu) \leqslant c_4 I_2(\mu)$$

for every $q \leqslant c_3 \sqrt{n}$.

Proof of Theorem 5.2.1. Assuming Theorem 5.2.3, we consider an isotropic log-concave probability measure μ on $\mathbb{R}^n$. From Markov's inequality, for any $q \geqslant 2$ we have
$$\mu(\{\|x\|_2 \geqslant e^3 I_q(\mu)\}) \leqslant e^{-3q}.$$
Then, from Borell's lemma - in the form of (2.4.2) - we get
$$\mu(\{\|x\|_2 \geqslant e^3 I_q(\mu) s\}) \leqslant (1 - e^{-3q}) \left(\frac{e^{-3q}}{1 - e^{-3q}}\right)^{(s+1)/2}$$
$$\leqslant e^{-qs}$$
for every $s \geqslant 1$. Choosing $q = c_3 \sqrt{n}$, and using (5.2.4), we get
$$\mu(\{\|x\|_2 \geqslant c_4 e^3 I_2(\mu) s\}) \leqslant \exp(-c_3 \sqrt{n} s)$$
for every $s \geqslant 1$. Since μ is isotropic, we have $I_2(\mu) = \sqrt{n}$. This proves the theorem. □

For the proof of Theorem 5.2.3 we need to introduce a number of parameters which allow a deeper study of the family of L_q-centroid bodies of μ. This is done in the next subsections.

5.2.2. Averages of norms on the sphere

Let C be a symmetric convex body in $\mathbb{R}^n$, and let $\|\cdot\|$ be the norm induced to $\mathbb{R}^n$ by C. For every $q \geqslant 1$ we define
$$M_q := M_q(C) = \left(\int_{S^{n-1}} \|\theta\|^q d\sigma(\theta)\right)^{1/q}.$$
Note that $M_1(C) = M(C)$. The parameters M_q were studied by Litvak, Milman and Schechtman in [**334**].

THEOREM 5.2.4 (Litvak-Milman-Schechtman). *Let C be a symmetric convex body in $\mathbb{R}^n$ and let $\|\cdot\|$ be the corresponding norm on $\mathbb{R}^n$. We denote by b the smallest constant for which $\|x\| \leqslant b\|x\|_2$ holds true for every $x \in \mathbb{R}^n$. Then,*
$$\max\left\{M_1, c_1 \frac{b\sqrt{q}}{\sqrt{n}}\right\} \leqslant M_q \leqslant \max\left\{2M_1, c_2 \frac{b\sqrt{q}}{\sqrt{n}}\right\}$$
for all $q \in [1, n]$, where $c_1, c_2 > 0$ are absolute constants.

Proof. The function $\|\cdot\| : S^{n-1} \to \mathbb{R}$ is Lipschitz continuous with constant b. By the spherical isoperimetric inequality it follows that

$$\sigma(\{x \in S^{n-1} : |\,\|x\| - M_1\,| > t\}) \leqslant 2\exp(-c_1 t^2 n/b^2)$$

for all $t > 0$. Then,

$$\int_{S^{n-1}} |\,\|x\| - M_1\,|^q d\sigma(x) \leqslant 2q \int_0^\infty t^{q-1} \exp(-c_1 t^2 n/b^2)\, dt$$

$$= \left(\frac{b}{\sqrt{c_1 n}}\right)^q 2q \int_0^\infty s^{q-1} \exp(-s^2)\, ds$$

$$\leqslant \left(C\frac{b\sqrt{q}}{\sqrt{n}}\right)^q,$$

for some absolute constant $C > 0$. The triangle inequality in $L^q(S^{n-1})$ implies that

(5.2.5) $$M_q - M_1 \leqslant \|\,\|x\| - M_1\,\|_q \leqslant C\frac{b\sqrt{q}}{\sqrt{n}}.$$

In other words,

$$M_q \leqslant 2\max\left\{M_1, C\frac{b\sqrt{q}}{\sqrt{n}}\right\}.$$

For the left hand side inequality we observe that C is contained in a strip of width $1/b$: there exists $z \in S^{n-1}$ with $\|z\| = b$ such that $C \subset \{y : |\langle y, z\rangle| \leqslant 1/b\}$. It follows that for every $t > 0$

$$tC \subset \{y : |\langle y, z\rangle| \leqslant t/b\}.$$

This can be also written in the form

$$\{x \in S^{n-1} : \|x\| \geqslant t\} \supset B_t := \{x \in S^{n-1} : |\langle x, z\rangle| \geqslant t/b\}$$

for every $t > 0$. A simple computation shows that if $t \leqslant b/3$ then

(5.2.6) $$\sigma(B_t) \geqslant \frac{c_2\sqrt{n}t}{b}\exp(-c_3 n t^2/b^2)$$

where $c_2, c_3 > 0$ are absolute constants. To see this we write

$$\sigma(B_t) = \frac{2}{I_{n-2}}\int_a^{\pi/2} \cos^{n-2}\theta\, d\theta$$

where $I_n = \int_{-\pi/2}^{\pi/2} \cos^n \theta\, d\theta$ and $a = \arcsin(t/b)$. We know that $I_n \leqslant \sqrt{\frac{\pi}{2n}}$, therefore

$$\sigma(B_t) \geqslant c\sqrt{n}\int_a^\gamma \cos^{n-2}\theta\, d\theta \geqslant (\gamma - a)c\sqrt{n}\cos^{n-2}\gamma$$

where $\gamma = \arcsin(2t/b)$. Assuming that $t \leqslant b/3$ and substituting we get (5.2.6). It follows that

$$M_q \geqslant t[\sigma(\{x \in S^{n-1} : \|x\| \geqslant t\})]^{1/q} \geqslant t[\sigma(B_t)]^{1/q}$$

$$\geqslant t\left(\frac{c_2\sqrt{n}t}{b}\right)^{1/q}\exp(-c_3 n t^2/(qb^2)).$$

for all $t \leqslant b/3$. Choosing $t = b\sqrt{q}/3\sqrt{n}$ we conclude the proof. □

The change of behavior of M_q occurs when $q \simeq n(M_1/b)^2$. This value is equivalent to the largest integer $k = k(C)$ for which the majority of k-dimensional sections of C are 4-Euclidean; this is the *critical dimension* of C which was introduced in Section 1.10b. Recall that the following asymptotic formula holds true:

FACT 5.2.5 (Milman-Schechtman). *There exist $c_1, c_2 > 0$ such that*
$$c_1 n \frac{M(C)^2}{b(C)^2} \leqslant k(C) \leqslant c_2 n \frac{M(C)^2}{b(C)^2}$$
for every symmetric convex body C in $\mathbb{R}^n$.

Note. Theorem 5.2.4 shows that $M_n \simeq b$. Since $M_q \leqslant b$ for all $q \geqslant 1$ and the function $q \mapsto M_q$ is clearly increasing, we conclude that $M_q \simeq b$ if $q \geqslant n$. In other words, we have a second change of behavior of M_q at the point $q = n$.

5.2.3. Mixed widths

Let C be a symmetric convex body in $\mathbb{R}^n$. For every $p \geqslant 1$ we define
$$w_p(C) = \left(\int_{S^{n-1}} h_C^p(\theta) \, d\sigma(\theta) \right)^{1/p}.$$
In particular, if μ is a log-concave probability measure on $\mathbb{R}^n$ then for every $p, q \geqslant 1$ we define
$$w_p(Z_q(\mu)) = \left(\int_{S^{n-1}} h_{Z_q(\mu)}^p(\theta) \, d\sigma(\theta) \right)^{1/p}.$$
Observe that $w_1(Z_q(\mu)) = w(Z_q(\mu))$ is the mean width of $Z_q(\mu)$.

The q-th moments of the Euclidean norm with respect to μ are related to the L_q-centroid bodies of μ through the following lemma.

LEMMA 5.2.6. *Let μ be a log-concave probability measure on $\mathbb{R}^n$. For every $q \geqslant 1$ we have that*
$$w_q(Z_q(\mu)) = a_{n,q} \sqrt{\frac{q}{q+n}} I_q(\mu)$$
where $a_{n,q} \simeq 1$.

Proof. Recall that from Lemma 3.2.16 we have
$$\left(\int_{S^{n-1}} |\langle x, \theta \rangle|^q d\sigma(\theta) \right)^{1/q} = a_{n,q} \frac{\sqrt{q}}{\sqrt{q+n}} \|x\|_2$$
for every $x \in \mathbb{R}^n$, where $a_{n,q} \simeq 1$. Since
$$w_q(Z_q(\mu)) = \left(\int_{S^{n-1}} \int_{\mathbb{R}^n} |\langle x, \theta \rangle|^q \, d\mu(x) \, d\sigma(\theta) \right)^{1/q},$$
the lemma follows. $\square$

REMARK 5.2.7. It is not hard to check that $a_{n,2} = \sqrt{(n+2)/(2n)}$ and
$$I_2(\mu) = \sqrt{n} w_2(Z_2(\mu)).$$

Let C be a symmetric convex body in $\mathbb{R}^n$. We may rewrite Theorem 5.2.4 in terms of $k(C)$ as follows.

THEOREM 5.2.8. *There exist $c_1, c_2, c_3 > 0$ such that for every symmetric convex body C in $\mathbb{R}^n$ we have:*
(i) *If $1 \leqslant q \leqslant k(C)$ then $M(C) \leqslant M_q(C) \leqslant c_1 M(C)$.*
(ii) *If $k(C) \leqslant q \leqslant n$ then $c_2 \sqrt{q/n}\, b(C) \leqslant M_q(C) \leqslant c_3 \sqrt{q/n}\, b(C)$.*

Let $k_*(C) = k(C^\circ)$. Observe that $M(C^\circ) = w(C)$ and $b(C^\circ) = R(C)$. So, we can translate Theorem 5.2.8 as follows:

THEOREM 5.2.9. *There exist $c_1, c_2, c_3 > 0$ such that for every symmetric convex body C in $\mathbb{R}^n$ we have:*

(i) *If $1 \leqslant q \leqslant k_*(C)$ then $w(C) \leqslant w_q(C) \leqslant c_1 w(C)$.*

(ii) *If $k_*(C) \leqslant q \leqslant n$ then $c_2 \sqrt{q/n} R(C) \leqslant w_q(C) \leqslant c_3 \sqrt{q/n} R(C)$.*

Also, by the definition of $k_*(C) := k(C^\circ)$ and the fact that $(C^\circ \cap F)^\circ = P_F(C)$, we get:

THEOREM 5.2.10. *There exist $c_1, c_2 > 0$ such that if $1 \leqslant k \leqslant k_*(C)$ then a random $F \in G_{n,k}$ satisfies*
$$c_1 w(C) B_F \subseteq P_F(C) \subseteq c_2 w(C) B_F$$
with probability greater than $1/2$.

5.2.4. The parameter q_*

DEFINITION 5.2.11. Let μ be a centered log-concave probability measure on $\mathbb{R}^n$. We define
$$q_*(\mu) = \max\{q \geqslant 1 : k_*(Z_q(\mu)) \geqslant q\}.$$

We will need a lower estimate for $q_*(\mu)$. The bound depends on the "ψ_α-behavior" of linear functionals with respect to μ. Recall that if μ is a centered log-concave probability measure on $\mathbb{R}^n$ and if $\alpha \in [1, 2]$ then we say that μ is a ψ_α-measure with constant b_α if

$$(5.2.7) \quad \left(\int_{\mathbb{R}^n} |\langle x, \theta \rangle|^q d\mu(x) \right)^{1/q} \leqslant b_\alpha q^{1/\alpha} \left(\int_{\mathbb{R}^n} |\langle x, \theta \rangle|^2 d\mu(x) \right)^{1/2}$$

for all $q \geqslant 2$ and all $\theta \in S^{n-1}$. Equivalently, if

$$(5.2.8) \quad Z_q(\mu) \subseteq b_\alpha q^{1/\alpha} Z_2(\mu)$$

for all $q \geqslant 2$. Observe that if μ is a ψ_α-measure with constant b_α, then $T(\mu) := \mu \circ T^{-1}$ is a ψ_α-measure with the same constant, for every $T \in GL(n)$. Also, from (5.2.7) we see that

$$(5.2.9) \quad R(Z_q(\mu)) \leqslant b_\alpha q^{1/\alpha} R(Z_2(\mu))$$

for all $q \geqslant 2$.

Finally, recall that from Theorem 2.4.6 we know that there exists an absolute constant $c > 0$ such that every centered log-concave probability measure μ on $\mathbb{R}^n$ is a ψ_1-measure with constant c.

The behavior of $I_q(\mu)$ is easily described in the range $q_*(\mu) \leqslant q \leqslant n$:

LEMMA 5.2.12. *There exist absolute constants $c_1, c_2 > 0$ such that if μ is a centered log-concave probability measure on $\mathbb{R}^n$ then, for every $n \geqslant q \geqslant q_*(\mu)$,*
$$c_1 R(Z_q(\mu)) \leqslant I_q(\mu) \leqslant c_2 R(Z_q(\mu)).$$

In particular, if μ is an isotropic ψ_α-measure with constant b_α then, for every $n \geqslant q \geqslant q_*(\mu)$,

$$(5.2.10) \quad I_q(\mu) \leqslant c_2 b_\alpha q^{1/\alpha}.$$

Proof. Let $n \geqslant q \geqslant q_*(\mu)$. By the definition of $q_*(\mu)$ we have $q \geqslant k_*(Z_q(\mu))$, and Theorem 5.2.9 (ii) shows that
$$c_3 \sqrt{q/n}\, R(Z_q(\mu)) \leqslant w_q(Z_q(\mu)) \leqslant c_4 \sqrt{q/n}\, R(Z_q(\mu)).$$
Now, from Lemma 5.2.6 we have that
$$w_q(Z_q(\mu)) \simeq \sqrt{q/n}\, I_q(\mu).$$
This proves the first assertion. For the second one, we use (5.2.9) and the fact that $R(Z_2(\mu)) = 1$ when μ is isotropic. $\square$

PROPOSITION 5.2.13 (Paouris). *There exists an absolute constant $c > 0$ with the following property: if μ is a centered log-concave probability measure on $\mathbb{R}^n$ which is a ψ_α-measure with constant b_α, then*
$$q_*(\mu) \geqslant c \frac{(k_*(Z_2(\mu)))^{\alpha/2}}{b_\alpha^\alpha}.$$
In particular, for every centered log-concave probability measure μ in $\mathbb{R}^n$ we have
$$q_*(\mu) \geqslant c \sqrt{k_*(Z_2(\mu))}.$$

Proof. Let $q_* := q_*(\mu)$. From Theorem 5.2.9 (i), Lemma 5.2.6, Hölder's inequality and Remark 5.2.7 we get
$$w(Z_{q_*}(\mu)) \geqslant c_1 w_{q_*}(Z_{q_*}(\mu)) = c_1 a_{n,q_*} \sqrt{\frac{q_*}{n+q_*}} I_{q_*}(\mu)$$
$$\geqslant c_1 a_{n,q_*} \sqrt{\frac{q_*}{n+q_*}} I_2(\mu)$$
$$= c_1 a_{n,q_*} \sqrt{\frac{q_*}{n+q_*}} \sqrt{n} w_2(Z_2(\mu))$$
(for the first steps observe that $q_* \simeq q_*+1 \geqslant k_*(Z_{q_*+1}(\mu)) \simeq k_*(Z_{q_*}(\mu))$). In other words,
$$(5.2.11) \qquad w(Z_{q_*}(\mu)) \geqslant c_2 \sqrt{q_*}\, w(Z_2(\mu)).$$
Since μ is a ψ_α-body with constant b_α, we have that
$$(5.2.12) \qquad R(Z_{q_*}(\mu)) \leqslant b_\alpha q_*^{1/\alpha} R(Z_2(\mu)).$$
Using the definition of q_*, Fact 5.2.5 and the inequalities (5.2.11) and (5.2.12), we write
$$q_* \geqslant c_3 k_*(Z_{q_*}(\mu)) \geqslant c_4 n \left(\frac{w(Z_{q_*}(\mu))}{R(Z_{q_*}(\mu))}\right)^2$$
$$\geqslant c_5 n \frac{c_2^2 q_*}{b_\alpha^2 q_*^{2/\alpha}} \frac{w^2(Z_2(\mu))}{R^2(Z_2(\mu))} = c_6 \frac{q_*^{1-2/\alpha}}{b_\alpha^2} k_*(Z_2(\mu)).$$
So, we get
$$q_*(\mu) \geqslant c \frac{[k_*(Z_2(\mu))]^{\alpha/2}}{b_\alpha^\alpha}.$$
The second assertion follows from the fact that every centered log-concave probability measure is a ψ_1-measure with (an absolute) constant $c > 0$. $\square$

Observe that if μ is isotropic then $k_*(Z_2(\mu)) = n$. So, we get the following:

COROLLARY 5.2.14. *There exists an absolute constant $c > 0$ with the following property: if μ is an isotropic log-concave measure on $\mathbb{R}^n$ which is a ψ_α-measure with constant b_α, then*

$$q_*(\mu) \geqslant \frac{cn^{\alpha/2}}{b_\alpha^\alpha}.$$

In particular, for every isotropic log-concave measure μ in $\mathbb{R}^n$ we have that

$$q_*(\mu) \geqslant c\sqrt{n}.$$

5.2.5. Proof of Paouris' inequality

We can now prove that if μ is an isotropic log-concave probability measure on $\mathbb{R}^n$ then $I_q(\mu)$ remains essentially constant as long as $q \leqslant \sqrt{n}$.

THEOREM 5.2.15 (Paouris). *There exist absolute constants $c_1, c_2 > 0$ with the following property: if μ is a centered log-concave probability measure on $\mathbb{R}^n$, then*

(5.2.13) $$I_q(\mu) \leqslant c_1 I_2(\mu)$$

for all $q \leqslant c_2 q_*(\mu)$. In particular, if μ is isotropic then (5.2.13) holds true for all $q \leqslant \sqrt{n}$.

Proof. From Corollary 5.2.14 we know that

$$q_*(\mu) = \max\{q \geqslant 2 : k_*(Z_q(\mu)) \geqslant q\} \geqslant c_2 \sqrt{n}$$

for some absolute constant $c_2 > 0$. For the rest of the proof we set $q = q_*(\mu)$. We may also assume that q is an integer. From Theorem 5.2.9 we have

$$I_q(\mu) \simeq \sqrt{n/q}\, w_q(Z_q(\mu)) \simeq \sqrt{n/q}\, w(Z_q(\mu)).$$

Therefore, the proof of the theorem will be complete if we show that

$$w(Z_q(\mu)) \leqslant c\, \sqrt{q/n}\, I_2(\mu).$$

We first observe that

$$\int_{G_{n,q}} \int \|P_F(x)\|_2^2 d\mu(x) d\nu_{n,q}(F) = \frac{q}{n} \int \|x\|_2^2 d\mu(x) = \frac{q}{n} I_2^2(\mu).$$

Taking also into account Theorem 5.2.10 we may find $F \in G_{n,q}$ which satisfies

(5.2.14) $$\int \|P_F(x)\|_2^2 d\mu(x) \leqslant \frac{cq}{n} I_2^2(\mu)$$

and

(5.2.15) $$P_F(Z_q(\mu)) \simeq w(Z_q(\mu)) B_F.$$

Then,

$$w(Z_q(\mu)) \simeq \left(\frac{|P_F(Z_q(\mu))|}{|B_2^q|}\right)^{1/q} \simeq \sqrt{q}\, |P_F(Z_q(\mu))|^{1/q}.$$

So, it is enough to show that

(5.2.16) $$|P_F(Z_q(\mu))|^{1/q} \leqslant \frac{C}{\sqrt{n}} I_2(\mu).$$

Now, we use Theorem 5.1.12: we have

(5.2.17) $$P_F(Z_q(\mu)) = Z_q(\pi_F(\mu)).$$

Recall that $\pi_F(\mu)$ is a q-dimensional centered log-concave probability measure. Then, from Theorem 5.1.13 we get

$$(5.2.18) \qquad |Z_q(\pi_F(\mu))|^{1/q} \simeq \frac{1}{\|\pi_F(\mu)\|_\infty^{1/q}} = \frac{|\det \mathrm{Cov}(\pi_F(\mu))|^{\frac{1}{2q}}}{L_{\pi_F(\mu)}}.$$

Using the arithmetic-geometric means inequality, the fact that $L_{\pi_F(\mu)} \geq c' > 0$ and (5.2.14) we see that

$$(5.2.19) \qquad |Z_q(\pi_F(\mu))|^{2/q} \leq c_1 \frac{\int \|x\|_2^2 d\pi_F \mu(x)}{q L_{\pi_F(\mu)}^2}$$

$$\leq \frac{c_2}{q} \int \|P_F(x)\|_2^2 \, d\mu(x) \leq \frac{c_3 I_2^2(\mu)}{n}.$$

This proves (5.2.16) and (5.2.13) follows. The second claim of the theorem clearly follows because $q_*(\mu) \geq c\sqrt{n}$ in the isotropic case. $\square$

We can actually describe completely the behavior of $I_q(K)$ in the range $2 \leq q \leq n$.

THEOREM 5.2.16. *Let μ be a centered log-concave measure on $\mathbb{R}^n$. For every $2 \leq q \leq n$,*

$$I_q(\mu) \simeq \max\{I_2(\mu), R(Z_q(\mu))\}.$$

Proof. It is clear that $I_2(\mu) \leq I_q(\mu)$ for all $q \geq 2$ and, for every $\theta \in S^{n-1}$,

$$h_{Z_q(\mu)}(\theta) = \left(\int_{\mathbb{R}^n} |\langle x, \theta \rangle|^q d\mu(x)\right)^{1/q} \leq \left(\int_{\mathbb{R}^n} \|x\|_2^q d\mu(x)\right)^{1/q} = I_q(\mu).$$

This shows that $R(Z_q(\mu)) \leq I_q(\mu)$ and hence

$$\max\{I_2(\mu), R(Z_q(\mu))\} \leq I_q(\mu).$$

For the reverse inequality we first note that, from Lemma 5.2.12, if $n \geq q \geq q_*(\mu)$ then $R(Z_q(\mu)) \simeq I_q(\mu)$. On the other hand, if $2 \leq q \leq q_*(\mu)$ then $I_q(\mu) \simeq I_2(\mu)$. It follows that

$$I_q(\mu) \leq c \max\{I_2(\mu), R(Z_q(\mu))\}$$

for every $2 \leq q \leq n$. $\square$

COROLLARY 5.2.17. *Let μ be an isotropic log-concave measure on $\mathbb{R}^n$ which is ψ_α-measure with constant b_α. Then,*

$$I_q(\mu) \leq c \max\{b_\alpha q^{1/\alpha}, \sqrt{n}\}$$

for every $2 \leq q \leq n$, where $c > 0$ is an absolute constant. In particular, for every isotropic log-concave measure μ in $\mathbb{R}^n$ we have that

$$I_q(\mu) \leq c_1 \max\{q, \sqrt{n}\}$$

for every $2 \leq q \leq n$, where $c_1 > 0$ is an absolute constant. $\square$

5.3. Small ball probability estimates

DEFINITION 5.3.1. Let μ be a centered log-concave probability measure on $\mathbb{R}^n$. We extend the definition of $I_q(\mu)$, allowing negative values of q, in the obvious way: for every $q \in (-n, \infty)$, $q \neq 0$, we define

$$I_q(\mu) := \left(\int_{\mathbb{R}^n} \|x\|_2^q d\mu(x)\right)^{1/q}.$$

The main result of this section is a second theorem of Paouris [414] which extends Theorem 5.2.3.

THEOREM 5.3.2 (Paouris). *Let μ be a centered log-concave probability measure on $\mathbb{R}^n$. For any integer $1 \leqslant k \leqslant q_*(\mu)$ we have*

$$I_{-k}(\mu) \simeq I_k(\mu).$$

In particular, the theorem shows that for all $k \leqslant q_*(\mu)$ one has $I_k(\mu) \leqslant cI_2(\mu)$, where $c > 0$ is a absolute constant. This was the statement of Theorem 5.2.3.

From Theorem 5.3.2 we can derive a small ball probability estimate for any isotropic log-concave measure μ on $\mathbb{R}^n$, using Markov's inequality and the fact that $q_*(\mu) \geqslant c\sqrt{n}$.

THEOREM 5.3.3 (Paouris). *Let μ be an isotropic log-concave measure on $\mathbb{R}^n$. Then, for any $0 < \varepsilon < \varepsilon_0$ we have*

$$\mu(\{x \in \mathbb{R}^n : \|x\|_2 < \varepsilon\sqrt{n}\}) \leqslant \varepsilon^{c\sqrt{n}},$$

where $\varepsilon_0, c > 0$ are absolute constants.

Proof. Let $1 \leqslant k \leqslant q_*(\mu)$. Then we can write

$$\mu(\{x \in \mathbb{R}^n : \|x\|_2 < \varepsilon I_2(\mu)\}) \leqslant \mu(\{x : \|x\|_2 < c_1 \varepsilon I_{-k}(\mu)\})$$
$$\leqslant (c_1 \varepsilon)^k \leqslant \varepsilon^{k/2},$$

for all $0 < \varepsilon < c_1^{-2}$ and every $k \leqslant q_*(\mu)$. Since, $q_*(\mu) \geqslant c_2\sqrt{n}$ we get the result with $\varepsilon_0 = c_1^{-2}$ and $c = c_2/2$. □

For the proof of Theorem 5.3.2 we need again to introduce some new parameters associated to μ and to establish some basic formulas which relate $I_q(\mu)$ with $I_{-q}(\mu)$. This is done in the next subsections.

5.3.1. Gaussian measure of dilates of symmetric convex bodies

In this subsection we describe the proof of the B-theorem of Cordero-Erausquin, Fradelizi and Maurey.

THEOREM 5.3.4 (Cordero-Fradelizi-Maurey). *Let K be a symmetric convex body in $\mathbb{R}^n$. Then, the function*

$$t \mapsto \gamma_n(e^t K)$$

is log-concave on $\mathbb{R}$.

Note. It is useful to also write the conclusion of Theorem 5.3.4 in the following form: if K is a symmetric convex body in $\mathbb{R}^n$ then

(5.3.1) $$\gamma_n(a^\lambda b^{1-\lambda} K) \geqslant \gamma_n(aK)^\lambda \gamma_n(bK)^{1-\lambda}$$

for all $a, b > 0$ and $\lambda \in (0, 1)$.

Proof. Given $(t_1, \ldots, t_n) \in \mathbb{R}^n$, we write $\Delta(t_1, \ldots, t_n)$ for the diagonal matrix with diagonal entries $t_1, \ldots, t_n$. If T is a linear operator on $\mathbb{R}^n$, we write $e^T K$ for the image of K under the operator e^T. We will prove that the function

$$(t_1, \ldots, t_n) \mapsto \gamma_n(e^{\Delta(t_1, \ldots, t_n)} K)$$

is log-concave on $\mathbb{R}^n$. The theorem follows if we restrict this result to n-tuples of the form $(t, \ldots, t)$.

We fix two n-tuples $(a_1, \ldots, a_n)$ and $(b_1, \ldots, b_n)$ in $\mathbb{R}^n$ and write A and B for the corresponding diagonal matrices. We define $f = f_{K,B,A} : \mathbb{R} \to \mathbb{R}$ with

$$f(t) = \gamma_n(e^{B+tA} K)$$

and we will show that f is log-concave; equivalently, we want to show that

(5.3.2) $$\frac{f''(t)}{f(t)} - \frac{[f'(t)]^2}{[f(t)]^2} \leqslant 0.$$

Since $f_{K,B,A}(s+t) = f_{K_1,0,A}(t)$ for all $s, t \in \mathbb{R}$, where $K_1 = e^{B+sA} K$, we may assume that $B = 0$ and $t = 0$. In other words, it is enough to check (5.3.2) for the function

$$f(t) = \gamma_n(e^{tA} K)$$

at the point $t = 0$. The theorem corresponds to the case $A = I$. With the change of variables $y = e^{tA} x$ and using the fact that for any diagonal matrix X we have $\det(e^X) = e^{\operatorname{tr}(X)}$, we get

$$f(t) = \gamma_n(e^{tA} K) = \frac{e^{t \cdot \operatorname{tr}(A)}}{(2\pi)^{n/2}} \int_K e^{-\|e^{tA} x\|_2^2 / 2} \, dx.$$

Observe that the function $t \mapsto e^{t \cdot \operatorname{tr}(A)}$ is log-affine. So, it is enough to check (5.3.2) for the function

$$g(t) = \int_K e^{-\|e^{tA} x\|_2^2 / 2} \, dx$$

at the point $t = 0$. The derivative of g is given by

$$g'(t) = -\int_K \langle e^{tA} x, A e^{tA} x \rangle e^{-\|e^{tA} x\|_2^2 / 2} \, dx,$$

and the second derivative of g is given by

$$g''(t) = \int_K \left(\langle e^{tA} x, A e^{tA} x \rangle^2 - 2 \langle e^{tA} x, A^2 e^{tA} x \rangle \right) e^{-\|e^{tA} x\|_2^2 / 2} \, dx.$$

We want to prove that

(5.3.3) $$\frac{g''(0)}{g(0)} - \frac{[g'(0)]^2}{[g(0)]^2} \leqslant 0.$$

This takes the form

$$\frac{\int_K \langle x, Ax \rangle^2 e^{-\|x\|_2^2/2} \, dx}{\int_K e^{-\|x\|_2^2/2} \, dx} - \left(\frac{\int_K \langle x, Ax \rangle e^{-\|x\|_2^2/2} \, dx}{\int_K e^{-\|x\|_2^2/2} \, dx} \right)^2$$

$$\leqslant 2 \frac{\int_K \langle x, A^2 x \rangle e^{-\|x\|_2^2/2} \, dx}{\int_K e^{-\|x\|_2^2/2} \, dx}.$$

Introducing the measure γ_K with density
$$d\gamma_K(x) = \frac{\mathbf{1}_K(x) e^{-\|x\|_2^2/2}\, dx}{\int_K e^{-\|y\|_2^2/2}\, dy},$$
which is the Gaussian measure restricted to K, we see that the previous estimate takes the compact form
$$\int [f_A(x)]^2\, d\gamma_K(x) - \left(\int f_A(x)\, d\gamma_K(x)\right)^2 \leqslant \frac{1}{2} \int \|\nabla f_A(x)\|_2^2\, d\gamma_K(x),$$
where $f_A(x) = \langle x, Ax \rangle$. That is, we want to verify a Poincaré inequality for the measure γ_K and the function f_A with constant $\frac{1}{2}$.

REMARK 5.3.5. The measure γ_K is log-concave with respect to γ_n: it belongs to the family of probability measures μ of the form $d\mu(x) = e^{-\varphi(x)}\, dx$ for some convex function φ on $\mathbb{R}^n$ which satisfies $(\operatorname{Hess}\varphi)(x) \geqslant I$ on the convex set $K = \{\varphi < \infty\}$. In fact, it is technically simpler to assume that φ is defined on the whole space $\mathbb{R}^n$ and not just on the convex set K. To this end, we consider γ_K as the pointwise limit of a sequence of densities of the form
$$e^{-\|x\|_2^2 - \psi(x)},$$
where ψ is constant on C and "big" outside K. We may assume that ψ is even when K is symmetric, and that $\operatorname{Hess}(\psi)$ is smooth and bounded on $\mathbb{R}^n$. Writing $\varphi(x) = \|x\|_2^2 + \psi(x)$ we work with a probability measure μ given by
$$(5.3.4) \qquad d\mu(x) = e^{-\varphi(x)}\, dx, \quad \varphi: \mathbb{R}^n \to \mathbb{R}, \quad \operatorname{Hess}\varphi \geqslant I.$$
In our situation, the function φ is furthermore even and the result we want to prove is now that, for
$$q(x) = \langle x, Ax \rangle - \int \langle x, Ax \rangle\, d\mu(x)$$
we have
$$\int q^2\, d\mu \leqslant \frac{1}{2} \int \|\nabla q\|_2^2\, d\mu.$$
It is a well known fact that probability measures μ of the form (5.3.4) satisfy the Poincaré inequality with constant 1:

FACT 5.3.6. For any probability measure μ of the form (5.3.4) and for any smooth function $g: \mathbb{R}^n \to \mathbb{R}$ with $g \in L_2(\mu)$ and $\int g\, d\mu = 0$ one has
$$(5.3.5) \qquad \int g^2\, d\mu \leqslant \int \|\nabla g\|_2^2\, d\mu.$$
Observe that a direct application of (5.3.5) would lead to an estimate inferior than then the one we need (by a factor of 2). To overcome this difficulty we will use in an essential way the additional assumption that $\int \nabla q\, d\mu = 0$. The next proposition completes the proof of Theorem 5.3.4.

PROPOSITION 5.3.7. *Let μ be a probability measure on $\mathbb{R}^n$, of the form $d\mu(x) = e^{-\varphi(x)}\, dx$, where $\varphi: \mathbb{R}^n \to \mathbb{R}$ is a convex function with $\operatorname{Hess}(\varphi) \geqslant I$. Then, for any smooth function $f \in L_2(\mu)$ with $\int f\, d\mu = 0$ and $\int \nabla f\, d\mu = 0$, one has*
$$\int f^2\, d\mu \leqslant \frac{1}{2} \int \|\nabla f\|_2^2\, d\mu.$$

We postpone the proof of Proposition 5.3.7 to Chapter 14 where we discuss (in more detail) Poincaré inequalities for log-concave probability measures.

5.3.2. The parameter d_*

DEFINITION 5.3.8. Let C be a symmetric convex body in $\mathbb{R}^n$. We define
$$d_*(C) = \min\left\{-\log \sigma\left(\left\{x \in S^{n-1} : h_C(x) \leqslant \frac{w(C)}{2}\right\}\right), n\right\}.$$

The parameter d_* was introduced by Klartag and Vershynin in [**288**], where it was also observed that $d_*(C)$ is always larger than $k_*(C)$:

PROPOSITION 5.3.9. *Let C be a symmetric convex body in $\mathbb{R}^n$. Then,*
$$d_*(C) \geqslant c k_*(C),$$
where $c > 0$ is an absolute constant.

Proof. From the spherical isoperimetric inequality it follows that
$$\sigma(\{x \in S^{n-1} : |h_C(x) - w(C)| > t w(C)\}) \leqslant \exp(-ct^2 k_*(C))$$
for every $t > 0$. Choosing $t = 1/2$ we get the result. □

LEMMA 5.3.10. *If K is a star body in $\mathbb{R}^n$ then*
$$\tfrac{1}{2}\sigma\left(S^{n-1} \cap \tfrac{1}{2}K\right) \leqslant \gamma_n(\sqrt{n}K) \leqslant \sigma(S^{n-1} \cap 2K) + e^{-cn}.$$

Proof. Write σ_r for the rotationally invariant probability measure on rS^{n-1}. For the left hand side inequality observe that, since K is star-shaped,
$$\begin{aligned}\gamma_n(\sqrt{n}K) &\geqslant \gamma_n(2\sqrt{n}B_2^n \cap \sqrt{n}K) \\ &\geqslant \gamma_n(2\sqrt{n}B_2^n)\sigma_{2\sqrt{n}}(2\sqrt{n}S^{n-1} \cap \sqrt{n}K) \\ &= \gamma_n(2\sqrt{n}B_2^n)\sigma\left(S^{n-1} \cap \tfrac{1}{2}K\right).\end{aligned}$$

From Markov's inequality we have
$$\gamma_n(\{x : \|x\|_2 \geqslant 2\sqrt{n}\}) \leqslant \frac{1}{4n}\int_{\mathbb{R}^n} \|x\|_2^2 d\gamma_n(x) = \frac{1}{4},$$
and hence
$$\gamma_n(2\sqrt{n}B_2^n) = 1 - \gamma_n(\{x : \|x\|_2 \geqslant 2\sqrt{n}\}) \geqslant \frac{3}{4}.$$
This shows that
$$\sigma\left(S^{n-1} \cap \tfrac{1}{2}K\right) \leqslant 2\gamma_n(\sqrt{n}K).$$
[In fact, starting from the observation that for every $0 < \lambda < 1/2$
$$\begin{aligned}\gamma_n(\{x : \|x\| \geqslant 2\sqrt{n}\}) &\leqslant e^{-2\lambda n}\int_{\mathbb{R}^n} e^{\lambda \|x\|_2^2} d\gamma_n(x) \\ &= e^{-2\lambda n}\frac{1}{(2\pi)^{n/2}}\int_{\mathbb{R}^n} e^{(2\lambda-1)\|x\|_2^2/2} dx \\ &= (1-2\lambda)^{-n/2} e^{-2\lambda n},\end{aligned}$$
and choosing $\lambda = 1/4$ one gets
$$\gamma_n(2\sqrt{n}B_2^n) \geqslant 1 - (2/e)^{n/2}$$

for $n \geq 6$.] Next, observe that
$$\sqrt{n}K \subseteq \left(\tfrac{1}{2}\sqrt{n}B_2^n\right) \cup C\left(\tfrac{1}{2}\sqrt{n}S^{n-1} \cap \sqrt{n}K\right)$$
where, for $A \subseteq \tfrac{1}{2}\sqrt{n}S^{n-1}$, we write $C(A)$ for the positive cone generated by A. It follows that
$$\gamma_n(\sqrt{n}K) \leq \gamma_n\left(\tfrac{1}{2}\sqrt{n}B_2^n\right) + \sigma_{\frac{\sqrt{n}}{2}}\left(\tfrac{1}{2}\sqrt{n}S^{n-1} \cap \sqrt{n}K\right).$$
Now,
$$\sigma_{\frac{\sqrt{n}}{2}}\left(\tfrac{1}{2}\sqrt{n}S^{n-1} \cap \sqrt{n}K\right) = \sigma(S^{n-1} \cap 2K),$$
and a direct computation shows that
$$\gamma_n\left(\rho\sqrt{n}B_2^n\right) \leq \left(\frac{\rho\sqrt{n}}{\sqrt{2\pi}}\right)^n |B_2^n|,$$
for all $0 < \rho \leq 1$. It follows that
$$\gamma_n\left(\tfrac{1}{2}\sqrt{n}B_2^n\right) \leq e^{-cn},$$
for some absolute constant $c > 0$. □

THEOREM 5.3.11 (Klartag-Vershynin). *For every $0 < \varepsilon < \tfrac{1}{2}$ we have*
$$\sigma(\{x \in S^{n-1} : h_C(x) < \varepsilon w(C)\}) < (c_1\varepsilon)^{c_2 d_*(C)} < (c_1\varepsilon)^{c_3 k_*(C)},$$
where $c_1, c_2, c_3 > 0$ are absolute constants.

Proof. Let h_C denote the support function of the symmetric convex body C. Let $m = \mathrm{med}(h_C)$ denote the median of h_C with respect to the measure σ on the unit sphere S^{n-1}. Markov's inequality shows that
$$\tfrac{1}{2}\mathrm{med}(h_C) \leq \int_{\{\theta : h_C(\theta) \geq m\}} h_C(\theta)\, d\sigma(\theta) \leq w(C).$$
On the other hand, it is known (and will be used in the end of the proof) that, conversely, $w(C) \leq c_0 m$ for some absolute constant $c_0 > 0$.

Set $L = m\sqrt{n}C^\circ$. According to Lemma 5.3.10, we have
$$(5.3.6) \qquad \gamma_n(2L) \geq \tfrac{1}{2}\sigma(S^{n-1} \cap mC^\circ) \geq \tfrac{1}{4},$$
by the definition of the Lévy mean. On the other hand, using Lemma 5.3.10 again, we have
$$(5.3.7) \qquad \gamma_n(\tfrac{1}{8}L) \leq \sigma\left(S^{n-1} \cap \tfrac{m}{4}C^\circ\right) + e^{-cn}$$
$$= \sigma\left(\left\{\theta \in S^{n-1} : h_C(\theta) \leq \tfrac{m}{4}\right\}\right) + e^{-cn}$$
$$\leq \sigma\left(\left\{\theta \in S^{n-1} : h_C(\theta) \leq \tfrac{w(C)}{2}\right\}\right) + e^{-cn}$$
$$\leq 2e^{-c_1 d_*(C)},$$
where $c_1 > 0$ is a suitable absolute constant (recall that $d_*(C) \leq n$). We may assume that $0 < \varepsilon < e^{-3}$. We apply the B-theorem for the body L, with $a = \varepsilon$, $b = 2$ and $\lambda = 3(\log \tfrac{1}{\varepsilon})^{-1}$. This gives
$$(5.3.8) \qquad \gamma_n(\varepsilon L)^{\frac{3}{\log(1/\varepsilon)}} \gamma_n(2L)^{1 - \frac{3}{\log(1/\varepsilon)}} \leq \gamma_n(\varepsilon^{\frac{3}{\log(1/\varepsilon)}} 2^{1 - \frac{1}{\log(1/\varepsilon)}} L) \leq \gamma_n(\tfrac{1}{8}L).$$

Combining (5.3.6), (5.3.7) and (5.3.8) we see that

$$\gamma_n(\varepsilon L) \leqslant 8e^{c_2 d_*(C) \log \varepsilon} \leqslant 8\varepsilon^{c_2 d_*(C)} \leqslant (c_3 \varepsilon)^{c_2 d_*(C)}.$$

According to Lemma 5.3.10 we can transfer this estimate to the spherical measure, and using the fact that $w(C) \leqslant c_0 m$ we finally obtain

$$\sigma\left(\{\theta \in S^{n-1} : h_C(\theta) \leqslant \varepsilon w(C)\}\right) \leqslant (c_4 \varepsilon)^{c_5 d_*(C)}$$

for $0 < \varepsilon < e^{-4}$. Adjusting the constants we get the theorem. □

Theorem 5.3.11 yields the following reverse Hölder inequalities.

COROLLARY 5.3.12. *Let C be a symmetric convex body in $\mathbb{R}^n$. Then, for every $1 \leqslant q \leqslant c_1 d_*(C)$,*

$$c_2 w(C) \leqslant \left(\int_{S^{n-1}} \frac{1}{h_C^q(x)} d\sigma(x)\right)^{-1/q} \leqslant c_3 w(C).$$

In other words, for every $1 \leqslant q \leqslant c_1 d_*(C)$ we have

$$w_{-q}(C) \simeq w(C).$$

Proof. The right hand side inequality follows easily from Hölder's inequality. For the left hand side inequality we use integration by parts and Theorem 5.3.11. □

In particular, since $d_*(C) \geqslant c k_*(C)$, we always have the following.

THEOREM 5.3.13 (Klartag-Vershynin). *Let C be a symmetric convex body in $\mathbb{R}^n$. Then, $w_q(C) \simeq w_{-q}(C)$ for all $1 \leqslant q \leqslant c k_*(C)$.*

Proof. From Theorem 5.2.8 we have $w_q(C) \simeq w(C)$ for all $q \leqslant k_*(C)$. From Theorem 5.3.12 we get $w_{-q}(C) \simeq w(C)$ for all $q \leqslant c k_*(C)$. Combining these two facts we get the result. □

5.3.3. Small ball probability estimates

We will first show that if f is a centered log-concave density on $\mathbb{R}^n$ then for every integer $1 \leqslant k < n$ we have

$$I_{-k}(f) \simeq \sqrt{n/k} w_{-k}(Z_k(f)).$$

The proof is based on two basic integral formulas. We formulate them as two separate propositions because they are of independent interest.

PROPOSITION 5.3.14 (Paouris). *Let f be a centered log-concave density on $\mathbb{R}^n$ and let $1 \leqslant k < n$ be a positive integer. Then,*

(5.3.9) $$I_{-k}(f) = c_{n,k} \left(\int_{G_{n,k}} \pi_F(f)(0) d\nu_{n,k}(F)\right)^{-1/k},$$

where

$$c_{n,k} = \left(\frac{(n-k)\omega_{n-k}}{n\omega_n}\right)^{1/k} \simeq \sqrt{n}.$$

Note. It is useful to note that in the case of a centered convex body K of volume 1 in $\mathbb{R}^n$ the statement of Proposition 5.3.14 takes the form

$$(5.3.10) \qquad I_{-k}(K) \simeq \sqrt{n} \left(\int_{G_{n,k}} |K \cap F^\perp| \, d\nu_{n,k}(F) \right)^{-1/k}$$

for all $1 \leqslant k < n$.

Proof. Let $1 \leqslant k < n$. Then, we have

$$\int_{G_{n,k}} \pi_F(f)(0) \, d\nu_{n,k}(F)$$

$$= \int_{G_{n,n-k}} \pi_{E^\perp}(f)(0) \, d\nu_{n,n-k}(E)$$

$$= \int_{G_{n,n-k}} \int_E f(y) \, dy \, d\nu_{n,n-k}(E)$$

$$= \int_{G_{n,n-k}} (n-k)\omega_{n-k} \int_{S_E} \int_0^\infty r^{n-k-1} f(r\theta) \, dr \, d\sigma_E(\theta) \, d\nu_{n,n-k}(E)$$

$$= \frac{(n-k)\omega_{n-k}}{n\omega_n} n\omega_n \int_{S^{n-1}} \int_0^\infty r^{n-k-1} f(r\theta) \, dr \, d\sigma(\theta)$$

$$= \frac{(n-k)\omega_{n-k}}{n\omega_n} \int_{\mathbb{R}^n} \|x\|_2^{-k} f(x) \, dx = \frac{(n-k)\omega_{n-k}}{n\omega_n} I_{-k}^{-k}(f).$$

It follows that

$$I_{-k}(f) = \left(\frac{(n-k)\omega_{n-k}}{n\omega_n} \right)^{1/k} \left(\int_{G_{n,k}} \pi_F(f)(0) \, d\nu_{n,k}(F) \right)^{-1/k}.$$

One can check $c_{n,k} = \left(\frac{(n-k)\omega_{n-k}}{n\omega_n} \right)^{1/k} \simeq \sqrt{n}$ and the proof is complete. $\square$

PROPOSITION 5.3.15 (Paouris). *Let C be a symmetric convex body in $\mathbb{R}^n$ and let $1 \leqslant k < n$ be a positive integer. Then,*

$$w_{-k}(C) \simeq \sqrt{k} \left(\int_{G_{n,k}} |P_F(C)|^{-1} d\nu_{n,k}(F) \right)^{-1/k}.$$

Proof. Using the Blaschke-Santaló and the reverse Santaló inequality, we write

$$w_{-k}^{-1}(C) = \left(\int_{S^{n-1}} \frac{1}{h_C^k(\theta)} d\sigma(\theta) \right)^{1/k}$$

$$= \left(\int_{G_{n,k}} \int_{S_F} \frac{1}{\|\theta\|_{(P_F C)^\circ}^k} d\sigma(\theta) d\nu_{n,k}(F) \right)^{1/k}$$

$$= \left(\int_{G_{n,k}} \frac{|(P_F(C))^\circ|}{|B_2^k|} d\nu_{n,k}(F) \right)^{1/k}$$

$$\simeq \left(\int_{G_{n,k}} \frac{|B_2^k|}{|P_F(C)|} d\nu_{n,k}(F) \right)^{1/k},$$

and the result follows. $\square$

Now, let f be a centered log-concave density on $\mathbb{R}^n$. Consider an integer $1 \leqslant k < n$ and let $F \in G_{n,k}$. Recall that, from Theorem 5.1.13, we have

$$\frac{1}{|P_F(Z_k(f))|^{1/k}} \simeq \pi_F(f)(0)^{1/k}.$$

Then, combining Proposition 5.3.14 and Proposition 5.3.15 we get the following.

THEOREM 5.3.16 (Paouris). *Let f be a centered log-concave density on $\mathbb{R}^n$. For every integer $1 \leqslant k < n$ we have*

$$w_{-k}(Z_k(f)) \simeq \sqrt{k}\left(\int_{G_{n,k}} \pi_F(f)(0) d\nu_{n,k}(F)\right)^{-\frac{1}{k}}$$

and

(5.3.11) $$I_{-k}(f) \simeq \sqrt{n/k}\, w_{-k}(Z_k(f)).$$

We are now ready to prove the main result of this section.

Proof of Theorem 5.3.2. Recall that, for every $1 \leqslant k < n$,

(5.3.12) $$w_k(Z_k(\mu)) \simeq \sqrt{k/n}\, I_k(\mu).$$

On the other hand, from (5.3.11) we see that

$$w_{-k}(Z_k(\mu)) \simeq \sqrt{k/n}\, I_{-k}(\mu).$$

Let $k_0 = \lfloor q_* \rfloor$, where $q_* = q_*(\mu)$. Then,

(5.3.13) $$k_*(Z_{k_0}(\mu)) \simeq k_*(Z_{q_*}(\mu)) \geqslant c_1 q_* \geqslant c_1 k_0.$$

From Theorem 5.3.13 we have

(5.3.14) $$w_{-k}(Z_{k_0}(\mu)) \simeq w_k(Z_{k_0}(\mu))$$

for all $1 \leqslant k \leqslant c_2 k_*(Z_{k_0}(\mu))$, and (5.3.13) shows that (5.3.14) holds true for all $k \leqslant c_3 q_*(\mu)$. Setting $k_1 = \lfloor c_3 q_*(\mu) \rfloor \simeq k_0$, and using the fact that $Z_{k_0}(\mu) \simeq Z_{k_1}(\mu)$, we get

(5.3.15) $$w_{-k_1}(Z_{k_1}(\mu)) \simeq w_{k_1}(Z_{k_1}(\mu)).$$

It is now clear that $I_{-k_1}(\mu) \simeq I_{k_1}(\mu)$ and since $k_1 \simeq q_*(\mu)$ we see that $I_q(\mu)$ is "constant" for $1 \leqslant |q| \leqslant c q_*(\mu)$. $\square$

5.4. A short proof of Paouris' deviation inequality

We close this chapter with a self-contained proof of the deviation inequality of Paouris which appears in [3]. Adamczak, Latała, Litvak, Oleszkiewicz, Pajor and Tomczak-Jaegermann use the language of (isotropic) log-concave random vectors. The exact statement of their result is the following.

THEOREM 5.4.1 (Adamczak-Latała-Litvak-Oleszkiewicz-Pajor-Tomczak). *There exists an absolute constant $C > 0$ such that if X is a log-concave random vector in $\mathbb{R}^n$, then for every $q \geqslant 1$ we have*

(5.4.1) $$(\mathbb{E}\|X\|_2^q)^{1/q} \leqslant C\left(\mathbb{E}\|X\|_2 + \max_{\theta \in S^{n-1}}(\mathbb{E}|\langle X,\theta\rangle|^q)^{1/q}\right)$$

Having this, we first recall the standard ψ_1-estimate

$$(\mathbb{E}|\langle X,\theta\rangle|^q)^{1/q} \leqslant Cq\left(\mathbb{E}|\langle X,\theta\rangle|^2\right)^{1/2},$$

which holds true for every log-concave measure and every $\theta \in S^{n-1}$. Taking into account the fact that if X is, additionally, isotropic then

$$\mathbb{E}\|X\|_2 \leqslant \left(\mathbb{E}\|X\|_2^2\right)^{1/2} = \sqrt{n},$$

we deduce from Theorem 5.4.1 that

$$(\mathbb{E}\|X\|_2^q)^{1/q} \leqslant C\left(\sqrt{n} + C'q\right)$$

for all $q \geqslant 1$. An immediate consequence of Markov's inequality for $q = t\sqrt{n}$ is the estimate of Paouris

$$\mathbb{P}\left(\|X\|_2 \geqslant ct\sqrt{n}\right) \leqslant e^{-t\sqrt{n}}$$

for every $t \geqslant 1$ and every isotropic log-concave random vector X in $\mathbb{R}^n$. Note that this is precisely the assertion of Theorem 5.2.1.

We start with some lemmas which will be used in the proof of Theorem 5.4.1.

LEMMA 5.4.2. *Let X be a random vector and let $\|\cdot\|$ be a norm on $\mathbb{R}^n$. Then, for every $q > 0$*

$$\min_{\theta \in S^{n-1}} (\mathbb{E}|\langle X,\theta\rangle|^q)^{1/q} \leqslant \frac{(\mathbb{E}\|X\|^q)^{1/q}}{\mathbb{E}\|X\|}\mathbb{E}\|X\|_2.$$

Proof. Since the norms $\|\cdot\|_2$ and $\|\cdot\|$ are equivalent on $\mathbb{R}^n$, there exists a largest $r > 0$ with the property that $r\|x\| \leqslant \|x\|_2$ for all $x \in \mathbb{R}^n$. By duality, we may find $\theta \in S^{n-1}$ such that $\|\theta\|_* = r$. Then, for every $x \in \mathbb{R}^n$ we have that $|\langle \theta, x\rangle| \leqslant r\|x\| \leqslant \|x\|_2$. This implies that

$$(\mathbb{E}|\langle X,\theta\rangle|^q)^{1/q} \leqslant r\left(\mathbb{E}\|X\|^q\right)^{1/q}$$

and

$$r\,\mathbb{E}\|X\| \leqslant \mathbb{E}\|X\|_2$$

Combining these two inequalities we get the lemma. $\square$

LEMMA 5.4.3. *Let X be a random vector in $\mathbb{R}^n$ with an even log-concave density f_X. Then, there exists a norm $\|\cdot\|$ on $\mathbb{R}^n$ such that*

$$(\mathbb{E}\|X\|^n)^{1/n} \leqslant 500\mathbb{E}\|X\|$$

Proof. Since f_X is even and log-concave, it attains its maximum at 0. We consider the set

$$K = \{x \in \mathbb{R}^n : f_X(x) \geqslant 25^{-n}f_X(0)\}$$

It is clear that K is a bounded symmetric convex set with non-empty interior. Therefore, it is the unit ball of a norm, which we denote by $\|\cdot\|$. Observe that

$$1 \geqslant \mathbb{P}(X \in K) = \int_K f_X \geqslant 25^{-n}f_X(0)|K|,$$

and hence

$$\mathbb{P}(\|X\| \leqslant 1/50) = \int_{K/50} f_X \leqslant 50^{-n}f_X(0)|K| \leqslant 2^{-n} \leqslant 1/2.$$

It follows that

(5.4.2) $$\mathbb{E}\|X\| \geqslant \frac{1}{50}\mathbb{P}(\|X\| > 1/50) \geqslant \frac{1}{100}.$$

Using the fact that f_X is log-concave, we see that for every $x \in \mathbb{R}^n \setminus K$ one has
$$f_{2X}(x) = 2^{-n} f_X(x/2) \geqslant 2^{-n} \sqrt{f_X(x) f_X(0)} \geqslant \left(\frac{5}{2}\right)^n f_X(x).$$

This implies that
$$\mathbb{E}\|X\|^n \leqslant 1 + \mathbb{E}\left(\|X\|^n \mathbf{1}_{\mathbb{R}^n \setminus K}\right) \leqslant 1 + (2/5)^n \mathbb{E}\|2X\|^n = 1 + (4/5)^n \mathbb{E}\|X\|^n,$$

which in turn gives $(\mathbb{E}\|X\|^n)^{1/n} \leqslant 5$. From (5.4.2) we get the result. $\square$

Let us also recall V. Milman's version of Dvoretzky theorem. Here and in the sequel G_n is a standard Gaussian random vector in $\mathbb{R}^n$; in particular, $\mathbb{E}(G_n) = 0$ and $\mathrm{Cov}(G_n) = I$.

THEOREM 5.4.4. *Let $\|\cdot\|$ be a norm on $\mathbb{R}^n$ and $b = \max\{\|t\| : t \in S^{n-1}\}$. For any $\varepsilon > 0$ there exists a constant $c(\varepsilon)$ such that for any integer $1 \leqslant q \leqslant c(\varepsilon) (\mathbb{E}\|G_n\|/b)^2$ there exists a subspace E of dimension q such that*
$$(1-\varepsilon)\|t\|_2 \leqslant \|t\| \frac{\mathbb{E}\|G_n\|_2}{\mathbb{E}\|G_n\|} \leqslant (1+\varepsilon)\|t\|_2$$

for all $t \in E$. Moreover the measure of $E \in G_{n,k}$ that satisfy the above inequality is greater than $1 - e^{-q}$.

Proof of Theorem 5.4.1. We first observe that it is enough to consider the case of symmetric log-concave random vectors. To see this, given a log-concave random vector X, we consider an independent copy X' of X and for every $q \geqslant 1$ we write
$$(\mathbb{E}\|X\|_2^q)^{1/q} \leqslant (\mathbb{E}\|X - \mathbb{E}(X)\|_2^q)^{1/q} + \|\mathbb{E}(X)\|_2$$
$$\leqslant (\mathbb{E}\|X - X'\|_2^q)^{1/q} + \mathbb{E}\|X\|_2.$$

Moreover, $X - X'$ is symmetric and log-concave, and
$$\mathbb{E}\|X - X'\|_2 \leqslant 2\mathbb{E}\|X\|_2.$$

So assuming that (5.4.1) holds true for $X - X'$, we conclude the same for X (with a possibly different constant C).

We proceed to examine the symmetric case. Let X be a symmetric log-concave random vector, and define $\|u\|' = (\mathbb{E}|\langle X, u\rangle|^q)^{1/q}$ for $u \in \mathbb{R}^m$. If G_n is a standard Gaussian random vector on $\mathbb{R}^n$, independent of X, invariance under orthogonal transformations shows that
$$\mathbb{E}|\langle X, G_n\rangle|^q = c_q^q \, \mathbb{E}\|X\|_2^q.$$

where
$$c_q = (\mathbb{E}\|G_1\|_2^q)^{1/q} \simeq \sqrt{q},$$

for $q \geqslant 1$. We now observe that
$$b^2 := \max_{\|u\|'_* \leqslant 1} \mathbb{E}|\langle G_n, u\rangle|^2 = \max_{\|u\|'_* \leqslant 1} \|u\|_2^2 = \left(\max_{\theta \in S^{n-1}} (\mathbb{E}|\langle X, \theta\rangle|^q)^{1/q}\right)^2.$$

From the Gaussian isoperimetric inequality it follows (see [**328**, Theorem 12.2]) that, for every $s \geqslant 0$,
$$\mathbb{P}(|f - \mathbb{E}(f)| \geqslant s) \leqslant \frac{2}{\sqrt{2\pi}} \int_{s/b}^{\infty} e^{-t^2/2} dt \leqslant 2 e^{-s^2/b^2}.$$

On the other hand,
$$\|f\|_q \leq \|f - \mathbb{E}(f)\|_q + \mathbb{E}(f)$$
and
$$\|f - \mathbb{E}(f)\|_q^q = \int_0^\infty q t^{q-1} \mathbb{P}(|f - \mathbb{E}(f)| > t) dt.$$
Therefore,
$$\|f - \mathbb{E}(f)\|_q^q \leq 2 \int_0^\infty q t^{q-1} e^{-\frac{t^2}{b^2}} dt = 2\Gamma\left(\frac{q}{2} + 1\right) b^q.$$
It follows (for $f = \|\cdot\|'$) that

(5.4.3)
$$(\mathbb{E}\|X\|_2^q)^{1/q} = c_q^{-1} (\mathbb{E}(\|G_n\|')^q)^{1/q}$$
$$\leq c_q^{-1} \left(\mathbb{E}\|G_n\|' + c_q \max_{\theta \in S^{n-1}} (\mathbb{E}|\langle X, \theta \rangle|^q)^{1/q} \right).$$

Fix a constant $c_0 = c(1/2)$ so that the assertion of Theorem 5.4.4 is satisfied with $\varepsilon = \frac{1}{2}$. We distinguish two cases:

Case 1. If $q \geq c_0 (\mathbb{E}\|G_n\|'/b)^2$ then using (5.4.3) we get
$$(\mathbb{E}\|X\|_2^q)^{1/q} \leq \frac{\sqrt{q} b}{c_q \sqrt{c_0}} + b \leq Cb.$$

Case 2. If $q \leq c_0 (\mathbb{E}\|G_n\|'/b)^2$ then there exists a subspace E of $\mathbb{R}^n$ of dimension $k \in [q, 2q]$ such that

(5.4.4)
$$\|t\|' \geq \frac{1}{2} \frac{\mathbb{E}\|G_n\|'}{\mathbb{E}\|G_n\|_2} \geq \frac{1}{2} \frac{\mathbb{E}\|G_n\|'}{\sqrt{n}},$$

for all $t \in E$. However by Lemmas 5.4.2 and 5.4.3 applied to $Y = P_E(X)$ we have that
$$\min_{\theta \in S^{n-1}} (\mathbb{E}|\langle X, \theta \rangle|^q)^{1/q} \leq 500 \mathbb{E}\|P_E(X)\|_2.$$
Hence for any $E \in G_{n,k}$ which satisfies (5.4.4), we get
$$\mathbb{E}\|G_n\|' \leq 1000 \sqrt{n} \mathbb{E}\|P_E(X)\|_2.$$
By the Cauchy-Scwharz inequality we have that
$$\int_{G_{n,k}} \|P_E x\|_2 \, d\nu_{n,k}(E) \leq \left(\int_{G_{n,k}} \|P_E x\|_2^2 \, d\nu_{n,k}(E) \right)^{1/2} = \sqrt{k/n} \|x\|_2,$$
hence
$$\int_{G_{n,k}} \mathbb{E}\|P_E(x)\|_2 \, d\nu_{n,k}(E) \leq \sqrt{k/n} \, \mathbb{E}\|x\|_2,$$
and using Markov's inequality we see that
$$\nu_{n,k}\left(\left\{ E \in G_{n,k} : \mathbb{E}\|P_E(X)\|_2 \leq 2\sqrt{k/n} \mathbb{E}\|X\|_2 \right\} \right) \geq \frac{1}{2}.$$
Therefore, there exists $E \in G_{n,k}$ such that
$$\mathbb{E}\|P_E(X)\|_2 \leq 2\sqrt{k/n} \, \mathbb{E}\|X\|_2$$
and (5.4.4) holds as well. For such an E we get
$$\mathbb{E}\|G_n\|' \leq 2000 \sqrt{k} \mathbb{E}\|X\|_2 \leq C\sqrt{q} \, \mathbb{E}\|X\|_2.$$
Using now (5.4.3) we complete the proof. $\square$

It is useful to rephrase Theorem 5.4.1 in the language of centroid bodies and compare the argument of [3] with the one of Paouris. The equivalent formulation is the following.

THEOREM 5.4.5. *Let μ be a centered log-concave probability measure on $\mathbb{R}^n$. For every $q \geqslant 1$,*
$$(5.4.5) \qquad I_q(\mu) \leqslant C\left(I_2(\mu) + R(Z_q(\mu))\right).$$

Note. This statement is a combination of Lemma 5.2.12 and of the main Theorem 5.2.15 of Paouris. Both results appeared in exactly this form in [**413**] (see Lemma 3.9 and Theorem 5.2). An advantage of the proof of Theorem 5.4.1 is probably that it is short and avoids the parameter $q_*(\mu)$. Nevertheless, the parameter $q_*(\mu)$ is useful and plays a rather important role as one can see from the next chapters of this book. The argument that we present below was communicated by E. Milman to one of the authors and provides a short route to Theorem 5.4.5 using Dvoretzky theorem and some basic tools of Paouris that we saw in the previous sections.

Proof. We start with the formula
$$(5.4.6) \qquad I_q(\mu) = c_{n,q} w_q(Z_q(\mu)),$$
where $c_{n,q} \simeq \max\{1, \sqrt{n/q}\}$. Since $w_q(Z_q(\mu)) \leqslant R(Z_q(\mu))$, we clearly have (5.4.5) when $q \geqslant n$, and hence in the sequel we may assume that q is an integer and $1 \leqslant q \leqslant n$.

Recall the result of Litvak, Milman and Schechtman (Theorem 5.2.4): we have
$$(5.4.7) \qquad w_q(Z_q(\mu)) \leqslant w(Z_q(\mu)) + c_1\sqrt{q/n}R(Z_q(\mu)).$$
Therefore, the theorem will follow if we show that
$$(5.4.8) \qquad w(Z_q(\mu)) \leqslant C\sqrt{q/n}(I_2(\mu) + R(Z_q(\mu))).$$
If $q \geqslant k_*(Z_q(\mu))$ then we have
$$(5.4.9) \qquad w(Z_q(\mu)) \leqslant c_2\sqrt{q/n}R(Z_q(\mu))$$
by the definition of $k_*(Z_q(\mu))$. If $q \leqslant k_*(Z_q(\mu))$ then Theorem 5.2.10 (Dvoretzky theorem for $Z_q(\mu)$) shows that a random $F \in G_{n,q}$ satisfies
$$\int \|P_F(x)\|_2^2 \, d\mu(x) \leqslant c_3(q/n)I_2^2(\mu)$$
and
$$(5.4.10) \qquad w(Z_q(\mu))B_F \subseteq c_4 P_F(Z_q(\mu)).$$
Since $P_F(Z_q(\mu)) = Z_q(\pi_F(\mu))$ (by Theorem 5.1.12) and $\pi_F(\mu)$ is a q-dimensional centered log-concave probability measure, from Theorem 5.1.13 we get
$$(5.4.11) \qquad \mathrm{v.rad}(Z_q(\pi_F(\mu))) \simeq \frac{\sqrt{q}}{\|\pi_F(\mu)\|_\infty^{1/q}} = \sqrt{q}\frac{|\det \mathrm{Cov}(\pi_F(\mu))|^{\frac{1}{2q}}}{L_{\pi_F(\mu)}}.$$
Using the the fact that $L_{\pi_F(\mu)} \geqslant c > 0$, we see that
$$(5.4.12) \qquad \mathrm{v.rad}(Z_q(\pi_F(\mu))) \leqslant c_5 \frac{\left(\int \|x\|_2^2 d\pi_F\mu(x)\right)^{1/2}}{L_{\pi_F(\mu)}}$$
$$\leqslant c_6 \left(\int \|P_F(x)\|_2^2 \, d\mu(x)\right)^{1/2} \leqslant c_7\sqrt{q/n}I_2(\mu).$$

Combining (5.4.10), (5.2.18) and (5.4.12) we have

(5.4.13) $$w(Z_q(\mu)) \leqslant c_8 \sqrt{q/n} I_2(\mu).$$

This completes the proof. □

Note that this argument avoids introducing the parameter $q_*(\mu)$. All the other main ideas and tools in Paouris' original argument are used in an essential way.

5.5. Further reading

5.5.1. The unconditional case

In the particular case of unconditional isotropic convex bodies, the inequality of Paouris had been previously proved by Bobkov and Nazarov. In fact, it was after this result that people started thinking whether an analogous estimate might be true in full generality. In this subsection we describe their results.

Let K be an isotropic unconditional convex body in $\mathbb{R}^n$ and let

$$K^+ = 2K \cap \mathbb{R}_+^n$$

be the normalized part of K in $\mathbb{R}_+^n = [0, +\infty)^n$ as in Section 4.1. Let $x = (x_1, \ldots, x_n) \in \mathbb{R}_+^n$. We write $x_1^*, \ldots, x_n^*$ for the coordinates of x in decreasing order. That is,

$$\max x_j = x_1^* \geqslant x_2^* \geqslant \cdots \geqslant x_n^* = \min x_j.$$

Let μ_K^+ denote the uniform distribution on K^+. Corollary 4.1.6 has the following consequence.

PROPOSITION 5.5.1. *Let K be an isotropic unconditional convex body in $\mathbb{R}^n$. Then,*

$$\mu_K^+(\{x \in \mathbb{R}_+^n : x_k^* \geqslant \alpha\}) \leqslant \binom{n}{k} e^{-ck\alpha}$$

for all $\alpha \geqslant 0$ and $1 \leqslant k \leqslant n$, where $c = 1/\sqrt{6}$.

PROOF. Let $1 \leqslant j_1 < \cdots < j_k \leqslant n$. From Corollary 4.1.6 we have

$$\mu_K^+(\{x \in \mathbb{R}_+^n : x_{j_1} \geqslant \alpha, \ldots, x_{j_k} \geqslant \alpha\}) \leqslant \exp(-ck\alpha).$$

Since

$$\{x \in \mathbb{R}_+^n : x_k^* \geqslant \alpha\} = \bigcup_{1 \leqslant j_1 < \cdots < j_k \leqslant n} \{x \in \mathbb{R}_+^n : x_{j_1} \geqslant \alpha, \ldots, x_{j_k} \geqslant \alpha\},$$

we get

$$\mu_K^+(\{x : x_k^* \geqslant \alpha\}) \leqslant \sum_{1 \leqslant j_1 < \cdots < j_k \leqslant n} \mu_K^+(\{x : x_{j_1} \geqslant \alpha, \ldots, x_{j_k} \geqslant \alpha\})$$

$$\leqslant \binom{n}{k} e^{-ck\alpha}$$

as claimed. □

THEOREM 5.5.2 (Bobkov-Nazarov). *Let K be an isotropic unconditional convex body in $\mathbb{R}^n$. Then, for every $t > 0$,*

$$|\{x \in K : \|x\|_1 \geqslant c_1(1+t)n\}| \leqslant n \exp\left(-2t \frac{n}{\log n + 1}\right),$$

where $c_1 = \sqrt{6}$.

Proof. Let $\alpha_1, \ldots, \alpha_n \geqslant 0$. From Proposition 5.5.1 we have

$$\left|\left\{x \in K : \|x\|_1 \geqslant \sum_{k=1}^n \alpha_k\right\}\right| = \mu_K^+\left(\left\{x \in \mathbb{R}_+^n : \sum_{k=1}^n x_k \geqslant 2\sum_{k=1}^n \alpha_k\right\}\right)$$

$$= \mu_K^+\left(\left\{x \in \mathbb{R}_+^n : \sum_{k=1}^n x_k^* \geqslant 2\sum_{k=1}^n \alpha_k\right\}\right)$$

$$\leqslant \sum_{k=1}^n \mu_K^+(\{x \in \mathbb{R}_+^n : x_k^* \geqslant 2\alpha_k\})$$

$$\leqslant \sum_{k=1}^n \binom{n}{k} \exp(-2ck\alpha_k),$$

where $c = 1/\sqrt{6}$. Since

$$\binom{n}{k} \leqslant \left(\frac{en}{k}\right)^k,$$

we get

(5.5.1) $$\left|\left\{x \in K : \|x\|_1 \geqslant \sum_{k=1}^n \frac{\alpha_k}{c}\right\}\right| \leqslant \sum_{k=1}^n \exp\left(-k\left(2\alpha_k - \log\frac{en}{k}\right)\right).$$

Given $t > 0$ we choose

$$\alpha_k = \frac{1}{2}\log\frac{en}{k} + \frac{t}{k}\frac{n}{\log n + 1}.$$

Since $\sum_{k=1}^n \frac{1}{k} \leqslant \log n + 1$ and $n! \geqslant (n/e)^n$, we get

$$\sum_{k=1}^n \alpha_k = \frac{1}{2}\log\left(\frac{n^n e^n}{n!}\right) + t\frac{n}{\log n + 1}\sum_{k=1}^n \frac{1}{k} \leqslant n + nt.$$

Going back to (5.5.1) we get

$$\left|\{x \in K : \|x\|_1 \geqslant \sqrt{6}(1+t)n\}\right| \leqslant n\exp\left(-2t\frac{n}{\log n + 1}\right)$$

for every $t > 0$. □

THEOREM 5.5.3 (Bobkov-Nazarov). *Let K be an isotropic unconditional convex body in $\mathbb{R}^n$. Then, for every $t \geqslant 4$,*

$$|\{x \in K : \|x\|_2 \geqslant c_2 t\sqrt{n}\}| \leqslant \exp\left(-\frac{t\sqrt{n}}{2}\right),$$

where $c_2 = \sqrt{6}$.

Proof. Let $\alpha_1, \ldots, \alpha_n \geqslant 0$. From Proposition 5.5.1 we have

$$\left|\left\{x \in K : \|x\|_2^2 \geqslant \sum_{k=1}^n \alpha_k^2\right\}\right| = \mu_K^+\left(\left\{x \in \mathbb{R}_+^n : \sum_{k=1}^n x_k^2 \geqslant 4\sum_{k=1}^n \alpha_k^2\right\}\right)$$

$$= \mu_K^+\left(\left\{x \in \mathbb{R}_+^n : \sum_{k=1}^n (x_k^*)^2 \geqslant 4\sum_{k=1}^n \alpha_k^2\right\}\right)$$

$$\leqslant \sum_{k=1}^n \mu_K^+(\{x \in \mathbb{R}_+^n : x_k^* \geqslant 2\alpha_k\})$$

$$\leqslant \sum_{k=1}^n \binom{n}{k} \exp(-2c\alpha_k).$$

This shows that

(5.5.2) $$\left|\left\{x \in K : \|x\|_2^2 \geq \sum_{k=1}^n \frac{\alpha_k^2}{c^2}\right\}\right| \leq \sum_{k=1}^n \exp\left(-k\left(2\alpha_k - \log\frac{en}{k}\right)\right),$$

where $c = 1/\sqrt{6}$. Given $t > 0$ we choose

$$\alpha_k = \frac{1}{2}\log\frac{en}{k} + t\frac{\sqrt{n}}{k}.$$

We check that if $t \geq 2$ then $\sum_{k=1}^n \alpha_k^2 \leq 4nt^2$, and going back to (5.5.2) we have

$$\left|\{x \in K : \|x\|_2 \geq 2\sqrt{6}t\sqrt{n}\}\right| \leq n\exp(-2t\sqrt{n}) \leq \exp(-t\sqrt{n})$$

for every $t \geq 2$. This proves the theorem. □

5.5.2. Small ball probability estimates in terms of width

In this subsection we prove a small ball probability estimate for a symmetric convex body K, which is due to Latała and Oleszkiewicz in terms of the inradius r of the body. Recall that

$$r(K) = \sup\{r > 0 : rB_2^n \subseteq K\}.$$

THEOREM 5.5.4 (Latała-Oleszkiewicz). *Let K be a symmetric convex body in $\mathbb{R}^n$ with inradius $r = r(K)$ and $\gamma_n(K) \leq 1/2$. For any $0 \leq s \leq 1/2$ we have*

(5.5.3) $$\gamma_n(sK) \leq (2s)^{r^2/4}\gamma_n(K).$$

REMARK 5.5.5. For the proof we will use Theorem 5.3.4 (the B-theorem) in the following form: if L is a symmetric convex body in $\mathbb{R}^n$ then the log-concavity of $t \mapsto \gamma_n(e^t L)$ implies that the function $u \mapsto [\gamma_n(uL)/\gamma_n(L)]^{\frac{1}{\log u}}$, $0 < u < 1$ is decreasing.

We start with a simple lemma.

LEMMA 5.5.6. *For any $u \in \mathbb{R}$ and any $t \geq 0$ we have*

(5.5.4) $$\gamma_1(u+t,\infty) \leq e^{-tu}\gamma_1(u,\infty).$$

Proof. We write

$$\int_{u+t}^\infty e^{-s^2/2}\,ds = \int_u^\infty e^{-(s+t)^2/2}\,ds \leq \int_u^\infty e^{-st}e^{-s^2/2}\,ds \leq e^{-ut}\int_u^\infty e^{-s^2/2}\,ds.$$

Since $\gamma_1(x,\infty) = \frac{1}{\sqrt{2\pi}}\int_x^\infty e^{-s^2/2}\,ds$, the result follows. □

Proof of Theorem 5.5.4. We first observe that the set $K^c \equiv \mathbb{R}^n \setminus K$ satisfies $\gamma_n(K^c) \geq 1/2$ and $\frac{1}{2}K \cap (K^c + \frac{r}{2}B_2^n) = \emptyset$. Applying the Gaussian isoperimetric inequality we have

$$\gamma_n(\tfrac{1}{2}K) \leq 1 - \gamma_n\left(K^c + \tfrac{r}{2}B_2^n\right) \leq \gamma_1\left(\tfrac{r}{2},\infty\right).$$

Therefore, we may define $u \geq r/2$ by the equation $\gamma_n(\frac{1}{2}K) = \gamma_1(u,\infty)$. Fix $0 < s < 1$. For any $s < t < 1$ we also have $\frac{t}{2}K \cap ((\frac{1}{2}K)^c + \frac{(1-t)r}{2}B_2^n) = \emptyset$, so arguing as before and using Lemma 5.5.6 we may write:

$$\gamma_n\left(\frac{t}{2}K\right) \leq \gamma_1\left(u + \frac{(1-t)r}{2},\infty\right)$$
$$\leq e^{-\frac{(1-t)ru}{2}}\gamma_1(u,\infty) \leq e^{-\frac{1-t}{4}r^2}\gamma_n\left(\frac{1}{2}K\right).$$

We rewrite the previous estimate as

$$\frac{\gamma_n(\frac{t}{2}K)}{\gamma_n(\frac{1}{2}K)} \leq e^{\frac{t-1}{4}r^2}$$

and observe that
$$\left[\frac{\gamma_n(\frac{t}{2}K)}{\gamma_n(\frac{1}{2}K)}\right]^{\frac{1}{\log t}} \geqslant e^{\frac{t-1}{\log t}\frac{r^2}{4}}$$
because $\log t < 0$. Now, since $s < t$, using Remark 5.5.5 for the body $L = \frac{1}{2}K$ we obtain
$$\left[\frac{\gamma_n(\frac{s}{2}K)}{\gamma_n(\frac{1}{2}K)}\right]^{\frac{1}{\log s}} \geqslant \left[\frac{\gamma_n(\frac{t}{2}K)}{\gamma_n(\frac{1}{2}K)}\right]^{\frac{1}{\log t}} \geqslant e^{\frac{t-1}{\log t}\frac{r^2}{4}}.$$
Letting $t \uparrow 1$ we conclude that
$$\left[\frac{\gamma_n(\frac{s}{2}K)}{\gamma_n(\frac{1}{2}K)}\right]^{\frac{1}{\log s}} \geqslant e^{\frac{r^2}{4}},$$
and hence
$$\gamma_n\left(\tfrac{s}{2}K\right) \leqslant e^{\frac{r^2}{4}\log s}\gamma_n\left(\tfrac{1}{2}K\right) = s^{r^2/4}\gamma_n\left(\tfrac{1}{2}K\right).$$
This is true for all $0 < s < 1$, therefore
$$\gamma_n(sK) \leqslant (2s)^{r^2/4}\gamma_n(K)$$
for all $s \in [0, 1/2]$. $\square$

5.5.3. Small ball probability and Dvoretzky theorem

The connection between the small ball probability estimate of Theorem 5.3.11 and Dvoretzky theorem was studied by Klartag and Vershynin in [**288**].

THEOREM 5.5.7 (Klartag-Vershynin). *Let K be a symmetric convex body in $\mathbb{R}^n$ and $1 \leqslant k \leqslant c_1 d_*(K)$. Then, with probability greater than $1 - e^{-c_2 k}$ a random $F \in G_{n,k}$ satisfies*
$$P_F(K) \supseteq c_3 w(K) B_F,$$
where $c_1, c_2, c_3 > 0$ are absolute constants.

From Theorem 5.2.10 we know that up to the critical dimension k_* the reverse inclusion also holds. Theorem 5.5.7 shows that for dimensions probably larger than k_*, namely up to d_*, the projection of the body still contains a large Euclidean ball. A computation for the example of the ℓ_1^n-ball shows that d_* can be much larger than k_*.

Theorem 5.5.7 is a direct consequence of the following:

THEOREM 5.5.8 (inradius of random projections). *Let K be a symmetric convex body in $\mathbb{R}^n$. If $1 \leqslant k \leqslant c_1 d_*(K)$, then*
$$(5.5.5) \qquad c_2 w(K) \leqslant \left(\int_{G_{n,k}} r(P_F(K))^{-k} d\nu_{n,k}(F)\right)^{-1/k} \leqslant c_3 w(K),$$
where $c_1, c_2, c_3 > 0$ are absolute constants.

We first prove a result of independent interest.

PROPOSITION 5.5.9 (dimension lift for the inradius of projections). *Let K be a convex body in $\mathbb{R}^n$ and let $1 \leqslant q < n$ be a positive integer. Then for any positive integer $1 \leqslant k < q/4$, we have*
$$(5.5.6) \qquad \left(\int_{G_{n,k}} r(P_F(K))^{-k} d\nu_{n,k}(F)\right)^{1/k} \leqslant Cw(K)/(w_{-q}(K))^2,$$
where $C > 0$ is an absolute constant.

Proposition 5.5.9 immediately implies Theorem 5.5.8:

Proof of Theorem 5.5.8. The right hand side inequality follows from Hölder's inequality: We may write

$$\left(\int_{G_{n,k}} r(P_F(K))^{-k}\,d\nu_{n,k}(F)\right)^{-1/k} \leqslant \int_{G_{n,k}} r(P_F(K))\,d\nu_{n,k}(F)$$

$$\leqslant \int_{G_{n,k}} w(P_F(K))\,d\nu_{n,k}(F) = w(K),$$

where we have used the invariance of the Haar measure and the fact that $r(P_F(K)) \leqslant h_K(\theta)$ for any $\theta \in S_F$.

For the left hand side inequality we employ the estimate of Proposition 5.5.9 and the definition of $d_*(K)$: For any $k \leqslant d_*(K)$ we have $w_{-k}(K) \geqslant c_2 w(K)$. □

Now, we turn to the proof of Proposition 5.5.9. We need a lemma.

LEMMA 5.5.10. *Let K be a symmetric convex body in $\mathbb{R}^n$ and let $1 \leqslant k < n$. Then,*

$$(5.5.7) \qquad w(K) \leqslant \left(\int_{G_{n,k}} w(P_F(K))^{2k}\,d\nu_{n,k}(F)\right)^{\frac{1}{2k}} \leqslant cw(K),$$

where $c > 0$ is a absolute constant.

Proof. The left hand side inequality follows immediately from Hölder's inequality and the fact that

$$w(K) = \int_{G_{n,k}} w(P_F(K))\,d\nu_{n,k}(F).$$

For the right hand side inequality we are going to prove the large deviation estimate

$$(5.5.8) \qquad \nu_{n,k}(\{F \in G_{n,k} : w(P_F(K)) \geqslant c_1 t w(K)\}) \leqslant e^{-c_2 t^2 k},$$

for $t > 1$. Then, the result follows if we use integration by parts.

Concentration of measure on the sphere shows that

$$\sigma(\{\theta \in S^{n-1} : h_K(\theta) > (1+\varepsilon)w(K)\}) \leqslant c_3 e^{-c_4 \varepsilon^2 k_*(K)} \leqslant e^{-c_5 \varepsilon^2},$$

for all $\varepsilon > 0$. This is equivalent to the fact that

$$(5.5.9) \qquad \int_{S^{n-1}} \exp(u h_K(\theta)/w(K))\,d\sigma(\theta) \leqslant e^{c_6 u^2},$$

for all $u > 1$. Now, we consider independent random points $\theta_1, \ldots, \theta_k$ which are uniformly distributed over the sphere. From (5.5.9), using independence and Hölder's inequality we get

$$(5.5.10) \qquad \mathbb{E}\left(\exp\left(\frac{u}{kw(K)} \sum_{i=1}^{k} h_K(\theta_i)\right)\right) \leqslant e^{c_6 u^2/k}$$

for all $u > 1$. Using Markov's inequality and choosing $u = u(t)$ in an optimal way we see that

$$(5.5.11) \qquad \mathbb{P}\left(\sum_{i=1}^{k} h_K(\theta_i) > c_7 tkw(K)\right) \leqslant e^{-kt^2},$$

for all $t > 1$, provided that $c_7 > 0$ is large enough.

We know that $F = \text{span}\{\theta_1, \ldots, \theta_k\}$ is uniformly distributed on $G_{n,k}$ almost surely. Then, we can write:

$\nu_{n,k}(F \in G_{n,k} : w(P_F(K)) > 2c_7 tw(K))$

$$\leqslant \frac{\mathbb{P}(\sum_{i=1}^{k} h_K(\theta_i) > c_7 tkw(K))}{\mathbb{P}(\sum_{i=1}^{k} h_K(\theta_i) > c_7 tkw(K) \mid w(P_F(K)) > 2c_7 tw(K))}.$$

As an upper bound for the numerator we use (5.5.11). For the denominator we may write:

$$\mathbb{P}\left(\sum_{i=1}^{k} h_K(\theta_i) > c_7 t k w(K) \;\Big|\; w(P_F(K)) > 2c_7 t w(K)\right)$$
$$\geqslant \mathbb{P}\left(\frac{1}{k}\sum_{i=1}^{k} h_K(\theta_i) > \frac{w(P_F(K))}{2} \;\Big|\; w(P_F(K)) > 2c_7 t w(K)\right)$$
$$= \frac{1}{\mathbb{P}(F \in G_{n,k} : w(P_F(K)) \geqslant 2c_7 t w(K))} \int_{\{F \in G_{n,k} : w(P_F) > 2c_7 t w(K)\}} \mathbb{P}(A_F)\, d\nu_{n,k}(F)$$
$$\geqslant \min_{F \in G_{n,k}} \mathbb{P}(A_F),$$

where A_F is the event:

$$A_F = \left\{ \frac{1}{k}\sum_{i=1}^{k} h_K(\theta_i) > \frac{w(P_F(K))}{2} \;\Big|\; \operatorname{span}\{\theta_1, \ldots, \theta_k\} = F \right\}.$$

In order to complete the proof it suffices to give a lower bound for $\mathbb{P}(A_F)$, uniformly in $F \in G_{n,k}$. To this end, fix $F \in G_{n,k}$. Note that if we condition on $F = \operatorname{span}\{\theta_1, \ldots, \theta_k\}$ then each of the vectors θ_i is uniformly distributed in S_F. Thus, we may write:

$$w(P_F(K)) = \mathbb{E}\left(\frac{1}{k}\sum_{i=1}^{k} h_K(\theta_i) \;\Big|\; \operatorname{span}\{\theta_1, \ldots, \theta_k\} = F\right)$$
$$\leqslant C\sqrt{k}\, w(P_F(K))\mathbb{P}(A_F) + \frac{w(P_F(K))}{2}(1 - \mathbb{P}(A_F)),$$

where we have used the fact that $h_K(\theta) \leqslant C\sqrt{\dim F}\, w(P_F(K)) = C\sqrt{k}\, w(P_F(K))$ for every $\theta \in S_F$. It follows that $\mathbb{P}(A_F) \geqslant c_8/\sqrt{k}$. Since F was arbitrary we obtain:

$$\nu_{n,k}(\{F \in G_{n,k} : w(P_F(K)) \geqslant c_7 t w(K)\}) \leqslant c_8^{-1}\sqrt{k}\, e^{-t^2 k} < e^{-c_9 t^2 k},$$

for all $t > 1$. $\square$

We will also need a geometric inequality due to Klartag [**271**].

THEOREM 5.5.11. *Let K be a symmetric convex body in $\mathbb{R}^n$. Then, for $1 \leqslant k \leqslant n-1$ we have*

(5.5.12) $$\left(\int_{G_{n,k}} R(K \cap E)^{n-k}\, \mathrm{v.rad}(K \cap E)^k\, d\nu_{n,k}(E)\right)^{1/n} \leqslant C \mathrm{v.rad}(K),$$

where $C > 0$ is an absolute constant.

Proof. Using polar coordinates we may write:

$$|K| = n\omega_n \int_0^\infty \int_{S^{n-1}} r^{n-1} \mathbf{1}_K(r\theta)\, d\sigma(\theta)\, dr$$
$$= \omega_n \int_{S^{n-1}} \rho_K(\theta)^n\, \sigma(\theta)$$
$$= \omega_n \int_{G_{n,k}} \int_{S_E} \rho_{K \cap E}(\theta)^n\, d\sigma_E(\theta)\, d\nu_{n,k}(E)$$
$$= \frac{n\omega_n}{k\omega_k} \int_{G_{n,k}} k\omega_k \left(\int_0^\infty \int_{S_E} r^{k-1} \|r\phi\|_2^{n-k} \mathbf{1}_{K \cap E}(r\phi)\, d\sigma_E(\phi)\, dr\right) d\nu_{n,k}(E)$$
$$= \frac{n\omega_n}{k\omega_k} \int_{G_{n,k}} \int_{K \cap E} \|y\|_2^{n-k}\, dy\, d\nu_{n,k}(E).$$

The next step is to give a lower bound for the integral $\int_{K \cap E} \|y\|_2^{n-k}\, dy$ in terms of $|K \cap E|$ and $R(K \cap E)$. We have the following:

Claim. Let K be a symmetric convex body in $\mathbb{R}^m$. Then, for any $0 < \varepsilon < 1$ we have

$$|\{x \in K : \|x\|_2 \geqslant \varepsilon R(K)\}| \geqslant (1-\varepsilon)^m |K|/2.$$

Proof of the Claim. Let $x_0 \in K$ such that $\|x_0\|_2 = R(K)$ and $v = x_0/\|x_0\|_2$. We consider the set K^+ defined as

$$K^+ := \{x \in K : \langle x, v\rangle \geqslant 0\}.$$

Since K is symmetric we have $|K^+| \geqslant |K|/2$. Note that

$$\{x \in K : \|x\|_2 \geqslant \varepsilon R(K)\} \supseteq \varepsilon x_0 + (1-\varepsilon)K^+.$$

Therefore,

$$|\{x \in K : \|x\|_2 \geqslant \varepsilon R(K)\}| \geqslant |\varepsilon x_0 + (1-\varepsilon)K^+| = (1-\varepsilon)^m |K^+| \geqslant (1-\varepsilon)^m |K|/2.$$

Applying this distributional inequality for $K_1 = K \cap E$ and $m = k$ we get:

$$\int_{K \cap E} \|y\|_2^{n-k}\, dy \geqslant |\{y \in K \cap E : \|y\|_2 \geqslant \varepsilon R(K \cap E)\}|(\varepsilon R(K \cap E))^{n-k}$$

$$\geqslant \frac{1}{2}(1-\varepsilon)^k \varepsilon^{n-k} |K \cap E| R(K \cap E)^{n-k}.$$

Inserting this estimate into the formula for $|K|$ we conclude that

$$\frac{|K|}{\omega_n} \geqslant \frac{n}{2k}(1-\varepsilon)^k \varepsilon^{n-k} \int_{G_{n,k}} \frac{|K \cap E|}{\omega_k} R(K \cap E)^{n-k}\, d\nu_{n,k}(E).$$

Taking n-th roots we see that

$$\left(\int_{G_{n,k}} R(K \cap E)^{n-k} \mathrm{v.rad}(K \cap E)^k\, d\nu_{n,k}(E)\right)^{1/n}$$

$$\leqslant \frac{2}{\varepsilon}\left(\frac{k}{n}\right)^{1/n}\left(\frac{\varepsilon}{1-\varepsilon}\right)^{k/n} \mathrm{v.rad}(K).$$

Choosing $\varepsilon = \frac{n}{n+k}$ we obtain the assertion of the theorem with $C = 4e$. $\square$

Proof of Proposition 5.5.9. Because of the monotonicity of moments it suffices to prove the assertion for $q = 4k$. Note that $r(P_F(K))^{-1} = R((P_F(K))^\circ) = R(K^\circ \cap F)$. We are going to apply to Theorem 5.5.11 to $K_1 = K^\circ \cap F$ for $F \in G_{n,q}$. We write

$$\int_{G_{n,k}} R(K^\circ \cap F)^{2k} \mathrm{v.rad}(K^\circ \cap F)^{2k}\, d\nu_{n,k}(F)$$

$$= \int_{G_{n,q}} \int_{E \in G_{F,k}} R(K^\circ \cap E)^{q-2k} \mathrm{v.rad}(K^\circ \cap E)^{q-2k}\, d\nu_{F,k}(E)\, d\nu_{n,q}(F)$$

$$\leqslant (4e)^q \int_{G_{n,q}} \mathrm{v.rad}(K^\circ \cap F)^q\, d\nu_{n,q}(F)$$

$$= (4e)^q \int_{G_{n,q}} \int_{S_F} h_{P_F(K)}^{-q}(\theta)\, d\sigma_F(\theta) d\nu_{n,q}(F)$$

$$= (4e)^q \int_{S^{n-1}} h_K^{-q}(\theta)\, d\sigma(\theta)$$

$$= (4e)^q w_{-q}^{-q}(K).$$

Now we apply Cauchy-Schwarz inequality and the previous estimate to get

$$\left(\int_{G_{n,k}} R(K^\circ \cap F)^k \, d\nu_{n,k}(F)\right)^{1/k}$$
$$\leqslant \left(\int_{G_{n,k}} R(K^\circ \cap F)^{2k} \mathrm{v.rad}(K^\circ \cap F)^{2k} \, d\nu_{n,k}(F)\right)^{\frac{1}{2k}}$$
$$\times \left(\int_{G_{n,k}} \mathrm{v.rad}(K^\circ \cap F)^{-2k} \, d\nu_{n,k}(F)\right)^{\frac{1}{2k}}$$
$$\leqslant (4e)^2 w_{-q}^{-2}(K) \left(\int_{G_{n,k}} \mathrm{v.rad}(K^\circ \cap F)^{-2k} \, d\nu_{n,k}(F)\right)^{\frac{1}{2k}}.$$

From Hölder's inequality we know that $1 \leqslant \mathrm{v.rad}(K^\circ \cap F) w(P_F(K))$ for any subspace F, and hence

$$\left(\int_{G_{n,k}} R(K^\circ \cap F)^k \, d\nu_{n,k}(F)\right)^{1/k} \leqslant \frac{(4e)^2}{w_{-q}(K)^2} \left(\int_{G_{n,k}} w(P_F(K))^{2k} \, d\nu_{n,k}\right)^{\frac{1}{2k}}.$$

An application of Lemma 5.5.10 completes the proof of the theorem. $\square$

5.6. Notes and references

L_q-centroid bodies

L_q-centroid bodies were introduced by Lutwak and Zhang in [**346**]. They defined a multiple of $Z_q(K)$, denoted by $\Gamma_q(K)$, choosing a normalization so that $\Gamma_q(B_2^n) = B_2^n$ for every q. Lutwak, Yang and Zhang [**347**] proved Theorem 5.1.4. A second proof was given by Campi and Gronchi in [**131**]. In our language this inequality takes the form $|Z_q(K)|^{1/n} \geqslant c\sqrt{q/n}$ for some absolute constant $c > 0$. Paouris initiated the study of L_q-centroid bodies from an asymptotic point of view in his PhD Thesis (the articles [**410**] and [**411**] come from that period).

The formulas of Proposition 5.1.5, Proposition 5.1.6 and Theorem 5.1.7 are standard consequences of the inclusion relations and the volume estimates that we obtained for the bodies $K_p(f)$ in Chapter 2. Their first main application is the formula

$$f(0)^{-1/n} \simeq |Z_n(f)|^{1/n}$$

of Proposition 5.1.8, which holds true for every centered log-concave density f on $\mathbb{R}^n$. This last result appears in the article [**414**] of Paouris.

Marginals and projections

Marginals of log-concave probability measures were studied by Klartag in [**275**]. The identity

$$P_F(Z_q(f)) = Z_q(\pi_F(f))$$

of Theorem 5.1.12 was observed in the setting of convex bodies by Paouris in [**412**] and [**413**]. Although it appears in disguised form in the article of Milman and Pajor [**384**], its importance was understood through its role in the proof of Paouris' deviation inequality. Soon after, it was observed that in the language of log-concave probability measures and their marginals, the identity takes the above much more elegant form; see [**414**].

Paouris' inequality

Theorem 5.2.1 is due to Paouris [**413**]; it was announced in [**412**]. The reduction of his deviation inequality to Theorem 5.2.3 had been observed earlier in his article [**411**].

Theorem 5.2.4 is due to Litvak, Milman and Schechtman [**334**]. It plays a key role in the approach of Paouris, who started from the formula $w_q(Z_q(\mu)) \simeq \sqrt{q/n} I_q(\mu)$, $2 \leqslant q \leqslant n$ of Lemma 5.2.6 and defined the parameter

$$q_*(\mu) = \max\{q \geqslant 2 : k_*(Z_q(\mu)) \geqslant q\}.$$

Using Theorem 5.2.4, Paouris proved in [**410**] (see also [**409**]) that $q_*(\mu) \geqslant c\sqrt{n}$ for every isotropic log-concave measure μ in $\mathbb{R}^n$. The main (and final) step in his argument is the inequality

$$w(Z_q(\mu)) \leqslant c\sqrt{q}, \quad 2 \leqslant q \leqslant q_*(\mu).$$

This was done in [**412**] and [**413**], when Proposition 5.1.8 and Theorem 5.1.12 were observed (see the proof of Theorem 5.2.15).

The upper bound $|Z_q(\mu)|^{1/n} \leqslant c\sqrt{q/n}$, $2 \leqslant q \leqslant n$, for the volume of the L_q-centroid body of an isotropic log-concave measure μ in $\mathbb{R}^n$ was proved by Paouris: Theorem 5.1.17 appears in [**413**].

In the particular case of unconditional isotropic convex bodies, the inequality of Paouris had been previously proved by Bobkov and Nazarov [**90**]. Their results (Theorem 5.5.2 and Theorem 5.5.3) are described in Section 5.4.1. The origin of the work of Bobkov and Nazarov is in the work of Schechtman, Zinn and Schmuckenschläger on the volume of the intersection of two L_p^n-balls (see [**456**], [**457**], [**458**] and [**462**]).

Let us also mention that, before Theorem 5.2.1, Guédon and Paouris had studied in [**241**] the case of the unit balls of the Schatten classes.

B-Theorem and the parameter d_*

Theorem 5.3.4 is due to Cordero-Erausquin, Fradelizi and Maurey [**142**]. It provides an affirmative answer to a question of Banaszczyk (see Latała [**307**]). In the same article, the question is generalized as follows: one says that a probability measure μ and a convex body K in $\mathbb{R}^n$ satisfy the B-theorem if the function $t \mapsto \mu(e^t K)$ is log-concave. This problem is studied in $\mathbb{C}^n$ and it is shown by complex interpolation that the B-theorem holds true for a more general class of sets and measures. It is also shown that the B-theorem is true for any unconditional log-concave probability measure μ provided that the convex body K is also unconditional.

The parameter d_* was introduced by Klartag and Vershynin in [**288**]. The main question in the article concerned the diameter of proportional random sections of high dimensional convex bodies (see Section 5.4.3). The results in [**288**] reveal a difference between the lower and the upper inclusions in Dvoretzky theorem: the expected upper bound for the diameter of random sections of a convex body sometimes continues to hold in much larger dimensions than the critical dimension of the body. An example is given by the cube; while $k(Q_n) \simeq \log n$, one has that $d(Q_n)$ is polynomial in n. B-theorem is the main tool for the small deviations estimate of Theorem 5.3.11. The use of the B-theorem in this context was observed by Latała and Oleszkiewicz in [**310**] in their proof of Theorem 5.5.4 (see Section 5.4.2).

Theorem 5.3.13 combines the result of Klartag and Vershynin with the one of Litvak, Milman and Schechtman, and plays a key role in Paouris' work on the negative moments $I_{-q}(\mu)$ of an isotropic log-concave probability measure μ.

Small ball probability estimates

The second main theorem of Paouris is from [**414**]. The formulas of Proposition 5.3.14 and Proposition 5.3.15 come from this work. Paouris combined them with Theorem 5.3.13

to obtain the formula
$$I_q(\mu) \simeq I_{-q}(\mu), \qquad 2 \leqslant q \leqslant q_*(\mu)$$
for any log-concave probability measure μ on $\mathbb{R}^n$. This extends Theorem 5.2.3 and leads to a second proof of Theorem 5.2.1. At the same time, it provides the small ball probability estimate of Theorem 5.3.3.

CHAPTER 6

Bodies with maximal isotropic constant

Recall that we write L_n for the maximum of all isotropic constants of convex bodies in $\mathbb{R}^n$,
$$L_n := \max\{L_K : K \text{ is isotropic in } \mathbb{R}^n\}.$$
We have seen that, if we want to study the magnitude of the isotropic constant of log-concave probability measures, we may do so by looking only at the class of convex bodies. Even more specifically, in this chapter we look at properties we can deduce for bodies with maximal isotropic constant, that is, isotropic constant equal to or very close to L_n. It turns out that the isotropic position of such bodies is closely related to their M-position (see Section 1.13.2 for characterizations of the M-position), and this enables one to establish several interesting facts: for example, Bourgain, Klartag and V. Milman obtained a reduction of the hyperplane conjecture to the question of boundedness of the isotropic constant of a restricted class of convex bodies, those that have volume ratio bounded by an absolute constant. Recall that the volume ratio of a centered convex body K in $\mathbb{R}^n$ is defined as

$$\mathrm{vr}(K) := \inf\left\{\left(\frac{|K|}{|\mathcal{E}|}\right)^{1/n} : \mathcal{E} \text{ is an origin symmetric ellipsoid inside } K\right\},$$

that it is an affine invariant and that it can be up to of order $\sqrt{n}$ for bodies K in $\mathbb{R}^n$. This reduction of the hyperplane conjecture is presented in Section 6.2.

In Sections 6.4 and 6.5, we give two more reductions of the conjecture to the study of two parameters that can be associated with any isotropic convex body. The first of these reductions, due to Dafnis and Paouris, shows that the hyperplane conjecture is directly related to the behavior of $I_{-p}(K)$ when p lies in the interval $[\sqrt{n}, n]$. If we consider the parameter

$$q_{-c}(K, \zeta) := \max\{p \geqslant 1 : I_2(K) \leqslant \zeta I_{-p}(K)\}, \qquad \zeta \geqslant 1,$$

then the hyperplane conjecture can be shown to be equivalent to the statement that there exist absolute constants $C, \xi > 0$ such that, for every isotropic convex body K in $\mathbb{R}^n$,

$$q_{-c}(K, \xi) \geqslant Cn.$$

Note also that the results of Chapter 5 show that $q_{-c}(K, \xi_0) \geqslant \sqrt{n}$ for some absolute constant $\xi_0 \geqslant 1$.

The final reduction in this chapter, due to Giannopoulos, Paouris and Vritsiou, is based on the parameter

$$I_1(K, Z_q^\circ(K)) = \int_K h_{Z_q(K)}(x)dx = \int_K \|\langle \cdot, x\rangle\|_{L_q(K)}dx.$$

It can be viewed as a continuation of Bourgain's approach and, roughly speaking, can be formulated as follows: given $q \geqslant 2$ and $\frac{1}{2} \leqslant s \leqslant 1$, an upper bound of the

213

form
$$I_1(K, Z_q^\circ(K)) \leqslant C_1 q^s \sqrt{n} L_K^2 \quad \text{for all bodies } K \text{ in isotropic position}$$
leads to the estimate
$$L_n \leqslant \frac{C_2 \sqrt[4]{n} \log^2 n}{q^{\frac{1-s}{2}}}.$$
Bourgain's estimate is (almost) recovered by choosing $q = 2$. However, the behavior of $I_1(K, Z_q^\circ(K))$ may allow us to use $s < 1$ along with large values of q. In Subsection 6.6.1 we discuss upper and lower bounds for this quantity. For every isotropic convex body K in $\mathbb{R}^n$, one has $I_1(K, Z_q^\circ(K)) \leqslant c_2 q \sqrt{n} L_K^2$. Any improvement of the exponent of q would lead to an estimate $L_n \leqslant Cn^\alpha$ with $\alpha < \frac{1}{4}$. It seems plausible that one could even have $I_1(K, Z_q^\circ(K)) \leqslant c\sqrt{qn} L_K^2$, at least when q is small, say $2 \leqslant q \ll \sqrt{n}$.

The proofs of these reductions rely heavily on the existence of convex bodies with maximal isotropic constant whose isotropic position is not only closely related to their M-position, but is also compatible with regular covering estimates; the existence of such isotropic bodies is proven in Section 6.3. Moreover, in Subsection 6.1.2 we see that L_n is essentially a monotone increasing function of the dimension n, in the sense that there exists an absolute constant C such that $L_k \leqslant CL_n$ for all dimensions $k < n$; this interesting property turns out to be key in establishing the existence of regular bodies, and hence all three reductions described above.

6.1. Symmetrization of isotropic convex bodies

6.1.1. Symmetrization and isotropic constant

Let K be a convex body in $\mathbb{R}^n$. Below we consider a way of symmetrizing K with respect to a given k-dimensional subspace E and a given convex body T in E; Bourgain, Klartag and V. Milman consider this symmetrization in [**107**], being mainly interested in its effect on the isotropic constant of K.

DEFINITION 6.1.1. Let K be a convex body in $\mathbb{R}^n$. Let E be a k-dimensional subspace of $\mathbb{R}^n$ and consider a centered convex body T in E. We define the (T, E)-*symmetrization* of K as the unique convex body $K(T, E)$ which has the following two properties:
   (i) If $y \in E^\perp$ then $|K \cap (y + E)| = |K(T, E) \cap (y + E)|$.
   (ii) If $y \in P_{E^\perp}(K)$, then $(K(T, E) - y) \cap E$ is homothetic to T and has its barycenter at the origin (of course it can be a singleton containing only the origin when $|K \cap (y + E)| = 0$).

The next lemma shows that $K(T, E)$ is indeed convex.

LEMMA 6.1.2. $K(T, E)$ *is a convex body.*

PROOF. We set $K_1 := K(T, E)$. It suffices to prove that, for every $y_1, y_2 \in P_{E^\perp}(K_1) = P_{E^\perp}(K)$ and every $\lambda \in (0, 1)$,
$$\lambda[K_1 \cap (y_1 + E)] + (1 - \lambda)[K_1 \cap (y_2 + E)] \subseteq K_1 \cap (\lambda y_1 + (1-\lambda)y_2 + E).$$
Since $(K_1 - y) \cap E$ is homothetic to T for every $y \in P_{E^\perp}(K)$, it is enough to check that
$$\lambda|K_1 \cap (y_1 + E)|^{\frac{1}{k}} + (1 - \lambda)|K_1 \cap (y_2 + E)|^{\frac{1}{k}} \leqslant |K_1 \cap (\lambda y_1 + (1-\lambda)y_2 + E)|^{\frac{1}{k}}.$$

6.1. SYMMETRIZATION OF ISOTROPIC CONVEX BODIES

By the definition of K_1 we have

$$|K_1 \cap (y+E)|^{\frac{1}{k}} = |K \cap (y+E)|^{\frac{1}{k}}$$

for every $y \in P_{E^\perp}(K)$, so the lemma follows from the Brunn-Minkowski inequality applied to K. □

The main result of this subsection compares the isotropic constants of K and $K(T, E)$.

THEOREM 6.1.3 (Bourgain-Klartag-V. Milman). *Let K be an isotropic convex body in $\mathbb{R}^n$. Let E be a k-dimensional subspace of $\mathbb{R}^n$ and let T be a convex body in E with volume 1 and center of mass at the origin. Then,*

$$\left(\frac{k+1}{n+1}\right)^{\frac{k}{n}} L_{K(T,E)} \leqslant L_K^{1-\frac{k}{n}} L_T^{\frac{k}{n}} |K \cap E|^{\frac{1}{n}} \leqslant \left(\frac{n+1}{k+1}\right)^{\frac{k}{n}} L_{K(T,E)}.$$

Before proving the theorem we need to introduce some terminology and to prove some auxiliary results.

DEFINITION 6.1.4. Let A be a centered convex body of volume 1 in $\mathbb{R}^n$. We say that $e \in S^{n-1}$ is an *axis of inertia* of A if it is an eigenvector of the operator $M_A(y) = \int_A \langle x, y \rangle x \, dx$. Since M_A is symmetric and positive definite, there exists an orthonormal basis formed by eigenvectors of M_A. Since $\det M_A = L_A^{2n}$, if $\{e_1, \ldots, e_n\}$ is such a basis, then it is clear that

$$L_A^{2n} = \prod_{i=1}^n \int_A \langle x, e_i \rangle^2 dx.$$

We will say that such a basis is a *basis of inertia axes* of A.

REMARK 6.1.5. Fubini's theorem shows that if $\theta \in \mathbb{R}^n$ then

(6.1.1) $$\int_{K_1} \langle x, \theta \rangle^2 = \int_{P_{E^\perp}(K)} \int_{(K_1 - y) \cap E} \langle y + z, \theta \rangle^2 \, dz \, dy.$$

If $\theta \in E^\perp$, then the inner integrand is independent of z, and since $|(K_1 - y) \cap E| = |(K - y) \cap E|$ for every $y \in P_{E^\perp}(K)$, we get

$$\int_{K_1} \langle x, \theta \rangle^2 = \int_K \langle x, \theta \rangle^2 dx.$$

If $\theta \in E$, then (6.1.1) gives

(6.1.2) $$\int_{K_1} \langle x, \theta \rangle^2 = \int_{P_{E^\perp}(K)} \int_{(K_1-y) \cap E} \langle z, \theta \rangle^2 \, dz \, dy$$
$$= \left(\int_{P_{E^\perp}(K)} |(K-y) \cap E|^{1+\frac{2}{k}} dy\right) \left(\int_T \langle z, \theta \rangle^2 dz\right)$$

(recall that T has volume 1 in the statement of Theorem 6.1.3). More generally, we have the following: if $\theta_1, \theta_2 \in E^\perp$ then

(6.1.3) $$\langle M_{K_1}(\theta_1), \theta_2 \rangle = \langle M_K(\theta_1), \theta_2 \rangle,$$

and if $\theta_1, \theta_2 \in E$ then

(6.1.4) $$\langle M_{K_1}(\theta_1), \theta_2 \rangle = c(K, E) \langle M_T(\theta_1), \theta_2 \rangle,$$

where
$$c(K, E) = \int_{P_{E^\perp}(K)} |K \cap (y + E)|^{1+\frac{2}{k}} dy.$$

LEMMA 6.1.6. *Let K be an isotropic convex body in $\mathbb{R}^n$. Let E be a k-dimensional subspace of $\mathbb{R}^n$ and let T be a centered convex body of volume 1 in E. If $e_1, \ldots, e_k$ are axes of inertia of T and $\{e_{k+1}, \ldots, e_n\}$ is any orthonormal basis of $E^\perp$, then $\{e_1, \ldots, e_n\}$ is a basis of inertia axes of $K(T, E)$.*

Proof. Set $K_1 := K(T, E)$. We first observe that the restriction of $M_{K(T,E)}$ onto $E^\perp$ is a multiple of the identity: from (6.1.3) we have
$$\langle M_{K_1}(\theta_1), \theta_2 \rangle = L_K^2 \langle \theta_1, \theta_2 \rangle$$
for all $\theta_1, \theta_2 \in E^\perp$, which shows that $M_{K_1}|_{E^\perp} = L_K^2 I_{E^\perp}$. Thus, any orthonormal basis $\{e_{k+1}, \ldots, e_n\}$ of $E^\perp$ consists of axes of inertia of K_1, and furthermore, combining that with the fact that M_{K_1} is symmetric, we have that E is invariant under M_{K_1}.

On the other hand, from (6.1.4) we see that
$$\langle M_{K_1}(e_i), \theta \rangle = c(K, E) \langle M_T(e_i), \theta \rangle$$
for every $\theta \in E$ and every $i = 1, \ldots, k$. Since $e_1, \ldots, e_k$ are eigenvectors of M_T, this shows that $e_1, \ldots, e_k$ are axes of inertia of K_1. $\square$

We also need two inequalities for concave functions. We first prove a variant of Theorem 2.2.2 (see [**182**]), which shows that the general inequality of the theorem, that compares the maximum of an arbitrary log-concave function f to the value at its barycenter, can be strengthened when f is p-concave for some $p > 0$ (where the latter means that $f^{1/p}$ is concave).

LEMMA 6.1.7. *Let C be a convex body in $\mathbb{R}^m$ and let $h : C \to \mathbb{R}$ be a non-negative concave function, not identically zero. Then,*
$$h\left(\frac{\int_C x h^p(x) dx}{\int_C h^p(x) dx}\right) \geqslant \frac{p+1}{m+p+1} \max_{x \in C} h(x).$$

Proof. From Jensen's inequality we have
$$h\left(\frac{\int_C x h^p(x) dx}{\int_C h^p(x) dx}\right) \geqslant \frac{\int_C h^{p+1}(x) dx}{\int_C h^p(x) dx}.$$
Define $\phi(t) := |\{x \in C : h(x) \geqslant t\}|$ and set $M = \max_{x \in C} h(x)$. By Fubini's theorem we have
$$\int_C h^p(x) dx = p \int_0^M t^{p-1} \phi(t) dt,$$
and thus, to prove the assertion of the lemma, it is enough to show that
$$(6.1.5) \qquad \int_0^M t^p \phi(t) dt \geqslant \frac{p}{m+p+1} M \int_0^M t^{p-1} \phi(t) dt.$$

First, we show that ϕ is m-concave. Observe that, if x, y are elements of C satisfying $h(x) \geqslant t$ and $h(y) \geqslant s$, then $h(\lambda x + (1-\lambda)y) \geqslant \lambda t + (1-\lambda)s$ because h is concave. This shows that $\lambda \{h \geqslant t\} + (1-\lambda)\{h \geqslant s\} \subseteq \{h \geqslant \lambda t + (1-\lambda)s\}$, and it implies, by the Brunn-Minkowski inequality and the definition of ϕ, that
$$\phi^{1/m}(\lambda t + (1-\lambda)s) \geqslant \lambda \phi^{1/m}(t) + (1-\lambda)\phi^{1/m}(s).$$

Next, set $r = p/(m+p+1)$ and consider a function $g : [0, M] \to \mathbb{R}^+$ such that $g^{1/m}$ is affine, $g(M) = 0$ and $g(rM) = \phi(rM)$. It follows that $g(t) = c(1 - t/M)^m$ for some $c > 0$ and for all $t \in [0, M]$. In addition, since $\phi^{1/m}$ is concave on $[0, M]$, $\phi(M) \geqslant g(M) = 0$ and $\phi(rM) = g(rM)$, we get that $\phi \leqslant g$ on $[0, rM]$ and $\phi \geqslant g$ on $[rM, M]$. Hence,

$$\int_0^M (t - rM)t^{p-1}\phi(t)\,dt \geqslant \int_0^M (t - rM)t^{p-1}g(t)\,dt.$$

Using the fact that $\int_0^1 t^a(1-t)^b dt = \Gamma(a+1)\Gamma(b+1)/\Gamma(a+b+2)$ for every $a, b > 0$, we see that the right hand side integral vanishes and this gives us (6.1.5). □

The second inequality is again for p-concave functions and compares moments of the function to appropriate powers of their maximum value.

PROPOSITION 6.1.8. *Let $f : \mathbb{R}^m \to [0, \infty)$ be a p-concave function with $\int f(x)dx = 1$. Then,*

$$\frac{(p+1)(p+2)}{(m+p+1)(m+p+2)}\|f\|_\infty^{2/p} \leqslant \int_{\mathbb{R}^m} f(x)^{1+2/p}dx \leqslant \|f\|_\infty^{2/p}.$$

Proof. The right hand side inequality is trivial: using the assumption that f is a density, we can write

$$\int_{\mathbb{R}^m} f(x)^{1+2/p}dx \leqslant \|f\|_\infty^{2/p} \int_{\mathbb{R}^m} f(x)dx = \|f\|_\infty^{2/p}.$$

For the left hand side inequality we use a similar argument to that of the previous lemma. Translating f if necessary, we may assume that $f(0) = \|f\|_\infty$. Integrating in polar coordinates we write

$$\int_{\mathbb{R}^m} f(x)^{1+2/p}dx = m\omega_m \int_{S^{m-1}} \int_0^\infty f(r\theta)^{1+2/p}r^{m-1}dr d\sigma(\theta).$$

Now, we fix $\theta \in S^{m-1}$ and we consider the function $g(r) := f(r\theta)$, $r \geqslant 0$. Then g has compact support, say contained in $[0, M]$ for some $M > 0$, and $g^{1/p}$ is concave.

Claim. Suppose b is such that

(6.1.6) $$\int_0^M g(r)r^{m-1}dr = \int_0^\infty \left(g(0)^{1/p} - br\right)_+^p r^{m-1}dr,$$

where $r_+ := \max\{r, 0\}$. Then, for every $k > 1$,

$$\int_0^M g(r)^k r^{m-1}dr \geqslant \int_0^\infty \left(g(0)^{1/p} - br\right)_+^{pk} r^{m-1}dr.$$

Proof of the Claim. Since the left hand side integral in (6.1.6) is finite, we must have $b > 0$. We set $h(r) = g(0)^{1/p} - g(r)^{1/p}$, and we have that h is a convex function and that $h(0) = 0$. This implies that the function $\tilde{h}(r) := \frac{h(r)}{r}$, $r > 0$, is increasing. In addition, given that

$$\int_0^\infty (g(0)^{1/p} - \tilde{h}(r)r)_+^p r^{m-1}dr = \int_0^M g(r)r^{m-1}dr = \int_0^\infty (g(0)^{1/p} - br)_+^p r^{m-1}dr,$$

it cannot be that $\tilde{h}(r)$ is always smaller or always larger than b. Therefore, there exists $r_0 \in [0, \infty)$ such that $\tilde{h} \leqslant b$ on $[0, r_0]$ and $\tilde{h} \geqslant b$ on $[r_0, \infty)$. Denote $(g(0)^{1/p} -$

$br)_+^p$ by $\phi(r)$. Then, to prove the claim, it suffices by Fubini's theorem to show that

$$\int_0^\infty \left(\int_0^{g(r)} s^{k-1}ds\right) r^{m-1}dr \geqslant \int_0^\infty \left(\int_0^{\phi(r)} s^{k-1}ds\right) r^{m-1}dr,$$

or equivalently that $\int_0^\infty \left(\int_{\phi(r)}^{g(r)} s^{k-1}ds\right) r^{m-1}dr \geqslant 0$. From the above we have that $(\phi(r)-g(r))(r-r_0) \geqslant 0$ for every r; since in addition ϕ^{k-1} is decreasing, we obtain

$$(6.1.7) \qquad \int_{\phi(r)}^{g(r)} s^{k-1}ds \geqslant \int_{\phi(r)}^{g(r)} \phi(r_0)^{k-1}ds$$

for $0 \leqslant r \leqslant r_0$ and

$$(6.1.8) \qquad \int_{g(r)}^{\phi(r)} s^{k-1}ds \leqslant \int_{g(r)}^{\phi(r)} \phi(r_0)^{k-1}ds$$

for $r \geqslant r_0$. From (6.1.8) and (6.1.7) we get

$$\int_0^\infty \left(\int_{\phi(r)}^{g(r)} s^{k-1}ds\right) r^{m-1}dr \geqslant \phi(r_0)^{k-1}\int_0^\infty (g(r)-\phi(r))r^{m-1}dr$$

and the right hand side integral vanishes according to assumption (6.1.6). $\square$

Carrying on with the proof of Proposition 6.1.8, we use the claim above with $k = 1 + 2/p$ and obtain

$$\int_0^\infty g(r)^{1+2/p}r^{m-1}dr \geqslant \int_0^\infty \left(g(0)^{1/p}-br\right)_+^{p+2}r^{m-1}dr,$$

where b is such that (6.1.6) holds. A simple calculation yields that

$$\frac{\int_0^\infty \left(g(0)^{1/p}-br\right)_+^{p+2}r^{m-1}dr}{\int_0^\infty \left(g(0)^{1/p}-br\right)_+^{p}r^{m-1}dr} = g(0)^{2/p}\frac{(p+1)(p+2)}{(m+p+1)(m+p+2)} = c_{m,p}\|f\|_\infty^{2/p}.$$

Combining (6.1.6) with the last two relations, we conclude that

$$\int_0^\infty g(r)^{1+2/p}r^{m-1}dr \geqslant c_{m,p}\|f\|_\infty^{2/p}\int_0^\infty \left(g(0)^{1/p}-br\right)_+^{p}r^{m-1}dr$$

$$= c_{m,p}\|f\|_\infty^{2/p}\int_0^\infty g(r)r^{m-1}dr.$$

We have thus proven that, for every $\theta \in S^{m-1}$,

$$\int_0^\infty f(r\theta)^{1+2/p}r^{m-1}dr \geqslant \frac{(p+1)(p+2)}{(m+p+1)(m+p+2)}\|f\|_\infty^{2/p}\int_0^\infty f(r\theta)r^{m-1}dr,$$

and integrating over S^{m-1} and using the fact that f is a density, we conclude the proof of the proposition. $\square$

Now, given a convex body K in $\mathbb{R}^n$ and a k-dimensional subspace E of $\mathbb{R}^n$, we can apply the above inequalities with the function $g(y) = |K \cap (E+y)|$ which is supported on the convex body $P_{E^\perp}(K)$, is non-negative and is k-concave by the Brunn-Minkowski inequality.

PROPOSITION 6.1.9 (Fradelizi). *Let K be a centered convex body in $\mathbb{R}^n$. Let $k < n$ and let E be a k-dimensional subspace of $\mathbb{R}^n$. Then,*

$$\max_{y \in E^\perp} |K \cap (E+y)| \leqslant \left(\frac{n+1}{k+1}\right)^k |K \cap E|.$$

Proof. Consider the function $g : P_{E^\perp}(K) \to \mathbb{R}$ with $g(y) = |K \cap (E+y)|$. Then, $h := g^{\frac{1}{k}}$ is non-negative and concave. Therefore, we may apply Lemma 6.1.7 with $C = P_{E^\perp}(K)$, $m = n - k$ and $p = k$ to write

$$h\left(\frac{\int xg(x)\,dx}{\int g(x)\,dx}\right) \geqslant \frac{k+1}{n+1} \max_{x \in C} h(x).$$

Since $\mathrm{bar}(g) = 0$, it follows that

$$|K \cap E|^{\frac{1}{k}} = h(0) = h\left(\frac{\int xg(x)\,dx}{\int g(x)\,dx}\right) \geqslant \frac{k+1}{n+1} \max_{x \in P_{E^\perp}(K)} h(x)$$
$$= \frac{k+1}{n+1} \max_{y \in E^\perp} |K \cap (E+y)|^{\frac{1}{k}},$$

which is the assertion of the proposition. $\square$

PROPOSITION 6.1.10. *Let K be a centered convex body of volume 1 in $\mathbb{R}^n$. Let E be a k-dimensional subspace of $\mathbb{R}^n$ and let T be a centered convex body of volume 1 in E. Then, for every $\theta \in E$,*

$$\left(\frac{k+1}{n+1}\right)^2 |K \cap E|^{\frac{2}{k}} \int_T \langle z,\theta \rangle^2 dz \leqslant \int_{K(T,E)} \langle x,\theta \rangle^2 dx$$

and

$$\int_{K(T,E)} \langle x,\theta \rangle^2 dx \leqslant \left(\frac{n+1}{k+1}\right)^2 |K \cap E|^{\frac{2}{k}} \int_T \langle z,\theta \rangle^2 dz.$$

Proof. From (6.1.2) we have

$$\int_{K(T,E)} \langle x,\theta \rangle^2 dx = \int_{P_{E^\perp}(K)} |K \cap (E+w)|^{1+\frac{2}{k}} dw \cdot \int_T \langle z,\theta \rangle^2 dz.$$

Let $g(y) = |K \cap (E+y)|$ on $P_{E^\perp}(K)$. By the Brunn-Minkowski inequality, $g^{1/k}$ is concave and $\int g(y)\,dy = |K| = 1$. Proposition 6.1.8 shows that

$$\frac{(k+1)(k+2)}{(n+1)(n+2)} \|g\|_\infty^{\frac{2}{k}} \int_T \langle z,\theta \rangle^2 dy \leqslant \int_{K(T,E)} \langle x,\theta \rangle^2 dx \leqslant \|g\|_\infty^{\frac{2}{k}} \int_T \langle z,\theta \rangle^2 dy.$$

Since K has its center of mass at the origin, Proposition 6.1.9 shows that

$$g(0) \leqslant \|g\|_\infty \leqslant \left(\frac{n+1}{k+1}\right)^k g(0).$$

Observing that $g(0) = |K \cap E|$ and

$$\frac{(k+1)(k+2)}{(n+1)(n+2)} \geqslant \left(\frac{k+1}{n+1}\right)^2,$$

we conclude the proof. $\square$

Proof of Theorem 6.1.3. We choose an orthonormal basis $\{e_1, \ldots, e_n\}$ as in Lemma 6.1.6. Then,

$$L_{K(T,E)}^{2n} = \prod_{i=1}^{n} \int_{K(T,E)} \langle x, e_i \rangle^2 dx = L_K^{2(n-k)} \prod_{i=1}^{k} \int_{K(T,E)} \langle x, e_i \rangle^2 dx.$$

By the first inequality of Proposition 6.1.10, this gives

$$L_{K(T,E)}^{2n} \geqslant \left(\frac{k+1}{n+1}\right)^{2k} L_K^{2(n-k)} |K \cap E|^2 \prod_{i=1}^{k} \int_T \langle x, e_i \rangle^2 dx,$$

and since $\{e_1, \ldots, e_k\}$ is a basis of axes of inertia of T,

$$L_{K(T,E)}^{2n} \geqslant \left(\frac{k+1}{n+1}\right)^{2k} L_K^{2(n-k)} L_T^{2k} |K \cap E|^2.$$

It follows that

$$L_{K(T,E)} \geqslant \left(\frac{k+1}{n+1}\right)^{k/n} L_K^{1-\frac{k}{n}} L_T^{\frac{k}{n}} |K \cap E|^{\frac{1}{n}}.$$

Using the second inequality of Proposition 6.1.10 and following the same argument, we get

$$L_{K(T,E)} \leqslant \left(\frac{n+1}{k+1}\right)^{k/n} L_K^{1-\frac{k}{n}} L_T^{\frac{k}{n}} |K \cap E|^{\frac{1}{n}}.$$

Finally, note that $[(k+1)/(n+1)]^{k/n} \geqslant c$ for all $k < n$, where $c > 0$ is an absolute constant. $\square$

6.1.2. Monotonicity with respect to the dimension

In this subsection we will see that

$$L_n = \sup\{L_K : K \text{ is an isotropic convex body in } \mathbb{R}^n\}$$

is essentially a monotone increasing function of the dimension n. From Lemma 3.2.5 we know that if K and T are isotropic convex bodies in $\mathbb{R}^n$ and $\mathbb{R}^m$ respectively, then

$$L_{K \times T} = L_K^{\frac{n}{n+m}} L_T^{\frac{m}{n+m}}.$$

Choosing K and T so that $L_K = L_n$ and $L_T = L_m$, we readily see that

$$L_{n+m}^{n+m} \geqslant L_n^n L_m^m$$

for all $n, m \in \mathbb{N}$. By induction we can then obtain the following

LEMMA 6.1.11. *Let $k, n \in \mathbb{N}$. If k divides n, then $L_k \leqslant L_n$.* $\square$

We will now prove that the sequence L_n is "monotone" in the following precise sense.

THEOREM 6.1.12 (Bourgain-Klartag-V. Milman). *There exists an absolute constant $C > 0$ such that: if $m, n \in \mathbb{N}$ and $m < n$ then $L_m \leqslant CL_n$.*

The proof will be based on Theorem 6.1.3 and on the following

PROPOSITION 6.1.13. *Let K be a centered convex body of volume 1 in $\mathbb{R}^n$. For every $k < n$ there exists a k-dimensional subspace E of $\mathbb{R}^n$ such that*

$$|K \cap E|^{1/n} \geqslant c,$$

where $c > 0$ is an absolute constant.

To make the proof easier to read, we first recall the following fact about the volume of sections with subspaces and projections onto subspaces of ellipsoids that will be used again in the sequel (the proof that follows is taken from [**287**] and [**149**]).

LEMMA 6.1.14. *Let $\mathcal{E}$ be an ellipsoid in $\mathbb{R}^n$, then $\mathcal{E} = T(B_2^n)$ for some $T \in GL(n)$. We denote the eigenvalues of the matrix $\sqrt{TT^*}$ by $\lambda_1 \geqslant \cdots \geqslant \lambda_n > 0$ (recall that TT^* is a symmetric, positive definite matrix). Then, for all $1 \leqslant k \leqslant n-1$,*

$$(6.1.9) \qquad \min_{F \in G_{n,k}} |\mathcal{E} \cap F| = \min_{F \in G_{n,k}} |P_F(\mathcal{E})| = \omega_k \prod_{i=n-k+1}^{n} \lambda_i,$$

and

$$(6.1.10) \qquad \max_{F \in G_{n,k}} |\mathcal{E} \cap F| = \max_{F \in G_{n,k}} |P_F(\mathcal{E})| = \omega_k \prod_{i=1}^{k} \lambda_i.$$

Proof. Using the left polar decomposition of T we can write $T = \sqrt{TT^*}\, U$ for some orthogonal matrix U and hence have $\mathcal{E} = \sqrt{TT^*}\, U(B_2^n) = \sqrt{TT^*}(B_2^n)$. Therefore, we may assume without loss of generality that $T = \sqrt{TT^*}$, and we can find an orthonormal basis $\{v_1, \ldots, v_n\}$ in $\mathbb{R}^n$ so that T is diagonal with respect to that basis. Let $F \in G_{n,k}$, then consider an orthonormal basis $\{u_1, \ldots, u_k\}$ in F. If V is a $(k \times n)$-matrix that allows us to change from $\{u_1, \ldots, u_k\}$ to $\{v_1, \ldots, v_n\}$, then it is easy to check that

$$\mathcal{E} \cap F = VTV^t(B_2^n \cap F),$$

and hence that

$$|\mathcal{E} \cap F| = \det(VTV^t)\,\omega_k.$$

Thus, our task reduces to showing that $\det(VTV^t) \geqslant \prod_{i=n-k+1}^{n} \lambda_i$ for every $(k \times n)$-matrix whose rows are orthonormal vectors.

We now use the Cauchy-Binet formula. Note that the sums below are over all subsets $A \subset \{1, \ldots, n\}$ with exactly k elements, and that for such A we write V_A to denote the submatrix obtained from V by considering only columns whose indices are in A. Then,

$$\det(VTV^t) = \sum_A \det(V_A T_A (V_A)^t) = \sum_A \left(\prod_{i \in A} \lambda_i\right) \det(V_A(V_A)^t)$$

$$\geqslant \left(\prod_{i=n-k+1}^{n} \lambda_i\right) \sum_A \det(V_A(V_A)^t) = \left(\prod_{i=n-k+1}^{n} \lambda_i\right) \det(VV^t)$$

$$= \prod_{i=n-k+1}^{n} \lambda_i.$$

Next, let $F_0 = \operatorname{span}\{v_{n-k+1}, \ldots, v_n\}$. Then, for every $F \in G_{n,k}$ we have

$$|P_{F_0}(\mathcal{E})| = |\mathcal{E} \cap F_0| = \omega_k \prod_{i=n-k+1}^{n} \lambda_i \leqslant |\mathcal{E} \cap F| \leqslant |P_F(\mathcal{E})|.$$

This shows that

$$\min_{F \in G_{n,k}} |P_F(\mathcal{E})| = |P_{F_0}(\mathcal{E})| = \omega_k \prod_{i=n-k+1}^{n} \lambda_i$$

and completes the proof of (6.1.9).

Observe now that $\mathcal{E}^\circ = (T^*)^{-1}(B_2^n) = T^{-1}(B_2^n)$ is also an ellipsoid; since the diagonal entries of T^{-1} are $\lambda_n^{-1} \geqslant \cdots \geqslant \lambda_1^{-1} > 0$, the same reasoning gives

$$(6.1.11) \qquad \min_{F \in G_{n,k}} |\mathcal{E}^\circ \cap F| = \min_{F \in G_{n,k}} |P_F(\mathcal{E}^\circ)| = \omega_k \left(\prod_{i=1}^k \lambda_i \right)^{-1}.$$

But $\mathcal{E}^\circ \cap F$ is an ellipsoid in the subspace F and $P_F(\mathcal{E})$ is its polar, hence, by the affine invariance of the product of volumes of a body and its polar, we get $|P_F(\mathcal{E})| \cdot |\mathcal{E}^\circ \cap F| = |B_2^n \cap F|^2 = \omega_k^2$ for every $F \in G_{n,k}$. This and (6.1.11) give (6.1.10). $\square$

Proof of Proposition 6.1.13. We will make use of the existence of an M-ellipsoid for K (see Section 1.13.2). There exists an ellipsoid $\mathcal{E}$ such that $|K| = |\mathcal{E}| = 1$ and

$$N(K, \mathcal{E}) \leqslant \exp(\beta n),$$

where $\beta > 0$ is an absolute constant.

If $\mathcal{E} = T(B_2^n)$ and $\lambda_1 \geqslant \cdots \geqslant \lambda_n > 0$ are the eigenvalues of the matrix $\sqrt{TT^*}$, then, as we saw before, there exists a $(n-k)$-dimensional "coordinate subspace" F_0 with respect to the axes of $\mathcal{E}$ such that

$$|P_{F_0}(\mathcal{E})| = \min_{F \in G_{n,n-k}} |P_F(\mathcal{E})| = \omega_{n-k} \prod_{i=k+1}^n \lambda_i.$$

Since

$$\omega_n \prod_{i=1}^n \lambda_i = |\mathcal{E}| = 1,$$

it is easy to check that $\prod_{i=k+1}^n \lambda_i \leqslant \omega_n^{-(n-k)/n}$, and hence

$$|P_{F_0}(\mathcal{E})| \leqslant \frac{\omega_{n-k}}{\omega_n^{(n-k)/n}} \leqslant \left(\frac{n}{n-k} \right)^{\frac{n-k}{2}}.$$

Since $N(P_{F_0}(K), P_{F_0}(\mathcal{E})) \leqslant N(K, \mathcal{E}) \leqslant \exp(\beta n)$, we get

$$(6.1.12) \qquad |P_{F_0}(K)|^{1/n} \leqslant e^\beta |P_{F_0}(\mathcal{E})|^{1/n} \leqslant e^\beta \left(\frac{n}{n-k} \right)^{\frac{n-k}{2n}} \leqslant C = C(\beta).$$

Define $E = F_0^\perp$. Using Proposition 6.1.9 and Fubini's theorem we get

$$(6.1.13) \quad 1 = |K| \leqslant \max_{y \in E^\perp} |K \cap (y+E)| |P_{F_0}(K)| \leqslant \left(\frac{n+1}{k+1} \right)^k |K \cap E| |P_{F_0}(K)|.$$

Then, (6.1.12) shows that

$$|K \cap E|^{1/n} \geqslant \frac{1}{C} \left(\frac{k+1}{n+1} \right)^{k/n} \geqslant c,$$

where $c > 0$ is an absolute constant. $\square$

Proof of Theorem 6.1.12. Let $m, n \in \mathbb{N}$ with $m < n$. Consider the largest integer s for which $2^s m \leqslant n$. Lemma 6.1.11 shows that

$$L_{2^s m} \geqslant L_m.$$

Set $k = 2^s m$. Let K be an isotropic convex body in $\mathbb{R}^n$ such that $L_K = L_n$. From Proposition 6.1.13 there exists a k-dimensional subspace E satisfying

(6.1.14) $$|K \cap E|^{1/n} \geqslant c.$$

Let T be an isotropic convex body in E such that $L_T = L_k$. Since K has extremal isotropic constant, using Theorem 6.1.3 we get

$$L_K \geqslant L_{K(T,E)} \geqslant L_K^{1-\frac{k}{n}} L_T^{\frac{k}{n}} |K \cap E|^{\frac{1}{n}} \left(\frac{k+1}{n+1}\right)^{\frac{k}{n}},$$

and (6.1.14) shows that

$$L_T \leqslant L_K \cdot \frac{n+1}{k+1} \left(\frac{1}{c}\right)^{\frac{n}{k}} \leqslant C L_K$$

because $n < 2k$. By the definition of K and T we have

$$L_m \leqslant L_k = L_T \leqslant C L_K = C L_n,$$

and the proof is complete. $\square$

6.2. Reduction to bounded volume ratio

Recall the definition of volume ratio: if K is a centered convex body in $\mathbb{R}^n$ then

$$\mathrm{vr}(K) = \inf\left\{ \left(\frac{|K|}{|\mathcal{E}|}\right)^{1/n} : \mathcal{E} \text{ is an origin symmetric ellipsoid inside } K \right\}.$$

For every $\alpha > 1$ define

$$L_n(\alpha) = \sup\{L_K : K \text{ is isotropic in } \mathbb{R}^n \text{ and } \mathrm{vr}(K) \leqslant \alpha\}.$$

In this section we will prove the following reduction of the slicing problem.

THEOREM 6.2.1 (Bourgain-Klartag-V. Milman). *There exist two absolute constants $c > 0$ and $\alpha > 1$ such that*

$$L_n \leqslant c[L_n(\alpha)]^4$$

for all $n \geqslant 1$.

In the largest part of this section we work with a symmetric isotropic convex body K in $\mathbb{R}^n$ such that

$$L_K = L'_n = \sup\{L_K : K \text{ is an isotropic symmetric convex body in } \mathbb{R}^n\}.$$

Recall that by Proposition 2.5.10 we have that $L'_n \simeq L_n$ and thus it suffices to prove Theorem 6.2.1 with L'_n instead of L_n. The reader will notice that the symmetry of K is not needed for the first two propositions that follow; this is clearly stated because they will be used again later on in their general form.

As a consequence of the monotonicity of the sequence $\{L_s\}$ we obtain the following property of K.

PROPOSITION 6.2.2. *Let K be an isotropic (not-necessarily symmetric) convex body in $\mathbb{R}^n$ with $L_K \geqslant L'_n$. For every s-codimensional subspace E of $\mathbb{R}^n$,*

$$|K \cap E|^{1/s} \leqslant C_1,$$

where $C_1 > 0$ is an absolute constant.

Proof. Let $E \in G_{n,n-s}$. Since K is isotropic, the measure μ_K with density $L_K^n \mathbf{1}_{\frac{K}{L_K}}$ is isotropic. Note that
$$Z_q(\mu_K) = \frac{1}{L_K} Z_q(K).$$
Then, from Proposition 5.1.12 we have
$$P_{E^\perp}(Z_k(K)) = L_K P_{E^\perp}(Z_k(\mu_K)) = L_K Z_k(\pi_{E^\perp}(\mu_K)).$$
From Theorem 5.1.13 we have
$$L_{\pi_{E^\perp}(\mu_K)} |Z_k(\pi_{E^\perp}(\mu_K))|^{1/s} \simeq 1$$
and, similarly, from Theorem 5.1.14 we have
$$|K \cap E|^{1/s} |P_{E^\perp}(Z_k(K))|^{1/s} \simeq 1.$$
Combining the above we get
$$(6.2.1) \qquad |K \cap E|^{1/s} \simeq \frac{L_{\pi_{E^\perp}(\mu_K)}}{L_K}.$$
On the other hand, by Theorem 6.1.12 we have
$$L_{\pi_{E^\perp}(\mu_K)} \leqslant L_s \leqslant C L_n \leqslant C_1 L'_n \leqslant C_1 L_K$$
for some absolute constants $C, C_1 > 0$, and the result follows. □

A key observation in the proof of Theorem 6.2.1 is that, given any M-ellipsoid of a body K as above, Proposition 6.2.2 imposes strong conditions on the axes of the ellipsoid.

PROPOSITION 6.2.3. *Let K be an isotropic (not-necessarily symmetric) convex body in $\mathbb{R}^n$ with $L_K \geqslant L'_n$. Let $\mathcal{E}$ be an ellipsoid with $|K| = |\mathcal{E}| = 1$ and $N(K, \mathcal{E}) \leqslant \exp(\beta n)$. If we write $\mathcal{E}$ in the form*
$$\mathcal{E} = \left\{ x \in \mathbb{R}^n : \sum_{i=1}^n \frac{\langle x, e_i \rangle^2}{n \lambda_i^2} \leqslant 1 \right\},$$
where $\{e_1, \ldots, e_n\}$ is an orthonormal basis of $\mathbb{R}^n$ and $\lambda_1 \leqslant \cdots \leqslant \lambda_n$, then
$$\lambda_{\lfloor n/2 \rfloor + 1} \leqslant C_2(\beta)$$
for some constant $C_2(\beta) \leqslant c \exp(2\beta)$.

Proof. Set $k = \lfloor n/2 \rfloor$ and let $E = \mathrm{span}\{e_1, \ldots, e_k\}$. From (6.1.13) we have
$$1 = |K| \leqslant \left(\frac{n+1}{n-k+1} \right)^{n-k} |K \cap E^\perp| \cdot |P_E(K)|.$$
Proposition 6.2.2 shows that
$$|K \cap E^\perp| \leqslant C_1^k.$$
Also, $P_E(K)$ is covered by at most $\exp(\beta n)$ translates of $P_E(\mathcal{E})$, so
$$|P_E(K)| \leqslant \exp(\beta n) \, \omega_k \, n^{k/2} \prod_{i=1}^k \lambda_i.$$
It follows that
$$1 \leqslant \left(\frac{n+1}{n-k+1} \right)^{n-k} \exp(\beta n) \, \omega_k \, n^{k/2} C_1^k \prod_{i=1}^k \lambda_i.$$

On the other hand, since $|\mathcal{E}| = 1$, we have
$$\prod_{i=1}^{n} \lambda_i = \frac{1}{n^{n/2}\omega_n},$$
and therefore
$$\lambda_{k+1}^{n-k} \leqslant \prod_{i=k+1}^{n} \lambda_i \leqslant \left(\frac{n+1}{n-k+1}\right)^{n-k} C_1^k \exp(\beta n) \frac{\omega_k n^{k/2}}{\omega_n n^{n/2}}.$$
This implies that
$$\lambda_{k+1} \leqslant C(\beta)$$
for some constant depending only on β. □

The last ingredient of the proof of Theorem 6.2.1 is the fact that every symmetric convex body which is in M-position has orthogonal projections of proportional dimension with bounded volume ratio. First recall the volume ratio theorem (Theorem 1.10.12): given a symmetric convex body K in $\mathbb{R}^n$ such that $B_2^n \subseteq K$ and $|K| = \alpha^n |B_2^n|$ for some $\alpha > 1$, we have for every $1 \leqslant k \leqslant n$ that a random subspace $E \in G_{n,k}$ satisfies
$$B_E \subseteq K \cap E \subseteq (c\alpha)^{\frac{n}{n-k}} B_E$$
with probability greater than $1 - e^{-n}$, where $c > 0$ is an absolute constant.

PROPOSITION 6.2.4. *Let K be a symmetric convex body in $\mathbb{R}^n$. Assume that $|K| = |B_2^n|$ and $N(K, B_2^n) \leqslant \exp(\beta n)$ for some constant $\beta > 0$. Then, for any $1 \leqslant k \leqslant n$, a random orthogonal projection $P_E(K)$ of K onto a k-dimensional subspace E of $\mathbb{R}^n$ has volume ratio bounded by a constant $C(\beta, k/n)$.*

Proof. Observing that
$$|B_2^n| = |K| \leqslant N(K, B_2^n) |K \cap B_2^n| \leqslant e^{\beta n} |K \cap B_2^n|,$$
and using the Blaschke-Santaló inequality we get
$$|\mathrm{conv}(K^\circ \cup B_2^n)|^{1/n} \leqslant C |B_2^n|^{1/n},$$
where C depends only on β. In other words, $W = \mathrm{conv}(K^\circ \cup B_2^n)$ has bounded volume ratio, hence we may apply Theorem 1.10.12 and deduce that
$$K^\circ \cap E \subseteq W \cap E \subseteq C(\beta)^{\frac{n}{n-k}} B_E$$
for a random $E \in G_{n,k}$. By duality, this implies that
$$P_E(K) \supseteq r B_E,$$
where $r = C(\beta)^{-\frac{n}{n-k}}$. Since
$$|P_E(K)| \leqslant N(P_E(K), B_E) |B_E| \leqslant N(K, B_2^n) |B_E| \leqslant \exp(\beta n) |B_E|,$$
this shows that
$$\bigl(|P_E(K)|/|rB_E|\bigr)^{1/k} \leqslant C(\beta, k/n) = \exp(\beta n/k) C(\beta)^{\frac{n}{n-k}},$$
whence we are done. □

Proof of Theorem 6.2.1. Recall that we have started with a symmetric isotropic convex body K with $L_K = L'_n$. There exists an absolute constant $\beta > 0$ and an ellipsoid $\mathcal{E}$ as in the statement of Proposition 6.2.3, that is, satisfying $|K| = |\mathcal{E}| = 1$

and $N(K, \mathcal{E}) \leqslant \exp(\beta n)$. Let $k = \lfloor n/2 \rfloor + 1$. If $F = \mathrm{span}\{e_1, \ldots, e_k\}$, from Proposition 6.2.2 and (6.1.13) we have

(6.2.2) $$|P_F(K)| \geqslant c_1^k.$$

Now,

(6.2.3) $$N(P_F(K), P_F(\mathcal{E})) \leqslant N(K, \mathcal{E}) \leqslant \exp(2\beta k)$$

and Proposition 6.2.3 shows that

(6.2.4) $$P_F(\mathcal{E}) \subseteq c\sqrt{k} B_F.$$

Let $\rho > 0$ be such that $|P_F(K)| = |\rho B_F|$. From (6.2.2)-(6.2.4) we see that $\rho \simeq_\beta \sqrt{k}$. This means that
$$\rho B_F \supseteq c_2 P_F(\mathcal{E}),$$
and hence (6.2.3) gives

(6.2.5) $$N(P_F(K), \rho B_F) \leqslant \exp(c_3 \beta k).$$

We apply Proposition 6.2.4 with the symmetric body $P_F(K)$ and we find a subspace E of F, with dimension $\dim E = s = \lfloor k/2 \rfloor + 1$, such that
$$\mathrm{vr}(P_E(K)) = \mathrm{vr}(P_E(P_F(K))) \leqslant C = C(\beta).$$

Actually, Proposition 6.2.4 shows that the body $P_E(K)$ contains a ball rB_E with $r = \rho C'(\beta)$, and that

(6.2.6) $$\frac{|P_E(K)|}{|rB_E|} \leqslant N(P_E(K), \rho B_E)(\rho/r)^s \leqslant (C''(\beta))^s.$$

Consider the $(\overline{B}_{E^\perp}, E^\perp)$-symmetrization of K, where $\overline{B}_{E^\perp}$ is the Euclidean ball of volume 1 in $E^\perp$. We denote this body by K_1.

Claim. $\mathrm{vr}(K_1) \leqslant \alpha$, where $\alpha > 1$ is an absolute constant.

Proof of the claim. We define the ellipsoid
$$\mathcal{L} = \{ax + by : a^2 + b^2 \leqslant 1, x \in rB_E, y \in |K \cap E^\perp|^{1/(n-s)} \overline{B}_{E^\perp}\}.$$
Note that $|K \cap E^\perp|^{1/(n-s)} \overline{B}_{E^\perp} = K_1 \cap E^\perp$. Then, it is not hard to check that
$$\frac{1}{\sqrt{2}} \mathcal{L} \subseteq \mathrm{conv}\{rB_E, K_1 \cap E^\perp\} \subseteq K_1.$$
On the other hand, $|\sqrt{2} \mathcal{L}| \geqslant |rB_E| \cdot |K_1 \cap E^\perp|$ and (6.2.6) shows that
$$|\mathcal{L}|^{1/n} \geqslant \frac{1}{\sqrt{2} C(\beta)^{s/n}} |P_E(K_1)|^{1/n} |K_1 \cap E^\perp|^{1/n} \geqslant \frac{1}{e\sqrt{2} C'(\beta)} |K_1|^{1/n}.$$
This shows that $\mathrm{vr}(K_1) \leqslant \alpha$, where $\alpha = 2e\, C'(\beta)$. $\square$

We now use Theorem 6.1.3. Since $s \geqslant n/4$, we have

(6.2.7) $$L_K^{\frac{1}{4}} L_{\overline{B}_{E^\perp}}^{\frac{3}{4}} |K \cap E^\perp|^{\frac{1}{n}} \leqslant c_1 L_{K_1}.$$

Moreover, from (6.2.5) we have $|P_E(K)| \leqslant \exp(c\beta n) |\rho B_E|$, which gives
$$|P_E(K)|^{1/n} \leqslant C''(\beta)$$
since $\rho \simeq_\beta \sqrt{k}$. This means that
$$|K \cap E^\perp|^{\frac{1}{n}} \geqslant c |P_E(K)|^{-\frac{1}{n}} \geqslant c(\beta).$$

Going back to (6.2.7) and taking into account the fact that $L_{\overline{B}_{E^\perp}} \geqslant c$, we get
$$L_K \leqslant c_1(\beta) L_{K_1}^4.$$
Since $L_K = L'_n$ and $L_{K_1} \leqslant L_n(\alpha)$, the proof is complete. $\square$

6.3. Regular isotropic convex bodies

In Subsection 1.13.2 we saw that, given any $0 < \alpha < 2$ and any symmetric convex body K, we can find a linear image $\tilde{K}$ of K such that
$$\max\{N(\tilde{K}, tB_2^n), N(B_2^n, t\tilde{K}), N(\tilde{K}^\circ, tB_2^n), N(B_2^n, t\tilde{K}^\circ)\} \leqslant \exp\left(\frac{c(\alpha)n}{t^\alpha}\right)$$
for every $t \geqslant 1$, where $c(\alpha)$ is a constant depending only on α and satisfies $c(\alpha) = O\big((2-\alpha)^{-\alpha/2}\big)$ as $\alpha \to 2$. Note that
$$|\tilde{K}| \leqslant e^n |c(\alpha)^{1/\alpha} B_2^n| \quad \text{and} \quad |B_2^n| \leqslant e^n |c(\alpha)^{1/\alpha} \tilde{K}|,$$
which implies that $e^{-1} c(\alpha)^{-1/\alpha} \leqslant r = \mathrm{v.rad}(\tilde{K}) \leqslant e c(\alpha)^{1/\alpha}$. It follows that
$$\max\{N(\tilde{K}, trB_2^n), N(rB_2^n, t\tilde{K}), N(\tilde{K}^\circ, trB_2^n), N(rB_2^n, t\tilde{K}^\circ)\} \leqslant \exp\left(\frac{e^\alpha [c(\alpha)]^2 n}{t^\alpha}\right)$$
for all $t \geqslant ec(\alpha)^{1/\alpha}$. As a consequence of that and the Rogers-Shephard inequality, we can state the following theorem for any convex body.

THEOREM 6.3.1 (Pisier). *For every $0 < \alpha < 2$ and every convex body K in $\mathbb{R}^n$, there exists an ellipsoid $\mathcal{E}_\alpha$ with $|\mathcal{E}_\alpha| = |K|$ such that, for every $t \geqslant 4ec(\alpha)^{1/\alpha} \geqslant 1$,*

(6.3.1) $$\log N(K, t\mathcal{E}_\alpha) \leqslant \frac{\kappa(\alpha)n}{t^\alpha},$$

where $\kappa(\alpha) \leqslant \kappa_1 (2-\alpha)^{-\alpha}$ and $\kappa_1 > 0$ is an absolute constant.

In the rest of this section we will abuse the classical terminology and we will call any ellipsoid $\mathcal{E}_\alpha$ which satisfies (6.3.1) an α-regular M-ellipsoid for K.

We will now show that, for every $1 \leqslant \alpha < 2$, there are isotropic convex bodies with maximal isotropic constant satisfying (6.3.1) for all sufficiently large t with some absolute constant $c''(\alpha) = O\big((2-\alpha)^{-2\alpha}\big)$. In particular, we have the following

THEOREM 6.3.2 (Dafnis-Paouris). *There exist absolute constants $\kappa, \tau > 1$ and $\delta > 0$ such that, for every $\alpha \in [1,2)$, we can find an isotropic convex body K_α in $\mathbb{R}^n$ with the following properties:*
  (i) $L_{K_\alpha} \geqslant \delta L_n$,
  (ii) *for every $t \geqslant \tau(2-\alpha)^{-3/2}$*

(6.3.2) $$\log N(K_\alpha, t\overline{B}_2^n) \leqslant \frac{\kappa n}{(2-\alpha)^{2\alpha} t^\alpha}.$$

Apart from the existence of α-regular M-ellipsoids for any convex body and Lemma 6.1.14, we will also use another property of ellipsoids (see e.g. [**510**]).

LEMMA 6.3.3. *Let $\mathcal{E}$ be an ellipsoid in $\mathbb{R}^{2m}$. If $\mathcal{E} = T(B_2^{2m})$, we denote the eigenvalues of the matrix $\sqrt{TT^*}$ by $\lambda_1 \geqslant \cdots \geqslant \lambda_{2m} > 0$. Then, there exists $F \in G_{2m,m}$ such that $P_F(\mathcal{E}) = \lambda_m B_F$.*

Proof. As in the proof of Lemma 6.1.14, we can assume that $T = \sqrt{TT^*}$, and we can find an orthonormal basis $\{v_1, \ldots, v_{2m}\}$ in $\mathbb{R}^{2m}$ so that T is diagonal with respect to that basis. We can also assume that $\lambda_1 > \cdots > \lambda_{2m} > 0$. Then

$$\mathcal{E}^\circ \cap v_{2m}^\perp = \left\{ x \in \mathbb{R}^{2m-1} : \sum_{i=1}^{2m-1} \lambda_i^2 \langle x, v_i \rangle^2 \leqslant 1 \right\}.$$

Since $\lambda_i > \lambda_m > \lambda_{2m-i}$ for every $i \leqslant m-1$, we can define $b_1, \ldots, b_{m-1} > 0$ by the equations

(6.3.3) $$\lambda_i^2 b_i^2 + \lambda_{2m-i}^2 = \lambda_m^2 (b_i^2 + 1).$$

Consider the subspace $F = \operatorname{span}\{u_1, \ldots, u_m\} \in G_{2m,m}$ where $u_m := v_m$ and

(6.3.4) $$u_i := \frac{b_i v_i + v_{2m-i}}{\sqrt{b_i^2 + 1}}, \quad i = 1, \ldots, m-1.$$

It is easy to check that $\{u_1, \ldots, u_m\}$ is an orthonormal basis for F and, using (6.3.3) and (6.3.4), we see that, for every $x \in F$,

$$\lambda_m^2 \|x\|_2^2 = \lambda_m^2 \sum_{i=1}^m \langle x, u_i \rangle^2 = \sum_{i=1}^{2m-1} \lambda_i^2 \langle x, v_i \rangle^2 = \|x\|_{\mathcal{E}^\circ}^2.$$

This shows that $\mathcal{E}^\circ \cap F = \lambda_m^{-1}(B_2^n \cap F)$ and therefore, by duality, $P_F(\mathcal{E}) = \lambda_m (B_2^n \cap F) = \lambda_m B_F$. □

In view of the last lemma, we choose to restrict ourselves to the cases that the dimension n is even, $n = 2m$ for some $m \geqslant 1$, and prove Theorem 6.3.2 for those. However, as we will see in Remark 6.3.6, it is not hard to then extend the theorem to all dimensions. Recall also that, when discussing upper bounds for L_n, we do not need to care about odd dimensions by the "monotonicity" of the sequence $\{L_n\}$.

We start with an isotropic convex body K_0 with $L_{K_0} \geqslant \delta_0 L_{2m}$, where $\delta_0 \in (0,1)$. Following the proof of Proposition 6.2.2 we get the following upper bound for the volume of sections of K_0.

LEMMA 6.3.4. *For every k-codimensional subspace E of $\mathbb{R}^{2m}$, $|K_0 \cap E|^{1/k} \leqslant c_1(\delta_0)$, where $c_1(\delta_0) > 0$ depends only on δ_0.* □

For subspaces $F \in G_{2m,m}$ we also have a lower bound of the same order.

LEMMA 6.3.5. *For every $F \in G_{2m,m}$ we have $|K_0 \cap F|^{1/m} \geqslant c_2(\delta_0)$, where $c_2(\delta_0) > 0$ depends only on δ_0.*

Proof. We consider an α-regular M-ellipsoid $\mathcal{E}_\alpha$ for K_0 (for the proof of this lemma we could have fixed $\alpha = 1$; however, we will later make use of the general argument again). Set $t_\alpha = \max\{4e[c(\alpha)]^{1/\alpha}, [\kappa(\alpha)]^{1/\alpha}\}$. Then,

(6.3.5) $$|P_H(K_0)| \leqslant N(K_0, t_\alpha \mathcal{E}_\alpha)|P_H(t_\alpha \mathcal{E}_\alpha)| \leqslant e^{2m}|P_H(t_\alpha \mathcal{E}_\alpha)|$$

for every $H \in G_{2m,m}$. We use the Rogers-Shephard inequality for both K_0 and $\mathcal{E}_\alpha$: since $|K_0| = |\mathcal{E}_\alpha| = 1$, we know that

(6.3.6) $$1 = c_1 \leqslant |K_0 \cap H|^{1/m} |P_{H^\perp}(K_0)|^{1/m} \leqslant c_2,$$

and similar estimates hold true for $\mathcal{E}_\alpha$ (see [**471**] or [**386**] for the left hand side inequality). Combining (6.3.6) with the conclusion of Lemma 6.3.4, we see that

6.3. REGULAR ISOTROPIC CONVEX BODIES

$$\min_{F \in G_{2m,m}} |P_{F^\perp}(K_0)|^{1/m} \geqslant c_3(\delta_0).$$ We then get from (6.3.5) that

$$\min_{F \in G_{2m,m}} |P_{F^\perp}(t_\alpha \mathcal{E}_\alpha)|^{1/m} \geqslant c_4(\delta_0).$$

Now, using (6.3.6) for $\mathcal{E}_\alpha$ we get $|\mathcal{E}_\alpha \cap F|^{1/m} \leqslant c_5(\delta_0) t_\alpha$ for every $F \in G_{2m,m}$. But from Lemma 6.1.14 we have that

(6.3.7) $$\max_{F \in G_{2m,m}} |P_F(\mathcal{E}_\alpha)|^{1/m} = \max_{F \in G_{2m,m}} |\mathcal{E}_\alpha \cap F|^{1/m} \leqslant c_5(\delta_0) t_\alpha.$$

Using (6.3.5) once again, we get $|P_F(K_0)|^{1/m} \leqslant c_6(\delta_0) t_\alpha^2$ for every $F \in G_{2m,m}$. Inserting this estimate into (6.3.6), we see that $|K_0 \cap F|^{1/m} \geqslant c_7(\delta_0)/t_\alpha^2$ for every $F \in G_{2m,m}$. We may choose $\alpha = 1$ now, and complete the proof with $c_2(\delta_0) = c_7(\delta_0)/t_1^2$. $\square$

In the proof of Theorem 6.3.2 we will often use the following facts (see Proposition 5.1.15): If $K \subset \mathbb{R}^n$ is isotropic, and if $1 \leqslant k < n$ and $F \in G_{n,k}$ then the body $\overline{K_{k+1}}(\pi_F(\mu_K))$ is almost isotropic and

(6.3.8) $$|K \cap F^\perp|^{1/k} \simeq \frac{L_{\overline{K_{k+1}}(\pi_F(\mu_K))}}{L_K},$$

where μ_K is the isotropic log-concave measure with density $L_K^n \mathbf{1}_{\frac{K}{L_K}}$. Also, for all $1 \leqslant q \leqslant k$,

(6.3.9) $$Z_q(\overline{K_{k+1}}(\pi_F(\mu_K))) \simeq |K \cap F^\perp|^{1/k} P_F(Z_q(K)).$$

Proof of Theorem 6.3.2. Let $\alpha \in [1, 2)$ and let $\mathcal{E}_\alpha$ be an α-regular M-ellipsoid for K_0 as in Theorem 6.3.1. Recall that $n = 2m$ and that $|\mathcal{E}_\alpha| = 1$. Also, if $\mathcal{E}_\alpha = T(B_2^n) = T(B_2^{2m})$, let $\lambda_1 \geqslant \cdots \geqslant \lambda_{2m} > 0$ be the eigenvalues of the matrix $\sqrt{TT^*}$; observe from Lemma 6.1.14 that

$$|B_2^m| \prod_{i=m+1}^{2m} \lambda_i = \min_{F \in G_{2m,m}} |P_F(\mathcal{E}_\alpha)| \leqslant \max_{F \in G_{2m,m}} |P_F(\mathcal{E}_\alpha)| = |B_2^m| \prod_{i=1}^{m} \lambda_i.$$

Using (6.3.5) and the conclusion of Lemma 6.3.5, we get

$$|B_2^m|^{1/m} \lambda_m \geqslant \min_{F \in G_{2m,m}} |P_F(\mathcal{E}_\alpha)|^{1/m} \geqslant \frac{e^{-2}}{t_\alpha} \min_{F \in G_{2m,m}} |P_F(K_0)|^{1/m}$$
$$\geqslant \frac{e^{-2}}{t_\alpha} \min_{F \in G_{2m,m}} |K_0 \cap F|^{1/m} \geqslant \frac{c_8(\delta_0)}{t_\alpha},$$

and hence

$$\lambda_m \geqslant \frac{c_9(\delta_0)}{t_\alpha} \sqrt{n}.$$

In a similar way, using (6.3.7), we see that $|B_2^m|^{1/m} \lambda_m \leqslant \max_{F \in G_{2m,m}} |P_F(\mathcal{E}_\alpha)|^{1/m} \leqslant c_5(\delta_0) t_\alpha$, and hence $\lambda_m \leqslant c_{10}(\delta_0) t_\alpha \sqrt{n}$. But from Lemma 6.3.3 we know that there exists a subspace $F_0 \in G_{2m,m}$ such that $P_{F_0}(\mathcal{E}_\alpha) = \lambda_m B_{F_0}$, therefore,

(6.3.10) $$\frac{c_9'(\delta_0)}{t_\alpha} \overline{B}_{F_0} \subseteq P_{F_0}(\mathcal{E}_\alpha) \subseteq c_{10}'(\delta_0) t_\alpha \overline{B}_{F_0}.$$

Let $W := \overline{K_{m+1}}(\pi_{F_0}(\mu_{K_0}))$ and $K_\alpha := W \times U(W)$, where $U \in O(2m)$ satisfies $U(F_0) = F_0^\perp$. From Proposition 2.5.12 we know that W is almost isotropic, which means that if $T : F_0 \to F_0$ is an invertible linear operator such that $T(W)$ is

isotropic, then $d_G(T(B_{F_0}), B_{F_0}) \leq C$ for some absolute constant C. Therefore, if we set $S := T \times UT(U^{-1}|_{F_0^\perp}) \in GL(2m)$ and write K_α for the convex body $W \times U(W)$ in $\mathbb{R}^n \equiv \mathbb{R}^{2m}$, then $S(K_\alpha)$ will be an isotropic body and in addition
$$d_G\big(S(B_{F_0} \times B_{F_0^\perp}), B_{F_0} \times B_{F_0^\perp}\big) \leq C,$$
by which it will also follow that $d_G(S(B_2^n), B_2^n) \leq C'$ since $d_G(B_2^n, B_{F_0} \times B_{F_0^\perp}) \leq \sqrt{2}$. But then, given that S preserves volumes since $T \in SL(F_0)$, we will obtain that $S(\overline{B}_2^n) \subseteq C''\overline{B}_2^n$ for some absolute constant C''. Hence, if we prove that the body $K_\alpha = W \times U(W)$ satisfies property (ii) of Theorem 6.3.2, that is, if we prove that
$$\log N(K_\alpha, t\overline{B}_2^n) \leq \frac{\kappa n}{(2-\alpha)^{2\alpha} t^\alpha} \quad \text{for every } t \geq \tau(2-\alpha)^{-3/2}$$
for some absolute constants κ, τ, we will have proven that the isotropic image $S(K_\alpha)$ of K_α satisfies (ii) as well (with slightly different constants κ and τ) since
$$N(S(K_\alpha), t\overline{B}_2^n) \leq N\Big(S(K_\alpha), \frac{t}{C''}S(\overline{B}_2^n)\Big) N\Big(\frac{1}{C''}S(\overline{B}_2^n), \overline{B}_2^n\Big)$$
$$= N\Big(S(K_\alpha), \frac{t}{C''}S(\overline{B}_2^n)\Big).$$
Noting also that $L_{K_\alpha} = L_{W \times U(W)} = L_W$ by Lemma 3.2.5 and because $L_{U(W)} = L_W$, we conclude that it suffices to show that K_α satisfies (i) and (ii).

Proof of (i): Since $L_{K_\alpha} = L_W$, from (6.3.8) and Lemma 6.3.5 we get
$$L_{K_\alpha} = L_W \simeq L_{K_0}|K_0 \cap F_0^\perp|^{1/m} \geq c_2(\delta_0)L_{K_0} \geq \delta L_n,$$
where $\delta \simeq \delta_0 c_2(\delta_0)$.

Proof of (ii): Using the fact that $N(A \times A, B \times B) \leq N(A, B)^2$ for any two nonempty sets A, B, and also the fact that $B_2^m \times B_2^m \subseteq \sqrt{2}B_2^{2m}$, we may write
$$N\big(K_\alpha, s\sqrt{2}\overline{B}_2^n\big) \leq N\big(W \times U(W), s(\overline{B}_{F_0} \times \overline{B}_{F_0^\perp})\big) \leq N\big(W, s\overline{B}_{F_0}\big)^2$$
for any $s > 0$. Now, recall that from (6.3.9) we have
$$Z_m(\overline{K_{m+1}}(\pi_{F_0}(\mu_{K_0}))) \simeq |K_0 \cap F_0^\perp|^{1/m} P_{F_0}(Z_m(K_0)),$$
therefore, using Lemmas 6.3.4, 6.3.5 and the fact that $\text{conv}(C, -C) \simeq Z_m(C)$ for every centered convex body C of volume 1 in F_0 or in $\mathbb{R}^n$, we get
$$\text{conv}(W, -W) \simeq Z_m(\overline{K_{m+1}}(\pi_{F_0}(\mu_{K_0}))) \simeq |K_0 \cap F_0^\perp|^{1/m} P_{F_0}(Z_m(K_0))$$
$$\simeq_{\delta_0} P_{F_0}(\text{conv}(K_0, -K_0)).$$
Recalling also (6.3.10), we see that for every $r > 0$
$$N(W, c'_{10}(\delta_0)t_\alpha r\overline{B}_{F_0}) \leq N\big(\text{conv}(W, -W), c'_{10}(\delta_0)t_\alpha r\overline{B}_{F_0}\big)$$
$$\leq N\big(\text{conv}(W, -W), rP_{F_0}(\mathcal{E}_\alpha)\big)$$
$$\leq N\big(c_{11}(\delta_0)P_{F_0}(\text{conv}(K_0, -K_0)), rP_{F_0}(\mathcal{E}_\alpha)\big)$$
$$\leq N\big(c_{11}(\delta_0)\text{conv}(K_0, -K_0), r\mathcal{E}_\alpha\big)$$
$$\leq N\big(K_0 - K_0, c_{12}(\delta_0)r(\mathcal{E}_\alpha - \mathcal{E}_\alpha)\big)$$
$$\leq N\big(K_0, c_{13}(\delta_0)r\mathcal{E}_\alpha\big)^2$$
(note that for the last two inequalities we have also used that $\mathcal{E}_\alpha$ is convex and symmetric, so $\mathcal{E}_\alpha - \mathcal{E}_\alpha = 2\mathcal{E}_\alpha$, that K_0 is convex and contains the origin, so

$\mathrm{conv}(K_0, -K_0) \subset K_0 - K_0$, as well as the fact that $N(A-A, B-B) \leqslant N(A,B)^2$. It follows that

$$(6.3.11) \qquad N(K_\alpha, t\overline{B}_2^n) \leqslant N\left(K_0, \frac{c_{13}(\delta_0)t}{\sqrt{2}c'_{10}(\delta_0)t_\alpha}\mathcal{E}_\alpha\right)^4$$

for every $t > 0$. Since $\mathcal{E}_\alpha$ is an α-regular M-ellipsoid for K_0, it remains to consider large enough $t \geqslant \tau(\delta_0, \alpha)$, where

$$\tau(\delta_0, \alpha) := 4\sqrt{2}c'_{10}(\delta_0)t_\alpha e[c(\alpha)]^{1/\alpha}/c_{13}(\delta_0),$$

to deduce from (6.3.1) and (6.3.11) that

$$\log N(K, t\overline{B}_2^n) \leqslant 4\log N\left(K_0, \frac{c_{14}(\delta_0)t}{t_\alpha}\mathcal{E}_\alpha\right) \leqslant \frac{4\kappa(\alpha)t_\alpha^\alpha}{c_{14}^\alpha(\delta_0)} \frac{n}{t^\alpha}.$$

Since $\tau(\delta_0, \alpha) \lesssim (2-\alpha)^{-3/2}$ and $\kappa(\alpha)t_\alpha^\alpha \lesssim (2-\alpha)^{-2\alpha}$, we conclude the proof. $\square$

REMARK 6.3.6. Now that, for every $\alpha \in [1,2)$, we have proven the existence of an isotropic body K in $\mathbb{R}^{2m}$ which has properties (i) and (ii) of Theorem 6.3.2, we can easily prove the existence of such bodies in $\mathbb{R}^{2m-1}$ as well: just note that for every subspace $F \in G_{2m,2m-1}$ we have that $cL_K \leqslant |K \cap F^\perp| \leqslant 2R(K)$. Combining this with the properties (6.3.8), (6.3.9) of the almost isotropic convex body $\overline{K_{2m}}(\pi_F(\mu_K))$ in the $(2m-1)$-dimensional subspace F, we get that

$$L_{\overline{K_{2m}}(\pi_F(\mu_K))} \simeq |K \cap F^\perp|^{1/(2m-1)} L_K \simeq L_K \geqslant \delta L_{2m} \geqslant c\delta L_{2m-1},$$

and also that

$$\overline{K_{2m}}(\pi_F(\mu_K)) \simeq Z_{2m-1}\left(\overline{K_{2m}}(\pi_F(\mu_K))\right)$$
$$\simeq |K \cap F^\perp|^{1/(2m-1)} P_F(Z_{2m-1}(K)) \simeq P_F(K).$$

Since $N(P_F(K), tB_F) = N(P_F(K), tP_F(B_2^{2m})) \leqslant N(K, tB_2^{2m})$ for every $t > 0$, we conclude that the body $\overline{K_{2m}}(\pi_F(\mu_K))$ will also satisfy properties (i) and (ii) of Theorem 6.3.2 with perhaps slightly different, but still independent of the dimension, constants κ, τ and δ.

6.4. Reduction to negative moments

In this section we describe the main result of the work of Dafnis and Paouris: they proved that a positive answer to the hyperplane conjecture is equivalent to some very strong small probability estimates for the Euclidean norm on isotropic convex bodies. Recall that, for $-n < p \leqslant \infty$, $p \neq 0$,

$$I_p(K) := \left(\int_K \|x\|_2^p dx\right)^{1/p}$$

and, given any $\zeta \geqslant 1$, consider the parameter

$$(6.4.1) \qquad q_{-c}(K, \zeta) := \max\{p \geqslant 1 : I_2(K) \leqslant \zeta I_{-p}(K)\}.$$

The results in this section reveal that the hyperplane conjecture is equivalent to the following statement:

> There exist absolute constants $C, \xi > 0$ such that, for every isotropic convex body K in $\mathbb{R}^n$,
> $$q_{-c}(K, \xi) \geqslant Cn.$$

Note that the results of Chapter 5 show that there exists a parameter $q_* := q_*(K)$ (related to the L_q-centroid bodies of K) with the following properties:

(i) $q_*(K) \geqslant c\sqrt{n}$,
(ii) $q_{-c}(K, \xi_0) \geqslant q_*(K)$ for some absolute constant $\xi_0 \geqslant 1$, and hence, $I_2(K) \leqslant \xi_0 I_{-q_*}(K)$.

What is not clear is the behavior of $I_{-p}(K)$ when p lies in the interval $[q_*, n]$.

The key idea in [149] is to start with an "extremal" isotropic convex body K in $\mathbb{R}^n$ with maximal isotropic constant $L_K \simeq L_n$ which is at the same time in α-regular M-position (the existence of such bodies was established in the previous section), and try taking advantage of the fact that small ball probability estimates are closely related to estimates for covering numbers. The key lemma is the following.

LEMMA 6.4.1. *Let K be a centered convex body of volume 1 in $\mathbb{R}^n$. Assume that, for some $s > 0$,*

(6.4.2) $$r_s := \log N(K, sB_2^n) < n.$$

Then,
$$I_{-r_s}(K) \leqslant 3es.$$

Proof. Let $z_0 \in \mathbb{R}^n$ be such that $|K \cap (-z_0 + sB_2^n)| \geqslant |K \cap (z + sB_2^n)|$ for every $z \in \mathbb{R}^n$. It follows that

(6.4.3) $$|(K + z_0) \cap sB_2^n| \cdot N(K, sB_2^n) \geqslant |K| = 1.$$

Let $q = r_s$. Then, using Markov's inequality, the definition of $I_{-q}(K + z_0)$ and (6.4.2), we get
$$|(K + z_0) \cap 3^{-1}I_{-q}(K + z_0)B_2^n| \leqslant 3^{-q} < e^{-q} = e^{-r_s} \leqslant \frac{1}{N(K, sB_2^n)}.$$

From (6.4.3) we obtain
$$|(K + z_0) \cap 3^{-1}I_{-q}(K + z_0)B_2^n| < |(K + z_0) \cap sB_2^n|,$$

and this implies
$$3^{-1}I_{-q}(K + z_0) \leqslant s.$$

Since K is centered, as an application of Theorem 2.2.2 we get that $I_{-k}(K + z) \geqslant \frac{1}{e}I_{-k}(K)$ for any $1 \leqslant k < n$ and $z \in \mathbb{R}^n$. To see this, we use Proposition 5.3.14 to write

$$I_{-k}(K + z) = c_{n,k}\left(\int_{G_{n,k}} |(K + z) \cap F^\perp| \, d\nu_{n,k}(F)\right)^{-1/k}$$

$$\geqslant \frac{c_{n,k}}{e}\left(\int_{G_{n,k}} |K \cap F^\perp| \, d\nu_{n,k}(F)\right)^{-1/k} = \frac{1}{e}I_{-k}(K).$$

This proves the lemma. □

THEOREM 6.4.2 (Dafnis-Paouris). *Assume that $q_{-c}(K, \zeta) \geqslant \beta n$ for some $\zeta \geqslant 1$, some $\beta \in (0, 1)$ and every isotropic convex body K in $\mathbb{R}^n$. Then,*

(6.4.4) $$L_n \leqslant \frac{C\zeta}{\sqrt{\beta}}\log^2\left(\frac{e}{\beta}\right),$$

where $C > 0$ is an absolute constant.

Proof. Set $\alpha := 2 - \log(e/\beta)^{-1}$ and with this α apply Theorem 6.3.2 to find an isotropic convex body K_α which satisfies its conclusion: for some absolute constants $\kappa, \tau \geqslant 1$ and $\delta > 0$ it holds that $L_{K_\alpha} \geqslant \delta L_n$ and

$$\log N(K_\alpha, t\sqrt{n}B_2^n) \leqslant \frac{\kappa n}{(2-\alpha)^{2\alpha} t^\alpha} \quad \text{for all } t \geqslant \tau \log^{3/2}\left(\frac{e}{\beta}\right).$$

We may clearly assume that $\tau^2 \leqslant e\kappa$ as well. We choose

$$t_1 = (e\kappa)^{1/\alpha} \frac{1}{\sqrt{\beta}} \log^2\left(\frac{e}{\beta}\right);$$

then $t_1^\alpha = e\kappa(2-\alpha)^{-2\alpha}(\sqrt{\beta})^{-\alpha}$ and, since $\tau \leqslant \sqrt{e\kappa} \leqslant (e\kappa)^{1/\alpha}$, we have that $t_1 \geqslant \tau(2-\alpha)^{-3/2} = \tau \log^{3/2}(e/\beta)$. Therefore,

$$r_1 := \log N(K_\alpha, t_1\sqrt{n}B_2^n) \leqslant \frac{\kappa n}{(2-\alpha)^{2\alpha} t_1^\alpha} \leqslant \frac{1}{e}(\sqrt{\beta})^\alpha n \leqslant \beta n,$$

and hence by Lemma 6.4.1 we obtain that

$$I_{-r_1}(K_\alpha) \leqslant 3et_1\sqrt{n}.$$

On the other hand, since $r_1 \leqslant \beta n$ and since $q_{-c}(K_\alpha, \zeta) \geqslant \beta n$, we have that

$$\sqrt{n}L_{K_\alpha} = I_2(K_\alpha) \leqslant \zeta I_{-r_1}(K_\alpha).$$

It follows that

$$L_{K_\alpha} \leqslant 3e\zeta t_1 = 3e\zeta(e\kappa)^{1/\alpha}\frac{1}{\sqrt{\beta}}\log^2\left(\frac{e}{\beta}\right) \leqslant \frac{3e^2\zeta\kappa}{\sqrt{\beta}}\log^2\left(\frac{e}{\beta}\right).$$

Since $L_{K_\alpha} \geqslant \delta L_n$, the result follows. $\square$

In the opposite direction, one can show that if the hyperplane conjecture is correct then there are absolute constants $\sigma, \xi > 0$ such that, for every isotropic convex body K in $\mathbb{R}^n$, one has $q_{-c}(K, \xi) \geqslant \sigma n$. This is an immediate consequence of the next theorem.

THEOREM 6.4.3 (Dafnis-Paouris). *There exists an absolute constant $C > 0$ such that, for every n and for every isotropic convex body K in $\mathbb{R}^n$,*

$$q_{-c}(K, CL_n) \geqslant n - 1.$$

Proof. We start with the formula

(6.4.5) $$I_{-s}(K) \simeq \sqrt{n}\left(\int_{G_{n,s}} |K \cap F^\perp|\, d\nu_{n,s}(F)\right)^{-1/s}.$$

Recall from Proposition 5.1.15 that

$$|K \cap F^\perp|^{1/s} \simeq \frac{L_{\overline{K_{s+1}}(\pi_F(\mu_K))}}{L_K}.$$

for every $F \in G_{n,s}$. Thus, we get

$$I_{-s}(K) \simeq \sqrt{n}\left(\int_{G_{n,s}} \left(\frac{L_{\overline{K_{s+1}}(\pi_F(f))}}{L_K}\right)^s d\nu_{n,s}(F)\right)^{-1/s}.$$

Now,

$$\int_{G_{n,s}} \left(\frac{L_{\overline{K_{s+1}}(\pi_F(f))}}{L_K}\right)^s d\nu_{n,s}(F) \leqslant \left(\frac{L_s}{L_K}\right)^s.$$

Therefore,
$$I_{-s}(K) \geqslant \frac{c_1\sqrt{n}L_K}{L_s} \geqslant \frac{c_2\sqrt{n}L_K}{L_n}$$
because $L_s \leqslant c_3 L_n$ for all integers $s \leqslant n-1$. Since $I_2(K) = \sqrt{n}L_K$, we get
$$q_{-c}(K,\delta) := \max\{p \geqslant 1 : I_2(K) \leqslant c_2^{-1}L_n I_{-p}(K)\} \geqslant n-1.$$
This is the claim of the theorem. $\square$

6.5. Reduction to $I_1(K, Z_q^\circ(K))$

Let K be a centered convex body of volume 1 in $\mathbb{R}^n$. We consider the parameter
$$I_1(K, Z_q^\circ(K)) = \int_K \|\langle \cdot, x\rangle\|_{L_q(K)} dx.$$
Generally, if K is a centered convex body of volume 1 in $\mathbb{R}^n$, then for every symmetric convex body C in $\mathbb{R}^n$ and for every $q \in (-n, \infty)$, $q \neq 0$, we define
$$I_q(K, C) := \left(\int_K \|x\|_C^q \, dx\right)^{1/q}.$$
The notation $I_1(K, Z_q^\circ(K))$ is then justified by the fact that $\|\langle \cdot, x\rangle\|_{L_q(K)}$ is the norm induced on $\mathbb{R}^n$ by the polar body $Z_q^\circ(K)$ of the L_q-centroid body of K.

The purpose of this section is to describe a work of Giannopoulos, Paouris and Vritsiou [213] which reduces the hyperplane conjecture to the study of the parameter $I_1(K, Z_q^\circ(K))$ when K belongs to the following subclass of isotropic convex bodies.

DEFINITION 6.5.1. Let $\kappa, \tau > 0$. We say that an isotropic convex body K in $\mathbb{R}^n$ is (κ, τ)-regular if
$$\log N(K, tB_2^n) \leqslant \frac{\kappa n^2 \log^4 n}{t^2} \text{ for all } t \geqslant \tau\sqrt{n}\log^{3/2} n.$$

Applying Theorem 6.3.2 with $\alpha = 2 - (\log n)^{-1}$, we already know that there are absolute constants $\kappa, \tau > 1$ and $\delta > 0$ such that, for every $n \in \mathbb{N}$, there exist (κ, τ)-regular isotropic convex bodies with maximal isotropic constant.

COROLLARY 6.5.2. For every $n \in \mathbb{N}$ we can find an isotropic convex body K in $\mathbb{R}^n$ with the following properties:
 (i) $L_K \geqslant \delta L_n$,
 (ii) $\log N(K, tB_2^n) \leqslant \kappa n^2 \log^4 n/t^2$ for all $t \geqslant \tau\sqrt{n}\log^{3/2} n$,
where $\kappa \geqslant \tau^2 \geqslant 1$ and $\delta > 0$ are absolute constants.

The main result of this section is the next reduction of the slicing problem.

THEOREM 6.5.3 (Giannopoulos-Paouris-Vritsiou). There exists an absolute constant $\rho \in (0,1)$ with the following property. Given $\kappa \geqslant \tau^2 \geqslant 1$, for every $n \geqslant n_0(\tau)$ and every (κ, τ)-regular isotropic convex body K in $\mathbb{R}^n$ we have: if
(6.5.1) $\qquad 2 \leqslant q \leqslant \rho^2 n \text{ and } I_1(K, Z_q^\circ(K)) \leqslant \rho n L_K^2,$
then
$$L_K^2 \leqslant C\kappa \sqrt{\frac{n}{q}} \log^4 n \, \max\left\{1, \frac{I_1(K, Z_q^\circ(K))}{\sqrt{qn}L_K^2}\right\},$$
where $C > 0$ is an absolute constant.

Observe that, for every isotropic convex body K in $\mathbb{R}^n$, we have that
$$I_1(K, Z_2^\circ(K)) \leqslant \sqrt{n} L_K^2 \leqslant \rho n L_K^2$$
if n is sufficiently large. But by Corollary 6.5.2 we know that, for some absolute constants $\kappa \geqslant \tau^2 \geqslant 1$ and $\delta > 0$, there exists a (κ, τ)-regular isotropic convex body K in $\mathbb{R}^n$ with $L_K \geqslant \delta L_n$. Therefore, Theorem 6.5.3 gives

(6.5.2) $$L_K^2 \leqslant C_1 \sqrt{n} \log^4 n,$$

which leads to a bound for L_n that is only logarithmically worse than Bourgain's and Klartag's bounds: $L_n \leqslant C_2 \sqrt[4]{n} \log^2 n$.

However, the behavior of $I_1(K, Z_q^\circ(K))$ may allow us to use much larger values of q. In Subsection 6.6.1 we discuss upper and lower bounds for this quantity. For every isotropic convex body K in $\mathbb{R}^n$ we have some simple general estimates:

(i) For every $2 \leqslant q \leqslant n$,
$$c_1 \max\left\{\sqrt{n} L_K^2, \sqrt{qn}, R(Z_q(K)) L_K\right\} \leqslant I_1(K, Z_q^\circ(K)) \leqslant c_2 q \sqrt{n} L_K^2.$$

(ii) If $2 \leqslant q \leqslant \sqrt{n}$, then
$$c_1 \max\left\{\sqrt{n} L_K^2, \sqrt{qn} L_K\right\} \leqslant I_1(K, Z_q^\circ(K)) \leqslant c_2 q \sqrt{n} L_K^2.$$

Any improvement of the exponent of q in the upper bound $I_1(K, Z_q^\circ(K)) \leqslant cq\sqrt{n} L_K^2$ would lead to an estimate $L_n \leqslant Cn^\alpha$ with $\alpha < \frac{1}{4}$. It seems plausible that one could even have $I_1(K, Z_q^\circ(K)) \leqslant c\sqrt{qn} L_K^2$, at least when q is small, say $2 \leqslant q \ll \sqrt{n}$. Some evidence is given by the following facts:

(iii) If K is an unconditional isotropic convex body in $\mathbb{R}^n$, then
$$c_1 \sqrt{qn} \leqslant I_1(K, Z_q^\circ(K)) \leqslant c_2 \sqrt{qn} \log n$$
for all $2 \leqslant q \leqslant n$.

(iv) If K is an isotropic convex body in $\mathbb{R}^n$ then, for every $2 \leqslant q \leqslant \sqrt{n}$, there exists a set $A_q \subseteq O(n)$ with $\nu(A_q) \geqslant 1 - e^{-q}$ such that $I_1(K, Z_q^\circ(U(K))) \leqslant c_3 \sqrt{qn} L_K^2$ for all $U \in A_q$.

The proofs of (i)-(iv) are given in Subsection 6.6.1.

6.5.1. Proof of the reduction

In this subsection we prove Theorem 6.5.3. We start with two auxiliary results. The first one provides an estimate for the L_q-norm of the maximum of N linear functionals on K.

LEMMA 6.5.4. *Let K be a convex body of volume 1 in $\mathbb{R}^n$, and consider any points $z_1, z_2, \ldots, z_N \in \mathbb{R}^n$. If $q \geqslant 1$ and $p \geqslant \max\{\log N, q\}$, then*

(6.5.3) $$\left(\int_K \max_{1 \leqslant i \leqslant N} |\langle x, z_i \rangle|^q dx\right)^{1/q} \leqslant \overline{\beta}_1 \max_{1 \leqslant i \leqslant N} h_{Z_p(K)}(z_i),$$

where $\overline{\beta}_1 > 0$ is an absolute constant.

Proof. Let $p \geqslant \max\{\log N, q\}$ and $\theta \in S^{n-1}$. Markov's inequality shows that
$$\left|\{x \in K : |\langle x, \theta \rangle| \geqslant e^3 h_{Z_p(K)}(\theta)\}\right| \leqslant e^{-3p}.$$

Since $x \mapsto |\langle x, \theta \rangle|$ is a seminorm, from Borell's lemma we get that

$$|\{x \in K : |\langle x, \theta \rangle| \geq e^3 t h_{Z_p(K)}(\theta)\}| \leq (1-e^{-3p})\left(\frac{e^{-3p}}{1-e^{-3p}}\right)^{\frac{t+1}{2}} \leq e^{-pt}$$

for every $t \geq 1$. We set $S := e^3 \max_{1 \leq i \leq N} h_{Z_p(K)}(z_i)$. Then, for every $t \geq 1$ we have that

$$|\{x \in K : \max_{1 \leq i \leq N} |\langle x, z_i \rangle| \geq St\}| \leq \sum_{i=1}^{N} |\{x \in K : |\langle x, z_i \rangle| \geq e^3 t h_{Z_p(K)}(z_i)\}|$$
$$\leq N e^{-pt}.$$

It follows that

$$\int_K \max_{1 \leq i \leq N} |\langle x, z_i \rangle|^q dx = q \int_0^\infty s^{q-1} |\{x \in K : \max_{1 \leq i \leq N} |\langle x, z_i \rangle| \geq s\}| ds$$
$$\leq S^q + q \int_S^\infty s^{q-1} |\{x \in K : \max_{1 \leq i \leq N} |\langle x, z_i \rangle| \geq s\}| ds$$
$$= S^q \left(1 + q \int_1^\infty t^{q-1} |\{x \in K : \max_{1 \leq i \leq N} |\langle x, z_i \rangle| \geq St\}| dt\right)$$
$$\leq S^q \left(1 + qN \int_1^\infty t^{q-1} e^{-pt} dt\right)$$
$$= S^q \left(1 + \frac{qN}{p^q} \int_p^\infty t^{q-1} e^{-t} dt\right)$$
$$\leq S^q \left(1 + \frac{qN}{p^q} e^{-p} p^q\right)$$
$$\leq (3S)^q,$$

where we have also used the fact that, for every $p \geq q \geq 1$,

$$\int_p^\infty t^{q-1} e^{-t} dt \leq e^{-p} p^q.$$

This finishes the proof (with $\overline{\beta}_1 = 3e^3$). $\square$

The second lemma concerns the L_q-centroid bodies of subsets of K.

LEMMA 6.5.5. *Let K be a convex body of volume 1 in $\mathbb{R}^n$ and let $1 \leq q, r \leq n$. There exists an absolute constant $\overline{\beta}_2 > 0$ such that if A is a convex subset of K with $|A| \geq 1 - e^{-\overline{\beta}_2 q}$, then*

(6.5.4) $$Z_p(K) \subseteq 2 Z_p(\overline{A})$$

for all $1 \leq p \leq q$. Also, for the opposite inclusion, it suffices to have $|A| \geq 2^{-\frac{r}{2}}$ to conclude that

(6.5.5) $$Z_p(\overline{A}) \subseteq 2 Z_p(K)$$

for all $r \leq p \leq n$.

Proof. Let $\theta \in S^{n-1}$. Note that

$$h_{Z_p(\overline{A})}(\theta) = \left(\int_{\overline{A}} |\langle x, \theta \rangle|^p dx\right)^{1/p} = \frac{1}{|A|^{\frac{1}{p}+\frac{1}{n}}} \left(\int_A |\langle x, \theta \rangle|^p dx\right)^{1/p}.$$

We first prove (6.5.5): since $A \subseteq K$ and assuming that $|A| \geqslant 2^{-\frac{r}{2}}$, we have

$$h_{Z_p(K)}(\theta) = \left(\int_K |\langle x, \theta \rangle|^p dx\right)^{1/p} \geqslant \left(\int_A |\langle x, \theta \rangle|^p dx\right)^{1/p}$$

$$\geqslant 2^{-\frac{r}{2p} - \frac{r}{2n}} \left(\int_{\overline{A}} |\langle x, \theta \rangle|^p dx\right)^{1/p} \geqslant \frac{1}{2} h_{Z_p(\overline{A})}(\theta)$$

for all $r \leqslant p \leqslant n$. On the other hand, assuming that $|A| \geqslant 1 - e^{-\overline{\beta}_2 q}$ and using the fact that $\|\langle \cdot, \theta \rangle\|_{2p} \leqslant c \|\langle \cdot, \theta \rangle\|_p$ for some absolute constant $c > 0$, we have

$$\int_K |\langle x, \theta \rangle|^p dx = \int_A |\langle x, \theta \rangle|^p dx + \int_{K \setminus A} |\langle x, \theta \rangle|^p dx$$

$$\leqslant |A|^{1+\frac{p}{n}} \int_{\overline{A}} |\langle x, \theta \rangle|^p dx + |K \setminus A|^{1/2} \left(\int_K |\langle x, \theta \rangle|^{2p} dx\right)^{1/2}$$

$$\leqslant \int_{\overline{A}} |\langle x, \theta \rangle|^p dx + e^{-\overline{\beta}_2 q/2} c^p \int_K |\langle x, \theta \rangle|^p dx$$

$$\leqslant \int_{\overline{A}} |\langle x, \theta \rangle|^p dx + \frac{1}{2} \int_K |\langle x, \theta \rangle|^p dx$$

for every $1 \leqslant p \leqslant q$, if $\overline{\beta}_2 > 0$ is chosen large enough. This proves (6.5.4). $\square$

Proof of Theorem 6.5.3. Let $\kappa \geqslant \tau^2 \geqslant 1$ and consider a (κ, τ)-regular isotropic convex body K in $\mathbb{R}^n$. Assume that the conditions (6.5.1) are also satisfied. We define a convex body W in $\mathbb{R}^n$, setting

$$W := \{x \in K : h_{Z_q(K)}(x) \leqslant C_1 I_1(K, Z_q^\circ(K))\},$$

where $C_1 = e^{2\overline{\beta}_2}$ and $\overline{\beta}_2 > 0$ is the constant which was defined in Lemma 6.5.5. From Markov's inequality we have that $|W| \geqslant 1 - e^{-2\overline{\beta}_2}$ and also trivially that $|W| \geqslant \frac{1}{2} \geqslant 2^{-q/2}$ (as long as $\overline{\beta}_2 \geqslant 1$). Then we set

$$K_1 := \overline{W}.$$

Applying both cases of Lemma 6.5.5 to the set W with $p = 2$, we see that

$$\frac{1}{2} Z_2(K_1) \subseteq Z_2(K) \subseteq 2 Z_2(K_1).$$

This implies that

$$\frac{1}{4} L_K^2 = \frac{1}{4} \int_K \langle x, \theta \rangle^2 dx \leqslant \int_{K_1} \langle x, \theta \rangle^2 dx \leqslant 4 \int_K \langle x, \theta \rangle^2 dx = 4 L_K^2$$

for every $\theta \in S^{n-1}$, and hence,

$$\frac{n L_K^2}{4} \leqslant \sum_{i=1}^n \int_{K_1} \langle x, e_i \rangle^2 dx = \int_{K_1} \|x\|_2^2 dx \leqslant 4 n L_K^2.$$

We also have

$$K_1 = |W|^{-1/n} W \subseteq 2W \subseteq 2K,$$

thus for every $x \in K_1$ we have $x/2 \in W$ and, using (6.5.5) of Lemma 6.5.5 (with $p = q$) we write

(6.5.6) $\qquad h_{Z_q(K_1)}(x) \leqslant 2 h_{Z_q(K)}(x) = 4 h_{Z_q(K)}(x/2) \leqslant 4 C_1 I_1(K, Z_q^\circ(K)).$

Finally,
$$\log \overline{N}(K_1, tB_2^n) \leqslant \log N(2K_1, tB_2^n) \leqslant \log N(4K, tB_2^n) \leqslant \frac{16\kappa n^2 \log^4 n}{t^2},$$
for all $t \geqslant 4\tau\sqrt{n}\log^{3/2} n$. We write
$$nL_K^2 \leqslant 4\int_{K_1} \|x\|_2^2 dx \leqslant 4\int_{K_1} \max_{z \in K_1} |\langle x, z\rangle| \, dx.$$
Observe now that for every $t \geqslant 4\tau\sqrt{n}\log^{3/2} n$ we can find $z_1, \ldots, z_{N_t} \in K_1$, with $|N_t| \leqslant \exp(16\kappa n^2 \log^4 n/t^2)$, such that $K_1 \subseteq \cup_{i=1}^{N_t}(z_i + tB_2^n)$. It follows that
$$\max_{z \in K_1} |\langle x, z\rangle| \leqslant \max_{1 \leqslant i \leqslant N_t} |\langle x, z_i\rangle| + \max_{w \in tB_2^n} |\langle x, w\rangle| = \max_{1 \leqslant i \leqslant N_t} |\langle x, z_i\rangle| + t\|x\|_2,$$
and hence
$$(6.5.7) \qquad nL_K^2 \leqslant 4\int_{K_1} \max_{1 \leqslant i \leqslant N_t} |\langle x, z_i\rangle| dx + 4t\int_{K_1} \|x\|_2 dx$$
$$\leqslant 4\int_{K_1} \max_{1 \leqslant i \leqslant N_t} |\langle x, z_i\rangle| dx + 8t\sqrt{n}L_K.$$

Recall also that by Borell's lemma (more precisely, from Proposition 5.1.2) we can find absolute constants $\beta_1, \beta_2 > 0$ so that
$$(6.5.8) \qquad Z_q(K) \subseteq \beta_1 q Z_1(K) \quad \text{and} \quad Z_q(K) \subseteq \beta_2 \frac{q}{p} Z_p(K)$$
for all $1 \leqslant p < q$. We choose
$$t_0^2 = 64 C_2 \kappa \max\left\{1, \frac{I_1(K, Z_q^\circ(K))}{\sqrt{qn}L_K^2}\right\} \frac{n^{3/2}}{\sqrt{q}} \log^4 n,$$
where $C_2 = 16 C_1 \beta_1 \overline{\beta}_1$ with $\overline{\beta}_1$ the constant from Lemma 6.5.4. With this choice of t_0, we have
$$(6.5.9) \qquad t_0^2 \geqslant 64 C_2 \kappa \sqrt{\frac{n}{q}} n \log^4 n \geqslant \frac{64 C_2 \kappa}{\rho} n \log^4 n$$
and
$$(6.5.10) \qquad t_0^2 \geqslant 64 C_2 \kappa \frac{I_1(K, Z_q^\circ(K))}{qL_K^2} n \log^4 n.$$
From (6.5.9) it is clear that
$$t_0^2 \geqslant 64 C_2 \kappa \frac{n \log^4 n}{\rho} \geqslant 16\tau^2 n \log^3 n,$$
provided that $n \geqslant n_0(\tau, \rho)$, so the above argument, leading up to (6.5.7), holds with $t = t_0$. We also set $p_0 := \frac{16\kappa n^2 \log^4 n}{t_0^2}$. Observe that $p_0 \geqslant q$: since q is such that $I_1(K, Z_q^\circ(K)) \leqslant \rho n L_K^2$, we have $\max\left\{1, \frac{I_1(K, Z_q^\circ(K))}{\sqrt{qn}L_K^2}\right\} \leqslant \rho\sqrt{n/q}$, and hence
$$t_0^2 \leqslant 64 C_2 \kappa \rho \frac{n^2 \log^4 n}{q}.$$
But then, if we choose $\rho < 1/(4C_2)$, we have
$$p_0 = \frac{16\kappa n^2 \log^4 n}{t_0^2} \geqslant \frac{16\kappa n^2 q \log^4 n}{64 C_2 \kappa \rho n^2 \log^4 n} = \frac{q}{4C_2 \rho} \geqslant q$$

as claimed. Therefore, using Lemma 6.5.4 with $q' = 1$, we can write

$$\int_{K_1} \max_{1 \leqslant i \leqslant N_{t_0}} |\langle x, z_i \rangle| dx \leqslant \overline{\beta}_1 \max_{1 \leqslant i \leqslant N_{t_0}} h_{Z_{p_0}(K_1)}(z_i) \leqslant \overline{\beta}_1 \beta_1 \frac{p_0}{q} \max_{1 \leqslant i \leqslant N_{t_0}} h_{Z_q(K_1)}(z_i).$$

Combining the above with (6.5.7), (6.5.6) and the definition of C_2, we get

(6.5.11) $$nL_K^2 \leqslant C_2 \frac{p_0}{q} I_1(K, Z_q^\circ(K)) + 8t_0 \sqrt{n} L_K.$$

Also, from (6.5.10) and the definition of p_0, we have

$$C_2 \frac{p_0}{q} I_1(K, Z_q^\circ(K)) = \frac{16 C_2 \kappa I_1(K, Z_q^\circ(K))}{qt_0^2} n^2 \log^4 n \leqslant \frac{1}{4} n L_K^2.$$

Therefore, (6.5.11) gives

$$nL_K^2 \leqslant C_3 t_0 \sqrt{n} L_K.$$

This shows that

$$L_K^2 \leqslant C_4 \frac{t_0^2}{n} = C\kappa \max\left\{1, \frac{I_1(K, Z_q^\circ(K))}{\sqrt{qn} L_K^2}\right\} \sqrt{\frac{n}{q}} \log^4 n,$$

as claimed. $\square$

6.6. Further reading

6.6.1. Simple estimates for $I_1(K, Z_q^\circ(K))$

In this section we give some upper and lower bounds for $I_1(K, Z_q^\circ(K))$ which hold true for every isotropic convex body K in $\mathbb{R}^n$ and any $1 \leqslant q \leqslant n$. In fact, our arguments are quite direct and make use of estimates for simple parameters of the bodies $Z_q(K)$, such as their radius or their volume, so that it is straightforward to reach analogous upper and lower bounds for $I_1(K, Z_q^\circ(M))$ in the more general case when K and M are not necessarily the same isotropic convex body.

Since $h_{Z_q(K)}(x) \leqslant R(Z_q(K)) \|x\|_2$, we have that

(6.6.1) $$I_1(K, Z_q^\circ(K)) \leqslant R(Z_q(K)) \int_K \|x\|_2 \, dx \leqslant R(Z_q(K)) \sqrt{n} L_K,$$

which, in combination with the fact that $R(Z_q(K)) \leqslant \beta_1 q L_K$ (a direct consequence of (6.5.8)), leads to the bound

(6.6.2) $$I_1(K, Z_q^\circ(K)) \leqslant \beta_1 q \sqrt{n} L_K^2.$$

More generally, we have that

(6.6.3) $$I_1(K, Z_q^\circ(M)) \leqslant R(Z_q(M)) \int_K \|x\|_2 \, dx \leqslant \beta_1 q \sqrt{n} L_K L_M.$$

However, in the case that M is an orthogonal transformation of K, the next lemma shows that the average of the quantity $I_1(K, Z_q^\circ(M))$ can be bounded much more effectively than in (6.6.3).

LEMMA 6.6.1. *Let K be a centered convex body of volume 1 in $\mathbb{R}^n$. For every $2 \leqslant q \leqslant n$,*

(6.6.4) $$\left(\int_{O(n)} I_1^q(K, Z_q^\circ(U(K))) \, d\nu(U)\right)^{1/q} \leqslant C\sqrt{q/n} I_q^2(K),$$

where $C > 0$ is an absolute constant.

Proof. We write

$$\int_{O(n)} I_1^q\bigl(K, Z_q^\circ(U(K))\bigr)\,d\nu(U) \leqslant \int_{O(n)} I_q^q\bigl(K, Z_q^\circ(U(K))\bigr)\,d\nu(U)$$

$$= \int_{O(n)} \int_K \int_{U(K)} |\langle x,y\rangle|^q dy\,dx\,d\nu(U)$$

$$= \int_K \int_K \int_{O(n)} |\langle x, Uy\rangle|^q d\nu(U)\,dy\,dx$$

$$= \int_K \int_K \|y\|_2^q \int_{S^{n-1}} |\langle x,\theta\rangle|^q d\sigma(\theta)\,dy\,dx$$

$$= c_{n,q}^q \int_K \int_K \|y\|_2^q \|x\|_2^q dy\,dx$$

$$= c_{n,q}^q I_q^{2q}(K),$$

where $c_{n,q} \simeq \sqrt{q/n}$. □

Recall that in the case that K is isotropic, one has from Theorem 5.2.16 that $I_q(K) \simeq \max\{\sqrt{n}L_K, R(Z_q(K))\}$. Then, Lemma 6.6.1 shows that, for every $2 \leqslant q \leqslant n$,

(6.6.5) $$\left(\int_{O(n)} I_1^q\bigl(K, Z_q^\circ(U(K))\bigr)\,d\nu(U)\right)^{1/q} \leqslant C_1 \max\{\sqrt{qn}, q^2\sqrt{q/n}\}\, L_K^2,$$

where $C_1 > 0$ is an absolute constant. Therefore, for every $2 \leqslant q \leqslant \sqrt{n}$, there exists a set $A_q \subseteq O(n)$ with $\nu(A_q) \geqslant 1 - e^{-q}$ such that $I_1\bigl(K, Z_q^\circ(U(K))\bigr) \leqslant C_2 \sqrt{qn}\, L_K^2$ for all $U \in A_q$.

We now pass to lower bounds; we will present three simple arguments. For the first one we do not have to assume that K or M are in isotropic position, only that they are centered and have volume 1: we use the following fact (see [**384**, Corollary 2.2.a]).

LEMMA 6.6.2. *Let $f : \mathbb{R}^n \to \mathbb{R}^+$ be a measurable function with $\|f\|_\infty = 1$. For every symmetric convex body C in $\mathbb{R}^n$, the function*

$$F(p) = \left(\frac{\int_{\mathbb{R}^n} \|x\|_C^p f(x)\,dx}{\int_C \|x\|_C^p\,dx}\right)^{\frac{1}{n+p}}$$

is increasing on $(-n, \infty)$.

Proof. First note that a direct computation gives

$$\int_C \|x\|_C^p\,dx = \frac{n}{n+p}|C|.$$

For every $p > q > -n$ and every $t > 0$ we may write

$$\int_{\mathbb{R}^n} \|x\|_C^p f(x)\,dx \geqslant t^{p-q} \int_{\mathbb{R}^n} \|x\|_C^q\,dx - \int_{tC}(t^{p-q}\|x\|_C^q - \|x\|_C^p)f(x)\,dx$$

$$\geqslant t^{p-q} \int_{\mathbb{R}^n} \|x\|_C^q\,dx - t^{p+n}\int_C (\|x\|_C^q - \|x\|_C^p)\,dx.$$

The optimal choice of t is

$$t = \left((q+n)\int_{\mathbb{R}^n}\|x\|_C^q f(x)\,dx\right)^{\frac{1}{n+q}}.$$

Inserting this value of t in the previous inequality we get $F(p) \geqslant F(q)$. □

Let K be any convex body in $\mathbb{R}^n$. Setting $f = \mathbf{1}_K$ in Lemma 6.6.2, we get $F(0) = (|K|/|C|)^{1/n}$ for every symmetric convex body C, and hence

(6.6.6) $$\left(\frac{n+p}{n}\frac{1}{|C|}\int_K \|x\|_C^p\,dx\right)^{\frac{1}{n+p}} \geqslant \left(\frac{|K|}{|C|}\right)^{1/n}$$

for all $p > 0$. This inequality can be rewritten as follows:

$$(6.6.7) \qquad \left(\frac{1}{|K|}\int_K \|x\|_C^p \, dx\right)^{1/p} \geqslant \left(\frac{n}{n+p}\right)^{1/p}\left(\frac{|K|}{|C|}\right)^{1/n}.$$

Applying (6.6.7) with $C = Z_q^\circ(M)$ and $p = 1$ we see that

$$I_1(K, Z_q^\circ(M)) = \int_K h_{Z_q(M)}(x)\, dx \geqslant \frac{n}{n+1}\frac{1}{|Z_q^\circ(M)|^{1/n}}.$$

Then, by the Blaschke-Santaló inequality, we get that

$$I_1(K, Z_q^\circ(M)) \geqslant c_1 n |Z_q(M)|^{1/n} \geqslant c_2 \sqrt{qn} L_M$$

for all $2 \leqslant q \leqslant \sqrt{n}$, because $|Z_q(M)|^{1/n} \geqslant c_3 \sqrt{q/n}\, L_M$ for this range of values of q by a result of Klartag and E. Milman (see Theorem 7.5.19 in Chapter 7). When $\sqrt{n} \leqslant q \leqslant n$, we have the weaker lower bound $|Z_q(M)|^{1/n} \geqslant c_4 \sqrt{q/n}$ of Lutwak, Yang and Zhang. It follows that $I_1(K, Z_q^\circ(M)) \geqslant c_5 \sqrt{qn}$ for this range of values of q.

For the second argument, we require that K is isotropic and we write

$$I_1(K, Z_q^\circ(M)) = \int_K h_{Z_q(M)}(x)\, dx = \int_K \max_{z \in Z_q(M)} |\langle x, z\rangle|\, dx$$
$$\geqslant \max_{z \in Z_q(M)} \int_K |\langle x, z\rangle|\, dx \geqslant c \max_{z \in Z_q(M)} \|z\|_2 L_K$$
$$= c\, R(Z_q(M)) L_K.$$

Finally, if M is isotropic as well, we can use Hölder's inequality to get

$$I_1(K, Z_q^\circ(M)) = \int_K h_{Z_q(M)}(x)\, dx$$
$$\geqslant \int_K h_{Z_2(M)}(x)\, dx = \int_K \|x\|_2 L_M\, dx \geqslant c\sqrt{n} L_K L_M.$$

All the estimates presented above are gathered in the next proposition.

PROPOSITION 6.6.3. *Let K and M be isotropic convex bodies in $\mathbb{R}^n$. For every $2 \leqslant q \leqslant n$,*

$$c_1 \max\left\{\sqrt{n} L_K L_M, \sqrt{qn}, R(Z_q(M)) L_K\right\} \leqslant I_1(K, Z_q^\circ(M)) \leqslant c_2 q \sqrt{n} L_K L_M.$$

In addition, if $2 \leqslant q \leqslant \sqrt{n}$ then

$$c_1 \max\left\{\sqrt{n} L_K L_M, \sqrt{qn} L_M\right\} \leqslant I_1(K, Z_q^\circ(M)) \leqslant c_2 q \sqrt{n} L_K L_M.$$

The situation is more or less clear in the unconditional case. Since $h_{Z_q(K)}(x) \leqslant c\sqrt{qn}\|x\|_\infty$ for every $q \geqslant 2$ by a result of Bobkov and Nazarov (see Theorem 8.6.6 in Chapter 8), we obtain the estimates

$$c_1 \sqrt{qn} \leqslant I_1(K, Z_q^\circ(K)) \leqslant c\sqrt{qn} \int_K \|x\|_\infty dx$$
$$= c\sqrt{qn} \int_K \max_{1 \leqslant i \leqslant n} |\langle x, e_i\rangle|\, dx \leqslant c_2 \sqrt{qn} \log n$$

for all $2 \leqslant q \leqslant n$, where we have also used Proposition 3.3.6 and the fact that $L_K \simeq 1$ in the last step. Note that the same estimates hold true for the quantity $I_1(K, Z_q^\circ(M))$ when M is too an unconditional isotropic convex body.

6.7. Notes and references

Symmetrization of isotropic convex bodies

The (T, E)-symmetrization of a convex body K in $\mathbb{R}^n$ (with respect to a subspace $E \in G_{n,k}$ and a centered convex body T in E) is studied by Bourgain, Klartag and V. Milman in [107] (and [106]) where it is shown that $L_{K(T,E)} \simeq L_K^{1-\frac{k}{n}} L_T^{\frac{k}{n}} |K \cap E|^{\frac{1}{n}}$ (Theorem 6.1.3).

Theorem 6.1.12 on the monotonicity of L_n with respect to the dimension appears in the same paper. Besides Theorem 6.1.3, Bourgain, Klartag and Milman make use of the existence of an M-ellipsoid in Proposition 6.1.13 and of some lemmas on the volume of sections and projections of ellipsoids that can be found in [287] and [149]. These ideas are exploited in the next sections of this chapter.

Reduction to bounded volume ratio

Theorem 6.2.1 appears in [107] as well. Proposition 6.2.2 which provides an upper bound on the volume radius of the sections of an isotropic convex body with maximal isotropic constant is used in the next sections.

Regular isotropic convex bodies

Theorem 6.3.2 on the existence of isotropic convex bodies with maximal isotropic constant and α-regular covering numbers appears in the work [149] of Dafnis and Paouris. It is stated, more generally, for a *coherent* class of probability measures (a class which is closed under the operations of taking marginals and products). The version of the theorem that is presented in the text is a variant of the one given in [213] and clarifies some minor issues about the dependence of the parameters on α as $\alpha \to 2^-$.

Reduction to negative moments

The results of this section are due to Dafnis and Paouris. They introduced the parameter $q_{-c}(\mu, \zeta) := \max\{p \geqslant 1 : I_2(\mu) \leqslant \zeta I_{-p}(\mu)\}$ in [149] and using Theorem 6.3.2 and the covering Lemma 6.4.1 they proved Theorem 6.4.2 and Theorem 6.4.3 (in the setting of coherent classes of log-concave probability measures). These results imply that a positive answer to the hyperplane conjecture is equivalent to the existence of two absolute constants $C, \xi > 0$ such that $q_{-c}(K, \xi) \geqslant Cn$ for every isotropic convex body K in $\mathbb{R}^n$.

Reduction to $I_1(K, Z_q^\circ(K))$

Theorem 6.5.3 is due to Giannopoulos, Paouris and Vritsiou [213]. It offers a reduction of the hyperplane conjecture to the study of the parameter $I_1(K, Z_q^\circ(K))$ in the sense that it immediately recovers a bound that is only slightly worse than Bourgain's and Klartag's bounds and leaves some hopes for improvements: an upper bound of the form $I_1(K, Z_q^\circ(K)) \leqslant C_1 q^s \sqrt{n} L_K^2$ *for some* $q \geqslant 2$ and $\frac{1}{2} \leqslant s \leqslant 1$ and *for all* isotropic convex bodies K in $\mathbb{R}^n$ leads to the estimate

$$L_n \leqslant \frac{C_2 \sqrt[4]{n} \log n}{q^{\frac{1-s}{2}}}.$$

The partial results of Section 6.6 on the behavior of $I_1(K, Z_q^\circ(K))$ for large values of q are also from [213].

CHAPTER 7

Logarithmic Laplace transform and the isomorphic slicing problem

In this chapter we first discuss Klartag's solution to the isomorphic slicing problem. Answering a question of V. Milman, Klartag originally proved that for every symmetric convex body K in $\mathbb{R}^n$ there exists a second symmetric convex body T in $\mathbb{R}^n$ which is close to K (with respect to the Banach-Mazur distance) and has bounded isotropic constant. More precisely, as we will see in Section 7.1, he defined an appropriate αn-concave function $f : K \to [0, \infty)$, with $\alpha \simeq \log n$, and he showed that the body $T = K_{n+2}(f)$ satisfies $d_G(K, T) \leqslant c \log n$ and $L_T \leqslant C$, where $c, C > 0$ are absolute constants. In a second work, Klartag used the logarithmic Laplace transform

$$\Lambda_\mu(\xi) = \log \left(\frac{1}{\mu(\mathbb{R}^n)} \int_{\mathbb{R}^n} e^{\langle \xi, x \rangle} d\mu(x) \right),$$

of an isotropic log-concave measure μ on $\mathbb{R}^n$ in order to find suitable candidates for a log-concave function $f : K \to [0, \infty)$ and thus, finally, to show that, if K is a convex body in $\mathbb{R}^n$, then, for every $\varepsilon \in (0, 1)$, there exist a centered convex body $T \subset \mathbb{R}^n$ and a point $x \in \mathbb{R}^n$ such that $\frac{1}{1+\varepsilon} T \subseteq K + x \subseteq (1 + \varepsilon) T$ and

$$L_T \leqslant \frac{C'}{\sqrt{\varepsilon}},$$

where $C' > 0$ is an absolute constant. Combining this fact with the deviation inequality of Paouris, Klartag was able to obtain the currently best known upper bound for the isotropic constant: for every log-concave probability measure μ on $\mathbb{R}^n$ one has

$$L_\mu \leqslant Cn^{1/4},$$

where $C > 0$ is an absolute constant. Another application of Klartag's second approach was observed by Giannopoulos, Paouris and Vritsiou; one can use it in order to give a purely convex geometric proof of the reverse Santaló inequality. We describe all these results in Sections 7.2, 7.3 and 7.4.

In the second part of the chapter, we describe an alternative approach to the hyperplane conjecture, that started with the work of Klartag and E. Milman and combines both the theory of the L_q-centroid bodies and the logarithmic Laplace transform method. A main result of it is that it provides lower bounds for the volume radius of the L_q-centroid bodies of an isotropic log-concave measure μ. Through a delicate analysis of the logarithmic Laplace transform of μ, Klartag and E. Milman show that

$$|Z_q(\mu)|^{1/n} \geqslant c_3 [\det \operatorname{Cov}(\mu)]^{\frac{1}{2n}} \sqrt{q/n} = c_3 \sqrt{q/n}$$

for all $q \leqslant \sqrt{n}$, where $c_3 > 0$ is an absolute constant. This leads again to the estimate $L_\mu \leqslant c\sqrt[4]{n}$. Vritsiou observed that the lower bound for $|Z_q(\mu)|^{1/n}$ might even hold for larger $q \in [1,n]$ and that, in fact, we may have

$$|Z_q(\mu)|^{1/n} \geqslant \frac{c}{\zeta}\sqrt{\frac{q}{n}}$$

for all $q \leqslant q^H_{-c}(\mu,\zeta)$, where $q^H_{-c}(\mu,\zeta)$ is the hereditary variant

$$q^H_{-c}(\mu,\zeta) := n \inf_k \inf_{E \in G_{n,k}} \frac{\lfloor q_{-c}(\pi_E\mu,\zeta) \rfloor}{k}$$

of $q_{-c}(\mu,\zeta)$ that was defined in Section 6.4 (recall that, given a k-dimensional subspace F of $\mathbb{R}^n$, $\pi_F\mu$ is the marginal measure of μ on F). In that case, we also have that

$$L_\mu \leqslant C\zeta \sqrt{\frac{n}{q^H_{-c}(\mu,\zeta)}}.$$

The results of Chapter 6 show that $q^H_{-c}(\mu,\zeta_0)$ is at least of the order of $\sqrt{n}$ (for some $\zeta_0 \simeq 1$) and may be much larger than $\sqrt{n}$: indeed, if the hyperplane conjecture is correct, one must have $q^H_{-c}(\mu,\zeta_1) \simeq n$ for some $\zeta_1 \simeq L_n \simeq 1$. Thus, using the parameter $q^H_{-c}(\mu,\cdot)$ permits one to extend the range of q with which the method of Klartag and E. Milman can be applied, and offers yet another reduction of the slicing problem.

7.1. Klartag's first approach to the isomorphic slicing problem

The main result of this section is due to Klartag [**272**].

THEOREM 7.1.1 (Klartag). *Let K be a symmetric convex body in $\mathbb{R}^n$. Then, there exists a symmetric convex body T in $\mathbb{R}^n$ such that*

$$d_G(K,T) \leqslant c \log n \quad \text{and} \quad L_T \leqslant C,$$

where $c, C > 0$ are absolute constants.

REMARKS 7.1.2. (i) The $\log n$ term in the bound for $d_G(K,T)$ comes from the fact that Klartag uses the MM^*-estimate. In fact, in the notation of Theorem 7.1.1, it is proven that $d_G(K,T) \leqslant cM(K)M^*(K)$.

(ii) As Klartag notes in [**272**], the slicing problem and its isomorphic version are related as follows: if we knew that for some $c_1, C_1 > 0$ and for any $n \geqslant 1$ and any isotropic convex body C in $\mathbb{R}^n$ there exists an *isotropic* convex body T with $d_G(C,T) \leqslant c_1$ and $L_T \leqslant C_1$, then we would be able to show that $L_K \leqslant C_2$ for any $n \geqslant 1$ and any isotropic convex body K in $\mathbb{R}^n$, where $C_2 > 0$ is an absolute constant.

The steps for the proof of Theorem 7.1.1 are roughly the following: Given a body K we will define an appropriate function $f : K \to [0,\infty)$ which will be α-concave with $\alpha > 0$ as small as possible. Then we will prove that Ball's body $T = K_{n+2}(f)$ is "close enough" to K, while $L_T \simeq L_f$ will be "well bounded".

7.1.1. Basic facts about s-concave functions and the bodies $K_p(f)$

Recall that a function $f : \mathbb{R}^n \to [0,\infty)$ is called s-concave (for some $s > 0$) if the function $f^{1/s}$ is concave on its support $\operatorname{supp}(f) = \{x : f(x) > 0\}$. Note that every s-concave function f is also log-concave.

LEMMA 7.1.3. *Let $f : \mathbb{R}^n \to [0,\infty)$ be an even function whose restriction to any line through the origin is s-concave. If $\operatorname{supp}(f) = \{x : f(x) > 0\}$ then, for every $s > n$,*
$$d_G(K_{n+2}(f), \operatorname{supp}(f)) \leqslant \frac{cs}{n},$$
where $c > 0$ is an absolute constant.

Proof. We may assume that $f(0) = 1$. We will compare the radial functions of the bodies $K_{n+2}(f)$ and $\operatorname{supp}(f)$.

Given $\theta \in S^{n-1}$ we set
$$M_\theta = \rho_{\operatorname{supp}(f)}(\theta) = \sup\{r > 0 : f(r\theta) > 0\}$$
and consider the restriction $f|_{\mathbb{R}\theta}$ of f onto $\mathbb{R}\theta$. This is an s-concave function, and since $f(0) = 1$, for every $0 \leqslant t \leqslant M_\theta$ we have
$$f(t\theta) \geqslant \left(1 - \frac{t}{M_\theta}\right)^s.$$

Then,
$$\rho_{K_{n+2}(f)}^{n+2}(\theta) = (n+2) \int_0^{M_\theta} f(t\theta) t^{n+1} dt$$
$$\geqslant (n+2) \int_0^{M_\theta} \left(1 - \frac{t}{M_\theta}\right)^s t^{n+1} dt = \frac{M_\theta^{n+2}}{\binom{n+s+2}{n+2}}.$$

On the other hand,
$$\rho_{K_{n+2}(f)}^{n+2}(\theta) \leqslant (n+2) \int_0^{M_\theta} t^{n+1} dt = M_\theta^{n+2}.$$

It follows that
$$M_\theta \geqslant \rho_{K_{n+2}(f)}(\theta) \geqslant \frac{M_\theta}{\binom{n+s+2}{n+2}^{\frac{1}{n+2}}} \geqslant \frac{nM_\theta}{cs},$$
because $\binom{n+s+2}{n+2} \leqslant \left(\frac{e(n+s+2)}{n+2}\right)^{n+2} \leqslant \left(\frac{cs}{n}\right)^{n+2}$ for $s \geqslant n$. $\square$

LEMMA 7.1.4. *Let $f : \mathbb{R}^n \to [0,\infty)$ be an even, s-concave function. Then, the body $T = K_{n+2}(f)$ is a symmetric convex body with*
$$c_1 L_f \leqslant L_T \leqslant c_2 L_f.$$

Proof. It is essentially the same as the proof of Proposition 2.5.9. Since f is even and s-concave, T is a symmetric convex body. Because f is even and s-concave, we also have $f(x) \leqslant f(0)$ for all $x \in \mathbb{R}^n$. Hence, $f(0) > 0$ otherwise we have nothing to prove. From Proposition 2.5.3 we have
$$\int_T \langle x, \theta \rangle^2 dx = \frac{1}{f(0)} \int_{\mathbb{R}^n} \langle x, \theta \rangle^2 f(x) \, dx,$$

for all $\theta \in S^{n-1}$. It follows that
$$|T|\operatorname{Cov}(T) = \frac{\int f}{f(0)}\operatorname{Cov}(f).$$

By the definition of the isotropic constant and Fradelizi's inequality (Theorem 2.2.2) we obtain
$$L_T \simeq \frac{1}{|T|^{\frac{1}{2}+\frac{1}{n}}}\left(\frac{1}{f(0)}\int_{\mathbb{R}^n} f(x)dx\right)^{\frac{1}{2}+\frac{1}{n}} L_f.$$

On the other hand, we know that
$$|T|^{\frac{1}{2}+\frac{1}{n}} = |K_{n+2}(f)|^{\frac{1}{2}+\frac{1}{n}} \simeq \left(\frac{1}{f(0)}\int_{\mathbb{R}^n} f(x)\,dx\right)^{\frac{1}{2}+\frac{1}{n}},$$

by Proposition 2.5.8 for $p=2$. $\square$

We summarize Lemma 7.1.3 and Lemma 7.1.4 in the next proposition.

PROPOSITION 7.1.5. *Let K be a symmetric convex body in $\mathbb{R}^n$, let $\alpha > 1$ and let $f : \mathbb{R}^n \to [0,\infty)$ be an αn-concave, even function with $\operatorname{supp}(f) = \operatorname{int}(K)$. Then, the symmetric convex body $T = K_{n+2}(f)$ satisfies $L_T \simeq L_f$ and*
$$d_{\mathrm{G}}(K,T) \leqslant c\alpha,$$
where $c > 0$ is an absolute constant. $\square$

7.1.2. Constructing a function on K

Let K be a symmetric convex body in $\mathbb{R}^n$. We turn to the construction of an αn-concave function f supported on K. In view of Proposition 7.1.5, if we obtain good estimates on α and L_f, this will yield our main theorem.

We write $\|\cdot\| = \|\cdot\|_K$ for the norm induced on $\mathbb{R}^n$ by K. We denote by m the median of $x \mapsto \|x\|$ with respect to the measure σ on S^{n-1} and define $F_K : K \to [0,\infty)$ as follows:
$$F_K(x) = \min\left\{t \in [0,1] : x \in (1-t)\left(K \cap \frac{1}{m}B_2^n\right) + tK\right\}.$$

Note that F_K is a convex function with $F_K \equiv 0$ on $K \cap \frac{1}{m}B_2^n$ and $F_K < 1$ on $\operatorname{int}(K)$. Thus, for every $\alpha > 1$ the function
$$f(x) = f_K(x) = (1 - F_K(x))^{\alpha n}\mathbf{1}_K(x)$$
is an αn-concave function with $\operatorname{supp}(f) = \operatorname{int}(K)$ and $f \equiv 1$ on $K \cap \frac{1}{m}B_2^n$.

The next lemma provides an upper bound for the expectation of f_K:

LEMMA 7.1.6. *Let K be a symmetric convex body in $\mathbb{R}^n$ and let*
$$\alpha = \lfloor cM(K)M^*(K) \rfloor.$$
Then,
$$\int_K (1 - F_K(x))^{\alpha n}\,dx \leqslant 2\left|K \cap \left(\frac{1}{m}B_2^n\right)\right|,$$
where $c > 0$ is an absolute constant.

7.1. ISOMORPHIC SLICING PROBLEM

Proof. We can write

$$\int_K (1 - F_K(x))^{\alpha n}\, dx = \alpha n \int_0^1 (1-s)^{\alpha n - 1} |\{x \in K : F_K(x) \leqslant s\}|\, ds$$

$$= \alpha n \int_0^1 (1-s)^{\alpha n - 1} \left|(1-s)\left(K \cap \frac{1}{m} B_2^n\right) + sK\right| ds.$$

Now we use Steiner's formula:

$$\left|(1-s)\left(K \cap \frac{1}{m} B_2^n\right) + sK\right| = \sum_{i=0}^n \binom{n}{i} V_i s^i (1-s)^{n-i},$$

where $V_i = V\left(K; i, (K \cap \frac{1}{m} B_2); n - i\right)$. It follows that

$$\int_K (1 - F_K(x))^{\alpha n}\, dx = \alpha n \sum_{i=0}^n \int_0^1 \binom{n}{i} V_i s^i (1-s)^{(\alpha+1)n - i - 1}\, ds.$$

Standard calculations with the Beta function show that

$$\int_K (1 - F_K(x))^{\alpha n}\, dx = \frac{\alpha}{\alpha + 1} \sum_{i=0}^n \frac{\binom{n}{i}}{\binom{(\alpha+1)n-1}{i}} V_i.$$

Using the elementary inequalities $(n/k)^k \leqslant \binom{n}{k} \leqslant (en/k)^k$ for $1 \leqslant k \leqslant n$ we can deduce the following:

$$\int_K (1 - F_K(x))^{\alpha n}\, dx = \frac{\alpha}{\alpha + 1} V_0 \left(1 + \sum_{i=1}^n \frac{\binom{n}{i}}{\binom{(\alpha+1)n-1}{i}} \frac{V_i}{V_0}\right)$$

$$\leqslant \frac{\alpha}{\alpha + 1} V_0 \left[1 + \sum_{i=1}^n \left(\frac{en}{(\alpha+1)n - 1}\right)^i \frac{V_i}{V_0}\right]$$

$$= \frac{\alpha}{\alpha + 1} V_0 \left[1 + \sum_{i=1}^n \left(\frac{en}{(1+\alpha)n - 1} \left(\frac{V_i}{V_0}\right)^{1/i}\right)^i\right].$$

Now, we use the Alexandrov-Fenchel inequalities to conclude that $(V_i/V_0)^{1/i}$ is non-increasing, i.e. for $1 \leqslant i \leqslant j$ one has

$$\left(\frac{V_i}{V_0}\right)^{1/i} \geqslant \left(\frac{V_j}{V_0}\right)^{1/j}.$$

In particular, if $\alpha + 1 > 4eV_1/V_0$ we have

$$\frac{en}{(1+\alpha)n - 1} \left(\frac{V_i}{V_0}\right)^{1/i} \leqslant \frac{2e}{1+\alpha} \frac{V_1}{V_0} \leqslant \frac{1}{2}.$$

Thus, we obtain

$$\int_K (1 - F_K(x))^{\alpha n}\, dx \leqslant \frac{\alpha}{\alpha + 1} V_0 \left(1 + \sum_{i=1}^n \frac{1}{2^i}\right) < 2V_0,$$

which proves the desired estimate as long as $\alpha + 1 > 4eV_1/V_0$.

We proceed to prove that $V_1/V_0 < \lfloor cM(K)M^*(K)\rfloor$ for a suitable absolute constant $c > 0$. First we estimate V_1 from above: Since $\frac{1}{m} B_2^n \cap K \subseteq \frac{1}{m} B_2^n$, by the monotonicity of the mixed volumes we get

$$V_1 = V\left(K; 1, \left(\frac{1}{m}B_2^n \cap K\right); n-1\right) \leqslant V\left(K; 1, \frac{1}{m}B_2^n; n-1\right)$$
$$= \frac{\omega_n}{m^{n-1}} M^*(K).$$

Next, we give a lower bound for V_0. Since m is the median of $\|\cdot\|$ on S^{n-1}, we have $\sigma(mK \cap S^{n-1}) \geqslant 1/2$ which implies that

$$V_0 = |K \cap \tfrac{1}{m}B_2^n| \geqslant \tfrac{1}{2}|\tfrac{1}{m}B_2^n| = \frac{\omega_n}{2m^n}.$$

Combining the above we conclude that

$$\frac{V_1}{V_0} \leqslant \frac{\omega_n}{m^{n-1}} M^*(K) \frac{2m^n}{\omega_n} = 2mM^*(K).$$

On the other hand,

$$M(K) = \int_{S^{n-1}} \|\theta\|_K \, d\sigma(\theta) \geqslant m\sigma(S^{n-1} \setminus (mK)) \geqslant \frac{m}{2}.$$

Therefore,

$$V_1/V_0 \leqslant 4M(K)M^*(K).$$

It is now clear that if we choose $\alpha = \lfloor cM(K)M^*(K) \rfloor$ (for a suitable absolute constant $c > 0$) we will then have $\alpha + 1 > 4eV_1/V_0$. □

The next proposition establishes a uniform upper bound for the isotropic constant of f_K when $\alpha \simeq M(K)M^*(K)$.

PROPOSITION 7.1.7. *Let K be a symmetric convex body in $\mathbb{R}^n$ and let $\alpha = cM(K)M^*(K)$. Then, for the function $f(x) = (1 - F_K(x))^{\alpha n}$ we have*

$$L_f \leqslant C,$$

where $C, c > 0$ are absolute constants.

Proof. We consider the probability measure μ_f whose density is proportional to f:

$$\mu_f(A) = \frac{\int_A f(x) \, dx}{\int_K f(x) \, dx}.$$

Since $f \equiv 1$ on $\tfrac{1}{m}B_2^n \cap K$, from Lemma 7.1.6 we get $\mu_f\left(\tfrac{1}{m}B_2^n \cap K\right) \geqslant 1/2$. In other words,

$$\mu_f(\{x : \|x\|_2 \leqslant 1/m\}) \geqslant \frac{1}{2},$$

which implies

$$\int \frac{1}{\sqrt{\|x\|_2}} \, d\mu_f(x) \geqslant \frac{\sqrt{m}}{2}.$$

Since f is log-concave, we know that the measure μ_f is log-concave, and hence Borell's lemma and Theorem 2.4.9 show that

$$(7.1.1) \quad \left(\int \|x\|_2^2 \, d\mu_f(x)\right)^{1/2} \leqslant c_1 \int \|x\|_2 \, d\mu_f(x)$$
$$\leqslant c_2 \left(\int \|x\|_2^{-1/2} \, d\mu_f(x)\right)^{-2} \leqslant c_2 \left(\frac{\sqrt{m}}{2}\right)^{-2}$$
$$= \frac{4c_2}{m} = \frac{c}{m}$$

for some absolute constant $c > 0$. Therefore, the arithmetic-geometric means inequality shows that

$$\left(\frac{\int_K f}{f(0)}\right)^{1/n} \sqrt{n} L_f = \sqrt{n}[\det \operatorname{Cov}(\mu_f)]^{\frac{1}{2n}} \leqslant \left(\int \|x\|_2^2 \, d\mu_f(x)\right)^{1/2} < \frac{c}{m}.$$

Since

$$\int_K f(x)\,dx \geqslant |\tfrac{1}{m}B_2^n \cap K| \geqslant \tfrac{1}{2}|\tfrac{1}{m}B_2^n|$$

and $f(0) = 1$, we see that $L_f \leqslant C$ for some absolute constant $C > 0$. □

Proof of Theorem 7.1.1. Let K be a symmetric convex body in $\mathbb{R}^n$. Let $K_1 = T(K)$ be a linear image of K for $T \in GL(n)$ satisfying

$$M(K_1)M^*(K_1) \leqslant c\log(1+d_K) \leqslant c\log n.$$

Define $F_{K_1} : K_1 \to [0,\infty)$ as before and set $f(x) = (1 - F_{K_1}(x))^{\alpha n}$ with $\alpha = cM(K_1)M^*(K_1)$. From Proposition 7.1.5 we see that the body $T = K_{n+2}(f)$ satisfies

$$d_G(K_1, T) \leqslant c_1 \alpha \leqslant c_2 M(K_1)M^*(K_1) \leqslant c_3 \log n.$$

Furthermore, Lemma 7.1.4 and Proposition 7.1.7 show that

$$L_T \simeq L_f \leqslant C.$$

On observing that $d(K,T) \leqslant d_G(K_1, T)$ we have the theorem. □

7.2. Logarithmic Laplace transform and convex perturbations

Starting with this section, we describe Klartag's second approach to the isomorphic slicing problem (see [**273**]) which is based on the logarithmic Laplace transform. The logarithmic Laplace transform of a Borel measure μ on $\mathbb{R}^n$ is defined by

$$\Lambda_\mu(\xi) = \log\left(\frac{1}{\mu(\mathbb{R}^n)} \int_{\mathbb{R}^n} e^{\langle \xi, x \rangle} \, d\mu(x)\right).$$

PROPOSITION 7.2.1. *If $\mu = \mu_K$ is the Lebesgue measure on some convex body K in $\mathbb{R}^n$, then*

$$(\nabla \Lambda_\mu)(\mathbb{R}^n) = \operatorname{int}(K).$$

If μ_ξ is the probability measure on $\mathbb{R}^n$ with density proportional to $e^{\langle \xi, x \rangle} \mathbf{1}_K(x)$, then

$$\operatorname{bar}(\mu_\xi) = \nabla \Lambda_\mu(\xi)$$

and

$$\operatorname{Hess}(\Lambda_\mu)(\xi) = \operatorname{Cov}(\mu_\xi).$$

Moreover, the map $\nabla \Lambda_\mu$ transports the measure ν with density $\det \operatorname{Hess}(\Lambda_\mu)(\xi)$ to μ. Equivalently, for every continuous non-negative function $\phi : \mathbb{R}^n \to \mathbb{R}$,

$$\int_K \phi(x)\,dx = \int_{\mathbb{R}^n} \phi(\nabla \Lambda_\mu(\xi)) \det \operatorname{Hess}(\Lambda_\mu)(\xi)\,d\xi = \int_{\mathbb{R}^n} \phi(\nabla \Lambda_\mu(\xi))\,d\nu(\xi).$$

Proof. Let $F = \Lambda_{\mu_K}$, that is

$$F(x) = \log\left(\frac{1}{|K|} \int_K e^{\langle x, y \rangle}\,dy\right).$$

Observe that F is a C^2-smooth, strictly convex function. Smoothness is clear, as we are integrating a smooth function on a convex body. The strict convexity follows from the Cauchy-Schwarz inequality. Differentiating under the integral sign we get

$$(7.2.1) \qquad \nabla F(\xi) = \frac{\int_K y e^{\langle \xi, y \rangle} \, dy}{\int_K e^{\langle \xi, z \rangle} \, dz} = \int_{\mathbb{R}^n} y \, d\mu_\xi(y) = \operatorname{bar}(\mu_\xi).$$

Since μ_ξ is supported on the compact convex set K, we have $\nabla F(\xi) = \operatorname{bar}(\mu_\xi) \in K$ for all $\xi \in \mathbb{R}^n$. This shows that $\nabla F(\mathbb{R}^n) \subseteq K$. To compute the Hessian we differentiate twice to get

$$\frac{\partial^2 F(\xi)}{\partial \xi_j \partial \xi_i} = \frac{\int_K y_i y_j e^{\langle \xi, y \rangle} \, dy}{\int_K e^{\langle \xi, y \rangle} \, dy} - \frac{\int_K y_i e^{\langle \xi, y \rangle} \, dy \int_K y_j e^{\langle \xi, y \rangle} \, dy}{\left(\int_K e^{\langle \xi, y \rangle} \, dy \right)^2}$$

$$= \int_{\mathbb{R}^n} y_i y_j \, d\mu_\xi(y) - \int_{\mathbb{R}^n} y_i \, d\mu_\xi(y) \int_{\mathbb{R}^n} y_j \, d\mu_\xi(y)$$

$$= \operatorname{Cov}(\mu_\xi)_{ij}$$

Now we prove that $\nabla F(\mathbb{R}^n) = \operatorname{int}(K)$. Let $z \in \operatorname{bd}(K)$ be an exposed point of K. There exist $u \in \mathbb{R}^n$ and $t \in \mathbb{R}$ such that $\langle u, z \rangle = t$ and for any $x \in K$, $x \neq z$ we have $\langle x, z \rangle < t$. Consider the measure μ_{ru} for large $r > 0$. Its density is proportional to $x \mapsto e^{r\langle u, x \rangle} \mathbf{1}_K(x)$ and it attains its unique maximum at z. Moreover, it is straightforward to verify that

$$\mu_{ru} \xrightarrow{w^*} \delta_z,$$

as $r \to \infty$, where δ_z is the Dirac mass at z. Therefore, by (7.2.1) we obtain

$$\nabla F(ru) \to z,$$

as $r \to \infty$. This shows that $z \in \overline{\nabla F(\mathbb{R}^n)}$. Since z was an arbitrary exposed point and since $\overline{\nabla F(\mathbb{R}^n)}$ is convex, from Straszewicz's theorem (see [**463**, Theorem 1.4.7] we get that $K \subseteq \overline{\nabla F(\mathbb{R}^n)}$. Moreover, $\nabla F(\mathbb{R}^n)$ is open and combining with the fact that $\nabla F(\mathbb{R}^n) \subseteq K \subseteq \overline{\nabla F(\mathbb{R}^n)}$ we conclude that $\nabla F(\mathbb{R}^n)$ is equal to the interior of K.

For the last assertion of the proposition note that since F is strictly convex, ∇F is one-to-one. So, for any continuous function $\phi : \mathbb{R}^n \to \mathbb{R}$, the change of variables $y = \nabla F(\xi)$ gives

$$\int_{\nabla F(\mathbb{R}^n)} \phi(y) \, dy = \int_{\mathbb{R}^n} \phi(\nabla F(\xi)) \det \operatorname{Hess} F(\xi) \, d\xi = \int_{\mathbb{R}^n} \phi(\nabla F(\xi)) \, d\nu(\xi).$$

This completes the proof. $\square$

The next lemma is the "log-concave analogue" of Lemma 7.1.3 and plays an important role in Klartag's work. The proof is simple if we take into account the properties of the bodies $K_p(f)$ that we discussed in Section 2.5 (more precisely, Lemma 2.5.2 and Proposition 2.5.3 (iv)).

LEMMA 7.2.2. *Let K be a convex body in $\mathbb{R}^n$, let $f : K \to (0, \infty)$ be a log-concave function such that*

$$M^{-n} \leqslant \inf_{x \in K} f(x) \leqslant \sup_{x \in K} f(x) \leqslant M^n$$

for some $M > 1$ and let $x_0 = \mathrm{bar}(f)$. Then, there exists a centered convex body T in $\mathbb{R}^n$ such that
$$\frac{1}{M} T \subseteq K - x_0 \subseteq MT$$
and
$$L_f \simeq L_T.$$

Proof. Since $x_0 = \mathrm{bar}(f)$, we have $x_0 \in K$. We define a function $g : K_1 = K - x_0 \to \mathbb{R}$ by $g(x) = f(x+x_0)$. Then, g is log-concave and $\mathrm{bar}(g) = 0$. We set $T = K_{n+1}(g)$. Then, T is a centered convex body, and since
$$\frac{1}{M^{n+1}} \leqslant \frac{1}{M^n} \leqslant \inf_{x \in K_1} g(x) \leqslant \sup_{x \in K_1} g(x) \leqslant M^n \leqslant M^{n+1},$$
from Lemma 2.5.2 and Proposition 2.5.3 (iv) we obtain
$$\frac{1}{M} T \subseteq K_1 \subseteq MT.$$
Note that $L_g = L_f$ because g is the composition of f with an affine function. From Proposition 2.5.12 we know that $L_T = L_{K_{n+1}(g)} \simeq L_g$, and hence the lemma is proved. □

7.3. Klartag's solution to the isomorphic slicing problem

The main result of Klartag in [**273**] is the next theorem.

THEOREM 7.3.1 (Klartag). *Let K be a convex body in $\mathbb{R}^n$. For every $\varepsilon \in (0,1)$ there exist a centered convex body $T \subset \mathbb{R}^n$ and a point $x \in \mathbb{R}^n$ such that*

(7.3.1) $$\frac{1}{1+\varepsilon} T \subseteq K + x \subseteq (1+\varepsilon) T$$

and

(7.3.2) $$L_T \leqslant \frac{C}{\sqrt{\varepsilon n s(K-K)^{1/n}}},$$

where $s(K - K) = |K - K| |(K - K)^\circ|$ and $C > 0$ is an absolute constant. In particular, we have
$$L_T \leqslant \frac{C'}{\sqrt{\varepsilon}},$$
where $C' > 0$ is an absolute constant.

Proof. We may assume that K is centered and that $|K - K| = 1$, because L_T and $s(K-K)$ are affine invariants.

Recall from Proposition 7.2.1 that if $\mu = \mu_K$ is the Lebesgue measure restricted on K, then the function $\nabla \Lambda_\mu$ transports the measure ν with density
$$\frac{d\nu}{d\xi} = \det \mathrm{Hess}\,(\Lambda_\mu)(\xi) \equiv \det \mathrm{Cov}(\mu_\xi)$$
to μ. This implies that
$$\nu(\mathbb{R}^n) = \int_{\mathbb{R}^n} \mathbf{1} \det \mathrm{Hess}\,(\Lambda_\mu)(\xi)\,d\xi = \int_K \mathbf{1}\,dx = |K| \leqslant |K-K| = 1.$$

Thus, for every $\varepsilon > 0$ we may write

$$(7.3.3) \quad |\varepsilon n(K-K)^\circ| \min_{\xi \in \varepsilon n(K-K)^\circ} \det \mathrm{Cov}(\mu_\xi) \leqslant$$

$$\leqslant \int_{\varepsilon n(K-K)^\circ} \det \mathrm{Cov}(\mu_\xi) \, d\xi = \nu(\varepsilon n(K-K)^\circ) \leqslant 1,$$

which means that there exists $\xi \in \varepsilon n(K-K)^\circ$ such that

$$\det \mathrm{Cov}(\mu_\xi) = \min_{\xi' \in \varepsilon n(K-K)^\circ} \det \mathrm{Cov}(\mu_{\xi'}) \leqslant |\varepsilon n(K-K)^\circ|^{-1}$$

$$= \left(\varepsilon n s(K-K)^{1/n}\right)^{-n}$$

(where the last equality holds because $|K - K| = 1$). Now, from the definition of μ_ξ and of the isotropic constant we have that

$$L_{\mu_\xi} = \left(\frac{\sup_{x \in K} e^{\langle \xi, x \rangle}}{\int_K e^{\langle \xi, x \rangle} dx}\right)^{\frac{1}{n}} [\det \mathrm{Cov}(\mu_\xi)]^{\frac{1}{2n}}.$$

Since $\xi \in \varepsilon n(K-K)^\circ$ and $K \cup (-K) \subset K - K$, we know that $|\langle \xi, x \rangle| \leqslant \varepsilon n$ for all $x \in K$, therefore $\sup_{x \in K} e^{\langle \xi, x \rangle} \leqslant \exp(\varepsilon n)$. On the other hand, since K is centered, from Jensen's inequality we have that

$$\frac{1}{|K|} \int_K e^{\langle \xi, x \rangle} dx \geqslant \exp\left(\frac{1}{|K|} \int_K \langle \xi, x \rangle \, dx\right) = 1,$$

which means that $\int_K e^{\langle \xi, x \rangle} dx \geqslant |K| \geqslant 4^{-n} |K-K|$ by the Rogers-Shephard inequality. Combining all these we get

$$L_{\mu_\xi} \leqslant \frac{4e^\varepsilon}{\sqrt{\varepsilon n s(K-K)^{1/n}}}.$$

Finally, we note that the function $f_\xi(x) = e^{\langle \xi, x \rangle} \mathbf{1}_K(x)$ (which is proportional to the density of μ_ξ) is obviously log-concave and satisfies

$$e^{-\varepsilon n} \leqslant \inf_{x \in \mathrm{supp}(f_\xi)} f_\xi(x) \leqslant \sup_{x \in \mathrm{supp}(f_\xi)} f_\xi(x) \leqslant e^{\varepsilon n}$$

(since $|\langle \xi, x \rangle| \leqslant \varepsilon n$ for all $x \in K$). Therefore, applying Lemma 7.2.2, we can find a centered convex body T_ξ in $\mathbb{R}^n$ such that

$$L_{T_\xi} \simeq L_{f_\xi} = L_{\mu_\xi} \leqslant \frac{4e^\varepsilon}{\sqrt{\varepsilon n s(K-K)^{1/n}}}$$

and

$$\frac{1}{e^\varepsilon} T_\xi \subseteq K - \mathrm{bar}(f_\xi) \subseteq e^\varepsilon T_\xi.$$

Since $e^{2\varepsilon} \leqslant 1 + c\varepsilon$ when $\varepsilon \in (0,1)$, the result follows. The, in particular case, follows immediately from the reverse Santaló inequality for the symmetric convex body $K - K$. $\square$

Using Theorem 7.3.1 and Paouris' deviation inequality, Klartag [**273**] improved Bourgain's upper bound for the isotropic constant by a logarithmic term.

THEOREM 7.3.2. *Let K be a centered convex body in $\mathbb{R}^n$. Then,*

$$L_K \leqslant Cn^{1/4}$$

where $C > 0$ is an absolute constant.

The following corollary follows immediately from Theorem 7.3.2.

COROLLARY 7.3.3. *Let $f : \mathbb{R}^n \to [0, \infty)$ be a centered log-concave function with finite, positive integral. Then,*
$$L_f \leqslant Cn^{1/4}$$
where $C > 0$ is an absolute constant.

Proof. Let $K = K_{n+1}(f)$. Since, f is a centered and log-concave function, K is a centered convex body in $\mathbb{R}^n$ and according to Theorem 7.3.2 we have $L_K \leqslant cn^{1/4}$. Taking into account the fact that $L_K \simeq L_f$ (by Proposition 2.5.12) we obtain the result. $\square$

For the proof of Theorem 7.3.2 we will need the following Lemma.

LEMMA 7.3.4. *Let K, T be two convex bodies in $\mathbb{R}^n$ and $t \geqslant 1$. Suppose that*
$$d_G(K, T) \leqslant \left(1 + \frac{t}{\sqrt{n}}\right)^2.$$
Then,
$$L_K \leqslant ctL_T,$$
where $c > 0$ is an absolute constant.

Proof. We may assume that $t < \sqrt{n}$, otherwise the conclusion of the lemma is trivial because of the known bound $L_K \leqslant c\sqrt{n}$. Since $d_G(K, T) \leqslant \left(1 + \frac{t}{\sqrt{n}}\right)^2$, there exist $x_0, y_0 \in \mathbb{R}^n$ such that

$$(7.3.4) \qquad \left(1 + \frac{t}{\sqrt{n}}\right)^{-1}(T + x_0) \subseteq (K + y_0) \subseteq \left(1 + \frac{t}{\sqrt{n}}\right)(T + x_0).$$

Applying an affine transformation to both T and K we may assume that T is isotropic. We set $K_1 = (1 + \frac{t}{\sqrt{n}})^{-1}(K + y_0) - x_0$. From (7.3.4) we have $K_1 \subseteq T$. Using (7.3.4) once again, we write

$$|K_1| = \left(1 + \frac{t}{\sqrt{n}}\right)^{-n}|K| \geqslant \left(1 + \frac{t}{\sqrt{n}}\right)^{-2n}|T| \geqslant e^{-2t\sqrt{n}}.$$

At this point we need to use Paouris' inequality: we have
$$|T \setminus (ct\sqrt{n}L_T B_2^n)| \leqslant e^{-4t\sqrt{n}},$$
for some absolute constant $c > 0$. Since $K_1 \subseteq T$, the last two estimates imply that
$$|K_1 \cap (ct\sqrt{n}L_T B_2^n)| \geqslant \frac{|K_1|}{2}.$$

Therefore, the median of the function $x \mapsto \|x\|_2$ on K_1, with respect to the uniform measure on K_1, is not larger than $ct\sqrt{n}L_T$. Since K_1 is convex, and hence the uniform measure on K_1 is a log-concave probability measure, using Theorem 2.4.9 as in the proof of Proposition 7.1.7 we obtain

$$\left(\frac{1}{|K_1|}\int_{K_1} \|x\|_2^2 \, dx\right)^{1/2} \leqslant Ct\sqrt{n}L_T,$$

for some absolute constant $C > 0$. It follows that

$$L_K \leqslant \frac{1}{\sqrt{n}}\left(\frac{1}{|K_1|^{1+\frac{2}{n}}}\int_{K_1}\|x\|_2^2 dx\right)^{1/2} \leqslant C\frac{tL_T}{|K_1|^{1/n}} < c'tL_T,$$

for some absolute constant $c' > 0$. This proves the lemma. $\square$

Proof of Theorem 7.3.2. Let $K \subset \mathbb{R}^n$ be a convex body and let $\varepsilon \in (0,1)$. According to Theorem 7.3.1 there exists a centered convex body T with
$$d_{\mathrm{G}}(K,T) \leqslant (1+\varepsilon)^2 \leqslant 1+3\varepsilon$$
and
$$L_T \leqslant \frac{C}{\sqrt{\varepsilon}}.$$
On the other hand, if we set $\varepsilon = \frac{1}{\sqrt{n}}$, Lemma 7.3.4 implies that
$$L_K \leqslant 3cL_T \leqslant 3c\frac{C}{\sqrt{\varepsilon}} = C'\sqrt[4]{n}.$$
This completes the proof. $\square$

7.4. Isotropic position and the reverse Santaló inequality

It was observed by Giannopoulos, Paouris and Vritsiou in [**214**] that one can use Klartag's approach in order to give a purely convex geometric proof of the reverse Santaló inequality. The proof consists of two steps.

7.4.1. Step 1: Lower bound involving the isotropic constant

We will need the following standard lemma.

LEMMA 7.4.1. *Let K be an isotropic convex body in $\mathbb{R}^n$. Then, for every $t > 0$ one has,*
$$\log N(B_2^n, tK^\circ) \leqslant \log N(B_2^n, t(K-K)^\circ) \leqslant \frac{c_2 n^{3/2} L_K}{t},$$
where $c_2 > 0$ is an absolute constant.

Proof. From Theorem 3.2.2 we know that
$$(7.4.1) \qquad N(K, tB_2^n) \leqslant 2\exp\left(\frac{6n^{3/2}L_K}{t}\right) \leqslant \exp\left(\frac{12n^{3/2}L_K}{t}\right)$$
for all $0 < t < (6/\log 2)n^{3/2}L_K$. Since $R(K) \leqslant (n+1)L_K$, we have
$$(7.4.2) \qquad N(K, tB_2^n) \leqslant \exp(c_3 n^{3/2} L_K/t)$$
for all $t > 0$. For every $t > 0$ we set $A(t) := \log N(K, tB_2^n)$ and $B(t) := \log N(B_2^n, tK^\circ)$. We use a well-known idea from [**492**] (see also [**318**, Section 3.3]). For any $t > 0$ we have $(t^2 K^\circ) \cap (4K) \subseteq 2tB_2^n$. Passing to the polar bodies we see that
$$B_2^n \subseteq \operatorname{conv}\left(\frac{t}{2}K^\circ, \frac{2}{t}K\right) \subseteq \frac{t}{2}K^\circ + \frac{2}{t}K.$$
We write
$$N(B_2^n, tK^\circ) \leqslant N\left(\frac{t}{2}K^\circ + \frac{2}{t}K, tK^\circ\right) = N\left(\frac{2}{t}K, \frac{t}{2}K^\circ\right)$$
$$\leqslant N\left(\frac{2}{t}K, \frac{1}{4}B_2^n\right) N\left(\frac{1}{4}B_2^n, \frac{t}{2}K^\circ\right)$$
$$= N\left(K, \frac{t}{8}B_2^n\right) N(B_2^n, 2tK^\circ).$$

7.4. ISOTROPIC POSITION AND THE REVERSE SANTALÓ INEQUALITY

Taking logarithms we get

$$B(t) \leqslant A(t/8) + B(2t),$$

for all $t > 0$. From (7.4.2) we have $tA(t) \leqslant c_3 n^{3/2} L_K$ for all $t > 0$ and this implies (after iterating) that

$$B(t) \leqslant B(2^m t) + \sum_{k=1}^{m} A(2^k t/16) \leqslant B(2^m t) + \sum_{k=1}^{m} \frac{c_4 n^{3/2} L_K}{2^k t},$$

for all $m \geqslant 1$. Note that $\lim_{s \to \infty} B(s) = 0$. Letting $m \to \infty$ we get

$$B(t) \leqslant \frac{c_5 n^{3/2} L_K}{t}$$

for all $t > 0$. Furthermore, arguing similarly with $N(K - K, tB_2^n)$ in place of $N(K, tB_2^n)$ we obtain:

(7.4.3) $\qquad \log N(B_2^n, tK^\circ) \leqslant \log N\big(B_2^n, t(K - K)^\circ\big) \leqslant \dfrac{c_6 n^{3/2} L_K}{t},$

where $c_6 > 0$ is an absolute constant. $\qquad \square$

Now we can prove the main result of this subsection:

PROPOSITION 7.4.2. *Let K be a convex body which contains the origin in its interior. Then,*

$$4|K|^{1/n} |nK^\circ|^{1/n} \geqslant |K - K|^{1/n} |n(K - K)^\circ|^{1/n} \geqslant \frac{c_1}{L_K},$$

where $c_1 > 0$ is an absolute constant.

Proof. We may assume that $|K| = 1$. From the Brunn-Minkowski inequality and the Rogers-Shephard inequality, we have $2 \leqslant |K - K|^{1/n} \leqslant 4$. Since $(K - K)^\circ \subseteq K^\circ$, we immediately see that

$$|K|^{1/n} |nK^\circ|^{1/n} \geqslant \frac{1}{4} |K - K|^{1/n} |n(K - K)^\circ|^{1/n},$$

so it remains to prove the second inequality. Since

$$|T(K) - T(K)| \big|(T(K) - T(K))^\circ\big| = |K - K| |(K - K)^\circ|$$

for any invertible affine transformation T of K, we may assume for the rest of the proof that K is isotropic. By Lemma 7.4.1 and the simple fact that $|A|/|B| \leqslant N(A, B)$ for any two convex bodies A, B we get that:

$$\frac{c_1}{t\sqrt{n}|(K - K)^\circ|^{1/n}} \leqslant \left(\frac{|B_2^n|}{|t(K - K)^\circ|}\right)^{1/n} \leqslant \exp(c_2 \sqrt{n} L_K/t)$$

for all $t > 0$. Choosing $t \simeq \sqrt{n} L_K$ we get

(7.4.4) $\qquad \dfrac{c_1}{nL_K |(K - K)^\circ|^{1/n}} \leqslant c_3.$

This, combined with the lower bound $|K - K|^{1/n} \geqslant 2$ that we have for the volume of $K - K$, completes the proof. $\qquad \square$

7.4.2. Step 2: Removing the isotropic constant

Combining Proposition 7.4.2 with Theorem 7.3.1 we can remove the isotropic constant L_K from the lower bound for $s(K)^{1/n}$ and conclude the reverse Santaló inequality.

THEOREM 7.4.3 (reverse Santaló inequality). *Let K be a convex body in $\mathbb{R}^n$ which contains the origin in its interior. Then,*
$$|K|^{1/n}|nK^\circ|^{1/n} \geqslant c_3,$$
where $c_3 > 0$ is an absolute constant.

Proof. Since $|K|^{1/n}|nK^\circ|^{1/n} \geqslant \tfrac{1}{4}|K-K|^{1/n}|n(K-K)^\circ|^{1/n}$, we may assume for the rest of the proof that K is symmetric. Using Theorem 7.3.1 with $\varepsilon = 1/2$, we find a convex body $T \subset \mathbb{R}^n$ and a point $x \in \mathbb{R}^n$ such that

(7.4.5)
$$\frac{2}{3}T \subseteq K + x \subseteq \frac{3}{2}T$$

and $L_T \leqslant c_0/\sqrt{ns(K)^{1/n}}$ for some absolute constant $c_0 > 0$. Proposition 7.4.2 shows that
$$|T-T|^{1/n}|n(T-T)^\circ|^{1/n} \geqslant \frac{c_1}{L_T},$$
where $c_1 > 0$ is an absolute constant too. Observe that $\tfrac{2}{3}(T-T) \subseteq K-K = 2K \subseteq \tfrac{3}{2}(T-T)$, and thus $K^\circ \supseteq \tfrac{4}{3}(T-T)^\circ$. Therefore, combining the above, we get
$$ns(K)^{1/n} = |nK^\circ|^{1/n}|K|^{1/n} \geqslant \frac{4}{9}|n(T-T)^\circ|^{1/n}|T-T|^{1/n}$$
$$\geqslant \frac{c_1'}{L_T} \geqslant c_2\sqrt{ns(K)^{1/n}},$$

and so it follows that
$$s(K)^{1/n} \geqslant \frac{c_3}{n}$$
with $c_3 = c_2^2$. This completes the proof. $\square$

7.5. Volume radius of the centroid bodies

7.5.1. The bodies $\Lambda_p(\mu)$

Klartag and E. Milman [**284**] further exploited the logarithmic Laplace transform to obtain additional information on the L_q-centroid bodies of an isotropic log-concave probability measure μ on $\mathbb{R}^n$ and an alternative proof of the bound $L_\mu = O(\sqrt[4]{n})$. Recall that the logarithmic Laplace transform of a Borel probability measure μ on $\mathbb{R}^n$ is defined by
$$\Lambda_\mu(\xi) = \log\left(\int_{\mathbb{R}^n} e^{\langle \xi, x \rangle} d\mu(x)\right).$$
It is easily checked that Λ_μ is convex and $\Lambda_\mu(0) = 0$. If $\operatorname{bar}(\mu) = 0$ then Jensen's inequality shows that
$$\Lambda_\mu(\xi) = \log\left(\int_{\mathbb{R}^n} e^{\langle \xi, x \rangle} d\mu(x)\right) \geqslant \int_{\mathbb{R}^n} \langle \xi, x \rangle d\mu(x) = 0$$
for all ξ; therefore, Λ_μ is a non-negative function. Further properties of Λ_μ in the log-concave case are described in the next proposition.

7.5. VOLUME RADIUS OF THE CENTROID BODIES

PROPOSITION 7.5.1. *Let μ be an n-dimensional log-concave probability measure. The set $A(\mu) = \{\Lambda_\mu < \infty\}$ is open and Λ_μ is C^∞ and strictly convex on $A(\mu)$. Moreover, for every $t \geqslant 0$ and $\alpha \geqslant 1$,*

$$\frac{1}{\alpha}\{\Lambda_\mu \leqslant \alpha t\} \subseteq \{\Lambda_\mu \leqslant t\} \subseteq \{\Lambda_\mu \leqslant \alpha t\}. \tag{7.5.1}$$

Proof. Let us write f for the log-concave density of μ. From Lemma 2.2.1 we know that there exist positive constants B, D, which depend on f, such that $f(x) \leqslant De^{-B\|x\|_2}$ for every $x \in \mathbb{R}^n$. This immediately shows that

$$\exp(\Lambda_\mu(\xi)) = \int_{\mathbb{R}^n} e^{\langle \xi, x \rangle} f(x)dx \leqslant \int_{\mathbb{R}^n} De^{\|\xi\|_2 \|x\|_2} e^{-B\|x\|_2} dx < +\infty$$

for every ξ with Euclidean norm less than B, which means that $A(\mu)$ is a neighborhood of the origin. In addition, for every $\xi_0 \in A(\mu)$ we have that

$$0 < \int_{\mathbb{R}^n} e^{\langle \xi_0, x \rangle} f(x)dx < \infty,$$

therefore, by Lemma 2.2.1 again, we know that for the log-concave function $\tilde{f}_{\xi_0}(x) = e^{\langle \xi_0, x \rangle} f(x)$ there exist constants $D_{\xi_0}, B_{\xi_0} > 0$ such that $\tilde{f}_{\xi_0}(x) \leqslant D_{\xi_0} e^{-B_{\xi_0} \|x\|_2}$ for every $x \in \mathbb{R}^n$. From this and the fact that for every $\xi \in \mathbb{R}^n$ we have

$$\int_{\mathbb{R}^n} e^{\langle \xi, x \rangle} f(x)dx = \int_{\mathbb{R}^n} e^{\langle \xi - \xi_0, x \rangle} e^{\langle \xi_0, x \rangle} f(x)dx = \int_{\mathbb{R}^n} e^{\langle \xi - \xi_0, x \rangle} \tilde{f}_{\xi_0}(x)dx,$$

it follows as before that $\Lambda_\mu(\xi) < \infty$ for every ξ with $\|\xi - \xi_0\|_2 < B_{\xi_0}$. Moreover, since $\tilde{f}_{\xi_0}$ decays exponentially, we may differentiate under the integral sign any finite number of times. We conclude that $A(\mu)$ is open and that Λ_μ is C^∞ on $A(\mu)$.

Strict convexity follows from the Cauchy-Schwarz inequality and the fact that, for any distinct points $\xi_1, \xi_2 \in A(\mu)$, the functions $e^{\langle \xi_1, x \rangle}$ and $e^{\langle \xi_2, x \rangle}$ cannot be written as multiples of one another μ-almost everywhere. Finally, the left hand side inclusion in (7.5.1) follows from Hölder's inequality: suppose that $\xi \in \{\Lambda_\mu \leqslant \alpha t\}$, then

$$\int_{\mathbb{R}^n} e^{\langle \xi, x \rangle} d\mu(x) \leqslant e^{\alpha t},$$

and thus

$$\int_{\mathbb{R}^n} \exp(\langle \tfrac{\xi}{\alpha}, x \rangle) d\mu(x) \leqslant \left(\int_{\mathbb{R}^n} e^{\langle \xi, x \rangle} d\mu(x) \right)^{\frac{1}{\alpha}} \leqslant e^t,$$

which is equivalent to writing that $\xi/\alpha \in \{\Lambda_\mu \leqslant t\}$. $\square$

DEFINITION 7.5.2. *For every $p > 0$ we define*

$$\Lambda_p(\mu) = \{\Lambda_\mu \leqslant p\} \cap (-\{\Lambda_\mu \leqslant p\}).$$

PROPOSITION 7.5.3. *Let μ be a log-concave probability measure with $\mathrm{bar}(\mu) = 0$. For every $p \geqslant 1$,*

$$\Lambda_p(\mu) \simeq pZ_p(\mu)^\circ.$$

Proof. First, suppose that $\xi \in \Lambda_p(\mu)$. Then,

$$\int_{\mathbb{R}^n} \exp(|\langle \xi, x \rangle|) d\mu(x) \leqslant \int_{\mathbb{R}^n} \exp(\langle \xi, x \rangle) d\mu(x) + \int_{\mathbb{R}^n} \exp(\langle -\xi, x \rangle) d\mu(x) \leqslant 2e^p.$$

Using the inequality $(et/p)^p \leqslant e^t$, valid for any $t \geqslant 0$, we see that:

$$h_{Z_p(\mu)}(\xi) = \left(\int_{\mathbb{R}^n} |\langle \xi, x\rangle|^p d\mu(x)\right)^{\frac{1}{p}} \leqslant \frac{p}{e}\left(\int_{\mathbb{R}^n} \exp(|\langle \xi, x\rangle|)d\mu(x)\right)^{1/p} \leqslant 2p.$$

Since $\xi \in \Lambda_p(\mu)$ was arbitrary, this amounts to $\Lambda_p(\mu) \subseteq 2p(Z_p(\mu))°$.

For the other inclusion, suppose $\xi \in \mathbb{R}^n$ is such that $h_{Z_p(\mu)}(\xi) \leqslant p$, that is:

$$(7.5.2) \qquad \left(\int_{\mathbb{R}^n} |\langle \xi, x\rangle|^p d\mu(x)\right)^{\frac{1}{p}} \leqslant p.$$

Write X for the random vector in $\mathbb{R}^n$ that is distributed according to μ. Then the function

$$\phi(t) = \mu(\langle X, \xi\rangle \geqslant t), \qquad t \in \mathbb{R},$$

is log-concave, according to the Prékopa-Leindler inequality. Furthermore, since the barycenter of μ is at the origin, we have $1/e \leqslant \phi(0) \leqslant 1 - 1/e$ by Grünbaum's lemma (Lemma 2.2.6). Using Markov's inequality, from (7.5.2) we get

$$\phi(3ep) \leqslant (3e)^{-p}.$$

Since ϕ is log-concave, then:

$$\mathbb{P}(\langle X, \xi\rangle \geqslant t) = \phi(t) \leqslant \phi(0)\left(\frac{\phi(3ep)}{\phi(0)}\right)^{\frac{t}{3ep}} \leqslant C\exp(-t/(3e))$$

for all $t \geqslant 3ep$. An identical bound holds for $\mathbb{P}(\langle X, \xi\rangle \leqslant -t)$, and combining the two, we obtain:

$$\mathbb{P}(|\langle X, \xi\rangle| \geqslant t) \leqslant C\exp(-t/(3e))$$

for all $t \geqslant 3ep$. Therefore:

$$\mathbb{E}\exp\left(\frac{|\langle \xi, X\rangle|}{6e}\right) = \frac{1}{6e}\int_0^\infty \exp\left(\frac{t}{6e}\right)\mathbb{P}(|\langle X, \xi\rangle| \geqslant t)dt$$

$$\leqslant \frac{1}{6e}\int_0^{3ep}\exp\left(\frac{t}{6e}\right)dt + C\int_{3ep}^\infty \exp(-t/(6e))dt$$

$$\leqslant \exp(\tilde{C}p).$$

Consequently:

$$\max\left\{\Lambda_\mu\left(\frac{1}{6e}\xi\right), \Lambda_\mu\left(-\frac{1}{6e}\xi\right)\right\} \leqslant \log \mathbb{E}\exp\left(\frac{|\langle \xi, X\rangle|}{6e}\right) \leqslant Cp,$$

for some $C \geqslant 1$, and using (7.5.1), this implies:

$$\max\left\{\Lambda_\mu\left(\frac{1}{6eC}\xi\right), \Lambda_\mu\left(-\frac{1}{6eC}\xi\right)\right\} \leqslant p$$

for any $\xi \in \mathbb{R}^n$ with $h_{Z_p(\mu)}(\xi) \leqslant p$. This shows that $p(Z_p(\mu))° \subseteq C'\Lambda_p(\mu)$, and the assertion follows. $\square$

LEMMA 7.5.4. *Let μ be a log-concave probability measure with* $\mathrm{bar}(\mu) = 0$. *For every $q, r > 0$,*

$$\nabla \Lambda_\mu\left(\frac{1}{2}\{\Lambda_\mu \leqslant q\}\right) \subseteq (q+r)\{\Lambda_\mu \leqslant r\}°.$$

Proof. Let $x \in \frac{1}{2}\{\Lambda_\mu \leqslant q\}$. Then, $\Lambda_\mu(2x) \leqslant q$. For every $z \in \{\Lambda_\mu \leqslant r\}$ we may write

$$\left\langle \nabla\Lambda_\mu(x), \frac{z}{2} \right\rangle \leqslant \Lambda_\mu(x) + \left\langle \nabla\Lambda_\mu(x), \frac{z}{2} \right\rangle \leqslant \Lambda_\mu\left(x + \frac{z}{2}\right)$$
$$\leqslant \frac{\Lambda_\mu(2x) + \Lambda_\mu(z)}{2} \leqslant \frac{q+r}{2},$$

using the fact that $\Lambda_\mu(x) \geqslant 0$ and the convexity of Λ_μ. Since

$$\langle \nabla\Lambda_\mu(x), z \rangle \leqslant q + r$$

for every $z \in \{\Lambda_\mu \leqslant r\}$ we see that $\nabla\Lambda_\mu(x) \in (q+r)\{\Lambda_\mu \leqslant r\}^\circ$. $\square$

COROLLARY 7.5.5. *Let μ be a log-concave probability measure with* $\mathrm{bar}(\mu) = 0$. *For every $p > 0$,*

$$\nabla\Lambda_\mu\left(\frac{1}{2}\Lambda_p(\mu)\right) \subseteq 2p\,\Lambda_p(\mu)^\circ.$$

Proof. We apply Lemma 7.5.4 with $q = r = p$. We have

$$\nabla\Lambda_\mu\left(\frac{1}{2}\Lambda_p(\mu)\right) \subseteq \nabla\Lambda_\mu\left(\frac{1}{2}\{\Lambda_\mu \leqslant p\}\right) \subseteq 2p\,\{\Lambda_\mu \leqslant p\}^\circ$$
$$\subseteq 2p\,\Lambda_p(\mu)^\circ,$$

because $\{\Lambda_\mu \leqslant p\} \supseteq \Lambda_p(\mu)$ implies that $\{\Lambda_\mu \leqslant p\}^\circ \subseteq \Lambda_p(\mu)^\circ$. $\square$

DEFINITION 7.5.6. *For every $p > 0$ we define*

$$\Psi_p = \left(\frac{1}{|\frac{1}{2}\Lambda_p(\mu)|}\int_{\frac{1}{2}\Lambda_p(\mu)} \det\mathrm{Hess}\,(\Lambda_\mu)(x)\,dx\right)^{1/n}.$$

PROPOSITION 7.5.7. *For every $p > 0$,*

$$|\Lambda_p(\mu)|^{1/n} \leqslant C\sqrt{\frac{p}{n}}\,\frac{1}{\sqrt{\Psi_p}}.$$

Proof. Using Corollary 7.5.5 and the change of variables $x = \nabla\Lambda_\mu(y)$, we write

$$|2p\Lambda_p(\mu)^\circ| \geqslant \left|\nabla\Lambda_\mu\left(\frac{1}{2}\Lambda_p(\mu)\right)\right| = \int_{\frac{1}{2}\Lambda_p(\mu)} \det\mathrm{Hess}\,(\Lambda_\mu)(y)\,dy$$
$$= \left|\frac{1}{2}\Lambda_p(\mu)\right|\Psi_p^n.$$

In other words,

$$|\Lambda_p(\mu)^\circ|^{1/n} \geqslant \frac{\Psi_p}{4p}|\Lambda_p(\mu)|^{1/n}.$$

From the Blaschke-Santaló inequality we have

$$|\Lambda_p(\mu)^\circ|^{1/n} \leqslant \frac{c}{n}\frac{1}{|\Lambda_p(\mu)|^{1/n}},$$

and hence,

$$|\Lambda_p(\mu)|^{2/n} \leqslant \frac{C^2 p}{n}\frac{1}{\Psi_p},$$

where $C^2 = 4c$. $\square$

7.5.2. A first formula for the volume radius of $Z_p(\mu)$

Let μ be a log-concave probability measure on $\mathbb{R}^n$ with density ρ. For every $\xi \in A(\mu) = \{\Lambda_\mu < \infty\}$ we set

$$\rho_\xi(x) = \frac{1}{Z_\xi} \rho(x) e^{\langle \xi, x \rangle},$$

where $Z_\xi > 0$ is chosen so that ρ_ξ becomes a probability density. Next, we set

$$b_\xi = \frac{1}{Z_\xi} \int_{\mathbb{R}^n} x \rho(x) e^{\langle \xi, x \rangle} dx$$

and we define a probability measure μ_ξ with density

$$\frac{1}{Z_\xi} \rho(x + b_\xi) e^{\langle \xi, x + b_\xi \rangle}.$$

LEMMA 7.5.8. *We have*

$$b_\xi = \nabla \Lambda_\mu(\xi) \quad \text{and} \quad \mathrm{Cov}(\mu_\xi) = \mathrm{Hess}\,(\Lambda_\mu)(\xi).$$

Proof. Both equalities follow from simple calculations (the proof is similar to that of Proposition 7.2.1): just observe that since the log-concave density $\rho(x) e^{\langle \xi^0, x \rangle}$ decays exponentially for every $\xi^0 \in A(\mu)$, we can differentiate twice under the integral sign. □

THEOREM 7.5.9 (Klartag-E. Milman). *Let μ be a log-concave probability measure on $\mathbb{R}^n$ with $\mathrm{bar}(\mu) = 0$. For every $1 \leqslant p \leqslant n$,*

$$|Z_p(\mu)|^{1/n} \simeq \sqrt{\frac{p}{n}} \inf_{\xi \in \frac{1}{2}\Lambda_p(\mu)} [\det \mathrm{Cov}(\mu_\xi)]^{\frac{1}{2n}}.$$

Proof of the lower bound. We combine Propositions 7.5.3 and 7.5.7. We have

$$|\Lambda_p(\mu)|^{1/n} \leqslant C \sqrt{\frac{p}{n}} \frac{1}{\sqrt{\Psi_p}}$$

and

$$\Lambda_p(\mu) \simeq p Z_p(\mu)^\circ.$$

Therefore, by the reverse Santaló inequality,

$$|Z_p(\mu)|^{1/n} \geqslant \frac{c_1}{n |Z_p(\mu)^\circ|^{1/n}} \geqslant \frac{c_2 p}{n |\Lambda_p(\mu)|^{1/n}}$$

$$\geqslant c_3 \frac{p}{n} \sqrt{\frac{n}{p}} \sqrt{\Psi_p} = c_3 \sqrt{\frac{p}{n}} \sqrt{\Psi_p}$$

$$= c_3 \sqrt{\frac{p}{n}} \left(\frac{1}{|\frac{1}{2}\Lambda_p(\mu)|} \int_{\frac{1}{2}\Lambda_p(\mu)} \det \mathrm{Hess}\,(\Lambda_\mu)(\xi) \, d\xi \right)^{\frac{1}{2n}}$$

$$\geqslant c_3 \sqrt{\frac{p}{n}} \inf_{\xi \in \frac{1}{2}\Lambda_p(\mu)} [\det \mathrm{Hess}\,(\Lambda_\mu)(\xi)]^{\frac{1}{2n}}$$

$$= c_3 \sqrt{\frac{p}{n}} \inf_{\xi \in \frac{1}{2}\Lambda_p(\mu)} [\det \mathrm{Cov}(\mu_\xi)]^{\frac{1}{2n}}.$$

For the proof of the upper bound we need the following.

7.5. VOLUME RADIUS OF THE CENTROID BODIES

PROPOSITION 7.5.10. *Let μ be a log-concave probability measure on $\mathbb{R}^n$ with* $\mathrm{bar}(\mu) = 0$. *For every $\xi \in \frac{1}{2}\Lambda_p(\mu)$,*

(7.5.3) $$Z_p(\mu_\xi) \simeq Z_p(\mu).$$

Proof. From Proposition 7.5.3, it is enough to show that, for every $\xi \in \frac{1}{2}\Lambda_p(\mu)$,
$$\Lambda_p(\mu_\xi) \simeq \Lambda_p(\mu).$$

We first observe that

(7.5.4) $$\Lambda_{\mu_\xi}(z) = \Lambda_\mu(z+\xi) - \Lambda_\mu(\xi) - \langle z, \nabla\Lambda_\mu(\xi)\rangle.$$

To see this, first note that
$$\log Z_\xi = \log \int_{\mathbb{R}^n} \rho(x) e^{\langle \xi, x\rangle} dx = \Lambda_\mu(\xi)$$

and
$$\langle z, b_\xi\rangle = \langle z, \nabla\Lambda_\mu(\xi)\rangle.$$

Then, write
$$\Lambda_{\mu_\xi}(z) = \log\left(\int_{\mathbb{R}^n} \frac{1}{Z_\xi} e^{\langle z,y\rangle} \rho_\xi(y) dy\right)$$
$$= \log\left(\int_{\mathbb{R}^n} e^{\langle z,y\rangle} e^{\langle \xi, y+b_\xi\rangle} \rho(y+b_\xi) dy\right) - \log Z_\xi$$
$$= \log\left(\int_{\mathbb{R}^n} e^{-\langle z, b_\xi\rangle} e^{\langle z, y+b_\xi\rangle} e^{\langle \xi, y+b_\xi\rangle} \rho(y+b_\xi) dy\right) - \Lambda_\mu(\xi)$$
$$= -\langle z, b_\xi\rangle + \log\left(\int_{\mathbb{R}^n} e^{\langle z+\xi, y+b_\xi\rangle} \rho(y+b_\xi) dy\right) - \Lambda_\mu(\xi)$$
$$= -\langle z, b_\xi\rangle + \Lambda_\mu(z+\xi) - \Lambda_\mu(\xi).$$

Claim. Let $D, p > 0$. If $\Lambda_\mu(2y) \leqslant Dp$ and $z \in \Lambda_p(\mu)$, then
$$\Lambda_\mu(z/2 + y) - \Lambda_\mu(y) - \langle z/2, \nabla\Lambda_\mu(y)\rangle \leqslant (D+1)p.$$

Proof of the Claim. We apply Lemma 7.5.4 with $q = Dp$ and $r = p$. We have $\Lambda_\mu(2y) \leqslant Dp$ and $\Lambda_\mu(-z) \leqslant p$. Therefore,
$$-\langle \nabla\Lambda_\mu(y), z\rangle \leqslant (D+1)p.$$

Then,
$$\Lambda_\mu(z/2+y) - \Lambda_\mu(y) - \langle z/2, \nabla\Lambda_\mu(y)\rangle \leqslant \Lambda_\mu(z/2+y) + \frac{(D+1)p}{2}$$
$$\leqslant \frac{\Lambda_\mu(z) + \Lambda_\mu(2y)}{2} + \frac{(D+1)p}{2}$$
$$\leqslant (D+1)p.$$

We can now continue the proof of Proposition 7.5.10:

(i) Assume that $z \in \Lambda_p(\mu)$. Note that $\Lambda_\mu(2\xi) \leqslant p$ because $\xi \in \frac{1}{2}\Lambda_p(\mu)$. The claim (with $D = 1$ and $y = \xi$) combined with (7.5.4) shows that
$$\Lambda_{\mu_\xi}(z/2) = \Lambda_\mu(z/2+\xi) - \Lambda_\mu(\xi) - \langle z/2, \nabla\Lambda_\mu(\xi)\rangle \leqslant 2p.$$

From Proposition 7.5.1 it follows that $\Lambda_{\mu_\xi}(z/4) \leqslant p$. By symmetry, the same argument applies to $-z$, and hence $z \in 4\Lambda_p(\mu_\xi)$. In other words,
$$\Lambda_p(\mu) \subseteq 4\Lambda_p(\mu_\xi).$$

(ii) Assume that $z \in \Lambda_p(\mu_\xi)$. From (7.5.4) we know that
$$\Lambda_{\mu_\xi}(-2\xi) = \Lambda_\mu(-\xi) - \Lambda_\mu(\xi) + 2\langle \xi, \nabla \Lambda_\mu(\xi)\rangle.$$
Note that
$$\Lambda_\mu(-\xi) \leqslant \frac{\Lambda_\mu(-2\xi) + \Lambda_\mu(0)}{2} \leqslant \frac{p+0}{2} = \frac{p}{2},$$
and similarly $\Lambda_\mu(\xi) \leqslant \frac{p}{2}$. From Lemma 7.5.4 we have
$$\langle \xi, \nabla \Lambda_\mu(\xi)\rangle \leqslant \frac{3p}{2}.$$
Since $\Lambda_\mu(\xi) \geqslant 0$ we conclude that
$$\Lambda_{\mu_\xi}(-2\xi) \leqslant \frac{7p}{2}.$$
Since $(\mu_\xi)_{-\xi} = \mu$, we may apply the argument from (i), using that $\Lambda_{\mu_\xi}(-2\xi) \leqslant Dp$ for $D = \frac{7}{2}$. We write
$$\Lambda_\mu(z/2) = \Lambda_{\mu_\xi}(z/2 - \xi) - \Lambda_{\mu_\xi}(-\xi) + \langle -z/2, \nabla \Lambda_{\mu_\xi}(-\xi)\rangle.$$
Using the facts that $\Lambda_{\mu_\xi}(z/2 - \xi) \leqslant \frac{1}{2}(\Lambda_{\mu_\xi}(z) + \Lambda_{\mu_\xi}(-2\xi)) \leqslant \frac{9p}{4}$, $\Lambda_{\mu_\xi}(-\xi) \geqslant 0$ and $\langle -z/2, \nabla \Lambda_{\mu_\xi}(-\xi)\rangle \leqslant \frac{9p}{4}$ (by a last application of Lemma 7.5.4 for the pair $-z, -\xi$) we see that $\Lambda_\mu(z/2) \leqslant 9p/2$, which shows that $\Lambda_\mu(z/9) \leqslant p$. Using the same argument for $-z$ we finally conclude that
$$\Lambda_p(\mu_\xi) \subseteq 9\Lambda_p(\mu),$$
and the result follows. $\square$

We will also use the known upper bound for $|Z_p(\mu)|^{1/n}$ (Theorem 5.1.17):

FACT 7.5.11. *Let ν be a log-concave probability measure on $\mathbb{R}^n$ with* $\mathrm{bar}(\nu) = 0$. *For every $2 \leqslant p \leqslant n$,*
$$|Z_p(\nu)|^{1/n} \leqslant C\sqrt{p}|Z_2(\nu)|^{1/n}.$$

Proof of the upper bound. Since
$$[\det \mathrm{Cov}(\mu_\xi)]^{\frac{1}{2n}} \simeq \sqrt{n}|Z_2(\mu_\xi)|^{1/n},$$
applying Fact 7.5.11 we get
$$\inf_{\xi \in \frac{1}{2}\Lambda_p(\mu)} |Z_p(\mu_\xi)|^{1/n} \leqslant C\sqrt{\frac{p}{n}} \inf_{\xi \in \frac{1}{2}\Lambda_p(\mu)} [\det \mathrm{Cov}(\mu_\xi)]^{\frac{1}{2n}}.$$
From Proposition 7.5.10 we know that $Z_p(\mu_\xi) \simeq Z_p(\mu)$ for all $\xi \in \frac{1}{2}\Lambda_p(\mu)$. It follows that
$$|Z_p(\mu)|^{1/n} \leqslant C\sqrt{\frac{p}{n}} \inf_{\xi \in \frac{1}{2}\Lambda_p(\mu)} [\det \mathrm{Cov}(\mu_\xi)]^{\frac{1}{2n}}.$$
This completes the proof of Theorem 7.5.9. $\square$

An immediate consequence of Theorem 7.5.9 is the following.

THEOREM 7.5.12 (Klartag-E. Milman). *Let μ be a log-concave probability on $\mathbb{R}^n$ with* $\mathrm{bar}(\mu) = 0$. *For every $1 \leqslant p \leqslant q \leqslant n$,*
$$\frac{|Z_p(\mu)|^{1/n}}{\sqrt{p}} \geqslant c\frac{|Z_q(\mu)|^{1/n}}{\sqrt{q}},$$
where $c > 0$ is an absolute constant.

Proof. Since $\Lambda_p(\mu) \subseteq \Lambda_q(\mu)$, we have
$$\inf_{\xi \in \frac{1}{2}\Lambda_p(\mu)} [\det \operatorname{Cov}(\mu_\xi)]^{\frac{1}{2n}} \geqslant \inf_{\xi \in \frac{1}{2}\Lambda_q(\mu)} [\det \operatorname{Cov}(\mu_\xi)]^{\frac{1}{2n}}.$$
Then, we apply the formula of Theorem 7.5.9. $\square$

REMARK 7.5.13. Another consequence of Theorem 7.5.9 is that if $x_0 \in \frac{1}{2}\Lambda_p(\mu)$ is such that
$$[\det \operatorname{Cov}(\mu_{x_0})]^{\frac{1}{2n}} \simeq \inf_{x \in \frac{1}{2}\Lambda_p(\mu)} [\det \operatorname{Cov}(\mu_x)]^{\frac{1}{2n}},$$
then, using (7.5.3) as well, we get that
$$|Z_p(\mu_{x_0})|^{1/n} \simeq \sqrt{\frac{p}{n}} [\det \operatorname{Cov}(\mu_{x_0})]^{\frac{1}{2n}}.$$
Naturally, the aim is to show a similar equivalence for the corresponding quantities of the measure μ instead of those of μ_{x_0}. To accomplish this, we need to be able to prove that
$$(7.5.5) \qquad \inf_{x \in \frac{1}{2}\Lambda_p(\mu)} [\det \operatorname{Cov}(\mu_x)]^{\frac{1}{2n}} \geqslant \frac{1}{A} [\det \operatorname{Cov}(\mu)]^{\frac{1}{2n}}$$
for as small a constant $A \geqslant 1$ and for as large an interval of $p \in [1, n]$ as possible. Observe that if we establish (7.5.5) for some p and $A \geqslant 1$, then we have by Theorem 7.5.9 that
$$|Z_p(\mu)|^{1/n} \geqslant \frac{c}{A} \sqrt{\frac{p}{n}} [\det \operatorname{Cov}(\mu)]^{\frac{1}{2n}},$$
and hence, by the definition of L_μ and by Corollary 5.1.9, we can conclude that
$$(7.5.6) \qquad L_\mu = \|\mu\|_\infty^{1/n} [\det \operatorname{Cov}(\mu)]^{\frac{1}{2n}} \leqslant c' \frac{[\det \operatorname{Cov}(\mu)]^{\frac{1}{2n}}}{|Z_n(\mu)|^{1/n}}$$
$$\leqslant c' \frac{[\det \operatorname{Cov}(\mu)]^{\frac{1}{2n}}}{|Z_p(\mu)|^{1/n}} \leqslant c'' A \sqrt{\frac{n}{p}},$$
where $c > 0$, c' and c'' are absolute constants (independent of the measure μ, the dimension n, or p and A).

Klartag and E. Milman defined in [**284**] a new hereditary parameter $q_\sharp^H(\mu)$ for isotropic measures μ, and gave a lower bound of the correct order for the volume radius of $Z_p(\mu)$ for every p up to that parameter.

THEOREM 7.5.14 (Klartag-E. Milman). *Let μ be an isotropic log-concave measure on $\mathbb{R}^n$. Then, for every $p \in [1, q_\sharp^H(\mu)]$, we have that*
$$|Z_p(\mu)|^{1/n} \geqslant c \sqrt{\frac{p}{n}},$$
where $c > 0$ is an absolute constant.

As we will see in the next subsection, for every isotropic log-concave measure μ on $\mathbb{R}^n$ which is a ψ_α-measure with constant b_α for some $\alpha \in [1, 2]$, one has
$$q_\sharp^H(\mu) \geqslant \frac{c'}{b_\alpha^\alpha} n^{\alpha/2}.$$
Using this fact, Klartag and E. Milman were able to give an upper bound for the isotropic constant of ψ_α-measures which also takes into account their ψ_α-constant;

in particular, they showed that the isotropic constant is less than an absolute multiple of the ψ_2-constant of a measure (improving on Bourgain's result which was comparing the isotropic constant to the ψ_2-constant in a non-linear way – see Theorem 3.4.1).

THEOREM 7.5.15 (Klartag-E. Milman). *Let μ be an isotropic log-concave measure on $\mathbb{R}^n$. If μ is a ψ_α-measure with constant b_α for some $\alpha \in [1,2]$, then*
$$L_\mu \leqslant C b_\alpha^{\alpha/2} n^{(2-\alpha)/4},$$
where C is an absolute constant. □

Note that since every log-concave probability measure is ψ_1 with some absolute constant b_1, the above theorem leads to the best currently known bound for the isotropic constant of an arbitrary measure μ on $\mathbb{R}^n$: $L_\mu \leqslant C' n^{1/4}$.

7.5.3. Volume radius of $Z_q(\mu)$ when $q \leqslant q_\sharp^H(\mu)$ and when $q \leqslant r_\sharp^H(\mu)$

In this section we describe the work of Vritsiou in [**501**]. For any $A \geqslant 1$ she introduced the parameter $r_\sharp(\mu, A)$ and modified the proof of Theorem 7.5.14 of Klartag and E. Milman to show that the lower bound $|Z_p(\mu)|^{1/n} \geqslant cA^{-1}\sqrt{p/n}$ continues to hold for all $p \leqslant r_\sharp(\mu, A)$. As we will see, both $q_\sharp(\mu)$ and $q_*(\mu)$ are dominated by $r_\sharp(\mu, A_0)$ for some absolute constant $A_0 \geqslant 1$. Thus, the results of [**501**] extend Theorem 7.5.14.

DEFINITION 7.5.16. For every n-dimensional (isotropic) log-concave measure, and for any $A \geqslant 1$, we define

(7.5.7) $r_\sharp(\mu, A) := \max\{1 \leqslant k \leqslant n-1 : \exists E \in G_{n,k} \text{ such that } L_{\pi_E\mu} \leqslant A\}$

(recall that we denote the marginal of μ with respect to a subspace E by $\pi_E\mu$). In other words, $r_\sharp(\mu, A)$ is the largest dimension $\leqslant n-1$ in which we can find at least one marginal of μ that has isotropic constant bounded above by A; as a convention, when μ is an 1-dimensional measure, we set $r_\sharp(\mu, A) = 1$ for every A. Furthermore, we define a "hereditary" variant of $r_\sharp(\mu, A)$, namely a related parameter that in a way controls the behavior of all marginals of μ with respect to $r_\sharp(\cdot, A)$: set

(7.5.8) $$r_\sharp^H(\mu, A) := n \inf_k \inf_{E \in G_{n,k}} \frac{r_\sharp(\pi_E\mu, A)}{k}.$$

In this section, we will show that if μ is an isotropic measure on $\mathbb{R}^n$ and $p \leqslant r_\sharp^H(\mu, A)$, then
$$|Z_p(\mu)|^{1/n} \geqslant \frac{c}{A}\sqrt{\frac{p}{n}} [\det \operatorname{Cov}(\mu)]^{\frac{1}{2n}} = \frac{c}{A}\sqrt{\frac{p}{n}}$$
for some absolute constant $c > 0$. As we remarked previously, to do so we have to show that
$$[\det \operatorname{Cov}(\mu_x)]^{\frac{1}{2n}} \geqslant \frac{c'}{A}$$
for every $x \in \frac{1}{2}\Lambda_p(\mu)$. The reason for introducing the hereditary parameter $r_\sharp^H(\mu, A)$ is that in order to compare $\det \operatorname{Cov}(\mu_x)$ with $\det \operatorname{Cov}(\mu)$ we need to compare the corresponding eigenvalues of each covariance matrix, taken in increasing order, one pair at a time; this requires that we have control over $r_\sharp(\pi_E\mu, A)$ of several marginals of μ of different dimensions. Turning to the details, we denote the eigenvalues of

Cov(μ_x) by $\lambda_1^x \leqslant \lambda_2^x \leqslant \cdots \leqslant \lambda_n^x$, and we write E_k for the k-dimensional subspace which is spanned by eigenvectors corresponding to the first k eigenvalues of Cov(μ_x). We start with the following lemma from [**284**].

LEMMA 7.5.17. *For every two integers $1 \leqslant s \leqslant k \leqslant n$ we have that*
$$\sqrt{\lambda_k^x} \geqslant c_1 \sup_{F \in G_{E_k,s}} |Z_s(\pi_F \mu_x)|^{1/s}, \tag{7.5.9}$$
where $c_1 > 0$ is an absolute constant.

Proof. Note that
$$\lambda_k^x = \max_{\theta \in S_{E_k}} \int_{E_k} \langle z, \theta \rangle^2 \, d\pi_{E_k}\mu_x(z) = \sup_{F \in G_{E_k,s}} \max_{\theta \in S_F} \int_F \langle z, \theta \rangle^2 \, d\pi_F \mu_x(z). \tag{7.5.10}$$

This is because, for every subspace F of E_k and every $\theta \in S_F \subseteq S_{E_k}$, we have that
$$\int_F \langle z, \theta \rangle^2 \, d\pi_F \mu_x(z) = \int_{\mathbb{R}^n} \langle z, \theta \rangle^2 \, d\mu_x(z) = \int_{E_k} \langle z, \theta \rangle^2 \, d\pi_{E_k}\mu_x(z),$$
while λ_k^x is the largest eigenvalue of Cov$(\pi_{E_k}\mu_x)$.

On the other hand, since μ_x is a centered, log-concave probability measure, which means that so are its s-dimensional marginals $\pi_F \mu_x$, we get from Corollary 5.1.9 that
$$|Z_s(\pi_F \mu_x)|^{1/s} \simeq \frac{1}{\|f_{\pi_F \mu_x}\|_\infty^{1/s}} = \frac{[\det \text{Cov}(\pi_F \mu_x)]^{\frac{1}{2s}}}{L_{\pi_F \mu_x}}. \tag{7.5.11}$$

Since $L_\nu \geqslant c$ for any isotropic measure ν, for some absolute constant $c > 0$, it follows that
$$|Z_s(\pi_F \mu_x)|^{1/s} \leqslant c'[\det \text{Cov}(\pi_F \mu_x)]^{\frac{1}{2s}} \leqslant c' \max_{\theta \in S_F} \sqrt{\int_F \langle z, \theta \rangle^2 \, d\pi_F \mu_x(z)}$$
for every $F \in G_{E_k,s}$, which combined with (7.5.10) gives us (7.5.9). $\square$

To bound the right hand side of (7.5.9) by an expression that involves det Cov(μ), we have to compare the volume of $Z_s(\pi_F \mu_x)$ to that of $Z_s(\pi_F \mu)$ (we are able to do that because of Proposition 7.5.10). The right choice of s is prompted by the following lemma.

LEMMA 7.5.18. *Recall that for some fixed $x \in \frac{1}{2}\Lambda_p(\mu)$ and every integer $k \leqslant n$, we denote by E_k the k-dimensional subspace which is spanned by eigenvectors corresponding to the first k eigenvalues of* Cov(μ_x). *For convenience, we also set* $s_k^x := r_\sharp(\pi_{E_k}\mu, A)$. *Then*
$$\sup_{F \in G_{E_k, s_k^x}} |Z_{s_k^x}(\pi_F \mu)|^{1/s_k^x} \geqslant \frac{c_2}{A} [\det \text{Cov}(\mu)]^{\frac{1}{2n}} = \frac{c_2}{A},$$
where $c_2 > 0$ is an absolute constant.

Proof. As in (7.5.11), we can write
$$|Z_{s_k^x}(\pi_F \mu)|^{1/s_k^x} \geqslant \frac{c_2}{\|f_{\pi_F \mu}\|_\infty^{1/s_k^x}} = \frac{c_2 [\det \text{Cov}(\pi_F \mu)]^{\frac{1}{2s_k^x}}}{L_{\pi_F \mu}}$$
for some absolute constant $c_2 > 0$ and for every $F \in G_{E_k, s_k^x}$. Remember that since μ is isotropic, $[\det \text{Cov}(\pi_F \mu)]^{1/(2s_k^x)} = [\det \text{Cov}(\mu)]^{1/(2n)} = 1$. Moreover, by the

definition of $s_k^x = r_\sharp(\pi_{E_k}\mu, A)$, there is at least one s_k^x-dimensional subspace of E_k, say F_0, such that the marginal $\pi_{F_0}(\pi_{E_k}\mu) \equiv \pi_{F_0}\mu$ has isotropic constant bounded above by A. Combining all of these, we get

$$\sup_{F \in G_{E_k, s_k^x}} |Z_{s_k^x}(\pi_F \mu)|^{1/s_k^x} \geqslant |Z_{s_k^x}(\pi_{F_0}\mu)|^{1/s_k^x} \geqslant \frac{c_2}{A}$$

as required. $\square$

Observe now that in order to compare $Z_{s_k^x}(\pi_F \mu_x)$ and $Z_{s_k^x}(\pi_F \mu)$ for every $F \in G_{E_k, s_k^x}$, we have two cases to consider:

(i) If $p \leqslant s_k^x = r_\sharp(\pi_{E_k}\mu, A)$, then by Proposition 7.5.10 (and the fact that $x \in \frac{1}{2}\Lambda_p(\mu) \subseteq \frac{1}{2}\Lambda_{s_k^x}(\mu)$) we have that $Z_{s_k^x}(\mu_x) \simeq Z_{s_k^x}(\mu)$, and therefore for every $F \in G_{E_k, s_k^x}$,

$$Z_{s_k^x}(\pi_F \mu_x) = P_F\big(Z_{s_k^x}(\mu_x)\big) \simeq P_F\big(Z_{s_k^x}(\mu)\big) = Z_{s_k^x}(\pi_F \mu)$$

as well.

(ii) If $s_k^x < p$, then we can write

$$Z_{s_k^x}(\pi_F \mu_x) \supseteq c_0 \frac{s_k^x}{p} Z_p(\pi_F \mu_x) \supseteq c_0' \frac{s_k^x}{p} Z_p(\pi_F \mu) \supseteq c_0' \frac{s_k^x}{p} Z_{s_k^x}(\pi_F \mu)$$

for some absolute constants $c_0, c_0' > 0$. We also recall that since

$$p \leqslant r_\sharp^H(\mu, A) = n \inf_k \inf_{E \in G_{n,k}} \frac{r_\sharp(\pi_E\mu, A)}{k} \leqslant \frac{n}{k} r_\sharp(\pi_{E_k}\mu, A),$$

it holds that $s_k^x/p = r_\sharp(\pi_{E_k}\mu, A)/p \geqslant k/n$.

To summarize the above, we see that in any case and for every $F \in G_{E_k, s_k^x}$,

$$(7.5.12) \qquad Z_{s_k^x}(\pi_F \mu_x) \supseteq c_0'' \min\left\{1, \frac{s_k^x}{p}\right\} Z_{s_k^x}(\pi_F \mu) \supseteq c_0'' \frac{k}{n} Z_{s_k^x}(\pi_F \mu),$$

where $c_0'' > 0$ is a small enough absolute constant. We now have everything we need to bound $|Z_p(\mu)|^{1/n}$ from below.

THEOREM 7.5.19 (Vritsiou). *Let μ be an n-dimensional isotropic measure and let $A \geqslant 1$. Then, for every $p \in [1, r_\sharp^H(\mu, A)]$, we have that*

$$|Z_p(\mu)|^{1/n} \geqslant \frac{c}{A}\sqrt{\frac{p}{n}},$$

where $c > 0$ is an absolute constant.

Proof. Combining Lemmas 7.5.17 and 7.5.18 with (7.5.12), we see that for every $p \in [1, r_\sharp^H(\mu, A)]$ and for every $x \in \frac{1}{2}\Lambda_p(\mu)$,

$$[\det \operatorname{Cov}(\mu_x)]^{1/2} = \prod_{k=1}^n \sqrt{\lambda_k^x} \geqslant \prod_{k=1}^n \frac{c}{A} \frac{k}{n} = \frac{c^n}{A^n} \frac{n!}{n^n}.$$

If we take n-th roots, the theorem then follows from Theorem 7.5.9. $\square$

REMARK 7.5.20. The parameter $r_\sharp^H(\mu, A)$ was introduced in [**501**] and the preceding argument generalizes the argument in [**284**] by Klartag and E. Milman, which made use of a different parameter for isotropic measures μ, and gave a lower bound of the correct order for the volume radius of $Z_p(\mu)$ for every p up to that parameter. To define the parameter of Klartag and Milman, let us consider the function $\Delta_\mu : [1, \infty) \to \mathbb{R}$ which maps q to

$$\Delta_\mu(q) = \mathrm{diam}(Z_q(\mu))$$

(this can be defined for every log-concave probability measure μ). For every positive constant c, we can set

$$q_{\sharp,c}(\mu) := \Delta_\mu^{-1}\big(c\sqrt{n}[\det \mathrm{Cov}(\mu)]^{\frac{1}{2n}}\big)$$
$$= \sup\big\{q \geqslant 1 \,|\, \mathrm{diam}(Z_q(\mu)) \leqslant c\sqrt{n}[\det \mathrm{Cov}(\mu)]^{\frac{1}{2n}}\big\}.$$

Also, if it happens that

$$\mathrm{diam}(Z_1(\mu)) \geqslant c\sqrt{n}[\det \mathrm{Cov}(\mu)]^{\frac{1}{2n}},$$

then we set $q_{\sharp,c}(\mu) = 1$. Klartag and Milman proved that there exists a small positive constant c_0 such that, if μ is isotropic and we write $q_\sharp(\mu) := q_{\sharp,c_0}(\mu)$, then $q_\sharp(\mu)$ is equivalent to Paouris' parameter

$$q_*(\mu) = \sup\{q \geqslant 1 \,|\, k_*(Z_q(\mu)) \geqslant q\}$$

in the sense that

$$q_*(\mu) \geqslant q_\sharp(\mu) \geqslant q_{*,c_0}(\mu) = \sup\{q \geqslant 1 \,|\, k_*(Z_q(\mu)) \geqslant c_0 q\}.$$

In addition, if c_0 is chosen to be sufficiently small, then we can ensure that for every $1 \leqslant q \leqslant q_\sharp(\mu)$,

$$k_*(Z_q(\mu)) \geqslant q \quad \text{and} \quad w(Z_q(\mu)) \geqslant c_1 \sqrt{q}[\det \mathrm{Cov}(\mu)]^{\frac{1}{2n}},$$

where $c_1 > 0$ is an absolute constant (the last inequality follows anyway from Paouris' theorem and the inequality $q_\sharp(\mu) \leqslant q_*(\mu)$). If we recall now the definition of the dual critical dimension $k_*(Z_q(\mu))$, we see that there exist many subspaces $F \in G_{n, \lceil q_\sharp(\mu) \rceil}$ such that

$$P_F(Z_{\lceil q_\sharp(\mu) \rceil}(\mu)) \simeq P_F(Z_{q_\sharp(\mu)}(\mu)) \simeq w(Z_{q_\sharp(\mu)}(\mu))B_F \supseteq c_1 \sqrt{q_\sharp(\mu)} B_F,$$

which means that for every such subspace

$$\big|Z_{\lceil q_\sharp(\mu) \rceil}(\pi_F \mu)\big|^{1/\lceil q_\sharp(\mu) \rceil} = \big|P_F(Z_{\lceil q_\sharp(\mu) \rceil}(\mu))\big|^{1/\lceil q_\sharp(\mu) \rceil} \geqslant c_1'$$

and that

(7.5.13) $$\sup_{F \in G_{n, \lceil q_\sharp(\mu) \rceil}} \big|Z_{\lceil q_\sharp(\mu) \rceil}(\pi_F \mu)\big|^{1/\lceil q_\sharp(\mu) \rceil} \geqslant c_1'$$

for some absolute constant c_1'. An analogous inequality holds for every marginal $\pi_E \mu$ of μ if we replace n by $\dim E$ and $q_\sharp(\mu)$ by $q_\sharp(\pi_E \mu)$; this shows that we have an analogue of Lemma 7.5.18 that allows us to work with the parameter $q_\sharp^H(\mu)$, which is a hereditary variant of $q_\sharp(\mu)$ defined as

$$q_\sharp^H(\mu) := n \inf_k \inf_{E \in G_{n,k}} \frac{q_\sharp(\pi_E \mu)}{k},$$

and deduce, in a similar way as above, the bound
$$|Z_p(\mu)|^{1/n} \geqslant c_1'' \sqrt{\frac{p}{n}}$$
for every $p \leqslant q_\sharp^H(\mu)$.

Since the parameters $q_\sharp(\mu)$ and $q_*(\mu)$ are equivalent for every isotropic measure μ in the sense that we described above, we have by Corollary 5.2.14 that, if μ is a ψ_α-measure with constant b_α for some $\alpha \in [1,2]$, then
$$q_\sharp(\mu) \geqslant \frac{c}{b_\alpha^\alpha} n^{\alpha/2}.$$
Consequently, since every marginal of μ is also a ψ_α-measure with constant b_α, we see that
$$q_\sharp^H(\mu) \geqslant \frac{c'}{b_\alpha^\alpha} n^{\alpha/2}.$$
Using this fact, Klartag and E. Milman were able to prove Theorem 7.5.15 in the way that we explained in Remark 7.5.13.

On the other hand, recall that for every isotropic measure μ and for every k-dimensional marginal $\pi_F \mu$ of μ, we have the relation
$$|Z_k(\pi_F \mu)|^{1/k} \simeq \frac{1}{\|\pi_F \mu\|_\infty^{1/k}} = \frac{1}{L_{\pi_F \mu}};$$
therefore, Remark 7.5.20 and, in particular, (7.5.13) imply that both $q_\sharp(\mu)$ and $q_*(\mu)$ are dominated by $r_\sharp(\mu, A_0)$ for some absolute constant $A_0 \geqslant 1$, thus showing that the use of the parameter $r_\sharp$ extends the range of validity of the above argument. In fact, $r_\sharp^H$ is (in a sense that we'll explain shortly) equivalent to a hereditary variant of the parameter of Dafnis and Paouris from [**149**] which is defined through the negative moments of the Euclidean norm with respect to an isotropic measure μ (in contrast with $q_*(\mu)$ which is related to the positive moments): recall that for every $\delta \geqslant 1$ and every isotropic log-concave measure μ, we write
$$q_{-c}(\mu, \delta) := \max\{1 \leqslant p \leqslant n-1 : I_{-p}(\mu) \geqslant \delta^{-1} I_2(\mu) = \delta^{-1} \sqrt{n}\}.$$
Then, if we set
$$q_{-c}^H(\mu, \delta) := n \inf_k \inf_{E \in G_{n,k}} \frac{\lfloor q_{-c}(\pi_E \mu, \delta) \rfloor}{k},$$
the following theorem comparing $r_\sharp^H$ and q_{-c}^H holds.

THEOREM 7.5.21. *There exist absolute constants $C_1, C_2 > 0$ such that for every isotropic measure μ on $\mathbb{R}^n$ and every $A \geqslant 1$,*
$$(7.5.14) \qquad r_\sharp^H(\mu, A) \leqslant q_{-c}^H(\mu, C_1 A) \leqslant r_\sharp^H(\mu, C_2 A).$$
By Theorem 7.5.19 and Remark 7.5.13, it then follows that
$$(7.5.15) \qquad L_\mu \leqslant CA \sqrt{\frac{n}{r_\sharp^H(\mu, A)}} \leqslant CA \sqrt{\frac{n}{q_{-c}^H\left(\mu, \frac{C_1}{C_2} A\right)}}$$
for every $A \geqslant C_2/C_1$.

The key step in proving Theorem 7.5.21 is showing the following consequence of Theorem 7.5.19.

7.5. VOLUME RADIUS OF THE CENTROID BODIES

LEMMA 7.5.22. *There exists a positive absolute constant C_1 such that, for every n-dimensional isotropic measure μ and every $A \geqslant 1$,*
$$r_\sharp^H(\mu, A) \leqslant \lfloor q_{-c}(\mu, C_1 A) \rfloor.$$
In other words, for every $p \leqslant \lceil r_\sharp^H(\mu, A) \rceil$ we have that
$$I_{-p}(\mu) \geqslant \frac{1}{C_1 A} I_2(\mu) = \frac{1}{C_1 A} \sqrt{n}.$$

Proof. Set $p_A := r_\sharp^H(\mu, A)$ and observe that
$$|Z_{\lceil p_A \rceil}(\mu)|^{1/n} \geqslant |Z_{p_A}(\mu)|^{1/n} \geqslant \frac{c'}{A} \sqrt{\frac{\lceil p_A \rceil}{n}}.$$
By Hölder's and Santaló's inequalities, this gives us that
$$w_{-\lceil p_A \rceil}(Z_{\lceil p_A \rceil}(\mu)) \geqslant w_{-n}(Z_{\lceil p_A \rceil}(\mu)) \geqslant \frac{|Z_{\lceil p_A \rceil}(\mu)|^{1/n}}{\omega_n^{1/n}} \geqslant \frac{c''}{A} \sqrt{\lceil p_A \rceil}.$$
Since $r_\sharp^H(\mu, A) \leqslant r_\sharp(\mu, A) \leqslant n-1$ by definition, we have $\lceil p_A \rceil \leqslant n-1$, and thus we can use (5.3.11) to conclude that
$$I_{-\lceil p_A \rceil}(\mu) \geqslant \frac{1}{C_1 A} \sqrt{n}$$
for some absolute constant $C_1 > 0$. This completes the proof. $\square$

Proof of Theorem 7.5.21. For the left hand side inequality of (7.5.14) we apply Lemma 7.5.22 for every marginal $\pi_E \mu$ of μ; we get that
$$r_\sharp^H(\pi_E \mu, A) \leqslant \lfloor q_{-c}(\pi_E \mu, C_1 A) \rfloor.$$
In addition, we observe that

(7.5.16) $\quad r_\sharp^H(\mu, A) = n \inf_k \inf_{F \in G_{n,k}} \frac{r_\sharp(\pi_F \mu, A)}{k}$

$$\leqslant n \inf_{s \leqslant \dim E} \inf_{F \in G_{E,s}} \frac{r_\sharp(\pi_F \mu, A)}{s} = \frac{n}{\dim E} r_\sharp^H(\pi_E \mu, A),$$

which means that for every integer k, for every subspace $E \in G_{n,k}$,
$$r_\sharp^H(\mu, A) \leqslant \frac{n}{k} r_\sharp^H(\pi_E \mu, A) \leqslant \frac{n}{k} \lfloor q_{-c}(\pi_E \mu, C_1 A) \rfloor,$$
or equivalently that $r_\sharp^H(\mu, A) \leqslant q_{-c}^H(\mu, C_1 A)$.

For the other inequality of (7.5.14) we recall the formula for the negative Euclidean moments which Proposition 5.3.14 gives us: if k is an integer such that
$$I_{-k}(\mu) \simeq \sqrt{n} \left(\int_{G_{n,k}} f_{\pi_E \mu}(0) \, d\nu_{n,k}(E) \right)^{-1/k} \geqslant \frac{1}{C_1 A} I_2(\mu) = \frac{1}{C_1 A} \sqrt{n},$$
namely if $k \leqslant \lfloor q_{-c}(\mu, C_1 A) \rfloor$, then there must exist at least one $E \in G_{n,k}$ such that $f_{\pi_E \mu}(0) \leqslant (C_1' A)^k$ for some absolute constant C_1' (depending only on C_1). Since $\pi_E \mu$ is isotropic, we have
$$L_{\pi_E \mu} = \|f_{\pi_E \mu}\|_\infty^{1/k} \leqslant e(f_{\pi_E \mu}(0))^{1/k} \leqslant C_2 A.$$
This means that
$$r_\sharp(\mu, C_2 A) \geqslant \lfloor q_{-c}(\mu, C_1 A) \rfloor,$$

and the same will hold for every marginal $\pi_F\mu$ of μ. The inequality now follows from the definitions of $r_\sharp^H(\mu, C_2 A)$ and $q_{-c}^H(\mu, C_1 A)$. □

7.6. Notes and references

Klartag's first approach to the isomorphic slicing problem

This section gives an account of Klartag's first approach to the isomorphic slicing problem in [**272**]. We describe the proof of his main result (Theorem 7.1.1) stating that if K is a symmetric convex body in $\mathbb{R}^n$ then there exists a symmetric convex body T in $\mathbb{R}^n$ such that $d(K,T) \leqslant c\log n$ and $L_T \leqslant C$ where $c, C > 0$ are absolute constants. Proposition 2.5.10 and a second proof of the reduction of the slicing problem to bodies with bounded volume ratio (Theorem 6.2.1) appear in the same article.

Logarithmic Laplace transform and convex perturbations

The main properties of the logarithmic Laplace transform that we discuss in this section are taken from [**273**].

Klartag's solution to the isomorphic slicing problem

Theorem 7.3.1 is due to Klartag [**273**] and provides a solution to the isomorphic slicing problem. Combined with Paouris' deviation inequality it leads to an improvement of Bourgain's upper bound for the isotropic constant by a logarithmic term: in the same article, Klartag showed that $L_K \leqslant Cn^{1/4}$ for every convex body in $\mathbb{R}^n$.

Isotropic position and the reverse Santaló inequality

The observation that one can use Klartag's approach in order to give a purely convex geometric proof of the reverse Santaló inequality is due to Giannopoulos, Paouris and Vritsiou [**214**].

Volume radius of $Z_p(\mu)$

The approach of this section and Theorem 7.5.15 are due by Klartag and E. Milman [**284**]. They defined the "hereditary" variant

$$(7.6.1) \qquad q_*^H(\mu) := n \inf_k \inf_{E \in G_{n,k}} \frac{q_*(\pi_E \mu)}{k},$$

of $q_*(\mu)$ and then, for every $q \leqslant q_*^H(\mu)$, they showed that $|Z_q(\mu)|^{1/n} \geqslant c_3\sqrt{q/n}$ where $c_3 > 0$ is an absolute constant. An immediate consequence of this inequality and of the fact that $q_*^H(\mu) \geqslant c\sqrt{n}$ is an alternative proof of the bound $L_\mu = O(\sqrt[4]{n})$ for the isotropic constant. Vritsiou [**501**] introduced the parameter $r_\sharp(\mu, A)$ and modified the proof of Theorem 7.5.14 to show that the lower bound $|Z_p(\mu)|^{1/n} \geqslant cA^{-1}\sqrt{p/n}$ continues to hold for all $p \leqslant r_\sharp(\mu, A)$. Both $q_\sharp(\mu)$ and $q_*(\mu)$ are dominated by $r_\sharp(\mu, A_0)$ for some absolute constant $A_0 \geqslant 1$. Thus, the results of [**501**] extend Theorem 7.5.14.

CHAPTER 8

Tail estimates for linear functionals

Let μ be a centered log-concave probability measure on $\mathbb{R}^n$. We say that a direction $\theta \in S^{n-1}$ is *sub-Gaussian* for μ with constant $b > 0$ if

$$\|\langle \cdot, \theta \rangle\|_{\psi_2} \leqslant b \|\langle \cdot, \theta \rangle\|_2.$$

This is equivalent to saying that θ is a ψ_2-direction with constant b.

The following question was originally posed by V. Milman in the framework of convex bodies: *Is it true that there exists an absolute constant $C > 0$ such that every centered convex body K of volume 1 has at least one sub-Gaussian direction with constant C?*

Note that, if θ is a sub-Gaussian direction for K with constant b, then, for every $T \in SL(n)$, $(T^*)^{-1}(\theta)/\|(T^*)^{-1}(\theta)\|_2$ is a sub-Gaussian direction, with the same constant, for $T(K)$. Therefore, in order to study this question we may assume that K is isotropic.

An affirmative answer was given for some special classes of convex bodies. The first such result is due to Bobkov and Nazarov and concerns the unconditional case: If K is an isotropic unconditional convex body in $\mathbb{R}^n$ then

$$\|\langle \cdot, \theta \rangle\|_{\psi_2} \leqslant c\sqrt{n} \|\theta\|_\infty$$

for every $\theta \in S^{n-1}$, where $c > 0$ is an absolute constant. This result implies, for example, that the diagonal direction is sub-Gaussian. Paouris proved that bodies of "small diameter" have sub-Gaussian directions with uniformly bounded constant: If K is an isotropic convex body in $\mathbb{R}^n$ and $K \subseteq (\gamma\sqrt{n}L_K)B_2^n$ for some $\gamma > 0$, then

$$\sigma\big(\{\theta \in S^{n-1} : \|\langle \cdot, \theta\rangle\|_{\psi_2} \geqslant c_1 \gamma t L_K\}\big) \leqslant \exp(-c_2\sqrt{n}t^2/\gamma)$$

for every $t \geqslant 1$, where $c_1, c_2 > 0$ are absolute constants.

In the general case, Klartag was the first to prove the existence of "almost sub-Gaussian" directions. More precisely, he proved that, for every log-concave probability measure μ on $\mathbb{R}^n$, there exists $\theta \in S^{n-1}$ such that

$$\mu\big(\{x : |\langle x, \theta\rangle| \geqslant ct\|\langle \cdot, \theta\rangle\|_2\}\big) \leqslant e^{-\frac{t^2}{[\log(t+1)]^{2\alpha}}}$$

for all $1 \leqslant t \leqslant \sqrt{n}\log^\alpha n$, where $\alpha = 3$. In the first part of this chapter we describe the best known estimate for the general case; it was proved by Giannopoulos, Paouris and Valettas that one can always have $\alpha = 1/2$. We work in the more general setting of centered log-concave probability measures μ on $\mathbb{R}^n$. A natural approach to the problem is to define the symmetric convex set $\Psi_2(\mu)$ with support function

$$h_{\Psi_2(\mu)}(\theta) := \sup\left\{\frac{\|\langle \cdot, \theta\rangle\|_{L_q(\mu)}}{\sqrt{q}} : 2 \leqslant q \leqslant n\right\}$$

271

and to estimate its volume. Observe that $\Psi_2(\mu)$ contains the ellipsoid $\frac{1}{\sqrt{2}}Z_2(\mu)$, where
$$h_{Z_2(\mu)}(\theta) := \|\langle \cdot, \theta\rangle\|_{L_2(\mu)}.$$
We would like to prove that, for any centered log-concave probability measure μ, $\Psi_2(\mu)$ has bounded volume ratio, and, more precisely, that

(∗) $$\left(\frac{|\Psi_2(\mu)|}{|Z_2(\mu)|}\right)^{1/n} \leqslant C,$$

where $C > 0$ is an absolute constant. This would imply that for some $\theta \in S^{n-1}$ we have
$$h_{\Psi_2(\mu)}(\theta) \leqslant Ch_{Z_2(\mu)}(\theta) = C\|\langle \cdot, \theta\rangle\|_2.$$
From Proposition 3.2.10 we know that, in the case $d\mu(x) = \mathbf{1}_K(x)dx$ of a centered convex body, for every $\theta \in S^{n-1}$ we have
$$\|\langle \cdot, \theta\rangle\|_{\psi_2} \simeq \sup\left\{\frac{\|\langle \cdot, \theta\rangle\|_q}{\sqrt{q}} : 2 \leqslant q \leqslant n\right\} = h_{\Psi_2(\mu)}(\theta).$$

Thus, the volume estimate (∗) would imply an affirmative answer to the question. One can prove that this volume estimate is almost true (see Giannopoulos, Paouris and Valettas [**210**] and [**212**]): for every centered log-concave probability measure μ on $\mathbb{R}^n$, one has
$$c_1 \leqslant \left(\frac{|\Psi_2(\mu)|}{|Z_2(\mu)|}\right)^{1/n} \leqslant c_2\sqrt{\log n},$$
where $c_1, c_2 > 0$ are absolute constants. The proof of this inequality is given in Section 8.2. As described above, an immediate consequence is the existence of at least one sub-Gaussian direction for μ with constant $b = O(\sqrt{\log n})$. Our main tool will be estimates for the covering numbers $N(Z_q(\mu), sB_2^n)$ that are presented in Section 8.1.

A natural, and probably even more interesting, question that arises is to study the distribution of the ψ_2-norm $\|\langle \cdot, \theta\rangle\|_{\psi_2}$ with respect to the probability measure σ on the sphere. In Section 8.3, we introduce the function
$$\psi_K(t) := \sigma\left(\{\theta \in S^{n-1} : \|\langle \cdot, \theta\rangle\|_{\psi_2} \leqslant ct\sqrt{\log n}L_K\}\right), \qquad t \geqslant 1,$$
for any isotropic convex body K in $\mathbb{R}^n$. We present some rather weak lower bounds for $\psi_K(t)$. The problem remains open; as we will see in the next chapter, it would be very useful to know "when $\psi_K(t)$ becomes greater than $1/2$".

In Section 8.4 we discuss the work of Paouris on the existence of super-Gaussian directions. Given a centered log-concave probability measure on $\mathbb{R}^n$, one can ask for "dual tail estimates". The question can be made precise as follows: is it true that there exists $\theta \in S^{n-1}$ for which
$$\mu(\{x \in \mathbb{R}^n : |\langle x, \theta\rangle| \geqslant t\mathbb{E}|\langle \cdot, \theta\rangle|\}) \geqslant e^{-r^2t^2}$$
for all $1 \leqslant t \leqslant \sqrt{n}/r$ and for some absolute constant $r > 0$? The smallest $r > 0$ for which the previous distributional inequality holds true is called the *super-Gaussian constant* of μ in the direction of θ and is denoted by $sg_\mu(\theta)$. Note that if $T \in GL(n)$ then $sg_{T(\mu)}(\theta) = sg_\mu(T^*\theta/\|T^*\theta\|_2)$ for all $\theta \in S^{n-1}$. Therefore, as in the sub-Gaussian case, we can always assume that μ is isotropic. The main result states that, if a measure μ satisfies the hyperplane conjecture, then a random direction

$\theta \in S^{n-1}$ is super-Gaussian for μ. More precisely, for any isotropic log-concave measure μ on $\mathbb{R}^n$, one has

$$\sigma(\{\theta \in S^{n-1} : sg_\mu(\theta) \geqslant ctL_\mu\}) \leqslant t^{-n}$$

for all $t > 1$. In particular, we have $\mathbb{E}(sg_\mu) \leqslant CL_\mu$, where $C > 0$ is an absolute constant.

Section 8.5 describes the ψ_α-behavior of a random marginal $\pi_F(\mu)$ of an isotropic log-concave measure μ on $\mathbb{R}^n$. Finally, in the last section of this chapter we present the results of Bobkov and Nazarov on the existence of sub-Gaussian directions for unconditional convex bodies. For this class, more precise information on the distribution of the ψ_2-norm is obtained. We also present an example of an isotropic convex body which is c-close to a Euclidean ball and has directions with ψ_2-constant of the order $\sqrt[4]{n}$.

8.1. Covering numbers of the centroid bodies

Our aim is to prove the following.

THEOREM 8.1.1 (Giannopoulos-Paouris-Valettas). *Let μ be an isotropic log-concave measure on $\mathbb{R}^n$ and let $1 \leqslant q \leqslant n$ and $t \geqslant 1$. Then,*

$$\log N\left(Z_q(\mu), c_1 t\sqrt{q} B_2^n\right) \leqslant c_2 \frac{n}{t^2} + c_3 \frac{\sqrt{qn}}{t},$$

where $c_1, c_2, c_3 > 0$ are absolute constants.

We present two proofs of Theorem 8.1.1. The first one is from [**210**] and exploits the idea of Talagrand's proof for the dual Sudakov inequality and the notion of the p-median. The second one appears in [**212**] and combines Steiner's formula and Kubota's integral representation of the quermassintegrals with a lemma on the covering numbers of projections of convex bodies.

8.1.1. First proof

Taking into account the duality of entropy numbers theorem of Artstein, Milman and Szarek (Theorem 1.8.6), in order to estimate $N(Z_q(\mu), sB_2^n)$ it suffices to estimate covering numbers of the form $N(B_2^n, sZ_q^\circ(\mu))$. To this end, we introduce the p-median of a symmetric convex body C.

DEFINITION 8.1.2 (p-median). Let C be a symmetric convex body in $\mathbb{R}^n$. For every $p > 0$ we define $m_p(C)$ as the unique positive number m_p for which $\gamma_n(m_p C) = 2^{-p}$.

REMARKS 8.1.3. (i) The function $p \mapsto m_p(C)$ is decreasing and convex; this follows from the fact that γ_n is a log-concave measure. Moreover, if we use the B-theorem of Cordero-Erausquin, Fradelizi and Maurey (Theorem 5.3.4) then we can show that it is a log-convex function.
(ii) The value $m_1(C)$ is the standard Lévy mean of the norm $\|\cdot\|_C$ with respect to the Gaussian measure.
(iii) We set $\beta_q := I_q(\gamma_n, C) = (\int_{\mathbb{R}^n} \|x\|_C^q \, d\gamma_n(x))^{1/q}$ for all $-n < q \neq 0$. From Markov's inequality we see that, for all $0 < p < n$,

$$\gamma_n\left(\frac{\beta_{-p}}{2} C\right) = \gamma_n\left(\left\{x : \|x\|_C \leqslant \frac{\beta_{-p}}{2}\right\}\right) \leqslant 2^{-p} = \gamma_n(m_p C),$$

and hence
$$I_{-p}(\gamma_n, C) \leqslant 2m_p(C). \tag{8.1.1}$$

(iv) If we also assume that the negative moments of $\|\cdot\|_C$ with respect to γ_n satisfy certain reverse Hölder inequalities then the quantities $I_{-p}(\gamma_n, C)$ and $m_p(C)$ are equivalent. To make this more precise, assume that there exists $\alpha > 2$ such that
$$I_{-p}(\gamma_n, C) \leqslant \alpha I_{-2p}(\gamma_n, C) \tag{8.1.2}$$
for all $1 \leqslant p < n/2$. Set $\beta_q = I_q(\gamma_n, C)$ as in (iii). Using the Paley-Zygmund inequality we write
$$\gamma_n(\{x : \|x\|_C \leqslant 2\beta_{-p}\}) \geqslant (1 - 2^{-p})^2 \frac{\beta_{-p}^{-2p}}{\beta_{-2p}^{-2p}} \geqslant (1 - 2^{-p})^2 \alpha^{-2p}$$
$$\geqslant (2\alpha)^{-2p} \geqslant \alpha^{-4p} \geqslant 2^{-8p \log \alpha}.$$

By the definition of $m_q(C)$ we get
$$m_{8p \log \alpha}(C) \leqslant 2I_{-p}(\gamma_n, C). \tag{8.1.3}$$

(v) Finally, observe that $I_{-p}(\gamma_n, C)$ can be expressed in terms of the negative moments of $\|\cdot\|_C$ with respect to σ. For every $1 \leqslant p \leqslant n-1$ we have
$$I_{-p}(\gamma_n, C) = c_{n,p} M_{-p}(C),$$
where $M_{-p}(C) = (\int_{S^{n-1}} \|\theta\|_C^{-p} d\sigma(\theta))^{-1/p}$ and $c_{n,p} \simeq \sqrt{n}$. Therefore, reverse Hölder inequalities for $I_{-p}(\gamma_n, C)$ correspond to reverse Hölder inequalities for $M_{-p}(C)$. The proof is direct, by integration in polar coordinates.

The next lemma is a variant of Talagrand's argument for the proof of the dual Sudakov inequality (Theorem 1.8.2).

LEMMA 8.1.4. *Let C be a symmetric convex body in $\mathbb{R}^n$. For every $p > 0$,*
$$\log N\left(B_2^n, \frac{m_p(C)}{\sqrt{p}} C\right) \leqslant 3p.$$

Proof. Let $t > 0$ (which will be suitably chosen). We consider a subset $\{z_1, \ldots, z_N\}$ of B_2^n which is maximal with respect to the condition "$i \neq j \Rightarrow \|z_i - z_j\|_C \geqslant t$". Then, $B_2^n \subseteq \bigcup_{j \leqslant N}(z_j + tC)$ and hence $N(B_2^n, tC) \leqslant N$. Moreover, the sets $z_j + \frac{t}{2}C$ have disjoint interiors and, for every $\lambda > 0$, the same is true for the sets $\lambda z_j + \frac{\lambda t}{2}C$. So, we get
$$1 \geqslant \gamma_n\left(\bigcup_{j=1}^N \left(\lambda z_j + \frac{\lambda t}{2}C\right)\right) = \sum_{j=1}^N \gamma_n\left(\lambda z_j + \frac{\lambda t}{2}C\right)$$
$$\geqslant \sum_{j=1}^N e^{-\lambda^2 \|z_j\|_2^2/2} \gamma_n\left(\frac{\lambda t}{2}C\right) \geqslant \sum_{j=1}^N e^{-\lambda^2/2} \gamma_n\left(\frac{\lambda t}{2}C\right)$$
$$= N e^{-\lambda^2/2} \gamma_n\left(\frac{\lambda t}{2}C\right),$$
where we have used the fact that $\gamma_n(x + A) \geqslant e^{-\|x\|_2^2/2} \gamma_n(A)$ for every symmetric convex set A and every x in $\mathbb{R}^n$, and the fact that $\|z_j\|_2 \leqslant 1$ for all j. This shows

that $N \leqslant e^{\lambda^2/2}(\gamma_n(\frac{\lambda t}{2}C))^{-1}$. Choosing $\lambda = 2m_p(C)/t$, we get
$$N \leqslant \exp\left(p + \frac{2m_p(C)^2}{t^2}\right),$$
and choosing $t = m_p(C)/\sqrt{p}$ we conclude the proof. $\square$

A main point in the argument leading to Theorem 8.1.1 is that we first derive this estimate for bodies with bounded isotropic constant. This is enough because we can show that convolution of log-concave probability measures interacts well with their L_q-centroid bodies; then, convolving with Gaussian measure provides a way to reduce many questions to the case of measures with bounded isotropic constant. We first recall some basic facts about convolution of measures.

DEFINITION 8.1.5 (convolution). Let $f, g : \mathbb{R}^n \to \mathbb{R}$ be two integrable functions. We define the *convolution* of f with g as the function $f * g : \mathbb{R}^n \to \mathbb{R}$ defined by
$$(f * g)(x) = \int_{\mathbb{R}^n} f(x - y)g(y)\,dy.$$

We can easily check that $f * g = g * f$. If f, g are densities then so is $f * g$, and if $f, g : \mathbb{R}^n \to \mathbb{R}^+$ are log-concave then $f * g$ is also log-concave; this is a direct consequence of the Prékopa-Leindler inequality. If f, g are centered (respectively, even), then $f * g$ is centered (respectively, even). Furthermore, if μ, ν are Borel probability measures on $\mathbb{R}^n$ which have densities f_μ, f_ν with respect to the Lebesgue measure on $\mathbb{R}^n$, we denote by $\mu * \nu$ the Borel probability measure with density $f_\mu * f_\nu$. Namely, for any Borel set A in $\mathbb{R}^n$ we have:
$$(\mu * \nu)(A) := \int_A (f_\mu * f_\nu)(x)\,dx = \int_{\mathbb{R}^n}\int_{\mathbb{R}^n} \mathbf{1}_A(y+z)\,f_\mu(y)f_\nu(z)\,dy\,dz.$$
One can then easily check that
$$\int_{\mathbb{R}^n} h(x)\,d(\mu * \nu)(x) = \int_{\mathbb{R}^n}\int_{\mathbb{R}^n} h(x+y)\,d\mu(x)\,d\nu(y)$$
for any Borel measurable function $h : \mathbb{R}^n \to \mathbb{R}^+$.

The next simple lemma describes the L_q-centroid body of the convolution of two log-concave probability measures.

LEMMA 8.1.6. *Let μ, ν be two log-concave probability measures on $\mathbb{R}^n$. Assume that at least one of them is symmetric. Then, for every $q \geqslant 1$ we have*
$$Z_q(\mu * \nu) \simeq Z_q(\mu) + Z_q(\nu).$$

Proof. It is enough to consider only even integer values of q. We shall prove the following: for every $k \in \mathbb{N}$,
$$\frac{1}{2}\big(Z_{2k}(\mu) + Z_{2k}(\nu)\big) \subseteq Z_{2k}(\mu * \nu) \subseteq Z_{2k}(\mu) + Z_{2k}(\nu).$$
For the left hand side inequality, given $\theta \in S^{n-1}$ we write
$$\begin{aligned}h^{2k}_{Z_{2k}(\mu*\nu)}(\theta) &= \int_{\mathbb{R}^n} \langle x, \theta\rangle^{2k}\,d(\mu*\nu)(x) \\ &= \int_{\mathbb{R}^n}\int_{\mathbb{R}^n} (\langle y,\theta\rangle + \langle z,\theta\rangle)^{2k}\,d\mu(z)\,d\nu(y) \\ &= \sum_{s=0}^{2k}\binom{2k}{s}\left(\int_{\mathbb{R}^n}\langle z,\theta\rangle^s d\mu(z)\right)\left(\int_{\mathbb{R}^n}\langle y,\theta\rangle^{2k-s}\,d\nu(y)\right).\end{aligned}$$

Since at least one of μ and ν is symmetric, for all odd s we have
$$\left(\int_{\mathbb{R}^n} \langle z, \theta \rangle^s \, d\mu(z)\right) \left(\int_{\mathbb{R}^n} \langle y, \theta \rangle^{2k-s} \, d\nu(y)\right) = 0,$$
and hence, all the terms in the above sum are non-negative. It follows that, for every $\theta \in S^{n-1}$,
$$h_{Z_{2k}(\mu*\nu)}^{2k}(\theta) \geqslant \left(\int_{\mathbb{R}^n} \langle z, \theta \rangle^{2k} \, d\mu(z) + \int_{\mathbb{R}^n} \langle y, \theta \rangle^{2k} d\nu(y)\right),$$
which shows that
$$h_{Z_{2k}(\mu*\nu)}(\theta) \geqslant \left(h_{Z_{2k}(\mu)}^{2k}(\theta) + h_{Z_{2k}(\nu)}^{2k}(\theta)\right)^{1/2k} \geqslant \frac{1}{2}\left(h_{Z_{2k}(\mu)}(\theta) + h_{Z_{2k}(\nu)}(\theta)\right).$$
For the right hand side inequality, observe that for all $0 \leqslant s \leqslant 2k$ we have
$$\int_{\mathbb{R}^n} |\langle y, \theta \rangle|^s \, d\mu(y) \leqslant \left(\int_{\mathbb{R}^n} |\langle y, \theta \rangle|^{2k} \, d\mu(y)\right)^{\frac{s}{2k}} = h_{Z_{2k}(\mu)}^s(\theta)$$
and similarly,
$$\int_{\mathbb{R}^n} |\langle z, \theta \rangle|^{2k-s} \, d\nu(z) \leqslant \left(\int_{\mathbb{R}^n} |\langle z, \theta \rangle|^{2k} \, d\nu(z)\right)^{\frac{2k-s}{2k}} = h_{Z_{2k}(\nu)}^{2k-s}(\theta),$$
which implies that
$$h_{Z_{2k}(\mu*\nu)}^{2k} \leqslant \sum_{s=0}^{2k} \binom{2k}{s} h_{Z_{2k}(\nu)}^s h_{Z_{2k}(\mu)}^{2k-s} = \left(h_{Z_{2k}(\mu)} + h_{Z_{2k}(\nu)}\right)^{2k}.$$
So, $Z_{2k}(\mu * \nu) \subseteq Z_{2k}(\mu) + Z_{2k}(\nu)$. $\square$

We consider the convolution of an isotropic log-concave measure with the standard Gaussian measure. Using Lemma 8.1.6 and Theorem 5.1.7 we have:

PROPOSITION 8.1.7. *Let μ be an isotropic log-concave measure on $\mathbb{R}^n$. Then, there exists a convex body $T = T_\mu$ in $\mathbb{R}^n$ with the following properties:*
(i) *T is almost isotropic with constant $c_1 > 0$ and $L_T \leqslant c_2$.*
(ii) *For every $q \geqslant 1$ we have*
$$c_3 Z_q(T) \subseteq Z_q(\mu) + \sqrt{q} B_2^n \subseteq c_4 Z_q(T),$$
where c_1, c_2, c_3, c_4 are absolute positive constants.

Proof. Consider the body $T := \overline{K_{n+1}}(\mu * \gamma_n)$ where γ_n is the standard n-dimensional Gaussian measure. Then, we can easily check that $Z_q(\gamma_n) = \alpha_q B_2^n$ with $\alpha_q \simeq \sqrt{q}$. We also have
$$f_{\mu*\gamma_n}(0) = \int_{\mathbb{R}^n} f_\mu(x) \, d\gamma_n(x) \leqslant \frac{1}{(2\pi)^{n/2}} \int_{\mathbb{R}^n} f_\mu(x) \, dx = (2\pi)^{-n/2}.$$
On the other hand, since
$$\int_{\mathbb{R}^n} \langle x, \theta \rangle^2 d(\mu * \gamma_n)(x) = \int_{\mathbb{R}^n} \langle x, \theta \rangle^2 \, d\mu(x) + \int_{\mathbb{R}^n} \langle x, \theta \rangle^2 \, d\gamma_n(x) = 2,$$
if we revisit the argument from Proposition 2.3.12 we see that
$$2n = \int \|x\|_2^2 f_{\mu*\gamma_n}(x) \, dx \geqslant (\omega_n \|f_{\mu*\gamma_n}\|_\infty)^{-2/n} \frac{n}{n+2}.$$

This shows that
$$[f_{\mu*\gamma_n}(0)]^{1/n} \simeq \|f_{\mu*\gamma_n}\|_\infty^{1/n} \simeq 1.$$
Using Theorem 5.1.7 we get
$$Z_q(T) \simeq [f_{\mu*\gamma_n}(0)]^{1/n} Z_q(\mu*\gamma_n) \simeq Z_q(\mu) + \sqrt{q} B_2^n,$$
where we have also used Lemma 8.1.6 and the fact that $Z_q(\gamma_n) \simeq \sqrt{q} B_2^n$. This already proves the second assertion. We also observe that $Z_2(T) \simeq B_2^n$ from which it easily follows (recall Lemma 3.3.4) that T is almost isotropic with an absolute constant $c_1 > 0$. Finally, note that
$$L_T \simeq L_{\mu*\gamma_n} = \|f_{\mu*\gamma_n}\|_\infty^{1/n} [\det \text{Cov}(\mu*\gamma_n)]^{\frac{1}{2n}} \leqslant c_2,$$
because $\text{Cov}(\mu*\gamma_n) = \text{Cov}(\mu) + \text{Cov}(\gamma_n) = 2I$. This proves the proposition. □

COROLLARY 8.1.8. *Let K be an isotropic convex body in $\mathbb{R}^n$. There exists a convex body K_1 with the following properties:*
(i) *K_1 is almost isotropic with constant $c_1 > 0$ and $L_{K_1} \leqslant c_2$.*
(ii) *For every $q \geqslant 1$ we have*
$$c_3 Z_q(K_1) \subseteq \frac{Z_q(K)}{L_K} + \sqrt{q} B_2^n \subseteq c_4 Z_q(K_1).$$

Proof. We apply the previous proposition for the measure μ with density $f_\mu = L_K^n \mathbf{1}_{\frac{K}{L_K}}$. □

Besides Lemma 8.1.4, for the proof of Theorem 8.1.1 we will use two basic facts from Chapter 5:

FACT 8.1.9. The inequality of Lutwak, Yang and Zhang (Proposition 5.1.16) provides a lower bound for the volume of the L_q-centroid body of a centered convex body K of volume 1 in $\mathbb{R}^n$. For every $1 \leqslant q \leqslant n$ we have
$$|Z_q(K)|^{1/n} \geqslant c_1 \sqrt{q/n},$$
where $c_1 > 0$ is an absolute constant.

FACT 8.1.10. From Theorem 5.3.16 (in particular, from (5.3.11)) we know that for any centered convex body K of volume 1 in $\mathbb{R}^n$ and every $1 \leqslant q \leqslant n-1$ we have
$$I_{-q}(K) \simeq \sqrt{n/q}\, w_{-q}(Z_q(K)).$$

Proof of Theorem 8.1.1. We first consider the case where μ is the uniform measure on an isotropic convex body K with bounded isotropic constant: we assume that $L_K \simeq 1$. We also assume that $1 \leqslant t \leqslant \sqrt{n/q}$.

We will use the fact that the r-means of the body $Z_q^\circ(K)$ stay constant and of the order of $\sqrt{qn}$ for $q \leqslant r \leqslant n$. To see this, we first show that $M_{-r}(Z_q^\circ(K)) = w_{-r}(Z_q(K))$ satisfy reverse Hölder inequalities: for every $1 \leqslant q \leqslant n$ we have
$$w_{-n}(Z_q(K)) = \left(\frac{|Z_q^\circ(K)|}{|B_2^n|}\right)^{-1/n} \geqslant \left(\frac{|Z_q(K)|}{|B_2^n|}\right)^{1/n} \geqslant c_2 \sqrt{q}$$
by the Blaschke-Santaló inequality and Fact 8.1.9. Moreover, if $1 \leqslant q \leqslant r \leqslant n$ it follows from Fact 8.1.10 and Hölder's inequality that
$$w_{-r}(Z_q(K)) \leqslant w_{-q}(Z_q(K)) \leqslant c_3 \sqrt{q/n} I_{-q}(K) \leqslant c_3 \sqrt{q/n}\, I_2(K)$$
$$= c_3 \sqrt{q} L_K,$$

and since L_K has been assumed bounded we get $w_{-r}(Z_q(K)) \leqslant c_4\sqrt{q}$. Combining the above we see that $w_{-r}(Z_q(K)) \simeq \sqrt{q}$ for all $q \leqslant r \leqslant n$. On the other hand, from Remark 8.1.3 (v) we have

$$I_{-r}(\gamma_n, Z_q^\circ(K)) \simeq \sqrt{n}w_{-r}(Z_q(K)).$$

In particular, for all $c_0 q \leqslant r \leqslant n$, where $c_0 = 8\log(\alpha_0)$ and α_0 is a constant with which the inverse inequalities for $I_{-s}(\gamma_n, Z_q^\circ(K))$ (needed to apply 8.1.3 (iv)) hold, we see that

$$m_r(Z_q^\circ(K)) \simeq I_{-r}(\gamma_n, Z_q^\circ(K)) \simeq \sqrt{n}w_{-r}(Z_q(K)) \simeq \sqrt{qn}.$$

From Lemma 8.1.4 it follows that

$$\log N\left(B_2^n, \frac{m_r(Z_q^\circ(K))}{\sqrt{r}} Z_q^\circ(K)\right) \leqslant c_2 r,$$

and hence

$$\log N\left(B_2^n, c_1\sqrt{n/r}\sqrt{q}Z_q^\circ(K)\right) \leqslant c_2 r.$$

Setting $t = \sqrt{n/r}$ we get for all $1 \leqslant t \leqslant \sqrt{n/q}$ that

$$\log N(B_2^n, c_1 t\sqrt{q}Z_q^\circ(K)) \leqslant c_2 \frac{n}{t^2}.$$

Using the duality of entropy numbers theorem we conclude that

$$\log N(Z_q(K), c_1' t\sqrt{q}B_2^n) \leqslant c_2' \frac{n}{t^2},$$

for all $1 \leqslant q \leqslant n$ and $1 \leqslant t \leqslant \sqrt{n/q}$, provided that K has bounded isotropic constant.

For the general case we use Corollary 8.1.8: There exists an almost isotropic body K_1 in $\mathbb{R}^n$ with absolute constant $c_1 > 0$ and $L_{K_1} \leqslant c_2$ such that

$$\frac{Z_q(K)}{L_K} + \sqrt{q}B_2^n \subseteq c_2 Z_q(K_1),$$

for $1 \leqslant q \leqslant n$. Therefore, we can write

$$\log N\left(\frac{Z_q(K)}{L_K}, t\sqrt{q}B_2^n\right) \leqslant \log N\left(\frac{Z_q(K)}{L_K} + \sqrt{q}B_2^n, t\sqrt{q}B_2^n\right)$$
$$\leqslant \log N(Z_q(K_1), c_3 t\sqrt{q}B_2^n) \leqslant c_4 \frac{n}{t^2},$$

applying the first part of the proof. This proves the theorem for $1 \leqslant t \leqslant \sqrt{n/q}$.

Next, let $t > \sqrt{n/q}$. We set $p = \sqrt{qn}/t$. Then, $1 \leqslant p < q$ and using Proposition 5.1.2 we can write

$$\log N(Z_q(K), t\sqrt{q}L_K B_2^n) \leqslant \log N\left(c\frac{q}{p} Z_p(K), t\sqrt{q}L_K B_2^n\right)$$
$$\leqslant \log N\left(Z_p(K), c't\sqrt{\frac{p}{q}}\sqrt{p}L_K B_2^n\right)$$
$$= \log N(Z_p(K), c's\sqrt{p}L_K B_2^n),$$

for $s := t\sqrt{p/q} = \sqrt{n/p}$. Then, the previous case shows that

$$\log N(Z_p(K), c's\sqrt{p}L_K B_2^n) \leqslant c_2 n/s^2 = c_2 p = c_2 \frac{\sqrt{qn}}{t},$$

and this completes the proof in the case of convex bodies.

Finally, using Theorem 5.1.7 we can extend the result to the setting of log-concave measures. □

The regularity of the covering numbers of $Z_q(\mu)$ can be also expressed as follows:

COROLLARY 8.1.11. *Let μ be an isotropic log-concave measure on $\mathbb{R}^n$ and let $1 \leqslant q \leqslant n$. Define $\beta \geqslant 1$ by the equation $q = n^{1/\beta}$ and set $\alpha := \min\{\beta, 2\}$. Then,*

$$N\left(Z_q(\mu), c_1 t \sqrt{q} B_2^n\right) \leqslant \exp\left(c_2 \frac{n}{t^\alpha}\right),$$

where $c_1, c_2 > 0$ are absolute constants.

Proof. Assume first that $\beta \geqslant 2$. Then, $q \leqslant \sqrt{n}$ and the results of Chapter 5 show that $w(Z_q(\mu)) \simeq \sqrt{q}$. In this case, our claim follows from Sudakov's inequality.

On the other hand, if $\beta \in [1, 2]$ then using the fact that $q^\beta = n$ we observe that for all $1 \leqslant t \leqslant \sqrt{q}$,

$$\frac{\sqrt{qn}}{t} \leqslant \frac{n}{t^\beta} = \frac{n}{t^\alpha}$$

and the result follows from Theorem 8.1.1. □

8.1.2. Second proof

Our second proof of Theorem 8.1.1 is from [**212**]; it is done in two steps. For the first step we recall Steiner's formula: if C is a convex body in $\mathbb{R}^n$ then

(8.1.4) $$|C + tB_2^n| = \sum_{k=0}^{n} \binom{n}{k} W_k(C) t^k,$$

for every $t > 0$, where $W_k(C)$ is the k-th quermassintegral of C. Recall also that, by Kubota's integral formula, $W_{n-k}(C)$ can be expressed in the form

$$W_{n-k}(C) = \frac{\omega_n}{\omega_k} \int_{G_{n,k}} |P_F(C)| \, d\nu_{n,k}(F).$$

The main idea is the following: using the elementary inequality

(8.1.5) $$|tB_2^n| N(C, 2tB_2^n) \leqslant |C + tB_2^n| \qquad (t > 0)$$

in order to estimate the covering number of $Z_q(\mu)$ by tB_2^n, it is enough to estimate the volume of the Minkowski sum $Z_q(\mu) + tB_2^n$. Because of Kubota's formula it is enough to estimate the volume of projections of $Z_q(\mu)$.

DEFINITION 8.1.12. Let μ be an isotropic log-concave measure on $\mathbb{R}^n$. For every $q \geqslant 1$ we define the *normalized L_q-centroid body* of μ by

$$K_q(\mu) := \frac{Z_q(\mu)}{\sqrt{q}}.$$

With this definition we have the following.

LEMMA 8.1.13. *Let μ be an isotropic log-concave probability measure on $\mathbb{R}^n$. Then, for every $1 \leqslant k, q \leqslant n$ and every $F \in G_{n,k}$ we have*

$$|P_F(K_q)|^{1/k} \leqslant c \max\{1, \sqrt{q/k}\} |B_2^k|^{1/k},$$

where $c > 0$ is an absolute constant.

Proof. We distinguish two cases:

(i) Assume that $1 \leqslant q \leqslant k$. Then,
$$|P_F(K_q)|^{1/k} = \frac{1}{\sqrt{q}}|P_F(Z_q)|^{1/k} = \frac{1}{\sqrt{q}}|Z_q(\pi_F(\mu))|^{1/k}.$$

Since $\pi_F(\mu)$ is an isotropic log-concave measure on the k-dimensional space F, from Theorem 5.1.17 we have
$$|Z_q(\pi_F(\mu))|^{1/k} \leqslant c_1\sqrt{q/k}$$
for all $1 \leqslant q \leqslant k$. This shows that in the first case we have
$$|P_F(K_q)|^{1/k} \leqslant c_2/\sqrt{k}.$$

(ii) Assume that $1 \leqslant k < q \leqslant n$. Then, we have $Z_q(\pi_F(\mu)) \subseteq c_3 \frac{q}{k} Z_k(\pi_F(\mu))$. Therefore,
$$|P_F(K_q)|^{1/k} \leqslant c_3 \frac{q}{k} \frac{1}{\sqrt{q}} |Z_k(\pi_F(\mu))|^{1/k} \leqslant c_4 \frac{\sqrt{q}}{k}.$$

These two cases, combined with the fact that $|B_2^k|^{1/k} \simeq \frac{1}{\sqrt{k}}$, complete the proof. $\square$

We are now able to prove a weak version of Theorem 8.1.1. This estimate was obtained by Giannopoulos, Pajor and Paouris in [**209**].

PROPOSITION 8.1.14. *Let μ be an isotropic log-concave measure on $\mathbb{R}^n$. Then, for every $1 \leqslant q \leqslant n$ and every $t > 0$ one has*
$$N(K_q(\mu), 2tB_2^n) \leqslant 2\exp\left(c_1\frac{n}{t} + c_2\frac{\sqrt{qn}}{\sqrt{t}}\right),$$
where $c_1, c_2 > 0$ are absolute constants.

Proof. Combining Kubota's formula with Lemma 8.1.13 we see that
$$W_{n-k}(K_q) \leqslant \omega_n \left(c\max\left\{1, \sqrt{q/k}\right\}\right)^k$$
for all $1 \leqslant k \leqslant n-1$. For any $t > 0$ we write
$$|K_q + tB_2^n| \leqslant \omega_n \sum_{k=0}^n \binom{n}{k} \left(c\max\{1, \sqrt{q/k}\}\right)^k t^{n-k}.$$
Taking into account (8.1.5) we get
$$N(K_q, 2tB_2^n) \leqslant \sum_{k \leqslant q} \binom{n}{k}\left(\frac{c\sqrt{q}}{t\sqrt{k}}\right)^k + \sum_{k > q} \binom{n}{k}\left(\frac{c}{t}\right)^k$$
$$\leqslant \sum_{k \leqslant q}\left(\frac{c_1 n\sqrt{q}}{k^{3/2}t}\right)^k + \sum_{k > q}\left(\frac{c_1 n}{kt}\right)^k,$$
where we have also used the inequality $\binom{n}{k} \leqslant (en/k)^k$. Observe that for every $1 \leqslant k \leqslant q$ we have
$$\left(\frac{c_1 n \sqrt{q}}{k^{3/2}t}\right)^k \leqslant \left(\frac{c_1 nq}{tk^2}\right)^k \leqslant \frac{(c_2\sqrt{qn}/\sqrt{t})^{2k}}{(2k)!}$$

while for $q \leq k \leq n$ we have
$$\left(\frac{c_1 n}{kt}\right)^k \leq \frac{(c_2 n/t)^k}{k!}.$$

We use these estimates, Taylor's expansion of the exponential function and the elementary inequality $u + v \leq 2uv$ for $u, v \geq 1$, to get
$$N(K_q, 2tB_2^n) \leq 2 \exp\left(C\frac{\sqrt{qn}}{\sqrt{t}} + C\frac{n}{t}\right),$$

for some absolute constant $C > 0$. □

For the second step of the proof we will need a small ball probability estimate from [**389**, Fact 3.2(c)]:

LEMMA 8.1.15. *Let $\theta \in S^{n-1}$, $1 \leq k \leq n-1$ and $r \geq e$. Then,*
$$\nu_{n,k}\left(\left\{F \in G_{n,k} : \|P_F(\theta)\|_2 \leq \frac{1}{r}\sqrt{\frac{k}{n}}\right\}\right) \leq \left(\frac{\sqrt{e}}{r}\right)^k.$$

Proof. Because of the invariance of the Haar measure we can write:
$$\nu_{n,k}(\{F \in G_{n,k} : \|P_F\theta\|_2 \leq \varepsilon\sqrt{k/n}\}) = \mu(\{U \in O(n) : \|P_k U\theta\|_2 \leq \varepsilon\sqrt{k/n}\})$$
$$= \sigma(\{u \in S^{n-1} : \|P_k(u)\|_2 \leq \varepsilon\sqrt{k/n}\}),$$

where P_k is the projection onto $\mathrm{span}\{e_i : 1 \leq i \leq k\}$. For any $A \subseteq S^{n-1}$ let $C(A)$ be the cone generated by A, i.e. $C(A) = \bigcup\{tu : u \in A, t \geq 0\}$. Note that the measure
$$\mu(A) := \gamma_n(C(A))$$
for $A \subseteq S^{n-1}$, is invariant under orthogonal transformations, hence by the uniqueness of the Haar measure we have $\mu \equiv \sigma$. Moreover, if
$$A_\varepsilon := \{u \in S^{n-1} : \|P_k(u)\|_2 \leq \varepsilon\sqrt{k/n}\},$$
then $\sigma(A_\varepsilon) = \gamma_n(C(A_\varepsilon))$ where
$$C(A_\varepsilon) = \{x \in \mathbb{R}^n : \|P_k(x)\|_2 \leq \varepsilon\sqrt{k/n}\|x\|_2\}\}.$$

Note that
$$C(A_\varepsilon) = \left\{x : (1-\gamma^2)\sum_{i \leq k} x_i^2 < \gamma^2 \sum_{i > k} x_i^2\right\},$$

where $\gamma = \varepsilon\sqrt{k/n}$. Therefore, integrating in polar coordinates we get:
$$\gamma_n(C(A_\varepsilon)) = (2\pi)^{-n/2}\int_{\{x : \gamma^2 \sum_{i>k} x_i^2 > (1-\gamma^2)\sum_{i\leq k} x_i^2\}} e^{-\|x\|_2^2/2}\,dx$$
$$= \frac{k(n-k)\omega_k\omega_{n-k}}{(2\pi)^{n/2}}\int_{r^2\gamma^2 > (1-\gamma^2)\rho^2} r^{n-k-1}\rho^{k-1}e^{-r^2/2}e^{-\rho^2/2}\,dr\,d\rho.$$

Applying the change of variables $r = \sqrt{2t(1-s)}$ and $\rho = \sqrt{2ts}$ we find

$$\gamma_n(C(A_\varepsilon)) = \frac{k(n-k)}{4\Gamma(\frac{k}{2}+1)\Gamma(\frac{n-k}{2}+1)} \int_0^\infty \int_0^{\gamma^2} t^{n/2-1} s^{\frac{k}{2}-1}(1-s)^{\frac{n-k}{2}-1} e^{-t}\, ds\, dt$$

$$= \frac{k(n-k)\Gamma(\frac{n}{2})}{4\Gamma(\frac{k}{2}+1)\Gamma(\frac{n-k}{2}+1)} \int_0^{\gamma^2} (1-s)^{\frac{n-k}{2}-1} s^{\frac{k}{2}-1}\, ds$$

$$= \frac{1}{B(\frac{k}{2}, \frac{n-k}{2})} \int_0^{\gamma^2} (1-s)^{\frac{n-k}{2}-1} s^{\frac{k}{2}-1}\, ds,$$

where B is the Beta function:

$$B(a,b) = \frac{\Gamma(a)\Gamma(b)}{\Gamma(a+b)} = \int_0^1 s^{a-1}(1-s)^{b-1}\, ds,$$

for $a, b > 0$. For any $\delta \in (0,1)$ and any $b \geqslant 1$ we have

$$B(a,b) \geqslant \int_0^\delta s^{a-1}(1-s)^{b-1}\, ds \geqslant (1-\delta)^{b-1} \int_0^\delta s^{a-1}\, ds = (1-\delta)^{b-1} \frac{\delta^a}{a}.$$

Applying this, for $a = \frac{k}{2}$ and $b = \frac{n-k}{2}$ with $1 \leqslant k \leqslant n-2$, we get:

$$\gamma_n(C(A_\varepsilon)) \leqslant \frac{k}{2\delta^{k/2}(1-\delta)^{(n-k)/2-1}} \int_0^{\gamma^2} (1-s)^{\frac{n-k}{2}-1} s^{\frac{k}{2}-1}\, ds$$

$$\leqslant \frac{\gamma^k}{\delta^{k/2}(1-\delta)^{(n-k)/2-1}}.$$

Choosing $\delta = k/n$ and taking into account the equation $\gamma = \varepsilon\sqrt{k/n}$ we obtain:

$$\gamma_n(C(A_\varepsilon)) \leqslant \frac{\varepsilon^k}{\left(\frac{n-k}{n}\right)^{\frac{n-k}{2}-1}},$$

from which the result follows. For $k = n-1$ we check that the small ball probability estimate is valid as well. □

Under the restriction $\log N(C, tB_2^n) \leqslant k$, Lemma 8.1.15 allows us to compare the covering numbers $N(C, tB_2^n)$ of a convex body C with the covering numbers of its random k-dimensional projections.

LEMMA 8.1.16. *Let C be a convex body in $\mathbb{R}^n$, let $r \geqslant e$, $s > 0$ and $1 \leqslant k \leqslant n-1$. If $N_s := N(C, sB_2^n)$, then there exists $\mathcal{F} \subseteq G_{n,k}$ such that $\nu_{n,k}(\mathcal{F}) \geqslant 1 - N_s^2 e^{k/2} r^{-k}$ and*

$$N\left(P_F(C), \frac{s}{2r}\sqrt{\frac{k}{n}} B_F\right) \geqslant N_s$$

for every $F \in \mathcal{F}$.

Proof. Let $N_s = N(C, sB_2^n)$. There exist $z_1, \ldots, z_{N_s} \in C$ such that $\|z_i - z_j\|_2 \geqslant s$ for all $1 \leqslant i \neq j \leqslant N_s$. Consider the set $\{w_m : 1 \leqslant m \leqslant n_s\}$ of all differences $w_m = z_i - z_j$ $(i \neq j)$. Note that $n_s \leqslant N_s^2$ and that $\|w_m\|_2 \geqslant s$ for every m. Lemma 8.1.15 shows that

$$\nu_{n,k}\left(\left\{F \in G_{n,k} : \|P_F(w_m)\|_2 \leqslant \frac{1}{r}\sqrt{\frac{k}{n}}\|w_m\|_2\right\}\right) \leqslant \left(\frac{\sqrt{e}}{r}\right)^k,$$

and hence
$$\nu_{n,k}\left(\left\{F : \|P_F(w_m)\|_2 \geq \frac{1}{r}\sqrt{\frac{k}{n}}\|w_m\|_2 \text{ for every } m\right\}\right) \geq 1 - N_s^2 e^{k/2} r^{-k}.$$

Let $\mathcal{F}$ be the subset of $G_{n,k}$ which is described by the previous relation. Then, for every $F \in \mathcal{F}$ and every $i \neq j$,
$$\|P_F(z_i) - P_F(z_j)\|_2 \geq \frac{1}{r}\sqrt{\frac{k}{n}}\|z_i - z_j\|_2 \geq \frac{s}{r}\sqrt{\frac{k}{n}}.$$

Since $P_F(z_i) \in P_F(C)$, it follows that
$$N\left(P_F(C), \frac{s}{2r}\sqrt{\frac{k}{n}}B_F\right) \geq N_s,$$

and this proves the lemma. $\square$

Second proof of Theorem 8.1.1. Let $1 \leq q \leq n$ and $N_t := N(K_q(\mu), tB_2^n)$. By Proposition 8.1.14 we may assume that $3 \leq N_t \leq e^{cn}$ and then, we choose $1 \leq k \leq n$ so that $\log N_t \leq k \leq 2\log N_t$. We distinguish two cases:

(i) Assume that $1 \leq t \leq \sqrt{n/q}$. We apply the previous lemma with $r = e^3$ and we see that with probability greater than $1 - N_t^2 e^{-5k/2} \geq 1 - e^{-k/2}$, a random subspace $F \in G_{n,k}$ satisfies
$$\frac{k}{2} \leq \log N_t \leq \log N\left(P_F(K_q(\mu)), c_1 t\sqrt{k/n}B_F\right),$$

where $c_1 > 0$ is an absolute constant.

If $\log N_t \leq q$ then we trivially have $\log N_t \leq n/t^2$ because $q \leq n/t^2$. So, we may assume that $\log N_t \geq q$ and in particular, $q \leq k$. We have
$$\frac{k}{2} \leq \log N\left(K_q(\pi_F(\mu)), ct\sqrt{k/n}B_F\right).$$

Note that $t\sqrt{k/n} \leq \sqrt{k/q} \leq k/q$. Then, applying Proposition 8.1.14 for the k-dimensional isotropic measure $\pi_F(\mu)$, we get
$$\frac{k}{2} \leq \frac{ck}{t\sqrt{k/n}} = c\frac{\sqrt{kn}}{t},$$

which shows that
$$\log N(K_q(\mu), tB_2^n) = \log N_t \leq k \leq c_2 \frac{n}{t^2},$$

with $c_2 = 4c^2$.

(ii) Assume that $t \geq \sqrt{n/q}$. We work as in the first proof of Theorem 8.1.1: we set $p = \sqrt{qn}/t \leq q$. From Proposition 5.1.3 we know that $Z_q(\mu) \subseteq \frac{c_0 q}{p}Z_p(\mu)$ for some absolute constant $c_0 > 1$. Then, we have
$$N\left(Z_q(\mu), t\sqrt{q}B_2^n\right) \leq N\left(\frac{c_0 q}{p}Z_p(\mu), t\sqrt{q}B_2^n\right)$$
$$\leq N\left(Z_p(\mu), \frac{t}{c_0}\sqrt{\frac{p}{q}}\sqrt{p}B_2^n\right)$$
$$= N\left(Z_p(\mu), \frac{1}{c_0}\sqrt{\frac{n}{p}}\sqrt{p}B_2^n\right).$$

Applying the result of case (i) for $Z_p(\mu)$ with $t = \frac{1}{c_0}\sqrt{n/p}$, we see that

$$N\left(Z_q(\mu), t\sqrt{q}B_2^n\right) \leqslant N\left(Z_p(\mu), \frac{1}{c_0}\sqrt{\frac{n}{p}}\sqrt{p}B_2^n\right)$$
$$\leqslant e^{cp} = \exp\left(\frac{c\sqrt{qn}}{t}\right),$$

and the proof is complete. □

8.2. Volume radius of the ψ_2-body

Let μ be an isotropic log-concave measure on $\mathbb{R}^n$. Recall the definition of $\Psi_2(\mu)$: it is the symmetric convex body with support function

$$h_{\Psi_2(\mu)}(\theta) = \sup_{1 \leqslant q \leqslant n} \frac{h_{Z_q(\mu)}(\theta)}{\sqrt{q}}.$$

From the definition, we have that $Z_q(\mu) \subseteq \sqrt{q}\Psi_2(\mu)$ for all $1 \leqslant q \leqslant n$. In particular, $B_2^n = Z_2(\mu) \subseteq \sqrt{2}\Psi_2(\mu)$, which implies that

$$|\Psi_2(\mu)|^{1/n} \geqslant \frac{c_1}{\sqrt{n}}.$$

Our aim is to give an upper bound for the volume of $\Psi_2(\mu)$.

THEOREM 8.2.1 (Giannopoulos-Paouris-Valettas). *Let μ be an isotropic log-concave measure on $\mathbb{R}^n$. Then,*

$$|\Psi_2(\mu)|^{1/n} \leqslant \frac{c_2\sqrt{\log n}}{\sqrt{n}}.$$

We start with the case where μ is the uniform measure on an isotropic convex body K in $\mathbb{R}^n$. As a first step, in the next subsection we assume that K has small diameter.

8.2.1. Isotropic convex bodies with small diameter

We assume that K is an isotropic convex body in $\mathbb{R}^n$ with radius $R(K) = \alpha\sqrt{n}L_K$, where $\alpha \geqslant 1$ is a positive constant. The main result of this subsection comes from [**410**] and shows that most directions $\theta \in S^{n-1}$ are "good ψ_2-directions" if α is uniformly bounded. The precise statement is as follows.

THEOREM 8.2.2. *Let K be an isotropic convex body in $\mathbb{R}^n$ with $R(K) = \alpha\sqrt{n}L_K$ for some $\alpha \geqslant 1$. Then,*

$$\sigma\left(\{\theta \in S^{n-1} : \|\langle \cdot, \theta\rangle\|_{\psi_2} \geqslant C\alpha t L_K\}\right) \leqslant \exp\left(-c\sqrt{n}t^2/\alpha\right)$$

for all $t \geqslant 1$, where $c, C > 0$ are absolute constants.

The main estimate of the theorem is a direct application of the concentration of measure on the Euclidean sphere: Let $f : S^{n-1} \to \mathbb{R}$ be an L-Lipschitz function and let $M = \mathrm{med}(f)$ be the Lévy median of f with respect to σ on S^{n-1}. Then, for any $\varepsilon > 0$ we have

(8.2.1) $\qquad \sigma(\{\theta \in S^{n-1} : |f(\theta) - M| > \varepsilon\}) \leqslant e^{-c'\varepsilon^2 n/L^2}.$

8.2. VOLUME RADIUS OF THE ψ_2-BODY

We use this result for the function $f : S^{n-1} \to \mathbb{R}$ defined by
$$f(\theta) = \|\langle \cdot, \theta \rangle\|_{\psi_2}.$$
For the Lipschitz constant of f we know that
$$|f(\theta) - f(\theta')| \leqslant \|\langle \cdot, \theta - \theta' \rangle\|_{\psi_2} \leqslant \|\theta - \theta'\|_2 \sqrt{\|\langle \cdot, u \rangle\|_{\psi_1} \|\langle \cdot, u \rangle\|_\infty}$$
$$\leqslant c'' \sqrt{R(K) L_K} \|\theta - \theta'\|_2,$$
where $u = \frac{\theta - \theta'}{\|\theta - \theta'\|_2}$, and since the diameter of K satisfies $R(K) \leqslant \alpha \sqrt{n} L_K$ we obtain
$$(8.2.2) \qquad |f(\theta) - f(\theta')| \leqslant c'' n^{1/4} \sqrt{\alpha} L_K \|\theta - \theta'\|_2$$
for all $\theta, \theta' \in S^{n-1}$. This shows that f is Lipschitz continuous with constant $L \leqslant c''(\alpha \sqrt{n} L_K^2)^{1/2}$. For the median of f we have the following estimates:

LEMMA 8.2.3. *Let K be an isotropic convex body in $\mathbb{R}^n$ and let $\alpha \geqslant 1$ such that $K \subseteq (\alpha \sqrt{n} L_K) B_2^n$. Then, for the function $f : S^{n-1} \to \mathbb{R}$ with $f(\theta) = \|\langle \cdot, \theta \rangle\|_{\psi_2}$ we have*
$$c_1 L_K \leqslant \mathrm{med}(f) \leqslant c_2 \alpha L_K,$$
where $c_1, c_2 > 0$ are absolute constants.

Proof. The lower bound is obvious: for any $\theta \in S^{n-1}$ we have
$$f(\theta) = \|\langle \cdot, \theta \rangle\|_{\psi_2} \geqslant c_1 \|\langle \cdot, \theta \rangle\|_2 = c_1 L_K,$$
because K is isotropic. For the upper bound we prove that
$$(8.2.3) \qquad \int_{S^{n-1}} \int_K e^{\left(\frac{|\langle x, \theta \rangle|}{\gamma} \right)^2} dx\, d\sigma(\theta) \leqslant 2,$$
where $\gamma = c_4 \alpha L_K$ and $c_4 > 0$ is an absolute constant. To see this, we write
$$\int_{S^{n-1}} \int_K \exp\left(\frac{\langle x, \theta \rangle^2}{\gamma^2} \right) dx\, d\sigma(\theta)$$
$$= 1 + \sum_{k=1}^\infty \frac{1}{k! \gamma^{2k}} \int_K \left(\int_{S^{n-1}} |\langle x, \theta \rangle|^{2k} d\sigma(\theta) \right) dx$$
$$\leqslant 1 + \sum_{k=1}^\infty \frac{c_5^{2k}}{k! \gamma^{2k}} \left(\frac{2k}{n+2k} \right)^k \int_K \|x\|_2^{2k} dx,$$
where we have used (3.2.6). Using the trivial estimate $\|x\|_2 \leqslant R(K)$ for all $x \in K$ we get:
$$\int_{S^{n-1}} \int_K \exp\left(\frac{\langle x, \theta \rangle^2}{\gamma^2} \right) dx\, d\sigma(\theta) \leqslant 1 + \sum_{k=1}^\infty \frac{(2c_5^2)^k}{k! \gamma^{2k}} \left(\frac{k}{n+2k} \right)^k [R(K)]^{2k}$$
$$\leqslant 1 + \sum_{k=1}^\infty \left(\frac{2e c_5^2 R(K)^2}{\gamma^2 (n+2k)} \right)^k.$$
Using again our assumption on the diameter of K we arrive at:
$$\int_{S^{n-1}} \int_K \exp\left(\frac{\langle x, \theta \rangle^2}{\gamma^2} \right) dx\, d\sigma(\theta) \leqslant 1 + \sum_{k=1}^\infty \left(\frac{2e c_5^2 \alpha^2 n L_K^2}{\gamma^2 (n+2k)} \right)^k$$
$$\leqslant 1 + \sum_{k=1}^\infty \left(\frac{2e c_5^2 \alpha^2 L_K^2}{\gamma^2} \right)^k.$$

Choosing $\gamma = 2e^{1/2}c_5\alpha L_K$ we get (8.2.3) with $c_4 = 2e^{1/2}c_5$.

Applying now Markov's inequality, we find a set $A \subseteq S^{n-1}$ with $\sigma(A) \geqslant 1/2$ such that for any $\theta \in A$ we have:

$$\int_K \exp\left(\frac{|\langle x, \theta \rangle|}{\gamma}\right)^2 dx \leqslant 4.$$

It follows that for any $\theta \in A$ and for any $k \in \mathbb{N}$ we have $\|\langle \cdot, \theta \rangle\|_{2k} \leqslant 3\gamma\sqrt{k}$, and hence that, for every $q \geqslant 1$,

$$\|\langle \cdot, \theta \rangle\|_q \leqslant 3\gamma\sqrt{q}.$$

Thus,

$$f(\theta) = \|\langle \cdot, \theta \rangle\|_{\psi_2} \leqslant c_6 \sup_{2 \leqslant q \leqslant n} \frac{\|\langle \cdot, \theta \rangle\|_q}{\sqrt{q}} \leqslant 3c_6\gamma$$

for any $\theta \in A$. In other words, we have

$$\sigma(\{\theta \in S^{n-1} : f(\theta) \leqslant 3c_6\gamma\}) \geqslant \sigma(A) \geqslant \frac{1}{2},$$

which shows that $\operatorname{med}(f) \leqslant 3c_6\gamma$. This proves the assertion of lemma with $c_2 = 3c_6c_4$. $\square$

Proof of Theorem 8.2.2. We use (8.2.1) for the function $f(\theta) = \|\langle \cdot, \theta \rangle\|_{\psi_2}$: for every $t \geqslant 1$ we can write

$$\sigma(\{\theta : f(\theta) \geqslant 2c_2 t\alpha L_K\}) \leqslant \sigma(\{\theta : f(\theta) - M \geqslant tM\}) \leqslant e^{-c't^2 n(\frac{M}{L})^2},$$

where we have used the upper estimate from Lemma 8.2.3. Then we use the lower estimate of Lemma 8.2.3 and the bound (8.2.2) for the Lipschitz constant of f to get

$$\frac{M}{L} \geqslant c_1 \frac{L_K}{c''\alpha^{1/2}n^{1/4}L_K} = c_7 n^{-1/4}\alpha^{-1/2}.$$

This proves the theorem with $c = c'/c_7^2$ and $C = 2c_2$. $\square$

REMARK 8.2.4. Choosing $t = 1$ in Theorem 8.2.2 we see that if $R(K) = \alpha\sqrt{n} L_K$ then

$$\|\langle \cdot, \theta \rangle\|_{\psi_2} \leqslant C\alpha L_K$$

with probability greater than $1 - \exp(-c\sqrt{n}/\alpha)$. Then, we easily check that

$$w(\Psi_2(K)) \leqslant C_2\alpha L_K,$$

where $C_2 > 0$ is an absolute constant. From Urysohn's inequality we immediately get:

THEOREM 8.2.5. *Let K be an isotropic convex body in $\mathbb{R}^n$ with $R(K) = \alpha\sqrt{n}L_K$. Then,*

$$\left(\frac{|\Psi_2(K)|}{|B_2^n|}\right)^{1/n} \leqslant w(\Psi_2(K)) \leqslant C\alpha L_K,$$

where $C > 0$ is an absolute constant.

8.2.2. Isotropic convex bodies

Next we consider the general case where μ is the uniform measure on an isotropic convex body K in $\mathbb{R}^n$. We know that

$$\Psi_2(K) = \operatorname{conv}\left\{\frac{Z_p(K)}{\sqrt{p}}, 1 \leqslant p \leqslant n\right\},$$

and using the fact that $Z_{2p}(K) \simeq Z_p(K)$, we may write

$$\Psi_2(K) \simeq \operatorname{conv}\left\{\frac{Z_p(K)}{\sqrt{p}}, p = 2^k, k = 1, \ldots, \log_2 n\right\}.$$

We set

$$m_1 := \log_2(\sqrt{n}), \quad m_2 := \log_2\left(\frac{n}{\log n}\right), \quad m_3 := \log_2 n = 2m_1,$$

and define the symmetric convex bodies $C_1, C_2, C_{2,1}, C_3$ and $C_{3,1}$ as follows:

$$C_1 := \operatorname{conv}\left\{\frac{Z_p(K)}{\sqrt{p}}, 1 \leqslant p \leqslant \sqrt{n}\right\},$$

$$C_2 := \operatorname{conv}\left\{\frac{Z_p(K)}{\sqrt{p}}, p = 2^k, m_1 \leqslant k \leqslant m_2\right\},$$

$$C_{2,1} := \operatorname{conv}\left\{\frac{Z_p(K)}{\sqrt{p}\sqrt{\log p}}, p = 2^k, m_1 \leqslant k \leqslant m_2\right\},$$

$$C_3 := \operatorname{conv}\left\{\frac{Z_p(K)}{\sqrt{p}}, p = 2^k, m_2 + 1 \leqslant k \leqslant m_3\right\},$$

$$C_{3,1} := \operatorname{conv}\left\{\frac{Z_p(K)}{\sqrt{p}\sqrt{\log p}}, p = 2^k, m_2 + 1 \leqslant k \leqslant m_3\right\}.$$

It is clear that

$$\Psi_2(K) \simeq \operatorname{conv}\{C_1, C_2, C_3\}.$$

We also define

$$V := \operatorname{conv}\{C_1, C_{2,1}, C_{3,1}\}.$$

Our next task is to give upper bounds for the covering numbers of the bodies C_1, C_2, $C_{2,1}$, C_3 and $C_{3,1}$ by multiples of the Euclidean unit ball.

(i) Covering numbers of C_1

We will use Theorem 8.2.5 and the next lemma.

LEMMA 8.2.6. *Let K be a centered convex body of volume 1 in $\mathbb{R}^n$ and let $1 \leqslant q \leqslant n$. Let A be a convex subset of K with volume $|A| \geqslant 1 - e^{-q}$. Then, for every $1 \leqslant p \leqslant c_1 q$, one has*

$$Z_p(K) \subseteq 2Z_p(\overline{A}),$$

where $c_1 > 0$ is an absolute constant.

Proof. Recall that there exists an absolute constant $c > 1$ such that $h_{Z_{2p}(K)}(\theta) \leqslant ch_{Z_p(K)}(\theta)$ for every $\theta \in S^{n-1}$ and $p \geqslant 1$.

We fix an absolute constant $c_1 > 0$ such that $e^{-q/2}c^{c_1 q} \leqslant \frac{1}{2}$. Then, working as in Lemma 6.5.5 we check that

$$\int_K |\langle x,\theta\rangle|^p dx \leqslant \int_{\overline{A}} |\langle x,\theta\rangle|^p dx + \frac{1}{2}\int_K |\langle x,\theta\rangle|^p dx.$$

for every $p \leqslant c_1 q$. □

PROPOSITION 8.2.7. *Let K be an isotropic convex body in $\mathbb{R}^n$. Then,*

$$\left(\frac{|C_1|}{|B_2^n|}\right)^{1/n} \leqslant w(C_1) \leqslant cL_K,$$

where $c > 0$ is an absolute constant. Moreover, for every $t \geqslant 1$,

$$N(C_1, c_1 t L_K B_2^n) \leqslant e^{\frac{c_2 n}{t^2}},$$

where $c_1, c_2 > 0$ are absolute constants.

Proof. From the inequality of Paouris we have $|K \cap s\sqrt{n}L_K B_2^n| \geqslant 1 - e^{-s\sqrt{n}}$ for all $s \geqslant c'$, where $c' > 0$ is an absolute constant. We set $s_0 = \max\{c_1^{-1}, c'\}$ where $c_1 > 0$ is the constant from Lemma 8.2.6. Let $A = K \cap s_0\sqrt{n}L_K B_2^n$. Then, $R(\overline{A}) \leqslant s_0\sqrt{n}L_K$ and $\overline{A}$ is almost isotropic. Also, from Lemma 8.2.6, for every $1 \leqslant p \leqslant \sqrt{n}$ we have $Z_p(K) \subseteq 2Z_p(\overline{A})$. It follows that

$$C_1 \subseteq 2C_1(\overline{A}) \subseteq 2\Psi_2(\overline{A}).$$

Now, the result follows from Theorem 8.2.5 and Sudakov's inequality. □

(ii) Covering numbers of the bodies C_2 and C_3

We start with a lemma on the covering numbers of a finite union of bounded sets.

LEMMA 8.2.8. *Let $A_1, \ldots, A_s$ be subsets of RB_2^k. For every $0 < t < R$,*

$$\overline{N}(\mathrm{conv}(A_1 \cup \cdots \cup A_s), 2tB_2^k) \leqslant \left(\frac{cR}{t}\right)^s \prod_{i=1}^s \overline{N}(A_i, tB_2^k).$$

Proof. Let $t > 0$ and let N_i be a subset of A_i respectively, with cardinality $|N_i| = \overline{N}(A_i, tB_2^k)$. Each $u \in \mathrm{conv}(A_1 \cup A_2 \cup \cdots \cup A_s)$ can be written in the form $z = \lambda_1 u_1 + \lambda_2 u_2 + \cdots + \lambda_s u_s$, where $u_i \in A_i$ and $\lambda_i \geqslant 0$ satisfy $\sum_{i=1}^s \lambda_i = 1$; so, we can see the s-tuples of the λ_i's as points in the unit sphere $S_{\ell_1^s} = \{\lambda = (\lambda_1, \ldots, \lambda_s) : \sum_{i=1}^s |\lambda_i| = 1\}$ of ℓ_1^s. We consider a subset Z of $S_{\ell_1^s}$, which is maximal with respect to the relation $\|y_i - y_j\|_1 \geqslant \frac{t}{R}$ ($i \neq j$). We know that $|Z| \leqslant (3R/t)^s$ for all $0 < t < R$.

Claim. The set

$$N := \{v = z_1 x_1 + z_2 x_2 + \cdots + z_s x_s : z = (z_i)_{i=1}^s \in Z, x_i \in N_i\},$$

is a $2t$-net of $\mathrm{conv}(A_1 \cup \cdots \cup A_s)$ with respect to B_2^k.

Indeed; let $u \in \mathrm{conv}(A_1 \cup \cdots \cup A_s)$. We can find $t_i \geqslant 0$ with $\sum t_i = 1$ and $u_i \in A_i$ such that $u = \sum_{i=1}^s t_i u_i$. There exists $z = (z_1, \ldots, z_s) \in Z$ such that $\sum_{i=1}^s |t_i - z_i| \leqslant t/R$. We can also find $x_i \in N_i$ with $\|u_i - x_i\|_2 \leqslant t$. Then, $\sum_{i=1}^s z_i x_i \in N$ and we have

$$\left\|u - \sum_{i=1}^s z_i x_i\right\|_2 \leqslant \left(\max_{1 \leqslant i \leqslant s} \|u_i\|_2\right) \sum_{i=1}^s |t_i - z_i| + \sum_{i=1}^s |z_i| \cdot \|u_i - x_i\|_2 \leqslant 2t.$$

Finally, observe that

$$\overline{N}(\operatorname{conv}(A_1 \cup \cdots \cup A_s), 2tB_2^k) \leqslant |N| \leqslant |Z| \prod_{i=1}^{s} \overline{N}(A_i, tB_2^k),$$

which proves the lemma. $\square$

PROPOSITION 8.2.9. *Let K be an isotropic convex body in $\mathbb{R}^n$. For every $t \geqslant 1$,*

$$\max\left\{N(C_2, c_1 t\sqrt{\log n} L_K B_2^n), N(C_3, c_2 t(\log \log n) L_K B_2^n)\right\} \leqslant e^{c_3 \frac{n}{t}}$$

and

$$\max\{N(C_{2,1}, c_1 L_K B_2^n), N(C_{3,1}, c_2 L_K B_2^n)\} \leqslant e^{c_3 n},$$

where $c_1, c_2, c_3 > 0$ are absolute constants.

Proof. We first consider the bodies C_2 and $C_{2,1}$. We set $s := m_2 - m_1$ and define

$$A_i := \frac{1}{2^{\frac{m_1+i}{2}}} Z_{2^{m_1+i}}(K) \text{ and } A_{i,1} := \frac{1}{2^{\frac{m_1+i}{2}} \sqrt{m_1+i}} Z_{2^{m_1+i}}(K),$$

for $i = 0, \ldots, s$. Observe that $\max\{R(A_i), R(A_{i,1})\} \leqslant \sqrt{n} L_K$ for all $0 \leqslant i \leqslant s$. From Theorem 8.1.1 we have that, for all $r \geqslant 1$,

$$\log N(A_i, cr L_K B_2^n) \leqslant \frac{c'n}{r^2} + \frac{c'n}{r\sqrt{\log n}}$$

and

$$\log N(A_{i,1}, cL_K B_2^n) \leqslant \frac{c'n}{m_1+i} + \frac{c'n}{\sqrt{m_1+i}\sqrt{\log n}} \leqslant \frac{c''n}{m_1+i}.$$

Using Lemma 8.2.8, we see that

$$\log N(C_2, 2cr L_K B_2^n) \leqslant \log^2 n + \frac{c'n \log n}{r^2} + \frac{c'n \log n}{r\sqrt{\log n}}.$$

Since $R(C_2) \leqslant \sqrt{n} L_K$, we assume that $1 \leqslant t \leqslant \sqrt{n}$. Then, $\log^2 n \leqslant \frac{n}{t}$. Setting $r = t\sqrt{\log n}$ we conclude that, for every $t \geqslant 1$,

$$\log N\left(C_2, 2ct\sqrt{\log n} L_K B_2^n\right) \leqslant \frac{3c'n}{t}.$$

In a similar way we see that

$$\log N(C_{2,1}, 2cL_K B_2^n) \leqslant \log^2 n + c''n \sum_{i=1}^{s} \frac{1}{m_1+i} \leqslant c'''n \sum_{j=m_1+1}^{2m_1} \frac{1}{j} \leqslant c'''n.$$

Now, we consider the bodies C_3 and $C_{3,1}$. We set $s := m_3 - m_2 = \log \log n$ and define

$$A_i := \frac{1}{2^{\frac{m_2+i}{2}}} Z_{2^{m_2+i}}(K), \ A_{i,1} := \frac{1}{2^{\frac{m_2+i}{2}} \sqrt{m_2+i}} Z_{2^{m_2+i}}(K),$$

for $i = 1, \ldots, s$. Note that $\max\{R(A_i), R(A_{i,1})\} \leqslant \sqrt{n} L_K$ for all $1 \leqslant i \leqslant s$. Theorem 8.1.1 shows that, for every $r \geqslant 1$,

$$\log N(A_i, cr L_K B_2^n) \leqslant \frac{c'n}{r} \text{ and } \log N(A_{i,1}, cL_K B_2^n) \leqslant \frac{c'n}{m_2+i} \leqslant \frac{c''n}{\log n}.$$

From Lemma 8.2.8 we get

$$\log N(C_3, 2cr L_K B_2^n) \leqslant \log^2 n + \frac{c'n(\log \log n)}{r}.$$

As before, we may assume that $1 \leqslant t \leqslant \sqrt{n}$, and hence, $\log^2 n \leqslant \frac{n}{t}$. Setting $r = t \log \log n$ we conclude that, for every $t \geqslant 1$,

$$\log N\left(C_3, 2ct(\log \log n) L_K B_2^n\right) \leqslant \frac{3c'n}{t}.$$

Also, by Lemma 8.2.8,

$$\log N\left(C_{3,1}, 2cL_K B_2^n\right) \leqslant c' \frac{n(\log \log n)}{\log n} \leqslant cn.$$

This completes the proof. □

PROPOSITION 8.2.10. *Let K be an isotropic convex body in $\mathbb{R}^n$. For every $t \geqslant 1$,*

$$N\left(\Psi_2(K), c_1 t \sqrt{\log n} L_K B_2^n\right) \leqslant e^{c_2 \frac{n}{t}}$$

and

$$N\left(V, c_3 L_K B_2^n\right) \leqslant e^{c_2 n},$$

where $c_1, c_2, c_3 > 0$ are absolute constants.

Proof. We apply Lemma 8.2.8 for $A_1 := C_1$, $A_2 := C_2$ and $A_3 := C_3$, and we use Proposition 8.2.7 and Proposition 8.2.9. We work in the same way for V. □

Proof of Theorem 8.2.1 in the case of isotropic convex bodies. The result is a direct consequence of Proposition 8.2.10 (for $t = 1$) and of the fact that for every pair of compact sets A and B of $\mathbb{R}^n$ we have $|A| \leqslant N(A, B)|B|$.

Note that the same argument shows that $|V|^{1/n} \leqslant cL_K |B_2^n|^{1/n}$. □

The existence of directions with small ψ_2-constant may be also expressed in the form of almost sub-Gaussian tail estimates.

THEOREM 8.2.11. *Let K be an isotropic convex body in $\mathbb{R}^n$. There exists $\theta \in S^{n-1}$ such that*

(8.2.4) $$|\{x \in K : |\langle x, \theta \rangle| \geqslant ctL_K\}| \leqslant \exp\left(-\frac{t^2}{\log(t+1)}\right)$$

for every $t \geqslant 1$, where $c > 0$ is an absolute constant.

Proof. We consider the symmetric convex bodies

$$V_1 := \operatorname{conv}\left\{\frac{Z_p(K)}{\sqrt{p}\sqrt{\log p}}, p \in [2, n]\right\} \text{ and } V_2 := \operatorname{conv}\left\{\frac{Z_p(K)}{\sqrt{p}\sqrt{\log p}}, p \geqslant 2\right\}.$$

Note that from Proposition 5.1.2 we have $V_1 \simeq V_2$. Then, $V_1 \subseteq cV$ and $|V_2|^{1/n} \leqslant cL_K |B_2^n|^{1/n}$. So, there exists $\theta \in S^{n-1}$ such that $h_{V_2}(\theta) \leqslant cL_K$. It follows that for every $p \geqslant 1$,

(8.2.5) $$h_{Z_p(K)}(\theta) \leqslant c\sqrt{p}\sqrt{\log p}\, L_K.$$

From Markov's inequality we have that, for every $p > 0$,

(8.2.6) $$|\{x \in K : |\langle x, \theta \rangle| \geqslant e h_{Z_p(K)}(\theta)\}| \leqslant e^{-p}.$$

Let $t \geqslant 1$. If we define p by the equation $\sqrt{p} = \frac{t}{\sqrt{\log(t+1)}}$, then (8.2.5) and (8.2.6) imply (8.2.4). □

8.2.3. Isotropic log-concave measures

We can finally give the proof of Theorem 8.2.1: we show that if μ is an isotropic log-concave measure on $\mathbb{R}^n$ then
$$|\Psi_2(\mu)|^{1/n} \leqslant \frac{c\sqrt{\log n}}{\sqrt{n}}.$$

Proof. Using Proposition 8.1.7, we establish the existence of a convex body $T = T_\mu$ in $\mathbb{R}^n$ of volume 1 such that $Z_2(T) \subseteq c_1 B_2^n$ and $Z_q(\mu) \subseteq c_2 Z_q(T)$ for all $1 \leqslant q \leqslant n$. Therefore, by definition of the Ψ_2 body we get:
$$\Psi_2(\mu) \subseteq c_2 \Psi_2(T).$$

Using the fact that Theorem 8.2.1 has been established for (isotropic) convex bodies, for any isotropic image T_1 of T we can write
$$|\Psi_2(T)|^{1/n} = |Z_2(T)|^{1/n} \left(\frac{|\Psi_2(T_1)|}{|Z_2(T_1)|}\right)^{1/n} \leqslant c_2 L_T^{-1} |\Psi_2(T_1)|^{1/n}$$
$$\leqslant c_2 C \frac{\sqrt{\log n}}{\sqrt{n}}.$$

The result follows. $\square$

In this setting, tail estimates take the following form.

THEOREM 8.2.12. *If μ is a centered log-concave probability measure on $\mathbb{R}^n$ then there exists $\theta \in S^{n-1}$ such that*
$$\mu(\{x \in \mathbb{R}^n : |\langle x, \theta \rangle| \geqslant ct \mathbb{E}|\langle \cdot, \theta \rangle|\}) \leqslant e^{-\frac{t^2}{\log(t+1)}}$$
for all $1 \leqslant t \leqslant \sqrt{n \log n}$, where $c > 0$ is an absolute constant.

Proof. Using (8.2.5) and Proposition 5.1.7 we see that there exists $\theta \in S^{n-1}$ such that for $p \leqslant n$
$$f_\mu(0)^{1/n} h_{Z_p(\mu)}(\theta) \leqslant c_1 h_{Z_p(\overline{K_{n+1}(\mu)})}(\theta) \leqslant c_2 \sqrt{p} \sqrt{\log p} \, h_{Z_2(\overline{K_{n+1}(\mu)})}(\theta)$$
$$\leqslant c_3 \sqrt{p} \sqrt{\log p} \, h_{Z_1(\overline{K_{n+1}(\mu)})}(\theta).$$

Again, Proposition 5.1.7 shows that
$$h_{Z_1(\overline{K_{n+1}(\mu)})}(\theta) \leqslant c_4 f_\mu(0)^{1/n} h_{Z_1(\mu)}(\theta) = c_4 f_\mu(0)^{1/n} \mathbb{E}|\langle \cdot, \theta \rangle|.$$

Combining the previous inequalities we arrive at
$$h_{Z_p(\mu)}(\theta) \leqslant c_5 \sqrt{p} \sqrt{\log p} \, \mathbb{E}|\langle \cdot, \theta \rangle|.$$

Then, we use Markov's inequality; we have
$$\mu\left(\{x \in \mathbb{R}^n : |\langle x, \theta \rangle| \geqslant e h_{Z_p(\mu)}(\theta)\}\right) \leqslant e^{-p},$$
and hence, given $1 \leqslant t \leqslant \sqrt{n \log n}$ we choose $\sqrt{p} = \frac{t}{\sqrt{\log(t+1)}}$ and the result follows.
$\square$

8.3. Distribution of the ψ_2-norm

In the previous section we discussed the question of the existence of sub-Gaussian directions for an isotropic convex body; we saw that one can always have a ψ_2-direction with constant $O(\sqrt{\log n})$. A natural question that arises is how many these directions are, in terms of the probability measure σ on the sphere.

DEFINITION 8.3.1. Let K be an isotropic convex body in $\mathbb{R}^n$. We introduce a function $\psi_K : [1, \infty) \to \mathbb{R}$ setting

$$\psi_K(t) := \sigma\left(\{\theta \in S^{n-1} : \|\langle \cdot, \theta \rangle\|_{\psi_2} \leqslant ct\sqrt{\log n} L_K\}\right).$$

It is clear that ψ_K is increasing. Also, from Remark 3.2.14 we know that $\|\langle \cdot, \theta \rangle\|_{\psi_2} \leqslant c\sqrt{n} L_K$ for all $\theta \in S^{n-1}$; it follows that $\psi_K(t) = 1$ if $t \geqslant c\sqrt{n/\log n}$. Thus, the question is to give lower bounds for $\psi_K(t)$ in the range $1 \leqslant t \leqslant c\sqrt{n/\log n}$.

In the next two subsections we obtain some rather weak lower bounds for $\psi_K(t)$. The problem remains open; as we will see in the next chapter, a significant improvement of these bounds would be very useful.

8.3.1. ψ_2-directions on subspaces

We first show that for every $1 \leqslant k \leqslant n$ and any $F \in G_{n,k}$ one can find $\theta \in S_F$ such that $\|\langle \cdot, \theta \rangle\|_{\psi_2} \leqslant g(n,k)$, where the behavior of the function $g(n,k)$ is naturally worse as k becomes smaller. More precisely, we have:

THEOREM 8.3.2. *Let K be an isotropic convex body in $\mathbb{R}^n$ and $\log^2 n \leqslant k \leqslant n$.*
(i) *If $\log^2 n \leqslant k \leqslant n/\log n$ then for every $F \in G_{n,k}$ there exists $\theta \in S_F$ such that*
(8.3.1) $$\|\langle \cdot, \theta \rangle\|_{\psi_2} \leqslant C\sqrt{n/k} L_K.$$
(ii) *If $n/\log n \leqslant k \leqslant n$ then for every $F \in G_{n,k}$ there exists $\theta \in S_F$ such that*
(8.3.2) $$\|\langle \cdot, \theta \rangle\|_{\psi_2} \leqslant C\sqrt{\log n} L_K,$$
where $C > 0$ is an absolute constant.

Note. In Chapter 9, Section 1 (see Remark 9.1.14) we give an argument that covers the remaining case $1 \leqslant k \leqslant \log^2 n$. One can show that for every $1 \leqslant k \leqslant \log^2 n$ and every $F \in G_{n,k}$ there exists $\theta \in S_F$ such that

$$\|\langle \cdot, \theta \rangle\|_{\psi_2} \leqslant C\sqrt{n/k}\sqrt{\log 2k} L_K,$$

where $C > 0$ is an absolute constant. In fact, for a random $F \in G_{n,k}$ the term $\sqrt{\log 2k}$ is not needed.

For the proof of Theorem 8.3.2 we first obtain estimates on the covering numbers of the projections of the L_q-centroid bodies of K, using Theorem 8.1.1. Recall the next fact on the covering numbers of K (see Theorem 3.2.2).

FACT 8.3.3. *Let K be an isotropic convex body in $\mathbb{R}^n$. Then, for every $t > 0$ one has*

$$N(K, t\sqrt{n} L_K B_2^n) \leqslant 2\exp(c_1 n/t).$$

Furthermore, one has

$$N(K - K, 2t\sqrt{n} L_K B_2^n) \leqslant 4\exp(c_2 n/t)$$

for every $t > 0$, where $c_1, c_2 > 0$ are absolute constants.

PROPOSITION 8.3.4. *Let K be an isotropic convex body in $\mathbb{R}^n$. For every $1 \leqslant q < k \leqslant n$, for every $F \in G_{n,k}$ and every $t \geqslant 1$, we have*

$$\log N\left(P_F(Z_q(K)), t\sqrt{q}L_K B_F\right) \leqslant \frac{c_1 k}{t^2} + \frac{c_2 \sqrt{qk}}{t},$$

where $c_1, c_2 > 0$ are absolute constants. Also, for every $k \leqslant q \leqslant n$, and for every $F \in G_{n,k}$ and $t \geqslant 1$,

$$\log N\left(P_F(Z_q(K)), t\sqrt{q}L_K B_F\right) \leqslant \frac{c_3 \sqrt{qk}}{t},$$

where $c_3 > 0$ is an absolute constant.

Proof. Since K is isotropic, the measure μ_K with density $L_K^n \mathbf{1}_{\frac{K}{L_K}}$ is isotropic. Note that

$$Z_q(\mu_K) = \frac{1}{L_K} Z_q(K).$$

Then, from Proposition 5.1.12 we have

$$P_F(Z_q(K)) = L_K P_F(Z_q(\mu_K)) = L_K Z_q(\pi_F(\mu_K))$$

for every $F \in G_{n,k}$.

Assume first that $1 \leqslant q \leqslant k$, and let $F \in G_{n,k}$. For any $t \geqslant 1$ we write

$$\log N\left(P_F(Z_q(K)), t\sqrt{q}L_K B_F\right) = \log N\left(L_K Z_q(\pi_F(\mu_K)), t\sqrt{q}L_K B_F\right)$$
$$= \log N\left(Z_q(\pi_F(\mu_K)), t\sqrt{q} B_F\right).$$

Since $\pi_F(\mu_K)$ is an isotropic log-concave probability measure on F, we can apply Theorem 8.1.1 to get

$$\log N\left(Z_q(\pi_F(\mu_K)), c_1 t\sqrt{q} B_F\right) \leqslant \frac{c_2 k}{t^2} + \frac{c_3 \sqrt{qk}}{t}$$

for all $t \geqslant 1$ (where $c_i > 0$ are absolute constants) and hence

$$\log N\left(P_F(Z_q(K)), c_1 t\sqrt{q}L_K B_F\right) \leqslant \frac{c_2 k}{t^2} + \frac{c_3 \sqrt{qk}}{t}$$

for all $t \geqslant 1$.

Next, we assume that $k \leqslant q \leqslant n$. For any $F \in G_{n,k}$ and $t \geqslant 1$, using Proposition 5.1.2 we can write

$$\log N\left(P_F(Z_q(K)), t\sqrt{q}L_K B_F\right) = \log N\left(Z_q(\pi_F(\mu_K)), t\sqrt{q} B_F\right)$$
$$\leqslant \log N\left(\frac{c_1 q}{k} Z_k(\pi_F(\mu_K)), \frac{t\sqrt{q}}{\sqrt{k}} \sqrt{k} B_F\right)$$
$$= \log N\left(Z_k(\pi_F(\mu_K)), \frac{c_2 t\sqrt{k}}{\sqrt{q}} \sqrt{k} B_F\right)$$
$$\leqslant c_3 \frac{\sqrt{qk}}{t}.$$

For the last inequality we consider the convex body $T := \overline{K_{k+2}}(\pi_F(\mu_K))$ and write

$$\log N\left(Z_k(\pi_F(\mu_K)), \frac{c_2 t\sqrt{k}}{\sqrt{q}} \sqrt{k} B_F\right) \simeq \log N\left(Z_k(T), \frac{c_3 t\sqrt{k}}{\sqrt{q}} \sqrt{k} L_T B_F\right),$$

using the fact that $L_{\pi_F(\mu_K)} \simeq L_T$ and

$$L_T Z_k(\pi_F(\mu_K)) \simeq Z_k(T)$$

by Theorem 5.1.7. Then, since $Z_k(T) \subseteq c_4(T-T)$ and T is almost isotropic, we may apply the second assertion of Fact 8.3.3 for T. This completes the proof. □

Using these estimates we can prove the existence of directions with relatively small ψ_2-norm on every subspace of $\mathbb{R}^n$. The dependence becomes better as the dimension increases.

Proof of Theorem 8.3.2. As in Definition 8.1.12, for every integer $q \geqslant 1$ we define the normalized L_q-centroid body K_q of K by $K_q = \frac{1}{\sqrt{q}L_K} Z_q(K)$, and then consider the convex body

$$Q = \operatorname{conv}\left(\bigcup_{i=1}^{\lfloor \log_2 n \rfloor} K_{2^i} \right).$$

Then, for every $F \in G_{n,k}$ we have

$$P_F(Q) = \operatorname{conv}\left(\bigcup_{i=1}^{\lfloor \log_2 n \rfloor} P_F(K_{2^i}) \right).$$

We will use Lemma 8.2.8 for the sets $A_i = P_F(K_{2^i})$. Note that $K_{2^i} \subseteq c_1 2^{i/2} B_2^n$, and hence, $N(A_i, tB_F) = 1$ if $c_1 2^{i/2} \leqslant t$. Also, $A_i \subseteq c_2 \sqrt{n} B_F$ for every i.

Using Proposition 8.3.4, for every $t \geqslant 1$ we can write

$$N(P_F(Q), 2tB_F) \leqslant (c_2\sqrt{n})^{\lfloor \log_2 n \rfloor} \left[\prod_{i=1}^{\lfloor \log_2 n \rfloor} N(P_F(K_{2^i}), tB_F) \right]$$

$$\leqslant e^{c_3 \log^2 n} \exp\left(C \sum_{i=1}^{\lfloor \log_2 n \rfloor} \frac{2^{i/2}\sqrt{k}}{t} + C \sum_{t^2 \leqslant 2^i \leqslant k} \frac{k}{t^2} \right)$$

$$\leqslant e^{c_3 \log^2 n} \exp\left(C \frac{\sqrt{kn}}{t} + C \frac{k}{t^2} \log\left(\frac{k}{t^2}\right) \right),$$

where the second term appears only if $k \geqslant ct^2$.

Now, we distinguish two cases:

(i) If $\log^2 n \leqslant k \leqslant \frac{n}{\log n}$ then we choose $t_0 = \sqrt{n/k}$. We observe that $\frac{\sqrt{kn}}{t_0} = k$ and

$$\frac{k}{t_0^2} \log\left(\frac{k}{t_0^2}\right) = \frac{k^2}{n} \log\left(\frac{k^2}{n}\right) \leqslant \frac{k}{\log n} \log\left(\frac{k^2}{n}\right) \leqslant k.$$

This gives $N(P_F(Q), \sqrt{n/k} B_F) \leqslant e^{ck}$. It follows that

$$|P_F(Q)| \leqslant |C\sqrt{n/k}\, B_F|.$$

So, there exists $\theta \in S_F$ such that

$$h_Q(\theta) = h_{P_F(Q)}(\theta) \leqslant C\sqrt{n/k},$$

which implies that

$$\|\langle \cdot, \theta \rangle\|_{2^i} \leqslant C\, 2^{i/2} \sqrt{n/k}\, L_K$$

for every $i = 1, 2, \ldots, \lfloor \log_2 n \rfloor$. Thus, we get the first part of the theorem.

(ii) If $\frac{n}{\log n} \leqslant k \leqslant n$ and $F \in G_{n,k}$ then we consider a subspace H of F with $\dim H = \lfloor \frac{n}{\log n} \rfloor$ and we apply part (i) to find $\theta \in S_H \subseteq S_F$ such that $\|\langle \cdot, \theta \rangle\|_{\psi_2} \leqslant C\sqrt{\log n}$. The result follows. $\square$

8.3.2. Distribution function of the ψ_2-norm

A simple argument, which is based on the estimates of the previous subsection, allows us to get a non-trivial but weak lower bound for $\psi_K(t)$.

THEOREM 8.3.5. *Let K be an isotropic convex body in $\mathbb{R}^n$. For every $t \geqslant 1$ we have*
$$\psi_K(t) \geqslant \exp(-cn/t^2),$$
where $c > 0$ is an absolute constant.

Recall that $\psi_K(t) = 1$ if $t \geqslant c\sqrt{n/\log n}$; so, the bound of Theorem 8.3.5 has a meaning only if $1 \leqslant t \leqslant c\sqrt{n/\log n}$. Actually, if $t \geqslant c\sqrt[4]{n/\log n}$ then we have a better estimate:

PROPOSITION 8.3.6. *Let K be an isotropic convex body in $\mathbb{R}^n$. For every $t \geqslant c_1 \sqrt[4]{n/\log n}$ we have*
$$\psi_K(t) \geqslant 1 - e^{-c_2 t^2 \log n},$$
where $c_1, c_2 > 0$ are absolute constants.

The proof of Proposition 8.3.6 is presented at the end of the section. For the proof of Theorem 8.3.5 we start with the following lemma.

LEMMA 8.3.7. *Let $1 \leqslant k \leqslant n$ and let A be a subset of S^{n-1} which satisfies $A \cap F \neq \emptyset$ for every $F \in G_{n,k}$. Then, for every $\varepsilon > 0$ we have*
$$\sigma(A_\varepsilon) \geqslant \frac{1}{2} \left(\frac{\varepsilon}{2} \right)^{k-1},$$
where
$$A_\varepsilon = \{ y \in S^{n-1} : \inf\{\|y - \theta\|_2 : \theta \in A\} \leqslant \varepsilon \}.$$

Proof. We write
$$\sigma(A_\varepsilon) = \int_{S^{n-1}} \mathbf{1}_{A_\varepsilon}(y) \, d\sigma(y) = \int_{G_{n,k}} \int_{S_F} \mathbf{1}_{A_\varepsilon}(y) \, d\sigma_F(y) \, d\nu_{n,k}(F),$$
and observe that, since $A \cap S_F \neq \emptyset$, the set $A_\varepsilon \cap S_F$ contains a cap $C_F(\varepsilon) = \{y \in S_F : \|y - \theta_0\|_2 \leqslant \varepsilon\}$ of Euclidean radius ε in S_F. It follows that
$$\int_{S_F} \mathbf{1}_{A_\varepsilon}(y) \, d\sigma_F(y) \geqslant \sigma_F(C_F(\varepsilon)) \geqslant \frac{1}{2} \left(\frac{\varepsilon}{2} \right)^{k-1},$$
from known estimates for caps (see e.g. [**39**]). $\square$

REMARK 8.3.8. The proof of the lemma shows that the strong assumption that $A \cap F \neq \emptyset$ for every $F \in G_{n,k}$ is not really needed if we want to estimate $\sigma(A_\varepsilon)$. We would have the same bound for $\sigma(A_\varepsilon)$ under the weaker assumption that $A \cap F \neq \emptyset$ for every F in a subset $\mathcal{F}_{n,k}$ of $G_{n,k}$ with measure $\nu_{n,k}(\mathcal{F}_{n,k}) \geqslant c^{-k}$.

THEOREM 8.3.9. *Let K be an isotropic convex body in $\mathbb{R}^n$. For every $\log^2 n \leqslant k \leqslant n$ there exists $\Theta_k \subseteq S^{n-1}$ such that*

$$\sigma(\Theta_k) \geqslant e^{-c_1 k \log k}$$

where $c_1 > 0$ is an absolute constant, and

$$\|\langle \cdot, y \rangle\|_{\psi_2} \leqslant C \max\left\{\sqrt{n/k}, \sqrt{\log n}\right\} L_K$$

for every $y \in \Theta_k$.

Proof. We fix $k \leqslant n/\log n$ and consider the set Θ_k of all $\theta \in S^{n-1}$ which satisfy (8.3.1). From Theorem 8.3.2 we have $\Theta_k \cap S_F \neq \emptyset$ for every $F \in G_{n,k}$. So, we can apply Lemma 8.3.7 with $\varepsilon = \frac{1}{\sqrt{k}}$. If $y \in (\Theta_k)_\varepsilon$ then there exists $\theta \in \Theta_k$ such that $\|y - \theta\|_2 \leqslant \varepsilon$, which gives

$$\|\langle \cdot, y - \theta \rangle\|_{\psi_2} \leqslant \|\langle \cdot, y - \theta \rangle\|_\infty^{1/2} \|\langle \cdot, y - \theta \rangle\|_{\psi_1}^{1/2} \leqslant c\varepsilon\sqrt{n}\, L_K,$$

if we use the fact that $\|\langle \cdot, \theta \rangle\|_{\psi_1} \leqslant c\|\langle \cdot, \theta \rangle\|_1 \leqslant cL_K$. It follows that

$$\|\langle \cdot, y \rangle\|_{\psi_2} \leqslant \|\langle \cdot, \theta \rangle\|_{\psi_2} + \|\langle \cdot, y - \theta \rangle\|_{\psi_2} \leqslant \|\langle \cdot, \theta \rangle\|_{\psi_2} + c\sqrt{n/k}\, L_K.$$

Since θ satisfies (8.3.1), we get the result for every $y \in \Theta_k$, with a different constant C. Finally, Lemma 8.3.7 shows that

$$\sigma(\Theta_k) \geqslant \frac{1}{2}\left(\frac{1}{2\sqrt{k}}\right)^{k-1} \geqslant e^{-c_1 k \log k},$$

which completes the proof for all k in this interval. A similar argument works for $k \geqslant n/\log n$: in this case, we apply Lemma 8.3.7 with $\varepsilon = \sqrt{\log n/n}$ and the measure estimate for Θ_k is the same. $\square$

Proof of Theorem 8.3.5. Because of Proposition 8.3.6 we may assume that $1 \leqslant t \leqslant \frac{\sqrt{n}}{(\log n)^{3/2}}$. Let k be the least integer for which $k \geqslant \frac{n}{t^2 \log n}$. Then,

$$\frac{n}{t^2} \simeq k \log n \geqslant k \log k,$$

and hence $e^{-c_1 k \log k} \geqslant e^{-c_2 n/t^2}$. Note also that $k \geqslant \log^2 n$, and hence, using the set Θ_k from Theorem 8.3.9 we see that

$$\psi_K(t) \geqslant \sigma(\Theta_k) \geqslant e^{-c_2 n/t^2}.$$

This proves our claim. $\square$

In order to obtain an improved estimate of $\psi_K(t)$ for large values of t, we use Theorem 8.3.5 to give upper bounds on the mean width of $\Psi_2(K)$. Then, a direct application of the isoperimetric inequality on the sphere will give the promised estimate. We start with the following:

PROPOSITION 8.3.10. *Let K be an isotropic convex body in $\mathbb{R}^n$ and let $1 \leqslant t \leqslant \frac{\sqrt{n}}{(\log n)^{3/2}}$. Then,*

$$w_{-\frac{n}{t^2}}(\Psi_2(K)) \leqslant ct\sqrt{\log n}\, L_K.$$

8.3. DISTRIBUTION OF THE ψ_2-NORM

Proof. From Markov's inequality we get

$$\sigma\left(\left\{\theta \in S^{n-1} : h_{\Psi_2(K)}(\theta) \leqslant \frac{1}{e}w_{-\frac{n}{t^2}}(\Psi_2(K))\right\}\right) \leqslant e^{-\frac{n}{t^2}}.$$

From Theorem 8.3.5 we know that

$$e^{-\frac{n}{t^2}} \leqslant \sigma\left(\{\theta \in S^{n-1} : h_{\Psi_2(K)}(\theta) \leqslant ct\sqrt{\log n}L_K\}\right)$$

for some absolute constant $c > 0$ and the result follows. $\square$

Using the result of Klartag-Vershynin on the negative moments of norms on the sphere (Corollary 5.3.12) we can give an upper bound for the mean width of the body $\Psi_2(K)$:

PROPOSITION 8.3.11. *Let K be an isotropic convex body in $\mathbb{R}^n$. Then,*

$$w(\Psi_2(K)) \leqslant c\sqrt[4]{n \log n}L_K.$$

Proof. Let $w := w(\Psi_2(K))$. Since $R(\Psi_2(K)) \leqslant c\sqrt{n}L_K$, using Proposition 5.3.9 we see that

$$d_*(\Psi_2(K)) \equiv d(\Psi_2^\circ(K)) \geqslant ck(\Psi_2^\circ(K)) \geqslant c\frac{w^2}{L_K^2}.$$

We choose t so that $\frac{n}{t^2} = c\frac{w^2}{L_K^2}$, i.e. $t = c_1\sqrt{n}L_K/w \geqslant 1$. If $t \geqslant \frac{\sqrt{n}}{(\log n)^{3/2}}$ then we immediately obtain the bound

$$w \leqslant c_1(\log n)^{3/2}L_K.$$

If $t \leqslant \frac{\sqrt{n}}{(\log n)^{3/2}}$ then we can apply Proposition 8.3.10 to get

$$w \leqslant cw_{-d_*}(\Psi_2(K)) \leqslant w_{-\frac{cw^2}{L_K^2}}(\Psi_2(K)) = w_{-\frac{n}{t^2}}(\Psi_2(K))$$

$$\leqslant c_2 t\sqrt{\log n}L_K \leqslant c_3 \frac{\sqrt{n}}{w}\sqrt{\log n}L_K^2,$$

which gives $w(\Psi_2(K)) \leqslant c\sqrt[4]{n \log n}L_K$. $\square$

Proof of Proposition 8.3.6. Since the function $h_{\Psi_2(K)}$ is Lipschitz continuous on the sphere, with constant $\sqrt{n}L_K$, we have

$$\sigma\left(\{\theta \in S^{n-1} : h_{\Psi_2(K)}(\theta) - w(\Psi_2(K)) \geqslant sw(\Psi_2(K))\}\right) \leqslant e^{-cns^2\left(\frac{w(\Psi_2(K))}{\sqrt{n}L_K}\right)^2}.$$

Let $u \geqslant 2w(\Psi_2(K))$. Then, $u = (1+s)w(\Psi_2(K))$ for some $s \geqslant 1$ and $sw(\Psi_2(K)) \geqslant u/2$. Therefore, we have

$$\sigma\left(\{\theta \in S^{n-1} : h_{\Psi_2(K)}(\theta) \geqslant u\}\right) \leqslant \exp\left(-cu^2/L_K^2\right).$$

If $t \geqslant c_1\sqrt[4]{n}/\sqrt[4]{\log n}$, Proposition 8.3.11 shows that $u = t\sqrt{\log n}L_K \geqslant 2w(\Psi_2(K))$. This completes the proof. $\square$

REMARK 8.3.12. Since the proof of Theorem 8.3.5 is now complete, we can remove the restriction $1 \leqslant t \leqslant \frac{\sqrt{n}}{(\log n)^{3/2}}$ from Proposition 8.3.10. Note also that, as can be seen from the proof, the estimate of Proposition 8.3.6 holds true for all $t \geqslant \frac{cw(\Psi_2(K))}{\sqrt{\log n}L_K}$. In the next chapter (see Theorem 9.1.13) we will see a stronger estimate for $w(\Psi_2(K))$. One can show that $w(\Psi_2(\mu)) \leqslant c\sqrt[4]{n}$.

8.4. Super-Gaussian directions

Given a centered log-concave probability measure on $\mathbb{R}^n$ one can ask for "dual tail estimates". The question can be made precise as follows: is it true that there exists $\theta \in S^{n-1}$ for which

$$\mu(\{x \in \mathbb{R}^n : |\langle x, \theta\rangle| \geq t\mathbb{E}|\langle \cdot, \theta\rangle|\}) \geq e^{-r^2 t^2}$$

for all $1 \leq t \leq \sqrt{n}/r$ and for some absolute constant $r > 0$?

The smallest $r > 0$ for which the previous distributional inequality holds is called the *super-Gaussian* constant of μ in the direction of θ and is denoted by $sg_\mu(\theta)$. Note that sg_μ, as the sub-Gaussian constant, satisfies $sg_{T(\mu)}(\theta) = sg_\mu(T^*\theta/\|T^*\theta\|_2)$ for all $T \in GL(n)$ and all $\theta \in S^{n-1}$. Therefore, as in the sub-Gaussian case we can always assume that μ is isotropic.

The next result of Paouris from [415] says that if a measure μ satisfies the hyperplane conjecture then a random direction $\theta \in S^{n-1}$ is super-Gaussian for μ with constant $O(1)$.

THEOREM 8.4.1. *Let μ be an isotropic log-concave measure on $\mathbb{R}^n$. Then,*

(8.4.1) $$\sigma(\{\theta \in S^{n-1} : sg_\mu(\theta) \geq ctL_\mu\}) \leq t^{-n},$$

for all $t > 1$. In particular, we have

(8.4.2) $$\mathbb{E}(sg_\mu) \leq CL_\mu,$$

where $C > 0$ is an absolute constant.

Proof. First, we need to check the analogue of the Lutwak-Yang-Zhang inequality (Proposition 5.1.16) in the setting of centered log-concave probability measures: one has

(8.4.3) $$|Z_p(\mu)|^{1/n} \geq c\|\mu\|_\infty^{-1/n}\sqrt{\frac{p}{n}},$$

for all $1 \leq p \leq n$, where $c > 0$ is an absolute constant. To see this, note that from Proposition 5.1.6 we have

$$|Z_p(\mu)|^{1/n} f_\mu(0)^{1/n} \geq c_1 |Z_p(\overline{K_{n+p}}(\mu))|^{1/n}$$

and by Proposition 5.1.16 we also have

$$|Z_p(\overline{K_{n+p}}(\mu))|^{1/n} \geq c_2\sqrt{p/n}.$$

Combining these two estimates, we get (8.4.3).

Using the fact that

$$\mathrm{v.rad}(Z_p^\circ(\mu)) = \left(\int_{S^{n-1}} h_{Z_p(\mu)}^{-n}(\theta)\, d\sigma(\theta)\right)^{1/n},$$

Markov's inequality, the Blaschke-Santaló inequality and (8.4.3) we see that

$$\sigma\left(\{\theta : h_{Z_p(\mu)}(\theta) \leq \alpha\}\right) \leq \alpha^n \mathrm{v.rad}(Z_p^\circ(\mu))^n$$
$$\leq \frac{\alpha^n}{[\mathrm{v.rad}(Z_p(\mu))]^n} \leq \left(\frac{\alpha\|\mu\|_\infty^{1/n}}{c\sqrt{p}}\right)^n.$$

Setting $\alpha = \frac{c\sqrt{p}}{et\|\mu\|_\infty^{1/n}}$, $t > 1$ we obtain

(8.4.4) $$\sigma\left(\left\{\theta \in S^{n-1} : h_{Z_p(\mu)}(\theta) \leqslant \frac{c\sqrt{p}}{et\|\mu\|_\infty^{1/n}}\right\}\right) \leqslant (et)^{-n}.$$

Since $h_{Z_p} \simeq h_{Z_{2p}}$ for all $p \geqslant 1$, using a standard "discretization" argument we conclude that:

$$\sigma\left(\left\{\theta : h_{Z_p(\mu)}(\theta) \geqslant c_3 \frac{\sqrt{p}}{t\|\mu\|_\infty^{1/n}}, \forall p \in [1, n]\right\}\right) \geqslant 1 - (\log_2 n)(et)^{-n} > 1 - t^{-n},$$

for all $t > 1$.

Claim. Let $t > 1$ and $\theta \in A_t := \left\{\theta : h_{Z_p(\mu)}(\theta) > \frac{c_3\sqrt{p}}{t\|\mu\|_\infty^{1/n}}, \forall p \in [1, n]\right\}$. Then,

$$\mu(\{x : |\langle x, \theta\rangle| \geqslant s\}) \geqslant e^{-c(s\|\mu\|_\infty^{1/n})^2},$$

for all $1 \leqslant s \leqslant \frac{\sqrt{n}}{t\|\mu\|_\infty^{1/n}}$.

Proof of the claim. Fix $t > 1$. Then, for all $1 \leqslant p \leqslant n$ we have

$$\mu\left(\left\{x : |\langle x, \theta\rangle| \geqslant \frac{c_4\sqrt{p}}{t\|\mu\|_\infty^{1/n}}\right\}\right) \geqslant \mu\left(\left\{x : |\langle x, \theta\rangle| \geqslant \frac{1}{2}h_{Z_p(\mu)}(\theta)\right\}\right) \geqslant e^{-c_5 p},$$

where in the last step we have used the Paley-Zygmund inequality and the fact that $h_{Z_{2p}} \simeq h_{Z_p}$ for all $1 \leqslant p \leqslant n$. Hence, given $1 \leqslant s \leqslant \frac{\sqrt{n}}{t\|\mu\|_\infty^{1/n}}$, we define p by the equation $p = (st\|\mu\|_\infty^{1/n})^2$ and we obtain:

(8.4.5) $$\mu(\{x : |\langle x, \theta\rangle| \geqslant c_4 s\}) \geqslant e^{-c_5 s^2(t\|\mu\|_\infty^{1/n})^2}$$

for all $1 \leqslant s \leqslant \sqrt{n}/(t\|\mu\|_\infty^{1/n})$. This completes the proof. $\square$

The claim shows that
$$\{\theta \in S^{n-1} : sg_\mu(\theta) \leqslant c_6 t\|\mu\|_\infty^{1/n}\} \supseteq A_t.$$

Using the measure estimate for A_t we conclude the first assertion of the theorem.

For the "in particular" case note that by the definition of sg_μ we trivially have $sg_\mu(\theta) \leqslant c\sqrt{n}$ for all $\theta \in S^{n-1}$. We also know that the set

$$A := \{\theta : sg_\mu(\theta) \leqslant cL_\mu\}$$

satisfies $\sigma(A) \geqslant 1 - e^{-n}$, therefore we can write:

$$\mathbb{E}(sg_\mu) = \int_A sg_\mu(\theta)\, d\sigma(\theta) + \int_{A^c} sg_\mu(\theta)\, d\sigma(\theta)$$
$$\leqslant cL_\mu + c_1\sqrt{n}e^{-c_2\sqrt{n}} \leqslant c'L_\mu,$$

taking into account the fact that $L_\mu \geqslant c_3 > 0$ for all μ. $\square$

The next result is in the same spirit as Theorem 3.4.1. The proof in this case is much simpler.

THEOREM 8.4.2. *Let μ be a centered log-concave probability measure on $\mathbb{R}^n$ and set $a_\mu = \sup_{\theta \in S^{n-1}} sg_\mu(\theta)$. Then,*

(8.4.6) $$L_\mu \leqslant ca_\mu,$$

where $c > 0$ is an absolute constant.

Proof. Recall that a_μ and L_μ are $GL(n)$-invariant, so we may assume that μ is isotropic.

Claim. We have that
$$|Z_n(\mu)|^{1/n} \geqslant \frac{c}{a_\mu},$$
where $c > 0$ is an absolute constant.

Indeed; if $r = sg_\mu(\theta)$ then we may write:
$$h^n_{Z_n(\mu)}(\theta) = n \int_0^\infty t^{n-1} \mu(\{x : |\langle x, \theta \rangle| \geqslant t\}) \, dt$$
$$\geqslant n c_1^n \int_0^{\sqrt{n}/r} t^{n-1} \mu(\{x : |\langle x, \theta \rangle| \geqslant t \|\langle \cdot, \theta \rangle\|_1\}) \, dt$$
$$\geqslant n c_1^n r^{-n} \int_0^{\sqrt{n}} t^{n-1} e^{-t^2} \, dt$$
$$\geqslant \frac{n c_1^n}{r^n} (c_2 \sqrt{n})^n.$$

This shows that $Z_n(\mu) \supseteq c_3 \frac{\sqrt{n}}{a_\mu} B_2^n$, which proves the claim.

On the other hand we have (from Proposition 5.1.8) that
$$\|\mu\|_\infty^{1/n} \cdot |Z_n(\mu)|^{1/n} \simeq 1.$$

Combining this with the claim we get the result. □

8.4.1. Super-Gaussian estimates for marginals

We can also prove that if μ is log-concave isotropic measure on $\mathbb{R}^n$, then random marginals of μ exhibit better super-gaussian behavior. This result comes from [415]. The precise statement is the following:

THEOREM 8.4.3. *Let μ be an isotropic log-concave probability measure on $\mathbb{R}^n$. Then, for every $1 \leqslant k \leqslant \sqrt{n}$ there exists $B_k \subseteq G_{n,k}$ with $\nu_{n,k}(B_k) \geqslant 1 - e^{-c\sqrt{n}}$ such that $a_{\pi_F \mu} = \sup_{\theta \in S_F} sg_{\pi_F \mu}(\theta) \leqslant c$ for all $F \in B_k$ where $c > 0$ is an absolute constant.*

Proof. The proof is very similar to that of Theorem 5.2.15. We define the following variant of $q_*(\mu)$:
$$(8.4.7) \qquad q_\natural(\mu) = \max\{q \leqslant n \mid k_*(Z_p(\mu)) \geqslant p \text{ for all } p \in [1, q]\}.$$

Note that the previous parameter is well defined since $k_*(Z_2(\mu)) = n$ and $k_*(Z_n(\mu)) \leqslant n$. (Actually, $q_\natural(\mu)$ is the "first root" of the equation $k_*(Z_q(\mu)) = q$ while $q_*(\mu)$ is the largest, hence $q_\natural(\mu) \leqslant q_*(\mu)$). For any $1 \leqslant p \leqslant q_\natural(\mu)$, arguing as in Proposition 5.2.13. we get:
$$k_*(Z_p(\mu)) \simeq n \left(\frac{w(Z_p(\mu))}{R(Z_p(\mu))}\right)^2 \geqslant c_1 n \left(\frac{w_p(Z_p(\mu))}{R(Z_p(\mu))}\right)^2,$$

where we have used Theorem 5.2.9. In addition, by Lemma 5.2.6, Hölder's inequality and the isotropic condition, we get
$$w_p(Z_p(\mu)) \geqslant c_2 \sqrt{\frac{p}{n}} I_p(\mu) \geqslant c_2 \sqrt{p},$$

for all $1 \leqslant p \leqslant n$. Taking into account that $R(Z_p(\mu)) \leqslant c_3 p$ for all p we conclude that:
$$k_*(Z_p(\mu)) \geqslant c_4 n/p$$
for all $p \leqslant q_\natural(\mu)$. In particular, we get
$$q_\natural \equiv q_\natural(\mu) = k_*(Z_{q_\natural}(\mu)) \geqslant c_4 n/q_\natural,$$
from which it follows that $q_\natural(\mu) \geqslant c_5 \sqrt{n}$ with $c_5 = c_4^{1/2}$. Let $1 \leqslant k \leqslant \sqrt{n}$. Fix $q \leqslant k$. Then, by Dvoretzky theorem we obtain a subset $B_{k,q}$ of $G_{n,k}$ with measure
$$\nu_{n,k}(B_{k,q}) \geqslant 1 - e^{-ck_*(Z_q)} \geqslant 1 - e^{-cn/q} \geqslant 1 - e^{-c\sqrt{n}},$$
with the property
$$P_F(Z_q(\mu)) \supseteq c_6 w(Z_q(\mu)) B_F \supseteq c_7 \sqrt{q} B_F$$
for every $F \in B_{k,q}$. Since, $Z_{2q} \simeq Z_q$ we conclude that there exists a set $B_k = \cap_j B_{k,2^j}$ for $j = 1, \ldots, \log_2 k$ with measure
$$\nu_{n,k}(B_k) \geqslant 1 - (\log_2 k) e^{-c\sqrt{n}} \geqslant 1 - e^{-c'\sqrt{n}},$$
such that for every $F \in B_k$ we have
$$Z_q(\pi_F(\mu)) \supseteq c\sqrt{q} B_F$$
for all $1 \leqslant q \leqslant k$. Using the Paley-Zygmund inequality as in the proof of Theorem 8.4.1 we conclude that
$$\pi_F(\mu)(\{x : |\langle x, \theta \rangle| \geqslant t\}) \geqslant e^{-c_8 t^2}$$
for all $1 \leqslant t \leqslant c_9 \sqrt{k}$. This shows that $\sup_{\theta \in S_F} sg_{\pi_F \mu}(\theta) \leqslant c_{10}$ for some absolute constant $c_{10} > 0$. $\square$

8.5. ψ_α-estimates for marginals of isotropic log-concave measures

The purpose of this section is to show, following [211] that a random marginal $\pi_F(\mu)$ of an isotropic log-concave measure μ on $\mathbb{R}^n$ exhibits better ψ_α-behavior.

For the study of marginals, we need a variant of the ψ_α norm. Starting from the well-known fact that $\|f\|_{\psi_\alpha} \simeq \sup\left\{\frac{\|f\|_q}{q^{1/\alpha}} : q \geqslant \alpha\right\}$ and recalling that if μ is the Lebesgue measure μ_K on an isotropic convex body K in $\mathbb{R}^n$ and if f is a linear functional then

(8.5.1) $$\|f\|_{\psi_\alpha} \simeq \sup_{q \geqslant \alpha} \frac{\|f\|_q}{q^{1/\alpha}} \simeq \sup_{\alpha \leqslant q \leqslant n} \frac{\|f\|_q}{q^{1/\alpha}},$$

we define
$$\|f\|_{\psi'_\alpha} := \sup_{\alpha \leqslant q \leqslant n} \frac{\|f\|_q}{q^{1/\alpha}}.$$

It is clear that $\|f\|_{\psi'_\alpha} \leqslant c\|f\|_{\psi_\alpha}$. In view of (8.5.1) this is a natural definition of a "ψ_α-norm" when one studies the behavior of linear functionals with respect to a log-concave measure on $\mathbb{R}^n$.

With this definition we prove the following:

THEOREM 8.5.1. *Let μ be an isotropic log-concave measure on $\mathbb{R}^n$.*

(i) If $k \leq \sqrt{n}$ then there exists $A_k \subseteq G_{n,k}$ with measure $\nu_{n,k}(A_k) > 1 - \exp(-c\sqrt{n})$ such that, for every $F \in A_k$, $\pi_F(\mu)$ is a ψ_2'-measure with constant C, where $C > 0$ is an absolute constant.

(ii) If $k = n^\delta$, $\frac{1}{2} < \delta < 1$ then there exists $A_k \subseteq G_{n,k}$ with measure $\nu_{n,k}(A_k) > 1 - \exp(-ck)$ such that, for every $F \in A_k$, $\pi_F(\mu)$ is a $\psi_{\alpha(\delta)}'$-measure with constant C, where $\alpha(\delta) = \frac{2\delta}{3\delta-1}$ and $C > 0$ is an absolute constant.

Proof. In the case $k \leq \sqrt{n}$ the proof is very similar to that of Theorem 8.4.3. For any $k \leq \sqrt{n}$ and $1 \leq q \leq k$, using the upper inclusion of Dvoretzky's theorem this time, we get

$$P_F(Z_q(\mu)) \subseteq 2w(Z_q(\mu))(B_2^n \cap F)$$

for all F in a subset $B_{k,q}$ of $G_{n,k}$ of measure

$$\nu_{n,k}(B_{k,q}) \geq 1 - e^{-c_1 k_*(Z_q(\mu))} \geq 1 - e^{-c_2\sqrt{n}}.$$

Applying this argument for $q = 2^i$, $i = 1, \ldots, \log_2 k$, and taking into account the fact that $Z_p(\mu) \subseteq Z_q(\mu) \subseteq cZ_p(\mu)$ if $p < q \leq 2p$, we conclude that there exists $B_k \subset G_{n,k}$ with $\nu_{n,k}(B_k) \geq 1 - e^{-c_3\sqrt{n}}$ such that, for every $F \in B_k$ and every $1 \leq q \leq k$,

$$Z_q(\pi_F(\mu)) = P_F(Z_q(\mu)) \subseteq 2w(Z_q(\mu))(B_2^n \cap F).$$

Since $w(Z_q(\mu)) \simeq \sqrt{q}$ for all $q \leq \sqrt{n}$, the last formula can be written in the form

$$h_{Z_q(\pi_F(\mu))}(\theta) \leq c_4\sqrt{q}$$

for all $F \in B_k$, $\theta \in S_F$ and $1 \leq q \leq k$. From the inequality

$$\sup_{1 \leq q \leq k} \frac{\|\langle \cdot, \theta \rangle\|_{L_q(\pi_F(\mu))}}{\sqrt{q}} = \sup_{1 \leq q \leq k} \frac{h_{Z_q(\pi_F(\mu))}(\theta)}{\sqrt{q}} \leq C, \quad \theta \in S_F$$

we immediately get the result.

For the second case we begin with a simple lemma.

LEMMA 8.5.2. *Let μ be a centered, log-concave probability measure on $\mathbb{R}^n$. Let $1 \leq k < n$ be a positive integer, $1 \leq p \leq k$ and $q \geq 1$. Then,*

$$(8.5.2) \qquad \left(\int_{G_{n,k}} R^p(Z_q(\pi_F(\mu)))\, d\nu_{n,k}(F)\right)^{1/p} \leq c w_k(Z_q(\mu)),$$

where $c > 0$ is an absolute constant.

Proof. Recall that $Z_q(\pi_F(\mu)) = P_F(Z_q(\mu))$ for any $F \in G_{n,k}$. Thus, $P_F(Z_q(\mu))$ is a k-dimensional convex body in F and by Theorem 5.2.9 we conclude that

$R(P_F(Z_q(\mu))) \simeq w_k(P_F(Z_q(\mu)))$. Hence we may write:

$$\left(\int_{G_{n,k}} R^k(Z_q(\pi_F(\mu))) \, d\nu_{n,k}(F)\right)^{1/k} \simeq \left(\int_{G_{n,k}} w_k^k(P_F(Z_q(\mu))) \, d\nu_{n,k}(F)\right)^{1/k}$$

$$= \left(\int_{G_{n,k}} \int_{S_F} h_{P_F(Z_q(\mu))}^k(\theta) \, d\sigma_F(\theta) \, d\nu_{n,k}(F)\right)^{1/k}$$

$$= \left(\int_{S^{n-1}} h_{Z_q(\mu)}^k(\theta) \, d\sigma(\theta)\right)^{1/k}$$

$$= w_k(Z_q(\mu)),$$

where we have used the invariance under $O(n)$ and the fact that $h_{P_F(C)}(\theta) = h_C(\theta)$ for any $\theta \in S_F$ and any convex body C. This proves the lemma. □

The next lemma gives upper bounds for the quantity $w_k(Z_q(\mu))$ when μ is an isotropic log-concave measure.

LEMMA 8.5.3. *Let μ be an isotropic log-concave measure on $\mathbb{R}^n$. Let $\delta \in (1/2, 1)$ and $k = n^\delta$ be a positive integer. For all $1 \leqslant q \leqslant k$ we have $w_k(Z_q(\mu)) \leqslant c_1 q^{1/\alpha}$, where $\alpha = \frac{2\delta}{3\delta - 1}$ and $c_1 > 0$ is an absolute constant.*

Proof. Fixing $1 \leqslant q \leqslant k$ we distinguish two cases:

1. If $k \leqslant k_*(Z_q(\mu))$. Then, by Theorem 5.2.9 we get

$$w_k(Z_q(\mu)) \simeq w(Z_q(\mu)) \leqslant w_q(Z_q(\mu)) \leqslant c_1 \sqrt{q/n} I_q(\mu),$$

where we have used Lemma 5.2.6. Since $q \leqslant k_*(Z_q(\mu))$, this implies that $q \leqslant q_*(\mu)$ and by Theorem 5.2.15 we conclude that $I_q(\mu) \leqslant c_2\sqrt{n}$. This proves that $w_k(Z_q(\mu)) \leqslant c_3\sqrt{q}$.

2. If $k > k_*(Z_q(\mu))$ then, using Theorem 5.2.9 again, we have

$$w_k(Z_q(\mu)) \leqslant c_4\sqrt{k/n} R(Z_q(\mu)) \leqslant c_5 q\sqrt{k/n},$$

where we have also used the fact that $R(Z_q(\mu)) \leqslant c_4' q$ for any isotropic log-concave measure μ. We are looking for the maximal value of α such that $q\sqrt{k/n} \leqslant q^{1/\alpha}$ or equivalently, $q^{1-1/\alpha} k^{1/2} \leqslant n^{1/2}$ for all $q \leqslant k$. Hence, it suffices to have $k^{3/2 - 1/\alpha} \leqslant n^{1/2}$ and since $k = n^\delta$ we get $(3/2 - 1/\alpha)\delta \leqslant 1/2$, from which it follows that $\alpha \leqslant \frac{2\delta}{3\delta - 1}$.

Combining the two cases we conclude that $w_k(Z_q(\mu)) \leqslant cq^{1/\alpha}$ where $\alpha = \alpha(\delta) = \frac{2\delta}{3\delta - 1}$ for all $1 \leqslant q \leqslant k = n^\delta$. This completes the proof. □

Proof of Theorem 8.5.1 (ii). We apply Markov's inequality for $q = 2^i$, $i = 1, \ldots, \log_2 k$ in Lemma 8.5.2, and taking into account the fact that $Z_p(\mu) \subseteq Z_q(\mu) \subseteq cZ_p(\mu)$ if $p < q \leqslant 2p$, we conclude that

$$\sup_{1 \leqslant q \leqslant k} \frac{R(Z_q(\pi_F(\mu)))}{w_k(Z_q(\mu))} \leqslant C,$$

where $C > 0$ is an absolute constant, for all F in a subset A_k of $G_{n,k}$ with measure $\nu_{n,k}(A_k) \geqslant 1 - (\log_2 k) e^{-2k} \geqslant 1 - e^{-k}$.

Next, we use the estimates from Lemma 8.5.3; for every $F \in A_k$ we have

$$\|\langle \cdot, \theta \rangle\|_{\psi'_{\alpha(\delta)}} = \sup_{1 \leqslant q \leqslant k} \frac{\|\langle \cdot, \theta \rangle\|_{L_q(\pi_F(\mu))}}{q^{1/\alpha(\delta)}} \leqslant C_1 \sup_{1 \leqslant q \leqslant k} \frac{R(Z_q(\pi_F(\mu)))}{w_k(Z_q(\mu))} \leqslant C_2$$

for all $\theta \in S_F$, where $C_2 > 0$ is an absolute constant. □

8.6. Further reading

In Section 8.6.1 we present some results on the existence of sub-Gaussian directions in the case of unconditional bodies. For this class, more precise information on the distribution of the ψ_2-norm had been obtained by Bobkov and Nazarov before the appearance of Klartag's general theorem on this question. In Section 8.6.2 we present an example of Paouris which shows that ψ_2-behavior is unstable with respect to the geometric distance; one can have an isotropic convex body which is c-close to a Euclidean ball and has (a small set of) directions with ψ_2-constant as large as $\sqrt[4]{n}$.

8.6.1. Linear functionals on unconditional convex bodies

We start with the example of B_1^n, the unit ball of ℓ_1^n, equipped with the uniform distribution μ_n. The density of μ_n is given by

$$\frac{d\mu_n(x)}{dx} = \frac{n!}{2^n} \mathbf{1}_{B_1^n}(x).$$

We also define $\Delta_n = \{x \in \mathbb{R}_+^n : x_1 + \cdots + x_n \leqslant 1\}$.

Let $\theta = (\theta_1, \ldots, \theta_n) \in \mathbb{R}^n$ and consider the linear functional $f_\theta(x) = \langle \theta, x \rangle = \theta_1 x_1 + \cdots + \theta_n x_n$.

LEMMA 8.6.1. *For every* $q \in \mathbb{N}$,

$$\int |f_\theta(x)|^{2q} d\mu_n(x) = \frac{n!(2q)!}{(n+2q)!} \sum \theta_1^{2q_1} \cdots \theta_n^{2q_n},$$

where the summation is over all non-negative integers $q_1, \ldots, q_n$ *with* $q_1 + \cdots + q_n = q$.

PROOF. Using the fact that $\|\cdot\|_1$ is unconditional, we write

(8.6.1) $$\int |f_\theta|^{2q} d\mu_n = \sum \frac{(2q)!}{(2q_1)! \cdots (2q_n)!} \theta_1^{2q_1} \cdots \theta_n^{2q_n} \int x_1^{2q_1} \cdots x_n^{2q_n} d\mu_n(x),$$

where the summation is over all non-negative integers $q_1, \ldots, q_n$ with $q_1 + \cdots + q_n = q$. Now, a simple computation shows that

$$\int x_1^{2q_1} \cdots x_n^{2q_n} d\mu_n(x) = n! \int_{\Delta_n} x_1^{2q_1} \cdots x_n^{2q_n} dx = \frac{n!(2q_1)! \cdots (2q_n)!}{(n+2q)!}.$$

This proves the lemma. □

THEOREM 8.6.2. *For every* $\theta \in \mathbb{R}^n$,

$$c_1 \|\theta\|_\infty \leqslant \sqrt{n} \|f_\theta\|_{L^{\psi_2}(\mu_n)} \leqslant c_2 \|\theta\|_\infty,$$

where $c_1 = 1/\sqrt{6}$ *and* $c_2 = 2\sqrt{2}$.

Proof. We set $\alpha = \sqrt{n}\|\theta\|_\infty$. Then, we have $\theta_1^{2q_1}\cdots\theta_n^{2q_n} \leq \alpha^{2q}/n^q$ for all $q_1,\ldots,q_n \geq 1$ with $q_1 + \cdots + q_n = q$. Since the sum in (8.6.1) consists of $\binom{n+q-1}{n-1}$ terms, we get

$$\text{(8.6.2)} \quad \int |f_\theta(x)|^{2q} d\mu_n(x) \leq \frac{n!(2q)!}{(n+2q)!}\frac{(n+q-1)!}{(n-1)!q!}\frac{\alpha^{2q}}{n^q}$$

$$= \frac{n}{(n+q)\cdots(n+2q)}\binom{2q}{q}\frac{q!\alpha^{2q}}{n^q}$$

$$\leq \frac{4^q q! \alpha^{2q}}{n^{2q}}.$$

Therefore, if $|t| < 1/(2\alpha)$, we have

$$\int \exp\left((tn f_\theta(x))^2\right)d\mu_n(x) = 1 + \sum_{q=1}^\infty \frac{t^{2q} n^{2q}}{q!}\int |f_\theta(x)|^{2q}d\mu_n(x)$$

$$\leq 1 + \sum_{q=1}^\infty (2t\alpha)^{2q} = \frac{1}{1-4t^2\alpha^2}.$$

If we choose $t = 1/(2\sqrt{2}\alpha)$, the last quantity becomes equal to 2. This shows that

$$n\|f_\theta\|_{L^{\psi_2}(\mu_n)} \leq 2\sqrt{2}\alpha = 2\sqrt{2}\sqrt{n}\|\theta\|_\infty.$$

We now turn to the lower bound. We may clearly assume that $\theta_j \geq 0$ for all $1 \leq j \leq n$, and Lemma 8.6.1 shows that for every $q \geq 1$ the function $F_{2q}(\theta) := \|f_\theta\|_{2q}$ is increasing with respect to each coordinate on $\mathbb{R}_+^n$. It follows that $F_{\psi_2}(\theta) := \|f_\theta\|_{L^{\psi_2}(\mu_n)}$ has the same property. So,

$$\text{(8.6.3)} \quad \|f_\theta\|_{L^{\psi_2}(\mu_n)} \geq \|\theta\|_\infty \|f_{e_1}\|_{L^{\psi_2}(\mu_n)}.$$

Since $g_1 := f_{e_1}/\|f_{e_1}\|_{L^{\psi_2}(\mu_n)}$ has ψ_2-norm equal to 1, we have

$$2 \geq \int \exp\left([g_1(x)]^2\right)d\mu_n(x) \geq \frac{1}{n!}\frac{1}{\|f_{e_1}\|_{L^{\psi_2}(\mu_n)}^{2n}}\int |f_{e_1}(x)|^{2n} d\mu_n(x).$$

Taking into account (8.6.2), we get

$$\|f_{e_1}\|_{L^{\psi_2}(\mu_n)}^{2n} \geq \frac{1}{2n!}\frac{n!(2n)!}{(3n)!} \geq \frac{1}{2(3n)^n}.$$

Then, (8.6.3) shows that

$$\sqrt{n}\|f_\theta\|_{L^{\psi_2}(\mu_n)} \geq \frac{\sqrt{n}}{\sqrt[2n]{2}\sqrt{3n}}\|\theta\|_\infty,$$

which gives the left hand side inequality of the theorem. $\square$

Note that the proof of Theorem 8.6.2 gives the following estimate for $\|f_\theta\|_{L^p(\mu_n)}$, $p \geq 1$.

PROPOSITION 8.6.3. *For every $\theta \in \mathbb{R}^n$ and every $p \geq 1$,*

$$\sqrt{n}\|f_\theta\|_p \leq c\sqrt{p}\|\theta\|_\infty,$$

where $c = 2\sqrt{2}$.

Proof. For $p = 2q$ where $q \in \mathbb{N}$ this follows immediately (with $c = 2$) from the estimate (8.6.2). It is then easily extended to all values of $p \geq 1$. $\square$

The estimates that we obtained for the ℓ_1^n-ball carry over to any isotropic unconditional convex body. The main tool is a comparison theorem of Bobkov and Nazarov that we discuss next. Recall that, from Theorem 4.1.5, there exists an absolute constant $C > 0$ with the following property: if K is an isotropic unconditional convex body in $\mathbb{R}^n$, then

$$\text{(8.6.4)} \quad |\{x \in K^+ : x_1 \geq \alpha_1, \ldots, x_n \geq \alpha_n\}| \leq \left(1 - \frac{\alpha_1 + \cdots + \alpha_n}{Cn}\right)^n$$

for all $(\alpha_1, \ldots, \alpha_n) \in K^+$. In particular, $K \subseteq (C/2)nB_1^n$ (the constant $C = \sqrt{6}$ works). We fix such a constant C and define $V = (C/2)nB_1^n$. We also denote by μ_T the uniform distribution on a convex body T.

We say that a function $F : \mathbb{R}^n \to \mathbb{R}^+$ belongs to the class $\mathcal{F}_n$ if it satisfies the following:

1. F is symmetric with respect to each coordinate: we have
$$F(x_1, \ldots, x_n) = F(\varepsilon_1 x_1, \ldots, \varepsilon_n x_n)$$
for every $x \in \mathbb{R}^n$ and all choices of signs ε_i.

2. There exists a positive Borel measure ν on $\mathbb{R}_+^n$ which is finite on all compact sets, such that
$$F(x) = \nu([0, x_1] \times \cdots \times [0, x_n])$$
for every $x = (x_1, \ldots, x_n) \in \mathbb{R}_+^n$. If F is absolutely continuous with respect to Lebesgue measure, this means that there exists a measurable function $q : \mathbb{R}_+^n \to \mathbb{R}^+$ such that
$$F(x) = \int_0^{x_1} \cdots \int_0^{x_n} q(z) dz$$
for every $x = (x_1, \ldots, x_n) \in \mathbb{R}_+^n$.

In this section we prove the following comparison theorem.

THEOREM 8.6.4 (Bobkov-Nazarov). *Let $F \in \mathcal{F}_n$. For every isotropic unconditional convex body K in $\mathbb{R}^n$ and every $t \geq 0$,*
$$\mu_K(\{x \in \mathbb{R}^n : F(x) \geq t\}) \leq \mu_V(\{x \in \mathbb{R}^n : F(x) \geq t\}).$$
In particular,
$$\int F(x) d\mu_K(x) \leq \int F(x) d\mu_V(x).$$

Proof. We will show that for every $\alpha_1, \ldots, \alpha_n \geq 0$,
$$\mu_K(|x_1| \geq \alpha_1, \ldots, |x_n| \geq \alpha_n) \leq \mu_V(|x_1| \geq \alpha_1, \ldots, |x_n| \geq \alpha_n).$$
We may assume that $\alpha = (\alpha_1, \ldots, \alpha_n)$ is in the interior of K (otherwise, the quantity on the left is equal to zero). Then, we also have $\alpha \in V$.

Observe that (8.6.4) is equivalent to
$$\mu_K(|x_1| \geq \alpha_1, \ldots, |x_n| \geq \alpha_n) \leq \left(1 - \frac{2(\alpha_1 + \cdots + \alpha_n)}{Cn}\right)^n.$$

Note that
$$|\{x \in bB_1^n : |x_i| \geq \alpha_i\}| = 2^n \left|\left\{x \in (\mathbb{R}^n)^+ : x_i \geq \alpha_i, \sum_{i=1}^n x_i \leq b\right\}\right|$$
$$= 2^n \left|\left\{y \in (\mathbb{R}^n)^+ : \sum_{i=1}^n y_i \leq b - \sum_{i=1}^n \alpha_i\right\}\right|$$
$$= \left|(b - \sum_{i=1}^n \alpha_i)B_1^n\right|$$

for every $b > 0$. Therefore,
$$\mu_V(|x_1| \geq \alpha_1, \ldots, |x_n| \geq \alpha_n) = \frac{|\{x \in V : |x_1| \geq \alpha_1, \ldots, |x_n| \geq \alpha_n\}|}{|V|}$$
$$= \frac{|[(C/2)n - (\alpha_1 + \cdots + \alpha_n)]B_1^n|}{|(C/2)nB_1^n|}$$
$$= \left(1 - \frac{2(\alpha_1 + \ldots + \alpha_n)}{Cn}\right)^n.$$

This proves the first claim. Observe that this is equivalent to the second claim for the function $F_\alpha = \mathbf{1}_{\{|x_1| \geqslant \alpha_1, \ldots, |x_n| \geqslant \alpha_n\}}$. Note that F_α corresponds to the Dirac measure ν_α which gives unit mass to the point α. Since the class $\{F_\alpha : \alpha \in \mathbb{R}_+^n\}$ coincides with the extreme points of the cone $\mathcal{F}_n$, the result follows. $\square$

We exploit the comparison theorem as follows. Let K be an isotropic unconditional convex body in $\mathbb{R}^n$. For every $\theta = (\theta_1, \ldots, \theta_n) \in \mathbb{R}^n$ consider the linear functional $f_\theta(x) = \theta_1 x_1 + \cdots + \theta_n x_n$. Note that for every $q \in \mathbb{N}$,

$$(8.6.5) \qquad \int |f_\theta(x)|^{2q} d\mu_K(x) = (2q)! \sum \frac{\theta_1^{2q_1} \cdots \theta_n^{2q_n}}{(2q_1)! \cdots (2q_n)!} \int x_1^{2q_1} \cdots x_n^{2q_n} d\mu_K(x),$$

where the summation is over all non-negative integers $q_1, \ldots, q_n$ with $q_1 + \cdots + q_n = q$. From Theorem 8.6.4 it is clear that

$$\int |x_1|^{p_1} \cdots |x_n|^{p_n} d\mu_K(x) \leqslant \int |x_1|^{p_1} \cdots |x_n|^{p_n} d\mu_V(x)$$

for all $p_1, \ldots, p_n \geqslant 0$. This gives immediately the following.

PROPOSITION 8.6.5. *Let K be an isotropic unconditional convex body in $\mathbb{R}^n$. For every $\theta \in \mathbb{R}^n$ and every integer $q \geqslant 1$,*

$$\int |f_\theta(x)|^{2q} d\mu_K(x) \leqslant \int |f_\theta(x)|^{2q} d\mu_V(x)$$

and

$$\|f_\theta\|_{L^{\psi_2}(\mu_K)} \leqslant \|f_\theta\|_{L^{\psi_2}(\mu_V)}.$$

Proof. The second inequality follows from the first if we take the Taylor expansion of $\exp((f_\theta/r)^2)$ with $r = \|f_\theta\|_{L^{\psi_2}(\mu_V)}$ and integrate with respect to μ_K. $\square$

Observe that

$$\|f_\theta\|_{L^{\psi_2}(\mu_V)} = \frac{Cn}{2} \|f_\theta\|_{L^{\psi_2}(\mu_n)}.$$

Then, the next theorem follows from Theorem 8.6.2.

THEOREM 8.6.6 (Bobkov-Nazarov). *Let K be an isotropic unconditional convex body in $\mathbb{R}^n$. For every $\theta \in \mathbb{R}^n$,*

$$\|f_\theta\|_{L^{\psi_2}(\mu_K)} \leqslant c\sqrt{n} \|\theta\|_\infty,$$

where $c > 0$ is an absolute constant. $\square$

8.6.2. Instability of the ψ_2-behavior: an example

In this section we show that examples of bad ψ_2-behavior may be given by very simple symmetric convex bodies of the form

$$(8.6.6) \qquad K = \{(x, t) \in \mathbb{R}^{n-1} \times \mathbb{R} : |t| \leqslant a, \|x\|_2 \leqslant f(|t|)\},$$

where $f : [0, R] \to \mathbb{R}^+$ is defined by $f(t) = a - bt$ for some $a > 0$ and $0 \leqslant b \leqslant a/R$ (this was observed by Paouris in [410]).

LEMMA 8.6.7. *There exist $a \simeq \sqrt{n}$ and $b \simeq 1/\sqrt{n}$ such that the symmetric convex body*

$$W = \{y = (x, t) : |t| \leqslant a, \|x\|_2 \leqslant a - b|t|\}$$

has volume 1 and satisfies

$$(8.6.7) \qquad c_1 \leqslant \int_W \langle y, \theta \rangle^2 dy \leqslant c_2$$

for every $\theta \in S^{n-1}$, where $c_1, c_2 > 0$ are absolute constants.

Proof. Let r be the solution of the equation $\omega_{n-1}r^{n-1} = 1$, and consider the body
$$K = \{(x,t) : |t| \leqslant r, \|x\|_2 \leqslant r - |t|/\sqrt{n}\}.$$
Then,
$$|K| = 2\omega_{n-1}r^{n-1} \cdot \sqrt{n}r \cdot \frac{1 - \left(1 - \frac{1}{\sqrt{n}}\right)^n}{n} \simeq 1,$$
since $r \simeq \sqrt{n}$. Let $s > 0$ be such that the body $W := sK$ has volume 1. Then, $s^n \simeq 1$, and W is of the form (8.6.6), where $R = a \simeq \sqrt{n}$ and $b = 1/\sqrt{n}$. Note that
$$\|\langle \cdot, e_n \rangle\|_1^{-1} \simeq |W \cap e_n^\perp| = \omega_{n-1}r^{n-1}s^{n-1} \simeq 1,$$
and hence
$$\text{(8.6.8)} \qquad \int_W \langle y, e_n \rangle^2 dy \simeq \|\langle \cdot, e_n \rangle\|_1^2 \simeq 1.$$

Since W is symmetric with respect to the coordinate subspaces, (8.6.7) will hold for every $\theta \in S^{n-1}$ if we check that
$$c_3 \leqslant \int_W \langle y, e_j \rangle^2 dy \leqslant c_4$$
for every $j = 1, \ldots, n-1$. To this end we estimate $|W \cap e_j^\perp|$. For the upper bound we use Hölder's inequality and the fact that $2\omega_{n-1}\int_0^a (a - bt)^{n-1} dt = |W| = 1$:
$$|W \cap e_j^\perp| = 2\omega_{n-2}\int_0^a (a - bt)^{n-2} dt$$
$$\leqslant 2\omega_{n-2}\left(\int_0^a (a - bt)^{n-1} dt\right)^{\frac{n-2}{n-1}} \sqrt[n-1]{a}$$
$$= 2\omega_{n-2}(2\omega_{n-1})^{-\frac{n-2}{n-1}} \sqrt[n-1]{a} \leqslant c_5 \sqrt[n-1]{a} \leqslant c_6$$
where $c_6 > 0$ is an absolute constant. For the lower bound we observe that
$$|W \cap e_j^\perp| = 2\omega_{n-2}\int_0^a (a - bt)^{n-2} dt \geqslant \frac{\omega_{n-2}}{a\omega_{n-1}} 2\omega_{n-1} \int_0^a (a - bt)^{n-1} dt$$
$$= \frac{\omega_{n-2}}{a\omega_{n-1}} \geqslant c_7 \frac{\sqrt{n}}{a} \geqslant c_8,$$
where $c_8 > 0$ is an absolute constant. It follows that
$$\int_W \langle y, e_j \rangle^2 dy \simeq |W \cap e_j^\perp|^{-2} \simeq 1.$$
and the lemma is proved. $\square$

Starting with W we may easily pass to a "similar" isotropic body.

THEOREM 8.6.8. *There exist $R_1 \simeq a_1 \simeq \sqrt{n}$ and $b_1 \simeq 1/\sqrt{n}$ such that the symmetric convex body*
$$Q = \{y = (x,t) : |t| \leqslant R_1, \|x\|_2 \leqslant a_1 - b_1|t|\}$$
is isotropic and has the following properties:
$$\text{(8.6.9)} \qquad c_1\sqrt{n}B_2^n \subseteq Q \subseteq c_2\sqrt{n}B_2^n$$
and
$$\text{(8.6.10)} \qquad \|\langle \cdot, e_n \rangle\|_{\psi_2} \geqslant c_3\sqrt[4]{n}$$
where $c_1, c_2, c_3 > 0$ are absolute constants.

Proof. Consider the body W of the previous lemma. There exists a diagonal operator $T = \operatorname{diag}(u, \ldots, u, v)$ such that $Q = T(W)$ is isotropic. From (8.6.7) we easily check (see the arguments in Section 3.3.1) that $u, v \simeq 1$. Then, Q can be written in the form (8.6.6) with $R_1 = av$, $a_1 = au$ and $b_1 = bu/v$.

We first prove (8.6.9). For every $y = (x, t) \in Q$ we have

$$\|y\|_2^2 = \|x\|_2^2 + t^2 \leqslant a_1^2 + R_1^2 \leqslant C^2 n,$$

where $C > 0$ is an absolute constant, because $a_1, R_1 \simeq \sqrt{n}$. This shows that $Q \subseteq C\sqrt{n}B_2^n$. On the other hand, the inradius of Q is equal to $\min\{R_1, d\}$, where d is the distance from $(0, 0)$ to the line $y = a_1 - b_1 t$ in $\mathbb{R}^2$. We have

$$d = \frac{a_1}{\sqrt{1 + b_1^2}} \simeq \sqrt{n},$$

and hence $Q \supseteq c\sqrt{n}B_2^n$ for some absolute constant $c > 0$.

Next we prove (8.6.10). For every $q \geqslant 1$ we have

$$\|\langle \cdot, e_n \rangle\|_q \leqslant c_4 q \|\langle \cdot, e_n \rangle\|_1 \leqslant c_5 q,$$

where $c_5 > 0$ is an absolute constant. Note that, by Markov's inequality,

(8.6.11) $$|\{y \in Q : |\langle y, e_n \rangle| \geqslant e\|\langle y, e_n \rangle\|_q\}| \leqslant e^{-q}$$

for every $q \geqslant 1$. For any $s > 0$ we compute

$$|\{y \in Q : |\langle y, e_n \rangle| \geqslant s\}| = 2\omega_{n-1} \int_s^{R_1} (a_1 - b_1 t)^{n-1} dt$$
$$= 2\omega_{n-1} \frac{a_1^n}{nb_1} \left(\left(1 - \frac{b_1 s}{a_1}\right)^n - \left(1 - \frac{b_1 R_1}{a_1}\right)^n\right).$$

Since $\omega_{n-1} a_1^{n-1} = |Q \cap e_n^\perp| \simeq 1$, we have

$$2\omega_{n-1} \frac{a_1^n}{nb_1} \simeq \frac{a_1}{nb_1} \geqslant c_6$$

for some absolute constant $c_6 > 0$. If $s \leqslant a_1/2b_1 (\simeq n)$, using the numerical inequality $1 - x \geqslant e^{-2x}$ for $x \in [0, 1/2]$, we obtain

$$\left(1 - \frac{b_1 s}{a_1}\right)^n \geqslant \exp(-2b_1 n s / a_1) \geqslant \exp(-c_7 s).$$

On the other hand,

$$\left(1 - \frac{b_1 R_1}{a_1}\right)^n \leqslant \exp(-b_1 R_1 n / a_1) \leqslant \exp(-c_8 \sqrt{n}).$$

If $s \leqslant \frac{c_8 \sqrt{n}}{2c_7}$ (and $n \gg 1$) then

$$\exp(-c_8 \sqrt{n}) \leqslant \frac{1}{2} \exp(-c_7 s).$$

It follows that if $s \leqslant c_9 \sqrt{n}$ then

(8.6.12) $$|\{y \in Q : |\langle y, e_n \rangle| \geqslant s\}| \geqslant \frac{c_6}{2} \exp(-c_7 s).$$

We choose $q_0 = (c_9/(ec_5))\sqrt{n}$. Since $\|\langle \cdot, e_n \rangle\|_q \leqslant c_5 q$, for every $q \leqslant q_0$ we may use (8.6.11) and (8.6.12) with $s = e\|\langle \cdot, e_n \rangle\|_q$ to get

$$\exp(-q) \geqslant \frac{c_6}{2} \exp(-c_7 e \|\langle \cdot, e_n \rangle\|_q),$$

which finally gives (if $q \gg 1$)

(8.6.13) $$c_7 e \|\langle \cdot, e_n \rangle\|_q \geqslant q + c_8 \geqslant \frac{q}{2}.$$

Thus we may conclude that

(8.6.14) $$\|\langle \cdot, e_n\rangle\|_{\psi_2} = \sup\left\{\frac{\|\langle \cdot, e_n\rangle\|_q}{\sqrt{q}} : q \geqslant 1\right\} \geqslant c_{10}\sqrt{q_0} = c_3 \sqrt[4]{n}$$

for some absolute constant $c_3 > 0$. □

REMARK 8.6.9. Theorem 8.6.8 shows that the simple ψ_2-estimate of the order of $\sqrt[4]{n}$ cannot be improved, even for bodies which have uniformly bounded geometric distance to a Euclidean ball.

We can actually give a complete description of the L_q-centroid bodies of Q.

THEOREM 8.6.10. *Let Q be the isotropic convex body in Theorem 8.6.8. There exists $q_1 \simeq \sqrt{n}$ such that:*

(i) *For every $1 \leqslant q \leqslant n$,*
$$w(Z_q(Q)) \simeq \sqrt{q}.$$
(ii) *If $1 \leqslant q \leqslant q_1$ then $R(Z_q(Q)) \simeq q$, and if $q \geqslant q_1$ then $R(Z_q(Q)) \simeq \sqrt{n}$.*
(iii) *If $1 \leqslant q \leqslant q_1$ then $k_*(Z_q(Q)) \simeq n/q$, and if $q_1 \leqslant q \leqslant n$ then $k_*(Z_q(Q)) \simeq q$.*

Proof. Let q_1 be the largest $q \geqslant 1$ for which $\|\langle \cdot, e_n\rangle\|_q \geqslant c_{10}q$ holds true (where c_{10} is the constant in (8.6.14)). Note that $q_1 \geqslant c\sqrt{n}$.

(i) Using the bound $|Z_q(K)|^{1/n} \geqslant c\sqrt{q/n}$ and Urysohn's inequality, for every $1 \leqslant q \leqslant n$ we have
$$w(Z_q(Q)) \geqslant c\sqrt{n}|Z_q(Q)|^{1/n} \geqslant c'\sqrt{q}.$$
On the other hand, since $R(Q) = O(\sqrt{n})$, we see that
$$w(Z_q(Q)) \leqslant w_q(Z_q(Q)) \leqslant c\sqrt{q/n}R(Q) \leqslant c''\sqrt{q}.$$

(ii) In the case $q \leqslant q_1$, (8.6.13) shows that
$$R(Z_q(Q)) \simeq q$$
(we have that $\|\langle \cdot, e_n\rangle\|_{q_1} \geqslant c_{10}q_1$, which gives that $R(Z_{q_1}(Q)) \simeq q_1$, and then we may use the inverse inclusions for the Z_p-bodies that we get from Borell's lemma to deduce the claim for all $q \leqslant q_1$). Since $R(Z_{q_1}(Q)) \simeq q_1$, it follows that if $q \geqslant q_1$ then
$$\sqrt{n} \lesssim q_1 \simeq R(Z_{q_1}(Q)) \leqslant R(Z_q(Q)) \leqslant R(Q) \simeq \sqrt{n}.$$

(iii) Since we have determined $R(Z_q(Q))$ and $w(Z_q(Q))$ for every value of $q \in [1,n]$, we can compute the parameter $k_*(Z_q(Q))$. □

8.7. Notes and references

Covering numbers of the centroid bodies

Theorem 8.1.1 is due to Giannopoulos, Paouris and Valettas. They presented two proofs of this result in [210] and [212]. An additional feature of the first proof is a reduction to the case of convex bodies with uniformly bounded isotropic constant. The convolution argument that serves this purpose is analogous to an essential ingredient in Klartag's work [276, Section 4]; Klartag's convolution with a Gaussian is substituted here by convolution with a Euclidean ball. A main step in the second proof that is presented in this section is a weaker estimate that had been previously obtained by Giannopoulos, Pajor and Paouris in [209].

Volume radius of the ψ_2-body

A logarithmic in the dimension bound on the volume radius of $\Psi_2(\mu)$ was first obtained by Klartag in [276] and then by Giannopoulos, Pajor and Paouris in [209]. Theorem 8.2.1 which establishes the best known estimate v.rad$(\Psi_2(\mu)) \leqslant c\sqrt{\log n}$ is proved in [210].

Distribution of the ψ_2-norm

The function $\psi_K(t) := \sigma\left(\{\theta \in S^{n-1} : \|\langle \cdot, \theta \rangle\|_{\psi_2} \leqslant ct\sqrt{\log n}L_K\}\right)$ is introduced in [**212**] and the lower bounds that are presented in this section come from the same article. Klartag [**276**] has obtained some information on the distribution of the ψ_2-norm, but for a different position of the body K. More precisely, he proved that if K is centered and has volume 1 then there exists $T \in SL(n)$ such that the body $K_1 = T(K)$ has the following property: there exists $A \subseteq S^{n-1}$ with measure $\sigma(A) \geqslant \frac{4}{5}$ such that, for every $\theta \in A$ and every $t \geqslant 1$,

$$|\{x \in K_1 : |\langle x, \theta \rangle| \geqslant ct\|\langle \cdot, \theta \rangle\|_1\}| \leqslant \exp\left(-\frac{ct^2}{\log^2 n \log^5(t+1)}\right).$$

In his result the position K_1 of K is chosen so that $\Psi_2(K_1)$ is in the ℓ-position. Note that in this case we have the estimate

$$w(\Psi_2(K_1)) \leqslant c_1 \log n \, [\text{v.rad}(\Psi_2(K_1))].$$

On the other hand, we may write

$$\int_{S^{n-1}} \frac{h_{\Psi_2(K_1)}(\theta)}{h_{Z_2(K_1)}(\theta)} d\sigma(\theta) \leqslant \left(\int_{S^{n-1}} h_{\Psi_2(K_1)}^2(\theta) d\sigma(\theta)\right)^{1/2} \cdot \left(\int_{S^{n-1}} h_{Z_2(K_1)}^{-n}(\theta) d\sigma(\theta)\right)^{1/n}$$

$$\leqslant c_2 w(\Psi_2(K_1)) \text{v.rad}(Z_2^\circ(K_1)) = c_2 \frac{w(\Psi_2(K_1))}{\text{v.rad}(Z_2(K_1))},$$

where we have used Cauchy-Schwarz inequality, Hölder's inequality, and the equivalence of L_1 and L_2 norms on S^{n-1}.

Combining the previous estimates we conclude that

$$\int_{S^{n-1}} \frac{h_{\Psi_2(K_1)}(\theta)}{h_{Z_2(K_1)}(\theta)} d\sigma(\theta) \leqslant c_3 \log n \left(\frac{|\Psi_2(K_1)|}{|Z_2(K_1)|}\right)^{1/n} \leqslant C(\log n)^{3/2},$$

where in the last step we have used Theorem 8.2.1. An application of Markov's inequality shows that for any $\delta \in (0,1)$ we may find $\Theta_\delta \subseteq S^{n-1}$ with measure $\sigma(\Theta_\delta) \geqslant 1 - \delta$ such that every $\theta \in \Theta_\delta$ is a ψ_2-direction for K_1 with constant $C\delta^{-1}(\log n)^{3/2}$.

It is worth mentioning that if one is not interested in the full range of t's for which sub-Gaussian tail estimates hold, then there are "many" directions which satisfy a sub-Gaussian tail estimate for t up to $\sqrt[4]{n}$. More precisely, it is proved in [**212**] that if μ is an isotropic log-concave measure on $\mathbb{R}^n$ then there exists $A \subseteq S^{n-1}$ with $\sigma(A) \geqslant 1 - e^{-c_1\sqrt{n}}$ such that, for any $\theta \in A$ we have

$$\mu(\{x : |\langle x, \theta \rangle| \geqslant c_2 t\}) \leqslant e^{-t^2},$$

for all $1 \leqslant t \leqslant c_3 \sqrt[4]{n}$, where $c_1, c_2, c_3 > 0$ are absolute constants. In particular, there exist "many" directions $\theta \in S^{n-1}$ such that

$$\|\langle \cdot, \theta \rangle\|_{\psi_2'(\mu)} \leqslant c\sqrt[4]{n},$$

where $c > 0$ is an absolute constant.

Super-Gaussian directions

All the results of this section are due to Paouris [**414**]. The proofs can be carried out in a setting much broader than the one of log-concave measures. This is somehow expected: as Klartag shows in [**280**], any non-degenerate n-dimensional measure has at least one direction which is "super-Gaussian" on a non-trivial interval. The question about super-Gaussian directions was also considered by Pivovarov [**432**], who gave an affirmative answer (up to a logarithmic in the dimension factor) for the class of 1-unconditional bodies.

Linear functionals on unconditional convex bodies

The results of this section are due to to Bobkov and Nazarov [**91**]. Their method yields even stronger results on the ψ_2-behavior of linear functionals on unconditional isotropic convex bodies. Let $C_n(\theta) = \|\theta\|_\infty \sqrt{n/\log n}$ for $\theta \in \mathbb{R}^n$ and $n \geqslant 2$. Note that, since the expectation of $\|\theta\|_\infty$ on S^{n-1} is of the order of $\sqrt{\log n/n}$, for a random $\theta \in S^{n-1}$ we have $C_n(\theta) \simeq 1$. It is proved in [**91**] that if K is an isotropic unconditional convex body in $\mathbb{R}^n$ then for every $\theta \in \mathbb{R}^n$ and every $p \geqslant 2$,

$$\|f_\theta\|_{L^p(\mu_K)} \leqslant c \max\{1, C_n(\theta)\} \sqrt{p \log p},$$

where $c > 0$ is an absolute constant.

A very precise description of the behavior of linear functionals on the ℓ_p^n-balls, $1 \leqslant p \leqslant \infty$, can be found in the article [**58**] of Barthe, Guédon, Mendelson and Naor.

ψ_α-estimates for marginals of isotropic log-concave measures

Theorem 8.5.1 appears in [**211**]. In the same article, the super-Gaussian behavior of marginals is also discussed. It is proved that if μ is an isotropic log-concave probability measure on $\mathbb{R}^n$ and if $k \leqslant \sqrt{n}$, then there exists $B_k \subseteq G_{n,k}$ with measure $\nu_{n,k}(B_k) > 1 - \exp(-c\sqrt{n})$ such that, for every $F \in B_k$, $\pi_F(\mu)$ is a super-Gaussian measure with constant c (where $c > 0$ is an absolute constant) in the sense that

$$\inf_{\theta \in S_F} \overline{sg}_{\pi_F(\mu)}(\theta) \geqslant c.$$

CHAPTER 9

M and M^*-estimates

In this chapter we discuss the question to obtain upper bounds for the basic parameters

$$M^*(K) = w(K) = \int_{S^{n-1}} h_K(x)\, d\sigma(x) \quad \text{and} \quad M(K) = \int_{S^{n-1}} \|x\|_K\, d\sigma(x),$$

of an isotropic convex body K in $\mathbb{R}^n$. Both questions are open and a best-possible answer seems highly non-trivial. We describe the available approaches and the so far known estimates.

A trivial upper bound for $w(K)$, which follows immediately from the inclusion $K \subseteq (n+1)L_K B_2^n$, is $w(K) \leqslant (n+1)L_K$. In Section 9.1 we present several proofs of the estimate

$$w(K) \leqslant Cn^{3/4} L_K,$$

where $C > 0$ is an absolute constant.

Regarding $M(K)$, a trivial upper bound, which follows immediately from the inclusion $K \supseteq cL_K B_2^n$, is $M(K) \leqslant (cL_K)^{-1}$. In Section 9.2 we will see that one can have a better estimate, at least in the symmetric case. For every isotropic symmetric convex body K in $\mathbb{R}^n$ one has

$$M(K) \leqslant \frac{C \log n}{n^{1/10} L_K},$$

where $C > 0$ is an absolute constant.

In fact, the same questions can be discussed for the centroid bodies $Z_q(\mu)$, $q \leqslant \sqrt{n}$, of an isotropic log-concave measure μ on $\mathbb{R}^n$. We describe an approach that utilizes the local structure of $Z_q(\mu)$. We first obtain a lower bound for the inradius of proportional projections of $Z_q(\mu)$ and using an "entropy extension" theorem we obtain regular entropy estimates for the covering numbers $N(\sqrt{q}B_2^n, tZ_q(\mu))$, $t \geqslant 1$. Although all these estimates are most probably not optimal, we can still conclude that $Z_q(\mu)$, with $q \leqslant n^{2/5}$, is an (almost) 1-regular convex body in the sense of Pisier's theorem. As a consequence of this fact one can also get an upper bound for the parameter $M(Z_q(\mu))$. For every $1 \leqslant q \leqslant n^{2/5}$,

$$M(Z_q(\mu)) = \int_{S^{n-1}} \|x\|_{Z_q(\mu)}\, d\sigma(x) \leqslant C \frac{(\log q)^{5/2}}{\sqrt[4]{q}}.$$

9.1. Mean width in the isotropic case

Let K be an isotropic convex body in $\mathbb{R}^n$. In this section we discuss the question to obtain an upper bound for

$$M^*(K) = w(K) = \int_{S^{n-1}} h_K(x)\, d\sigma(x).$$

The next theorem provides the best known upper bound for $w(K)$, which, however, does not seem to be optimal.

THEOREM 9.1.1. *Let K be an isotropic convex body in $\mathbb{R}^n$. Then,*
$$w(K) \leqslant Cn^{3/4}L_K,$$
where $C > 0$ is an absolute constant.

Below, we describe several proofs of Theorem 9.1.1.

9.1.1. Approach through covering numbers

Let K be an isotropic convex body in $\mathbb{R}^n$. The first proof of the bound $w(K) \leqslant Cn^{3/4}L_K$ is due to Hartzoulaki [249] and uses the known bounds for the covering numbers of K (Theorem 3.2.2) and the Dudley-Fernique decomposition (see Section 3.3.3). Recall that for every $t > 0$ we have
$$N(K, tB_2^n) \leqslant 2\exp\left(\frac{6n^{3/2}L_K}{t}\right).$$
We also know that if $N_j = N(K, R/2^j B_2^n)$ then there exist $Z_j \subseteq \frac{3R}{2^j}B_2^n$ with cardinality $|Z_j| \leqslant N_j N_{j-1}$, $j = 1, 2, \ldots$, which satisfy the following: for every $y \in K$ and $m \in \mathbb{N}$ we can find $z_j \in Z_j$ and $w_m \in (R/2^m)B_2^n$ such that $y = z_1 + \cdots + z_m + w_m$. Consequently, for every $x \in \mathbb{R}^n$ we have
$$\max_{y \in K}|\langle y, x\rangle| \leqslant \sum_{j=1}^m \max_{z \in Z_j}|\langle x, z\rangle| + \max_{w \in R/2^m B_2^n}|\langle x, w\rangle|$$
$$= \sum_{j=1}^m \max_{z \in Z_j}|\langle x, z\rangle| + \frac{R}{2^m}\|x\|_2.$$
Note that
$$\log|Z_j| \leqslant \log N(K, (R/2^j)B_2^n) + \log N(K, (R/2^{j-1})B_2^n) \leqslant \frac{c_1 2^j}{R}n^{3/2}L_K.$$
Integration in polar coordinates shows that
$$w(K) \simeq \frac{1}{\sqrt{n}}\int_{\mathbb{R}^n} h_K(x)\,d\gamma_n(x).$$
We write
$$\int_{\mathbb{R}^n} h_K(x)\,d\gamma_n(x) \leqslant \sum_{j=1}^m \int_{\mathbb{R}^n}\max_{z\in Z_j}|\langle x,z\rangle|\,d\gamma_n(x) + \frac{R}{2^m}\int_{\mathbb{R}^n}\|x\|_2\,d\gamma_n(x)$$
$$\leqslant \sum_{j=1}^m \sqrt{\log|Z_j|}\max_{z\in Z_j}\|z\|_2 + \frac{R}{2^m}\sqrt{n}$$
$$\leqslant \sum_{j=1}^m \frac{3R}{2^j}c_2 n^{3/4}\sqrt{L_K}\frac{2^{j/2}}{\sqrt{R}} + \frac{R}{2^m}\sqrt{n}$$
$$\leqslant c_3 n^{3/4}\sqrt{R}\sqrt{L_K} + \frac{R}{2^m}\sqrt{n}.$$
Choosing m with $R \simeq 2^m$ and using the fact that $R(K) \leqslant c_4 n L_K$ we get the result.
$\square$

9.1.2. A bound through L_q-centroid bodies

We know that for every convex body K of volume 1 in $\mathbb{R}^n$, and for every $1 \leqslant q \leqslant n$,
$$I_q(K) \simeq \sqrt{n/q}\, w_q(Z_q(K)).$$
From Theorem 5.2.3 we also have $I_q(K) \leqslant CI_2(K)$ for all $2 \leqslant q \leqslant q_*(K)$. Then, Hölder's inequality implies that
$$w(Z_q(K)) \leqslant w_q(Z_q(K)) \leqslant c_1 \sqrt{q/n}\, I_2(K)$$
for all $2 \leqslant q \leqslant q_*(K)$. Assuming that K is in the isotropic position we have $I_2(K) = \sqrt{n} L_K$, and hence
$$w(Z_q(K)) \leqslant c_1 \sqrt{q} L_K$$
for all $2 \leqslant q \leqslant q_*(K)$. Recall that for an isotropic convex body K in $\mathbb{R}^n$ we have $q_*(K) \geqslant c\sqrt{n}$ (Corollary 5.2.14). Finally, recall that if $1 \leqslant p < q$ then
$$Z_q(K) \subseteq c_2 \frac{q}{p} Z_p(K),$$
and also $Z_n(K) \supseteq c_3 \mathrm{conv}(K, -K)$.

Using the above, we get a very simple proof of the bound for $w(K)$. For every $q \leqslant q_*(K)$,
$$w(K) \leqslant c_4 w(Z_n(K)) \leqslant c_4 c_2 \frac{n}{q} w(Z_q(K)) \leqslant c_4 c_2 c_1 \frac{n}{\sqrt{q}} L_K.$$
Since $q_*(K) \geqslant c\sqrt{n}$, choosing $q \simeq \sqrt{n}$ we get the result.

9.1.3. The method of random polytopes

In this subsection we describe an approach which is due to Pivovarov [**432**]. Using the large deviation estimate of Paouris for the Euclidean norm, he constructs a random polytope inside K which has small mean width. Then, he exploits the estimate of the next lemma in order to compare the mean width of this random polytope with the mean width of the body.

LEMMA 9.1.2. *Let C be a centered convex body in $\mathbb{R}^n$. Let $\theta \in S^{n-1}$ and $\varepsilon \in (0,1)$. Then,*
$$|\{x \in C : \langle x, \theta \rangle \geqslant \varepsilon h_C(\theta)\}| \geqslant e^{-1}(1-\varepsilon)^n |C|.$$

Proof. Let $x_0 \in C$ satisfy $h_C(\theta) = \langle x_0, \theta \rangle$. We observe that
$$\{x \in C : \langle x, \theta \rangle \geqslant \varepsilon h_C(\theta)\} \supseteq \{x \in C : \langle x, \theta \rangle \geqslant \varepsilon \langle x_0, \theta \rangle\} \supseteq \varepsilon x_0 + (1-\varepsilon) C^+,$$
where $C^+ = \{x \in C : \langle x, \theta \rangle \geqslant 0\}$. Thus, we get
$$|\{x \in C : \langle x, \theta \rangle \geqslant \varepsilon h_C(\theta)\}| \geqslant |\varepsilon x_0 + (1-\varepsilon) C^+| = (1-\varepsilon)^n |C^+|.$$
From Grünbaum's Lemma 2.2.6 we know that $|C^+| \geqslant \frac{1}{e}|C|$ and the result follows. $\square$

We consider N random points $x_1, \ldots, x_N$ independently and uniformly distributed in K and consider the random polytope $K_N := \mathrm{conv}\{x_1, \ldots, x_N\} \subseteq K$. From Paouris' deviation inequality we have

(9.1.1) $$\mathbb{P}\left(\max_{1 \leqslant j \leqslant N} \|x_j\|_2 \geqslant Ct\sqrt{n} L_K\right) \leqslant Ne^{-t\sqrt{n}}$$

for every $t \geqslant 1$. If we assume that $N < e^{t\sqrt{n}}$ we get the following.

LEMMA 9.1.3. *Let $t \geqslant 1$ and $N < e^{t\sqrt{n}}$. Then, the random polytope K_N satisfies*
$$w(K_N) \leqslant C_1 t \sqrt{\log N} L_K,$$
with probability greater than $1 - Ne^{-t\sqrt{n}}$.

Proof. We know that for every $z_1, \ldots, z_N$ one has

(9.1.2) $$\int_{S^{n-1}} \max_{1 \leqslant j \leqslant N} |\langle z_j, \theta \rangle| \, d\sigma(\theta) \leqslant c_1 \frac{\sqrt{\log N}}{\sqrt{n}} \max_{1 \leqslant j \leqslant N} \|z_j\|_2.$$

From (9.1.1) we see that with probability greater than $1 - Ne^{-t\sqrt{n}}$ we have $\|x_j\|_2 \leqslant Ct\sqrt{n}L_K$ for all $j = 1, \ldots, N$. Since $h_{K_N}(\theta) = \max_{j \leqslant N} \langle x_j, \theta \rangle$, it follows that

$$w(K_N) = \int_{S^{n-1}} \max_{1 \leqslant j \leqslant N} \langle x_j, \theta \rangle \, d\sigma(\theta) \leqslant \int_{S^{n-1}} \max_{1 \leqslant j \leqslant N} |\langle x_j, \theta \rangle| \, d\sigma(\theta)$$
$$\leqslant c_1 \frac{\sqrt{\log N}}{\sqrt{n}} \max_{1 \leqslant j \leqslant N} \|x_j\|_2 \leqslant c_1 C t \sqrt{\log N} L_K,$$

making use of (9.1.2). □

On the other hand, using Lemma 9.1.2 we get:

LEMMA 9.1.4. *Let K be a convex body of volume 1 in $\mathbb{R}^n$ and let $x_1, \ldots, x_N$ be independent random points which are uniformly distributed in K. Then, for the random polytope $K_N = \mathrm{conv}\{x_1, \ldots, x_N\}$ we have: for every $\theta \in S^{n-1}$ and $\varepsilon \in (0, 1)$,*
$$\mathbb{P}\left(h_{K_N}(\theta) \leqslant \varepsilon h_K(\theta)\right) \leqslant \exp(-Nv(\varepsilon, \theta)),$$
where $v(\varepsilon, \theta) = |\{x \in K : \langle x, \theta \rangle \geqslant \varepsilon h_K(\theta)\}|$.

Proof. We write
$$\mathbb{P}\left(h_{K_N}(\theta) \leqslant \varepsilon h_K(\theta)\right) = \mathbb{P}\left(\max_{1 \leqslant j \leqslant N} \langle x_j, \theta \rangle \leqslant \varepsilon h_K(\theta)\right)$$
$$= |\{x \in K : \langle x, \theta \rangle \leqslant \varepsilon h_K(\theta)\}|^N$$
$$= (1 - v(\varepsilon, \theta))^N \leqslant e^{-Nv(\varepsilon, \theta)}.$$
□

Upper bound for $w(K)$. Let K be an isotropic convex body in $\mathbb{R}^n$. Fix $t > 1$ and $\varepsilon \in (0, 1/2)$ which will be suitably chosen. We consider $N < e^{t\sqrt{n}}$ and independent, uniform random points $x_1, \ldots, x_N$ from K. Then, the following conditions are satisfied:

(i) $w(K_N) \leqslant C_1 t \sqrt{\log N} L_K$ with probability greater than $1 - Ne^{-t\sqrt{n}}$,
(ii) For fixed $\theta \in S^{n-1}$ we have $\mathbb{P}(h_{K_N}(\theta) \leqslant \varepsilon h_K(\theta)) \leqslant \exp(-Nv(\varepsilon, \theta))$.

Also, recall that $c_1 L_K B_2^n \subseteq K \subseteq c_0 n L_K B_2^n$ or, equivalently, $c_1 L_K \leqslant h_K(\theta) \leqslant c_0 n L_K$ for every $\theta \in S^{n-1}$. We fix $\delta \in (0, 1)$ and consider a δ-net $\mathcal{N}$ for S^{n-1} with cardinality $|\mathcal{N}| \leqslant (3/\delta)^n$.

CLAIM 9.1.5. *If $0 < \delta \leqslant \frac{c_1 \varepsilon}{2c_0 n}$ then we have*
$$\mathbb{P}\left(\exists \theta \in S^{n-1} : h_{K_N}(\theta) \leqslant \frac{\varepsilon}{2} h_K(\theta)\right) \leqslant \mathbb{P}(\exists z \in \mathcal{N} : h_{K_N}(z) \leqslant \varepsilon h_K(z)).$$

Proof of the Claim. Assume that there exists $\theta \in S^{n-1}$ such that $h_{K_N}(\theta) \leqslant \frac{\varepsilon}{2} h_K(\theta)$. We choose $z \in \mathcal{N}$ such that $\|z - \theta\|_2 < \delta$. Then,

$$h_{K_N}(z) \leqslant h_{K_N}(\theta) + h_{K_N}(z - \theta) \leqslant \frac{\varepsilon}{2} h_K(\theta) + h_K(z - \theta)$$
$$\leqslant \frac{\varepsilon}{2} h_K(z) + c_0 n \delta L_K \leqslant \frac{\varepsilon}{2} h_K(z) + \frac{c_0}{c_1} n \delta h_K(z).$$

Since $0 < \delta \leqslant \frac{c_1 \varepsilon}{2 c_0 n}$, the claim follows. □

Let $0 < \delta \leqslant \frac{c_1 \varepsilon}{2 c_0 n}$. Using the claim, we see that

$$\mathbb{P}\left(\exists \theta \in S^{n-1} : h_{K_N}(\theta) \leqslant \frac{\varepsilon}{2} h_K(\theta)\right) \leqslant |\mathcal{N}| \exp(-N v(\varepsilon))$$
$$\leqslant (3/\delta)^n \exp(-N v(\varepsilon)),$$

where

$$v(\varepsilon) = \inf_{z \in \mathcal{N}} v(\varepsilon, z) \geqslant e^{-1}(1 - \varepsilon)^n \geqslant e^{-3\varepsilon n}$$

by (ii) and Lemma 9.1.2, provided that $\frac{1}{n} < \varepsilon < \frac{1}{2}$.

If we establish that this last probability is small, we get $h_K(\theta) \leqslant \frac{2}{\varepsilon} h_{K_N}(\theta)$ for every $\theta \in S^{n-1}$. In this case, as long as we also have $N \geqslant n e^{3\varepsilon n} \log(3/\delta)$, we can write:

$$w(K) = \int_{S^{n-1}} h_K(\theta) \, d\sigma(\theta) \leqslant \frac{2}{\varepsilon} \int_{S^{n-1}} h_{K_N}(\theta) \, d\sigma(\theta)$$
$$\leqslant \frac{2}{\varepsilon} C_1 t \sqrt{\log N} L_K.$$

Choose $\delta = \frac{c_1 \varepsilon}{2 c_0 n}$. Then, it suffices to have $N \geqslant n e^{4\varepsilon n} \log(\frac{6 c_0 n}{c_1 \varepsilon})$ and $\frac{1}{n} < \varepsilon < \frac{1}{2}$. On the other hand, we must have $N < e^{t\sqrt{n}}$ and this yields the additional restriction $\varepsilon \gtrsim \frac{1}{t\sqrt{n}}$. We choose $\varepsilon = 1/\sqrt{n}$. Then, we have $n e^{4\varepsilon n} \log(\frac{6 c_0 n}{c_1 \varepsilon}) < e^{c_2 \sqrt{n}}$ for some absolute constant $c_2 > 1$. We choose $N = e^{c_2 \sqrt{n}}$ and $t = 2 c_2 > 1$, and this leads to the bound $w(K) \leqslant C_2 n^{3/4} L_K$, with $C_2 = 4 c_2^{3/2} C_1$. □

9.1.4. ψ_2-directions and mean width

Our next proof of Theorem 9.1.1 is related to the distribution of the norm $\|\langle \cdot, \theta \rangle\|_{\psi_2}$ on the sphere. We consider an isotropic convex body K in $\mathbb{R}^n$ and for every $2 \leqslant q \leqslant n$ we define

$$k_*(q) = n \left(\frac{w(Z_q(K))}{R(Z_q(K))}\right)^2.$$

Since $\|\langle \cdot, \theta \rangle\|_q \leqslant c q L_K$ for every $\theta \in S^{n-1}$, we have $R(Z_q(K)) \leqslant c q L_K$. Therefore,

$$w(Z_q(K)) \leqslant c q L_K \frac{\sqrt{k_*(q)}}{\sqrt{n}}.$$

From Proposition 5.1.2 we see that

$$w(K) \simeq w(Z_n(K)) \leqslant \frac{c n}{q} w(Z_q(K)) \leqslant c \sqrt{n} \sqrt{k_*(q)} \, L_K.$$

We define

$$\rho_* = \rho_*(K) := \min_{2 \leqslant q \leqslant n} k_*(q).$$

Since q was arbitrary, we get:

PROPOSITION 9.1.6. *For every isotropic convex body K in $\mathbb{R}^n$ we have*
$$w(K) \leqslant c\sqrt{n}\sqrt{\rho_*(K)}\, L_K.$$

Our next observation is the following: for every $q \geqslant 1$ the function $\theta \mapsto \|\langle \cdot, \theta \rangle\|_q$ is Lipschitz continuous with constant $R(Z_q(K))$; it follows that

$$\sigma\left(|\,\|\langle \cdot, \theta\rangle\|_q - w(Z_q(K))\,| \geqslant \frac{w(Z_q(K))}{2}\right) \leqslant \exp(-ck_*(q)) \leqslant \exp(-2c_1\rho_*)$$

where $c_1 > 0$ is an absolute constant. We assume that $\log n \leqslant e^{c\rho_*}$. Then,
$$\|\langle \cdot, \theta\rangle\|_q \simeq w(Z_q(K))$$
for all θ in a subset A_q of S^{n-1} with measure $\sigma(A_q) \geqslant 1 - \exp(-c_1\rho_*)$. Choosing $q_i = 2^i$, $i \leqslant \log_2 n$ and setting $A = \bigcap A_{q_i}$, we have the following:

LEMMA 9.1.7. *For every isotropic convex body K in $\mathbb{R}^n$ with $\rho_*(K) \geqslant C\log\log n$ we can find $A \subset S^{n-1}$ with $\sigma(A) \geqslant 1 - e^{-c\rho_*}$ such that*
$$\|\langle \cdot, \theta\rangle\|_q \simeq w(Z_q(K))$$
for every $\theta \in A$ and every $2 \leqslant q \leqslant n$. In particular,
$$\|\langle \cdot, \theta\rangle\|_{\psi_2} \simeq \max_{2 \leqslant q \leqslant n} \frac{w(Z_q(K))}{\sqrt{q}}$$
for all $\theta \in A$.

Lemma 9.1.7 shows that if the parameter $\rho_*(K)$ is large and the norm $\|\langle \cdot, \theta\rangle\|_{\psi_2}$ is well-bounded on a relatively large subset of the sphere then a similar upper bound holds for almost all directions. As a consequence we obtain a good upper bound for the mean width of the body K.

PROPOSITION 9.1.8. *Let K be an isotropic convex body in $\mathbb{R}^n$ which satisfies the next two conditions:*
  (i) $\rho_* := \rho_*(K) \geqslant C \log \log n$.
  (ii) *For some $b_n > 0$ we have $\|\langle \cdot, \theta\rangle\|_{\psi_2} \leqslant b_n L_K$ for all θ in a set $B \subseteq S^{n-1}$ with $\sigma(B) > e^{-c\rho_*}$.*

Then,
$$\|\langle \cdot, \theta\rangle\|_{\psi_2} \leqslant Cb_n L_K \tag{9.1.3}$$
for all θ in a set $A \subseteq S^{n-1}$ with $\sigma(A) > 1 - e^{-c\rho_*}$. Also,
$$w(Z_q(K)) \leqslant c\sqrt{q}\, b_n\, L_K \tag{9.1.4}$$
for all $2 \leqslant q \leqslant n$ and
$$w(K) \leqslant C_1 \sqrt{n}\, b_n\, L_K. \tag{9.1.5}$$

Proof. We can find $u \in A \cap B$, where A is the set from Lemma 9.1.7. Since $u \in B$, we have
$$\|\langle \cdot, u\rangle\|_q \leqslant c\sqrt{q}\|\langle \cdot, u\rangle\|_{\psi_2} \leqslant C_1\sqrt{q}\, b_n L_K$$
for every $2 \leqslant q \leqslant n$, and since u is also in A, this implies that
$$w(Z_q(K)) \leqslant C_2\sqrt{q}\, b_n L_K$$
for every $2 \leqslant q \leqslant n$. Choosing $q = n$ we get (9.1.5) (recall that $w(K) \simeq w(Z_n(K))$). From Lemma 9.1.7 we have that if $\theta \in A$ then
$$\|\langle \cdot, \theta\rangle\|_q \leqslant cw(Z_q(K)) \leqslant C_3\sqrt{q}\, b_n L_K$$

for every $2 \leqslant q \leqslant n$. Thus, we have checked (9.1.4)

Finally, for every $\theta \in A$ we get
$$\|\langle \cdot, \theta \rangle\|_{\psi_2} \simeq \max_{2 \leqslant q \leqslant n} \frac{\|\langle \cdot, \theta \rangle\|_q}{\sqrt{q}} \leqslant C b_n L_K.$$

This completes the proof. □

Propositions 9.1.6 and 9.1.8 provide a dichotomy: if the parameter $\rho_*(K)$ is small, then we can use Proposition 9.1.6 to give an upper bound for $w(K)$. If $\rho_*(K)$ is large, we can combine Proposition 9.1.8 with our estimate for $\psi_K(t)$ from Theorem 8.3.5 to get
$$\psi_K(t) \geqslant e^{-c_1 n/t^2} \geqslant e^{-c\rho_*},$$
with $t \simeq \sqrt{n/\rho_*}$. This results in the estimate
$$w(K) \leqslant C\sqrt{n \log n} \sqrt{n/\rho^*} L_K.$$

Combining the above, we get one more general upper bound for the mean width of K.

THEOREM 9.1.9. *For every isotropic convex body K in $\mathbb{R}^n$ we have*
$$w(K) \leqslant C\sqrt{n} \min\left\{\sqrt{\rho_*}, \sqrt{n \log n / \rho_*}\right\} L_K,$$
where $c > 0$ is an absolute constant.

Note that this recovers the bound $w(K) \leqslant C n^{3/4} \sqrt[4]{\log n} L_K$, which is slightly worse than the best known. However, with this argument, one would be able to obtain a better result if one could establish better lower bounds for the function $\psi_K(t)$.

We conclude this section with an argument based on Theorem 5.3.12 which allows us to remove the logarithmic term from the estimate in Theorem 9.1.9. We need a lemma.

LEMMA 9.1.10. *Let K be an isotropic convex body in $\mathbb{R}^n$. For every $1 \leqslant q \leqslant n$ we have*
$$w(Z_q(K)) \leqslant c\sqrt{q} L_K \left(1 + \sqrt{\frac{q}{d_*(Z_q(K))}}\right),$$
where $c > 0$ is an absolute constant.

Proof. We distinguish two cases:

(i) Let $d_*(Z_q(K)) \geqslant q$. Then by Theorem 5.3.12 we obtain:
$$w(Z_q(K)) \leqslant c_1 w_{-q}(Z_q(K)) \leqslant c_2 \sqrt{q/n} I_{-q}(K) \leqslant c_2 \sqrt{q/n} I_2(K) = c_2 \sqrt{q} L_K,$$
where we have also used (5.3.11).

(ii) Let $d_*(Z_q(K)) < q$. We consider $t > 1$ such that $q/t = d_*(Z_q(K))$. Then, applying Theorem 5.3.12 once again, we get:
$$w(Z_q(K)) \leqslant c_3 w_{-q/t}(Z_q(K)) \leqslant c_4 t w_{-q/t}(Z_{q/t}(K))$$
$$\leqslant c_5 t \sqrt{q/(tn)} I_{-q/t}(K) \leqslant c_5 \sqrt{qt} L_K.$$

Substituting the value of t into this last estimate we get
$$w(Z_q(K)) \leqslant c_5 \frac{q}{\sqrt{d_*(Z_q(K))}} L_K.$$

Taking into account both cases we have
$$w(Z_q(K)) \leqslant c\sqrt{q}L_K \max\left\{1, \sqrt{\frac{q}{d_*(Z_q(K))}}\right\}.$$
This proves the lemma. □

THEOREM 9.1.11. *Let K be an isotropic convex body in $\mathbb{R}^n$. Then, we have*
(9.1.6) $$w(K) \leqslant c\sqrt{n}\min\left\{\sqrt{\rho_*}, \sqrt{n/\sigma_*}\right\}L_K,$$
where $\sigma_* = \max\{2 \leqslant q \leqslant n : d_*(Z_q(K)) \geqslant q\}$. In particular, we have that
$$w(K) \leqslant cn^{3/4}L_K,$$
where $c > 0$ is an absolute constant.

Proof. We have already proved that $w(K) \leqslant c\sqrt{n}\sqrt{\rho_*}L_K$. Thus, we need to prove that
$$w(K) \leqslant c\frac{n}{\sqrt{\sigma_*}}L_K.$$
We use the previous lemma: For any $1 \leqslant q \leqslant n$ we can write
(9.1.7) $$w(K) \leqslant c_1\frac{n}{q}w(Z_q(K)) \leqslant c_2\frac{n}{q}\left(\sqrt{q} + \frac{q}{\sqrt{d_*(Z_q(K))}}\right)L_K$$
$$\leqslant c_2 n\left(\frac{1}{\sqrt{q}} + \frac{1}{\sqrt{d_*(Z_q(K))}}\right)L_K.$$

Note that σ_* is well defined because $d_*(Z_2(K)) = n$ and $d_*(Z_n(K)) \leqslant n$, and satisfies $d_*(Z_{\sigma_*}(K)) = \sigma_*$ because the function $q \mapsto d_*(Z_q(K))$ is continuous. Therefore, setting $q = \sigma_*$ in (9.1.7) we get the result.

The bound for $w(K)$ follows easily, once we observe that
$$\sigma_* = d_*(Z_{\sigma_*}(K)) \geqslant c_3 k_*(Z_{\sigma_*}(K)) \geqslant c_4 k_*(\sigma_*) \geqslant c_4\rho_*.$$
Thus, the proof of the theorem is complete. □

9.1.5. Mean width of $\Psi_2(\mu)$

Recall that for any log-concave probability measure μ on $\mathbb{R}^n$, the Ψ_2-body of μ is the symmetric convex body whose support function is
$$h_{\Psi_2(\mu)}(\theta) = \sup_{2\leqslant q\leqslant n}\frac{\|\langle\cdot,\theta\rangle\|_q}{\sqrt{q}}, \ \theta \in S^{n-1}.$$

Using Theorem 8.5.1 for $k \simeq \sqrt{n}$ we can give an upper estimate for the mean width of $\Psi_2(\mu)$. We begin with the following simple lemma.

LEMMA 9.1.12. *Let μ be a log-concave probability measure on $\mathbb{R}^n$. Then, for any positive integer $1 \leqslant k < n$ and any $F \in G_{n,k}$ we have:*
(9.1.8) $$P_F(\Psi_2(\mu)) \subseteq C\sqrt{\frac{n}{k}}\Psi_2(\pi_F(\mu)),$$
where $C > 0$ is an absolute constant.

Proof. Let $\theta \in S_F$. Then, we may write:

$$h_{P_F(\Psi_2(\mu))}(\theta) = h_{\Psi_2(\mu)}(\theta) = \sup_{2 \leqslant q \leqslant n} \frac{h_{Z_q(\mu)}(\theta)}{\sqrt{q}}$$

$$\leqslant \sup_{2 \leqslant q \leqslant k} \frac{h_{Z_q(\pi_F(\mu))}(\theta)}{\sqrt{q}} + \sup_{k \leqslant q \leqslant n} \frac{h_{Z_q(\pi_F(\mu))}(\theta)}{\sqrt{q}}$$

$$\leqslant \sup_{2 \leqslant q \leqslant k} \frac{h_{Z_q(\pi_F(\mu))}(\theta)}{\sqrt{q}} + c\sqrt{\frac{n}{k}} \frac{h_{Z_k(\pi_F(\mu))}(\theta)}{\sqrt{k}},$$

where we have used the fact that $h_{Z_q(\mu)}(\theta) = h_{Z_q(\pi_F(\mu))}(\theta)$ for any $\theta \in S_F$ and that $h_{Z_q(\mu)} \leqslant c\frac{q}{k} h_{Z_k(\mu)}$ for $k \leqslant q$. It follows that

$$h_{P_F(\Psi_2(\mu))}(\theta) \leqslant C\sqrt{\frac{n}{k}} \sup_{2 \leqslant q \leqslant k} \frac{h_{Z_q(\pi_F(\mu))}(\theta)}{\sqrt{q}}.$$

Since $\theta \in S_F$ was arbitrary, the result follows. $\square$

Using also Theorem 8.5.1 we get:

THEOREM 9.1.13. *Let μ be an isotropic log-concave probability measure on $\mathbb{R}^n$. Then,*

(9.1.9) $$w(\Psi_2(\mu)) \leqslant c\sqrt[4]{n}.$$

In particular, if K is an isotropic convex body in $\mathbb{R}^n$ then

(9.1.10) $$w(K) \leqslant cn^{3/4} L_K,$$

where $c > 0$ is an absolute constant.

Proof. Using the previous lemma we can write

$$w(\Psi_2(\mu)) = \int_{G_{n,k}} w(P_F(\Psi_2(\mu))) \, d\nu_{n,k}(F) \leqslant C\sqrt{\frac{n}{k}} \int_{G_{n,k}} w(\Psi_2(\pi_F(\mu))) \, d\nu_{n,k}.$$

By Theorem 8.5.1, for $k \leqslant \sqrt{n}$ we deduce the existence of a subset A_k of $G_{n,k}$ such that $\nu_{n,k}(A_k) \geqslant 1 - e^{-c\sqrt{n}}$ and $\Psi_2(\pi_F(\mu)) \subseteq c_1 B_F$ for all $F \in G_{n,k}$. On the other hand, for every $F \in G_{n,k}$, since $\pi_F(\mu)$ is an isotropic log-concave probability measure on F we also have that $\Psi_2(\pi_F(\mu)) \subseteq c_2 \sqrt{\dim F} B_F$. Thus, we can write:

$$\int_{G_{n,k}} w(\Psi_2(\pi_F(\mu))) \, d\nu_{n,k}(F) = \int_{A_k} w(\Psi_2(\pi_F(\mu))) \, d\nu_{n,k}(F)$$

$$+ \int_{A_k^c} w(\Psi_2(\pi_F(\mu))) \, d\nu_{n,k}(F)$$

$$\leqslant c_1(1 - e^{-c\sqrt{n}}) + c_2\sqrt{k} e^{-c\sqrt{n}} \leqslant c_3.$$

Finally, we conclude that

$$w(\Psi_2(\mu)) \leqslant C\sqrt{\frac{n}{k}} \int_{G_{n,k}} w(\Psi_2(\pi_F(\mu))) \, d\nu_{n,k}(F) \leqslant C'\sqrt{\frac{n}{k}}.$$

The result follows if we choose $k = \lfloor \sqrt{n} \rfloor$. The "in particular" case follows immediately if we consider the isotropic measure μ with density function $f_\mu = L_K^n \mathbf{1}_{\frac{K}{L_K}}$, and then observe that $\Psi_2(\mu) = L_K^{-1} \Psi_2(K)$ and take into account the fact that $w(K) \simeq w(Z_n(K)) \leqslant c\sqrt{n} w(\Psi_2(K))$. $\square$

REMARK 9.1.14. Applying Lemma 9.1.12 we can cover the case $1 \leqslant k \leqslant \log^2 n$ in Theorem 8.3.2. One can show that if K is an isotropic convex body in $\mathbb{R}^n$ then for every $1 \leqslant k \leqslant \log^2 n$ and every $F \in G_{n,k}$ there exists $\theta \in S_F$ such that

$$\|\langle \cdot, \theta \rangle\|_{\psi_2} \leqslant C\sqrt{n/k}\sqrt{\log 2k}\, L_K,$$

where $C > 0$ is an absolute constant. In fact, for a random $F \in G_{n,k}$ the term $\sqrt{\log 2k}$ is not needed.

For the proof, let $1 \leqslant k \leqslant \log^2 n$ and $F \in G_{n,k}$. Consider the measure $\nu = \pi_F(\mu_K)$. Recall that $\overline{K}_{k+2}(\nu)$ is almost isotropic, therefore Theorem 8.3.2 shows that there exists $\theta \in S_F$ such that

$$h_{\Psi_2(\overline{K}_{k+2}(\nu))}(\theta) \leqslant c_1 \sqrt{\log 2k} L_{\overline{K}_{k+2}(\nu)}.$$

Then, Lemma 9.1.12 shows that

$$\begin{aligned}\|\langle \cdot, \theta \rangle\|_{\psi_2} &\simeq h_{\Psi_2(K)}(\theta) = h_{P_F(\Psi_2(K))}(\theta) \\ &\leqslant c\sqrt{n/k}\frac{L_K}{L_{\overline{K}_{k+2}(\nu)}} h_{\Psi_2(\overline{K}_{k+2}(\nu))}(\theta) \leqslant C\sqrt{n/k}\sqrt{\log 2k}\, L_K.\end{aligned}$$

In fact, since $k \leqslant \log^2 n \leqslant q_*(K)$, for a random $F \in G_{n,k}$ we know that $\overline{K}_{k+2}(\nu)$ is a ψ_2-body (see 8.5), and hence, $h_{\Psi_2(\overline{K}_{k+2}(\nu))}(\theta) \leqslant c_2 L_{\overline{K}_{k+2}(\nu)}$ for all $\theta \in S_F$. Using this estimate we may remove the $\sqrt{\log 2k}$-term for a random $F \in G_{n,k}$.

9.2. Estimates for $M(K)$ in the isotropic case

In this section we discuss the dual question to obtain an upper bound for

$$M(K) = \int_{S^{n-1}} \|x\|_K d\sigma(x),$$

where K is an isotropic convex body in $\mathbb{R}^n$. Since $K \supseteq cL_K B_2^n$ we readily see that $M(K) \leqslant (cL_K)^{-1}$. The example of the normalized ℓ_∞^n ball shows that the best one could hope is $M(K) \leqslant C\sqrt{\log n/n}$. Note that trying to obtain a bound of the form $M(K) \leqslant n^{-\delta} L_K^{-1}$ goes towards the direction of giving an upper bound for L_K. Indeed, assume that we have $M(K) \leqslant Cn^{-\delta}L_K^{-1}$; applying Hölder's inequality and taking into account the fact that an isotropic convex body K has volume 1, we check that $M(K) \geqslant cn^{-1/2}$ and hence we conclude that $L_K \leqslant C_1 n^{\frac{1}{2}-\delta}$, for some absolute constant $C_1 > 0$.

Paouris and Valettas proved that for every isotropic symmetric convex body K in $\mathbb{R}^n$ one has

$$(9.2.1) \qquad M(K) \leqslant \frac{C(\log n)^{1/3}}{n^{1/12}L_K},$$

where $C > 0$ is an absolute constant. Afterwards, Giannopoulos, Stavrakakis, Tsolomitis and Vritsiou extended this result to the case of L_q-centroid bodies of an isotropic, log-concave probability measure μ on $\mathbb{R}^n$. In [**216**] it is proved that if μ is an isotropic log-concave measure on $\mathbb{R}^n$, then:

(i) For every $1 \leqslant q \leqslant n^{3/7}$,

$$(9.2.2) \qquad M(Z_q(\mu)) \leqslant C\frac{(\log q)^{5/6}}{\sqrt[6]{q}}.$$

(ii) For every q that satisfies $L_n \log q \leqslant q \leqslant \sqrt{L_n}\, n^{3/4}$,

$$M(Z_q(\mu)) \leqslant C \frac{\sqrt[3]{L_n}(\log q)^{5/6}}{\sqrt[6]{q}} \tag{9.2.3}$$

(iii) For every isotropic symmetric convex body K in $\mathbb{R}^n$,

$$M(K) \leqslant C \frac{\sqrt[4]{L_n}(\log n)^{5/6}}{L_K \sqrt[8]{n}}.$$

Recall that if K is an isotropic symmetric convex body in $\mathbb{R}^n$ then we have $K \supseteq Z_q(K)$ for all $1 \leqslant q \leqslant n$. Thus, in order to bound $M(K)$ from above it suffices to bound $M(Z_q(K))$ for values of q that are of the order of n (or "as large as possible"). One may use e.g. (9.2.3) to obtain some estimates for $M(K)$; however, these are weaker than (9.2.1).

The method to obtain the above results is the following. First we obtain a lower bound for the inradius of projections of the body $Z_q(\mu)$ onto subspaces of arbitrarily small codimension εn; this is a function of ε and q. Then, combining this bound with an entropy extension result (e.g. Proposition 9.2.7 below) we obtain regular entropy estimates for the covering numbers $N(\sqrt{q} B_2^n, t Z_q(\mu))$. Then, a standard application of Dudley-Fernique decomposition leads to M-estimates for $Z_q(\mu)$.

In this section we present an improved version of the ideas behind these results. In particular, we get

THEOREM 9.2.1. *Let K be an isotropic symmetric convex body in $\mathbb{R}^n$. Then,*

$$M(K) \leqslant C \frac{\log n}{n^{1/10} L_K}, \tag{9.2.4}$$

where $C > 0$ is an absolute constant.

Most probably, (9.2.4) is not optimal. It is an interesting question to establish the best possible dependence on the dimension.

9.2.1. Projections of the centroid bodies

Let μ be an isotropic log-concave measure on $\mathbb{R}^n$. We will give lower bounds for the inradius of proportional projections of $Z_q(\mu)$ and $Z_q^\circ(\mu)$. Let $1 \leqslant k \leqslant n-1$ and consider a random subspace $F \in G_{n,k}$. An upper bound for the radius of $Z_q(\mu) \cap F$, and hence a lower bound for the inradius of $P_F(Z_q^\circ(\mu))$, follows from the low M^*-estimate. Recall that $w(Z_q(\mu)) \simeq \sqrt{q}$ if $2 \leqslant q \leqslant q_*(\mu)$. Therefore, we get:

PROPOSITION 9.2.2. *Let μ be an isotropic log-concave measure on $\mathbb{R}^n$. If $2 \leqslant q \leqslant q_*(\mu)$ and if $\varepsilon \in (0,1)$ and $k = \lfloor (1-\varepsilon)n \rfloor$, then a subspace $F \in G_{n,k}$ satisfies*

$$R(Z_q(\mu) \cap F) \leqslant \frac{c_1 \sqrt{q}}{\sqrt{\varepsilon}} \quad \text{or equivalently} \quad P_F(Z_q^\circ(\mu)) \supseteq \frac{c_2 \sqrt{\varepsilon}}{\sqrt{q}} B_F \tag{9.2.5}$$

with probability greater than $1 - c_3 \exp(-c_4 \varepsilon n)$, where c_i are absolute constants.

Our aim is to provide analogous upper bounds for $R(Z_q^\circ(\mu) \cap F)$, $F \in G_{n,k}$. Some information is provided by the next two propositions.

PROPOSITION 9.2.3 (version for "small" q). Let μ be an isotropic log-concave measure on $\mathbb{R}^n$. Let $q \geqslant 2$ and $0 < \varepsilon < 1$. Assume that

$$(9.2.6) \qquad \varepsilon \geqslant C_1 \max\left\{\frac{\log q}{q^2}, \frac{q^2}{n}\right\}$$

for some large enough absolute constant $C_1 > 0$. Then, there exist $k \geqslant (1-\varepsilon)n$ and $F \in G_{n,k}$ such that

$$P_F(Z_q(\mu)) \supseteq \frac{c\varepsilon}{|\log \varepsilon|} \frac{\sqrt{q}}{\log q} B_F.$$

This proposition is complemented by the next result for "large" values of q.

PROPOSITION 9.2.4 (version for "large" q). Let μ be an isotropic log-concave measure on $\mathbb{R}^n$. Let $q \geqslant 2$ and $0 < \varepsilon < 1$. Assume that

$$(9.2.7) \qquad \varepsilon \geqslant C_1 \max\left\{\frac{\log q}{q^2}, \frac{q}{n}\right\}$$

for some large enough absolute constant $C_1 > 0$. Then, there exist $k \geqslant (1-\varepsilon)n$ and $F \in G_{n,k}$ such that

$$P_F(Z_q(\mu)) \supseteq \frac{c_1 \varepsilon}{|\log \varepsilon| L_{\varepsilon n}} \frac{\sqrt{q}}{\log q} B_F \supseteq \frac{c_2 \varepsilon^{\frac{3}{4}}}{\sqrt[4]{n} |\log \varepsilon|} \frac{\sqrt{q}}{\log q} B_F.$$

The proof of the two propositions depends on our knowledge about the volume of the L_q-centroid bodies of μ. Recall that for every $1 \leqslant m \leqslant n$ and any $H \in G_{n,m}$ we have

$$P_H(Z_q(\mu)) = Z_q(\pi_H(\mu)).$$

From Chapter 7 we know that if ν is an isotropic log-concave measure on $\mathbb{R}^m$, then

$$|Z_q(\nu)|^{1/m} \geqslant c_1 \sqrt{q/m}$$

for all $q \leqslant \sqrt{m}$. It follows that

$$(9.2.8) \qquad \left|P_H(Z_q(\mu))\right|^{1/m} \geqslant c_1 \sqrt{q/m}$$

for all $H \in G_{n,m}$ and all $q \leqslant \sqrt{m}$. For larger values of q we can still use the bound

$$(9.2.9) \qquad |Z_q(\mu)|^{1/n} \geqslant c_2 \sqrt{q/n}\, L_\mu^{-1}$$

for the volume of the L_q-centroid bodies of μ. The lack of a "uniform" (in q) formula for $|Z_q(\mu)|^{1/n}$ is the reason why we consider "small" and "large" values of q separately.

A second tool that is needed for the proof is the existence of α-regular M-ellipsoids for $Z_q(\mu)$. We know that for every $0 < \alpha < 2$ there exists an ellipsoid $\mathcal{E}_\alpha$ such that

$$\max\{N(Z_q(\mu), t\mathcal{E}_\alpha), N(\mathcal{E}_\alpha, tZ_q(\mu))\} \leqslant \exp\left(\frac{cn}{(2-\alpha)^{\alpha/2} t^\alpha}\right)$$

for every $t \geq 1$, where $c > 0$ is an absolute constant. We fix $\overline{\alpha} = 2 - \frac{1}{\log q}$ and set $\mathcal{E} = \mathcal{E}_{\overline{\alpha}}$. Then,

$$(9.2.10) \quad \max\left\{N(Z_q(\mu), t\mathcal{E}), N(\mathcal{E}, tZ_q(\mu))\right\} \leq \exp\left(\frac{ct^{1/\log q}(\log q)n}{t^2}\right)$$

$$\leq \exp\left(\frac{c(\log q)n}{t^2}\right)$$

for all $1 \leq t \leq q$.

Finally, we use the following result (that was suggested by E. Milman) on the diameter of sections of a convex body that satisfies regular entropy estimates; it is due to V. Milman and can be essentially found in [**382**].

LEMMA 9.2.5. *Let C be a symmetric convex body in $\mathbb{R}^m$ and let $\mathcal{E}$ be an ellipsoid such that*

$$(9.2.11) \quad \log N(C, t\mathcal{E}) \leq \frac{\gamma m}{t^2}$$

for all $1 \leq t \leq \beta$ and some constant $\gamma \geq 1$. Then, for every integer $\frac{\gamma m}{\beta^2} \leq k < m$ there exists $F \in G_{m,k}$ such that

$$(9.2.12) \quad C \cap F^\perp \subseteq c_2\sqrt{\gamma}\sqrt{\frac{m}{k}}\log\left(\frac{m}{k}\right)\mathcal{E} \cap F^\perp,$$

where $c_1, c_2 > 0$ are absolute constants.

Proof. Without loss of generality we may assume that $\mathcal{E} = B_2^m$. Otherwise, we write $\mathcal{E} = T(B_2^m)$ for some $T \in GL(m)$ and observe that $N(T^{-1}(C), tB_2^m) = N(C, t\mathcal{E})$ and that $T^{-1}(C) \cap E^\perp \subseteq \rho B_{E^\perp}$ if and only if $C \cap F^\perp \subseteq \rho \mathcal{E} \cap F^\perp$, where $F = (T(E^\perp))^\perp \in G_{m,\dim E}$.

Fix $k \geq \gamma m/\beta^2$. We define $t_j = 2^{-(j-1)}\sqrt{\gamma m/k}$ for $j = 1, \ldots, s$ as long as $2^{s-1} \leq \sqrt{\gamma m/k}$. Note that $t_j \leq \beta$ for all j by the restrictions on k.

For any $1 \leq j \leq s$ we consider a t_j-net N_j of C with cardinality $|N_j| \leq \exp(4^{j-1}k)$. In particular, we have $|N_1| \leq e^k$.

Step 1. Fix $\theta \in S^{m-1}$ and $\gamma m/\beta^2 \leq k < m$. We know that for any $\delta \in (0, e^{-2})$ one has,

$$\nu_{m,k}(\{F \in G_{m,k} : \|P_F\theta\|_2 < \delta\sqrt{k/m}\}) < \delta^k.$$

Applying this for the net N_1 we see that

$$\nu_{m,k}(\{F \in G_{m,k} : \exists u \in N_1 \text{ with } \|P_F u\|_2 < \delta\sqrt{k/m}\|u\|_2\}) \leq \delta^k e^k.$$

Choosing $\delta = e^{-3}$ we obtain

$$(9.2.13) \quad \nu_{m,k}(\{F \in G_{m,k} : \forall u \in N_1, \|P_F u\|_2 \geq e^{-3}\sqrt{k/m}\|u\|_2\}) \geq 1 - e^{-2k}.$$

Step 2. Fix $\theta \in S^{m-1}$ and $1 \leq k < m$. For any $\varepsilon > 0$ we have

$$\nu_{m,k}(\{F \in G_{m,k} : \|P_F(\theta)\|_2 \geq (1+\varepsilon)\sqrt{k/m}\}) \leq e^{-c_3\varepsilon^2 k}.$$

Set $Z_j = N_j - N_{j+1}$. For any fixed $1 \leq j < s$ and $\varepsilon \geq 1$ we obtain

$$\nu_{m,k}(\{F \in G_{m,k} : \exists z \in Z_j \text{ with } \|P_F(z)\|_2 \geq 2\varepsilon\sqrt{k/m}\|z\|_2)$$

$$\leq e^{-c_3\varepsilon^2 k}|N_{j+1}|^2 < \exp(-c_3\varepsilon^2 k + 4^j k).$$

Choose $\varepsilon = \varepsilon_j \simeq 2^j$. Then, we get

(9.2.14) $\quad \nu_{m,k}(\{F: \forall 1 \leqslant j < s, \forall z \in Z_j, \|P_F(z)\|_2 \leqslant c_4 2^j \sqrt{k/m}\|z\|_2\})$
$$\geqslant 1 - \sum_{j<s} e^{-c_5 4^j k} \geqslant 1 - e^{-c_6 k}.$$

Thus, there exist "many" $F \in G_{m,k}$ such that we simultaneously have:
  (i) For all $u \in N_1$, $\|P_F(u)\|_2 \geqslant e^{-3}\sqrt{k/m}\|u\|_2$.
  (ii) For all $1 \leqslant j < s$ and for all $z \in Z_j$, $\|P_F(z)\|_2 \leqslant c_4 2^j \sqrt{k/m}\|z\|_2$.

Consider such a subspace F, and let $x \in C \cap F^\perp$. For any $1 \leqslant j \leqslant s$ there exists $z_j \in N_j$ such that $\|x - z_j\|_2 \leqslant t_j$, and hence $\|z_j - z_{j+1}\|_2 \leqslant 2t_j$ for $1 \leqslant j < s$. So, we may write

$$\|x\|_2 \leqslant \|x - z_1\|_2 + \|z_1\| \leqslant t_1 + c_7\sqrt{m/k}\|P_F(z_1)\|_2$$

$$\leqslant t_1 + c_7\sqrt{m/k}\left(\sum_{j<s}\|P_F(z_j - z_{j+1})\|_2 + \|P_F(z_s - x)\|_2\right)$$

$$\leqslant c_8 \sum_{1 \leqslant j < s} 2^j t_j + c_9\sqrt{m/k}\, t_s$$

$$\leqslant c_{10} s \sqrt{\frac{\gamma m}{k}} + c_{11}\sqrt{\frac{m}{k}}\sqrt{\frac{\gamma m}{k}}\frac{1}{2^s}.$$

Note that $\sqrt{m/k} \leqslant \sqrt{\gamma m/k}$, so s can be chosen as the maximal integer with $2^s \leqslant \sqrt{m/k}$. Then $s \simeq \log(m/k)$ and the result follows. $\square$

Proof of Proposition 9.2.3. Consider an ellipsoid $\mathcal{E}$ that satisfies (9.2.10). Let $0 < \lambda_1 \leqslant \cdots \leqslant \lambda_n$ be the axes of $\mathcal{E}$, and let $\{u_1, \ldots, u_n\}$ be an orthonormal basis which corresponds to the λ_j. For every $1 \leqslant m, s \leqslant n$ we set

$$H_m := \text{span}\{u_1, \ldots, u_m\} \quad \text{and} \quad F_s = \text{span}\{u_{s+1}, \ldots, u_n\}.$$

We have

$$N\Big(P_{H_m}\big(Z_q(\mu)\big), t P_{H_m}(\mathcal{E})\Big) \leqslant N\big(Z_q(\mu), t\mathcal{E}\big) \leqslant e^{c(\log q)n/t^2},$$

and hence

(9.2.15) $\quad \big|P_{H_m}\big(Z_q(\mu)\big)\big|^{1/m} \leqslant |B_{H_m}|^{1/m}(t\lambda_m)\exp\left(\frac{c(\log q)n}{t^2 m}\right)$

for all $1 \leqslant t \leqslant q$. Assuming that

(9.2.16) $\quad m \geqslant \max\left\{q^2, \frac{n \log q}{q^2}\right\}$

we may choose $t = (n(\log q)/m)^{1/2}$ and apply (9.2.8) to get

(9.2.17) $\quad \lambda_m \geqslant \left(\frac{m}{n \log q}\right)^{1/2} \sqrt{q}.$

Let $\varepsilon \in (0,1)$ satisfy (9.2.6). We will use the bound of (9.2.17) with $m = \lceil \frac{\varepsilon n}{2} \rceil$, and one can check that m satisfies (9.2.16), because of the assumption that $\varepsilon \geqslant C_1 \log q/q^2$.

We set $s = \lfloor \frac{\varepsilon n}{2} \rfloor$. Then, we have $m \leqslant s$ and

$$N\Big(P_{F_s}(\mathcal{E}), t P_{F_s}\big(Z_q(\mu)\big)\Big) \leqslant N\big(\mathcal{E}, t Z_q(\mu)\big) \leqslant e^{cn \log q/t^2} \leqslant e^{2c(n-s)\log q/t^2},$$

for every $1 \leqslant t \leqslant q$.

We now use the duality of entropy theorem to write
$$N(Z_q^\circ(\mu) \cap F_s, t\mathcal{E}^\circ \cap F_s) \leqslant N\Big(P_{F_s}(\mathcal{E}), at P_{F_s}(Z_q(\mu))\Big)^b \leqslant e^{c_1(n-s)\log q/t^2}.$$

We apply Lemma 9.2.5 for the body $Z_q^\circ(\mu) \cap F_s$ (with $\gamma = c_1 \log q$) to find a subspace F of F_s, of dimension $k \geqslant (1-\varepsilon/2)(n-s) \geqslant (1-\varepsilon)n$, such that
$$Z_q^\circ(\mu) \cap F \subseteq C\sqrt{\log q}\,\frac{|\log \varepsilon|}{\sqrt{\varepsilon}}\,\mathcal{E}^\circ \cap F,$$
and hence
$$(9.2.18) \qquad P_F(Z_q(\mu)) \supseteq \frac{c\sqrt{\varepsilon}}{|\log \varepsilon|\sqrt{\log q}} P_F(\mathcal{E}).$$

The restriction for this is $\frac{\varepsilon}{2}(n-s) \geqslant c_1 n \frac{\log q}{q^2}$, which is satisfied because of (9.2.6). From (9.2.17) we have
$$\mathcal{E} \cap F_s \supseteq \lambda_{s+1} B_{F_s} \supseteq \frac{c\sqrt{\varepsilon}}{\sqrt{\log q}} \sqrt{q}\, B_{F_s},$$
provided that $q \leqslant \sqrt{\varepsilon n}$. This is also satisfied because of (9.2.6). Then,
$$P_F(\mathcal{E}) = P_F\big(P_{F_s}(\mathcal{E})\big) = P_F(\mathcal{E} \cap F_s) \supseteq \frac{c\sqrt{\varepsilon}\sqrt{q}}{\sqrt{\log q}} P_F(B_{F_s})$$
$$= \frac{c\sqrt{\varepsilon}\sqrt{q}}{\sqrt{\log q}} B_F.$$

Combining this fact with (9.2.18) we conclude the proof. □

The proof of Proposition 9.2.4 is more or less the same:

Sketch of the proof of Proposition 9.2.4. This time we use the estimate
$$(9.2.19) \qquad \big|P_H(Z_q(\mu))\big|^{1/m} \geqslant \frac{c_1}{L_m}\sqrt{q/m}$$
for all $H \in G_{n,m}$ and all $q \leqslant m$. We define H_m, F_s as in the proof of Proposition 9.2.3 and we consider an ellipsoid $\mathcal{E}$ that satisfies (9.2.10). Assuming that $q \leqslant m$, from (9.2.19) and (9.2.15) we get
$$\lambda_m \geqslant \frac{1}{L_m}\left(\frac{m}{(\log q)n}\right)^{1/2}\sqrt{q}.$$

Next, fix some $\varepsilon \in (0,1)$ that satisfies (9.2.7), set $s = \lfloor \frac{\varepsilon n}{2} \rfloor$ and consider any $q \leqslant \varepsilon n/2$. As previously, we find a subspace F of F_s, of dimension $k \geqslant (1 - \varepsilon/2)(n-s) \geqslant (1-\varepsilon)n$, such that
$$P_F(Z_q(\mu)) \supseteq \frac{c\sqrt{\varepsilon}}{|\log \varepsilon|\sqrt{\log q}} P_F(\mathcal{E})$$
and
$$P_F(\mathcal{E}) = P_F\big(P_{F_s}(\mathcal{E})\big) \supseteq P_F(\lambda_{s+1} B_{F_s}) \supseteq \frac{c\sqrt{\varepsilon}\sqrt{q}}{L_s \sqrt{\log q}} B_F.$$

Since $s \simeq \varepsilon n$ and $L_{\varepsilon n} \leqslant C\sqrt[4]{\varepsilon n}$, the result follows. □

Recall that if K is a symmetric convex body of volume 1 in $\mathbb{R}^n$, then
$$P_F(K) \supseteq P_F(Z_q(K))$$

for all $q \geq 1$. Also, the measure μ_K with density $L_K^n \mathbf{1}_{K/L_K}$ is isotropic and $Z_q(K) = L_K Z_q(\mu_K)$. Choosing $q = C_1^{-1}\varepsilon n$ and applying Proposition 9.2.4 with $\mu = \mu_K$ we get:

COROLLARY 9.2.6. *Let K be an isotropic symmetric convex body in $\mathbb{R}^n$. For every $0 < \varepsilon < 1$ with $\varepsilon \geq \frac{C\sqrt[3]{\log n}}{n^{2/3}}$ there exist $k \geq (1-\varepsilon)n$ and $F \in G_{n,k}$ such that*

$$P_F(K) \supseteq L_K P_F(Z_q(\mu_K)) \supseteq \frac{c_2 \varepsilon^{\frac{5}{4}}}{|\log \varepsilon|} \frac{\sqrt[4]{n}}{\log n} L_K B_F.$$

9.2.2. Dual covering numbers of the centroid bodies

In Chapter 8 we saw that if μ is an isotropic log-concave measure on $\mathbb{R}^n$ then, for any $1 \leq q \leq n$ and $t \geq 1$,

$$(9.2.20) \qquad \log N\big(Z_q(\mu), c_1 t \sqrt{q} B_2^n\big) \leq c_2 \frac{n}{t^2} + c_3 \frac{\sqrt{qn}}{t},$$

where $c_1, c_2, c_3 > 0$ are absolute constants. Using the results of the previous section and the next entropy extension result one can obtain regular entropy estimates for the dual covering numbers.

PROPOSITION 9.2.7. *Let K be a symmetric convex body in $\mathbb{R}^n$ and assume that $B_2^n \subseteq \rho K$ for some $\rho \geq 1$. Let W be a subspace of $\mathbb{R}^n$ with $\dim W = m$ and $P_{W^\perp}(K) \supseteq B_{W^\perp}$. Then, we have*

$$N(B_2^n, 4K) \leq \overline{N}(B_2^n, 2K) \leq (3\rho)^m.$$

Proof. The result will follow by the next two claims.

Claim 1. For any $x \in B_2^n$ there exists $w_x \in (1+\rho)(K \cap W)$ such that

$$x \in w_x + K.$$

Consider $x \in B_2^n$. Then, there exists $y_x \in K$ such that $x = P_{W^\perp}(y_x) + P_W(x)$. Setting $w_x := P_W(x - y_x) \in W$ we observe that $x - w_x = y_x \in K$ and

$$\|w_x\|_K = \|x - y_x\|_K \leq \rho \|x\|_2 + \|y_x\|_K \leq \rho + 1,$$

where we have used the fact that $B_2^n \subseteq \rho K$.

Claim 2. There exists a finite subset $\mathcal{N}$ of $(1+\rho)(K \cap W)$, with cardinality at most $(2+\rho)^m$, satisfying the property: for any $w \in (1+\rho)(K \cap W)$ there exists $z_w \in \mathcal{N}$ such that $w \in z_w + K \cap W$.

Set $C := (K \cap W)$ and consider a 1-net $\mathcal{N}$ of $(1+\rho)C$ with respect to C. Then, standard volumetric arguments show that there exists such a net with cardinality

$$N((1+\rho)C, C) \leq \frac{|(1+\rho)C + C|}{|C|} \leq (2+\rho)^m.$$

Combining the above we get the result. $\square$

Using also Proposition 9.2.3, we obtain some estimates for the covering numbers $N(\sqrt{q}B_2^n, tZ_q(\mu))$.

9.2. ESTIMATES FOR $M(K)$ IN THE ISOTROPIC CASE

PROPOSITION 9.2.8. *Let μ be an isotropic log-concave measure on $\mathbb{R}^n$. Assume that $2 \leqslant q \leqslant \sqrt{n}$. Then, for any*

$$1 \leqslant t \leqslant \min\left\{\sqrt{q}, c_1 \frac{n \log q}{q^2}\right\}$$

we have

$$(9.2.21) \qquad \log N\left(\sqrt{q} B_2^n, t Z_q(\mu)\right) \leqslant c_2 \frac{n \log^2 q \log t}{t},$$

and, by the duality of entropy theorem,

$$(9.2.22) \qquad \log N\left(\sqrt{q} Z_q^\circ(\mu), t B_2^n\right) \leqslant c_2 \frac{n \log^2 q \log t}{t},$$

where $c_1, c_2 > 0$ are absolute constants.

Proof. Note that, since $B_2^n \subseteq Z_q(\mu)$, the interesting range for t is up to $\sqrt{q}$. Given some $\varepsilon \in (0,1)$ satisfying (9.2.6) and $k = (1-\varepsilon)n$, Proposition 9.2.3 yields the existence of some $F \in G_{n,k}$ such that

$$P_{F^\perp}(Z_q(\mu)) \supseteq c_3 \frac{\varepsilon}{|\log \varepsilon|} \frac{\sqrt{q}}{\log q} B_{F^\perp}.$$

We shall apply Proposition 9.2.7 for the body $K := \frac{t}{4\sqrt{q}} Z_q(\mu)$; note that $\rho = \frac{4\sqrt{q}}{t} > 1$. We know that

$$P_{F^\perp}(K) \supseteq c_4 \frac{\varepsilon t}{|\log \varepsilon| \log q} B_{F^\perp}.$$

We choose $\varepsilon \simeq \frac{\log q \log t}{t}$ and assume that $t \geqslant c_5 \log q$ so that $c_4 \frac{\varepsilon t}{|\log \varepsilon| \log q} \geqslant 1$. Then, we may apply Proposition 9.2.7 to get

$$(9.2.23) \qquad \log N(\sqrt{q} B_2^n, t Z_q(\mu)) \leqslant \varepsilon n \log\left(\frac{12\sqrt{q}}{t}\right)$$

as long as $c_5 \log q \leqslant t \leqslant c_6 n \log q / q^2$.

The restriction $t \leqslant c_6 n \log q / q^2$ is coming up from (9.2.6) and from the choice of ε. Note that, if e.g. $q \leqslant n^{2/5}$, then this allows us to consider any t up to $\sqrt{q}$.

With this choice of ε, we get from (9.2.23) that

$$(9.2.24) \qquad \log N\left(\sqrt{q} B_2^n, t Z_q(\mu)\right) \leqslant c_7 \frac{n \log^2 q \log t}{t}.$$

This proves (9.2.21) provided that $c_5 \log q \leqslant t \leqslant c_6 n (\log q)/q^2$.

When $1 \leqslant t \leqslant c_5 \log q$, we use the inequality

$$N\left(\sqrt{q} B_2^n, t Z_q(\mu)\right) \leqslant N\left(\sqrt{q} B_2^n, c_5 (\log q) Z_q(\mu)\right) N\left(c_5 \log q Z_q(\mu), t Z_q(\mu)\right).$$

Observe that the second covering number is less than $\left(1 + \frac{c_5 \log q}{t}\right)^n$; this completes the proof. □

Note. Although all these estimates are most probably not optimal, we can still conclude that $Z_q(\mu)$, with $q \leqslant n^{2/5}$, is (almost) 1-regular in the terminology of Pisier's theorem.

REMARK 9.2.9. In Section 3.2 we saw that if K is an isotropic convex body in $\mathbb{R}^n$ then
$$\log N(K, t\sqrt{n} L_K B_2^n) \leqslant \frac{cn}{t}$$
for all $t > 0$. In the symmetric case, using Corollary 9.2.6 and arguing as before we obtain a dual estimate.

PROPOSITION 9.2.10. *Let K be an isotropic symmetric convex body in $\mathbb{R}^n$. Then, for any $c_1 \sqrt[4]{n} \log n \leqslant t \leqslant \sqrt{n}$ we have*
$$\max\{\log N(\sqrt{n} L_K B_2^n, tK), \log N(\sqrt{n} L_K K^\circ, t B_2^n)\} \leqslant \frac{c_2 n^{6/5} \log n}{t^{4/5}},$$
where $c_1, c_2 > 0$ are absolute constants.

9.2.3. M-estimates for $Z_q(\mu)$

Using the results of the previous subsections and the method of Section 9.1.1, we can get an upper bound for the parameter
$$M(Z_q(\mu)) = \int_{S^{n-1}} \|x\|_{Z_q(\mu)} \, d\sigma(x).$$
For example, we have

THEOREM 9.2.11 (version for small q). *Let μ be an isotropic log-concave measure on $\mathbb{R}^n$. For every $1 \leqslant q \leqslant cn^{2/5} \log n$,*

(9.2.25) $$M(Z_q(\mu)) \leqslant C \frac{(\log q)^{5/2}}{\sqrt[4]{q}},$$

where $C, c > 0$ are absolute constants.

One can also check that, for every $\sqrt{n} \leqslant q \leqslant c(n \log(n/q) \log q)^{4/5}$,

(9.2.26) $$M(Z_q(\mu)) \leqslant C \frac{n^{1/6} (\log q)^{4/3}}{\sqrt[3]{q}},$$

where $c, C > 0$ are absolute constants. Finally, using Proposition 9.2.10 we obtain Theorem 9.2.1: if K is an isotropic symmetric convex body in $\mathbb{R}^n$ then
$$M(K) \leqslant C \frac{\log n}{n^{1/10} L_K},$$
where $C > 0$ is an absolute constant.

9.3. Further reading

9.3.1. Estimate for $M(K)$ in terms of the type–2 constant

In this section we present a result due to E. Milman [366] which provides an upper bound for $M(K)$, when K is an isotropic symmetric convex body in $\mathbb{R}^n$ with type-2 constant $T_2(X_K)$. Recall that if X_K is the normed space with unit ball K, we write $T_{2,k}(X_K)$ for the best constant $T > 0$ such that
$$\mathbb{E} \left\| \sum_{i=1}^k \varepsilon_i x_i \right\|_K \leqslant T \left(\sum_{i=1}^k \|x_i\|_K^2 \right)^{1/2}$$
for all $x_1, \ldots, x_k \in X$. Then, the type-2 constant of X_K is defined as $T_2(X_K) := \sup_k T_{2,k}(X_K)$.

9.3. FURTHER READING

LEMMA 9.3.1. *Let $\lambda_j > 0$ and $v_j \in \mathbb{R}^n$, $j = 1, \ldots, N$, such that $\mu = \sum_{j=1}^N \lambda_j \delta_{v_j}$ is an isotropic measure. Let $\{g_j\}_{j=1}^N$ be independent real-valued standard Gaussian random variables. Then,*

$$G_\mu = \sum_{j=1}^N g_j \sqrt{\lambda_j} v_j$$

is a standard Gaussian random vector.

Proof. We check that $\mathbb{E}(\langle G_\mu, x\rangle \langle G_\mu, y\rangle) = \langle x, y\rangle$ for all $x, y \in \mathbb{R}^n$, which shows that $\operatorname{Cov}(G_\mu) = I$. □

PROPOSITION 9.3.2 (E. Milman). *Let μ be an isotropic, compactly supported isotropic measure on $\mathbb{R}^n$. Then, for any symmetric convex body K in $\mathbb{R}^n$ we have:*

(9.3.1) $$I_2(\mu, K) \geqslant \sqrt{n}\frac{M_2(K)}{T_2(X_K)},$$

where $I_2^2(\mu, K) = \int_{\mathbb{R}^n} \|x\|_K^2 d\mu(x)$.

Proof. Assume first that μ is a discrete isotropic measure $\mu = \sum_{j=1}^N \lambda_j \delta_{v_j}$ with support $\{v_1, \ldots, v_N\}$. The previous lemma shows that

$$\mathbb{E}\|G_\mu\|_K^2 = (2\pi)^{-n/2} \int_{\mathbb{R}^n} \|x\|_K^2 \, d\gamma_n(x) = nM_2^2(K).$$

By the definition of $T_{2,N}(X_K)$ we may write

$$M_2(K) = \frac{1}{\sqrt{n}}(\mathbb{E}\|G_\mu\|^2)^{1/2} \leqslant \frac{cT_{2,N}(X_K)}{\sqrt{n}}\left(\sum_{j=1}^N \lambda_j \|v_j\|_K^2\right)^{1/2}.$$

Since

$$I_2^2(\mu, K) = \int_{\mathbb{R}^n} \|x\|_K^2 \, d\mu(x) = \sum_{j=1}^N \lambda_j \|v_j\|_K^2,$$

we get the result in the discrete case.

For the general case we approximate the measure μ (in the w^*-topology) by measures of the form $\mu_\varepsilon = \sum_{j=1}^{N_\varepsilon} \lambda_j^\varepsilon \delta_{v_j^\varepsilon}$ (the details can be found in [**366**]). □

THEOREM 9.3.3 (E. Milman). *Let K be an isotropic symmetric convex body in $\mathbb{R}^n$. Then,*

(9.3.2) $$M(K) \leqslant C\frac{T_2(X_K)}{\sqrt{n}L_K},$$

where $C > 0$ is an absolute constant. In particular, we get

$$L_K \leqslant C' T_2(X_K).$$

Proof. An application of the previous proposition for the isotropic measure μ_K gives

(9.3.3) $$M(K) \leqslant M_2(K) \leqslant T_2(X_K)\frac{I_2(\mu_K, K)}{\sqrt{n}} = T_2(X_K)\frac{I_2(K, K)}{L_K\sqrt{n}}$$

and the result follows once we recall that $I_2^2(K, K) = \frac{n}{n+2}|K|$. □

9.4. Notes and references

Mean width in the isotropic case

The first proof of the upper bound $w(K) \leqslant cn^{3/4}L_K$ appeared in the PhD Thesis of Hartzoulaki [249]. It combines the entropy estimate of Theorem 3.2.2 with an idea of Giannopoulos and Milman from [203, Theorem 5.6] on the mean width of convex bodies whose covering numbers by a ball are α-regular. It is an observation of Paouris that Theorem 5.2.3 (which implies that $w(Z_{q_*}(K)) \leqslant c\sqrt{q_*}L_K$) and the fact that $q_* \geqslant c'\sqrt{n}$ allow us to choose $q \simeq \sqrt{n}$ in the standard inequality $w(K) \leqslant c_1 w(Z_n(K)) \leqslant (c_2 n/q)w(Z_q(K))$, $2 \leqslant q \leqslant n$ and to obtain yet another proof of the same bound. The approach of Pivovarov through random polytopes is from [432]. It leads to some interesting questions regarding the behavior of the volume $v(\varepsilon, \theta) = |\{x \in K : \langle x, \theta \rangle \geqslant \varepsilon h_K(\theta)\}|$ of "caps" of an isotropic convex body in a random direction that might eventually lead to a better estimate (see [432] for a discussion). The parameter ρ_* was introduced in [212], and Theorem 9.1.9 which relates the question to the distribution of the ψ_2-norm of linear functionals appears in the same article. The upper bound $w(\Psi_2(\mu)) \leqslant c\sqrt[4]{n}$ for the mean width of $\Psi_2(\mu)$ (given by Theorem 9.1.13) is also from [212].

Estimating $M(K)$ in isotropic position

Although the question to obtain upper bounds for $M(K)$ in the isotropic position is natural, it had not been studied until recently. The first non-trivial estimate $M(K) \leqslant \frac{C(\log n)^{1/3}}{n^{1/12}L_K}$ on this problem was obtained by Paouris and Valettas (unpublished). The argument was based on a lower bound for the inradius of projections of isotropic convex bodies of arbitrarily small codimension; the proof used some ideas from Bourgain's argument for the isotropic constant of ψ_2-bodies. A similar result with cubic dependence on ε appears in [61]. Also, in [287], under the additional assumption that $L_n \leqslant C$ for all $n \geqslant 1$, it is proved that for every isotropic convex body K in $\mathbb{R}^n$ and any $0 < \varepsilon < 1$ there exists $F \in G_{n, \lfloor(1-\varepsilon)n\rfloor}$ so that

$$P_F(K) \supseteq c\varepsilon^3 \sqrt{n}B_F.$$

The argument that we use for Proposition 9.2.3 and Proposition 9.2.4 is from [216]. Some main ideas of the proof come from [287]. Note that these results guarantee the existence of *one* $\lfloor(1-\varepsilon)n\rfloor$-dimensional projection of $Z_q(\mu)$ with "large" inradius. However, it is proved in [208] that, for every fixed proportion $\mu \in (0,1)$ and every $0 < s < 1/(2-\mu)$, the maximal inradius of $\lfloor \mu n \rfloor$-dimensional projections and the random inradius of $\lfloor s\mu n \rfloor$-dimensional projections of a symmetric convex body C in $\mathbb{R}^n$ are comparable up to a constant depending on μ and s. See also [496] for a similar result, and especially [335] where it is shown that if C is a symmetric convex body in $\mathbb{R}^n$, and if $1 \leqslant k < m < n$ and $\mu = \frac{n-k}{n-m}$, then assuming that $R(C \cap F) \leqslant r$ for some $F \in G_{n,m}$ we have that a random subspace $E \in G_{n,k}$ satisfies

$$R(C \cap E) \leqslant r\left(c\sqrt{\tfrac{n}{n-m}}\right)^{\frac{\mu}{\mu-1}}$$

with probability greater than $1 - 2e^{-(n-k)/2}$, where $c > 0$ is an absolute constant. Using this fact one can obtain versions of the results of this section concerning random proportional projections of $Z_q(\mu)$.

The idea to combine the information on the inradius of projections of arbitrarily small codimension with "entropy extension" in order to obtain regular entropy estimates is new. In our presentation we use a variant (Proposition 9.2.7) of a lemma of Vershynin and Rudelson from [496]. The interested reader will find more on entropy extension results in [333].

CHAPTER 10

Approximating the covariance matrix

The question that we discuss in this chapter was posed by Kannan, Lovász and Simonovits and has its origin in the problem of finding a fast algorithm for the computation of the volume of a given convex body. It can be stated in the more general setting of probability measures as follows: we consider an isotropic log-concave random vector X in $\mathbb{R}^n$ and a sequence $\{X_i\}$ of independent copies of X. By the law of large numbers we know that the empirical covariance matrix $\frac{1}{N}\sum_{i=1}^{N} X_i \otimes X_i$ converges to the identity matrix $I = \mathbb{E}(X \otimes X)$ as $N \to \infty$. We are interested in the rate of this convergence: given $0 < \varepsilon < 1$, we ask for the minimal value of N for which

$$\left\| \frac{1}{N} \sum_{i=1}^{N} X_i \otimes X_i - I \right\| \leqslant \varepsilon$$

with probability close to 1, say greater than $1 - \varepsilon$.

In the language of isotropic convex bodies, we consider N independent random points $x_1, \ldots, x_N$ uniformly distributed in an isotropic convex body K in $\mathbb{R}^n$, and then the task becomes to find N_0, as small as possible, for which the following holds true: if $N \geqslant N_0$, then with probability greater than $1 - \varepsilon$ one must have

$$(1-\varepsilon) L_K^2 \leqslant \frac{1}{N} \sum_{i=1}^{N} \langle x_i, \theta \rangle^2 \leqslant (1+\varepsilon) L_K^2$$

for every $\theta \in S^{n-1}$. Kannan, Lovász and Simonovits (see [**267**]) proved that one can take $N_0 = C(\varepsilon) n^2$ for some constant $C(\varepsilon) > 0$ depending only on ε. This was improved to $N_0 = C(\varepsilon) n (\log n)^3$ by Bourgain and to $N_0 = C(\varepsilon) n (\log n)^2$ by Rudelson. One can actually check (see [**205**]) that this last estimate can be recovered by Bourgain's argument too if we incorporate Alesker's theorem into it. It was finally proved by Adamczak, Litvak, Pajor and Tomczak-Jaegermann that the best estimate for N_0 is $C(\varepsilon) n$. We describe the solution of the problem in Section 10.1. We outline Bourgain's and Rudelson's arguments in Section 10.2.

The study of spectral properties of random matrices with independent rows or columns that have log-concave distribution is a direction where much progress has been achieved. The underlying principle is that, because of high-dimension and of the log-concavity assumption, such matrices behave in more or less the same way as matrices with independent entries. We do not cover this topic in detail since it would have taken us far from the central theme of this book. For more information on the non-asymptotic theory of random matrices, the interested reader may consult the article [**448**] of Rudelson and Vershynin, and Vershynin's lecture notes [**497**].

10.1. Optimal estimate

The question of Kannan, Lovász and Simonovits was answered in an optimal way by Adamczak, Litvak, Pajor and Tomczak-Jaegermann in [1]. Their estimates are summarized in the next theorem.

THEOREM 10.1.1 (Adamczak-Litvak-Pajor-Tomczak). *Let $X_1, \ldots, X_N$, be i.i.d. isotropic log-concave random vectors in $\mathbb{R}^n$. For every $\varepsilon \in (0,1)$ and $t \geqslant 1$ there exists a constant $C(\varepsilon, t) > 0$ such that: if $N \geqslant C(\varepsilon, t)n$ then with probability greater than $1 - e^{-ct\sqrt{n}}$ we have*

$$(10.1.1) \qquad \left\| \frac{1}{N} \sum_{i=1}^{N} X_i \otimes X_i - I \right\|_{\ell_2^n \to \ell_2^n} \leqslant \varepsilon,$$

where $c > 0$ is an absolute constant. The argument shows that we can choose

$$C(\varepsilon, t) \simeq t^4 \varepsilon^{-2} \log^2 \left(2t^2 \varepsilon^{-2} \right).$$

Note that (10.1.1) can be written in the form

$$\max_{\theta \in S^{n-1}} \left| \frac{1}{N} \sum_{i=1}^{N} \left(\langle X_i, \theta \rangle^2 - \mathbb{E} \langle X_i, \theta \rangle^2 \right) \right| \leqslant \varepsilon$$

because one has $\|M\| = \max_{\theta \in S^{n-1}} \langle M\theta, \theta \rangle$ for any symmetric positive definite matrix M, and $\mathbb{E}\langle X_i, \theta \rangle^2 = \|\theta\|_2^2$.

REMARK 10.1.2. The proof of Theorem 10.1.1 is based on Theorem 10.1.4 which we discuss in the next subsection and which requires that $N \leqslant e^{\sqrt{n}}$ because of the use of Paouris' theorem. This does not create any problems, because we can use the following argument for larger values of N. Assume that Theorem 10.1.1 has been proved for $N \leqslant e^{\sqrt{n}}$ and let $N > e^{\sqrt{n}}$. If $X_i = (X_{i1}, \ldots, X_{in})$ are the random vectors in the theorem, we consider the smallest integer m for which $N \leqslant e^{\sqrt{m}}$ (note that $m > n$) and we define new random vectors $Y_i = (Y_{i1}, \ldots, Y_{im})$ in $\mathbb{R}^m$ setting $Y_{ij} = X_{ij}$ if $j \leqslant n$ and $Y_{ij} = g_{ij}$ if $j > n$, where g_{ij} are independent standard Gaussian random variables. Observe that $Y_1, \ldots, Y_m$ are i.i.d. isotropic log-concave random vectors in $\mathbb{R}^m$ and $N \leqslant e^{\sqrt{m}}$, therefore we can apply the theorem for them. If we identify $\theta = (\theta_1, \ldots, \theta_n) \in S^{n-1}$ with $z = (z_1, \ldots, z_m) \in S^{m-1}$, where $z_j = \theta_j$ if $j \leqslant n$ and $z_j = 0$ if $j > n$, then we have

$$\max_{\theta \in S^{n-1}} \left| \frac{1}{N} \sum_{i=1}^{N} \left(\langle X_i, \theta \rangle^2 - \mathbb{E}(\langle X_i, \theta \rangle^2) \right) \right|$$

$$\leqslant \max_{z \in S^{m-1}} \left| \frac{1}{N} \sum_{i=1}^{N} \left(\langle Y_i, z \rangle^2 - \mathbb{E}(\langle Y_i, z \rangle^2) \right) \right| \leqslant \varepsilon$$

with probability even closer to 1. This shows that, for the proof of the theorem we may assume without loss of generality that $N \leqslant e^{\sqrt{n}}$.

10.1.1. The main technical step

DEFINITION 10.1.3. Let $X_1, \ldots, X_N$ be i.i.d random vectors in $\mathbb{R}^n$. We consider the $n \times N$ random matrix A whose columns are the X_i's. Viewing A as an operator

$A : l_2^N \to l_2^n$, we denote its norm by $\|A\|$. For every $1 \leq m \leq N$ we define
$$A_m = \max_{\substack{F \subset \{1,\dots,N\} \\ |F| \leq m}} \|A|_{\mathbb{R}^F}\| = \max_{\substack{z \in S^{N-1} \\ |\operatorname{supp}(z)| \leq m}} \|Az\|_2.$$

Note that A_m is increasing in m.

The main technical step for the proof of Theorem 10.1.1 is an upper bound for A_m.

THEOREM 10.1.4 (Adamczak-Litvak-Pajor-Tomczak). *Let n, N be positive integers with $n \leq N \leq e^{\sqrt{n}}$ and consider N i.i.d. isotropic log-concave random vectors $X_1, \dots, X_N$ in $\mathbb{R}^n$. For every $t \geq 1$,*

(10.1.2) $$\mathbb{P}\left(\exists m \leq N \;:\; A_m \geq Ct\left(\sqrt{n} + \sqrt{m}\log\frac{2N}{m}\right)\right) \leq e^{-ct\sqrt{n}},$$

where $C, c > 0$ are absolute constants.

REMARK 10.1.5. The term $\sqrt{n} + \sqrt{m}\log\frac{2N}{m}$ in (10.1.2) can be replaced by $\sqrt{n} + \sqrt{m}\log\frac{2N}{\max\{n,m\}}$ because if $m < n$ then we have

$$\sqrt{n} + \sqrt{m}\log\frac{2N}{m} = \sqrt{n} + \sqrt{m}\log\frac{n}{m} + \sqrt{m}\log\frac{2N}{n}$$
$$\leq 2\sqrt{n} + \sqrt{m}\log\frac{2N}{n}.$$

Note that in the case $m = N$ we obtain an upper bound for the operator norm $\|A\|$.

COROLLARY 10.1.6. *Let n, N be positive integers with $n \leq N \leq e^{\sqrt{n}}$ and consider N i.i.d. isotropic log-concave random vectors $X_1, \dots, X_N$ in $\mathbb{R}^n$. Let A be the $n \times N$ random matrix whose columns are the X_j's. For every $t \geq 1$ one has*

$$\|A : \ell_2^N \to \ell_2^n\| \leq Ct(\sqrt{n} + \sqrt{N})$$

with probability greater than $1 - \exp(-ct\sqrt{n})$, where $C, c > 0$ are absolute constants.

Another consequence of Theorem 10.1.4 is an improvement of the main technical step in Bourgain's argument (Theorem 10.2.2).

COROLLARY 10.1.7. *Let n, N be positive integers with $n \leq N \leq e^{\sqrt{n}}$ and consider N i.i.d. isotropic log-concave random vectors $X_1, \dots, X_N$ in $\mathbb{R}^n$. Then,*

$$\mathbb{P}\left(\exists E \subseteq \{1,\dots,N\} \;:\; \Big\|\sum_{i \in E} X_i\Big\|_2 \geq Ct\left(\sqrt{n|E|} + |E|\log\frac{2N}{|E|}\right)\right) \leq e^{-ct\sqrt{n}},$$

where $C, c > 0$ are absolute constants.

Proof. Let $E \subseteq \{1,\dots,N\}$ with $|E| = m$. Consider the vector $z \in S^{N-1}$ with coordinates $z_i = \frac{1}{\sqrt{m}}$ if $i \in E$ and $z_i = 0$ otherwise. Then,

$$\Big\|\sum_{i \in E} X_i\Big\|_2 = \sqrt{m}\|Az\|_2 \leq \sqrt{m}A_m,$$

and the result follows from Theorem 10.1.4. $\square$

For the proof of Theorem 10.1.4 we need a number of auxiliary results that we gather in the next remark.

REMARK 10.1.8. Let n, N be positive integers with $n \leqslant N \leqslant e^{\sqrt{n}}$ and consider N i.i.d. isotropic log-concave random vectors $X_1, \ldots, X_N$ in $\mathbb{R}^n$.

(i) As a standard consequence of Borell's lemma we know that there exists an absolute constant $\varrho > 0$ such that

$$\|\langle X_i, \theta \rangle\|_{\psi_1} \leqslant \varrho \tag{10.1.3}$$

for every $\theta \in S^{n-1}$ and for all $i \leqslant N$. We fix this constant in the rest of this section.

(ii) As an immediate consequence of Paouris' inequality we have that, for every $N \leqslant e^{\sqrt{n}}$ and for every $t \geqslant 1$,

$$\max_{1 \leqslant i \leqslant N} \|X_i\|_2 \leqslant Ct\sqrt{n}, \tag{10.1.4}$$

with probability greater than $1 - e^{-t\sqrt{n}}$, where $C > 0$ is an absolute constant. This is clear by the union bound, since

$$\mathbb{P}\{\|X_i\|_2 \geqslant C_1 t\sqrt{n}\} \leqslant e^{-t\sqrt{n}} \tag{10.1.5}$$

for every $i \leqslant N$ and $t \geqslant 1$, where $C_1 > 0$ is an absolute constant.

(iii) For any $x_1, \ldots, x_N \in \mathbb{R}^n$ there exists $E \subset \{1, \ldots, N\}$ such that

$$\sum_{i \neq j} \langle x_i, x_j \rangle \leqslant 4 \sum_{i \in E} \sum_{j \in E^c} \langle x_i, x_j \rangle. \tag{10.1.6}$$

To see this, consider independent random variables $\delta_1, \ldots, \delta_N$ that take the values 0 or 1 with probability $1/2$ and observe that

$$\mathbb{E}\Big\langle \sum_{i=1}^N \delta_i x_i, \sum_{j=1}^N (1-\delta_j) x_j \Big\rangle = \frac{1}{4} \sum_{i \neq j} \langle x_i, x_j \rangle.$$

(iv) Given $E \subset \{1, \ldots, N\}$ with $|E| = m$ and $\varepsilon, \alpha \in (0, 1]$, we can choose an ε-net $\mathcal{N}$ of $B_2^N \cap \alpha B_\infty^N \cap \mathbb{R}^E$, with respect to the Euclidean metric, that has cardinality

$$|\mathcal{N}| \leqslant \left(\frac{3}{\varepsilon}\right)^m. \tag{10.1.7}$$

In what follows, for any E that appears in our arguments we will be fixing an ε-net of $B_2^N \cap \alpha B_\infty^N \cap \mathbb{R}^E$ that satisfies (10.1.7); this will be denoted by $\mathcal{N}(E, \varepsilon, \alpha)$.

We also need the following two lemmas.

LEMMA 10.1.9. *Let $X_1, \ldots, X_N$ be i.i.d. isotropic log-concave random vectors in $\mathbb{R}^n$. Then, if $m \leqslant N$, for all $\varepsilon, \alpha \in (0, 1]$ and $L \geqslant 2m \log \frac{24eN}{m\varepsilon}$ we have*

$$\mathbb{P}\left(\max_{F, E, z} \sum_{i \in E} \left|\left\langle z_i X_i, \sum_{j \in F \setminus E} z_j X_j \right\rangle\right| > \varrho \alpha L A_m\right) \leqslant e^{-L/2},$$

where $\max_{F,E,z}$ denotes the triple maximum over all $F \subset \{1, \ldots, N\}$ with $|F| \leqslant m$, all $E \subseteq F$ and all $z \in \mathcal{N}(F, \varepsilon, \alpha)$.

Proof. We denote by Ω the underlying probability space. For every $F \subseteq \{1, \ldots, N\}$ with $|F| \leqslant m$, and every $E \subseteq F$ and $z \in \mathcal{N}(F, \varepsilon, \alpha)$, we define

$$\Omega(F, E, z) = \left\{\omega \in \Omega : \sum_{i \in E}\left|\left\langle z_i X_i(\omega), \sum_{j \in F \setminus E} z_j X_j(\omega)\right\rangle\right| > \varrho \alpha L A_m\right\}.$$

We fix (F, E, z) and set $Y = \sum_{j \in F \setminus E} z_j X_j$. Then, since the X_i's are independent, we observe that Y is independent from X_i, $i \in E$, and by the definition of A_m we have $\|Y\|_2 \leqslant A_m$. We also observe that if $\omega \in \Omega(F, E, z)$ then $Y(\omega) \neq 0$ (otherwise, $\langle z_i X_i, Y \rangle = 0$ for all $i \in E$ and the condition in the definition of $\Omega(F, E, z)$ cannot be satisfied). Then, using also the fact that $\|z\|_\infty \leqslant \alpha$, we get

$$\sum_{i \in E} |\langle z_i X_i, Y \rangle| \leqslant \sum_{i \in E} |z_i| |\langle X_i, Y \rangle| \leqslant \alpha \sum_{i \in E} |\langle X_i, Y \rangle|$$

$$= \alpha \|Y\|_2 \sum_{i \in E} \left| \left\langle X_i, \frac{Y}{\|Y\|_2} \right\rangle \right| \leqslant \alpha A_m \sum_{i \in E} \left| \left\langle X_i, \frac{Y}{\|Y\|_2} \right\rangle \right|.$$

Since $A_m \geqslant \|Y\|_2 > 0$ on $\Omega(F, E, z)$, we have

$$\mathbb{P}(\Omega(F, E, z)) \leqslant \mathbb{P}\left(\sum_{i \in E} \left| \left\langle X_i, \frac{Y}{\|Y\|_2} \right\rangle \right| > \varrho L \right).$$

Using Markov's inequality and the assumption $\|\langle X_i, \theta \rangle\|_{\psi_1} \leqslant \varrho$ for all $1 \leqslant i \leqslant N$ and $\theta \in S^{n-1}$, we can write

$$\mathbb{P}\left(\sum_{i \in E} \left| \left\langle X_i, \frac{Y}{\|Y\|_2} \right\rangle \right| > \varrho L \right) \leqslant e^{-L} \mathbb{E} \exp\left(\frac{1}{\varrho} \sum_{i \in E} \left| \left\langle X_i, \frac{Y}{\|Y\|_2} \right\rangle \right| \right)$$

$$\leqslant 2^{|E|} e^{-L} \leqslant 2^m e^{-L}.$$

Finally, taking the union over all (F, E, z) and taking into account (10.1.7) we get

$$\mathbb{P}\left(\max_{F, E, z} \sum_{i \in E} \left| \left\langle z_i X_i, \sum_{j \in F \setminus E} z_j X_j \right\rangle \right| > \varrho \alpha L A_m \right)$$

$$= \mathbb{P}\left(\max_{|F| \leqslant m} \max_{E \subseteq F} \max_{z \in \mathcal{N}(F, \varepsilon, \alpha)} \sum_{i \in E} \left| \left\langle z_i X_i, \sum_{j \in F \setminus E} z_j X_j \right\rangle \right| > \varrho \alpha L A_m \right)$$

$$\leqslant \left[\left(\sum_{k=1}^m \binom{N}{k} 2^k \right) \left(\frac{3}{\varepsilon} \right)^m \right] \max_{F, E, z} \mathbb{P}(\Omega(F, E, z))$$

$$\leqslant \left(\sum_{k=1}^m \left(\frac{2eN}{k} \right)^k \right) \left(\frac{3}{\varepsilon} \right)^m 2^m e^{-L} \leqslant \left(\frac{4eN}{m} \right)^m \left(\frac{6}{\varepsilon} \right)^m e^{-L}$$

$$= \exp\left(m \log \frac{24 eN}{m \varepsilon} - L \right),$$

using that $\left(\frac{2eN}{k} \right)^k$ is an increasing function of k, which implies

$$\sum_{k=1}^m \left(\frac{2eN}{k} \right)^k \leqslant m(2eN/m)^m \leqslant (4eN/m)^m.$$

Our assumption for L completes the proof. $\square$

The second lemma is of the same type.

LEMMA 10.1.10. *Let $X_1, \ldots, X_N$ be i.i.d. isotropic log-concave random vectors in $\mathbb{R}^n$. Let $1 \leqslant k, m \leqslant N$, and let $\varepsilon, \alpha \in (0, 1]$ and $\beta, L > 0$. We write $B(m, \beta)$ for* the set of all $x \in \beta B_2^N$ with $|\mathrm{supp}(x)| \leqslant m$. We also consider a subset $\mathcal{B}$ of $B(m,\beta)$ with cardinality equal to M. Then,

$$\mathbb{P}\left(\max_{F,x,z} \sum_{i\in F}\left|\left\langle z_i X_i, \sum_{j\in F^c} x_j X_j \right\rangle\right| > \varrho\alpha\beta L A_m \right) \leqslant M\left(\frac{12eN}{k\varepsilon}\right)^k e^{-L},$$

where $\max_{F,x,z}$ is the triple maximum over all $F \subset \{1,\ldots,N\}$ with $|F| \leqslant k$, all $x \in \mathcal{B}$ and all $z \in \mathcal{N}(F,\varepsilon,\alpha)$.

Proof. For every $F \subseteq \{1,\ldots,N\}$ with $|F| \leqslant k$, and every $x \in \mathcal{B}$ and $z \in \mathcal{N}(F,\varepsilon,\alpha)$, we define

$$\Omega(F,x,z) = \left\{\omega \in \Omega : \sum_{i\in F}\left|\left\langle z_i X_i(\omega), \sum_{j\in F^c} x_j X_j(\omega) \right\rangle\right| > \varrho\alpha\beta L A_m \right\}.$$

We fix (F,x,z) and we set $Y = \sum_{j\in F^c} x_j X_j$. Since the X_i's are independent, we have that Y is independent from X_i, $i \in F$. Moreover, $\|Y\|_2 \leqslant \beta A_m$. We also note that, as in the previous lemma, if $\omega \in \Omega(F,E,z)$ then $Y(\omega) \neq 0$. Then, using also the fact that $\|z\|_\infty \leqslant \alpha$, we get

$$\sum_{i\in F}|\langle z_i X_i, Y\rangle| \leqslant \sum_{i\in F}|z_i||\langle X_i, Y\rangle| \leqslant \alpha \sum_{i\in F}|\langle X_i, Y\rangle|$$

$$= \alpha\|Y\|_2 \sum_{i\in F}\left|\left\langle X_i, \frac{Y}{\|Y\|_2}\right\rangle\right| \leqslant \alpha\beta A_m \sum_{i\in F}\left|\left\langle X_i, \frac{Y}{\|Y\|_2}\right\rangle\right|.$$

Since $A_m \geqslant \beta^{-1}\|Y\|_2 > 0$, we have

$$\mathbb{P}(\Omega(F,x,z)) \leqslant \mathbb{P}\left(\sum_{i\in F}\left|\left\langle X_i, \frac{Y}{\|Y\|_2}\right\rangle\right| > \varrho L\right).$$

Then,

$$\mathbb{P}\left(\sum_{i\in F}\left|\left\langle X_i, \frac{Y}{\|Y\|_2}\right\rangle\right| > \varrho L\right) \leqslant e^{-L}\mathbb{E}\exp\left(\frac{1}{\varrho}\sum_{i\in F}\left|\left\langle X_i, \frac{Y}{\|Y\|_2}\right\rangle\right|\right)$$

$$\leqslant 2^{|F|}e^{-L} \leqslant 2^k e^{-L}.$$

Finally, taking the union and using (10.1.7) we conclude that

$$\mathbb{P}\left(\max_{F,x,z}\sum_{i\in F}\left|\left\langle z_i X_i, \sum_{j\in F^c} x_j X_j\right\rangle\right| > \varrho\alpha\beta L A_m\right)$$

$$\leqslant M\left(\sum_{l=1}^k \binom{N}{l}\right)\left(\frac{3}{\varepsilon}\right)^k \max_{F,x,z}\mathbb{P}(\Omega(F,x,z))$$

$$\leqslant M\left(\frac{2eN}{k}\right)^k\left(\frac{3}{\varepsilon}\right)^k 2^k e^{-L},$$

which proves the lemma. $\square$

Proof of Theorem 10.1.4. The assumption $N \leqslant e^{\sqrt{n}}$ and the union bound show that it is enough to check that, for t large enough and for any fixed $m \leqslant N$,

$$\mathbb{P}\left(A_m \geqslant Ct\left(\sqrt{n} + \sqrt{m}\log\frac{2N}{m}\right)\right) \leqslant e^{-ct\sqrt{n}}.$$

10.1. OPTIMAL ESTIMATE

We will define a finite set $\mathcal{M} \subset \mathbb{R}^n$ so that we can estimate $\max_{x \in \mathcal{M}} \|Ax\|_2$ with high probability and, at the same time, we can approximate by $\mathcal{M}$ any point from B_2^N with at most m non-zero coordinates.

Note that if for some $x \in S^{N-1}$ we know that $|\mathrm{supp}(x)| \sim s$ and $\|x\|_\infty \leqslant \frac{1}{\sqrt{s}}$ for some $s \geqslant 1$ then, using (10.1.6) and Lemma 10.1.9 we can estimate $\|Ax\|_2$ with probability close to 1. The set $\mathcal{M}$ will consist of sums of vectors with disjoint supports, that roughly satisfy these two conditions.

To make this precise, we first define $\mathcal{M}$ distinguishing two cases:

Case 1. Assume that
$$m \log \frac{96eN}{m} \leqslant \sqrt{n}.$$
Then, we set
$$\mathcal{M} = \bigcup_{\substack{E \subset \{1,\ldots,N\} \\ |E|=m}} \mathcal{N}(E, 1/4, 1).$$

Case 2. If $m \log \frac{96eN}{m} > \sqrt{n}$ then we consider the least integer l for which
$$\frac{m}{2^l} \log \frac{96 \cdot 2^l eN}{m} \leqslant \sqrt{n},$$
and we fix positive integers $a_0, a_1, \ldots, a_l$ so that the next three conditions are satisfied:

(i) $\sum_{k=0}^l a_k = m$.
(ii) $a_k \leqslant m \cdot 2^{-k+1}$ for all $1 \leqslant k \leqslant l$.
(iii) $a_0 \leqslant m \cdot 2^{-l}$.

Then, we set $\mathcal{M} = \mathcal{M}_0 \cap 2B_2^N$, where $\mathcal{M}_0$ is the set of all vectors of the form $x = \sum_{k=0}^l x_k$, where the x_i's have disjoint supports and
$$x_0 \in \bigcup_{\substack{E \subset \{1,\ldots,N\} \\ |E| \leqslant a_0}} \mathcal{N}(E, 1/4, 1), \quad x_k \in \bigcup_{\substack{E \subset \{1,\ldots,N\} \\ |E| \leqslant a_k}} \mathcal{N}\left(E, 2^{-k}, \sqrt{2^k/m}\right), \ 1 \leqslant k \leqslant l.$$

Observe that if $x \in \mathcal{M}$ then $|\mathrm{supp}(x)| \leqslant \sum_{k=0}^l a_k = m$ and $\|x\|_2 \leqslant 2$.

We will give the proof for Case 2. The first case can be handled in a similar way; it is in fact simpler because of the simpler form of $\mathcal{M}$.

Case 2 (proof). We fix $x \in \mathcal{M}$ and write $x = \sum_{k=0}^l x_k$. We denote the support of x_k by F_k, and the coordinates of x by $x(i)$, $1 \leqslant i \leqslant N$. Then, we have

$$\|Ax\|_2^2 = \left\langle \sum_{1 \leqslant i \leqslant N} x(i) X_i, \sum_{1 \leqslant j \leqslant N} x(j) X_j \right\rangle$$
$$= \sum_{1 \leqslant i \leqslant N} x(i)^2 \|X_i\|_2^2 + \sum_{i \neq j} \langle x(i) X_i, x(j) X_j \rangle$$
$$\leqslant 2 \max_i \|X_i\|_2^2 + D_x \leqslant 2 \max \left\{ 2 \max_i \|X_i\|_2^2, D_x \right\},$$

where
$$D_x = \sum_{i \neq j} \langle x(i) X_i, x(j) X_j \rangle.$$

From (10.1.4) we know that $\max_i \|X_i\|_2 \leqslant C_0 t\sqrt{n}$ with probability greater than $1 - e^{-t\sqrt{n}}$, and we would like to have a similar estimate for D_x. We split D_x into two parts, corresponding to the construction of x: we set

$$D'_x = \sum_{k=0}^{l} \sum_{\substack{i,j \in F_k \\ i \neq j}} \langle x(i)X_i, x(j)X_j \rangle,$$

and

$$D''_x = \sum_{k=0}^{l} \sum_{\substack{i \in F_k \\ j \in F_k^c}} \langle x(i)X_i, x(j)X_j \rangle = 2\sum_{k=1}^{l} \sum_{i \in F_k} \sum_{r \in G_k} \left\langle x(i)X_i, \sum_{j \in F_r} x(j)X_j \right\rangle,$$

where $G_k = \{0, k+1, k+2, \ldots, l\}$. Note that

$$D_x = D'_x + D''_x.$$

In order to estimate D'_x, D''_x we will use (10.1.6) and Lemma 10.1.9. More precisely, from (10.1.6) we see that for every k there exists a subset F'_k of F_k such that

$$D'_x \leqslant 4 \sum_{k=0}^{l} \sum_{\substack{i \in F'_k \\ j \in F_k \setminus F'_k}} \langle x(i)X_i, x(j)X_j \rangle$$

$$\leqslant 4 \max_{\substack{F \subset \{1,\ldots,N\} \\ |F| \leqslant m/2^l}} \max_{E \subset F} \max_{v \in \mathcal{N}(F, 1/4, 1)} \sum_{i \in E} \left| \left\langle v_i X_i, \sum_{j \in F \setminus E} v_j X_j \right\rangle \right|$$

$$+ 4 \sum_{k=1}^{l} \max_{\substack{F \subset \{1,\ldots,N\} \\ |F| \leqslant 2m/2^k}} \max_{E \subset F} \max_{v \in \mathcal{N}(F, 2^{-k}, \sqrt{2^k/m})} \sum_{i \in E} \left| \left\langle v_i X_i, \sum_{j \in F \setminus E} v_j X_j \right\rangle \right|.$$

Using now Lemma 10.1.9 with $L = 2t\sqrt{n}$, $\varepsilon = 1/4$, $\alpha = 1$ for the first term and $L = \frac{8mt}{2^k} \log \frac{24eN \cdot 4^k}{m}$, $\varepsilon = 2^{-k}$, $\alpha = \sqrt{2^k/m}$ for the second one, we get

$$\mathbb{P}\left(\max_{x \in \mathcal{M}} D'_x > 8\varrho t A_m \sqrt{n} + 2\varrho t A_m \sum_{k=1}^{l} \sqrt{\frac{2^k}{m}} \frac{8m}{2^k} \log \frac{24 \cdot 4^k eN}{m} \right)$$

$$\leqslant \exp(-t\sqrt{n}) + \sum_{k=1}^{l} \exp\left(-t \frac{2m}{2^k} \log \frac{24 \cdot 4^k eN}{m} \right)$$

$$\leqslant \exp(-t\sqrt{n}) + l \exp\left(-t \frac{2m}{2^l} \log \frac{24 \cdot 4^l eN}{m} \right).$$

By the choice of l and the fact that $l \leqslant 2\sqrt{n}$ (because $m \leqslant N \leqslant e^{\sqrt{n}}$) we see that there exists an absolute constant $C > 0$ such that

$$\mathbb{P}\left(\max_{x \in \mathcal{M}} D'_x > t A_m \left(8\varrho \sqrt{n} + C\varrho \sqrt{m} \log \frac{2N}{m} \right) \right)$$

$$\leqslant \exp(-t\sqrt{n}) + l \exp(-t\sqrt{n}) \leqslant (2\sqrt{n} + 1) \exp(-t\sqrt{n}).$$

We estimate D''_x in a similar way. For every $1 \leqslant k \leqslant l$, we consider the set $\mathcal{M}_k = \mathcal{M}'_k \cap 2B_2^N$, where $\mathcal{M}'_k$ is the set of all vectors of the form $x = x_0 +$

$\sum_{s=k+1}^{l} x_s$, where the x_i's have disjoint supports for $s \geqslant k+1$, and

$$x_0 \in \bigcup_{\substack{E \subset \{1,\ldots,N\} \\ |E| \leqslant a_0}} \mathcal{N}(E, 1/4, 1), \ x_s \in \bigcup_{\substack{E \subset \{1,\ldots,N\} \\ |E| \leqslant a_s}} \mathcal{N}\left(E, 2^{-s}, \sqrt{2^s/m}\right), \ s \geqslant k+1.$$

We have $\mathcal{M}_k \subset 2B_2^N$ and from (10.1.7) we get

$$|\mathcal{M}_k| \leqslant \left(\frac{3}{1/4}\right)^{a_0} \binom{N}{a_0} \prod_{s=k+1}^{l} (3 \cdot 2^s)^{a_s} \binom{N}{a_s}$$

$$\leqslant \left(\frac{12eN}{a_0}\right)^{a_0} \prod_{s=k+1}^{l} \left(\frac{3 \cdot 2^s eN}{a_s}\right)^{a_s}$$

$$\leqslant \exp\left(\frac{m}{2^l} \log \frac{12 \cdot 2^l eN}{m} + \sum_{s=k+1}^{l} \frac{2m}{2^s} \log \frac{3 \cdot 4^s eN}{2m}\right)$$

$$\leqslant \exp\left(\sum_{s=k+1}^{l+1} \frac{2m}{2^s} \log \frac{3 \cdot 4^s eN}{2m}\right)$$

$$\leqslant \exp\left(\frac{m}{2^k}\left(\log \frac{6 \cdot 4^k eN}{m} \sum_{s=0}^{l-k} \frac{1}{2^s} + \log 4 \sum_{s=1}^{l-k} \frac{s}{2^s}\right)\right)$$

$$\leqslant \exp\left(\frac{4m}{2^k} \log \frac{6 \cdot 4^k eN}{m}\right).$$

Moreover, we observe that

$$D_x'' = 2 \sum_{k=1}^{l} \sum_{i \in F_k} \left\langle x(i) X_i, \sum_{r \in G_k} \sum_{j \in F_r} x(j) X_j \right\rangle$$

$$\leqslant 2 \sum_{k=1}^{l} \max_{\substack{F \subset \{1,\ldots,N\} \\ |F| \leqslant 2m/2^k}} \max_{u \in \mathcal{N}(F, 2^{-k}, \sqrt{2^k/m})} \max_{v \in \mathcal{M}_k} \left|\sum_{i \in F} \left\langle u_i X_i, \sum_{j \in F^c} v_j X_j \right\rangle\right|.$$

Applying Lemma 10.1.10 for every term, with $L_k = \frac{12m}{2^k} t \log \frac{12 \cdot 4^k eN}{m}$ and

$$\varepsilon_k = 2^{-k}, \ \alpha_k = \sqrt{2^k/m}, \ \beta = 2, \ \mathcal{B}_k = \mathcal{M}_k,$$

we obtain

$$\mathbb{P}\left(D_x'' > 96 \varrho t A_m \sum_{k=1}^{l} \sqrt{\frac{2^k}{m} \frac{m}{2^k} \log \frac{12 \cdot 4^k eN}{m}}\right)$$

$$\leqslant \sum_{k=1}^{l} \exp\left(\frac{4m}{2^k} \log \frac{6 \cdot 4^k eN}{m} + \frac{2m}{2^k} \log \frac{6 \cdot 4^k eN}{m} - t \frac{12m}{2^k} \log \frac{12 \cdot 4^k eN}{m}\right)$$

$$\leqslant \sum_{k=1}^{l} \exp\left(-t \frac{6m}{2^k} \log \frac{12 \cdot 4^k eN}{m}\right) \leqslant l \exp\left(-t \frac{6m}{2^l} \log \frac{12 \cdot 4^l eN}{m}\right).$$

As in the case of D_x', we check that

$$\mathbb{P}\left(\max_{x \in \mathcal{M}} D_x'' > 3C \varrho A_m t \sqrt{m} \log \frac{2N}{m}\right) \leqslant 2\sqrt{n} \exp\left(-t\sqrt{n}\right),$$

where $C > 0$ is an absolute constant. Since $D_x = D'_x + D''_x$, we have
(10.1.8)
$$\mathbb{P}\left(\max_{x \in \mathcal{M}} D_x > A_m t \left(8\varrho\sqrt{n} + 4C\varrho\sqrt{m}\log\frac{2N}{m}\right)\right) \leqslant (4\sqrt{n} + 1)\exp\left(-t\sqrt{n}\right).$$

In order to complete the proof, it remains to show that we can approximate by a vector from $\mathcal{M}$ any vector from B_2^N whose support has cardinality less than or equal to m.

We consider $z \in S^{N-1}$ with $|\operatorname{supp}(z)| \leqslant m$ and denote its coordinates by z_i, $1 \leqslant i \leqslant N$. We can find $n_1, \ldots, n_m$ so that $|z_{n_1}| \geqslant |z_{n_2}| \geqslant \cdots \geqslant |z_{n_N}|$, and $z_{n_i} = 0$ if $i > m$. If $m \log \frac{96eN}{M} \leqslant \sqrt{n}$, then we can consider only the support of z, which we denote by E_0. Otherwise, we set

$$E_0 = \{n_i\}_{1 \leqslant i \leqslant m/2^l}$$

and

$$E_1 = \{n_i\}_{m/2 < i \leqslant m}, \; E_2 = \{n_i\}_{m/4 < i \leqslant m/2}, \ldots, E_l = \{n_i\}_{m/2^l < i \leqslant m/2^{l-1}},$$

where l is the least integer satisfying

$$\frac{m}{2^l} \log \frac{96 \cdot 2^l eN}{m} \leqslant \sqrt{n}.$$

It is clear that

$$a_0 := |E_0| \leqslant m/2^l, \; a_k := |E_k| \leqslant m/2^k + 1 \leqslant m/2^{k-1}, \; 1 \leqslant k \leqslant l,$$

and $\sum_{i=0}^{l} a_i = m$. Finally, note that for every $k \geqslant 1$ we have

$$\|P_{E_k} z\|_\infty \leqslant |z_{n_s}| \leqslant \sqrt{2^k/m},$$

where $s = \lfloor m/2^k \rfloor$. So, for every $k \geqslant 1$ the vector $P_{E_k} z$ can be approximated by a vector in the net $\mathcal{N}\left(E_k, 2^{-k}, \sqrt{2^k/m}\right)$ and the vector $P_{E_0} z$ can be approximated by a vector in the net $\mathcal{N}(E_0, 1/4, 1)$. Then, we can find $x \in \mathcal{M}$ of the form $x = \sum_{k=0}^{l} x_k$, such that

$$\|z - x\|_2^2 \leqslant \sum_{k=0}^{l} \|P_{E_k} z - x_k\|_2^2 \leqslant 2^{-4} + \sum_{k=1}^{l} 2^{-2k} < 4/10.$$

Moreover, we have $\operatorname{supp}(x) = \operatorname{supp}(z)$, therefore the vector $w = z - x$ satisfies $|\operatorname{supp}(w)| \leqslant m$. Considering all $z \in S^{N-1}$ with $|\operatorname{supp}(z)| \leqslant m$, we get

$$A_m = \max_{\substack{z \in S^{N-1} \\ |\operatorname{supp}(z)| \leqslant m}} \|Az\|_2 \leqslant \max_{x \in \mathcal{M}} \|Ax\|_2 + \sqrt{4/10} \max_{\substack{w \in S^{N-1} \\ |\operatorname{supp}(w)| \leqslant m}} \|Aw\|_2$$
$$= \max_{x \in \mathcal{M}} \|Ax\|_2 + \sqrt{4/10} A_m,$$

and hence

$$A_m \leqslant 3 \max_{x \in \mathcal{M}} \|Ax\|_2.$$

This completes the proof of the theorem because

$$A_m^2 \leqslant 9 \max_{x \in \mathcal{M}} \|Ax\|_2^2 \leqslant 9 \max\{4 \max_i \|X_i\|_2^2, 2 \max_{x \in \mathcal{M}} D_x\},$$

and using (10.1.4) and (10.1.8) we finally get

$$\mathbb{P}\left(A_m \leqslant Ct\varrho\left(\sqrt{n} + \sqrt{m}\log\frac{2N}{m}\right)\right) \geqslant 1 - c'\sqrt{n}e^{-c_1 t\sqrt{n}}$$
$$\geqslant 1 - \exp(-ct\sqrt{n}),$$

where $C, c > 0$ are absolute constants. □

10.1.2. Proof of Theorem 10.1.1

Let $X_1, \ldots, X_N$ be i.i.d. isotropic log-concave random vectors in $\mathbb{R}^n$. Our aim is to estimate

$$\max_{\theta \in S^{n-1}} \left| \frac{1}{N} \sum_{i=1}^{N} (\langle X_i, \theta \rangle^2 - 1) \right|.$$

The next lemma will allow us to work with a finite set of points θ in S^{n-1}.

LEMMA 10.1.11. *Let $x_1, \ldots, x_N \in \mathbb{R}^n$. Let $\gamma, \varepsilon \in (0,1)$ and let $\mathcal{N}$ be a $\gamma\varepsilon$-net for S^{n-1}. If*

$$\max_{\theta \in \mathcal{N}} \left| \frac{1}{N} \sum_{i=1}^{N} (\langle x_i, \theta \rangle^2 - 1) \right| \leqslant \varepsilon,$$

then

$$\max_{\theta \in S^{n-1}} \left| \frac{1}{N} \sum_{i=1}^{N} (\langle x_i, \theta \rangle^2 - 1) \right| \leqslant c_1(\gamma)\varepsilon,$$

where $c_1(\gamma) > 0$ is a constant depending only on γ.

Proof. We define a semi-norm $\|\cdot\|$ on $\mathbb{R}^n$ by

$$\|y\| = \left(\frac{1}{N} \sum_{i=1}^{N} \langle x_i, y \rangle^2\right)^{1/2}.$$

Then we have

$$1 - \varepsilon \leqslant \sqrt{1-\varepsilon} \leqslant \max_{\theta \in \mathcal{N}} \|\theta\| \leqslant \sqrt{1+\varepsilon} \leqslant 1 + \varepsilon/2,$$

and, using the fact that $\mathcal{N}$ is a $\gamma\varepsilon$-net together with the triangle inequality and the homogeneity of $\|\cdot\|$, we get

$$\max_{\theta \in S^{n-1}} \|\theta\| \leqslant \frac{1+\varepsilon/2}{1-\gamma\varepsilon} = 1 + \frac{2\gamma+1}{2(1-\gamma\varepsilon)}\varepsilon \leqslant 1 + \frac{2\gamma+1}{2(1-\gamma)}\varepsilon.$$

For the lower bound, we consider any $\theta \in S^{n-1}$ and find $\theta_1 \in \mathcal{N}$ such that $\|\theta - \theta_1\|_2 \leqslant \gamma\varepsilon$. Then,

$$\|\theta\| \geqslant \|\theta_1\| - \|\theta - \theta_1\| \geqslant (1-\varepsilon) - \gamma\varepsilon\left(1 + \frac{2\gamma+1}{2(1-\gamma)}\varepsilon\right) \geqslant 1 - \frac{2\gamma+3}{2(1-\gamma)}\varepsilon.$$

This shows that $|\|\theta\| - 1| \leqslant c_1(\gamma)\varepsilon$, where

$$c_1(\gamma) = \frac{2\gamma+3}{2(1-\gamma)}.$$

Since $t \mapsto t^2$ is Lipschitz with constant $2(1 + c_1(\gamma))$ on the interval $[0, 1 + c_1(\gamma)]$, we finally get

$$\max_{\theta \in S^{n-1}} \left| \frac{1}{N} \sum_{i=1}^{N} (\langle x_i, \theta \rangle^2 - 1) \right| \leqslant c_2(\gamma)\varepsilon,$$

where $c_2(\gamma) = 2c_1(\gamma)(1 + c_1(\gamma)) = \frac{5(2\gamma+3)}{2(1-\gamma)^2}$. □

For the proof of Theorem 10.1.1 we also need the next Bernstein-type inequality.

LEMMA 10.1.12. *Let $Z_1, \ldots, Z_N$ be i.i.d. random variables with $\mathbb{E}(Z_i) = 0$, $\|Z_i\|_\infty \leqslant a$ and $\mathrm{Var}(Z_i) = \mathbb{E}(Z_i^2) = \sigma^2$. Then,*

$$\mathbb{P}(Z_1 + \cdots + Z_N \geqslant sN) \leqslant \exp\left(-\frac{3N}{8} \min\left\{\frac{s^2}{\sigma^2}, \frac{s}{a}\right\}\right)$$

for all $s > 0$.

Proof. Let $s > 0$. From Markov's inequality and by independence we have

$$\mathbb{P}\left(\sum_{j=1}^{N} Z_j \geqslant sN\right) \leqslant e^{-\lambda s} \mathbb{E} \exp\left(\frac{\lambda}{N} \sum_{j=1}^{N} Z_j\right) = e^{-\lambda s} \prod_{j=1}^{N} \mathbb{E} \exp(\lambda Z_j/N)$$

for every $\lambda > 0$. Next, observe that

$$\mathbb{E} \exp(\lambda Z_j/N) = 1 + \sum_{k=2}^{\infty} \frac{\lambda^k \mathbb{E}(Z_j^k)}{N^k k!} \leqslant 1 + \mathbb{E}(Z_j^2) \sum_{k=2}^{\infty} \frac{\lambda^k a^{k-2}}{N^k k!}$$

$$= 1 + \frac{\sigma^2}{a^2}\left(e^{\frac{\lambda a}{N}} - \frac{\lambda a}{N} - 1\right).$$

Since $e^u \geqslant 1 + u$ for all $u \in \mathbb{R}$, we get

$$\prod_{j=1}^{N} \mathbb{E} \exp(\lambda Z_j/N) \leqslant \exp\left(\frac{\sigma^2 N}{a^2}\left(e^{\frac{\lambda a}{N}} - \frac{\lambda a}{N} - 1\right)\right).$$

This shows that

$$\mathbb{P}\left(\sum_{j=1}^{N} Z_j \geqslant sN\right) \leqslant \exp\left(\frac{\sigma^2 N}{a^2}\left(e^{\frac{\lambda a}{N}} - \frac{\lambda a}{N} - 1\right) - \lambda s\right)$$

for all $\lambda > 0$. We choose λ so that $\exp(\lambda a/N) = 1 + sa/\sigma^2$. This leads to the bound

$$\mathbb{P}\left(\sum_{j=1}^{N} Z_j \geqslant sN\right) \leqslant \exp\left(-\frac{\sigma^2 N}{a^2} F\left(\frac{as}{\sigma^2}\right)\right),$$

where $F(r) = (1+r)\log(1+r) - r$, $r > 0$. One can check that $F(r) \geqslant r^2/(2 + 2r/3)$, and this implies the assertion of the lemma, because $F(r) \geqslant \frac{3r}{8}$ if $r \geqslant 1$ and $F(r) \geqslant \frac{3r^2}{8}$ if $0 < r < 1$. □

Proof of Theorem 10.1.1. We fix $\gamma, \eta \in (0,1)$ and a $\gamma\eta$-net $\mathcal{N}$ of S^{n-1} with cardinality $|\mathcal{N}| \leqslant (3/\gamma\eta)^n$. We shall estimate

$$\max_{\theta \in \mathcal{N}} \left| \frac{1}{N} \sum_{i=1}^{N} \left(\langle X_i, \theta \rangle^2 - \mathbb{E}(\langle X_i, \theta \rangle^2)\right) \right|.$$

We fix $B > 1$ that will be specified later, and write

(10.1.9) $\quad \max_{\theta \in \mathcal{N}} \left| \frac{1}{N} \sum_{i=1}^{N} \left(\langle X_i, \theta \rangle^2 - \mathbb{E}(\langle X_i, \theta \rangle^2) \right) \right|$

$$\leq \frac{1}{N} \max_{\theta \in \mathcal{N}} \left| \sum_{i=1}^{N} \left((|\langle X_i, \theta \rangle| \wedge B)^2 - \mathbb{E}\left((|\langle X_i, \theta \rangle| \wedge B)^2 \right) \right) \right|$$

$$+ \max_{\theta \in \mathcal{N}} \frac{1}{N} \sum_{i=1}^{N} \left(|\langle X_i, \theta \rangle|^2 - B^2 \right) \mathbf{1}_{\{|\langle X_i, \theta \rangle| \geq B\}}$$

$$+ \max_{\theta \in \mathcal{N}} \frac{1}{N} \mathbb{E} \left(\sum_{i=1}^{N} \left(|\langle X_i, \theta \rangle|^2 - B^2 \right) \mathbf{1}_{\{|\langle X_i, \theta \rangle| \geq B\}} \right),$$

where we write $f \wedge g(x) := \min\{f(x), g(x)\}$ for any pair of functions f and g. We will estimate each term separately.

In order to control the first term of the right hand side in (10.1.9), for fixed $\theta \in S^{n-1}$ we apply Bernstein's inequality (Lemma 10.1.12) for the random variables

$$Z_i = (|\langle X_i, \theta \rangle| \wedge B)^2 - \mathbb{E}\left((|\langle X_i, \theta \rangle| \wedge B)^2 \right).$$

Note that $\|Z_i\|_\infty \leq B^2$ and

$$\text{Var}(Z_i) \leq \mathbb{E}\left(\langle X_i, \theta \rangle^4 \right) \leq C_1$$

for some absolute constant $C_1 > 0$. If we choose $s = tB\sqrt{n/N}$ we get

$$\mathbb{P} \left(\left| \frac{1}{N} \sum_{i=1}^{N} \left((|\langle X_i, \theta \rangle| \wedge B)^2 - \mathbb{E}\left((|\langle X_i, \theta \rangle| \wedge B)^2 \right) \right) \right| \geq tB\sqrt{n/N} \right)$$

$$\leq \exp\left(-c_1 \min(t^2 B^2 n, t\sqrt{Nn}/B) \right),$$

where $c_1 > 0$ is an absolute constant. It follows that

(10.1.10) $\quad \max_{\theta \in \mathcal{N}} \left| \frac{1}{N} \sum_{i=1}^{N} \left((|\langle X_i, \theta \rangle| \wedge B)^2 - \mathbb{E}\left((|\langle X_i, \theta \rangle| \wedge B)^2 \right) \right) \right| \leq tB\sqrt{n/N}$

with probability greater than

(10.1.11) $\quad 1 - \exp\left(n \log\left(\frac{3}{\gamma \eta} \right) - c_1 \min(t^2 B^2 n, t\sqrt{Nn}/B) \right).$

For the third term in (10.1.9), using the Cauchy-Schwarz inequality and the fact that $\|\langle \cdot, \theta \rangle\|_{\psi_1} \leq \varrho$, we get

(10.1.12) $\quad \mathbb{E}\left(\langle X_i, \theta \rangle^2 \mathbf{1}_{\{|\langle X_i, \theta \rangle| \geq B\}} \right) \leq \left(\mathbb{E}|\langle X_i, \theta \rangle|^4 \right)^{1/2} \left(\mathbb{P}(|\langle X_i, \theta \rangle| \geq B) \right)^{1/2}$

$$\leq C_2 \exp\left(-\frac{B}{2\varrho} \right).$$

For the second term in (10.1.9) we will use Theorem 10.1.4; taking into account Remark 10.1.5 as well, we know that if $t \geq 1$ then with probability greater than $1 - \exp(-ct\sqrt{n})$ we have that for every $m \leq N$ and for every $z \in S^{N-1}$ with $|\text{supp}(z)| = m$,

$$\left\| \sum_{i=1}^{N} z_i X_i \right\|_2 \leq C_3 t \left(\sqrt{n} + \sqrt{m} \log \frac{2N}{n} \right).$$

Then, for every $E \subset \{1, \ldots, N\}$ we have

$$(10.1.13) \quad \max_{\theta \in S^{n-1}} \left(\sum_{i \in E} |\langle X_i, \theta \rangle|^2 \right)^{1/2} \leqslant \max \left\{ \left\| \sum_{i=1}^{N} z_i X_i \right\|_2 \, \middle| \, z \in S^{N-1} \cap \mathbb{R}^E \right\}$$

$$\leqslant C_3 t \left(\sqrt{n} + \sqrt{|E|} \log \frac{2N}{n} \right).$$

Let $\theta \in \mathcal{N}$. We write

$$\sum_{i=1}^{N} |\langle X_i, \theta \rangle|^2 \mathbf{1}_{\{|\langle X_i, \theta \rangle| \geqslant B\}} = \sum_{i \in E_B(\theta)} \langle X_i, \theta \rangle^2,$$

where $E_B(\theta) = \{i \leqslant N : |\langle X_i, \theta \rangle| \geqslant B\}$. Using (10.1.13) we see that

$$(10.1.14) \quad B|E_B(\theta)|^{1/2} \leqslant \left(\sum_{i \in E_B(\theta)} \langle X_i, \theta \rangle^2 \right)^{1/2} \leqslant C_3 t \left(\sqrt{n} + \sqrt{|E_B(\theta)|} \log \frac{2N}{n} \right).$$

We choose

$$(10.1.15) \quad B = 2C_3 t \log \frac{2N}{n}.$$

Then, (10.1.14) shows that

$$(10.1.16) \quad |E_B(\theta)| \leqslant 4 C_3^2 t^2 n B^{-2} = \frac{n}{\log^2(2N/n)}.$$

Therefore, for this choice of B we have

$$(10.1.17) \quad \max_{\theta \in \mathcal{N}} \left(\sum_{i=1}^{N} |\langle X_i, \theta \rangle|^2 \mathbf{1}_{\{|\langle X_i, \theta \rangle| \geqslant B\}} \right) \leqslant 4 C_3^2 t^2 n.$$

for all $\theta \in \mathcal{N}$, with probability greater than $1 - \exp(-c_2 t \sqrt{n})$. Combining (10.1.10), (10.1.12) and (10.1.17) we get

$$(10.1.18) \quad \max_{\theta \in \mathcal{N}} \left| \frac{1}{N} \sum_{i=1}^{N} \left(\langle X_i, \theta \rangle^2 - \mathbb{E}(\langle X_i, \theta \rangle^2) \right) \right|$$

$$\leqslant 2 C_3 t^2 \log \left(\frac{2N}{n} \right) \sqrt{\frac{n}{N}} + 4 C_3^2 t^2 \frac{n}{N} + C_2 \left(\frac{n}{2N} \right)^{\frac{C_3 t}{\varrho}}$$

with probability greater than
$$(10.1.19)$$
$$1 - \exp(-c_2 t \sqrt{n}) - \exp\left(n \log\left(\frac{3}{\gamma \eta}\right) - c_3 \min\left(t^4 n \log^2(2N/n), \frac{\sqrt{Nn}}{\log(2N/n)} \right) \right).$$

Given $\varepsilon \in (0, 1)$ we choose the parameters γ and η as follows:

$$\gamma = \frac{1}{2} \quad \text{and} \quad \eta = \frac{\varepsilon}{40}.$$

We also set

$$C(\varepsilon, t) := C_0 t^4 \varepsilon^{-2} \log^2 \left(2 t^2 \varepsilon^{-2} \right),$$

where $C > 0$ is a (large enough) absolute constant. If $N \geqslant C(\varepsilon,t)n$ then we see that the lower bound (for the probability) in (10.1.19) is greater than $1 - \exp(-ct\sqrt{n})$ and that (with the same probability) we have

$$(10.1.20) \qquad \max_{\theta \in \mathcal{N}} \left| \frac{1}{N} \sum_{i=1}^{N} \left(\langle X_i, \theta \rangle^2 - \mathbb{E}(\langle X_i, \theta \rangle^2) \right) \right| \leqslant \frac{\varepsilon}{40}.$$

We complete the proof using Lemma 10.1.11 in order to pass from $\mathcal{N}$ to the sphere S^{n-1}. □

10.1.3. Uniform appoximation of moments of higher order

Using the same strategy, Adamczak, Litvak, Pajor and Tomczak-Jaegermann actually prove in [**1**] a more general theorem concerning uniform approximation of moments of one-dimensional marginals of an isotropic log-concave measure by the corresponding empirical sums.

THEOREM 10.1.13 (Adamczak-Litvak-Pajor-Tomczak). *Let $X_1, \ldots, X_N$ be i.i.d. isotropic log-concave random vectors in $\mathbb{R}^n$. For every $p \geqslant 2$, and for every $\varepsilon \in (0,1)$ and $t \geqslant 1$, there exists a constant $C(\varepsilon, t, p) > 0$ such that: if $N \geqslant C(\varepsilon, t, p)n^{p/2}$ then with probability greater than $1 - e^{-c_p t \sqrt{n}}$ one has*

$$\max_{\theta \in S^{n-1}} \left| \frac{1}{N} \sum_{i=1}^{N} \left(|\langle X_i, \theta \rangle|^p - \mathbb{E}\left(|\langle X_i, \theta \rangle^p|\right) \right) \right| \leqslant \varepsilon,$$

where $c_p > 0$ is a constant depending only on p. We can choose

$$C(\varepsilon, t, p) = C_p t^{2p} \varepsilon^{-2} \log^{2p-2}\left(2t^2 \varepsilon^{-2}\right),$$

where $C_p > 0$ depends only on p.

In fact, they obtain a stronger statement.

THEOREM 10.1.14. *Let $X_1, \ldots, X_N$ be i.i.d. isotropic log-concave random vectors in $\mathbb{R}^n$. If $n \leqslant N \leqslant e^{\sqrt{n}}$ then, for any $p \geqslant 2$ and $s, t \geqslant 1$, one has*

$$\max_{\theta \in S^{n-1}} \left| \frac{1}{N} \sum_{i=1}^{N} \left(|\langle X_i, \theta \rangle|^p - \mathbb{E}(|\langle X_i, \theta \rangle^p|) \right) \right|$$
$$\leqslant C^{p-1} t s^{p-1} p \log^{p-1}\left(\frac{2N}{n}\right) \sqrt{\frac{n}{N}} + \frac{C^p s^p n^{p/2}}{N} + C^p p^p \left(\frac{n}{2N}\right)^s$$

with probability greater than

$$1 - \exp(-cs\sqrt{n}) - \exp(-c_p \min(u, v)),$$

where $u = t^2 s^{2p-2} n \log^{2p-2}(2N/n)$, $v = ts^{-1}\sqrt{Nn}/\log(2N/n)$, $C, c > 0$ are absolute constants and $c_p > 0$ depends only on p.

We briefly describe the additional probabilistic tools that are needed for the proof. Below, we write $\varepsilon_1, \ldots, \varepsilon_N$ for a sequence of independent Rademacher random variables. The next theorem can be found in the book of Ledoux and Talagrand (see [**318**, Theorem 4.12]).

THEOREM 10.1.15. *Let $F : \mathbb{R}^+ \mapsto \mathbb{R}^+$ be a convex increasing function and let $\phi_i : \mathbb{R} \mapsto \mathbb{R}$, $1 \leq i \leq N$, be 1-Lipschitz functions with $\phi_i(0) = 0$. Then, for every bounded $T \subset \mathbb{R}^n$, one has*

$$\mathbb{E}\left(F\left(\frac{1}{2}\max_{t \in T}\left|\sum_{i=1}^{N}\varepsilon_i\phi_i(t_i)\right|\right)\right) \leq \mathbb{E}\left(F\left(\frac{1}{2}\max_{t \in T}\left|\sum_{i=1}^{N}\varepsilon_i t_i\right|\right)\right).$$

COROLLARY 10.1.16. *Let $\mathcal{F}$ be a family of functions which are uniformly bounded by B. Then, for every N-tuple of independent random variables $X_1, \ldots, X_N$ and for every $p \geq 1$ we have*

$$\mathbb{E}\max_{f \in \mathcal{F}}\left|\sum_{i=1}^{N}(|f(X_i)|^p - \mathbb{E}|f(X_i)|^p)\right| \leq 4pB^{p-1}\mathbb{E}\max_{f \in \mathcal{F}}\left|\sum_{i=1}^{N}\varepsilon_i f(X_i)\right|.$$

Proof. Applying Theorem 10.1.15 with $F(x) = x$ we get

$$\mathbb{E}\left(\frac{1}{2}\max_{t \in T}\left|\sum_{i=1}^{N}\varepsilon_i\phi_i(t_i)\right|\right) \leq \mathbb{E}\left(\frac{1}{2}\max_{t \in T}\left|\sum_{i=1}^{N}\varepsilon_i t_i\right|\right).$$

Symmetrization shows that

$$\frac{1}{4}\mathbb{E}\max_{t \in T}\left|\sum_{i=1}^{N}(\phi_i(t_i) - \mathbb{E}\phi_i(t_i))\right| \leq \mathbb{E}\left(\frac{1}{2}\max_{t \in T}\left|\sum_{i=1}^{N}\varepsilon_i\phi_i(t_i)\right|\right).$$

Therefore,

$$\frac{1}{4}\mathbb{E}\max_{t \in T}\left|\sum_{i=1}^{N}(\phi_i(t_i) - \mathbb{E}\phi_i(t_i))\right| \leq \mathbb{E}\left(\frac{1}{2}\max_{t \in T}\left|\sum_{i=1}^{N}\varepsilon_i t_i\right|\right).$$

Applying the last inequality with $\phi_i(s) = \frac{|s|^p \wedge B^p}{pB^{p-1}}$ we get the result. □

The next tool is a version of Talagrand's concentration inequality for the supremum of a bounded empirical process (see [**316**, Section 7.3] for a proof).

THEOREM 10.1.17. *Let $X_1, \ldots, X_N$ be independent random variables with values in the measurable space $(\mathcal{S}, \mathcal{B})$ and let $\mathcal{F}$ be a countable family of measurable functions $f : \mathcal{S} \mapsto [-a, a]$, such that for every i, $\mathbb{E}f(X_i) = 0$. We consider the random variable*

$$Z = \sup_{f \in \mathcal{F}}\sum_{i=1}^{N}f(X_i).$$

and define

$$\sigma^2 = \sup_{f \in \mathcal{F}}\sum_{i=1}^{N}\mathbb{E}f(X_i)^2.$$

Then, for every $t \geq 0$ we have

$$\mathbb{P}(Z \geq \mathbb{E}Z + t) \leq \exp\left(-\frac{t^2}{2(\sigma^2 + 2a\mathbb{E}Z) + 3at}\right).$$

Proof of Theorem 10.1.14. Using Corollary 10.1.16 with a constant $B > 1$ which will be suitably chosen, and then using Theorem 10.1.15 and the fact that $t \mapsto |t| \wedge B$ is 1-Lipschitz, we write

$$\mathbb{E} \max_{\theta \in S^{n-1}} \left| \sum_{i=1}^{N} \left((|\langle X_i, \theta \rangle| \wedge B)^p - \mathbb{E}(|\langle X_i, \theta \rangle| \wedge B)^p \right) \right|$$

$$\leqslant 4pB^{p-1} \, \mathbb{E} \max_{\theta \in S^{n-1}} \left| \sum_{i=1}^{N} \varepsilon_i (|\langle X_i, \theta \rangle| \wedge B) \right|$$

$$\leqslant 8pB^{p-1} \, \mathbb{E} \max_{\theta \in S^{n-1}} \left| \sum_{i=1}^{N} \varepsilon_i \langle X_i, \theta \rangle \right| \leqslant 8pB^{p-1} \, \mathbb{E} \left\| \sum_{i=1}^{N} \varepsilon_i X_i \right\|_2$$

$$\leqslant 8pB^{p-1} \sqrt{Nn}.$$

Since the random vectors X_i are isotropic and log-concave, from the ψ_1-behavior of linear functionals we see that $\mathbb{E}(|\langle X_i, \theta \rangle| \wedge B)^{2p} \leqslant C^{2p} p^{2p}$, therefore using Theorem 10.1.17 we see that for every $t \geqslant 1$ one has

$$(10.1.21) \quad \max_{\theta \in S^{n-1}} \left| \sum_{i=1}^{N} \left((|\langle X_i, \theta \rangle| \wedge B)^p - \mathbb{E}(|\langle X_i, \theta \rangle| \wedge B)^p \right) \right| \leqslant 16tpB^{p-1}\sqrt{Nn}$$

with probability greater than

$$(10.1.22) \quad 1 - \exp\left(-\frac{64 B^{2p-2} t^2 Nn}{2N C^{2p} p^{2p} + 32pB^{2p-1}\sqrt{Nn} + 24pB^{2p-1} t\sqrt{Nn}} \right)$$

$$\geqslant 1 - \exp\left(-c_p \min\left(t^2 n B^{2p-2}, t\sqrt{Nn}/B \right) \right).$$

Next, we write

$$\max_{\theta \in S^{n-1}} \left| \frac{1}{N} \sum_{i=1}^{N} (\langle X_i, \theta \rangle^p - \mathbb{E}\langle X_i, \theta \rangle^p) \right|$$

$$\leqslant \frac{1}{N} \max_{\theta \in S^{n-1}} \left| \sum_{i=1}^{N} ((|\langle X_i, \theta \rangle| \wedge B)^p - \mathbb{E}(|\langle X_i, \theta \rangle| \wedge B)^p) \right|$$

$$+ \max_{\theta \in S^{n-1}} \frac{1}{N} \sum_{i=1}^{N} (|\langle X_i, \theta \rangle|^p - B^p) \mathbf{1}_{\{|\langle X_i, \theta \rangle| \geqslant B\}}$$

$$+ \max_{\theta \in S^{n-1}} \frac{1}{N} \mathbb{E}\left(\sum_{i=1}^{N} (|\langle X_i, \theta \rangle|^p - B^p) \mathbf{1}_{\{|\langle X_i, \theta \rangle| \geqslant B\}} \right)$$

and we estimate the other two terms using the same more or less ideas as in the proof of Theorem 10.1.1 (see [**1**] for the details). $\square$

10.2. Further reading

10.2.1. Bourgain's argument

In this section we sketch Bourgain's original argument from [**103**] in the setting of isotropic convex bodies. The approach of Adamczak, Litvak, Pajor and Tomczak-Jaegermann uses the same strategy. We chose to add Paouris' inequality in the argument; this saves a log n-term in Bourgain's final estimate. Nevertheless, the weaker inequality of Alesker would have served the same purpose.

THEOREM 10.2.1 (Bourgain). *Let K be an isotropic convex body in $\mathbb{R}^n$ and let $\delta, \varepsilon \in (0,1)$. If $N \geqslant \kappa(\delta,\varepsilon)\varepsilon^{-2}n(\log n)^2$, then N random points $x_1,\ldots,x_N$ which are chosen independently and uniformly from K satisfy with probability greater than $1-\delta$ the following: For every $\theta \in S^{n-1}$,*

$$(10.2.1) \qquad (1-\varepsilon)L_K^2 \leqslant \frac{1}{N}\sum_{j=1}^N \langle x_j,\theta\rangle^2 \leqslant (1+\varepsilon)L_K^2.$$

Note. In this subsection, we use the notation $\kappa(\delta), \kappa(\delta,\varepsilon)$ etc. for a constant which depends logarithmically on $\delta,\varepsilon \in (0,1)$ (for example, $\kappa(\delta) = O(\log(2/\delta))$).

We may assume that $N \leqslant e^{\sqrt{n}}$ (see Remark 10.1.2). An important step for the proof of Theorem 10.2.1 is done in the next theorem.

THEOREM 10.2.2 (Bourgain). *Let $\delta \in (0,1)$ and let $x_1,\ldots,x_N$ be random points independently and uniformly distributed in K. With probability greater than $1-\delta$ we have*

$$\|x_j\|_2 \leqslant c_1\sqrt{n}L_K$$

for all $j \in \{1,\ldots,N\}$, and

$$\left\|\sum_{i \in E} x_i\right\|_2 \leqslant c_2 L_K\sqrt{|E|}\sqrt{n} + \kappa_3(\delta)L_K(\log N)|E|$$

for all $E \subseteq \{1,\ldots,N\}$, where $\kappa_3(\delta) = O\left(\log\frac{2}{\delta}\right)$.

Note. The main technical step in the work of Adamczak, Litvak, Pajor and Tomczak-Jaegermann is exactly a stronger version of this statement; they prove that

$$\mathbb{P}\left(\exists E \subseteq \{1,\ldots,N\} : \left\|\sum_{i \in E} x_i\right\|_2 \geqslant CtL_K\left(\sqrt{n|E|} + |E|\log\frac{2N}{|E|}\right)\right) \leqslant e^{-ct\sqrt{n}}$$

for every $t \geqslant 1$, where $C, c > 0$ are absolute constants (see Corollary 10.1.7).

Next, Bourgain uses Lemma 10.1.12: Consider a ζ-net $\mathcal{N}$ for S^{n-1}, with cardinality $|\mathcal{N}| \leqslant (3/\zeta)^n$. For every $\theta \in \mathcal{N}$ we define a random variable $f_\theta : K \to \mathbb{R}$ by

$$f_\theta(x) = \frac{1}{L_K^2}\langle x,\theta\rangle^2 \mathbf{1}_{\{z \in K : |\langle z,\theta\rangle| \leqslant B\}}(x).$$

If $B \geqslant \kappa_1(\varepsilon)L_K$, then we have

$$(10.2.2) \qquad 1 - \mathbb{E}(f_\theta) = \frac{1}{L_K^2}\int_{\{z \in K : |\langle z,\theta\rangle| > B\}} \langle x,\theta\rangle^2 dx \leqslant \frac{\varepsilon}{4}.$$

Applying Lemma 10.1.12 to $f_\theta - \mathbb{E}(f_\theta)$ one can check that if $N \geqslant \kappa_2(\delta,\zeta)\varepsilon^{-2}n(B/L_K)^2$ the following holds true.

LEMMA 10.2.3. *Let K be an isotropic convex body in $\mathbb{R}^n$. Fix $\delta,\zeta,\varepsilon \in (0,1)$. If*

$$B \geqslant \kappa_1(\varepsilon)L_K \quad \text{and} \quad N \geqslant \kappa_2(\delta,\zeta)\varepsilon^{-2}n(B/L_K)^2,$$

then N random points $x_1,\ldots,x_N$ that are chosen independently and uniformly from K satisfy with probability greater than $1-\delta$ the following: for all θ in a ζ-net for S^{n-1},

$$\left(1 - \frac{\varepsilon}{2}\right)L_K^2 \leqslant \frac{1}{N}\sum_{j \in T_B(\theta)} \langle x_j,\theta\rangle^2 \leqslant \left(1 + \frac{\varepsilon}{2}\right)L_K^2$$

where $T_B(\theta) = \{j : |\langle x_j,\theta\rangle| \leqslant B\}$.

Proof of Theorem 10.2.1. We may assume that $x_1, \ldots, x_N$ satisfy the conclusion of Theorem 10.2.2 with probability greater than $1 - \frac{\delta}{2}$. We fix $\zeta \in (0,1)$ (which will be later chosen of the order of ε). Let
$$B = 4\kappa_3(\delta) L_K \log N,$$
where $\kappa_3(\delta)$ is the constant in Theorem 10.2.2. If $N \geqslant \kappa_4(\varepsilon, \delta)$ and

(10.2.3) $$N \geqslant \kappa_5(\delta, \zeta) \varepsilon^{-2} n \cdot \log^2 N,$$

then we know that with probability greater than $1 - \frac{\delta}{2}$, the random points $x_1, \ldots, x_N \in K$ satisfy

(10.2.4) $$\left(1 - \frac{\varepsilon}{2}\right) L_K^2 \leqslant \frac{1}{N} \sum_{j \in T_B(\theta)} \langle x_j, \theta \rangle^2 \leqslant \left(1 + \frac{\varepsilon}{2}\right) L_K^2$$

for all θ in a ζ-net $\mathcal{N}$ of S^{n-1}.

For every $\beta \geqslant B$ and every $\theta \in S^{n-1}$ we define
$$E_\beta(\theta) = \{j \leqslant N : |\langle x_j, \theta \rangle| > \beta\}.$$

Using Theorem 10.2.2 we can estimate the cardinality of $E_\beta(\theta)$ as follows:
$$\beta |E_\beta(\theta)| \leqslant \sum_{j \in E_\beta(\theta)} |\langle x_j, \theta \rangle| \leqslant \max_{\varepsilon_j = \pm 1} \left\| \sum_{j \in E_\beta(\theta)} \varepsilon_j x_j \right\|_2 \leqslant 2 \max_{F \subseteq E_\beta(\theta)} \left\| \sum_{j \in F} x_j \right\|_2$$
$$\leqslant 2 c_2 L_K \sqrt{n} \sqrt{|E_\beta(\theta)|} + 2\kappa_3(\delta) L_K (\log N) |E_\beta(\theta)|$$
$$\leqslant 2 c_2 L_K \sqrt{n} \sqrt{|E_\beta(\theta)|} + \frac{\beta}{2} |E_\beta(\theta)|.$$

This gives: for every $\theta \in S^{n-1}$,
$$\beta^2 |E_\beta(\theta)| \leqslant 16 c_2^2 L_K^2 n.$$

It follows that
$$\sum_{j \in E_B(\theta)} \langle x_j, \theta \rangle^2 = \sum_{k=0}^{k_0-1} \sum_{j \in E_{2^k B}(\theta) \setminus E_{2^{k+1} B}(\theta)} \langle x_j, \theta \rangle^2$$
$$\leqslant \sum_{k=0}^{k_0-1} |E_{2^k B}(\theta)| (2^{k+1} B)^2 \leqslant c k_0 n L_K^2,$$

where the summation is over all non-empty E, and k_0 is the least integer for which $R(K) \leqslant 2^{k_0} B$. Since K is isotropic, we have $R(K) \leqslant (n+1) L_K$. So, recalling the definition of B, we get
$$k_0 \leqslant c \log \left(\frac{n L_K}{B} \right) \leqslant c' \log n.$$

Now, if N satisfies (10.2.3), we have
$$\frac{1}{N} \sum_{j \in E_B(\theta)} \langle x_j, \theta \rangle^2 \leqslant \frac{c L_K^2 n \log n}{N} < \frac{\varepsilon L_K^2}{20}.$$

Combining with (10.2.4) we conclude the proof for every $\theta \in \mathcal{N}$. Finally, choosing $\zeta = \varepsilon/10$ and using a standard successive approximation argument (see, for example, Lemma 10.1.11 in the next section) we get a similar estimate for every $\theta \in S^{n-1}$. □

10.2.2. Rudelson's argument

Next, we present Rudelson's approach (see [**447**]) to the problem; it is based on the next theorem.

THEOREM 10.2.4 (Rudelson). *Let X be an isotropic log-concave random vector in $\mathbb{R}^n$, let $N \geqslant cn \log n$ and let $X_1, \ldots, X_N$ be independent copies of X. Then,*

$$\mathbb{E} \left\| \frac{1}{N} \sum_{i=1}^{N} X_i \otimes X_i - I \right\| \leqslant C \frac{\sqrt{\log n}}{\sqrt{N}} \cdot \left(\mathbb{E} \|X\|_2^{\log N} \right)^{1/\log N}.$$

where $C > 0$ is an absolute constant.

From Theorem 10.2.4 and Paouris' theorem we get Rudelson's estimate for the Kannan-Lovász-Simonovits question.

THEOREM 10.2.5. *Let $\varepsilon \in (0,1)$ and let K be an isotropic convex body in $\mathbb{R}^n$. Consider the isotropic log-concave measure μ_K with density $L_K^n \mathbf{1}_{\frac{K}{L_K}}$. If*

$$N \geqslant C \frac{n}{\varepsilon^2} \cdot \log \frac{n}{\varepsilon^2}$$

and $X_1, \ldots, X_N$ are independent random vectors with distribution μ_K, then

$$\mathbb{E} \left\| \frac{1}{N} \sum_{i=1}^{N} X_i \otimes X_i - I \right\| \leqslant \varepsilon.$$

Proof. From Theorem 5.4.1 and the standard ψ_1 estimates for linear functionals, we get that if $N \leqslant e^{\sqrt{n}}$ then

$$\left(\mathbb{E} \|X\|_2^{\log N} \right)^{1/\log N} \leqslant C'(\sqrt{n} + C'' \log N) \leqslant C_1 \max\{\sqrt{n}, \log N\}.$$

Applying Theorem 10.2.4 we get

$$\mathbb{E} \left\| \frac{1}{N} \sum_{i=1}^{N} X_i \otimes X_i - I \right\| \leqslant C \frac{\sqrt{\log n}}{\sqrt{N}} \cdot C_1 \max\{\sqrt{n}, \log N\} \leqslant \varepsilon$$

if we assume that $N \geqslant C_2 \frac{n}{\varepsilon^2} \cdot \log \frac{n}{\varepsilon}$, where $C_2 > 0$ is a large enough absolute constant. $\square$

For the proof of Theorem 10.2.4, Rudelson originally constructed a suitable majorizing measure in order to prove the next fact.

LEMMA 10.2.6. *Let $X_1, \ldots, X_N$ be random vectors in $\mathbb{R}^n$ and let $\varepsilon_1, \ldots, \varepsilon_N$ be symmetric Bernoulli random variables. Then,*

(10.2.5) $$\mathbb{E} \left\| \sum_{i=1}^{N} \varepsilon_i X_i \otimes X_i \right\| \leqslant C \sqrt{\log n} \max_{i \leqslant N} \|X_i\|_2 \cdot \left\| \sum_{i=1}^{N} X_i \otimes X_i \right\|^{1/2},$$

where $C > 0$ is an absolute constant.

Then, he simplified his argument, using a non-commutative version of Khintchine's inequality due to Lust-Piquard and Pisier from [**341**]: recall that for every $1 \leqslant p \leqslant \infty$ the p-Schatten norm of an $n \times n$ matrix A is the ℓ_p norm of the sequence of eigenvalues $s_i(A)$ of $\sqrt{A^*A}$. In other words,

$$\|A\|_{C_p^n} = \|(s_i(A))_{i=1}^n\|_p = \left(\sum_{i=1}^{n} s_i(A)^p \right)^{1/p}.$$

In the case $p = \infty$, this norm coincides with the operator norm of A, i.e. $\|A\| = \max_{i \leqslant n} s_i(A)$. It follows that for $p = \log n$ one has

(10.2.6) $$\|A\|_{C_p^n} \leqslant \|A\| \leqslant e \|A\|_{C_p^n}.$$

The inequality of Lust-Piquard and Pisier is the following statement.

THEOREM 10.2.7. *We consider N self-adjoint $n \times n$ matrices $A_1, \ldots, A_N$ and the symmetric Bernoulli random variables $\varepsilon_1, \ldots, \varepsilon_N$. Then, for every $2 \leqslant p < \infty$, we have*

$$\left\| \left(\sum_{i=1}^{N} A_i^2 \right)^{1/2} \right\|_{C_p^n} \leqslant \left(\mathbb{E} \left\| \sum_{i=1}^{N} \varepsilon_i A_i \right\|_{C_p^n}^p \right)^{1/p} \leqslant C\sqrt{p} \left\| \left(\sum_{i=1}^{N} A_i^2 \right)^{1/2} \right\|_{C_p^n}.$$

Using (10.2.6) and Theorem 10.2.7 for $p = \log n$ we get

(10.2.7) $$\mathbb{E} \left\| \sum_{i=1}^{N} \varepsilon_i A_i \right\| \leqslant C\sqrt{\log n} \left\| \left(\sum_{i=1}^{N} A_i^2 \right)^{1/2} \right\|.$$

Then, one can give a simple proof of Rudelson's lemma:

Proof of Lemma 10.2.6. Replacing $\mathbb{R}^n$ with $\mathrm{span}\{X_1, \ldots, X_N\}$, we may assume that $n \leqslant N$. Then, the result follows from (10.2.7): note that

$$\left\| \left(\sum_{i=1}^{N} (X_i \otimes X_i)^2 \right)^{1/2} \right\| = \left\| \sum_{i=1}^{N} \|X_i\|_2^2 \, X_i \otimes X_i \right\|^{1/2} \leqslant \max_{i \leqslant N} \|X_i\|_2 \left\| \sum_{i=1}^{N} X_i \otimes X_i \right\|^{1/2}.$$

$\square$

Proof of Theorem 10.2.4. We first use symmetrization. Indeed, from the inequality

$$\mathbb{E} \left\| \sum_{i=1}^{N} (X_i - \mathbb{E}(X_i)) \right\| \leqslant 2\mathbb{E} \left\| \sum_{i=1}^{N} \varepsilon_i X_i \right\|,$$

we get

$$\mathbb{E} \left\| \frac{1}{N} \sum_{i=1}^{N} X_i \otimes X_i - I \right\| = \mathbb{E} \left\| \frac{1}{N} \sum_{i=1}^{N} (X_i \otimes X_i - \mathbb{E}(X_i \otimes X_i)) \right\|$$

$$\leqslant \frac{2}{N} \mathbb{E} \left\| \sum_{i=1}^{N} \varepsilon_i X_i \otimes X_i \right\|,$$

where the last expectation is with respect to all X_i and all signs ε_i. Now, Lemma 10.2.6 shows that

$$\mathbb{E} \left\| \sum_{i=1}^{N} \varepsilon_i X_i \otimes X_i \right\| \leqslant C\sqrt{\log n} \, \mathbb{E} \left(\max_{i \leqslant N} \|X_i\|_2 \cdot \left\| \sum_{i=1}^{N} X_i \otimes X_i \right\|^{1/2} \right).$$

From the Cauchy-Schwarz inequality we get

$$\mathbb{E} \left\| \sum_{i=1}^{N} \varepsilon_i X_i \otimes X_i \right\| \leqslant C\sqrt{\log n} \left(\mathbb{E} \max_{i \leqslant N} \|X_i\|_2^2 \right)^{1/2} \left(\mathbb{E} \left\| \sum_{i=1}^{N} X_i \otimes X_i \right\| \right)^{1/2}.$$

Next, using Hölder's inequality we write

$$\left(\mathbb{E} \max_{i \leqslant N} \|X_i\|_2^2 \right)^{1/2} \leqslant \left(\mathbb{E} \left(\sum_{i=1}^{N} \|X_i\|_2^{\log N} \right)^{2/\log N} \right)^{1/2}$$

$$\leqslant N^{1/\log N} \cdot \left(\mathbb{E} \|X\|_2^{\log N} \right)^{1/\log N}.$$

If we denote $E := \mathbb{E} \left\| \frac{1}{N} \sum_{i=1}^{N} X_i \otimes X_i - I \right\|$, then combining the above we obtain

$$E \leqslant C \frac{\sqrt{\log n}}{\sqrt{N}} \left(\mathbb{E} \|X\|_2^{\log N} \right)^{1/\log N} \sqrt{E+1}.$$

If
$$C\frac{\sqrt{\log n}}{\sqrt{N}}\left(\mathbb{E}\|X\|_2^{\log N}\right)^{1/\log N} < 1,$$
then we conclude that
$$E \leqslant 2C\frac{\sqrt{\log n}}{\sqrt{N}}\left(\mathbb{E}\|X\|_2^{\log N}\right)^{1/\log N}$$
as claimed. $\square$

10.3. Notes and references

History of the question

Kannan, Lovász and Simonovits proved in [**267**] that if X is an isotropic log-concave random vector in $\mathbb{R}^n$ and $\{X_i\}$ is a sequence of independent copies of X then for every $0 < \varepsilon < 1$ and $N \geqslant N_0 = C(\varepsilon)n^2$ one has

$$\left\|\frac{1}{N}\sum_{i=1}^N X_i \otimes X_i - I\right\| \leqslant \varepsilon$$

with probability greater than $1 - \varepsilon$, where $C(\varepsilon) > 0$ is a constant depending only on ε.

The estimate on N_0 was improved to $N_0 \simeq C(\varepsilon)n(\log n)^3$ by Bourgain in [**103**]. We sketch the argument of Bourgain in Section 10.2.1. Shortly after, Rudelson [**447**] improved this result to $N_0 \simeq C(\varepsilon)n(\log n)^2$; Rudelson's approach is sketched in Section 10.2.2.

One should mention that Rudelson's estimate can be recovered if we incorporate Alesker's deviation inequality into Bourgain's argument. This was noted in [**205**] where Bourgain's ideas were used to prove the next general fact: If μ is an isotropic log-concave measure on $\mathbb{R}^n$ which satisfies a ψ_α-estimate with constant C_α for some $\alpha \in [1,2]$, and if $p > 2$ and $N \geqslant c_2^p(\delta)h_p(n\log n)^{p/2}$, where $h_p \leqslant \min\{(p-2)^{-1}, \log n\}$, then N random points $x_1, \ldots, x_N$ distributed according to μ satisfy with probability $> 1 - \delta$

$$c_1 L_\mu \leqslant \left(\frac{1}{N}\sum_{j=1}^N |\langle x_j, y\rangle|^p\right)^{1/p} \leqslant c_2 C_\alpha p^{1/\alpha} L_\mu$$

for all $y \in S^{n-1}$. Using Rudelson's approach and constructing an appropriate majorizing measure, Guédon and Rudelson [**242**] proved a much stronger result: for any isotropic log-concave measure μ on $\mathbb{R}^n$ and for any $\varepsilon \in (0,1)$ and $p \geqslant 2$ one only needs $N = \lfloor C_p \varepsilon^{-2} n^{p/2} \log n \rfloor$ sample points so that for any $t > 0$, with probability greater than $1 - c\exp(-(t/C_p'\varepsilon)^{1/p})$ we have

$$(1-t)\mathbb{E}_\mu |\langle x, y\rangle|^p \leqslant \frac{1}{N}\sum_{j=1}^N |\langle x_j, y\rangle|^p \leqslant (1+t)\mathbb{E}_\mu |\langle x, y\rangle|^p$$

for all $y \in \mathbb{R}^n$, where $C_p, C_p' > 0$ are constants depending only on p. In other words, the empirical p-th moments of the linear functionals are almost isometrically the same as the exact ones.

A main tool in Rudelson's argument is Theorem 10.2.4 which naturally relates the problem with the behavior of $q \mapsto I_q(\mu)$. This was observed by Giannopoulos, Hartzoulaki and Tsolomitis in [**202**]; they proved that for any $\varepsilon \in (0,1)$ and $\rho > 2$ there exists $n_0(\rho)$ such that if K is an isotropic unconditional convex body in $\mathbb{R}^n$, $n \geqslant n_0(\rho)$, and if $N \geqslant c\varepsilon^{-\rho}n\log n$, where $c > 0$ is an absolute constant, then N independent random points $x_1, \ldots, x_N$ uniformly distributed in K satisfy with probability greater than $1 - \varepsilon$

$$(1-\varepsilon)L_K^2 \leqslant \frac{1}{N}\sum_{i=1}^N \langle x_i, \theta\rangle^2 \leqslant (1+\varepsilon)L_K^2$$

for every $\theta \in S^{n-1}$. The proof of this fact employed the estimate
$$\mathbb{E}_\mu \max_{i \leq N} \|x_i\|_2^{2q} \leq (cn)^q$$
for N polynomial in n and for large values of q. This was available by the results of Bobkov and Nazarov in the unconditional case (see Chapter 5). As an immediate consequence of his deviation inequality, Paouris observed in [412] and [413] that for every isotropic convex body K in $\mathbb{R}^n$, and more generally for every isotropic log-concave measure μ on $\mathbb{R}^n$ we have ε-approximation of the identity operator with $N \simeq C(\varepsilon) n \log n$.

Optimal estimate

Aubrun's article [25] is the first one where a linear dependence of the sample size N on the dimension n is achieved. He considered the unconditional case and using methods from random matrix theory he improved the result of [202] as follows: if X is an isotropic unconditional log-concave random vector in $\mathbb{R}^n$ then, for any $\varepsilon \in (0,1)$ and $N \geq C\varepsilon^{-2} n$ one has with probability larger than $1 - C\exp(-cn^{1/5})$
$$(1-\varepsilon)\|y\|_2^2 \leq \frac{1}{N} \sum_{j=1}^N \langle X_j, y\rangle^2 \leq (1+\varepsilon)\|y\|_2^2$$
for all $y \in \mathbb{R}^n$. The general case was settled by Adamczak, Litvak, Pajor and Tomczak-Jaegermann in [1]. This work occupies most of this chapter.

We chose to discuss only the case of log-concave distributions. Various extensions to larger classes as well as alternative approaches have appeared after [1]. Srivastava and Vershynin [472] established the optimal bound for every distribution whose k-dimensional marginals have uniformly bounded $2+\varepsilon$ moments outside a sphere of radius $O(\sqrt{k})$. Their argument is based on randomization of the deterministic spectral sparsification technique of Batson, Spielman and Srivastava [64], and provides an alternative approach to the Kannan-Lovasz-Simonovits problem. Mendelson and Paouris presented in [360] and [361] an approach that works for measures on $\mathbb{R}^n$ that are supported in a relatively small ball and whose linear functionals are uniformly bounded in L_p for some $p > 8$.

CHAPTER 11

Random polytopes in isotropic convex bodies

Let K be an isotropic convex body in $\mathbb{R}^n$. For every $N \geqslant n$ we consider N independent random points $x_1, \ldots, x_N$ uniformly distributed in K and define the random polytope $K_N := \mathrm{conv}\{\pm x_1, \ldots, \pm x_N\}$; if $N \geqslant n+1$, we also set $C_N := \mathrm{conv}\{x_1, \ldots, x_N\}$. In this chapter we discuss some basic questions about the geometry of these random polytopes:

(i) To determine the asymptotic behavior of the "volume radius" $|K_N|^{1/n}$ and $|C_N|^{1/n}$.
(ii) To understand the typical "asymptotic shape" of K_N and C_N.
(iii) To estimate the isotropic constant of K_N and C_N.

The same questions can be formulated and studied in the more general setting of N i.i.d. isotropic log-concave random vectors $X_1, \ldots, X_N$.

In Section 11.1 we obtain some information on the first question. Using the classical method of Steiner symmetrization we show that the expected "volume radius" of K_N is minimized when $K = B(n)$, the Euclidean ball of volume 1 in $\mathbb{R}^n$. Then, we estimate $\mathbb{E}|K_N|^{1/n}$ in this particular case to obtain the lower bound

$$\mathbb{E}|K_N|^{1/n} \geqslant c \min\left\{\frac{\sqrt{\log(2N/n)}}{\sqrt{n}}, 1\right\}$$

for $N \geqslant c_1 n$, where $c_1 > 0$ is an absolute constant. Using a different method, due to Pivovarov, we obtain in Section 11.2 sharp results, with high probability, in the cases that N is linear in the dimension: if $\gamma > 1$ and $n \leqslant N \leqslant \gamma n$ then, with probability greater than $1 - \exp(-c_1 n)$, we have

$$|K_N|^{1/n} \geqslant \frac{c_2(\gamma) L_K \sqrt{\log(1 + N/n)}}{\sqrt{n}},$$

where $c_1, c_2 > 0$ are absolute constants and $c_2(\gamma) = c_2/\sqrt{\log(1+\gamma)}$.

In Section 11.3 we discuss the asymptotic shape of K_N. A general, and rather precise, description was obtained by Dafnis, Giannopoulos and Tsolomitis (see [**147**] and [**148**]). The main idea comes from a previous study (of the special case) of symmetric random ± 1-polytopes (see Section 11.5 for a brief discussion of related results and ideas). Roughly speaking, given any isotropic log-concave measure μ on $\mathbb{R}^n$ and any $cn \leqslant N \leqslant e^n$, the random polytope K_N defined by N i.i.d. random points $X_1, \ldots, X_N$ which are distributed according to μ satisfies, with high probability, the next two conditions:

(i) $K_N \supseteq c\, Z_{\log(N/n)}(\mu)$.
(ii) For every $\alpha > 1$ and $q \geqslant 1$,
$$\mathbb{E}\left[\sigma(\{\theta : h_{K_N}(\theta) \geqslant \alpha h_{Z_q(\mu)}(\theta)\})\right] \leqslant N\alpha^{-q}.$$

Using the inclusion (i) and the known lower bounds for the volume of the L_q-centroid bodies (from Chapters 5 and 7) we can show that, if K is a convex body of volume 1 in $\mathbb{R}^n$ and if $x_1, \ldots, x_N$ are independent random points uniformly distributed in K, then for $n \leqslant N \leqslant e^{\sqrt{n}}$ we have

$$|K_N|^{1/n} \geqslant c_1 L_K \frac{\sqrt{\log(2N/n)}}{\sqrt{n}},$$

while in the range $e^{\sqrt{n}} \leqslant N \leqslant e^n$ we have

$$|K_N|^{1/n} \geqslant c_1 \frac{\sqrt{\log(2N/n)}}{\sqrt{n}}$$

with probability exponentially close to 1. On the other hand, the weak reverse inclusion (ii) is sufficient for some sharp upper bounds: for all $n \leqslant N \leqslant \exp(n)$ we have

$$\mathbb{E}\left[w(K_N)\right] \leqslant C\, w(Z_{\log N}(\mu)),$$

and

$$|K_N|^{1/n} \leqslant C \frac{\sqrt{\log(2N/n)}}{\sqrt{n}}$$

with probability greater than $1 - \frac{1}{N}$, where $C > 0$ is an absolute constant.

In Section 11.4 we discuss the isotropic constant of K_N. The question can be made precise as follows: We are given a convex body K of volume 1 in $\mathbb{R}^n$ and for every $N > n$ we consider N independent random points $x_1, \ldots, x_N$ uniformly distributed in K. Is it true that, with probability tending to 1 as $n \to \infty$, the isotropic constant of the random polytope $K_N := \operatorname{conv}\{\pm x_1, \ldots, \pm x_N\}$ is bounded by CL_K where $C > 0$ is a constant independent of K, n and N? We describe a method which was initiated by Klartag and Kozma in [**282**]; they showed that, if $N > n$ and if $G_1, \ldots, G_N$ are independent standard Gaussian random vectors in $\mathbb{R}^n$, then the isotropic constant of the random polytopes

$$K_N := \operatorname{conv}\{\pm G_1, \ldots, \pm G_N\} \quad \text{and} \quad C_N := \operatorname{conv}\{G_1, \ldots, G_N\}$$

is bounded by an absolute constant $C > 0$ with probability greater than $1 - Ce^{-cn}$. As we will see, variants of the argument of [**282**] also work in the cases that the vertices x_j of K_N are distributed according to the uniform measure on an isotropic convex body that is either ψ_2 (with constant b) or unconditional. The general case remains open.

11.1. Lower bound for the expected volume radius

Let μ be an isotropic log-concave measure on $\mathbb{R}^n$. For every $p > 0$ and $N \geqslant n+1$ we consider the quantity

$$\mathbb{E}_p(\mu, N) = \left(\mathbb{E}\left|C(x_1, \ldots, x_N)\right|^p\right)^{1/pn},$$

where

$$C(x_1, \ldots, x_N) = C_N = \operatorname{conv}\{x_1, \ldots, x_N\}.$$

In this section we consider the case $\mu = \mu_K$ (the Lebesgue measure on K) where K is a convex body of volume 1 in $\mathbb{R}^n$ and we give a lower bound for the expected "volume radius" $\mathbb{E}_{1/n}(\mu_K, N)$ of the random polytope C_N using the method of comparison with the Euclidean ball. The idea is to use Steiner symmetrization and it has been used, for example, by Groemer (see [**225**], [**226**] and [**227**]) in his work on Sylvester's problem. For our purposes we need to further exploit this method; this was done in [**217**] (see also [**250**]).

THEOREM 11.1.1. *Let K be a convex body of volume 1 in $\mathbb{R}^n$. We write $B(n)$ for the centered ball of volume 1 in $\mathbb{R}^n$. Then,*

$$\mathbb{E}_p(\mu_K, N) \geqslant \mathbb{E}_p(\mu_{B(n)}, N)$$

for every $p > 0$. In particular, the expected "volume radius" $\mathbb{E}_{1/n}(\mu_K, N)$ of a random N-tope in K is minimal when $K = B(n)$.

The method we use is to show that $\mathbb{E}_p(\mu_K, N)$ decreases under Steiner symmetrization: Let H be an $(n-1)$-dimensional subspace of $\mathbb{R}^n$. We identify H with $\mathbb{R}^{n-1}$ and write $x = (y, t)$, $y \in H$, $t \in \mathbb{R}$, for a point $x \in \mathbb{R}^n$. If K is a convex body in $\mathbb{R}^n$ with $|K| = 1$ and $P(K)$ is the orthogonal projection of K onto H, then

$$\mathbb{E}_p^{pn}(\mu_K, N) = \int_{P(K)} \cdots \int_{P(K)} M_{p,K}(y_1, \ldots, y_N) dy_N \cdots dy_1$$

where

$$M_{p,K}(y_1, \ldots, y_N) = \int_{\ell(K, y_1)} \cdots \int_{\ell(K, y_N)} |C((y_1, t_1), \ldots, (y_N, t_N))|^p dt_N \cdots dt_1$$

and $\ell(K, y) = \{t \in \mathbb{R} : (y, t) \in K\}$.

We fix $y_1, \ldots, y_N \in H$ and consider the function $F_Y : \mathbb{R}^N \to \mathbb{R}$ defined by

$$F_Y(t_1, \ldots, t_N) = |C((y_1, t_1), \ldots, (y_N, t_N))|,$$

where $Y = (y_1, \ldots, y_N)$. The key observation is the following lemma from [**226**] (the basic idea of its proof comes from a work of Macbeath [**348**]).

LEMMA 11.1.2. *For any $y_1, \ldots, y_N \in P(K)$, the function F_Y is convex.*

Proof. We set $A := \text{conv}\{y_1, \ldots, y_N\}$. For any $\vec{t} = (t_1, \ldots, t_N) \in \mathbb{R}^N$ we define two functions $f_{\vec{t}}, g_{\vec{t}} : A \to \mathbb{R}$ by

$$f_{\vec{t}}(y) = \min\{u : (y, u) \in C((y_1, t_1), \ldots, (y_N, t_N))\}$$

and

$$g_{\vec{t}}(y) = \max\{u : (y, u) \in C((y_1, t_1), \ldots, (y_N, t_N))\}.$$

Then, it is easy to check that $f_{\vec{t}}$ is convex, $g_{\vec{t}}$ is concave, and

$$C((y_1, t_1), \ldots, (y_N, t_N)) = \{(y, u) : y \in A, f_{\vec{t}}(y) \leqslant u \leqslant g_{\vec{t}}(y)\}.$$

Now, let $\vec{t}, \vec{s} \in \mathbb{R}^N$ and define

$$D := \left\{(y, u) : y \in A, \frac{f_{\vec{t}}(y) + f_{\vec{s}}(y)}{2} \leqslant u \leqslant \frac{g_{\vec{t}}(y) + g_{\vec{s}}(y)}{2}\right\}.$$

Observe that, by Fubini's theorem,

$$|D| = \frac{1}{2}|C((y_1, t_1), \ldots, (y_N, t_N))| + \frac{1}{2}|C((y_1, s_1), \ldots, (y_N, s_N))|.$$

On the other hand, by convexity, we easily check that

$$C\left(\left(y_1, \tfrac{t_1+s_1}{2}\right), \ldots, \left(y_N, \tfrac{t_N+s_N}{2}\right)\right) \subseteq D.$$

It follows that

$$F_Y\left(\tfrac{t_1+s_1}{2},\ldots,\tfrac{t_N+s_N}{2}\right) = \left|C\left((y_1,\tfrac{t_1+s_1}{2}),\ldots,(y_N,\tfrac{t_N+s_N}{2})\right)\right|$$
$$\leqslant \frac{1}{2}|C(\{(y_i,t_i):i\leqslant N\})| + \frac{1}{2}|C(\{(y_i,s_i):i\leqslant N\})|$$
$$= \frac{1}{2}F_Y(t_1,\ldots,t_N) + \frac{1}{2}F_Y(s_1,\ldots,s_N).$$

This shows that F_Y is a convex function. $\square$

We now also fix $r_1,\ldots,r_N > 0$ and define

$$Q = \{U = (u_1,\ldots,u_N) : |u_i| \leqslant r_i,\ i=1,\ldots,N\}.$$

For every N-tuple $W = (w_1,\ldots,w_N) \in \mathbb{R}^N$ we set

$$G_W(u_1,\ldots,u_N) = F_Y(w_1+u_1,\ldots,w_N+u_N),$$

and write

$$G_W(U) = F_Y(W+U).$$

This is the volume of the polytope which is generated by the points (y_i, w_i+u_i). Finally, for every $W \in \mathbb{R}^N$ and $\alpha > 0$, we define

$$A(W,\alpha) = \{U \in Q : G_W(U) \leqslant \alpha\}.$$

With this notation, we have:

LEMMA 11.1.3. *Let $\alpha > 0$ and $\lambda \in (0,1)$. If $W, W' \in \mathbb{R}^N$, then*

$$|A(\lambda W + (1-\lambda)W', \alpha)| \geqslant |A(W,\alpha)|^\lambda |A(W',\alpha)|^{1-\lambda}.$$

Proof. Let $U \in A(W,\alpha)$ and $U' \in A(W',\alpha)$. Then, using the convexity of F_Y we see that

$$G_{\lambda W+(1-\lambda)W'}(\lambda U+(1-\lambda)U') = F_Y(\lambda(W+U)+(1-\lambda)(W'+U'))$$
$$\leqslant \lambda F_Y(W+U) + (1-\lambda)F_Y(W'+U')$$
$$= \lambda G_W(U) + (1-\lambda)G_{W'}(U')$$
$$\leqslant \alpha.$$

Therefore,

$$A(\lambda W+(1-\lambda)W', \alpha) \supseteq \lambda A(W,\alpha) + (1-\lambda)A(W',\alpha)$$

and the result follows from the Brunn-Minkowski inequality. $\square$

Observe that the polytopes $C((y_i, w_i+u_i)_{i\leqslant N})$ and $C((y_i, -w_i-u_i)_{i\leqslant N})$ have the same volume since they are reflections of each other with respect to H. It follows that

$$A(-W,\alpha) = -A(W,\alpha)$$

for every $\alpha > 0$. Taking $W' = -W$ and $\lambda = 1/2$ in Lemma 11.1.3 we obtain the following:

LEMMA 11.1.4. *Let $y_1,\ldots,y_N \in H$. For every $W \in \mathbb{R}^N$ and every $\alpha > 0$,*

$$|A(O,\alpha)| \geqslant |A(W,\alpha)|,$$

where O is the origin in $\mathbb{R}^N$. $\square$

For every $y \in P(K)$, we denote by $w(y)$ the midpoint and by $2r(y)$ the length of $\ell(K,y)$. Let $S_H(K)$ be the Steiner symmetrization of K with respect to $H^\perp$. By definition, $P(S_H(K)) = P(K) = P$ and for every $y \in P$ the midpoint and length of $\ell(S_H(K),y)$ are $w'(y) = 0$ and $2r'(y) = 2r(y)$ respectively.

LEMMA 11.1.5. *Let $y_1, \ldots, y_N \in P(K) = P(S_H(K))$. Then,*
$$M_{p,K}(y_1, \ldots, y_N) \geqslant M_{p,S_H(K)}(y_1, \ldots, y_N)$$
for every $p > 0$.

Proof. Let $r_i = r(y_i)$ and $w_i = w(y_i)$, $i = 1, \ldots, N$. In the notation of the previous lemmas, we have
$$M_{p,K}(y_1, \ldots, y_N) = \int_Q [G_W(U)]^p dU$$
$$= \int_0^\infty |\{U \in Q : G_W(U) \geqslant t^{1/p}\}| dt$$
$$= \int_0^\infty \left(|Q| - |A(W, t^{1/p})|\right) dt,$$
where $W = (w_1, \ldots, w_N)$. By the definition of $S_H(K)$,
$$M_{p,S_H(K)}(y_1, \ldots, y_N) = \int_Q [G_O(U)]^p dU = \int_0^\infty \left(|Q| - |A(O, t^{1/p})|\right) dt,$$
and the result follows from Lemma 11.1.4. $\square$

It is now clear that $\mathbb{E}_p(\mu_K, N)$ decreases under Steiner symmetrization.

THEOREM 11.1.6. *Let K be a convex body of volume 1 in $\mathbb{R}^n$ and let $H \in G_{n,n-1}$. If $S_H(K)$ is the Steiner symmetrization of K with respect to H, then*
$$\mathbb{E}_p(\mu_{S_H(K)}, N) \leqslant \mathbb{E}_p(\mu_K, N)$$
for every $p > 0$.

Proof. We may assume that $H = \mathbb{R}^{n-1}$. Since $P(S_H(K)) = P(K)$, Lemma 11.1.5 shows that
$$\mathbb{E}_p^{pn}(\mu_K, N) = \int_{P(K)} \cdots \int_{P(K)} M_{p,K}(y_1, \ldots, y_N) dy_N \cdots dy_1$$
$$\geqslant \int_{P(S_H(K))} \cdots \int_{P(S_H(K))} M_{p,S_H(K)}(y_1, \ldots, y_N) dy_N \cdots dy_1$$
$$= \mathbb{E}_p^{pn}(\mu_{S_H(K)}, N).$$
$\square$

Proof of Theorem 11.1.1. Since the ball $B(n)$ of volume 1 is the Hausdorff limit of a sequence of successive Steiner symmetrizations of K, Theorem 11.1.6 shows that the expected "volume radius" is minimal in the case of $B(n)$. $\square$

The argument shows that a more general fact holds true.

THEOREM 11.1.7. *Let K be a convex body of volume 1 in $\mathbb{R}^n$. Then,*

$$\int_K \cdots \int_K f(|C(x_1,\ldots,x_N)|)dx_N\cdots dx_1$$
$$\geqslant \int_{B(n)} \cdots \int_{B(n)} f(|C(x_1,\ldots,x_N)|)dx_N\cdots dx_1$$

for every increasing function $f:[0,\infty)\to[0,\infty)$. □

It is now clear that, for our original question to give a lower bound for the expected "volume radius" of C_N, it is enough to consider the case of $B(n)$. We will show that if $N\geqslant c_1 n$, where $c_1>0$ is an absolute constant, then the convex hull of N random points from $K=B(n)$ contains a ball of radius $c\sqrt{\log(2N/n)}/\sqrt{n}$.

PROPOSITION 11.1.8. *There exist $c_1,c_2>0$ which satisfy the following: Let $B(n)$ be the centered ball of volume 1 in $\mathbb{R}^n$. If $N\geqslant c_1 n$ and $x_1,\ldots,x_N$ are independent random points uniformly distributed in $B(n)$, then*

$$\operatorname{conv}\{x_1,\ldots,x_N\}\supseteq c_2\min\left\{\frac{\sqrt{\log(N/n)}}{\sqrt{n}},1\right\}B(n)$$

with probability greater than $1-\exp(-n)$.

Proof. Let r_n be the radius of $B(n)$ and let $\alpha\in(0,1)$ be a constant which will be suitably chosen. Consider the random polytope $C_N:=\operatorname{conv}\{x_1,\ldots,x_N\}$. With probability equal to one, C_N has non-empty interior and, for every $J=\{j_1,\ldots,j_n\}\subset\{1,\ldots,N\}$, the points $x_{j_1},\ldots,x_{j_n}$ are affinely independent. Write H_J for the affine subspace determined by $x_{j_1},\ldots,x_{j_n}$ and H_J^+, H_J^- for the two closed halfspaces whose bounding hyperplane is H_J.

If $\alpha B(n)\not\subseteq C_N$, then there exists $x\in\alpha B(n)\setminus C_N$, and hence, there is a facet of C_N defining some affine subspace H_J as above that satisfies the following: either $x\in H_J^-$ and $C_N\subset H_J^+$, or $x\in H_J^+$ and $C_N\subset H_J^-$. Observe that, for every J, the probability of each of these two events is bounded by

$$\bigl(\mu_{B(n)}(\{\langle x,e_1\rangle\leqslant\alpha r_n\})\bigr)^{N-n}.$$

It follows that

(11.1.1) $\operatorname{Prob}(\alpha B(n)\not\subseteq C_N)\leqslant 2\binom{N}{n}\bigl(\mu_{B(n)}(\{\langle x,e_1\rangle\leqslant\alpha r_n\})\bigr)^{N-n}.$

Note that if $\frac{1}{\sqrt{n}}\simeq\frac{\omega_n}{\omega_{n-1}}\leqslant\alpha\leqslant\frac{1}{4}$ then

$$\mu_{B(n)}\bigl(\{\langle x,e_1\rangle\geqslant\alpha r_n\}\bigr)=\omega_{n-1}r_n^n\int_\alpha^1(1-t^2)^{(n-1)/2}dt$$
$$\geqslant\omega_{n-1}r_n^n\int_\alpha^{2\alpha}(1-t^2)^{(n-1)/2}dt$$
$$\geqslant\frac{\omega_{n-1}}{\omega_n}\alpha(1-4\alpha^2)^{(n-1)/2}$$
$$\geqslant\exp\bigl(-4(n-1)\alpha^2\bigr)\geqslant\exp(-4\alpha^2 n).$$

Going back to (11.1.1) we get

$$\operatorname{Prob}(\alpha B(n) \not\subseteq C_N) \leqslant 2 \binom{N}{n} \left(1 - \exp(-4\alpha^2 n)\right)^{N-n}$$

$$\leqslant \left(\frac{2eN}{n}\right)^n \exp\left(-(N-n)e^{-4\alpha^2 n}\right).$$

A simple computation shows that this probability is smaller than $\exp(-n)$ if $\alpha \simeq \min\left\{\sqrt{\log(N/n)}/\sqrt{n}, 1\right\}$, for all $N \geqslant c_1 n$, where $c_1 > 0$ is a (large enough) absolute constant (for this choice of α the restriction $\frac{\omega_n}{\omega_{n-1}} \leqslant \alpha \leqslant \frac{1}{4}$ is also satisfied). This completes the proof. $\square$

Combining the above we have:

THEOREM 11.1.9. *Let K be a convex body of volume 1 in $\mathbb{R}^n$ and let $x_1, \ldots, x_N$ be independent random points uniformly distributed in K. If $N \geqslant c_1 n$ then*

$$\mathbb{E}|K_N|^{1/n} \geqslant \mathbb{E}|C_N|^{1/n} \geqslant c_2 \min\left\{\frac{\sqrt{\log(2N/n)}}{\sqrt{n}}, 1\right\},$$

where $c_1 > 1$ and $c_2 > 0$ are absolute constants.

There are two questions which arise naturally from the above discussion. The first one is to obtain a lower bound, and also an upper bound, for the volume radius of C_N (or K_N) that holds true not only in expectation but also with probability close to 1; Proposition 11.1.8 shows that this is the case if $K = B(n)$. The second one is if it is possible to extend these results in the remaining case $n+1 \leqslant N \leqslant c_1 n$. We address both questions in the next sections.

11.2. Linear number of points

We will work in the more general setting of an isotropic log-concave measure μ on $\mathbb{R}^n$. Let $\{X_i\}_{i=1}^\infty$ be a sequence of independent random vectors with distribution μ. Recall that, for every $p > 0$ and $N \geqslant n+1$,

$$\mathbb{E}_p(\mu, N) = \left(\mathbb{E}|C(X_1, \ldots, X_N)|^p\right)^{1/pn}.$$

In Section 3.5.1 we saw that

$$\mathbb{E}|\operatorname{conv}(X_1, \ldots, X_n)|^2 = \frac{1}{n!}$$

and

$$\mathbb{E}|\operatorname{conv}(X_1, \ldots, X_n)|^2 \leqslant \mathbb{E}_2^{2n}(\mu, n+1) \leqslant (n+1)^2 \mathbb{E}|\operatorname{conv}(X_1, \ldots, X_n)|^2.$$

It follows that

$$\frac{c_1}{\sqrt{n}} \leqslant \mathbb{E}_2(\mu, n+1) \leqslant \frac{c_2}{\sqrt{n}}$$

where $c_1, c_2 > 0$ are absolute constants. From Markov's inequality we get:

PROPOSITION 11.2.1. *Let $X_1, \ldots, X_{n+1}$ be i.i.d. isotropic log-concave random vectors in $\mathbb{R}^n$. Then, with probability greater than $1 - e^{-c_1 n}$,*

$$|C_{n+1}|^{1/n} = |\operatorname{conv}\{X_1, \ldots, X_{n+1}\}|^{1/n} \leqslant \frac{C_1}{\sqrt{n}}$$

and

$$|K_n|^{1/n} = |\text{conv}\{\pm X_1, \ldots, \pm X_n\}|^{1/n} \leqslant \frac{C_1}{\sqrt{n}},$$

where $c_1, C_1 > 0$ are absolute constants.

Proof. For the symmetric case observe that $K_N = T(B_1^n)$ where $T(e_i) = X_i$, $i = 1, \ldots, n$, and hence

$$|K_N| = \frac{2^n}{n!}|\det(X_1, \ldots, X_N)|$$

and

$$\mathbb{E}|K_N|^2 = 4^n \, \mathbb{E}|\text{conv}(X_1, \ldots, X_n)|^2 = \frac{4^n}{n!}.$$

Then, we use Markov's inequality. $\square$

We will show that a similar lower bound holds with high probability (the argument below is due to Pivovarov [**433**]).

PROPOSITION 11.2.2. *Let $X_1, \ldots, X_n$ be i.i.d. isotropic log-concave random vectors in $\mathbb{R}^n$. Then, with probability greater than $1 - e^{-c_2 n}$,*

$$|K_n|^{1/n} = |\text{conv}\{\pm X_1, \ldots, \pm X_n\}|^{1/n} \geqslant \frac{c_3}{\sqrt{n}},$$

where $c_2, c_3 > 0$ are absolute constants.

For the proof we need the next lemma.

LEMMA 11.2.3. *Let X be an isotropic log-concave random vector in $\mathbb{R}^n$. For every $1 \leqslant k \leqslant n$ and $F \in G_{n,k}$ we consider the random variable $Y := \frac{\|P_F(X)\|_2}{\sqrt{k}}$. Then,*

$$\mathbb{E}|Y|^{-1/2} \leqslant C_2,$$

where $C_2 > 0$ is an absolute constant.

Proof. Let μ be the distribution of X. Using the equivalence of "small" positive and negative moments of $\|x\|_2$ for the isotropic measure $\pi_F(\mu)$ we write

$$\int_{\mathbb{R}^n} \|P_F(x)\|_2^{-1/2} d\mu(x) = \int_F \|x\|_2^{-1/2} d\pi_F(\mu)(x)$$

$$\leqslant C_2 \left(\int_F \|x\|_2^2 d\pi_F(\mu)(x) \right)^{-1/4}$$

$$= C_2 k^{-1/4}$$

and the lemma follows. $\square$

Proof of Proposition 11.2.2. We set $E_0 = \{0\}$ and $E_k = \text{span}\{X_1, \ldots, X_k\}$ for $1 \leqslant k \leqslant n-1$. Let T be the linear map defined by $T(e_i) = X_i$. Then

$$|K_n| = \frac{2^n}{n!}|\det(X_1, \ldots, X_n)| = \frac{2^n}{n!}|\det T|.$$

For every $k = 1, \ldots, n$ we set

$$Y_k = \frac{\|P_{E_{k-1}^\perp}(X_k)\|_2}{\sqrt{n-k+1}}.$$

If we write $\mathbb{E}_k$ for the expectation with respect to X_k, for fixed $X_1, \ldots, X_{k-1}$, then Lemma 11.2.3 shows that
$$\mathbb{E}_k(Y_k^{-1/2}) \leqslant C_2$$
(note that $Y_k > 0$ with probability 1). Integrating first with respect to X_n, then with respect to X_{n-1} etc., we get
$$\mathbb{E}\Big(\prod_{k=1}^n Y_k^{-1/2}\Big) \leqslant C_2^n.$$

Then,
$$\mathbb{P}\left(|\det T| \leqslant (eC_2)^{-2n}\sqrt{n!}\right) = \mathbb{P}\left(\prod_{k=1}^n \|P_{E_{k-1}^\perp}(X_k)\|_2 \leqslant (eC_2)^{-2n}\sqrt{n!}\right)$$
$$= \mathbb{P}\left(\prod_{k=1}^n Y_k \leqslant (eC_2)^{-2n}\right)$$
$$= \mathbb{P}\left(\prod_{k=1}^n Y_k^{-1/2} \geqslant (eC_2)^n\right) \leqslant e^{-n}$$

from Markov's inequality. $\square$

COROLLARY 11.2.4. *Let $X_1, \ldots, X_n$ be i.i.d. isotropic log-concave random vectors in $\mathbb{R}^n$. Then,*
$$\mathbb{E}\,|\det(X_1, \ldots, X_n)|^{1/n} \geqslant c_4\sqrt{n},$$
where $c_4 > 0$ is an absolute constant. $\square$

The analogue of Proposition 11.2.2 for a random simplex $\operatorname{conv}\{X_1, \ldots, X_{n+1}\}$ is also true.

PROPOSITION 11.2.5. *Let $X_1, \ldots, X_{n+1}$ be i.i.d. isotropic log-concave random vectors in $\mathbb{R}^n$. Then, with probability greater than $1 - e^{-c_5 n}$,*
$$|C_{n+1}|^{1/n} = |\operatorname{conv}\{X_1, \ldots, X_{n+1}\}|^{1/n} \geqslant \frac{c_6}{\sqrt{n}},$$
where $c_5, c_6 > 0$ are absolute constants.

Proof. First observe that from Borell's lemma
$$\mathbb{P}\left(\|X_i\|_2 \leqslant C_3 n \text{ for all } 1 \leqslant i \leqslant n+1\}\right) \geqslant 1 - e^{-c_7 n}.$$
So, we may assume that $\|X_i\|_2 \leqslant C_3 n$. We define $Y_i = X_i - X_1$, $i = 2, \ldots, n+1$ and consider the symmetric random polytope $K'_{n+1} = \operatorname{conv}\{\pm Y_2, \ldots, \pm Y_{n+1}\}$. By the Rogers-Shephard inequality we have
$$|C_{n+1}| = |\operatorname{conv}\{0, Y_2, \ldots, Y_{n+1}\}| \geqslant 4^{-n}|K'_{n+1}|,$$
and hence, it remains to estimate $|K'_{n+1}|$ from below. With probability one, $X_2, \ldots, X_{n+1}$ are linearly independent. Let $v \in \mathbb{R}^n$ be such that $\langle v, X_i \rangle = 1$, $2 \leqslant i \leqslant n+1$. Consider the linear map $T : \mathbb{R}^n \to \mathbb{R}^n$ defined by $T(x) = x - \langle x, v \rangle X_1$. Then, $T(X_i) = X_i - X_1$, $2 \leqslant i \leqslant n+1$, and $K'_{n+1} = T(A_n)$, where $A_n = \operatorname{conv}\{\pm X_2, \ldots, \pm X_{n+1}\}$. Therefore,
$$|K'_{n+1}| = |\det T| \cdot |A_n|.$$

Since $\|X_i\|_2 \leqslant C_3 n$ for all i, we have $\|v\|_2 \geqslant c_8/n$ by the Cauchy-Schwarz inequality. Observe that $\det T = 1 - \langle v, X_1 \rangle$ (note that $1 - \langle v, X_1 \rangle$ is an eigenvalue of T with eigenvector X_1, and the remaining $(n-1)$ eigenvalues of T are equal to 1). This implies that

$$\mathrm{Prob}(|\det T| < 2^{-n}) = \mathbb{E}_v \left[\mu(\{x \in \mathbb{R}^n : |\langle v, x \rangle - 1| < 2^{-n}\}) \right]$$
$$= \mu\left(\left\{ x \in \mathbb{R}^n : 1 - \frac{1}{2^n} < \langle x, v \rangle < 1 + \frac{1}{2^n} \right\} \right),$$

where μ is the distribution of the X_i's. Using the fact that

(11.2.1) $\qquad \mu(\{x \in \mathbb{R}^n : s < \langle x, \theta \rangle < t\}) \leqslant c_9 |t - s|$

for all $t > s$ in $\mathbb{R}$ and $\theta \in S^{n-1}$, we get that

$$\mu\left(\left\{ x \in \mathbb{R}^n : 1 - \frac{1}{2^n} < \langle x, v \rangle < 1 + \frac{1}{2^n} \right\} \right) \frac{2 c_9}{2^n \|v\|_2} \leqslant \exp(-c_{10} n)$$

(because $\|v\|_2 \geqslant c_8/n$). We have already seen that, with probability greater than $1 - \exp(-c_2 n)$, the volume of A_n is larger than $(c_3/\sqrt{n})^n$. Since we also have $|\det T| \geqslant 2^{-n}$, the proof is complete. $\qquad \square$

Note. A way to justify (11.2.1) is the following. Let f be the density of μ. If $\theta \in S^{n-1}$ and $G : \mathbb{R} \to \mathbb{R}^+$ is the function

$$G(t) := \int_{\{x : \langle x, \theta \rangle \leqslant t\}} f(x) dx,$$

then $g = G'$ is isotropic and log-concave on $\mathbb{R}$, and hence $\sup(g) \leqslant c_9$ for some absolute constant $c_9 > 0$. Therefore,

$$\mu(\{x \in \mathbb{R}^n : s < \langle x, \theta \rangle < t\}) = G(t) - G(s) \leqslant c_9 |t - s|$$

for all $t > s$.

Using Proposition 11.2.2 and Proposition 11.2.5 we get:

THEOREM 11.2.6. *Let K be a convex body of volume 1 in $\mathbb{R}^n$ and let $x_1, \ldots, x_N$ be independent random points uniformly distributed in K. Let $\gamma > 1$. If $n + 1 \leqslant N \leqslant \gamma n$ then, with probability greater than $1 - \exp(-c_1 n)$ we have*

$$|C_N|^{1/n} \geqslant \frac{c_2(\gamma) L_K \sqrt{\log(1 + N/n)}}{\sqrt{n}},$$

and if $n \leqslant N \leqslant \gamma n$ then, with probability greater than $1 - \exp(-c_1 n)$ we have

$$|K_N|^{1/n} \geqslant \frac{c_2(\gamma) L_K \sqrt{\log(1 + N/n)}}{\sqrt{n}},$$

where $c_1, c_2 > 0$ are absolute constants and $c_2(\gamma) = c_2/\sqrt{\log(1 + \gamma)}$.

Proof. We may assume that K is isotropic. Given $x_1, \ldots, x_N$ we set $C_{n+1} = \mathrm{conv}\{x_1, \ldots, x_{n+1}\}$ and $K_n = \mathrm{conv}\{\pm x_1, \ldots, \pm x_n\}$. Then, applying Proposition 11.2.2 and Proposition 11.2.5 for the isotropic measure $L_K^n \mathbf{1}_{\frac{1}{L_K} \mu_K}$ we see that with probability greater than $1 - e^{-c_1 n}$ we have

$$|K_N|^{1/n} \geqslant |K_n|^{1/n} \geqslant \frac{c_2 L_K}{\sqrt{n}} \geqslant \frac{c_2(\gamma) L_K \sqrt{\log(1+N/n)}}{\sqrt{n}}$$

and

$$|C_N|^{1/n} \geqslant |C_{n+1}|^{1/n} \geqslant \frac{c_2 L_K}{\sqrt{n}} \geqslant \frac{c_2(\gamma) L_K \sqrt{\log(1+N/n)}}{\sqrt{n}}$$

where $c_1, c_2 > 0$ are absolute constants and $c_2(\gamma) = c_2/\sqrt{\log(1+\gamma)}$. $\square$

11.3. Asymptotic shape

11.3.1. The main idea

The main idea for the description of the asymptotic shape of the random polytope K_N comes from the study of the behavior of symmetric random ± 1-polytopes; these are the absolute convex hulls of random subsets of the discrete cube $E_2^n = \{-1, 1\}^n$. The natural way to define these random polytopes is to fix $N > n$ and to consider the convex hull

$$K_{n,N} = \text{conv}\{\pm \vec{X}_1, \ldots, \pm \vec{X}_N\}$$

of N independent random points $\vec{X}_1, \ldots, \vec{X}_N$, uniformly distributed over E_2^n. This class of random polytopes was first studied by Giannopoulos and Hartzoulaki in [**201**] where it was proved that a random $K_{n,N}$ has the largest possible volume among all ± 1-polytopes with N vertices, at every scale of n and N. This is a consequence of the following fact: If $n \geqslant n_0$ and if $N \geqslant n(\log n)^2$, then

(11.3.1) $$K_{n,N} \supseteq c\left(\sqrt{\log(N/n)} B_2^n \cap B_\infty^n\right)$$

with probability greater than $1 - e^{-n}$, where $c > 0$ is an absolute constant.

In [**336**], Litvak, Pajor, Rudelson, and Tomczak–Jaegermann worked in a more general setting which contains the previous Bernoulli model and the Gaussian model; let $K_{n,N}$ be the absolute convex hull of the rows of the random matrix $\Gamma_{n,N} = (\xi_{ij})_{1 \leqslant i \leqslant N, 1 \leqslant j \leqslant n}$, where ξ_{ij} are independent symmetric random variables satisfying the conditions $\|\xi_{ij}\|_{L^2} \geqslant 1$ and $\|\xi_{ij}\|_{L^{\psi_2}} \leqslant \rho$ for some $\rho \geqslant 1$. For this larger class of random polytopes, the estimates from [**201**] were generalized and improved in two ways: the article [**336**] provides estimates for all $N \geqslant (1+\delta)n$, where $\delta > 0$ can be as small as $1/\log n$, and establishes the following inclusion: for every $0 < \beta < 1$,

(11.3.2) $$K_{n,N} \supseteq c(\rho)\left(\sqrt{\beta \log(N/n)} B_2^n \cap B_\infty^n\right)$$

with probability greater than $1 - \exp(-c_1 n^\beta N^{1-\beta}) - \exp(-c_2 N)$. The proof in [**336**] is based on a lower bound of the order of $\sqrt{N}$ for the smallest singular value of the random matrix $\Gamma_{n,N}$, with probability greater than $1 - \exp(-cN)$.

In a sense, both works correspond to the study of the size of a random polytope $K_N = \text{conv}\{\pm x_1, \ldots, \pm x_N\}$ spanned by N independent random points $x_1, \ldots, x_N$ uniformly distributed in the unit cube $Q_n := [-1/2, 1/2]^n$. Dafnis, Giannopoulos and Tsolomitis showed in [**147**] that the estimates (11.3.1) and (11.3.2) can be rewritten in terms of the L_q-centroid bodies of Q_n; the connection comes from the following observation.

REMARK 11.3.1. For any $x \in \mathbb{R}^n$ and $t > 0$, define
$$K_{1,2}(x,t) := \inf\{\|u\|_1 + t\|x-u\|_2 : u \in \mathbb{R}^n\}.$$
If we write $(x_j^*)_{j \leqslant n}$ for the decreasing rearrangement of $(|x_j|)_{j \leqslant n}$ we have Holmstedt's approximation formula
$$\frac{1}{c}K_{1,2}(x,t) \leqslant \sum_{j=1}^{\lfloor t^2 \rfloor} x_j^* + t\left(\sum_{j=\lfloor t^2 \rfloor+1}^{n} (x_j^*)^2\right)^{1/2} \leqslant K_{1,2}(x,t)$$
where $c > 0$ is an absolute constant (see [255]). Now, for any $\alpha \geqslant 1$ define $C(\alpha) = \alpha B_2^n \cap B_\infty^n$. Then,
$$h_{C(\alpha)}(\theta) = K_{1,2}(\theta, \alpha)$$
for every $\theta \in S^{n-1}$. On the other hand,
$$\|\langle \cdot, \theta \rangle\|_{L^q(Q_n)} \simeq \sum_{j \leqslant q} \theta_j^* + \sqrt{q}\left(\sum_{q < j \leqslant n} (\theta_j^*)^2\right)^{1/2}$$
for every $q \geqslant 1$ (see, for example, [58]). In other words,
$$C(\sqrt{q}) \simeq Z_q(Q_n).$$
This shows that (11.3.2) or (11.3.1) can be written in the form
(11.3.3) $\qquad K_{n,N} \supseteq c(\rho) Z_{\beta \log(N/n)}(Q_n).$

Starting from this observation, the idea is to compare a random polytope $K_N = \text{conv}\{\pm x_1, \ldots, \pm x_N\}$ spanned by N independent random points $x_1, \ldots, x_N$ uniformly distributed in an isotropic convex body K with the L_q-centroid body $Z_q(K)$ of K for a suitable value $q = q(N,n) \simeq \log(N/n)$. As we will see, an analogue of (11.3.3) holds true in full generality.

11.3.2. Comparison with centroid bodies

We will work in the more general setting of isotropic log-concave random vectors. The main result on the asymptotic shape of K_N is the next theorem from [147] which asserts that with high probability K_N contains the centroid body $Z_{\log(N/n)}(\mu)$ of μ.

THEOREM 11.3.2 (Dafnis-Giannopoulos-Tsolomitis). *Let $0 < \beta \leqslant \frac{1}{2}$ and $\gamma > 1$. Let $X_1, \ldots, X_N$ be i.i.d. isotropic log-concave random vectors in $\mathbb{R}^n$, with $N \geqslant c\gamma n$ where $c > 1$ is an absolute constant. Then, for all $q \leqslant c_2 \beta \log(N/n)$ we have that*
$$K_N \supseteq c_1 Z_q(\mu)$$
with probability greater than
$$1 - \exp(-c_3 N^{1-\beta} n^\beta) - \mathbb{P}(\|\Gamma : \ell_2^n \to \ell_2^N\| \geqslant \gamma\sqrt{N}) \geqslant 1 - \exp(-c_0 \gamma \sqrt{N}),$$
where $\Gamma : \ell_2^n \to \ell_2^N$ is the random operator $\Gamma(y) = (\langle X_1, y \rangle, \ldots \langle X_N, y \rangle)$.

For the proof of Theorem 11.3.2 we need a number of lemmas.

LEMMA 11.3.3. *Let $0 < t < 1$ and $q \geqslant 1$. For every $\theta \in S^{n-1}$ one has*
$$\mathbb{P}(\{|\langle X, \theta \rangle| \geqslant t\|\langle \cdot, \theta \rangle\|_q\}) \geqslant \frac{(1-t^q)^2}{C^q}.$$

Proof. We apply the Paley-Zygmund inequality

$$\mathbb{P}\left(g(X) \geqslant t^q \mathbb{E}\left(g(X)\right)\right) \geqslant (1-t^q)^2 \frac{[\mathbb{E}\left(g(X)\right)]^2}{\mathbb{E}\left(g(X)^2\right)}$$

for the function $g(x) = |\langle x, \theta \rangle|^q$. Since

$$\mathbb{E}\left(g(X)^2\right) = \mathbb{E}\,|\langle X, \theta \rangle|^{2q} \leqslant C^q \left(\mathbb{E}\,|\langle X, \theta \rangle|^q\right)^2 = C^q \left[\mathbb{E}\left(g(X)\right)\right]^2$$

for some absolute constant $C > 0$, the lemma follows. □

LEMMA 11.3.4. *For every* $\sigma \subseteq \{1, \ldots, N\}$ *and any* $\theta \in S^{n-1}$ *and* $q \geqslant 1$ *one has*

(11.3.4) $$\mathbb{P}\left(\{\max_{j \in \sigma} |\langle X_j, \theta \rangle| \leqslant \frac{1}{2} \|\langle \cdot, \theta \rangle\|_q\}\right) \leqslant \exp\left(-|\sigma|/(4C^q)\right),$$

where $C > 0$ *is an absolute constant.*

Proof. Applying Lemma 11.3.3 with $t = 1/2$ we see that

(11.3.5) $$\mathbb{P}\left(\max_{j \in \sigma} |\langle X_j, \theta \rangle| \leqslant \frac{1}{2} \|\langle \cdot, \theta \rangle\|_q\right) = \prod_{j \in \sigma} \mathbb{P}\left(|\langle X_j, \theta \rangle| \leqslant \frac{1}{2} \|\langle \cdot, \theta \rangle\|_q\right)$$

$$\leqslant \left(1 - \frac{1}{4C^q}\right)^{|\sigma|}$$

$$\leqslant \exp\left(-|\sigma|/(4C^q)\right),$$

since $1 - v < e^{-v}$ for every $v > 0$. □

Proof of Theorem 11.3.2. Let $\Gamma : \ell_2^n \to \ell_2^N$ be the random operator defined by

$$\Gamma(y) = (\langle X_1, y \rangle, \ldots, \langle X_N, y \rangle).$$

The argument that follows is a modification of an idea from [**336**]. Define $m = \lfloor 8(N/n)^{2\beta} \rfloor$ and $k = \lfloor N/m \rfloor$. Fix a partition $\sigma_1, \ldots, \sigma_k$ of $\{1, \ldots, N\}$ with $m \leqslant |\sigma_i|$ for all $i = 1, \ldots, k$ and define the norm

$$\|u\|_0 = \frac{1}{k} \sum_{i=1}^k \|P_{\sigma_i}(u)\|_\infty$$

on $\mathbb{R}^N$. Since

$$h_{K_N}(z) = \max_{1 \leqslant j \leqslant N} |\langle X_j, z \rangle| \geqslant \|P_{\sigma_i} \Gamma(z)\|_\infty$$

for all $z \in \mathbb{R}^n$ and $i = 1, \ldots, k$, we observe that

$$h_{K_N}(z) \geqslant \|\Gamma(z)\|_0.$$

So, it suffices to prove that $\|\Gamma(z)\|_0 \geqslant c \|\langle \cdot, z \rangle\|_q$ for all $z \in \mathbb{R}^n$ and for as large a q as possible with probability greater than $1 - \exp(-f(n, \gamma))$, where f is a suitable function of $N \geqslant \gamma n$ and n. To do so, we first prove an estimate of this form for each individual point, and then we use a net argument to pass from a finite collection of points to the whole space. If $z \in \mathbb{R}^n$ and $\|\Gamma(z)\|_0 < \frac{1}{4} \|\langle \cdot, z \rangle\|_q$, then, Markov's inequality implies that there exists $I \subset \{1, \ldots, k\}$ with $|I| > k/2$ such that $\|P_{\sigma_i}\Gamma(z)\|_\infty < \frac{1}{2}\|\langle\cdot,z\rangle\|_q$, for all $i\in I$. It follows that for fixed $z\in\mathbb{R}^n$ we have

(11.3.6) $\mathbb{P}\left(\|\Gamma(z)\|_0 < \frac{1}{4}\|\langle\cdot,z\rangle\|_q\right)$

$$\leqslant \sum_{|I|=\lfloor(k+1)/2\rfloor} \mathbb{P}\left(\|P_{\sigma_i}\Gamma(z)\|_\infty < \frac{1}{2}\|\langle\cdot,z\rangle\|_q, \text{ for all } i\in I\right)$$

$$\leqslant \sum_{|I|=\lfloor(k+1)/2\rfloor} \prod_{i\in I} \mathbb{P}\left(\|P_{\sigma_i}\Gamma(z)\|_\infty < \frac{1}{2}\|\langle\cdot,z\rangle\|_q\right)$$

$$\leqslant \sum_{|I|=\lfloor(k+1)/2\rfloor} \prod_{i\in I} \exp\left(-|\sigma_i|/(4C^q)\right)$$

$$\leqslant \binom{k}{\lfloor(k+1)/2\rfloor} \exp\left(-c_1 km/C^q\right)$$

$$\leqslant \exp\left(k\log 2 - c_1 km/C^q\right).$$

Choosing

$$q = \frac{\beta}{\log C}\log(N/n)$$

we see that

$$\mathbb{P}\left(\|\Gamma(z)\|_0 < \frac{1}{4}\|\langle\cdot,z\rangle\|_q\right) \leqslant \exp\left(-c_2 N^{1-\beta}n^\beta\right).$$

Next, let $S = \{z : \|\langle\cdot,z\rangle\|_q/2 = 1\}$ and consider a δ-net U of S (with respect to $\|\langle\cdot,z\rangle\|_q/2$) of cardinality $|U| \leqslant (3/\delta)^n$. For every $u\in U$ we have

$$\mathbb{P}\left(\|\Gamma(u)\|_0 < \frac{1}{2}\right) \leqslant \exp\left(-c_2 N^{1-\beta}n^\beta\right),$$

and hence,

$$\mathbb{P}\left(\bigcup_{u\in U}\left\{\|\Gamma(u)\|_0 < \frac{1}{2}\right\}\right) \leqslant \exp\left(n\log(3/\delta) - c_2 N^{1-\beta}n^\beta\right).$$

Fix $\gamma > 1$ and set

$$\Omega_\gamma = \{\Gamma : \|\Gamma : \ell_2^n \to \ell_2^N\| \leqslant \gamma\sqrt{N}\}.$$

Since $Z_q(\mu) \supseteq B_2^n$ for all $q \geqslant 2$, we have

$$\|\Gamma(z)\|_0 \leqslant \frac{1}{\sqrt{k}}\|\Gamma(z)\|_2 \leqslant \gamma\sqrt{N/k}\|z\|_2 \leqslant \gamma\sqrt{N/k}\|\langle\cdot,z\rangle\|_q$$

for all $z\in\mathbb{R}^n$ and all Γ in Ω_γ.

This allows us to pass the estimates we have for elements of the net U to all elements in S (and then by homogeneity to all points in $\mathbb{R}^n$). Indeed, let $z\in S$. There exists $u\in U$ such that $\frac{1}{2}\|\langle\cdot,z-u\rangle\|_q < \delta$, which implies that

$$\|\Gamma(u)\|_0 \leqslant \|\Gamma(z)\|_0 + 2\gamma\delta\sqrt{N/k}$$

on Ω_γ. Now, choose $\delta = \sqrt{k/N}/(8\gamma)$. Then,

(11.3.7) $\quad \mathbb{P}(\{\Gamma \in \Omega_\gamma : \exists z \in \mathbb{R}^n : \|\Gamma(z)\|_0 \leqslant \|\langle \cdot, z\rangle\|_q/8\})$
$$= \mathbb{P}(\{\Gamma \in \Omega_\gamma : \exists z \in S : \|\Gamma(z)\|_0 \leqslant 1/4\})$$
$$\leqslant \mathbb{P}(\{\Gamma \in \Omega_\gamma : \exists u \in U : \|\Gamma(u)\|_0 \leqslant 1/2\})$$
$$\leqslant \exp\left(n\log(12\gamma\sqrt{N/k}) - c_2 N^{1-\beta}n^\beta\right)$$
$$\leqslant \exp\left(-c_3 N^{1-\beta}n^\beta\right)$$

provided that N is large enough. Since $h_{K_N}(z) \geqslant \|\Gamma(z)\|_0$ for every $z \in \mathbb{R}^n$, we get that $K_N \supseteq cZ_q(\mu)$ with probability greater than $1 - \exp\left(-c_3 N^{1-\beta}n^\beta\right) - \mathbb{P}(\|\Gamma : \ell_2^n \to \ell_2^N\| \geqslant \gamma\sqrt{N})$.

We now analyze the restriction for N; we need $n\log(12\gamma\sqrt{N/k}) \leqslant \frac{c_2}{2}N^{1-\beta}n^\beta$ for some suitable constant $C > 0$. Assuming

(11.3.8) $\qquad\qquad\qquad N \geqslant c_4 \gamma n,$

and since $0 < \beta \leqslant \frac{1}{2}$, using the definitions of k and m we see that it is enough to guarantee
$$\log(N/n) \leqslant C\sqrt{N/n},$$
which is valid if $N/n \geqslant c_6$ for a suitable absolute constant $c_6 > 0$. We get the result taking (11.3.8) into account.

It remains to estimate the probability
$$\mathbb{P}(\Omega_\gamma) = \mathbb{P}(\|\Gamma : \ell_2^n \to \ell_2^N\| \geqslant \gamma\sqrt{N}).$$
The best known bound can be extracted from Theorem 10.1.4: one has
$$\mathbb{P}(\|\Gamma : \ell_2^n \to \ell_2^N\| \geqslant \gamma\sqrt{N}) \leqslant \exp(-c_0\gamma\sqrt{N})$$
for all $N \geqslant \gamma n$. Assuming that $\beta \leqslant 1/2$, one gets
$$K_N \supseteq c_1 Z_q(\mu) \text{ for all } q \leqslant c_2 \beta \log(N/n),$$
with probability greater than $1 - \exp(-c\sqrt{N})$. $\square$

REMARK 11.3.5. Paouris and Werner [**419**] have studied the relation between the family of L_q-centroid bodies and the family of floating bodies of a convex body K of volume 1 in $\mathbb{R}^n$. Given $\delta \in (0, \frac{1}{2}]$, the floating body $K_{(\delta)}$ of K is the intersection of all halfspaces whose defining hyperplanes cut off a set of volume δ from K. It was observed in [**384**] that $K_{(\delta)}$ is isomorphic to an ellipsoid as long as δ stays away from 0. In [**419**] it is proved that
$$c_1 Z_{\log(1/\delta)}(K) \subseteq K_{(\delta)} \subseteq c_2 Z_{\log(1/\delta)}(K)$$
where $c_1, c_2 > 0$ are absolute constants. From Theorem 11.3.2 it follows that if K is isotropic and if, for example, $N \geqslant n^2$ then
$$K_N \supseteq c_3 K_{(1/N)}$$
with probability greater than $1 - o_n(1)$, where $c_3 > 0$ is an absolute constant. This fact should be compared with the following well-known result from [**47**]: for any convex body K in $\mathbb{R}^n$ one has $c|K_{(1/N)}|^{1/n} \leqslant \mathbb{E}|K_N|^{1/n} \leqslant c_n|K_{(1/N)}|^{1/n}$ (where the constant on the left is absolute and the right hand side inequality holds true with a constant c_n depending on the dimension, for N large enough; the critical value of N is exponential in n).

We will use the inclusion of Theorem 11.3.2 to give a lower bound for $|K_N|^{1/n}$. From the results of Paouris (Chapter 5) and Klartag and E. Milman (Chapter 7) we know that if μ is an isotropic log-concave measure on $\mathbb{R}^n$ then, for every $1 \leqslant q \leqslant \sqrt{n}$,
$$|Z_q(\mu)|^{1/n} \simeq \sqrt{q/n}.$$
Thus, using also Proposition 11.2.2 for the "small" values of N, we immediately get:

THEOREM 11.3.6. *Let $n \leqslant N \leqslant e^{\sqrt{n}}$ and $X_1, \ldots, X_N$ be i.i.d. isotropic log-concave random vectors in $\mathbb{R}^n$. Then,*

$$(11.3.9) \qquad |K_N|^{1/n} \geqslant c_1 \frac{\sqrt{\log(2N/n)}}{\sqrt{n}}$$

with probability greater than $1 - \exp(-c_2\sqrt{N})$, where $c_1, c_2 > 0$ are absolute constants. $\square$

In order to emphasize the fact that this improves Theorem 11.1.9, we state it again in the setting of convex bodies.

THEOREM 11.3.7. *Let K be a convex body of volume 1 in $\mathbb{R}^n$ and let $x_1, \ldots, x_N$ be independent random points uniformly distributed in K. If $n \leqslant N \leqslant e^{\sqrt{n}}$ then, with probability greater than $1 - \exp(-c_2\sqrt{N})$ we have*

$$(11.3.10) \qquad |K_N|^{1/n} \geqslant c_1 L_K \frac{\sqrt{\log(2N/n)}}{\sqrt{n}},$$

where $c_1, c_2 > 0$ are absolute constants. $\square$

Note. The estimate of Klartag and E. Milman is no longer available in the range $e^{\sqrt{n}} \leqslant N \leqslant e^n$. Nevertheless, using the lower bound
$$|Z_q(K)|^{1/n} \geqslant |Z_q(\overline{B_2^n})|^{1/n} \geqslant c\sqrt{q/n}$$
of Proposition 5.1.16, we have the weaker estimate

$$(11.3.11) \qquad |K_N|^{1/n} \geqslant c_1 \frac{\sqrt{\log(2N/n)}}{\sqrt{n}}$$

with probability exponentially close to 1.

11.3.3. Weak reverse inclusion

It is natural to ask whether the inclusion given by Theorem 11.3.2 is sharp. Let us consider the case of N independent random points $x_1, \ldots, x_N$ uniformly distributed in an isotropic convex body K. It is not hard to see that we cannot expect a reverse inclusion of the form $K_N \subseteq c\, Z_q(K)$ with probability close to 1, unless q is of the order of n. Observe that, for any $\alpha > 0$,

$$\mathbb{P}\big(K_N \subseteq \alpha Z_q(K)\big) = \mathbb{P}(x_1, x_2, \ldots, x_N \in \alpha Z_q(K))$$
$$= \Big(\mathbb{P}\big(x \in \alpha Z_q(K)\big)\Big)^N$$
$$\leqslant |\alpha Z_q(K)|^N.$$

From Theorem 5.1.17 we know that for every $q \leqslant n$ the volume of $Z_q(K)$ is bounded by $(c\sqrt{q/n}L_K)^n$. This implies that

(11.3.12) $$\mathbb{P}\big(K_N \subseteq \alpha Z_q(K)\big) \leqslant (c\alpha\sqrt{q/n}L_K)^{nN},$$

where $c > 0$ is an absolute constant. Assume that K has bounded isotropic constant and we want to keep $\alpha \simeq 1$. Then, (11.3.12) shows that, independently from the value of N, we have to choose q of the order of n so that it might be possible to show that $\mathbb{P}\big(K_N \subseteq \alpha Z_q(K)\big)$ is really close to 1. Actually, if $q \sim n$ then this is always the case, because $Z_n(K) \supseteq c\operatorname{conv}\{K, -K\}$.

However, we can describe a "weak reverse inclusion" starting from the following simple observation.

LEMMA 11.3.8. *Let $X_1, \ldots, X_N$ be i.i.d. isotropic log-concave random vectors in $\mathbb{R}^n$. Fix $\alpha > 1$ and $q \geqslant 1$. Then, for every $\theta \in S^{n-1}$ one has*
$$\mathbb{P}\left(h_{K_N}(\theta) \geqslant \alpha h_{Z_q(\mu)}(\theta)\right) \leqslant N\alpha^{-q}.$$

Proof. Markov's inequality shows that
$$\mathbb{P}\left(|\langle X, \theta\rangle| \geqslant \alpha \|\langle \cdot, \theta\rangle\|_q\right) \leqslant \alpha^{-q}.$$

Then,
$$\mathbb{P}\left(h_{K_N}(\theta) \geqslant \alpha h_{Z_q(\mu)}(\theta)\right) = \mathbb{P}\left(\max_{1 \leqslant j \leqslant N} |\langle X_j, \theta\rangle| \geqslant \alpha \|\langle \cdot, \theta\rangle\|_q\right)$$
$$\leqslant N\mathbb{P}\left(|\langle X, \theta\rangle| \geqslant \alpha \|\langle \cdot, \theta\rangle\|_q\right),$$

and the result follows. $\square$

LEMMA 11.3.9. *Let $X_1, \ldots, X_N$ be i.i.d. isotropic log-concave random vectors in $\mathbb{R}^n$. For every $\alpha > 1$ and $q \geqslant 1$ one has*
$$\mathbb{E}\left[\sigma(\{\theta : h_{K_N}(\theta) \geqslant \alpha h_{Z_q(\mu)}(\theta)\})\right] \leqslant N\alpha^{-q}.$$

Proof. Immediate: observe that
$$\mathbb{E}\left[\sigma(\{\theta : h_{K_N}(\theta) \geqslant \alpha h_{Z_q(\mu)}(\theta)\})\right] = \int_{S^{n-1}} \mathbb{P}\left(h_{K_N}(\theta) \geqslant \alpha h_{Z_q(\mu)}(\theta)\right) d\sigma(\theta)$$

by Fubini's theorem. $\square$

The estimate of Lemma 11.3.9 is already enough to show that if $q \geqslant 2\log N$ then, on the average, $h_{K_N}(\theta) \leqslant ch_{Z_q(\mu)}(\theta)$ with probability greater than $1 - N^{-1}$. In particular, the mean width of a random K_N is bounded by the mean width of $Z_{\log N}(\mu)$:

PROPOSITION 11.3.10. *Let $n \leqslant N \leqslant \exp(n)$ and $X_1, \ldots, X_N$ be i.i.d. isotropic log-concave random vectors in $\mathbb{R}^n$. Then,*
$$\mathbb{E}\left[w(K_N)\right] \leqslant c\,w(Z_{\log N}(\mu)),$$

where $c > 0$ is an absolute constant.

Proof. Let $q \geqslant 1$. We write
$$w(K_N) \leqslant \int_{A_N} h_{K_N}(\theta)\, d\sigma(\theta) + \sigma(A_N^c)R(K_N),$$

where $A_N = \{\theta : h_{K_N}(\theta) \leqslant eh_{Z_q(\mu)}(\theta)\}$. Then,
$$w(K_N) \leqslant e\int_{A_N} h_{Z_q(\mu)}(\theta)\, d\sigma(\theta) + \sigma(A_N^c)R(K_N),$$

and hence, by Lemma 11.3.9,
$$\mathbb{E}\, w(K_N) \leqslant ew(Z_q(\mu)) + cNe^{-q}\mathbb{E}\left(R(K_N)\right).$$
Since $R(K_N) \leqslant c_1\sqrt{n}w(K_N)$, we get
$$\mathbb{E}\left(w(K_N)\right) \leqslant ew(Z_q(\mu)) + c_2N\sqrt{n}e^{-q}\mathbb{E}\left(w(K_N)\right).$$
Choosing $q = c_3 \log N$ we can clearly have $c_2 N\sqrt{n}e^{-q} \leqslant 1/2$. Then,
$$\mathbb{E}\left(w(K_N)\right) \leqslant 2ew(Z_{c_3 \log N}(\mu)).$$
Using the fact that $Z_{c_3 \log N}(\mu) \subseteq c_4 Z_{\log N}(\mu)$, and hence
$$w(Z_{c_3 \log N}(\mu)) \leqslant c_4\, w(Z_{\log N}(\mu)),$$
we get the assertion of the proposition. $\square$

Recall that, for every $1 \leqslant k \leqslant n$, the normalized quermassintegrals of a convex body C in $\mathbb{R}^n$ are defined by
$$Q_k(C) := \left(\frac{W_{n-k}(C)}{\omega_n}\right)^{1/k} = \left(\frac{1}{\omega_k}\int_{G_{n,k}}|P_F(C)|\,d\nu_{n,k}(F)\right)^{1/k},$$
where the second equality is a consequence of Kubota's formula. Note that $Q_1(C) = w(C)$. In the case $N \leqslant e^{\sqrt{n}}$ our results allow us to determine the expectation of $Q_k(K_N)$ for all values of k (see [148]).

THEOREM 11.3.11. *Let $X_1, \ldots, X_N$ be i.i.d. isotropic log-concave random vectors in $\mathbb{R}^n$. If $n \leqslant N \leqslant \exp(\sqrt{n})$ then for every $1 \leqslant k \leqslant n$ we have*
$$c_1\sqrt{\log(2N/n)} \leqslant \mathbb{E}\left[Q_k(K_N)\right] \leqslant c_2\sqrt{\log N},$$
where $c_1, c_2 > 0$ are absolute constants.

Proof. Using the fact that $Q_k(\cdot)$ is decreasing in k, we immediately get
$$\mathbb{E}\left[Q_k(K_N)\right] \geqslant \mathbb{E}\left[Q_n(K_N)\right] = \mathbb{E}\left(\frac{|K_N|}{\omega_n}\right)^{1/n}.$$
From Theorem 11.3.2 we know that
$$\mathbb{E}\left(\frac{|K_N|}{\omega_n}\right)^{1/n} \geqslant c_1\sqrt{\log(2N/n)},$$
where $c_1 > 0$ is an absolute constant. On the other hand, Proposition 11.3.10 states that
$$\mathbb{E}\left[Q_1(K_N)\right] = \mathbb{E}\left[w(K_N)\right] \leqslant c\, w(Z_{\log N}(\mu)),$$
where $c > 0$ is an absolute constant. Therefore,
$$\mathbb{E}\left[Q_k(K_N)\right] \leqslant \mathbb{E}\left[Q_1(K_N)\right] \leqslant c\, w(Z_{\log N}(\mu)),$$
for all $1 \leqslant k \leqslant n$. Assuming that $\log N \leqslant \sqrt{n}$, we know (from Chapter 5, Section 2) that
$$w(Z_{\log N}(K)) \leqslant c'\sqrt{\log N}.$$
It follows that
$$\mathbb{E}\left[Q_k(K_N)\right] \leqslant c_2\sqrt{\log N},$$
where $c_2 > 0$ is an absolute constant. $\square$

11.3.4. Upper bound for the volume radius

In this subsection we further exploit the simple estimate of Lemma 11.3.8 to obtain a sharp upper bound for the "volume radius" $|K_N|^{1/n}$ of K_N in the full range of values of N.

THEOREM 11.3.12 (Dafnis-Giannopoulos-Tsolomitis). *Let $X_1, \ldots, X_N$ be i.i.d. isotropic log-concave random vectors in $\mathbb{R}^n$. For every $n \leqslant N \leqslant e^n$, one has*

$$|K_N|^{1/n} \leqslant C \frac{\sqrt{\log(2N/n)}}{\sqrt{n}}$$

with probability greater than $1 - \frac{1}{N}$, where $C > 0$ is an absolute constant.

We will use a number of observations which follow from the results of Chapter 5.

LEMMA 11.3.13. *For any symmetric convex body A in $\mathbb{R}^n$ and any $1 \leqslant q \leqslant n$,*

$$|A|^{1/n} \leqslant \frac{c_1 w_{-q}(A)}{\sqrt{n}},$$

where $c_1 > 0$ is an absolute constant.

Proof. An application of Hölder's inequality shows that

$$\left(\frac{|A^\circ|}{|B_2^n|}\right)^{1/n} = \left(\int_{S^{n-1}} \frac{1}{h_A^n(\theta)} d\sigma(\theta)\right)^{1/n} \geqslant \left(\int_{S^{n-1}} \frac{1}{h_A^q(\theta)} d\sigma(\theta)\right)^{1/q}$$
$$= \frac{1}{w_{-q}(A)}.$$

From the Blaschke-Santaló inequality it follows that

$$|A|^{1/n} \leqslant |B_2^n|^{2/n} |A^\circ|^{-1/n} \leqslant |B_2^n|^{1/n} w_{-q}(A) \leqslant \frac{c_1 w_{-q}(A)}{\sqrt{n}}$$

because $|B_2^n|^{1/n} \simeq 1/\sqrt{n}$. $\square$

Next, recall that if μ is an isotropic log-concave measure on $\mathbb{R}^n$ then, for any $1 \leqslant q \leqslant n-1$,

$$w_{-q}(Z_q(\mu)) \simeq \frac{\sqrt{q}}{\sqrt{n}} I_{-q}(\mu)$$

where

$$I_p(\mu) = \left(\int_{\mathbb{R}^n} \|x\|_2^p \, d\mu(x)\right)^{1/p}, \qquad 0 \neq p > -n.$$

LEMMA 11.3.14. *Let $X_1, \ldots, X_N$ be i.i.d. isotropic log-concave random vectors in $\mathbb{R}^n$. Let $n \leqslant N \leqslant e^n$ and set $q = 2\log(2N)$. Then,*

(11.3.13) $$w_{-q}(K_N) \leqslant c_4 w_{-q/2}(Z_q(\mu))$$

with probability greater than $1 - e^{-q}$.

Proof. With probability greater than $1 - e^{-cn}$ we have $\|X_i\|_2 \leqslant c_1 n$ for all $1 \leqslant i \leqslant N$. So, we may assume that $K_N \subseteq c_1 n B_2^n$. We write

$$[w_{-q/2}(Z_q(\mu))]^{-q} = \left(\int_{S^{n-1}} \frac{1}{h_{Z_q(\mu)}^{q/2}(\theta)} d\sigma(\theta)\right)^2$$

$$\leqslant \left(\int_{S^{n-1}} \frac{1}{h_{K_N}^q(\theta)} d\sigma(\theta)\right) \left(\int_{S^{n-1}} \frac{h_{K_N}^q(\theta)}{h_{Z_q(\mu)}^q(\theta)} d\sigma(\theta)\right).$$

Since $K_N \subseteq c_1 n B_2^n$ and $Z_q(\mu) \supseteq Z_2(\mu) \supseteq B_2^n$, we have $h_{K_N}(\theta) \leqslant c_1 n h_{Z_q(\mu)}(\theta)$ for all $\theta \in S^{n-1}$. Therefore,

$$(11.3.14) \qquad \int_{S^{n-1}} \frac{h_{K_N}^q(\theta)}{h_{Z_q(\mu)}^q(\theta)} d\sigma(\theta) = \int_0^{c_1 n} q t^{q-1} \left[\sigma\left(\theta : h_{K_N}(\theta) \geqslant t h_{Z_q(\mu)}(\theta)\right)\right] dt.$$

Taking expectations in (11.3.14) and using Lemma 11.3.9, we see that, for every $\alpha > 1$,

$$\mathbb{E}\left[\int_{S^{n-1}} \frac{h_{K_N}^q(\theta)}{h_{Z_q(K)}^q(\theta)} d\sigma(\theta)\right] \leqslant \alpha^q + \int_\alpha^{c_1 n} q t^{q-1} N t^{-q} dt$$

$$= \alpha^q + qN \log\left(\frac{c_1 n}{\alpha}\right).$$

Choosing $\alpha = 2e$ and using the fact that $e^q = (2N)^2 \geqslant qN \log\left(\frac{c_1 n}{2e}\right)$ we see that

$$\mathbb{E}\left[\int_{S^{n-1}} \frac{h_{K_N}^q(\theta)}{h_{Z_q(\mu)}^q(\theta)} d\sigma(\theta)\right] \leqslant c_2^q$$

where $c_2 > 0$ is an absolute constant. Then, Markov's inequality implies that

$$\int_{S^{n-1}} \frac{h_{K_N}^q(\theta)}{h_{Z_q(\mu)}^q(\theta)} d\sigma(\theta) \leqslant (c_2 e)^q$$

with probability greater than $1 - e^{-q}$. Thus, we conclude that $[w_{-q/2}(Z_q(\mu))]^{-q} \leqslant c_3^q [w_{-q}(K_N)]^{-q}$ and the lemma is proved. $\square$

Proof of Theorem 11.3.12. We define $q := 2 \log(2N)$. From Lemma 11.3.13 we have

$$|K_N|^{1/n} \leqslant \frac{c_1}{\sqrt{n}} w_{-q}(K_N).$$

Now, (11.3.13) shows that

$$|K_N|^{1/n} \leqslant \frac{c_5}{\sqrt{n}} w_{-q/2}(Z_q(\mu))$$

with probability greater than $1 - e^{-q}$. Since $Z_q(\mu) \subseteq c Z_{q/2}(\mu)$, we can write

$$w_{-q/2}(Z_q(\mu)) \leqslant c_6 w_{-q/2}(Z_{q/2}(\mu)) \leqslant \frac{c_7 \sqrt{q}}{\sqrt{n}} I_{-q/2}(\mu).$$

Since μ is isotropic, we have $I_{-q/2}(\mu) \leqslant I_2(\mu) = \sqrt{n}$, which implies

$$w_{-q/2}(Z_q(\mu)) \leqslant c_7 \sqrt{q}.$$

Putting everything together, we have
$$|K_N|^{1/n} \leqslant \frac{c\sqrt{q}}{\sqrt{n}} \simeq \frac{\sqrt{\log N}}{\sqrt{n}},$$
with probability greater than $1 - e^{-q} \geqslant 1 - \frac{1}{N}$. This completes the proof. $\square$

11.4. Isotropic constant

The study of the isotropic constant of random convex bodies was initiated by Klartag and Kozma in [**282**] with the case of Gaussian random polytopes. They proved that if $N > n$ and if $G_1, \ldots, G_N$ are independent standard Gaussian random vectors in $\mathbb{R}^n$, then the isotropic constant of the random polytopes
$$K_N := \mathrm{conv}\{\pm G_1, \ldots, \pm G_N\} \quad \text{and} \quad C_N := \mathrm{conv}\{G_1, \ldots, G_N\}$$
is bounded by an absolute constant $C > 0$ with probability greater than $1 - Ce^{-cn}$. The argument of [**282**] works for other classes of random polytopes with vertices which have independent coordinates (for example, if the vertices are uniformly distributed in the cube $Q_n := [-1/2, 1/2]^n$ or in the discrete cube $E_2^n := \{-1, 1\}^n$). Alonso-Gutiérrez (see [**9**]) has obtained a positive answer in the situation where K_N or C_N is spanned by N random points uniformly distributed on the Euclidean sphere S_2^{n-1}. In this section we discuss the following problem:

QUESTION 11.4.1. Let K be a convex body in $\mathbb{R}^n$. For every $N > n$ consider N independent random points $x_1, \ldots, x_N$ uniformly distributed in K and consider the random polytopes $K_N := \mathrm{conv}\{\pm x_1, \ldots, \pm x_N\}$ and $C_N := \mathrm{conv}\{x_1, \ldots, x_N\}$.

> Is it true that, with probability tending to 1 as $n \to \infty$, one has $L_{K_N} \leqslant CL_K$ and $L_{C_N} \leqslant CL_K$ where $C > 0$ is a constant independent from K, n and N?

In view of the lower bound $|K_N|^{1/n} \geqslant c\frac{\sqrt{\log(N/n)}}{\sqrt{n}} L_K$ in the range $n \leqslant N \leqslant e^{\sqrt{n}}$ one is even tempted to ask if $L_{K_N} \leqslant C$ where $C > 0$ is a constant independent from K, n and N, at least for that range of n and N.

11.4.1. ψ_2-case: the approach of Klartag and Kozma

In all the results that we mentioned in the introduction of this section, the distribution of the vertices of the random polytope C_N or K_N is a ψ_2-measure. In this subsection we assume that K is a ψ_2-body with constant b and for simplicity we restrict our discussion to the case of K_N. Our starting point is the same as in the approach of Klartag and Kozma.

FACT 11.4.2. Let D be a centered convex body in $\mathbb{R}^n$. Then,
$$|D|^{2/n} n L_D^2 \leqslant \frac{1}{|D|} \int_D \|x\|_2^2 \, dx.$$

In view of Fact 11.4.2, in order to prove that $K_N := \mathrm{conv}\{\pm x_1, \ldots, \pm x_N\}$ has bounded isotropic constant with probability close to 1, it suffices to give a lower bound for the "volume radius" $|K_N|^{1/n}$ and an upper bound for the expected value of $\|\cdot\|_2^2$ on K_N. Observe that the problem is affinely invariant, and hence, we may assume that K is an isotropic convex body.

The results of the previous section give the following lower bound for the "volume radius" of K_N:

PROPOSITION 11.4.3. *Let K be a convex body of volume 1 in $\mathbb{R}^n$ and let $x_1, \ldots, x_N$ be independent random points uniformly distributed in K. If $n \leqslant N \leqslant \exp(\sqrt{n})$ then with probability greater than $1 - e^{-c_1 n}$ we have*

$$|K_N|^{1/n} \geqslant c_2 \frac{\sqrt{\log(2N/n)}}{\sqrt{n}} L_K,$$

and if $\exp(\sqrt{n}) \leqslant N \leqslant \exp(n)$ then with probability greater than $1 - e^{-c_1 n}$ we have

$$|K_N|^{1/n} \geqslant c_2 \frac{\sqrt{\log(2N/n)}}{\sqrt{n}},$$

where $c_1, c_2 > 0$ are absolute constants.

Upper bound for the expectation of $\|\cdot\|_2^2$

We write $\mathcal{F}(K_N)$ for the family of facets of K_N. We denote by $[y_1, \ldots, y_n]$ the convex hull of $y_1, \ldots, y_n$. Observe that, with probability equal to 1, all the facets of K_N or C_N are simplices. Also, if $F = [y_1, \ldots, y_n]$ is a facet of K_N then we must have $y_j = \varepsilon_j x_{i_j}$ and $i_j \neq i_s$ for all $1 \leqslant j \neq s \leqslant n$. In other words, x_i and $-x_i$ cannot belong to the same facet of K_N.

The next lemma reduces the computation of the expectation of $\|x\|_2^2$ on K_N to a similar problem on the facets of K_N.

LEMMA 11.4.4. *Let $F_1, \ldots, F_M$ be the facets of K_N. Then,*

$$\frac{1}{|K_N|} \int_{K_N} \|x\|_2^2 dx \leqslant \frac{n}{n+2} \max_{1 \leqslant s \leqslant M} \frac{1}{|F_s|} \int_{F_s} \|u\|_2^2 du.$$

Proof. Every point $x \neq 0$ in K_N can be uniquely represented in the form $x = ty$, where $t \in [0, 1]$ and $y \in \mathrm{bd}(K_N)$. Integrating with respect to these coordinates we see that

$$\frac{1}{|K_N|} \int_{K_N} \|x\|_2^2 dx = \frac{1}{|K_N|} \int_{\mathrm{bd}(K_N)} \int_0^1 \|ty\|_2^2 t^{n-1} \langle y, \nu(y) \rangle \, dt \, dy,$$

where $\nu(y)$ is the unit outward normal to $\mathrm{bd}(K_N)$ at y (note that this is uniquely determined almost everywhere). If $y \in F_s$ for some $s = 1, \ldots, M$, then $\langle y, \nu(y) \rangle = d(0, F_s)$, where $d(0, F_s)$ is the Euclidean distance from 0 to the affine subspace determined by F_s. It follows that

$$\frac{1}{|K_N|} \int_{K_N} \|x\|_2^2 dx = \frac{1}{|K_N|} \sum_{s=1}^M \frac{d(0, F_s)}{n+2} \int_{F_s} \|u\|_2^2 du.$$

On the other hand,

$$|K_N| = \frac{1}{n} \sum_{s=1}^M d(0, F_s) |F_s|.$$

Combining the above we easily get the lemma. $\square$

LEMMA 11.4.5. *Let $y_1, \ldots, y_n \in \mathbb{R}^n$ and define $F = [y_1, \ldots, y_n]$. Then,*

$$\frac{1}{|F|} \int_F \|u\|_2^2 du \leqslant \frac{2}{n(n+1)} \max_{\varepsilon_j = \pm 1} \|\varepsilon_1 y_1 + \cdots + \varepsilon_n y_n\|_2^2.$$

Proof. We write $F = T(\Delta^{n-1})$ where $\Delta^{n-1} = [e_1, \ldots, e_n]$ and $T_{ij} = \langle y_j, e_i \rangle =: y_{ji}$. Assume that $\det T \neq 0$. It follows that

$$\frac{1}{|F|} \int_F \|u\|_2^2 du = \frac{1}{|\Delta^{n-1}|} \int_{\Delta^{n-1}} \|Tu\|_2^2 du$$

$$= \frac{1}{|\Delta^{n-1}|} \int_{\Delta^{n-1}} \sum_{i=1}^n \left(\sum_{j=1}^n y_{ji} u_j \right)^2 du.$$

Direct computation shows that

$$\frac{1}{|\Delta^{n-1}|} \int_{\Delta^{n-1}} u_{j_1} u_{j_2} = \frac{1 + \delta_{j_1, j_2}}{n(n+1)}.$$

Therefore, we get

$$\frac{1}{|F|} \int_F \|u\|_2^2 du = \frac{1}{n(n+1)} \sum_{i=1}^n \left(\sum_{j=1}^n y_{ji}^2 + \left(\sum_{j=1}^n y_{ji} \right)^2 \right).$$

Since

$$\sum_{i=1}^n \left(\sum_{j=1}^n y_{ji}^2 \right) = \operatorname{Ave}_{\varepsilon_j = \pm 1} \sum_{i=1}^n \left(\sum_{j=1}^n \varepsilon_j y_{ji} \right)^2$$

$$= \operatorname{Ave}_{\varepsilon_j = \pm 1} \|\varepsilon_1 y_1 + \cdots + \varepsilon_n y_n\|_2^2$$

and

$$\sum_{i=1}^n \left(\sum_{j=1}^n y_{ji} \right)^2 = \|y_1 + \cdots + y_n\|_2^2,$$

the proof is complete. $\square$

We will make use of the following Bernstein-type inequality (for a proof see e.g. [**109**] and [**110**]).

LEMMA 11.4.6. *Let $g_1, \ldots, g_m$ be independent random variables with $\mathbb{E}(g_j) = 0$ on some probability space (Ω, μ). Assume that $\|g_j\|_{\psi_2} \leqslant b$ for all $j \leqslant m$ and some constant $\alpha > 0$. Then,*

$$\operatorname{Prob}\left(\left| \sum_{j=1}^m g_j \right| > \alpha m \right) \leqslant 2 \exp(-\alpha^2 m / 8b^2)$$

for every $\alpha > 0$. $\square$

PROPOSITION 11.4.7. *Let K be an isotropic convex body in $\mathbb{R}^n$. Assume that $\|\langle \cdot, \theta \rangle\|_{\psi_2} \leqslant b L_K$ for all $\theta \in S^{n-1}$. Fix $N > n$ and let $x_1, \ldots, x_N$ be independent random points uniformly distributed in K. Then, with probability greater than $1 - \exp(-cn \log(2N/n))$ we have*

$$\max_{\varepsilon_j = \pm 1} \|\varepsilon_1 x_{i_1} + \cdots + \varepsilon_n x_{i_n}\|_2 \leqslant C b L_K n \sqrt{\log(2N/n)}$$

for all $\{i_1, \ldots, i_n\} \subseteq \{1, \ldots, N\}$.

Proof. We first fix a subset $\{y_1, \ldots, y_n\}$ of $\{\pm x_1, \ldots, \pm x_N\}$ and also fix $\theta \in S^{n-1}$ and a choice of signs $\varepsilon_j = \pm 1$. We apply Lemma 11.4.6 (with $m = n$) to the random variables $g_j(y_1, \ldots, y_n) = \langle \varepsilon_j y_j, \theta \rangle$ on $\Omega = K^n$. Note that $\mathbb{E}(g_j) = 0$ because K is centered. Also, by our assumption we know that $\|g_j\|_{\psi_2} \leqslant bL_K$. Therefore,

$$\mathrm{Prob}\big(\{|\langle \varepsilon_1 y_1 + \cdots + \varepsilon_n y_n, \theta \rangle| > \alpha b L_K n\}\big) \leqslant 2\exp(-c_1 \alpha^2 n)$$

for every $\alpha > 0$. Consider a 1/2-net $\mathcal{N}$ for S^{n-1} with cardinality $|\mathcal{N}| \leqslant 5^n$. If $\alpha \geqslant C_1$ then, with probability greater than $1 - \exp(-c_2 \alpha^2 n)$ we have

$$|\langle \varepsilon_1 y_1 + \cdots + \varepsilon_n y_n, \theta \rangle| \leqslant \alpha b L_K n$$

for every $\theta \in \mathcal{N}$ and every choice of signs $\varepsilon_j = \pm 1$. Using a standard successive approximation argument, and taking into account all 2^n possible choices of signs $\varepsilon_j = \pm 1$, we get that, with probability greater than $1 - \exp(-c_2 n \log(2N/n))$, for every $\theta \in S^{n-1}$ and every choice of signs $\varepsilon_j = \pm 1$ we have

$$|\langle \varepsilon_1 y_1 + \cdots + \varepsilon_n y_n, \theta \rangle| \leqslant C b L_K n \sqrt{\log(2N/n)},$$

and hence

$$\max_{\varepsilon_j = \pm 1} \|\varepsilon_1 y_1 + \cdots + \varepsilon_n y_n\|_2 \leqslant C b L_K n \sqrt{\log(2N/n)}.$$

Since the number of subsets $\{y_1, \ldots, y_n\}$ of $\{\pm x_1, \ldots, \pm x_N\}$ is bounded by $(2eN/n)^n$, choosing $C > 0$ large enough we immediately get that, with high probability,

$$\max_{\varepsilon_j = \pm 1} \|\varepsilon_1 y_1 + \cdots + \varepsilon_n y_n\|_2 \leqslant C b L_K n \sqrt{\log(2N/n)}$$

for all $\{i_1, \ldots, i_n\} \subseteq \{1, \ldots, N\}$. $\square$

Combining Proposition 11.4.7 and Lemmas 11.4.4 and 11.4.5 we obtain:

PROPOSITION 11.4.8. *Let K be an isotropic convex body in $\mathbb{R}^n$. Assume that $\|\langle \cdot, \theta \rangle\|_{\psi_2} \leqslant bL_K$ for all $\theta \in S^{n-1}$. Fix $N > n$ and let $x_1, \ldots, x_N$ be independent random points uniformly distributed in K. Then, with probability greater than $1 - \exp(-cn \log(2N/n))$ we have*

$$\frac{1}{|K_N|} \int_{K_N} \|x\|_2^2 dx \leqslant C b^2 L_K^2 \log(2N/n),$$

where $C > 0$ is an absolute constant.

We are ready to prove our main result. We will use the fact that if K is a ψ_2-body with constant b then $L_K \leqslant Cb$ (see Chapter 7).

THEOREM 11.4.9 (Klartag-Kozma). *Let K be an isotropic convex body in $\mathbb{R}^n$. Assume that $\|\langle \cdot, \theta \rangle\|_{\psi_2} \leqslant bL_K$ for all $\theta \in S^{n-1}$. Fix $N > n$ and let $x_1, \ldots, x_N$ be independent random points uniformly distributed in K. Then, with probability greater than $1 - \exp(-cn)$ we have*

$$L_{K_N} \leqslant Cb^2.$$

Proof. From Fact 11.4.2 we know that

$$|K_N|^{2/n} n L_{K_N}^2 \leqslant \frac{1}{|K_N|} \int_{K_N} \|x\|_2^2 \, dx.$$

From Proposition 11.4.3, with probability greater than $1 - e^{-c_1 n}$ we have

$$|K_N|^{2/n} \geqslant c_2 \frac{\log(2N/n)}{n},$$

and hence
$$L_{K_N}^2 \leqslant \frac{C}{\log(2N/n)} \frac{1}{|K_N|} \int_{K_N} \|x\|_2^2\, dx.$$
Now, Proposition 11.4.8 shows that with probability greater than $1 - e^{-c_1 n}$ we have
$$L_{K_N}^2 \leqslant C_2 b^2 L_K^2 \leqslant C_3 b^4,$$
which proves the theorem. $\square$

REMARK 11.4.10. If we assume that $N \leqslant e^{\sqrt{n}}$ then we obtain the stronger estimate
$$L_{K_N} \leqslant Cb$$
because for a random K_N we can use the lower bound $|K_N|^{2/n} \geqslant c_2 \frac{\log(2N/n)}{n} L_K^2$.

11.5. Further reading

11.5.1. Isotropic constant of random polytopes: the unconditional case

In the case of unconditional convex bodies one has an affirmative answer to Question 11.4.1, although these are not necessarily ψ_2-bodies.

THEOREM 11.5.1 (Dafnis-Giannopoulos-Guédon). *Let K be an isotropic unconditional convex body in $\mathbb{R}^n$. For every $N \geqslant n$ consider N independent random points $x_1, \ldots, x_N$ uniformly distributed in K. Then, with probability greater than $1 - C_1 \exp(-cn)$ the random polytope $K_N := \operatorname{conv}\{\pm x_1, \ldots, \pm x_N\}$ has isotropic constant bounded by an absolute constant $C > 0$.*

The method of proof (see [**146**]) is similar to that of Klartag and Kozma; the additional tool is the estimate of Bobkov and Nazarov for the ψ_2-norm of linear functionals on isotropic unconditional convex bodies. The starting point is now a stronger estimate for L_D in terms of the expectation of the ℓ_1^n-norm on D (see Section 4.2.1; this is a result from [**384**]).

LEMMA 11.5.2. *Let D be a symmetric convex body in $\mathbb{R}^n$. Then,*
$$|D|^{1/n} n L_D \leqslant c \frac{1}{|D|} \int_D \|x\|_1\, dx,$$
where $c > 0$ is an absolute constant.

Recall that for every $\theta \in \mathbb{R}^n$ we have
$$\|\langle \cdot, \theta \rangle\|_{\psi_2} \leqslant c\sqrt{n}\|\theta\|_\infty,$$
where $c > 0$ is an absolute constant. Now, let $y_1, \ldots, y_n$ be independent random points uniformly distributed in K. We fix $\theta \in \mathbb{R}^n$ with $\|\theta\|_\infty = 1$ and a choice of signs $\varepsilon_j = \pm 1$, and apply Lemma 11.4.6 (with $m = n$) to the random variables $g_j(y_1, \ldots, y_n) = \langle \varepsilon_j y_j, \theta \rangle$ on $\Omega = K^n$ to get
$$\operatorname{Prob}\{|\langle \varepsilon_1 y_1 + \cdots + \varepsilon_n y_n, \theta \rangle| > \alpha n\} \leqslant 2\exp(-c\alpha^2)$$
for every $\alpha > 0$. Consider a $1/2$-net $\mathcal{N}$ for S_∞^n with cardinality $|\mathcal{N}| \leqslant 5^n$. Choosing $\alpha = C\sqrt{n}\sqrt{\log(2N/n)}$ where $C > 0$ is a large enough absolute constant, we see that, with probability greater than $1 - \exp(-c_1 n \log(2N/n))$ we have
$$|\langle \varepsilon_1 y_1 + \cdots + \varepsilon_n y_n, \theta \rangle| \leqslant C n^{3/2} \sqrt{\log(2N/n)}$$
for every $\theta \in \mathcal{N}$ and every choice of signs $\varepsilon_j = \pm 1$. Using a standard successive approximation argument, and taking into account all 2^n possible choices of signs $\varepsilon_j = \pm 1$, we get that with probability greater than $1 - \exp(-c_2 n \log(2N/n))$ we have
$$|\langle \varepsilon_1 y_1 + \cdots + \varepsilon_n y_n, \theta \rangle| \leqslant C n^{3/2} \sqrt{\log(2N/n)}$$

for every $\theta \in S_\infty^n$ and every choice of signs $\varepsilon_j = \pm 1$, and hence
$$\max_{\varepsilon_j=\pm 1} \|\varepsilon_1 y_1 + \cdots + \varepsilon_n y_n\|_1 \leqslant Cn^{3/2}\sqrt{\log(2N/n)}.$$

Now, let $N \geqslant n$ and let $x_1, \ldots, x_N$ be independent random points uniformly distributed in K. Since the number of subsets $\{y_1, \ldots, y_n\}$ of $\{\pm x_1, \ldots, \pm x_N\}$ is bounded by $(2eN/n)^n$, we immediately get the following.

PROPOSITION 11.5.3. *Let K be an isotropic unconditional convex body in $\mathbb{R}^n$. Fix $N \geqslant n$ and let $x_1, \ldots, x_N$ be independent random points uniformly distributed in K. Then, with probability greater than $1 - \exp(-cn\log(2N/n))$ we have*
$$\max_{\varepsilon_j=\pm 1} \|\varepsilon_1 x_{i_1} + \cdots + \varepsilon_n x_{i_n}\|_1 \leqslant Cn^{3/2}\sqrt{\log(2N/n)}$$
for all $\{i_1, \ldots, i_n\} \subseteq \{1, \ldots, N\}$.

The next lemma reduces the computation of the expectation of $\|x\|_1$ on K_N to a similar problem on the facets of K_N.

LEMMA 11.5.4. *Let $F_1, \ldots, F_m$ be the facets of K_N. Then,*
$$\frac{1}{|K_N|}\int_{K_N}\|x\|_1 dx \leqslant \max_{1 \leqslant s \leqslant m} \frac{1}{|F_s|}\int_{F_s}\|u\|_1 du.$$

Proof. As in the proof of Lemma 11.4.4 we check that
$$\frac{1}{|K_N|}\int_{K_N}\|x\|_1 dx = \frac{1}{|K_N|}\sum_{s=1}^m \frac{d(0, F_s)}{n+1}\int_{F_s}\|u\|_1 du,$$
where $d(0, F_s)$ is the Euclidean distance from 0 to the affine subspace determined by F_s. Since
$$|K_N| = \frac{1}{n}\sum_{s=1}^m d(0, F_s)|F_s|,$$
the result follows. □

Let $y_1, \ldots, y_n \in \mathbb{R}^n$ and define $F = \operatorname{conv}\{y_1, \ldots, y_n\}$. Then, $F = T(\Delta^{n-1})$ where $\Delta^{n-1} = \operatorname{conv}\{e_1, \ldots, e_n\}$ and $T_{ij} = \langle y_j, e_i \rangle =: y_{ji}$. Assume that $\det T \neq 0$. It follows that
$$\frac{1}{|F|}\int_F \|u\|_1 du = \frac{1}{|\Delta^{n-1}|}\int_{\Delta^{n-1}}\|Tu\|_1 du$$
$$= \frac{1}{|\Delta^{n-1}|}\int_{\Delta^{n-1}}\sum_{i=1}^n\Big|\sum_{j=1}^n y_{ji}u_j\Big| du$$
$$= \sum_{i=1}^n \frac{1}{|\Delta^{n-1}|}\int_{\Delta^{n-1}}\Big|\sum_{j=1}^n y_{ji}u_j\Big| du$$
$$\leqslant \sum_{i=1}^n \left(\frac{1}{|\Delta^{n-1}|}\int_{\Delta^{n-1}}\Big(\sum_{j=1}^n y_{ji}u_j\Big)^2 du\right)^{1/2}.$$

Using the fact that
$$\frac{1}{|\Delta^{n-1}|}\int_{\Delta^{n-1}} u_{j_1}u_{j_2} = \frac{1 + \delta_{j_1, j_2}}{n(n+1)},$$
we see that
$$\frac{1}{|F|}\int_F \|u\|_1 du \leqslant \frac{1}{\sqrt{n(n+1)}}\sum_{i=1}^n\left(\sum_{j=1}^n y_{ji}^2 + \Big(\sum_{j=1}^n y_{ji}\Big)^2\right)^{1/2}$$
$$\leqslant \frac{1}{n}\sum_{i=1}^n\left[\Big(\sum_{j=1}^n y_{ji}^2\Big)^{1/2} + \Big|\sum_{j=1}^n y_{ji}\Big|\right].$$

From the classical Khintchine inequality (see [**484**] for the best constant $\sqrt{2}$) we know that
$$\Big(\sum_{j=1}^n y_{ji}^2\Big)^{1/2} \leqslant \sqrt{2}\mathbb{E}_{\epsilon_j=\pm 1}\Big|\sum_{j=1}^n \varepsilon_j y_{ji}\Big|.$$
Therefore,
$$\sum_{i=1}^n \Big[\Big(\sum_{j=1}^n y_{ji}^2\Big)^{1/2} + \Big|\sum_{j=1}^n y_{ji}\Big|\Big] \leqslant \sqrt{2}\mathbb{E}_{\epsilon_j=\pm 1}\Big\|\sum_{j=1}^n \varepsilon_j y_j\Big\|_1 + \Big\|\sum_{j=1}^n y_j\Big\|_1.$$
This shows that
$$\frac{1}{|F|}\int_F \|u\|_1 du \leqslant \frac{\sqrt{2}+1}{n}\max_{\varepsilon_j=\pm 1}\|\varepsilon_1 y_1 + \cdots + \varepsilon_n y_n\|_1.$$
Then, Proposition 11.5.3 and Lemma 11.5.4 immediately imply our upper bound:

PROPOSITION 11.5.5. *Let K be an isotropic 1-unconditional convex body in $\mathbb{R}^n$. Fix $N \geqslant n$ and let $x_1, \ldots, x_N$ be independent random points uniformly distributed in K. Then, with probability greater than $1 - \exp(-cn\log(2N/n))$ we have*
$$\frac{1}{|K_N|}\int_{K_N} \|x\|_1 dx \leqslant C\sqrt{n}\sqrt{\log(2N/n)}$$
where $C > 0$ is an absolute constant. □

We are ready to prove our main result. Recall that if K is an unconditional convex body then $L_K \leqslant C$.

THEOREM 11.5.6. *Let K be an unconditional isotropic convex body in $\mathbb{R}^n$. Fix $N \geqslant n$ and let $x_1, \ldots, x_N$ be independent random points uniformly distributed in K. Then, with probability greater than $1 - \exp(-c_1 n)$ we have*
$$L_{K_N} \leqslant C.$$

Proof. From Lemma 11.5.2 we know that
$$|K_N|^{1/n} n L_{K_N} \leqslant \frac{1}{|K_N|}\int_{K_N} \|x\|_1 dx.$$
From Proposition 11.4.3, with probability greater than $1 - e^{-c_1 n}$ we have
$$|K_N|^{1/n} \geqslant c_2 \frac{\sqrt{\log(2N/n)}}{\sqrt{n}},$$
and hence
$$\sqrt{n} L_{K_N} \leqslant \frac{C}{\sqrt{\log(2N/n)}} \frac{1}{|K_N|}\int_{K_N} \|x\|_1 dx.$$
Now, Proposition 11.5.5 shows that with probability greater than $1 - e^{-c_1 n}$ we have $L_{K_N} \leqslant C$. □

11.5.2. Geometry of 0-1 polytopes

The point of view of our results on the "asymptotic shape" of random polytopes has its origin in analogous, and naturally much more precise, results on the asymptotic shape of random 0/1 polytopes or, equivalently, ± 1-polytopes, which we describe in this subsection.

By definition, a ± 1-polytope is the convex hull of a subset of the vertices of $[-1,1]^n$. In order to define random ± 1-polytopes with a prescribed number of vertices, we consider n independent and identically distributed ± 1 random variables $X_1, \ldots, X_n$, defined on some probability space $(\Omega, \mathcal{F}, \mathbb{P})$, with distribution
$$\mathbb{P}(X=1) = \mathbb{P}(X=-1) = \tfrac{1}{2},$$

we set $\vec{X} = (X_1, \ldots, X_n)$ and, for a fixed N satisfying $n < N \leqslant 2^n$, we consider N independent copies $\vec{X}_1, \ldots, \vec{X}_N$ of $\vec{X}$. This procedure defines the random ± 1-polytope

(11.5.1) $$K_N = \mathrm{conv}\{\vec{X}_1, \ldots, \vec{X}_N\}.$$

Note that K_N has at most N vertices.

Dyer, Füredi and McDiarmid established in [163] a sharp threshold for the expected volume of these random ± 1 polytopes.

THEOREM 11.5.7 (Dyer-Füredi-McDiarmid). *Let $\kappa = 2/\sqrt{e}$ and consider the random polytope K_N defined in (11.5.1). For every $\varepsilon \in (0,1)$,*

$$\limsup_{n \to \infty} \left\{ 2^{-n} \mathbb{E}|K_N| : N \leqslant (\kappa - \varepsilon)^n \right\} = 0$$

and

$$\liminf_{n \to \infty} \left\{ 2^{-n} \mathbb{E}|K_N| : N \geqslant (\kappa + \varepsilon)^n \right\} = 1.$$

Their strategy, which we describe below, was later used for the study of a question of Fukuda and Ziegler on the maximal possible number of facets of a ± 1-polytope in $\mathbb{R}^n$. In general, if P is a polytope in $\mathbb{R}^n$, we write $f_{n-1}(P)$ for the number of its facets. Let

$$g(n) := \max\left\{ f_{n-1}(P_n) : P_n \text{ is a } 0/1 \text{ polytope in } \mathbb{R}^n \right\}.$$

Fukuda and Ziegler (see [186] and [509]) asked what the behavior of $g(n)$ is as $n \to \infty$. The best known upper bound to date is

$$g(n) \leqslant 30(n-2)!$$

(for n large enough), which is established by Fleiner, Kaibel and Rote in [176]. Regarding lower bounds, a major breakthrough in this direction was made by Bárány and Pór in [48]; they proved that

$$g(n) \geqslant \left(\frac{cn}{\log n} \right)^{n/4},$$

where $c > 0$ is an absolute constant. Later, Gatzouras, Giannopoulos and Markoulakis showed in [195] and [196] that the exponent $n/4$ can in fact be improved to $n/2$:

THEOREM 11.5.8 (Gatzouras-Giannopoulos-Markoulakis). *There exists a constant $c > 0$ such that*

$$g(n) \geqslant \left(\frac{cn}{\log n} \right)^{n/2}.$$

It is interesting to compare this estimate with the known bounds for the expected number of facets of the convex hull $P_{N,n}$ of N independent random points which are uniformly distributed on the sphere S^{n-1}. In [124] it is shown that there exist two constants $c_1, c_2 > 0$, such that

$$\left(c_1 \log \frac{N}{n} \right)^{n/2} \leqslant \mathbb{E}[f_{n-1}(P_{N,n})] \leqslant \left(c_2 \log \frac{N}{n} \right)^{n/2}$$

for all n and N satisfying $2n \leqslant N \leqslant 2^n$. In the case of ± 1 polytopes, N can be as large as 2^n, therefore one might conjecture that $g(n)$ is of the order of $n^{n/2}$. Theorem 11.5.8 gives a lower bound which is "practically of this order".

We pass to a brief description of the ideas behind the above results. In order to determine the threshold $N(n) = (2/\sqrt{e})^n$, Dyer, Füredi and McDiarmid introduced two families of convex subsets of the cube $C = [-1,1]^n$. For every $\vec{x} \in C$, set

$$q(\vec{x}) := \inf\left\{ \mathrm{Prob}(\vec{X} \in H) : \vec{x} \in H, H \text{ a closed halfspace} \right\}.$$

If $\beta > 0$ then the β-center of C is defined by

$$Q^\beta = \{\vec{x} \in C : q(\vec{x}) \geqslant \exp(-\beta n)\};$$

it is easily checked that Q^β is a convex polytope.

Next, consider the function $f : (-1, 1) \to \mathbb{R}$ with
$$f(x) = \tfrac{1}{2}(1+x)\log(1+x) + \tfrac{1}{2}(1-x)\log(1-x),$$
extend it to a continuous function on $[-1, 1]$ by setting $f(\pm 1) = \log 2$, and for every $\vec{x} = (x_1, \ldots, x_n) \in C$ set
$$F(\vec{x}) = \frac{1}{n} \sum_{i=1}^{n} f(x_i).$$

The next lemma was proved in [**163**].

LEMMA 11.5.9. *For every $\vec{x} \in (-1,1)^n$ we have $q(\vec{x}) \leqslant \exp(-nF(\vec{x}))$.*

The second family of subsets of C introduced in [**163**] is as follows: for every $\beta > 0$, set
$$F^{\beta} = \{\vec{x} \in C : F(\vec{x}) \leqslant \beta\}.$$
Since f is a strictly convex function on $(-1, 1)$, it is clear that F^β is convex. Lemma 11.5.9 and the definition of Q^β show that if $\vec{x} \in Q^\beta \cap (-1,1)^n$ then $F(\vec{x}) \leqslant \beta$. In other words, we have the following.

LEMMA 11.5.10. $Q^\beta \cap (-1,1)^n \subseteq F^\beta$ *for every $\beta > 0$.* □

Observe that as $\beta \to \log 2$, both Q^β and F^β approach C. The main technical fact is that the two families are very close, in the following sense:

THEOREM 11.5.11. *There exist $\gamma \in \left(0, \tfrac{1}{10}\right)$ and $n_0 = n_0(\gamma) \in \mathbb{N}$ with the following property: If $n \geqslant n_0$ and $4 \log n / n \leqslant \beta < \log 2$, then*
$$F^{\beta-\varepsilon} \cap \gamma C \subseteq Q^\beta$$
for some $\varepsilon \leqslant 3 \log n / n$.

Theorem 11.5.11 was proved in [**195**] and strengthens a previous estimate from [**48**]. Now, fix $n^8 \leqslant N \leqslant 2^n$ and define $\alpha = (\log N)/n$. The family (Q^β) is related to the random polytope K_N through a lemma from [**163**]: If n is sufficiently large, one has that
$$\text{Prob}\big(K_N \supseteq Q^{\alpha-\varepsilon}\big) > 1 - 2^{-(n-1)}$$
for some $\varepsilon \leqslant 3 \log n / n$.

Combining this fact with Theorem 11.5.11 one gets the following.

LEMMA 11.5.12. *Let $n^8 \leqslant N \leqslant 2^n$ and $n \geqslant n_0(\gamma)$. Then,*
$$\text{Prob}\big(K_N \supseteq F^{\alpha-\varepsilon} \cap \gamma C\big) > 1 - 2^{-(n-1)}$$
for some $\varepsilon \leqslant 6 \log n / n$.

Bárány and Pór proved that K_N is weakly sandwiched between $F^{\alpha-\varepsilon} \cap \gamma C$ and $F^{\alpha+\delta}$ in the sense that $K_N \supseteq F^{\alpha-\varepsilon} \cap \gamma C$ and most of the surface area of $F^{\alpha+\delta} \cap \gamma C$ is outside K_N for small positive values of δ (the estimate for δ given below is checked in [**195**]).

LEMMA 11.5.13. *If $n \geqslant n_0$ and $\alpha < \log 2 - 12n^{-1}$, then*
$$\text{Prob}\big(\, |\text{bd}(F^{\alpha+\delta}) \cap \gamma C \cap K_N| \geqslant \tfrac{1}{2} |\text{bd}(F^{\alpha+\delta}) \cap \gamma C|\, \big) \leqslant \tfrac{1}{100}.$$
for some $\delta \leqslant 6/n$.

The last ingredient is the following geometric lemma from [**48**].

LEMMA 11.5.14. *Let $\gamma \in \left(0, \tfrac{1}{10}\right)$ and assume that $\beta + \zeta < \log 2$. Then,*
$$|\text{bd}(F^{\beta+\zeta}) \cap \gamma C \cap H| \leqslant (3\zeta n)^{(n-1)/2} |S^{n-1}|$$
for every closed halfspace H whose interior is disjoint from $F^\beta \cap \gamma C$.

The main idea for the final step of the proof is that for a random K_N and for each half-space H_A which is defined by a facet A of K_N and has interior disjoint from K_N, we also have that H_A has interior disjoint from $F^{\alpha-\varepsilon} \cap \gamma C$ and hence cuts a small amount (independent from A) of the surface of $\mathrm{bd}(F^{\alpha+\delta}) \cap \gamma C$. Since the surface area of $\mathrm{bd}(F^{\alpha+\delta}) \cap \gamma C$ is mostly outside K_N we see that the number of facets of K_N must be large, depending on the total surface of $\mathrm{bd}(F^{\alpha+\delta}) \cap \gamma C$.

Under some restrictions on the range of values of N, one thus obtains a lower bound for the expected number of facets $\mathbb{E}[f_{n-1}(K_N)]$, for each fixed N. In particular, one has

THEOREM 11.5.15 (Gatzouras-Giannopoulos-Markoulakis). *There exist two positive constants a and b such that: for all sufficiently large n, and all N satisfying $n^a \leqslant N \leqslant \exp(bn)$, there exists a 0/1 polytope K_N in $\mathbb{R}^n$ with*

$$f_{n-1}(K_N) \geqslant \left(\frac{\log N}{a \log n}\right)^{n/2}.$$

It is clear that the lower bound for $g(n)$ follows: one only has to choose $N = \lfloor \exp(bn) \rfloor$.

A general "large deviations approach" to the geometry of polytopes spanned by random points with independent coordinates was developed by Gatzouras and Giannopoulos in [**193**] and [**194**]. They considered an even Borel probability measure μ on the real line with $\mathbb{E}_\mu(x^2) = 1$ and the property that

$$(11.5.2) \qquad \int_{\mathbb{R}} e^{tx} \, d\mu(x) = \mathbb{E}(e^{tX}) < \infty \qquad \text{for all } t \text{ in an open interval.}$$

The condition (11.5.2) ensures that X has finite moments of all orders. If $X_1, \ldots, X_n$ are independent and identically distributed random variables, defined on the product space $(\Omega^n, \mathcal{F}^{\otimes n}, P^n)$, each with distribution μ, one can set $\vec{X} = (X_1, \ldots, X_n)$ and, for a fixed N satisfying $N > n$, consider N independent copies $\vec{X}_1, \ldots, \vec{X}_N$ of $\vec{X}$, defined on the product space $(\Omega^{nN}, \mathcal{F}^{\otimes nN}, \mathrm{Prob})$. This procedure defines the random polytope

$$(11.5.3) \qquad K_N := \mathrm{conv}\{\vec{X}_1, \ldots, \vec{X}_N\}.$$

In this more general setting one can then establish very precise analogues of most of the statements that we discussed about ± 1-polytopes. In order to give a flavor of this idea, let us consider the case where μ is supported in some interval $[-\alpha, \alpha]$. Then, the random polytope K_N is contained in $[-\alpha, \alpha]^n$ almost surely.

Let $\varphi(t) := \mathbb{E}(e^{tX})$, $t \in \mathbb{R}$, denote the moment generating function of X, and let $\psi(t) := \log \varphi(t)$ be its cumulant generating function (or logarithmic moment generating function). By Hölder's inequality, ψ is a convex function on $\mathbb{R}$. Consider the Legendre transform λ of ψ; this is the function

$$\lambda(x) := \sup\{tx - \psi(t) : t \in \mathbb{R}\}.$$

Define

$$\kappa = \kappa(\mu) := \frac{1}{2\alpha} \int_{-\alpha}^{\alpha} \lambda(x) \, dx.$$

For a large class of distributions μ one can establish the following threshold for the expected volume of K_N.

THEOREM 11.5.16 (Gatzouras-Giannopoulos). *Let μ be an even Borel probability measure, supported in some interval $[-\alpha, \alpha]$, and assume that $0 < \kappa(\mu) < \infty$. Then*

$$\lim_{n \to \infty} \sup\left\{(2\alpha)^{-n} \mathbb{E}(|K_N|) : N \leqslant \exp((\kappa - \varepsilon)n)\right\} = 0$$

holds for every $\varepsilon > 0$. Furthermore

$$\lim_{n \to \infty} \inf\left\{(2\alpha)^{-n} \mathbb{E}(|K_N|) : N \geqslant \exp((\kappa + \varepsilon)n)\right\} = 1.$$

holds for every $\varepsilon > 0$, whenever the distribution μ satisfies

(11.5.4) $$\lim_{x\uparrow\alpha} \frac{-\log P(X \geqslant x)}{\lambda(x)} = 1.$$

One can see that if we exclude trivial cases then $\kappa(\mu) > 0$. Furthermore, the proof shows that when $\kappa(\mu) = \infty$ then in fact $\sup\{(2\alpha)^{-n}\mathbb{E}(|K_N|) : N \leqslant e^{rn}\} \to 0$ as $n \to \infty$, for any $r > 0$. Condition (11.5.4) holds for a large class of compactly supported distributions.

11.6. Notes and references

Lower bound for the expected volume radius

Theorem 11.1.1 is due to Giannopoulos and Tsolomitis and appears in [**217**]. The idea of Steiner symmetrization in this context goes back to Blaschke who used it to show that the expected area of a random triangle inside a convex region of area 1 is minimized in the case where the region is a disc. It has been used by Groemer in [**225**], [**226**] and [**227**] in his work on Sylvester's problem for higher dimensions and arbitrary number of points. In order to obtain the full statement of Theorem 11.1.1 one has to further exploit this method. Hartzoulaki and Paouris used the same ideas in [**250**] for a version of Theorem 11.1.1 in which volume is replaced by other quermassintegrals.

Lemma 11.1.2 is from [**226**] (the basic idea of its proof comes from a work of Macbeath [**348**]).

The proof of Proposition 11.1.8 is from [**202**]. The idea comes from the article [**163**] of Dyer, Füredi and McDiarmid. Proposition 11.2.2 is due to Pivovarov [**433**].

Asymptotic shape

The results of this section were proved by Dafnis, Giannopoulos and Tsolomitis. They have their origin in the study of the behavior of symmetric random ± 1-polytopes by Giannopoulos-Hartzoulaki in [**201**] and by Litvak, Pajor, Rudelson, and Tomczak–Jaegermann in [**336**] who worked in a more general setting and improved the estimates from [**201**] covering all $N \geqslant (1+\delta)n$, where $\delta > 0$ can be as small as $1/\log n$, and establishing the inclusion
$$K_{n,N} \supseteq c\left(\sqrt{\beta \log(N/n)} B_2^n \cap B_\infty^n\right)$$
with probability greater than $1 - \exp(-c_1 n^\beta N^{1-\beta}) - \exp(-c_2 N)$ for every $0 < \beta < 1$. Dafnis, Giannopoulos and Tsolomitis observed in [**147**] that this estimate can be written in the form
$$K_{n,N} \supseteq c Z_{\beta \log(N/n)}(Q_n)$$
and obtained the inclusion $K_N \supseteq Z_q(K)$ for $q = q(N,n) \simeq \log(N/n)$ (of Theorem 11.3.2) for the random polytope $K_N = \text{conv}\{\pm x_1, \ldots, \pm x_N\}$ spanned by N independent random points $x_1, \ldots, x_N$ uniformly distributed in an isotropic convex body K. The method of proof is a modification of ideas from [**336**].

One of the main results in [**147**] is Theorem 11.3.12 which provides the sharp upper bound
$$|K_N|^{1/n} \leqslant C \frac{\sqrt{\log(2N/n)}}{\sqrt{n}} L_K$$
for the volume of K_N with probability greater than $1 - \frac{1}{N}$ (see [**217**] and [**202**] for previous weaker results). It combines the information on the asymptotic shape of K_N with the formula $w_{-q}(Z_q(\mu)) \simeq \frac{\sqrt{q}}{\sqrt{n}} I_{-q}(\mu)$ of Paouris (see Chapter 5).

Theorem 11.3.11 appears in [**148**] where the same line of thought is continued. Subsequent works of Alonso-Gutiérrez and Prochno (and Dafnis, Hernandez-Cifre) combining the approach and the results of this chapter with additional probabilistic tools, determine other geometric parameters of K_N (see, for example, [**15**], [**16**] and [**14**]).

Small-ball probabilities

Paouris and Pivovarov introduced in [**417**] a general way of generating random convex sets. Let $\mu_1, \ldots, \mu_N$, $N \geqslant n$, be probability measures on $\mathbb{R}^n$ which are absolutely continuous with respect to Lebesgue measure and assume that their densities f_i are uniformly bounded: $\|f_i\|_\infty \leqslant 1$ for all $i = 1, \ldots N$. If x_i are independent random points with distribution μ_i, then the random matrix $[x_1 \cdots x_N]$ with columns x_i acts on any convex body C in $\mathbb{R}^N$ producing the random convex set

$$[x_1 \cdots x_N]C := \Big\{ \sum_{i=1}^n c_i x_i : c = (c_1, \ldots, c_N) \in C \Big\}.$$

The choice $C = B_1^N$ corresponds to the random polytope $K_N = \operatorname{conv}\{\pm x_1, \ldots, \pm x_N\}$ that we discuss in this chapter. The main result of [**417**] states that the expected volume

$$\mathcal{F}_C(\mu_1, \ldots, \mu_N) = \int_{\mathbb{R}^n} \cdots \int_{\mathbb{R}^n} |[x_1, \ldots, x_N]C| \, d\mu_N(x_N) \cdots d\mu_1(x_1)$$

of $[x_1 \cdots x_N]C$ is minimized when $\mu_1 = \cdots = \mu_N = \mu_{\overline{B}_2^n}$. From this fact one can deduce Groemer's inequality that we described in Section 11.1 as well as the Lutwak-Zhang-Yang inequalities for L_q-centroid bodies and Orlicz bodies, the Bourgain-Meyer-Milman-Pajor random zonotope inequality (see [**111**]) etc. The proof exploits Lemma 11.1.2 but it employs rearrangement inequalities instead of Steiner symmetrization.

A second work of Paouris and Pivovarov (see [**418**]) provides estimates for the probability

$$\operatorname{Prob}_{\otimes \mu_i}\Big(|[x_1, \ldots, x_N]C|^{1/n} \leqslant \varepsilon\Big)$$

for small $\varepsilon > 0$. Using the techniques of [**417**] they first show that the extremal case is when $\mu_1 = \cdots = \mu_N = \mu_{\overline{B}_2^n}$. Then, instead of working with the ball, they transfer the problem to the standard Gaussian measure and they use the Gaussian representation of the intrinsic volumes of C to obtain information on the small ball behavior of $|[x_1, \ldots, x_N]C|$. This approach leads to precise estimates in the cases $C = B_1^N$ or $C = B_\infty^N$ that correspond to convex hulls and Minkowski sums of line segments generated by independent random points.

Isotropic constant

The first class of random polytopes K_N in $\mathbb{R}^n$ for which uniform bounds (independent of n and N) for the isotropic constant were established was the class of Gaussian random polytopes. Klartag and Kozma proved in [**282**] that if $N > n$ and if $G_1, \ldots, G_N$ are independent standard Gaussian random vectors in $\mathbb{R}^n$, then the isotropic constant of the random polytopes $K_N := \operatorname{conv}\{\pm G_1, \ldots, \pm G_N\}$ and $C_N := \operatorname{conv}\{G_1, \ldots, G_N\}$ is bounded by an absolute constant $C > 0$ with probability greater than $1 - Ce^{-cn}$. So far, there is essentially no other successful approach to this question; one can see that the same idea works in the case that the vertices x_j of K_N are distributed according to the uniform measure on an isotropic ψ_2-convex body, leading to a bound depending on its ψ_2-constant b. We chose to present the proof of this result (see Theorem 11.4.9). With a similar method, Alonso-Gutiérrez [**9**] has obtained a uniform bound in the situation where K_N or C_N is spanned by N random points uniformly distributed on the Euclidean sphere S_2^{n-1}. The unconditional case was studied by Dafnis, Giannopoulos and Guédon: Theorem 11.5.1 appears in [**146**].

CHAPTER 12

Central limit problem and the thin shell conjecture

The *central limit problem* is the question of identifying those high-dimensional distributions that have approximately Gaussian marginals. A typical example is given by the random vector $X = (X_1, \ldots, X_n)$ which is distributed uniformly in the cube $Q(n) = [-\sqrt{3}, \sqrt{3}]^n$ (the normalization is so that $\mathrm{Var}(X_j^2) = 1$ for all $1 \leqslant j \leqslant n$). It is well-known that, if the θ_j's satisfy e.g. Lindeberg's condition, then the distribution of

$$\langle X, \theta \rangle = \sum_{j=1}^{n} \theta_j X_j$$

is approximately Gaussian. A second example is given by the ball $D(n) = \sqrt{n+2} B_2^n$. Let $X = (X_1, \ldots, X_n)$ be a random vector which is uniformly distributed in $D(n)$. Here, the random variables X_j are no longer independent. Based on Maxwell's observation that, if n is large enough, then

$$\sigma\left(\{\theta \in S^{n-1} : \theta_j \leqslant t\}\right) \simeq \sqrt{\frac{n}{2\pi}} \int_{-\infty}^{t} \exp(-s^2 n/2)\, ds$$

for all $t \in [-1, 1]$, as well as the symmetry of $D(n)$, one can check that the distribution of $\langle X, \theta \rangle$ is close to the standard normal distribution for any $\theta \in S^{n-1}$.

Let us now assume that μ is an isotropic Borel probability measure on $\mathbb{R}^n$, i.e. normalized so that

$$\mathbb{E}_\mu(x_j) = 0 \quad \text{and} \quad \mathbb{E}_\mu(x_i x_j) = \delta_{ij}, \qquad i, j = 1, \ldots, n.$$

In this chapter we will see that if μ satisfies a *thin shell bound* then the question has an affirmative answer. A more precise version of this statement reads as follows: if

$$\mu\left(\left|\frac{\|x\|_2}{\sqrt{n}} - 1\right| \geqslant \varepsilon\right) \leqslant \varepsilon$$

for some $\varepsilon \in (0, 1/2)$, then, for all directions θ in a subset A of S^{n-1} with $\sigma(A) \geqslant 1 - \exp(-c_1 \sqrt{n})$, we have

$$|\mathbb{P}\left(\langle X, \theta \rangle \leqslant t\right) - \Phi(t)| \leqslant c_2(\varepsilon + n^{-\alpha}) \qquad \text{for all } t \in \mathbb{R},$$

where $\Phi(t)$ is the standard Gaussian distribution function and $c_1, c_2, \alpha > 0$ are absolute constants.

In Section 12.1 we describe the proof of a version of this statement following Bobkov [78]. It is useful to describe here some of the main ideas of the argument. First, one can consider a random vector θ which is uniformly distributed on S^{n-1}, and, given $t \in \mathbb{R}$, define $F_t : S^{n-1} \to \mathbb{R}$ by

$$F_t(\theta) = \mu\left(\langle x, \theta \rangle \leqslant t\right)$$

and observe that
$$F(t) := \int_{S^{n-1}} F_t(\theta)\,d\sigma(\theta) = \mu\left(\|x\|_2 \theta_1 \leqslant t\right).$$
Using our thin shell bound, we see that $\|x\|_2 \theta_1$ is close to a standard normal random variable, and hence
$$\int_{S^{n-1}} F_t(\theta)\,d\sigma(\theta) \simeq \Phi(t).$$
The proof would be complete if we were able to show that, for most $\theta \in S^{n-1}$,
$$F_t(\theta) \simeq F(t).$$
Here, we use the deviation estimates for Lipschitz functions on the sphere to show that F_t is almost constant and approximately equal to its expectation on a large part of S^{n-1}. This proves our claim for a fixed value of t and we conclude the proof by a discretization argument.

The approach that we described is quite general; of course, we have assumed that the dimension is large enough but we have not assumed independence of the coordinate functions $x \mapsto x_j$ and we have not made any symmetry assumptions about μ. Such statements have appeared more than once in the literature (see works of Sudakov [**482**], Diaconis and Freedman [**157**], von Weizsäker [**502**]), but it was the work of Anttila, Ball and Perissinaki [**17**] that made it widely known in the context of isotropic convex bodies or, more generally, log-concave distributions. In Section 12.2 we discuss this last work in detail.

The main question now becomes how to identify those high-dimensional distributions for which a thin shell bound holds true. It is not hard to construct simple examples of isotropic distributions for which this is not the case; as Klartag notes in [**281**], if we write σ_t for the uniform probability measure on the sphere tS^{n-1} then, for $t_1 = \sqrt{n}/2$ and $t_2 = \sqrt{7n}/2$, the isotropic measure
$$\mu = \frac{\sigma_{t_1} + \sigma_{t_2}}{2}$$
does not have Gaussian marginals (which means that it does not satisfy a thin shell bound). As we will see, the assumption of log-concavity guarantees a thin shell bound, and hence an affirmative answer to the central limit problem. In fact, the following quantitative conjecture has been proposed: *There exists an absolute constant $C > 0$ such that, for any $n \geqslant 1$ and any isotropic log-concave random vector X in $\mathbb{R}^n$, one has*
$$\sigma_X^2 := \mathbb{E}\left(\|X\|_2 - \sqrt{n}\right)^2 \leqslant C^2.$$
A variant of σ_X has been introduced by Bobkov and Koldobsky in [**85**]: they define
$$\overline{\sigma}_X^2 = \frac{n\mathrm{Var}(\|X\|_2^2)}{\left(\mathbb{E}\|X\|_2^2\right)^2} = \frac{\mathrm{Var}(\|X\|_2^2)}{n}$$
and they ask if
$$\overline{\sigma}_X^2 \leqslant C^2$$
for some absolute constant $C > 0$ and all isotropic log-concave random vectors. The two questions are equivalent; we briefly discuss the parameter $\overline{\sigma}_X$ in Section 12.3, where we also show that it is uniformly bounded for the class of uniform measures on the ℓ_p^n-balls. This was the first class of measures for which the thin shell conjecture

was confirmed through the "subindependence of coordinates" theorem of Ball and Perissinaki.

In Section 12.4 we give an account of Klartag's positive answer to the thin shell conjecture for the class of unconditional isotropic log-concave random vectors, which is one of the special cases for which this question was fully verified. Klartag proved that if K is an unconditional isotropic convex body in $\mathbb{R}^n$ then

$$\sigma_K^2 := \mathbb{E}_{\mu_K}\left(\|x\|_2 - \sqrt{n}\right)^2 \leqslant C^2,$$

where $C \leqslant 4$ is an absolute positive constant.

In Section 12.5 we present a result of Eldan and Klartag which shows that the thin shell conjecture is stronger than the hyperplane conjecture: there exists an absolute constant $C > 0$ such that

$$L_n \leqslant C\sigma_n$$

for every $n \geqslant 1$, where

$$\sigma_n^2 := \sup_X \mathbb{E}(\|X\|_2 - \sqrt{n})^2,$$

is the supremum of σ_X^2 over all isotropic log-concave random vectors X in $\mathbb{R}^n$.

The next chapter is devoted to a complete proof of the currently best known estimate for the thin shell conjecture.

12.1. From the thin shell estimate to Gaussian marginals

In this section, following Bobkov [78] we consider any isotropic Borel probability measure μ on $\mathbb{R}^n$. That is, we assume that μ has second moments and that

$$\mathbb{E}_\mu(x) = 0 \quad \text{and} \quad \mathbb{E}_\mu(x_i x_j) = \delta_{ij}, \qquad i,j = 1,\ldots,n.$$

Given $\theta \in S^{n-1}$ we set

$$F_\theta(t) = \mu(\{x : \langle x, \theta \rangle \leqslant t\}), \qquad t \in \mathbb{R}$$

and the average distribution function

$$F(t) = \int_{S^{n-1}} F_\theta(t)\, d\sigma(\theta).$$

Our first task will be to estimate the *Lévy distance* $L(F_\theta, F)$ of F_θ and F. Given two distribution functions F and G, their Lévy distance is defined as follows:

$$L(F,G) = \min\{\eta \geqslant 0 : F(t-\eta) - \eta \leqslant G(t) \leqslant F(t+\eta) + \eta \text{ for all } t \in \mathbb{R}\}.$$

Note that $L(F,G) \leqslant \|F - G\|_\infty \leqslant 1$.

THEOREM 12.1.1 (Bobkov). *For every $\delta \in (0,1)$ we have*

$$\sigma(\{\theta : L(F_\theta, F) \geqslant \delta\}) \leqslant c\delta^{-3/2} \exp(-\delta^4 n/8)$$

where $c > 0$ is an absolute constant.

Proof. We will use the following observation. If $g : \mathbb{R} \to \mathbb{R}$ is a Lipschitz continuous function with $\|g\|_{\text{Lip}} \leqslant C$, then using the fact that $\mathbb{E}_\mu(\langle x, y \rangle^2) = \|y\|_2^2$ we see that the function $f(y) = \mathbb{E}_\mu(g(\langle x, y \rangle))$ satisfies

$$|f(y) - f(z)| \leqslant \mathbb{E}_\mu\big(|g(\langle x, y \rangle) - g(\langle x, z \rangle)|\big) \leqslant C\, \mathbb{E}_\mu\big(|\langle x, y - z \rangle|\big)$$
$$\leqslant C\,\|y - z\|_2;$$

therefore, $\|f\|_{\text{Lip}} \leq C$. From the spherical isoperimetric inequality it follows that
$$(12.1.1) \qquad \sigma(\{\theta : |f(\theta) - \mathbb{E}_\sigma(f)| \geq \delta\}) \leq 2\exp(-(n-1)\delta^2/2C^2)$$
for all $\delta > 0$. For any $a \in \mathbb{R}$ we consider the function
$$g_a(t) := \begin{cases} 1 & \text{if } t < a, \\ 1 - \frac{2(t-a)}{\delta} & \text{if } a \leq t \leq a + \delta/2, \\ 0 & \text{if } t > a + \delta/2. \end{cases}$$
Then, $\|g_a\|_{\text{Lip}} = \frac{2}{\delta}$, which implies that the function $f_a(y) = \mathbb{E}_\mu(g_a(\langle x, y \rangle))$ satisfies
$$(12.1.2) \qquad \sigma(\{\theta : |f_a(\theta) - \mathbb{E}_\sigma(f_a)| \geq \delta\}) \leq 2\exp(-(n-1)\delta^4/8)$$
for all $\delta > 0$. Since $\mathbf{1}_{(-\infty,a]} \leq g_a \leq \mathbf{1}_{(-\infty,a+\delta/2]}$, we have
$$(12.1.3) \qquad F_\theta(a) \leq f_a(\theta) \leq F_\theta(a + \delta/2)$$
for all $\theta \in S^{n-1}$. Integrating on S^{n-1} we get
$$(12.1.4) \qquad F(a) \leq \mathbb{E}_\sigma(f_a) \leq F(a + \delta/2).$$
We fix $\delta \in (0, 1)$ and define
$$S_a = \{\theta \in S^{n-1} : |f_a(\theta) - \mathbb{E}_\sigma(f_a)| < \delta\}.$$
We also consider an increasing sequence of real numbers
$$-1/\sqrt{\delta} = a_0 < a_1 < \cdots < a_N = 1/\sqrt{\delta} - \delta$$
satisfying $a_k - a_{k-1} \leq \delta/2$ for all $1 \leq k \leq N$. Let $\theta \in \bigcap_{k=1}^N S_{a_k}$. If $a_{k-1} \leq t < a_k$ then
$$F_\theta(t) \leq F_\theta(a_k) \leq f_{a_k}(\theta) \leq \mathbb{E}_\sigma(f_{a_k}) + \delta \leq F(a_k + \delta/2) + \delta \leq F(t + \delta) + \delta.$$
In the same way we check that $F(t) \leq F_\theta(t + \delta) + \delta$.

Next, observe that since $\int t^2 dF_\theta(t) = \int t^2 dF(t) = \mathbb{E}_\mu(\langle x, \theta \rangle^2) = 1$, Markov's inequality implies
$$(1 - F_\theta(1/\sqrt{\delta})) + F_\theta(-1/\sqrt{\delta}) \leq \delta$$
and
$$(1 - F(1/\sqrt{\delta})) + F(-1/\sqrt{\delta}) \leq \delta.$$
It follows that
$$F_\theta(1/\sqrt{\delta}) \geq 1 - \delta \quad \text{and} \quad F_\theta(-1/\sqrt{\delta}) \leq \delta$$
for all $\theta \in S^{n-1}$, and also
$$F(1/\sqrt{\delta}) \geq 1 - \delta \quad \text{and} \quad F(-1/\sqrt{\delta}) \leq \delta.$$
This guarantees that $F_\theta(t) \leq F(t+\delta)+\delta$ and $F(t) \leq F_\theta(t+\delta)+\delta$ for all $t \notin [a_0, a_N]$ as well.

Combining the above we conclude that
$$\{\theta : L(F_\theta, F) \leq \delta\} \supseteq \bigcap_{k=1}^N S_{a_k},$$
and taking into account (12.1.2) we obtain
$$\sigma(\{\theta : L(F_\theta, F) > \delta\}) \leq \sum_{k=1}^N (1 - \sigma(S_{a_k})) \leq 2N\exp(-(n-1)\delta^4/8).$$

Since $a_N - a_0 \simeq \frac{2}{\sqrt{\delta}}$ it is clear that we can choose the sequence $\{a_k\}$ so that $a_k - a_{k-1} \leqslant \delta/2$ and $N\delta \simeq 1/\sqrt{\delta}$, therefore we may assume that $N \simeq \delta^{-3/2}$. □

The next lemma compares the Lévy distance with the Kantorovich-Rubinstein distance

$$\kappa(F, G) = \|F - G\|_{L_1} = \int_{-\infty}^{\infty} |F(t) - G(t)|\, dt.$$

LEMMA 12.1.2. *Let F and G be two distribution functions such that*

$$\int_{-\infty}^{\infty} t^2 dF(t) = \int_{-\infty}^{\infty} t^2 dG(t) = 1.$$

Then,

$$\kappa(F, G) \leqslant c\sqrt{L(F, G)}$$

where $c > 0$ is an absolute constant.

Proof. We set $\delta = L(F, G)$. Then, $G(t - \delta) - \delta \leqslant F(t) \leqslant G(t + \delta) + \delta$ for all $t \in \mathbb{R}$, which shows that

$$\begin{aligned}
F(t) - G(t) &= F(t) - F(t - \delta) + F(t - \delta) - G(t) \\
&\leqslant F(t) - F(t - \delta) + G(t) + \delta - G(t) \\
&= F(t) - F(t - \delta) + \delta,
\end{aligned}$$

and similarly, $G(t) - F(t) \leqslant G(t) - G(t - \delta) + \delta$. Therefore,

$$|F(t) - G(t)| \leqslant (F(t) - F(t - \delta)) + (G(t) - G(t - \delta)) + \delta$$

for all $t \in \mathbb{R}$. Then, for any $M > 0$ we have

$$\int_{-M}^{M} |F(t) - G(t)|\, dt \leqslant 2(M + 1)\delta.$$

On the other hand, $F(-t) + (1 - F(t)) \leqslant 1/t^2$ and $G(-t) + (1 - G(t)) \leqslant 1/t^2$ for all $|t| > M$. Therefore,

$$\int_{-\infty}^{-M} |F(t) - G(t)|\, dt + \int_{M}^{\infty} |F(t) - G(t)|\, dt \leqslant 2/M.$$

Combining the above we get

$$\kappa(F, G) \leqslant 2(M + 1)\delta + 2/M,$$

and making the optimal choice for M we see that $\kappa(F, G) \leqslant 4\sqrt{\delta} + 2\delta \leqslant 6\sqrt{\delta} = 6\sqrt{L(F, G)}$. □

In our case, Lemma 12.1.2 shows that $\kappa(F_\theta, F) \leqslant 6\sqrt{L(F_\theta, F)}$ and then Theorem 12.1.1 implies the next

THEOREM 12.1.3. *For every $\delta \in (0, 1)$ we have*

$$\sigma(\{\theta : \kappa(F_\theta, F) \geqslant \delta\}) \leqslant c_1 \delta^{-3} \exp(-c_2 \delta^8 n)$$

where $c_1, c_2 > 0$ are absolute constants.

Our next observation is that the Lévy distance $L(F, G)$ of two distribution functions can be also compared with the uniform distance $\|F - G\|_\infty$ under some assumption on the boundedness of the derivative of, say, G:

LEMMA 12.1.4. *Let F and G be two distribution functions. If $\|G'\|_\infty < \infty$ then*
$$\|F - G\|_\infty \leqslant (1 + \|G'\|_\infty) L(F, G).$$

Proof. We set $\delta = L(F, G)$ and observe that, for every $t \in \mathbb{R}$,
$$F(t) - G(t) \leqslant (F(t) - G(t + \delta)) + |G(t + \delta) - G(t)| \leqslant \delta + \|G'\|_\infty \delta$$
and
$$G(t) - F(t) \leqslant (G(t) - G(t - \delta)) + (G(t - \delta) - F(t)) \leqslant \|G'\|_\infty \delta + \delta.$$
This shows that $|F(t) - G(t)| \leqslant (1 + \|G'\|_\infty)\delta$. □

In our setting, we can use Lemma 12.1.4 starting from the equation
$$F(t) = (\mu \otimes \sigma)(\{(x, \theta) : \|x\|_2 \theta_1 \leqslant t\}),$$
where θ_1 is the first coordinate of $\theta \in S^{n-1}$. We write Φ_n and φ_n for the distribution function and the density of $\sqrt{n}\theta_1$ respectively. Then, for all $t \in \mathbb{R}$ we have
$$F(t) = \mathbb{E}_\mu\left(\Phi_n\left(\frac{\sqrt{n}t}{\|x\|_2}\right)\right) \text{ and } F'(t) = \mathbb{E}_\mu\left(\frac{\sqrt{n}}{\|x\|_2}\varphi_n\left(\frac{\sqrt{n}t}{\|x\|_2}\right)\right).$$

Since φ_n attains its maximum at zero, we see that
$$\|F'\|_\infty = \varphi_n(0)\mathbb{E}_\mu\left(\frac{\sqrt{n}}{\|x\|_2}\right).$$
It is also easy to check that
$$\varphi_n(0) = \left(2\sqrt{n}\int_0^{\pi/2} \cos^{n-2} s\, ds\right)^{-1} \leqslant \frac{1}{2}$$
for all $n \geqslant 1$. This shows that $\|F'\|_\infty \leqslant \mathbb{E}_\mu\left(\frac{\sqrt{n}}{2\|x\|_2}\right)$ and from Lemma 12.1.4 we get
$$\|F_\theta - F\|_\infty \leqslant c\mathbb{E}_\mu\left(\frac{\sqrt{n}}{\|x\|_2}\right) L(F_\theta, F).$$

As a consequence of Theorem 12.1.1 we have

THEOREM 12.1.5. *For every $\delta > 0$ we have*
$$\sigma\left(\left\{\theta : \|F_\theta - F\|_\infty \geqslant c_1 \delta \mathbb{E}_\mu\left(\frac{\sqrt{n}}{\|x\|_2}\right)\right\}\right) \leqslant c_2 \delta^{-3/2} \exp(-c_3 \delta^4 n)$$
where $c_1, c_2, c_3 > 0$ are absolute constants.

Now, we are able to show that most of the random variables $x \mapsto \langle x, \theta \rangle$ are very close to a standard Gaussian random variable under the following general hypothesis which states that the Euclidean norm concentrates near the value $\sqrt{n}$.

DEFINITION 12.1.6 (concentration hypothesis). Let $0 < \varepsilon_n < \frac{1}{2}$. We say that μ satisfies the *concentration hypothesis* with constant ε_n if
$$\mu\left(\left|\frac{\|x\|_2}{\sqrt{n}} - 1\right| \geqslant \varepsilon_n\right) \leqslant \varepsilon_n.$$

12.1. FROM THE THIN SHELL ESTIMATE TO GAUSSIAN MARGINALS

THEOREM 12.1.7 (Bobkov). *Let μ be an isotropic probability measure on $\mathbb{R}^n$. Assume that*

$$\mu\left(\left|\frac{\|x\|_2}{\sqrt{n}} - 1\right| \geq \varepsilon_n\right) \leq \varepsilon_n$$

for some $0 < \varepsilon_n < 1/3$. Then, for every $\delta > 0$ we have

$$\sigma\left(\left\{\theta : \sup_{t \in \mathbb{R}} |F_\theta(t) - \Phi(t)| \geq 2\delta + \frac{6}{\sqrt{n}} + 4\varepsilon_n\right\}\right) \leq c_1 \delta^{-3/2} \exp(-c_2 \delta^4 n)$$

where Φ is the standard Gaussian distribution function and $c_1, c_2 > 0$ are absolute constants.

Proof. We shall use the following two claims (their proof is given at the end of this section):

(i) For every $r > 0$ and $t \in \mathbb{R}$,

(12.1.5) $$|\Phi_n(rt) - \Phi_n(t)| \leq |r - 1|.$$

(ii) For all $n \geq 1$,

(12.1.6) $$\|\Phi_n - \Phi\|_\infty \leq \frac{4}{\sqrt{n}}.$$

For any $x \in \mathbb{R}^n$ with $\left|\frac{\|x\|_2}{\sqrt{n}} - 1\right| \leq \varepsilon_n$ and for any $t \in \mathbb{R}$, using (12.1.5) we write

(12.1.7) $$\left|\Phi_n\left(\frac{t\sqrt{n}}{\|x\|_2}\right) - \Phi_n(t)\right| \leq \left|\frac{\sqrt{n}}{\|x\|_2} - 1\right| \leq \frac{\varepsilon_n}{1 - \varepsilon_n} \leq \frac{3\varepsilon_n}{2}.$$

This implies that

(12.1.8) $$\left|\Phi_n\left(\frac{t\sqrt{n}}{\|x\|_2}\right) - \Phi(t)\right| \leq \|\Phi_n - \Phi\|_\infty + \frac{3\varepsilon_n}{2}.$$

For any $x \in \mathbb{R}^n$ with $\left|\frac{\|x\|_2}{\sqrt{n}} - 1\right| \geq \varepsilon_n$ and for any $t \in \mathbb{R}$, we use the trivial bound

(12.1.9) $$\left|\Phi_n\left(\frac{t\sqrt{n}}{\|x\|_2}\right) - \Phi(t)\right| \leq 1.$$

Recall that

$$F(t) = \mathbb{E}_\mu\left(\Phi_n\left(\frac{\sqrt{n}t}{\|x\|_2}\right)\right).$$

Therefore, using the concentration hypothesis, from (12.1.8) and (12.1.9) we get

(12.1.10) $$\|F - \Phi\|_\infty \leq \mathbb{E}_\mu\left(\sup_t \left|\Phi_n\left(\frac{t\sqrt{n}}{\|x\|_2}\right) - \Phi(t)\right|\right) \leq \|\Phi_n - \Phi\|_\infty + \frac{5\varepsilon_n}{2},$$

and (12.1.6) shows that

(12.1.11) $$L(F, \Phi) \leq \|F - \Phi\|_\infty \leq \frac{4}{\sqrt{n}} + \frac{5\varepsilon_n}{2}.$$

Then, we can apply Theorem 12.1.1 to get

$$L(F_\theta, \Phi) \leq \delta + \frac{4}{\sqrt{n}} + \frac{5\varepsilon_n}{2}$$

with probability greater than $1 - c\delta^{-3/2}\exp(-\delta^4 n/8)$, where $c > 0$ is an absolute constant. Since $\|\Phi'\|_\infty = \frac{1}{\sqrt{2\pi}}$, Lemma 12.1.4 shows that

$$\|F_\theta - \Phi\|_\infty \leq \left(1 + \frac{1}{\sqrt{2\pi}}\right)\left(\delta + \frac{4}{\sqrt{n}} + \frac{5\varepsilon_n}{2}\right) \leq 2\delta + \frac{6}{\sqrt{n}} + 4\varepsilon_n$$

with probability greater than $1 - c\delta^{-3/2}\exp(-\delta^4 n/8)$ as claimed. □

Proof of (12.1.5). Note that $\Phi'_n = \varphi_n$ is even, continuous and decreasing on $[0, \infty)$. For $r \geq 1$ we need to show that $\Phi_n(rt) - \Phi_n(t) \leq r - 1$. This holds true as an equality at $r = 1$, and since the left hand side is a concave function of r we only need to show that $\Phi'_n(1) = t\phi_n(t) \leq 1$; this is easily checked: in fact, we have

$$t\phi_n(t) \leq \int_0^t \phi_n(u)\,du \leq \frac{1}{2}.$$

For $0 \leq r \leq 1$ we need to show that $\Phi_n(t) - \Phi_n(rt) \leq 1 - r$. This holds true at $r = 0$ and $r = 1$, and since the left hand side is a convex function of r the claim follows. □

Proof of (12.1.6). We may assume that $n \geq 16$, otherwise there is nothing to prove. Let $G = (g_1, \ldots, g_n)$ be a standard Gaussian random vector in $\mathbb{R}^n$. Observe that

$$\Phi_n(t) = \mathbb{P}\left(g_1 \leq \frac{\|G\|_2 t}{\sqrt{n}}\right) \quad \text{and} \quad \Phi(t) = \mathbb{P}(g_1 \leq t).$$

Therefore,

(12.1.12) $$|\Phi_n(t) - \Phi(t)| = \mathbb{P}\left(t \leq g_1 \leq \frac{\|G\|_2 t}{\sqrt{n}}\right) + \mathbb{P}\left(\frac{\|G\|_2 t}{\sqrt{n}} \leq g_1 \leq t\right).$$

We distinguish two cases:

(i) If $t \geq \sqrt{n}$ then the second term of the right hand side is equal to zero and we have

(12.1.13) $$|\Phi_n(t) - \Phi(t)| = \mathbb{P}\left(t \leq g_1 \leq \frac{\|G\|_2 t}{\sqrt{n}}\right) \leq 1 - \Phi(\sqrt{n}) < \frac{1}{\sqrt{n}}.$$

(ii) If $t < \sqrt{n}$ then (12.1.12) takes the form

(12.1.14) $$|\Phi_n(t) - \Phi(t)| = \mathbb{E}\left|\Phi\left(\frac{\|\tilde{G}\|_2 t}{\sqrt{n - t^2}}\right) - \Phi(t)\right|,$$

where $\tilde{G} = (g_2, \ldots, g_n)$. Note that

$$\left|\Phi\left(\frac{\|\tilde{G}\|_2 t}{\sqrt{n - t^2}}\right) - \Phi(t)\right| \leq \frac{1}{2}e^{-t^2/2} + \frac{1}{2}e^{-t^2\|\tilde{G}\|_2^2/2(n-t^2)},$$

and hence

$$|\Phi_n(t) - \Phi(t)| \leq \frac{1}{2}e^{-t^2/2} + \frac{1}{2}\left(\mathbb{E}e^{-t^2 g_2^2/2(n-t^2)}\right)^{n-1}$$

$$\leq \frac{1}{2}e^{-t^2/2} + \frac{1}{2}\left(1 - \frac{t^2}{n}\right)^{\frac{n-1}{2}}.$$

If $t^2 \geq \sqrt{n}/2$ then the right hand side is bounded by

$$\frac{1}{2}e^{-\sqrt{n}/4} + \frac{1}{2}\left(1 - \frac{1}{2\sqrt{n}}\right)^{\frac{n-1}{2}} \leq \frac{2}{\sqrt{n}}.$$

Finally, if $t^2 < \sqrt{n}/2$ then we use (12.1.5) to write

$$\left|\Phi\left(\frac{\|\tilde{G}\|_2 t}{\sqrt{n - t^2}}\right) - \Phi(t)\right| \leq \frac{|\|\tilde{G}\|_2 - \sqrt{n - t^2}|}{\sqrt{n - t^2}} \leq \frac{|\|\tilde{G}\|_2^2 - (n - t^2)|}{n - t^2}$$

$$\leq \frac{|\|\tilde{G}\|_2^2 - (n - 1)|}{n - t^2} + \frac{|t^2 - 1|}{n - t^2} \leq \frac{2|\|\tilde{G}\|_2^2 - (n - 1)|}{n} + \frac{1}{\sqrt{n}},$$

using that $t^2 < \sqrt{n}/2 \leqslant n/2$. Since
$$\mathbb{E}\,|\,\|\tilde{G}\|_2^2 - (n-1)| \leqslant \sqrt{\operatorname{Var}(\|\tilde{G}\|_2^2)} = \sqrt{2n},$$
the claim follows. □

12.2. The log-concave case

In this section we describe the work of Anttila, Ball and Perissinaki. We consider an isotropic symmetric convex body K in $\mathbb{R}^n$, which we view as a probability space with the Lebesgue measure μ_K on K. For every $\theta \in S^{n-1}$ we consider the random variable $X_\theta(x) = \langle x, \theta \rangle$. Since K is isotropic, we have
$$\mathbb{E}(X_\theta) = 0 \quad \text{and} \quad \operatorname{Var}(X_\theta) = L_K^2$$
for every $\theta \in S^{n-1}$. We will show that, under the concentration hypothesis with some constant ε, most of these random variables have to be very close to a Gaussian random variable γ with mean 0 and variance L_K^2.

We denote by $g(s)$ the density of γ and for simplicity we write $g_\theta(s)$ for the density of X_θ. Note that
$$g_\theta(s) = f_{K,\theta}(s) = |K \cap \{\langle x, \theta \rangle = s\}|$$
and
$$g(s) = \frac{1}{\sqrt{2\pi}L_K} \exp\left(-\frac{s^2}{2L_K^2}\right).$$

THEOREM 12.2.1 (Anttila-Ball-Perissinaki). *Let K be an isotropic symmetric convex body in $\mathbb{R}^n$ which satisfies the concentration hypothesis with constant ε for some $0 < \varepsilon < \frac{1}{2}$. Then, for every $\delta > 0$,*
$$\sigma\left(\left\{\theta : \left|\int_{-t}^{t} g_\theta(s)\,ds - \int_{-t}^{t} g(s)\,ds\right| \leqslant \delta + 4\varepsilon + \frac{c_1}{\sqrt{n}} \text{ for every } t \in \mathbb{R}\right\}\right)$$
$$\geqslant 1 - n\,e^{-c_2\delta^2 n},$$
where $c_1, c_2 > 0$ are absolute constants.

The dependence on δ is better than the one in the previous section; this is due to the log-concavity of the uniform measure on K (that will be used in the proof). The argument is divided into three steps. We first consider the average function
$$A_K(t) = \int_{S^{n-1}} \int_{-t}^{t} g_\theta(s)\,ds\,d\sigma(\theta)$$
and show that $A_K(t)$ is close to $\int_{-t}^{t} g(s)\,ds$ for every $t > 0$:

THEOREM 12.2.2 (first step). *Let K be an isotropic convex body in $\mathbb{R}^n$. If K satisfies the concentration hypothesis with constant ε then*
$$\left|A_K(t) - \int_{-t}^{t} g(s)\,ds\right| \leqslant 4\varepsilon + \frac{c_1}{\sqrt{n}}$$
for every $t > 0$.

The proof of Theorem 12.2.2 has many similarities with the one that we presented in the previous section; in particular, convexity does not play an essential role in this step. So, we concentrate in the second step where we use the estimate of Theorem 12.2.2 for the average $A_K(t)$ to obtain a similar estimate for "most directions" $\theta \in S^{n-1}$. More precisely, we prove:

THEOREM 12.2.3 (second step). *Let K be an isotropic symmetric convex body in $\mathbb{R}^n$. If K satisfies the concentration hypothesis with constant ε then, for every $t > 0$ and $\delta > 0$,*

$$\sigma\left(\left\{\theta : \left|\int_{-t}^{t} g_\theta(s)ds - \int_{-t}^{t} g(s)ds\right| \geqslant \delta + 4\varepsilon + \frac{c_3}{\sqrt{n}}\right\}\right) \leqslant 2e^{-c_4\delta^2 n},$$

where $c_3, c_4 > 0$ are absolute constants.

Note the concentration hypothesis is used in the previous step; here, we use log-concavity. The idea is to show that $\int_{-t}^{t} g_\theta(s)ds$ is the (restriction on S^{n-1} of the) radial function of a symmetric convex body in $\mathbb{R}^n$ and then use the spherical isoperimetric inequality in the context of Lipschitz continuous functions on the sphere. We make use of a classical inequality of Busemann.

THEOREM 12.2.4 (Busemann). *Let K be a symmetric convex body in $\mathbb{R}^n$. We define $F : \mathbb{R}^n \to \mathbb{R}$ with $F(0) = 0$ and*

$$F(x) = \frac{\|x\|_2}{|K \cap x^\perp|}$$

for all $x \neq 0$. Then, F is a norm.

Proof. We only need to check the triangle inequality. It suffices to prove that if x and y are two linearly independent vectors in $\mathbb{R}^n$ then

$$F(\lambda x + \mu y) \leqslant \lambda F(x) + \mu F(y)$$

for all $\lambda, \mu > 0$ with $\lambda + \mu = 1$. We set $E = \operatorname{span}\{x, y\}$ and define $f : E \to \mathbb{R}$ by $f(w) = |K \cap (w + E^\perp)|$. Note that f is even and log-concave. Applying Theorem 2.5.5 with $p = 1$ we get
(12.2.1)
$$\left(\int_0^\infty f(r(\lambda x + \mu y))\, dr\right)^{-1} \leqslant \lambda \left(\int_0^\infty f(rx)\, dr\right)^{-1} + \mu \left(\int_0^\infty f(ry)\, dr\right)^{-1}.$$

Note that for every $w \in E \setminus \{0\}$ we have

$$\int_0^\infty f(rw)\, dr = \frac{1}{\|w\|_2} \int_0^\infty f\left(\frac{rw}{\|w\|_2}\right) dr$$

$$= \frac{1}{\|w\|_2} \int_0^\infty \left|K \cap \left(\frac{rw}{\|w\|_2} + E^\perp\right)\right| dr$$

$$= \frac{1}{2\|w\|_2} |K \cap \operatorname{span}\{w, E^\perp\}|$$

$$= \frac{|K \cap w^\perp|}{2\|w\|_2} = \frac{1}{2F(w)}.$$

Going back to (12.2.1) we readily see that $F(\lambda x + \mu y) \leqslant \lambda F(x) + \mu F(y)$. $\square$

12.2. THE LOG-CONCAVE CASE

PROPOSITION 12.2.5. *Let K be a symmetric convex body in $\mathbb{R}^n$. Fix $t > 0$ and define*
$$\|x\|_t = \frac{\|x\|_2}{\int_{-t}^{t} g_{\frac{x}{\|x\|_2}}(s) ds}.$$
Then, $\|\cdot\|_t$ is a norm on $\mathbb{R}^n$.

Proof. Recall that $g_\theta(s) = |K \cap \{\langle x, \theta \rangle = s\}|$ for every $\theta \in S^{n-1}$. We define
$$v(x, t) = \int_{-t}^{t} g_{\frac{x}{\|x\|_2}}(s) ds$$
and prove that for all $x, y \in \mathbb{R}^n$,
$$\frac{1}{2}\left(\frac{\|x\|_2}{v(x,t)} + \frac{\|y\|_2}{v(y,t)}\right) \geq \frac{\left\|\frac{x+y}{2}\right\|_2}{v\left(\frac{x+y}{2}, t\right)}.$$

We may clearly assume that x and y are linearly independent. Consider the convex body $K' = K \times [-1, 1]$ in $\mathbb{R}^{n+1}$. Theorem 12.2.4 shows that $\frac{\|\theta\|_2}{|K' \cap \theta^\perp|}$ defines a norm on $\mathbb{R}^{n+1}/\{0\}$. That is,
$$\frac{1}{2}\left(\frac{\|\theta\|_2}{|K' \cap \theta^\perp|} + \frac{\|\phi\|_2}{|K' \cap \phi^\perp|}\right) \geq \frac{\left\|\frac{\theta+\phi}{2}\right\|_2}{|K' \cap \left(\frac{\theta+\phi}{2}\right)^\perp|}$$
for all linearly independent $\theta, \phi \in \mathbb{R}^{n+1}$.

Let $r \in (0, 1)$ be defined by the equation $tr = \sqrt{1 - r^2}$. We observe that if $z \in \mathbb{R}^n \setminus \{0\}$ and
$$u(z) = \left(r \frac{z}{\|z\|_2}, \sqrt{1 - r^2}\right),$$
then the projection of $K' \cap u(z)^\perp$ onto the first n coordinates is $\{w \in K : |\langle w, z \rangle| \leq t\|z\|_2\}$. It follows that
$$v(z, t) = \sqrt{1 - r^2}|K' \cap u(z)^\perp|.$$

We define $\eta(z) = \|z\|_2 u(z)$. Then, $|K' \cap u(z)^\perp| = |K' \cap \eta(z)^\perp|$ and $\|\eta(z)\|_2 = \|z\|_2$. If we set $\theta = \eta(x)$ and $\phi = \eta(y)$, Busemann's inequality shows that

(12.2.2) $$\frac{1}{2}\left(\frac{\|x\|_2}{v(x,t)} + \frac{\|y\|_2}{v(y,t)}\right) \geq \frac{1}{\sqrt{1-r^2}} \frac{\left\|\frac{\theta+\phi}{2}\right\|_2}{|K' \cap \left(\frac{\theta+\phi}{2}\right)^\perp|}.$$

Observe that
$$\frac{\theta + \phi}{2} = \left(r \frac{x+y}{2}, \sqrt{1-r^2}\frac{\|x\|_2 + \|y\|_2}{2}\right)$$
$$= \frac{\|x+y\|_2}{2}\left(r \frac{x+y}{\|x+y\|_2}, \sqrt{1-r^2}\frac{\|x\|_2 + \|y\|_2}{\|x+y\|_2}\right).$$

Then, the projection of $K' \cap \left(\frac{\theta+\phi}{2}\right)^\perp$ onto the first n coordinates is a strip perpendicular to $\frac{x+y}{2}$, with width
$$s = \frac{\|x\|_2 + \|y\|_2}{\|x+y\|_2} t,$$

and this gives
$$\frac{v\left(\frac{x+y}{2},s\right)}{|K'\cap\left(\frac{\theta+\phi}{2}\right)^\perp)|}=\frac{\sqrt{1-r^2}}{2}\frac{\|x\|_2+\|y\|_2}{\left\|\frac{\theta+\phi}{2}\right\|_2}.$$

So, (12.2.2) takes the form

(12.2.3) $$\frac{1}{2}\left(\frac{\|x\|_2}{v(x,t)}+\frac{\|y\|_2}{v(y,t)}\right)\geq\frac{1}{2}\frac{\|x\|_2+\|y\|_2}{v\left(\frac{x+y}{2},s\right)}.$$

Observe that if $a\geq 1$, then for every $z\in\mathbb{R}^n$ we have
$$v(z,at)\leq a\,v(z,t)$$
(this follows from the fact that $g_\theta(as)\leq g_\theta(s)$ for all $\theta\in S^{n-1}$ and $s>0$). It follows that
$$\frac{\|x\|_2+\|y\|_2}{2v\left(\frac{x+y}{2},s\right)}=\frac{\|x\|_2+\|y\|_2}{2v\left(\frac{x+y}{2},\frac{\|x\|_2+\|y\|_2}{\|x+y\|_2}t\right)}\geq\frac{\|x\|_2+\|y\|_2}{2\frac{\|x\|_2+\|y\|_2}{\|x+y\|_2}v\left(\frac{x+y}{2},t\right)}=\frac{\left\|\frac{x+y}{2}\right\|_2}{v\left(\frac{x+y}{2},t\right)}.$$

Going back to (12.2.3) we conclude the proof. □

LEMMA 12.2.6. *Let K be a symmetric convex body of volume 1 in $\mathbb{R}^n$. For every $\theta\in S^{n-1}$ and every $t>0$,*
$$\int_t^\infty g_\theta(s)ds\leq\frac{1}{2}e^{-2g_\theta(0)t}.$$

Proof. Consider the function
$$H(t)=\int_t^\infty g_\theta(s)ds=\int_0^\infty \mathbf{1}_{[t,\infty)}(s)g_\theta(s)ds.$$

Using the fact that g_θ is log-concave and applying the Prékopa-Leindler inequality, we may easily check that H is log-concave. It follows that
$$(\log H)(t)-(\log H)(0)\leq (\log H)'(0)t$$
for every $t>0$. Observe that $H(0)=1/2$ by the symmetry of K, and
$$(\log H)'(0)=-\frac{g_\theta(0)}{H(0)}=-2g_\theta(0).$$

It follows that
$$H(t)\leq H(0)\exp\left((\log H)'(0)t\right)=\frac{1}{2}\exp(-2g_\theta(0)t),$$
as stated in the lemma. □

LEMMA 12.2.7. *Let K be an isotropic symmetric convex body in $\mathbb{R}^n$. For every $t>0$, the norm*
$$\|x\|_t=\frac{\|x\|_2}{\int_{-t}^t g_{\frac{x}{\|x\|_2}}(s)\,ds}$$
satisfies
$$a\|x\|_2\leq\|x\|_t\leq b\|x\|_2$$
for every $x\in\mathbb{R}^n$, where a,b are two positive constants such that $a\geq 1$ and $b/a\leq c$ for some absolute constant $c>0$.

Proof. Since K is isotropic, we know that $g_\theta(0) \simeq L_K^{-1}$ for every $\theta \in S^{n-1}$. Then, by the symmetry of K we have
$$\int_{-t}^{t} g_\theta(s)\,ds \leqslant \min\{2t\, g_\theta(0), 1\} \leqslant \min\left\{\frac{c_1 t}{L_K}, 1\right\}.$$
Also, Lemma 12.2.6 shows that
$$\int_{-t}^{t} g_\theta(s)\,ds = 1 - 2\int_{t}^{\infty} g_\theta(s)\,ds \geqslant 1 - e^{-2g_\theta(0)t} \geqslant 1 - e^{-\frac{c_2 t}{L_K}}.$$
We easily check that
$$1 - e^{-\frac{c_2 t}{L_K}} \geqslant \frac{c_2 t}{2 L_K}$$
if $c_2 t \leqslant L_K$. In any case,
$$\int_{-t}^{t} g_\theta(s)\,ds \geqslant \min\left\{\frac{c_3 t}{L_K}, 1 - e^{-1}\right\}.$$
In other words
$$a := \max\left\{\frac{L_K}{c_4 t}, 1\right\} \leqslant \|\theta\|_t \leqslant b := \max\left\{\frac{L_K}{c_3 t}, \frac{e}{e-1}\right\}$$
for every $\theta \in S^{n-1}$. Finally, observe that $a \geqslant 1$ and b/a is bounded independently of t and L_K. $\square$

Proof of Theorem 12.2.3. Let $t > 0$ and $\delta > 0$ be fixed. Recall that (as in (12.1.1)) by the spherical isoperimetric inequality, for every C-Lipschitz function $f: S^{n-1} \to \mathbb{R}$ and any $\delta > 0$ we have
$$\sigma(\{\theta : |f(\theta) - \mathbb{E}_\sigma(f)| \geqslant \delta\}) \leqslant 2\exp(-(n-1)\delta^2/2C^2).$$
We shall apply this to the function $f(\theta) = \int_{-t}^{t} g_\theta(s)\,ds$. Observe that
$$\left|\int_{-t}^{t} g_\theta(s)\,ds - \int_{-t}^{t} g_\phi(s)\,ds\right| = \left|\frac{1}{\|\theta\|_t} - \frac{1}{\|\phi\|_t}\right|$$
$$\leqslant \frac{\|\theta - \phi\|_t}{\|\theta\|_t \|\phi\|_t} \leqslant \frac{b}{a^2}\|\theta - \phi\|_2 \leqslant c\|\theta - \phi\|_2,$$
where c is the absolute constant in Lemma 12.2.7. Also, note that $\mathbb{E}(f) = A_K(t)$. It follows that
$$\sigma\left(\left\{\theta : \left|\int_{-t}^{t} g_\theta(s)\,ds - A_K(t)\right| \geqslant \delta\right\}\right) \leqslant 2\exp\left(-\frac{(n-1)\delta^2}{2c^2}\right).$$
Combining this with Theorem 12.2.2 we get
$$\sigma\left(\left\{\theta : \left|\int_{-t}^{t} g_\theta(s)ds - \int_{-t}^{t} g(s)ds\right| \geqslant \delta + 4\varepsilon + \frac{c_1}{\sqrt{n}}\right\}\right) \leqslant 2\exp\left(-c_2 \delta^2 n\right),$$
where $c_2 = 1/(2c^2)$. $\square$

The end of the proof of Theorem 12.2.1 is quite similar to that of Theorem 12.1.1. First, fix some $\theta \in S^{n-1}$. Since
$$g_\theta(s) \leqslant g_\theta(0) \leqslant \frac{c_1}{L_K}$$

and
$$g(s) = \frac{1}{\sqrt{2\pi}L_K}\exp(-s^2/(2L_K^2)) \leqslant \frac{1}{\sqrt{2\pi}L_K}$$
for every $s > 0$, the function
$$H(t) = \left|\int_{-t}^{t} g_\theta(s)ds - \int_{-t}^{t} g(s)ds\right|$$
is Lipschitz continuous with constant $d \leqslant \frac{c_2}{L_K}$, where $c_2 > 0$ is an absolute constant.

Also, there is an absolute constant $c_3 > 0$ such that $H(t) \leqslant 1/\sqrt{n}$ for every $t \geqslant c_3 L_K \log n$. This is a consequence of the equality
$$H(t) = 2\left|\int_t^\infty g_\theta(s)ds - \int_t^\infty g(s)ds\right|$$
and of Lemma 12.2.6: if $c_3 > 0$ is chosen large enough, when $t \geqslant c_3 L_K \log n$ we have
$$\max\left\{\int_t^\infty g_\theta(s)ds, \int_t^\infty g(s)ds\right\} < \frac{1}{2\sqrt{n}}.$$
Define $t_k = k\alpha$, where $\alpha = L_K/\sqrt{n}$ and $k = 1,\ldots,k_0 = [c_3\sqrt{n}\log n] + 1$. From Theorem 12.2.3, for every $\delta > 0$ we have
$$\sigma(A) \leqslant 2c_3\sqrt{n}(\log n)e^{-c_6\delta^2 n},$$
where
$$A = \left\{\theta : \exists k \leqslant k_0 \text{ s.t. } \left|\int_{-t_k}^{t_k} g_\theta(s)ds - \int_{-t_k}^{t_k} g(s)ds\right| \geqslant \delta + 4\varepsilon + \frac{c_5}{\sqrt{n}}\right\}$$
and $c_5, c_6 > 0$ are absolute constants. If θ is not in A, then
$$H(t_k) \leqslant \delta + 4\varepsilon + \frac{c_5}{\sqrt{n}}$$
for all $k = 1,\ldots,k_0$. Since H is $\frac{c_2}{L_K}$-Lipschitz, we get a similar estimate for $H(t)$, $t \in [0, c_3 L_K \log n]$. Finally, if $t > c_3 L_K \log n$, we know that $H(t) < 1/\sqrt{n}$. This proves Theorem 12.2.1.

12.3. The thin shell conjecture

Let K be an isotropic convex body in $\mathbb{R}^n$. In this section we study two parameters of K. The first one is defined by
$$\overline{\sigma}_K^2 = \frac{\mathrm{Var}(\|x\|_2^2)}{nL_K^4} = \frac{n\mathrm{Var}(\|x\|_2^2)}{\left(\mathbb{E}\|x\|_2^2\right)^2}.$$
Note that the last expression is invariant under homotheties, and hence, easier to compute. The parameter $\overline{\sigma}_K$ was introduced in [**85**] by Bobkov and Koldobsky, who also asked if it is uniformly bounded.

Variance hypothesis. *There exists an absolute constant $C > 0$ such that $\overline{\sigma}_K^2 \leqslant C$ for every isotropic convex body.*

More generally, for any isotropic log-concave random vector X in $\mathbb{R}^n$ we define
$$\overline{\sigma}_X^2 = \frac{1}{n}\mathrm{Var}(\|X\|_2^2) \quad \text{and} \quad \sigma_X^2 = \mathbb{E}(\|X\|_2 - \sqrt{n})^2.$$

12.3. THE THIN SHELL CONJECTURE

We also set
$$\overline{\sigma}_n^2 = \sup_X \overline{\sigma}_X^2 \quad \text{and} \quad \sigma_n^2 = \sup_X \sigma_X^2,$$
where the supremum is over all isotropic log-concave random vectors X in $\mathbb{R}^n$. The next lemma shows that these two parameters are equivalent.

LEMMA 12.3.1. *For every $n \geq 1$ one has*
$$(12.3.1) \qquad \sigma_n^2 \leq \overline{\sigma}_n^2 \leq C\sigma_n^2,$$
where $C > 0$ is an absolute constant.

Proof. We first observe that, for any X,
$$\mathbb{E}(\|X\|_2 - \sqrt{n})^2 \leq \frac{1}{n}\mathbb{E}\big[(\|X\|_2 - \sqrt{n})^2(\|X\|_2 + \sqrt{n})^2\big]$$
$$= \frac{1}{n}\mathbb{E}(\|X\|_2^2 - n)^2.$$
This proves that $\sigma_X^2 \leq \overline{\sigma}_X^2$, and the left hand side inequality in (12.3.1) follows. For the right hand side inequality, we use the bound
$$\mathbb{E}\big(\|X\|_2^4 \mathbf{1}_{\{\|X\|_2 \geq C\sqrt{n}\}}\big) \leq C\exp(-\sqrt{n}),$$
which follows, after an application of the Cauchy-Schwarz inequality, from the deviation inequality of Paouris (see Chapter 5) and the fact that, by Borell's lemma, $\mathbb{E}(\|X\|_2^k) \leq C(k/2)^k[\mathbb{E}(\|X\|_2^2)]^{k/2} = C(k/2)^k n^{k/2}$ for any $k \geq 2$. Since
$$\|X\|_2^2 - n = (\|X\|_2 - \sqrt{n})(\|X\|_2 + \sqrt{n}),$$
we may write
$$\mathbb{E}(\|X\|_2^2 - n)^2 = \mathbb{E}([\|X\|_2^2 - n)^2 \mathbf{1}_{\{\|X\|_2 \leq C\sqrt{n}\}}] + \mathbb{E}[(\|X\|_2^2 - n)^2 \mathbf{1}_{\{\|X\|_2 \geq C\sqrt{n}\}}]$$
$$\leq (C+1)^2 n\mathbb{E}(\|X\|_2 - \sqrt{n})^2 + \mathbb{E}[\|X\|_2^4 \mathbf{1}_{\{\|X\|_2 \geq C\sqrt{n}\}}],$$
and the right hand side inequality in (12.3.1) follows. $\square$

Lemma 12.3.1 shows that the variance hypothesis is equivalent to the question if σ_n is bounded:

CONJECTURE 12.3.2 (thin shell conjecture). *There exists an absolute constant $C > 0$ such that, for any $n \geq 1$ and any isotropic log-concave measure μ on $\mathbb{R}^n$, one has*
$$\sigma_\mu^2 := \int_{\mathbb{R}^n} \big(\|x\|_2 - \sqrt{n}\big)^2 d\mu(x) \leq C^2.$$

A simple computation shows that if $K = B_2^n$ then
$$\mathbb{E}\|x\|_2^4 = \frac{n}{n+4} \quad \text{and} \quad \mathbb{E}\|x\|_2^2 = \frac{n}{n+2}.$$

Therefore,
$$(12.3.2) \qquad \overline{\sigma}_{B_2^n}^2 = n\left(\frac{\mathbb{E}\|x\|_2^4}{(\mathbb{E}\|x\|_2^2)^2} - 1\right) = \frac{4}{n+4}.$$

Actually, the minimum of $\overline{\sigma}_K$ over all K is attained at the Euclidean ball as the next theorem shows.

THEOREM 12.3.3. *Let K be an isotropic convex body in $\mathbb{R}^n$. Then,*
$$\overline{\sigma}_K \geq \overline{\sigma}_{B_2^n}.$$

Proof. Let x be uniformly distributed in K. The distribution function $F(r) = |\{x \in K : \|x\|_2 \leqslant r\}|$ has density

$$F'(r) = n\omega_n r^{n-1} \sigma(r^{-1}K)$$

for $r > 0$. We define $q(r) = n\omega_n \sigma(r^{-1}K)$. Observe that q is increasing and can be assumed absolutely continuous. Therefore, we can write q in the form

$$q(r) = n \int_r^\infty \frac{p(s)}{s^n} ds,$$

where $p : (0, +\infty) \to \mathbb{R}$ is a non-negative measurable function. Then, Fubini's theorem shows that

$$\int_0^\infty p(s)ds = n \int_0^\infty \frac{p(s)}{s^n} \left(\int_0^s r^{n-1} dr \right) ds = \int_0^\infty r^{n-1} q(r) dr = 1,$$

which means that p is the density of some positive random variable ξ.

Also, for every $\alpha > -n$,

$$\mathbb{E}\|x\|_2^\alpha = \int_0^\infty r^{\alpha+n-1} q(r) dr = \frac{n}{n+\alpha} \int_0^\infty s^\alpha p(s) ds = \frac{n}{n+\alpha} \mathbb{E}(\xi^\alpha).$$

We can now compute

$$\operatorname{Var}(\|x\|_2^2) = \frac{n}{n+4} \mathbb{E}(\xi^4) - \left(\frac{n}{n+2} \mathbb{E}(\xi^2) \right)^2$$

$$= \frac{4n}{(n+4)(n+2)^2} \left(\mathbb{E}(\xi^2) \right)^2 + \frac{n}{n+4} \operatorname{Var}(\xi^2)$$

$$\geqslant \frac{4n}{(n+4)(n+2)^2} \left(\mathbb{E}(\xi^2) \right)^2.$$

It follows that

$$\overline{\sigma}_K^2 = n \frac{\operatorname{Var}(\|x\|_2^2)}{\left(\mathbb{E}\|x\|_2^2 \right)^2} \geqslant n \frac{\frac{4n}{(n+4)(n+2)^2} \left(\mathbb{E}(\xi^2) \right)^2}{\left(\frac{n}{n+2} \mathbb{E}(\xi^2) \right)^2} = \frac{4}{n+4},$$

and the theorem follows from (12.3.2). $\square$

Note. Simple computations show that

$$\overline{\sigma}_{B_1^n}^2 = 1 - \frac{2(n+1)}{(n+3)(n+4)} \to 1 \text{ as } n \to \infty$$

and

$$\overline{\sigma}_{B_\infty^n} = \frac{4}{5} \text{ for every } n.$$

In the next subsection we show that $\overline{\sigma}_K^2$ is uniformly bounded for all ℓ_p^n, $p \in [1, \infty]$.

12.3.1. The variance hypothesis for p-balls

Let $1 \leqslant p \leqslant \infty$ and let $r_{p,n} > 0$ be a constant such that $|r_{p,n} B_p^n| = 1$. We write $L_{p,n}$ for the isotropic constant of B_p^n and $\mu_{p,n}$ for the Lebesgue measure on $r_{p,n} B_p^n$. As the next theorem shows, most of the volume of the normalized ℓ_p^n-ball lies in a very thin spherical shell around the radius $\sqrt{n} L_{p,n}$:

12.3. THE THIN SHELL CONJECTURE

THEOREM 12.3.4. *Let* $1 \leqslant p \leqslant \infty$. *For every* $t > 0$,

$$\mu_{p,n}\left(\left|\frac{\|x\|_2^2}{n} - L_{p,n}^2\right| \geqslant t\right) \leqslant \frac{CL_{p,n}^4}{nt^2},$$

where $C > 0$ *is an absolute constant.*

The proof is based on the fact that normalized ℓ_p^n-balls have the following subindependence property.

THEOREM 12.3.5 (Ball-Perissinaki). *Let* $K := r_{p,n}B_p^n$ *and* $P := \mu_{p,n}$. *If* $t_1, \ldots, t_n$ *are non-negative numbers, then*

$$P\left(\bigcap_{i=1}^n \{|x_i| \geqslant t_i\}\right) \leqslant \prod_{i=1}^n P(\{|x_i| \geqslant t_i\}).$$

Proof. The theorem will follow by induction if we show that

$$P\left(\bigcap_{i=1}^n \{|x_i| \geqslant t_i\}\right) \leqslant P\left(|x_1| \geqslant t_1\right) P\left(\bigcap_{i=2}^n \{|x_i| \geqslant t_i\}\right).$$

Set

$$S = \bigcap_{i=2}^n \{|x_i| \geqslant t_i\}.$$

Then, we need to prove that

$$\frac{|K \cap S \cap \{|x_1| \geqslant t_1\}|}{|K|} \leqslant \frac{|K \cap \{|x_1| \geqslant t_1\}|}{|K|} \cdot \frac{|K \cap S|}{|K|}.$$

We will apply the following simple fact: if μ is a positive measure on $[0,1]$ and $f : [0,1] \to \mathbb{R}$ is increasing, then

(12.3.3) $$\mu([0,1]) \int_0^s f \, d\mu \leqslant \mu([0,s]) \int_0^1 f \, d\mu$$

for all $s \in [0,1]$. If

$$f(u) = \frac{|K \cap S \cap \{|x_1| = 1 - u\}|}{|K \cap \{|x_1| = 1 - u\}|},$$

it is not hard to check that f is increasing. Let μ be the probability measure with density

$$g(u) = \frac{|K \cap \{|x_1| = 1 - u\}|}{|K|}.$$

Then,

$$\int_0^1 f(u) \, d\mu = \int_0^1 \frac{|K \cap S \cap \{|x_1| = 1 - u\}|}{|K \cap \{|x_1| = 1 - u\}|} \frac{|K \cap \{|x_1| = 1 - u\}|}{|K|} du$$

$$= \frac{|K \cap S|}{|K|}$$

and

$$\int_0^{1-t_1} f(u) \, d\mu = \frac{|K \cap S \cap \{|x_1| \geqslant t_1\}|}{|K|}.$$

Since

$$\mu([0, 1 - t_1]) = \frac{|K \cap \{|x_1| \geqslant t_1\}|}{|K|},$$

applying (12.3.3) for $s = 1 - t_1$ we get the result. □

Theorem 12.3.5 immediately implies an anti-correlation inequality for the coordinate functions.

COROLLARY 12.3.6. *Let* $K := r_{p,n} B_p^n$. *Then,*
$$\int_K x_i^2 x_j^2 dx \leqslant \int_K x_i^2 dx \cdot \int_K x_j^2 dx$$
for all $i \neq j$ *in* $\{1, \ldots, n\}$.

Proof. We write
$$\int_K x_i^2 x_j^2 dx = 4 \int_{K \cap \{x_i \geqslant 0, x_j \geqslant 0\}} x_i^2 x_j^2 dx$$
$$= 4 \int_0^\infty \int_0^\infty 4 t_i t_j P(x_i \geqslant t_i, x_j \geqslant t_j) dt_i dt_j$$
$$\leqslant 4 \int_0^\infty \int_0^\infty 4 t_i t_j P(x_i \geqslant t_i) P(x_j \geqslant t_j) dt_i dt_j$$
$$= 4 \left(\int_0^\infty 2 t_i P(x_i \geqslant t_i) dt_i \right) \left(\int_0^\infty 2 t_j P(x_j \geqslant t_j) dt_j \right)$$
$$= 4 \int_{K \cap \{x_i \geqslant 0\}} x_i^2 dx \int_{K \cap \{x_j \geqslant 0\}} x_j^2 dx$$
$$= \int_K x_i^2 dx \int_K x_j^2 dx.$$

Proof of Theorem 12.3.4. From the Cauchy-Schwarz inequality we have
$$n^2 L_{p,n}^4 = \left(\int_K \|x\|_2^2 dx \right)^2 \leqslant \int_K \|x\|_2^4 dx.$$

On the other hand, using Corollary 12.3.6 we have
$$\int_K \|x\|_2^4 dx = \int_K \left(\sum_{i=1}^n x_i^2 \right)^2 dx = \sum_{i=1}^n \int_K x_i^4 dx + \sum_{i \neq j} \int_K x_i^2 x_j^2 dx$$
$$\leqslant n \int_K x_1^4 dx + \sum_{i \neq j} \int_K x_i^2 dx \int_K x_j^2 dx$$
$$= n \int_K x_1^4 dx + n(n-1) L_{p,n}^4.$$

Since
$$\int_K x_1^4 dx \leqslant C \left(\int_K x_1^2 dx \right)^2 = C L_{p,n}^4$$
for some absolute constant $C > 0$, we get
$$L_{p,n}^4 \leqslant \frac{1}{n^2} \int_K \|x\|_2^4 dx \leqslant \left(1 + \frac{C}{n} \right) L_{p,n}^4.$$

This implies that
$$\int_K \left(\frac{\|x\|_2^2}{n} - L_{p,n}^2 \right)^2 dx = \frac{1}{n^2} \int_K \|x\|_2^4 dx - L_{p,n}^4 \leqslant \frac{C}{n} L_{p,n}^4.$$

Then, Chebyshev's inequality gives

$$t^2 \mu_{p,n}\left(\left|\frac{\|x\|_2^2}{n} - L_{p,n}^2\right| \geqslant t\right) \leqslant \int_K \left(\frac{\|x\|_2^2}{n} - L_{p,n}^2\right)^2 dx \leqslant \frac{C}{n} L_{p,n}^4$$

for every $t > 0$, which is exactly the assertion of the theorem. □

COROLLARY 12.3.7. *For every $t > 0$,*

$$\mu_{p,n}\left(\left|\frac{\|x\|_2}{\sqrt{n}} - L_{p,n}\right| \geqslant t\right) \leqslant \frac{CL_{p,n}^2}{nt^2}.$$

Proof. Let $t > 0$. We have

$$\mu_{p,n}\left(\left|\,\|x\|_2 - \sqrt{n}L_{p,n}\,\right| \geqslant t\sqrt{n}\right) \leqslant \mu_{p,n}\left(\left|\,\|x\|_2^2 - nL_{p,n}^2\,\right| \geqslant tnL_{p,n}\right)$$
$$\leqslant \frac{CL_{p,n}^4}{t^2 n L_{p,n}^2} = \frac{CL_{p,n}^2}{t^2 n}$$

by Theorem 12.3.4. □

PROPOSITION 12.3.8. *There exists an absolute constant $C > 0$ such that $\overline{\sigma}_{B_p^n}^2 \leqslant C$ for every n and every $p \in [1, \infty]$.*

Proof. In the proof of Theorem 12.3.4 we saw that

$$n^2 L_{p,n}^4 \leqslant \int_K \|x\|_2^4 dx \leqslant (n^2 + Cn) L_{p,n}^4$$

for some absolute constant $C > 0$. Then,

$$\overline{\sigma}_{B_p^n}^2 = n\left(\frac{\mathbb{E}\|x\|_2^4}{n^2 L_{p,n}^4} - 1\right) \leqslant C$$

for all p and n. □

12.4. The thin shell conjecture in the unconditional case

Let $X = (X_1, \ldots, X_n)$ be a random vector, uniformly distributed in an unconditional convex body K in $\mathbb{R}^n$. We also assume that X is isotropic; equivalently, we have $\mathbb{E}(X_i^2) = 1$ for all $i = 1, \ldots, n$ (note that $\mathbb{E}(X_i) = 0$ and $\mathbb{E}(X_i X_j) = 0$ for $i \neq j$ are automatically satisfied). Our aim is to show that X satisfies the thin shell conjecture.

THEOREM 12.4.1 (Klartag). *For every unconditional isotropic convex body K in $\mathbb{R}^n$ we have*

$$\sigma_K^2 := \mathbb{E}_{\mu_K}\left(\|x\|_2 - \sqrt{n}\right)^2 \leqslant C^2,$$

where $C \leqslant 4$ is an absolute positive constant.

Theorem 12.4.1 was proved by Klartag in [**279**]. In the next two subsections we describe the ingredients of the proof, which is then given in Section 12.4.3.

12.4.1. A general upper bound for the variance

The proof of Theorem 12.4.1 is based on an analysis of the Neumann Laplacian on convex bodies (see Hörmander [**256**] and Helffer-Sjöstrand [**251**]). We list some of its properties which will be useful in this section. In what follows, we assume that K is a convex body of volume 1 in $\mathbb{R}^n$ with C^∞-smooth boundary. We will say that a function $\varphi : K \to \mathbb{R}$ belongs to the class $C^\infty(K)$ if it has derivatives of all orders and these are bounded in the interior of K. Then, the boundary values of φ and its derivatives are well-defined and they are C^∞-smooth on $\mathrm{bd}(K)$.

We denote by $\mathcal{D}$ the class of all $C^\infty(K)$-smooth functions $u : K \to \mathbb{R}$ which satisfy
$$\langle \nabla u(x), \nu(x) \rangle = 0$$
for all $x \in \mathrm{bd}(K)$, where $\nu(x)$ is the outer normal vector at the point $x \in \mathrm{bd}(K)$. We will use Stokes theorem in the form
$$\int_K \langle \nabla u, \nabla v \rangle = -\int_K (\Delta u) v + \int_{\mathrm{bd}(K)} \langle v \nabla u, \nu \rangle = -\int_K (\Delta u) v$$
for all $v \in C^\infty(K)$ and $u \in \mathcal{D}$.

For every function $u \in C^\infty(K)$ we define
$$\|u\|_{H^{-1}(K)} = \sup \left\{ \int_K \varphi u : \varphi \in C^\infty(K), \int_K \|\nabla \varphi\|_2^2 \leqslant 1 \right\}.$$
Note that $\|u\|_{H^{-1}(K)} = \infty$ if $\int_K u \neq 0$ (if $\int_K \|\nabla \varphi\|_2^2 \leqslant 1$ then the same is true for $\varphi_1 = \varphi + \alpha$ for any $\alpha \in \mathbb{R}$). We write $\partial_i f$ for the partial derivative of f with respect to the i-th coordinate. Finally, for every $f \in L^2(K)$ we set
$$\mathrm{Var}_K(f) = \int_K (f(x) - \mathbb{E}_{\mu_K}(f))^2 dx,$$
where $\mathbb{E}_{\mu_K}(f) = \int_K f$.

We fix the function $\rho : K \to \mathbb{R}$ defined by $\rho(x) = -\mathrm{dist}(x, \mathrm{bd}(K))$. This is a C^∞-smooth convex function with bounded derivatives of all orders in a neighborhood of $\mathrm{bd}(K)$, and also, $\rho(x) \leqslant 0$ if $x \in K$ and
$$\rho(x) = 0 \quad \text{and} \quad \|\nabla \rho(x)\|_2 = 1 \quad \text{if } x \in \mathrm{bd}(K).$$
Observe that $\nabla \rho(x) = \nu(x)$ for every $x \in \mathrm{bd}(K)$.

Using integration by parts we obtain the next lemma (see Lichnerowicz [**325**], Hörmander [**256**] and Kadlec [**264**] for related results).

LEMMA 12.4.2. *Let $u \in \mathcal{D}$ and set $f = -\Delta u$. Then,*
$$\int_K f^2 = \sum_{i=1}^n \int_K \|\nabla \partial_i u\|_2^2 + \int_{\mathrm{bd}(K)} \langle (\mathrm{Hess}\,\rho)(\nabla u), \nabla u \rangle.$$

Proof. By the definition of $\mathcal{D}$ we have that the function $x \mapsto \langle \nabla u(x), \nabla \rho(x) \rangle$ vanishes on the boundary of K. Moreover, the vector ∇u is tangential to the boundary of K, and hence the derivative of $x \mapsto \langle \nabla u(x), \nabla \rho(x) \rangle$ in the direction of ∇u vanishes on $\mathrm{bd}(K)$. This means that
$$\langle \nabla u(x), \nabla (\langle \nabla u(x), \nabla \rho(x) \rangle) \rangle = 0 \text{ for all } x \in \mathrm{bd}(K).$$
Equivalently, we can write this in the form
$$(12.4.1) \qquad \langle (\mathrm{Hess}\,u)(\nabla \rho), \nabla u \rangle + \langle (\mathrm{Hess}\,\rho)(\nabla u), \nabla u \rangle = 0 \text{ for all } x \in \mathrm{bd}(K).$$

Stokes theorem implies that

$$\int_K f^2 = \int_K (\Delta u)^2 = -\int_K \langle \nabla(\Delta u), \nabla u\rangle + \int_{\mathrm{bd}(K)} \langle (\Delta u \nabla u), \nabla \rho\rangle.$$

The integral on the boundary of K is zero; so, one more application of Stokes theorem gives

$$\int_K f^2 = -\sum_{i=1}^n \int_K \partial_i u \Delta(\partial_i u) = \sum_{i=1}^n \int_K \|\nabla \partial_i u\|_2^2 - \int_{\mathrm{bd}(K)} \sum_{i=1}^n \langle \partial_i u \nabla \partial_i u, \nabla \rho\rangle.$$

On the other hand,

$$\sum_{i=1}^n \langle (\partial_i u \nabla \partial_i u), \nabla \rho\rangle = \langle (\mathrm{Hess}\, u)(\nabla \rho), \nabla u\rangle.$$

Therefore, using also (12.4.1) we get

$$\int_K f^2 = \sum_{i=1}^n \int_K \|\nabla \partial_i u\|_2^2 + \int_{\mathrm{bd}(K)} \langle (\mathrm{Hess}\,\rho)(\nabla u), \nabla u\rangle,$$

and this proves our claim. $\square$

LEMMA 12.4.3. *Let K be a convex body in $\mathbb{R}^n$ with C^∞ smooth boundary. If $f : K \to \mathbb{R}$ is a $C^\infty(K)$-smooth function then*

$$\mathrm{Var}_K(f) \leqslant \sum_{i=1}^n \|\partial_i f\|_{H^{-1}(K)}^2.$$

Proof. We may assume that $\int_K f = 0$. Then, there exists a function $u \in \mathcal{D}$ (see e.g. [**181**, Chapter 7]) such that

$$f = -\Delta u.$$

From Stokes theorem we get

$$\int_K f^2 = -\int_K f \Delta u = \int_K \langle \nabla f, \nabla u\rangle - \int_{\mathrm{bd}(K)} \langle f \nabla u, \nu\rangle = \sum_{i=1}^n \int_K \partial_i(f)\,\partial_i u,$$

where the integral on the boundary vanishes because $u \in \mathcal{D}$. Then, using the definition of the $H^{-1}(K)$-norm and the Cauchy-Schwarz inequality, we obtain

$$\int_K f^2 = \sum_{i=1}^n \int_K \partial_i(f)\,\partial_i u \leqslant \sum_{i=1}^n \|\partial_i f\|_{H^{-1}(K)} \cdot \sqrt{\int_K \|\nabla \partial_i u\|_2^2}$$

$$\leqslant \sqrt{\sum_{i=1}^n \|\partial_i f\|_{H^{-1}(K)}^2} \cdot \sqrt{\sum_{i=1}^n \int_K \|\nabla \partial_i u\|_2^2}.$$

From Lemma 12.4.2 we have

(12.4.2) $$\sum_{i=1}^n \int_K \|\nabla \partial_i u\|_2^2 \leqslant \int_K f^2,$$

because the Hessian of the convex function ρ is positive semidefinite. Thus, the last two inequalities prove the lemma. $\square$

12.4.2. Transportation of measure

Let μ_1, μ_2 be two finite Borel measures on $\mathbb{R}^n$, and let $T : \mathbb{R}^n \to \mathbb{R}^n$ be a measurable map which transports μ_1 to μ_2; that is,

$$\mu_2(A) = \mu_1(T^{-1}(A))$$

for every Borel subset A of $\mathbb{R}^n$. This is equivalent to the fact that $\int (\varphi \circ T) \, d\mu_1 = \int \varphi \, d\mu_2$ for any bounded measurable function φ. Observe that the function

$$x \mapsto (x, Tx)$$

transports the measure μ_1 to a measure γ on $\mathbb{R}^n \times \mathbb{R}^n$, which has marginals μ_1 and μ_2. The L^2-*Wasserstein distance* of μ_1 and μ_2 is defined by

$$W_2(\mu_1, \mu_2) = \inf_\gamma \left(\iint_{\mathbb{R}^n \times \mathbb{R}^n} \|x - y\|_2^2 d\gamma(x, y) \right)^{1/2},$$

where the infimum is over all measures γ with marginals μ_1 and μ_2. If there is no such measure γ, we set $W_2(\mu_1, \mu_2) = \infty$.

Let μ be a finite, compactly supported Borel measure on $\mathbb{R}^n$. For every C^∞-smooth function $u : \mathbb{R}^n \to \mathbb{R}$ with $\int u \, d\mu = 0$ we set

$$\|u\|_{H^{-1}(\mu)} = \sup \left\{ \int_{\mathbb{R}^n} u \varphi \, d\mu : \varphi \in C^\infty(\mathbb{R}^n), \int_{\mathbb{R}^n} \|\nabla \varphi\|_2^2 d\mu \leqslant 1 \right\}.$$

This definition clearly agrees with our previous definition of $\|\cdot\|_{H^{-1}(K)}$; we have $\|u\|_{H^{-1}(\mu_K)} = \|u\|_{H^{-1}(K)}$, where μ_K is the Lebesgue measure on K. We will need the following theorem that extends an observation of Brenier [**120**] (the proof that is given below is from [**498**, Section 7.6]).

THEOREM 12.4.4. *Let μ be a finite, compactly supported Borel measure on $\mathbb{R}^n$ and let $h : \mathbb{R}^n \to \mathbb{R}$ be a bounded measurable function whose integral is equal to zero. For small enough $\varepsilon > 0$, we consider the measure μ_ε whose density with respect to μ is the non-negative function $1 + \varepsilon h$. Then,*

$$\|h\|_{H^{-1}(\mu)} \leqslant \liminf_{\varepsilon \to 0^+} \frac{W_2(\mu, \mu_\varepsilon)}{\varepsilon}.$$

Proof. It suffices to prove that

$$(12.4.3) \qquad \int_{\mathbb{R}^n} h \varphi \, d\mu \leqslant \left(\int_{\mathbb{R}^n} \|\nabla \varphi\|_2^2 d\mu \right)^{1/2} \liminf_{\varepsilon \to 0^+} \frac{W_2(\mu, \mu_\varepsilon)}{\varepsilon}$$

for every C^∞-smooth function $\varphi : \mathbb{R}^n \to \mathbb{R}^n$. We may additionally assume that φ has compact support, because μ is compactly supported. If φ satisfies the above, then its second derivatives are bounded; so, there exists $R = R(\varphi) > 0$ such that

$$(12.4.4) \qquad |\varphi(y) - \varphi(x)| \leqslant \|\nabla \varphi(x)\|_2 \|x - y\|_2 + R\|x - y\|_2^2$$

for all $x, y \in \mathbb{R}^n$. Let $\varepsilon > 0$ satisfy $\varepsilon \|h\|_\infty < 1$ and consider any measure γ with marginals μ and μ_ε (note that μ_ε is a non-negative measure by the restriction on ε). Then,

$$\int_{\mathbb{R}^n} h \varphi \, d\mu = \frac{1}{\varepsilon} \int_{\mathbb{R}^n} \varphi \, d(\mu_\varepsilon - \mu) = \frac{1}{\varepsilon} \iint_{\mathbb{R}^n \times \mathbb{R}^n} (\varphi(y) - \varphi(x)) \, d\gamma(x, y).$$

We set

$$W_2^\gamma(\mu, \mu_\varepsilon) = \left(\iint_{\mathbb{R}^n \times \mathbb{R}^n} \|x - y\|_2^2 \, d\gamma(x, y) \right)^{1/2}.$$

12.4. THE THIN SHELL CONJECTURE IN THE UNCONDITIONAL CASE

Then, (12.4.4) and the Cauchy-Schwarz inequality imply

$$\int_{\mathbb{R}^n} h\varphi \, d\mu \leq \frac{1}{\varepsilon} \iint_{\mathbb{R}^n \times \mathbb{R}^n} \|\nabla\varphi(x)\|_2 \|x-y\|_2 \, d\gamma(x,y)$$
$$+ \frac{R}{\varepsilon} \iint_{\mathbb{R}^n \times \mathbb{R}^n} \|x-y\|_2^2 \, d\gamma(x,y)$$
$$\leq \frac{1}{\varepsilon} \left(\int_{\mathbb{R}^n} \|\nabla\varphi\|_2^2 d\mu \right)^{1/2} W_2^\gamma(\mu,\mu_\varepsilon) + \frac{R}{\varepsilon} \left(W_2^\gamma(\mu,\mu_\varepsilon) \right)^2.$$

Taking the infimum over all possible γ, we obtain

$$(12.4.5) \qquad \int_{\mathbb{R}^n} h\varphi \, d\mu \leq \frac{1}{\varepsilon} \left(\int_{\mathbb{R}^n} \|\nabla\varphi\|_2^2 d\mu \right)^{1/2} W_2(\mu,\mu_\varepsilon) + \frac{R}{\varepsilon} \left(W_2(\mu,\mu_\varepsilon) \right)^2.$$

Note that R depends only on φ and we may assume that $\liminf_{\varepsilon \to 0^+} \frac{W_2(\mu,\mu_\varepsilon)}{\varepsilon} < \infty$, otherwise the theorem is trivially true. Then,

$$\liminf_{\varepsilon \to 0^+} \frac{\left(W_2(\mu,\mu_\varepsilon)\right)^2}{\varepsilon} = \liminf_{\varepsilon \to 0^+} \varepsilon \left(\frac{W_2(\mu,\mu_\varepsilon)}{\varepsilon} \right)^2 = 0,$$

and letting $\varepsilon \to 0^+$ in (12.4.5) we obtain (12.4.3). $\square$

Next, we consider the standard orthonormal basis $\{e_1, \ldots, e_n\}$ of $\mathbb{R}^n$, a convex body K of volume 1 in $\mathbb{R}^n$, and we fix $x \in K$. We denote by $x + \mathbb{R}e_i$ the line through x in the direction of e_i. The intersection of this line with K is a closed line segment or a point. We denote the endpoints of this line segment by $\mathfrak{B}_i^-(x)$ and $\mathfrak{B}_i^+(x)$, where $\langle \mathfrak{B}_i^-(x), e_i \rangle \leq \langle \mathfrak{B}_i^+(x), e_i \rangle$. Then,

$$K \cap (x + \mathbb{R}e_i) = \left[\mathfrak{B}_i^-(x), \mathfrak{B}_i^+(x) \right].$$

For $i = 1, \ldots, n$ we consider the projections

$$\pi_i(x_1, \ldots, x_n) = (x_1, \ldots, x_{i-1}, x_{i+1}, \ldots, x_n).$$

Then, $\pi_i(K)$ is a convex body in $\mathbb{R}^{n-1}$, and for every $y \in \pi_i(K)$ we define $q_i^-(y) \in \mathbb{R}$ as the smallest, and $q_i^+(y)$ as the largest, i-th coordinate among all points of K with $\pi_i(x) = y$. With this notation, we have the next lemma.

LEMMA 12.4.5. *Let K be a convex body of volume 1 in $\mathbb{R}^n$ with C^∞-smooth boundary. We fix $i = 1, \ldots, n$ and consider a $C^\infty(K)$-smooth function $\Psi : K \to \mathbb{R}$ such that, for every $x \in K$,*

$$(12.4.6) \qquad \Psi(\mathfrak{B}_i^-(x)) = \Psi(\mathfrak{B}_i^+(x)).$$

For small enough $\varepsilon > 0$ we denote by μ_ε the measure whose density with respect to μ is $1 + \varepsilon \partial_i \Psi$. Then, we have that

$$\liminf_{\varepsilon \to 0^+} \frac{W_2(\mu, \mu_\varepsilon)}{\varepsilon} \leq \left(\int_K \left[\Psi(x) - \Psi(\mathfrak{B}_i^+(x)) \right]^2 dx \right)^{1/2}.$$

Proof. Without loss of generality we assume that $i = 1$. For small enough $\varepsilon > 0$, the function $1 + \varepsilon \partial_1 \Psi$ is positive on K, which implies that μ_ε is a non-negative measure.

We fix such an $\varepsilon > 0$ and we write $x = (t, y)$, where $y = (x_2, \ldots, x_n) \in \mathbb{R}^{n-1}$. We also fix a point $y \in \pi_1(K)$ and we write $p = q_1^-(y)$, $q = q_1^+(y)$. Then, (12.4.6) implies

$$\int_p^q (1 + \varepsilon \partial_1 \Psi(t,y)) dt = q - p + \varepsilon \Psi(t,y)\big|_{t=p}^q = q - p.$$

It follows that the densities $t \mapsto 1$ and $t \mapsto 1+\varepsilon\partial_1\Psi(t,y)$ have the same integral on the interval $[p,q]$. Therefore, there exists a unique function $T = T^y : [p,q] \to [p,q]$ which transports the measure with density $1+\varepsilon\partial_1\Psi(t,y)$ to the Lebesgue measure on $[p,q]$ and satisfies the relation

$$\int_p^{x_1} (1+\varepsilon\partial_1\Psi(t,y))dt = \int_p^{T(x_1)} \mathbf{1}\, dt.$$

Then, for every $x_1 \in [p,q]$ we have

$$T(x_1) = x_1 + \varepsilon\left[\Psi(x_1,y) - \Psi(p,y)\right].$$

This implies that

$$\int_p^q |T(t) - t|^2 (1+\varepsilon\partial_1\Psi(t,y))dt = \varepsilon^2 \int_p^q \left[\Psi(t,y) - \Psi(p,y)\right]^2 dt + \varepsilon^3 R$$

where $|R|$ is bounded independently of ε and y.

Now, for every $(x_1,y) \in K$ we define $S(x_1,y) = (T^y(x_1), y)$. This is a well-defined one-to-one and continuous map. Moreover, Fubini's theorem shows that for every continuous function $\varphi : K \to \mathbb{R}$ one has

$$\int_K \varphi(S(x))d\mu_\varepsilon(x) = \int_{\pi(K)} \left[\int_p^q \varphi(T^y(x_1),y)(1+\varepsilon\partial_1\Psi)dx_1\right]dy$$

$$= \int_{\pi(K)} \left[\int_p^q \varphi(x_1,y)dx_1\right]dy = \int_K \varphi(x)d\mu(x).$$

It follows that S transports μ_ε to μ. Therefore,

$$W_2(\mu,\mu_\varepsilon)^2 \leqslant \int_K \|S(x) - x\|_2^2 d\mu_\varepsilon(x) = \varepsilon^2 \int_K \left[\Psi(x) - \Psi(\mathfrak{B}_1^-(x))\right]^2 dt + \varepsilon^3 R',$$

where $|R'|$ does not depend on ε. Dividing by ε^2 and letting $\varepsilon \to 0^+$ we conclude the proof. $\square$

12.4.3. Thin shell estimate in the unconditional case

Next, we restrict ourselves to the unconditional case.

PROPOSITION 12.4.6. *Let K be an unconditional convex body of volume 1 in $\mathbb{R}^n$. If $\Psi : K \to \mathbb{R}$ is an unconditional continuous function, then*

$$\operatorname{Var}_K(\Psi) \leqslant \sum_{i=1}^n \int_K \left(\Psi(x) - \Psi(\mathfrak{B}_i^+(x))\right)^2 dx.$$

Proof. We may assume that K has C^∞-smooth boundary and that Ψ is a $C^\infty(K)$-smooth function. Then, from Lemma 12.4.3 we have that

$$\operatorname{Var}_K(\Psi) \leqslant \sum_{i=1}^n \|\partial_i\Psi\|_{H^{-1}(K)}^2.$$

Using the symmetries of Ψ we see that $\int_K \partial_i \Psi = 0$ for any $i = 1, \ldots, n$. So, we may apply Theorem 12.4.4. Combining this with Lemma 12.4.5 we get

$$\|\partial_i\Psi\|_{H^{-1}(K)}^2 \leqslant \int_K \left(\Psi(x) - \Psi(\mathfrak{B}_i^+(x))\right)^2 dx.$$

Adding these inequalities we get the result. $\square$

12.4. THE THIN SHELL CONJECTURE IN THE UNCONDITIONAL CASE 413

COROLLARY 12.4.7. *Let $f_1, \ldots, f_n : \mathbb{R} \to \mathbb{R}$ be even continuous functions and let $\Psi(x_1, \ldots, x_n) = \sum_{i=1}^{n} f_i(x_i)$. Then,*

$$\operatorname{Var}_K(\Psi) \leqslant \sum_{i=1}^{n} \int_K \sup_{s,t \in J_i(x)} (f_i(s) - f_i(t))^2 dx,$$

where $J_i(x)$ is an interval in $\mathbb{R}$ which is symmetric with respect to the origin and has the same length as $[\mathfrak{B}_i^-(x), \mathfrak{B}_i^+(x)]$.

LEMMA 12.4.8. *Let $X = (X_1, \ldots, X_n) \in \mathbb{R}^n$ be a random vector with an unconditional log-concave density. If $p_1, \ldots, p_n > 0$ and $a_1, \ldots, a_n \geqslant 0$, then*

$$(12.4.7) \qquad \operatorname{Var}\left(\sum_{i=1}^{n} a_i |X_i|^{p_i}\right) \leqslant \sum_{i=1}^{n} \frac{2p_i^2}{p_i + 1} a_i^2 \mathbb{E}\left(|X_i|^{2p_i}\right).$$

Proof. We first assume that X is uniformly distributed in an unconditional convex body K of volume 1 in $\mathbb{R}^n$. Then, for $x = (x_1, \ldots, x_n) \in \mathbb{R}^n$ we set

$$\Psi(x_1, \ldots, x_n) = \sum_{i=1}^{n} a_i |x_i|^{p_i},$$

and the assertion of the lemma becomes

$$\operatorname{Var}_K(\Psi) \leqslant \sum_{i=1}^{n} \frac{2p_i^2}{p_i + 1} \int_K a_i^2 |x_i|^{2p_i} dx_1 \cdots dx_n.$$

By Proposition 12.4.6, it is enough to show that

$$\int_K \left(\Psi(x) - \Psi(\mathfrak{B}_i^+(x))\right)^2 dx \leqslant \frac{2p_i^2}{p_i + 1} \int_K a_i^2 |x_i|^{2p_i} dx_1 \cdots dx_n.$$

We fix $i \leqslant n$ and we prove this last equality using Fubini's theorem. For fixed

$$x' = (x_1, \ldots, x_{i-1}, x_{i+1}, \ldots, x_n) \in \pi_i(K)$$

we set $r = q_i^+(x') \geqslant 0$. Then, we need to check that

$$\int_{-r}^{r} \left[\sum_{j=1}^{n} a_j |x_j|^{p_j} - \left(a_i r^{p_i} + \sum_{j \neq i} a_j |x_j|^{p_j}\right)\right]^2 dx_i \leqslant \frac{2p_i^2}{p_i + 1} \int_{-r}^{r} a_i^2 |x_i|^{2p_i} dx_i.$$

This reduces to

$$\int_{-r}^{r} (a_i |x_i|^{p_i} - a_i r^{p_i})^2 dx_i \leqslant \frac{2p_i^2}{p_i + 1} \int_{-r}^{r} (a_i |x_i|^{p_i})^2 dx_i,$$

which is easily verified.

For the general case, let $f : \mathbb{R}^n \to [0, \infty)$ be the unconditional log-concave density of X. First, let us make the additional assumption that f is s-concave for some integer $s \geqslant 1$ and let us write $N = n + s$. We consider the unconditional convex body K in $\mathbb{R}^N$ which is defined as follows:

$$K = \left\{(x, y) : x \in \mathbb{R}^n, \ y \in \mathbb{R}^s, \ \|y\|_2 \leqslant \omega_s^{-\frac{1}{s}} f^{1/s}(x)\right\},$$

where ω_s is the volume of the s-dimensional Euclidean unit ball. From our result in the first case, it follows that the lemma also holds true in the case where the density f is s-concave.

Finally, if $f = e^{-\psi}$ is the unconditional log-concave density of X, then for every $s > 0$ the function

$$x \mapsto \left(1 - \frac{\psi(x)}{s}\right)^s_+$$

with $x_+ = \max\{x, 0\}$, is unconditional and s-concave. Normalizing we get a density that converges weakly to $e^{-\psi}$ (and $\|\cdot\|_{L_1}$-uniformly on $\mathbb{R}^n$) as $s \to \infty$. Therefore, the general case follows from our result in the s-concave case. □

We are now able to prove the main theorem.

THEOREM 12.4.9. *Let $X = (X_1, \ldots, X_n)$ be a random vector in $\mathbb{R}^n$, which has an unconditional log-concave density and satisfies $\mathbb{E}(X_i^2) = 1$ for all $i = 1, \ldots, n$. If $a_1, \ldots, a_n \geqslant 0$ then*

$$\mathrm{Var}\left(\sum_{i=1}^n a_i X_i^2\right) \leqslant C' \sum_{i=1}^n a_i^2,$$

where $C' \leqslant 16$ is an absolute positive constant. In particular,

(12.4.8) $$\mathbb{E}\left(\|X\|_2 - \sqrt{n}\right)^2 \leqslant C^2,$$

where $C \leqslant 4$ is an absolute positive constant. Moreover, for every $p \geqslant 1$ we have

$$\sqrt{\mathrm{Var}(\|X\|_p)} \leqslant C_p \cdot n^{1/p - 1/2},$$

where $C_p > 0$ is a constant depending only on p.

Proof. By the Prékopa-Leindler inequality, the random variable X_i has an even log-concave density for every i. From (12.4.7) and the equivalence of $\|X_i\|_{L_4}$ and $\|X_i\|_{L_2}$, we have

$$\mathrm{Var}\left(\sum_{i=1}^n a_i X_i^2\right) \leqslant \frac{8}{3} \sum_{i=1}^n a_i^2 \mathbb{E}|X_i|^4 \leqslant 16 \sum a_i^2 \left(\mathbb{E}|X_i|^2\right)^2 = 16 \sum_{i=1}^n a_i^2.$$

Moreover, setting $a_i = 1$ for all i, we see that

$$\mathbb{E}\left[\left(\|X\|_2 - \sqrt{n}\right)^2\right] \leqslant \frac{1}{n} \mathbb{E}\left[\left(\|X\|_2 - \sqrt{n}\right)^2 \left(\|X\|_2 + \sqrt{n}\right)^2\right]$$

$$= \frac{1}{n} \mathbb{E}\left(\|X\|_2^2 - n\right)^2 \leqslant 16,$$

as claimed.

For the last assertion, we set $E_p = \mathbb{E}\|X\|_p^p$. From (12.4.7) and the equivalence of $\|X_i\|_{L_p}$ and $\|X_i\|_{L_2}$, we have

$$\mathbb{E}\left(\|X\|_p^p - E_p\right)^2 = \mathrm{Var}\left(\sum_{i=1}^n |X_i|^p\right) \leqslant 2^{1-p} p \Gamma(2p+1) n.$$

Now, for $p \geqslant 2$ we obtain $\mathbb{E}|X_i|^p \geqslant \left(\mathbb{E}(X_i^2)\right)^{p/2} = 1$, while for $1 \leqslant p \leqslant 2$

$$\mathbb{E}|X_i|^p \geqslant (\mathbb{E}|X_i|)^p \geqslant 2^{-p/2} \left(\mathbb{E}X_i^2\right)^{p/2} = 2^{-p/2} \geqslant 2^{-1/2};$$

the above show that $E_p = \sum_i \mathbb{E}|X_i|^p \geqslant n/\sqrt{2}$. Therefore,

$$\mathrm{Var}(\|X\|_p) \leqslant \mathbb{E}\left(\|X\|_p - E_p^{1/p}\right)^2 \leqslant E_p^{-\frac{2(p-1)}{p}} \mathbb{E}\left(\|X\|_p^p - E_p\right)^2 \leqslant C_p n^{2/p - 1},$$

and the proof is complete. □

REMARK 12.4.10. Theorem 12.4.9 has some interesting consequences. From (12.4.8) we get

$$\mathbb{P}\left(\|X\|_2 \leqslant \sqrt{n}-8\right) \leqslant \frac{1}{4} \quad \text{and} \quad \mathbb{P}\left(\|X\|_2 \leqslant \sqrt{n}+8\right) \geqslant \frac{3}{4}.$$

From the B-theorem (Theorem 5.3.4) we have that the function $s \mapsto \mathbb{P}\left(\|X\|_2 \leqslant e^s\right)$ is log-concave, and hence, for every $t > 0$,

$$\mathbb{P}\left(\|X\|_2 \leqslant (\sqrt{n}-8)\cdot\left(\frac{\sqrt{n}-8}{\sqrt{n}+8}\right)^t\right) \leqslant \frac{1}{4\cdot 3^t}.$$

After some elementary calculations we see that for n large enough and $1 \ll s \leqslant \sqrt{n}$ we have

$$\mathbb{P}\left(\|X\|_2 \leqslant \sqrt{n}-s\right) \leqslant C\exp(-\alpha(s)),$$

where $\alpha(s) = \log\left(1+\frac{s-8}{\sqrt{n}-8}\right)/\log\left(1+\frac{16}{\sqrt{n}-8}\right) \geqslant c_1 s$, and hence

$$\mathbb{P}\left(\|X\|_2 \leqslant \sqrt{n}-s\right) \leqslant C\exp(-c_2 s),$$

for all $0 \leqslant s \leqslant \sqrt{n}$.

12.5. Thin shell conjecture and the hyperplane conjecture

Recall that $\sigma_n^2 := \sup_X \mathbb{E}(\|X\|_2 - \sqrt{n})^2$, where the supremum is taken over all isotropic log-concave random vectors X in $\mathbb{R}^n$. The thin shell conjecture (see Section 12.3) asks whether

$$\sigma_n \leqslant C$$

for all n, where $C > 0$ is an absolute constant. In this section we present a result of Eldan and Klartag showing that the thin shell conjecture is stronger than the hyperplane conjecture.

THEOREM 12.5.1 (Eldan-Klartag). *There exists an absolute constant $C > 0$ such that*

$$L_n \leqslant C\sigma_n$$

for every $n \geqslant 1$.

For the proof, Eldan and Klartag introduce in [**166**] a third parameter as follows.

DEFINITION 12.5.2. For every $n \geqslant 1$ we define

$$\underline{\sigma}_n = \frac{1}{\sqrt{n}}\sup_X \left\|\mathbb{E}X\|X\|_2^2\right\|_2 = \frac{1}{\sqrt{n}}\sup_X \sup_{\theta \in S^{n-1}} \mathbb{E}\left(\langle X,\theta\rangle\|X\|_2^2\right),$$

where the supremum is over all isotropic log-concave random vectors X in $\mathbb{R}^n$.

LEMMA 12.5.3. *For every $n \geqslant 1$ one has*

(12.5.1) $$2 \leqslant \underline{\sigma}_n \leqslant c_1 \sigma_n,$$

where $c_1 > 0$ is an absolute constant.

Proof. We first observe that, since $\mathbb{E}(X) = 0$, one has

$$\mathbb{E}\left(\langle X, \theta\rangle \|X\|_2^2\right) = \mathbb{E}\left(\langle X, \theta\rangle(\|X\|_2^2 - n)\right)$$
$$\leqslant \sqrt{\mathbb{E}\langle X, \theta\rangle^2 \mathbb{E}(\|X\|_2^2 - n)^2} \leqslant C\sqrt{n}\sigma_n,$$

using also the Cauchy-Schwarz inequality, the fact that $\mathbb{E}\langle X, \theta\rangle^2 = 1$ by the isotropicity of X, and Lemma 12.3.1. Finally, we can show that $\underline{\sigma}_n \geqslant 2$ by considering the isotropic log-concave random vector $Y = (Y_1, \ldots, Y_n)$ whose coordinates $Y_1, \ldots, Y_n$ are independent random variables with density $\mathbf{1}_{[-1,\infty)} e^{-x-1}$. A simple computation shows that

$$\mathbb{E}\left(\frac{\sum_{j=1}^n Y_j}{\sqrt{n}} \|Y\|_2^2\right) = \sqrt{n} \int_{-1}^{\infty} x^3 e^{-x-1} dx = 2\sqrt{n}$$

and the result follows. □

Let us now recall the definition and properties of the logarithmic Laplace transform of a compactly supported Borel probability measure μ on $\mathbb{R}^n$, which we assume is not degenerate (its support is not contained in a hyperplane) and hence can be associated with a convex body

$$K := \text{int}\left[\text{conv}(\text{supp}(\mu))\right].$$

The logarithmic Laplace transform

$$\Lambda_\mu(\xi) = \log\left(\int_{\mathbb{R}^n} e^{\langle \xi, x\rangle} d\mu(x)\right)$$

of μ is a strictly convex C^∞-function on $\mathbb{R}^n$. For every $\xi \in \mathbb{R}^n$ we consider the probability measure μ_ξ whose density $d\mu_\xi/d\mu$ is proportional to the function $x \mapsto \exp(\langle \xi, x\rangle)$. It is easily checked, by differentiation, that

$$\nabla \Lambda_\mu(\xi) = \text{bar}(\mu_\xi) \quad \text{and} \quad \text{Hess}\Lambda_\mu(\xi) = \text{Cov}(\mu_\xi).$$

Following the proof of Proposition 7.2.1 we check that $\nabla \Lambda_\mu(\mathbb{R}^n) = K$. Using the fact that Λ_μ is strictly convex, and hence $\text{Hess}\Lambda_\mu$ is positive-definite everywhere, we may now apply change of variables to get the next lemma:

LEMMA 12.5.4. *With the above notation one has*

$$|K| = |\nabla \Lambda_\mu(\mathbb{R}^n)| = \int_{\mathbb{R}^n} \det(\text{Hess}\Lambda_\mu(\xi))\, d\xi.$$

Note that the fact that we restrict ourselves to compactly supported isotropic log-concave measures does not affect the generality of Theorem 12.5.1 since, by Proposition 2.5.12, we know that

$$L_n = \sup\{L_\mu : \mu \text{ is a finite (non-degenerate) log-concave measure on } \mathbb{R}^n\}$$
$$\leqslant C \sup\{L_\mu : \mu \text{ is uniformly distributed on a convex body in } \mathbb{R}^n\}.$$

Taking into account Lemma 12.3.1, in order to prove Theorem 12.5.1, it is enough to establish the next

THEOREM 12.5.5. *Let μ be a compactly supported isotropic log-concave measure on $\mathbb{R}^n$. Then,*

$$L_\mu \leqslant C\underline{\sigma}_n,$$

where $C > 0$ is an absolute constant.

12.5. THIN SHELL CONJECTURE AND THE HYPERPLANE CONJECTURE

Plan of the proof. We set $p = n/(\underline{\sigma}_n)^2$. By Lemma 12.5.3, we know that $\underline{\sigma}_n \leqslant c_1 \sigma_n$, and thus there can be only few small n for which p might be less than 1, and for those n we have $L_n \lesssim \sqrt{n} \leqslant \underline{\sigma}_n$, so there is nothing more to prove. For all other n we have $p = n/(\underline{\sigma}_n)^2 \in [1, n]$ and thus, recalling Remark 7.5.13, one can check that Theorem 12.5.5 would follow from a lower bound for the "volume radius" of $Z_p(\mu)$ of the form

$$|Z_p(\mu)|^{1/n} \geqslant c\sqrt{\frac{p}{n}},$$

with $c > 0$ an absolute constant; reemploying the notation and arguments of Chapter 7, we see that, in order to prove such a bound, we have to show that

(12.5.2) $\qquad [\det \mathrm{Cov}(\mu_x)]^{\frac{1}{2n}} \gtrsim [\det \mathrm{Cov}(\mu)]^{\frac{1}{2n}} = 1,$

for every $x \in \frac{1}{2}\Lambda_p(\mu)$. $\qquad\square$

To do so, Eldan and Klartag introduce the notion of a log-concave Riemannian package.

DEFINITION 12.5.6. A Riemannian package of dimension n is a quadruple $X = (U, g, \Psi, x_0)$ where U is an open subset of $\mathbb{R}^n$, g is a Riemannian metric on U, $x_0 \in U$ and $\Psi : U \to \mathbb{R}$ is a function with $\Psi(x_0) = 0$.

One can associate a Riemannian metric with any compactly supported Borel probability measure. Furthermore, given that any such measure corresponds to a convex body, in the way that we saw above, and vice versa, we can also think of the procedure as associating a Riemannian metric with any convex body or, more precisely, with any affine equivalence class of convex bodies (since the construction we will now describe does not distinguish between two bodies that belong to the same affine class). We begin by considering a compactly supported (non-degenerate) Borel probability measure μ on $\mathbb{R}^n$, and for every $\xi \in \mathbb{R}^n$ we define

(12.5.3) $\qquad g_\mu(\xi)(u, v) := \langle \mathrm{Cov}(\mu_\xi) u, v \rangle, \qquad u, v \in \mathbb{R}^n.$

Then, $g_\mu(\xi)$ is a positive-definite bilinear form inducing a Riemannian metric on $\mathbb{R}^n$. We also set

(12.5.4) $\qquad \Psi_\mu(\xi) = \log \dfrac{\det \mathrm{Hess}\Lambda_\mu(\xi)}{\det \mathrm{Hess}\Lambda_\mu(0)} = \log \dfrac{\det \mathrm{Cov}(\mu_\xi)}{\det \mathrm{Cov}(\mu)}, \qquad \xi \in \mathbb{R}^n.$

Then $X_\mu = (\mathbb{R}^n, g_\mu, \Psi_\mu, 0)$ is an n-dimensional Riemannian package according to Definition 12.5.6.

Moreover, if $X = (U, g, \Psi, x_0)$ and $Y = (V, h, \Phi, y_0)$ are two Riemannian packages, we say that a function $\varphi : U \to V$ is an isomorphism of X and Y if the following conditions are satisfied:

(i) φ is a Riemannian isometry between (U, g) and (V, h).
(ii) $\varphi(x_0) = y_0$.
(iii) $\Phi(\varphi(x)) = \Psi(x)$ for all $x \in U$.

We then call X and Y isomorphic, and we write $X \cong Y$.

Finally, we say that an n-dimensional Riemannian package is *log-concave* if it is isomorphic to a Riemannian package X_μ, where μ is a non-degenerate compactly supported log-concave probability measure on $\mathbb{R}^n$.

It is useful to also describe a dual construction of the Riemannian package that we associated with a measure μ, a construction that makes the connection with convex bodies more clear. We consider the Legendre transform

$$\Lambda_\mu^*(x) = \sup_{\xi \in \mathbb{R}^n} \left(\langle \xi, x \rangle - \Lambda_\mu(\xi)\right), \qquad \xi \in \mathbb{R}^n$$

and recall that $\Lambda_\mu^* : K \to \mathbb{R}^n$ is a strictly convex C^∞ function and that $\nabla \Lambda_\mu^* : K \to \mathbb{R}^n$ is the inverse map of $\nabla \Lambda_\mu : \mathbb{R}^n \to K$ (recall that $K = \text{int}\,[\text{conv}(\text{supp}(\mu))]$). We define

$$\Phi(x) = \log \frac{\det \text{Hess}\Lambda_\mu^*(\text{bar}(\mu))}{\det \text{Hess}\Lambda_\mu^*(x)}, \qquad x \in K$$

and, for every $x \in K$, we set

$$h(x)(u,v) := \langle [\text{Hess}\Lambda_\mu^*](x) u, v \rangle, \qquad u, v \in \mathbb{R}^n.$$

Then, h is a Riemannian metric on K and

$$[\text{Hess}\Lambda_\mu(\xi)]^{-1} = [\text{Hess}\Lambda_\mu^*](\nabla \Lambda_\mu(\xi)), \qquad \xi \in \mathbb{R}^n.$$

From this identity it follows that the Riemannian package $\tilde{X}_\mu = (K, h, \Phi, \text{bar}(\mu))$ is isomorphic to the Riemannian package $X_\mu = (\mathbb{R}^n, g_\mu, \Psi_\mu, 0)$ under the isomorphism $\xi \to \nabla \Lambda_\mu(\xi)$.

Next, let φ be a smooth real-valued function on a Riemannian manifold (M, g). We will write $\nabla_g \varphi(x_0)$ for the gradient of φ at the point $x_0 \in M$ which we compute with respect to the Riemannian metric g; this belongs to the tangent space $T_{x_0}(M)$ of M at x_0. On the other hand, we will write $\nabla \varphi(x_0)$ for the usual gradient of a function $\varphi : \mathbb{R}^n \to \mathbb{R}$ at a point $x_0 \in \mathbb{R}^n$. Also, the length of a tangent vector $v \in T_{x_0}(M)$ with respect to the metric g is given by $\|v\|_g := \sqrt{g_{x_0}(v, v)}$.

LEMMA 12.5.7. *Let $X = (U, g, \Psi, \xi_0)$ be a log-concave n-dimensional Riemannian package. Then, for every $\xi \in U$ one has*

$$\|\nabla_g \Psi(\xi)\|_g \leqslant \sqrt{n} \underline{\sigma}_n.$$

The proof of Lemma 12.5.7 is given in the appendix of this section.

LEMMA 12.5.8. *Let μ be a non-degenerate compactly supported probability measure on $\mathbb{R}^n$. Let Λ_μ denote the logarithmic Laplace transform of μ and let $X_\mu = (\mathbb{R}^n, g_\mu, \Psi_\mu, 0)$ be the corresponding Riemannian package. Then, for all $\xi, \eta \in \mathbb{R}^n$ one has*

$$d(\xi, \eta) \leqslant \sqrt{\Lambda_\mu(2\xi - \eta) - \Lambda_\mu(\eta) - 2\langle \nabla \Lambda_\mu(\eta), \xi - \eta \rangle},$$

where $d(\xi, \eta)$ is the Riemannian distance between ξ and η with respect to the Riemannian metric g_μ. If $\text{bar}(\mu) = 0$ we get

$$d(\xi, 0) \leqslant \sqrt{\Lambda_\mu(2\xi)}.$$

Proof. In the case $\xi = \eta$ the first claim is obvious. Assuming that $\xi \neq \eta$, we need to define a path from η to ξ with Riemannian length less than or equal to the right hand side of our inequality. We set $\theta = (\xi - \eta)/\|\xi - \eta\|_2$ and $R = \|\xi - \eta\|_2$. If we consider the interval

$$\gamma(t) = \eta + t\theta, \qquad 0 \leqslant t \leqslant R$$

12.5. THIN SHELL CONJECTURE AND THE HYPERPLANE CONJECTURE

then this path connects η and ξ, and has Riemannian length equal to

$$\int_0^R \sqrt{g_\mu(\gamma(t))(\theta,\theta)}\,dt = \int_0^R \sqrt{[\partial_{\theta\theta}\Lambda_\mu](\eta+t\theta)}\,dt$$

$$= \int_0^R \sqrt{\frac{d^2\Lambda_\mu(\eta+t\theta)}{dt^2}}\,dt$$

$$\leqslant \sqrt{\int_0^{2R}(2R-t)\frac{d^2\Lambda_\mu(\eta+t\theta)}{dt^2}\,dt\int_0^R \frac{dt}{2R-t}},$$

by the Cauchy-Schwarz inequality. We have

$$\int_0^R \frac{dt}{2R-t} = \log 2 \leqslant 1.$$

Also, using Taylor's theorem we can write

$$\int_0^R (2R-t)\frac{d^2\Lambda_\mu(\eta+t\theta)}{dt^2}\,dt = \Lambda_\mu(\eta+2R\theta) - [\Lambda_\mu(\eta) + 2R\langle\theta,\nabla\Lambda_\mu(\eta)\rangle].$$

This proves the first inequality. The second claim follows from the first one, because $\Lambda_\mu(0)=0$ and $\nabla\Lambda_\mu(0)=0$ if we assume that $\text{bar}(\mu)=0$. $\square$

Proof of Theorem 12.5.5. Let $X=(\mathbb{R}^n, g_\mu, \Psi_\mu, 0)$ be the Riemannian package corresponding to μ. By Lemma 12.5.8, for any $x \in \frac{1}{2}\Lambda_p(\mu)$ we have that

$$d(0,x) \leqslant \sqrt{\Lambda(2x)} \leqslant \sqrt{p} = \sqrt{n}/\underline{\sigma}_n,$$

and thus, by Lemma 12.5.7 and by the mean value theorem we see that

$$\Psi_\mu(0) - \Psi_\mu(x) \leqslant \sqrt{n}\underline{\sigma}_n\, d(0,x) \leqslant n$$

for any $x \in \frac{1}{2}\Lambda_p(\mu)$. Since $\Psi_\mu(0)=0$ and

$$\Psi_\mu(x) = \log\,[\det\text{Hess}\Lambda_\mu(x)] = \log\,[\det\text{Cov}(\mu_x)],$$

we get

$$\det\text{Cov}(\mu_x) \geqslant e^{-n}$$

for all $x \in \frac{1}{2}\Lambda_p(\mu)$, which implies the theorem. $\square$

REMARK 12.5.9. We add an observation of Eldan and Klartag regarding the quantity $\big\|\mathbb{E}X\|X\|_2^2\big\|_2$ when X is an isotropic log-concave random vector in $\mathbb{R}^n$.

PROPOSITION 12.5.10. *Let X be an isotropic log-concave random vector in $\mathbb{R}^n$. Then,*

$$\big\|\mathbb{E}X\|X\|_2^2\big\|_2^2 \leqslant Cn^3 \int_{S^{n-1}} \left(\mathbb{E}\langle X,\theta\rangle^3\right)^2 d\sigma(\theta),$$

where $C>0$ is an absolute constant.

Proof. We define $F(\theta) := \mathbb{E}\left(\langle X,\theta\rangle^3\right)$. Note that $F(\theta)$ is a homogeneous polynomial of degree 3, with Laplacian

$$\Delta F(\theta) = 6\mathbb{E}\left(\langle X,\theta\rangle\|X\|_2^2\right).$$

We set $v = \mathbb{E}\left(X\|X\|_2^2\right)$. Then, the function

$$\theta \mapsto F(\theta) - \frac{6}{2n+4}\|\theta\|_2^2\langle\theta,v\rangle$$

is a homogeneous harmonic polynomial of degree 3. This means that we can decompose the restriction of F on S^{n-1} into spherical harmonics as follows:

$$F(\theta) = \frac{6}{2n+4}\langle\theta, v\rangle + \left(F(\theta) - \frac{6}{2n+1}\langle\theta, v\rangle\right).$$

Using the orthogonality of spherical harmonics of different degrees, we get

$$\int_{S^{n-1}} F^2(\theta) d\sigma(\theta) \geqslant \frac{36}{(2n+4)^2} \int_{S^{n-1}} \langle\theta, v\rangle^2 d\sigma(\theta)$$
$$= \frac{36}{n(2n+4)^2}\|v\|_2^2,$$

as claimed. □

Proposition 12.5.10 and Theorem 12.5.5 show that, if we could prove that $|\mathbb{E}\langle X, \theta\rangle^3| \leqslant C/n$ for most $\theta \in S^{n-1}$, this would imply a positive answer to the hyperplane conjecture.

12.5.1. Appendix: proof of Lemma 12.5.7

First, we need to establish a series of cases in which certain Riemannian packages are isomorphic.

LEMMA 12.5.11. *Let μ and ν be two non-degenerate compactly supported measures on $\mathbb{R}^n$ such that*

$$\nu = T(\mu)$$

for some linear map $T : \mathbb{R}^n \to \mathbb{R}^n$. Then, $X_\mu \cong X_\nu$.

Proof. One can check that T^* is the required isometry between the Riemannian manifolds $(\mathbb{R}^n, g_\nu)$ and $(\mathbb{R}^n, g_\mu)$. □

REMARK 12.5.12. Another way to convince ourselves that the two packages are isomorphic is to observe that the construction of a Riemannian package can be done in a more abstract way. Consider any n-dimensional vector space V, its dual space V^*, and a compactly supported Borel probability measure μ on V whose support is not contained in a proper affine subspace of V (in the case of the lemma, $V = \mathbb{R}^n$ with some fixed Euclidean structure). We can define the logarithmic Laplace transform of μ as a function $\Lambda : V^* \to \mathbb{R}$ given by

$$\xi \in V^* \to \Lambda(\xi) := \log\left(\int_V \exp(\xi(x)) d\mu(x)\right).$$

The family of probability measures μ_ξ, $\xi \in V^*$, on V can also be well-defined in the same manner. Given $\xi \in V^*$ and $\eta, \zeta \in T_\xi V^* \equiv V^*$, we set

$$g_\xi(\eta, \zeta) := \int_V \eta(x)\zeta(x) d\mu_\xi(x) - \left(\int_V \eta(x) d\mu_\xi(x)\right)\left(\int_V \zeta(x) d\mu_\xi(x)\right).$$

This definition of g_ξ is consistent with (12.5.3). Moreover, there exists a linear operator $A_\xi : V^* \to V^*$, which is self-adjoint and positive-definite with respect to g_0, and satisfies

$$g_\xi(\eta, \zeta) = g_0(A_\xi \eta, \zeta)$$

for all $\eta, \zeta \in V^*$. So, we can define

$$\Psi(\xi) = \log \det A_\xi,$$

and this definition is also consistent with (12.5.4). So, $X_\mu = (V^*, g, \Psi, 0)$ is the Riemannian package corresponding to μ. However, in this situation, the Riemannian package X_ν corresponding to ν is also constructed from the same data, and hence the two packages must be isomorphic.

Lemma 12.5.13. *Let μ and ν be two non-degenerate compactly supported measures on $\mathbb{R}^n$ such that*
$$\nu = T(\mu)$$
for some affine map $T : \mathbb{R}^n \to \mathbb{R}^n$. Then, $X_\mu \cong X_\nu$.

Proof. Note that the only difference from the previous lemma is that here we allow T to be a combination of linear maps and translations. But because of Lemma 12.5.11, we may only deal with the case in which T is a translation of the form
$$T(x) = x + x_0$$
for some $x_0 \in \mathbb{R}^n$. Then,
$$\Lambda_\nu(\xi) = \langle \xi, x_0 \rangle + \Lambda_\mu(\xi), \qquad \xi \in \mathbb{R}^n$$
and, taking second derivatives, we see that $g_\mu = g_\nu$ and $\Psi_\mu = \Psi_\nu$. Thus, $X_\mu = (\mathbb{R}^n, g_\mu, \Psi_\mu, 0)$ and $X_\nu = (\mathbb{R}^n, g_\nu, \Psi_\nu, 0)$ are isomorphic. $\square$

Lemma 12.5.14. *Let $X = (U, g, \Psi, \xi_0)$ be a log-concave Riemannian package. Fix $\xi_1 \in U$ and define*
$$\tilde{\Psi}(\xi) = \Psi(\xi) - \Psi(\xi_1), \qquad \xi \in U.$$
Then, $Y = (U, g, \tilde{\Psi}, \xi_1)$ is an n-dimensional log-concave Riemannian package.

Proof. Let μ be a compactly supported log-concave probability measure on $\mathbb{R}^n$ such that the corresponding Riemannian package $X_\mu = (\mathbb{R}^n, g_\mu, \Psi_\mu, 0)$ is isomorphic to X. We can identify ξ_1 with some point in $\mathbb{R}^n$, which we denote by ξ_1 as well. Then, we may assume that
$$\tilde{\Psi}(\xi) = \Psi(\xi) - \Psi(\xi_1), \qquad \xi \in \mathbb{R}^n.$$
The claim of the lemma becomes now that $Y = (\mathbb{R}^n, g_\mu, \tilde{\Psi}, \xi_1)$ is log-concave.

Recall that μ_{ξ_1} is the compactly supported probability measure on $\mathbb{R}^n$ whose density with respect to μ is proportional to the function $x \mapsto \exp(\langle \xi_1, x \rangle)$. We set $\nu = \mu_{\xi_1}$ and we observe that ν is log-concave and satisfies
$$\Lambda_\nu(\xi) = \Lambda_\mu(\xi + \xi_1) - \Lambda_\mu(\xi_1), \qquad \xi \in \mathbb{R}^n.$$
We will show that Y is isomorphic to $X_\nu = (\mathbb{R}^n, g_\nu, \Psi_\nu, 0)$. To this end, we will prove that the translation
$$\varphi(\xi) = \xi + \xi_1, \qquad \xi \in \mathbb{R}^n,$$
is an isomorphism between X_ν and Y. Using the relation between Λ_ν and Λ_μ we check that
$$\mathrm{Hess}\Lambda_\nu(\xi) = \mathrm{Hess}\Lambda_\mu(\xi + \xi_1), \qquad \xi \in \mathbb{R}^n.$$
So, φ is a Riemannian isometry between $(\mathbb{R}^n, g_\nu)$ and $(\mathbb{R}^n, g_\mu)$, which satisfies $\varphi(0) = \xi_1$. The equality of Hessians implies that $\tilde{\Psi}(\varphi(\xi)) = \Psi_\nu(\xi)$ for all $\xi \in \mathbb{R}^n$. This shows that φ is an isomorphism. $\square$

Proof of Lemma 12.5.7. We first assume that $\xi = \xi_0$. We will show that
$$\|\nabla_g \Psi(\xi_0)\|_g \leqslant \sqrt{n}\underline{\sigma}_n$$
for every n-dimensional log-concave Riemannian package $X = (U, g, \Psi, \xi_0)$. We can identify X with $X_\mu = (\mathbb{R}^n, g_\mu, \Psi_\mu, 0)$ for some compactly supported log-concave probability measure μ on $\mathbb{R}^n$. By applying an affine map we may also assume that μ is isotropic. Then, our claim becomes
$$\|\nabla_{g_\mu} \Psi_\mu(0)\|_{g_\mu} \leqslant \sqrt{n}\underline{\sigma}_n.$$
Since μ is isotropic, we have
$$\mathrm{Hess}\Lambda_\mu(0) = \mathrm{Cov}(\mu) = I.$$
Therefore, we need to show that
$$\|\nabla \Psi_\mu(0)\|_2 \leqslant \sqrt{n}\underline{\sigma}_n.$$

This is equivalent to
$$\partial_\theta \log \frac{\det \mathrm{Hess}\Lambda_\mu(\xi)}{\det \mathrm{Hess}\Lambda_\mu(0)}\Big|_{\xi=0} \leqslant \sqrt{n}\underline{\sigma}_n$$
for all $\theta \in S^{n-1}$. We check that $\partial_\theta \log \det \mathrm{Hess}\Lambda_\mu(\xi)$ is equal to the trace of the matrix
$$(\mathrm{Hess}\Lambda_\mu(\xi))^{-1}(\mathrm{Hess}\partial_\theta \Lambda_\mu(\xi)).$$
Since μ is isotropic, this gives
$$\partial_\theta \log \frac{\det \mathrm{Hess}\Lambda_\mu(\xi)}{\det \mathrm{Hess}\Lambda_\mu(0)}\Big|_{\xi=0} = \Delta \partial_\theta \Lambda_\mu(0)$$
$$= \int_{\mathbb{R}^n} \langle x, \theta \rangle \|x\|_2^2 d\mu(x)$$
$$\leqslant \sqrt{n}\underline{\sigma}_n,$$
by the definition of $\underline{\sigma}_n$, where Δ is the standard Laplacian on $\mathbb{R}^n$. This proves the lemma in the case that $\xi = \xi_0$.

For the general case we employ Lemma 12.5.14. If $\xi \neq \xi_0$, we consider the log-concave Riemannian package $Y = (U, g, \tilde{\Psi}, \xi)$. Applying the previous result for Y we get
$$\|\nabla_g \Psi(\xi)\|_g = \|\nabla_g \tilde{\Psi}(\xi)\|_g \leqslant \sqrt{n}\underline{\sigma}_n,$$
and now the proof is complete. $\square$

12.6. Notes and references

From the thin shell estimate to Gaussian marginals

Using the spherical isoperimetric inequality, Sudakov showed in [**482**] that for every $\delta > 0$ there exists $n(\delta)$ such that if $n \geqslant n(\delta)$ and if μ is a Borel probability measure on $\mathbb{R}^n$ that satisfies $\mathbb{E}_\mu(\langle x, \theta \rangle^2) \leqslant 1$ for all $\theta \in S^{n-1}$ then we may find $A_\delta \subset S^{n-1}$ with $\sigma(A_\delta) \geqslant 1 - \delta$ and such that $\kappa(F_\theta, F) \leqslant \delta$ for all $\theta \in A_\delta$. Variants of the problem were studied by Diaconis and Freedman in [**157**], and by von Weizsäker in [**502**] who considered weighted sums $\langle x, G \rangle$ with Gaussian coefficients and obtained a similar result for a different metric κ^*. Our presentation in Section 12.1 follows Bobkov's article [**78**].

The log-concave case

The case where μ_K is the Lebesgue measure on an isotropic symmetric convex body K in $\mathbb{R}^n$ was studied by Anttila, Ball and Perissinaki in [**17**] who made use of the log-concavity of μ_K and worked with the uniform metric (see Section 12.2; Theorem 12.2.4 is due to Busemann [**126**]).

Around the same time, the same problem was studied by Brehm and Voigt [**117**], Voigt [**500**], Brehm, Vogt and Voigt [**118**], Brehm, Hinow, Vogt and Voigt [**119**], who made sharp computations for specific convex bodies.

Bobkov offers in [**78**] a stronger version of Theorem 12.1.7 for an isotropic measure μ on $\mathbb{R}^n$ which is also log-concave. For any $\delta > 0$ one has
$$\sigma\left(\left\{\theta : \sup_{t \in \mathbb{R}} e^{c|t|}|F_\theta(t) - F(t)| \geqslant \delta\right\}\right) \leqslant C\sqrt{n}\log n \exp(-c\delta^2 n),$$
where $c, C > 0$ are absolute constants. Other contributions related to the central limit problem for convex sets (before or soon after Klartag's first general proof of the thin shell estimate) were made by Koldobsky and Lifshits [**293**], Naor and Romik [**396**], Paouris [**411**], Sodin [**469**], E. Meckes and M. W. Meckes [**357**] and [**358**], E. Milman [**368**], Wojtaszczyk [**503**], Bastero and Bernués [**63**].

The thin shell conjecture

Theorem 12.3.3 is from [**85**] where Bobkov and Koldoky introduced the parameter $\overline{\sigma}_K$ and stated the variance hypothesis, which is an equivalent form of the thin shell conjecture.

Theorem 12.3.5 on the subindependence of coordinate slabs for the ℓ_p^n-balls, $p \geqslant 1$, is due to Ball and Perissinaki [**45**]. The simpler proof that we present in the text is from [**17**] where Anttila, Ball and Perissinaki confirmed the variance hypothesis in this case (see Proposition 12.3.8). Pilipczuk and Wojtaszczyk introduced in [**424**] the so-called negative association property, which is stronger than the sub-independence of coordinate slabs, and they proved it for generalized Orlicz balls; this implies a strong concentration property and moment comparison inequalities for this class of bodies (a simpler proof was then given in [**505**]).

The thin shell conjecture in the unconditional case

The thin shell conjecture for the class of unconditional log-concave measures was established by Klartag in [**279**]. He also obtained the Berry-Esseen type theorem

$$\sup_{a \leqslant b}\left|\mu\left(\left\{x : a \leqslant \sum_{i=1}^n \theta_i x_i \leqslant b\right\}\right) - \frac{1}{\sqrt{2\pi}}\int_a^b e^{-t^2/2}dt\right| \leqslant C\sum_{i=1}^n \theta_i^4$$

for all $\theta = (\theta_1, \ldots, \theta_n) \in S^{n-1}$.

Klartag's approach was extended by Barthe and Cordero-Erausquin in [**56**]. They adapted Klartag's techniques to provide variants and improvements of the Brascamp-Lieb variance inequality which takes into account the invariance properties of the underlying measure and they applied them to spectral gap estimates for log-concave measures with many symmetries and to non-interacting conservative spin systems.

Thin shell conjecture and the hyperplane conjecture

The inequality $L_n \leqslant C\sigma_n$ was proved by Eldan and Klartag in [**166**]. It should be compared with a result of Ball and Nguyen [**43**] that will be presented in Chapter 16 and states that the (stronger) Kannan-Lovász-Simonovits conjecture implies the hyperplane conjecture.

In connection with Remark 12.5.9 we mention another work of Eldan and Klartag [**167**], where it is shown that stability estimates for the Brunn-Minkowski inequality are closely related to the thin shell conjecture.

CHAPTER 13

The thin shell estimate

In this chapter we present a proof of the currently best known estimate for the thin shell problem, which is due to Guédon and E. Milman.

In Chapter 12 we discussed the ε-concentration hypothesis of Anttila, Ball and Perissinaki: "There exists a sequence $\varepsilon_n \downarrow 0$ such that, if K is an isotropic symmetric convex body in $\mathbb{R}^n$, then

$$\mathbb{P}\left(\left|\frac{\|x\|_2}{\sqrt{n}L_K} - 1\right| \geqslant \varepsilon_n\right) \leqslant \varepsilon_n,$$

where L_K is the isotropic constant of K". We saw that this type of statement implies that one-dimensional marginal distributions of uniform measures on isotropic convex bodies are approximately Gaussian with probability close to one.

The ε-hypothesis was studied by many authors and it was verified in some special cases. In [**277**] Bo'az Klartag, in a breakthrough work, gave a positive answer in full generality. Originally, Klartag showed that, if X is an isotropic log-concave measure then, for all $0 < \varepsilon \leqslant 1$ one has

$$\mathbb{P}\left(\left|\frac{\|X\|_2}{\sqrt{n}} - 1\right| \geqslant \varepsilon\right) \leqslant Cn^{-c\varepsilon^2}$$

where $c, C > 0$ are absolute constants. This establishes the ε-hypothesis with $\varepsilon_n \simeq \sqrt{\frac{\log \log n}{\log n}}$. Soon after, Fleury, Guédon and Paouris obtained in [**180**] a slightly weaker result; a main ingredient in their approach was the proof of the fact that if X is an isotropic log-concave random vector in $\mathbb{R}^n$, then for every $1 < q \leqslant c(\log n)^{1/3}$ one has

$$(\mathbb{E}\|X\|_2^q)^{1/q} \leqslant \left(1 + \frac{cq}{(\log n)^{1/3}}\right)(\mathbb{E}\|X\|_2^2)^{1/2}.$$

This allowed them to verify the ε-hypothesis with $\varepsilon_n \simeq (\log \log n)^2/(\log n)^{1/6}$.

Subsequently, Klartag obtained in [**278**] power-type estimates for the ε-hypothesis by showing that

$$\mathbb{E}\left(\frac{\|X\|_2^2}{n} - 1\right)^2 \leqslant \frac{C}{n^\alpha}$$

with some $\alpha \simeq 1/5$. Let us give a short description of his main ideas. Assuming that the density of X is log-concave and *radially symmetric*, one can write the density of $\|X\|_2$ in the form $n\omega_n t^{n-1} f(t)$, $t > 0$. The function $t^{n-1} f(t)$ attains its maximum at some point t_0, and using the log-concavity of f one can conclude that

$$t^{n-1} f(t) \leqslant t_0^{n-1} f(t_0) \exp(-c(t-t_0)^2)$$

whenever $|t - t_0| \leqslant c\sqrt{n}$. This implies that $\|X\|_2$ is highly concentrated around its mean and proves the desired power-type estimate with $\alpha \simeq 1$. The next idea is to reduce the general case to the radial one using concentration of measure arguments. For any subspace E of $\mathbb{R}^n$, let f_E denote the density of the isotropic log-concave random vector $P_E(X)$. Fixing $k \simeq n^b$ for some small $b > 0$ and fixing $r > 0$, Klartag showed that the map

$$(E, \theta) \mapsto \log f_E(r\theta),$$

defined on $G_{n,k} \times S_E$, is more or less equal to a Lipschitz function. Concentration of measure on the Grassmannian $G_{n,k}$ allows one to show that this function is *almost constant*. This leads to the fact that, for most $E \in G_{n,k}$ and *for all* $\theta \in S_E$, the function $f_E(r\theta)$ has more or less the same value. Consequently, for most $E \in G_{n,k}$, the function f_E is approximately radial. Applying the "radial result", one sees that

$$\mathbb{E}\left(\frac{\|P_E(X)\|_2^2}{\sqrt{k}} - 1\right)^2 \leqslant \frac{C}{k}$$

for most $E \in G_{n,k}$. Since $\|X\|_2 \simeq \sqrt{n/k}\|P_E(X)\|_2$ for a random E uniformly distributed in $G_{n,k}$ and independent of X, we can conclude the proof.

Building on the ideas of Klartag (but also on his previous work joint with Guédon and Paouris) Fleury later obtained a stronger thin shell estimate in [**177**]: if X is an isotropic log-concave random vector in $\mathbb{R}^n$, then for every $2 \leqslant q \leqslant c_1 \sqrt[4]{n}$ one has

$$(\mathbb{E}\|X\|_2^q)^{1/q} \leqslant \left(1 + \frac{c_2 q}{\sqrt[4]{n}}\right)(\mathbb{E}\|X\|_2^2)^{1/2},$$

where $c_1, c_2 > 0$ are absolute constants; as a consequence of this fact,

$$\mathbb{P}\left(1 - \frac{t}{n^{1/8}} \leqslant \frac{\|X\|_2}{\sqrt{n}} \leqslant 1 + \frac{t}{n^{1/8}}\right) \geqslant 1 - C_1 e^{-c_3 t}$$

and

$$\mathbb{P}\left(\frac{\|X\|_2}{\sqrt{n}} \geqslant 1 + \frac{t}{n^{1/8}}\right) \leqslant C_2 e^{-c_4 t^2}$$

for every $0 \leqslant t \leqslant n^{1/8}$, where $C_1, C_2, c_3, c_4 > 0$ are absolute constants. We present Fleury's argument in Section 13.1.

Section 13.2 is devoted to the work of Guédon and E. Milman [**240**] that provides the best known estimates for the problem. They showed that, if X is an isotropic log-concave random vector in $\mathbb{R}^n$ and if $1 \leqslant |p - 2| \leqslant c_1 n^{1/6}$, then one has

$$1 - C\frac{|p-2|}{n^{1/3}} \leqslant \frac{(\mathbb{E}\|X\|_2^p)^{1/p}}{(\mathbb{E}\|X\|_2^2)^{1/2}} \leqslant 1 + C\frac{|p-2|}{n^{1/3}},$$

while, if $c_1 n^{1/6} \leqslant |p - 2| \leqslant c_2 \sqrt{n}$, then

$$1 - C\frac{\sqrt{|p-2|}}{n^{1/4}} \leqslant \frac{(\mathbb{E}\|X\|_2^p)^{1/p}}{(\mathbb{E}\|X\|_2^2)^{1/2}} \leqslant 1 + C\frac{\sqrt{|p-2|}}{n^{1/4}}.$$

This leads to the thin shell estimate

$$\mathbb{P}\left(\big|\|X\|_2 - \sqrt{n}\big| \geqslant t\sqrt{n}\right) \leqslant C \exp(-c\sqrt{n} \min\{t^3, t\})$$

for every $t > 0$, where $C, c > 0$ are absolute constants. In particular,

$$\sqrt{\operatorname{Var}(\|X\|_2)} \leqslant C n^{1/3}.$$

We discuss the proofs in detail.

13.1. The method of proof and Fleury's estimate

In this section we describe the thin shell estimate of Fleury in [**177**]; some of the main ideas and tools were previously developed by Klartag.

13.1.1. The function $h_{k,p}$

We start with an isotropic log-concave random vector X in $\mathbb{R}^n$ and write μ for the distribution of X and f for its density.

DEFINITION 13.1.1. Let $1 \leqslant k < n$. We fix $E_0 \in G_{n,k}$ and $\theta_0 \in S_{E_0}$ and for any $p \geqslant -k+1$ we define a function $h_{k,p} : SO(n) \to \mathbb{R}$ by

$$(13.1.1) \qquad h_{k,p}(U) = k\omega_k \int_0^\infty t^{p+k-1} \pi_{U(E_0)} f(tU(\theta_0))\, dt.$$

Our first step will be to rewrite the ratio $\left(\mathbb{E}\|X\|_2^p\right)^{1/p} / \left(\mathbb{E}\|X\|_2^2\right)^{1/2}$ in terms of the functions $h_{k,s}$ on $SO(n)$. In the next section we will also deal with negative values of p. The restrictions on p and k will finally be $|p| \leqslant \frac{n-1}{2}$ and $|p| \leqslant \frac{k-1}{2}$. So, in this subsection we slightly deviate from Fleury's presentation. For the moment, the reader may check the next lemmas for $p \geqslant 2$.

LEMMA 13.1.2. *Let $0 \neq |p| \leqslant \frac{n-1}{2}$ and let k be an integer in $[2,n]$ such that $|p| \leqslant \frac{k-1}{2}$. Then,*

$$\mathbb{E}\|X\|_2^p = c_{n,k,p} \mathbb{E}_{Y,F} \|P_F(Y)\|_2^p = c_{n,k,p} \mathbb{E}_U [h_{k,p}(U)],$$

where

$$c_{n,k,p} = \frac{\Gamma\left(\frac{p+n}{2}\right)\Gamma\left(\frac{k}{2}\right)}{\Gamma\left(\frac{n}{2}\right)\Gamma\left(\frac{p+k}{2}\right)}.$$

Proof. There exists a constant $c_{n,k,p} > 0$ such that

$$\|x\|_2^p = c_{n,k,p} \int_{G_{n,k}} \|P_F(x)\|_2^p d\nu_{n,k}(F)$$

for all $x \in \mathbb{R}^n$. Therefore,

$$\frac{\mathbb{E}\|X\|_2^p}{\mathbb{E}\|G_n\|_2^p} = \frac{\mathbb{E}_{X,F}\|P_F(X)\|_2^p}{\mathbb{E}_{X,F}\|P_F(G_n)\|_2^p} = \frac{\mathbb{E}_{X,F}\|P_F(X)\|_2^p}{\mathbb{E}\|G_k\|_2^p},$$

where G_s is a standard Gaussian random vector in $\mathbb{R}^s$. Then we check that

$$\mathbb{E}\|G_s\|_2^p = \frac{2^{p/2}\Gamma\left(\frac{p+s}{2}\right)}{\Gamma\left(\frac{s}{2}\right)},$$

and hence

$$(13.1.2) \qquad \mathbb{E}\|X\|_2^p = \frac{\Gamma\left(\frac{p+n}{2}\right)\Gamma\left(\frac{k}{2}\right)}{\Gamma\left(\frac{n}{2}\right)\Gamma\left(\frac{p+k}{2}\right)} \mathbb{E}_{X,F}\|P_F(X)\|_2^p.$$

Recall from (13.1.1) that

$$h_{k,p}(U) = k\omega_k \int_0^\infty t^{p+k-1} \pi_{U(E_0)} f(tU(\theta_0))\, dt$$

for some fixed $E_0 \in G_{n,k}$ and some fixed $\theta_0 \in S_{E_0}$. We compute $\mathbb{E}_{X,F}\|P_F(X)\|_2^p$ integrating in polar coordinates on $F \in G_{n,k}$ and then using the $SO(n)$-invariance of the Haar measures on $G_{n,k}$, S_F and $SO(n)$ we see that

(13.1.3) $$\mathbb{E}_{X,F}\|P_F(X)\|_2^p = \mathbb{E}_U[h_{k,p}(U)].$$

The result follows. □

Note. Setting $p = 0$ in the last equation we see that

(13.1.4) $$\mathbb{E}_U[h_{k,0}(U)] = 1.$$

Fleury uses the formula of Lemma 13.1.2 to get

LEMMA 13.1.3. *For any $1 \leqslant k \leqslant n$ and $p \geqslant 2$, we have*

(13.1.5) $$\frac{(\mathbb{E}\|X\|_2^p)^{1/p}}{(\mathbb{E}\|X\|_2^2)^{1/2}} \leqslant \frac{(\mathbb{E}_{X,F}\|P_F(X)\|_2^p)^{1/p}}{(\mathbb{E}_{X,F}\|P_F(X)\|_2^2)^{1/2}}.$$

Proof. We need to show that

(13.1.6) $$\frac{c_{n,k,p}}{c_{n,k,2}^{p/2}} \leqslant 1.$$

We consider the function

$$s \mapsto J(s) = \frac{\mathbb{E}\|G_s\|_2^p}{(\mathbb{E}\|G_s\|_2^2)^{p/2}} = \frac{\Gamma\left(\frac{s+p}{2}\right)\Gamma\left(\frac{s}{2}\right)^{\frac{p}{2}-1}}{\Gamma\left(\frac{s+2}{2}\right)^{\frac{p}{2}}}.$$

Using the fact that the function $(\log \circ \Gamma)'$ is concave one can check that J is a decreasing function of s. On observing that

$$\frac{c_{n,k,p}}{c_{n,k,2}^{p/2}} = \frac{J(n)}{J(k)},$$

we verify (13.1.6). □

13.1.2. Log-Lipschitz constant of $h_{k,p}$

Let X be an isotropic log-concave random vector in $\mathbb{R}^n$. We write μ for the distribution of X and f for its density. Lemma 13.1.2 and (13.1.4) show that

(13.1.7) $$\frac{(\mathbb{E}\|X\|_2^p)^{1/p}}{(\mathbb{E}\|X\|_2^2)^{1/2}} \leqslant \frac{(\mathbb{E}_U[h_{k,p}(U)])^{1/p}(\mathbb{E}_U[h_{k,0}(U)])^{1/2-1/p}}{(\mathbb{E}_U[h_{k,2}(U)])^{1/2}}$$

for all $p \geqslant 2$.

In order to bound the right hand side we will use the logarithmic Sobolev inequality for $SO(n)$ and to this end we would like to have control on the log-Lipschitz constant $\|\log h_{k,s}\|_{\text{Lip}}$ for different values of s. It will be convenient to consider the isotropic log-concave random vector $Y = \frac{X+G_n}{\sqrt{2}}$ where G_n is a standard Gaussian random vector in $\mathbb{R}^n$ which is independent from X. We write ν for the distribution of Y and g for its density.

This idea was used by Klartag and Fleury who considered a truncated version of the function $h_{k,p}$; we fix $E_0 \in G_{n,k}$ and $\theta_0 \in S_{E_0}$, and for a fixed $\delta > 0$ (that will be suitably chosen) and any $p \geqslant 0$ we define $\tilde{h}_{k,p} : SO(n) \to [0, +\infty)$ by

(13.1.8) $$\tilde{h}_{k,p}(U) = \int_0^{\delta\sqrt{k}} t^{p+k-1}[\pi_{U(E_0)}(g)](tU(\theta_0))dt.$$

Klartag obtained the following estimate for the Lipschitz constant of $\log \tilde{h}_{k,p}$:

PROPOSITION 13.1.4 (Klartag). *Let $E_0 \in G_{n,k}$ and $x_0 \in E_0$ with $\|x_0\|_2 \leqslant \delta\sqrt{k}$. For the function $M : SO(n) \to \mathbb{R}$ defined by*
$$M(U) = \log \pi_{U(F_0)} g(U(x_0))$$
one has $\|M\|_{\mathrm{Lip}} \leqslant Ck^2$. Consequently, for every $p \geqslant 0$,
$$\|\log \tilde{h}_{k,p}\|_{\mathrm{Lip}} \leqslant Ck^2.$$

We will not provide the proof of Proposition 13.1.4 since, as we will see in the next section, Guédon and E. Milman replaced it with a stronger one. It is now clear that in order to continue with the original plan of proof, we need to have a variant of 13.1.7 where the functions $h_{k,s}$ will be replaced by the functions $\tilde{h}_{k,s}$ (for a suitable choice of δ). This is accomplished in the next proposition.

PROPOSITION 13.1.5. *There exists an absolute constant $\delta > 0$ such that*
$$\frac{(\mathbb{E}\|Y\|_2^p)^{1/p}}{(\mathbb{E}\|Y\|_2^2)^{1/2}} \leqslant \left(1 + \overline{C}_1 e^{-\overline{c}_2 \min\{k, \sqrt{n}\}}\right) \frac{(\mathbb{E}\tilde{h}_{k,p}(U))^{1/p}(\mathbb{E}\tilde{h}_{k,0}(U))^{1/2-1/p}}{(\mathbb{E}\tilde{h}_{k,2}(U))^{1/2}},$$
for all $2 \leqslant p \leqslant \overline{c}_1 \min\{k, \sqrt{n}\}$, where $\overline{C}_1, \overline{c}_1, \overline{c}_2 > 0$ are absolute constants.

For the proof we use the fact that for any $1 \leqslant k \leqslant n$ and any $y \neq 0$ in $\mathbb{R}^n$ one has
$$(13.1.9) \qquad \nu_{n,k}\left(\left\{F \in G_{n,k} : \|P_F(y)\|_2 \geqslant 2s\sqrt{k/n}\|y\|_2\right\}\right) \leqslant C_2 e^{-c_1 s^2 k}$$
for every $s \geqslant 1$. This is a standard consequence of the spherical isoperimetric inequality. We also use Paouris' inequality: for every $s \geqslant 1$ we also have
$$(13.1.10) \qquad \mathbb{P}(\|Y\|_2 \geqslant C_4 s\sqrt{n}) \leqslant C_3 e^{-c_2 s\sqrt{n}}.$$
Here, $C_2, C_3, C_4 > 1$ and $c_1, c_2 > 0$ are absolute constants.

LEMMA 13.1.6. *Let $\delta = 2C_4$. Then, for all $1 \leqslant k < n$ and $2 \leqslant p \leqslant \overline{c}\min\{k, \sqrt{n}\}$, we have*
$$\mathbb{E}_Y\left(\int_{G_{n,k}} \|P_F(Y)\|_2^p d\nu_{n,k}(F)\right)$$
$$\leqslant \left(1 + C_1 e^{-c_1 \min\{k, \sqrt{n}\}}\right) \mathbb{E}_Y\left(\int_{G_{n,k}} \|P_F(Y)\|_2^p \mathbf{1}_{\{\|P_F(Y)\|_2 \leqslant \delta\sqrt{k}\}} d\nu_{n,k}(F)\right),$$
where $C_1 > 1$ and $c_1, \overline{c} > 0$ are absolute constants.

Proof. We write
$$\mathbb{P}_{Y,F}\left(\{\|P_F(Y)\|_2 \geqslant \delta\sqrt{k}\}\right)$$
$$= \mathbb{E}_Y\left(\mathbb{P}_F(\{\|P_F(Y)\|_2 \geqslant \delta\sqrt{k}\})\right)$$
$$= \mathbb{E}_Y\left[\left(\mathbb{P}_F(\{\|P_F(Y)\|_2 \geqslant \delta\sqrt{k}\})\right) \mathbf{1}_{\{\|Y\|_2 \leqslant \frac{\delta}{2}\sqrt{n}\}}\right]$$
$$+ \mathbb{E}_Y\left[\left(\mathbb{P}_F(\{\|P_F(Y)\|_2 \geqslant \delta\sqrt{k}\})\right) \mathbf{1}_{\{\|Y\|_2 \geqslant \frac{\delta}{2}\sqrt{n}\}}\right]$$
$$\leqslant \mathbb{E}_Y\left(\mathbb{P}_F\left(\left\{\|P_F(Y)\|_2 \geqslant 2\|Y\|_2\sqrt{k/n}\right\}\right)\right) + \mathbb{P}_Y\left(\|Y\|_2 \geqslant \delta\sqrt{n}/2\right).$$

By the choice of δ we have $\delta\sqrt{n}/2 = C_4\sqrt{n}$. Using (13.1.9) and (13.1.10) (with $s = 1$) we get

(13.1.11) $\quad \mathbb{P}_{Y,F}\left(\{\|P_F(Y)\|_2 \geqslant \delta\sqrt{k}\}\right) \leqslant C_2 e^{-c_1 k} + C_3 e^{-c_2\sqrt{n}} \leqslant C_5 e^{-c_3 \min\{k,\sqrt{n}\}}.$

Since Y is log-concave, we know that

$$(\mathbb{E}\|Y\|_2^{2p})^{1/2p} \leqslant C_6 (\mathbb{E}\|Y\|_2^p)^{1/p}.$$

Then, using (13.1.2) we see that

$$\left(\mathbb{E}_Y\left(\int_{G_{n,k}} \|P_F(Y)\|_2^{2p} d\nu_{n,k}(F)\right)\right)^{\frac{1}{2p}} \leqslant C_7 \left(\mathbb{E}_Y\left(\int_{G_{n,k}} \|P_F(Y)\|_2^p d\nu_{n,k}(F)\right)\right)^{\frac{1}{p}}.$$

An application of the Cauchy-Schwarz inequality gives

$$\mathbb{E}_Y \left(\int_{G_{n,k}} \|P_F(Y)\|_2^p \mathbf{1}_{\{\|P_F(Y)\|_2 \geqslant \delta\sqrt{k}\}} d\nu_{n,k}(F)\right)$$

$$\leqslant \left(\mathbb{E}_Y \left(\int_{G_{n,k}} \|P_F(Y)\|_2^{2p} d\nu_{n,k}(F)\right)\right)^{\frac{1}{2}} \cdot \mathbb{P}_{Y,F}(\{\|P_F(Y)\|_2 \geqslant \delta\sqrt{k}\})^{\frac{1}{2}}$$

$$\leqslant C_8^p e^{-c_3 \min\{k,\sqrt{n}\}} \mathbb{E}_Y \left(\int_{G_{n,k}} \|P_F(Y)\|_2^p d\nu_{n,k}(F)\right)$$

$$\leqslant e^{-c_4 \min\{k,\sqrt{n}\}} \mathbb{E}_Y \left(\int_{G_{n,k}} \|P_F(Y)\|_2^p d\nu_{n,k}(F)\right),$$

provided that $p \leqslant \bar{c}\min\{k, \sqrt{n}\}$ for a small enough absolute constant $\bar{c} > 0$, and the lemma follows. $\square$

Proof of Proposition 13.1.5. We may assume that k is large enough (otherwise, we anyway have $\bar{c}_1 \min\{k, \sqrt{n}\} < 2$ by adjusting the constant $\bar{c}_1$, and then there is nothing to prove). From (13.1.11) we know that, for $\delta = 2C_4$,

$$\mathbb{P}_{Y,F}\left(\{\|P_F(Y)\|_2 \geqslant \delta\sqrt{k}\}\right) \leqslant C_5 e^{-c_3 \min\{k,\sqrt{n}\}},$$

and hence

$$(1 + 2C_5 e^{-c_3' \min\{k,\sqrt{n}\}})\mathbb{P}_{Y,F}\left(\{\|P_F(Y)\|_2 \leqslant \delta\sqrt{k}\}\right)^{1/2-1/p} \geqslant 1$$

Using also Lemma 13.1.3 and Lemma 13.1.6 we write

$$\frac{(\mathbb{E}\|Y\|_2^p)^{1/p}}{(\mathbb{E}\|Y\|_2^2)^{1/2}} \leqslant \frac{(\mathbb{E}_{F,Y}\|P_F(Y)\|_2^p)^{1/p}}{(\mathbb{E}_{F,Y}\|P_F(Y)\|_2^2)^{1/2}} \leqslant (1 + C_9 e^{-c_4 \min\{k,\sqrt{n}\}})$$

$$\times \frac{(\mathbb{E}_{Y,F}(\|P_F(Y)\|_2^p \mathbf{1}_{\{\|P_F(Y)\|_2 \leqslant \delta\sqrt{k}\}}))^{1/p} \mathbb{P}_{Y,F}(\|P_F(Y)\|_2 \leqslant \delta\sqrt{k})^{1/2-1/p}}{(\mathbb{E}_{Y,F}(\|P_F(Y)\|_2^2 \mathbf{1}_{\{\|P_F(Y)\|_2 \leqslant \delta\sqrt{k}\}}))^{1/2}}.$$

To finish the proof notice that, for $q = 0, 2$ and p, one has

$$\mathbb{E}\left[\|P_F(Y)\|_2^q \mathbf{1}_{\{\|P_F(Y)\|_2 \leqslant \delta\sqrt{k}\}}\right] = \mathbb{E}_F\left[k\omega_k \int_{S_F} \int_0^{\delta\sqrt{k}} t^{k+q-1} \pi_F g(t\theta)\, dt\, d\sigma(\theta)\right]$$

$$= \mathbb{E}_U(\tilde{h}_{k,q}(U))$$

where the last equality follows by integration in polar coordinates on $F \in G_{n,k}$ and the $SO(n)$-invariance of the Haar measures on $G_{n,k}$, S_F and $SO(n)$ as in Lemma 13.1.2. □

13.1.3. Logarithmic Sobolev inequality

Having Proposition 13.1.5 Fleury is then using the logarithmic Sobolev inequality for $SO(n)$ (equipped with the geodesic distance d).

THEOREM 13.1.7. *For every Lipschitz continuous function $F : SO(n) \to \mathbb{R}$ one has*
$$\mathrm{Ent}(F^2) \leqslant \frac{C}{n} \mathbb{E} |\nabla F(U)|^2,$$
where
$$\mathrm{Ent}(F^2) = \mathbb{E}\left[F^2 \log(F^2)\right] - \mathbb{E}\left(F^2\right) \log(\mathbb{E}\left(F^2\right))$$
and
$$|\nabla F(U)| = \limsup_{d(V,U) \to 0} \frac{|F(V) - F(U)|}{d(V,U)}.$$

Fleury uses Theorem 13.1.7 in order to compare the moments of the function $\tilde{h}_{k,p}(U)$. More generally, if $h : SO(n) \to \mathbb{R}^+$ is a function with $\|\log h\|_{\mathrm{Lip}} \leqslant A$ then, applying the log-Sobolev inequality for $F = h^{q/2}$ we see that, for any $q > 0$,

$$\frac{d}{dq}\left[\log\left((\mathbb{E}\,(h^q))^{1/q}\right)\right] = \frac{1}{q^2} \frac{\mathrm{Ent}\,(h^q)}{\mathbb{E}\,(h^q)} \leqslant \frac{C}{n} \frac{\mathbb{E}\,[h^{q-2}|\nabla h|^2]}{\mathbb{E}\,(h^q)}$$
$$= \frac{C}{n} \frac{\mathbb{E}\,[h^q |\nabla(\log h)|^2]}{\mathbb{E}\,(h^q)} \leqslant \frac{CA^2}{n},$$

because $\nabla h = h(\nabla(\log h))$. Integrating this, we get

(13.1.12) $$(\mathbb{E}\,|h|^q)^{1/q} \leqslant \exp\left(\frac{CA^2}{n}(q - r)\right)(\mathbb{E}\,|h|^r)^{1/r}$$

for all $q > r > 0$.

PROPOSITION 13.1.8. *Let $4 \leqslant p \leqslant k$ and set $A = \max\{\|\log \tilde{h}_{k,p}\|_{\mathrm{Lip}}, \|\log \tilde{h}_{k,0}\|_{\mathrm{Lip}}\}$. If $A \leqslant \bar{c}\sqrt{n}$ then*

$$(\mathbb{E}\,(\tilde{h}_{k,p}))^{1/p}(\mathbb{E}\,(\tilde{h}_{k,0}))^{1/2 - 1/p} \leqslant \left(1 + \frac{CA^2}{pn} + \frac{Cp}{k}\right)(\mathbb{E}\,(\tilde{h}_{k,2}))^{1/2}.$$

For the proof of Proposition 13.1.8 we employ Theorem 2.2.5 which we state here again as a lemma.

LEMMA 13.1.9. *Let $\phi : \mathbb{R}^+ \to \mathbb{R}^+$ be an integrable log-concave function. Then, the function $\Psi : (0, \infty) \to \mathbb{R}$ defined by*
$$\Psi(p) = \frac{1}{\Gamma(p)} \int_0^\infty t^{p-1} \phi(t)\, dt$$
is log-concave.

Proof of Proposition 13.1.8. We fix $p \geqslant 4$. Applying Lemma 13.1.9 for the log-concave function
$$\phi(t) = \pi_{U(F_0)} g(tU(\theta_0)) \cdot \mathbf{1}_{[0, \delta\sqrt{k}]}(t),$$

for every $U \in SO(n)$ we have
$$\frac{1}{\Gamma(k+2)}\tilde{h}_{k,2}(U) \geqslant \Big(\frac{1}{\Gamma(k+p)}\tilde{h}_{k,p}(U)\Big)^{2/p}\Big(\frac{1}{\Gamma(k)}\tilde{h}_{k,0}(U)\Big)^{1-2/p}.$$
Simple computations with the Gamma function show that
$$\frac{[\Gamma(k+p)]^{2/p}[\Gamma(k)]^{1-2/p}}{\Gamma(k+2)} \leqslant \Big(\frac{k+p}{k+2}\Big)^2 \leqslant 1 + \frac{3p}{k}.$$
Therefore,
$$(13.1.13) \qquad [\tilde{h}_{k,p}(U)]^{2/p}[\tilde{h}_{k,0}(U)]^{1-2/p} \leqslant \Big(1 + \frac{3p}{k}\Big)\tilde{h}_{k,2}(U).$$
We apply (13.1.12) four times:

(i) If $h = \tilde{h}_{k,p}$ and $q = 1$, $r = 2/p$, we get
$$(13.1.14) \qquad \mathbb{E}\,(\tilde{h}_{k,p}^{2/p}) \geqslant e^{-\frac{C_1 A^2}{pn}}(\mathbb{E}\,(\tilde{h}_{k,p}))^{2/p}.$$

(ii) If $h = \tilde{h}_{k,0}$ and $q = 1$, $r = 1 - 2/p$, we get
$$(13.1.15) \qquad \mathbb{E}\,(\tilde{h}_{k,0}^{1-2/p}) \geqslant e^{-\frac{C_2 A^2}{pn}}(\mathbb{E}\,(\tilde{h}_{k,0}))^{1-2/p}.$$

(iii) If $h = \tilde{h}_{k,p}$ and $q = 4/p$, $r = 2/p$, we get
$$(13.1.16) \qquad \mathbb{E}\,(\tilde{h}_{k,p}^{4/p}) \leqslant e^{\frac{C_3 A^2}{p^2 n}}(\mathbb{E}\,(\tilde{h}_{k,p}^{2/p}))^2$$

(iv) If $h = \tilde{h}_{k,0}$ and $q = 2(1 - 2/p)$, $r = 1 - 2/p$, we get
$$(13.1.17) \qquad \mathbb{E}\,(\tilde{h}_{k,0}^{2(1-2/p)}) \leqslant e^{\frac{C_4 A^2}{n}}(\mathbb{E}\,(\tilde{h}_{k,0}^{1-2/p}))^2.$$

Denoting by $\mathrm{Cov}(Z_1, Z_2)$ the covariance of two random variables Z_1, Z_2, from (13.1.16) and (13.1.17) we see that if $A \leqslant \bar{c}\sqrt{n}$ for some small enough absolute constant $\bar{c} > 0$, then
$$|\mathrm{Cov}(\tilde{h}_{k,p}^{2/p}, \tilde{h}_{k,0}^{1-2/p})| \leqslant \sqrt{\mathrm{Var}(\tilde{h}_{k,p}^{2/p})}\sqrt{\mathrm{Var}(\tilde{h}_{k,0}^{1-2/p})}$$
$$\leqslant \Big(e^{\frac{C_3 A^2}{np^2}} - 1\Big)^{1/2}\Big(e^{\frac{C_4 A^2}{n}} - 1\Big)^{1/2}\mathbb{E}\,(\tilde{h}_{k,p}^{2/p})\mathbb{E}\,(\tilde{h}_{k,0}^{1-2/p})$$
$$\leqslant \frac{C_5 A^2}{pn}\mathbb{E}\,(\tilde{h}_{k,p}^{2/p})\mathbb{E}\,(\tilde{h}_{k,0}^{1-2/p}).$$

Then, if we integrate (13.1.13), taking into account our bound for $|\mathrm{Cov}(\tilde{h}_{k,p}^{2/p}, \tilde{h}_{k,0}^{1-2/p})|$ and using (13.1.14) and (13.1.15) in the last step, we see that
$$\Big(1 + \frac{3p}{k}\Big)\mathbb{E}_U(\tilde{h}_{k,2}) \geqslant \mathbb{E}_U[\tilde{h}_{k,p}^{2/p}\tilde{h}_{k,0}^{1-2/p}]$$
$$= \mathbb{E}_U(\tilde{h}_{k,p}^{2/p})\mathbb{E}_U(\tilde{h}_{k,0}^{1-2/p}) + \mathrm{Cov}(\tilde{h}_{k,p}^{2/p}, \tilde{h}_{k,0}^{1-2/p})$$
$$\geqslant \Big(1 - \frac{C_5 A^2}{pn}\Big)\mathbb{E}_U(\tilde{h}_{k,p}^{2/p})\mathbb{E}_U(\tilde{h}_{k,0}^{1-2/p})$$
$$\geqslant \Big(1 - \frac{C_6 A^2}{pn}\Big)(\mathbb{E}\,(\tilde{h}_{k,p}))^{2/p}(\mathbb{E}_U(\tilde{h}_{k,0}))^{1-2/p},$$

and the proposition follows because
$$1 + \frac{3p}{k} \leqslant \left(1 - \frac{C_6 A^2}{pn}\right)\left(1 + \frac{CA^2}{pn} + \frac{Cp}{k}\right)$$
if we choose $C > 0$ large enough and $\bar{c} > 0$ small enough. $\square$

13.1.4. Fleury's estimate

The precise statement of Fleury's theorem is the following.

THEOREM 13.1.10 (Fleury). *Let X be an isotropic log-concave random vector in $\mathbb{R}^n$. For every $2 \leqslant p \leqslant c_1 \sqrt[4]{n}$ one has*

(13.1.18) $$(\mathbb{E}\|X\|_2^2)^{1/2} \leqslant (\mathbb{E}\|X\|_2^p)^{1/p} \leqslant \left(1 + \frac{c_2 p}{\sqrt[4]{n}}\right)(\mathbb{E}\|X\|_2^2)^{1/2},$$

where $c_1, c_2 > 0$ are absolute constants. As a consequence, one has

(13.1.19) $$\mathbb{P}\left(1 - \frac{t}{n^{1/8}} \leqslant \frac{\|X\|_2}{\sqrt{n}} \leqslant 1 + \frac{t}{n^{1/8}}\right) \geqslant 1 - C_1 e^{-c_3 t}$$

and

(13.1.20) $$\mathbb{P}\left(\frac{\|X\|_2}{\sqrt{n}} \geqslant 1 + \frac{t}{n^{1/8}}\right) \leqslant C_2 e^{-c_4 t^2}$$

for every $0 \leqslant t \leqslant n^{1/8}$, where $C_1, C_2, c_3, c_4 > 0$ are absolute constants.

Proof. Recall that $Y = \frac{X + G_n}{\sqrt{2}}$. Combining Proposition 13.1.5 with Proposition 13.1.8, and taking into account the estimate of Proposition 13.1.4 for the Lipschitz constant of $\log \tilde{h}_{k,p}$, we see that if $k \leqslant \sqrt[4]{n}$ then for every integer $4 \leqslant p \leqslant k/2$ we have

(13.1.21) $$(\mathbb{E}\|Y\|_2^p)^{1/p} \leqslant \left(1 + \frac{c_1 p}{k} + \frac{c_2 k^4}{pn}\right)(\mathbb{E}\|Y\|_2^2)^{1/2}.$$

The theorem will follow by the next lemmas:

LEMMA 13.1.11. *For all $k \leqslant \sqrt[4]{n}$ and $\sqrt{k} \leqslant p \leqslant k/2$ we have*

(13.1.22) $$(\mathbb{E}\|X\|_2^p)^{1/p} \leqslant \left(1 + \frac{c_3 p}{k}\right)(\mathbb{E}\|X\|_2^2)^{1/2}$$

Proof. For every integer $p \geqslant 2$, using the symmetry and independence of G_n, the convexity of the function $t \mapsto t^p$ and the Cauchy-Schwarz inequality, we write

$$\begin{aligned}
\mathbb{E}\|Y\|_2^{2p} &= \mathbb{E}\left(\frac{\|X + G_n\|_2^2}{2}\right)^p \\
&= \frac{1}{2}\mathbb{E}\left(\left(\frac{\|X + G_n\|_2^2}{2}\right)^p + \left(\frac{\|X - G_n\|_2^2}{2}\right)^p\right) \\
&\geqslant \mathbb{E}\left(\frac{\|X\|_2^2 + \|G_n\|_2^2}{2}\right)^p \\
&\geqslant \mathbb{E}\left(\|X\|_2^p \|G_n\|_2^p\right) = \mathbb{E}\|X\|_2^p \, \mathbb{E}\|G_n\|_2^p \\
&\geqslant \mathbb{E}\|X\|_2^p \left(\mathbb{E}\|G_n\|_2^2\right)^{p/2} = n^{p/2}\mathbb{E}\|X\|_2^p.
\end{aligned}$$

Since $\mathbb{E}\|X\|_2^2 = \mathbb{E}\|Y\|_2^2 = n$, this implies that

$$(13.1.23) \quad \frac{(\mathbb{E}\|X\|_2^p)^{1/p}}{(\mathbb{E}\|X\|_2^2)^{1/2}} \leqslant \left(\frac{(\mathbb{E}\|Y\|_2^{2p})^{1/(2p)}}{(\mathbb{E}\|Y\|_2^2)^{1/2}}\right)^2 \leqslant \left(1 + \frac{2c_1 p}{k} + \frac{c_2 k^4}{2pn}\right)^2$$

$$\leqslant \left(1 + \frac{2c_1 p}{k} + \frac{c_2}{2p}\right)^2 \leqslant \left(1 + \frac{2c_1 p}{k} + \frac{c_2 p}{2k}\right)^2$$

$$\leqslant 1 + \frac{c_3 p}{k}$$

for some absolute constant $c_3 > 0$ (recall that $k^4 \leqslant n$ and note that $p/k \geqslant 1/p$ by the restrictions on k and p). $\square$

LEMMA 13.1.12. *If $\sqrt[8]{n} \leqslant p \leqslant c_4 \sqrt[4]{n}$ then*

$$(\mathbb{E}\|X\|_2^p)^{1/p} \leqslant \left(1 + \frac{c_5 p}{\sqrt[4]{n}}\right)(\mathbb{E}\|X\|_2^2)^{1/2},$$

$c_4, c_5 > 0$ are absolute constants.

Proof. We choose $k = \lfloor \sqrt[4]{n} \rfloor$ and apply Lemma 13.1.11. $\square$

LEMMA 13.1.13. *For every $0 \leqslant t \leqslant n^{1/8}$,*

$$\mathbb{P}\left(\frac{\|X\|_2}{\sqrt{n}} \geqslant 1 + \frac{t}{n^{1/8}}\right) \leqslant C_2 e^{-c_5 t^2}.$$

Proof. We set $k = \lfloor \sqrt[4]{n} \rfloor$. Let $t > 0$ with $2 \leqslant t \leqslant \sqrt{k}$. Choose $p = \frac{t\sqrt{k}}{2}$ and observe that $\sqrt{k} \leqslant p \leqslant k/2$. From Markov's inequality,

$$\mathbb{P}\left(\|X\|_2 \geqslant \left(1 + \frac{t}{\sqrt{k}}\right)(\mathbb{E}\|X\|_2^p)^{1/p}\right) \leqslant \left(1 + \frac{t}{\sqrt{k}}\right)^{-p}$$

$$\leqslant \exp\left(-\frac{pt}{2\sqrt{k}}\right) = \exp\left(-\frac{t^2}{4}\right),$$

using the inequality $e^{-a/2}(1+a) \geqslant 1$ for $a \in [0,1]$. From (13.1.22) we get

$$(13.1.24)$$

$$\mathbb{P}\left(\|X\|_2 \geqslant \left(1 + \frac{C_3 t}{\sqrt{k}}\right)(\mathbb{E}\|X\|_2^2)^{1/2}\right) \leqslant \mathbb{P}\left(\|X\|_2 \geqslant \left(1 + \frac{t}{\sqrt{k}}\right)(\mathbb{E}\|X\|_2^p)^{1/p}\right)$$

$$\leqslant \exp\left(-\frac{t^2}{4}\right)$$

where $C_3 > 0$ is an absolute constant such that

$$\left(1 + \frac{t}{\sqrt{k}}\right)\left(1 + \frac{c_3 p}{k}\right) \leqslant 1 + \frac{C_3 t}{\sqrt{k}}$$

for our choice of $p\,(= t\sqrt{k}/2)$. $\square$

LEMMA 13.1.14. *If $0 \leqslant t \leqslant \sqrt[8]{n}$ then*

$$\mathbb{P}\left(1 - \frac{t}{n^{1/8}} \leqslant \frac{\|X\|_2}{\sqrt{n}} \leqslant 1 + \frac{t}{n^{1/8}}\right) \geqslant 1 - C_1 e^{-c_6 t}.$$

Proof. We first observe that, because of (13.1.22), we can choose $k \simeq \sqrt[4]{n}$ and $p \simeq \sqrt{k}$ so that
$$\operatorname{Var}(\|X\|_2^p) \leqslant \frac{1}{16}(\mathbb{E}\|X\|_2^p)^2.$$
Then, from Markov's inequality we get
$$\frac{1}{4} \geqslant \mathbb{P}\Big(\big|\|X\|_2^p - (\mathbb{E}\|X\|_2^p)\big| \geqslant \frac{1}{2}\mathbb{E}\|X\|_2^p\Big) \geqslant \mathbb{P}\Big(\|X\|_2 \leqslant \frac{1}{2^{1/p}}(\mathbb{E}\|X\|_2^p)^{1/p}\Big)$$
$$\geqslant \mathbb{P}\Big(\|X\|_2 \leqslant \Big(1 - \frac{c_7}{\sqrt{k}}\Big)(\mathbb{E}\|X\|_2^2)^{1/2}\Big).$$
On the other hand, by (13.1.24) we have
$$\mathbb{P}\Big(\|X\|_2 \leqslant \Big(1 + \frac{c_8 t}{\sqrt{k}}\Big)(\mathbb{E}\|X\|_2^2)^{1/2}\Big) \geqslant \frac{3}{4}.$$
We may assume that $t > 1$, otherwise (13.1.19) is trivially true. We define $c_9 = \max\{c_7, c_8\}$ and setting $\lambda = \frac{2}{1+t}$ we write
$$1 - \frac{c_9}{\sqrt{k}} = \lambda\Big(1 - \frac{c_9 t}{\sqrt{k}}\Big) + (1-\lambda)\Big(1 + \frac{c_9}{\sqrt{k}}\Big).$$
The function $w : [0, \infty) \to [0,1]$ with $u \mapsto w(u) = \mathbb{P}(\|X\|_2 \leqslant u)$ is log-concave, and hence
$$\mathbb{P}\Big(\|X\|_2 \leqslant \Big(1 - \frac{c_9}{\sqrt{k}}\Big)(\mathbb{E}\|X\|_2^2)^{1/2}\Big) \geqslant \mathbb{P}\Big(\|X\|_2 \leqslant \Big(1 - \frac{c_9 t}{\sqrt{k}}\Big)(\mathbb{E}\|X\|_2^2)^{1/2}\Big)^\lambda$$
$$\times \mathbb{P}\Big(\|X\|_2 \leqslant \Big(1 + \frac{c_9}{\sqrt{k}}\Big)(\mathbb{E}\|X\|_2^2)^{1/2}\Big)^{1-\lambda}.$$
It follows that
$$\mathbb{P}\Big(\|X\|_2 \leqslant \Big(1 - \frac{c_9 t}{\sqrt{k}}\Big)(\mathbb{E}\|X\|_2^2)^{1/2}\Big) \leqslant \Big(\frac{1}{4}\Big)^{1/\lambda}\Big(\frac{4}{3}\Big)^{1/\lambda - 1} \leqslant \Big(\frac{1}{3}\Big)^{1/\lambda} \leqslant e^{-t/2}.$$
Combining this estimate with Lemma 13.1.13 we get the result. $\square$

LEMMA 13.1.15. *For every* $2 \leqslant p \leqslant \sqrt[8]{n}$ *we have*
$$(\mathbb{E}\|X\|_2^2)^{1/2} \leqslant (\mathbb{E}\|X\|_2^p)^{1/p} \leqslant \Big(1 + \frac{c_2 p}{\sqrt[4]{n}}\Big)(\mathbb{E}\|X\|_2^2)^{1/2}.$$

Proof. Combining Lemma 13.1.13 with Paouris' inequality (13.1.10) we see that
$$(13.1.25) \quad \mathbb{P}\big(\big|\|X\|_2 - \sqrt{n}\big| \geqslant s\sqrt{n}\big) \leqslant Ce^{-c(n^{1/8}s)^2}\mathbf{1}_{\{s \leqslant 1\}}(s) + Ce^{-c\sqrt{n}s}\mathbf{1}_{\{s \geqslant 1\}}(s)$$
for all $s > 0$. We consider the function
$$F(x) = \frac{\|x\|_2^2}{n} - 1.$$
We write
$$\mathbb{E}|F(X)|^p = p\int_0^\infty t^{p-1}\mathbb{P}(\{|F(X)| \geqslant t\})\,dt$$
and using (13.1.25) we see that for every $2 \leqslant p \leqslant \sqrt[8]{n}$
$$(\mathbb{E}|F(X)|^p)^{1/p} \leqslant C_1 \frac{\sqrt{p}}{n^{1/8}} + C_2 \frac{p^2}{n} \leqslant C_3 \frac{\sqrt{p}}{n^{1/8}}.$$

Note that $\mathbb{E}(F(X)) = 0$. Therefore, for any integer $2 \leqslant p \leqslant \sqrt[8]{n}$ we get

$$\frac{\mathbb{E}\|X\|_2^{2p}}{(\mathbb{E}\|X\|_2^2)^p} = \mathbb{E}(1 + F(X))^p = 1 + \sum_{k=2}^{p} \binom{p}{k} \mathbb{E}|F(X)|^k$$

$$\leqslant 1 + \sum_{k=2}^{p} \left(\frac{C_4 p}{\sqrt{k} n^{1/8}}\right)^k \leqslant 1 + 2 \max_{2 \leqslant k \leqslant p} \left(\frac{C_5 p}{\sqrt{k} n^{1/8}}\right)^k.$$

Since $p \leqslant \sqrt[8]{n}$ we can check that $g(t) = \left(\frac{C_5 p}{\sqrt{t} n^{1/8}}\right)^t$ is decreasing on $[2, p]$. Therefore,

$$\frac{\mathbb{E}\|X\|_2^{2p}}{(\mathbb{E}\|X\|_2^2)^p} \leqslant 1 + 2g(2) \leqslant 1 + \frac{Cp^2}{\sqrt[4]{n}}.$$

Since $\left(1 + \frac{Cp^2}{\sqrt[4]{n}}\right)^{\frac{1}{2p}} \leqslant 1 + \frac{C'p}{\sqrt[4]{n}}$, the lemma follows. □

13.2. The thin shell estimate of Guédon and E. Milman

In this section we provide a proof of the best known result on the problem which is due to Guédon and E. Milman [**240**].

THEOREM 13.2.1 (Guédon-E. Milman). *Let X be an isotropic log-concave random vector in $\mathbb{R}^n$. Then,*

(13.2.1) $$\mathbb{P}\left(\left|\|X\|_2 - \sqrt{n}\right| \geqslant t\sqrt{n}\right) \leqslant C \exp(-c\sqrt{n} \min\{t^3, t\})$$

for every $t > 0$, where $C, c > 0$ are absolute constants. In particular,

(13.2.2) $$\sqrt{\mathrm{Var}(\|X\|_2)} \leqslant Cn^{1/3}.$$

Theorem 13.2.1 implies a large deviation estimate which complements Paouris' theorem.

THEOREM 13.2.2 (Guédon-E. Milman). *Let X be an isotropic log-concave random vector in $\mathbb{R}^n$. Then,*

(13.2.3) $$\mathbb{P}\left(\|X\|_2 \geqslant (1+t)\sqrt{n}\right) \leqslant \exp(-c\sqrt{n} \min\{t^3, t\})$$

for every $t \geqslant 0$, where $c > 0$ is an absolute constant.

Theorem 13.2.1 also implies a small ball probability estimate.

THEOREM 13.2.3 (Guédon-E. Milman). *Let X be an isotropic log-concave random vector in $\mathbb{R}^n$. Then,*

(13.2.4) $$\mathbb{P}\left(\|X\|_2 \leqslant (1-t)\sqrt{n}\right) \leqslant \exp\left(-c_1 \sqrt{n} \min\left\{t^3, \log\frac{c_2}{1-t}\right\}\right)$$

for every $t \in (0, 1)$, where $c_1, c_2 > 0$ are absolute constants.

All these results will follow from the following main technical theorem.

THEOREM 13.2.4. *Let X be an isotropic log-concave random vector in $\mathbb{R}^n$. Then, if $1 \leqslant |p - 2| \leqslant c_1 n^{1/6}$ one has*

(13.2.5) $$1 - C\frac{|p-2|}{n^{1/3}} \leqslant \frac{(\mathbb{E}\|X\|_2^p)^{1/p}}{(\mathbb{E}\|X\|_2^2)^{1/2}} \leqslant 1 + C\frac{|p-2|}{n^{1/3}},$$

and if $c_1 n^{1/6} \leqslant |p-2| \leqslant c_2 \sqrt{n}$ then

(13.2.6) $$1 - C\frac{\sqrt{|p-2|}}{n^{1/4}} \leqslant \frac{(\mathbb{E}\|X\|_2^p)^{1/p}}{(\mathbb{E}\|X\|_2^2)^{1/2}} \leqslant 1 + C\frac{\sqrt{|p-2|}}{n^{1/4}}.$$

Observe that the concentration estimates that are provided by Theorems 13.2.1, 13.2.2 and 13.2.3 recover the large deviations inequality of Paouris, improve the thin shell estimate of Fleury and interpolate continuously between all scales of t. One can also check that if X has a better ψ_α behavior (for example, if X is ψ_2) then the proofs can be easily modified to yield better bounds. As the proof shows, the estimates are better if one assumes that the distribution of X has super-Gaussian marginals. This can be assumed, without loss of generality, if one replaces X by an isotropic random vector Y whose distribution is the convolution of the distribution of X with a Gaussian measure. Klartag and E. Milman introduced in [285] the idea to use the convolution of the distribution of X with a random orthogonal image of itself. This leads to improved small ball estimates; we present this technique and incorporate it in our presentation.

13.2.1. A variant of the L_p-centroid bodies

A main technical step in the proof of Theorem 13.2.4 is an upper bound for the log-Lipschitz constant $A_{k,p}$ of the function $h_{k,p}$. The estimate of Guédon and E. Milman, which is currently the best known, is that if $p \geqslant -k+1$ then

$$A_{k,p} \leqslant C \left[\max\{k,p\}\right]^{3/2}.$$

For the proof of this inequality, Guédon and E. Milman use a variant of the L_q-centroid bodies, the so-called *one-sided L_q-centroid bodies*.

DEFINITION 13.2.5. Let $f : \mathbb{R}^m \to \mathbb{R}^+$ be a log-concave density. For every $p \geqslant 1$ we define a convex body $Z_p^+(f)$ with support function

$$h_{Z_p^+(f)}(y) = \left(2 \int_{\mathbb{R}^m} \langle x, y \rangle_+^p f(x) dx\right)^{1/p},$$

where $a_+ = \max\{a, 0\}$. When f is even, it is clear that $Z_p^+(f) = Z_p(f)$. In any case, we easily verify that

$$Z_p^+(f) \subseteq 2^{1/p} Z_p(f).$$

We also introduce a variant of the bodies $K_p(f)$; the reader will notice that the difference between the definition below and the definition in Section 2.5 (see (2.5.1)) is not essential.

DEFINITION 13.2.6. For every log-concave function $f : \mathbb{R}^m \to \mathbb{R}^+$ with $0 < \int f < \infty$ and $f(0) > 0$, and for every $p > 0$, we define

$$\|x\| = \|x\|_{K_p(f)} := \left(\int_0^\infty p r^{p-1} f(rx) dr\right)^{-1/p}.$$

We easily check that the function $\|\cdot\|$ satisfies: $\|x\| \geqslant 0$ with equality if and only if $x = 0$, $\|x+y\| \leqslant \|x\| + \|y\|$, and $\|\lambda x\| = \lambda \|x\|$ for all $\lambda \geqslant 0$ and all $x, y \in \mathbb{R}^m$. For the proof of the triangle inequality see Theorem 2.5.5. In other words, $\|\cdot\|_{K_p(f)}$

has all the properties of a norm with the exception that it may fail to be even. We set
$$\|x\|_{\widehat{K}_p(f)} := \max\left\{\|x\|_{K_p(f)}, \|-x\|_{K_p(f)}\right\}.$$
So, the unit ball of this norm is the body
$$\widehat{K}_p(f) = K_p(f) \cap (-K_p(f)).$$
By the triangle inequality we have
$$\left|\|x\|_{K_p(f)} - \|y\|_{K_p(f)}\right| \leqslant \|x-y\|_{\widehat{K}_p(f)} \tag{13.2.7}$$
for all $x, y \in \mathbb{R}^m$.

From the symmetry of B_2^m it follows that if $c_1 B_2^m \subseteq K \subseteq c_2 B_2^m$ then $c_1 B_2^m \subseteq K \cap (-K) \subseteq c_2 B_2^m$. Applying this observation to $K_p(f)$ we see that
$$\frac{\|x\|_{\widehat{K}_p(f)}}{\|y\|_{K_p(f)}} \leqslant d_G(K_p(f), B_2^m) \frac{\|x\|_2}{\|y\|_2} \tag{13.2.8}$$
for all $x, y \in \mathbb{R}^m$.

We need the following theorem which compares $K_{m+p}(f)$ to $Z_p^+(K_{m+p}(f))$.

THEOREM 13.2.7. *Let $f : \mathbb{R}^m \to \mathbb{R}^+$ be a log-concave density. For every $p \geqslant 1$ we have*
$$c_1 Z_p^+(K_{m+p}(f)) \subseteq |K_{m+p}(f)|^{1/p} K_{m+p}(f)$$
$$\subseteq c_2 \left(\frac{\Gamma(m+p+1)}{\Gamma(m)\Gamma(p+1)}\right)^{1/p} Z_p^+(K_{m+p}(f)),$$
where $c_1, c_2 > 0$ are absolute constants.

For the proof of the theorem we will need a series of lemmas. First, recall the inequality
$$e^{-m\left(\frac{1}{p}-\frac{1}{q}\right)} \frac{K_p(f)}{f(0)^{1/p}} \subseteq \frac{K_q(f)}{f(0)^{1/q}} \subseteq \frac{\Gamma(q+1)^{1/q}}{\Gamma(p+1)^{1/p}} \frac{K_p(f)}{f(0)^{1/p}}, \tag{13.2.9}$$
which holds true for every centered log-concave density f and for all $q \geqslant p > 0$; this is exactly equivalent to (2.5.3).

In the sequel we also use the notation
$$H_\theta^+ = \{x \in \mathbb{R}^m : \langle x, \theta \rangle \geqslant 0\}.$$

LEMMA 13.2.8. *Let K be a convex body in $\mathbb{R}^m$. We fix $\theta \in S^{m-1}$ and define $f_\theta = \pi_\theta(\mathbf{1}_K)$ where π_θ denotes the marginal with respect to $\mathrm{span}\{\theta\}$. Then,*
$$\left(\frac{f_\theta(0)}{\|f_\theta\|_\infty}\right)^{1/p} \left(\frac{\Gamma(m)\Gamma(p+1)}{\Gamma(m+p+1)}\right)^{1/p} h_K(\theta) \leqslant \frac{h_{Z_p^+(K)}(\theta)}{(2|K \cap H_\theta^+|)^{1/p}} \leqslant h_K(\theta).$$

Proof. The right hand side inequality is simple: we write
$$h_{Z_p^+(K)}(\theta) = \left(2\int_0^{h_K(\theta)} t^p f_\theta(t)\, dt\right)^{1/p}$$
$$\leqslant \left(2\int_0^{h_K(\theta)} f_\theta(t)\, dt\right)^{1/p} h_K(\theta) = (2|K \cap H_\theta^+|)^{1/p} h_K(\theta).$$

For the left hand side inequality we repeat the proof of Lemma 3.2.8. We have

$$f_\theta(t) \geqslant \left(1 - \frac{t}{h_K(\theta)}\right)^{m-1} f_\theta(0)$$

for all $t \in [0, h_K(\theta)]$. Therefore,

$$h^p_{Z_p^+(K)}(\theta) = 2\int_0^{h_K(\theta)} t^p f_\theta(t)dt \geqslant 2\int_0^{h_K(\theta)} t^p \left(1 - \frac{t}{h_K(\theta)}\right)^{m-1} f_\theta(0)dt$$

$$= 2f_\theta(0)h_K^{p+1}(\theta)\int_0^1 s^p(1-s)^{m-1}ds$$

$$= \frac{\Gamma(m)\Gamma(p+1)}{\Gamma(p+m+1)} 2f_\theta(0)h_K^{p+1}(\theta).$$

On observing that

$$2f_\theta(0)h_K(\theta) = \frac{f_\theta(0)}{\|f_\theta\|_\infty} 2\|f_\theta\|_\infty h_K(\theta) \geqslant \frac{f_\theta(0)}{\|f_\theta\|_\infty}(2|K \cap H_\theta^+|),$$

we get the result. $\square$

LEMMA 13.2.9. *Let f be a log-concave density on $\mathbb{R}$ and let $\varepsilon \in (0,1)$. If*

$$\varepsilon \leqslant \int_0^\infty f(x)dx \leqslant 1 - \varepsilon,$$

then

$$f(0) \geqslant \varepsilon\|f\|_\infty.$$

Proof. We define

$$F(x) = \int_{-\infty}^x f(t)dt \quad \text{and} \quad G(x) = 1 - F(x) = \int_x^\infty f(t)dt.$$

Using the Prékopa-Leindler inequality we check that F and G are log-concave. Therefore, the functions f/F and $-f/G$ are decreasing. It follows that

$$f(x) \leqslant f(y)\max\left\{\frac{F(x)}{F(y)}, \frac{G(x)}{G(y)}\right\}$$

for every $x, y \in \mathbb{R}$. The assumption of the lemma implies that $F(0) \geqslant \varepsilon$ and $G(0) \geqslant \varepsilon$. Setting $y = 0$ we get the result. $\square$

LEMMA 13.2.10. *Let $f : \mathbb{R}^m \to \mathbb{R}$ be a log-concave function with barycenter at 0. For every $p \geqslant 1$ and every $\theta \in S^{n-1}$ one has*

$$\left(\frac{|K_{m+p}(f) \cap H_\theta^+|}{|K_{m+p}(f)|}\right)^{1/p} \geqslant c,$$

where $c > 0$ is an absolute constant.

Proof. We may assume that $f(0) = 1$ and that f is a density. We will use the identity

$$(13.2.10) \qquad |K \cap H_\theta^+| = \omega_m \int_{S^{m-1} \cap H_\theta^+} \|u\|_K^{-m} d\sigma(u),$$

which is easily checked, by integration in polar coordinates, for every convex body K which contains 0 (we denote by $\|\cdot\|_K$ the Minkowski functional of K). Note that, since f is a density, the same formula shows that

$$|K_m(f)| = \int_{\mathbb{R}^m} f(x)dx = 1.$$

From (13.2.9) we see that, for every $u \in S^{m-1}$,

$$e^{-\frac{mp}{m+p}} \|u\|_{K_m(f)}^{-m} \leqslant \|u\|_{K_{m+p}(f)}^{-m} \leqslant \frac{\Gamma(m+p+1)^{\frac{m}{m+p}}}{\Gamma(m+1)} \|u\|_{K_m(f)}^{-m}.$$

Inserting these inequalities into (13.2.10) and using Stirling's formula we see that

$$(13.2.11) \qquad e^{-p} \leqslant \frac{|K_{m+p}(f) \cap H_\theta^+|}{|K_m(f) \cap H_\theta^+|} \leqslant C^p$$

for every $\theta \in S^{m-1}$. Using once again (13.2.10), the definition of $K_m(f)$ and polar coordinates, we see that

$$|K_m(f) \cap H_\theta^+| = \int_{H_\theta^+} f(x)dx = \mathbb{P}(W_1 \geqslant 0),$$

where W_1 is the random variable $\pi_\theta(f)$ on $\mathbb{R}$. Since W_1 is log-concave and centered, Grünbaum's lemma (Lemma 2.2.6) in dimension 1 shows that

$$(13.2.12) \qquad \frac{|K_m(f) \cap H_\theta^+|}{|K_m(f)|} \geqslant \frac{1}{e}.$$

Using (13.2.11) and the fact that $|K_m(f)| = 1$ we see that

$$|K_{m+p}(f)| = |K_{m+p}(f) \cap H_\theta^+| + |K_{m+p}(f) \cap H_{-\theta}^+|$$
$$\leqslant C^p(|K_m(f) \cap H_\theta^+| + |K_m(f) \cap H_{-\theta}^+|) = C^p.$$

Taking also into account (13.2.12) we get

$$\frac{|K_{m+p}(f) \cap H_\theta^+|}{|K_{m+p}(f)|} = \frac{|K_{m+p}(f) \cap H_\theta^+|}{|K_m(f) \cap H_\theta^+|} \frac{|K_m(f) \cap H_\theta^+|}{|K_{m+p}(f)|} \geqslant e^{-p} \frac{e^{-1}}{C^p}.$$

This concludes the proof. □

Proof of Theorem 13.2.7. We apply Lemma 13.2.8 for $K = K_{m+p}(f)$ and using Lemma 13.2.10 we see that, for every $\theta \in S^{m-1}$,

$$c_3 \left(\frac{f_\theta(0)}{\|f_\theta\|_\infty}\right)^{1/p} \left(\frac{\Gamma(m)\Gamma(p+1)}{\Gamma(m+p+1)}\right)^{1/p} \leqslant |K_{m+p}(f)|^{-1/p} \frac{h_{Z_p^+(K_{m+p}(f))}(\theta)}{h_{K_{m+p}(f)}(\theta)} \leqslant c_4.$$

Let $f_\theta := \pi_\theta(\mathbf{1}_{K_{m+p}(f)})$. Lemma 13.2.10 shows that

$$c_5^p \leqslant \int_0^\infty f_\theta(x)dx \leqslant 1 - c_5^p.$$

for some absolute constant $c_5 \in (0,1)$; employing Lemma 13.2.9 we get
$$\min_{\theta \in S^{m-1}} \left(\frac{f_\theta(0)}{\|f_\theta\|_\infty} \right)^{1/p} \geqslant c_6 > 0,$$
and hence
$$c_7 \left(\frac{\Gamma(m)\Gamma(p+1)}{\Gamma(m+p+1)} \right)^{1/p} K_{m+p}(f) \subseteq |K_{m+p}(f)|^{-1/p} Z_p^+(K_{m+p}(f)) \subseteq c_8 K_{m+p}(f).$$
This proves the theorem. □

Our main result in this subsection provides an estimate for the distance of $K_{m+p}(f)$ from the Euclidean ball.

THEOREM 13.2.11. *Let $f : \mathbb{R}^m \to \mathbb{R}$ be a centered log-concave density. For any $p \geqslant -m+1$, one has*
$$d_G(K_{m+p}(f), B_2^m) \leqslant C \max\left\{ \frac{m}{m+p}, 1 \right\} d_G(Z_{\max\{p,m\}}^+(f), B_2^m),$$
where $C > 0$ is an absolute constant.

Proof. We use Theorem 13.2.7 to obtain
$$c_1 Z_p^+(K_{m+p}(f)) \subseteq |K_{m+p}(f)|^{1/p} K_{m+p}(f)$$
$$\subseteq c_2 \left(\frac{\Gamma(m+p+1)}{\Gamma(m)\Gamma(p+1)} \right)^{1/p} Z_p^+(K_{m+p}(f)),$$
where $c_1, c_2 > 0$ are absolute constants. On the other hand, integration in polar coordinates shows that
$$Z_p^+(K_{m+p}(f)) = Z_p^+(f),$$
and hence, combining the above and using Stirling's formula we get
$$d_G(K_{m+p}(f), B_2^m) \leqslant c_3 \frac{m+p}{p} \operatorname{dist}(Z_p^+(f), B_2^m).$$
It follows that if $p \geqslant m$ then
$$d_G(K_{m+p}(f), B_2^m) \leqslant 2c_3 d_G(Z_{\max\{p,m\}}^+(f), B_2^m)$$
$$\leqslant 2c_3 \max\left\{ \frac{m}{m+p}, 1 \right\} d_G(Z_{\max\{p,m\}}^+(f), B_2^m),$$
and the assertion of the theorem is satisfied with $C = 2c_3$. In the case $p < m$ we use (13.2.9) to write
$$d_G(K_{m+p}(f), B_2^m) \leqslant C_1 \frac{m+q}{m+p} \operatorname{dist}(K_{m+q}(f), B_2^m)$$
$$\leqslant C_2 \frac{m+q}{m+p} \frac{m+q}{q} d_G(Z_q^+(f), B_2^m)$$
for all $q \geqslant \max\{p, 1\}$, and setting $q = m$ we conclude the proof. □

REMARK 13.2.12. Let X be an isotropic random vector in $\mathbb{R}^n$ with a log-concave density f. We consider the random vector $Y = (X + G_n)/\sqrt{2}$ and denote its density by g. Then, we have an analogue of Proposition 8.1.7: for every $q \geqslant 2$,
$$c_1 \sqrt{q}\, B_2^n \subseteq Z_q^+(g) \subseteq c_2 q\, B_2^n,$$

where $c_1, c_2 > 0$ are absolute constants. In particular,

(13.2.13) $$d_G(Z_q^+(g), B_2^n) \leqslant c_3\sqrt{q}.$$

Let us sketch a proof of (13.2.13): we fix $\theta \in S^{n-1}$ and we define $Y_1 = P_\theta(Y)$, $X_1 = P_\theta(X)$ and $G_1 = P_\theta(G_n)$, where P_θ denotes the orthogonal projection onto span$\{\theta\}$. Then,

$$h_{Z_q^+(g)}^q(\theta) = 2\mathbb{E}(Y_1)_+^q = \frac{2}{2^{q/2}}\mathbb{E}(X_1 + G_1)_+^q \geqslant \frac{2}{2^{q/2}}\mathbb{E}[(G_1)_+^q]\mathbb{P}(X_1 \geqslant 0).$$

From Grünbaum's lemma we have $\mathbb{P}(X_1 \geqslant 0) \geqslant 1/e$, and this implies that

$$h_{Z_q^+(g)}^q(\theta) \geqslant \frac{1}{e2^{q/2}}\mathbb{E}|G_1|^q,$$

if we also take into account the symmetry of G_1. Using the fact that

$$c_1\sqrt{q} \leqslant \left(\mathbb{E}|G_1|^q\right)^{1/q} \leqslant c_2\sqrt{q}$$

for all $q \geqslant 1$, we get the left hand side inclusion. Next, we observe that

$$\frac{1}{2}h_{Z_q^+(g)}^q(\theta) \leqslant h_{Z_q(g)}^q(\theta) = \mathbb{E}|Y_1|^q = \mathbb{E}\left|\frac{X_1+G_1}{\sqrt{2}}\right|^q \leqslant \frac{2^{q-1}}{2^{q/2}}\mathbb{E}\left(|X_1|^q + |G_1|^q\right).$$

Since

$$\left(\mathbb{E}|X_1|^q\right)^{1/q} \leqslant cq\left(\mathbb{E}|X_1|^2\right)^{1/2} = cq,$$

the right hand side inclusion follows as well.

13.2.2. The log-Lipschitz constant of $h_{k,p}$

We write $M_{k,l}$ for the set of all $k \times l$ matrices with real entries and set $M_n := M_{n,n}$. We consider

$$SO(n) = \{U \in M_n : U^*U = I, \det(U) = 1\}$$

equipped with the standard invariant Riemannian metric d. We fix an orthonormal basis on $\mathbb{R}^n$ and taking the derivative of the equation $U^*U = I$ we see that the tangent space $T_I(SO(n))$ at the identity $I \in SO(n)$ can be identified with the set of all anti-symmetric matrices

$$\{B \in M_n : B^* + B = 0\}.$$

For every $B \in T_I(SO(n))$ we set

$$|B|^2 = \langle B, B \rangle := d_I(B, B) = \frac{1}{2}\|B\|_{\mathrm{HS}}^2,$$

where

$$\|A\|_{\mathrm{HS}}^2 = \mathrm{tr}(A^*A) = \sum_{i=1}^{k}\sum_{j=1}^{l} a_{ij}^2$$

for every $A = (a_{ij}) \in M_{k,l}$.

Recall that Y is an isotropic log-concave random vector in $\mathbb{R}^n$. We denote by g the density of Y, we fix $k, p, E_0 \in G_{n,k}$ and $\theta_0 \in S_{E_0}$, and we define

$$h_{k,p}(U) = k\omega_k \int_0^\infty t^{p+k-1}\pi_{U(E_0)}g(tU(\theta_0))\,dt.$$

13.2. THE THIN SHELL ESTIMATE OF GUÉDON AND E. MILMAN

THEOREM 13.2.13 (Guédon-E. Milman). *For the log-Lipschitz constant $A_{k,p}$ of $h_{k,p}$ we have*

$$A_{k,p} \leqslant C \max\{k,p\}\, d_{\mathrm{G}}(Z^+_{\max\{k,p\}}(g), B_2^n),$$

where $C > 0$ is an absolute constant.

Proof. We may assume that $2 \leqslant k \leqslant n/2$; this restriction makes the presentation simpler and does not create any problem to our arguments later on. Since $E_0 \in G_{n,k}$ is arbitrary, by the symmetry and transitivity of $SO(n)$ it is enough to estimate $|\nabla_{U_0} \log h_{k,p}|$ at $U_0 = I$. We consider an orthonormal basis $\{\theta_0, e_2, \ldots, e_k\}$ of E_0 and we extend it to an orthonormal basis $\{\theta_0, e_2, \ldots, e_n\}$ of $\mathbb{R}^n$.

The anti-symmetric matrix $M = \nabla_I \log h_{k,p} \in T_I(SO(n))$ can be written in the form

$$M = \begin{pmatrix} M_1 & M_2 \\ -M_2^* & 0 \end{pmatrix}, \qquad M_1 = \begin{pmatrix} 0 & V_1 \\ -V_1^* & 0 \end{pmatrix}, \qquad M_2 = \begin{pmatrix} V_2 \\ V_3 \end{pmatrix}$$

where $M_1 \in M_{k,k}$, $M_2 \in M_{k,n-k}$, $V_1 \in M_{1,k-1}$, $V_2 \in M_{1,n-k}$ and $V_3 \in M_{k-1,n-k}$. It suffices to observe that the lower $(n-k) \times (n-k)$ submatrix of M is 0 because every rotation of $E_0^\perp$ leaves $\pi_{U(E_0)}(g)$, and hence $h_{k,p}$, unaltered. Also, the lower $(k-1) \times (k-1)$ submatrix of M_1 is 0 because every rotation which fixes θ_0 and acts invariantly on E_0 leaves $h_{k,p}$ unaltered. It follows that

$$|\nabla_I \log h_{k,p}|^2 = \|V_1\|_{\mathrm{HS}}^2 + \|V_2\|_{\mathrm{HS}}^2 + \|V_3\|_{\mathrm{HS}}^2.$$

We will analyze these three terms separately.

For every $i = 1, 2, 3$ we write T_i for the subspace of $T_I(SO(n))$ consisting of those matrices M (as above) for which $V_j = 0$ if $j \neq i$. For every $B \in T_i$, a *type-i movement* is a geodesic $s \mapsto U_s := \exp_I(sB)$ in $SO(n)$. It is clear that $\frac{d}{ds} U_s\big|_{s=0} = B$, and hence

$$\frac{d}{ds} \log h_{k,p}(U_s)\big|_{s=0} = \langle \nabla_I \log h_{k,p}, B \rangle.$$

Then,

$$\|V_i\|_{\mathrm{HS}} = \sup \left\{ \frac{\langle \nabla_I \log h_{k,p}, B \rangle}{|B|} : 0 \neq B \in T_i \right\},$$

so we need to give an upper bound for the derivative of $\log h_{k,p}$ induced by each type-i movement.

The case of type-1 movement

Let $\{U_s\}$ be a type-1 movement generated by some $B \in T_1$ with $|B| = 1$. We write

$$\xi_0 = \frac{d}{ds} U_s(\theta_0)\big|_{s=0} \in T_{\theta_0} S(\mathbb{R}^n).$$

Using the natural embedding $T_\theta S(\mathbb{R}^n) \subset T_\theta \mathbb{R}^n \simeq \mathbb{R}^n$ we see that U_s is a rotation in the $\{\theta_0, \xi_0\}$-plane and that ξ_0 is orthogonal to θ_0 in E_0, so $U_s(E_0) = E_0$. From the assumption that $|B| = 1$ it follows that $\|\xi_0\|_2 = 1$. By the definition of $h_{k,p}$ we then have

$$h_{k,p}(U_s) = k\omega_k \int_0^\infty t^{p+k-1} \pi_{E_0} g(tU_s(\theta_0))\, dt = c_{p,k} \|U_s(\theta_0)\|_{K_{k+p}(\pi_{E_0} g)}^{-(k+p)},$$

where $c_{p,k} = k\omega_k/(k+p)$. It follows that

$$|\langle \nabla_I \log h_{k,p}, B \rangle| = \left| \frac{d}{ds} \log h_{k,p}(U_s) \right|_{s=0} \right|$$

$$= (k+p) \left| \frac{d}{ds} \log \|U_s(\theta_0)\|_{K_{k+p}(\pi_{E_0} g)} \right|_{s=0} \right|.$$

From the triangle inequality (13.2.7) we have

$$\left| \frac{d}{ds} \|U_s(\theta_0)\|_{K_{k+p}(\pi_{E_0} g)} \right| \leq \left\| \frac{d}{ds} U_s(\theta_0) \right\|_{\widehat{K}_{k+p}(\pi_{E_0}(g))},$$

and using (13.2.8) we conclude that

$$|\langle \nabla_I \log h_{k,p}, B \rangle| \leq (k+p) \frac{\|\xi_0\|_{\widehat{K}_{k+p}(\pi_{E_0} g)}}{\|\theta_0\|_{K_{k+p}(\pi_{E_0} g)}}$$

$$\leq (k+p) d_G(K_{k+p}(\pi_{E_0} g), B_{E_0}).$$

The case of type-2 movement

Let $\{U_s\}$ be a type-1 movement generated by some $B \in T_2$ with $|B| = 1$. We set $\theta_s := U_s(\theta_0)$ and

$$\xi_s := \frac{d}{ds} \theta_s \in T_{\theta_s} S(\mathbb{R}^n).$$

Note that $\xi_0 \in E_0^\perp$ and that U_s is a rotation in the $\{\theta_0, \xi_0\} = \{\theta_s, \xi_s\}$-plane. From the assumption that $|B| = 1$ it follows that $\|\xi_0\|_2 = 1$. We write E^1 for the orthogonal complement of θ_0 in E_0. Then, U_s rotates E_0 into

$$E_s := U_s(E_0) = E^1 \oplus \text{span}\{\theta_s\},$$

and hence U_s leaves $H := E_0 \oplus \text{span}\{\xi_0\} = E_s \oplus \text{span}\{\xi_s\} \in G_{n,k+1}$ invariant. Therefore,

$$h_{k,p}(U_s) = k\omega_k \int_0^\infty \int_{-\infty}^\infty t^{p+k-1} \pi_H g(t\theta_s + r\xi_s) dr dt.$$

With the change of variables $r = vt$ we get

$$h_{k,p}(U_s) = k\omega_k \int_0^\infty \int_{-\infty}^\infty t^{p+k} \pi_H g(t(\theta_s + v\xi_s)) dv dt$$

$$= c_{p,k} \int_{-\infty}^\infty \|\theta_s + v\xi_s\|_{K_{k+p+1}(\pi_H g)}^{-(k+p+1)} dv,$$

where $c_{p,k} = k\omega_k/(k+p+1)$. Using the fact that $\frac{d}{ds}\xi_s = -\theta_s$, together with (13.2.7) and (13.2.8) for $\|\cdot\|_{K_{k+p+1}(\pi_H g)}$, we write

$$|\langle \nabla_I \log h_{k,p}, B \rangle| = \left| \frac{d}{ds} \log h_{k,p}(U_s) \right|_{s=0} \right|$$

$$\leq (k+p+1) \sup_{v \in \mathbb{R}} \frac{\|\xi_0 - v\theta_0\|_{\widehat{K}_{k+p+1}(\pi_H g)}}{\|\theta_0 + v\xi_0\|_{K_{k+p+1}(\pi_H g)}}$$

$$\leq (k+p+1) d_G(K_{k+p+1}(\pi_H g), B_H) \sup_{v \in \mathbb{R}} \frac{\|\xi_0 - v\theta_0\|_2}{\|\theta_0 + v\xi_0\|_2}$$

$$= (k+p+1) d_G(K_{k+p+1}(\pi_H g), B_H),$$

where we have also used the fact that $\theta_0 \perp \xi_0$.

The case of type-3 movement

We finally examine type-3 movements; these act on subspaces of dimension $(k-1)(n-k)$. Let $0 \neq B \in T_3$ which generates a type-3 movement $\{U_s\}$, and set $e_{s,j} := U_s(e_j)$ and $f_j := \frac{d}{ds}e_{s,j}\big|_{s=0}$, $j = 2, \ldots, k$. Then, we have that $U_s(\theta_0) = \theta_0$ and $f_j \in E_0^\perp$ for all j. We define $F_0 := \text{span}\{f_2, \ldots, f_k\}$ and, by a small perturbation of B if needed, we assume that $\dim(F_0) = k - 1$. We set $H = E_0 \oplus F_0 \in G_{n,2k-1}$ and we observe that H is invariant under U_s, because U_s is an isometry which acts as the identity on the orthogonal complement. It follows that $H = E_s \oplus F_s$, where $E_s := U_s(E_0)$ and $F_s := U_s(F_0)$. Therefore,

$$h_{k,p}(U_s) = k\omega_k \int_0^\infty \int_{F_s} t^{p+k-1} \pi_H g(t\theta_0 + y) dy dt.$$

With the change of variables $y = zt$ we get

$$h_{k,p}(U_s) = k\omega_k \int_0^\infty \int_{F_s} t^{p+2k-2} \pi_H g(t(\theta_0 + z)) dz dt$$

$$= c_{p,k} \int_{F_s} \|\theta_0 + z\|_{K_{2k+p-1}(\pi_H g)}^{-(2k+p-1)} dz,$$

where $c_{p,k} = k\omega_k/(2k+p-1)$. Using the fact that U_s is orthogonal, we rewrite this in the form

$$h_{k,p}(U_s) = c_{p,k} \int_{F_0} \|\theta_0 + U_s(z)\|_{K_{2k+p-1}(\pi_H g)}^{-(2k+p-1)} dz.$$

Then, the triangle inequality (13.2.7) for $\|\cdot\|_{K_{2k+p-1}(\pi_H g)}$ gives

$$|\langle \nabla_I \log h_{k,p}, B\rangle| = \left|\frac{d}{ds} \log h_{k,p}(U_s)\big|_{s=0}\right|$$

$$\leqslant (2k+p-1) \sup_{z \in F_0} \frac{\|Bz\|_{\widehat{K}_{2k+p-1}(\pi_H g)}}{\|\theta_0 + z\|_{K_{2k+p-1}(\pi_H g)}},$$

and (13.2.8) implies that

$$\frac{|\langle \nabla_I \log h_{k,p}, B\rangle|}{(2k+p-1)d_G(K_{2k+p-1}(\pi_H g), B_H)} \leqslant \sup_{z \in F_0} \frac{\|Bz\|_2}{\|\theta_0 + z\|_2}$$

$$\leqslant \|B\|_{\text{op}} \sup_{z \in F_0} \frac{\|z\|_2}{\sqrt{1 + \|z\|_2^2}}$$

$$\leqslant \frac{\|B\|_{\text{HS}}}{\sqrt{2}} = |B|,$$

where we have used the facts that $\theta_0 \perp F_0$ and $\|B\|_{\text{op}} \leqslant \|B\|_{\text{HS}}/\sqrt{2}$ for any antisymmetric matrix (which is an immediate consequence of the Cauchy-Schwarz inequality).

Proof of Theorem 13.2.13. Recall that

$$|\nabla_I \log h_{k,p}|^2 = \|V_1\|_{\text{HS}}^2 + \|V_2\|_{\text{HS}}^2 + \|V_3\|_{\text{HS}}^2.$$

The theorem immediately follows if we apply Theorem 13.2.11 (with f a suitable marginal density of g) for each type-i movement: we saw that

$$\|V_1\|_{\text{HS}} \leqslant (k+p) d_G(K_{k+p}(\pi_E g), B_E)$$

for some $E \in G_{n,k}$,
$$\|V_2\|_{\mathrm{HS}} \leqslant (k+p+1)d_{\mathrm{G}}(K_{k+p+1}(\pi_E g), B_E)$$
for some $E \in G_{n,k+1}$, and
$$\|V_3\|_{\mathrm{HS}} \leqslant (2k+p-1)d_{\mathrm{G}}(K_{2k+p-1}(\pi_E g), B_E)$$
for some $E \in G_{n,2k-1}$. $\square$

13.2.3. Estimates for the moments of the Euclidean norm

We pass to the proof of Theorem 13.2.4. In order to compare $(\mathbb{E}\|Y\|_2^p)^{1/p}$ and $(\mathbb{E}\|Y\|_2^2)^{1/2}$ we prove an estimate for the derivative of $p \mapsto \log((\mathbb{E}\|Y\|_2^p)^{1/p})$.

THEOREM 13.2.14. *Let $0 \neq |p| \leqslant \frac{n-1}{2}$ and let k be an integer in $[2,n]$ such that $|p| \leqslant \frac{k-1}{2}$. Then,*

$$(13.2.14) \qquad \frac{d}{dp}\log((\mathbb{E}\|Y\|_2^p)^{1/p}) \leqslant C\left(\frac{\max\{A_{k,p}^2, A_{k,0}^2\}}{p^2 n} + \frac{1}{k}\right)$$

where $C > 0$ is an absolute constant.

PROOF. Recall that from Lemma 13.1.2 we have
$$\mathbb{E}\|Y\|_2^p = c_{n,k,p}\mathbb{E}_U[h_{k,p}(U)],$$
where
$$c_{n,k,p} = \frac{\Gamma\left(\frac{p+n}{2}\right)\Gamma\left(\frac{k}{2}\right)}{\Gamma\left(\frac{n}{2}\right)\Gamma\left(\frac{p+k}{2}\right)}.$$

The first step of the proof is done in the next

LEMMA 13.2.15. *Let $0 \neq |p| \leqslant \frac{n-1}{2}$ and let k be an integer in $[2,n]$ such that $|p| \leqslant \frac{k-1}{2}$. Then,*

$$\frac{d}{dp}\log\left(\mathbb{E}\|Y\|_2^p\right)^{1/p} \leqslant \frac{c}{p^2 n}(2A_{k,p}^2 + 3A_{k,0}^2) + \frac{d}{dp}\left(\frac{1}{p}\log\frac{\Gamma(k+p)}{\Gamma(k)}\right)$$
$$+ \frac{d}{dp}\left(\frac{1}{p}\log\frac{\Gamma\left(\frac{p+n}{2}\right)\Gamma\left(\frac{k}{2}\right)}{\Gamma\left(\frac{n}{2}\right)\Gamma\left(\frac{p+k}{2}\right)}\right),$$

where $c > 0$ is an absolute constant.

PROOF. We first write
$$\frac{d}{dp}\log\left(\mathbb{E}\|Y\|_2^p\right)^{1/p}$$
$$= \frac{d}{dp}\log\left((\mathbb{E}_U[h_{k,p}(U)])^{1/p}\right) + \frac{d}{dp}\left(\frac{1}{p}\log\frac{\Gamma\left(\frac{p+n}{2}\right)\Gamma\left(\frac{k}{2}\right)}{\Gamma\left(\frac{n}{2}\right)\Gamma\left(\frac{p+k}{2}\right)}\right).$$

For every $U \in SO(n)$ we define a (not necessarily probability) measure μ_U on $\mathbb{R}^+$ with density $k\omega_k t^{k-1}\pi_{U(F_0)}g(tU(\theta_0))$. Next, we define the probability measure $\mu_{k,p} := \mathbb{E}_U \mu_U$ on $[0,\infty)$. With these definitions we have

$$(13.2.15) \qquad h_{k,p}(U) = \mathbb{E}_{\mu_U}(t^p) \quad \text{and} \quad \mathbb{E}_U[h_{k,p}(U)] = \mathbb{E}_U(\mathbb{E}_{\mu_U}(t^p)) = \mathbb{E}_{\mu_{k,p}}(t^p).$$

13.2. THE THIN SHELL ESTIMATE OF GUÉDON AND E. MILMAN

We agree that for every measure space (Ω, μ) and every measurable function $f : \Omega \to [0, \infty)$ we will write

$$\mathbb{E}_\mu(f) = \int_\Omega f \, d\mu \quad \text{and} \quad \operatorname{Ent}_\mu(f) = \mathbb{E}_\mu(f \log f) - \mathbb{E}_\mu(f) \log(\mathbb{E}_\mu(f)).$$

With this notation we have

$$(13.2.16) \qquad \frac{d}{dp} \log\left((\mathbb{E}_\mu f^p)^{1/p}\right) = \frac{1}{p^2} \frac{\operatorname{Ent}_\mu(f^p)}{\mathbb{E}_\mu(f^p)}.$$

In particular,

$$(13.2.17) \qquad \frac{d}{dp} \log\left((\mathbb{E}_U[h_{k,p}(U)])^{1/p}\right) = \frac{1}{p^2} \frac{\operatorname{Ent}_{\mu_{k,p}}(t^p)}{\mathbb{E}_{\mu_{k,p}}(t^p)} = \frac{1}{p^2} \frac{\operatorname{Ent}_{\mu_{k,p}}(t^p)}{\mathbb{E}_U h_{k,p}(U)}.$$

Using the equation $\mathbb{E}_{\mu_{k,p}}(t^p) = \mathbb{E}_U(\mathbb{E}_{\mu_U}(t^p))$ and the definition of Ent we see that the numerator can be written in the form

$$(13.2.18) \qquad \operatorname{Ent}_{\mu_{k,p}}(t^p) = \mathbb{E}_U\left(\operatorname{Ent}_{\mu_U}(t^p)\right) + \operatorname{Ent}_U\left(\mathbb{E}_{\mu_U}(t^p)\right)$$
$$= \mathbb{E}_U\left(\operatorname{Ent}_{\mu_U}(t^p)\right) + \operatorname{Ent}_U\left(h_{k,p}(U)\right).$$

For the second term in (13.2.18) we use the log-Sobolev inequality on $SO(n)$ (Theorem 13.1.7) to write

$$(13.2.19) \qquad \frac{1}{p^2} \frac{\operatorname{Ent}_U h_{k,p}(U)}{\mathbb{E}_U h_{k,p}(U)} \leqslant \frac{c}{p^2 n} \frac{\mathbb{E}_U[|\nabla(\log h_{k,p})(U)|^2 h_{k,p}(U)]}{\mathbb{E}_U h_{k,p}(U)} \leqslant \frac{c A_{k,p}^2}{p^2 n},$$

where $A_{k,p}$ is the log-Lipschitz constant of $U \mapsto h_{k,p}(U)$. For the first term in (13.2.18), using (13.2.16) we first write

$$\frac{1}{p^2} \frac{\operatorname{Ent}_{\mu_U}(t^p)}{\mathbb{E}_{\mu_U}(t^p)} = \frac{d}{dp}\left[\log\left((\mathbb{E}_{\mu_U}(t^p))^{1/p}\right)\right]$$
$$= \frac{d}{dp}\left[\frac{1}{p}\left(\log \frac{h_{k,p}(U)}{\Gamma(k+p)} - \log \frac{h_{k,0}(U)}{\Gamma(k)} + \log \frac{\Gamma(k+p)}{\Gamma(k)} + \log h_{k,0}(U)\right)\right].$$

Next, using Lemma 13.1.9 we see that the function $p \mapsto \log(h_{k,p}(U)/\Gamma(k+p))$ is concave on $[-k+1, \infty)$, and hence

$$\frac{d}{dp}\left[\frac{1}{p}\left(\log \frac{h_{k,p}(U)}{\Gamma(k+p)} - \log \frac{h_{k,0}(U)}{\Gamma(k)}\right)\right] \leqslant 0.$$

This implies that

$$\frac{1}{p^2} \frac{\operatorname{Ent}_{\mu_U}(t^p)}{\mathbb{E}_{\mu_U}(t^p)} \leqslant \frac{d}{dp}\left(\frac{1}{p} \log \frac{\Gamma(k+p)}{\Gamma(k)}\right) - \frac{1}{p^2} \log h_{k,0}(U).$$

Integrating this with respect to $U \in O(n)$ and recalling that $h_{k,p}(U) = \mathbb{E}_{\mu_U}(t^p)$ we have

$$\frac{1}{p^2} \mathbb{E}_U(\operatorname{Ent}_{\mu_U}(t^p)) \leqslant \mathbb{E}_U(\mathbb{E}_{\mu_U}(t^p)) \frac{d}{dp}\left(\frac{1}{p} \log \frac{\Gamma(k+p)}{\Gamma(k)}\right)$$
$$+ \frac{1}{p^2} \mathbb{E}_U\left(\mathbb{E}_{\mu_U}(t^p) \log(1/h_{k,0}(U))\right)$$
$$= \mathbb{E}_U[h_{k,p}(U)] \frac{d}{dp}\left(\frac{1}{p} \log \frac{\Gamma(k+p)}{\Gamma(k)}\right)$$
$$+ \frac{1}{p^2} \mathbb{E}_U\left[\log(1/h_{k,0}(U)) h_{k,p}(U)\right]$$

and hence,
(13.2.20)
$$\frac{1}{p^2}\frac{\mathbb{E}_U(\mathrm{Ent}_{\mu_U}(t^p))}{\mathbb{E}_U(\mathbb{E}_{\mu_U}(t^p))} \leqslant \frac{d}{dp}\left(\frac{1}{p}\log\frac{\Gamma(k+p)}{\Gamma(k)}\right) + \frac{1}{p^2}\frac{\mathbb{E}_U[\log(1/h_{k,0}(U))h_{k,p}(U)]}{\mathbb{E}_U(h_{k,p}(U))}.$$

We now use Jensen's inequality and the Cauchy-Schwarz inequality to bound the second term as follows:

$$\frac{\mathbb{E}_U[\log(1/h_{k,0}(U))h_{k,p}(U)]}{\mathbb{E}_U[h_{k,p}(U)]} \leqslant \log\left(\frac{\mathbb{E}_U\left(\frac{h_{k,p}(U)}{h_{k,0}(U)}\right)}{\mathbb{E}_U[h_{k,p}(U)]}\right)$$

$$\leqslant \log\left(\frac{(\mathbb{E}_U[h_{k,p}(U)^2])^{1/2}}{\mathbb{E}_U[h_{k,p}(U)]}(\mathbb{E}_U[h_{k,0}(U)^{-2}])^{1/2}\right).$$

Then, we use the reverse Hölder inequality

$$(\mathbb{E}_U(f^q))^{1/q} \leqslant \exp\left(C\frac{\|\log f\|_{\mathrm{Lip}}^2}{n}(q-r)\right)(\mathbb{E}_U(f^r))^{1/r} \qquad (q > r > 0)$$

(this is the same as (13.1.12) in Fleury's argument) to handle the various moments that appear above. Setting $\|f\|_q := (\mathbb{E}_U|f(U)|^q)^{1/q}$ we have

$$\|h_{k,p}\|_2 \leqslant \exp\left(\frac{CA_{k,p}^2}{n}\right)\|h_{k,p}\|_1$$

and

$$\|h_{k,0}^{-1}\|_2 \leqslant \exp\left(\frac{2CA_{k,0}^2}{n}\right)\|h_{k,0}^{-1}\|_0 = \exp\left(\frac{2CA_{k,0}^2}{n}\right)\frac{1}{\|h_{k,0}\|_0}$$

$$\leqslant \exp\left(\frac{3CA_{k,0}^2}{n}\right)\frac{1}{\|h_{k,0}\|_1}.$$

Since
$$\|h_{k,0}\|_1 = \mathbb{E}_U[h_{k,0}(U)] = \mathbb{E}_{\mu_{k,0}}(1) = 1,$$
we get
(13.2.21)
$$\frac{1}{p^2}\frac{\mathbb{E}_U[\log(1/h_{k,0}(U))h_{k,p}(U)]}{\mathbb{E}_U[h_{k,p}(U)]} \leqslant \frac{C}{p^2 n}(A_{k,p}^2 + 3A_{k,0}^2).$$

Inserting (13.2.19), (13.2.20) and (13.2.21) into (13.2.17) we conclude the proof. □

We are now ready to complete the proof of Theorem 13.2.14 by analyzing the three terms in the bound of Lemma 13.2.15. Using the fact that the function $p \mapsto \frac{d}{dp}\log\Gamma(p)$ is concave, we easily check that

(13.2.22)
$$\frac{d}{dp}\left(\frac{1}{p}\log\frac{\Gamma\left(\frac{p+n}{2}\right)\Gamma\left(\frac{k}{2}\right)}{\Gamma\left(\frac{n}{2}\right)\Gamma\left(\frac{p+k}{2}\right)}\right) \leqslant 0.$$

For the second term we observe first that if $q \neq 0$ has the same sign as p and at the same time satisfies $k + p + q > 0$ then, from Jensen's inequality, we may write

$$\frac{d}{dp}\left(\frac{1}{p}\log\frac{\Gamma(k+p)}{\Gamma(k)}\right) = \frac{1}{pq}\frac{\int_0^\infty \log(t^q)t^{p+k-1}\exp(-t)dt}{\Gamma(p+k)} - \frac{1}{p^2}\log\frac{\Gamma(k+p)}{\Gamma(k)}$$

$$\leqslant \frac{1}{pq}\log\frac{\Gamma(k+p+q)}{\Gamma(k+p)} - \frac{1}{p^2}\log\frac{\Gamma(k+p)}{\Gamma(k)}$$

$$= \frac{1}{p}\log\left(\frac{\Gamma(k+p+q)^{1/q}}{\Gamma(k+p)^{1/q}}\frac{\Gamma(k)^{1/p}}{\Gamma(k+p)^{1/p}}\right).$$

We choose $q = (p+k-1)\frac{p}{k-1}$. Then, q satisfies the above, and using Stirling's formula and the fact that $p \geqslant -\frac{k-1}{2}$ we see that

$$(13.2.23) \qquad \frac{d}{dp}\left(\frac{1}{p}\log\frac{\Gamma(k+p)}{\Gamma(k)}\right) \leqslant \frac{C}{k}.$$

Going back to Lemma 13.2.15 we get

$$(13.2.24) \qquad \frac{d}{dp}\log((\mathbb{E}\|Y\|_2^p)^{1/p}) \leqslant C\left(\frac{\max\{A_{k,p}^2, A_{k,0}^2\}}{p^2 n} + \frac{1}{k}\right)$$

as claimed. $\square$

Next, we prove a version of Theorem 13.2.4 under some regularity assumption on the growth of the L_q-centroid bodies of the distribution of Y. Using the results of Section 13.2.5, in the final subsection we will see how this additional hypothesis can be removed.

THEOREM 13.2.16. *Let Y be an isotropic log-concave random vector in $\mathbb{R}^n$ such that*

$$(13.2.25) \qquad d_G(Z_q^+(g), B_2^n) \leqslant \beta\sqrt{q}$$

for some $\beta \geqslant 1$ and for all $q \geqslant 2$, where g is the density of Y. Then, for all $|p| \leqslant \bar{c}\sqrt{n}$ one has

$$(13.2.26) \qquad 1 - \overline{C}\beta^2\frac{\sqrt{|p-2|}}{n^{1/4}} \leqslant \frac{(\mathbb{E}\|Y\|_2^p)^{1/p}}{(\mathbb{E}\|Y\|_2^2)^{1/2}} \leqslant 1 + \overline{C}\beta^2\frac{\sqrt{|p-2|}}{n^{1/4}},$$

where $\bar{c}, \overline{C} > 0$ are absolute constants.

Proof. From Theorem 13.2.13 we have

$$A_{k,p} \leqslant C_1 \max\{k,p\}\, d_G(Z_{\max\{k,p\}}^+(g), B_2^n),$$

where $C_1 > 0$ is an absolute constant. We have assumed that $|p| \leqslant \frac{k-1}{2}$, and then (13.2.25) gives

$$\max\{A_{k,p}, A_{k,0}\} \leqslant C_1\beta k^{3/2}.$$

Therefore, Theorem 13.2.14 implies that

$$(13.2.27) \qquad \frac{d}{dp}\log((\mathbb{E}\|Y\|_2^p)^{1/p}) \leqslant C_2\beta^2\left(\frac{k^3}{p^2 n} + \frac{1}{k}\right)$$

for every $|p| \leqslant \frac{n-1}{2}$ and every $k \in [2,n]$ with $k \geqslant 2|p|+1$. We optimize this bound with respect to k: the optimal choice is

$$k = \lceil \sqrt{|p|}\sqrt[4]{n}\rceil,$$

which satisfies our restrictions provided that e.g. $|p| \in \left[\frac{4}{\sqrt{n}}, \frac{\sqrt{n}}{64}\right]$. For all these values of p we get
$$\frac{d}{dp} \log((\mathbb{E}\|Y\|_2^p)^{1/p}) \leq \frac{C_3 \beta^2}{\sqrt{|p|} \sqrt[4]{n}}.$$

In what follows we set $p_0 := 4/\sqrt{n}$. If $2 < p \leq \sqrt{n}/64$, integrating with respect to p we see that
$$\log((\mathbb{E}\|Y\|_2^p)^{1/p}) - \log((\mathbb{E}\|Y\|_2^2)^{1/2}) \leq \frac{C_3 \beta^2}{\sqrt[4]{n}} \int_2^p \frac{1}{\sqrt{p}} dp$$
$$= \frac{2C_3 \beta^2 (\sqrt{p} - \sqrt{2})}{\sqrt[4]{n}} \leq \frac{2C_3 \beta^2 \sqrt{|p-2|}}{\sqrt[4]{n}}.$$

Similarly, for $p \in [p_0, 2]$ we have
$$\log((\mathbb{E}\|Y\|_2^2)^{1/2}) - \log((\mathbb{E}\|Y\|_2^p)^{1/p}) \leq \frac{C_3 \beta^2}{\sqrt[4]{n}} \int_p^2 \frac{1}{\sqrt{p}} dp$$
$$= \frac{2C_3 \beta^2 (\sqrt{2} - \sqrt{p})}{\sqrt[4]{n}} \leq \frac{2C_3 \beta^2 \sqrt{|p-2|}}{\sqrt[4]{n}}.$$

In other words, for all $p \in [p_0, \sqrt{n}/64]$ we have

(13.2.28) $\quad 1 - C_5 \beta^2 \frac{\sqrt{|p-2|}}{\sqrt[4]{n}} \leq \exp\left(-C_4 \beta^2 \frac{\sqrt{|p-2|}}{\sqrt[4]{n}}\right)$

$$\leq \frac{(\mathbb{E}\|Y\|_2^p)^{1/p}}{(\mathbb{E}\|Y\|_2^2)^{1/2}}$$

$$\leq \exp\left(C_4 \beta^2 \frac{\sqrt{|p-2|}}{\sqrt[4]{n}}\right) \leq 1 + C_5 \beta^2 \frac{\sqrt{|p-2|}}{\sqrt[4]{n}}.$$

For the interval $p \in [-\sqrt{n}/64, -p_0]$ we work in the same way and we first prove a similar estimate with 2 replaced by $-p_0$. We have
$$\log((\mathbb{E}\|Y\|_2^{-p_0})^{-1/p_0}) - \log((\mathbb{E}\|Y\|_2^p)^{1/p}) \leq \frac{C_3 \beta^2}{\sqrt[4]{n}} \int_p^{-p_0} \frac{1}{\sqrt{|p|}} dp$$
$$= \frac{2C_3 \beta^2 (\sqrt{|p|} - \sqrt{p_0})}{\sqrt[4]{n}} \leq \frac{2C_3 \beta^2 \sqrt{|p| - p_0}}{\sqrt[4]{n}}.$$

Therefore,

(13.2.29) $\quad \frac{(\mathbb{E}\|Y\|_2^p)^{1/p}}{(\mathbb{E}\|Y\|_2^{-p_0})^{-1/p_0}} \geq \exp\left(-C_4 \frac{\sqrt{|p|-p_0}}{\sqrt[4]{n}}\right) \geq 1 - C_5 \beta^2 \frac{\sqrt{|p|-p_0}}{\sqrt[4]{n}}.$

It remains to study the moments of $\|Y\|_2$ in the range $p \in [-p_0, p_0]$. We set $k_0 = \lceil \sqrt{|p_0|} \sqrt[4]{n} \rceil$ (note that $k_0 \simeq 1$) and follow Fleury's argument for the proof of Proposition 13.1.8. First, using Lemma 13.1.9 we see that
$$h_{k_0, p_0}^{1/2}(U) h_{k_0, -p_0}^{1/2}(U) \leq \frac{(\Gamma(k_0 + p_0)\Gamma(k_0 - p_0))^{1/2}}{\Gamma(k_0)} h_{k_0, 0}(U)$$
$$\leq (1 + C_6 p_0^2) h_{k_0, 0}(U).$$

Then, taking expectation with respect to U we get

$$(1 + C_6 p_0^2) \geqslant \mathbb{E}_U[h_{k_0,p_0}^{1/2}(U)] \mathbb{E}_U[h_{k_0,-p_0}^{1/2}(U)] + \mathrm{Cov}_U\left(h_{k_0,p_0}^{1/2}(U), h_{k_0,-p_0}^{1/2}(U)\right)$$

$$\geqslant \mathbb{E}_U[h_{k_0,p_0}^{1/2}(U)] \mathbb{E}_U[h_{k_0,-p_0}^{1/2}(U)] - \sqrt{\mathrm{Var}_U\left(h_{k_0,p_0}^{1/2}(U)\right) \mathrm{Var}\left(h_{k_0,-p_0}^{1/2}(U)\right)}$$

$$= \mathbb{E}_U[h_{k_0,p_0}^{1/2}(U)] \mathbb{E}_U[h_{k_0,-p_0}^{1/2}(U)] - \left(\mathbb{E}_U[h_{k_0,p_0}(U)] - \left(\mathbb{E}_U[h_{k_0,p_0}^{1/2}(U)]\right)^2\right)^{1/2}$$

$$\times \left(\mathbb{E}_U[h_{k_0,-p_0}(U)] - \left(\mathbb{E}_U[h_{k_0,-p_0}^{1/2}(U)]\right)^2\right)^{1/2}.$$

Using the log-Sobolev inequality for h_{k_0,p_0} and $h_{k_0,-p_0}$ we see that

$$(1 + C_6 p_0^2) \geqslant \left(\mathbb{E}_U[h_{k_0,p_0}(U)] \mathbb{E}_U[h_{k_0,-p_0}(U)]\right)^{1/2}$$

$$\times \left(\exp\left(-\frac{C_7}{2} \frac{A_{k_0,p_0}^2 + A_{k_0,-p_0}^2}{n}\right) - C_7 \frac{A_{k_0,p_0} A_{k_0,-p_0}}{n}\right).$$

Using the estimate $\max\{A_{k_0,p_0}, A_{k_0,-p_0}\} \leqslant C_8 \beta k_0^{3/2}$ we conclude that

$$\frac{(\mathbb{E}_U[h_{k_0,p_0}(U)])^{1/p_0}}{(\mathbb{E}_U[h_{k_0,-p_0}(U)])^{-1/p_0}} \leqslant \left(1 + \frac{C_9 \beta^2}{n}\right)^{2/p_0} \leqslant 1 + \frac{C_{10} \beta^2}{\sqrt{n}},$$

and hence

$$(13.2.30) \qquad \frac{(\mathbb{E}\|Y\|_2^{p_0})^{1/p_0}}{(\mathbb{E}\|Y\|_2^{-p_0})^{-1/p_0}} \leqslant \frac{(\mathbb{E}_U[h_{k_0,p_0}(U)])^{1/p_0}}{(\mathbb{E}_U[h_{k_0,-p_0}(U)])^{-1/p_0}} \leqslant 1 + \frac{C_{10} \beta^2}{\sqrt{n}}.$$

Theorem 13.2.16 follows from (13.2.28), (13.2.29) and (13.2.30). □

REMARK 13.2.17. Let X be an isotropic random vector in $\mathbb{R}^n$ with a log-concave density f. In Remark 13.2.12 we saw that if we consider the random vector $Y = (X + G_n)/\sqrt{2}$ and denote its density by g then

$$(13.2.31) \qquad d_G(Z_q^+(g), B_2^n) \leqslant c\sqrt{q}$$

for every $q \geqslant 2$, where $c > 0$ is an absolute constant. Therefore, Theorem 13.2.16 immediately implies the next theorem.

THEOREM 13.2.18. *Let X be an isotropic log-concave random vector in $\mathbb{R}^n$ and let $Y = (X + G_n)/\sqrt{2}$. Then, for all $|p| \leqslant \bar{c}\sqrt{n}$ one has*

$$(13.2.32) \qquad 1 - \overline{C} \frac{\sqrt{|p-2|}}{n^{1/4}} \leqslant \frac{(\mathbb{E}\|Y\|_2^p)^{1/p}}{(\mathbb{E}\|Y\|_2^2)^{1/2}} \leqslant 1 + \overline{C} \frac{\sqrt{|p-2|}}{n^{1/4}},$$

where $\bar{c}, \overline{C} > 0$ are absolute constants.

13.2.4. Deviation estimates

Next we obtain the deviation estimates of Theorems 13.2.2 and 13.2.3 for Y.

THEOREM 13.2.19. *Let Y be an isotropic log-concave random vector in $\mathbb{R}^n$ such that*

$$(13.2.33) \qquad d_G(Z_q^+(g), B_2^n) \leqslant \beta\sqrt{q}$$

for some $\beta \geqslant 1$ and for all $q \geqslant 2$, where g is the density of Y. Then,

$$(13.2.34) \qquad \mathbb{P}\left(\|Y\|_2 \geqslant (1+t)\sqrt{n}\right) \leqslant \exp(-c_1(\beta)\sqrt{n} \min\{t^3, t\})$$

for all $t \geq 0$, and

(13.2.35) $\quad \mathbb{P}\left(\|Y\|_2 \leq (1-t)\sqrt{n}\right) \leq C_1 \exp\left(-c_2(\beta)\sqrt{n} \max\left\{t^3, \log \frac{c_3}{1-t}\right\}\right)$

for all $t \in (0,1)$.

Proof. We fix the values of the constants $\overline{c}, \overline{C}$ from Theorem 13.2.16 and set
$$\varepsilon_n := \frac{2\sqrt{2\overline{C}}\beta^2}{\sqrt[4]{n}}$$

Observe that there exists a constant $t_0 \in (0,1]$ such that for every $t \in (\varepsilon_n, t_0]$ we can find $p_1 \in (4, \sqrt{n}/64]$ and $p_2 \in [-\sqrt{n}/64, 0)$ satisfying

(13.2.36) $\quad t = 2\overline{C}\beta^2 \dfrac{\sqrt{p_1 - 2}}{\sqrt[4]{n}} \quad \text{and} \quad t = 2\overline{C}\beta^2 \dfrac{\sqrt{|p_2 - 2|}}{\sqrt[4]{n}}.$

Therefore, we have

(13.2.37) $\quad \left(1 - \tfrac{t}{2}\right)\sqrt{n} \leq \left(\mathbb{E}\|Y\|_2^{p_2}\right)^{1/p_2} \leq \left(\mathbb{E}\|Y\|_2^{p_1}\right)^{1/p_1} \leq \left(1 + \tfrac{t}{2}\right)\sqrt{n}.$

Proof of (13.2.34). Using the elementary inequality
$$\frac{1+t}{1+t/2} \geq 1 + \frac{t}{3} \quad (0 \leq t \leq 1)$$
and applying Markov's inequality we see that
$$\mathbb{P}\left(\|Y\|_2 \geq (1+t)\sqrt{n}\right) \leq \mathbb{P}\left(\|Y\|_2 \geq (1+t/3)(\mathbb{E}\|Y\|_2^{p_1})^{1/p_1}\right)$$
$$\leq (1+t/3)^{-p_1} \leq \exp(-p_1 t/4).$$

Solving the first equation (13.2.36) for p_1 in terms of t, we obtain
$$\mathbb{P}\left(\|Y\|_2 \geq (1+t)\sqrt{n}\right) \leq \exp(-c_1 \sqrt{n} t^3/\beta^4)$$
for all $t \in [\varepsilon_n, t_0]$. On the other hand, it is clear that
$$\mathbb{P}\left(\|Y\|_2 \geq (1+t)\sqrt{n}\right) \leq \frac{1}{(1+t)^2} \leq 1 - \frac{t}{2} \leq \exp(-t/2)$$
for all $t \in [0, \varepsilon_n]$, which extends the estimate (13.2.34) to the whole interval $[0, t_0]$ (it suffices to replace the constant c_1 by a different absolute constant $c_2 > 0$). In the case $t \geq t_0$ we apply Borell's lemma (as in Section 5.2.1) to see that
$$\mathbb{P}\left(\|Y\|_2 \geq (1+t)\sqrt{n}\right) \leq \exp(-c_3 \sqrt{n} t),$$
and this completes the proof of (13.2.34).

Proof of (13.2.35). We first write
$$\mathbb{P}\left(\|Y\|_2 \leq (1-t)\sqrt{n}\right) \leq \mathbb{P}\left(\|Y\|_2 \leq (1-t/2)(\mathbb{E}\|Y\|_2^{p_2})^{1/p_2}\right)$$
$$\leq (1-t/2)^{-p_2} \leq \exp(p_2 t/2).$$

Solving the second equation (13.2.36) for p_2 in terms of t, we obtain
$$\mathbb{P}\left(\|Y\|_2 \leq (1-t)\sqrt{n}\right) \leq C_2 \exp(-c\sqrt{n} t^3/\beta^4)$$
for all $t \in [\varepsilon_n, t_0]$. It is simple to extend this estimate to the interval $t \in (0, t_0]$ by adjusting the constant C_2. Finally, observe that if we set $p_3 = p_2(t_0)$ then
$$\left(\mathbb{E}\|Y\|_2^{p_3}\right)^{1/p_3} \geq \frac{1}{2}\sqrt{n},$$

and hence, for all $0 < \varepsilon < 1/2$ we have

$$\mathbb{P}(\|Y\|_2 \leqslant \varepsilon\sqrt{n}) \leqslant \mathbb{P}(\|Y\|_2 \leqslant 2\varepsilon(\mathbb{E}\|Y\|_2^{p_3})^{1/p_3})$$
$$\leqslant (2\varepsilon)^{-p_3} = \exp\left(-c_3(\beta)\sqrt{n}\log\left(\tfrac{1}{2\varepsilon}\right)\right).$$

Changing the constants, if needed, we obtain (13.2.35). □

As in the previous section, starting with an arbitrary isotropic log-concave random vector X and setting $Y = \frac{X+G_n}{\sqrt{2}}$ we immediately get:

THEOREM 13.2.20. *Let X be an isotropic log-concave random vector in $\mathbb{R}^n$ and let $Y = (X + G_n)/\sqrt{2}$. Then,*

(13.2.38) $\qquad \mathbb{P}(\|Y\|_2 \geqslant (1+t)\sqrt{n}) \leqslant \exp(-c_1\sqrt{n}\min\{t^3, t\})$

for all $t \geqslant 0$, and

(13.2.39) $\qquad \mathbb{P}(\|Y\|_2 \leqslant (1-t)\sqrt{n}) \leqslant C_1 \exp\left(-c_2\sqrt{n}\max\left\{t^3, \log\frac{c_3}{1-t}\right\}\right)$

for all $t \in (0, 1)$.

13.2.5. Inner regularization

Having Theorem 13.2.20 one only needs to transfer the estimates (13.2.38) and (13.2.39) from $Y = \frac{X+G_n}{\sqrt{2}}$ to X. This step can be done (see Klartag, and then Fleury, Guédon-E. Milman) for (13.2.38). Klartag and E. Milman introduced in [285] a way of *thickening from inside* the distribution of an isotropic log-concave random vector which serves the same purpose very efficiently for both (13.2.38) and (13.2.39). More precisely, they proved the following

THEOREM 13.2.21 (Klartag-E. Milman). *Let X be an isotropic log-concave random vector in $\mathbb{R}^n$ and let X' denote an independent copy of X. For any $U \in O(n)$ we consider the random vectors*

$$Y_\pm^U := \frac{X \pm U(X')}{\sqrt{2}}.$$

Then:
(i) *For any $U \in O(n)$ and $s > 0$ we have*

$$\mathbb{P}(\|X\|_2 \geqslant s) \leqslant \left(2\max\{\mathbb{P}(\|Y_+^U\|_2 \geqslant s), \mathbb{P}(\|Y_-^U\|_2 \geqslant s)\}\right)^{1/2}$$

and

$$\mathbb{P}(\|X\|_2 \leqslant s) \leqslant \left(2\max\{\mathbb{P}(\|Y_+^U\|_2 \leqslant s), \mathbb{P}(\|Y_-^U\|_2 \leqslant s)\}\right)^{1/2}.$$

(ii) *For any $U \in O(n)$ and any $p \geqslant 2$ one has*

$$Z_p^+(Y_\pm^U) \subseteq CpB_2^n.$$

(iii) *With probability greater than $1 - \exp(-cn)$ a random $U \in O(n)$ satisfies*

$$Z_p^+(Y_\pm^U) \supseteq c_1\sqrt{p}B_2^n$$

for all $2 \leqslant p \leqslant c_2\sqrt{n}$.

The main step for the proof of Theorem 13.2.21 is the next

THEOREM 13.2.22. *Let X be an isotropic log-concave random vector in $\mathbb{R}^n$ and let X' denote an independent copy of X. Let U be uniformly distributed in $O(n)$ and set*
$$Y = \frac{X + U(X')}{\sqrt{2}}.$$
Then, for all $2 \leqslant p \leqslant c_1\sqrt{n}$,
$$\nu(\{U : Z_p^+(Y) \supseteq c_2\sqrt{p}B_2^n\}) \geqslant 1 - \exp(-c_3 n),$$
where $c_1, c_2, c_3 > 0$ are absolute constants.

We start with the proof of Theorem 13.2.22. Our first lemma compares $Z_p^+(Y)$ with $Z_p^+(X) + U(Z_p^+(X))$:

LEMMA 13.2.23. *For any $U \in O(n)$ we have*
$$Z_p^+(Y) \supseteq \frac{1}{2\sqrt{2}e^{1/p}}(Z_p^+(X) + U(Z_p^+(X))).$$

Proof. Fix $\theta \in S^{n-1}$ and write
$$Y_1 = P_\theta(Y), \quad X_1 = P_\theta(X) \quad \text{and} \quad X_1' = P_\theta(U(X')),$$
where P_θ is the orthogonal projection onto $\text{span}\{\theta\}$. Since all these random variables are centered and log-concave, using Grünbaum's lemma as in Remark 13.2.12 we write
$$h^p_{Z_p^+(Y)}(\theta) = 2\mathbb{E}\,(Y_1)_+^p = \frac{2}{2^{p/2}}\mathbb{E}\,(X_1 + X_1')_+^p$$
$$\geqslant \frac{2}{2^{p/2}}\mathbb{E}\,(X_1)_+^p \mathbb{P}(X_1' \geqslant 0) \geqslant \frac{1}{e 2^{p/2}}\,2\mathbb{E}\,(X_1)_+^p.$$
Exchanging the roles of X_1 and X_1' we get
$$h^p_{Z_p^+(Y)}(\theta) \geqslant \frac{1}{e 2^{p/2}}\max\left\{h^p_{Z_p^+(X)}(\theta), h^p_{Z_p^+(U(X'))}(\theta)\right\},$$
and hence
$$h_{Z_p^+(Y)}(\theta) \geqslant \frac{1}{\sqrt{2}e^{1/p}}\frac{h_{Z_p^+(X)}(\theta) + h_{Z_p^+(U(X'))}(\theta)}{2}.$$
Observe that $Z_p^+(U(X')) = U(Z_p^+(X')) = U(Z_p^+(X))$ to conclude the proof. □

LEMMA 13.2.24. *Let K be a centered convex body in $\mathbb{R}^n$. Assume that*
$$N(K, B_2^n) \leqslant \exp(\alpha_1 n) \quad \text{and} \quad \text{v.rad}(K) \geqslant r_1 > 0.$$
Then,
$$N(K^\circ, B_2^n) \leqslant \exp(\alpha_2 n),$$
where $\alpha_2 \leqslant \alpha_1 + \log(c/r_1)$, and $c > 0$ is an absolute constant.

Proof. We consider the body $K_s = K \cap -K$. By the König-Milman theorem (or the duality of entropy theorem, see Section 1.8.3) we know that
$$N(K^\circ, B_2^n) \leqslant N(K_s^\circ, B_2^n) \leqslant c_1^n N(B_2^n, K_s).$$
Therefore,
$$N(K^\circ, B_2^n) \leqslant c_1^n \frac{|B_2^n + K_s/2|}{|K_s/2|} \leqslant c_1^n N(K_s/2, B_2^n)\frac{|B_2^n|}{|K_s/2|}.$$

Since $|K_s| = |K \cap -K| \geqslant 2^{-n}|K|$, we finally get
$$N(K^\circ, B_2^n) \leqslant (4c_1)^n N(K, B_2^n)(\mathrm{v.rad}(K))^{-n} \leqslant (8c_1/r_1)^n \exp(\alpha_1 n),$$
which proves the lemma. □

LEMMA 13.2.25. *Let A be a compact set in $\mathbb{R}^n$, $n \geqslant 2$, such that $N(A, B_2^n) \leqslant \exp(\alpha_1 n)$. Then,*
$$\nu(\{U \in O(n) : A \cap U(A) \subseteq \alpha_3 B_2^n\}) \geqslant 1 - \exp(-\alpha_2 n),$$
where $\alpha_2 = \alpha_1 + (\log 2)/2$ and $\alpha_3 = c_1 \exp(6\alpha_1)$ for some absolute constant $c_1 > 0$.

Proof. If we set $N = \lfloor \exp(\alpha_1 n) \rfloor$, there exist $x_1, \ldots, x_N \in \mathbb{R}^n$ such that $A \subseteq \bigcup_{i=1}^{N}(x_i + B_2^n)$. We define $R = 4c\exp(6\alpha_1)$, where $c > 0$ will be chosen large enough, and without loss of generality, we assume that $x_1, \ldots, x_M$ are the points of the set $\{x_1, \ldots, x_N\}$ that do not belong to RB_2^n. For every $i = 1, \ldots, M$, the cone $C_i = \{t(x_i + B_2^n) : t \geqslant 0\}$ intersects S^{n-1} at a spherical cap B_{r_i} of radius $r_i \leqslant 1/R$. By rotational invariance, for all $1 \leqslant i, j \leqslant M$ we have
$$\nu(\{U \in O(n) : U(x_i + B_2^n) \cap (x_j + B_2^n) \neq \emptyset\}) \leqslant \sigma(B_{2/R}).$$
Assuming that $2/R < (2c)^{-1}$ we have
$$\sigma(B_{2/R}) \leqslant (2c/R)^{n-1}.$$
Therefore,
$$\nu(\{U \in O(n) : A \cap U(A) \subseteq (R+1)B_2^n\})$$
$$\geqslant \nu\left(\bigcap_{i,j=1}^{M} \{U \in O(n) : U(x_i + B_2^n) \cap (x_j + B_2^n) = \emptyset\}\right)$$
$$\geqslant 1 - M^2(2c/R)^{n-1}.$$
Since $M \leqslant \exp(2\alpha_1(n-1))$, we get the result with $c_1 = 5c$. □

For the proof of Theorem 13.2.22 we will also use the following facts:

(i) For any (log-concave) probability density g on $\mathbb{R}^n$ and any $p \geqslant 1$ one has
$$Z_p^+(g) \subseteq 2^{1/p} Z_p(g) \subseteq Z_p^+(g) - Z_p^+(g).$$
We have already seen the first inclusion, and the second follows from
$$h_{Z_p^+(g) - Z_p^+(g)}(\theta) = h_{Z_p^+(g)}(\theta) + h_{Z_p^+(g)}(-\theta) \geqslant 2^{1/p} h_{Z_p(g)}(\theta),$$
where the last inequality follows from the definitions and the elementary inequality $a^{1/p} + b^{1/p} \geqslant (a+b)^{1/p}$ for $a, b \geqslant 0$.

(ii) For all $2 \leqslant p \leqslant \sqrt{n}$ one has
$$w(Z_p(X)) \leqslant c\sqrt{p},$$
where $c > 0$ is an absolute constant. This follows from the fact that $q^*(\mu) \geqslant c_1 \sqrt{n}$, where μ is the distribution of X, and Paouris' inequality (see Chapter 5).

(iii) For all $2 \leqslant p \leqslant \sqrt{n}$ one has
$$\mathrm{v.rad}(Z_p(X)) \geqslant c_2 \sqrt{p},$$
where $c_2 > 0$ is an absolute constant. This is one of the main results of Klartag and E. Milman in [**284**] (see Chapter 7).

Proof of Theorem 13.2.22. Let $2 \leqslant p \leqslant c\sqrt{n}$, where $0 < c < 1$ is a small enough constant. From Sudakov's inequality we have

$$N(Z_p^+(X), \sqrt{p}B_2^n) \leqslant N(2^{1/p}Z_p(X), \sqrt{p}B_2^n) \leqslant \exp\left(\frac{C_1 n w^2(2^{1/p}Z_p(X))}{p}\right)$$
$$\leqslant \exp(c_3 n).$$

By the Rogers-Shephard inequality,

$$2^{n/p}|Z_p(X)| \leqslant |Z_p^+(X) - Z_p^+(X)| \leqslant 4^n |Z_p^+(X)|,$$

and hence

$$\mathrm{v.rad}(Z_p^+(X)) \geqslant c_4 \sqrt{p}.$$

Applying Lemma 13.2.24 we get

$$N(\sqrt{p}(Z_p^+(X))^\circ, B_2^n) \leqslant \exp(c_5 n).$$

Then, Lemma 13.2.25 shows that for $c_6 = c_5 + (\log 2)/2$ we can find $c_7 > 0$ such that

$$\nu\left(\left\{U \in O(n) : (Z_p^+(X))^\circ \cap U((Z_p^+(X))^\circ) \subseteq \frac{c_7}{\sqrt{p}} B_2^n\right\}\right) \geqslant 1 - \exp(-c_6 n).$$

By duality,

$$\nu\left(\{U \in O(n) : Z_p^+(X) + U(Z_p^+(X)) \supseteq c_7^{-1} \sqrt{p} B_2^n\}\right)$$
$$\geqslant \nu\left(\{U \in O(n) : \mathrm{conv}(Z_p^+(X) \cup U(Z_p^+(X))) \supseteq c_7^{-1} \sqrt{p} B_2^n\}\right)$$
$$\geqslant 1 - \exp(-c_6 n).$$

The result now follows from Lemma 13.2.23. $\square$

Note. A proof of the inclusion $Z_p^+(X) + U(Z_p^+(X)) \supseteq c\sqrt{p}B_2^n$ can be also given using the results of Chapter 9; see [**116**].

We can now prove Theorem 13.2.21.

Proof of Theorem 13.2.21. (i) Let $U \in O(n)$. We write

$$2\max\left\{\mathbb{P}\left(\left\|\frac{X+U(X')}{\sqrt{2}}\right\|_2 \leqslant s\right), \mathbb{P}\left(\left\|\frac{X-U(X')}{\sqrt{2}}\right\|_2 \leqslant s\right)\right\}$$
$$\geqslant \mathbb{P}\left(\left\|\frac{X+U(X')}{\sqrt{2}}\right\|_2 \leqslant s\right) + \mathbb{P}\left(\left\|\frac{X-U(X')}{\sqrt{2}}\right\|_2 \leqslant s\right)$$
$$= \mathbb{P}\left(\frac{\|X\|_2^2 + \|X'\|_2^2}{2} + \langle X, U(X')\rangle \leqslant s^2\right)$$
$$+ \mathbb{P}\left(\frac{\|X\|_2^2 + \|X'\|_2^2}{2} - \langle X, U(X')\rangle \leqslant s^2\right)$$
$$\geqslant \mathbb{P}(\|X\|_2 \leqslant s \text{ and } \|X'\|_2 \leqslant s \text{ and } \langle X, U(X')\rangle \leqslant 0)$$
$$+ \mathbb{P}(\|X\|_2 \leqslant s \text{ and } \|X'\|_2 \leqslant s \text{ and } \langle X, U(X')\rangle > 0)$$
$$= \mathbb{P}(\|X\|_2 \leqslant s \text{ and } \|X'\|_2 \leqslant s) = [\mathbb{P}(\|X\|_2 \leqslant s)]^2.$$

In the same way we check that

$$2\max\left\{\mathbb{P}\left(\left\|\frac{X+U(X')}{\sqrt{2}}\right\|_2 \geqslant s\right), \mathbb{P}\left(\left\|\frac{X-U(X')}{\sqrt{2}}\right\|_2 \geqslant s\right)\right\} \geqslant [\mathbb{P}(\|X\|_2 \geqslant s)]^2.$$

This proves the first assertion of the theorem.

(ii) We fix $\theta \in S^{n-1}$ and set $Y_1 = P_\theta(Y_+^U)$, $X_1 = P_\theta(X)$ and $X_2 = P_\theta(U(X'))$, where P_θ is the orthogonal projection onto span$\{\theta\}$. Then,

$$h_{Z_p(Y_+^U)}(\theta) = \left(\mathbb{E}|Y_1|^p\right)^{1/p} = \left(\mathbb{E}\left|\frac{X_1+X_2}{\sqrt{2}}\right|^p\right)^{1/p}$$
$$\leqslant \frac{1}{\sqrt{2}}\left((\mathbb{E}|X_1|^p)^{1/p} + (\mathbb{E}|X_2|^p)^{1/p}\right)$$
$$= \frac{1}{\sqrt{2}}\left(h_{Z_p(X)}(\theta) + h_{Z_p(U(X))}(\theta)\right).$$

Using the fact that $Z_p^+(g) \subseteq 2^{1/p} Z_p(g)$ we get

$$Z_p^+(Y_+^U) \subseteq 2^{1/p} Z_p(Y_+^U) \subseteq \frac{2^{1/p}}{\sqrt{2}}(Z_p(X) + U(Z_p(X))).$$

Since $Z_p(X) \subseteq CpB_2^n$, we get the second assertion of the theorem for Y_+^U (and we work in the same way for Y_-^U).

(iii) Let $p_i = 2^i$, where i is a positive integer. From Theorem 13.2.22 we know that there exists $c_1 > 0$ such that if $2 \leqslant p_i \leqslant c_1\sqrt{n}$ then a random $U \in O(n)$ satisfies

$$Z_{p_i}^+(Y_\pm^U) \supseteq c_2\sqrt{p_i}\, B_2^n$$

with probability greater than $1 - \exp(-c_3 n)$. Using the fact that $Z_{2p}(g) \subseteq cZ_p(g)$, we easily see that

$$Z_p^+(Y_\pm^U) \supseteq c_4\sqrt{p}\, B_2^n$$

for all $2 \leqslant p \leqslant c_1\sqrt{n}$, with probability greater than $1 - (\log n)\exp(-c_3 n)$. This proves the last assertion of the theorem. $\square$

13.2.6. Proof of the main results

We are now ready to prove Theorems 13.2.1, 13.2.2 and 13.2.3.

THEOREM 13.2.26. *Let X be an isotropic log-concave random vector in $\mathbb{R}^n$. Then,*

(13.2.40) $$\mathbb{P}\left(\,|\,\|X\|_2 - \sqrt{n}\,|\geqslant t\sqrt{n}\right) \leqslant C\exp(-c\sqrt{n}\min\{t^3, t\})$$

for all $t \geqslant 0$, and

(13.2.41) $$\mathbb{P}(\|X\|_2 \leqslant \varepsilon\sqrt{n}) \leqslant (C\varepsilon)^{c\sqrt{n}}$$

for all $0 \leqslant \varepsilon \leqslant 1/C$.

Proof of Theorem 13.2.26. We fix $U \in O(n)$ which satisfies the assertions (i), (ii) and (iii) in Theorem 13.2.21. Since

$$c_1\sqrt{p}B_2^n \subseteq Z_p^+(Y_\pm^U) \subseteq c_2 p B_2^n$$

for all $2 \leqslant p \leqslant c_2\sqrt{n}$, we see that Y_+^U and Y_-^U satisfy (13.2.40) and (13.2.41). Now, the first assertion of Theorem 13.2.21 shows that the same holds true for X. $\square$

13.3. Notes and references

The thin shell estimate

The ε-hypothesis was studied by many authors and it was verified in some special cases. In [**277**] Bo'az Klartag first gave a positive answer in full generality. In fact, he established normal approximation for multidimensional marginal distributions. Originally, Klartag showed that, if X is an isotropic log-concave measure then, for all $0 < \varepsilon \leqslant 1$ one has

$$\mathbb{P}\left(\left|\frac{\|X\|_2}{\sqrt{n}} - 1\right| \geqslant \varepsilon\right) \leqslant C n^{-c\varepsilon^2}$$

where $c, C > 0$ are absolute constants. This establishes the ε-hypothesis with $\varepsilon_n \simeq \sqrt{\frac{\log \log n}{\log n}}$. Subsequently, Klartag obtained in [**278**] power-type estimates for the ε-hypothesis by showing that

$$\mathbb{E}\left(\frac{\|X\|_2^2}{n} - 1\right)^2 \leqslant \frac{C}{n^\alpha}$$

with some $\alpha \simeq 1/5$. In both works, the basic idea is to show that a typical multi-dimensional marginal of an isotropic log-concave measure is approximately spherically-symmetric. A main difference in [**278**] is the use of concentration inequalities on the orthogonal group. These play an important role in the works of Fleury and Guédon and E. Milman that we present in this chapter.

Soon after Klartag's first proof of a thin shell estimate, Fleury, Guédon and Paouris obtained in [**180**] a slightly weaker result; they showed that the ε-hypothesis is satisfied for any isotropic convex body K in $\mathbb{R}^n$ with $\varepsilon_n \simeq (\log \log n)^2/(\log n)^{1/6}$. One of their main steps was to show that $I_q(K)$ stays "almost equal" to $I_2(K)$ for large values of q; they proved that

$$I_q(K) \leqslant \left(1 + \frac{cq}{(\log n)^{1/3}}\right) I_2(K)$$

for every $1 < q \leqslant c(\log n)^{1/3}$.

Building on the ideas of Klartag (but also on his joint work with Guédon and Paouris) Fleury [**177**] proved that if X is an isotropic log-concave random vector in $\mathbb{R}^n$, then for every $2 \leqslant q \leqslant c_1 \sqrt[4]{n}$ one has

$$(\mathbb{E}\|X\|_2^q)^{1/q} \leqslant \left(1 + \frac{c_2 q}{\sqrt[4]{n}}\right) (\mathbb{E}\|X\|_2^2)^{1/2},$$

where $c_1, c_2 > 0$ are absolute constants and obtained the thin shell estimates that we present in Section 13.1. Fleury's articles [**178**] and [**179**] are also related to the thin shell and KLS-conjectures. The logarithmic Sobolev inequality is due to Gross (see [**235**] and [**236**]).

Section 13.2 is devoted to the work of Guédon and E. Milman [**240**] that provides the best known estimates for the problem:

$$\mathbb{P}\left(\left|\|X\|_2 - \sqrt{n}\right| \geqslant t\sqrt{n}\right) \leqslant C \exp(-c\sqrt{n} \min\{t^3, t\})$$

for every $t > 0$, where $C, c > 0$ are absolute constants. As the proof shows, the estimates are better if one assumes that the distribution of X has super-Gaussian marginals. This can be assumed, without loss of generality, if one replaces X by an isotropic random vector Y whose distribution is the convolution of the distribution of X with a Gaussian measure. Klartag and E. Milman introduced in [**285**] the idea to use the convolution of the distribution of X with a random orthogonal image of itself. This leads to improved small ball estimates; we present this technique and incorporate it in our presentation.

Central limit theorems for convex bodies

Combining his thin shell estimate with Sodin's moderate deviation estimates from [**469**], Klartag proved in [**278**] that if X is an isotropic log-concave random vector in $\mathbb{R}^n$ then the density f_θ of $\langle X, \theta \rangle$ satisfies

$$\int_{-\infty}^{\infty} |f_\theta(t) - \gamma(t)|\, dt \leqslant \frac{1}{n^\kappa}$$

and

$$\sup_{|t| \leqslant n^\kappa} \left| \frac{f_\theta(t)}{\gamma(t)} - 1 \right| \leqslant \frac{1}{n^\kappa},$$

for all θ in a subset A of S^{n-1} with measure $\sigma(A) \geqslant 1 - c_1 \exp(-c_2 \sqrt{n})$, where γ is the density of a standard Gaussian random variable, and c_1, c_2, κ are absolute constants. Multidimensional versions of the first result appear already in [**277**]. A generalization of the second result to higher dimensions appears in [**165**]: Eldan and Klartag showed that if $1 \leqslant k \leqslant n^{a_1}$ then there exists a subset $\mathcal{A}_{n,k}$ of $G_{n,k}$ with measure $\nu_{n,k}(\mathcal{A}_{n,k}) \geqslant 1 - c_1 \exp(-n^{a_2})$ such that

$$\sup_{\|x\|_2 \leqslant n^{a_3}} \left| \frac{f_E(x)}{\gamma_E(x)} - 1 \right| \leqslant \frac{C}{n^{a_4}}$$

for every $E \in \mathcal{A}_{n,k}$, where f_E is the density of $P_E(X)$, γ is the standard Gaussian density in E, and $c_1, C, a_1, a_2, a_3, a_4 > 0$ are absolute constants.

CHAPTER 14

Kannan-Lovász-Simonovits conjecture

In this Chapter we discuss the Kannan-Lovász-Simonovits conjecture. Recall from the Introduction that the KLS-conjecture asserts that there exists an absolute constant $c > 0$ such that

$$\mathrm{Is}_n := \min\{\mathrm{Is}_\mu : \mu \text{ is an isotropic log-concave measure on } \mathbb{R}^n\} \geqslant c,$$

where the *Cheeger constant* Is_μ of a log-concave probability measure μ is defined as the best constant $\kappa \geqslant 0$ such that

$$\mu^+(A) \geqslant \kappa \min\{\mu(A), 1 - \mu(A)\}$$

for every Borel subset A of $\mathbb{R}^n$, and where

$$\mu^+(A) := \liminf_{t \to 0^+} \frac{\mu(A_t) - \mu(A)}{t}$$

is the Minkowski content of A. In the first two sections of this chapter we introduce four isoperimetric constants (the Cheeger constant Is_μ, the Poincaré constant Poin_μ, the exponential concentration constant Exp_μ and the first moment concentration constant FM_μ) associated with a Borel probability measure μ on $\mathbb{R}^n$ and we discuss their relation. Complementing classical results of Maz'ya, Cheeger, Gromov, V. Milman, Buser, Ledoux and others, a theorem of E. Milman establishes the equivalence of all four constants in the log-concave setting: one has

$$\mathrm{Is}_\mu \simeq \sqrt{\mathrm{Poin}_\mu} \simeq \mathrm{Exp}_\mu \simeq \mathrm{FM}_\mu$$

for every log-concave probability measure μ, where $a \simeq b$ means that $c_1 a \leqslant b \leqslant c_2 a$ for some absolute constants $c_1, c_2 > 0$. We outline the proofs of these facts in Section 14.1 and Section 14.2.

As an application, E. Milman established stability of the Cheeger constant of convex bodies under perturbations. Loosely speaking, if K and T are two convex bodies in $\mathbb{R}^n$ and if $|K| \simeq |T| \simeq |K \cap T|$, then

$$\mathrm{Is}_K \simeq \mathrm{Is}_T.$$

We give the precise statement and present the ideas of the proof in Section 14.3.

In the second part of this chapter we discuss the KLS-conjecture. Kannan, Lovász and Simonovits conjectured that $\mathrm{Is}_\mu \geqslant c$ for every isotropic log-concave measure μ on $\mathbb{R}^n$ and they obtained the lower bound $\sqrt{n}\mathrm{Is}_\mu \geqslant c$, where $c > 0$ is an absolute constant. In Section 14.4 we describe a number of proofs of this inequality (besides the original argument of Kannan, Lovász and Simonovits we discuss the ones of Bobkov and E. Milman). In Section 14.7 we also describe subsequent work of Bobkov which shows that $\sqrt[4]{n}\sqrt{\sigma_\mu}\mathrm{Is}_\mu \geqslant c$; this provides a direct link between the KLS-conjecture and the thin shell conjecture. Combined with the thin shell estimate of Guédon and E. Milman, this result leads to the bound $n^{5/12}\mathrm{Is}_\mu \geqslant c$.

In Section 14.5 we present Klartag's logarithmic in the dimension lower bound for the Poincaré constant $\operatorname{Poin}_K := \operatorname{Poin}_{\mu_K}$ of an unconditional isotropic convex body K in $\mathbb{R}^n$; one has

$$\operatorname{Is}_K \simeq \sqrt{\operatorname{Poin}_K} \geqslant \frac{c}{\log n},$$

where $c > 0$ is an absolute positive constant.

In Section 14.6 we describe a result of Eldan which, again, connects the thin shell conjecture with the KLS-conjecture: there exists an absolute constant $C > 0$ such that

$$\frac{1}{\operatorname{Is}_n^2} \leqslant C \log n \sum_{k=1}^n \frac{\sigma_k^2}{k}.$$

This, combined with the thin shell estimate of Guédon and E. Milman, gives the best known bound for Is_n: $\operatorname{Is}_n^{-1} \leqslant C n^{1/3} \log n$.

14.1. Isoperimetric constants for log-concave probability measures

Let μ be a Borel probability measure on $\mathbb{R}^n$. For every Borel subset A of $\mathbb{R}^n$ we define its *Minkowski content* with respect to μ as follows:

$$(14.1.1) \qquad \mu^+(A) = \liminf_{t \to 0^+} \frac{\mu(A_t) - \mu(A)}{t},$$

where $A_t = \{x : d(x, A) < t\}$ is the t-extension of A with respect to the Euclidean metric. In this section we recall various *isoperimetric constants* which provide information on the interplay between the measure μ and the underlying Euclidean metric. Of course, we are particularly interested in the case of log-concave probability measures.

14.1.1. Cheeger constant

DEFINITION 14.1.1 (Cheeger constant). We say that μ satisfies *Cheeger's inequality* with constant $\kappa \geqslant 0$ if

$$(14.1.2) \qquad \mu^+(A) \geqslant \kappa \min\{\mu(A), 1 - \mu(A)\}$$

for every Borel subset A of $\mathbb{R}^n$. The *Cheeger constant* Is_μ of μ is the best constant $\kappa \geqslant 0$ for which (14.1.2) is satisfied for all A.

We also define a function $I_\mu : [0, 1] \to [0, +\infty]$ as follows:

$$I_\mu(t) = \inf\{\mu^+(A) : A \text{ Borel}, \mu(A) = t\}.$$

This function is called the *isoperimetric profile* of μ. Note that

$$(14.1.3) \qquad \operatorname{Is}_\mu = \inf_{0 < t \leqslant 1/2} \frac{\min\{I_\mu(t), I_\mu(1-t)\}}{t}.$$

The next theorem gives an equivalent description of the Cheeger constant.

THEOREM 14.1.2 (Rothaus, Cheeger, Maz'ya). *Let μ be a Borel probability measure on $\mathbb{R}^n$ that satisfies Cheeger's inequality. Then, $\alpha_1 \leqslant \operatorname{Is}_\mu \leqslant 2\alpha_1$, where α_1 is the largest constant with the following property: for every integrable, locally Lipschitz function $f : \mathbb{R}^n \to \mathbb{R}$,*

$$(14.1.4) \qquad \alpha_1 \int_{\mathbb{R}^n} |f(x) - \mathbb{E}_\mu(f)| \, d\mu(x) \leqslant \int_{\mathbb{R}^n} \|\nabla f\|_2 \, d\mu(x).$$

14.1. ISOPERIMETRIC CONSTANTS

Recall that f is called locally Lipschitz if for every $x \in \mathbb{R}^n$ there exists $r > 0$ such that the restriction of f onto $B(x,r) := \{y : \|y - x\|_2 < r\}$ is Lipschitz, i.e. $\|\nabla f\|_2$ is bounded on $B(x,r)$, where

$$(14.1.5) \qquad \|\nabla f(x)\|_2 = \limsup_{y \to x} \frac{|f(y) - f(x)|}{\|y - x\|_2}.$$

If f is continuous then $\|\nabla f\|_2$ is Borel measurable, and if f is differentiable at x then $\|\nabla f(x)\|_2$ is the usual length of the gradient of f at x. Rademacher's theorem shows that if f is locally Lipschitz then $\|\nabla f(x)\|_2$ is finite, and f is differentiable almost everywhere with respect to the Lebesgue measure; therefore, the definition in (14.1.5) creates no confusion in the absolutely continuous case.

The proof of Theorem 14.1.2 makes use of the co-area formula.

THEOREM 14.1.3 (co-area formula). *Let μ be a Borel probability measure on $\mathbb{R}^n$. For every locally Lipschitz function $f : \mathbb{R}^n \to \mathbb{R}$ we have*

$$(14.1.6) \qquad \int_{\mathbb{R}^n} \|\nabla f(x)\|_2 \, d\mu(x) \geqslant \int_0^\infty \mu^+(\{x : |f(x)| > s\}) \, ds.$$

Proof. For any $t > 0$ we define $f_t : \mathbb{R}^n \to \mathbb{R}$ by $f_t(x) = \sup\{|f(y)| : y \in B(x,t)\}$. Note that f_t is measurable. Observe that for any $x \in \mathbb{R}^n$ we have

$$\limsup_{t \to 0} \frac{f_t(x) - |f(x)|}{t} \leqslant \limsup_{y \to x} \frac{|f(y)| - |f(x)|}{\|y - x\|_2}$$

$$\leqslant \limsup_{y \to x} \frac{|f(y) - f(x)|}{\|y - x\|_2} = \|\nabla f(x)\|_2.$$

Using Fatou's lemma we see that

$$\int_{\mathbb{R}^n} \|\nabla f(x)\|_2 d\mu(x) \geqslant \int_{\mathbb{R}^n} \limsup_{t \to 0} \frac{f_t(x) - |f(x)|}{t} d\mu(x)$$

$$\geqslant \limsup_{t \to 0} \int_{\mathbb{R}^n} \frac{f_t(x) - |f(x)|}{t} d\mu(x)$$

$$\geqslant \liminf_{t \to 0} \int_{\mathbb{R}^n} \frac{f_t(x) - |f(x)|}{t} d\mu(x)$$

$$= \liminf_{t \to 0} \int_0^\infty \frac{1}{t} \bigl(\mu(\{f_t > s\}) - \mu(\{|f| > s\})\bigr) ds.$$

For any $s > 0$ we set $A(s) = \{|f| > s\}$. One can check that $\{f_t > s\} = (A(s))_t$ and hence, by Fatou's lemma again, we may write

$$\int_{\mathbb{R}^n} \|\nabla f(x)\|_2 d\mu(x) \geqslant \liminf_{t \to 0} \int_0^\infty \frac{\mu((A(s))_t) - \mu(A(s))}{t} ds$$

$$\geqslant \int_0^\infty \liminf_{t \to 0} \frac{\mu((A(s))_t) - \mu(A(s))}{t} ds$$

$$= \int_0^\infty \mu^+(A(s)) \, ds$$

as claimed. $\square$

Proof of Theorem 14.1.2. We set $\kappa = \mathrm{Is}_\mu$. Let $f : \mathbb{R}^n \to \mathbb{R}$ be an integrable, locally Lipschitz function. It suffices to prove the desired inequality for approximations of f of the form $(f + n) \cdot \mathbf{1}_{\{f > -n\}} - n$, since, by the continuity of f, every set

$\{f > -n\}$ is open and so $\nabla((f+n) \cdot \mathbf{1}_{\{f>-n\}})$ coincides with $\nabla(f+n) = \nabla f$ on this set, while $\|\nabla((f+n) \cdot \mathbf{1}_{\{f>-n\}})\|_2 \leqslant \|\nabla f\|_2$ on the set $\{f \leqslant -n\}$. Thus we may assume that f is bounded from below and hence, by adding a suitable constant, that $f > 0$. The co-area formula shows that

$$(14.1.7) \qquad \int_{\mathbb{R}^n} \|\nabla f(x)\|_2 d\mu(x) \geqslant \int_0^\infty \mu^+(\{x : f(x) > s\}) \, ds$$

$$\geqslant \kappa \int_0^\infty \min\{\mu(A(s)), 1 - \mu(A(s))\} \, ds,$$

where $A(s) = \{f > s\}$. Using the facts that

$$\|\mathbf{1}_B - \mathbb{E}_\mu(\mathbf{1}_B)\|_1 = 2\mu(B)(1 - \mu(B))$$

for every Borel subset B of $\mathbb{R}^n$ and the simple identity

$$\mathbb{E}_\mu(f(g - \mathbb{E}_\mu(g))) = \mathbb{E}_\mu(g(f - \mathbb{E}_\mu(f))),$$

we may write

$$\int_{\mathbb{R}^n} \|\nabla f(x)\|_2 d\mu(x) \geqslant \kappa \int_0^\infty \mu(A(s))(1 - \mu(A(s))) \, ds$$

$$= \frac{\kappa}{2} \int_0^\infty \|\mathbf{1}_{A(s)} - \mathbb{E}_\mu(\mathbf{1}_{A(s)})\|_1 \, ds$$

$$\geqslant \frac{\kappa}{2} \sup\left\{ \int_0^\infty \int_{\mathbb{R}^n} (\mathbf{1}_{A(s)} - \mathbb{E}_\mu(\mathbf{1}_{A(s)})) g \, d\mu \, ds \;\middle|\; \|g\|_\infty \leqslant 1 \right\}$$

$$= \frac{\kappa}{2} \sup\left\{ \int_0^\infty \int_{\mathbb{R}^n} \mathbf{1}_{A(s)} (g - \mathbb{E}_\mu(g)) \, d\mu \, ds \;\middle|\; \|g\|_\infty \leqslant 1 \right\}$$

$$= \frac{\kappa}{2} \sup\left\{ \int_{\mathbb{R}^n} f(g - \mathbb{E}_\mu(g)) \, d\mu \;\middle|\; \|g\|_\infty \leqslant 1 \right\}$$

$$= \frac{\kappa}{2} \sup\left\{ \int_{\mathbb{R}^n} g(f - \mathbb{E}_\mu(f)) \, d\mu \;\middle|\; \|g\|_\infty \leqslant 1 \right\}$$

$$= \frac{\kappa}{2} \|f - \mathbb{E}_\mu(f)\|_1.$$

This shows that $\kappa \leqslant 2\alpha_1$.

For the inverse inequality we consider any closed subset A of $\mathbb{R}^n$ and for small $\varepsilon > 0$ we define the function

$$f_\varepsilon(x) = \max\left\{0, 1 - \frac{d(x, A_{\varepsilon^2})}{\varepsilon - \varepsilon^2}\right\}.$$

Then, $0 \leqslant f_\varepsilon \leqslant 1$, $f_\varepsilon \equiv 1$ on $A_{\varepsilon^2} \supseteq A$, $f \equiv 0$ on $\{x : d(x, A) > \varepsilon\}$, and $f_\varepsilon \longrightarrow \mathbf{1}_A$ as $\varepsilon \to 0$. Finally, f_ε is Lipschitz: we have

$$|f_\varepsilon(x) - f_\varepsilon(y)| \leqslant \frac{1}{\varepsilon(1 - \varepsilon)} \left| d(x, A_{\varepsilon^2}) - d(y, A_{\varepsilon^2}) \right| \leqslant \frac{\|x - y\|_2}{\varepsilon(1 - \varepsilon)},$$

therefore $\|\nabla f_\varepsilon(x)\|_2 \leqslant (\varepsilon - \varepsilon^2)^{-1}$. Since $\nabla f_\varepsilon(x) = 0$ on $C = \{x : d(x, A) > \varepsilon\} \cup \{x : d(x, A) < \varepsilon^2\}$, we get

$$\int_{\mathbb{R}^n} \|\nabla f_\varepsilon(x)\|_2 d\mu(x) \leqslant \int_{\mathbb{R}^n \setminus C} \|\nabla f_\varepsilon(x)\|_2 d\mu(x)$$

$$\leqslant \frac{1}{1 - \varepsilon} \frac{\mu(A_\varepsilon) - \mu(A)}{\varepsilon} - \frac{\varepsilon}{1 - \varepsilon} \frac{\mu(A_{\varepsilon^2}) - \mu(A)}{\varepsilon^2},$$

and since we have assumed (14.1.4) we get

$$\alpha_1 \int_{\mathbb{R}^n} |f_\varepsilon(x) - \mathbb{E}_\mu(f_\varepsilon)| \, d\mu(x) \leqslant \frac{1}{1-\varepsilon} \frac{\mu(A_\varepsilon) - \mu(A)}{\varepsilon} - \frac{\varepsilon}{1-\varepsilon} \frac{\mu(A_{\varepsilon^2}) - \mu(A)}{\varepsilon^2}.$$

Letting $\varepsilon \to 0^+$ we see that

$$\mu^+(A) \geqslant \alpha_1 \|\mathbf{1}_A - \mathbb{E}_\mu(\mathbf{1}_A)\|_1 = 2\alpha_1 \mu(A)(1 - \mu(A)),$$

and the result follows. □

14.1.2. Poincaré constant

DEFINITION 14.1.4 (Poincaré constant). We say that μ satisfies the *Poincaré inequality* with constant $\kappa > 0$ if

$$(14.1.8) \qquad \kappa \operatorname{Var}_\mu(f) \leqslant \int_{\mathbb{R}^n} \|\nabla f\|_2^2 \, d\mu,$$

for all locally Lipschitz functions f on $\mathbb{R}^n$ that are square integrable. The *Poincaré constant* (or *spectral gap*) Poin_μ of μ is the best constant $\kappa > 0$ for which (14.1.8) is satisfied.

The classical Poincaré inequality is closely related to the eigenvalues of the Laplace-Beltrami operator

$$\Delta(f) = \operatorname{div}(\nabla f).$$

It is known that the eigenvalues of $-\Delta$ are non-negative and form a discrete set, hence, they can be written in ascending order $0 < \lambda_1 \leqslant \lambda_2 \leqslant \cdots$, since Δ vanishes only for constant functions. In the case where μ is a probability measure with density $e^{-\varphi(x)}$ and φ is a C^1-function on $\mathbb{R}^n$, the Laplace-Beltrami operator is defined by

$$\mathcal{L}_\mu f = \Delta f - \langle \nabla f, \nabla \varphi \rangle.$$

One can check that the following Green's formula holds:

$$(14.1.9) \qquad \int_{\mathbb{R}^n} (\mathcal{L}_\mu f) g \, d\mu = -\int_{\mathbb{R}^n} \langle \nabla f, \nabla g \rangle \, d\mu,$$

for all smooth bounded functions $f, g \in C_b^\infty(\mathbb{R}^n)$. Indeed; using Green's theorem we may write:

$$\int_{\mathbb{R}^n} g(\Delta f - \langle \nabla f, \nabla \varphi \rangle) e^{-\varphi} = \int_{\mathbb{R}^n} g \Delta f e^{-\varphi} - \int_{\mathbb{R}^n} g \langle \nabla f, \nabla \varphi \rangle e^{-\varphi}$$

$$= -\int_{\mathbb{R}^n} \langle \nabla f, \nabla(g e^{-\varphi}) \rangle - \int_{\mathbb{R}^n} g \langle \nabla f, \nabla \varphi \rangle e^{-\varphi}$$

$$= -\int_{\mathbb{R}^n} \langle \nabla f, \nabla g \rangle e^{-\varphi}.$$

The next theorem shows that $\operatorname{Poin}_\mu = \lambda_1$, where λ_1 is the first eigenvalue of the differential operator $-\mathcal{L}_\mu$.

THEOREM 14.1.5. *Let μ be a Borel probability measure with density $e^{-\varphi(x)}$, where φ is a C^1-function on $\mathbb{R}^n$. Then,*

$$(14.1.10) \qquad \lambda_1 \operatorname{Var}_\mu(f) \leqslant \int_{\mathbb{R}^n} \|\nabla f\|_2^2 \, d\mu,$$

for all smooth functions f with compact support on $\mathbb{R}^n$. We have equality if f is an eigenfunction that corresponds to λ_1. Therefore, $\mathrm{Poin}_\mu = \lambda_1$.

Proof. Consider $C_b^\infty(\mathbb{R}^n)$ as a subspace of $L_2(\mu)$ with the inner product $\langle f, g \rangle = \int_{\mathbb{R}^n} fg \, d\mu$. The operator $-\mathcal{L}_\mu$ is self-adjoint and positive, and there exists an orthonormal basis $\{\phi_j\}$ of $L_2(\mu)$ which consists of eigenfunctions corresponding to the eigenvalues λ_j. From Parseval's identity, if $f \in L_2(\mu)$ then we have

$$f = \sum_{j=1}^\infty \langle f, \phi_j \rangle \phi_j \quad \text{and} \quad \|f\|_{L_2(\mu)}^2 = \sum_{j=1}^\infty \langle f, \phi_j \rangle^2.$$

Consider the *energy*

$$\mathcal{E}(f, g) = \int_{\mathbb{R}^n} \langle \nabla f, \nabla g \rangle \, d\mu.$$

Note that $\mathcal{E}(f) := \mathcal{E}(f, f) = \int_{\mathbb{R}^n} \|\nabla f\|_2^2 d\mu$. Using (14.1.9) we see that, for every $s \geqslant 1$,

$$\begin{aligned}
0 &\leqslant \mathcal{E}\left(f - \sum_{j=1}^s \langle f, \phi_j \rangle \phi_j, f - \sum_{j=1}^s \langle f, \phi_j \rangle \phi_j\right) \\
&= \mathcal{E}(f, f) - 2\sum_{j=1}^s \langle f, \phi_j \rangle \mathcal{E}(f, \phi_j) + \sum_{j,k=1}^s \langle f, \phi_j \rangle \langle f, \phi_k \rangle \mathcal{E}(\phi_j, \phi_k) \\
&= \mathcal{E}(f, f) + 2\sum_{j=1}^s \langle f, \phi_j \rangle \langle f, \mathcal{L}_\mu \phi_j \rangle - \sum_{j,k=1}^s \langle f, \phi_j \rangle \langle f, \phi_k \rangle \langle \phi_j, \mathcal{L}_\mu \phi_k \rangle \\
&= \mathcal{E}(f, f) - 2\sum_{j=1}^s \lambda_j \langle f, \phi_j \rangle^2 + \sum_{j=1}^s \lambda_j \langle f, \phi_j \rangle^2.
\end{aligned}$$

In other words,

$$\sum_{j=1}^s \lambda_j \langle f, \phi_j \rangle^2 \leqslant \mathcal{E}(f)$$

for every s, and this gives

$$\lambda_1 \|f\|_{L_2(\mu)}^2 = \lambda_1 \sum_{j=1}^\infty \langle f, \phi_j \rangle^2 \leqslant \sum_{j=1}^\infty \lambda_j \langle f, \phi_j \rangle^2 \leqslant \mathcal{E}(f).$$

Note that $\mathcal{E}(f) = \mathcal{E}(f - \mathbb{E}_\mu(f))$; so,

$$\mathrm{Var}_\mu(f) \leqslant \frac{1}{\lambda_1} \mathcal{E}(f) = \frac{1}{\lambda_1} \int_{\mathbb{R}^n} \|\nabla f\|_2^2 d\mu.$$

This shows that the Poincaré inequality is satisfied with $\kappa = \lambda_1$. □

It was proved by Maz'ya (see [**355**], [**356**]) and independently by Cheeger (see [**140**]) that the Poincaré constant of μ is bounded by its Cheeger constant; in other words, the Poincaré inequality follows if the isoperimetric inequality (14.1.2) is satisfied. More precisely, one has

THEOREM 14.1.6 (Maz'ya, Cheeger). *Let μ be a Borel probability measure on $\mathbb{R}^n$ that satisfies Cheeger's inequality. Then,*

(14.1.11) $$\mathrm{Poin}_\mu \geqslant \frac{\mathrm{Is}_\mu^2}{4}.$$

Proof. Let $\kappa = \mathrm{Is}_\mu$. By the co-area formula and the definition of the Cheeger constant, for every positive smooth function g we have

$$(14.1.12) \qquad \kappa \int_0^\infty \min\{\mu(\{g \geqslant s\}), 1 - \mu(\{g \geqslant s\})\} \, ds \leqslant \int_0^\infty \mu^+(\{g \geqslant s\}) \, ds$$

$$\leqslant \int_{\mathbb{R}^n} \|\nabla g\|_2 d\mu.$$

Let f be a smooth function and set $m = \mathrm{med}(f)$. Then, we have $\mu(\{f \geqslant m\}) \geqslant \frac{1}{2}$ and $\mu(\{f \leqslant m\}) \geqslant \frac{1}{2}$. We set $f^+ = \max\{f - m, 0\}$ and $f^- = -\min\{f - m, 0\}$. Then, $f - m = f^+ - f^-$ and by the definition of m we have

$$\mu(\{(f^+)^2 \geqslant s\}) \leqslant \frac{1}{2} \quad \text{and} \quad \mu(\{(f^-)^2 \geqslant s\}) \leqslant \frac{1}{2}$$

for all $s > 0$. Using (14.1.12) with $g = (f^+)^2$ and $g = (f^-)^2$ and applying integration by parts we see that

$$\kappa \int_{\mathbb{R}^n} |f - m|^2 d\mu = \kappa \int_{\mathbb{R}^n} (f^+)^2 d\mu + \kappa \int_{\mathbb{R}^n} (f^-)^2 d\mu$$

$$= \kappa \int_0^\infty \mu(\{(f^+)^2 \geqslant s\}) \, ds + \kappa \int_0^\infty \mu(\{(f^-)^2 \geqslant s\}) \, ds$$

$$\leqslant \int_{\mathbb{R}^n} \|\nabla((f^+)^2)\|_2 d\mu + \int_{\mathbb{R}^n} \|\nabla((f^-)^2)\|_2 d\mu$$

$$= \int_{\mathbb{R}^n} (\|\nabla((f^+)^2)\|_2 + \|\nabla((f^-)^2)\|_2) \, d\mu.$$

Note that

$$\|\nabla((f^+)^2)\|_2 + \|\nabla((f^-)^2)\|_2 \leqslant 2|f - m| \, \|\nabla f\|_2.$$

So, an application of the Cauchy-Schwarz inequality shows that

$$\kappa \int_{\mathbb{R}^n} |f - m|^2 d\mu \leqslant 2 \left(\int_{\mathbb{R}^n} |f - m|^2 d\mu \right)^{1/2} \left(\int_{\mathbb{R}^n} \|\nabla f\|_2^2 d\mu \right)^{1/2}.$$

It follows that

$$\frac{\kappa^2}{4} \int_{\mathbb{R}^n} |f - m|^2 d\mu \leqslant \int_{\mathbb{R}^n} \|\nabla f\|_2^2 d\mu.$$

Since

$$\int_{\mathbb{R}^n} |f - \mathbb{E}_\mu(f)|^2 d\mu = \min_{\alpha \in \mathbb{R}} \int_{\mathbb{R}^n} |f - \alpha|^2 d\mu \leqslant \int_{\mathbb{R}^n} |f - m|^2 d\mu,$$

we get the theorem. $\square$

In the case of log-concave probability measures, Buser [**130**] (see also Ledoux [**312**]) proved that, conversely, the Cheeger constant of a measure μ is bounded by its Poincaré constant. More precisely, we have

THEOREM 14.1.7 (Buser, Ledoux). *Let μ be a log-concave probability measure on $\mathbb{R}^n$. Then,*

$$\mathrm{Poin}_\mu \leqslant C^2 \mathrm{Is}_\mu^2,$$

where $C > 0$ is an absolute constant.

We sketch a proof using semigroup methods (see the lecture notes [315] of Ledoux for a presentation of the theory of Markov diffusion generators). We assume that $d\mu = e^{-\varphi(x)} dx$ for some C^2 convex function φ on $\mathbb{R}^n$ and consider the semigroup of operators with generator the Laplace-Beltrami operator,

$$\mathcal{L}_\mu(f) = \Delta f - \langle \nabla \varphi, \nabla f \rangle.$$

This is well-defined on $C_b^\infty(\mathbb{R}^n)$ and it is known that there exists a unique semigroup $(P_t)_{t \geq 0}$ of bounded linear operators on $L_2(\mu)$ that satisfy

$$\mathcal{L}_\mu f = \lim_{t \to 0} \frac{P_t f - f}{t}$$

and

$$\frac{d}{dt}(P_t f) = \mathcal{L}_\mu P_t f = P_t \mathcal{L}_\mu f$$

for any $f \in C_b^\infty(\mathbb{R}^n)$. The next basic properties are easily verified: $P_0 f = f$, $P_{t+s} f = P_t(P_s f)$, $[P_t(fg)]^2 \leq P_t(f^2) P_t(g^2)$ for all $f, g \in C_b^\infty(\mathbb{R}^n)$ and $t, s \geq 0$. Moreover, for every $f \in C_b^\infty(\mathbb{R}^n)$ and every $p \geq 1$,

(14.1.13) $$|P_t(f)|^p \leq P_t(|f|^p).$$

The measure μ is time reversible and invariant with respect to $(P_t)_{t \geq 0}$: this means that for any $f, g \in C_b^\infty(\mathbb{R}^n)$ one has

(14.1.14) $$\int_{\mathbb{R}^n} f P_t g \, d\mu = \int_{\mathbb{R}^n} g P_t f \, d\mu$$

and

(14.1.15) $$\int_{\mathbb{R}^n} P_t f \, d\mu = \int_{\mathbb{R}^n} f \, d\mu$$

for all $t \geq 0$ respectively.

We define a symmetric bilinear form Γ by the equation $\Gamma(f, g) = \langle \nabla f, \nabla g \rangle$ for all $f, g \in C_b^\infty(\mathbb{R}^n)$. We also set $\Gamma(f) = \Gamma(f, f) = \|\nabla f\|_2^2$. Note that

$$2\Gamma(f, g) = \mathcal{L}_\mu(fg) - f\mathcal{L}_\mu(g) - g\mathcal{L}_\mu(f) = \lim_{t \to 0} \frac{1}{t}[P_t(fg) - P_t(f)P_t(g)]$$

for all $f, g \in C_b^\infty(\mathbb{R}^n)$. One can check that

$$\Gamma(h(f), g) = h'(f)\Gamma(f, g)$$

and

$$\Gamma(fg, h) = f\Gamma(g, h) + g\Gamma(f, h)$$

for all $f, g, h \in C_b^\infty(\mathbb{R}^n)$.

Next, we define the (iterated) operator Γ_2 by replacing the product of two functions by Γ: we set

$$2\Gamma_2(f, g) = \mathcal{L}_\mu \Gamma(f, g) - \Gamma(f, \mathcal{L}_\mu g) - \Gamma(g, \mathcal{L}_\mu f)$$

for all $f, g \in C_b^\infty(\mathbb{R}^n)$. A direct computation shows that

$$\Gamma_2(f) := \Gamma_2(f, f) = \langle (\text{Hess}\varphi)(\nabla f), \nabla f \rangle + \|\text{Hess} f\|_2^2 \geq \|\text{Hess} f\|_2^2,$$

where we have used that φ is convex, and hence $\text{Hess}\,\varphi$ is positive semi-definite. We conclude that

$$\Gamma_2(f) \geq 0,$$

a property that will be useful in the sequel.

THEOREM 14.1.8 (Bakry-Ledoux). *For any $t \geqslant 0$ and any $f \in C_b^\infty(\mathbb{R}^n)$ we have the pointwise estimate*
$$2t\|\nabla P_t(f)\|_2^2 \leqslant P_t(f^2) - (P_t(f))^2.$$

We need the next lemma.

LEMMA 14.1.9. *For every $f \in C_b^\infty(\mathbb{R}^n)$ and every $t \geqslant 0$ one has*
$$\Gamma P_t f \leqslant P_t \Gamma f.$$

Proof. Observe that the function F defined by $F(s) = P_s(\Gamma(P_{t-s}f))$ on $[0,t]$ is increasing: we have
$$F'(s) = P_s \mathcal{L}_\mu \Gamma(P_{t-s}f) - 2P_s \Gamma(P_{t-s}f, \mathcal{L}_\mu P_{t-s}f)$$
$$= P_s(\Gamma_2(P_{t-s}(f)))$$
by the definition of Γ_2. Since $h = \Gamma_2(P_{t-s}(f)) \geqslant 0$, we get $F'(s) = P_s(h) \geqslant 0$ by (14.1.13).

Since F is increasing, from $F(0) \leqslant F(t)$ we get $\Gamma P_t(f) \leqslant P_t \Gamma f$. □

Proof of Theorem 14.1.8. We write
$$P_t(f^2) - (P_t f)^2 = \int_0^t \frac{d}{ds} P_s((P_{t-s}f)^2) \, ds.$$

Differentiation and the definition of Γ show that
$$P_t(f^2) - (P_t f)^2 = 2 \int_0^t P_s(\Gamma(P_{t-s}f)) \, ds.$$

Since $P_s(\Gamma(P_{t-s}f)) \geqslant \Gamma P_s(P_{t-s}f) = \Gamma P_t f$, we get
$$P_t(f^2) - (P_t f)^2 \geqslant 2 \int_0^t \Gamma(P_t f) \, ds = 2t \Gamma(P_t f)$$
as claimed. □

From Theorem 14.1.8 we can see that, for any $2 \leqslant q \leqslant \infty$,

(14.1.16) $$\big\| \|\nabla P_t(f)\|_2 \big\|_{L_q(\mu)} \leqslant \frac{1}{\sqrt{2t}} \|f\|_{L_q(\mu)}.$$

To see this we use the fact that
$$(P_t(f^2))^{q/2} \leqslant P_t((f^2)^{q/2}) = P_t(|f|^q)$$
and Theorem 14.1.8 in the form $\|\nabla P_t f\|_2^2 \leqslant \frac{1}{2t} P_t(f^2)$. Then, we write
$$\left(\int_{\mathbb{R}^n} \|\nabla P_t f\|_2^q \, d\mu \right)^{1/q} = \left(\int_{\mathbb{R}^n} (\|\nabla P_t f\|_2^2)^{q/2} \, d\mu \right)^{1/q}$$
$$\leqslant \frac{1}{\sqrt{2t}} \left(\int_{\mathbb{R}^n} [P_t(f^2)]^{q/2} \, d\mu \right)^{1/q}$$
$$\leqslant \frac{1}{\sqrt{2t}} \left(\int_{\mathbb{R}^n} P_t(|f|^q) \, d\mu \right)^{1/q}$$
$$= \frac{1}{\sqrt{2t}} \left(\int_{\mathbb{R}^n} |f|^q \, d\mu \right)^{1/q}.$$

Using this with $q = \infty$ we get:

COROLLARY 14.1.10 (Ledoux). *For any $t \geqslant 0$,*
$$\|f - P_t(f)\|_{L_1(\mu)} \leqslant \sqrt{2t}\big\| \|\nabla f\|_2 \big\|_{L_1(\mu)}.$$

Proof. Consider $g \in C_b^\infty(\mathbb{R}^n)$ with $\|g\|_\infty = 1$ and write
$$\int_{\mathbb{R}^n} g(f - P_t f)\, d\mu = -\int_{\mathbb{R}^n} g\Big(\int_0^t (\mathcal{L}_\mu P_s f)\, ds\Big)\, d\mu$$
$$= \int_0^t \Big(-\int_{\mathbb{R}^n} (g\mathcal{L}_\mu P_s f)\, d\mu\Big)\, ds.$$

We know that
$$-\int_{\mathbb{R}^n} g(\mathcal{L}_\mu P_s f)\, d\mu = -\int_{\mathbb{R}^n} g(P_s \mathcal{L}_\mu f)\, d\mu = -\int_{\mathbb{R}^n} (P_s g)(\mathcal{L}_\mu f)\, d\mu$$
$$= \int_{\mathbb{R}^n} \langle \nabla P_s(g), \nabla f \rangle\, d\mu \leqslant \big\| \|\nabla P_s(g)\|_2 \big\|_{L_\infty(\mu)} \int_{\mathbb{R}^n} \|\nabla f\|_2 d\mu.$$

Therefore,
$$\int_0^t \Big(-\int_{\mathbb{R}^n} g\mathcal{L}_\mu P_s f\, d\mu\Big) ds \leqslant \Big(\int_0^t \big\| \|\nabla P_s(g)\|_2 \big\|_{L_\infty(\mu)} ds\Big) \int_{\mathbb{R}^n} \|\nabla f\|_2 d\mu$$
$$\leqslant \Big(\int_0^t \frac{1}{\sqrt{2s}} \|g\|_\infty ds\Big) \int_{\mathbb{R}^n} \|\nabla f\|_2 d\mu$$
$$= \sqrt{2t} \|g\|_\infty \int_{\mathbb{R}^n} \|\nabla f\|_2 d\mu$$
$$= \sqrt{2t} \int_{\mathbb{R}^n} \|\nabla f\|_2 d\mu,$$

and the result follows. $\square$

Applying the inequality of Corollary 14.1.10 to smooth functions that approximate the indicator function of open sets with smooth boundary we can give a proof of Theorem 14.1.7. We will use the next lemma:

LEMMA 14.1.11. *For any f with $\mathbb{E}_\mu(f) = 0$ one has*
$$\|P_t f\|_{L_2(\mu)} \leqslant e^{-\lambda_1 t} \|f\|_{L_2(\mu)}$$
for all $t \geqslant 0$.

Proof. Differentiating the function $G(t) = e^{2\lambda_1 t} \|P_t f\|_{L_2(\mu)}^2$ we get
$$G'(t) = 2e^{2\lambda_1 t}\Big(\lambda_1 \|P_t f\|_{L_2(\mu)}^2 - \int_{\mathbb{R}^n} \|\nabla P_t f\|_2^2\, d\mu\Big) \leqslant 0,$$
using basic properties of the semigroup $(P_t)_{t \geqslant 0}$ and the fact that μ satisfies the Poincaré inequality with constant λ_1. $\square$

Proof of Theorem 14.1.7. Let A be an open subset of $\mathbb{R}^n$ with smooth boundary. From Corollary 14.1.10 and properties (14.1.13)–(14.1.15) of the operators P_t we get
$$\sqrt{2t}\mu^+(A) \geqslant \int_A (1 - P_t(\mathbf{1}_A))\, d\mu + \int_{A^c} P_t(\mathbf{1}_A)\, d\mu$$
$$= 2\Big(\mu(A) - \int_A P_t(\mathbf{1}_A)\, d\mu\Big) = 2\Big(\mu(A) - \|P_{t/2}(\mathbf{1}_A)\|_{L_2(\mu)}^2\Big)$$

for every $t \geq 0$. Using Lemma 14.1.11 and the fact that $P_t(a) = a$ for all constant functions a, we may write

$$\|P_{t/2}(\mathbf{1}_A)\|_{L_2(\mu)}^2 = [\mu(A)]^2 + \|P_{t/2}(\mathbf{1}_A - \mathbb{E}_\mu(\mathbf{1}_A))\|_{L_2(\mu)}^2$$
$$\leq [\mu(A)]^2 + e^{-\lambda_1 t}\|\mathbf{1}_A - \mathbb{E}_\mu(\mathbf{1}_A)\|_{L_2(\mu)}^2.$$

Combining the above we get

$$\sqrt{2t}\mu^+(A) \geq 2\mu(A)(1 - \mu(A))(1 - e^{-\lambda_1 t}) \geq (1 - e^{-\lambda_1 t})\min\{\mu(A), 1 - \mu(A)\}$$

for all $t > 0$. This shows that

$$\mathrm{Is}_\mu \geq \frac{1}{\sqrt{2}} \sup_{t>0} \frac{1 - e^{-\lambda_1 t}}{\sqrt{t}},$$

and choosing $t = 1/\lambda_1$ we see that $\mathrm{Is}_\mu \geq \frac{1-e^{-1}}{\sqrt{2}}\sqrt{\lambda_1}$. □

14.1.3. Exponential concentration

DEFINITION 14.1.12 (exponential concentration constant). We say that μ satisfies an *exponential inequality* with constant $\kappa > 0$ if

(14.1.17) $$\mu(\{x : |f(x) - \mathbb{E}_\mu(f)| \geq t\}) \leq e^{1-\kappa t}$$

for all $t > 0$ and all integrable 1-Lipschitz functions f. The *exponential concentration constant* Exp_μ of μ is the best constant $\kappa > 0$ for which (14.1.17) is satisfied. We also agree to write $\mathrm{Exp}_\mu(f)$ for the best constant $\kappa > 0$ for which (14.1.17) is satisfied by f. From the definition one can check that

(14.1.18) $$\frac{1}{\mathrm{Exp}_\mu(f)} \simeq \|f - \mathbb{E}_\mu(f)\|_{L_{\psi_1}(\mu)}.$$

Gromov and Milman showed in [**232**] that exponential concentration follows from the Poincaré inequality. There exists an absolute constant $c > 0$ such that

(14.1.19) $$c\sqrt{\mathrm{Poin}_\mu} \leq \mathrm{Exp}_\mu.$$

This follows from an estimate for the concentration function of μ in terms of its Poincaré constant.

THEOREM 14.1.13 (Gromov-V. Milman). *Let μ be a Borel probability measure on $\mathbb{R}^n$ that satisfies the Poincaré inequality with constant κ. Then, the concentration function α_μ of μ satisfies*

$$\alpha_\mu(t) \leq \exp\left(-t\sqrt{\kappa}/4\right)$$

for all $t > 0$.

We give an argument for Theorem 14.1.13 that uses the notion of the *expansion coefficient* of μ.

DEFINITION 14.1.14. Let (X, d, μ) be a metric probability space. The expansion coefficient of μ is defined for every $\varepsilon > 0$ as follows:

$$\mathrm{Exp}_\mu(\varepsilon) = \sup\{s \geq 1 : \mu(B_\varepsilon) \geq s\mu(B) \text{ for all } B \subseteq X \text{ with } \mu(B_\varepsilon) \leq 1/2\}.$$

One has the following general fact.

PROPOSITION 14.1.15. *Assume that for some $\varepsilon > 0$ we have $\mathrm{Exp}_\mu(\varepsilon) \geq s > 1$. Then, for every $t > 0$ we have*
$$\alpha_\mu(t) \leq \frac{s}{2} s^{-t/\varepsilon}.$$

Proof. Let $A \subseteq X$ with $\mu(A) \geq \frac{1}{2}$ and set $B = X \setminus A_t$. There exists $k \geq 0$ such that $k\varepsilon \leq t < (k+1)\varepsilon$. Note that for any $1 \leq j \leq k$ we have $(X \setminus A_{j\varepsilon})_\varepsilon \cap A_{(j-1)\varepsilon} = \emptyset$ which gives that $(X \setminus A_{j\varepsilon})_\varepsilon \subseteq X \setminus A_{(j-1)\varepsilon}$. By the definition of the expansion coefficient and the assumption $\mathrm{Exp}_\mu(\varepsilon) \geq s$, we have

$$\mu(X \setminus A_t) \leq \mu(X \setminus A_{k\varepsilon}) \leq \frac{1}{s}\mu(X \setminus A_{(k-1)\varepsilon}) \leq \frac{1}{s^2}\mu(X \setminus A_{(k-2)\varepsilon})$$
$$\leq \cdots \leq \frac{1}{s^k}\mu(A^c) \leq \frac{1}{2} s^{-k} \leq \frac{1}{2} s^{-(\frac{t}{\varepsilon}-1)}$$

since $t < (k+1)\varepsilon$. $\square$

Proof of Theorem 14.1.13. Let A and B be Borel subsets of $\mathbb{R}^n$ with $\mathrm{dist}(A,B) \geq t$. We set $a = \mu(A) > 0$, $b = \mu(B) > 0$. Next, we define $f : \mathbb{R}^n \to \mathbb{R}$ by

$$f(x) = \frac{1}{a} - \frac{1}{t}\left(\frac{1}{a} + \frac{1}{b}\right)\min\{t, d(x,A)\}.$$

Observe that $f(x) = 1/a$ on A, $f(x) = -1/b$ on B and

$$\|\nabla f(x)\|_2 \leq \frac{1}{t}\left(\frac{1}{a} + \frac{1}{b}\right) \quad \text{for all } x \notin A \cup B,$$

while $\|\nabla f(x)\|_2 = 0$ on $A \cup B$. Consequently,

$$\int_{\mathbb{R}^n} \|\nabla f\|_2^2 d\mu \leq \frac{1}{t^2}\left(\frac{1}{a} + \frac{1}{b}\right)^2 (1-a-b).$$

On the other hand, if $m = \mathbb{E}_\mu(f)$ we have

$$\mathrm{Var}_\mu(f) \geq \int_A (f-m)^2 d\mu + \int_B (f-m)^2 d\mu$$
$$\geq a\left(\frac{1}{a} - m\right)^2 + b\left(-\frac{1}{b} - m\right)^2$$
$$\geq \frac{1}{a} + \frac{1}{b}.$$

From the Poincaré inequality we get

$$\left(\frac{1}{a} + \frac{1}{b}\right) \leq \frac{1}{\kappa t^2}\left(\frac{1}{a} + \frac{1}{b}\right)^2 (1-a-b),$$

and hence

$$\kappa t^2 \leq \frac{a+b}{ab}(1-a-b) \leq \frac{1-a-b}{ab} \leq \frac{1-a}{ab}.$$

This implies that

$$\mu(A)\mu(B) \leq \frac{1}{\kappa t^2}\mu(X \setminus A).$$

Consider $\varepsilon > 0$ and B a Borel subset in $\mathbb{R}^n$ with $\mu(B_\varepsilon) \leq 1/2$. Setting $A := X \setminus B_\varepsilon$ we readily see that $\mathrm{dist}(A,B) \geq \varepsilon$ and $\mu(A) \geq 1/2$, and the previous estimates show that

$$\mu(B) \leq \frac{2}{\varepsilon^2 \kappa}\mu(B_\varepsilon).$$

This shows that
$$\operatorname{Exp}_\mu(\varepsilon) \geqslant \frac{\kappa\varepsilon^2}{2}.$$

We choose $\varepsilon = 2/\sqrt{\kappa}$. Then, Proposition 14.1.15 gives
$$\alpha_\mu(t) \leqslant 2^{-t\sqrt{\kappa}/2} < \exp(-t\sqrt{\kappa}/4)$$
as claimed. $\square$

Having this bound for $\alpha_\mu(t)$ we may apply Theorem 1.7.2:

THEOREM 14.1.16. *Let μ be a Borel probability measure on $\mathbb{R}^n$ that satisfies the Poincaré inequality with constant κ. Then, for every 1-Lipschitz function $f : \mathbb{R}^n \to \mathbb{R}$ and every $t > 0$ we have*
$$\mu\left(\{x \in X : |f(x) - \operatorname{med}(f)| > t\}\right) \leqslant 2\exp\left(-t\sqrt{\kappa}/4\right),$$
where $\operatorname{med}(f)$ is a Lévy mean of f.

A standard argument (see [**387**, Appendix V]) shows that we may replace the median by expectation in Theorem 14.1.16. Then, the definition of Exp_μ gives:

THEOREM 14.1.17. *Let μ be a Borel probability measure on $\mathbb{R}^n$ that satisfies the Poincaré inequality with constant κ. Then,*
$$\operatorname{Exp}_\mu \geqslant c\sqrt{\kappa},$$
where $c > 0$ is an absolute constant. Therefore, $\operatorname{Exp}_\mu \geqslant c\sqrt{\operatorname{Poin}_\mu}$.

In the next section we will see that this inequality can be reversed.

14.2. Equivalence of the isoperimetric constants

E. Milman introduced in [**369**] a weaker notion of concentration property as follows.

DEFINITION 14.2.1 (first moment concentration). Let μ be a Borel probability measure on $\mathbb{R}^n$. We say that μ satisfies *first moment concentration* with constant κ if
$$(14.2.1) \qquad \|f - \mathbb{E}_\mu(f)\|_{L_1(\mu)} \leqslant \frac{1}{\kappa}$$
for every integrable 1-Lipschitz function $f : \mathbb{R}^n \to \mathbb{R}$. The *first moment constant* FM_μ of μ is the best constant $\kappa > 0$ for which (14.2.1) is satisfied.

Assume that μ satisfies exponential concentration with constant κ. Then, for any integrable 1-Lipschitz function $f : \mathbb{R}^n \to \mathbb{R}$ we have
$$\|f - \mathbb{E}_\mu(f)\|_{L_1(\mu)} = \int_0^\infty \mu(\{x : |f(x) - \mathbb{E}_\mu(f)| \geqslant t\})\, dt$$
$$\leqslant e\int_0^\infty e^{-\kappa t} dt = \frac{e}{\kappa}.$$

This shows that $\operatorname{FM}_\mu \geqslant e^{-1}\operatorname{Exp}_\mu$. Summarizing, so far we have seen that for any Borel probability measure μ on $\mathbb{R}^n$ one has:
$$\operatorname{Is}_\mu \lesssim \sqrt{\operatorname{Poin}_\mu} \lesssim \operatorname{Exp}_\mu \lesssim \operatorname{FM}_\mu.$$

Although first moment concentration is weaker than exponential concentration, in the setting of log-concave probability measures the four constants that we defined are completely equivalent. More precisely, E. Milman proved the next

THEOREM 14.2.2 (E. Milman). *For every log-concave probability measure μ on $\mathbb{R}^n$ one has*

(14.2.2) $$\mathrm{Is}_\mu \simeq \sqrt{\mathrm{Poin}_\mu} \simeq \mathrm{Exp}_\mu \simeq \mathrm{FM}_\mu,$$

where $a \simeq b$ means that $c_1 a \leqslant b \leqslant c_2 a$ for some absolute constants $c_1, c_2 > 0$.

A first observation towards Theorem 14.2.2 is that the various isoperimetric constants that we discussed in the previous section can have equivalent descriptions through *generalized Poincaré inequalities*.

DEFINITION 14.2.3. Let $1 \leqslant p, q \leqslant \infty$ and let μ be a Borel probability measure on $\mathbb{R}^n$. We will say that μ satisfies a (p,q)-Poincaré inequality with constant κ if

(14.2.3) $$\kappa \|f - \mathbb{E}_\mu(f)\|_{L_p(\mu)} \leqslant \big\| \|\nabla f\|_2 \big\|_{L_q(\mu)}$$

for every integrable, locally Lipschitz function f. We will write $C_\mu^{p,q}$ for the best constant κ with which μ satisfies (14.2.3).

With this terminology we have $\sqrt{\mathrm{Poin}_\mu} = C_\mu^{2,2}$, $\mathrm{FM}_\mu = C_\mu^{1,\infty}$ and $\mathrm{Is}_\mu \simeq C_\mu^{1,1}$ according to Theorem 14.1.2.

REMARK 14.2.4. Recall that by Lemma 2.4.10, for any Orlicz function Φ that is strictly increasing on $\mathbb{R}^+$ and for every $f \in L_\Phi(\mu)$ we have

(14.2.4) $$\frac{1}{2}\|f - \mathbb{E}_\mu(f)\|_{L_\Phi(\mu)} \leqslant \|f - \mathrm{med}_\mu(f)\|_{L_\Phi(\mu)} \leqslant 3\|f - \mathbb{E}_\mu(f)\|_{L_\Phi(\mu)}.$$

We say that μ satisfies a (p,q)-Poincaré inequality with constant κ with respect to the median if

$$\kappa \|f - \mathrm{med}_\mu(f)\|_{L_p(\mu)} \leqslant \big\| \|\nabla f\|_2 \big\|_{L_q(\mu)}$$

for every integrable, locally Lipschitz function f. We will write $C_\mu^{p,q,L}$ for the best constant κ with which μ satisfies the above inequality. Note that

(14.2.5) $$\frac{1}{3} C_\mu^{p,q} \leqslant C_\mu^{p,q,L} \leqslant 2 C_\mu^{p,q}$$

for every $p, q \geqslant 1$.

THEOREM 14.2.5. *Let $1 \leqslant p \leqslant \infty$, $2 \leqslant q \leqslant \infty$ and set $r = \frac{1}{p} + 1 - \frac{1}{q}$. Assume that μ is a log-concave probability measure on $\mathbb{R}^n$ that satisfies a (p,q)-Poincaré inequality. Then,*

(14.2.6) $$\min\{I(t), I(1-t)\} \geqslant \frac{C_\mu^{p,q}}{32} t^r \quad \textit{for all } t \in [0, 1/2].$$

Proof. Note that since $p \geqslant 1$ and $q \geqslant 2$ we have $\frac{1}{2} \leqslant r \leqslant 2$. Let A be an open subset of $\mathbb{R}^n$ with smooth boundary. As in the proof of Theorem 14.1.7 we apply

Corollary 14.1.10 to get

$$\sqrt{2t}\mu^+(A) \geqslant \int_A (1 - P_t(\mathbf{1}_A))\,d\mu + \int_{A^c} P_t(\mathbf{1}_A)\,d\mu$$
$$= 2\left(\mu(A) - \int_A P_t(\mathbf{1}_A)\,d\mu\right)$$
$$= 2\left(\mu(A)(1-\mu(A)) - \int_{\mathbb{R}^n}(P_t(\mathbf{1}_A) - \mu(A))(\mathbf{1}_A - \mu(A))\,d\mu\right).$$

Using Hölder's inequality, the definition of the constant $C_\mu^{p,q}$ and the invariance of μ with respect to $(P_t)_{t\geqslant 0}$, we get

$$\int_{\mathbb{R}^n}(P_t(\mathbf{1}_A) - \mu(A))(\mathbf{1}_A - \mu(A))\,d\mu \leqslant \|P_t(\mathbf{1}_A) - \mu(A)\|_{L_p(\mu)}\|\mathbf{1}_A - \mu(A)\|_{L_{p^*}(\mu)}$$
$$\leqslant (C_\mu^{p,q})^{-1}\big\|\|\nabla P_t(\mathbf{1}_A)\|_2\big\|_{L_q(\mu)}\|\mathbf{1}_A - \mu(A)\|_{L_{p^*}(\mu)},$$

where p^* is the conjugate exponent of p. Combining this with (14.1.16) we arrive at the inequality

(14.2.7) $\sqrt{2t}\mu^+(A) \geqslant$
$$2\left(\mu(A)(1-\mu(A)) - \frac{1}{\sqrt{2t}C_\mu^{p,q}}\|\mathbf{1}_A - \mu(A)\|_{L_q(\mu)}\|\mathbf{1}_A - \mu(A)\|_{L_{p^*}(\mu)}\right)$$

that holds for all $t > 0$. We may now optimize in t. Using the rough estimate

$$\|\mathbf{1}_A - \mu(A)\|_{L_s(\mu)} \leqslant 2[\mu(A)(1-\mu(A))]^{1/s}$$

that is valid for all $s \geqslant 1$, and choosing

$$t = \frac{32}{(C_\mu^{p,q})^2}[\mu(A)(1-\mu(A))]^{2(1/q-1/p)},$$

we obtain

$$\mu^+(A) \geqslant \frac{C_\mu^{p,q}}{8}[\mu(A)(1-\mu(A))]^r \geqslant \frac{C_\mu^{p,q}}{8\cdot 2^r}\min\{\mu(A)^r, (1-\mu(A))^r\},$$

where $r = 1 - 1/q + 1/p$. Since $\frac{1}{2} \leqslant r \leqslant 2$, we have $2^r \leqslant 4$; this completes the proof. $\square$

The proof of the equivalence $\mathrm{FM}_\mu \simeq \mathrm{Is}_\mu$ requires one more (deep) result (see the notes and references of this chapter for historical remarks).

THEOREM 14.2.6. *Let μ be a log-concave probability measure on $\mathbb{R}^n$. Then, the isoperimetric profile I_μ of μ is concave on $(0,1)$, and for every $t \in (0,1)$ we have $I(t) = I(1-t)$. As a consequence,*

$$\mathrm{Is}_\mu := \inf_{0<t<1}\frac{I(t)}{\min\{t, 1-t\}} = \inf_{0<t\leqslant 1/2}\frac{I(t)}{t} = 2I(1/2),$$

which means that we can calculate the Cheeger constant of a log-concave probability measure μ by looking only at Borel sets A with $\mu(A) = 1/2$.

With these tools one can complete the proof of

THEOREM 14.2.7 (E. Milman). *Let μ be a log-concave probability measure on $\mathbb{R}^n$. Then,*

$$\mathrm{FM}_\mu = C_\mu^{1,\infty} \leqslant c\, C_\mu^{1,1} \simeq \mathrm{Is}_\mu,$$

where $c > 0$ is an absolute constant.

Proof. From Theorem 14.2.5 (with $p = 1$, $q = \infty$ and $r = 2$) we know that

$$\min\{I(t), I(1-t)\} \geqslant c_1 \, C_\mu^{1,\infty} t^2 \quad \text{for all } t \in [0, 1/2],$$

where $c_1 > 0$ is an absolute constant. The concavity of the isoperimetric profile and the symmetry around $1/2$ imply

$$\frac{\min\{I(t), I(1-t)\}}{t} = \frac{I(t)}{t} \geqslant 2I(1/2) \quad \text{for all } t \in (0, 1/2],$$

and hence

$$\min\{I(t), I(1-t)\} \geqslant \frac{c_1}{2} C_\mu^{1,\infty} t \quad \text{for all } t \in [0, 1/2].$$

Recall from (14.1.3) that

$$\text{Is}_\mu = \inf_{0 < t \leqslant 1/2} \frac{\min\{I_\mu(t), I_\mu(1-t)\}}{t}.$$

It follows that $\text{Is}_\mu \geqslant c C_\mu^{1,\infty}$, and the proof is complete. $\square$

REMARK 14.2.8. In fact, E. Milman proved that there exists a single form of 1-Lipschitz functions that determine the order of all the above constants: one can "compute" the Poincaré constant of a log-concave probability measure μ on $\mathbb{R}^n$ just by testing functions of the form $x \mapsto d(x, A)$.

THEOREM 14.2.9 (E. Milman). *Let μ be a log-concave probability measure on $\mathbb{R}^n$. Then,*

$$(14.2.8) \qquad \text{Is}_\mu \simeq \inf\left\{\frac{1}{\int d(x, A) d\mu(x)} : \mu(A) \geqslant \frac{1}{2}\right\}.$$

Proof. Let A be a Borel subset of $\mathbb{R}^n$ with $\mu(A) \geqslant 1/2$. Then, if we write g for the function $x \mapsto d(x, A)$, we have that g is 1-Lipschitz and that $\text{med}_\mu(g) = 0$, so by Theorem 14.1.2 and (14.2.4),

$$\frac{1}{2} \text{Is}_\mu \leqslant \frac{\int \|\nabla g\|_2 d\mu}{\int g \, d\mu} \leqslant \frac{1}{\int g \, d\mu}.$$

Conversely, we use Theorem 14.2.2 and Remark 14.2.4 to write

$$\text{Is}_\mu \geqslant c \, C_\mu^{1,\infty,L} = \inf \frac{c}{\int |f| \, d\mu},$$

where the infimum is over all 1-Lipschitz functions with $\text{med}_\mu(f) = 0$. Choosing such a function f and setting $A_1 = \{f \leqslant 0\}$, $A_2 = \{f \geqslant 0\}$, we have that $\mu(A_i) \geqslant 1/2$, $i = 1, 2$, and by the continuity of f that $f(x) = 0$ for every $x \in \text{bd}(A_1) \cup \text{bd}(A_2)$. Given also that f is 1-Lipschitz, we see that

$$\int_{\mathbb{R}^n} |f| \, d\mu \leqslant \int_{\mathbb{R}^n \setminus A_1} d(x, \text{bd}(A_1)) \, d\mu(x) + \int_{\mathbb{R}^n \setminus A_2} d(x, \text{bd}(A_2)) \, d\mu(x)$$

$$= \int_{\mathbb{R}^n} d(x, A_1) \, d\mu(x) + \int_{\mathbb{R}^n} d(x, A_2) \, d\mu(x)$$

whence the conclusion follows. $\square$

14.3. Stability of the Cheeger constant

Theorem 14.2.2 leads to some very useful results on the stability of the Cheeger constant of convex bodies under perturbations. In a few words, E. Milman proved in [**369**] that if K and T are two convex bodies in $\mathbb{R}^n$ then

(14.3.1) $$|K| \simeq |T| \simeq |K \cap T| \quad \text{implies} \quad \text{Is}_K \simeq \text{Is}_T.$$

The results that we discuss in this section can be useful when trying to estimate the Cheeger constant by reducing the problem to some class of convex bodies with additional properties. In fact, in the last two sections of this chapter we will see some of their applications.

The next theorem provides a precise formulation of (14.3.1).

THEOREM 14.3.1 (E. Milman). *Let K and T be two convex bodies in $\mathbb{R}^n$. If*
$$|K \cap T| \geqslant \alpha_K |K| \quad \text{and} \quad |K \cap T| \geqslant \alpha_T |T|,$$
then
$$\text{Is}_K \geqslant \frac{c \alpha_K^2}{\log(1 + 1/\alpha_T)} \text{Is}_T.$$

For the proof we need a series of lemmas.

DEFINITION 14.3.2. Let μ be a log-concave probability measure on $\mathbb{R}^n$. For any Borel function $f : \mathbb{R}^n \to \mathbb{R}$ and any $\delta \in (0,1)$, the *δ-percentile* of f with respect to μ is defined by
$$Q_{\mu,\delta}(f) := \inf\{q \in \mathbb{R} : \mu(\{f \leqslant q\}) \geqslant \delta\}.$$

REMARKS 14.3.3. The next simple, though useful, properties of the functional $f \mapsto Q_{\mu,\delta}(f)$ can be easily verified:
  (i) Note that $Q_{\mu,1/2}(f) = \text{med}_\mu(f)$.
  (ii) For any constant $a \in \mathbb{R}$ we have $Q_{\mu,\delta}(f+a) = Q_{\mu,\delta}(f) + a$.
  (iii) The functional is monotone, i.e. if $f \leqslant g$, then $Q_{\mu,\delta}(f) \leqslant Q_{\mu,\delta}(g)$.
  (iv) The previous two properties and a simple application of the triangle inequality yield that for any constant $b \in \mathbb{R}$, one has
  $$Q_{\mu,\delta}(|f|) \leqslant Q_{\mu,\delta}(|f-b|) + b.$$
  (v) Markov's inequality shows that
  $$\mu(|f| \leqslant t\|f\|_{\psi_1(\mu)}) \geqslant 1 - e^{-t}.$$
  For any $\delta \in (0,1)$ choose $t = t_\delta = \log(1/\delta)$ to conclude that
  $$Q_{\mu,1-\delta}(|f|) \leqslant \log\left(\frac{1}{\delta}\right) \cdot \|f\|_{\psi_1(\mu)}.$$

LEMMA 14.3.4. *Let μ be a log-concave probability measure on $\mathbb{R}^n$ and let f be a 1-Lipschitz function with either of the following two properties:*
  (i) $\mathbb{E}_\mu(f) = 0$ *and* $\|f\|_{L_1(\mu)} \geqslant (3\text{FM}_\mu)^{-1}$,
  (ii) $\text{med}_\mu(f) = 0$ *and* $\|f\|_{L_1(\mu)} \geqslant (6\text{FM}_\mu)^{-1}$.

Then,

(14.3.2) $$\|f\|_{\psi_1} \leqslant C \|f\|_{L_1(\mu)}.$$

Consequently,

(14.3.3) $$Q_{\mu,1-\varepsilon_0}(|f|) \geqslant \frac{\|f\|_{L_1(\mu)}}{2},$$

for some absolute constants $C > 0$ and $0 < \varepsilon_0 < 1$.

Proof. Using Lemma 2.4.10 and Theorem 14.2.2 we write

$$\|f\|_{\psi_1} \simeq \|f - \mathbb{E}_\mu(f)\|_{\psi_1} \simeq \frac{1}{\mathrm{Exp}_\mu(f)} \leqslant \frac{1}{\mathrm{Exp}_\mu} \leqslant \frac{C}{\mathrm{FM}_\mu} \leqslant 6C\|f\|_{L_1(\mu)}.$$

It follows that

$$\|f\|_{L_2(\mu)} \leqslant 2\|f\|_{\psi_1} \leqslant C_1\|f\|_{L_1(\mu)}$$

for some absolute constant $C_1 > 0$. This also allows us to derive (14.3.3) as a consequence of the Paley-Zygmund inequality: for any $\gamma \in (0,1)$ we have $Q_{\mu,1-\varepsilon(\gamma)}(|f|) \geqslant \gamma\|f\|_{L_1(\mu)}$, where $\varepsilon(\gamma) = (1-\gamma)^2/C_1^2$. $\square$

LEMMA 14.3.5. *Let K and T be two convex bodies in $\mathbb{R}^n$ with $T \subseteq K$. If $|T| \geqslant \alpha|K|$ then*

(14.3.4) $$\mathrm{FM}_T \geqslant \frac{c}{\log(1+1/\alpha)}\mathrm{Exp}_K.$$

Proof. Write μ_T and μ_K for the uniform probability measures on T and K respectively. Let g be a 1-Lipschitz function on T with $\mathrm{med}_{\mu_T}(g) = 0$ and the property that

$$\int |g| \, d\mu_T \geqslant \frac{1}{4\mathrm{FM}_T}.$$

Since T is convex, we may clearly extend g to a 1-Lipschitz function on K; for example, we may set $f(x) = g(P_T(x))$, where $P_T(x)$ is the metric projection on T, i.e. is the unique y in T so that $d(x,y) = \mathrm{dist}(x,T)$. It is well known that this function is a contraction (see [**463**]), hence it is clear that the aforementioned extension preserves the Lipschitz constant of g. Without loss of generality we assume that $\mathbb{E}_{\mu_K}(f) \geqslant 0$ (otherwise we may work with $-g$ and $-f$). Note that we can estimate $\mathbb{E}_{\mu_K}(f)$ as follows:

(14.3.5) $$\frac{\alpha}{2} \leqslant \mu_K(\{f \leqslant 0\}) \leqslant \mu_K(\{|f - \mathbb{E}_{\mu_K}(f)| \geqslant \mathbb{E}_{\mu_K}(f)\})$$
$$\leqslant e \cdot \exp\left(-\mathrm{Exp}_K \mathbb{E}_{\mu_K}(f)\right).$$

From Lemma 14.3.4 it follows that

$$\frac{1}{2}\|g\|_{L_1(\mu_T)} \leqslant Q_{\mu_T,1-\varepsilon_0}(|g|),$$

where $\varepsilon_0 > 0$ is an absolute constant. Using this inequality, the assumption that $|T| \geqslant \alpha|K|$, Remark 14.3.3 and (14.3.5) we may write

$$\frac{1}{8\mathrm{FM}_T} \leqslant \frac{1}{2}\|g\|_{L_1(\mu_T)} \leqslant Q_{\mu_T,1-\varepsilon_0}(|g|) \leqslant Q_{\mu_K,1-\varepsilon_0\alpha}(|f|)$$
$$\leqslant Q_{\mu_K,1-\varepsilon_0\alpha}(|f - \mathbb{E}_{\mu_K}(f)|) + \mathbb{E}_{\mu_K}(f)$$
$$\leqslant C_1 \log\left(1 + \frac{1}{\varepsilon_0\alpha}\right)\|f - \mathbb{E}_{\mu_K}(f)\|_{L_{\psi_1}(\mu_K)} + \frac{\log(2e/\alpha)}{\mathrm{Exp}_K}$$
$$\leqslant C_2 \frac{\log(1+1/\alpha)}{\mathrm{Exp}_K},$$

where $C_1, C_2 > 0$ are absolute constants. This concludes the proof. $\square$

14.3. STABILITY OF THE CHEEGER CONSTANT

LEMMA 14.3.6. *Let K and T be two convex bodies in $\mathbb{R}^n$ with $T \subseteq K$. If $|T| \geqslant \alpha |K|$ then*

$$\mathrm{Is}_K \geqslant \alpha^2 \mathrm{Is}_T. \tag{14.3.6}$$

Proof. We first assume that $|T| = p|K|$ for some $p \in (1/2, 1]$. For any Borel set $A \subset K$ with $|A| = |K|/2$ we have that

$$1 - \frac{1}{2p} \leqslant \mu_T(A \cap T) \leqslant \frac{1}{2p}$$

and

$$\mu_K^+(A) \geqslant \frac{|T|}{|K|} \mu_T^+(A \cap T) \geqslant p \, \mathrm{Is}_T \, \min\{\mu_T(A \cap T), 1 - \mu_T(A \cap T)\}.$$

Combining the above we get

$$\mathrm{Is}_K = \inf\{2\mu_K^+(A) : \mu_K(A) = 1/2\} \geqslant (2p - 1)\mathrm{Is}_T.$$

Setting $\alpha_0 := |T|/|K| \geqslant \alpha$ and iterating this procedure with a sequence of convex bodies $T = T_0 \subseteq T_1 \subseteq \cdots \subseteq T_k = K$ such that $|T_i| = p|T_{i+1}|$ for every $i \leqslant k-1$, for some $p = \alpha_0^{1/k} \in (1/2, 1]$, we see that

$$2\mathrm{Is}_K \geqslant (2p-1)^k \mathrm{Is}_T = (2p-1)^{\frac{\log \alpha_0}{\log p}} \mathrm{Is}_T = \alpha_0^{\frac{\log(2p-1)}{\log p}} \mathrm{Is}_T.$$

Letting $p = \alpha_0^{1/k} \to 1^-$ we get the result. $\square$

Proof of Theorem 14.3.1. Applying Lemma 14.3.6, Theorem 14.2.2 and Lemma 14.3.5 we can write

$$\mathrm{Is}_K \geqslant \alpha_K^2 \mathrm{Is}_{K \cap T} \geqslant c_1 \alpha_K^2 \mathrm{FM}_{K \cap T} \tag{14.3.7}$$

$$\geqslant c_2 \frac{\alpha_K^2}{\log(1 + 1/\alpha_T)} \mathrm{Exp}_T \geqslant c_3 \frac{\alpha_K^2}{\log(1 + 1/\alpha_T)} \mathrm{Is}_T$$

for some absolute constants $c_i > 0$. By symmetry, an analogous inequality holds if we interchange the roles of K and T. $\square$

REMARK 14.3.7. In [369] it is also proved that an analogue of Theorem 14.3.1 holds true in the context of log-concave probability measures. Recall that the total variation distance of μ_1 and μ_2 is defined by

$$d_{\mathrm{TV}}(\mu_1, \mu_2) = \frac{1}{2} \int_{\mathbb{R}^n} |f_{\mu_1}(x) - f_{\mu_2}(x)| \, dx,$$

where f_{μ_i} is the density of μ_i. One can show that if μ_1, μ_2 are two log-concave probability measures on $\mathbb{R}^n$ and if

$$d_{\mathrm{TV}}(\mu_1, \mu_2) \leqslant 1 - \varepsilon$$

for some $\varepsilon \in (0, 1)$, then

$$c(\varepsilon)^{-1} \mathrm{Is}_{\mu_2} \leqslant \mathrm{Is}_{\mu_1} \leqslant c(\varepsilon) \mathrm{Is}_{\mu_2}$$

where $c(\varepsilon) \simeq \varepsilon^2 / \log(1 + 1/\varepsilon)$.

14.4. The conjecture and the first lower bounds

We come now to the conjecture of Kannan, Lovász and Simonovits. In [**266**] they define the *isoperimetric coefficient* of a convex body K in $\mathbb{R}^n$ as the largest number $\psi(K)$ with the property that for every measurable subset A of K one has

$$\mu_K^+(A) \geqslant \psi(K) \frac{|A| |K \setminus A|}{|K|}, \tag{14.4.1}$$

where μ_K is the normalized Lebesgue measure on K. Their interest in this parameter arose in connection with randomized volume algorithms. In fact, there was a literature on the subject before [**266**], and the available lower bounds for $\psi(K)$ were of the order of $1/\mathrm{diam}(K)$. Note that instead of (14.4.1) one could have asked for the bound

$$\mu_K^+(A) \geqslant \psi(K) \frac{\min\{|A|, |K \setminus A|\}}{|K|}, \tag{14.4.2}$$

because the quantities on the right hand sides are always within a factor of 2 to each other. Thus, $\psi(K) \simeq \mathrm{Is}_{\mu_K}$.

The main result in [**266**] is a lower bound for $\psi(K)$ in terms of the quantity

$$M_1(K) = \frac{1}{|K|} \int_K \|x - \mathrm{bar}(K)\|_2 \, dx.$$

THEOREM 14.4.1 (Kannan-Lovász-Simonovits). *For every convex body K in $\mathbb{R}^n$ one has*

$$\psi(K) \geqslant \frac{\ln 2}{M_1(K)}. \tag{14.4.3}$$

Note that, if K is isotropic, then $M_1(K) = I_1(K)$. Also, in that case $M_1(K) \leqslant \sqrt{n} L_K$, so (14.4.3) takes the form

$$\psi(K) \geqslant \frac{c}{\sqrt{n} L_K}, \tag{14.4.4}$$

where $c > 0$ is an absolute constant.

The conjecture of Kannan, Lovász and Simonovits is that a much stronger bound is valid. In the setting of centered convex bodies, they conjecture that

$$\psi(K) \simeq \frac{1}{\sqrt{\alpha(K)}}, \tag{14.4.5}$$

where $\alpha(K)$ is the largest eigenvalue of the matrix of inertia $M_{ij} := \int_K x_i x_j dx$. It was actually proved in [**266**] that one always has

$$\psi(K) \leqslant \frac{10}{\sqrt{\alpha(K)}},$$

therefore the question is about the lower bound. Since $M = L_K^2 I$ when K is in the isotropic position, we clearly have $\alpha(K) = L_K$ in this case. So, the *KLS-conjecture* may be formulated as follows:

CONJECTURE 14.4.2 (KLS-conjecture for isotropic bodies). *If K is an isotropic convex body in $\mathbb{R}^n$, then*

$$\psi(K) \simeq \frac{1}{L_K}.$$

Recall that, when K is an isotropic convex body in $\mathbb{R}^n$, then $1/L_K$ is (approximately) equal to the $(n-1)$-dimensional volume of the section of K with any hyperplane passing through the origin: $|K \cap \theta^\perp| \simeq 1/L_K$ for every $\theta \in S^{n-1}$. Since $|K \cap \theta^\perp|$ is the Minkowski content of the intersection of K with $\{x : \langle x, \theta \rangle \geqslant 0\}$ or $\{x \in K : \langle x, \theta \rangle \leqslant 0\}$, the KLS-conjecture can be re-read as saying that, when K is isotropic, then the half-spaces are approximate isoperimetric minimizers for the measure μ_K.

Taking into account the different normalization that we have chosen in the definition of isotropic log-concave measures, we can restate the KLS-conjecture in this more general setting as follows:

CONJECTURE 14.4.3 (KLS-conjecture for isotropic log-concave measures). *If μ is an isotropic log-concave measure on $\mathbb{R}^n$, then*
$$\mathrm{Is}_\mu \geqslant c$$
for some absolute constant $c > 0$.

14.4.1. Localization lemma and the bound of Kannan, Lovász and Simonovits

In this subsection we sketch the original proof of Theorem 14.4.1 that was based on the localization lemma of Lovász and Simonovits (see Section 2.6.1).

Sketch of the proof of Theorem 14.4.1. We first fix $\varepsilon > 0$ small enough and consider the intersection of the $\varepsilon/2$-extension of $\mathrm{bd}(A)$ with K; we write K_3 for the closure of this set. In the definition (14.4.1) of the isoperimetric coefficient we replace $\mathrm{bd}(A)$ by K_3, and we replace A by $K_1 = A \setminus K_3$ and $K \setminus A$ by $K_2 = (K \setminus A) \setminus K_3$. The theorem will follow if we prove the next claim and then let $\varepsilon \to 0^+$.

CLAIM 14.4.4. *Let K be a convex body and let $K = K_1 \cup K_2 \cup K_3$ be a decomposition of K into three measurable sets such that $\mathrm{dist}(K_1, K_2) = \varepsilon > 0$. Then,*

$$(14.4.6) \qquad |K_1||K_2| \leqslant \frac{M_1(K)}{\varepsilon \ln 2} |K||K_3|.$$

For the proof of Claim 14.4.4 we may assume that K_1 and K_2 are closed. We may also assume that $\mathrm{bar}(K) = 0$. We define $f_i = \mathbf{1}_{K_i}$, $i = 1, 2, 3$, and $f_4(x) = \frac{\|x\|_2}{\varepsilon \ln 2}$. Then, we need to show that

$$\int_K f_1(x) dx \int_K f_2(x) dx \leqslant \int_K f_3(x) dx \int_K f_4(x) dx,$$

which by Theorem 2.6.5 reduces to the following one-dimensional problem: let $[a, b]$ be a closed interval in $\mathbb{R}$ with $a \leqslant 0 \leqslant b$ and let $u \in [a, b]$. If $[a, b] = J_1 \cup J_2 \cup J_3$ is a partition of $[a, b]$ into three measurable sets, such that $\mathrm{dist}(J_1, J_2) \geqslant \varepsilon$, then

$$\int_{J_1} e^t dt \int_{J_2} e^t dt \leqslant \frac{1}{\varepsilon \ln 2} \int_{J_3} e^t dt \int_a^b |t - u| e^t dt.$$

This is indeed verified with elementary (but delicate) arguments.

14.4.2. Bobkov's argument

Bobkov's argument in [**76**] works for an arbitrary log-concave probability measure μ on $\mathbb{R}^n$. In what follows, we assume that μ has a density (with respect to the Lebesgue measure)
$$d\mu(x) = e^{-\varphi(x)}dx,$$
where $\varphi : \mathbb{R}^n \to \mathbb{R}$ is a convex function.

THEOREM 14.4.5. *Let μ be a log-concave probability measure on $\mathbb{R}^n$. Then,*
$$\mathrm{Is}_\mu \geqslant \frac{c}{\|f\|_{L_2(\mu)}},$$
where $f(x) = \|x - \mathrm{bar}(\mu)\|_2$ and $c > 0$ is an absolute constant.

This is a consequence of the next inequality which relates the Minkowski content of a set A with the distribution of the Euclidean norm with respect to μ.

THEOREM 14.4.6. *Let μ be a log-concave probability measure on $\mathbb{R}^n$. For every Borel subset A of $\mathbb{R}^n$ and for every $r > 0$ we have*

(14.4.7) $\quad 2r\mu^+(A) \geqslant \mu(A) \log \dfrac{1}{\mu(A)} + (1-\mu(A)) \log \dfrac{1}{1-\mu(A)} + \log \mu(rB_2^n).$

For the proof of Theorem 14.4.6 we need its functional analogue. Recall that, for every non-negative measurable function $f : \mathbb{R}^n \to \mathbb{R}^+$ with $\int f \log(1+f) < \infty$, the entropy of f with respect to μ is

(14.4.8) $\quad\quad\quad \mathrm{Ent}_\mu(f) = \displaystyle\int_{\mathbb{R}^n} f \log f \, d\mu - \int_{\mathbb{R}^n} f \, d\mu \log \int_{\mathbb{R}^n} f \, d\mu.$

Note that the entropy is non-negative (by Jensen's inequality) and homogeneous of degree 1.

PROPOSITION 14.4.7. *Let μ be a centered log-concave probability measure on $\mathbb{R}^n$. For every locally Lipschitz function $f : \mathbb{R}^n \to [0,1]$ and for every $r > 0$ we have*
$$\mathrm{Ent}_\mu(f) + \mathrm{Ent}_\mu(1-f) + \log \mu(rB_2^n) \leqslant 2r \int_{\mathbb{R}^n} \|\nabla f\|_2 \, d\mu.$$

Proof. We may assume that f is smooth, constant outside a compact set, with $0 < f(x) < 1$ for all $x \in \mathbb{R}^n$.

Given $t, s \in (0,1)$ with $t+s = 1$, we set
$$f_t(z) = \sup\left\{ f\left(z + \frac{s}{t}(z-x)\right) : x \in rB_2^n \right\};$$
we then define $w, u, v : \mathbb{R}^n \to \mathbb{R}^+$ by
$$u(x) \equiv u_t(x) = (f(x))^{1/t} e^{-\varphi(x)}$$
$$v(y) = \mathbf{1}_{rB_2^n}(y) e^{-\varphi(y)}$$
$$w(z) \equiv w_t(x) = f_t(z) e^{-\varphi(z)}.$$
The functions $w = w_t, u = u_t, v$ satisfy
$$w(tx+sy) \geqslant (u(x))^t (v(y))^s,$$

for all $x, y \in \mathbb{R}^n$. An application of the Prékopa-Leindler inequality gives
$$\int_{\mathbb{R}^n} f_t \, d\mu \geqslant \left(\int_{\mathbb{R}^n} f^{1/t} \, d\mu \right)^t [\mu(rB_2^n)]^s.$$
From Taylor's theorem we see that, when $s = 1 - t > 0$ is sufficiently small,
$$f_t(z) = f(z) + [r \|\nabla f\|_2 + \langle \nabla f(z), z \rangle] s + O(s^2),$$
uniformly in $z \in \mathbb{R}^n$. On the other hand,
$$\left(\int_{\mathbb{R}^n} f^{1/t} d\mu \right)^t = \int_{\mathbb{R}^n} f \, d\mu + s \operatorname{Ent}_\mu(f) + O(s^2).$$
To see this, we define $h(t) = (\int_{\mathbb{R}^n} f^{1/t} d\mu)^t = \exp(t \log \int_{\mathbb{R}^n} f^{1/t} d\mu)$ and write
$$h(t) = h(1) + h'(1)(t-1) + O((t-1)^2) = h(1) - h'(1)s + O(s^2),$$
where
$$h'(t) = \left(\int_{\mathbb{R}^n} f^{1/t} d\mu \right)^t \left(\log \int_{\mathbb{R}^n} f^{1/t} d\mu - \frac{\int_{\mathbb{R}^n} f^{1/t} \log f \, d\mu}{t \int_{\mathbb{R}^n} f^{1/t} d\mu} \right),$$
which implies that $h'(1) = -\operatorname{Ent}_\mu(f)$. Combining the above and letting $s \to 0^+$ we get
$$\operatorname{Ent}_\mu(f) + \log \mu(rB_2^n) \int_{\mathbb{R}^n} f \, d\mu \leqslant r \int_{\mathbb{R}^n} \|\nabla f\|_2 d\mu + \int_{\mathbb{R}^n} \langle \nabla f(x), x \rangle \, d\mu(x).$$
Applying the same reasoning to $1 - f$ and adding the two inequalities we get the result. $\square$

Proof of Theorem 14.4.6. We may approximate the indicator function $\mathbf{1}_A$ of A by Lipschitz functions with values in $[0, 1]$, as in the proof of Theorem 14.1.2, and then use Proposition 14.4.7 for them. $\square$

Proof of Theorem 14.4.5. We may assume that μ is centered, and hence $f(x) = \|x\|_2$. Let A be a Borel subset of $\mathbb{R}^n$ with $\mu(A) = p \in (0, 1)$. From Theorem 14.4.6 we know that for any $r > 0$,
$$(14.4.9) \qquad \mu^+(A) \geqslant \frac{1}{2r} \left[p \log \frac{1}{p} + (1-p) \log \frac{1}{1-p} + \log \mu(rB_2^n) \right].$$
We choose $r_0 > 0$ so that $\mu(r_0 B_2^n) = \frac{2}{3}$. Then, Borell's lemma shows that for every $t > 1$ we have
$$1 - \mu(tr_0 B_2^n) \leqslant \frac{1}{3} 2^{-\frac{t-1}{2}}.$$
Using the inequality $-\ln(1-x) \leqslant \frac{x}{1-x}$ for $0 < x < 1$, we get
$$-\log \mu(tr_0 B_2^n) \leqslant -\log \left(1 - \frac{1}{3} 2^{-\frac{t-1}{2}} \right) \leqslant 2^{-\frac{t+1}{2}}.$$
Then, we apply (14.4.9) with $r = tr_0$ to get
$$(14.4.10) \qquad \mu^+(A) \geqslant \frac{1}{2tr_0} \left[p \log(1/p) + (1-p) \log(1/(1-p)) + \log \mu(tr_0 B_2^n) \right]$$
$$\geqslant \frac{1}{2tr_0} \left[p \log(1/p) + (1-p) \log(1/(1-p)) - 2^{-\frac{t+1}{2}} \right].$$
We may clearly assume that $0 < p \leqslant 1/2$. We apply (14.4.10) with $t = 3 \log(1/p) \geqslant 1$. Observe that we also have
$$(1-p) \log(1/(1-p)) \geqslant 2^{-\frac{t+1}{2}}.$$

We check this by considering the function

$$g(p) = (1-p)\log\frac{1}{1-p} - 2^{-\frac{t+1}{2}} = (1-p)\log\frac{1}{1-p} - \frac{1}{\sqrt{2}}p^{3\log 2/2}$$

on $[0, 1/2]$. We have $g(0) = 0$ and we can see that g is concave. Moreover, the inequality $g(1/2) \geqslant 0$ is equivalent to the inequality

$$\frac{\log 2}{2} \geqslant 2^{-\frac{3}{4\log 2}},$$

which can be verified by direct calculation. It follows that

$$g(p) \geqslant 0 \text{ for all } p \in [0, 1/2].$$

Going back to (14.4.10) we have

$$(14.4.11) \qquad \mu^+(A) \geqslant \frac{1}{2tr_0}p\log\frac{1}{p} = \frac{p}{6r_0} = \frac{1}{6r_0}\min\{\mu(A), 1-\mu(A)\}.$$

A similar argument works if we assume that $p \geqslant 1/2$, and it leads to the same estimate.

It remains to estimate r_0. Recall that $f(x) = \|x\|_2$. Since $\mu(\{x : \|x\|_2 \geqslant \sqrt{3}\|f\|_{L_2(\mu)}\}) \leqslant \frac{1}{3}$ by Markov's inequality, the choice of r_0 implies that

$$r_0 \leqslant \sqrt{3}\|f\|_{L_2(\mu)}.$$

Then, (14.4.11) gives

$$\mu^+(A) \geqslant \frac{1}{6\sqrt{3}\|f\|_{L_2(\mu)}}\min\{\mu(A), 1-\mu(A)\}.$$

In other words, $\text{Is}_\mu \geqslant c/\|f\|_{L_2(\mu)}$, with $c = (6\sqrt{3})^{-1}$. $\square$

The function $f(x) = \|x - \text{bar}(\mu)\|_2$ satisfies $\|f\|_{L_2(\mu)} = \sqrt{n}$ in the isotropic case. Therefore, the result that we described takes then the following form.

THEOREM 14.4.8. *Let μ be an isotropic log-concave measure on $\mathbb{R}^n$. Then,*

$$\sqrt{n}\text{Is}_\mu \geqslant c,$$

where $c > 0$ is an absolute constant.

14.4.3. A third approach

A different approach which leads to the same lower bound was offered by E. Milman in [**369**]. The proof is very simple if we take into account Theorem 14.2.2.

THEOREM 14.4.9. *Let μ be a log-concave probability measure in $\mathbb{R}^n$. Then,*

$$\text{Is}_\mu \geqslant \frac{1}{2\inf_{z\in\mathbb{R}^n}\|f_z\|_{L_2(\mu)}},$$

where $f_z(x) = \|x - z\|_2$.

Proof. In view of Theorem 14.2.2 instead of estimating Is_μ from below we are going to estimate FM_μ. Let $f : \mathbb{R}^n \to \mathbb{R}$ be a 1-Lipschitz function and fix $z \in \mathbb{R}^n$. Then, we may write

$$\int_{\mathbb{R}^n} |f - \mathbb{E}_\mu(f)|\, d\mu \leqslant \int_{\mathbb{R}^n} |f(x) - f(z)|\, d\mu(x) + |f(z) - \mathbb{E}_\mu(f)|$$

$$\leqslant 2 \int_{\mathbb{R}^n} |f(x) - f(z)|\, d\mu(x) \leqslant 2 \int_{\mathbb{R}^n} \|x - z\|_2\, d\mu(x).$$

Therefore,

$$\mathrm{FM}_\mu \geqslant \frac{1}{2} \sup_{z \in \mathbb{R}^n} \frac{1}{\int \|x - z\|_2\, d\mu(x)}.$$

This proves the theorem. □

14.5. Poincaré constant in the unconditional case

Klartag offers in [279] a logarithmic in the dimension lower bound for the Poincaré constant $\mathrm{Poin}_K := \mathrm{Poin}_{\mu_K}$ of an unconditional isotropic convex body K in $\mathbb{R}^n$.

THEOREM 14.5.1 (Klartag). *For every unconditional isotropic convex body K in $\mathbb{R}^n$ we have*

$$\mathrm{Is}_K \simeq \sqrt{\mathrm{Poin}_K} \geqslant \frac{c}{\log n},$$

where $c > 0$ is an absolute positive constant.

The proof makes use of E. Milman's Theorem 14.3.1, which allows us to exploit the next special case.

PROPOSITION 14.5.2. *Let K be an unconditional convex body in $\mathbb{R}^n$. Assume that K has C^∞-smooth boundary and that $K \subseteq [-R, R]^n$ for some $R > 0$. Then,*

$$\mathrm{Poin}_K \geqslant \frac{\pi^2}{R^2}.$$

Proof. For every $x = (x_1, \ldots, x_n) \in \mathbb{R}^n$ and any $1 \leqslant i \leqslant n$ we define $\sigma_i(x) = (x_1, \ldots, x_{i-1}, -x_i, x_{i+1}, \ldots, x_n)$. We also set $\sigma_i(\varphi)(x) = \varphi(\sigma_i(x))$ for any function $\varphi : K \to \mathbb{R}$.

Claim. There exist $1 \leqslant i \leqslant n$ and a non-zero Neumann eigenfunction φ corresponding to $\lambda_1 = \mathrm{Poin}_K$, such that

$$\sigma_i(\varphi) = -\varphi.$$

Proof of the Claim. We write $h \in E_\lambda$ if h is a Neumann eigenfunction φ corresponding to λ. We first observe that, since K is unconditional, if f is an eigenfunction of λ_1 then the same is true for $\sigma_i(f)$, $i = 1, \ldots, n$. We start with an eigenfunction $f_0 \in E_{\lambda_1}$ and we define

$$f_i = f_{i-1} + \sigma_i(f_{i-1}), \qquad i = 1, \ldots, n.$$

Then, $f_i \in E_{\lambda_1}$ for all $0 \leqslant i \leqslant n$. If $f_i \equiv 0$ for some i then we consider the minimal i with this property, and since $f_{i-1} \not\equiv 0$ and $\sigma_i(f_{i-1}) = -f_{i-1}$, our claim is proved.

Otherwise, we have that $g = f_n$ is a non-zero eigenfunction in E_{λ_1}. Note that $\sigma_i(g) = g$, and hence $\sigma_i(\partial_i g) = -\partial_i g$ for all $i = 1, \ldots, n$. It follows that

$$\int_K \nabla g = 0.$$

Now, we can see that $\partial_1 g, \ldots, \partial_n g$ belong to E_{λ_1}: using (12.4.2) we check that

$$\lambda_1^2 \int_K g^2 = \int_K |\Delta g|^2 \geqslant \sum_{i=1}^n \int_K \|\nabla \partial_i g\|_2^2 \geqslant \lambda_1 \sum_{i=1}^n \int_K (\partial_i g)^2$$
$$= \lambda_1 \int_K \|\nabla g\|_2^2 = \lambda_1^2 \int_K g^2,$$

and this shows that we must have equality everywhere. On the other hand,

$$\int_K \|\nabla g\|_2^2 > 0,$$

and hence, there exists $1 \leqslant i \leqslant n$ such that $\partial_i g \not\equiv 0$. The function $\partial_i g$ then satisfies our claim. $\square$

We can now continue the proof of Proposition 14.5.2. We first check that for every $0 < r \leqslant R$ and any smooth odd function $\psi : [-r, r] \to \mathbb{R}$,

(14.5.1) $$\frac{\pi^2}{R^2} \int_{-r}^r \psi^2(x) dx \leqslant \frac{\pi^2}{r^2} \int_{-r}^r [\psi'(x)]^2 dx.$$

If φ is a non-zero eigenfunction of λ_1 that satisfies the claim, using Fubini's theorem we get

$$\frac{\pi^2}{R^2} \int_K \varphi^2 \leqslant \int_K |\partial_i \varphi|^2 \leqslant \int_K \|\nabla \varphi\|_2^2 = \lambda_1 \int_K \varphi^2.$$

This shows that $\lambda_1 \geqslant \pi^2/R^2$. $\square$

Proof of Theorem 14.5.1. We set $R = C \log n$, for some large enough absolute constant $C > 0$, and consider the unconditional convex body

$$T = K \cap [-R, R]^n.$$

From Proposition 14.5.2 we see that

$$\text{Poin}_T \geqslant \frac{\pi^2}{R^2} \geqslant \frac{c_1}{\log^2 n}.$$

Using ψ_1-tail estimates for the functionals $\langle \cdot, e_i \rangle$ in an elementary way, we check that

$$|T| \geqslant \left(1 - \frac{1}{n}\right) |K|.$$

Then, from Theorem 14.3.1 we get

$$\text{Is}_K \geqslant c_2 \text{Is}_T \geqslant c_3 \sqrt{\text{Poin}_T} \geqslant \frac{c}{\log n},$$

where $c > 0$ is an absolute constant. $\square$

14.6. KLS-conjecture and the thin shell conjecture

In this section we describe Eldan's "approximate reduction" of the KLS-conjecture to the thin shell conjecture. Recall that the KLS-conjecture asks if $\text{Is}_n \geqslant c$, where

$$\text{Is}_n = \inf_\mu \inf_A \frac{\mu^+(A)}{\mu(A)},$$

and the infimum is over all isotropic log-concave measures μ on $\mathbb{R}^n$ and all Borel sets $A \subset \mathbb{R}^n$ with $0 < \mu(A) \leqslant 1/2$, and $\mu^+(A)$ denotes the Minkowski content of A. Alternatively,
$$\mathrm{Is}_n^{-2} \simeq \sup_{\mu} \mathrm{Poin}_\mu^{-1} = \sup_{\mu} \sup_{\varphi} \frac{\int \varphi^2 d\mu}{\int \|\nabla \varphi\|_2^2 d\mu},$$
where the supremum is over all isotropic log-concave measures μ on $\mathbb{R}^n$ and all smooth functions φ with $\int_{\mathbb{R}^n} \varphi \, d\mu = 0$. Eldan (see [**164**]) proved the following.

THEOREM 14.6.1 (Eldan). *There exists an absolute constant $C > 0$ such that*
$$\mathrm{Is}_\mu^{-2} \leqslant C \log n \sum_{k=1}^n \frac{\sigma_k^2}{k}$$
for every isotropic log-concave measure μ on $\mathbb{R}^n$. Assuming the thin shell conjecture, we have
$$\mathrm{Is}_n := \inf_\mu \mathrm{Is}_\mu \geqslant c/\log n.$$

Note that using the thin shell estimate of Guédon and E. Milman one may conclude that $\mathrm{Is}_n^{-1} \leqslant C\sqrt[3]{n}\sqrt{\log n}$. Theorem 14.6.1 will be a consequence of a bound involving the constant
$$K_n^2 := \sup_X \sup_{\theta \in S^{n-1}} \sum_{i,j=1}^n [\mathbb{E}(X_i X_j \langle X, \theta \rangle)]^2,$$
where the supremum is over all isotropic log-concave random vectors X in $\mathbb{R}^n$. Equivalently, K_n can be expressed as
$$K_n := \sup_\mu \left\| \int_{\mathbb{R}^n} x_1 x \otimes x \, d\mu(x) \right\|_{\mathrm{HS}},$$
where the supremum is over all isotropic log-concave measures μ on $\mathbb{R}^n$. The relation between K_n and σ_n is given by the next lemma.

LEMMA 14.6.2. *There exists an absolute constant $C > 0$ such that*
$$K_n^2 \leqslant C \sum_{k=1}^n \frac{\sigma_k^2}{k}.$$

Proof. Let X be an isotropic log-concave random vector in $\mathbb{R}^n$. We fix $\theta \in S^{n-1}$ and set
$$A = \mathbb{E}[X \otimes X \langle X, \theta \rangle].$$
We need to show that
$$\|A\|_{\mathrm{HS}}^2 \leqslant C \sum_{k=1}^n \frac{\sigma_k^2}{k}.$$
Let $k \leqslant n$ and let $E \in G_{n,k}$. We set $Y = \|P_E(X)\|_2 - \sqrt{k}$. Since $P_E(X)$ is isotropic and k-dimensional, we have
$$\mathbb{E}\|P_E(X)\|_2^2 = k \quad \text{and} \quad \mathrm{Var}(Y) \leqslant \mathbb{E}(Y^2) \leqslant \sigma_k^2$$
from the definition of σ_k. From Lemma 12.3.1 we see that
$$\mathrm{Var}(\|P_E(X)\|_2^2) \leqslant Ck\sigma_k^2$$

for some sufficiently large absolute constant $C > 0$. Applying the Cauchy-Schwarz inequality and using the fact that

$$\mathbb{E}(\langle X, \theta \rangle) = 0 \quad \text{and} \quad \text{Var}(|\langle X, \theta \rangle|) = 1$$

we get

$$\mathbb{E}[\langle X, \theta \rangle \|P(X)\|_2^2] \leqslant \sqrt{\text{Var}(|\langle X, \theta \rangle|) \text{Var}(\|P(X)\|_2^2)} \leqslant C\sqrt{k}\sigma_k,$$

or equivalently

$$\text{tr}(P_E A P_E) \leqslant C\sqrt{k}\sigma_k.$$

We write $\lambda_1, \ldots, \lambda_s$ for the non-negative eigenvalues of A in decreasing order, and $r_1, \ldots, r_{n-s}$ for the negative eigenvalues of A. Let $k \leqslant s$ and let E_k be the k-dimensional subspace which corresponds to the eigenvalues $\lambda_1, \ldots, \lambda_k$. From the trace inequality above, it follows that the matrix $P_{E_k} A P_{E_k}$ has at least one eigenvalue which is smaller than $C\sigma_k/\sqrt{k}$. Thus, we get

$$\lambda_k^2 \leqslant C' \frac{\sigma_k^2}{k}, \qquad k = 1, \ldots, s.$$

In the same way we check that $r_k^2 \leqslant C'\sigma_k^2/k$ for all $k = 1, \ldots, n-s$. This implies that

$$\|A\|_{\text{HS}}^2 = \sum_{k=1}^{s} \lambda_k^2 + \sum_{k=1}^{n-s} r_k^2 \leqslant 2C' \sum_{k=1}^{n} \frac{\sigma_k^2}{k},$$

so the proof is complete. $\square$

In view of Lemma 14.6.2, Theorem 14.6.1 will clearly follow from the next fact.

THEOREM 14.6.3. *There exists an absolute constant $C > 0$ such that*

$$\text{Is}_n^{-2} \leqslant C K_n^2 \log n.$$

For the proof of Theorem 14.6.3 we need to prove the following fact.

THEOREM 14.6.4. *Let μ be an isotropic log-concave measure on $\mathbb{R}^n$ and let E be a measurable subset of $\mathbb{R}^n$. If $\mu(E) = 1/2$, then*

$$\mu(E_{D/\delta} \setminus E) \geqslant c$$

for some absolute constants $D, c > 0$, where $\delta = \frac{1}{K_n \sqrt{\log n}}$ and $E_{D/\delta}$ is the D/δ-extension of E.

Once we have proved Theorem 14.6.4, we can employ the next result of E. Milman (see [**371**] and [**372**]).

THEOREM 14.6.5 (E. Milman). *Let μ be a log-concave probability measure on $\mathbb{R}^n$. Assume that there exist $0 < \lambda < 1/2$ and $\eta > 0$ such that for every measurable subset E of $\mathbb{R}^n$ with $\mu(E) \geqslant 1/2$ one has $\mu(E_\eta) \geqslant 1 - \lambda$. Then, for every measurable subset E of $\mathbb{R}^n$ with $0 < \mu(E) \leqslant 1/2$,*

$$\frac{\mu^+(E)}{\mu(E)} \geqslant \frac{1 - 2\lambda}{\eta}.$$

Proof of Theorem 14.6.3. Let μ be an isotropic log-concave measure on $\mathbb{R}^n$. It is clear that if we apply Theorem 14.6.5 with $\lambda = \frac{1}{2} - c$ and $\eta = D/\delta$ we get

$$\mathrm{Is}_\mu = \inf \left\{ \frac{\mu^+(E)}{\mu(E)} : 0 < \mu(E) \leqslant \frac{1}{2} \right\} \geqslant \frac{2c\delta}{D} = \frac{c'}{K_n \sqrt{\log n}},$$

where $c' > 0$ is an absolute constant. Since μ was arbitrary, we obtain the assertion of Theorem 14.6.3:

$$\mathrm{Is}_n^{-2} \leqslant C K_n^2 \log n$$

for some absolute constant $C > 0$. $\square$

The next three subsections are devoted to the proof of Theorem 14.6.4.

14.6.1. A stochastic localization scheme

We refer to [**159**] and [**401**] for definitions and background on semimartingales and stochastic integration. For a continuous time stochastic process, we write dX_t for the differential of X_t and $[X]_t$ for the quadratic variation of X_t. The quadratic variation of a pair (X_t, Y_t) of continuous time stochastic processes X_t, Y_t will be denoted by $[X, Y]_t$.

Note also that, given a matrix B, we will write $\|B\|_2$ for the spectral norm of B, that is, the largest singular value of B (or equivalently, the square root of the largest eigenvalue of the positive semi-definite matrix B^*B); below we will use this notation only for symmetric, positive semi-definite matrices B, in which case $\|B\|_2$ is just the largest eigenvalue of B.

We consider an isotropic random vector X in $\mathbb{R}^n$ with log-concave density f. By standard concentration estimates for log-concave isotropic densities that we saw in Chapter 2, we may assume that

(14.6.1) $$\mathrm{supp}(f) \subseteq nB_2^n.$$

If B is an $n \times n$ matrix and $u \in \mathbb{R}^n$ we define $V_f(u, B)$ by

$$V_f(u, B) = \int_{\mathbb{R}^n} e^{\langle u, x \rangle - \frac{1}{2}\langle Bx, x \rangle} f(x)\, dx.$$

We also define

$$a_f(u, B) = \frac{1}{V_f(u, B)} \int_{\mathbb{R}^n} x e^{\langle u, x \rangle - \frac{1}{2}\langle Bx, x \rangle} f(x)\, dx$$

and

$$A_f(u, B) = \frac{1}{V_f(u, B)} \int_{\mathbb{R}^n} [(x - a_f(u, B)) \otimes (x - a_f(u, B))] e^{\langle u, x \rangle - \frac{1}{2}\langle Bx, x \rangle} f(x)\, dx.$$

The assumption that $\mathrm{supp}(f) \subseteq nB_2^n$ guarantees that V_f, a_f and A_f are smooth functions of u and B.

Let W_t be a standard Wiener process in $\mathbb{R}^n$. We consider the system of stochastic differential equations

(14.6.2) $$u_0 = 0, \quad du_t = A_f^{-1/2}(u_t, B_t) dW_t + A_f^{-1}(u_t, B_t) a_f(u_t, B_t)\, dt$$
$$B_0 = 0, \quad dB_t = A_f^{-1}(u_t, B_t)\, dt.$$

Since the functions A_f, a_f are smooth and $A_f(u, B)$ is positive definite for all u, B, we may conclude the existence and uniqueness of a solution in some interval $[0, t_0]$ (see [**401**]) where t_0 is an almost surely strictly positive random variable; this solution is a new stochastic process that is adapted to the filtration of the standard Wiener process. Note that $[B]_t = 0$.

We construct a 1-parameter family of functions $\Gamma_t(f)$ as follows. We define

(14.6.3) $$F_t(x) = V_f^{-1}(u_t, B_t) e^{\langle u_t, x \rangle - \frac{1}{2}\langle B_t x, x \rangle}$$

and then we set
$$\Gamma_t(f)(x) = f(x) F_t(x).$$

We agree to write
$$a_t := a_f(u_t, B_t), \quad A_t := A_f(u_t, B_t), \quad V_t := V_f(u_t, B_t), \quad f_t := \Gamma_t(f).$$

The purpose of this construction is justified by the subsequent theorem, Theorem 14.6.6. Remember that in Section 14.2 we saw that certain concentration estimates imply isoperimetric inequalities; these implications become optimal and even clearer in Theorem 14.6.5. This explains why, in order to establish an estimate as in Theorem 14.6.3, we try to establish good concentration estimates for the log-concave measure defined by the density f (namely Theorem 14.6.4). But for densities of the form (14.6.3), such concentration estimates are not hard to obtain, as can be seen by Theorem 14.6.6 which combines known results of Brascamp-Lieb [**114**] and Gromov-V. Milman [**232**].

Observe now that, as will be proved below, the stochastic process f_t is a martingale (adapted to the filtration of W_t), so we also have
$$\int_A f(x) dx = \mathbb{E}\left[\int_A f_t(x) dx\right]$$
for every $t > 0$ and every Borel set A, where $\mathbb{E}$ here denotes conditional expectation. Combining the above, we see that, after showing Theorem 14.6.6, the proof of Theorem 14.6.4 will be a matter of how to pass back to usable concentration estimates for the original density f.

THEOREM 14.6.6. *Let $\phi : \mathbb{R}^n \to \mathbb{R}$ be a convex function and let B be a positive definite $n \times n$ matrix. Assume that μ is a probability measure with*
$$d\mu(x) = Z e^{-\phi(x) - \frac{1}{2}\langle Bx, x \rangle} dx,$$
and $\mathrm{bar}(\mu) = 0$.

(i) *For any Borel subset A of $\mathbb{R}^n$ with $\mu(A) \geqslant 1/10$ we have*

(14.6.4) $$\mu(A_{D/\sqrt{\lambda_1}}) \geqslant \frac{95}{100},$$

where $\lambda_1 = (\|B^{-1}\|_2)^{-1}$ is the smallest eigenvalue of B, $A_{D/\sqrt{\lambda_1}}$ is the $D/\sqrt{\lambda_1}$-extension of A, and $D > 0$ is a suitably chosen absolute constant.

(ii) *For every $\theta \in S^{n-1}$ we have*
$$\int_{\mathbb{R}^n} \langle x, \theta \rangle^2 d\mu(x) \leqslant 3D^2/\lambda_1.$$

Proof. Let ρ denote the density of μ and write E for the complement of $A_{D/\sqrt{\lambda_1}}$, where $D > 0$ will be suitably chosen. We define
$$f(x) = \rho(x) \mathbf{1}_A \quad \text{and} \quad g(x) = \rho(x) \mathbf{1}_E.$$

Let $x \in A$ and $y \in E$. Since $\|x-y\|_2 \geqslant D/\sqrt{\lambda_1}$, and since $\langle Bu, u\rangle \geqslant \lambda_1 \|u\|_2^2$ for every vector $u \in \mathbb{R}^n$, we may write
$$2\langle Bx,x\rangle + 2\langle By,y\rangle = \langle B(x+y), x+y\rangle + \langle B(x-y), x-y\rangle$$
$$\geqslant \langle B(x+y), x+y\rangle + D^2$$
which shows that
$$e^{-D^2/8} \exp\left(-\frac{1}{2}\left\langle B\frac{x+y}{2}, \frac{x+y}{2}\right\rangle\right) \geqslant \left(e^{-\frac{1}{2}\langle Bx,x\rangle} e^{-\frac{1}{2}\langle By,y\rangle}\right)^{1/2}.$$
Using also the assumption that ϕ is convex, we conclude that
$$e^{-D^2/8} \rho\left(\frac{x+y}{2}\right) \geqslant \sqrt{f(x)g(y)}.$$
Applying the Prékopa-Leindler inequality we see that
$$\mu(A)\mu(B) = \int_{\mathbb{R}^n} f(x)dx \cdot \int_{\mathbb{R}^n} g(x)dx \leqslant e^{-D^2/4}.$$
This shows that
$$\mu(A_{D/\sqrt{\lambda_1}}) \geqslant 1 - \frac{1}{\mu(A)} e^{-D^2/4} \geqslant 1 - 10 e^{-D^2/4},$$
and choosing D large enough we obtain (i).

For the second claim we observe that
$$(14.6.5) \quad \int_{\mathbb{R}^n} \langle x,\theta\rangle^2 d\mu(x) = \int_0^{+\infty} 2t\mu(\{x: |\langle x,\theta\rangle|\geqslant t\})\,dt$$
$$= \int_0^{+\infty} 2t\big(\mu(\{x: \langle x,\theta\rangle\geqslant t\}) + \mu(\{x: \langle x,\theta\rangle\leqslant -t\})\big)\,dt,$$
so we define
$$g(t) = \mu(\{x: \langle x,\theta\rangle\geqslant t\})$$
and set $A = \{x: \langle x,\theta\rangle \geqslant 0\}$. By Grünbaum's lemma (see Lemma 2.2.6, since $\mathrm{bar}(\mu) = 0$ we have that $1/e \leqslant \mu(A) \leqslant 1 - 1/e$. Therefore, we can apply (i) for both the set A and $\mathbb{R}^n \setminus A$, whence we get
$$\max\{g(D/\sqrt{\lambda_1}), 1 - g(-D/\sqrt{\lambda_1})\} \leqslant 1/20.$$
Since g is log-concave, it follows that
$$g(tD/\sqrt{\lambda_1}) \leqslant 20^{-t/2}$$
for all $t > 1$, and similarly that
$$1 - g(-tD/\sqrt{\lambda_1}) \leqslant 20^{-t/2}$$
for all $t > 1$. These two estimates and (14.6.5) clearly imply (ii). $\square$

REMARK 14.6.7. From the proof it can be seen that the lower bound $95/100$ in (14.6.4) is not important and we can replace it with any positive number smaller than 1 at the expense of choosing a larger absolute constant D. In other words, for every $\varepsilon \in (0,1)$ there is an absolute constant $D > 0$ such that, if A is a Borel set with $\mu(A) \geqslant 1/10$, then $\mu(A_{D/\sqrt{\lambda_1}}) \geqslant 1 - \varepsilon$. We will use this observation later on to deduce Theorem 14.6.3 from the results we have seen in the previous sections of this chapter.

Let us now look at the basic properties of the processes F_t and $\Gamma_t(f)$ that we need; these are included in the next two lemmas.

LEMMA 14.6.8. *The function F_t satisfies the equations:*

$$F_0(x) = 1$$
$$dF_t(x) = \langle x - a_t, A_t^{-1/2} dW_t \rangle F_t(x)$$
$$a_t = \int_{\mathbb{R}^n} x f(x) F_t(x) dx$$
$$A_t = \int_{\mathbb{R}^n} [(x - a_t) \otimes (x - a_t)] f(x) F_t(x) dx$$

for all $x \in \mathbb{R}^n$ and all $t \in [0, t_0]$. The last two equations show that a_t and A_t are the barycenter and the covariance matrix of f_t.

Proof. We will show that $dF_t(x) = \langle x - a_t, A_t^{-1/2} dW_t \rangle F_t(x)$ for any $x \in \mathbb{R}^n$. We define

$$G_t(x) = V_t F_t(x) = e^{\langle u_t, x \rangle - \frac{1}{2} \langle B_t x, x \rangle}.$$

If we write $Q_t(x)$ for the quadratic variation of the process $\langle x, u_t \rangle$ then

$$d\langle x, u_t \rangle = \langle A_t^{-1/2} x, dW_t + A_t^{-1/2} a_t dt \rangle,$$

and hence

$$dQ_t(x) = \langle A_t^{-1} x, x \rangle dt.$$

Using Itô's formula we write

$$dG_t(x) = \left(\langle x, du_t \rangle - \frac{1}{2} \langle dB_t x, x \rangle + \frac{1}{2} dQ_t(x) \right) G_t(x)$$
$$= \left(\langle x, A_t^{-1/2} dW_t + A_t^{-1} a_t dt \rangle - \frac{1}{2} \langle A_t^{-1} x, x \rangle dt + \frac{1}{2} \langle A_t^{-1} x, x \rangle dt \right) G_t(x)$$
$$= \langle x, A_t^{-1/2} dW_t + A_t^{-1} a_t dt \rangle G_t(x).$$

We also have

$$dV_t(x) = d\left(\int_{\mathbb{R}^n} e^{\langle u_t, x \rangle - \frac{1}{2} \langle B_t x, x \rangle} f(x) dx \right)$$
$$= \int_{\mathbb{R}^n} dG_t(x) f(x) dx$$
$$= \int_{\mathbb{R}^n} \langle x, A_t^{-1/2} dW_t + A_t^{-1} a_t dt \rangle G_t(x) f(x) dx$$
$$= V_t \langle a_t, A_t^{-1/2} dW_t + A_t^{-1} a_t dt \rangle.$$

From Itô's formula we get

$$dV_t^{-1} = -\frac{dV_t}{V_t^2} + \frac{d[V]_t}{V_t^3}$$
$$= -V_t^{-1} \langle a_t, A_t^{-1/2} dW_t + A_t^{-1} a_t dt \rangle + V_t^{-1} \langle A_t^{-1} a_t, a_t \rangle.$$

Applying Itô's formula again we get

$$dF_t(x) = d(V_t^{-1}G_t(x)) = G_t(x)dV_t^{-1} + V_t^{-1}dG_t(x) + d[V^{-1}, G(x)]_t$$
$$= -V_t^{-1}\left\langle a_t, A_t^{-1/2}dW_t + A_t^{-1}a_t dt\right\rangle G_t(x) + V_t^{-1}\langle A_t^{-1}a_t, a_t\rangle G_t(x)$$
$$+ V_t^{-1}\langle x, A_t^{-1/2}dW_t + A_t^{-1}a_t dt\rangle G_t(x) - \langle A_t^{-1/2}a_t, A_t^{-1/2}x\rangle V_t^{-1}G_t(x)dt$$
$$= \langle A_t^{-1/2}dW_t, x - a_t\rangle F_t(x).$$

The other two equations can be verified directly. □

This process may be viewed as a continuous version of an iterative scheme: each time, we normalize our density so that it will become isotropic, and then we multiply it by a linear function which is equal to 1 at the barycenter and whose gradient has a random direction distributed uniformly on the ellipsoid of inertia.

Next, we analyze the basic properties of $\Gamma_t(f)$.

LEMMA 14.6.9. *The process $\Gamma_t(f)$ has the following properties:*
  (i) *The function $\Gamma_t(f)$ is almost surely well defined, finite and log-concave for every $t > 0$.*
  (ii) *For every $t > 0$,*
$$\int_{\mathbb{R}^n} f_t(x)dx = 1.$$
  (iii) *One has the semigroup property*
$$\Gamma_{s+t}(f) \sim \frac{1}{\sqrt{\det A_s}}\Gamma_t(\sqrt{\det A_s}\Gamma_s(f) \circ L^{-1}) \circ L,$$
  where
$$L(x) = A_s^{-1/2}(x - a_s).$$
  (iv) *For every $x \in \mathbb{R}^n$ the process $f_t(x)$ is a martingale.*

Proof. For the first claim we define $t_0 = \inf\{t : \det A_t = 0\}$ and we show that $t_0 = \infty$ (note that $t_0 > 0$ by continuity). The fact that $f_t := \Gamma_t(f)$ is log-concave follows from (14.6.3).

We first show that the second and the third claim hold true for all $t < t_0$. Since we have assumed that $\text{supp}(f) \subseteq nB_2^n$, whenever $\Gamma_t(f)$ is well defined, we have that $0 < V_f(u_t, B_t) < \infty$, and hence the second claim follows from the definition of $\Gamma_t(f)$.

For the third claim, we fix $0 < s < t_0 - t$ and we set

(14.6.6) $$L(x) = A_s^{-1/2}(x - a_s).$$

Then, the function
$$g(x) = \sqrt{\det A_s} f_s(L^{-1}(x))$$
is an isotropic density. We have
$$d\Gamma_t(g)(x)\,|_{t=0} = g(x)\langle x, dW_t\rangle$$
$$= \sqrt{\det A_t} f_s(L^{-1}(x))\langle L(L^{-1}(x)), dW_t\rangle$$
$$= \sqrt{\det A_s} f_s(L^{-1}(x))\langle L^{-1}(x) - a_s, A_s^{-1/2}dW_t\rangle.$$

On the other hand,
$$df_s(L^{-1}(x)) = f_s(L^{-1}(x))\langle L^{-1}(x) - a_s, A_s^{-1/2}dW_s\rangle,$$

or equivalently,
$$d\Gamma_t(\sqrt{\det A_s}\Gamma_s(f) \circ L^{-1})\big|_{t=0} \sim \sqrt{\det A_s}d\Gamma_t(f) \circ L^{-1}\big|_{t=s},$$
and this proves the third claim.

Finally, we show that $t_0 = \infty$. We need the next lemma.

LEMMA 14.6.10. *For every $n \geq 1$ there exists a constant $c(n) > 0$ such that, if f is an isotropic log-concave density in $\mathbb{R}^n$ satisfying (14.6.1), then*
$$\mathbb{P}\big(A_f(u_t, B_t) \geq c(n)I \text{ for all } t \in [0, c(n)]\big) \geq c(n).$$

Assuming the lemma, we define
$$s_1 = \min\{t : \|A_t^{-1}\|_2 = 1/c(n)\},$$
where $c(n)$ is the constant from Lemma 14.6.10. Note that s_1 is well defined and positive with probability 1. At time s_1, we can define L_1 as in (14.6.6) and continue the process on the function $f \circ L_1^{-1}$. We do the same thing every time we have $\|A_t^{-1}\|_2 = 1/c(n)$. This produces hitting times s_i which, by Lemma 14.6.10 satisfy
$$\mathbb{P}(s_{i+1} - s_i > c(n) \mid s_1, s_2, \ldots, s_i) > c(n).$$
This implies that, almost surely, $s_{i+1} - s_i > c(n)$ for infinitely many values of i, and hence $\lim s_i = +\infty$ almost surely, which shows that $t_0 = \infty$.

The fourth claim is an immediate consequence of Lemma 14.6.8. □

Summarizing the results of this subsection, we see that, from (14.6.3) and Theorem 14.6.6 (ii),
$$(14.6.7) \qquad A_t \leq 3D^2\|B_t^{-1}\|_2 I \leq 3D^2 \left(\int_0^t \frac{ds}{\|A_s\|_2}\right)^{-1} I$$
for all $t > 0$. Since we have assumed that $\mathrm{supp}(\mu) \subseteq nB_2^n$, we can crudely bound A_t by $n^2 I$, and going back to (14.6.7) we get
$$A_t \leq \frac{3D^2 n^2}{t} I,$$
which improves upon the inequality $A_t \leq n^2 I$ for large t, $t > 3D^2$. However, we need a better dependence on t for small values of t, and this will be the purpose of the next subsection.

14.6.2. Analysis of the matrix A_t

The next proposition provides more accurate estimates for the covariance matrix A_t of f_t.

PROPOSITION 14.6.11. *There exist absolute constants $C, c > 0$ with the following property: if $f : \mathbb{R}^n \to \mathbb{R}^+$ is an isotropic log-concave density and if A_t is the covariance matrix of $\Gamma_t(f)$, then:*

(i) *For the event*
$$F := \{\|A_t\|_2 < CK_n^2(\log n)e^{-ct} \text{ for all } t > 0\},$$
we have
$$\mathbb{P}(F) \geq 1 - n^{-10}.$$

(ii) *Assuming that the event F happens, we also have that, for every $t > \frac{1}{K_n^2 \log n}$,*
$$\|B_t^{-1}\|_2 \leqslant C K_n^2 \log n,$$
where $C > 0$ is an absolute constant.

We start with an identity for the differential of the process A_t.

LEMMA 14.6.12. *We have*
$$dA_t = \int_{\mathbb{R}^n} (x - a_t) \otimes (x - a_t) \langle x - a_t, A_t^{-1/2} dW_t \rangle f_t(x) dx - A_t dt.$$

Proof. Using Lemma 14.6.8 and Itô's formula we write

$$(14.6.8) \quad dA_t = d\left(\int_{\mathbb{R}^n} (x - a_t) \otimes (x - a_t) f_t(x) dx \right)$$
$$= \int_{\mathbb{R}^n} (x - a_t) \otimes (x - a_t) df_t(x) dx - 2 \int_{\mathbb{R}^n} da_t \otimes (x - a_t) f_t(x) dx$$
$$- 2 \int_{\mathbb{R}^n} (x - a_t) \otimes d[a_t, f_t(x)]_t dx + d[a_t, a_t] \int_{\mathbb{R}^n} f_t(x) dx.$$

We look at each term separately. The second term is
$$\int_{\mathbb{R}^n} da_t \otimes (x - a_t) f_t(x) dx = da_t \otimes \int_{\mathbb{R}^n} (x - a_t) f_t dt = 0.$$

Note that
$$(14.6.9) \quad da_t = d\left(\int_{\mathbb{R}^n} x f(x) F_t(x) dx \right)$$
$$= \int_{\mathbb{R}^n} x f(x) F_t(x) \langle x - a_t, A_t^{-1/2} dW_t \rangle dx$$
$$= \left(\int_{\mathbb{R}^n} (x - a_t) \otimes (x - a_t) f_t(x) dx \right) (A_t^{-1/2} dW_t)$$
$$= A_t^{1/2} dW_t,$$

with the third equality following from the definition of a_t, and more precisely from the fact that
$$\int_{\mathbb{R}^n} a_t f(x) F_t(x) \langle x - a_t, A_t^{-1/2} dW_t \rangle = 0.$$

From (14.6.9) we obtain $da_t = A_t^{1/2} dW_t$, and hence
$$(14.6.10) \quad d[a_t, a_t]_t = A_t dt$$
and
$$d[a_t, f_t(x)] = f_t(x) A_t^{1/2} A_t^{-1/2} x dt = f_t(x) x dt.$$
This implies that
$$(14.6.11) \quad \int_{\mathbb{R}^n} (x - a_t) \otimes d[a_t, f_t(x)]_t dx = \int_{\mathbb{R}^n} (x - a_t) \otimes x f_t(x) dx dt$$
$$= \int_{\mathbb{R}^n} (x - a_t) \otimes (x - a_t) f_t(x) dx dt = A_t dt.$$

Inserting (14.6.10) and (14.6.11) into (14.6.8) we get

$$dA_t = \int_{\mathbb{R}^n} (x - a_t) \otimes (x - a_t) df_t(x) dx - A_t dt,$$

which is the assertion of the lemma. $\square$

Note that the term $A_t dt$ is positive definite, therefore the eigenvalues of A_t decrease with t. We define

$$\tilde{A}_t = A_t + \int_0^t A_s ds.$$

Then, $\tilde{A}_0 = A_0 = I$ and

(14.6.12) $$d\tilde{A}_t = \int_{\mathbb{R}^n} (x - a_t) \otimes (x - a_t) \langle x - a_t, A_t^{-1/2} dW_t \rangle f_t(x) dx.$$

It is clear that $A_t \leqslant \tilde{A}_t$ for all $t > 0$. Our next task will be to bound $\|\tilde{A}_t\|_2$, and hence $\|A_t\|_2$.

Given $t > 0$ we consider an orthonormal basis $v_1, \ldots, v_n$ with respect to which $\tilde{A}_t$ is diagonal, and we write $\alpha_{ij} = \langle v_i, \tilde{A}_t v_j \rangle$ for the entries of $\tilde{A}_t$ with respect to this basis. Then, we can write (14.6.12) in the form

$$d\alpha_{ij} = \int_{\mathbb{R}^n} \langle x, v_i \rangle \langle x, v_j \rangle \langle A_t^{-1/2} x, dW_t \rangle f_t(x + a_t) dx.$$

We set

(14.6.13) $$\xi_{ij} = \frac{1}{\sqrt{\alpha_{ii}\alpha_{jj}}} \int_{\mathbb{R}^n} \langle x, v_i \rangle \langle x, v_j \rangle A_t^{-1/2} x f_t(x + a_t) dx.$$

Then, we have

(14.6.14) $$d\alpha_{ij} = \sqrt{\alpha_{ii}\alpha_{jj}} \langle \xi_{ij}, dW_t \rangle$$

and

$$\frac{d}{dt}[\alpha_{ij}]_t = \alpha_{ii}\alpha_{jj}\|\xi_{ij}\|_2^2.$$

The next lemma provides some bounds for $\|\xi_{ij}\|_2$.

LEMMA 14.6.13. *The vectors ξ_{ij} satisfy the following:*

(i) *For all $1 \leqslant i \leqslant n$ one has $\|\xi_{ii}\|_2 \leqslant C$, where $C > 0$ is an absolute constant.*

(ii) *For all $1 \leqslant i \leqslant n$ one has*

$$\sum_{j=1}^n \|\xi_{ij}\|_2^2 \leqslant K_n^2.$$

Proof. Since $A_t^{1/2} v_i = \sqrt{\alpha_{ii}} v_i$, we have

$$\xi_{ij} = \int_{\mathbb{R}^n} \langle A_t^{-1/2} x, v_i \rangle \langle A_t^{-1/2} x, v_j \rangle A_t^{-1/2} x f_t(x + a_t) dx.$$

We define $\tilde{f}_t(x) = \sqrt{\det A_t} f_t(A_t^{1/2} x + a_t)$. If we set $y = A_t^{-1/2} x$, the previous equation takes the form

(14.6.15) $$\xi_{ij} = \int_{\mathbb{R}^n} \langle y, v_i \rangle \langle y, v_j \rangle y \tilde{f}_t(y) dy.$$

Observe that $\tilde{f}_t$ is isotropic. Using the Cauchy-Schwarz inequality we write

$$
(14.6.16) \qquad \|\xi_{ii}\|_2 = \left\| \int_{\mathbb{R}^n} \langle y, v_i \rangle^2 y \tilde{f}_t(y) dy \right\|_2
$$

$$
= \int_{\mathbb{R}^n} \langle y, v_i \rangle^2 \left\langle y, \frac{\xi_{ii}}{\|\xi_{ii}\|_2} \right\rangle \tilde{f}_t(y) dy
$$

$$
\leqslant \left(\int_{\mathbb{R}^n} \langle y, v_i \rangle^4 \tilde{f}_t(y) dy \right)^{1/2} \left(\int_{\mathbb{R}^n} \left\langle y, \frac{\xi_{ii}}{\|\xi_{ii}\|_2} \right\rangle^2 \tilde{f}_t(y) dy \right)^{1/2}.
$$

This leads to (i) if we use the fact that $\|\langle \cdot, v_i \rangle\|_{L^4(\tilde{f}_t)} \leqslant c \|\langle \cdot, v_i \rangle\|_{L^2(\tilde{f}_t)}$ by the log-concavity of $\tilde{f}_t$.

Now, by the definition of K_n we see that, for every $1 \leqslant i \leqslant n$,

$$
\sum_{j=1}^n \|\xi_{ij}\|_2^2 = \sum_{j=1}^n \sum_{k=1}^n \left| \int_{\mathbb{R}^n} \langle y, v_i \rangle \langle y, v_j \rangle \langle y, v_k \rangle \tilde{f}_t(y) dy \right|^2
$$

$$
= \left\| \int_{\mathbb{R}^n} y \otimes y \langle y, v_i \rangle \tilde{f}_t(y) dy \right\|_{\mathrm{HS}}^2
$$

$$
\leqslant K_n^2.
$$

This completes the proof. $\square$

Proof of Proposition 14.6.11. We fix a positive integer p, which will be suitably chosen, and we define

$$
(14.6.17) \qquad S_t := \mathrm{tr}(\tilde{A}_t^p).
$$

Observe that S_t is a smooth function with respect to the entries α_{ij} of $\tilde{A}_t$. Assuming that the basis $\{v_1, \ldots, v_n\}$ is fixed, these are Itô processes, and thus, S_t is an Itô process too. We fix $t > 0$ and we compute dS_t. If we consider the set M of all $(p+1)$-tuples $(j_1, \ldots, j_{p+1})$ with $1 \leqslant j_i \leqslant n$ and $j_1 = j_{p+1}$, then we have

$$
(14.6.18) \qquad S_t = \sum_{(j_1, \ldots, j_{p+1}) \in M} \alpha_{j_1 j_2} \alpha_{j_2 j_3} \cdots \alpha_{j_p j_{p+1}}.
$$

Since $\mathrm{tr}(\tilde{A}_t^p)$ does not depend on the choice of our coordinate system, we may choose the latter so that $\tilde{A}_t$ will be diagonal. Therefore, we may assume that $\alpha_{ij} = 0$ whenever $i \neq j$ and that (14.6.14) is satisfied. In other words, we calculate the differential of S_t using a basis that depends on t, but given a specific value of t we compute the differential at that time t with respect to a fixed orthonormal basis. This implies that $d(\alpha_{j_1 j_2} \alpha_{j_2 j_3} \cdots \alpha_{j_p j_{p+1}})$ is zero unless there are at most two indices i_1, i_2 such that $j_{i_1} \neq j_{i_1+1}$ and $j_{i_2} \neq j_{i_2+1}$. Therefore, there are two types of terms with non-zero differential. The first type contains only diagonal entries and has the form α_{ii}^p. In this case, using (14.6.14) we write

$$
(14.6.19) \qquad d\alpha_{ii}^p = p \alpha_{ii}^{p-1} d\alpha_{ii} + \frac{p(p-1)}{2} \alpha_{ii}^{p-2} d[\alpha_{ij}]_t
$$

$$
= p \alpha_{ii}^p \langle \xi_{ii}, dW_t \rangle + p(p-1) \alpha_{ii}^p \|\xi_{ii}\|_2^2 dt.
$$

The second type contains exactly two off-diagonal entries and, if we take into account the symmetry of $\tilde{A}_t$ and the restriction $j_1 = j_{p+1}$, it has the form

$$
\alpha_{ii}^{k_1} \alpha_{ij} \alpha_{jj}^{k_2} \alpha_{ji} \alpha_{ii}^{k_3} = \alpha_{ii}^k \alpha_{jj}^{p-k-2} \alpha_{ij}^2,
$$

where $i \neq j$ and $0 \leq k \leq p - 2$. Since $\alpha_{ij} = 0$, we get

$$d(\alpha_{ii}^k \alpha_{jj}^{p-k-2} \alpha_{ij}^2) = \alpha_{ii}^k \alpha_{jj}^{p-k-2}(2\alpha_{ij} d\alpha_{ij} + d[\alpha_{ij}]_t)$$
$$= \alpha_{ii}^{k+1} \alpha_{jj}^{p-k-1} \|\xi_{ij}\|_2^2 dt.$$

We may assume that $\alpha_{11} \geq \alpha_{22} \geq \cdots \geq \alpha_{nn}$, therefore, for all $i < j$ and for all k, we will have

(14.6.20) $$d(\alpha_{ii}^k \alpha_{jj}^{p-k-2} \alpha_{ij}^2) \leq \alpha_{ii}^p \|\xi_{ij}\|_2^2 dt.$$

Coming back to (14.6.18) we see that for every $1 \leq i \leq n$ the sum on the right hand side contains exactly one term of the first type and $\binom{p}{2}$ terms of the second type. Using the estimates (14.6.19), (14.6.20) and Lemma 14.6.13 we finally get

$$dS_t \leq \sum_{i=1}^n \left(p\alpha_{ii}^p \langle \xi_{ii}, dW_t \rangle + p(p-1)\alpha_{ii}^p \|\xi_{ii}\|_2^2 dt\right) + \sum_{i<j} p(p-1)\alpha_{ii}^p \|\xi_{ij}\|_2^2$$
$$\leq \sum_{i=1}^n p\alpha_{ii}^p \langle \xi_{ii}, dW_t \rangle + p^2 \sum_{i=1}^n \alpha_{ii}^p \sum_{j=1}^n \|\xi_{ij}\|_2^2 dt$$
$$\leq \sum_{i=1}^n p\alpha_{ii}^p \langle \xi_{ii}, dW_t \rangle + p^2 S_t K_n^2 dt.$$

By a well-known property of Itô processes, there exists a unique decomposition $S_t = M_t + E_t$, where M_t is a local martingale and E_t is an adapted process of locally bounded variation. From the previous inequality we have

(14.6.21) $$dE_t \leq p^2 K_n^2 S_t dt$$

and

$$\frac{d[S]_t}{dt} = \left\| \sum_{i=1}^n p\alpha_{ii}^p \xi_{ii} \right\|_2^2.$$

From Lemma 14.6.13 (i) we get

(14.6.22) $$\frac{d[S]_t}{dt} \leq Cp^2 S_t^2.$$

Now, we write $\log S_t$ in the form $\log S_t = Y_t + Z_t$, where Y_t is a local martingale with $Y_0 = 0$ and Z_t is an adapted process of locally bounded variation. From Itô's formula and (14.6.22) we have

(14.6.23) $$\frac{d[Y]_t}{dt} = \frac{1}{S_t^2} \frac{d[S]_t}{dt} \leq Cp^2.$$

We use the fact that there exists a standard Wiener process $\tilde{W}_t$ such that Y_t has the same distribution as $\tilde{W}_{[Y]_t}$. Applying the reflection principle we get

$$\mathbb{P}\left(\max_{t \in [0,p]} \tilde{W}_t \geq tp\right) = 2\mathbb{P}(\tilde{W}_p \geq tp) \leq Ce^{-\frac{1}{2}t^2 p}.$$

Choosing $t = C_1$, where C_1 is a large enough absolute positive constant, we have

$$\mathbb{P}\left(\max_{t \in [0,p]} \tilde{W}_t \geq C_1 p\right) \leq e^{-10p}.$$

From (14.6.23) it follows that
$$\mathbb{P}\left(\max_{t\in[0,1/p]} Y_t \geqslant C_2 p\right) \leqslant e^{-10p}$$
for some absolute constant $C_2 > 0$. Using now Itô's formula and (14.6.21) we get
$$\frac{dZ_t}{dt} = \frac{1}{S_t}\frac{dE_t}{dt} - \frac{1}{2S_t^2}\frac{d[S]_t}{dt} \leqslant K_n^2 p^2.$$
Assuming that $K_n \geqslant 1$, from the last two inequalities we see that there exists an absolute constant $C > 0$ such that
$$\mathbb{P}\left(\max_{0\leqslant t\leqslant \frac{1}{K_n^2 p}} \log S_t - \log n > Cp\right) < e^{-10p}.$$
Choosing $p = \lceil \log n \rceil$ we get
$$\mathbb{P}\left(\max_{0\leqslant t\leqslant \frac{1}{K_n^2 \log n}} S_t^{1/\lceil \log n \rceil} > C_1\right) < n^{-10},$$
for some absolute constant $C_1 > 0$. We define
$$F = \left\{\max_{0\leqslant t\leqslant \frac{1}{K_n^2 \log n}} S_t^{1/\lceil \log n \rceil} \leqslant C_1\right\}.$$
Whenever F holds, we have
$$\|A_t\|_2 \leqslant \|\tilde{A}_t\|_2 \leqslant C_1$$
for all $0 \leqslant t \leqslant \frac{1}{K_n^2 \log n}$. Since

(14.6.24) $$A_t \leqslant 3D^2 \|B_t^{-1}\|_2 I \leqslant 3D^2 \left(\int_0^t \frac{ds}{\|A_s\|_2}\right)^{-1} I,$$

where $B_t = \int_0^t A_s^{-1} ds$, from (14.6.24) we get
$$\frac{d}{dt} B_t = A_t^{-1} \geqslant \frac{1}{3D^2 \|B_t^{-1}\|_2} I,$$
therefore,
$$\frac{d}{dt}\frac{1}{\|B_t^{-1}\|_2} \geqslant \frac{1}{3D^2 \|B_t^{-1}\|_2}.$$
By the definition of B_t it follows that, whenever F holds,
$$\frac{1}{\|B_{\delta^2}^{-1}\|_2} \geqslant C\delta^2$$
where $\delta^2 = \frac{1}{K_n^2 \log n}$. Then,

(14.6.25) $$B_t \geqslant c\delta^2 e^{(t-\delta^2)/3D^2} I$$

for all $t > \delta^2$, which gives
$$A_t \leqslant C\delta^{-2} e^{(\delta^2-t)/3D^2} I.$$
This establishes the first claim of the proposition, while the second one follows immediately from (14.6.25). □

Combining Theorem 14.6.6 and Proposition 14.6.11 (ii) we get the following corollary.

COROLLARY 14.6.14. *There exist absolute constants $C, c_0, c_1, D > 0$ such that, whenever the event*
$$F := \{\|A_t\|_2 < CK_n^2(\log n)e^{-c_0 t} \text{ for all } t > 0\}$$
holds, then we also have
$$\int_{E_{D/\delta}} f_t(x)dx \geq c_1$$
for every measurable subset E of $\mathbb{R}^n$ with
$$\int_E f_t(x)dx \geq \frac{1}{10}$$
and every $t > \delta^2$, where $\delta^2 = \frac{1}{K_n^2 \log n}$ and $E_{D/\delta}$ is the (D/δ)-extension of E.

REMARK 14.6.15. Taking Remark 14.6.7 into account too, we see that c_1 can be chosen as close to 1 as we want, as long as we allow D to increase accordingly. Since the dependence of D (and hence of C and c_0) on c_1 is explicit and known from the start, this has no consequence to the other calculations.

14.6.3. Reduction of the KLS-conjecture to the thin shell conjecture

We are now ready to prove Theorem 14.6.4. Let f be an isotropic log-concave density in $\mathbb{R}^n$ and let E be a measurable subset of $\mathbb{R}^n$. We assume that
$$\int_E f(x)dx = \frac{1}{2},$$
and we will show that
$$\int_{E_{D/\delta} \setminus E} f(x)dx \geq c$$
for some absolute constants $D, c > 0$, where $\delta = \frac{1}{K_n \sqrt{\log n}}$ and $E_{D/\delta}$ is the D/δ-extension of E.

We consider the family $f_t := \Gamma_t(f)$ and we fix $t > 0$. As we have seen, the process $\{f_t(x)\}$ is a martingale, and so we have
$$\int_{E_{D/\delta}} f(x)dx = \mathbb{E}\left[\int_{E_{D/\delta}} f_t(x)dx\right].$$

But now, from Corollary 14.6.14 it is clear that, if we want to bound the right hand side from below, it suffices to bound $\int_E f_t(x)dx$ away from 0, provided that t is not too small. We define
$$g(t) = \int_E f_t(x)dx,$$
and we need to show the following.

LEMMA 14.6.16. *There exists $T > 0$ such that, for every $t \in [0, T]$,*
$$(14.6.26) \qquad \mathbb{P}\left(g(t) \geq \frac{1}{10}\right) \geq \frac{1}{2}.$$

REMARK 14.6.17. Note that T here is a constant, not a random variable. In addition, as will become clear from the following proof, the lower bound $1/2$ for the probability in (14.6.26) is not important, and can be replaced by any positive constant $c < 1$ by reducing T accordingly.

Proof. Making the substitution $y = A_t^{-1/2}(x - a_t)$ we write

$$dg(t) = \int_E f_t(x)\langle x - a_t, A_t^{-1/2} dW_t\rangle dx$$

$$= \sqrt{\det A_t} \int_{A_t^{-1/2}(E-a_t)} f_t(A_t^{1/2}y + a_t)\langle y, dW_t\rangle dy$$

$$= \Big\langle \sqrt{\det A_t} \int_{A_t^{-1/2}(E-a_t)} f_t(A_t^{1/2}y + a_t) y \, dy, dW_t \Big\rangle.$$

We define

$$\tilde{f}_t = \sqrt{\det A_t}\, f_t(A_t^{1/2}y + a_t) \quad \text{and} \quad E_t = A_t^{-1/2}(E - a_t).$$

In this notation we have

(14.6.27) $$dg(t) = \Big\langle \int_{E_t} y \tilde{f}_t(y) dy, dW_t \Big\rangle.$$

In the case that $\int_{E_t} y \tilde{f}_t(y) dy \neq 0$ we set

$$\theta = \frac{\int_{E_t} y \tilde{f}_t(y) dy}{\|\int_{E_t} y \tilde{f}_t(y) dy\|_2}.$$

Since $\tilde{f}_t$ is isotropic, we have

$$\Big\| \int_{E_t} y \tilde{f}_t(y) dy \Big\|_2 = \int_{E_t} \langle y, \theta\rangle \tilde{f}_t(y) dy$$

$$\leq \int_{E_t} |\langle y, \theta\rangle| \tilde{f}_t(y) dy \leq \Big(\int_{E_t} \langle y, \theta\rangle^2 \tilde{f}_t(y) dy\Big)^{1/2} \leq 1.$$

Therefore, in any case $\Big\|\int_{E_t} y \tilde{f}_t(y) dy\Big\|_2 \leq 1$, and hence

$$\frac{d}{dt}[g]_t \leq 1$$

for all $t > 0$. We define $h(t) = (g(t) - 1/2)^2$. By Itô's formula we get

$$dh(t) = 2(g(t) - 1/2) dg(t) + d[g]_t.$$

Combining the above we see that

$$\mathbb{E}\left[(g(t) - 1/2)^2\right] \leq t,$$

and the lemma follows from Markov's inequality. $\square$

Proof of Theorem 14.6.4. Let T be the constant from Lemma 14.6.16. We set

$$G = \Big\{g(T) \geq \frac{1}{10}\Big\} \cap F,$$

where

$$F := \{\|A_t\|_2 < CK_n^2(\log n)e^{-ct} \text{ for all } t > 0\}.$$

From Lemma 14.6.16 and the fact that $\mathbb{P}(F) \geqslant 1 - n^{-10}$, we see that
$$\mathbb{P}(G) \geqslant \frac{2}{5}$$
for all $n \geqslant 2$. Since
$$\int_{E_{D/\delta}} f(x)dx = \mathbb{E}\left[\int_{E_{D/\delta}} f_T(x)dx\right],$$
Corollary 14.6.14 shows that there exist absolute constants $c_0, D > 0$ such that

(14.6.28)
$$\int_{E_{D/\delta}} f(x)dx = \mathbb{E}\left[\int_{E_{D/\delta}} f_T(x)dx\right]$$
$$\geqslant \mathbb{P}(G)\mathbb{E}\left[\int_{E_{D/\delta}} f_T(x)dx \bigg| G\right] \geqslant c_0.$$

This completes the proof. □

Proof of Theorem 14.6.3 using the results of Sections 14.2 and 14.3. By Remarks 14.6.17, 14.6.15 and Proposition 14.6.11 (ii) we can see that the following holds: given $\varepsilon \in (0,1)$ we can choose D first, and then large enough $n \geqslant n_0$ and a small constant T so that we will simultaneously have that

(i) whenever the event $F = \{\|A_t\|_2 < C(D)K_n^2(\log n)e^{-c_0(D)t}$ for all $t > 0\}$ will hold, it will follow that
$$\int_{E_{D/\delta}} f_t(x)dx \geqslant 1 - \varepsilon_1$$
for every measurable subset E of $\mathbb{R}^n$ with
$$\int_E f_t(x)dx \geqslant \frac{1}{10}$$
and every $t > \delta^2$, where $\delta^2 = \frac{1}{K_n^2 \log n}$;

(ii) for every measurable subset E of $\mathbb{R}^n$ with
$$\int_E f(x)dx = \frac{1}{2}$$
we will have
$$\mathbb{P}\left(\int_E f_T(x)dx \geqslant \frac{1}{10}\right) \geqslant 1 - \varepsilon_2;$$

(iii) the event F will hold with probability
$$\mathbb{P}(F) \geqslant 1 - n^{-10} \geqslant 1 - \varepsilon_3$$
(we also need n to be large enough so that $T > \delta^2$), where ε_i, $i = 1, 2, 3$, are such that
$$(1-\varepsilon_1)(1-\varepsilon_2-\varepsilon_3) \geqslant 1 - \varepsilon.$$
It is then easy to see that (14.6.28) would now give

(14.6.29)
$$\int_{E_{D/\delta}} f(x)dx \geqslant 1 - \varepsilon$$

for every measurable subset E with $\int_E f(x) = 1/2$ (and hence for every measurable subset E with $\int_E f(x) > 1/2$).

We will use this observation with ε equal to the constant ε_0 from Lemma 14.3.4 and a measurable subset E such that $\int_E f(x) \geqslant 1/2$ and

$$\int_{\mathbb{R}^n} d(x,E) f(x)\, dx \geqslant \frac{1}{6\mathrm{FM}_\mu},$$

where μ is the isotropic measure whose density is f (the existence of such measurable subsets follows for example from the second part of the argument proving Theorem 14.2.9. By the previous analysis we see that

$$\mu(\{d(x,E) \leqslant D/\delta\}) \geqslant 1 - \varepsilon_0,$$

which implies that

$$Q_{1-\varepsilon_0}(d(\cdot, E)) \leqslant D/\delta.$$

But by Lemma 14.3.4 and Theorem 14.2.2 we also have that

$$Q_{1-\varepsilon_0}(d(\cdot, E)) \geqslant \frac{1}{2} \int_{\mathbb{R}^n} d(x,E)\, d\mu(x) \geqslant \frac{1}{12\mathrm{FM}_\mu} \geqslant \frac{c}{\mathrm{Is}_\mu}.$$

The assertion of Theorem 14.6.3 now follows. $\square$

Appendix: Proof of Lemma 14.6.10

Lemma 14.6.10 states that for every $n \geqslant 1$ there exists a constant $c(n) > 0$ such that, if f is an isotropic log-concave density in $\mathbb{R}^n$ satisfying $\mathrm{supp}(f) \subseteq n B_2^n$, then

$$\mathbb{P}\big(A_f(u_t, B_t) \geqslant c(n) I \text{ for all } t \in [0, c(n)]\big) \geqslant c(n).$$

All the constants that appear in the proof will depend on n: Since f is isotropic and log-concave, we may find two constants $c_1, c_2 > 0$ such that $f(x) \geqslant c_1$ whenever $\|x\|_2 \leqslant c_2$ (see e.g. [**339**, Theorem 5.14]). We define

$$g(x) = c_1 \mathbf{1}_{\{x : \|x\|_2 \leqslant c_2\}}(x).$$

By uniform concentration estimates for isotropic log-concave densities that we can obtain trough Lemma 2.2.1, or simply by our assumption that $\mathrm{supp}(f) \subseteq n B_2^n$ and by the fact that $f(x) \leqslant L_n^n$ for all x, we can find constants $c_3, c_4 > 0$ such that

$$\int_{\mathbb{R}^n} e^{\langle x, y \rangle} f(x)\, dx \leqslant c_3$$

whenever $\|y\|_2 \leqslant c_4$. Therefore, if $\|u\|_2 \leqslant c_4$ and B is positive semi-definite, then

$$V_f(u,B) = \int_{\mathbb{R}^n} e^{\langle u, x\rangle - \frac{1}{2}\langle Bx, x\rangle} f(x)\, dx \leqslant c_3.$$

If in addition $B \leqslant I$, we get

$$(14.6.30) \quad A_f(u,B) \geqslant \frac{1}{c_3} \int_{\mathbb{R}^n} (x - a_f(u,B)) \otimes (x - a_f(u,B)) e^{-c_4 \|x\|_2 - \frac{1}{2}\|x\|_2^2} g(x)\, dx$$

$$\geqslant \frac{c_1}{c_3} \int_{\{x : \|x\|_2 \leqslant c_2\}} x \otimes x\, e^{-c_4 \|x\|_2 - \frac{1}{2}\|x\|_2^2}\, dx$$

$$= c_5 I,$$

for some constant $c_5 > 0$. We define the stopping times

$$T_1 = \inf\{t > 0 : \|u_t\|_2 \geqslant c_4\}, \quad T_2 = \inf\{t > 0 : B_t \geqslant I\} \text{ and } T = \min\{T_1, T_2\}.$$

From (14.6.30) we see that $A_t \geqslant c_5 I$ for all $t \in [0, T]$. Therefore, the proof of the lemma will be complete if we show that

$$\mathbb{P}(T > c) \geqslant c$$

for some $c > 0$ depending on n.

We set $E = \{T_2 \leq T_1\}$. Assume that E holds. From (14.6.30) we have

(14.6.31)
$$A_t \geq c_5 I$$

for all $t \in [0, T_2]$ and from $\frac{d}{dt} B_t = A_t^{-1}$ we get that

$$B_t \leq c_5^{-1} t I$$

for all $t \in [0, T_2]$. Taking $t = T_2$ we conclude that: if E holds, then $T \geq c_5$. It remains to show that

$$\mathbb{P}(T_1 > c) \geq c$$

for some $c > 0$ depending on n. We can also assume that $\mathbb{P}(E) \leq 1/10$ (otherwise we would have $\mathbb{P}(T \geq c_5) \geq \mathbb{P}(E) \geq 1/10$ and we would be done).

We recall the definition (14.6.2) and using Itô's formula we get

(14.6.32)
$$d\|u_t\|_2^2 = 2\langle u_t, A_t^{-1/2} dW_t\rangle + 2\langle A_t^{-1} a_t, u_t\rangle dt + \|A_t^{-1/2}\|_{\text{HS}}^2 dt.$$

We define a process e_t by the equations

$$e_0 = 0 \quad \text{and} \quad de_t = 2\langle u_t, A_t^{-1/2} dW_t\rangle.$$

Note that, from (14.6.31), if $t \in [0, T]$ then

(14.6.33)
$$[e]_t = 4\int_0^t \langle A_s^{-1/2} u_s, A_s^{-1/2} u_s\rangle \, ds \leq 4 c_4^2 c_5^{-1} t.$$

Now, we use the fact (Dambis, Dubins-Schwartz theorem) that there exists a standard Wiener process $\tilde{W}_t$ such that e_t has the same distribution as $\tilde{W}_{[e]_t}$. It is known that there exists a constant $c_6 > 0$ such that

(14.6.34)
$$\mathbb{P}(F) \geq \frac{9}{10}, \quad \text{where } F = \left\{\max_{s \in [0, c_6]} \tilde{W}_s \leq \frac{c_4^2}{2}\right\}.$$

We set $\delta = \min\left\{T, \frac{c_5 c_6}{4 c_4^2}\right\}$. From (14.6.33) we see that

(14.6.35)
$$F \subseteq \left\{\max_{t \in [0, \delta]} e_t \leq \frac{c_4^2}{2}\right\}.$$

Applying (14.6.31) again, and assuming (14.6.1) (which implies that $\|a_t\|_2 \leq n$ for any time t), we get

(14.6.36)
$$\int_0^t \left(\|A_s^{-1/2}\|_{\text{HS}}^2 + 2|\langle A_s^{-1} a_s, u_s\rangle|\right) ds \leq n c_5^{-1}(1 + c_4) t \leq c_7 t$$

for all $t \in [0, T]$, where $c_7 > 0$ is a constant depending on n. Inserting (14.6.35) and (14.6.36) into (14.6.32) we see that if F holds, then

$$\|u_t\|_2^2 \leq \frac{c_4^2}{2} + c_7 t$$

for all $t \in [0, \delta]$. Assuming that $\delta = T_1$, we get

$$c_4^2 = \|u_{T_1}\|_2^2 \leq \frac{c_4^2}{2} + c_7 T_1,$$

which implies that $T \geq \frac{c_4^2}{2 c_7}$ (since we now consider the event $T_1 \leq T_2$). So, if $F \cap E^c$ holds, we have

$$T = T_1 \geq \min\left\{\frac{c_4^2}{2 c_7}, \frac{c_5 c_6}{4 c_4^2}\right\}.$$

Since $\mathbb{P}(F \cap E^c) \geq \frac{8}{10}$, the proof is complete. $\square$

14.7. Further reading

14.7.1. A Poincaré-type inequality

The next Poincaré-type inequality was used for the proof of the B-theorem in Chapter 5.

PROPOSITION 14.7.1. *Let μ be a probability measure on $\mathbb{R}^n$, of the form $d\mu(x) = e^{-\varphi(x)} dx$, where $\varphi : \mathbb{R}^n \to \mathbb{R}$ is a convex function with $\mathrm{Hess}\,(\varphi) \geqslant I$. Then, for any smooth function $f \in L_2(\mu)$ with $\int f\, d\mu = 0$ and $\int \nabla f\, d\mu = 0$, one has*

$$\int f^2 \, d\mu \leqslant \frac{1}{2} \int \|\nabla f\|_2^2 \, d\mu.$$

Note. Recall from Fact 5.3.6 that for any smooth function $g : \mathbb{R}^n \to \mathbb{R}$ with $g \in L_2(\mu)$ and $\int g\, d\mu = 0$ one has

$$\int g^2 \, d\mu \leqslant \int \|\nabla g\|_2^2 \, d\mu. \tag{14.7.1}$$

To see this, note that $\mathrm{Hess}(\varphi) \geqslant I$ implies that $\Gamma_2(f) \geqslant \Gamma(f)$. This allows us to prove a refinement of Lemma 14.1.9: For every $f \in C_b^\infty(\mathbb{R}^n)$ and every $t \geqslant 0$ one has

$$\|\nabla(P_t f)\|_2^2 \leqslant e^{-2t} P_t(\|\nabla f\|_2^2). \tag{14.7.2}$$

The proof follows the same lines as in Lemma 14.1.9; one differentiates the function $F(s) = e^{-2s} P_s(\Gamma P_{s-t} f)$ to check that $\Gamma(P_t f) \leqslant e^{-2t} P_t(\Gamma f)$, which is equivalent to (14.7.2).

Then, using basic properties of the semigroup $(P_t)_{t \geqslant 0}$ we may write

$$\mathrm{Var}_\mu(f) = \int_{\mathbb{R}^n} \int_0^\infty -\frac{d}{dt}(P_t f)^2 \, dt \, d\mu$$
$$= -2 \int_0^\infty \int_{\mathbb{R}^n} P_t f \mathcal{L}_\mu(P_t f) \, d\mu \, dt$$
$$= 2 \int_0^\infty \int_{\mathbb{R}^n} \|\nabla(P_t f)\|_2^2 \, d\mu \, dt$$
$$\leqslant 2 \int_0^\infty e^{-2t} \int_{\mathbb{R}^n} P_t(\|\nabla f\|_2^2) \, d\mu \, dt = \int_{\mathbb{R}^n} \|\nabla f\|_2^2 \, d\mu,$$

where in the last inequality we used (14.7.2). $\square$

Proof of Proposition 5.3.7. We consider the Laplace-Beltrami operator $\mathcal{L}_\mu u = \Delta u - \langle \nabla \varphi, \nabla u \rangle$, $u \in C_b^\infty(\mathbb{R}^n)$. Recall the following properties of $\mathcal{L}_\mu$:

(i) If $u, \nabla v \in L_2(\mu)$ are compactly supported and $u \in C^2$, then

$$\int v \mathcal{L}_\mu u \, d\mu = -\int \langle \nabla u, \nabla v \rangle \, d\mu.$$

(ii) $\nabla(\mathcal{L}_\mu u) = \mathcal{L}_\mu(\nabla u) - \mathrm{Hess}\varphi(\nabla u)$.

Combining the previous two formulas we obtain

$$\int (\mathcal{L}_\mu u)^2 \, d\mu = -\int \langle \nabla u, \nabla(\mathcal{L}_\mu u) \rangle \, d\mu \tag{14.7.3}$$
$$= -\int \langle \nabla u, \mathcal{L}_\mu(\nabla u) \rangle \, d\mu + \int \langle (\mathrm{Hess}\,\varphi)(\nabla u), \nabla u \rangle \, d\mu$$
$$= \int \left(\|\mathrm{Hess}(u)\|_{\mathrm{HS}}^2 + \langle (\mathrm{Hess}\,\varphi)(\nabla u), \nabla u \rangle \right) d\mu.$$

Let $f \in L_2(\mu)$ with $\int f\, d\mu = 0$ and $\int \nabla f\, d\mu = 0$. Assume also that $\nabla f \in L_2(\mu)$. We start with the observation that, since $\int f\, d\mu = 0$, we have

$$\min \left\{ \int (g-f)^2 \, d\mu \; : \; g \in L_2(\mu), \int g \, d\mu = 0 \right\} = 0.$$

We will also use the following classical fact:

FACT 14.7.2. *The space $\{\mathcal{L}_\mu u : u \in C^2 \text{ smooth with compact support}\}$ is $L_2(\mu)$-dense in the space $\{g \in L_2(\mu) : \int g\, d\mu = 0\}$ (for a proof see [142]).*

As a consequence we get

$$\inf\left\{\int (\mathcal{L}_\mu u - f)^2\, d\mu : u \in C^2 \text{ smooth, compactly supported}\right\} = 0.$$

To prove the result it suffices to prove that for every such function u we have

$$\int \left((\mathcal{L}_\mu u - f)^2 - f^2 + \frac{1}{2}\|\nabla f\|_2^2\right) d\mu \geqslant 0,$$

or equivalently,

$$\int \left((\mathcal{L}_\mu u)^2 - 2f\mathcal{L}_\mu u + \frac{1}{2}\|\nabla f\|_2^2\right) d\mu \geqslant 0.$$

From (14.7.3), from the fact that $\langle(\text{Hess}\,\varphi)(\nabla u), \nabla u\rangle \geqslant \|\nabla u\|_2^2$ and from (i), we see that it is enough to prove that

$$\int \left(\|\text{Hess}(u)\|_{\text{HS}}^2 + \|\nabla u\|_2^2 + 2\langle\nabla u, \nabla f\rangle + \frac{1}{2}\|\nabla f\|_2^2\right) d\mu \geqslant 0.$$

The last inequality can be rewritten as

$$\int \left(\|\text{Hess}(u)\|_{\text{HS}}^2 - \|\nabla u\|_2^2 + \frac{1}{2}\|2\nabla u + \nabla f\|_2^2\right) d\mu \geqslant 0.$$

Denoting $\alpha := \int \nabla u\, d\mu$, introducing a function u_0 such that $\nabla u_0 = \nabla u - \alpha$ and using the assumption that $\int \nabla f\, d\mu = 0$, we see that the last inequality is equivalent to

$$\|\alpha\|_2^2 + \int \left(\|\text{Hess}(u_0)\|_{\text{HS}}^2 - \|\nabla u_0\|_2^2 + \frac{1}{2}\|2\nabla u_0 + \nabla f\|_2^2\right) d\mu \geqslant 0.$$

Therefore, it is enough to prove that

$$\int \left(\|\text{Hess}(u_0)\|_{\text{HS}}^2 - \|\nabla u_0\|_2^2\right) d\mu \geqslant 0.$$

Using (14.7.1) for the functions $\frac{\partial u_0(x)}{\partial x_j}$ for $j = 1, \ldots, n$ and summing over all $j = 1, \ldots, n$ we conclude the last inequality, and the result follows. □

14.7.2. Bobkov's bound for the isoperimetric constant

Bobkov obtained in [80] a sharper form of Theorem 14.4.5, in which the L_2-norm of $f(x) = \|x - \text{bar}(\mu)\|_2$ is replaced by the quantity $[\text{Var}_\mu(\|x\|_2^2)]^{1/4}$.

THEOREM 14.7.3 (Bobkov). *Let μ be a log-concave probability measure on $\mathbb{R}^n$. Then,*

$$\text{Is}_\mu \geqslant \frac{c}{[\text{Var}_\mu(\|x\|_2^2)]^{1/4}},$$

where $c > 0$ is an absolute constant.

Note that, by Borell's lemma,

$$[\text{Var}_\mu(\|x\|_2^2)]^{1/4} \leqslant [\mathbb{E}_\mu(\|x\|_2^4)]^{1/4} \leqslant C\mathbb{E}_\mu(\|x\|_2)$$

and this shows that the estimate of Theorem 14.7.3 is always stronger than that of Theorem 14.4.5. In fact, one can check that, in some cases, the difference between the two bounds may be essential. For example, if $B := \overline{B}_2^n$ is the Euclidean ball of volume 1, then $\text{Var}_{\mu_B}(\|x\|_2^2) \simeq 1/n^2$, and hence Theorem 14.7.3 implies that $\text{Is}_{\mu_B} \geqslant c\sqrt{n}$, which is the correct dependence on the dimension in this case.

Recall also that, by the definition of $\overline{\sigma}_K$ in Section 12.3, if K is an isotropic convex body in $\mathbb{R}^n$ then $nL_K^4 \overline{\sigma}_K^2 = \text{Var}(\|x\|_2^2)$. Therefore, we get

THEOREM 14.7.4. *Let K be an isotropic convex body in $\mathbb{R}^n$. Then,*
$$\sqrt[4]{n} L_K \sqrt{\sigma_K} \operatorname{Is}_{\mu_K} \geqslant c,$$
where $c > 0$ is an absolute constant.

Note. Combining this fact with the thin shell estimate of Guédon and E. Milman (see Chapter 13) we get
$$n^{5/12} \operatorname{Is}_n \geqslant c,$$
where $c > 0$ is an absolute constant. As we saw in Section 14.6, Eldan has substantially improved this estimate in [164].

Proof of Theorem 14.7.3. The starting point for the proof is a variant of Claim 14.4.4 for log-concave probability measures which reads as follows.

FACT 14.7.5. *Let $g : \mathbb{R}^n \to \mathbb{R}^+$ be a non-negative continuous function and let A, B be open subsets of $\mathbb{R}^n$ with $\operatorname{dist}(A, B) = \varepsilon > 0$ and $C = \mathbb{R}^n \setminus (A \cup B)$. If the inequality*

$$(14.7.4) \qquad \mu(A)\mu(B) \leqslant \frac{\mu(C)}{\varepsilon} \int_{\mathbb{R}^n} g \, d\mu$$

holds true for every one-dimensional log-concave probability measure μ, then it holds true for every log-concave probability measure μ on $\mathbb{R}^n$.

Note that if we let $\varepsilon \to 0^+$ then (14.7.4) gives

$$(14.7.5) \qquad \mu(A)\mu(B) \leqslant \mu^+(C) \int g \, d\mu.$$

In fact, one can check that if (14.7.5) holds true for all A, B and C then the same is true for (14.7.4). Using the inequality $2 \min\{\mu(A), \mu(B)\} \geqslant 2\mu(A)\mu(B) \geqslant \min\{\mu(A), \mu(B)\}$ that holds for all Borel subsets A, B of $\mathbb{R}^n$ satisfying $\mu(A) + \mu(B) = 1$ and the fact that

$$\operatorname{Is}_\mu := \inf \left\{ \frac{\mu^+(A)}{\min\{\mu(A), 1 - \mu(A)\}} : A \subset \mathbb{R}^n \text{ Borel}, 0 < \mu(A) < 1 \right\},$$

we get:

FACT 14.7.6. *Let $g : \mathbb{R}^n \to \mathbb{R}^+$ be a non-negative continuous function. If the inequality*

$$(14.7.6) \qquad \operatorname{Is}_\mu^{-1} \leqslant \int g \, d\mu$$

holds true for every one-dimensional log-concave probability measure μ, then

$$(14.7.7) \qquad \operatorname{Is}_\mu^{-1} \leqslant 2 \int_{\mathbb{R}^n} g \, d\mu$$

for every log-concave probability measure μ on $\mathbb{R}^n$.

We continue the proof of Theorem 14.7.3 as follows. From [76] it is known that if ξ is a random variable with log-concave distribution μ on $\mathbb{R}$ then

$$(14.7.8) \qquad \frac{1}{\sqrt{2}} \sqrt{\operatorname{Var}(\xi)} \leqslant \operatorname{Is}_\mu^{-1} \leqslant \sqrt{3} \sqrt{\operatorname{Var}(\xi)}.$$

Consider a one-dimensional log-concave probability measure μ on $\mathbb{R}^n$. This is then the distribution of a random vector $u + \xi\theta$, where $u \in \mathbb{R}^n$ and $\theta \in S^{n-1}$ are fixed and orthogonal to each other, and ξ is a random variable with log-concave distribution on the real line. It follows that μ satisfies (14.7.8) as well. From Fact 14.7.6, if we show that

$$(14.7.9) \qquad \sqrt{\operatorname{Var}(\xi)} \leqslant \mathbb{E}(g(u + \xi\theta))$$

for some continuous function $g : \mathbb{R}^n \to \mathbb{R}^+$ and for all u, θ and ξ as above, we will conclude that

$$(14.7.10) \qquad \operatorname{Is}_\mu^{-1} \leqslant 2\sqrt{3} \int_{\mathbb{R}^n} g \, d\mu$$

for all log-concave probability measures μ on $\mathbb{R}^n$.

We consider the function
$$g(x) = C\sqrt{|\,\|x\|_2^2 - \alpha|},$$
where $\alpha \in \mathbb{R}$ and $C > 0$ are to be chosen. The fact that $\mathbb{E}(g(u+\xi\theta)) = C\mathbb{E}\sqrt{|\,\|u\|_2^2 + \xi^2 - \alpha|}$ satisfies (14.7.9) follows from the Khintchine-type inequalities for polynomials that were presented in Section 2.6.2. More precisely, we use the fact that $\|F\|_2 \leqslant c\|F\|_0 \leqslant c\|F\|_{1/2}$ for any polynomial F of degree 2. Therefore,

$$(14.7.11) \qquad \operatorname{Var}(\xi^2)^{1/2} \leqslant \left(\mathbb{E}|\,\|u\|_2^2 + \xi^2 - \alpha|^2\right)^{1/2} \leqslant c\left(\mathbb{E}\sqrt{|\,\|u\|_2^2 + \xi^2 - \alpha|}\right)^2.$$

Note also that if $A^2 = \mathbb{E}(\xi^2)$ then
$$\operatorname{Var}(\xi^2)^{1/2} \geqslant \|\xi^2 - A^2\|_0 = \|\xi - A\|_0 \|\xi + A\|_0$$
$$\geqslant \frac{1}{c^2}\|\xi - A\|_2 \|\xi + A\|_2 \geqslant \frac{1}{c^2}\operatorname{Var}(\xi).$$

Because of (14.7.11) we conclude that
$$\sqrt{\operatorname{Var}(\xi)} \leqslant C\mathbb{E}\sqrt{|\,\|u\|_2^2 + \xi^2 - \alpha|},$$
which is exactly (14.7.9). Thus, (14.7.10) also holds.

Finally, using Hölder's inequality we check that
$$\int g\,d\mu = C\mathbb{E}_\mu\sqrt{|\,\|x\|_2^2 - \alpha|} \leqslant C\left(\mathbb{E}_\mu|\,\|x\|_2^2 - \alpha|^2\right)^{1/4},$$
and then we can choose $\alpha = \mathbb{E}_\mu(\|x\|_2^2)$ to minimize the right hand side. We conclude that
$$(14.7.12) \qquad \operatorname{Is}_\mu^{-1} \leqslant 2\sqrt{3}\,C(\operatorname{Var}_\mu(\|x\|_2^2))^{1/4}$$
for every log-concave probability measure μ on $\mathbb{R}^n$. $\square$

E. Milman proposed in [**369**] a different (more geometric) approach to the above bound, using his results on the stability of the Cheeger constant.

THEOREM 14.7.7. *Let μ be a log-concave probability measure in $\mathbb{R}^n$. Then,*
$$\operatorname{Is}_\mu \geqslant c \sup_{z\in\mathbb{R}^n} \frac{1}{\sqrt[4]{\operatorname{Var}_\mu(\|x-z\|_2^2)}},$$
where $c > 0$ is an absolute constant.

Sketch of the proof. Let us assume that $z = 0$ and set $t := \mathbb{E}_\mu(\|x\|_2)$ and $\sigma := \sqrt{\operatorname{Var}_\mu(\|x\|_2)}$. We consider the ball $B = rB_2^n$ with $r = t + 2\sigma$. We apply Chebyshev's inequality
$$\mu(\{x \in \mathbb{R}^n : |\,\|x\|_2 - t| \geqslant \varepsilon\}) \leqslant \frac{\operatorname{Var}_\mu(\|x\|_2)}{\varepsilon^2}$$
(with $\varepsilon = 2\sigma$) to obtain $\mu(B) \geqslant 3/4$. Now we employ the variant of Theorem 14.3.1 from Remark 14.3.7 to get $\operatorname{Is}_\mu \simeq \operatorname{Is}_\nu$ where $\nu = \frac{\mu|_B}{\mu(B)}$, taking into account that $d_{\mathrm{TV}}(\mu,\nu) \leqslant 2/3$. Hence, it suffices to bound Is_ν from below. We distinguish two cases (see [**369**] for the details):

(i) If $t \leqslant 2\sigma$ then we may show that
$$\int_{\mathbb{R}^n} \|x\|_2\,d\nu(x) = \frac{1}{\mu(B)}\int_B \|x\|_2\,d\mu(x) \leqslant 4t \leqslant 4\sqrt[4]{\operatorname{Var}_\mu(\|x\|_2^2)},$$
and the result follows from Theorem 14.4.5.

(ii) If $t > 2\sigma$ then we use the fact (due to Kannan-Lovász-Simonovits) that for any log-concave probability measure ν on $\mathbb{R}^n$ which is supported on a convex body K we have

$$(14.7.13) \qquad \mathrm{Is}_\nu \geqslant c \left(\int \chi_K(x)\, d\nu(x) \right)^{-1},$$

where $\chi_K(x)$ denotes the longest symmetric interval that is contained in K and is centered at x. In the case of a ball of radius r we can easily see that

$$\chi_{rB_2^n}(x) = 2\sqrt{r^2 - \|x\|_2^2}.$$

Inserting this into (14.7.13) we can show that $\mathrm{Is}_\nu \geqslant c_1/\sqrt{t\sigma}$.

Note that $\sqrt{t\sigma} \leqslant \sqrt[4]{\mathrm{Var}_\mu(\|x\|_2^2)}$; this proves the theorem. $\square$

14.8. Notes and references

Isoperimetric constants for log-concave probability measures

Theorem 14.1.2 can be found in [**84**]; it follows from works of Cheeger, Maz'ya, Rothaus and Ledoux. Theorem 14.1.6 was proved by Maz'ya (see [**355**], [**356**]) and independently by Cheeger (see [**140**]). The inverse inequality of Theorem 14.1.7 is due to Buser [**130**] (see also Ledoux [**312**]).

The books of Chavel [**138**], [**139**] and the survey [**137**] are very useful references on isoperimetric inequalities. The survey [**315**] of Ledoux presents the abstract theory of the geometry of Markov diffusion generators. Theorem 14.1.8 is due to Bakry and Ledoux [**29**]; it extends a main step in Ledoux's proof of Buser's theorem. It can be viewed as a dimension free form of the Li-Yau parabolic gradient inequality [**324**]. Lemma 14.1.11 is standard.

Theorem 14.1.13 was proved by Gromov and Milman in [**232**].

Concavity of the isoperimetric profile

The concavity of the isoperimetric profile on a convex domain was first obtained by Sternberg and Zumbrun in [**479**]. They showed that if $n \geqslant 2$ and if K is a convex body in $\mathbb{R}^n$, then I_{μ_K} is concave on $[0, 1]$, where μ_K is the uniform probability measure on K. Kuwert later noted in [**303**] that $I_{\mu_K}^{n/(n-1)}$ is also concave on $[0, 1]$. This is precisely the correct power to use. Kuwert's result was extended to convex domains in Riemannian manifolds with non-negative Ricci curvature by Bayle and Rosales [**65**]. E. Milman showed in [**369**] that the concavity of I_μ remains valid for log-concave measures on $\mathbb{R}^n$. The one-dimensional case had been settled by Bobkov.

However, all one needs for transferring concentration to isoperimetry is the fact that $I(t)/t$ is non-increasing on $[0, 1]$. In fact, on convex domains on manifolds with non-negative Ricci curvature, one has that $[I_\mu(t)]^{n/(n-1)}/t$ is non-increasing. This was noted by E. Milman in [**371**]. The best possible estimate, transferring concentration to isoperimetry can be found in E. Milman's [**372**, Theorem 2.1].

Equivalence of the isoperimetric constants and stability of the Cheeger constant

E. Milman introduced first moment concentration in [**369**] and proved Theorem 14.2.2: For every log-concave probability measure μ on $\mathbb{R}^n$ one has

$$\mathrm{Is}_\mu \simeq \sqrt{\mathrm{Poin}_\mu} \simeq \mathrm{Exp}_\mu \simeq \mathrm{FM}_\mu.$$

One of the main applications that he had in mind was Theorem 14.3.1 on the stability of the Cheeger constant. All the results of Sections 14.2 and 14.3 come from his work.

Lower bounds

The conjecture of Kannan, Lovász and Simonovits was stated in [**266**] in connection with randomized volume algorithms. Early lower bounds for Is_μ, that provide some estimates depending on the dimension, can be found in works of Payne and Weinberger [**421**], Li and Yau [**324**]. The main result in [**266**] is Theorem 14.4.1. The other two proofs that we present in Section 14.4 are due to Bobkov [**76**] and E. Milman [**369**].

Theorem 14.7.3 was proved by Bobkov [**80**]. E. Milman offers a different proof in [**369**].

Sodin proved in [**470**] that Is_μ is uniformly bounded for the ℓ_p^n-balls, $1 \leqslant p \leqslant 2$. The case $p \geqslant 2$ was settled by Latała and Wojtaszczyk [**311**]. See also [**55**], [**79**], [**83**], [**258**], [**259**].

Theorem 14.5.1 which establishes the lower bound $c/\log n$ for the Poincaré constant $\mathrm{Poin}_K := \mathrm{Poin}_{\mu_K}$ of an unconditional isotropic convex body K in $\mathbb{R}^n$, is due to Klartag [**279**].

KLS-conjecture and the thin shell conjecture

Theorem 14.6.1 is due to Eldan [**164**]. He showed that there exists an absolute constant $C > 0$ such that
$$\mathrm{Is}_\mu^{-2} \leqslant C \log n \sum_{k=1}^n \frac{\sigma_k^2}{k}$$
for every isotropic log-concave measure μ on $\mathbb{R}^n$. This reduces the KLS-conjecture to the thin shell conjecture up to a factor of $\log n$. Also, combined with the results of Chapter 13, Eldan's theorem shows that $\mathrm{Is}_n^{-1} \leqslant C \sqrt[3]{n}\sqrt{\log n}$. This is the best known general estimate.

CHAPTER 15

Infimum convolution inequalities and concentration

This chapter is devoted to a probabilistic approach and related conjectures of Latała and Wojtaszczyk on the geometry of log-concave measures. The starting point is an *infimum convolution inequality* that was first introduced by Maurey who gave a simple proof of Talagrand's two level concentration inequality for the product exponential measure. In general, if μ is a probability measure and φ is a non-negative measurable function on $\mathbb{R}^n$, we say that the pair (μ, φ) has *property* (τ) if, for every bounded measurable function f on $\mathbb{R}^n$,

$$\left(\int_{\mathbb{R}^n} e^{f \Box \varphi} d\mu \right) \left(\int_{\mathbb{R}^n} e^{-f} d\mu \right) \leqslant 1,$$

where $f \Box \varphi$ is the infimum convolution of f and φ, defined by

$$(f \Box \varphi)(x) = \inf \left\{ f(x - y) + \varphi(y) : y \in \mathbb{R}^n \right\}.$$

Property (τ) for a pair (μ, φ) is directly related to the concentration properties of the measure μ since it implies that, for every measurable $A \subseteq \mathbb{R}^n$ and every $t > 0$, we have

$$\mu \left(x \notin A + B_\varphi(t) \right) \leqslant (\mu(A))^{-1} e^{-t},$$

where $B_\varphi(t) = \{\varphi \leqslant t\}$. Therefore, given a measure μ it is natural to look for the optimal cost function φ such that (μ, φ) has property (τ). The first main observation is that, if we restrict ourselves to even probability measures μ and convex cost functions φ, then the largest possible cost function cannot be larger than the Cramer transform Λ_μ^* of μ; this is the Legendre transform of the logarithmic Laplace transform of μ. It is conjectured that, when μ is a log-concave probability measure, then the pair $(\mu, \Lambda_\mu^*(\frac{\cdot}{\beta}))$ always has property (τ) for some absolute constant $\beta > 0$; in that case, we say that μ has the infimum convolution property. A detailed analysis shows that if μ has the infimum convolution property then μ satisfies the following concentration inequality: for every $p \geqslant 2$ and every Borel subset A of $\mathbb{R}^n$,

$$\mu(A) \geqslant \frac{1}{2} \quad \text{implies} \quad 1 - \mu(A + \beta Z_p(\mu)) \leqslant e^{-p}(1 - \mu(A)).$$

In fact, under the assumption that μ satisfies Cheeger's inequality

$$\mu^+(A) \geqslant \frac{1}{\gamma} \min\{\mu(A), 1 - \mu(A)\}$$

with constant $1/\gamma$, the converse is also true up to a constant depending only on γ.

Assuming that a log-concave probability measure μ satisfies the above concentration inequality, one can establish a very strong form of comparison between weak and strong moments: for any norm $\|\cdot\|$ on $\mathbb{R}^n$ and every $p \geqslant 2$,

$$\left(\int_{\mathbb{R}^n} \bigl|\|x\| - \mathrm{med}(\|x\|)\bigr|^p d\mu(x)\right)^{1/p} \leqslant C\beta \sup_{\|u\|_* \leqslant 1} \left(\int_{\mathbb{R}^n} |\langle u, x\rangle|^p d\mu(x)\right)^{1/p},$$

where $\|\cdot\|_*$ denotes the dual norm of $\|\cdot\|$. This would imply an affirmative answer to most of the conjectures addressed in this book: among them, both the hyperplane and the thin shell conjectures.

An affirmative answer has been given for some rather restricted classes of measures: even log-concave product measures, uniform distributions on ℓ_p^n-balls and rotationally invariant log-concave measures. In the last section of the chapter we present one such result: if μ is an unconditional log-concave probability measure on $\mathbb{R}^n$, then for any norm $\|\cdot\|$ on $\mathbb{R}^n$ and for all $p \geqslant 1$,

$$\left(\int_{\mathbb{R}^n} \|x\|^p d\mu(x)\right)^{1/p} \leqslant C_1 \int_{\mathbb{R}^n} \|x\| d\nu(x) + C_2 \sup_{\|y\|_* \leqslant 1} \left(\int_{\mathbb{R}^n} |\langle y, x\rangle|^p d\mu(x)\right)^{1/p},$$

where ν is the product exponential measure with density $d\nu(x) = \frac{1}{2^n} e^{-\|x\|_1} dx$ and $C_1, C_2 > 0$ are absolute constants.

We also discuss the analogue of Paouris' large deviation inequality when the Euclidean norm is replaced by the ℓ_r-norm for some $1 \leqslant r \leqslant \infty$. Latała proved that for any isotropic log-concave measure μ on $\mathbb{R}^n$ one has that, if $1 \leqslant r \leqslant 2$, then

$$\mu(\{x : \|x\|_r \geqslant t\}) \leqslant \exp\left(\frac{1}{c_1} tn^{\frac{1}{2}-\frac{1}{r}}\right)$$

for all $t \geqslant c_1 n^{1/r}$, and if $r \geqslant 2$ then

$$\mu(\{\|x\|_r \geqslant t\}) \leqslant \exp\left(-\frac{1}{c_2} t\right)$$

for all $t \geqslant c_2 r n^{1/r}$. His very interesting argument is based on the study of order statistics: he shows that, if μ is an isotropic log-concave measure on $\mathbb{R}^n$, then for every $k = 1, \ldots, n$ and for every $t \geqslant C \log(en/k)$ one has

$$\mu\left(\{x : x_k^* \geqslant t\}\right) \leqslant \exp(-\sqrt{k} t/C),$$

where $(x_1^*, \ldots, x_n^*)$ is the decreasing rearrangement of $(|x_1|, \ldots, |x_n|)$.

15.1. Property (τ)

DEFINITION 15.1.1 (infimum convolution). Let f and g be (Borel) measurable functions on $\mathbb{R}^n$. We denote by $f \square g$ the *infimum convolution* of f and g, defined by

$$(f \square g)(x) = \inf\{f(x-y) + g(y) : y \in \mathbb{R}^n\}.$$

If μ is a probability measure on $\mathbb{R}^n$ and φ is a non-negative measurable function on $\mathbb{R}^n$, we say that the pair (μ, φ) has *property* (τ) if, for every bounded measurable function f on $\mathbb{R}^n$ we have

$$\left(\int_{\mathbb{R}^n} e^{f \square \varphi} d\mu\right)\left(\int_{\mathbb{R}^n} e^{-f} d\mu\right) \leqslant 1.$$

From the definition of property (τ) one can deduce a number of first basic properties.

LEMMA 15.1.2. *If (μ_i, φ_i) has property (τ) in $\mathbb{R}^{n_i}$, $i = 1, 2$, then $(\mu_1 \otimes \mu_2, \varphi)$ has property (τ) in $\mathbb{R}^{n_1} \times \mathbb{R}^{n_2}$, where $\varphi(x_1, x_2) = \varphi_1(x_1) + \varphi_2(x_2)$.*

Proof. Let $f : \mathbb{R}^{n_1} \times \mathbb{R}^{n_2} \to \mathbb{R}$ be a bounded measurable function. We set

$$\psi(y) = -\log\left(\int_{\mathbb{R}^{n_1}} e^{-f(x,y)} d\mu_1(x)\right)$$

and $f^y(x) := f(x, y)$. Then, using property (τ) for the pair (μ_1, φ_1) we see that, for all $y, y_1 \in \mathbb{R}^{n_2}$,

$$\int_{\mathbb{R}^{n_1}} e^{f \square \varphi(x,y)} d\mu_1(x) \leqslant \int_{\mathbb{R}^{n_1}} e^{f^{y_1} \square \varphi_1(x) + \varphi_2(y-y_1)} d\mu_1(x) \leqslant e^{\psi(y_1) + \varphi_2(y-y_1)}.$$

It follows that

$$\int_{\mathbb{R}^{n_1}} e^{f \square \varphi(x,y)} d\mu_1(x) \leqslant e^{\psi \square \varphi_2(y)}.$$

Then, applying property (τ) for the pair (μ_2, φ_2), we see that

$$\int e^{f \square \varphi} d(\mu_1 \otimes \mu_2) \leqslant \int_{\mathbb{R}^{n_2}} e^{\psi \square \varphi_2(y)} d\mu_2(y) \leqslant \left(\int_{\mathbb{R}^{n_2}} e^{-\psi(y)} d\mu_2(y)\right)^{-1}$$

$$\leqslant \left(\int e^{-f} d\mu_1 \otimes d\mu_2\right)^{-1}.$$

It follows that $(\mu_1 \otimes \mu_2, \varphi)$ has property (τ). $\square$

Recall that if μ_1 and μ_2 are probability measures on $\mathbb{R}^n$ then their convolution $\mu_1 * \mu_2$ is defined through

$$\int_{\mathbb{R}^n} h \, d(\mu_1 * \mu_2) = \int_{\mathbb{R}^n} \int_{\mathbb{R}^n} h(x+y) d\mu_1(x) d\mu_2(y).$$

Note also that if we assume that μ_1 and μ_2 are absolutely continuous with respect to Lebesgue measure with densities f_1 and f_2, then $\mu_1 * \mu_2$ is the measure with density $f_1 * f_2$.

LEMMA 15.1.3. *If (μ_i, φ_i) has property (τ) in $\mathbb{R}^n$, $i = 1, 2$, then the pair $(\mu_1 * \mu_2, \varphi_1 \square \varphi_2)$ has property (τ) in $\mathbb{R}^n$.*

Proof. The proof is analogous to the one of Lemma 15.1.2; we just give a brief description. Let g be a bounded measurable function on $\mathbb{R}^n$. For any $x \in \mathbb{R}^n$ we define $g_x : \mathbb{R}^n \to \mathbb{R}$ by $g_x(y) := g(x+y)$. Then, for any x we have that g_x is a bounded measurable function on $\mathbb{R}^n$. Define $h : \mathbb{R}^n \to \mathbb{R}$ by

$$e^{-h(x)} = \int e^{-g_x} d\mu_2.$$

Using the fact that (μ_2, φ_2) has property (τ) we get:

$$\int e^{\varphi_2 \square g_x} d\mu_2 \leqslant e^{h(x)},$$

and this in turn yields that

$$(h \square \varphi_1)(x) \geqslant \log \int \exp([g \square (\varphi_1 \square \varphi_2)](x+y)) \, d\mu_2(y).$$

Combining this with the fact that (μ_1, φ_1) has property (τ) we get the result. $\square$

LEMMA 15.1.4. *Assume that (μ_1, φ_1) has property (τ) in $\mathbb{R}^{n_1}$. Let $\varphi_2 : \mathbb{R}^{n_2} \to \mathbb{R}$ be a non-negative measurable function and $\psi : \mathbb{R}^{n_1} \to \mathbb{R}^{n_2}$ be a function which satisfies $\varphi_2(\psi(x) - \psi(y)) \leqslant \varphi_1(x - y)$ for all $x, y \in \mathbb{R}^{n_1}$. Let μ_2 be the probability measure $\psi(\mu_1)$ on $\mathbb{R}^{n_2}$, that is, $\mu_2(A) = \mu_1(\psi^{-1}(A))$. Then, (μ_2, φ_2) has property (τ) in $\mathbb{R}^{n_2}$.*

Proof. We first check that

$$[(f \circ \psi) \Box \varphi_1] \geqslant [(f \Box \varphi_2) \circ \psi]$$

for every bounded measurable function $f : \mathbb{R}^{n_2} \to \mathbb{R}$. From $\mu_2 = \psi(\mu_1)$ it follows that

$$\int_{\mathbb{R}^{n_2}} e^{(f \Box \varphi_2)(y)} d\mu_2(y) = \int_{\mathbb{R}^{n_1}} e^{((f \Box \varphi_2) \circ \psi)(x)} d\mu_1(x)$$
$$\leqslant \int_{\mathbb{R}^{n_1}} e^{((f \circ \psi) \Box \varphi_1)(x)} d\mu_1(x)$$
$$\leqslant \left(\int_{\mathbb{R}^{n_1}} e^{-(f \circ \psi)(x)} d\mu_1(x) \right)^{-1}$$
$$= \left(\int_{\mathbb{R}^{n_2}} e^{-f(y)} d\mu_2(y) \right)^{-1}.$$

This shows that (μ_2, φ_2) has property (τ). $\square$

Property (τ) for a pair (μ, φ) is directly related to the concentration properties of the measure μ as the next proposition shows.

PROPOSITION 15.1.5. *Assume that (μ, φ) has property (τ) in $\mathbb{R}^n$. Then, for every measurable $A \subseteq \mathbb{R}^n$ and every $t > 0$, we have*

$$\mu(x \notin A + B_\varphi(t)) \leqslant (\mu(A))^{-1} e^{-t},$$

where $B_\varphi(t) = \{\varphi \leqslant t\}$.

Proof. For every $n \geqslant t$ we consider the function $f_{A,n}(x) = n \mathbf{1}_{A^c}(x)$. Observe that if $x \notin A + B_\varphi(t)$, then $(f_{A,n} \Box \varphi)(x) \geqslant t$. Indeed,

$$(f_{A,n} \Box \varphi)(x) = \inf_z \{f_{A,n}(z) + \varphi(x - z)\}.$$

If $z \in A$, then $f_{A,n}(z) = 0$ and, since $x \notin A + B_\varphi(t)$, we have $\varphi(x - z) \geqslant t$. Therefore,

$$f_{A,n}(z) + \varphi(x - z) = \varphi(x - z) \geqslant t.$$

On the other hand, if $z \notin A$ then $f_{A,n}(z) = n$ and $\varphi(x - z) \geqslant 0$, which gives

$$f_{A,n}(z) + \varphi(x - z) \geqslant n \geqslant t.$$

Now, from property (τ) we get

$$\int_{\mathbb{R}^n} e^{f_{A,n} \Box \varphi} d\mu \leqslant \left(\int_{\mathbb{R}^n} e^{-f_{A,n}} d\mu \right)^{-1} = \left(\int_A e^{-f_{A,n}} d\mu + \int_{\mathbb{R}^n \setminus A} e^{-f_{A,n}} d\mu \right)^{-1}$$
$$= \left(\mu(A) + e^{-n}(1 - \mu(A)) \right)^{-1} \leqslant 1/\mu(A).$$

From Markov's inequality,
$$e^t\mu(x \notin A + B_\varphi(t)) \leq e^t\mu(x : (f_{A,n}\Box\varphi)(x) \geq t)$$
$$\leq \int_{\mathbb{R}^n} e^{f_{A,n}\Box\varphi} d\mu \leq (\mu(A))^{-1}.$$

Therefore, $\mu(x \notin A + B_\varphi(t)) \leq (\mu(A))^{-1}e^{-t}$, as claimed. $\square$

The next proposition offers a sharper estimate of the same type.

PROPOSITION 15.1.6. *Assume that (μ, φ) has property (τ) in $\mathbb{R}^n$. Then, for every measurable $A \subseteq \mathbb{R}^n$ and every $t > 0$, we have*

(15.1.1) $$\mu(A + B_\varphi(t)) \geq \frac{e^t \mu(A)}{(e^t - 1)\mu(A) + 1}.$$

In particular, for all $t > 0$,

(15.1.2) $\quad\quad \mu(A) > 0 \quad \text{implies} \quad \mu(A + B_\varphi(t)) \geq \min\{e^{t/2}\mu(A), 1/2\},$

(15.1.3) $\quad\quad \mu(A) \geq 1/2 \quad \text{implies} \quad 1 - \mu(A + B_\varphi(t)) \leq e^{-t/2}(1 - \mu(A))$

and

(15.1.4) $\quad\quad \mu(A) = \nu(-\infty, x] \quad \text{implies} \quad \mu(A + B_\varphi(t)) \geq \nu(-\infty, x + t/2],$

where ν is the symmetric exponential measure with density $\frac{1}{2}e^{-|x|}$.

Proof. We set $f(x) = t\mathbf{1}_{A^c}$. If $x \notin A + B_\varphi(t)$ then we have $f\Box\varphi(x) = \inf_y(f(y) + \varphi(x-y)) \geq t$, because either $y \notin A$, and then $f(y) = t$, or $y \in A$, and then $\varphi(x-y) \geq t$ by the assumption that $x \notin A + B_\varphi(t)$. Applying property (τ) for f we get
$$1 \geq \int_{\mathbb{R}^n} e^{f\Box\varphi(x)} d\mu(x) \int_{\mathbb{R}^n} e^{-f(x)} d\mu(x)$$
$$\geq \left[\mu(A + B_\varphi(t)) + e^t(1 - \mu(A + B_\varphi(t)))\right] \left[\mu(A) + e^{-t}(1 - \mu(A))\right].$$

Then, we solve for $\mu(A + B_\varphi(t))$ to get (15.1.1).

Next, we note that the function $f_t(p) := e^t p/((e^t - 1)p + 1)$ is increasing in p. Thus, if $p > e^{-t/2}/2$ we see that $f_t(p) \geq f_t(e^{-t/2}/2) \geq 1/2$. Using the fact that if $p \leq e^{-t/2}/2$ then
$$(e^t - 1)p + 1 \leq e^{t/2} + 1 - \frac{1}{2}(e^{t/2} + e^{-t/2}) < e^{t/2},$$
we get $f_t(p) > e^{t/2}p$ in this case. This implies (15.1.2).

Assuming that $p \geq 1/2$, we simply observe that
$$1 - f_t(p) = \frac{1-p}{(e^t - 1)p + 1} \leq \frac{1-p}{(e^t + 1)/2} < e^{-t/2}(1-p)$$
to get (15.1.3).

For the last assertion we set $F(x) = \nu(-\infty, x]$ and $g_t(p) = F(F^{-1}(p) + t)$. The previous calculations show that if $t, p > 0$ and if $F^{-1}(p) + t/2 \leq 0$ or $F^{-1}(p) \geq 0$, then $f_t(p) \geq g_{t/2}(p)$. Using the observation that $g_{t+s} = g_t \circ g_s$ and $f_{t+s} = f_t \circ f_s$, we can then see that $f_t(p) \geq g_{t/2}(p)$ for all $t, p > 0$. Therefore, (15.1.4) follows from (15.1.1). $\square$

15.1.1. Talagrand's concentration inequality for the product exponential measure

In this subsection we describe an application of property (τ). We present Maurey's proof [**354**] of Talagrand's concentration inequality [**489**] for the product exponential measure.

THEOREM 15.1.7 (Talagrand). *For every measurable $A \subseteq \mathbb{R}^n$ and every $t > 0$,*

$$\nu_n(x \notin A + 6\sqrt{t}B_2^n + 9tB_1^n) \leqslant \frac{1}{\nu_n(A)}e^{-t},$$

where $\nu_n = \nu \otimes \cdots \otimes \nu$ is the product exponential measure with density

$$d\nu_n(x) = \frac{1}{2^n}e^{-\|x\|_1}dx.$$

We start with the one-dimensional case. Let $\psi : \mathbb{R} \to \mathbb{R}$ be defined by

$$\psi(t) = \begin{cases} t^2/18, & \text{if } |t| \leqslant 2 \\ 2(|t|-1)/9, & \text{if } |t| > 2. \end{cases}$$

Note that ψ is even, convex and continuously differentiable. We also consider the probability measure μ_e on $\mathbb{R}$ with density $\mathbf{1}_{(0,+\infty)}(x)e^{-x}$.

PROPOSITION 15.1.8. *The pair (μ_e, ψ) has property (τ).*

Proof. Let f be a bounded continuous function on $(0, +\infty)$. We write λ for the function $f\square\psi$ and we set

$$I_0 = \int_0^{+\infty} e^{-f(x)-x}dx \quad \text{and} \quad I_1 = \int_0^{+\infty} e^{\lambda(y)-y}dy.$$

For every $t \in (0,1)$ we define $x(t)$ and $y(t)$ through the equations

$$\int_0^{x(t)} e^{-f(x)-x}dx = tI_0 \quad \text{and} \quad \int_0^{y(t)} e^{\lambda(y)-y}dy = tI_1.$$

By their definition, $x(t)$ and $y(t)$ are differentiable, with

$$x'(t) = I_0 e^{f(x(t))+x(t)} \quad \text{and} \quad y'(t) = I_1 e^{-\lambda(y(t))+y(t)}.$$

We have

$$\lambda(y(t)) = \inf_{y \in \mathbb{R}}\{f(y) + \psi(y(t)-y)\} \leqslant f(x(t)) + \psi(y(t)-x(t)).$$

Therefore,

$$y'(t) \geqslant I_1 e^{-f(x(t))-\psi(y(t)-x(t))+y(t)}.$$

We define

$$z(t) = \frac{x(t)+y(t)}{2} - \psi(y(t)-x(t)).$$

Then,

$$z'(t) = \frac{x'(t)+y'(t)}{2} - \psi'(y(t)-x(t))(y'(t)-x'(t))$$
$$= \left(\frac{1}{2} + \psi'(y(t)-x(t))\right)x'(t) + \left(\frac{1}{2} - \psi'(y(t)-x(t))\right)y'(t).$$

We easily check that $|\psi'| \leqslant 1/2$ on $\mathbb{R}$, and hence $z(t)$ is increasing.

We write x, y for $x(t), y(t)$. Using the inequality
$$\frac{1}{2}\left(ua + \frac{v}{a}\right) \geq \sqrt{uv}, \quad u, v, a > 0$$
with $a = \exp(f(x))$, we get
$$z'(t) \geq (1 + 2\psi'(y-x))I_0 e^x \frac{e^{f(x)}}{2} + (1 - 2\psi'(y-x))I_1 e^{-\psi(y-x)+y}\frac{e^{-f(x)}}{2}$$
$$\geq \sqrt{1 - 4(\psi'(y-x))^2}\sqrt{I_0 I_1} e^{(x+y)/2 - \psi(y-x)/2}$$
$$= \sqrt{1 - 4(\psi'(y-x))^2}\sqrt{I_0 I_1} e^{(x+y)/2 - \psi(y-x)} e^{\psi(y-x)/2}$$
$$= \sqrt{1 - 4(\psi'(y-x))^2}\sqrt{I_0 I_1} e^{z(t)} e^{\psi(y-x)/2}.$$

Claim. For every s,
$$\left(1 - 4(\psi'(s))^2\right) e^{\psi(s)} \geq 1.$$

Proof of the Claim. Since ψ is even, it suffices to prove the inequality for $s \geq 0$. On $[2, +\infty)$ we have $\psi'(s) = 2/9$ and ψ is increasing. This shows that if the inequality holds at $s = 2$, then it will hold true for all $s \geq 2$. We ask
$$\left(1 - 4(2/9)^2\right) e^{2/9} \geq 1.$$
or equivalently, $e^{2/9} \geq 81/65$. This inequality is valid, because
$$e^{2/9} \geq 1 + \frac{2}{9} + \frac{1}{2}\left(\frac{2}{9}\right)^2 = \frac{101}{81} > \frac{81}{65}.$$

For $s \in [0, 2]$ we have $\psi'(s) = s/9$, and hence we need to check that $e^{-s^2/18} \leq 1 - 4s^2/81$. It is then sufficient to check that the function
$$r(u) = 1 - \frac{4u}{81} - e^{-u/18}$$
is non-negative on $[0, 4]$. Differentiation shows that r is concave, so it is enough to check the values $r(0)$ and $r(4)$. But, $r(0) = 0$ and $r(4) \geq 0$ reduces to $e^{2/9} \geq 81/65$ which has been already checked. □

From the claim and the previous inequality we see that
$$z'(t) \geq \sqrt{I_0 I_1} e^{z(t)},$$
and hence
$$\left(-e^{-z(t)}\right)' \geq \sqrt{I_0 I_1}.$$
Integrating on $[0, 1]$ and using the fact that $z(0) = 0$, we get
$$1 \geq e^{-z(0)} - e^{-z(1)} = \int_0^1 \left(-e^{-z(t)}\right)' dt \geq \sqrt{I_0 I_1}.$$
In other words,
$$\left(\int_0^\infty e^{f \Box \psi} d\mu_e\right)\left(\int_0^\infty e^{-f} d\mu_e\right) = I_0 I_1 \leq 1.$$
Since f was arbitrary, the pair (μ_e, ψ) has property (τ). □

We now consider the symmetric image μ'_e of μ_e on $(-\infty, 0)$, with density $\mathbf{1}_{(-\infty,0)}(x)e^x$. By symmetry, (μ'_e, ψ) has property (τ). If ν is the symmetric exponential probability measure on $\mathbb{R}$ with density $\frac{1}{2}e^{-|x|}$, we easily check that

$$\nu = \mu_e * \mu'_e.$$

From Lemma 15.1.3, the pair $(\nu, \psi\square\psi)$ has property (τ). Taking into account the definition of ψ, we see that $\xi := \psi\square\psi$ is given by

$$\xi(t) = \begin{cases} t^2/36, & \text{if } |t| \leq 4 \\ 2(|t|-2)/9, & \text{if } |t| > 4. \end{cases}$$

Now, we consider the product measure $\nu_n = \nu \otimes \cdots \otimes \nu$ (n times) on $\mathbb{R}^n$. If we define the function $\xi_n : \mathbb{R}^n \to \mathbb{R}$ with

$$\xi_n(x_1, \ldots, x_n) = \sum_{i=1}^{n} \xi(x_i),$$

Lemma 15.1.2 implies the following.

THEOREM 15.1.9. *The pair (ν_n, ξ_n) has property (τ) on $\mathbb{R}^n$.* □

From Theorem 15.1.9 and Proposition 15.1.5 we see that for every measurable $A \subseteq \mathbb{R}^n$ and every $t > 0$,

$$\nu_n(x \notin A + \{\xi_n < t\}) \leq \frac{1}{\nu_n(A)} e^{-t}.$$

We are now ready for the proof of Talagrand's inequality.

Proof of Theorem 15.1.7. It suffices to prove that

$$\{\xi_n \leq t\} \subseteq 6\sqrt{t}B_2^n + 9tB_1^n.$$

Let $x \in \mathbb{R}^n$ with $\xi_n(x) \leq t$. We define y and z in $\mathbb{R}^n$ as follows: $y_i = x_i$ if $|x_i| \leq 4$ and $y_i = 0$ otherwise, $z_i = x_i$ if $|x_i| > 4$ and $z_i = 0$ otherwise. Clearly,

$$x = y + z.$$

Observe that

$$\|y\|_2^2 = \sum_{\{i:|x_i|\leq 4\}} x_i^2 = 36 \sum_{\{i:|x_i|\leq 4\}} \xi(x_i) \leq 36\xi_n(x) \leq 36t,$$

and hence $y \in 6\sqrt{t}B_2^n$. Also, if $|x_i| > 4$, then

$$\xi(x_i) = \frac{2}{9}(|x_i| - 2) \geq \frac{2}{9}\left(|x_i| - \frac{|x_i|}{2}\right) = \frac{|x_i|}{9},$$

and hence

$$\|z\|_1 = \sum_{\{i:|x_i|>4\}} |x_i| \leq 9 \sum_{\{i:|x_i|>4\}} \xi(x_i) \leq 9\xi_n(x) \leq 9t,$$

which shows that $z \in 9tB_1^n$. □

15.1.2. Property (τ) in Gauss space

Let γ be the standard Gaussian probability measure on $\mathbb{R}$, with density $\frac{1}{\sqrt{2\pi}}e^{-x^2/2}$, and let $\gamma_n = \gamma \otimes \cdots \otimes \gamma$ be the product measure on $\mathbb{R}^n$. Property (τ) can be used to establish Gaussian concentration.

THEOREM 15.1.10. *The pair $(\gamma_n, \|x\|_2^2/4)$ has property (τ).*

Proof. From Lemma 15.1.2 it is enough to show that the pair $(\gamma, x^2/4)$ has property (τ) on $\mathbb{R}$. Nevertheless, we will give a direct proof for the n-dimensional case, using the Prékopa-Leindler inequality.

Let f be a bounded measurable function on $\mathbb{R}^n$. We define $\varphi(y) = \|y\|_2^2/4$ and $\psi = f \square \varphi$. If

$$m(x) = f(x) + \frac{\|x\|_2^2}{2}, \; g(y) = -\psi(y) + \frac{\|y\|_2^2}{2} \text{ and } h(z) = \frac{\|z\|_2^2}{2},$$

then we easily check that

$$h\left(\frac{x+y}{2}\right) \leqslant \frac{m(x)+g(y)}{2}.$$

So,

$$\left(\int_{\mathbb{R}^n} e^{-m(x)}dx\right)\left(\int_{\mathbb{R}^n} e^{-g(x)}dx\right) \leqslant \left(\int_{\mathbb{R}^n} e^{-h(x)}dx\right)^2.$$

In other words,

$$\left(\int_{\mathbb{R}^n} e^{-f}d\gamma_n\right)\left(\int_{\mathbb{R}^n} e^{f\square\varphi}d\gamma_n\right) \leqslant 1$$

which proves the theorem. $\square$

As an application, we give a proof of a result of Pisier on the concentration of Lipschitz functions with respect to the measure γ_n.

THEOREM 15.1.11. *Let $f : \mathbb{R}^n \to \mathbb{R}$ be a Lipschitz function with constant 1 with respect to the Euclidean norm. Then, for every $t > 0$,*

$$\int_{\mathbb{R}^n}\int_{\mathbb{R}^n} \exp\left(t(f(x)-f(y))\right) d\gamma_n(x)\, d\gamma_n(y) \leqslant e^{t^2}.$$

Proof. We fix $t > 0$, consider the function $\varphi(y) = \|y\|_2^2/4$ and define

$$\psi_t = (tf)\square\varphi.$$

Let $x \in \mathbb{R}^n$ and $y = y(x,t) \in \mathbb{R}^n$ such that

$$\psi_t(x) = tf(y) + \frac{\|x-y\|_2^2}{4}.$$

Since $\|f\|_{\text{Lip}} \leqslant 1$, we get

(15.1.5) $\qquad \psi_t(x) \geqslant tf(x) - t\|x-y\|_2 + \frac{\|x-y\|_2^2}{4} \geqslant tf(x) - t^2.$

From Theorem 15.1.10,

$$\left(\int_{\mathbb{R}^n} e^{\psi_t}d\gamma_n\right)\left(\int_{\mathbb{R}^n} e^{-tf}d\gamma_n\right) \leqslant 1.$$

Using (15.1.5) we get

$$\int_{\mathbb{R}^n} e^{tf} d\gamma_n \cdot \int_{\mathbb{R}^n} e^{-tf} d\gamma_n \leqslant e^{t^2},$$

or equivalently,

$$\iint \exp\left(t(f(x) - f(y))\right) d\gamma_n(x)\, d\gamma_n(y) \leqslant e^{t^2}$$

as claimed. $\square$

COROLLARY 15.1.12. *Let $f : \mathbb{R}^n \to \mathbb{R}$ be a Lipschitz function with $\|f\|_{\mathrm{Lip}} \leqslant 1$. Then,*

$$\gamma_n\left(\left\{x : \left|f(x) - \int_{\mathbb{R}^n} f d\gamma_n\right| \geqslant s\right\}\right) \leqslant 2e^{-s^2/4}$$

for every $s > 0$.

Proof. Let $s > 0$. From Theorem 15.1.11 and Jensen's inequality we obtain

$$\mathbb{E}\left(\exp\left(t(f - \mathbb{E}(f))\right)\right) \leqslant e^{t^2}.$$

for every $t > 0$. This shows that

$$\gamma_n\left(x : f(x) - \mathbb{E}f \geqslant s\right) \leqslant \exp\left(t^2 - ts\right)$$

for every $t > 0$. Minimizing with respect to t and applying the same argument to $-f$, we conclude the proof. $\square$

15.1.3. Concentration and property (τ)

Proposition 15.1.5 and 15.1.6 show that property (τ) implies concentration. Here, we show that, conversely, one can establish property (τ) from concentration estimates.

PROPOSITION 15.1.13. *Let φ be an even convex function with $\varphi(0) = 0$. Let μ be a Borel probability measure on $\mathbb{R}^n$ and assume that there exists $\gamma > 1$ such that for every $t > 0$ and every Borel subset A of $\mathbb{R}^n$*

(15.1.6) $\quad \mu(A) = \nu(-\infty, x] \quad \textit{implies} \quad \mu(A + \gamma B_\varphi(t)) \geqslant \nu(-\infty, x + \max\{t, \sqrt{t}\}],$

where $B_\varphi(t) = \{\varphi \leqslant t\}$ (compare with (15.1.4) in Proposition 15.1.6). Then, the pair $(\mu, \varphi(\frac{\cdot}{36\gamma}))$ has property (τ).

Proof. Let $f : \mathbb{R}^n \to \mathbb{R}$. For any measurable function h on $\mathbb{R}^k$ and $t \in \mathbb{R}$ we set

$$A(h, t) := \{x \in \mathbb{R}^k : h(x) < t\}.$$

Let g be an increasing right-continuous function on $\mathbb{R}$ such that

$$\mu(A(f, t)) = \nu(A(g, t)).$$

Then, the distribution of g with respect to ν is the same as the distribution of f with respect to μ. Therefore,

$$\int_{\mathbb{R}^n} e^{-f(x)} d\mu(x) = \int_{\mathbb{R}} e^{-g(x)} d\nu(x).$$

The proof will be complete if we show that

$$\int_{\mathbb{R}^n} e^{f \square \varphi(\frac{\cdot}{36\gamma})} d\mu \leqslant \int_{\mathbb{R}} e^{g \square \xi} d\nu,$$

where $\xi = \xi_1$ is the cost function for ν in Theorem 15.1.9. In fact, we will check that for every $t \in \mathbb{R}$ one has
$$\mu\left(A\left(f\Box\varphi(\tfrac{\cdot}{36\gamma}), t\right)\right) \geq \nu(A(g\Box\xi, t)).$$
Since g is increasing, one can easily check that the set $A(g\Box\xi, t)$ is a half-line, and hence it is enough to prove that

(15.1.7) $\quad g(x_1) + \xi(x_2) < t \quad$ implies $\quad \mu\left(A\left(f\Box\varphi(\tfrac{\cdot}{36\gamma}), t\right)\right) \geq \nu(-\infty, x_1 + x_2].$

We fix x_1 and x_2 with $g(x_1) + \xi(x_2) < t$ and take $s_1 > g(x_1)$ and $s_2 = \xi(x_2)$ with $s_1 + s_2 < t$. We define $A := A(f, s_1)$; then,
$$\mu(A) = \nu(A(g, s_1)) \geq \nu(-\infty, x_1].$$
By the definition of ξ it easily follows that $x_2 \leq \max\{6\sqrt{s_2}, 9s_2\}$, and our assumption (15.1.6) shows that
$$\mu(A + \gamma B_\varphi(36s_2)) \geq \nu(-\infty, x_1 + x_2].$$
Since $\gamma > 1$ and φ satisfies $\varphi(\lambda x) \leq \lambda\varphi(x)$ for all $0 < \lambda < 1$, one can check that $\gamma B_\varphi(t) \subseteq B_{\varphi(\frac{\cdot}{\gamma})}(t)$ for all $t > 0$, and hence
$$A(f, s_1) + \gamma B_\varphi(36s_2) \subseteq A(f, s_1) + B_{\varphi(\frac{\cdot}{\gamma})}(36s_2)$$
$$\subseteq A(f, s_1) + B_{\varphi(\frac{\cdot}{36\gamma})}(s_2) \subseteq A\left(f\Box\varphi(\tfrac{\cdot}{36\gamma}), s_1 + s_2\right).$$
This proves (15.1.7). $\qquad\square$

The next lemma gives the relation between the inequalities (15.1.2) and (15.1.3).

LEMMA 15.1.14. *For any Borel subset K of $\mathbb{R}^n$ and any $\gamma > 1$ the following two conditions are equivalent:*
  (i) *For every Borel subset A of $\mathbb{R}^n$ with $\mu(A) > 0$ one has $\mu(A + K) > \min\{\gamma\mu(A), \tfrac{1}{2}\}$.*
  (ii) *For every Borel subset A_1 of $\mathbb{R}^n$ with $\mu(A_1) \geq \tfrac{1}{2}$ one has $1 - \mu(A_1 - K) < \tfrac{1}{\gamma}(1 - \mu(A_1))$.*

Proof. (i)$\Rightarrow$(ii) We assume that $\mu(A_1) \geq 1/2$ but $1 - \mu(A_1 - K) \geq \tfrac{1}{\gamma}(1 - \mu(A_1))$. We set $A := \mathbb{R}^n \setminus (A_1 - K)$. Then,
$$(A + K) \cap A_1 = \emptyset,$$
which implies $\mu(A + K) \leq 1/2$. On the other hand,
$$\mu(A + K) \leq 1 - \mu(A_1) \leq \gamma(1 - \mu(A_1 - K)) = \gamma\mu(A).$$
Therefore, $\mu(A + K) \leq \min\{\gamma\mu(A), 1/2\}$, which is a contradiction.

(ii)$\Rightarrow$(i). Let A be a Borel subset of $\mathbb{R}^n$ such that $\mu(A) > 0$ and $\mu(A + K) \leq \min\{\gamma\mu(A), 1/2\}$. We set $A_1 := \mathbb{R}^n \setminus (A + K)$. Then, $\mu(A_1) \geq 1/2$. Moreover, $(A_1 - K) \cap A = \emptyset$, and hence
$$1 - \mu(A_1 - K) \geq \mu(A) \geq \frac{1}{\gamma}\mu(A + K) = \frac{1}{\gamma}(1 - \mu(A_1)),$$
which is a contradiction. $\qquad\square$

This leads us to the next proposition:

PROPOSITION 15.1.15. *Let $t > 0$ and let K be a symmetric convex set in $\mathbb{R}^n$ such that, for every Borel subset A of $\mathbb{R}^n$,*

$$\mu(A) > 0 \quad \text{implies} \quad \mu(A + K) > \min\{e^t \mu(A), 1/2\}.$$

Then, for any Borel set A we have that

$$\mu(A) = \nu(-\infty, x] \quad \text{implies} \quad \mu(A + 2K) > \nu(-\infty, x + t].$$

Proof. We fix a Borel subset A of $\mathbb{R}^n$ with $\mu(A) = \nu(-\infty, x]$ and we distinguish three cases:

(i) $x \geqslant 0$. Then we may apply Lemma 15.1.14 with $\gamma = e^t$ to get

$$\mu(A + K) > 1 - e^{-t}(1 - \mu(A)) = \nu(-\infty, x + t].$$

(ii) $-t \leqslant x \leqslant 0$. Then $e^t \mu(A) \geqslant 1/2$, and from our assumption we get

$$\mu(A + K) \geqslant \min\{e^t \mu(A), 1/2\} = 1/2 = \nu(-\infty, 0].$$

Then, by the previous case (for $A + K$ instead of A and $x = 0$), we obtain

$$\mu(A + 2K) = \mu((A + K) + K) > \nu(-\infty, t] \geqslant \nu(-\infty, x + t].$$

(iii) $x \leqslant -t$. Note that $A + 2K = (A+K) + K \supseteq A + K$ and $e^t \mu(A) = \nu(-\infty, x+t] \leqslant 1/2$. Thus, our assumption implies that

$$\mu(A + K) > e^t \mu(A) = \nu(-\infty, x + t].$$

This concludes the proof. $\square$

The main result of this subsection provides some sufficient conditions for a pair (μ, φ) to have property (τ), in the case where the cost function φ is convex and even. Recall from Chapter 14 that a probability measure μ on $\mathbb{R}^n$ satisfies *Cheeger's inequality with constant κ* if for every Borel subset A of $\mathbb{R}^n$,

$$(15.1.8) \quad \mu^+(A) := \liminf_{t \to 0^+} \frac{\mu(A + tB_2^n) - \mu(A)}{t} \geqslant \kappa \min\{\mu(A), 1 - \mu(A)\}.$$

We also know (see Section 14.1) that Cheeger's inequality implies exponential concentration: more precisely,

$$\mu(A) = \nu(-\infty, x] \quad \Rightarrow \quad \mu(A + tB_2^n) \geqslant \nu(-\infty, x + \kappa t].$$

THEOREM 15.1.16. *Let $\varphi : \mathbb{R}^n \to \mathbb{R}$ be an even convex function with $\varphi(0) = 0$ and*

$$(15.1.9) \quad \min\{1, \varphi(x)\} \leqslant (\alpha \|x\|_2)^2$$

for all $x \in \mathbb{R}^n$. We assume that the measure μ satisfies Cheeger's inequality (15.1.8) with constant κ and that there exists $\gamma > 1$ such that

$$(15.1.10) \quad \mu(A) > 0 \quad \text{implies} \quad \mu(A + \gamma B_\varphi(t)) > \min\{e^t \mu(A), 1/2\}$$

for all $t \geqslant 1$. Then, $(\mu, \varphi(\cdot/C))$ has property (τ), where $C = 36 \max\{2\gamma, \alpha/\kappa\}$.

Proof. From Proposition 15.1.13 it is enough to show that

$$(15.1.11) \quad \mu(A) = \nu(-\infty, x] \quad \text{implies} \quad \mu(A + \beta B_\varphi(t)) \geqslant \nu(-\infty, x + \max\{t, \sqrt{t}\}]$$

for all $t > 0$ and a suitable constant $\beta > 1$.

Assume first that $t < 1$. From (15.1.9) we see that $\alpha B_\varphi(t) \supseteq \sqrt{t} B_2^n$ for all $t < 1$. Then, Cheeger's inequality implies that (15.1.11) holds for all $t < 1$ with $\beta = \alpha/\kappa$.

15.2. Infimum convolution conjecture

On the other hand, for $t \geqslant 1$, Proposition 15.1.15 shows that our assumption (15.1.10) implies (15.1.11) with $\beta = 2\gamma$.

Therefore, (15.1.10) holds for all $t \geqslant 0$ with $\beta = \max\{2\gamma, \alpha/\kappa\}$ and the assertion follows from Proposition 15.1.13. $\square$

15.2. Infimum convolution conjecture

A direct observation is that if (μ, ϕ) has property (τ) and $\phi_1 \leqslant \phi$ then (μ, ϕ_1) has property (τ) too. So, it is natural to ask which are the best pairs (μ, φ) having property (τ). The right way to put this question is to fix the measure μ and to ask for the best possible cost function. To do this, we will use again the logarithmic Laplace transform of a measure μ; we will also need the following definition.

DEFINITION 15.2.1. Let $f : \mathbb{R}^n \to (-\infty, +\infty]$. The *Legendre transform* $\mathcal{L}(f)$ of f is defined by
$$\mathcal{L}(f)(x) := \sup_{y \in \mathbb{R}^n} \{\langle x, y\rangle - f(y)\}.$$

Below, we list a few basic properties that one can verify directly from the definition.
  (i) The Legendre transform of any function is a convex function.
  (ii) If f is convex and lower semi-continuous, then $\mathcal{L}(\mathcal{L}(f)) = f$, otherwise $\mathcal{L}(\mathcal{L}(f)) \leqslant f$.
  (iii) If $f \geqslant g$ then $\mathcal{L}(f) \leqslant \mathcal{L}(g)$.
  (iv) The Legendre transform satisfies $\mathcal{L}(cf)(x) = c\mathcal{L}(f)(x/c)$ and if $g(x) = f(x/c)$, then $\mathcal{L}(g)(x) = \mathcal{L}(f)(cx)$, where $c > 0$.

DEFINITION 15.2.2. Let μ be a probability measure on $\mathbb{R}^n$. We define
$$M_\mu(v) := \int_{\mathbb{R}^n} e^{\langle v, x\rangle} d\mu(x) = \exp(\Lambda_\mu(v))$$
where
$$\Lambda_\mu(v) = \log\left(\int_{\mathbb{R}^n} e^{\langle v, x\rangle} d\mu(x)\right)$$
is the logarithmic Laplace transform of μ. We also define
$$\Lambda_\mu^*(v) := \mathcal{L}(\Lambda_\mu)(v) = \sup_{u \in \mathbb{R}^n} \left\{\langle v, u\rangle - \log \int_{\mathbb{R}^n} e^{\langle u, x\rangle} d\mu(x)\right\}.$$

The function Λ_μ^* is called the *Cramer transform* of μ and plays a crucial role in the theory of large deviations. Often in this theory, the "extremal" functions are linear functionals. Therefore, it is interesting to see what happens if we take $f(x) = \langle x, y\rangle$ in the definition of property (τ). This idea leads to the next results.

PROPOSITION 15.2.3. Let μ be an even probability measure on $\mathbb{R}^n$ and let φ be a convex cost function such that (μ, φ) has property (τ). Then,
$$\varphi(v) \leqslant 2\Lambda_\mu^*(v/2) \leqslant \Lambda_\mu^*(v).$$

Proof. We choose $f(x) = \langle x, v\rangle$. Then,
$$f \square \varphi(x) = \inf_y \{f(x-y) + \varphi(y)\} = \inf_y \{\langle x - y, v\rangle + \varphi(y)\}$$
$$= \langle x, v\rangle - \mathcal{L}\varphi(v).$$

Using property (τ) and the fact that μ is even we write

$$1 \geqslant \int_{\mathbb{R}^n} e^{f\square\varphi} d\mu \int_{\mathbb{R}^n} e^{-f} d\mu = e^{-\mathcal{L}\varphi(v)} \int_{\mathbb{R}^n} e^{\langle x,v\rangle} d\mu \int_{\mathbb{R}^n} e^{-\langle x,v\rangle} d\mu$$
$$= e^{-\mathcal{L}\varphi(v)} M_\mu^2(v).$$

It follows that $\mathcal{L}\varphi(v) \geqslant 2\Lambda_\mu(v)$, and applying the Legendre transform on both sides we see that $\varphi(v) = \mathcal{L}\mathcal{L}\varphi(v) \leqslant 2\Lambda_\mu^*(v/2)$. The inequality $2\Lambda_\mu^*(v/2) \leqslant \Lambda_\mu^*(v)$ follows from the convexity of Λ_μ^*. □

DEFINITION 15.2.4 (infimum convolution property). An even probability measure μ has the *infimum convolution property* with constant β if the pair $(\mu, \Lambda_\mu^*(\frac{\cdot}{\beta}))$ has property (τ). In this case we say that μ satisfies **IC**(β).

As an immediate consequence of Lemma 15.1.2 and of the additive properties of Λ_μ^* we see that the **IC**-inequality behaves well with respect to products.

PROPOSITION 15.2.5. *Let μ_i be even probability measures on $\mathbb{R}^{n_i}$, $1 \leqslant i \leqslant k$. If μ_i satisfies* **IC**(β_i), *then $\mu = \otimes_{i=1}^k \mu_i$ satisfies* **IC**(β) *with $\beta = \max_i \beta_i$.*

Proof. We observe that

$$\Lambda_\mu(x_1,\ldots,x_k) = \sum_{i=1}^k \Lambda_{\mu_i}(x_i) \quad \text{and} \quad \Lambda_\mu^*(x_1,\ldots,x_k) = \sum_{i=1}^k \Lambda_{\mu_i}^*(x_i).$$

Since **IC**(β) implies **IC**(β_0) for all $\beta_0 \geqslant \beta$, the result immediately follows from Lemma 15.1.2. □

It is useful to note that property **IC**(β) is invariant under linear transformations.

PROPOSITION 15.2.6. *Let $T: \mathbb{R}^n \to \mathbb{R}^k$ be a linear map and let μ be a probability measure on $\mathbb{R}^n$ which satisfies* **IC**(β). *Then, the probability measure $\mu \circ T^{-1}$ on $\mathbb{R}^k$ satisfies* **IC**(β).

For the proof we first establish an equivalent description of the property **IC**(β). In the next lemma, for any $v = (v_0, v_1, \ldots, v_n)$ in $\mathbb{R}^{n+1}$ we write $\overline{v}$ for the vector $(v_1, v_2, \ldots, v_n) \in \mathbb{R}^n$.

LEMMA 15.2.7. *A probability measure μ on $\mathbb{R}^n$ satisfies* **IC**(β) *if and only if for any nonempty $V \subseteq \mathbb{R}^{n+1}$ and any bounded measurable function f on $\mathbb{R}^n$,*

$$(15.2.1) \qquad \int_{\mathbb{R}^n} e^{f\square\psi_V} d\mu \int_{\mathbb{R}^n} e^{-f} d\mu \leqslant \sup_{v \in V}\left\{e^{v_0} \int_{\mathbb{R}^n} e^{\beta\langle x,\overline{v}\rangle} d\mu(x)\right\},$$

where

$$\psi_V(x) := \sup_{v \in V}\{v_0 + \langle x, \overline{v}\rangle\}.$$

Proof. We set $V = \{(v_0, \overline{v}) : v_0 = -\Lambda_\mu(\beta\overline{v})\}$. Then, the right hand side is equal to 1 and $\psi_V(x) = \Lambda_\mu^*(x/\beta)$, so if μ satisfies (15.2.1) for this set V, it also satisfies **IC**(β). Conversely, let us assume that μ satisfies **IC**(β) and let V be a non-empty set. If the supremum on the right hand side is infinite then the inequality is obvious, so we may assume that it is equal to some $s < \infty$. This means that for any $(v_0, \overline{v}) \in V$ we have $v_0 + \Lambda_\mu(\beta\overline{v}) \leqslant \log s$, that is $v_0 \leqslant \log s - \Lambda_\mu(\beta\overline{v})$. It follows that

$$\psi_V(x) = \sup_{v \in V}\{v_0 + \langle x, \overline{v}\rangle\} \leqslant \log s + \sup_{v \in V}\{\langle x, \overline{v}\rangle - \Lambda_\mu(\beta\overline{v})\}$$
$$\leqslant \log s + \sup_{\overline{v} \in \mathbb{R}^n}\{\langle x, \overline{v}\rangle - \Lambda_\mu(\beta\overline{v})\} = \log s + \Lambda_\mu^*(x/\beta).$$

Since μ satisfies $\mathbf{IC}(\beta)$, we readily see that the left hand side of (15.2.1) is less than s. $\square$

Proof of Proposition 15.2.6. Let $V \subseteq \mathbb{R} \times \mathbb{R}^k$ and let $f : \mathbb{R}^k \to \mathbb{R}$. We define $\tilde{f} : \mathbb{R}^n \to \mathbb{R}$ by $\tilde{f}(x) := f(T(x))$ and set $\tilde{V} := \{(v_0, T^*(\overline{v})) : (v_0, \overline{v}) \in V\}$. It is easy to check that $\psi_V(T(x)) = \psi_{\tilde{V}}(x)$ and $(f\square\psi_V)(T(x)) \leqslant \tilde{f}\square\psi_{\tilde{V}}(x)$, and hence

$$\int_{\mathbb{R}^k} e^{f\square\psi_V} d(\mu \circ T^{-1}) \leqslant \int_{\mathbb{R}^n} e^{\tilde{f}\square\psi_{\tilde{V}}} d\mu$$

and

$$\int_{\mathbb{R}^k} e^{-f} d(\mu \circ T^{-1}) = \int_{\mathbb{R}^n} e^{-\tilde{f}} d\mu.$$

Note that by the definition of $\tilde{V}$ we have:

$$\sup_{v \in V}\left\{e^{v_0} \int_{\mathbb{R}^k} e^{\beta\langle x, \overline{v}\rangle} d(\mu \circ T^{-1})\right\} = \sup_{v \in \tilde{V}}\left\{e^{v_0} \int_{\mathbb{R}^n} e^{\beta\langle x, \overline{v}\rangle} d\mu\right\}.$$

From Lemma 15.2.7 we know that (15.2.1) is satisfied by $\mu, \tilde{f}$ and $\tilde{V}$, and the previous relations show that (15.2.1) is satisfied by $\mu \circ T^{-1}, f$ and V. Applying Lemma 15.2.7, in the opposite direction this time, we get the result. $\square$

Next, we show that every even log-concave measure on $\mathbb{R}$ satisfies $\mathbf{IC}(\beta)$ for some absolute constant $\beta > 0$. We first show that the exponential measure has this property.

PROPOSITION 15.2.8. *For any $x \in \mathbb{R}$,*

$$\frac{1}{5}\min(x^2, |x|) \leqslant \Lambda_\nu^*(x) \leqslant \min(x^2, |x|).$$

In particular, ν satisfies $\mathbf{IC}(9)$.

Proof. A direct computation shows that $\Lambda_\nu(x) = -\log(1-x^2)$ for $|x| < 1$ (and $= +\infty$ for $|x| \geqslant 1$) and

$$\Lambda_\nu^*(x) = \sqrt{1+x^2} - 1 - \log\left(\frac{\sqrt{1+x^2}+1}{2}\right)$$

for all x. Since $a/2 \leqslant a - \log(1 + a/2) \leqslant a$ for all $a \geqslant 0$, we get

$$\frac{1}{2}(\sqrt{1+x^2} - 1) \leqslant \Lambda_\nu^*(x) \leqslant \sqrt{1+x^2} - 1.$$

Then, we observe that

$$\min(|x|, x^2) \geqslant \sqrt{1+x^2} - 1 = \frac{x^2}{\sqrt{1+x^2}+1} \geqslant \frac{1}{\sqrt{2}+1}\min(|x|, x^2).$$

Now, Theorem 15.1.9 and the fact that $\xi(x) \geqslant \min((x/9)^2, |x|/9)$ show that ν satisfies $\mathbf{IC}(9)$. $\square$

THEOREM 15.2.9. *Every even log-concave probability measure μ on $\mathbb{R}$ satisfies* **IC**(96).

Proof. Let μ be an even log-concave probability measure on $\mathbb{R}$. By Proposition 15.2.6 we may assume that μ is isotropic. We denote the density of μ by $g(x)$ and we write
$$\mu[x, \infty) = e^{-h(x)}.$$
From Hensley's inequality (see Theorem 2.2.3) we know that
$$g(0) = g(0)\left(\int_{\mathbb{R}} x^2 g(x)dx\right)^{1/2} \geqslant \frac{1}{2\sqrt{3}} \geqslant \frac{1}{8}.$$
Let $T : \mathbb{R} \to \mathbb{R}$ be a function such that
$$\nu(-\infty, x) = \mu(-\infty, T(x)).$$
Then, $\mu = \nu \circ T^{-1}$, T is odd and concave on $[0, \infty)$. In particular,
$$|T(x) - T(y)| \leqslant 2|T(x-y)|$$
for all $x, y \in \mathbb{R}$. Note that $T'(0) = 1/(2g(0)) \leqslant 4$. Since T is concave, we get $T(x) \leqslant 4x$ for $x \geqslant 0$. Moreover, for $x \geqslant 0$ we have that $h(T(x)) = x + \log 2$. We define
$$\tilde{h}(x) = \begin{cases} x^2, & \text{if } |x| \leqslant 2/3 \\ \max\{4/9, h(|x|)\}, & \text{if } |x| > 2/3. \end{cases}$$
We will show that $(\mu, \tilde{h}(\frac{\cdot}{48}))$ has property (τ). To this end, we first observe that $\tilde{h}((T(x) - T(y))/96) \leqslant \tilde{h}(T(|x-y|)/48)$. In view of Lemma 15.1.4 we need to check that
$$(15.2.2) \qquad \tilde{h}\left(\frac{T(x)}{48}\right) \leqslant \xi(x)$$
for $x \geqslant 0$, where $\xi(x)$ is as in Theorem 15.1.9. We distinguish two cases.

(i) If $T(x) \leqslant 32$ then
$$\tilde{h}\left(\frac{T(x)}{48}\right) = \left(\frac{T(x)}{48}\right)^2 \leqslant \min\left\{\frac{4}{9}, \left(\frac{x}{6}\right)^2\right\} \leqslant \xi(x).$$

(ii) If $T(x) \geqslant 32$ then $x \geqslant 8$ and
$$\tilde{h}\left(\frac{T(x)}{48}\right) = \max\left\{\frac{4}{9}, h\left(\frac{T(x)}{48}\right)\right\} \leqslant \max\left\{\frac{4}{9}, \frac{h(T(x))}{48} + \frac{47\log 2}{48}\right\}$$
$$= \max\left\{\frac{4}{9}, \frac{x}{48} + \log 2\right\} \leqslant \frac{x}{9} \leqslant \xi(x).$$

This proves (15.2.2) in both cases. To conclude the proof we need to show that $\Lambda_\mu^*(x) \leqslant \tilde{h}(x)$. For $|x| \leqslant 2/3$ it is enough to observe that
$$\Lambda_\mu^*(x) = \min\{1, \Lambda_\mu^*(x)\} \leqslant x^2 = \tilde{h}(x),$$

where the first equality and the inequality in the middle are a special case of Proposition 15.3.3 (see the next section). For the case $|x| > 2/3$ note that

$$\Lambda_\mu(t) \geqslant tx + \log \mu[x, \infty) = tx - h(x)$$

for all $x, t \geqslant 0$, which implies that

$$\Lambda_\mu^*(x) = \Lambda_\mu^*(|x|) = \sup_{t \geqslant 0}\{t|x| - \Lambda_\mu(t)\} \leqslant h(|x|) \leqslant \tilde{h}(x).$$

This completes the proof. $\square$

From Proposition 15.2.5 we immediately get

COROLLARY 15.2.10. *Every even log-concave product probability measure on $\mathbb{R}^n$ satisfies* **IC**(96).

In fact, one is tempted to conjecture something much more general.

CONJECTURE 15.2.11 (infimum convolution conjecture). *There exists an absolute constant $\beta > 0$ such that every even log-concave probability measure μ on $\mathbb{R}^n$ satisfies* **IC**(β).

15.3. Concentration inequalities

In this section we discuss concentration inequalities which follow from the infimum convolution property. L_q-centroid bodies enter this discussion in a natural way.

DEFINITION 15.3.1. Let μ be a probability measure on $\mathbb{R}^n$. For every $p \geqslant 1$ we define

$$M_p(\mu) := \left\{ v \in \mathbb{R}^n : \int_{\mathbb{R}^n} |\langle v, x \rangle|^p d\mu(x) \leqslant 1 \right\}.$$

Note that

$$Z_p(\mu) := (M_p(\mu))^\circ = \left\{ x \in \mathbb{R}^n : |\langle v, x \rangle|^p \leqslant \int_{\mathbb{R}^n} |\langle v, y \rangle|^p d\mu(y) \text{ for all } v \in \mathbb{R}^n \right\}.$$

For every $p > 0$ we also set

$$B_p(\mu) := \{v \in \mathbb{R}^n : \Lambda_\mu^*(v) \leqslant p\}.$$

The next two results describe the geometry of $B_p(\mu)$ for large and small values of p respectively.

PROPOSITION 15.3.2. *Let μ be an even probability measure μ on $\mathbb{R}^n$. For every $p \geqslant 1$ we have*

$$Z_p(\mu) \subseteq 2^{1/p} e B_p(\mu).$$

Proof. Given $v \in Z_p(\mu)$, we need to show that $\Lambda_\mu^*(v/(2^{1/p}e)) \leqslant p$, or equivalently

$$\frac{\langle u, v \rangle}{2^{1/p}e} - \Lambda_\mu(u) \leqslant p$$

for all $u \in \mathbb{R}^n$. We fix $u \in \mathbb{R}^n$ with

$$\int_{\mathbb{R}^n} |\langle u, x \rangle|^p d\mu(x) = \beta^p.$$

Then, $u/\beta \in M_p(\mu)$. Since $Z_p(\mu) = (M_p(\mu))^\circ$, we have $\langle u/\beta, v \rangle \leqslant 1$. We distinguish two cases:

(i) If $\beta \leqslant 2^{1/p}ep$ then, using the fact that
$$\Lambda_\mu(u) \geqslant \int_{\mathbb{R}^n} \langle u, x \rangle d\mu(x) = 0,$$
we get
$$\frac{\langle u, v \rangle}{2^{1/p}e} - \Lambda_\mu(u) \leqslant \frac{\beta}{2^{1/p}e} \langle u/\beta, v \rangle \leqslant p \cdot 1.$$

(ii) If $\beta > 2^{1/p}ep$ then we have
$$\int_{\mathbb{R}^n} e^{\langle u, x \rangle} d\mu(x) \geqslant \int_{\mathbb{R}^n} \left| e^{\langle u, x \rangle/p} \right|^p \mathbf{1}_{\{\langle u, x \rangle \geqslant 0\}} d\mu(x)$$
$$\geqslant \int_{\mathbb{R}^n} \left| \frac{\langle u, x \rangle}{p} \right|^p \mathbf{1}_{\{\langle u, x \rangle \geqslant 0\}} d\mu(x)$$
$$\geqslant \frac{1}{2} \int_{\mathbb{R}^n} \left| \frac{\langle u, x \rangle}{p} \right|^p d\mu(x),$$

therefore
$$\int_{\mathbb{R}^n} e^{2^{1/p}ep\langle u, x \rangle/\beta} d\mu(x) \geqslant \frac{1}{2} \int_{\mathbb{R}^n} \left| \frac{2^{1/p}e\langle u, x \rangle}{\beta} \right|^p d\mu(x) = e^p.$$

So, $\Lambda_\mu(2^{1/p}epu/\beta) \geqslant p$, which implies that
$$\Lambda_\mu(u) \geqslant \frac{\beta}{2^{1/p}ep} \Lambda_\mu(2^{1/p}epu/\beta) \geqslant \frac{\beta}{2^{1/p}e}.$$

Therefore,
$$\frac{\langle u, v \rangle}{2^{1/p}e} - \Lambda_\mu(u) \leqslant \frac{\beta}{2^{1/p}e} \langle u/\beta, v \rangle - \frac{\beta}{2^{1/p}e} \leqslant 0$$
because $\langle u/\beta, v \rangle \leqslant 1$, and the result follows. $\square$

PROPOSITION 15.3.3. *If μ is an even isotropic probability measure on $\mathbb{R}^n$, then*
$$\min\{1, \Lambda_\mu^*(u)\} \leqslant \|u\|_2^2$$
for all u. In particular,
$$\sqrt{p} B_2^n \subseteq B_p(\mu)$$
for all $p \in (0, 1)$.

Proof. Since μ is even and isotropic, we have
$$\int_{\mathbb{R}^n} e^{\langle u, x \rangle} d\mu(x) = 1 + \sum_{k=1}^\infty \frac{1}{(2k)!} \int_{\mathbb{R}^n} \langle u, x \rangle^{2k} d\mu(x) \geqslant 1 + \sum_{k=1}^\infty \frac{\|u\|_2^{2k}}{(2k)!}$$
$$= \cosh(\|u\|_2).$$

Then, for $\|u\|_2 < 1$,
$$\Lambda_\mu^*(u) \leqslant \mathcal{L}(\log \cosh)(\|u\|_2)$$
$$= \frac{1}{2} \left[(1 + \|u\|_2) \log(1 + \|u\|_2) + (1 - \|u\|_2) \log(1 - \|u\|_2) \right] \leqslant \|u\|_2^2,$$

where in the end we used the elementary inequality $\log(1+x) \leqslant x$ for $x > -1$. $\square$

In order to establish reverse inclusions we need some notion of regularity on the growth of moments of linear functionals, which we introduce in the next subsection. By Borell's lemma this regularity is certainly satisfied in the log-concave case.

15.3.1. α-regular measures

DEFINITION 15.3.4. We say that a measure μ on $\mathbb{R}^n$ is α-regular if for any $p \geqslant q \geqslant 2$ and every $v \in \mathbb{R}^n$,

$$\left(\int_{\mathbb{R}^n} |\langle v,x\rangle|^p d\mu(x)\right)^{1/p} \leqslant \alpha \frac{p}{q} \left(\int_{\mathbb{R}^n} |\langle v,x\rangle|^q d\mu(x)\right)^{1/q}.$$

REMARK 15.3.5. For all $p \geqslant q$ we have $M_p(\mu) \subseteq M_q(\mu)$ and $Z_q(\mu) \subseteq Z_p(\mu)$. If the measure μ is α-regular, then $M_q(\mu) \subseteq \alpha\frac{p}{q}M_p(\mu)$ and $Z_p(\mu) \subseteq \alpha\frac{p}{q}Z_q(\mu)$ for all $p \geqslant q \geqslant 2$. Moreover, for every even measure μ we have $\Lambda_\mu^*(0) = 0$, and the convexity of Λ_μ^* implies that $B_q(\mu) \subseteq B_p(\mu) \subseteq \frac{p}{q}B_q(\mu)$ for all $p \geqslant q > 0$.

Recall that, by Borell's lemma, every log-concave probability measure is c-regular. The next proposition shows that, in fact, one may assume that $c = 1$.

PROPOSITION 15.3.6. *Every even log-concave measure is 1-regular.*

Proof. Let μ be an even log-concave measure on $\mathbb{R}^n$. We need to show that, for every $u \in \mathbb{R}^n$,

$$(\mathbb{E}_\mu |\langle x,u\rangle|^p)^{1/p} \leqslant \frac{p}{q}(\mathbb{E}_\mu |\langle x,u\rangle|^q)^{1/q}$$

for all $p \geqslant q \geqslant 2$. Barlow, Marshall and Proschan have proved that

$$(\mathbb{E}_\mu |\langle x,u\rangle|^p)^{1/p} \leqslant \frac{\Gamma(p+1))^{1/p}}{\Gamma(q+1))^{1/q}}(\mathbb{E}_\mu |\langle x,u\rangle|^q)^{1/q},$$

so it is enough to show that the function $f(x) := \frac{1}{x}(\Gamma(x+1))^{1/x}$ is decreasing on $[2,\infty)$. We use a precise form of Stirling's formula: one has

$$\Gamma(x+1) = x\Gamma(x) = \sqrt{2\pi}x^{x+1/2}e^{-x+\theta(x)},$$

where $\theta(x) = \int_0^\infty \arctan(t/x)(e^{2\pi t}-1)^{-1}dt$ is a decreasing function. Therefore,

$$\log f(x) = \frac{\theta(x)}{x} + \frac{\log(2\pi x)}{2x} - 1$$

is decreasing on $[2,\infty)$. $\square$

The next results complement Proposition 15.3.2 and Proposition 15.3.3 under the α-regularity assumption.

PROPOSITION 15.3.7. *If μ is α-regular for some $\alpha \geqslant 1$, then for any $p \geqslant 2$ we have*

$$B_p(\mu) \subseteq 4e\alpha Z_p(\mu).$$

Proof. We first check that if $u \in M_p(\mu)$ then

$$\Lambda_\mu\left(\frac{pu}{2e\alpha}\right) \leqslant p.$$

We fix $u \in M_p(\mu)$ and set $\tilde{u} := \frac{pu}{2e\alpha}$. Then,

$$\left(\int_{\mathbb{R}^n} |\langle \tilde{u},x\rangle|^k d\mu(x)\right)^{1/k} = \frac{p}{2e\alpha}\left(\int_{\mathbb{R}^n} |\langle u,x\rangle|^k d\mu(x)\right)^{1/k},$$

which is bounded by $\frac{p}{2e\alpha}$ if $k \leqslant p$ and by $\frac{k}{2e}$ if $k > p$. It follows that

$$\int_{\mathbb{R}^n} e^{\langle \tilde{u}, x \rangle} d\mu(x) \leqslant \int_{\mathbb{R}^n} e^{|\langle \tilde{u}, x \rangle|} d\mu(x) = \sum_{k=0}^{\infty} \frac{1}{k!} \int_{\mathbb{R}^n} |\langle \tilde{u}, x \rangle|^k d\mu(x)$$

$$\leqslant \sum_{k \leqslant p} \frac{1}{k!} \left| \frac{p}{2e\alpha} \right|^k + \sum_{k > p} \frac{1}{k!} \left| \frac{k}{2e} \right|^k \leqslant e^{\frac{p}{2e\alpha}} + 1 \leqslant e^p$$

and the claim follows.

Now, let $v \notin 4e\alpha Z_p(\mu)$. We can find $u \in M_p(\mu)$ such that $\langle v, u \rangle > 4e\alpha$ and then

$$\Lambda_\mu^*(v) \geqslant \left\langle v, \frac{pu}{2e\alpha} \right\rangle - \Lambda_\mu \left(\frac{pu}{2e\alpha} \right) > \frac{p}{2e\alpha} 4e\alpha - p = p.$$

Therefore, $v \notin B_p(\mu)$. $\square$

PROPOSITION 15.3.8. *If μ is even, isotropic and α-regular for some $\alpha \geqslant 1$, then*

$$\Lambda_\mu^*(u) \geqslant \min \left\{ \frac{\|u\|_2}{2\alpha e}, \frac{\|u\|_2^2}{2\alpha^2 e^2} \right\}.$$

In particular,

$$B_p(\mu) \subseteq \max\{2\alpha ep, \alpha e\sqrt{2p}\} B_2^n$$

for all $p > 0$.

Proof. Since μ is even, isotropic and α-regular, for every $v \in \mathbb{R}^n$ we have

$$\int_{\mathbb{R}^n} e^{\langle v, x \rangle} d\mu(x) = \sum_{k=0}^{\infty} \frac{1}{(2k)!} \int_{\mathbb{R}^n} \langle v, x \rangle^{2k} d\mu(x) \leqslant 1 + \frac{\|v\|_2^2}{2} + \sum_{k=2}^{\infty} \int_{\mathbb{R}^n} \frac{(\alpha k \|v\|_2)^{2k}}{(2k)!}$$

$$\leqslant 1 + \frac{\|v\|_2^2}{2} + \sum_{k=2}^{\infty} \left(\frac{\alpha e \|v\|_2}{2} \right)^{2k}.$$

Assuming that $\alpha e \|v\|_2 \leqslant 1$, we get

$$\int_{\mathbb{R}^n} e^{\langle v, x \rangle} d\mu(x) \leqslant 1 + \frac{\|v\|_2^2}{2} + \frac{4}{3} \left(\frac{\alpha e \|v\|_2}{2} \right)^4$$

$$\leqslant 1 + \frac{\alpha^2 e^2 \|v\|_2^2}{2} + \frac{(\alpha e \|v\|_2)^4}{8} \leqslant e^{\alpha^2 e^2 \|v\|_2^2 / 2},$$

so $\Lambda_\mu(v) \leqslant \alpha^2 e^2 \|v\|_2^2 / 2$ in this case.

Let $u \in \mathbb{R}^n$. If $\|u\|_2 \leqslant \alpha e$, we may write $\Lambda_\mu^*(u) \geqslant \langle u, u/(\alpha^2 e^2) \rangle - \Lambda_\mu(u/(\alpha^2 e^2))$ and, since $\Lambda_\mu(u/(\alpha^2 e^2)) \leqslant \|u\|_2^2/(2\alpha^2 e^2)$ in this case, we see that

$$\Lambda_\mu^*(u) \geqslant \frac{\|u\|_2^2}{2\alpha^2 e^2}.$$

If $\|u\|_2 \geqslant \alpha e$, we write $\Lambda_\mu^*(u) \geqslant \langle u, u/(\alpha e \|u\|_2) \rangle - \Lambda_\mu(u/(\alpha e \|u\|_2))$ and, since $\Lambda_\mu(u/(\alpha e \|u\|_2)) \leqslant 1/2 \leqslant \|u\|_2/(2\alpha e)$ in this case, one can check that

$$\Lambda_\mu^*(u) \geqslant \frac{\|u\|_2}{2\alpha e}.$$

This proves the proposition. $\square$

DEFINITION 15.3.9. We say that a measure μ satisfies the concentration inequality with constant β - and we write $\mathbf{CI}(\beta)$ - if for every $p \geqslant 2$ and every Borel subset A of $\mathbb{R}^n$,

(15.3.1) $\quad\quad\quad \mu(A) \geqslant \dfrac{1}{2} \quad \text{implies} \quad 1 - \mu(A + \beta Z_p(\mu)) \leqslant e^{-p}(1 - \mu(A)).$

Note that $Z_p(\mu)$ is in a sense the best convex symmetric set that one can hope to use in an implication like (15.3.1). This can be seen from the next proposition.

PROPOSITION 15.3.10. *Let μ be an even α-regular probability measure on $\mathbb{R}^n$. Assume that K is a convex set such that, for any half-space A with $\mu(A) \geqslant 1/2$ we have*

$$1 - \mu(A + K) \leqslant e^{-p}/2.$$

Then,

$$K \supseteq c(\alpha) Z_p(\mu)$$

for all $p \geqslant p(\alpha)$, where $c(\alpha)$ and $p(\alpha)$ depend only on α.

Proof. We fix $v \in \mathbb{R}^n$ and set $A = \{x : \langle v, x \rangle < 0\}$. Then,

$$A + K = \{x : \langle v, x \rangle < a(v)\},$$

where $a(v) = \sup_{x \in K} \langle x, v \rangle$.

Then,

$$\mu(\{x : |\langle x, v \rangle| \geqslant a(v)\}) = 2\mu(\{x : \langle x, v \rangle \geqslant a(v)\}) = 2(1 - \mu(A + K)) \leqslant e^{-p}.$$

Since μ is α-regular, we have $\|\langle \cdot, v \rangle\|_p \leqslant \alpha(p/q)\|\langle \cdot, v, \rangle\|_q$ for all $p \geqslant q \geqslant 2$. Using the Paley-Zygmund inequality we see that for all $q \geqslant 2$,

$$\mu(\{x : |\langle x, v \rangle| \geqslant \|\langle \cdot, v \rangle\|_q/2\}) = \mu(\{x : |\langle x, v \rangle|^q \geqslant 2^{-q}\|\langle \cdot, v \rangle\|_q^q\})$$

$$\geqslant (1 - 2^{-q})^2 \frac{\|\langle \cdot, v \rangle\|_q^{2q}}{\|\langle \cdot, v \rangle\|_{2q}^{2q}}$$

$$\geqslant \frac{9}{16}(2\alpha)^{-2q} > (3\alpha)^{-2q}.$$

Thus, if $p \geqslant p(\alpha) = \max\{8\log(3\alpha), 2\log 4\}$ and $c(\alpha) = (8\alpha \log(3\alpha))^{-1}$, we get

$$\mu(\{x : |\langle x, v \rangle| \geqslant c(\alpha)\|\langle \cdot, v \rangle\|_p\}) \geqslant \mu\left(\left\{x : |\langle x, v \rangle| \geqslant \frac{1}{2}\|\langle \cdot, v \rangle\|_{p/(4\log(3\alpha))}\right\}\right)$$

$$> (3\alpha)^{-p/(2\log(3\alpha))}$$

$$= e^{-p/2} \geqslant 2e^{-p}.$$

It follows that

$$c(\alpha)\|\langle \cdot, v \rangle\|_p = c(\alpha)\left(\int_{\mathbb{R}^n} |\langle v, x \rangle|^p \, d\mu(x)\right)^{1/p} \leqslant a(v)$$

and this shows that $c(\alpha) Z_p(\mu) \subseteq K$. $\square$

The next theorem shows that the infimum convolution property implies the concentration inequality property. In fact, under the assumption that Cheeger's inequality

(15.3.2) $\quad\quad\quad \mu^+(A) \geqslant \dfrac{1}{\gamma} \min\{\mu(A), 1 - \mu(A)\}$

is satisfied with constant $1/\gamma$, then the two properties are equivalent up to a constant depending on γ.

THEOREM 15.3.11. *Let μ be an even α-regular isotropic probability measure, where $\alpha \geqslant 1$.*
  (i) *If μ satisfies $\mathbf{IC}(\beta)$ then μ satisfies $\mathbf{CI}(8e\alpha\beta)$.*
  (ii) *If μ satisfies $\mathbf{CI}(\beta)$, and it also satisfies Cheeger's inequality (15.3.2), then μ satisfies $\mathbf{IC}(36 \max\{6e\beta, \gamma\})$.*

Proof. (i) We assume that μ satisfies $\mathbf{IC}(\beta)$. From Remark 15.3.5, Proposition 15.1.6 and the definition of $B_p(\mu)$ we get
$$\mu(A + 2\beta B_p(\mu)) \geqslant \mu(A + \beta B_{2p}(\mu)) \geqslant 1 - e^{-p}(1 - \mu(A)).$$
From Proposition 15.3.7 it follows that
$$\mu(A + 8e\alpha\beta Z_p(\mu)) \geqslant 1 - e^{-p}(1 - \mu(A)).$$
This shows that μ satisfies $\mathbf{CI}(8e\alpha\beta)$.

(ii) We assume that μ satisfies $\mathbf{CI}(\beta)$. Using Remark 15.3.5 and Proposition 15.3.2 we see that if $\mu(A) \geqslant 1/2$ and $p \geqslant 1$ then
$$e^{-p}(1 - \mu(A)) > e^{-2p}(1 - \mu(A)) \geqslant 1 - \mu(A + \beta Z_{2p}(\mu))$$
$$\geqslant 1 - \mu(A + e2^{1/2p}\beta B_{2p}(\mu)) \geqslant 1 - \mu(A + 3e\beta B_p(\mu)).$$

Then, Lemma 15.1.14 shows that (15.1.10) holds true with $\gamma = 3e\beta$. We also have that Λ_μ^* is even, convex, and $\Lambda_\mu^*(0) = 0$. Finally, from Proposition 15.3.3 we have $\min\{1, \Lambda_\mu^*(u)\} \leqslant \|u\|_2^2$. Using Theorem 15.1.16 we get the result. □

Next, we show that $\mathbf{CI}$ implies exponential concentration (see Definition 14.1.12) for isotropic measures.

PROPOSITION 15.3.12. *Let μ be an isotropic log-concave probability measure on $\mathbb{R}^n$ which satisfies $\mathbf{CI}(\beta)$. If f is a 1-Lipschitz function with respect to the standard Euclidean norm then*

(15.3.3) $$\mu(\{x \in \mathbb{R}^n : |f(x) - \operatorname{med}(f)| > t\}) \leqslant e^{1-t/\beta_1},$$

where $\beta_1 = 4e^2\beta$. We also have

(15.3.4) $$\mu(\{x \in \mathbb{R}^n : |f(x) - \mathbb{E}_\mu(f)| > t\}) \leqslant e^{1-t/\beta_2},$$

where $\beta_2 = 8e^3\beta$.

Proof. Let $A_t = \{x \in \mathbb{R}^n : f(x) - \operatorname{med}(f) > t\}$ and $A = \{x : f(x) \leqslant \operatorname{med}(f)\}$. We have $\mu(A) \geqslant 1/2$, and $\mathbf{CI}(\beta)$ implies that
$$1 - \mu(A + \beta Z_p(\mu)) \leqslant e^{-p}(1 - \mu(A)) \leqslant e^{-p}/2.$$
Let $p \geqslant 1$. From Propositions 15.3.2 and 15.3.8 we have
$$Z_p(\mu) \subseteq 2eB_p(\mu) \subseteq 4e^2 p B_2^n.$$
We set $t = 4\beta e^2 p$ (then, $t \geqslant 4\beta e^2 = \beta_1$). Since f is 1-Lipschitz, we have $A_t \cap (A + tB_2^n) = \emptyset$, thus
$$\mu(A_t) \leqslant 1 - \mu(A + tB_2^n) \leqslant 1 - \mu(A + \beta Z_p(\mu)) \leqslant e^{-t/\beta_1}/2.$$
In the same way we show that if $t > \beta_1$ then
$$\mu(\{x : f(x) - \operatorname{med}(f) < -t\}) \leqslant e^{-t/\beta_1}/2,$$

and this proves the result for $t \geqslant \beta_1$. If $t \leqslant \beta_1$, then we obviously have $\mu(A_t) \leqslant 1 \leqslant e^{1-t/\beta_1}$. Integration by parts shows that

$$|\mathbb{E}_\mu(f) - \mathrm{med}(f)| \leqslant \int_0^\infty \mu(\{x : |f(x) - \mathrm{med}(f)| \geqslant t\})\, dt \leqslant e\beta_1,$$

and then considering the cases $t \geqslant 2e\beta_1$ and $t < 2e\beta_1$ we complete the proof. □

In Chapter 14 we saw that in the context of log-concave measures exponential concentration is equivalent to Cheeger's inequality and the Poincaré inequality. This allows us to prove that for log-concave probability measures the properties **IC** and **CI** are equivalent and with the additional assumption of isotropicity they imply the Kannan-Lovász-Simonovits conjecture.

THEOREM 15.3.13 (Latała-Wojtaszczyk). *Let μ be a log-concave probability measure on $\mathbb{R}^n$. Then:*
  (i) *If μ satisfies $\mathbf{IC}(\beta)$, then μ satisfies $\mathbf{CI}(\beta_0)$ with $\beta_0 \simeq \beta$.*
  (ii) *If μ satisfies $\mathbf{CI}(\beta_0)$, then μ satisfies $\mathbf{IC}(\beta)$ with $\beta \simeq \beta_0$.*
  (iii) *If μ satisfies either $\mathbf{IC}(\beta)$ or $\mathbf{CI}(\beta)$ and is in addition isotropic, then it satisfies Cheeger's inequality (15.3.2) with $\gamma \simeq \beta$.*

Proof. We can find an affine map T such that $\mu \circ T^{-1}$ is isotropic. Since $Z_p(\mu \circ T^{-1}) = T(Z_p(\mu))$ we see that $\mathbf{CI}(\beta)$ is affinely invariant. We also know from Proposition 15.2.6 that $\mathbf{IC}(\beta)$ is affinely invariant, and hence for the proof we may assume μ is isotropic. Log-concavity implies that μ is 1-regular as well.

Then, (i) is a direct consequence of Theorem 15.3.11. For (iii) we may assume, because of (i), that μ satisfies $\mathbf{CI}(\beta)$. Then, by Proposition 15.3.12 we have that $\mathrm{Exp}_\mu \simeq 1/\beta$, and the claim follows.

For (ii) we can use Theorem 15.3.11 (ii) again, because (iii) shows that μ satisfies Cheeger's inequality. □

15.4. Comparison of weak and strong moments

In this section we show that the **CI** property of a probability measure μ implies a very strong form of comparison of weak and strong moments of any norm with respect to μ.

PROPOSITION 15.4.1. *Let μ be a probability measure on $\mathbb{R}^n$ which is α-regular and satisfies $\mathbf{CI}(\beta)$. Then, for any norm $\|\cdot\|$ on $\mathbb{R}^n$ and every $p \geqslant 2$,*

$$\left(\int_{\mathbb{R}^n} |\,\|x\| - \mathrm{med}(\|x\|)|^p\, d\mu\right)^{1/p} \leqslant 2\alpha\beta \sup_{\|u\|_* \leqslant 1} \left(\int_{\mathbb{R}^n} |\langle u, x\rangle|^p d\mu\right)^{1/p},$$

where $\|\cdot\|_*$ denotes the dual norm of $\|\cdot\|$.

Proof. For every $p \geqslant 2$ we define

$$m_p := \sup_{\|u\|_* \leqslant 1} \left(\int_{\mathbb{R}^n} |\langle u, x\rangle|^p d\mu\right)^{1/p}.$$

We set

$$M := \mathrm{med}(\|x\|), \quad A := \{x : \|x\| \leqslant M\} \quad \text{and} \quad \tilde{A} := \{x : \|x\| \geqslant M\}.$$

Then, $\mu(A) \geq 1/2$ and $\mu(\tilde{A}) \geq 1/2$, so by **CI**(β) and Remark 15.3.5 we get: for every $t \geq p$,
$$1 - \mu\left(A + \beta\frac{\alpha t}{p}Z_p(\mu)\right) \leq \frac{1}{2}e^{-t} \quad \text{and} \quad 1 - \mu\left(\tilde{A} + \beta\frac{\alpha t}{p}Z_p(\mu)\right) \leq \frac{1}{2}e^{-t}.$$

Given $y \in Z_p(\mu)$, there exists $u \in \mathbb{R}^n$ with $\|u\|_* \leq 1$ such that
$$\|y\| = \langle u, y \rangle \leq \left(\int_{\mathbb{R}^n} |\langle u, x \rangle|^p d\mu(x)\right)^{1/p} \leq m_p,$$
and hence
$$\|x\| \leq M + tm_p$$
for all $x \in A + tZ_p(\mu)$. Then, for every $t \geq p$ we have that
$$\mu\left(\left\{x : \|x\| \geq M + \frac{\alpha\beta t}{p}m_p\right\}\right) \leq 1 - \mu\left(A + \beta\frac{\alpha t}{p}Z_p(\mu)\right) \leq \frac{1}{2}e^{-t}.$$

In the same way we show that $\|x\| \geq M - tm_p$ for all $x \in \tilde{A} + tZ_p(\mu)$ and $\mu(\{x : \|x\| \leq M - \alpha\beta tm_p/p\}) \leq e^{-t}/2$, therefore
$$\mu\left(x : |\|x\| - M| \geq \frac{\alpha\beta t}{p}m_p\right) \leq e^{-t} \quad \text{for all } t \geq p.$$

Integration by parts gives
$$\left(\int_{\mathbb{R}^n} |\|x\| - M|^p d\mu\right)^{1/p}$$
$$\leq \frac{\alpha\beta m_p}{p}\left[p + \left(p\int_p^\infty t^{p-1}\mu\left(\left\{x : |\|x\| - M| \geq \frac{\alpha\beta t}{p}m_p\right\}\right) dt\right)^{1/p}\right]$$
$$\leq \frac{\alpha\beta m_p}{p}\left[p + \left(p\int_p^\infty t^{p-1}e^{-t}dt\right)^{1/p}\right]$$
$$\leq \alpha\beta m_p\left(1 + \frac{\Gamma(p+1)^{1/p}}{p}\right) \leq 2\alpha\beta m_p,$$
which proves our claim. $\square$

REMARK 15.4.2. Under the assumptions of Proposition 15.4.1, by the triangle inequality we get (for $\gamma = 4\alpha\beta$) that if $p \geq q \geq 2$ then
(15.4.1)
$$\left(\int_{\mathbb{R}^n} \left|\|x\| - \left(\int_{\mathbb{R}^n}\|y\|^q d\mu(y)\right)^{1/q}\right|^p d\mu(x)\right)^{1/p} \leq \gamma \sup_{\|u\|_* \leq 1}\left(\int_{\mathbb{R}^n}|\langle u,x\rangle|^p d\mu\right)^{1/p}.$$

This leads us to the next definition.

DEFINITION 15.4.3. We say that a probability measure μ on $\mathbb{R}^n$ has *comparable weak and strong moments with constant* γ if (15.4.1) holds for any norm $\|\cdot\|$ on $\mathbb{R}^n$.

The next conjecture is wide open (see the next section for a few partial results).

CONJECTURE 15.4.4 (weak and strong moments). *There exists an absolute constant $\gamma_0 > 0$ such that every even log-concave probability measure on $\mathbb{R}^n$ has comparable weak and strong moments with constant γ_0.*

PROPOSITION 15.4.5. *Let μ be an isotropic probability measure on $\mathbb{R}^n$ which has comparable weak and strong moments with constant γ. Then:*

(i) $\int_{\mathbb{R}^n} |\|x\|_2 - \sqrt{n}|^2 d\mu(x) \leqslant \gamma^2$.

(ii) *If μ is also α-regular then for all $p > 2$,*

$$\left(\int_{\mathbb{R}^n} \|x\|_2^p d\mu\right)^{1/p} \leqslant \sqrt{n} + \frac{\gamma\alpha}{2}p.$$

Proof. Observe that $\int_{\mathbb{R}^n} \|x\|_2^2 d\mu = n$ and $\|u\|_2^* = \|u\|_2$. Hence, (i) follows directly from (15.4.1) with $p = q = 2$. Moreover, (15.4.1) with $q = 2$ implies

$$\left(\int_{\mathbb{R}^n} \|x\|_2^p d\mu\right)^{1/p} \leqslant \sqrt{n} + \gamma \sup_{\|u\|_2 \leqslant 1} \left(\int_{\mathbb{R}^n} |\langle u, x\rangle|^p d\mu\right)^{1/p} \leqslant \sqrt{n} + \frac{\gamma\alpha}{2}p$$

by the α-regularity and isotropicity of μ. □

REMARK 15.4.6. Recall that property (i) plays a crucial role in the proofs of the thin shell estimate. Also, Paouris' inequality states that the moments of the Euclidean norm for even isotropic log-concave measures are bounded by $C(p + \sqrt{n})$. Thus, Conjecture 15.4.4 would imply both the central limit theorem and Paouris' concentration inequality.

15.5. Further reading

15.5.1. Tail estimates for order statistics of isotropic log-concave vectors

Let μ be an isotropic log-concave measure on $\mathbb{R}^n$. The starting point for this section is the question if there exists an analogue of Paouris' large deviation inequality when the Euclidean norm is replaced by the ℓ_r-norm for some $1 \leqslant r \leqslant \infty$. A first observation is that

$$\mu(\{x : \|x\|_r \geqslant tn^{1/r}\}) \leqslant \frac{1}{t^r n} \int_{\mathbb{R}^n} \|x\|_r^r d\mu(x) = \frac{1}{t^r n} \sum_{i=1}^n \int_{\mathbb{R}^n} |x_i|^r d\mu(x)$$

by Markov's inequality, and

$$\left(\int_{\mathbb{R}^n} |x_i|^r d\mu(x)\right)^{1/r} \leqslant c_1 r \left(\int_{\mathbb{R}^n} |x_i|^2 d\mu(x)\right)^{1/2} = c_1 r,$$

because μ is log-concave and isotropic. It follows that

$$\mu(\{x : \|x\|_r \geqslant 2c_1 r n^{1/r}\}) \leqslant \frac{1}{2^r} \leqslant \frac{1}{2}.$$

Next, observe that if $x \in tB_2^n$ then we have $\|x\|_r \leqslant tn^{\frac{1}{r} - \frac{1}{2}}$ if $1 \leqslant r \leqslant 2$ and $\|x\|_r \leqslant t$ if $r \geqslant 2$. Therefore, if we make the additional assumption that μ satisfies exponential concentration with constant α, then we have: in the case $1 \leqslant r \leqslant 2$,

$$\mu(\{x : \|x\|_r \geqslant 2n^{1/r} + \alpha t n^{\frac{1}{r} - \frac{1}{2}}\}) \leqslant e^{-t}$$

for all $t > 0$, while in the case $r \geqslant 2$,

$$\mu(\{x : \|x\|_r \geqslant 2c_1 r n^{1/r} + \alpha t\}) \leqslant e^{-t}$$

for all $t > 0$. It is then natural to expect the following:

Let μ be an isotropic log-concave measure on $\mathbb{R}^n$. If $1 \leqslant r \leqslant 2$ then

$$\mu(\{x : \|x\|_r \geqslant t\}) \leqslant \exp\left(-\frac{1}{c_1} t n^{\frac{1}{2} - \frac{1}{r}}\right) \qquad (15.5.1)$$

for all $t \geqslant c_1 n^{1/r}$, and if $r \geqslant 2$ then

(15.5.2) $$\mu(\{\|x\|_r \geqslant t\}) \leqslant \exp\left(-\frac{1}{c_2}t\right)$$

for all $t \geqslant c_2 r n^{1/r}$.

Note that (15.5.1) is equivalent to

(15.5.3) $$\left(\int_{\mathbb{R}^n} \|x\|_r^p d\mu(x)\right)^{1/p} \leqslant c_3(n^{1/r} + pn^{\frac{1}{r}-\frac{1}{2}})$$

for all $p \geqslant 2$ and $1 \leqslant r \leqslant 2$, while (15.5.2) is equivalent to

(15.5.4) $$\left(\int_{\mathbb{R}^n} \|x\|_r^p d\mu(x)\right)^{1/p} \leqslant c_4(rn^{1/r} + p)$$

for all $p \geqslant 2$ and $2 \leqslant r < \infty$. In fact (15.5.1) and its equivalent form (15.5.3) follow from Paouris' inequality and Hölder's inequality. So, we concentrate on the case $r > 2$.

The approach of Latała in [308] is based on the study of order statistics. Recall that if $x = (x_1, \ldots, x_n) \in \mathbb{R}^n$ then we denote by $x_1^* \geqslant x_2^* \geqslant \cdots \geqslant x_n^*$ the decreasing rearrangement of $|x_1|, |x_2|, \ldots, |x_n|$. An immediate but useful observation is that

(15.5.5) $$\|x\|_r^r = \sum_{i=1}^n |x_i|^r = \sum_{i=1}^n |x_i^*|^r \simeq \sum_{k=0}^{\log_2 n} 2^k |x_{2^k}^*|^r.$$

The starting point is the next observation.

PROPOSITION 15.5.1. *Let μ be an isotropic log-concave measure on $\mathbb{R}^n$ which satisfies exponential concentration with constant α. Then, for any $1 \leqslant k \leqslant n$,*

(15.5.6) $$\mu(\{x : x_k^* \geqslant t\}) \leqslant \exp(-\sqrt{k}t/(3\alpha))$$

for all $t \geqslant 8\alpha \log\left(\frac{en}{k}\right)$.

Proof. We have $\mu(\{x : |x_i| \leqslant 2\}) \geqslant \frac{1}{2}$ and hence $\mu(\{x : |x_i| \geqslant 2+t\}) \leqslant e^{-t/\alpha}$ for all $t > 0$. For every $t > 0$ we define

(15.5.7) $$A(t) := \{x : \mathrm{card}(\{i : |x_i| \geqslant t\}) < k/2\}.$$

Note that if $t \geqslant 4\alpha \log\left(\frac{en}{k}\right)$ then $\mu(A(t)) \geqslant \frac{1}{2}$, because

$$1 - \mu(A(t)) = \mu\left(\sum_{i=1}^n \mathbf{1}_{\{|x_i|\geqslant t\}} \geqslant \frac{k}{2}\right) \leqslant \frac{2}{k}\int_{\mathbb{R}^n} \sum_{i=1}^n \mathbf{1}_{\{|x_i|\geqslant t\}}(x)\, d\mu(x)$$
$$\leqslant \frac{2n}{k} e^{-t/2\alpha} \leqslant \frac{2n}{k}\left(\frac{en}{k}\right)^{-2} \leqslant \frac{1}{2}.$$

We set $D = A(4\alpha \log(en/k))$ and observe that if $z = x + y \in D + \sqrt{k}sB_2^n$ then less than $k/2$ of the $|x_i|$'s are greater than $4\alpha \log(en/k)$ and less than $k/2$ of the $|y_i|$'s are greater than $\sqrt{2}s$, which implies that

(15.5.8) $$\mu\left(\left\{x : x_k^* \geqslant 4\alpha \log\left(\frac{en}{k}\right) + \sqrt{2}s\right\}\right) \leqslant 1 - \mu(D + \sqrt{k}sB_2^n) \leqslant \exp(-\sqrt{k}s/\alpha).$$

The result follows. $\square$

We will show that a similar estimate holds true without appealing to the exponential concentration property (and that α in the above estimates can be replaced by an absolute constant $C > 0$).

THEOREM 15.5.2 (Latała). *Let μ be an isotropic log-concave measure on $\mathbb{R}^n$. For every $k = 1, \ldots, n$ and for every $t \geqslant C \log\left(\frac{en}{k}\right)$ one has*

$$\mu(\{x : x_k^* \geqslant t\}) \leqslant \exp(-\sqrt{k}t/C),$$

where $(x_1^*, \ldots, x_n^*)$ is the decreasing rearrangement of $(|x_1|, \ldots, |x_n|)$.

For every $t > 0$ we consider the function
$$f_t(x) = \sum_{i=1}^{n} \mathbf{1}_{\{x_i \geq t\}}(x) = \text{card}(\{i \leq n : x_i \geq t\}).$$
The theorem will follow by the next proposition.

PROPOSITION 15.5.3. *Let μ be an isotropic log-concave probability measure on $\mathbb{R}^n$. For every $p \geq 1$ and $t \geq C \log\left(\frac{nt^2}{p^2}\right)$ we have*
$$\|f_t\|_{L^p(\mu)} \leq \left(\frac{Cp}{t}\right)^2.$$

We start with the next lemma which makes essential use of Paouris' inequality.

LEMMA 15.5.4. *Let μ be an isotropic log-concave measure on $\mathbb{R}^n$. If A is a convex subset of $\mathbb{R}^n$ with $0 < \mu(A) \leq 1/e$ then*

(15.5.9) $$\sum_{i=1}^{n} \mu(A \cap \{x_i \geq t\}) \leq C_1 \mu(A) \left[\frac{\log^2(\mu(A))}{t^2} + n e^{-t/C_1}\right]$$

for all $t \geq C_1$, where $C_1 > 0$ is an absolute constant. Moreover, if $1 \leq u \leq t/C_2$ then

(15.5.10) $$\text{card}(\{i \leq n : \mu(A \cap \{x_i \geq t\}) \geq e^{-u}\mu(A)\}) \leq \frac{C_2^2 u^2}{t^2} \log^2(\mu(A)),$$

where $C_2 > 0$ is an absolute constant.

Proof. We define a new measure ν on $\mathbb{R}^n$, setting
$$\nu(B) = \frac{\mu(A \cap B)}{\mu(A)}$$
for every Borel subset B of $\mathbb{R}^n$. Since A is convex, we have that ν is a log-concave probability measure but it may no longer be isotropic. Since the statement of the lemma is invariant under a permutation of the coordinates, we may assume that

(15.5.11) $$\int x_1^2 d\nu(x) \geq \int x_2^2 d\nu(x) \geq \cdots \geq \int x_n^2 d\nu(x).$$

Note also that, for every Borel subset B of $\mathbb{R}^n$,

(15.5.12) $$\mu(B) \geq \mu(A \cap B) = \nu(B)\mu(A).$$

Let $\alpha > 0$. We define
$$m = m(\alpha) := \text{card}\left(\left\{i \leq n : \int x_i^2 d\nu(x) \geq \alpha\right\}\right).$$

From (15.5.11) we have

(15.5.13) $$\int x_i^2 d\nu(x) \geq \alpha, \quad i = 1, \ldots, m$$

and hence

(15.5.14) $$\int \left(\sum_{i=1}^{m} x_i^2\right) d\nu(x) \geq \alpha m.$$

By the Paley-Zygmund inequality,

(15.5.15) $$\nu\left(\left\{x : \sum_{i=1}^{m} x_i^2 \geq \frac{1}{2}\int \left(\sum_{i=1}^{m} x_i^2\right) d\nu(x)\right\}\right) \geq \frac{1}{4} \frac{\left(\int \left(\sum_{i=1}^{m} x_i^2\right) d\nu(x)\right)^2}{\int \left(\sum_{i=1}^{m} x_i^2\right)^2 d\nu(x)}.$$

Since $f(x) = \left(\sum_{i=1}^m x_i^2\right)^{1/2}$ is a seminorm, we have $\|f\|_{L^4(\mu)} \leqslant c_1 \|f\|_{L^2(\mu)}$, where $c_1 > 0$ is an absolute constant. From (15.5.14) and (15.5.15) we conclude that

$$\nu\left(\left\{x : \sum_{i=1}^m x_i^2 \geqslant \frac{\alpha m}{2}\right\}\right) \geqslant \nu\left(\left\{x : \sum_{i=1}^m x_i^2 \geqslant \frac{1}{2}\int\left(\sum_{i=1}^m x_i^2\right) d\nu(x)\right\}\right)$$

$$\geqslant \frac{1}{4c_1^4} =: \frac{1}{C}.$$

It follows from (15.5.12) that

$$\mu(A) \leqslant C\mu\left(\left\{x : \sum_{i=1}^m x_i^2 \geqslant \frac{\alpha m}{2}\right\}\right).$$

Applying the deviation estimate of Paouris to the measure $\pi_{\mathbb{R}^m}(\mu)$ we get

$$\mu\left(\left\{x : \sum_{i=1}^m x_i^2 \geqslant \frac{\alpha m}{2}\right\}\right) \leqslant \exp\left(-\frac{1}{C_3}\sqrt{\alpha m}\right),$$

provided that $\alpha \geqslant C_3$, where $C_3 > 0$ is an absolute constant. This implies that

(15.5.16) $\qquad m = m(\alpha) \leqslant \frac{C_3^2}{\alpha} \log^2\left(\frac{C}{\mu(A)}\right) \leqslant \frac{C_4}{\alpha} \log^2(\mu(A)),$

for all $\alpha \geqslant C_3$ (taking into account the fact that $\mu(A) \leqslant 1/e$ and choosing $C_3 > 0$ large enough).

Proof of (15.5.10). By the log-concavity of ν we have

$$\frac{\mu(A \cap \{x_i \geqslant t\})}{\mu(A)} = \nu(\{x_i \geqslant t\}) \leqslant c_2 \exp(-t/\|x_i\|_{L^2(\nu)}).$$

Let $1 \leqslant u \leqslant t/C_2$. Then, assuming that $\mu(A \cap \{x_i \geqslant t\}) \geqslant e^{-u}\mu(A)$, we get

$$c_2 e^u \geqslant \exp(t/\|x_i\|_{L^2(\nu)}),$$

which implies that

$$\int x_i^2 d\nu(x) \geqslant \frac{c_3^2 t^2}{u^2}.$$

From (15.5.16) we conclude that

$$\mathrm{card}\left(\left\{i \leqslant n : \frac{\mu(A \cap \{x_i \geqslant t\})}{\mu(A)} \geqslant e^{-u}\right\}\right) \leqslant \mathrm{card}\left(\left\{i \leqslant n : \int x_i^2 d\nu(x) \geqslant \frac{c_3^2 t^2}{u^2}\right\}\right)$$

$$= m\left(\frac{c_3^2 t^2}{u^2}\right) \leqslant \frac{C_2 u^2}{t^2} \log^2(\mu(A)),$$

if $1 \leqslant u \leqslant t/C_2$, where $C_2 > 0$ is an absolute constant.

Proof of (15.5.9). We assume that $t \geqslant \sqrt{C_3}$ and define an integer $k_0 \geqslant 0$ so that $2^{k_0} \leqslant t/\sqrt{C_3} \leqslant 2^{k_0+1}$. Then, we define

$$I_0 = \left\{i \leqslant n : \int x_i^2 d\nu(x) \geqslant t^2\right\}, \quad I_{k_0+1} = \left\{i \leqslant n : \int x_i^2 d\nu(x) < \frac{t^2}{4^{k_0}}\right\}$$

and

$$I_j = \left\{i \leqslant n : \frac{t^2}{4^j} \leqslant \int x_i^2 d\nu(x) < \frac{t^2}{4^{j-1}}\right\}, \quad j = 1, 2, \ldots, k_0.$$

From (15.5.16) we see that, for every $j = 0, 1, \ldots, k_0$ we have

$$\mathrm{card}(I_j) \leqslant \frac{C_4 4^j}{t^2} \log^2(\mu(A)).$$

For $j = k_0 + 1$ we will use the trivial bound $\mathrm{card}(I_{k_0+1}) \leqslant n$. Observe that, for every $i \in I_j$, $j = 1, \ldots, k_0 + 1$, we have

$$\nu(\{x_i \geqslant t\}) \leqslant \nu(\{x_i \geqslant 2^{j-1}\|x_i\|_{L^2(\nu)}\}) \leqslant c_4 e^{-2^j/C}.$$

It follows that

$$\sum_{i=1}^n \nu(\{x_i \geqslant t\}) = \sum_{j=0}^{k_0+1} \sum_{i \in I_j} \nu(\{x_i \geqslant t\})$$

$$\leqslant \operatorname{card}(I_0) + c_4 \sum_{j=1}^{k_0} \operatorname{card}(I_j) e^{-2^j/C} + c_4 \operatorname{card}(I_{k_0+1}) e^{-2^{k_0+1}/C}$$

$$\leqslant C_4 \frac{\log^2(\mu(A))}{t^2} + \frac{C_4 \log^2(\mu(A))}{t^2} \sum_{j=1}^{k_0} c_4 4^j e^{-2^j/C} + c_4 n e^{-t/(\sqrt{C_3}C)}$$

$$\leqslant C_5 \left[\frac{\log^2(\mu(A))}{t^2} + n e^{-t/C_5} \right],$$

because $\sum_{j=1}^{k_0} 4^j e^{-2^j/C} \leqslant c_5$. By the definition of ν we get

$$\sum_{i=1}^n \mu(A \cap \{x_i \geqslant t\}) \leqslant \mu(A) \sum_{i=1}^n \nu(\{x_i \geqslant t\}) \leqslant C_5 \mu(A) \left[\frac{\log^2(\mu(A))}{t^2} + n e^{-t/C_5} \right],$$

which is (15.5.9). $\square$

Proof of Proposition 15.5.3. Since $f_t(x) \leqslant n$ for all x, the statement is trivially true if $t\sqrt{n} \leqslant Cp$. So, we may assume that $t\sqrt{n} \geqslant 10p$. Increasing the constants C_1 and C_2 of Lemma 15.5.4 we may assume that

(15.5.17) $$\mu(\{\langle x, \theta\rangle \geqslant t\}) \leqslant e^{-t/C_m}$$

for all $\theta \in S^{n-1}$ and for all $t \geqslant C_m$, $m = 1, 2$.

We fix $p \geqslant 1$ and $t \geqslant C \log\left(\frac{nt^2}{p^2}\right)$. If $C > 0$ is chosen large enough, we may assume that $t \geqslant \max\{C_1, 4C_2\}$ and

$$t^2 n e^{-t/C_1} \leqslant p^2.$$

We choose $\ell = 2^k$ such that $p \leqslant \ell \leqslant 2p$. We will prove that

$$\|f_t\|_{L^\ell(\mu)} \leqslant \left(\frac{C\ell}{t}\right)^2,$$

and the result will follow from Hölder's inequality and the fact that $\ell \leqslant 2p$. Our assumption on p now takes the form $t\sqrt{n} \geqslant 5\ell$.

We set $B_\emptyset = \mathbb{R}^n$ and, for any $1 \leqslant i_1, \ldots, i_s \leqslant n$,

$$B_{i_1,\ldots,i_s} = \{x : x_{i_1} \geqslant t, \ldots, x_{i_s} \geqslant t\}.$$

We write

$$m(\ell) = \|f_t\|_{L^\ell(\mu)}^\ell = \int \left(\sum_{i=1}^n \mathbf{1}_{\{x_i \geqslant t\}}(x)\right)^\ell = \sum_{i_1,\ldots,i_\ell=1}^n \mu(B_{i_1,\ldots,i_\ell}).$$

We need to show that

$$m(\ell) \leqslant \left(\frac{C\ell}{t}\right)^{2\ell}.$$

We will break the sum $m(\ell)$ into several parts. We first define an integer $j_1 \geqslant 2$ so that

$$2^{j_1-2} < \log\left(\frac{nt^2}{\ell^2}\right) \leqslant 2^{j_1-1}.$$

Then, we set
$$I_0 = \{(i_1, \ldots, i_\ell) : \mu(B_{i_1,\ldots,i_\ell}) > e^{-\ell}\}$$
$$I_j = \{(i_1, \ldots, i_\ell) : e^{-2^j \ell} < \mu(B_{i_1,\ldots,i_\ell}) \leqslant e^{-2^{j-1}\ell}\}, \qquad 0 < j < j_1$$
$$I_{j_1} = \{(i_1, \ldots, i_\ell) : \mu(B_{i_1,\ldots,i_\ell}) \leqslant e^{-2^{j_1-1}\ell}\}.$$

Finally, we write
$$m(\ell) = \sum_{j=0}^{j_1} m_j(\ell),$$
where
$$m_j(\ell) = \sum_{(i_1,\ldots,i_\ell) \in I_j} \mu(B_{i_1,\ldots,i_\ell}).$$

Bound for $m_{j_1}(\ell)$. We simply observe that
$$e^{2^{j_1-1}\ell} \geqslant \left(\frac{nt^2}{\ell^2}\right)^\ell$$
by the definition of j_1, and hence
$$m_{j_1}(\ell) \leqslant \operatorname{card}(I_{j_1}) e^{-2^{j_1-1}\ell} \leqslant n^\ell \left(\frac{\ell^2}{nt^2}\right)^\ell = \left(\frac{\ell}{t}\right)^{2\ell}.$$

Bound for $m_0(\ell)$. For every $1 \leqslant s \leqslant \ell$ and $I \subset \{1, \ldots, n\}^\ell$ we set
$$P_s(I) = \{(i_1, \ldots, i_s) : (i_1, \ldots, i_\ell) \in I \text{ for some } i_{s+1}, \ldots, i_\ell\}.$$
Using Lemma 15.5.4 we see that
$$\sum_{P_{s+1}(I_0)} \mu(B_{i_1,\ldots,i_{s+1}}) \leqslant \sum_{P_s(I_0)} \sum_{i_{s+1}=1}^n \mu(B_{i_1,\ldots,i_s} \cap \{x_{i_{s+1}} \geqslant t\})$$
$$\leqslant C_1 \sum_{P_s(I_0)} \mu(B_{i_1,\ldots,i_s}) \left[\frac{\log^2(\mu(B_{i_1,\ldots,i_s}))}{t^2} + ne^{-t/C_1}\right].$$

Now, we use the fact that if $(i_1, \ldots, i_s) \in P_s(I_0)$ then, for some $(i_1, \ldots, i_\ell) \in I_0$ we have $\mu(B_{i_1,\ldots,i_s}) \geqslant \mu(B_{i_1,\ldots,i_\ell}) \geqslant e^{-\ell}$, which implies that $\frac{\log^2(\mu(B_{i_1,\ldots,i_s}))}{t^2} \leqslant \frac{\ell^2}{t^2}$. We also have $nt^2 e^{-t/C_1} \leqslant p^2 \leqslant 4\ell^2$, and hence
$$ne^{-t/C_1} \leqslant \frac{4\ell^2}{t^2}.$$
Using these estimates, we finally have
$$\sum_{P_{s+1}(I_0)} \mu(B_{i_1,\ldots,i_{s+1}}) \leqslant \frac{5C_1 \ell^2}{t^2} \sum_{P_s(I_0)} \mu(B_{i_1,\ldots,i_{s+1}}).$$
By induction, we get
$$m_0(\ell) = \sum_{(i_1,\ldots,i_\ell) \in I_0} \mu(B_{i_1,\ldots,i_\ell}) \leqslant \left(\frac{5C_1 \ell^2}{t^2}\right)^{\ell-1} \sum_{i_1 \in P_1(I_0)} \mu(B_{i_1})$$
$$\leqslant \left(\frac{5C_1 \ell^2}{t^2}\right)^{\ell-1} ne^{-t/C_1} \leqslant \left(\frac{C\ell}{t}\right)^{2\ell}.$$

Bound for $m_j(\ell)$, $0 < j < j_1$. We first prove a combinatorial lemma.

LEMMA 15.5.5. *Let $\ell_0 \geqslant \ell_1 \geqslant \cdots \geqslant \ell_s$ be a sequence of positive integers. Consider the family*
$$\mathcal{F} = \{f : \{1, \ldots, \ell_0\} \to \{0, 1, \ldots, s\} : \operatorname{card}(\{r : f(r) \geqslant i\}) \leqslant \ell_i \text{ for all } 1 \leqslant i \leqslant s\}.$$
Then,
$$\operatorname{card}(\mathcal{F}) \leqslant \prod_{i=1}^{s} \left(\frac{e\ell_{i-1}}{\ell_i}\right)^{\ell_i}.$$

Proof. Every function $f : \{1, \ldots, \ell_0\} \to \{0, 1, \ldots, s\}$ is completely determined by the sets $A_i = \{r : f(r) \geqslant i\}$, $i = 1, \ldots, s$, and $A_0 = \{1, \ldots, \ell_0\}$. Suppose that the set A_{i-1} is already chosen and $\operatorname{card}(A_{i-1}) = a_{i-1} \leqslant \ell_{i-1}$. Then, we can choose a set $A_i \subseteq A_{i-1}$ of cardinality $\leqslant \ell_i$ in
$$\sum_{j=0}^{\ell_i} \binom{a_{i-1}}{j} \leqslant \sum_{j=0}^{\ell_i} \binom{\ell_{i-1}}{j} \leqslant \left(\frac{e\ell_{i-1}}{\ell_i}\right)^{\ell_i}$$
ways. By induction, we get the result. $\square$

We fix $0 < j < j_1$ and define r_1 so that
$$2^{r_1} < \frac{t}{C_2} \leqslant 2^{r_1+1}.$$

Note that $r_1 > j_1$ provided that we choose C large enough with respect to C_2, e.g. $C \geqslant 8C_2$. For every $(i_1, \ldots, i_\ell) \in I_j$ we define a function $f_{i_1, \ldots, i_\ell} : \{1, \ldots, \ell\} \to \{j, j+1, \ldots, r_1\}$ setting:

(i) $f_{i_1, \ldots, i_\ell}(s) = j$ if $\frac{\mu(B_{i_1, \ldots, i_s})}{\mu(B_{i_1, \ldots, i_{s-1}})} \geqslant \exp(-2^{j+1})$.

(ii) $f_{i_1, \ldots, i_\ell}(s) = r$ if $\exp(-2^{r+1}) \leqslant \frac{\mu(B_{i_1, \ldots, i_s})}{\mu(B_{i_1, \ldots, i_{s-1}})} < \exp(-2^r)$, $j < r < r_1$.

(iii) $f_{i_1, \ldots, i_\ell}(s) = r_1$ if $\frac{\mu(B_{i_1, \ldots, i_s})}{\mu(B_{i_1, \ldots, i_{s-1}})} < \exp(-2^{r_1})$.

We first observe that, by the definition of r_1, for every i_1 we have
$$\mu(\{x_{i_1} \geqslant t\}) \leqslant e^{-t/C_2} \leqslant \mu(B_\emptyset) \exp(-2^{r_1}),$$
which implies that
$$f_{i_1, \ldots, i_\ell}(1) = r_1$$
for all $(i_1, \ldots, i_\ell)$.

We define
$$\mathcal{F}_j = \{f_{i_1, \ldots, i_\ell} : (i_1, \ldots, i_\ell) \in I_j\}.$$

Claim. We have
$$\operatorname{card}(\mathcal{F}_j) \leqslant e^{2\ell}.$$

Proof of the Claim. If $f = f_{i_1, \ldots, i_\ell} \in \mathcal{F}_j$ and if $r > j$, then
$$\exp(-2^j \ell) < \mu(B_{i_1, \ldots, i_\ell}) < \exp(-2^r \operatorname{card}(\{s : f(s) \geqslant r\})),$$
which implies that
(15.5.18) $$\operatorname{card}(\{s : f(s) \geqslant r\}) \leqslant 2^{j-r} \ell =: \ell_r.$$
The last inequality holds for $r = j$ as well. Since $\ell_{r-1}/\ell_r = 2$ and $\sum_{r=j+1}^{r_1} \ell_r \leqslant \ell$, Lemma 15.5.5 shows that
$$\operatorname{card}(\mathcal{F}_j) \leqslant \prod_{r=j+1}^{r_1} \left(\frac{e\ell_{r-1}}{\ell_r}\right)^{\ell_r} \leqslant e^{2\ell}.$$

Claim. If $f \in \mathcal{F}_j$ and $I_j(f) = \{(i_1, \ldots, i_\ell) \in I_j : f_{i_1, \ldots, i_\ell} = f\}$, then
$$\operatorname{card}(I_j(f)) \leqslant \left(\frac{C\ell^2}{t^2}\right)^\ell \exp\left(\frac{3}{8} 2^j \ell\right).$$

Proof of the Claim. We define
$$n_r := \operatorname{card}(\{1 \leqslant s \leqslant \ell : f(s) = r\}), \qquad r = j, j+1, \ldots, r_1.$$
Clearly,
$$n_j + n_{j+1} + \cdots + n_{r_1} = \ell.$$
Note that, if $i_1, \ldots, i_{s-1}$ are fixed and if $f(s) = r < r_1$ then $s \geqslant 2$, hence $\mu(B_{i_1,\ldots,i_{s-1}}) \leqslant \mu(x_{i_1} \geqslant t) \leqslant 1/e$, and so applying Lemma 15.5.4 with $u = 2^{r+1} \leqslant t/C_2$, and also taking into account the definition of I_j (which provides a lower bound for $\mu(B_{i_1,\ldots,i_{s-1}})$), we see that i_s can take at most
$$\frac{4C_2 2^{2r}}{t^2} \log^2(\mu(B_{i_1,\ldots,i_{s-1}})) \leqslant \frac{4C_2 2^{2(r+j)} \ell^2}{t^2} \leqslant \frac{4C_2 \ell^2}{t^2} e^{2(r+j)} =: m_r$$
values. This shows that
$$\operatorname{card}(I_j(f)) \leqslant n^{n_{r_1}} \prod_{r=j}^{r_1-1} m_r^{n_r} = n^{n_{r_1}} \left(\frac{4C_2 \ell^2}{t^2}\right)^{\ell - n_{r_1}} \exp\left(\sum_{r=j}^{r_1-1} 2(r+j) n_r\right).$$
From (15.5.18) we have
$$n_r \leqslant \ell_r = 2^{j-r} \ell,$$
and hence
$$\sum_{r=j}^{r_1-1} 2(r+j) n_r \leqslant 2^{j+2} \ell \sum_{r=j}^{\infty} r 2^{-r} \leqslant (C + 2^{j-2}) \ell.$$
Since $t \geqslant C \log\left(\frac{nt^2}{4\ell^2}\right)$, if we choose C large enough we have
$$n_{r_1} \leqslant 2^{j-r_1} \ell \leqslant \frac{2C_2}{t} 2^j \ell \leqslant \frac{2^{j-3} \ell}{\log(nt^2/(4\ell^2))}.$$
Therefore,
$$\operatorname{card}(I_j(f)) \leqslant \left(\frac{C\ell^2}{t^2}\right)^{\ell} \left(\frac{nt^2}{4\ell^2}\right)^{n_{r_1}} e^{2^{j-2}\ell} \leqslant \left(\frac{C\ell^2}{t^2}\right)^{\ell} \exp\left(\frac{3}{8} 2^j \ell\right),$$
and the claim is proved. □

Combining the two Claims we get
$$\operatorname{card}(I_j) \leqslant \operatorname{card}(\mathcal{F}_j) \left(\frac{C\ell^2}{t^2}\right)^{\ell} \exp\left(\frac{3}{8} 2^j \ell\right) \leqslant \left(\frac{C\ell^2}{t^2}\right)^{\ell} \exp\left(\left(2 + \frac{3}{8} 2^j\right) \ell\right),$$
and hence
$$m_j(\ell) \leqslant \operatorname{card}(I_j) \exp(-2^{j-1} \ell) \leqslant \left(\frac{C\ell^2}{t^2}\right)^{\ell} \exp(-2^{j-3} \ell).$$
Adding all our estimates, we get
$$m(\ell) = m_0(\ell) + m_{j_1}(\ell) + \sum_{j=1}^{j_1-1} m_j(\ell) \leqslant \left(\frac{\ell}{t}\right)^{2\ell} \left(C^{\ell} + 1 + \sum_{j=1}^{\infty} C^{\ell} e^{-2^{j-3}\ell}\right),$$
that is,
$$m(\ell) \leqslant \left(\frac{C\ell}{t}\right)^{2\ell}.$$
This proves Proposition 15.5.3. □

Proof of Theorem 15.5.2. Let $t > 0$. Observe that
$$\{x : x_k^* \geqslant t\} \subseteq \{x : f_t(x) \geqslant k/2\} \cup \{x : f_t(-x) \geqslant k/2\}.$$
It follows that
$$\mu(\{x : x_k^* \geqslant t\}) \leqslant \left(\frac{2}{k}\right)^p \left(\|f_t\|_{L^p(\mu)}^p + \|f_t\|_{L^p(-\mu)}^p\right) \leqslant 2 \left(\frac{C\sqrt{2}p}{t\sqrt{k}}\right)^{2p}$$

for all $p \geq 1$ which satisfy $t \geq C \log\left(\frac{nt^2}{p^2}\right)$. Then, we may choose an absolute constant $C_1 > 0$ so that if $t \geq C_1 \log\left(\frac{en}{k}\right)$ we are allowed to choose $p = \frac{1}{eC\sqrt{2}} t\sqrt{k}$, and the result follows. □

We are now ready to prove versions of (15.5.2) and (15.5.4) with constants depending on r.

THEOREM 15.5.6. *Let μ be an isotropic log-concave measure on $\mathbb{R}^n$. For every $r > 2$ we have*

(15.5.19) $$\mu(\{x : \|x\|_r \geq t\}) \leq \exp\left(-\frac{1}{C}\left(\frac{r-2}{r}\right)^{1/r} t\right)$$

for all $t \geq C\left(rn^{1/r} + \left(\frac{r}{r-2}\right)^{1/r} \log n\right)$. Equivalently,

(15.5.20) $$\left(\int \|x\|_r^p d\mu(x)\right)^{1/p} \leq C\left(rn^{1/r} + \left(\frac{r}{r-2}\right)^{1/r} (\log n + p)\right)$$

for all $p \geq 2$.

Proof. We set $s = \lfloor \log_2 n \rfloor$. Then,

$$\|x\|_r^r = \sum_{i=1}^n |x_i^*|^r \leq \sum_{k=0}^s 2^k |x_{2^k}^*|^r.$$

From Theorem 15.5.2 we have

(15.5.21) $$\mu\left(\left\{x : |x_k^*|^r \geq C_3^r \log^r\left(\frac{en}{k}\right) + t^r\right\}\right) \leq \exp\left(-\frac{1}{C}\sqrt{k}t\right)$$

for all $t > 0$. Since

$$\sum_{k=0}^s 2^k \log^r(en2^{-k}) \leq Cn \sum_{j=1}^\infty j^r 2^{-j} \leq (Cr)^r n,$$

for all $t_1, \ldots, t_k \geq 0$ we get

$$\mu\left(\left\{x : \|x\|_r \geq C\left(rn^{1/r} + \left(\sum_{k=0}^s t_k\right)^{1/r}\right)\right\}\right) \leq \mu\left(\sum_{k=0}^s y_k \geq \sum_{k=0}^s t_k\right),$$

where

$$y_k := 2^k(|x_{2^k}^*|^r - C_3^r \log^r(en2^{-k})).$$

Using (15.5.21) we write

$$\mu\left(\left\{x : \|x\|_r \geq C\left(rn^{1/r} + \left(\sum_{k=0}^s t_k\right)^{1/r}\right)\right\}\right) \leq \sum_{k=0}^s \mu(y_k \geq t_k)$$
$$\leq \sum_{k=0}^s \exp\left(-\frac{1}{C} 2^{\frac{k}{2} - \frac{k}{r}} t_k^{1/r}\right).$$

Given $t > 0$ we now choose t_k through the equation $t = 2^{\frac{k}{2} - \frac{k}{r}} t_k^{1/r}$. Then,

$$\sum_{k=0}^s t_k = t^r \sum_{k=0}^s 2^{\frac{k(2-r)}{2}} \leq t^r (1 - 2^{\frac{2-r}{2}})^{-1} \leq Ct^r \frac{r}{r-2}.$$

This implies that

$$\mu\left(\left\{x : \|x\|_r \geq C\left(rn^{1/r} + t\left(\frac{r}{r-2}\right)^{1/r}\right)\right\}\right) \leq (\log_2 n + 1)\exp(-t/C),$$

and the theorem follows. □

Note. If $r \geq 2 + \delta$ for some $\delta > 0$, Theorem 15.5.6 shows that

$$\mu(\{x : \|x\|_r \geq t\}) \leq \exp\left(-\frac{1}{C_1(\delta)}t\right)$$

for all $t \geq C_1(\delta)rn^{1/r}$, and

$$\left(\int \|x\|_r^p d\mu(x)\right)^{1/p} \leq C_2(\delta)(rn^{1/r} + p)$$

for all $p \geq 2$, where $C_1(\delta), C_2(\delta) \leq C(1 + \delta^{-1/2})$.

15.5.2. Comparison of moments: the unconditional case

In this section we discuss the problem of comparison of moments in the unconditional case. We consider an unconditional isotropic log-concave measure μ on $\mathbb{R}^n$ and we ask if, for any norm $\|\cdot\|$ on $\mathbb{R}^n$ and for all $p \geq 1$,

$$\left(\int \|x\|^p d\mu(x)\right)^{1/p} \leq C_1 \int \|x\| d\mu(x) + C_2 \sup_{\|y\|_* \leq 1} \left(\int |\langle y, x \rangle|^p d\mu(x)\right)^{1/p},$$

where $C_1, C_2 > 0$ are absolute constants. We will describe the next (almost optimal) result of Latała from [**309**].

THEOREM 15.5.7 (Latała). *Let μ be an unconditional log-concave probability measure on $\mathbb{R}^n$. For any norm $\|\cdot\|$ on $\mathbb{R}^n$ and for all $p \geq 1$,*

$$\left(\int \|x\|^p d\mu(x)\right)^{1/p} \leq C_1 \int \|x\| d\nu_n(x) + C_2 \sup_{\|y\|_* \leq 1} \left(\int |\langle y, x \rangle|^p d\mu(x)\right)^{1/p},$$

where ν is the product exponential measure with density $d\nu_n(x) = \frac{1}{2^n}e^{-\|x\|_1}dx$ and $C_1, C_2 > 0$ are absolute constants.

The proof is based on a result of Talagrand which characterizes the boundedness of empirical processes. Given a probability measure μ on $\mathbb{R}^n$ with finite p-moments for all $p > 0$, for any non-empty subset T of $\mathbb{R}^n$ we define

$$\gamma_\mu(T) = \inf \sup_{t \in T} \sum_{n=0}^{\infty} \|\langle \pi_{n+1}(t) - \pi_n(t), \cdot \rangle\|_{L^{2^n}(\mu)},$$

where the infimum is over all families of subsets $T_n \subset T$ and functions $\pi_n : T \to T_n$, $n \geq 0$, which satisfy:

(i) $\text{card}(T_0) = 1$, $\text{card}(T_n) \leq 2^{2^n}$ for all $n \geq 1$, and
(ii) $\lim_{n \to \infty} \pi_n(t) = t$ for all $t \in T$.

Talagrand proved in [**490**] that if ν_n is the product exponential measure on $\mathbb{R}^n$ then

(15.5.22) $$\gamma_{\nu_n}(A) \leq C \int \sup_{y \in A} \langle y, x \rangle d\nu_n(x)$$

for every symmetric $A \subseteq \mathbb{R}^n$.

Proof of Theorem 15.5.7. Let $A := K^\circ = \{x \in \mathbb{R}^n : \|x\|_* \leq 1\}$ be the polar body of the unit ball of $\|\cdot\|$. Then, $\|x\| = \max_{y \in A}\langle y, x \rangle$ for all $x \in \mathbb{R}^n$. Fix a family of subsets $A_n \subset A$ and functions $\pi_n : A \to A_n$, $n \geq 0$, that satisfy (i) and (ii).

Let $p \geq 1$. We choose $n_0 \geq 1$ such that $2^{n_0-1} < 2p \leq 2^{n_0}$. We have

$$\|x\| = \max_{y \in A}\langle y, x \rangle \leq \max_{y \in A}|\langle \pi_{n_0}(y), x \rangle| + \max_{y \in A}\sum_{n=n_0}^{\infty} |\langle \pi_{n+1}(y) - \pi_n(y), x \rangle|.$$

Then,

$$\left(\int \|x\|^p d\mu(x)\right)^{1/p} = \left(\int \max_{y\in A} |\langle y, x\rangle|^p d\mu(x)\right)^{1/p}$$

$$\leqslant \left(\int \max_{y\in A} |\langle \pi_{n_0}(y), x\rangle|^p d\mu(x)\right)^{1/p}$$

(15.5.23)
$$+ \left(\int \max_{y\in A} \left(\sum_{n=n_0}^{\infty} |\langle \pi_{n+1}(y) - \pi_n(y), x\rangle|\right)^p d\mu(x)\right)^{1/p}.$$

For the first term we write

(15.5.24) $\left(\int \max_{y\in A} |\langle \pi_{n_0}(y), x\rangle|^p d\mu(x)\right)^{1/p} \leqslant \left(\int \sum_{y\in A_{n_0}} |\langle y, x\rangle|^p d\mu(x)\right)^{1/p}$

$$\leqslant |A_{n_0}|^{1/p} \max_{z\in A_{n_0}} \left(\int |\langle z, x\rangle|^p d\mu(x)\right)^{1/p}$$

$$\leqslant 16 \max_{z\in A_{n_0}} \left(\int |\langle z, x\rangle|^p d\mu(x)\right)^{1/p}$$

$$\leqslant 16 \max_{\|z\|_*\leqslant 1} \left(\int |\langle z, x\rangle|^p d\mu(x)\right)^{1/p}.$$

Let

$$R(t) = \Big\{x : \max_{y\in A} \sum_{n=n_0}^{\infty} |\langle \pi_{n+1}(y) - \pi_n(y), x\rangle|$$

$$\geqslant t \max_{y\in A} \sum_{n=n_0}^{\infty} \|\langle \pi_{n+1}(y) - \pi_n(y), \cdot\rangle\|_{L^{2^n}(\mu)} \Big\}.$$

From Markov's inequality, for all $t \geqslant 16$ we have

$$\mu(R(t)) \leqslant \mu\left(\bigcup_{n=n_0}^{\infty} \bigcup_{y\in A} \{x : |\langle \pi_{n+1}(y) - \pi_n(y), x\rangle| \geqslant t \|\langle \pi_{n+1}(y) - \pi_n(y), \cdot\rangle\|_{L^{2^n}(\mu)}\}\right)$$

$$\leqslant \sum_{n=n_0}^{\infty} \sum_{z\in A_n - A_{n+1}} \mu\left(\{x : |\langle z, x\rangle| \geqslant t \|\langle z, \cdot\rangle\|_{L^{2^n}(\mu)}\}\right)$$

$$\leqslant \sum_{n=n_0}^{\infty} |A_n|\cdot|A_{n+1}| t^{-2^n}$$

$$\leqslant \sum_{n=n_0}^{\infty} \left(\frac{8}{t}\right)^{2^n} \leqslant \left(\frac{8}{t}\right)^{2^{n_0}} + \sum_{k=2^{n_0}+1}^{\infty} \left(\frac{8}{t}\right)^k \leqslant 2\left(\frac{8}{t}\right)^{2^{n_0}} \leqslant 2\left(\frac{8}{t}\right)^{2p}.$$

Then,

$$\left(\int \max_{y\in A} \left(\sum_{n=n_0}^{\infty} |\langle \pi_{n+1}(y) - \pi_n(y), x\rangle|\right)^p d\mu(x)\right)^{1/p}$$

$$\leqslant \max_{y\in A} \sum_{n=n_0}^{\infty} \|\langle \pi_{n+1}(y) - \pi_n(y), \cdot\rangle\|_{L^{2^n}(\mu)} \left[16 + \left(2p\int_0^{\infty} t^{p-1}\left(\frac{8}{16+t}\right)^{2p}\right)^{1/p}\right]$$

$$\leqslant 32 \max_{y\in A} \sum_{n=n_0}^{\infty} \|\langle \pi_{n+1}(y) - \pi_n(y), \cdot\rangle\|_{L^{2^n}(\mu)}.$$

Since μ is unconditional and isotropic, the comparison theorem of Bobkov and Nazarov (see Chapter 8) shows that

$$\|\langle \pi_{n+1}(y) - \pi_n(y), \cdot \rangle\|_{L^{2^n}(\mu)} \leqslant C \|\langle \pi_{n+1}(y) - \pi_n(y), \cdot \rangle\|_{L^{2^n}(\nu_n)}$$

for every $n \geqslant n_0$. It follows from (15.5.22) that

$$\max_{y \in A} \sum_{n=n_0}^{\infty} \|\langle \pi_{n+1}(y) - \pi_n(y), \cdot \rangle\|_{L^{2^n}(\mu)} \leqslant C \int \|x\| d\nu_n(x).$$

Combining this with (15.5.23) and (15.5.24) we get the result. □

REMARK 15.5.8. Observe that for every $y \in \mathbb{R}^n$ we have $\|\langle y, \cdot \rangle\|_{L^q(\mu)} \leqslant C \|\langle y, \cdot \rangle\|_{L^q(\nu_n)}$. Since

$$\left(\int \|x\|^p d\nu(x) \right)^{1/p} \simeq \int \|x\| d\nu(x) + \sup_{\|y\|_* \leqslant 1} \left(\int |\langle y, x \rangle|^p d\nu_n(x) \right)^{1/p},$$

it follows that

$$\left(\int \|x\|^p d\mu(x) \right)^{1/p} \leqslant C \left(\int \|x\|^p d\nu_n(x) \right)^{1/p}.$$

REMARK 15.5.9. Observe that for every norm $\|\cdot\|$ on $\mathbb{R}^n$ we have

$$\int \|x\| d\nu_n(x) = \int_{E_2^n} \int \left\| \sum_{i=1}^n \varepsilon_i |x_i| e_i \right\| d\nu_n(x) d\varepsilon$$

$$\leqslant \int \max_{1 \leqslant i \leqslant n} |x_i| d\nu_n(x) \cdot \int_{E_2^n} \left\| \sum_{i=1}^n \varepsilon_i e_i \right\| d\varepsilon \leqslant C \log n \int_{E_2^n} \left\| \sum_{i=1}^n \varepsilon_i e_i \right\| d\varepsilon.$$

Since μ is unconditional, from Jensen's inequality we also have

$$\int \|x\| d\mu(x) = \int_{E_2^n} \int \left\| \sum_{i=1}^n \varepsilon_i |x_i| e_i \right\| d\mu(x) d\varepsilon$$

$$\geqslant \int_{E_2^n} \left\| \sum_{i=1}^n (\varepsilon_i \mathbb{E}_\mu(|x_i|)) e_i \right\| d\varepsilon$$

$$\geqslant c \int_{E_2^n} \left\| \sum_{i=1}^n \varepsilon_i e_i \right\| d\varepsilon.$$

Combining the above we get the following:

THEOREM 15.5.10. *Let $\|\cdot\|$ be a norm on $\mathbb{R}^n$. Then, for every unconditional and isotropic log-concave probability measure μ on $\mathbb{R}^n$ and every $p \geqslant 1$, we have*

$$\left(\int \|x\|^p d\mu(x) \right)^{1/p} \leqslant C \left[(\log n) \int \|x\| d\mu(x) + \sup_{\|y\|_* \leqslant 1} \left(\int |\langle y, x \rangle|^p d\mu(x) \right)^{1/p} \right].$$

15.6. Notes and references

Infimum convolution and concentration inequalities

Property (τ) was introduced by Maurey in [**354**], who emphasized its relation with concentration properties of product measures and gave a short proof for Talagrand's concentration inequality (see [**489**]) for the product exponential measure. We follow Maurey's proof of Theorem 15.1.7 in Section 15.1.1. The applications of property (τ) to concentration in Gauss space come from the same paper. See also the articles of Gozlan [**224**] and Lehec [**319**]..

Infimum convolution conjecture

In Sections 15.2, 15.3 and 15.4 we follow the presentation of Latała and Wojtaszczyk from [**311**]. The conjectures that are stated in this chapter are wide open: they have been verified only in some special cases, i.e. for product symmetric log-concave measures, uniform distributions on ℓ_p^n-balls, rotationally invariant log-concave measures. It is an open problem whether these properties hold (even up to logarithmic factors) in the unconditional case.

Tail estimates for order statistics of isotropic log-concave vectors

The results of Section 15.5.1 are due to Latała [**308**]. They form the basis of subsequent work of Adamczak, Latała, Litvak, Pajor and Tomczak-Jaegermann [**2**] who established tail estimates for the Euclidean norms of projections of sums of independent log-concave random vectors and uniform versions of these in the form of tail estimates for operator norms of matrices and their sub-matrices in the setting of a log-concave ensemble, leading to uniform bounds for the operator norm of sub-matrices (of a given size) of matrices with independent isotropic log-concave random rows.

Comparison of moments: the unconditional case

Theorem 15.5.7 is due to Latała [**309**]. He also shows that if $2 \leqslant q < \infty$ and if $F = (\mathbb{R}^n, \|\cdot\|)$ has cotype-q constant bounded by β then for any unconditional log-concave probability measure μ on $\mathbb{R}^n$ and for all $p \geqslant 1$ we have

$$\left(\int \|x\|^p d\mu(x)\right)^{1/p} \leqslant C(q,\beta)\left(\int \|x\| d\mu(x) + \sup_{\|y\|_* \leqslant 1}\left(\int |\langle y,x\rangle|^p d\mu(x)\right)^{1/p}\right),$$

where $C(q,\beta) > 0$ is a constant depending only on q and β.

CHAPTER 16

Information theory and the hyperplane conjecture

Let $(\Omega, \mathcal{A}, P)$ be a probability space. The *Shannon entropy* of a non-negative measurable function $f : \Omega \to \mathbb{R}$, with the property that $\int f \log(1+f) < +\infty$, is defined as $\mathrm{Ent}(f) = -\int_\Omega f \log f + \int_\Omega f \, \log \int_\Omega f$. In the case of an isotropic random vector X with density f, the entropy of X takes the form

$$\mathrm{Ent}(X) = -\int_{\mathbb{R}^n} f \log f.$$

Recall also that a random vector X in $\mathbb{R}^n$ with density f is said to satisfy the Poincaré inequality with constant κ if, for every smooth function h with $\int_{\mathbb{R}^n} fh = 0$, one has

$$\kappa \int_{\mathbb{R}^n} fh^2 \leqslant \int_{\mathbb{R}^n} f \|\nabla h\|_2^2.$$

It is known that among isotropic random vectors, the standard Gaussian random vector G has the largest entropy. If X is an isotropic random vector in $\mathbb{R}^n$, then the gap $\mathrm{Ent}(G) - \mathrm{Ent}(X)$ may be viewed as a measure of the "distance" between X and G. An inequality of Pinsker [**425**], Csiszár and Kullback (see [**425**], [**145**] and [**301**]) states that if f and g are the densities of X and G then

$$\left(\int_{\mathbb{R}^n} |f - g|\right)^2 \leqslant 2(\mathrm{Ent}(G) - \mathrm{Ent}(X)).$$

If the X_i are independent copies of a random vector X, then the normalized sums

$$Y_n = \frac{X_1 + \cdots + X_n}{\sqrt{n}}$$

approach the standard Gaussian distribution as $n \to \infty$. The Shannon-Stam inequality (see [**464**], [**465**] and [**473**]) asserts that $\mathrm{Ent}(\frac{X+Y}{\sqrt{2}}) \geqslant \mathrm{Ent}(X)$, that is, the entropy of the normalized sum of two independent copies of a random vector is larger than that of the original.

Ball observed that comparing the entropy gap $\mathrm{Ent}\left(\frac{X+Y}{\sqrt{2}}\right) - \mathrm{Ent}(X)$ with $\mathrm{Ent}(G) - \mathrm{Ent}(X)$ provides a link between the KLS-conjecture and the hyperplane conjecture. As we will see in Section 16.1, if X is an isotropic log-concave random vector in $\mathbb{R}^n$ and if

$$\mathrm{Ent}\left(\frac{X+Y}{\sqrt{2}}\right) - \mathrm{Ent}(X) \geqslant \delta(\mathrm{Ent}(G) - \mathrm{Ent}(X))$$

for some $\delta > 0$ and an independent copy Y of X, then the isotropic constant L_X of X satisfies

$$L_X \leqslant e^{1+\frac{2}{\delta}}.$$

Ball and Nguyen proved in [**43**] that if X is an isotropic log-concave random vector in $\mathbb{R}^n$ and its density f satisfies the Poincaré inequality with constant $\kappa > 0$, then

$$\mathrm{Ent}\left(\frac{X+Y}{\sqrt{2}}\right) - \mathrm{Ent}(X) \geqslant \frac{\kappa}{4(1+\kappa)}\left(\mathrm{Ent}(G) - \mathrm{Ent}(X)\right),$$

where Y is an independent copy of X. The proof is presented in Section 16.3. Recall that in Chapter 12 we saw that $L_N \leqslant C\sigma_n$; in other words, the KLS-conjecture implies the hyperplane conjecture, in the sense that $L_n \leqslant C\sigma_n$. From the theorem of Ball and Nguyen one can conclude that, for each individual isotropic log-concave random vector X, a bound for the Poincaré constant implies a bound for the isotropic constant: if the density of X satisfies the Poincaré inequality with constant $\kappa > 0$, then

$$L_X \leqslant e^{1+\frac{8(1+\kappa)}{\kappa}}.$$

In other words, the KLS-conjecture "strongly" implies the hyperplane conjecture.

In the last section of this chapter we discuss the monotonicity of entropy theorem of Artstein-Avidan, Ball, Barthe and Naor [**22**] that extends the Shannon-Stam inequality in the case of variance 1 random variables. Given a random variable with variance 1, they showed that $\mathrm{Ent}(Y_n)$ increases with n: one has

$$\mathrm{Ent}\left(\frac{X_1+\cdots+X_n}{\sqrt{n}}\right) \leqslant \mathrm{Ent}\left(\frac{X_1+\cdots+X_n+X_{n+1}}{\sqrt{n+1}}\right)$$

for all $n \geqslant 1$.

16.1. Entropy gap and the isotropic constant

Recall that the isotropic constant of an isotropic log-concave random vector X in $\mathbb{R}^n$ with density f is defined by

$$L_X := L_f = \left(\sup_{x \in \mathbb{R}^n} f(x)\right)^{\frac{1}{n}}$$

and satisfies

$$L_X \leqslant e f(0)^{1/n}.$$

The next theorem relates L_X to entropy.

THEOREM 16.1.1 (Ball). *Let X be an isotropic log-concave random vector in $\mathbb{R}^n$. We assume that there exists a constant $\delta \in (0,1)$ such that*

(16.1.1) $$\mathrm{Ent}\left(\frac{X+Y}{\sqrt{2}}\right) - \mathrm{Ent}(X) \geqslant \delta(\mathrm{Ent}(G) - \mathrm{Ent}(X))$$

for an independent copy Y of X. Then,

$$L_X \leqslant e^{1+\frac{2}{\delta}}.$$

Proof. We will use the fact that

(16.1.2) $$-\log f(0) \leqslant \mathrm{Ent}(X) \leqslant -\log f(0) + n.$$

To see this, we write $f = e^{-\varphi}$, where φ is a convex function on $\mathbb{R}^n$, and using Jensen's inequality and the fact that $\int_{\mathbb{R}^n} x f(x) dx = 0$ we write

$$-\log f(0) = \varphi(0) \leqslant \int_{\mathbb{R}^n} \varphi(x) f(x) dx = \mathrm{Ent}(X).$$

Next, using the convexity of φ and integration by parts we get

$$\text{Ent}(X) - \varphi(0) = \int_{\mathbb{R}^n} f(x)\left(\varphi(x) - \varphi(0)\right) dx \leqslant \int_{\mathbb{R}^n} f(x)\langle\nabla\varphi(x), x\rangle dx = n.$$

This proves (16.1.2).

For the proof of Theorem 16.1.1 we will first show that

(16.1.3) $$\text{Ent}\left(\frac{X+Y}{\sqrt{2}}\right) \leqslant -\log f(0) + 2n.$$

Recall that the density of $\frac{X+Y}{\sqrt{2}}$ is given by

$$h(x) = 2^{n/2} \int_{\mathbb{R}^n} f(x-y)f(y) dy.$$

Assuming first that f is even, and using the log-concavity of f, we see that

$$h(0) = 2^{n/2} \int_{\mathbb{R}^n} f^2(y) dy \geqslant 2^{n/2} \int_{\mathbb{R}^n} f(2y)f(0) dy = 2^{-n/2} f(0).$$

Then, (16.1.2) gives

$$\text{Ent}\left(\frac{X+Y}{\sqrt{2}}\right) \leqslant \frac{n}{2}\log 2 - \log f(0) + n \leqslant -\log f(0) + \frac{3}{2}n.$$

In the general case, where f is not necessarily symmetric, we consider independent copies Y, X', Y' of X and observe that $\frac{X-X'}{\sqrt{2}}$ and $\frac{Y-Y'}{\sqrt{2}}$ are independent symmetric log-concave random vectors in $\mathbb{R}^n$ with density $g(x) = 2^{n/2}\int_{\mathbb{R}^n} f(x+y)f(y)\,dy$. The argument that we used above shows that $g(0) \geqslant 2^{-n/2} f(0)$. Using the Shannon-Stam inequality (see the introduction of this chapter) and the bound that we obtained in the symmetric case, we have that

$$\text{Ent}\left(\frac{X+Y}{\sqrt{2}}\right) \leqslant \text{Ent}\left(\frac{X+Y-X'-Y'}{\sqrt{4}}\right) = \text{Ent}\left(\frac{\frac{X-X'}{\sqrt{2}} + \frac{Y-Y'}{\sqrt{2}}}{\sqrt{2}}\right)$$
$$\leqslant -\log g(0) + \frac{3}{2}n$$
$$\leqslant -\log f(0) + 2n.$$

This proves (16.1.3). Now, we use our assumption about the entropy to write

(16.1.4) $$(1-\delta)\text{Ent}(X) \leqslant \text{Ent}\left(\frac{X+Y}{\sqrt{2}}\right) - \delta\text{Ent}(G) \leqslant \text{Ent}\left(\frac{X+Y}{\sqrt{2}}\right),$$

(using also the fact that $\text{Ent}(G) \geqslant 0$). Combining (16.1.4) with (16.1.2) we immediately get

$$(1-\delta)(-\log f(0)) \leqslant -\log f(0) + 2n,$$

which is exactly the assertion of the theorem. $\square$

In the next section we will see that if the density f of X satisfies the Poincaré inequality with constant κ then (16.1.1) is satisfied for some constant $\delta = \delta(\kappa)$ depending only on κ. The precise statement is the following:

THEOREM 16.1.2 (Ball-Nguyen). *Let X be an isotropic log-concave random vector in $\mathbb{R}^n$. We assume that its density f satisfies the Poincaré inequality with constant $\kappa > 0$. If Y is an independent copy of X then*

$$\operatorname{Ent}\left(\frac{X+Y}{\sqrt{2}}\right) - \operatorname{Ent}(X) \geqslant \frac{\kappa}{4(1+\kappa)}\left(\operatorname{Ent}(G) - \operatorname{Ent}(X)\right).$$

where G is a standard Gaussian random vector in $\mathbb{R}^n$.

Taking into account Theorem 16.1.1 we obtain

THEOREM 16.1.3 (Ball-Nguyen). *Let X be an isotropic log-concave random vector in $\mathbb{R}^n$. We assume that its density f satisfies the Poincaré inequality with constant $\kappa > 0$. Then,*

$$L_X \leqslant e^{1+\frac{8(1+\kappa)}{\kappa}}.$$

It is known that if the density f of an isotropic log-concave random vector X in $\mathbb{R}^n$ satisfies the Poincaré inequality with some constant κ then $\kappa \leqslant 1$. Thus, we can write the estimate of Theorem 16.1.3 in the simpler form $L_X \leqslant e^{17/\kappa}$. In particular, Theorem 16.1.3 shows that the KLS-conjecture implies the hyperplane conjecture, and in fact this implication is valid for each individual vector irrespective of whether either one of the conjectures holds in general, for all isotropic vectors.

16.2. Entropy jumps for log-concave random vectors with spectral gap

In this section we present the proof of Theorem 16.1.2 following Ball and Nguyen [**43**].

Given a random vector X in $\mathbb{R}^n$ with sufficiently smooth density f, instead of entropy we prefer to work with the *Fisher information* of X which is defined by

$$(16.2.1) \qquad J(X) = J(f) = \int_{\mathbb{R}^n} \frac{\|\nabla f\|_2^2}{f}.$$

16.2.1. Fisher information of isotropic log-concave random vectors

The next proposition shows that among all isotropic random vectors a standard Gaussian random vector has minimal Fisher information.

PROPOSITION 16.2.1. *Let X be an isotropic random vector in $\mathbb{R}^n$. Then $J(X) \geqslant J(G)$, where G is a standard Gaussian random vector in $\mathbb{R}^n$.*

Proof. We first compute $J(G)$. If we write g for the density of G, then

$$J(G) = \int_{\mathbb{R}^n} \frac{\|\nabla g\|^2}{g} = \frac{1}{(2\pi)^{n/2}} \int_{\mathbb{R}^n} \|x\|_2^2 e^{-\|x\|_2^2/2} dx = n.$$

We write f for the density of X. Then,

$$0 \leqslant \int_{\mathbb{R}^n} \left\|\frac{\nabla f(x)}{f(x)} + x\right\|_2^2 f(x) dx = J(X) + 2\int_{\mathbb{R}^n} \langle \nabla f(x), x\rangle dx + \int_{\mathbb{R}^n} \|x\|_2^2 f(x) dx$$

$$= J(X) + 2\int_{\mathbb{R}^n} \sum_{i=1}^n x_i \frac{\partial f(x)}{\partial x_i} dx + n$$

$$= J(X) - 2n\int_{\mathbb{R}^n} f(x) dx + n = J(X) - J(G),$$

where, for the last equality, we used integration by parts. $\square$

It is known (see [**28**] or [**50**]) that the Fisher information of X is related with the derivative of the entropy along the Ornstein-Uhlenbeck semigroup associated with X, which can be defined as follows. If X is a random vector with density f then, for any $t \geqslant 0$, we consider the random vector

$$X_t = e^{-t}X + \sqrt{1 - e^{-2t}}G,$$

with density f_t. The generator L of the semigroup $\{f_t\}_{t \geqslant 0}$ is defined by the equation

$$(16.2.2) \qquad \frac{\partial}{\partial t}f_t(x) = L(f_t)(x) := \Delta f_t(x) + \operatorname{div}(xf_t(x))$$

It is known that starting at time $t = 0$ with a smooth density f, the density f_t is strictly positive for $t > 0$; it is also C^∞ on $\mathbb{R}^n$, and f_t as well as its derivatives decrease exponentially to 0 at ∞ (see [**135**] for a complete proof of these assertions). The next fact is immediate:

PROPOSITION 16.2.2. *Let X be an isotropic log-concave random vector in $\mathbb{R}^n$ and let G be a standard Gaussian random vector. Then,*

$$\operatorname{Ent}(G) - \operatorname{Ent}(X) = \int_0^\infty (J(f_t) - n)dt.$$

Proof. Observe that by Green's theorem and (16.2.2) we may write:

$$\frac{\partial}{\partial t}\operatorname{Ent}(f_t) = -\frac{\partial}{\partial t}\int_{\mathbb{R}^n} f_t \log f_t$$

$$= -\int_{\mathbb{R}^n} \frac{\partial f_t}{\partial t} \log f_t - \int_{\mathbb{R}^n} f_t \frac{\partial(\log f_t)}{\partial t}$$

$$= -\int_{\mathbb{R}^n} L(f_t) \log f_t - \int_{\mathbb{R}^n} \frac{\partial f_t}{\partial t}$$

$$= -\int_{\mathbb{R}^n} \Delta f_t \log f_t - \int_{\mathbb{R}^n} \operatorname{div}(xf_t(x)) \log f_t(x)\, dx$$

$$= J(f_t) - \sum_{i=1}^n \int_{\mathbb{R}^n} \frac{\partial}{\partial x_i}(x_i f_t(x)) \log f_t(x)\, dx,$$

where in the last equality we have applied integration by parts. Applying integrations by parts consecutively to the second summand we get:

$$\int_{\mathbb{R}^n} \frac{\partial}{\partial x_i}(x_i f_t(x)) \log f_t\, dx = -\int_{\mathbb{R}^n} x_i \frac{\partial f_t}{\partial x_i} = \int_{\mathbb{R}^n} f_t.$$

Integration with respect to t completes the proof. $\square$

The next proposition describes the behavior of f_t with respect to the Poincaré inequality.

PROPOSITION 16.2.3. *Let X be an isotropic random vector which satisfies the Poincaré inequality with constant $\kappa > 0$. Then, for every $t > 0$ we have that f_t satisfies the Poincaré inequality with the same constant.*

Sketch of the proof. A simple computation shows that if X_1 and X_2 are independent random vectors in $\mathbb{R}^n$ and if they satisfy the Poincaré inequality

$$\operatorname{Var}[h(X_i)] \leqslant \alpha_i \mathbb{E}\|\nabla h(X_i)\|_2^2, \qquad i = 0, 1$$

for every smooth function h on $\mathbb{R}^n$ (with constants α_i) then
$$\mathrm{Var}[h(\sqrt{\lambda}X_1 + \sqrt{1-\lambda}X_2)] \leqslant (\lambda\alpha_1 + (1-\lambda)\alpha_2)\mathbb{E}\|\nabla h(\sqrt{\lambda}X_1 + \sqrt{1-\lambda}X_2)\|_2^2$$
for every smooth function h on $\mathbb{R}^n$. We apply this fact with $X_1 = X$, $X_2 = G$ and $\lambda = e^{-2t}$ (and we see that actually the constant $\lambda\alpha_1 + (1-\lambda)\alpha_2$ gets better, $\leqslant \alpha_1$, since the Poincaré constant of the standard Gaussian measure is 1 while that of any other isotropic measure is at most 1 as can be seen by 14.1.8 applied to the function $x \mapsto \langle x, e_1 \rangle$). □

Let X, Y be two independent random vectors and let
$$X_t = e^{-t}X + \sqrt{1-e^{-2t}}G_1 \quad \text{and} \quad Y_t = e^{-t}Y + \sqrt{1-e^{-2t}}G_2$$
be their evolutes along the Ornstein-Uhlenbeck semigroup, where G_1, G_2 are independent standard Gaussian random vectors, independent from X and Y. Then,
$$\frac{X_t + Y_t}{\sqrt{2}} = e^{-t}\frac{X+Y}{\sqrt{2}} + \sqrt{1-e^{-2t}}G,$$
where $G = \frac{G_1+G_2}{\sqrt{2}}$. Since we may assume that the density of X has compact support, the density of each X_t may be assumed smooth enough and satisfies all the integrability properties that are required for the arguments below.

LEMMA 16.2.4. *Let X be an isotropic log-concave random vector in $\mathbb{R}^n$. Then, f_t is log-concave, i.e. it has the form $f_t = e^{-\varphi_t}$ for some convex function φ_t on $\mathbb{R}^n$. Moreover,*
$$\frac{\partial}{\partial t}J(f_t) = 2J(f_t) - 2\mathrm{tr}\left(\int_{\mathbb{R}^n} f_t\,(\mathrm{Hess}\,\varphi_t)^2\right).$$

Proof. The fact that f_t is log-concave follows directly from the Prékopa-Leindler inequality. For the second claim, we apply integration by parts to get
$$(16.2.3) \qquad J(f_t) = \int_{\mathbb{R}^n} \frac{\|\nabla f_t\|_2^2}{f_t} = -\int_{\mathbb{R}^n} f_t \Delta \log f_t = \mathrm{tr}\left(\int_{\mathbb{R}^n} f_t\,\mathrm{Hess}\,\varphi_t\right).$$

For notational convenience we write $\partial_j(f_t)$ for the partial derivative of f_t with respect to x_j. Note that
$$J(f_t) = \int_{\mathbb{R}^n} \frac{\|\nabla f_t\|_2^2}{f_t} = \sum_{i=1}^n \int_{\mathbb{R}^n} \frac{(\partial_i(f_t))^2}{f_t}.$$

Using also (16.2.2) we compute
$$\frac{\partial}{\partial t}J(f_t) = \sum_{i=1}^n \int_{\mathbb{R}^n} 2\frac{\partial_i(f_t) \cdot \partial_i \partial_t f_t}{f_t} - \sum_{i=1}^n \int_{\mathbb{R}^n} \partial_t f_t \left(\frac{\partial_i f_t}{f_t}\right)^2$$
$$= \sum_{i=1}^n \int_{\mathbb{R}^n} -2\partial_i(\varphi_t) \cdot \partial_i \left(\sum_{j=1}^n \partial_j\left((\partial_j \varphi_t + x_j)f_t\right)\right)$$
$$- \sum_{i=1}^n \int_{\mathbb{R}^n} \left(\sum_{j=1}^n \partial_j\left((-\partial_j \varphi_t + x_j)f_t\right)\right)(\partial_i \varphi_t)^2.$$

Denoting by A, B the last two terms, and using (16.2.3) and integration by parts, we get

$$A = 2\sum_{i,j=1}^{n} \int_{\mathbb{R}^n} (\partial_{ij}\varphi_t) \cdot \partial_i\left((-\partial_j\varphi_t + x_j)f_t)\right)$$

$$= 2J(f_t) - 2\mathrm{tr}\left(\int_{\mathbb{R}^n} f_t\left(\mathrm{Hess}\,\varphi_t\right)^2\right) + 2\sum_{i,j=1}^{n}\int_{\mathbb{R}^n} f_t(\partial_{ij}\varphi_t)(\partial_i\varphi_t)(\partial_j\varphi_t - x_j).$$

Integrating by parts once again, we see that

$$B = -\sum_{i=1}^{n}\int_{\mathbb{R}^n}\left(\sum_{j=1}^{n}\partial_j\left((-\partial_j\varphi_t + x_j)f_t\right)\right)(\partial_i\varphi_t)^2$$

$$= 2\sum_{i,j=1}^{n}\int_{\mathbb{R}^n} f_t(-\partial_j\varphi_t + x_j)(\partial_i\varphi_t)(\partial_{ij}\varphi_t)$$

$$= -2\sum_{i,j=1}^{n}\int_{\mathbb{R}^n} f_t(\partial_{ij}\varphi_t)(\partial_i\varphi_t)(\partial_j\varphi_t - x_j).$$

Adding these two equalities we obtain the result. □

LEMMA 16.2.5. *Let X be an isotropic log-concave random vector in $\mathbb{R}^n$ and let G be a standard Gaussian random vector. Then,*

$$2\int_0^{\infty} e^{-2t}(J(f_t) - n)dt \geqslant \mathrm{Ent}(G) - \mathrm{Ent}(X).$$

Proof. From Lemma 16.2.4 we have

$$\frac{\partial}{\partial t}(J(f_t) - n) = -2(J(f_t) - n) - 2\mathrm{tr}\left(\int_{\mathbb{R}^n} f_t\left(\mathrm{Hess}\,\varphi_t - I\right)^2\right)$$

$$\leqslant -2(J(f_t) - n).$$

Integrating this inequality over $(t, +\infty)$ for every $t > 0$ we get

$$J(f_t) - n \geqslant 2(\mathrm{Ent}(G) - \mathrm{Ent}(X_t))$$

for all $t > 0$, and an integration by parts shows that

$$\int_0^{\infty} e^{-2t}(J(f_t) - n)dt \geqslant \mathrm{Ent}(G) - \mathrm{Ent}(X) - \int_0^{\infty} e^{-2t}(J(f_t) - n)dt,$$

which is the assertion of the lemma. □

16.2.2. Fisher information for marginals

According to Lemma 16.2.4, the derivative $\frac{\partial}{\partial t}J(f_t)$ of $J(f_t)$ depends on the quantity $\mathrm{tr}\left(\int_{\mathbb{R}^n} f_t\left(\mathrm{Hess}\,\varphi_t\right)^2\right)$. The next lemma describes the behavior of this quantity for marginal densities.

LEMMA 16.2.6. *Let $w = e^{-\phi} : \mathbb{R}^n \longrightarrow \mathbb{R}^+$ be a positive smooth function and let E be a subspace of $\mathbb{R}^n$. We consider the function*

$$h(x) = \pi_E w(x) = \int_{E^\perp} w(x+y)dy = \int_{E^\perp} e^{-\phi(x+y)}dy.$$

and we write it in the form $h(x) = e^{-\psi(x)}$. If P_E denotes the orthogonal projection onto E then for every $x \in E$ we have

$$(16.2.4) \qquad h(x)\text{Hess}\psi(x) \leq \int_{E^\perp} w(x+y) P_E(\text{Hess}\phi(x+y)) P_E dy$$

in the operator sense. Moreover, if $\text{Hess}\psi(x) \geq 0$, then

$$(16.2.5) \quad \text{tr}\left((\text{Hess}(\psi(x)))^2 h(x)\right) \leq \int_{E^\perp} \text{tr}\left((P_E(\text{Hess}\phi(x+y))P_E)^2\right) w(x+y) dy.$$

Proof. From the definition of h we have the identities

$$\nabla h(x) = \int_{E^\perp} P_E(\nabla w(x+y)) dy$$

and

$$\text{Hess } h(x) = \int_{E^\perp} P_E(\text{Hess } w(x+y)) P_E dy.$$

Note also that

$$h(x)\text{Hess } \psi(x) = h(x)\text{Hess } (-\log h)(x) = \frac{\nabla h(x) \otimes \nabla h(x)}{h(x)} - \text{Hess } h(x).$$

Substituting these into the first claim of the lemma, we see that it is enough to prove that

$$\frac{\nabla h(x) \otimes \nabla h(x)}{h(x)} \leq \int_{E^\perp} \frac{P_E \nabla w(x+y) \otimes P_E \nabla w(x+y)}{w(x+y)} dy.$$

This last inequality can be written in the form

$$\int_{E^\perp} P_E(\nabla w(x+y)) dy \otimes \int_{E^\perp} P_E(\nabla w(x+y)) dy$$
$$\leq \int_{E^\perp} \frac{P_E \nabla w(x+y) \otimes P_E \nabla w(x+y)}{w(x+y)} dy \cdot \int_{E^\perp} w(x+y) dy,$$

which is true by the Cauchy-Schwarz inequality.

For the second claim, we first use the observation that if A, B are symmetric operators with $A \geq B$, then

$$\text{tr}(AH) \geq \text{tr}(BH)$$

for every operator $H \geq 0$. Therefore, if $\text{Hess } \psi(x) \geq 0$, we get

$$\text{tr}\left[(\text{Hess } \psi(x))^2 h(x)\right] \leq \int_{E^\perp} \text{tr}\left[P_E(\text{Hess}\phi(x+y))P_E \text{Hess}\psi(x)\right] w(x+y) dy.$$

Applying the Cauchy-Schwarz inequality we see that the right hand side of the last inequality is less than or equal to

$$\left(\int_{E^\perp} \text{tr}\left[(P_E \text{Hess } \phi(x+y) P_E)^2\right] w(x+y) dy\right)^{1/2} \left(\int_{E^\perp} \text{tr}\left[(\text{Hess}\psi(x))^2\right] w(x+y) dy\right)^{1/2}.$$

On the other hand, the second integral is equal to $\text{tr}\left[(\text{Hess } \psi(x))^2 h(x)\right]$, and this gives

$$\text{tr}\left[(\text{Hess } \psi(x))^2 h(x)\right] \leq \int_{E^\perp} \text{tr}\left[(P_E \text{Hess } \phi(x+y) P_E)^2\right] w(x+y) dy$$

which completes the proof. $\square$

Using Lemma 16.2.6 we obtain the main inequality that will be used in the proof of Theorem 16.1.2.

THEOREM 16.2.7. *Let X be a log-concave random vector in $\mathbb{R}^n$, with a smooth density of the form $f = e^{-\varphi}$, where φ is a convex function on $\mathbb{R}^n$, and let Y be an independent copy of X. We denote by $h = e^{-\psi}$ the density of the random vector $\frac{X+Y}{\sqrt{2}}$ and we set*

$$K = \operatorname{tr}\left[\int_{\mathbb{R}^n} (\operatorname{Hess} \varphi)^2 f\right],$$

$$K_2 = \operatorname{tr}\left[\int_{\mathbb{R}^n} (\operatorname{Hess} \psi)^2 h\right],$$

$$M = \operatorname{tr}\left[\left(\int_{\mathbb{R}^n} (\operatorname{Hess} \varphi) f\right)^2\right].$$

Then,

$$K_2 \leqslant \frac{K+M}{2}.$$

Proof. From the Prékopa-Leindler inequality it follows that h is log-concave, and hence $\operatorname{Hess} \psi \geqslant 0$. We denote by $w(x,y) = f(x)f(y)$ the density of the random vector (X,Y) in $\mathbb{R}^{2n}$. We set

$$e_i = \left(0, \ldots, \frac{1}{\sqrt{2}}, 0, \ldots, 0, \frac{1}{\sqrt{2}}, 0, \ldots\right)$$

where the non-zero entries are the i-th and the $(n+i)$-th. We write E for the subspace of $\mathbb{R}^{2n}$ which is spanned by the orthonormal basis $\{e_1, \ldots, e_n\}$. Integrating (16.2.5) we see that

$$K_2 \leqslant \int_{\mathbb{R}^{2n}} w(x,y) \operatorname{tr}\left[\left[\langle\operatorname{Hess}(-\log w)(x,y)e_i, e_j\rangle\right]_{i,j}^2\right] dxdy$$

$$= \int_{\mathbb{R}^{2n}} f(x)f(y) \sum_{i,j=1}^n \langle\operatorname{Hess}(-\log w)(x,y)e_i, e_j\rangle^2 dxdy.$$

On the other hand,

$$\langle\operatorname{Hess}(-\log w)(x,y)e_i, e_j\rangle^2 = \frac{1}{4}(\partial_{ji}\varphi(x) + \partial_{ji}\varphi(y))^2.$$

Therefore,

$$K_2 \leqslant \frac{1}{4}\sum_{i,j=1}^n \int_{\mathbb{R}^{2n}} f(x)f(y)\left((\partial_{ji}\varphi(x))^2 + 2\partial_{ji}\varphi(x)\partial_{ji}\varphi(y) + (\partial_{ji}\varphi(y))^2\right) dxdy$$

$$= \frac{1}{2}\sum_{i,j=1}^n \int_{\mathbb{R}^n} f(\partial_{ji}\varphi)^2 + \frac{1}{2}\sum_{i,j=1}^n \left(\int_{\mathbb{R}^n} f\partial_{ji}\varphi\right)^2$$

$$= \frac{K+M}{2}$$

as claimed. $\square$

Proof of Theorem 16.1.2. We consider the log-concave vectors X_t with density $f_t = e^{-\varphi_t}$. From (16.2.3) we know that
$$J(t) := J(X_t) = \operatorname{tr}\left(\int_{\mathbb{R}^n} f_t \operatorname{Hess} \varphi_t\right).$$

We set
$$K(t) := \operatorname{tr}\left(\int_{\mathbb{R}^n} f_t \left(\operatorname{Hess} \varphi_t\right)^2\right) = -\frac{1}{2} e^{2t} \frac{\partial}{\partial t}\left(e^{-2t} J(t)\right),$$
where the equality follows from Lemma 16.2.4. If $Z_t = \frac{X_t + Y_t}{\sqrt{2}}$ and if $h_t = e^{-\psi_t}$ is the log-concave density of Z_t then we also have
$$J_2(t) := J(Z_t) = \operatorname{tr}\left(\int_{\mathbb{R}^n} h_t \operatorname{Hess} \psi_t\right)$$
and we set
$$K_2(t) := \operatorname{tr}\left(\int_{\mathbb{R}^n} h_t \left(\operatorname{Hess} \psi_t\right)^2\right) = -\frac{1}{2} e^{2t} \frac{\partial}{\partial t}\left(e^{-2t} J_2(t)\right).$$

From Theorem 16.2.7 we get

(16.2.6) $$K_2(t) \leqslant \frac{K(t) + M(t)}{2} = K(t) - \frac{K(t) - M(t)}{2},$$

where
$$M(t) := \operatorname{tr}\left[\left(\int_{\mathbb{R}^n} \left(\operatorname{Hess} \varphi_t\right) f\right)^2\right].$$

We rewrite (16.2.6) in the form

(16.2.7) $$\frac{\partial}{\partial t}\left(e^{-2t}(J_2(t) - J(t))\right) \geqslant e^{-2t}(K(t) - M(t)).$$

According to Proposition 16.2.3, f_t satisfies the Poincaré inequality with constant κ. We apply the Poincaré inequality for the density $f_t = e^{-\varphi_t}$ and the functions
$$s_i(x) = \partial_i \varphi_t(x) - \sum_{j=1}^n x_j \int_{\mathbb{R}^n} (\partial_{ij} \varphi_t) f_t,$$
which satisfy $\int s_i f_t = 0$. Adding the inequalities $\int_{\mathbb{R}^n} \|\nabla s_i\|_2^2 f_t \geqslant \kappa \int_{\mathbb{R}^n} s_i^2 f_t$, which follow from the Poincaré inequality, we get
$$\operatorname{tr}\left(\int_{\mathbb{R}^n} f_t \left(\operatorname{Hess} \varphi_t\right)^2\right) - \operatorname{tr}\left(\int_{\mathbb{R}^n} f_t \operatorname{Hess} \varphi_t\right)^2$$
$$\geqslant \kappa \left(\operatorname{tr}\left(\int_{\mathbb{R}^n} f_t \operatorname{Hess} \varphi_t\right)^2 - \operatorname{tr}\left(\int_{\mathbb{R}^n} f_t \operatorname{Hess} \varphi_t\right)\right).$$

We rewrite this in the form $K(t) - M(t) \geqslant \kappa \left(M(t) - J(t)\right)$, or equivalently
$$K(t) - M(t) \geqslant \frac{\kappa}{\kappa + 1} \left(K(t) - J(t)\right).$$

Then, from (16.2.7) we get
$$\frac{\partial}{\partial t}\left(e^{-2t}(J_2(t) - J(t))\right) \geqslant \frac{\kappa}{\kappa + 1} e^{-2t}(K(t) - J(t)),$$

which we integrate from t to ∞, to arrive at
$$J(t) - J_2(t) \geqslant \frac{\kappa}{\kappa+1} e^{2t} \int_t^\infty e^{-2s} \left(K(s) - J(s)\right) ds.$$
Finally, from Proposition 16.2.3 and Proposition 16.2.2 we have that
$$\begin{aligned}
\operatorname{Ent}\left(\frac{X+Y}{\sqrt{2}}\right) - \operatorname{Ent}(X) &= \int_0^\infty (J(t) - J_2(t)) dt \\
&\geqslant \frac{\kappa}{\kappa+1} \int_0^\infty e^{2t} \int_t^\infty e^{-2s} \left(K(s) - J(s)\right) ds dt \\
&= \frac{\kappa}{2(1+\kappa)} \int_0^\infty (1 - e^{-2t})(K(t) - J(t)) dt \\
&= \frac{\kappa}{2(1+\kappa)} \int_0^\infty (1 - e^{-2t})\left(-\frac{1}{2}\frac{\partial}{\partial t}(J(t) - n)\right) dt \\
&= \frac{\kappa}{2(1+\kappa)} \int_0^\infty e^{-2t}(J(t) - n) dt.
\end{aligned}$$
Combining this inequality with Lemma 16.2.4 we get the result. □

16.3. Further reading

16.3.1. Monotonicity of entropy

Let X be a random variable with variance 1. For every $n \geqslant 2$ we consider the normalized sums
$$Y_n = \frac{X_1 + \cdots + X_n}{\sqrt{n}},$$
where X_i are independent copies of X. Recall that $\operatorname{Ent}(Y_2) \geqslant \operatorname{Ent}(Y_1)$ according to the Shannon-Stam inequality. Inductively, it follows that
$$\operatorname{Ent}(Y_{2^k}) \geqslant \operatorname{Ent}(Y_{2^{k-1}}).$$
Here we describe a proof of the conjecture that $\operatorname{Ent}(Y_n)$ increases with n. This fact was explicitly stated by Lieb and it was proved to hold true by Artstein, Ball, Barthe and Naor in [**22**].

THEOREM 16.3.1 (Artstein-Ball-Barthe-Naor). *Let $\{X_j\}_{j=1}^\infty$ be a sequence of independent and identically distributed square-integrable random variables. Then,*
$$\operatorname{Ent}\left(\frac{X_1 + \cdots + X_n}{\sqrt{n}}\right) \leqslant \operatorname{Ent}\left(\frac{X_1 + \cdots + X_n + X_{n+1}}{\sqrt{n+1}}\right).$$

In the same work, a non-identically distributed version of Theorem 16.3.1 was also obtained.

THEOREM 16.3.2. *Let $X_1, X_2, \ldots, X_{n+1}$ be independent random variables and let $(a_1, \ldots, a_{n+1}) \in S^n$ be a unit vector. Then,*
$$\operatorname{Ent}\left(\sum_{j=1}^{n+1} a_j X_j\right) \geqslant \sum_{j=1}^{n+1} \frac{1 - a_j^2}{n} \operatorname{Ent}\left(\frac{1}{\sqrt{1-a_j^2}} \sum_{i \neq j} a_i X_i\right).$$

In particular,
$$\operatorname{Ent}\left(\frac{X_1 + \cdots + X_n + X_{n+1}}{\sqrt{n+1}}\right) \geqslant \frac{1}{n+1} \sum_{j=1}^{n+1} \operatorname{Ent}\left(\frac{1}{\sqrt{n}} \sum_{i \neq j} X_i\right).$$

One of the main ingredients in the proof of Theorem 16.3.2 is the next formula for the Fisher information.

THEOREM 16.3.3. *Let $w : \mathbb{R}^n \to (0, \infty)$ be a C^2 density on $\mathbb{R}^n$ with*

$$\int \frac{\|w\|_2^2}{w} \cdot \int \|\operatorname{Hess}(w)\| < \infty.$$

Let e be a unit vector and let h be the marginal density in the direction of e, defined by

$$h(t) = \int_{te+e^\perp} w.$$

Then, the Fisher information of h satisfies

$$(16.3.1) \qquad J(h) \leqslant \int_{\mathbb{R}^n} \left(\frac{\operatorname{div}(pw)}{w}\right)^2 w$$

for every continuously differentiable vector field $p : \mathbb{R}^n \to \mathbb{R}^n$ with $\int \|p\|_2 w < \infty$ and the property that $\langle p(x), e \rangle = 1$ for every x.

Proof. Our assumptions for w show that

$$(16.3.2) \qquad h'(t) = \int_{te+e^\perp} \partial_e(w)$$

for all t. Assuming that $\int_{\mathbb{R}^n} \left(\frac{\operatorname{div}(pw)}{w}\right)^2 w$ is finite we have that $\operatorname{div}(pw)$ is integrable on $\mathbb{R}^n$ and hence on almost every hyperplane perpendicular to e. If $\int \|p\|_2 w$ is also finite, then on almost all of these hyperplanes the integral of the $(n-1)$-dimensional divergence of pw in the hyperplane is equal to zero by the Gauss-Green Theorem. Since the component of p in the direction of e is always 1, we have

$$h'(t) = \int_{te+e^\perp} \operatorname{div}(pw)$$

for almost every t. Therefore, applying also Jensen's inequality, we get

$$J(h) = \int \frac{h'(t)^2}{h(t)} dt = \int \frac{(\int \operatorname{div}(pw))^2}{\int w} \leqslant \int_{\mathbb{R}^n} \frac{(\operatorname{div}(pw))^2}{w},$$

as claimed. $\square$

Note. One can show that if w satisfies $\int \|x\|_2^2 w(x)\, dx < \infty$, then equality in (16.3.1) is attained for a suitable vector field p.

Proof of Theorem 16.3.2. Let f_i be the density of X_i and consider the product density

$$w(x_1, \ldots, x_{n+1}) = f_1(x_1) \cdots f_{n+1}(x_{n+1}).$$

The density of $\sum_{i=1}^{n+1} a_i X_i$ is the marginal of w in the direction of $(a_1, \ldots, a_{n+1}) \in S^n$. We will show that if w satisfies the conditions of Theorem 16.3.3, then for every unit vector $\vec{a} = (a_1, \ldots, a_{n+1}) \in S^n$ and every $b_1, \ldots, b_{n+1} \in \mathbb{R}$ satisfying

$$\sum_{j=1}^{n+1} b_j \sqrt{1 - a_j^2} = 1$$

we have

$$(16.3.3) \qquad J\left(\sum_{j=1}^{n+1} a_j X_j\right) \leqslant n \sum_{j=1}^{n+1} b_j^2 J\left(\frac{1}{\sqrt{1-a_j^2}} \sum_{i \neq j} a_i X_i\right).$$

Then, we choose $b_j = \frac{1}{n}\sqrt{1-a_j^2}$, we apply (16.3.3) to the Ornstein-Uhlenbeck evolutes $X_i^{(t)}$ of the X_i's, and finally integrate with respect to $t \in (0, \infty)$ to get Theorem 16.3.2.

In order to prove (16.3.3), for every j we set
$$\vec{a}_j = \frac{1}{\sqrt{1-a_j^2}}(a_1, \ldots, a_{j-1}, 0, a_{j+1}, \ldots, a_n).$$

Note that $\vec{a}_j$ is also a unit vector. Using Theorem 16.3.3 we find a vector field $p_j : \mathbb{R}^{n+1} \to \mathbb{R}^{n+1}$ such that $\langle p_j, \vec{a}_j \rangle = 1$ and
$$J\left(\frac{1}{\sqrt{1-a_j^2}} \sum_{i \neq j} a_i X_i\right) = \int_{\mathbb{R}^n} \left(\frac{\mathrm{div}(p_j w)}{w}\right)^2 w.$$

Moreover, we may assume that p_j does not depend on the coordinate x_j, and that the j-th coordinate of p_j is identically 0, since we may restrict to n dimensions, use Theorem 16.3.3, and then artificially add the j-th coordinate keeping the same conclusions.

We define a vector field $p : \mathbb{R}^{n+1} \to \mathbb{R}^{n+1}$ given by $p = \sum_{j=1}^{n+1} b_j p_j$. Since
$$\sum_{j=1}^{n+1} b_j \sqrt{1-a_j^2} = 1,$$
we have $\langle p, \vec{a} \rangle = 1$. By Theorem 16.3.3,
$$J\left(\sum_{i=1}^{n+1} a_i X_i\right) \leqslant \int_{\mathbb{R}^n} \left(\frac{\mathrm{div}(pw)}{w}\right)^2 w = \int_{\mathbb{R}^n} \left(\sum_{j=1}^{n+1} b_j \frac{\mathrm{div}(p_j w)}{w}\right)^2 w.$$

We set
$$y_j = b_j \frac{\mathrm{div}(wp_j)}{w}.$$

Our aim is to show that
$$\|y_1 + \cdots + y_{n+1}\|^2 \leqslant n(\|y_1\|^2 + \cdots + \|y_{n+1}\|^2)$$
in the Hilbert space $L^2(w)$ with weight w. A simple application of the Cauchy-Schwarz inequality would give a coefficient of $n+1$ instead of n on the right hand side. However, our y_i have additional properties. Define $T_1 : L^2(w) \to L^2(w)$ by
$$(T_1 \phi)(x) = \int \phi(u, x_2, \ldots, x_{n+1}) f_1(u) du.$$

Note that $T_1(\phi)$ is independent of the first coordinate. In a similar way, define T_i by integrating out the i-th coordinate against f_i. Then the operators T_i are commuting orthogonal projections on the Hilbert space $L^2(w)$. Moreover, for each i, $T_i(y_i) = y_i$ because y_i is already independent of the i-th coordinate, and for each j we have $T_1 \circ \cdots \circ T_{n+1}(y_j) = 0$ because we integrate a divergence. These properties ensure the slightly stronger inequality that we need. We summarize it in the following lemma which completes the proof of Theorem 16.3.1.

LEMMA 16.3.4. *Let $T_1, \ldots, T_m$ be m commuting orthogonal projections in a Hilbert space H. Let $y_1, \ldots, y_m \in H$ such that $T_1 \circ \cdots \circ T_m(y_j) = 0$ for every $1 \leqslant j \leqslant m$. Then,*
$$\|T_1(y_1) + \cdots + T_m(y_m)\|^2 \leqslant (m-1)(\|y_1\|^2 + \cdots + \|y_m\|^2).$$

Proof. Since the projections are commuting, we can write $H = \oplus_{\epsilon \in \{0,1\}^m} H_\epsilon$, where $H_\epsilon = \{x : T_i(x) = \epsilon_i x, 1 \leqslant i \leqslant m\}$. This is an orthogonal decomposition, so for every $\phi \in H$ we have
$$\|\phi\|^2 = \sum_{\epsilon \in \{0,1\}^m} \|\phi_\epsilon\|^2.$$

We decompose each y_i separately in the form $y_i = \sum_{\epsilon \in \{0,1\}^m} y_\epsilon^i$. The condition in the statement of the lemma implies that $y_{(1,\ldots,1)}^i = 0$ for each i. Therefore,

$$T_1(y_1) + \cdots + T_m(y_m) = \sum_{i=1}^m \sum_{\epsilon \in \{0,1\}^m} T_i(y_\epsilon^i) = \sum_{\epsilon \in \{0,1\}^m} \sum_{\epsilon_i = 1} y_\epsilon^i.$$

Now, we can write

$$\|T_1(y_1) + \cdots + T_m(y_m)\|^2 = \sum_{\epsilon \in \{0,1\}^m} \left\| \sum_{\epsilon_i = 1} y_\epsilon^i \right\|^2.$$

Every vector on the right hand side is a sum of at most $m - 1$ summands, since the only vector with m ones does not contribute anything to the sum. Thus we can complete the proof of the lemma using the Cauchy-Schwarz inequality:

$$\|T_1(y_1) + \cdots + T_m(y_m)\|^2 \leqslant \sum_{\epsilon \in \{0,1\}^m} (m-1) \sum_{\epsilon_i = 1} \|y_\epsilon^i\|^2$$
$$= (m-1)(\|y_1\|^2 + \cdots + \|y_m\|^2). \qquad \square$$

REMARK 16.3.5. An analogous result to Theorem 16.3.1 is also valid in the case where X is a random vector with density f. In this case the Fisher information is a matrix, which in the sufficiently smooth case can be written in the form

$$[J(f)]_{ij} = \int \frac{\partial f}{\partial x_i} \cdot \frac{\partial f}{\partial x_j} \frac{1}{f}.$$

Theorem 16.3.3 generalizes in the following way. If h is the (vector) marginal of f on the subspace E, for every $x \in E$ we define $h(x) = \int_{x+E^\perp} f$. Then, for every unit vector e in E

$$\langle J(h)(e), e \rangle = \inf_p \int \left(\frac{\mathrm{div}(fp)}{f} \right)^2 f,$$

where the infimum is taken over all $p : \mathbb{R}^n \to \mathbb{R}^n$ for which the orthogonal projection of p into E is constantly e. The argument is exactly the same as in the one-dimensional case.

16.4. Notes and references

Entropy gap and the isotropic constant

Theorem 16.1.1 appears in [43]. It was presented by Ball in a series of lectures in 2006, as part of a more general program proposing a probabilistic viewpoint on the geometry of high-dimensional convex bodies.

Bobkov and Madiman have also investigated the connections of information theory with the hyperplane conjecture (see [87], [88] and [89]). They provide a formulation of this conjecture in information-theoretic terms; more precisely, they show that the hyperplane conjecture is equivalent to the assertion that all log-concave probability measures are at most a bounded distance away from Gaussian, where distance is measured by relative entropy per coordinate.

The analogies between the classical Brunn-Minkowski theory and information theoretic inequalities have been understood and studied for a number of years. We do not cover this topic; the interested reader can get a feeling on this subject from the book of Cover and Thomas [144] and the articles of Costa and Cover [143], Dembo, Cover and Thomas [156].

Entropy jumps for log-concave random vectors with spectral gap

Theorem 16.1.2 is due to Ball and Nguyen. It is a multidimensional analogue of previous results of Ball, Barthe and Naor [42] who had proved that, if X is a random variable with variance 1 and density f that satisfies the Poincaré inequality with constant κ, then

$$\mathrm{Ent}(Y_2) - \mathrm{Ent}(X) \geqslant \frac{\kappa}{2(1+\kappa)}(\mathrm{Ent}(G) - \mathrm{Ent}(X)).$$

An extension of this result was obtained by Arstein-Avidan, Ball, Barthe and Naor [23]: If $X_1, \ldots, X_n$ are independent copies of a random variable X with variance 1 and density f that satisfies the Poincaré inequality with constant c, then for every $a \in \mathbb{R}^n$ with $\sum_{i=1}^n a_i^2 = 1$ we have that

$$\mathrm{Ent}(G) - \mathrm{Ent}\left(\sum_{i=1}^n a_i X_i\right) \leqslant \frac{2\|a\|_4^4}{c + (2-c)\|a\|_4^4}\left(\mathrm{Ent}(G) - \mathrm{Ent}(X)\right).$$

Monotonicity of entropy

Theorem 16.3.1 is due to Artstein-Avidan, Ball, Barthe and Naor [22]. It answers a problem that was known for several years and was formally stated as a conjecture by Lieb in [326].

Bibliography

[1] R. Adamczak, A. E. Litvak, A. Pajor and N. Tomczak-Jaegermann, *Quantitative estimates of the convergence of the empirical covariance matrix in log-concave ensembles*, J. Amer. Math. Soc. **23** (2010), No. 2, 535-561.

[2] R. Adamczak, R. Latała, A. E. Litvak, A. Pajor and N. Tomczak-Jaegermann, *Tail estimates for norms of sums of log-concave random vectors*, Preprint.

[3] R. Adamczak, R. Latała, A. E. Litvak, K. Oleszkiewicz, A. Pajor and N. Tomczak-Jaegermann, *A short proof of Paouris' inequality*, Can. Math. Bul. (to appear).

[4] J. M. Aldaz, *The weak type $(1,1)$ bounds for the maximal function associated to cubes grow to infinity with the dimension*, Annals of Math. **173** (2011), 1013-1023.

[5] A. D. Alexandrov, *On the theory of mixed volumes of convex bodies II: New inequalities between mixed volumes and their applications* (in Russian), Mat. Sb. N.S. **2** (1937), 1205-1238.

[6] A. D. Alexandrov, *On the theory of mixed volumes of convex bodies IV: Mixed discriminants and mixed volumes* (in Russian), Mat. Sb. N.S. **3** (1938), 227-251.

[7] A. D. Alexandrov, *Convex polyhedra*, Gosudarstv. Izdat. Techn.-Teor. Lit., Moscow-Leningrad 1950, Academie-Verlag, Berlin 1958, Springer-Verlag, Berlin 2005.

[8] S. Alesker, *ψ_2-estimate for the Euclidean norm on a convex body in isotropic position*, Geom. Aspects of Funct. Analysis (Lindenstrauss-Milman eds.), Oper. Theory Adv. Appl. **77** (1995), 1-4.

[9] D. Alonso-Gutiérrez, *On the isotropy constant of random convex sets*, Proc. Amer. Math. Soc. **136** (2008), 3293-3300.

[10] D. Alonso-Gutiérrez, *On an extension of the Blaschke-Santaló inequality and the hyperplane conjecture*, J. Math. Anal. Appl. **344** (2008), 292-300.

[11] D. Alonso-Gutiérrez, *A remark on the isotropy constant of polytopes*, Proc. Amer. Math. Soc. **139** (2011), 2565-2569.

[12] D. Alonso-Gutiérrez, J. Bastero, J. Bernués and G. Paouris, *High dimensional random sections of isotropic convex bodies*, J. Math. Appl. **361** (2010), 431-439.

[13] D. Alonso-Gutiérrez, J. Bastero, J. Bernues and P. Wolff, *On the isotropy constant of projections of polytopes*, Journal of Functional Analysis, **258** (2010), 1452-1465.

[14] D. Alonso-Gutiérrez, N. Dafnis, M. A. Hernandez Cifre and J. Prochno, *On mean outer radii of random polytopes*, Indiana Univ. Math. J. (to appear).

[15] D. Alonso-Gutiérrez and J. Prochno, *Estimating support functions of random polytopes via Orlicz rorms*, Discret. Comput. Geom., DOI 10.1007/s00454-012-9468-7.

[16] D. Alonso-Gutiérrez and J. Prochno, *On the Gaussian behavior of marginals and the mean width of random polytopes*, Proc. Amer. Math. Soc. (to appear).

[17] M. Anttila, K. M. Ball and E. Perissinaki, *The central limit problem for convex bodies*, Trans. Amer. Math. Soc. **355** (2003), 4723-4735.

[18] T. W. Anderson, *The integral of a symmetric unimodal function over a symmetric convex set and some probability inequalities*, Proc. Amer. Math. Soc. **6** (1955), 170-176.

[19] A. Aomoto, *Jacobi polynomials associated with Selberg integrals*, SIAM J. Math. Anal. **19** (1987), 545-549.

[20] J. Arias-de-Reyna, K. M. Ball and R. Villa, *Concentration of the distance in finite-dimensional normed spaces*, Mathematika **45** (1998), 245-252.

[21] S. Artstein, *Proportional concentration phenomena on the sphere*, Israel J. Math. **132** (2002), 337-358.

[22] S. Artstein–Avidan, K. M. Ball, F. Barthe and A. Naor, *Solution of Shannon's problem on the monotonicity of entropy*, J. Amer. Math. Soc. **17** (2004), 975-982.

[23] S. Artstein-Avidan, K. M. Ball, F. Barthe and A. Naor, *On the rate of convergence in the entropic central limit theorem*, Probab. Theory Related Fields **129** (2004), 381-390.

[24] S. Artstein, V. D. Milman and S. J. Szarek, *Duality of metric entropy*, Annals of Math., **159** (2004), no. 3, 1313-1328.

[25] G. Aubrun, *Sampling convex bodies: a random matrix approach*, Proc. Amer. Math. Soc. **135** (2007), 1293-1303.

[26] G. Aubrun, *Maximal inequality for high dimensional cubes*, Confluentes Mathematici **1** (2009), 169-179.

[27] A. Badrikian and S. Chevet, *Mesures cylindriques*, in Espaces de Wiener et Fonctions Aléatoires Gaussiennes, Lecture Notes in Mathematics **379** (1974), Springer.

[28] D. Bakry and M. Emery, *Diffusions hypercontractives*, Séminaire de Probabilités XIX, Lecture Notes in Math. **1123** (1985), 179-206.

[29] D. Bakry and M. Ledoux, *Lévy-Gromov's isoperimetric inequality for an infinite-dimensional diffusion generator*, Invent. Math. **123** (1996), 259-281.

[30] K. M. Ball, *Isometric problems in ℓ_p and sections of convex sets*, Ph.D. Dissertation, Trinity College, Cambridge (1986).

[31] K. M. Ball, *Cube slicing in $\mathbb{R}^n$*, Proc. Amer. Math. Soc. **97** (1986), 465-473.

[32] K. M. Ball, *Logarithmically concave functions and sections of convex sets in $\mathbb{R}^n$*, Studia Math. **88** (1988), 69-84.

[33] K. M. Ball, *Some remarks on the geometry of convex sets*, Geometric aspects of functional analysis (1986/87), Lecture Notes in Math. **1317**, Springer, Berlin (1988), 224-231.

[34] K. M. Ball, *Volumes of sections of cubes and related problems*, Lecture Notes in Mathematics **1376**, Springer, Berlin (1989), 251-260.

[35] K. M. Ball, *Normed spaces with a weak Gordon-Lewis property*, Lecture Notes in Mathematics **1470**, Springer, Berlin (1991), 36-47.

[36] K. M. Ball, *Shadows of convex bodies*, Trans. Amer. Math. Soc. **327** (1991), 891-901.

[37] K. M. Ball, *Volume ratios and a reverse isoperimetric inequality*, J. London Math. Soc. (2) **44** (1991), 351-359.

[38] K. M. Ball, *Ellipsoids of maximal volume in convex bodies*, Geom. Dedicata **41** (1992), 241-250.

[39] K. M. Ball, *An elementary introduction to modern convex geometry*, Flavors of Geometry, Math. Sci. Res. Inst. Publ. **31**, Cambridge Univ. Press (1997).

[40] K. M. Ball, *A remark on the slicing problem*, Geometric aspects of functional analysis, Lecture Notes in Math. **1745**, Springer, Berlin (2000), 21-26.

[41] K. M. Ball, *Convex geometry and functional analysis*, Handbook of the geometry of Banach spaces, Vol. **I**, North-Holland, Amsterdam, (2001), 161-194.

[42] K. M. Ball, F. Barthe and A. Naor, *Entropy jumps in the presence of a spectral gap*, Duke Math. J. **119** (2003), 41-63.

[43] K. M. Ball and V. H. Nguyen, *Entropy jumps for isotropic log-concave random vectors and spectral gap*, Studia Math. **213** (2012), 81-96.

[44] K. M. Ball and A. Pajor, *Convex bodies with few faces*, Proc. Amer. Math. Soc. **110** (1990), no. 1, 225-231.

[45] K. M. Ball and E. Perissinaki, *The subindependence of coordinate slabs for the ℓ_p^n-balls*, Israel J. Math. **107** (1998), 289-299.

[46] I. Bárány and Z. Füredi, *Approximation of the sphere by polytopes having few vertices*, Proc. Amer. Math. Soc. **102** (1988), 651-659.

[47] I. Bárány and D. G. Larman, *Convex bodies, economic cap coverings, random polytopes*, Mathematika **35** (1988), 274–291.

[48] I. Bárány and A. Pór, *On 0 − 1 polytopes with many facets*, Adv. Math. **161** (2001), 209–228.

[49] R. E. Barlow, A. W. Marshall and F. Proschan, *Properties of probability distributions with monotone hazard rate*, Ann. Math. Statist. **34** (1963), 375-389.

[50] A. R. Barron, *Entropy and the central limit theorem*, Ann. Probab. **14** (1986), 336-342.

[51] F. Barthe, *Inégalités fonctionelles et géométriques obtenues par transport des mesures*, Thèse de Doctorat de Mathématiques, Université de Marne-la-Vallée (1997).

[52] F. Barthe, *Inégalités de Brascamp-Lieb et convexité*, C. R. Acad. Sci. Paris Ser. I Math. **324** (1997), no. 8, 885-888.

[53] F. Barthe, *On a reverse form of the Brascamp-Lieb inequality*, Invent. Math. **134** (1998), 335-361.

[54] F. Barthe, *Log-concave and spherical models in isoperimetry*, Geom. Funct. Anal. **12** (2002), 32-55.

[55] F. Barthe, *Un théorème de la limite centrale pour les ensembles convexes*, Séminaire Bourbaki, Vol. 2008/2009, 997-1011.

[56] F. Barthe and D. Cordero-Erausquin, *Invariances in variance estimates*, Proc. London Math. Soc. **106** (2013) 33-64.

[57] F. Barthe, M. Fradelizi and B. Maurey, *A short solution to the Busemann-Petty problem*, Positivity **3** (1999), no. 1, 95-100.

[58] F. Barthe, O. Guédon, S. Mendelson and A. Naor, *A probabilistic approach to the geometry of the ℓ_p^n-ball*, Ann. Prob. **33** (2005), 480-513.

[59] F. Barthe and A. Koldobsky, *Extremal slabs in the cube and the Laplace transform*, Adv. Math. **174** (2003), 89-114.

[60] F. Barthe and A. Naor, *Hyperplane projections of the unit ball of ℓ_p^n*, Discrete Comput. Geom. **27** (2002), 215-226.

[61] J. Bastero, *Upper bounds for the volume and diameter of m-dimensional sections of convex bodies*, Proc. Amer. Math. Soc. **135** (2007), 1851-1859.

[62] J. Bastero and J. Bernués, *Gaussian behaviour and average of marginals for convex bodies*, Extracta Math. **22** (2007), 115-126.

[63] J. Bastero and J. Bernués, *Asymptotic behavior of averages of k-dimensional marginals of measures on $\mathbb{R}^n$*, Studia Math. **190** (2009), 1-31.

[64] J. Batson, D. Spielman and N. Srivastava, *Twice-Ramanujan Sparsifiers*, STOC 2009, SICOMP special issue (2012).

[65] V. Bayle and C. Rosales, *Some isoperimetric comparison theorems for convex bodies in Riemannian manifolds*, Indiana Univ. Math. J. **54** (2005), 1371-1394.

[66] E. F. Beckenbach and R. Bellman, *Inequalities*, Springer-Verlag (1971).

[67] L. Berwald, *Verallgemeinerung eines Mittelwertsatzes von J. Favard für positive konkave Funktionen*, Acta Math. **79** (1947), 17-37.

[68] W. Blaschke, *Kreis und Kugel*, Leipzig (1916).

[69] W. Blaschke, *Über affine Geometrie VII: Neue Extremaigenschaften von Ellipse und Ellipsoid*, Ber. Vergh. Sächs. Akad. Wiss. Leipzig, Math.-Phys. Kl. **69** (1917), 306-318, Ges. Werke **3**, 246-258.

[70] W. Blaschke, *Lösung des "Vierpunktproblems" von Sylvester aus der Theorie der geometrischen Wahrscheinlichkeiten*, Ber. Verh. sachs. Acad. Wiss., Math. Phys. Kl. **69** (1917), 436-453.

[71] W. Blaschke, *Vorlesungen über Differentialgeometrie II*, Springer, Berlin (1923).

[72] N. M. Blachman, *The convolution inequality for entropy powers*, IEEE Trans. Info. Theory **2** (1965), 267-271.

[73] S. G. Bobkov, *Extremal properties of half-spaces for log-concave distributions*, Ann. Probab. **24** (1996), 35-48.

[74] S. G. Bobkov, *A functional form of the isoperimetric inequality for the Gaussian measure*, J. Funct. Anal. **135** (1996), 39-49.

[75] S. G. Bobkov, *An isoperimetric inequality on the discrete cube and an elementary proof of the isoperimetric inequality in Gauss space*, Ann. Probab. **25** (1997), 206–214.

[76] S. G. Bobkov, *Isoperimetric and analytic inequalities for log-concave probability measures*, Ann. Prob. **27** (1999), 1903-1921.

[77] S. G. Bobkov, *Remarks on the growth of L^p-norms of polynomials*, Geom. Aspects of Funct. Analysis (Milman-Schechtman eds.), Lecture Notes in Math. **1745** (2000), 27-35.

[78] S. G. Bobkov, *On concentration of distributions of random weighted sums*, Ann. Probab. **31** (2003), 195-215.

[79] S. G. Bobkov, *Spectral gap and concentration for some spherically symmetric probability measures*, Geom. Aspects of Funct. Analysis, Lecture Notes in Math. **1807**, Springer, Berlin (2003), 37-43.

[80] S. G. Bobkov, *On isoperimetric constants for log-concave probability distributions*, Geometric aspects of functional analysis, Lecture Notes in Math., **1910**, Springer, Berlin (2007), 81-88.

[81] S. G. Bobkov, *Large deviations and isoperimetry over convex probability measures with heavy tails*, Electronic Journal of Probability **12** (2007), 1072-1100.

[82] S. G. Bobkov, *Convex bodies and norms associated to convex measures*, Probab. Theory Related Fields **147** (2010), 303-332.

[83] S. G. Bobkov, *Gaussian concentration for a class of spherically invariant measures*, Problems in mathematical analysis 46, J. Math. Sci. **167** (2010), 326-339.

[84] S. G. Bobkov and C. Houdré, *Some connections between isoperimetric and Sobolev-type inequalities*, Mem. Amer. Math. Soc. **129** (1997).

[85] S. G. Bobkov and A. Koldobsky, *On the central limit property of convex bodies*, Geom. Aspects of Funct. Analysis (Milman-Schechtman eds.), Lecture Notes in Math. **1807** (2003), 44-52.

[86] S. G. Bobkov and M. Ledoux, *From Brunn-Minkowski to Brascamp-Lieb and to logarithmic Sobolev inequalities*, Geom. Funct. Anal. **10** (2000), 1028–1052.

[87] S. G. Bobkov and M. Madiman, *The entropy per coordinate of a random vector is highly constrained under convexity conditions*, IEEE Transactions on Information Theory, vol. **57** (2011), no. 8, 4940-4954.

[88] S. G. Bobkov and M. Madiman, *Reverse Brunn-Minkowski and reverse entropy power inequalities for convex measures*, J. Math. Sciences (New York), vol. **179** (2011), no. 1, 2-6. Translated from: Problems in Math. Analysis **61** (2011), 5-8.

[89] S. G. Bobkov and M. Madiman, *Concentration of the information in data with log-concave distributions*, Ann. Probab. **39** (2011), 1528-1543.

[90] S. G. Bobkov and F. L. Nazarov, *On convex bodies and log-concave probability measures with unconditional basis*, Geom. Aspects of Funct. Analysis (Milman-Schechtman eds.), Lecture Notes in Math. **1807** (2003), 53-69.

[91] S. G. Bobkov and F. L. Nazarov, *Large deviations of typical linear functionals on a convex body with unconditional basis*, Stochastic Inequalities and Applications, Progr. Probab. **56**, Birkhäuser, Basel (2003), 3-13.

[92] V. Bogachev, *Gaussian measures*, Amer. Math. Soc. (1998).

[93] E. D. Bolker, *A class of convex bodies*, Trans. Amer. Math. Soc. **145** (1969), 323-345.

[94] T. Bonnesen and W. Fenchel, *Theorie der konvexen Körper*, Springer, Berlin, 1934. Reprint: Chelsea Publ. Co., New York, 1948. English translation: BCS Associates, Moscow, Idaho, 1987.

[95] C. Borell, *Complements of Lyapunov's inequality*, Math. Ann. **205** (1973), 323-331.

[96] C. Borell, *Convex measures on locally convex spaces*, Ark. Mat. **12** (1974), 239-252.

[97] C. Borell, *Convex set functions in d-space*, Period. Math. Hungar. **6** (1975), 111-136.

[98] C. Borell, *The Brunn-Minkowski inequality in Gauss space*, Inventiones Math. **30** (1975), 207-216.

[99] J. Bourgain, *On high dimensional maximal functions associated to convex bodies*, Amer. J. Math. **108** (1986), 1467-1476.

[100] J. Bourgain, *On the L^p-bounds for maximal functions associated to convex bodies*, Israel J. Math. **54** (1986), 257-265.

[101] J. Bourgain, *On the distribution of polynomials on high dimensional convex sets*, Lecture Notes in Mathematics **1469**, Springer, Berlin (1991), 127-137.

[102] J. Bourgain, *On the Busemann-Petty problem for perturbations of the ball*, Geom. Funct. Anal. **1** (1991), no. 1, 1-13.

[103] J. Bourgain, *Random points in isotropic convex bodies*, in *Convex Geometric Analysis* (Berkeley, CA, 1996) Math. Sci. Res. Inst. Publ. **34** (1999), 53-58.

[104] J. Bourgain, *On the isotropy constant problem for ψ_2-bodies*, Geom. Aspects of Funct. Analysis (Milman-Schechtman eds.), Lecture Notes in Math. **1807** (2003), 114-121.

[105] J. Bourgain, *On the Hardy-Littlewood maximal function for the cube*, Israel J. Math. (to appear).

[106] J. Bourgain, B. Klartag and V. D. Milman, *A reduction of the slicing problem to finite volume ratio bodies*, C. R. Acad. Sci. Paris, Ser. I **336** (2003), no. 4, 331-334.

[107] J. Bourgain, B. Klartag and V. D. Milman, *Symmetrization and isotropic constants of convex bodies*, Geom. Aspects of Funct. Analysis, Lecture Notes in Math. **1850** (2004), 101-115.

[108] J. Bourgain and J. Lindenstrauss, *Projection bodies*, Geometric aspects of functional analysis (1986/87), Lecture Notes in Mathematics **1317**, Springer, Berlin (1988), 250-270.

[109] J. Bourgain, J. Lindenstrauss and V. D. Milman, *Minkowski sums and symmetrizations*, Geom. Aspects of Funct. Analysis (Lindenstrauss-Milman eds.), Lecture Notes in Math. **1317** (1988), 44-74.

[110] J. Bourgain, J. Lindenstrauss and V. D. Milman, *Approximation of zonoids by zonotopes*, Acta Math. **162** (1989), no. 1-2, 73-141.

[111] J. Bourgain, M. Meyer, V. D. Milman and A. Pajor, *On a geometric inequality*, Geometric aspects of functional analysis (1986/87), Lecture Notes in Math., **1317**, Springer, Berlin (1988), 271-282.

[112] J. Bourgain and V. D. Milman, *New volume ratio properties for convex symmetric bodies in $\mathbb{R}^n$*, Invent. Math. **88** (1987), 319-340.

[113] H. J. Brascamp and E. H. Lieb, *Best constants in Young's inequality, its converse and its generalization to more than three functions*, Adv. in Math. **20** (1976), 151-173.

[114] H. J. Brascamp and E. H. Lieb, *On extensions of the Brunn-Minkowski and Prékopa-Leindler theorems, including inequalities for log-concave functions, and with an application to the diffusion equation*, J. Funct. Anal. **22** (1976), 366-389.

[115] H. J. Brascamp, E. H. Lieb and J. M. Luttinger, *A general rearrangement inequality for multiple integrals*, J. Funct. Anal. **17** (1974), 227-237.

[116] S. Brazitikos and P. Stavrakakis, *On the intersection of random rotations of a symmetric convex body*, Preprint.

[117] U. Brehm and J. Voigt, *Asymptotics of cross sections for convex bodies*, Beiträge Algebra Geom. **41** (2000), 437-454.

[118] U. Brehm, H. Vogt and J. Voigt, *Permanence of moment estimates for p-products of convex bodies*, Studia Math. **150** (2002), 243-260.

[119] U. Brehm, P. Hinow, H. Vogt and J. Voigt, *Moment inequalities and central limit properties of isotropic convex bodies*, Math. Zeitsch. **240** (2002), 37-51.

[120] Y. Brenier, *Décomposition polaire et réarrangement monotone des champs de vecteurs*, C. R. Acad. Sci. Paris Sér. I Math., **305** (1987), 805-808.

[121] H. Brezis, *Functional analysis, Sobolev spaces and partial differential equations*, Universitext, Springer, New York (2011).

[122] H. Brunn, *Über Ovale und Eiflächen*, Dissertation, München, 1887.

[123] H. Brunn, *Über Curven ohne Wendepunkte*, Habilitationsschrift, München, 1887.

[124] C. Buchta, J. Müller and R. F. Tichy, *Stochastical approximation of convex bodies*, Math. Ann. **271** (1985), 225-235.

[125] Y. D. Burago and V. A. Zalgaller, *Geometric Inequalities*, Springer Series in Soviet Mathematics, Springer-Verlag, Berlin-New York (1988).

[126] H. Busemann, *A theorem on convex bodies of the Brunn-Minkowski type*, Proc. Nat. Acad. Sci. U.S.A **35** (1949), 27-31.

[127] H. Busemann, *Volume in terms of concurrent cross-sections*, Pacific J. Math. **3** (1953), 1-12.

[128] H. Busemann and C. M. Petty, *Problems on convex bodies*, Math. Scand. **4** (1956), 88-94.

[129] H. Busemann, *Volumes and areas of cross-sections*, Amer. Math. Monthly **67** (1960), 248-250.

[130] P. Buser, *A note on the isoperimetric constant*, Ann. Sci. École Norm. Sup. **15** (1982), 213-230.

[131] S. Campi and P. Gronchi, *The L^p-Busemann-Petty centroid inequality*, Adv. in Math. **167** (2002), 128-141.

[132] A. Carbery, *Radial Fourier multipliers and associated maximal functions*, Recent Progress in Fourier Analysis, North-Holland Math. Studies **111** (1985), 49-56.

[133] A. Carbery and J. Wright, *Distributional and L_q-norm inequalities for polynomials over convex bodies in $\mathbb{R}^n$*, Math. Res. Lett. **8** (2001), no. 3, 233-248.

[134] B. Carl and A. Pajor, *Gelfand numbers of operators with values in a Hilbert space*, Invent. Math. **94** (1988), 479-504.

[135] E. A. Carlen and A. Soffer, *Entropy production by block variable summation and central limit theorem*, Commun. Math. Phys. **140** (1991), 339-371.

[136] G. D. Chakerian, *Inequalities for the difference body of a convex body*, Proc. Amer. Math. Soc. **18** (1967), 879-884.

[137] I. Chavel, *The Laplacian on Riemannian manifolds*, Spectral theory and geometry, London Math. Soc., Lecture Note Ser. **273**, Cambridge University Press, Cambridge (1999), 30-75.

[138] I. Chavel, *Isoperimetric inequalities - Differential geometric and analytic perspectives*, Cambridge Tracts in Mathematics **145**, Cambridge University Press, Cambridge (2001).

[139] I. Chavel, *Riemannian geometry - A modern introduction*, Second edition, Cambridge Studies in Advanced Mathematics **98**, Cambridge University Press, Cambridge (2006).

[140] J. Cheeger, *A lower bound for the smallest eigenvalue of the Laplacian*, Problems in Analysis (Papers dedicated to Salomon Bochner, 1969), Princeton Univ. Press, Princeton (1970), 195-199.

[141] D. Cordero-Erausquin, *Some applications of mass transport to Gaussian-type inequalities*, Arch. Rational Mech. Anal. **161** (2002), 257-269.

[142] D. Cordero-Erausquin, M. Fradelizi and B. Maurey, *The (B)-conjecture for the Gaussian measure of dilates of symmetric convex sets and related problems*, J. Funct. Anal. **214** (2004), 410-427.

[143] M. Costa and T. Cover, *On the similarity of the entropy power inequality and the Brunn-Minkowski inequality*, IEEE Trans. Inform. Theory **30** (1984), 837-839.

[144] T. Cover and J. Thomas, *Elements of Information Theory*, New York: J. Wiley (1991).

[145] I. Csiszár, *Informationstheoretische Konvergenzbegriffe im Raum der Wahrscheinlichkeitsverteilungen*, Magyar Tud. Akad. Mat. Kutató Int. Közl. **7** (1962), 137-158.

[146] N. Dafnis, A. Giannopoulos and O. Guédon, *On the isotropic constant of random polytopes*, Advances in Geometry **10** (2010), 311-321.

[147] N. Dafnis, A. Giannopoulos and A. Tsolomitis, *Asymptotic shape of a random polytope in a convex body*, J. Funct. Anal. **257** (2009), 2820-2839.

[148] N. Dafnis, A. Giannopoulos and A. Tsolomitis, *Quermassintegrals and asymptotic shape of a random polytope in an isotropic convex body*, Michigan Math. Journal (to appear).

[149] N. Dafnis and G. Paouris, *Small ball probability estimates, ψ_2-behavior and the hyperplane conjecture*, J. Funct. Anal. **258** (2010), 1933-1964.

[150] N. Dafnis and G. Paouris, *Estimates for the affine and dual affine quermassintegrals of convex bodies*, Illinois J. of Math. (to appear).

[151] L. Dalla and D. G. Larman, *Volumes of a random polytope in a convex set*, Applied geometry and discrete mathematics. Discrete Math. Theoret. Comput. Sci. **4** (Amer. Math. Soc.) (1991), 175-180.

[152] S. Dar, *Remarks on Bourgain's problem on slicing of convex bodies*, in Geometric Aspects of Functional Analysis, Operator Theory: Advances and Applications **77** (1995), 61-66.

[153] S. Dar, *On the isotropic constant of non-symmetric convex bodies*, Israel J. Math. **97** (1997), 151-156.

[154] S. Dar, *Isotropic constants of Schatten class spaces*, Convex Geometric Analysis, MSRI Publications **34** (1998), 77-80.

[155] S. Das Gupta, *Brunn-Minkowski inequality and its aftermath*, J. Multivariate Anal. **10** (1980), 296-318.

[156] A. Dembo, T. Cover, and J. Thomas, *Information-theoretic inequalities*, IEEE Trans. Inform. Theory **37** (1991), 1501-1518.

[157] P. Diaconis and D. Freedman, *Asymptotics of graphical projection pursuit*, Ann. of Stat. **12** (1984), 793-815.

[158] R. M. Dudley, *The sizes of compact subsets of Hilbert space and continuity of Gaussian processes*, J. Funct. Anal. **1** (1967), 290-330.

[159] R. Durrett, *Stochastic calculus: A practical introduction*, Cambridge University Press (2003).

[160] A. Dvoretzky, *A theorem on convex bodies and applications to Banach spaces*, Proc. Nat. Acad. Sci. U.S.A **45** (1959), 223-226.

[161] A. Dvoretzky, *Some results on convex bodies and Banach spaces*, in Proc. Sympos. Linear Spaces, Jerusalem (1961), 123-161.

[162] A. Dvoretzky and C. A. Rogers, *Absolute and unconditional convergence in normed linear spaces*, Proc. Nat. Acad. Sci., U.S.A **36** (1950), 192-197.

[163] M. E. Dyer, Z. Füredi and C. McDiarmid, *Volumes spanned by random points in the hypercube*, Random Structures Algorithms **3** (1992), 91-106.

[164] R. Eldan, *Thin shell implies spectral gap up to polylog via a stochastic localization scheme* (preprint).

[165] R. Eldan and B. Klartag, *Pointwise estimates for marginals of convex bodies*, J. Funct. Anal. **254** (2008), 2275-2293.

[166] R. Eldan and B. Klartag, *Approximately Gaussian marginals and the hyperplane conjecture* Concentration, functional inequalities and isoperimetry, Contemp. Math. **545**, Amer. Math. Soc., Providence, RI (2011), 55-68.
[167] R. Eldan and B. Klartag, *Dimensionality and the stability of the Brunn-Minkowski inequality*, Ann. Sc. Norm. Super. Pisa (to appear).
[168] A. Erhard, *Symétrisation dans l'espace de Gauss*, Math. Scand. **53** (1983), 281-301.
[169] W. Feller, *An Introduction to Probability and its Applications* Vol. I, 3rd ed., Wiley, New York (1968).
[170] W. Feller, *An Introduction to Probability and its Applications* Vol. II, 2nd ed., Wiley, New York (1971).
[171] W. Fenchel, *Inégalités quadratiques entre les volumes mixtes des corps convexes*, C.R. Acad. Sci. Paris **203** (1936), 647-650.
[172] W. Fenchel and B. Jessen, *Mengenfunktionen und konvexe Körper*, Danske Vid. Selskab. Mat.-fys. Medd. **16** (1938), 1-31.
[173] X. Fernique, *Régularité des trajectoires des fonctions aléatoires Gaussiennes*, Ecole d'Eté de St. Flour IV, 1974, Lecture Notes in Mathematics **480** (1975), 1-96.
[174] T. Figiel, J. Lindenstrauss and V. D. Milman, *The dimension of almost spherical sections of convex bodies*, Acta Math. **139** (1977), 53-94.
[175] T. Figiel and N. Tomczak-Jaegermann, *Projections onto Hilbertian subspaces of Banach spaces*, Israel J. Math. **33** (1979), 155-171.
[176] T. Fleiner, V. Kaibel and G. Rote, *Upper bounds on the maximal number of faces of 0/1 polytopes*, European J. Combin. **21** (2000), 121-130.
[177] B. Fleury, *Concentration in a thin Euclidean shell for log-concave measures*, J. Funct. Anal. **259** (2010), 832-841.
[178] B. Fleury, *Between Paouris concentration inequality and variance conjecture*, Ann. Inst. Henri Poincaré Probab. Stat. **46** (2010), 299-312.
[179] B. Fleury, *Poincaré inequality in mean value for Gaussian polytopes*, Probability Theory and Related Fields, Vol. **152**, Issue 1-2, (2012), 141-178.
[180] B. Fleury, O. Guédon and G. Paouris, *A stability result for mean width of L_p-centroid bodies*, Adv. Math. **214**, 2 (2007), 865-877.
[181] G. B. Folland, *Introduction to Partial Differential Equations*, Mathematical Notes, Princeton University Press, Princeton, NJ (1976).
[182] M. Fradelizi, *Sections of convex bodies through their centroid*, Arch. Math. **69** (1997), 515-522.
[183] M. Fradelizi and O. Guédon, *The extreme points of subsets of s-concave probabilities and a geometric localization theorem*, Discrete Comput. Geom. **31** (2004), 327-335.
[184] M. Fradelizi and O. Guédon, *A generalized localization theorem and geometric inequalities for convex bodies*, Adv. in Math. **204** (2006), 509-529.
[185] A. Frieze and R. Kannan, *Log-Sobolev inequalities and sampling from log-concave distributions*, Ann. Appl. Probab. **9** (1999), 14-26.
[186] K. Fukuda, *Frequently Asked Questions in Polyhedral Computation*, available at (http://www.ifor.math.ethz.ch/staff/fukuda/polyfaq/polyfaq.html).
[187] R. J. Gardner, *Intersection bodies and the Busemann-Petty problem*, Trans. Amer. Math. Soc. **342** (1994), 435-445.
[188] R. J. Gardner, *On the Busemann-Petty problem concerning central sections of centrally symmetric convex bodies*, Bull. Amer. Math. Soc. **30** (1994), 222-226.
[189] R. J. Gardner, *A positive answer to the Busemann-Petty problem in three dimensions*, Annals of Math. **140** (1994), 435-447.
[190] R. J. Gardner, *The Brunn-Minkowski inequality*, Bull. Amer. Math. Soc. (N.S.) **39** (2002), 355-405.
[191] R. J. Gardner, *Geometric Tomography*, Second Edition Encyclopedia of Mathematics and its Applications **58**, Cambridge University Press, Cambridge (2006).
[192] R. J. Gardner, A. Koldobsky, and T. Schlumprecht, *A complete analytic solution to the Busemann-Petty problem*, Annals of Math. **149** (1999), 691-703.
[193] D. Gatzouras and A. Giannopoulos, *A large deviations approach to the geometry of random polytopes*, Mathematika **53** (2006), 173-210.
[194] D. Gatzouras and A. Giannopoulos, *Threshold for the volume spanned by random points with independent coordinates*, Israel J. Math. **169** (2009), 125-153.

[195] D. Gatzouras, A. Giannopoulos and N. Markoulakis, *Lower bound for the maximal number of facets of a 0/1 polytope*, Discrete Comput. Geom. **34** (2005), 331-349.

[196] D. Gatzouras, A. A. Giannopoulos and N. Markoulakis, *On the maximal number of facets of 0/1 polytopes*, Geom. Aspects of Funct. analysis, Lecture Notes in Math. **1910**, Springer, Berlin (2007), 117-125.

[197] A. Giannopoulos, *A note on a problem of H. Busemann and C.M. Petty concerning sections of symmetric convex bodies*, Mathematika **37** (1990), 239-244.

[198] A. Giannopoulos, *On the mean value of the area of a random polygon in a plane convex body*, Mathematika **39** (1992), 279-290.

[199] A. Giannopoulos, *Problems on convex bodies*, PhD Thesis (June 1993), University of Crete.

[200] A. Giannopoulos, *Notes on isotropic convex bodies*, Lecture Notes, Warsaw 2003, available at http://users.uoa.gr/~apgiannop/.

[201] A. Giannopoulos and M. Hartzoulaki, *Random spaces generated by vertices of the cube*, Discrete Comput. Geom. **28** (2002), 255-273.

[202] A. Giannopoulos, M. Hartzoulaki and A. Tsolomitis, *Random points in isotropic unconditional convex bodies*, J. London Math. Soc. **72** (2005), 779-798.

[203] A. Giannopoulos and V. D. Milman, *Mean width and diameter of proportional sections of a symmetric convex body*, J. Reine Angew. Math. **497** (1998), 113-139.

[204] A. Giannopoulos and V. D. Milman, *Extremal problems and isotropic positions of convex bodies*, Israel J. Math. **117** (2000), 29-60.

[205] A. Giannopoulos and V. D. Milman, *Concentration property on probability spaces*, Adv. in Math. **156** (2000), 77-106.

[206] A. Giannopoulos and V. D. Milman, *Euclidean structure in finite dimensional normed spaces*, Handbook of the Geometry of Banach spaces (Lindenstrauss-Johnson eds), Elsevier (2001), 707-779.

[207] A. Giannopoulos and V. D. Milman, *Asymptotic convex geometry: short overview*, Different faces of geometry, 87-162, Int. Math. Ser. **3**, Kluwer/Plenum, New York, 2004.

[208] A. Giannopoulos, V. D. Milman and A. Tsolomitis, *Asymptotic formulas for the diameter of sections of symmetric convex bodies*, Journal of Functional Analysis **223** (2005), 86-108.

[209] A. Giannopoulos, A. Pajor and G. Paouris, *A note on subgaussian estimates for linear functionals on convex bodies*, Proc. Amer. Math. Soc. **135** (2007), 2599-2606.

[210] A. Giannopoulos, G. Paouris and P. Valettas, *On the existence of subgaussian directions for log-concave measures*, Contemporary Mathematics **545** (2011), 103-122.

[211] A. Giannopoulos, G. Paouris and P. Valettas, ψ_α-*estimates for marginals of log-concave probability measures*, Proc. Amer. Math. Soc. **140** (2012), 1297-1308.

[212] A. Giannopoulos, G. Paouris and P. Valettas, *On the distribution of the ψ_2-norm of linear functionals on isotropic convex bodies*, in Geom. Aspects of Funct. Analysis, Lecture Notes in Mathematics **2050** (2012), 227-254.

[213] A. Giannopoulos, G. Paouris and B-H. Vritsiou, *A remark on the slicing problem*, Journal of Functional Analysis **262** (2012), 1062-1086.

[214] A. Giannopoulos, G. Paouris and B-H. Vritsiou, *The isotropic position and the reverse Santaló inequality*, Israel J. Math., doi: 10.1007/s11856-012-0173-2.

[215] A. Giannopoulos and M. Papadimitrakis, *Isotropic surface area measures*, Mathematika **46** (1999), 1-13.

[216] A. Giannopoulos, P. Stavrakakis, A. Tsolomitis and B-H. Vritsiou, *Geometry of the L_q-centroid bodies of an isotropic log-concave measure*, Trans. Amer. Math. Soc. (to appear).

[217] A. Giannopoulos and A. Tsolomitis, *On the volume radius of a random polytope in a convex body*, Math. Proc. Cambridge Phil. Soc. **134** (2003), 13-21.

[218] M. Giertz, *A note on a problem of Busemann*, Math. Scand. **25** (1969), 145-148.

[219] D. Gilbarg and N. S. Trudinger, *Elliptic Partial Differential Equations of Second Order*, 2nd ed., Grundlehren Math. Wiss. **224**, Springer, Berlin (1983).

[220] E. D. Gluskin, *Extremal properties of orthogonal parallelepipeds and their applications to the geometry of Banach spaces*, Mat. Sb. (N.S.) **136** (1988), 85-96.

[221] P. R. Goodey and W. Weil, *Zonoids and generalizations*, in Handbook of Convex Geometry (P. M. Gruber and J.M. Wills, Eds.) North-Holland (1993).

[222] Y. Gordon, *On Milman's inequality and random subspaces which escape through a mesh in $\mathbb{R}^n$*, Lecture Notes in Mathematics **1317** (1988), 84-106.

[223] Y. Gordon, M. Meyer and S. Reisner, *Zonoids with minimal volume product - a new proof*, Proc. Amer. Math. Soc. **104** (1988), 273-276.

[224] N. Gozlan, *Characterization of Talagrand's like transportation-cost nequalities on the real line*, J. Funct. Anal. **250** (2007), 400-425.

[225] H. Groemer, *On some mean values associated with a randomly selected simplex in a convex set*, Pacific J. Math. **45** (1973), 525-533.

[226] H. Groemer, *On the mean value of the volume of a random polytope in a convex set*, Arch. Math. **25** (1974), 86-90.

[227] H. Groemer, *On the average size of polytopes in a convex set*, Geom. Dedicata **13** (1982), 47-62.

[228] H. Groemer, *Geometric applications of Fourier series and spherical harmonics*, Encyclopedia of Mathematics and its Applications, **61**, Cambridge University Press, Cambridge (1996).

[229] M. Gromov, *Dimension, nonlinear spectra and width*, Geom. Aspects of Funct. Analysis, Lecture Notes in Math. **1317**, Springer, Berlin (1988), 132-184.

[230] M. Gromov, *Convex sets and Kähler manifolds*, in "Advances in Differential Geometry and Topology", World Scientific Publishing, Teaneck NJ (1990), 1-38.

[231] M. Gromov, *Metric Structures for Riemannian and Non-Riemannian Spaces*, based on "Structures métriques des variétés Riemanniennes" (L. LaFontaine, P. Pansu. eds.), English translation by Sean M. Bates, Birkhäuser, Boston-Basel-Berlin, 1999 (with Appendices by M. Katz, P. Pansu and S. Semmes).

[232] M. Gromov and V. D. Milman, *A topological application of the isoperimetric inequality*, Amer. J. Math. **105** (1983), 843-854.

[233] M. Gromov and V. D. Milman, *Brunn theorem and a concentration of volume phenomenon for symmetric convex bodies*, GAFA Seminar Notes, Tel Aviv University (1984).

[234] M. Gromov and V. D. Milman, *Generalization of the spherical isoperimetric inequality to uniformly convex Banach spaces*, Compos. Math. **62** (1987), 263-282.

[235] L. Gross, *Logarithmic Sobolev inequalities*, Amer. J. Math. **97** (1975), 1061-1083.

[236] L. Gross, *Logarithmic Sobolev inequalities and contractivity properties of semigroups*, Dirichlet forms, Varenna 1992, Lecture Notes in Math. **1563**, Springer, Berlin (1993), 54-88.

[237] P. M. Gruber, *Convex and Discrete Geometry*, Grundlehren Math. Wiss. **336**, Springer, Heidelberg (2007).

[238] B. Grünbaum, *Partitions of mass-distributions and of convex bodies by hyperplanes*, Pacific J. Math. **10** (1960), 1257-1261.

[239] O. Guédon, *Kahane-Khinchine type inequalities for negative exponent*, Mathematika **46** (1999), 165-173.

[240] O. Guédon and E. Milman, *Interpolating thin shell and sharp large-deviation estimates for isotropic log-concave measures*, Geom. Funct. Anal. **21** (2011), 1043-1068.

[241] O. Guédon and G. Paouris, *Concentration of mass on the Schatten classes*, Ann. Inst. H. Poincare Probab. Statist. **43** (2007), 87-99.

[242] O. Guédon and M. Rudelson, L_p-*moments of random vectors via majorizing measures*, Adv. Math. **208** (2007), no. 2, 798-823.

[243] U. Haagerup, *The best constants in the Khintchine inequality*, Studia Math. **70** (1982), 231-283.

[244] H. Hadwiger and D. Ohmann, *Brunn-Minkowskischer Satz und Isoperimetrie*, Math. Z. **66** (1956), 1-8.

[245] H. Hadwiger, *Radialpotenzintegrale zentralsymmetrischer Rotationskörper und Ungleichheitsaussagen Busemannscher Art*, Math. Scand. **23** (1968), 193-200.

[246] G. H. Hardy, J. E. Littlewood and G. Pólya, *Inequalities*, 2nd ed. London: Cambridge University Press (1964).

[247] G. Hargé, *A particular case of correlation inequality for the Gaussian measure*, Ann. Probab. **27** (1999), 1939-1951.

[248] L. H. Harper, *Optimal numberings and isoperimetric problems on graphs*, J. Combin. Theory **1** (1966), 385-393.

[249] M. Hartzoulaki, *Probabilistic methods in the theory of convex bodies*, Ph.D. Thesis (March 2003), University of Crete.

[250] M. Hartzoulaki and G. Paouris, *Quermassintegrals of a random polytope in a convex body*, Arch. Math. **80** (2003), 430-438.

[251] B. Helffer and J. Sjöstrand, *On the correlation for Kac-like models in the convex case*, J. Statist. Phys. **74** (1994), 349-409.

[252] D. Hensley, *Slicing the cube in $\mathbb{R}^n$ and probability (bounds for the measure of a central cube slice in $\mathbb{R}^n$ by probability methods)*, Proc. Amer. Math. Soc. **73**(1979), 95-100.

[253] D. Hensley, *Slicing convex bodies: bounds for slice area in terms of the body's covariance*, Proc. Amer. Math. Soc. **79** (1980), 619-625.

[254] R. Henstock and A. M. Macbeath, *On the measure of sum-sets, I: The theorems of Brunn, Minkowski and Lusternik*, Proc. London Math. Soc. **3** (1953), 182-194.

[255] T. Holmstedt, *Interpolation of quasi-normed spaces*, Math. Scand. **26** (1970), 177-199.

[256] L. Hörmander, *L^2-estimates and existence theorems for the $\bar{\partial}$ operator*, Acta Math. **113** (1965), 89-152.

[257] L. Hörmander, *Notions of Convexity*, Progress in Math. **127**, Birkhäuser, Boston-Basel-Berlin (1994).

[258] N. Huet, *Spectral gap for some invariant log-concave probability measures*, Mathematika **57** (2011), 51-62.

[259] N. Huet, *Isoperimetry for spherically symmetric log-concave probability measures*, Rev. Mat. Iberoam. **27** (2011), 93-122.

[260] F. John, *Extremum problems with inequalities as subsidiary conditions*, Courant Anniversary Volume, Interscience, New York (1948), 187-204.

[261] M. Junge, *On the hyperplane conjecture for quotient spaces of L_p*, Forum Math. **6** (1994), 617-635.

[262] M. Junge, *Proportional subspaces of spaces with unconditional basis have good volume properties*, in Geometric Aspects of Functional Analysis, Operator Theory: Advances and Applications **77** (1995), 121-129.

[263] W. B. Johnson and J. Lindenstrauss, *Extensions of Lipschitz mappings into a Hilbert space*, in Conference in modern analysis and probability (New Haven, Conn.) (1982), 189-206.

[264] J. Kadlec, *The regularity of the solution of the Poisson problem in a domain whose boundary is similar to that of a convex domain*, Czechoslovak Math. J. **89** (1964), 386-393.

[265] J.-P. Kahane, *Some Random Series of Functions*, Cambridge Studies in Advanced Mathematics **5**, Cambridge Univ. Press, Cambridge (1985).

[266] R. Kannan, L. Lovász and M. Simonovits, *Isoperimetric problems for convex bodies and a localization lemma*, Discrete Comput. Geom. **13** (1995), 541-559.

[267] R. Kannan, L. Lovász and M. Simonovits, *Random walks and $O^*(n^5)$ volume algorithm for convex bodies*, Random Structures Algorithms II **1** (1997), 1-50.

[268] M. Kanter, *Unimodality and dominance for symmetric random vectors*, Trans. Amer. Math. Sot. **229** (1977), 65-85.

[269] B. S. Kashin, *Sections of some finite-dimensional sets and classes of smooth functions*, Izv. Akad. Nauk. SSSR Ser. Mat. **41** (1977), 334-351.

[270] A. Khintchine, *Über dyadische Brüche*, Math. Z. **18** (1923), 109-116.

[271] B. Klartag, *A geometric inequality and a low M-estimate*, Proc. Amer. Math. Soc. **132** (2004), 2619-2628.

[272] B. Klartag, *An isomorphic version of the slicing problem*, J. Funct. Anal. **218** (2005), 372-394.

[273] B. Klartag, *On convex perturbations with a bounded isotropic constant*, Geom. Funct. Analysis **16** (2006), 1274-1290.

[274] B. Klartag, *Isomorphic and almost-isometric problems in high-dimensional convex geometry*, International Congress of Mathematicians. Vol. II, Eur. Math. Soc., Zurich (2006), 1547-1562.

[275] B. Klartag, *Marginals of geometric inequalities*, Geom. Aspects of Funct. Analysis, Lecture Notes in Math. **1910**, Springer, Berlin (2007), 133-166.

[276] B. Klartag, *Uniform almost sub-gaussian estimates for linear functionals on convex sets*, Algebra i Analiz (St. Petersburg Math. Journal) **19** (2007), 109-148.

[277] B. Klartag, *A central limit theorem for convex sets*, Invent. Math. **168** (2007), 91-131.

[278] B. Klartag, *Power-law estimates for the central limit theorem for convex sets*, J. Funct. Anal. **245** (2007), 284-310.

[279] B. Klartag, *A Berry-Esseen type inequality for convex bodies with an unconditional basis*, Probab. Theory Related Fields **145** (2009), 1-33.

[280] B. Klartag, *On nearly radial marginals of high-dimensional probability measures*, J. Eur. Math. Soc, Vol. **12**, (2010), 723-754.

[281] B. Klartag, *High-dimensional distributions with convexity properties*, European Congress of Mathematics, Eur. Math. Soc., Zurich (2010), 401-417.

[282] B. Klartag and G. Kozma, *On the hyperplane conjecture for random convex sets*, Israel J. Math. **170** (2009), 253-268.

[283] B. Klartag and E. Milman, *On volume distribution in 2-convex bodies*, Israel J. Math. **164** (2008), 221-249.

[284] B. Klartag and E. Milman, *Centroid Bodies and the Logarithmic Laplace Transform - A Unified Approach*, J. Funct. Anal. **262** (2012), 10-34.

[285] B. Klartag and E. Milman, *Inner regularization of log-concave measures and small-ball estimates*, in Geom. Aspects of Funct. Analysis, Lecture Notes in Math. **2050**, 267-278.

[286] B. Klartag and V. D. Milman, *Geometry of log-concave functions and measures*, Geom. Dedicata **112** (2005), 169-182.

[287] B. Klartag and V. D. Milman, *Rapid Steiner Symmetrization of most of a convex body and the slicing problem*, Combin. Probab. Comput. **14** (2005), 829-843.

[288] B. Klartag and R. Vershynin, *Small ball probability and Dvoretzky theorem*, Israel J. Math. **157** (2007), 193-207.

[289] H. Knothe, *Contributions to the theory of convex bodies*, Michigan Math. J. **4** (1957), 39-52.

[290] A. Koldobsky, *An application of the Fourier transform to sections of star bodies*, Israel J. Math. **106** (1998), 157-164.

[291] A. Koldobsky, *Intersection bodies, positive definite distributions and the Busemann-Petty problem*, Amer. J. Math. **120** (1998), 827-840.

[292] A. Koldobsky, *Fourier analysis in convex geometry*, Mathematical Surveys and Monographs **116**, American Mathematical Society, Providence, RI, 2005.

[293] A. Koldobsky and M. Lifshits, *Average volume of sections of star bodies*, Geom. Aspects of Funct. Analysis (Milman-Schechtman eds.), Lecture Notes in Math. **1745** (2000), 119-146.

[294] A. Koldobsky, D. Ryabogin and A. Zvavitch, *Fourier analytic methods in the study of projections and sections of convex bodies*, Fourier analysis and convexity, 119-130, Appl. Numer. Harmon. Anal., Birkhäuser Boston, Boston, MA, 2004.

[295] A. Koldobsky, D. Ryabogin and A. Zvavitch, *Projections of convex bodies and the Fourier transform*, Israel J. Math. **139** (2004), 361-380.

[296] A. Koldobsky and V. Yaskin, *The interface between convex geometry and harmonic analysis*, CBMS Regional Conference Series in Mathematics, AMS, **108** (2008).

[297] A. Koldobsky and M. Zymonopoulou, *Extremal sections of complex ℓ_p-balls, $0 < p \leqslant 2$*, Studia Math. **159** (2003), 185-194.

[298] H. König, M. Meyer and A. Pajor, *The isotropy constants of the Schatten classes are bounded*, Math. Ann. **312** (1998), 773-783.

[299] H. König and V. D. Milman, *On the covering numbers of convex bodies*, Geometric Aspects of Functional Analysis (Lindenstrauss-Milman eds.), Lecture Notes in Math. **1267**, Springer (1987), 82-95.

[300] M. A. Krasnosel'skii and J. B. Rutickii, *Convex functions and Orlicz spaces*, Translated from the first Russian edition by L. F. Boron, P. Noordhoff Ltd., Groningen (1961).

[301] S. Kullback, *A lower bound for discrimination information in terms of variation*, IEEE Trans. Info. Theory **4** (1967), 126-127.

[302] G. Kuperberg, *From the Mahler conjecture to Gauss linking integrals*, Geom. Funct. Anal. **18** (2008), 870-892.

[303] E. Kuwert, *Note on the isoperimetric profile of a convex body*, Geometric Analysis and Nonlinear Partial Differential Equations, Springer, Berlin (2003), 195-200.

[304] S. Kwapien, *A theorem on the Rademacher series with vector coefficients*, Proc. Int. Conf. on Probability in Banach Spaces, Lecture Notes in Math. **526**, Springer, 1976.

[305] D. G. Larman and C. A. Rogers, *The existence of a centrally symmetric convex body with central sections that are unexpectedly small*, Mathematika **22** (1975), 164-175.

[306] R. Latała, *On the equivalence between geometric and arithmetic means for log-concave measures*, Convex geometric analysis (Berkeley, CA, 1996), Math. Sci. Res. Inst. Publ., **34**, Cambridge Univ. Press, Cambridge (1999), 123-127.

[307] R. Latała, *On some inequalities for Gaussian measures*, Proceedings of the International Congress of Mathematicians, Beijing, Vol. II, Higher Ed. Press, Beijing (2002), 813-822.

[308] R. Latała, *Order statistics and concentration of ℓ_r-norms for log-concave vectors*, J. Funct. Anal. **261** (2011), 681-696.

[309] R. Latała, *Weak and strong moments of random vectors*, Marcinkiewicz centenary volume, Banach Center Publ. **95**, Polish Acad. Sci. Inst. Math., Warsaw (2011), 115-121.

[310] R. Latała and K. Oleszkiewicz, *Small ball probability estimates in terms of widths*, Studia Math. **169** (2005), 305-314.

[311] R. Latała and J. O. Wojtaszczyk, *On the infimum convolution inequality*, Studia Math. **189** (2008), 147-187.

[312] M. Ledoux, *A simple analytic proof of an inequality by P. Buser*, Proc. Am. Math. Soc. **121** (1994), 951-959.

[313] M. Ledoux, *Isoperimetry and Gaussian Analysis*, Ecole d'Eté de Probabilités de St.-Flour 1994, Lecture Notes in Math. **1709** (1996), 165-294.

[314] M. Ledoux, *Concentration of measure and logarithmic Sobolev inequalities*, Séminaire de probabilités XXXIII, Lecture Notes in Math. **1709**, Springer, Berlin (1999), 120-216.

[315] M. Ledoux, *The geometry of Markov diffusion generators*, Ann. Fac. Sci. Toulouse Math. **9** (2000), 305-366.

[316] M. Ledoux, *The concentration of measure phenomenon*, Mathematical Surveys and Monographs **89**, American Mathematical Society, Providence, RI (2001).

[317] M. Ledoux, *Spectral gap, logarithmic Sobolev constant, and geometric bounds*, Surveys in Differential Geometry, vol. IX, Int. Press, Somerville (2004), 219-240.

[318] M. Ledoux and M. Talagrand, *Probability in Banach spaces*, Ergeb. Math. Grenzgeb., 3. Folge, Vol. **23** Springer, Berlin (1991).

[319] J. Lehec, *The symmetric property (τ) for the Gaussian measure*, Ann. Fac. Sci. Toulouse Math. **17** (2008), 357-370.

[320] L. Leindler, *On a certain converse of Hölder's inequality II*, Acta. Sci. Math. Szeged **33** (1972), 217-223.

[321] P. Lévy, *Problèmes Concrets d'Analyse Fonctionelle*, Gauthier-Villars, Paris (1951).

[322] D. R. Lewis, *Finite dimensional subspaces of L_p*, Studia Math. **63** (1978), 207-212.

[323] D. R. Lewis, *Ellipsoids defined by Banach ideal norms*, Mathematika **26** (1979), 18-29.

[324] P. Li and S. T. Yau, *On the parabolic kernel of the Schrödinger operator*, Acta Math. **156** (1986), 153-201.

[325] A. Lichnerowicz, *Géométrie des groupes de transformations*, Travaux et Recherches Mathématiques, III. Dunod, Paris, 1958.

[326] E. H. Lieb, *Proof of an entropy conjecture of Wehrl*, Comm. Math. Phys. **62** (1978), 35-41.

[327] E. H. Lieb, *Gaussian kernels have only Gaussian maximizers*, Inventiones Mathematicae **102** (1990), 179-208.

[328] M. A. Lifshits, *Gaussian random functions*, Mathematics and its Applications **322**, Kluwer Academic Publishers, Dordrecht, 1995.

[329] J. Lindenstrauss, *Almost spherical sections, their existence and their applications*, Jber. Deutsch. Math.-Vereinig., Jubiläumstagung (Teubner, Stuttgart), (1990), 39-61.

[330] J. Lindenstrauss and V. D. Milman, *The Local Theory of Normed Spaces and its Applications to Convexity*, Handbook of Convex Geometry (edited by P.M. Gruber and J.M. Wills), Elsevier 1993, 1149-1220.

[331] J. Lindenstrauss and L. Tzafriri, *Classical Banach Spaces I & II*, Springer Verlag, (1977-1979).

[332] J. E. Littlewood, *On bounded bilinear forms in an infinite number of variables*, Q. J. Math. (Oxford) **1** (1930), 164-174.

[333] A. E. Litvak, V. D. Milman, A. Pajor and N. Tomczak-Jeagermann, *Entropy extension*, (Russian) Funktsional. Anal. i Prilozhen. **40** (2006), no. 4, 65-71, 112; translation in Funct. Anal. Appl. **40** (2006), no. 4, 298-303.

[334] A. Litvak, V. D. Milman and G. Schechtman, *Averages of norms and quasi-norms*, Math. Ann. **312** (1998), 95-124.

[335] A. E. Litvak, A. Pajor and N. Tomczak-Jaegermann, *Diameters of sections and coverings of convex bodies*, J. Funct. Anal. **231** (2006), no. 2, 438-457.

[336] A. E. Litvak, A. Pajor, M. Rudelson and N. Tomczak-Jaegermann, *Smallest singular value of random matrices and geometry of random polytopes*, Adv. Math. **195** (2005), no. 2, 491-523.

[337] L. H. Loomis and H. Whitney, *An inequality related to the isoperimetric inequality*, Bull. Amer. Math. Soc. **55** (1949), 961-962.

[338] L. Lovász and M. Simonovits, *Random walks in a convex body and an improved volume algorithm*, Random Structures and Algorithms **4** (1993), 359-412.

[339] L. Lovász and S. Vempala, *The geometry of logconcave functions and sampling algorithms*, Random Structures Algorithms **30** (2007), 307-358.

[340] G. J. Lozanovskii, *On some Banach lattices*, Siberian J. Math. **10** (1969), 419-431.

[341] F. Lust-Piquard and G. Pisier, *Non-commutative Khintchine and Paley inequalities*, Arkiv för Mat. **29** (1991), 241-260.

[342] L. Lusternik, *Die Brunn-Minkowskische Ungleichung für beliebige messbare Mengen*, Doklady Akad. SSSR **3** (1935), 55-58.

[343] E. Lutwak, *Dual mixed volumes*, Pacific J. Math. **58** (1975), 531-538.

[344] E. Lutwak, *On some affine isoperimetric inequalities*, J. Differential Geom. **23** (1986), 1-13.

[345] E. Lutwak, *Intersection bodies and dual mixed volumes*, Advances in Math. **71** (1988), 232-261.

[346] E. Lutwak and G. Zhang, *Blaschke-Santaló inequalities*, J. Differential Geom. **47** (1997), 1-16.

[347] E. Lutwak, D. Yang and G. Zhang, L^p *affine isoperimetric inequalities*, J. Differential Geom. **56** (2000), 111-132.

[348] A. M. Macbeath, *An extremal property of the hypersphere*, Proc. Cambridge Phil. Soc. **47** (1951), 245-247.

[349] K. Mahler, *Ein Minimalproblem für konvexe Polygone*, Mathematika B (Zutphen) **7** (1938), 118-127.

[350] K. Mahler, *Übertragungsprinzip für konvexe Körper*, Casopis Pest. Mat. Fys. **68** (1939), 93-102.

[351] E. Markessinis, G. Paouris and Ch. Saroglou, *Comparing the M-position with some classical positions of convex bodies*, Math. Proc. Cambridge Philos. Soc. **152** (2012), 131-152.

[352] E. Markessinis and P. Valettas, *Distances between classical positions of centrally symmetric convex bodies*, Houston Journal of Mathematics (to appear).

[353] A. W. Marshall, I. Olkin and F. Proschan, *Monotonicity of ratios of means and other applications of majorization*, Inequalities edited by O. Shisha, Acad. Press, New York, London (1967), 177-190.

[354] B. Maurey, *Some deviation inequalities*, Geom. Funct. Anal. **1** (1991), 188-197.

[355] V. G. Maz'ya, *The negative spectrum of the higher-dimensional Schrödinger operator*, Dokl. Akad. Nauk SSSR **144** (1962), 721-722.

[356] V. G. Maz'ya, *On the solvability of the Neumann problem*, Dokl. Akad. Nauk SSSR **147** (1962), 294-296.

[357] E. Meckes and M. W. Meckes, *The central limit problem for random vectors with symmetries*, J. Theoret. Probab. **20** (2007), 697-720.

[358] M. W. Meckes, *Gaussian marginals of convex bodies with symmetries*, Beiträge Algebra Geom. **50** (2009), 101-118.

[359] M. L. Mehta, *Random matrices*, Second Edition, Academic Press, Boston, 1991.

[360] S. Mendelson and G. Paouris, *On generic chaining and the smallest singular value of random matrices with heavy tails*, J. Funct. Anal. **262** (2012), 3775-3811.

[361] S. Mendelson and G. Paouris, *On the singular values of random matrices*, Journal of the European Mathematical Society, to appear.

[362] M. Meyer, *Une characterisation volumique de certains éspaces normés*, Israel J. Math. **55** (1986), 317-326.

[363] M. Meyer and A. Pajor, *Sections of the unit ball of* ℓ_p^n, J. Funct. Anal. **80** (1988), 109-123.

[364] M. Meyer and A. Pajor, *On Santaló's inequality*, in Geom. Aspects of Funct. Analysis (eds. J. Lindenstrauss, V. D. Milman), Lecture Notes in Math. **1376**, Springer, Berlin, 1989, 261-263.

[365] M. Meyer and A. Pajor, *On the Blaschke–Santaló inequality*, Arch. Math. **55** (1990), 82-93.

[366] E. Milman, *Dual mixed volumes and the slicing problem*, Adv. Math. **207** (2006), 566-598.

[367] E. Milman, *Uniform tail-decay of Lipschitz functions implies Cheeger's isoperimetric inquality under convexity assumptions*, C. R. Math. Acad. Sci. Paris **346** (2008), 989-994.

[368] E. Milman, *On Gaussian marginals of uniformly convex bodies*, J. Theoret. Probab. **22** (2009), no. 1, 256-278.

[369] E. Milman, *On the role of convexity in isoperimetry, spectral gap and concentration*, Invent. Math. **177** (2009), no. 1, 1-43.

[370] E. Milman, *On the role of convexity in functional and isoperimetric inequalities*, Proc. Lond. Math. Soc. **99** (2009), no. 1, 32-66.

[371] E. Milman, *Isoperimetric and concentration inequalities: equivalence under curvature lower bound*, Duke Math. J. **154** (2010), no. 2, 207-239.

[372] E. Milman, *Isoperimetric bounds on convex manifolds*, in Proceedings of the Workshop on "Concentration, Functional Inequalities and Isoperimetry", Contemporary Math. **545** (2011), 195-208.

[373] E. Milman and S. Sodin, *An isoperimetric inequality for uniformly log-concave measures and uniformly convex bodies*, J. Funct. Anal. **254** (2008), no. 5, 1235-1268.

[374] V. D. Milman, *New proof of the theorem of Dvoretzky on sections of convex bodies*, Funct. Anal. Appl. **5** (1971), 28-37.

[375] V. D. Milman, *Geometrical inequalities and mixed volumes in the Local Theory of Banach spaces*, Astérisque **131** (1985), 373-400.

[376] V. D. Milman, *Random subspaces of proportional dimension of finite dimensional normed spaces: approach through the isoperimetric inequality*, Lecture Notes in Mathematics **1166** (1985), 106-115.

[377] V. D. Milman, *Almost Euclidean quotient spaces of subspaces of finite dimensional normed spaces*, Proc. Amer. Math. Soc. **94** (1985), 445-449.

[378] V. D. Milman, *Inegalité de Brunn-Minkowski inverse et applications à la théorie locale des espaces normés*, C.R. Acad. Sci. Paris **302** (1986), 25-28.

[379] V. D. Milman, *The concentration phenomenon and linear structure of finite-dimensional normed spaces*, Proceedings of the ICM, Berkeley (1986), 961-975.

[380] V. D. Milman, *Isomorphic symmetrization and geometric inequalities*, Lecture Notes in Mathematics **1317** (1988), 107-131.

[381] V. D. Milman, *The heritage of P. Lévy in geometrical functional analysis*, Astérisque **157-158** (1988), 273-302.

[382] V. D. Milman, *A note on a low M^*-estimate*, in "Geometry of Banach spaces, Proceedings of a conference held in Strobl, Austria, 1989" (P.F. Muller and W. Schachermayer, Eds.), LMS Lecture Note Series, Vol. 158, Cambridge University Press (1990), 219-229.

[383] V. D. Milman, *Dvoretzky's theorem - Thirty years later*, Geom. Functional Anal. **2** (1992), 455-479.

[384] V. D. Milman and A. Pajor, *Isotropic position and inertia ellipsoids and zonoids of the unit ball of a normed n-dimensional space*, Lecture Notes in Mathematics **1376**, Springer, Berlin (1989), 64-104.

[385] V. D. Milman and A. Pajor, *Cas limites dans les inégalités du type de Khinchine et applications géométriques*, C.R. Acad. Sci. Paris **308** (1989), 91-96.

[386] V. D. Milman and A. Pajor, *Entropy and asymptotic geometry of non-symmetric convex bodies*, Adv. Math. **152** (2000), no. 2, 314-335.

[387] V. D. Milman and G. Schechtman, *Asymptotic Theory of Finite Dimensional Normed Spaces*, Lecture Notes in Mathematics **1200**, Springer, Berlin (1986).

[388] V. D. Milman and G. Schechtman, *Global versus Local asymptotic theories of finite-dimensional normed spaces*, Duke Math. Journal **90** (1997), 73-93.

[389] V. D. Milman and S. J. Szarek, *A geometric lemma and duality of entropy numbers*, in Geom. Aspects of Funct. Analysis, Lecture Notes in Mathematics **1745**, Springer, Berlin, (2000), 191-222.

[390] H. Minkowski, *Volumen und Oberfläche*, Math. Ann. **57** (1903), 447-495.

[391] H. Minkowski, *Geometrie der Zahlen*, Teubner, Leipzig, 1910.

[392] H. Minkowski, *Gesammelte Abhandlungen*, vol. II, Teubner Leipzig (1911), 131-229.

[393] S. J. Montgomery-Smith, *The distribution of Rademacher sums*, Proc. Amer. Math. Soc. **109**, (1990), 517–522.

[394] C. Müller, *Spherical harmonics*, Lecture Notes in Math. **17**, Springer-Verlag, Berlin- New York (1966).

[395] , D. Müller, *A geometric bound for maximal functions associated to convex bodies*, Pacific J. Math. **142** (1990), 297-312.

[396] A. Naor and D. Romik, *Projecting the surface measure of the sphere of ℓ_p^n*, Ann. Inst. H. Poincaré Probab. Statist. **39** (2003), 241-261.

[397] A. Naor and T. Tao, *Random martingales and localization of maximal inequalities*, J. Funct. Analysis **259** (2010), 731-779.

[398] F. L. Nazarov, *The Hörmander proof of the Bourgain-Milman theorem*, Geom. Aspects of Funct. Analysis, Lecture Notes in Math. **2050**, Springer, Heidelberg (2012), 335-343.

[399] F. L. Nazarov and A. N. Podkorytov, *Ball, Haagerup, and distribution functions*, in Complex analysis, operators, and related topics, Oper. Theory Adv. Appl. **113**, Birkhauser, Basel (2000), 247-267.

[400] F. L. Nazarov, M. Sodin and A. Volberg, *The geometric Kannan-Lovsz-Simonovits lemma, dimension-free estimates for the distribution of the values of polynomials, and the distribution of the zeros of random analytic functions*, Algebra i Analiz, **14** (2002), 214-234.

[401] B. Oksendal, *Stochastic Differential Equations: An Introduction with Applications*, Berlin, Springer (2003).

[402] K. Oleszkiewicz, *On p-pseudostable random variables, Rosenthal spaces and ℓ_p^n ball slicing*, Geometric Aspects of Funct. Analysis, Lecture Notes in Math. **1807** (2003), 188-210.

[403] K. Oleszkiewicz and A. Pelczynski, *Polydisc slicing in $\mathbb{C}^n$*, Studia Math. **142** (2000), 281-294.

[404] A. Pajor and N. Tomczak-Jaegermann, *Remarques sur les nombres d'entropie d'un opérateur et de son transposé*, C.R. Acad. Sci. Paris **301** (1985), 743-746.

[405] A. Pajor and N. Tomczak-Jaegermann, *Subspaces of small codimension of finite dimensional Banach spaces*, Proc. Amer. Math. Soc. **97** (1986), 637-642.

[406] R. E. A. C. Paley and A. Zygmund, *A note on analytic functions in the unit circle*, Proc. Camb. Philos. Soc. **28** (1932), 266-272.

[407] G. Paouris, *On the isotropic constant of non-symmetric convex bodies*, Geom. Aspects of Funct. Analysis (Milman-Schechtman eds.), Lecture Notes in Math. **1745** (2000), 239-243.

[408] G. Paouris, *Ψ_2-estimates for linear functionals on zonoids*, Geom. Aspects of Funct. Analysis (Milman-Schechtman eds.), Lecture Notes in Math. **1807** (2003), 211-222.

[409] G. Paouris, *Volume inequalities for sections and projections of convex bodies*, PhD Thesis (January 2004), University of Crete.

[410] G. Paouris, *On the ψ_2-behavior of linear functionals on isotropic convex bodies*, Studia Math. **168** (2005), 285-299.

[411] G. Paouris, *Concentration of mass and central limit properties of isotropic convex bodies*, Proc. Amer. Math. Soc. **133** (2005), 565-575.

[412] G. Paouris, *Concentration of mass on isotropic convex bodies*, C. R. Math. Acad. Sci. Paris **342** (2006), 179-182.

[413] G. Paouris, *Concentration of mass in convex bodies*, Geom. Funct. Analysis **16** (2006), 1021-1049.

[414] G. Paouris, *Small ball probability estimates for log-concave measures*, Trans. Amer. Math. Soc. **364** (2012), 287-308.

[415] G. Paouris, *On the existence of supergaussian directions on convex bodies*, Mathematika **58** (2012), 389-408.

[416] G. Paouris, *On the isotropic constant of marginals*, Studia Mathematica (to appear).

[417] G. Paouris and P. Pivovarov, *A probabilistic take on isoperimetric-type inequalities*, Adv. Math. **230** (2012), 1402-1422.

[418] G. Paouris and P. Pivovarov, *Small-ball probabilities for the volume of random convex sets*, Discrete Comput. Geom. **49** (2013), 601-646.

[419] G. Paouris and E. Werner, *Relative entropy of cone measures and L_p-centroid bodies*, Proc. Lond. Math. Soc. **104** (2012), 253-286.

[420] M. Papadimitrakis, *On the Busemann-Petty problem about convex, centrally symmetric bodies in $\mathbb{R}^n$*, Mathematika **39** (1992), 258-266.

[421] L. E. Payne and H. F. Weinberger, *An optimal Poincaré inequality for convex domains*, Arch. Ration. Mech. Anal. **5** (1960), 286-292.

[422] C. M. Petty, *Surface area of a convex body under affine transformations*, Proc. Amer. Math. Soc. **12** (1961), 824-828.

[423] S. K. Pichorides, *On a recent result of A. Giannopoulos on sections of convex symmetric bodies in $\mathbb{R}^n$*, Sém. Anal. Harm. Orsay 1989-90, 83-86.

[424] M. Pilipczuk and J. O. Wojtaszczyk, *The negative association property for the absolute values of random variables equidistributed on a generalized Orlicz ball*, Positivity **12** (2008), 421-474.

[425] M. S. Pinsker, *Information and Information Stability of Random Variables and Processes*, Holden-Day, San Francisco (1964).

[426] G. Pisier, *Sur les espaces de Banach K-convexes*, Séminaire d'Analyse Fonctionnelle 79/80, Ecole Polytechniques, Palaiseau, Exp. 11 (1980).

[427] G. Pisier, *Holomorphic semi-groups and the geometry of Banach spaces*, Ann. of Math. **115** (1982), 375-392.

[428] G. Pisier, *Probabilistic methods in the geometry of Banach spaces*, Lecture Notes in Mathematics **1206** (1986), 167-241.

[429] G. Pisier, *A new approach to several results of V. Milman*, J. Reine Angew. Math. **393** (1989), 115-131.

[430] G. Pisier, *The Volume of Convex Bodies and Banach Space Geometry*, Cambridge Tracts in Mathematics **94** (1989).

[431] L. Pitt, *A Gaussian correlation inequality for symmetric convex sets*, Ann. Probability **5** (1977), 470-474.

[432] P. Pivovarov, *On the volume of caps and bounding the mean-width of an isotropic convex body*, Math. Proc. Cambridge Philos. Soc. **149** (2010), 317-331.

[433] P. Pivovarov, *On determinants and the volume of random polytopes in isotropic convex bodies*, Geometriae Dedicata **149**, (2010), 45-58.

[434] A. Prékopa, *Logarithmic concave measures with applications to stochastic programming*, Acta. Sci. Math. Szeged **32** (1971), 301-316.

[435] A. Prékopa, *On logarithmic concave measures and functions* Acta Sci. Math. Szeged **34** (1973), 335-343.

[436] M. M. Rao and Z. D. Ren, *Theory of Orlicz spaces*, Monographs and Textbooks in Pure and Applied Mathematics, **146** (1991), Marcel Dekker Inc., New York.

[437] M. M. Rao and Z. D. Ren, *Applications of Orlicz spaces*, Monographs and Textbooks in Pure and Applied Mathematics, **250** (2002), Marcel Dekker Inc., New York.

[438] S. Reisner, *Random polytopes and the volume-product of symmetric convex bodies*, Math. Scand. **57** (1985), 386-392.

[439] S. Reisner, *Zonoids with minimal volume product*, Math. Z. **192** (1986), 339-346.

[440] S. Reisner, *Minimal volume-product in Banach spaces with a 1-unconditional basis*, J. London Math. Soc. **36** (1987), 126-136.

[441] Y. Rinott, *On convexity of measures*, Ann. Probab. **4** (1976), 1020-1026.

[442] R. T. Rockafellar, *Convex analysis*, Princeton Mathematical Series **28**, Princeton University Press, Princeton, NJ (1970).

[443] C. A. Rogers and G. C. Shephard, *The difference body of a convex body*, Arch. Math. (Basel) **8** (1957), 220-233.

[444] C. A. Rogers and G. C. Shephard, *Convex bodies associated with a given convex body*, J. London Soc. **33** (1958), 270-281.

[445] C. A. Rogers and G. C. Shephard, *Some extremal problems for convex bodies*, Mathematika **5** (1958), 93-102.

[446] L. Rotem, *On the mean width of log-concave functions*, Geom. Aspects of Funct. Analysis, Lecture Notes in Math. **2050**, Springer, Heidelberg (2012), 355-372.

[447] M. Rudelson, *Random vectors in the isotropic position*, J. Funct. Anal. **164** (1999), no. 1, 60-72.

[448] M. Rudelson and R. Vershynin, *Non-asymptotic theory of random matrices: extreme singular values*, Proceedings of the International Congress of Mathematicians, Volume **III**, Hindustan Book Agency, New Delhi (2010), 1576-1602.

[449] J. Saint-Raymond, *Sur le volume des corps convexes symétriques*, Initiation Seminar on Analysis: G. Choquet-M. Rogalski-J. Saint-Raymond, 20th Year: 1980/1981, Publ. Math. Univ. Pierre et Marie Curie, vol. **46**, Univ. Paris VI, Paris 1981, Exp. No. 11, 25 (French).

[450] J. Saint-Raymond, *Le volume des ideaux d'operateurs classiques*, Studia Math. **80** (1984), 63-75.

[451] L. A. Santaló, *Un invariante afin para los cuerpos convexos del espacio de n dimensiones*, Portugaliae Math. **8** (1949), 155-161.

[452] Ch. Saroglou, *Characterizations of extremals for some functionals on convex bodies*, Canad. J. Math. **62** (2010), 1404-1418.

[453] Ch. Saroglou, *Minimal surface area position of a convex body is not always an M-position*, Israel. J. Math. **30**, (2012), 1-15.

[454] G. Schechtman, *Concentration results and applications*, Handbook of the Geometry of Banach Spaces (Johnson-Lindenstrauss eds.), Vol. 2, Elsevier (2003), 1603-1634.

[455] G. Schechtman, T. Schlumprecht and J. Zinn, *On the gaussian measure of the intersection of symmetric convex sets*, Ann. Probab **26** (1998), 346-357.

[456] G. Schechtman and M. Schmuckenschläger, *Another remark on the volume of the intersection of two L_p^n balls*, Geom. Aspects of Funct. Analysis (1989-90), Lecture Notes in Math., **1469**, Springer, Berlin (1991), 174-178.

[457] G. Schechtman and J. Zinn, *On the volume of the intersection of two L_p^n balls*, Proc. Am. Math. Soc. **110** (1990), 217-224.

[458] G. Schechtman and J. Zinn, *Concentration on the ℓ_p^n ball*, Geom. Aspects of Funct. Analysis, Lecture Notes in Math., **1745**, Springer, Berlin (2000), 245-256.

[459] E. Schmidt, *Die Brunn-Minkowski Ungleichung*, Math. Nachr. **1** (1948), 81-157.

[460] M. Schmuckenschläger, *Die Isotropiekonstante der Einheitskugel eines n-dimensionalen Banachraumes mit einer 1-unbedingten Basis*, Arch. Math. (Basel) **58** (1992), 376-383.

[461] M. Schmuckenschläger, *A concentration of measure phenomenon on uniformly convex bodies*, Geometric aspects of functional analysis, Oper. Theory Adv. Appl., Birkhäuser, Basel, **77** (1995), 275-287.

[462] M. Schmuckenschläger, *Volume of intersections and sections of the unit ball of ℓ_p^n*, Proc. Amer. Math. Soc. **126** (1998), 1527-1530.

[463] R. Schneider, *Convex Bodies: The Brunn-Minkowski Theory*, Encyclopedia of Mathematics and its Applications **44**, Cambridge University Press, Cambridge (1993).

[464] C. E. Shannon, *A mathematical theory of communication*, Bell System Tech. J. **27** (1948), 379-423.

[465] C. E. Shannon and W. Weaver, *The mathematical theory of communication*, University of Illinois Press, Urbana, IL (1949).

[466] S. Sherman, *A theorem on convex sets with applications*, Ann. Math. Statist. **26** (1955), 763-767.

[467] Z. Sidák, *On multivariate normal probabilities of rectangles: their dependence on correlation*, Ann. Math. Statist. **39** (1968), 1425-1434.

[468] D. Slepian, *The one-sided barrier problem for Gaussian noise*, Bell System Technical Journal (1962), 463-501.

[469] S. Sodin, *Tail-sensitive Gaussian asymptotics for marginals of concentrated measures in high dimension*, Geom. Aspects of Funct. Analysis (Milman-Schechtman eds.), Lecture Notes in Mathematics **1910**, Springer, Berlin (2007), 271-295.

[470] S. Sodin, *An isoperimetric inequality on the ℓ_p balls*, Ann. Inst. Henri Poincaré Probab. Stat. **44** (2008), 362-373.

[471] J. Spingarn, *An inequality for sections and projections of a convex set*, Proc. Amer. Math. Soc. **118** (1993), 1219–1224.

[472] N. Srivastava and R. Vershynin, *Covariance estimation for distributions with $2+\epsilon$ moments*, Annals of Probability (to appear).

[473] A. Stam, *Some inequalities satisfied by the quantities of information of Fisher and Shannon*, Information and Control **2** (1959), 101-112.

[474] P. Stavrakakis and P. Valettas, *On the geometry of log-concave probability measures with bounded log-Sobolev constant*, Proc. of the Asymptotic Geom. Anal. Programme, Fields Institute Comm. **68** (2013).

[475] E. Stein, *The development of square functions in the work of A. Zygmund*, Bull. Amer. Math. Soc. **7** (1982), 359-376.

[476] E. Stein, *Three variations on the theme of maximal functions*, Recent Progress in Fourier Analysis, North-Holland Math. Studies **111** (1985), 229-244.

[477] E. Stein and J. O. Strömberg, *Behavior of maximal functions in $\mathbb{R}^n$ for large n*, Ark. Mat. **21** (1983), 259-269.

[478] J. Steiner, *Über parallele Flächen*, Monatsber. Preuss. Akad. Wiss. Berlin (1840), 114-18.

[479] P. Sternberg and K. Zumbrun, *On the connectivity of boundaries of sets minimizing perimeter subject to a volume constraint*, Commun. Anal. Geom. **7** (1999), 199-220.

[480] D. W. Stroock, *Probability Theory, an Analytic View*, Cambridge University Press, Cambridge (1993).

[481] V. N. Sudakov, *Gaussian random processes and measures of solid angles in Hilbert spaces*, Soviet Math. Dokl. **12** (1971), 412-415.

[482] V. N. Sudakov, *Typical distributions of linear functionals in finite-dimensional spaces of high dimension*, Soviet Math. Dokl. **19** (1978), 1578-1582.

[483] V. N. Sudakov and B. S. Tsirelson, *Extremal properties of half-spaces for spherically invariant measures*, J. Soviet. Math. **9** (1978), 9-18; translated from Zap. Nauch. Sem. L.O.M.I. **41** (1974), 14-24.

[484] S. J. Szarek, *On the best constants in the Khinchin inequality*, Studia Math. **58** (1976), no. 2, 197-208.

[485] S. J. Szarek, *On Kashin's almost Euclidean orthogonal decomposition of ℓ_1^n*, Bull. Acad. Polon. Sci. **26** (1978), no. 8, 691-694.

[486] S. J. Szarek and N. Tomczak-Jaegermann, *On nearly Euclidean decompositions of some classes of Banach spaces*, Compositio Math. **40** (1980), 367-385.

[487] M. Talagrand, *Regularity of Gaussian processes*, Acta Math. **159** (1987), 99-147.

[488] M. Talagrand, *An isoperimetric theorem on the cube and the Kintchine-Kahane inequalities*, Proc. Amer. Math. Soc. **104** (1988), 905-909.

[489] M. Talagrand, *A new isoperimetric inequality and the concentration of measure phenomenon*, Geometric aspects of functional analysis (1989–90), Lecture Notes in Math. **1469**, Springer, Berlin (1991), 94-124.

[490] M. Talagrand, *The supremum of some canonical processes*, Amer. J. Math. **116** (1994), 283-325.

[491] M. Talagrand, *The generic chaining - Upper and lower bounds of stochastic processes*, Springer Monographs in Mathematics, Springer-Verlag, Berlin, 2005.

[492] N. Tomczak-Jaegermann, *Dualité des nombres d'entropie pour des opérateurs à valeurs dans un espace de Hilbert*, C.R. Acad. Sci. Paris **305** (1987), 299-301.

[493] N. Tomczak-Jaegermann, *Banach-Mazur Distances and Finite Dimensional Operator Ideals*, Pitman Monographs **38** (1989), Pitman, London.

[494] P. S. Urysohn, *Mittlere Breite und Volumen der konvexen Körper im n-dimensionalen Raume*, Mat. Sb. SSSR **31** (1924), 313-319.

[495] J. D. Vaaler, *A geometric inequality with applications to linear forms*, Pacific J. Math. **83** (1979), 543-553.

[496] R. Vershynin, *Isoperimetry of waists and local versus global asymptotic convex geometries* (with an appendix by M. Rudelson and R. Vershynin), Duke Math. Journal **131** (2006), no. 1, 1-16.

[497] R. Vershynin, *Introduction to the non-asymptotic analysis of random matrices*, Compressed sensing, Cambridge Univ. Press, Cambridge (2012), 210-268.

[498] C. Villani, *Topics in Optimal Transportation*, Graduate Texts in Mathematics **58**, Amer. Math. Soc. (2003).

[499] H. Vogt and J. A. Voigt, *A monotonicity property of the Γ-function*, J. Inequal. Pure Appl. Math. **3**, no. 5, Article 73 (2002).

[500] J. Voigt, *A concentration of mass property for isotropic convex bodies in high dimensions*, Israel J. Math. **115** (2000), 235-251.

[501] B-H. Vritsiou, *Further unifying two approaches to the hyperplane conjecture*, International Math. Research Notices, doi: 10.1093/imrn/rns263.

[502] H. von Weizsäker, *Sudakov's typical marginals, random linear functionals and a conditional central limit theorem*, Probab. Theory Relat. Fields **107** (1997), 313-324.

[503] J. O. Wojtaszczyk, *The square negative correlation property for generalized Orlicz balls*, Geom. Aspects of Funct. Analysis, Lecture Notes in Math., **1910**, Springer, Berlin (2007), 305-313.

[504] J. O. Wojtaszczyk, *A simpler proof of the negative association property for absolute values of measures tied to generalized Orlicz balls*, Bull. Pol. Acad. Sci. Math. **57** (2009), 41-56.
[505] J. O. Wojtaszczyk, *No return to convexity*, Studia Math. **199** (2010), 227-239.
[506] G. Zhang, *Centered bodies and dual mixed volumes*, Trans. Amer. Math. Soc. **345** (1994), 771-801.
[507] G. Zhang, *Intersection bodies and the Busemann-Petty inequalities in* $\mathbb{R}^4$, Annals of Math. **140** (1994), 331-346.
[508] G. Zhang, *A positive solution to the Busemann-Petty problem in* $\mathbb{R}^4$, Annals of Math. **149** (1999), 535-543.
[509] G. M. Ziegler, *Lectures on 0/1 polytopes*, in "Polytopes-Combinatorics and Computation" (G. Kalai and G. M. Ziegler, Eds.),pp. 1-44, DMV Seminars, Birkhäuser, Basel, 2000.
[510] C. Zong, *Strange phenomena in convex and discrete geometry*, Universitext, Springer (2003).

Subject Index

(T, E)-symmetrization, 214
(κ, τ)-regular body, 234
(τ)-property, 512
$0 - 1$-polytope, 383
B-theorem, 190
$I_q(K, C)$, 234
$I_q(\mu)$, 182
$K_p(f)$
    Ball's bodies, 84
    convexity of, 87
    volume, 90
L_n
    monotonicity, 220
L_q-Rogers-Shephard inequality, 179
L_q-affine isoperimetric inequality, 175
L_q-centroid body, 174
    volume, 181
M-ellipsoid, 56
M-position, 57
    of order α, 57
MM^*-estimate, 52
$QS(X)$, 54
$Z_p^+(f)$, 437
Z_q-projection formula, 179
$\Lambda_p(\mu)$, 257
α-regular measure, 529
β-center, 384
ℓ-norm, 50
γ-concave
    function, 66
$\phi(K)$
    functional, 133
ψ_2-body, 284
ψ_α
    body, 114
    estimate, 79
    measure, 79
ψ_α-norm, 78
ε-concentration hypothesis, 394
d_*-parameter, 193
j-th area measure, 14
p-median, 273
q_{-c}-parameter, 231
q_*-parameter, 186
$r_\sharp^H$-parameter, 264
s-concave
    measure, 66

Alexandrov
    inequalities for quermassintegrals, 15
    uniqueness theorem, 16
Alexandrov-Fenchel inequality, 15
asymptotic shape, 357
axis
    of inertia, 215

Ball
    normalized version of the Brascamp-Lieb inequality, 24
    reverse isoperimetric inequality, 24
Barthe
    reverse Brascamp-Lieb inequality, 22
barycenter, 2
    of a function, 64
    of a measure, 64
basis
    of inertia axes, 215
Binet ellipsoid, 107
Blaschke
    formula, 129
    selection theorem, 3
Blaschke-Santaló inequality, 11
Borell lemma, 10, 80
Bourgain
    upper bound for ψ_2-bodies, 123
    upper bound for the isotropic constant, 117
Bourgain-Milman inequality, 55
Brascamp-Lieb inequality, 22
Brunn concavity principle, 4
Brunn-Minkowski inequality, 4
Busemann formula, 131
Busemann-Petty problem, 132

Cauchy formula, 16
Cauchy-Binet formula, 157
central limit problem, 389

centroid bodies
  covering numbers, 273
centroid body, 174
  normalized, 279
Cheeger constant, 462
Cheeger inequality, 462
comparison
  weak and strong moments, 533
concentration
  exponential, 27
  first moment, 473
  normal, 26
  of measure, 25
concentration function, 26
concentration inequality, 531
conjecture
  Gaussian correlation, 159
  hyperplane, 104
  infimum convolution, 527
  isotropic constant, 104
  Kannan-Lovász-Simonovits, 480
  Mahler, 55
  random simplex, 128
  thin shell, 403
  weak and strong moments, 534
constant
  Cheeger, 462
  exponential concentration, 471
  isoperimetric, 462
  isotropic, 75
  log-Lipschitz, 443
  Poincaré, 465
  super-Gaussian, 272
contact point, 17
convex
  body, 2
  set, 2
convex body, 2
  2-convex, 147
  almost isotropic, 92
  barycenter, 2
  centered, 2
  isotropic, 72
  mean width, 12
  mixed width, 185
  polar, 2
  position, 16
  small diameter, 117
  symmetric, 2
  unconditional, 139
  width, 2
convolution, 275
covariance matrix, 76
covering number, 34
Cramer transform, 523

difference body, 8
direction
  super-Gaussian, 272
discrete cube, 26
distance
  Banach-Mazur, 3
  geometric, 3
  Hausdorff, 3
  Wasserstein, 410
dual
  norm, 3
  space, 3
duality of entropy
  theorem, 37
Dudley-Fernique decomposition, 121
Dvoretzky theorem, 43
Dvoretzky-Rogers lemma, 20

ellipsoid, 17
  M-ellipsoid, 56
  Binet, 107
  maximal volume, 17
  minimal volume, 17
empirical process
  boundedness, 544
entropy
  duality conjecture, 37
  duality theorem, 37
  monotonicity, 559
Euclidean
  norm, 1
  unit ball, 1
  unit sphere, 1
exponential
  inequality, 471
exponential concentration
  constant, 471

first moment concentration, 473
Fisher information, 552
formula
  Z_q-projection, 179
  Blashke, 129
  Busemann, 131
  Cauchy, 16
  Cauchy-Binet, 157
  Fourier inversion, 170
  Holmstedt, 368
  Kubota, 13
  Saint-Raymond, 151
  Steiner, 13
Fourier
  inversion formula, 170
  transform, 163
Fradelizi inequality, 67
function
  γ-concave, 66
  ψ_K, 292
  barycenter, 64
  centered, 64
  concentration, 26

indicator, 7
isotropic, 76
Lipschitz, 27
logarithmically concave, 64
radial, 2
support, 2
Walsh, 51
functional
  $\phi(K)$, 133

Gauss space, 26
Gaussian
  correlation conjecture, 159
  isoperimetric inequality, 31
  measure, 26
Grünbaum lemma, 71
Grassmann manifold, 2

Haar measure, 2
Hausdorff
  limit, 16
Hausdorff metric, 3
hyperplane conjecture, 104
hyperplane sections, 105
hypothesis
  ε-concentration, 394
  variance, 402

indicator function, 7
inequality
  L_q-Rogers-Shephard, 179
  Alexandrov-Fenchel, 15
  Barthe, 22
  Blaschke-Santaló, 11
  Bobkov-Nazarov, 202
  Borell, 69
  Brascamp-Lieb, 22
  Brunn-Minkowski, 4
  Busemann, 398
  Cheeger, 462, 522
  concentration, 531
  dual Sudakov, 35
  Dudley, 40
  exponential, 471
  Fradelizi, 67
  Hadamard, 145
  Kahane-Khintchine, 34
  Khintchine, 33
  log-Sobolev, 431
  Loomis-Whitney, 140
  Lyapunov, 69
  MacLaurin, 157
  Minkowski, 13
  non-commutative Khintchine, 352
  Poincaré, 192, 465
  Prékopa-Leindler, 5
  Rogers-Shephard, 8
  Sudakov, 35
  Talagrand, 516
  Urysohn, 12
  Vaaler, 161
  Young, 22
inertia
  axis, 215
  moments, 105
inertia matrix, 76
infimum convolution, 512
  property, 524
infimum convolution conjecture, 527
information
  Fisher, 552
isomorphic slicing problem, 244
isoperimetric
  coefficient, 480
  constant, 462
  problem, 29
isoperimetric inequality
  discrete cube, 32
  for convex bodies, 10
  in Gauss space, 31
  spherical, 29
isotropic
  condition, 73
  constant of a convex body, 75
  constant of a measure, 76
  convex body, 72
  function, 76
  measure, 17, 75
  position, 73
  random vector, 78
isotropic constant
  monotonicity, 220
  stability, 117
isotropic convex body
  (κ, τ)-regular, 234
  circumradius, 108
  covering numbers, 109
  inradius, 108
  with small diameter, 284

John
  representation of the identity, 18
  theorem, 19

k(X)
  critical dimension, 45
Kahane-Khintchine inequality, 34
Kannan-Lovasz-Simonovits conjecture, 480
Kashin decomposition, 50
Khintchine inequality, 33
Klartag
  bound for the isotropic constant, 252
  isomorphic slicing problem, 251
KLS-conjecture, 480
Knothe map, 7
Kubota formula, 13

Löwner position, 20

Lévy family, 26
Lévy mean, 27
Legendre transform, 523
lemma
   Arias de Reyna-Ball-Villa, 30
   Borell, 10, 80
   Dvoretzky-Rogers, 20
   Grünbaum, 71
   Lewis, 50
   Lovász-Simonovits, 94, 481
   Lozanovskii, 146
   Sidák, 158
   Slepian, 40
Lipschitz function, 27
localization lemma, 94, 481
log-concave
   density, 64
   function, 64
   measure, 64
log-Lipschitz constant, 443
logarithmic Laplace transform, 249
Loomis-Whitney inequality, 140

Mahler conjecture, 55
majorizing measure theorem, 42
manifold
   Grassmann, 2
map
   Knothe, 7
   Minkowski, 16
marginal, 178
matrix
   covariance, 76
   inertia, 76
mean width, 12
measure
   α-regular, 529
   s-concave, 66
   barycenter, 64
   centered, 64
   concentration, 25
   convolution, 275, 513
   even, 64
   Gaussian, 26
   Haar, 2
   isotropic, 17, 75
   logarithmically concave, 64
   majorizing, 42
   marginal, 178
   mixed area, 14
   peakedness, 161
   product exponential, 516
   surface area, 14
   symmetric, 64
   symmetric exponential, 515
metric
   Hausdorff, 3
metric probability space, 25

Milman
   M-ellipsoid, 56
   isomorphic symmetrization, 55
   low M^*-estimate, 52
   quotient of subspace theorem, 54
   reverse Brunn-Minkowski inequality, 57
   version of Dvoretzky theorem, 44
minimal
   mean width, 21
   surface, 21
   surface invariant, 21
Minkowski
   content, 10, 29, 462
   existence theorem, 15
   inequality for mixed volumes, 13
   map, 16
   sum, 2
   theorem on mixed volumes, 12
mixed
   area measure, 14
   volume, 13
mixed width, 185
monotonicity
   of L_n, 220
movement
   type-i, 443

Neumann Laplacian, 408
norm
   ψ_α-norm, 78
   Orlicz, 78
   Rademacher projection, 52
   trace dual, 50
   unconditional, 139
   unitarily invariant, 151

order statistics, 536
Orlicz norm, 78
Ornstein-Uhlenbeck semigroup, 553

packing number, 34
Paouris
   deviation inequality, 182
   small ball estimate, 190
parameter
   d_*, 193
   q_{-c}, 231
   q_{-c}^H, 268
   q_*, 186
   $q_{\sharp,c}$, 267
   $q_\sharp^H$, 267
   $r_\sharp^H$, 264
Pisier
   α-regular M-position, 57
   norm of the Rademacher projection, 52
Poincaré constant, 465
Poincaré inequality, 192, 465
polar body, 2
polynomials

Khintchine type inequalities, 97
polytope, 15
   $0-1$, 383
   random, 357
polytopes
   with few facets, 160
   with few vertices, 156
position
   M-position, 57
   isotropic, 19, 73
   John, 17
   Löwner, 20
   Lewis, 149
   minimal mean width, 21
   minimal surface, 21
   of a convex body, 16
Prékopa-Leindler inequality, 5
principle
   Brunn, 4
problem
   Busemann-Petty, 132
   Sylvester, 128
process
   empirical, 348
   Gaussian, 38
   sub-Gaussian, 38
projection body, 16
property (τ), 512

quermassintegrals, 13
   Alexandrov inequalities, 15
   normalized, 14

Rademacher
   functions, 33
   projection, 51
radial function, 2
Radon transform, 163
   spherical, 163
random
   polytope, 357
random polytopes
   asymptotic shape, 368
   isotropic constant, 377
   volume radius, 375
random vector
   isotropic, 78
reverse
   Brascamp-Lieb inequality, 22
   Brunn-Minkowski inequality, 57
   isoperimetric inequality, 24
   Santaló inequality, 55, 254
   Urysohn inequality, 52
Riemannian
   manifold, 418
   metric, 417
   package, 417
   package, isomorphism, 417
   package, log-concave, 417

Rogers-Shephard ineqality, 8

Saint-Raymond
   formula, 151
Schatten class, 150
semigroup
   Ornstein-Uhlenbeck, 553
separated set, 34
set
   convex, 2
   separated, 34
   star-shaped, 2
Sidák lemma, 158
Slepian lemma, 40
slicing problem, 104
   reduction to $I_1(K, Z_q^\circ(K))$, 234
   reduction to $q_{-c}(K, \delta)$, 231
   reduction to bounded volume ratio, 223
space
   normed, 3
spectral gap, 465
spherical
   cone, 46
   isoperimetric inequality, 29
   symmetrization, 30
star body, 2
Steiner
   formula, 13
   symmetrization, 4
sub-Gausian
   direction, 271
subindependence of coordinate slabs, 405
Sudakov inequality, 35
super-Gaussian
   constant, 272
   direction, 272
support function, 2
surface area, 10
   measure, 14
Sylvester's problem, 128
symmetrization
   (T, E), 214
   isomorphic, 55
   spherical, 30
   Steiner, 4, 358

Talagrand
   comparison theorem, 42
   isoperimetric inequality for the discrete cube, 32
   majorizing measure theorem, 42
theorem
   Adamczak-Litvak-Pajor-Tomczak, 334, 335, 347
   Alesker, 115
   Alexandrov, 16
   Anttila-Ball-Perissinaki, 397
   Artstein-Ball-Barthe-Naor, 559

Artstein-Milman-Szarek, 37
Bakry-Ledoux, 469
Ball, 24, 87, 91, 166, 169, 550
Ball-Nguyen, 552
Ball-Perissinaki, 405
Blaschke, 3
Bobkov, 391, 395, 506
Bobkov-Nazarov, 142, 202, 203, 306, 307
Borell, 64, 66
Bourgain, 97, 118, 121, 123, 350
Bourgain-Klartag-Milman, 215, 220, 223
Bourgain-Milman, 55
Busemann, 398
Carl-Pajor, 158
Cordero-Fradelizi-Maurey, 190
Dafnis-Giannopoulos-Guédon, 381
Dafnis-Giannopoulos-Tsolomitis, 368, 375
Dafnis-Paouris, 227, 232, 233
Dudley-Sudakov, 39
Dvoretzky, 43
Dyer-Füredi-McDiarmid, 384
Eldan, 487
Eldan-Klartag, 415
Figiel-Tomczak, 51
Fleury, 433
Fradelizi-Guédon, 94
Gatzouras-Giannopoulos, 386
Gatzouras-Giannopoulos-Markoulakis, 384, 386
Giannopoulos-Milman, 21
Giannopoulos-Paouris-Valettas, 273, 284
Giannopoulos-Paouris-Vritsiou, 234
Gluskin, 158
Gromov-V. Milman, 471
Guédon-E. Milman, 436, 443
John, 18, 19
König-Meyer-Pajor, 155
König-Milman, 37
Kannan-Lovász-Simonovits, 480
Kanter, 161
Kashin, 48
Klartag, 91, 207, 244, 251, 407, 485
Klartag-E. Milman, 260, 262, 453
Klartag-Kozma, 380
Klartag-Vershynin, 194, 205
Koldobsky, 164, 165
Latała, 82
Latała-Oleszkiewicz, 204
Litvak-Milman-Schechtman, 183
Lutwak-Yang-Zhang, 175, 181
Meyer-Pajor, 161
Milman, E., 474–477, 488
Milman, V. D., 44, 52, 54, 56
Milman-Pajor, 9
Milman-Schechtman, 45
Minkowski, 12, 15
Pajor-Tomczak, 35

Paouris, 179, 181–183, 187, 188, 190
Petty, 21
Pisier, 52, 57, 227
Rudelson, 352
Szarek-Tomczak, 48
Talagrand, 32, 42
thin shell
  conjecture, 403
transform
  Cramer, 523
  Fourier, 163
  Legendre, 523
  Radon, 163
type-i movement, 443

unconditional convex body, 139
unitarily invariant norm, 151
Urysohn inequality, 12

variance hypothesis, 402
volume, 1
  mixed, 13
  of L_q-centroid bodies, 181
  radius, 9
  ratio, 17
  sections of the cube, 166
volume product, 11
volume ratio, 47
  outer, 144
  theorem, 48
  theorem, global form, 49
  uniformly bounded, 47

Walsh functions, 51
Wasserstein distance, 410
weak and strong moments comparison, 533
Wiener process, 489

zonoid, 16
  Löwner position, 149
  Lewis position, 149
  minimal mean width position, 150

Author Index

Adamczak, R., 197, 334, 335, 347, 355, 547, 565
Aldaz, J. M., 134, 565
Alesker, S., 115, 135, 183, 354, 565
Alexandrov, A. D., 14–16, 58, 565
Alonso-Gutiérrez, D., 134, 137, 155, 156, 158, 171, 377, 387, 388, 565
Anderson, T. W., 171, 565
Anttila, M., 397, 422, 423, 565
Aomoto, A., 171, 565
Arias de Reyna, J., 30, 60, 171, 565
Artstein-Avidan, S., 37, 60, 273, 559, 563, 565
Aubrun, G., 134, 355, 566

Bárány, I., 171, 371, 384, 385, 566
Badrikian, A., 61, 566
Bakry, D., 469, 509, 566
Ball, K. M., 18, 24, 30, 58–60, 84, 100, 101, 134, 136, 137, 149, 155, 166, 169, 171, 172, 397, 405, 422, 423, 550, 552, 559, 562, 563, 565, 566
Banaszczyk, W., 210
Barlow, R. E., 100, 566
Barron, A. R., 566
Barthe, F., 22, 23, 59, 60, 101, 137, 172, 312, 368, 423, 510, 559, 563, 565–567
Bastero, J., 156, 332, 422, 565, 567
Batson, J., 355, 567
Bayle, V., 509, 567
Beckenbach, E. F., 567
Bellman, R., 567
Bernués, J., 156, 422, 565, 567
Berwald, L., 567
Blachman, N. M., 567
Blaschke, W., 3, 11, 58, 135, 387, 567
Bobkov, S. G., 60, 71, 101, 141, 143, 170, 171, 202, 203, 210, 241, 306, 307, 312, 381, 391, 395, 402, 422, 423, 506, 509, 510, 562, 567, 568
Bogachev, V., 1, 568
Bolker, E. D., 149, 171, 568
Bonnesen, T., 58, 568

Borell, C., 10, 31, 59, 60, 65, 66, 69, 100, 568
Bourgain, J., 55, 59, 61, 97, 101, 111, 117, 118, 121, 123, 134–136, 171, 214, 215, 220, 223, 242, 349, 354, 379, 568
Brascamp, H. J., 22, 59, 569
Brazitikos, S., 456, 569
Brehm, U., 422, 569
Brenier, Y., 410, 569
Brezis, H., 1, 569
Brunn, H., 4, 58, 569
Buchta, C., 384, 569
Burago, Y. D., 58, 569
Busemann, H., 131, 132, 136, 398, 422, 569
Buser, P., 467, 509, 569

Campi, S., 209, 569
Carbery, A., 101, 134, 569
Carl, B., 158, 171, 569
Carlen, E. A., 569
Cauchy, A. L., 16
Chakerian, G. D., 59, 569
Chavel, I., 509, 569
Cheeger, J., 462, 466, 509, 570
Chevet, S., 61, 566
Cordero-Erausquin, D., 171, 190, 210, 273, 567, 570
Costa, M., 562, 570
Cover, T., 562, 570
Csiszár, I., 570

Dafnis, N., xx, 221, 227, 232, 233, 242, 367, 368, 374, 375, 381, 387, 388, 565, 570
Dalla, L., 135, 570
Dar, S., 117, 120, 121, 135, 155, 156, 171, 570
Das Gupta, S., 59, 570
Dembo, A., 562, 570
Diaconis, P., 390, 422, 570
Dudley, R. M., 39, 40, 61, 570
Durrett, R., 570
Dvoretzky, A., 20, 43, 59, 61, 570
Dyer, M. E., 384, 385, 387, 570

Eldan, R., 415, 423, 459, 487, 510, 570
Emery, M., 566
Erhard, A., 60, 571

Füredi, Z., 171, 384, 385, 387, 566, 570
Feller, W., 1, 571
Fenchel, W., 15, 58, 568, 571
Fernique, X., 61, 571
Figiel, T., 51, 60, 61, 571
Fleiner, T., 384, 571
Fleury, B., 433, 571
Folland, G. B., 571
Fradelizi, M., 67, 94, 100, 101, 137, 190, 210, 216, 219, 273, 567, 570, 571
Freedman, D., 390, 422, 570
Frieze, A., 571
Fukuda, K., 384, 571

Gardner, R. J., 58, 136, 137, 571
Gatzouras, D., xx, 384–386, 571
Giannopoulos, A., 17, 21, 58, 59, 100, 135–137, 171, 234, 242, 254, 270, 273, 279, 280, 284, 301, 310, 322, 332, 354, 359, 367, 368, 374, 375, 381, 384–388, 570–572
Giertz, M., 136, 572
Gilbarg, D., 572
Gluskin, E. D., 158, 171, 572
Goodey, P. R., 171, 572
Gordon, Y., 53, 61, 572
Gozlan, N., 546, 573
Grünbaum, B., 71, 100, 573
Groemer, H., 58, 135, 163, 359, 387, 573
Gromov, M., 60, 100, 171, 471, 573
Gronchi, P., 209, 569
Gross, L., 458, 573
Grothendieck, A., 43, 59
Gruber, P. M., 1, 58, 573
Guédon, O., 94, 101, 210, 312, 354, 368, 381, 388, 436, 443, 567, 570, 571, 573

Hörmander, L., 408, 574
Haagerup, U., 34, 60, 573
Hadwiger, H., 58, 136, 573
Hardy, G. H., 69, 573
Hargé, G., 171, 573
Harper, L. H., 32, 60, 573
Hartzoulaki, M., xx, 109, 135, 314, 332, 354, 359, 367, 387, 572, 573
Helffer, B., 408, 574
Hensley, D., 100, 106, 574
Henstock, R., 58, 574
Hernandez Cifre, M. A., 387, 565
Hinow, P., 422, 569
Hioni, L., xx
Holmstedt, T., 368, 574
Houdré, C., 509, 568
Huet, N., 510, 574

Jessen, B., 58, 571
John, F., 18, 19, 59, 574
Johnson, W. B., 574
Junge, M., 155, 171, 574

König, H., 37, 60, 150, 155, 171, 575
Kadlec, J., 408, 574
Kahane, J.-P., 34, 60, 574
Kaibel, V., 384, 571
Kannan, R., 101, 109, 135, 334, 354, 510, 571, 574
Kanter, M., 161, 171, 574
Kashin, B., 48, 50, 61, 574
Khintchine, A., 34, 60, 574
Klartag, B., 91, 100, 111, 135, 147, 171, 193–195, 205, 207, 209, 210, 214, 215, 220, 221, 223, 241, 242, 244, 249, 251, 256, 260, 262–264, 270, 310, 311, 332, 372, 377, 380, 388, 407, 415, 423, 429, 453, 459, 485, 510, 568, 571, 574, 575
Knothe, H., 7, 59, 575
Koldobsky, A., 137, 163, 172, 402, 422, 423, 567, 571, 575
Kozma, G., 377, 380, 388, 575
Krasnosel'skii, M. A., 100, 575
Kubota, T., 13
Kullback, S., 575
Kuperberg, G., 61, 137, 575
Kuwert, E., 509, 575
Kwapien, S., 60, 575

Lévy, P., 30, 60, 576
Larman, D. G., 135, 136, 371, 566, 570, 575
Latała, R., 82, 101, 197, 204, 210, 510, 533, 536, 544, 547, 565, 576
Ledoux, M., 60, 171, 254, 347, 348, 467, 469, 509, 566, 576
Lehec, J., 546, 576
Leindler, L., 5, 59, 576
Lewis, D. R., 61, 149, 171, 576
Li, P., 509, 510, 576
Lichnerowicz, A., 408, 576
Lieb, E. H., 22, 59, 563, 569, 576
Lifshits, M. A., 199, 422, 575, 576
Lindenstrauss, J., 58, 60, 100, 171, 379, 568, 571, 574, 576
Littlewood, J. E., 60, 69, 573, 576
Litvak, A. E., 183, 197, 210, 332, 334, 335, 347, 355, 367, 369, 387, 547, 565, 576
Loomis, L. H., 140, 170, 577
Lovász, L., 101, 109, 135, 334, 354, 510, 574, 577
Lozanovskii, G. J., 146, 577
Lust-Piquard, F., 352, 577
Lusternik, L., 58, 577
Luttinger, J. M., 59, 569
Lutwak, E., 136, 175, 181, 209, 241, 577

Müller, C., 579

Müller, D., 134, 579
Müller, J., 384, 569
Macbeath, A. M., 58, 359, 387, 574, 577
Madiman, M., 562, 568
Mahler, K., 59, 61, 577
Markessinis, E., xx, 59, 577
Markoulakis, N., xx, 384–386, 572
Marshall, A. W., 100, 566, 577
Maurey, B., 32, 60, 137, 190, 210, 273, 516, 546, 567, 570, 577
Maz'ya, V. G., 462, 466, 509, 577
McDiarmid, C., 384, 385, 387, 570
Meckes, E., 422, 577
Meckes, M. W., 422, 577
Mehta, M. L., 171, 577
Mendelson, S., 312, 355, 368, 567, 577
Meyer, M., 11, 59, 61, 150, 155, 161, 162, 171, 569, 573, 575, 577
Milman, E., 135, 147, 155, 171, 201, 241, 256, 260, 262–264, 270, 325, 330, 372, 422, 436, 443, 453, 474, 475, 477, 488, 573, 575, 578
Milman, V. D., 1, 9, 17, 21, 37, 43–45, 52, 54–56, 58–61, 100, 101, 105, 109, 111, 134, 135, 170, 171, 183, 185, 209, 210, 214, 215, 220, 221, 223, 228, 240, 242, 273, 281, 325, 332, 354, 371, 379, 471, 566, 568, 569, 571–573, 576, 578
Milman,E., 476
Minkowski, H., 4, 12, 13, 15, 58, 578
Montgomery-Smith, S. J., 579

Naor, A., 134, 172, 312, 368, 422, 559, 563, 565–567, 579
Nazarov, F., 61, 101, 141, 143, 167, 170, 172, 202, 203, 210, 241, 306, 307, 312, 381, 568, 579
Nguyen, V. H., 423, 552, 563, 566

Ohmann, D., 58, 573
Oksendal, B., 579
Oleszkiewicz, K., 172, 197, 204, 210, 565, 576, 579
Olkin, I., 100, 577

Pólya, G., 69, 573
Pór, A., 384, 385, 566
Pajor, A., 9, 11, 35, 53, 59–61, 100, 101, 105, 109, 134, 135, 150, 155, 158, 161, 162, 170, 171, 197, 209, 228, 240, 280, 310, 332, 334, 335, 347, 355, 367, 369, 371, 387, 547, 565, 566, 569, 572, 575–577, 579
Paley, R. E. A. C., 274, 579
Paouris, G., xx, 59, 101, 113, 175, 177, 179, 181–183, 187, 188, 190, 195–197, 201, 209, 210, 221, 227, 232–234, 242, 254, 270, 273, 279, 280, 284, 298, 300, 301, 310, 311, 322, 332, 355, 359, 371, 372, 387, 388, 422, 565, 570–573, 577, 579
Papadimitrakis, M., 59, 136, 171, 572, 579
Payne, L. E., 510, 580
Pelczynski, A., 172, 579
Perissinaki, E., xx, 397, 405, 422, 423, 565, 566
Petty, C. M., 22, 59, 132, 136, 569, 580
Pichorides, S. K., 136, 580
Pilipczuk, M., 423, 580
Pinsker, M. S., 580
Pisier, G., 1, 52, 57–59, 61, 227, 352, 577, 580
Pitt, L., 159, 171, 580
Pivovarov, P., 311, 315, 332, 364, 387, 388, 580
Podkorytov, A. N., 167, 172, 579
Prékopa, A., 5, 59, 580
Prochno, J., 387, 565
Proschan, F., 100, 566, 577

Rao, M. M., 100, 580
Reisner, S., 61, 573, 580
Ren, Z. D., 100, 580
Rinott, Y., 171, 580
Rockafellar, R. T., 1, 580
Rogers, C. A., 8, 20, 43, 59, 136, 570, 575, 580
Romik, D., 422, 579
Rosales, C., 509, 567
Rote, G., 384, 571
Rotem, L., 100, 580
Rothaus, O. S., 462, 509
Rudelson, M., 332, 333, 352, 354, 367, 369, 387, 573, 577, 580
Rutickii, J. B., 100, 575
Ryabogin, D., 172, 575

Saint-Raymond, J., 61, 151, 152, 171, 581
Santaló, L. A., 11, 59, 581
Saroglou, C., 59, 135, 577, 581
Schechtman, G., 1, 45, 58, 60, 61, 101, 171, 183, 185, 210, 576, 578, 581
Schlumprecht, T., 171, 571, 581
Schmidt, E., 60, 581
Schmuckenschläger, M., 170, 171, 210, 581
Schneider, R., 1, 58, 250, 581
Shannon, C. E., 581
Shephard, G. C., 8, 59, 580
Sherman, S., 171, 581
Sidák, Z., 158, 171, 581
Simonovits, M., 101, 135, 334, 354, 510, 574, 577
Sjöstrand, J., 408, 574
Slepian, D., 40, 61, 581
Sodin, M., 101, 579
Sodin, S., 422, 459, 510, 578, 581
Soffer, A., 569
Spielman, D., 355, 567

Spingarn, J., 228, 581
Srivastava, N., 355, 567, 581
Stam, A., 581
Stavrakakis, P., xx, 322, 332, 456, 569, 572, 581
Stein, E., 134, 581
Steiner, J., 13, 58, 582
Sternberg, P., 509, 582
Strömberg, J. O., 134, 582
Stroock, D. W., 60, 582
Sudakov, V. N., 31, 35, 39, 40, 60, 61, 390, 422, 582
Szarek, S. J., 34, 37, 48, 60, 61, 273, 281, 383, 566, 578, 582

Talagrand, M., 32, 35, 42, 43, 60, 61, 117, 135, 254, 347, 348, 516, 544, 576, 582
Tao, T., 134, 579
Thomas, J., 562, 570
Tichy, R. F., 384, 569
Tomczak-Jaegermann, N., 1, 35, 36, 48, 51, 53, 58, 60, 61, 197, 254, 332, 334, 335, 347, 355, 367, 369, 387, 547, 565, 571, 576, 579, 582
Trudinger, N. S., 572
Tsirelson, B. S., 31, 60, 582
Tsolomitis, A., xx, 135, 322, 332, 354, 359, 367, 368, 374, 375, 387, 570, 572
Tzafriri, L., 100, 171, 576

Urysohn, P. S., 12, 59, 61, 582

Vaaler, J. D., 161, 171, 582
Valettas, P., 59, 273, 279, 284, 301, 310, 322, 332, 572, 581
Vempala, S., 577
Vershynin, R., 193–195, 205, 210, 332, 333, 355, 575, 581, 582
Villa, R., 30, 60, 171, 565
Villani, C., 410, 582
Vogt, H., 90, 422, 569, 582
Voigt, J. A., 90, 422, 569, 582
Volberg, A., 101, 579
von Weizsäker, H., 390, 582
Vritsiou, B.-H., 234, 242, 254, 266, 270, 322, 332, 572, 582

Weaver, W., 581
Weil, W., 171, 572
Weinberger, H. F., 510, 580
Werner, E., 371, 579
Whitney, H., 140, 170, 577
Wojtaszczyk, J., 422, 423, 510, 533, 547, 576, 582
Wolff, P., 156, 565
Wright, J., 101, 569

Yang, D., 175, 181, 209, 241, 577
Yaskin, V., 172, 575

Yau, S. T., 509, 510, 576

Zalgaller, V. A., 58, 569
Zhang, G., 136, 175, 181, 209, 241, 577, 583
Ziegler, G., 384, 583
Zinn, J., 171, 210, 581
Zong, C., 583
Zumbrun, K., 509, 582
Zvavitch, A., 172, 575
Zygmund, A., 274, 579
Zymonopoulou, M., 172, 575

Published Titles in This Series

196 Silouanos Brazitikos, Apostolos Giannopoulos, Petros Valettas, and Beatrice-Helen Vritsiou, Geometry of Isotropic Convex Bodies, 2014

195 Ching-Li Chai, Brian Conrad, and Frans Oort, Complex Multiplication and Lifting Problems, 2013

194 Samuel Herrmann, Peter Imkeller, Ilya Pavlyukevich, and Dierk Peithmann, Stochastic Resonance, 2014

193 Robert Rumely, Capacity Theory with Local Rationality, 2013

192 Messoud Efendiev, Attractors for Degenerate Parabolic Type Equations, 2013

191 Grégory Berhuy and Frédérique Oggier, An Introduction to Central Simple Algebras and Their Applications to Wireless Communication, 2013

190 Aleksandr Pukhlikov, Birationally Rigid Varieties, 2013

189 Alberto Elduque and Mikhail Kochetov, Gradings on Simple Lie Algebras, 2013

188 David Lannes, The Water Waves Problem, 2013

187 Nassif Ghoussoub and Amir Moradifam, Functional Inequalities: New Perspectives and New Applications, 2013

186 Gregory Berkolaiko and Peter Kuchment, Introduction to Quantum Graphs, 2013

185 Patrick Iglesias-Zemmour, Diffeology, 2013

184 Frederick W. Gehring and Kari Hag, The Ubiquitous Quasidisk, 2012

183 Gershon Kresin and Vladimir Maz'ya, Maximum Principles and Sharp Constants for Solutions of Elliptic and Parabolic Systems, 2012

182 Neil A. Watson, Introduction to Heat Potential Theory, 2012

181 Graham J. Leuschke and Roger Wiegand, Cohen-Macaulay Representations, 2012

180 Martin W. Liebeck and Gary M. Seitz, Unipotent and Nilpotent Classes in Simple Algebraic Groups and Lie Algebras, 2012

179 Stephen D. Smith, Subgroup complexes, 2011

178 Helmut Brass and Knut Petras, Quadrature Theory, 2011

177 Alexei Myasnikov, Vladimir Shpilrain, and Alexander Ushakov, Non-commutative Cryptography and Complexity of Group-theoretic Problems, 2011

176 Peter E. Kloeden and Martin Rasmussen, Nonautonomous Dynamical Systems, 2011

175 Warwick de Launey and Dane Flannery, Algebraic Design Theory, 2011

174 Lawrence S. Levy and J. Chris Robson, Hereditary Noetherian Prime Rings and Idealizers, 2011

173 Sariel Har-Peled, Geometric Approximation Algorithms, 2011

172 Michael Aschbacher, Richard Lyons, Stephen D. Smith, and Ronald Solomon, The Classification of Finite Simple Groups, 2011

171 Leonid Pastur and Mariya Shcherbina, Eigenvalue Distribution of Large Random Matrices, 2011

170 Kevin Costello, Renormalization and Effective Field Theory, 2011

169 Robert R. Bruner and J. P. C. Greenlees, Connective Real K-Theory of Finite Groups, 2010

168 Michiel Hazewinkel, Nadiya Gubareni, and V. V. Kirichenko, Algebras, Rings and Modules, 2010

167 Michael Gekhtman, Michael Shapiro, and Alek Vainshtein, Cluster Algebras and Poisson Geometry, 2010

166 Kyung Bai Lee and Frank Raymond, Seifert Fiberings, 2010

165 Fuensanta Andreu-Vaillo, José M. Mazón, Julio D. Rossi, and J. Julián Toledo-Melero, Nonlocal Diffusion Problems, 2010

164 Vladimir I. Bogachev, Differentiable Measures and the Malliavin Calculus, 2010

163 Bennett Chow, Sun-Chin Chu, David Glickenstein, Christine Guenther, James Isenberg, Tom Ivey, Dan Knopf, Peng Lu, Feng Luo, and Lei Ni, The Ricci Flow: Techniques and Applications: Part III: Geometric-Analytic Aspects, 2010